# Food, Pharmaceutical and Bioengineering Division 2015



Core Programming Area at the 2015 AIChE Meeting

Salt Lake City, Utah, USA
8-13 November 2015

Volume 1 of 2

ISBN: 978-1-5108-1860-6

**Printed from e-media with permission by:**

Curran Associates, Inc.
57 Morehouse Lane
Red Hook, NY 12571

Some format issues inherent in the e-media version may also appear in this print version.



Printed by Curran Associates, Inc. (2016)

For permission requests, please contact AIChE
at the address below.

AIChE
120 Wall Street, FL 23
New York, NY 10005-4020

Phone: (800) 242-4363
Fax: (203) 775-5177

www.aiche.org

**Additional copies of this publication are available from:**

Curran Associates, Inc.
57 Morehouse Lane
Red Hook, NY 12571 USA
Phone: 845-758-0400
Fax: 845-758-2634
Email: curran@proceedings.com
Web: www.proceedings.com

# TABLE OF CONTENTS

## VOLUME 1

## VOLUME 2

# 422871 Rational Engineering of an or-Gate CAR for Adoptive T-Cell Therapy Against B-Cell Malignancies

**Abstract:**

Chimeric antigen receptor (CAR)-based immunotherapy for treatment of cancer has shown great promise in recent years for the treatment of cancers resistant to surgery, chemotherapy and radiation. The most clinically successful CAR, targeting the B-cell antigen CD19, has been shown to be effective against both acute and chronic B-cell malignancies. However, the CD19 CAR can be susceptible to mutational escape as demonstrated by clinical trials in which multiple patients treated with the single-input CD19-targeting CAR relapsed due to the emergence of CD19-negative leukemic cells. As a solution to this challenge, receptors that trigger robust T-cell activation in response to multiple antigens can significantly reduce the probability of antigen escape and increase the robustness of CAR–T-cell therapy. Here, we present the development of OR-gate CARs targeting CD19 and CD20, two clinically relevant antigens associated with B-cell malignancies. Bispecific CARs were rationally designed and optimized based on the structural requirements of CD19 and CD20 CARs. Primary human T cells expressing CD19-OR-CD20 CARs demonstrate robust cytokine production and efficient lysis of Burkitt's lymphoma cells, including tumor cells that have lost CD19 expression due to CRISPR/Cas9-guided genome engineering. To our knowledge, this is the first demonstration of a bi-specific CAR that effectively targets a clinically relevant pair of tumor antigens, and ongoing work aims to optimize the technology for clinical translation.

# 430583 Yeast Artificial Antigen Presenting Cells (yAPCs): A Robust Experimental System for Modulating T-Cell Activation

**Abstract:**

T-cell mediated immune response is triggered by T cell receptor (TCR) binding of agonist peptide-major histocompatibility complex (pMHC) on the surface of antigen presenting cells (APCs). Despite being among the most important interactions in the immune system, the exact mechanism relating pMHC-TCR binding to T cell activation is not fully understood. Recent studies using glass-supported lipid bilayers have shown that just a few agonist pMHC can initiate TCR triggering, and that the extent of T-cell triggering is influenced by the number of agonist pMHC. Furthermore, T cells can discriminate between a single agonist pMHC among thousands of endogenous pMHC. In spite of this incredible specificity and sensitivity for a specific pMHC ligand, recent kinetic studies using fluorescence resonance energy transfer (FRET) have predicted that *in situ* TCR generally exhibit relatively low affinity for cognate pMHC ($K_D$ values ranging from 1-100 μM). Several theories have been proposed to explain this seemingly paradoxical nature of TCR triggering, including triggering of pre-formed TCR microclusters, force-induced receptor deformation, TCR pseudodimerization with agonist and endogenous pMHC, and a kinetic-segregation model. With the aim to better understand how specific and relatively weak pMHC-TCR binding produces T cell activation, we have approached the problem of TCR triggering by investigating how local pMHC valency and overall pMHC density contribute to T-cell activation using yeast artificial antigen presenting cells (yAPCs).

Conventional strategies to investigate TCR triggering are often acellular, planar, and tend to rely on lithography and other expensive fabrication techniques. While these systems are well-suited for quantitative analysis of cell-cell signaling, they often lack the ability to control pMHC clustering and restrict stimulatory pMHC-TCR interactions to a single plane. To address these shortcomings, we have engineered a robust cellular system that allows for precise control over local pMHC valency and overall pMHC density by engineering yAPCs expressing surface-displayed multivalent protein scaffolds. These scaffolds are capable of binding recombinant human pMHC proteins with high affinity, and presenting them to T cells in precise stoichiometric arrangements. The engineered yAPCs presenting pMHC at different local and absolute levels have been used to selectively engage antigen-specific T cells and induce a dose-dependent cytokine response in T cell hybridoma recognizing the influenza epitope $HA_{306-318}$. This system provides a unique perspective on the study of TCR triggering and can be used to elucidate the role that pMHC valency plays in T cell activation. While this system is currently focused on multivalent pMHC presentation, on-going experiments are exploring the inclusion of costimulatory and adhesion molecules to emulate natural T cell activation and contribute to existing models of TCR triggering.

# 431655 Nanoallergens: A Nanoparticle Platform for Identifying Crucial Allergy Epitopes

**Abstract:**

Allergy is a disorder of the immune system which is elicited by the allergen-mediated-clustering of the immunoglobulin E (IgE) molecules on the surface of mast cells and basophils. Allergenic diseases are a major cause of illness and disability in the US, and there is broad consensus that the prevalence of type I allergy is increasing worldwide. The majority of the current therapies used to treat allergy related adverse immunologic responses involve non-specific suppression of parts of the immune system. More targeted therapies would improve the effectiveness of allergy treatment, but would also require a more thorough assessment of antigenic regions (also known as epitopes) on allergen proteins. During allergic reactions, allergen specific IgE molecules bind multivalently to an allergen protein, causing the release of inflammatory cytokines. A possible allergy inhibitor design would be a mimetic of these IgE binding epitopes, which competitively block IgEs from binding allergen proteins, thus preventing allergic reactions. Therefore, we have engineered a new method for determining the most antigenic (allergic reaction inducing) epitopes on allergens using allergen mimetics we call nanoallergens. Nanoallergens are liposomal nanoparticles that have been modified to display linear peptide as epitope mimetics in a multivalent fashion and trigger allergic reactions. Using a method developed in our laboratory, we can synthesize nanoallergens to display several epitope peptides, allowing us to systematically determine a single allergen epitope's contribution to the overall antigenicity of an allergen protein. Using nanoallergens, we demonstrate the importance of a single highly antigenic allergy epitope in the major peanut allergen ara h 2. We also present several other, lower antigenic epitopes that are important for overall degranulation response, which are vastly underestimated by more conventional epitope characterization techniques. This novel method is a versatile tool that can be used for any allergen and will likely provide important information for future inhibitor design.

# A Mechanistic Insight into Pollen Grains as an Oral Vaccine Delivery System

Md Jasim Uddin and Harvinder Singh Gill

Department of Chemical Engineering
Texas Tech University
Lubbock, TX
2015 AIChE Annual Meeting
November 8, 2015

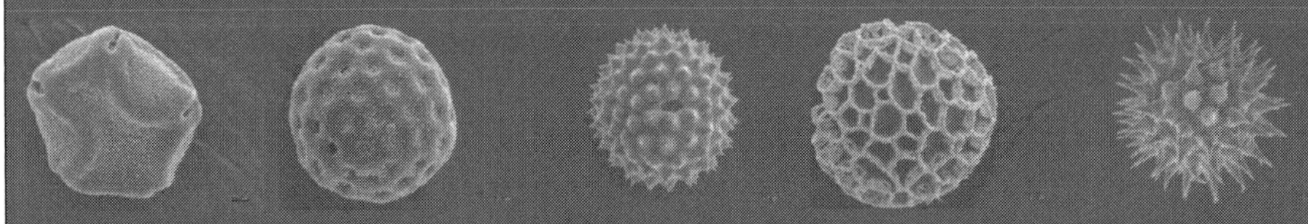

# Mucosal Vaccines: An Unmet Need

- Mucosal surfaces cover 400 $m^2$ area
- 70% of infectious agents enter by mucosal routes
- Mucosal vaccination can induce secretory IgA
- Advantages of an oral vaccine:
  - -Needle free vaccine
  - -Low cost

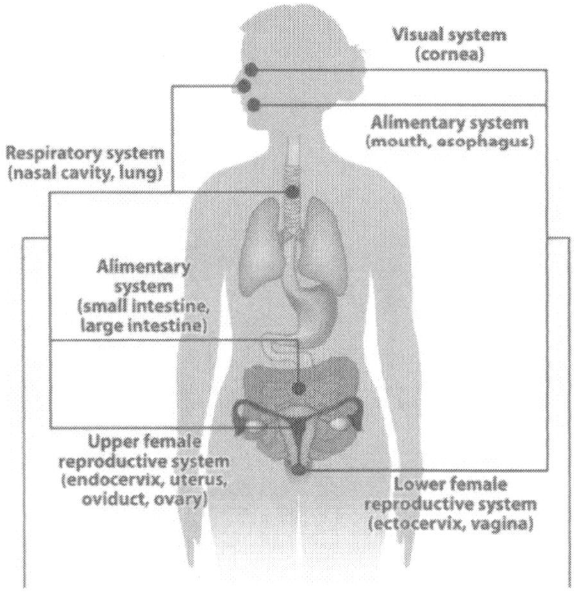


Woodrow et al, (2012) *Annu. Rev. Biomed. Eng.*, 17-46

2

## Research Motivation

❑Oral vaccination requires protection against enzymatic degradation and an adjuvant to enhance immunogenicity

❑Our lab developed pollen grains as novel microcapsules for oral vaccine delivery. In vivo study with pollen grains showed that antigen-specific systemic and mucosal antibodies can be induced.

**How do pollens enhance the immune response?**

Atwe et al, (2014) *Journal of Controlled Release*,45-52
Area Plenary: Bionanotechnology II, November 9, 253B (Salt Palace Convention Center)

3

## Research Questions

1. How do pollens affect permeability of epithelial cells?
2. Are pollens toxic to the cells?
3. What are the interactions between pollen grains and antigen presenting cells (APCs) such as macrophages, dendritic cells?
4. Are pollen capable of activating APCs?

4

# Q1: Effect of Pollen on TJs of Epithelial Cells

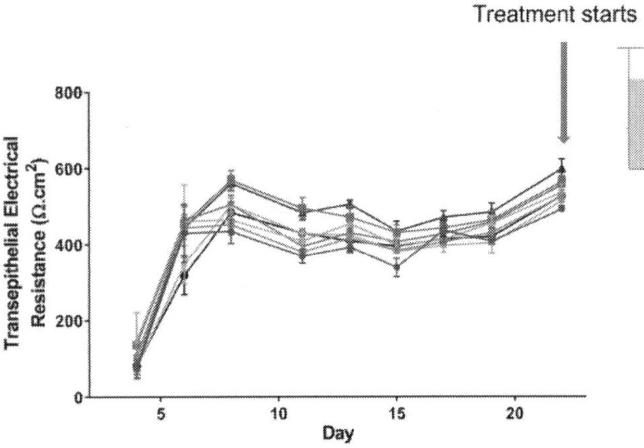


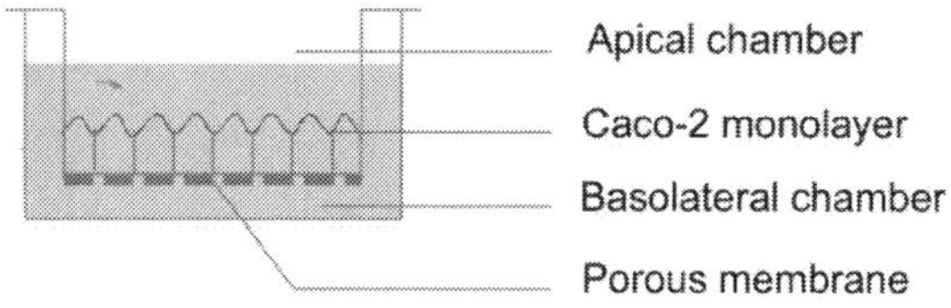


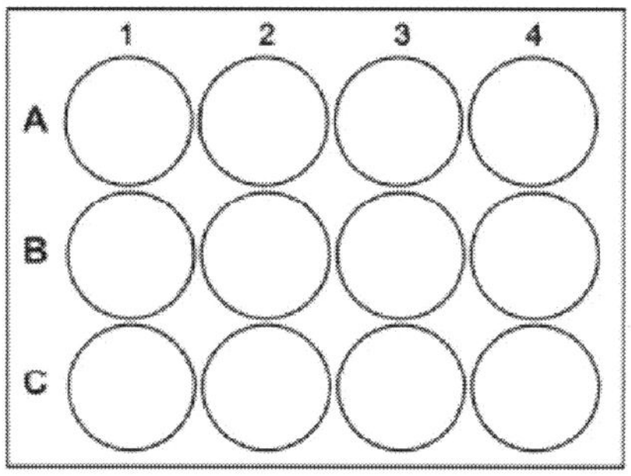


5

# Q1: Effect of Pollen on TJs of Epithelial Cells

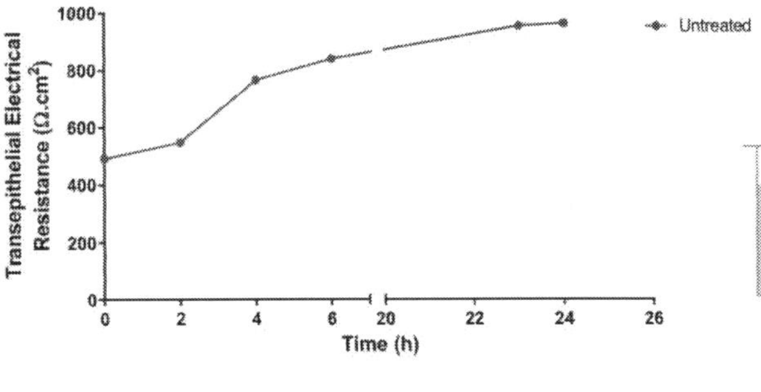


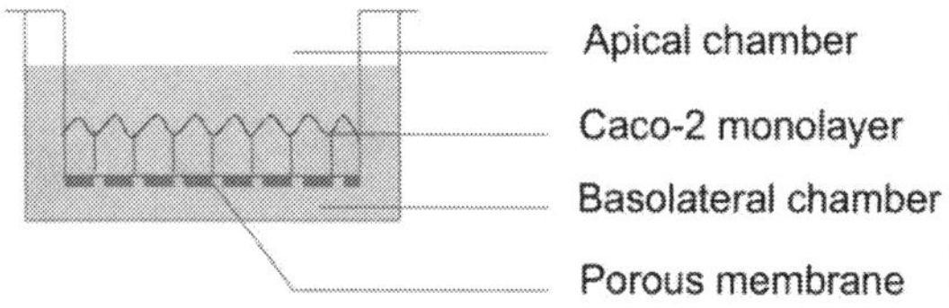


6

# Q1: Effect of Pollen on TJs of Epithelial Cells

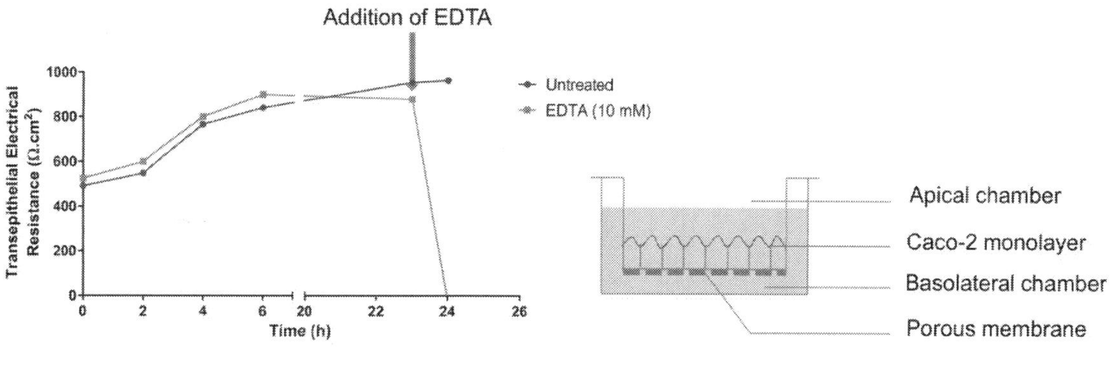


7

# Q1: Effect of Pollen on TJs of Epithelial Cells

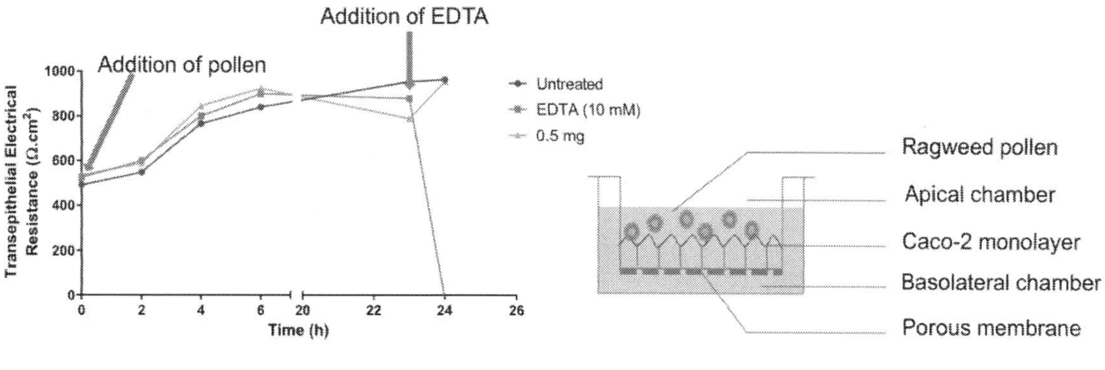


8

# Q1: Effect of Pollen on TJs of Epithelial Cells

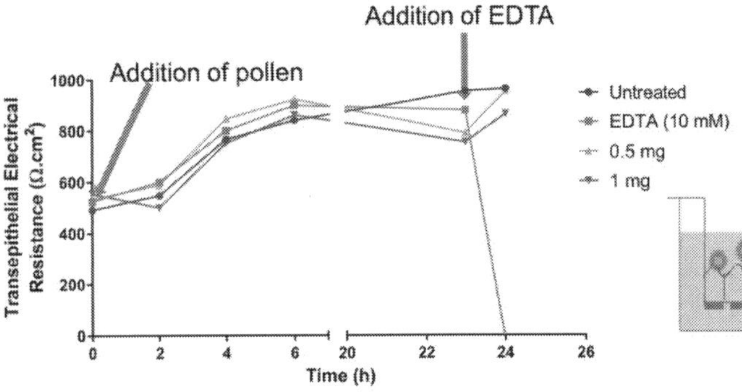


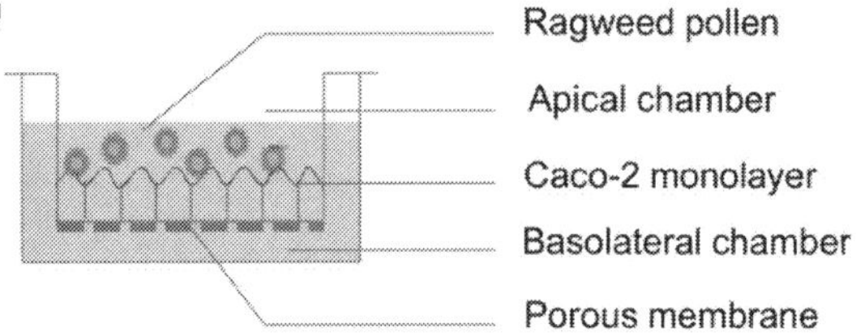


9

# Q1: Effect of Pollen on TJs of Epithelial Cells

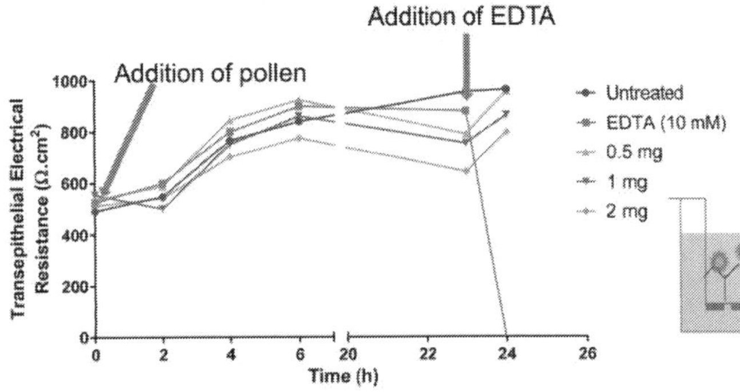


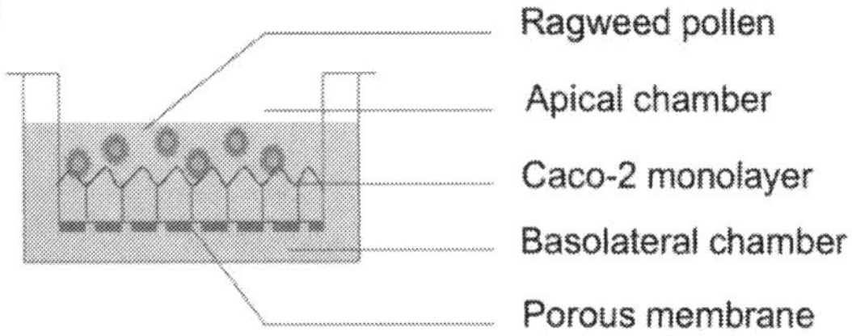


10

# Q1: Effect of Pollen on TJs of Epithelial Cells

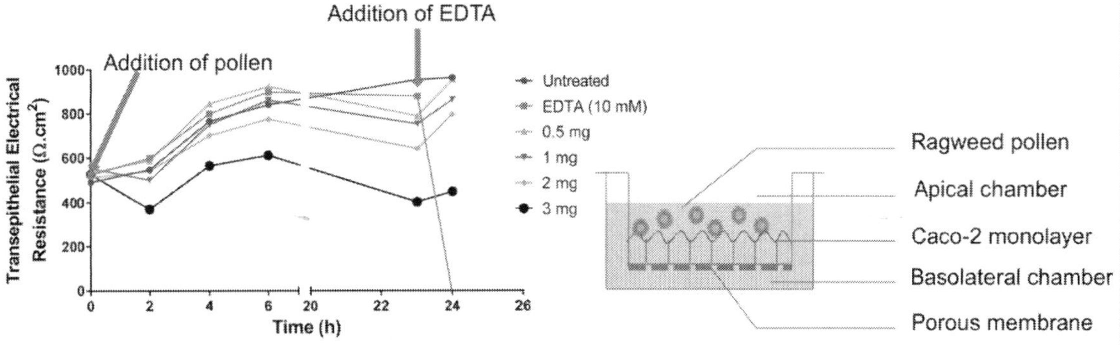


11

# Q1: Effect of Pollen on TJs of Epithelial Cells

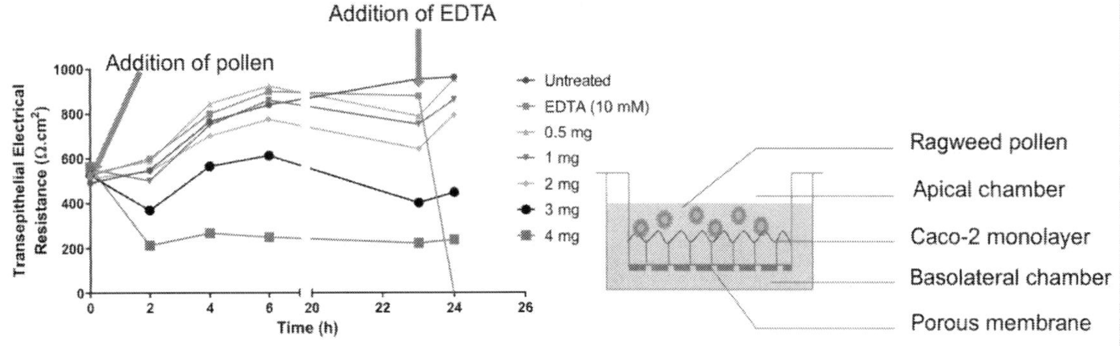


12

# Q1: Effect of Pollen on TJs of Epithelial Cells

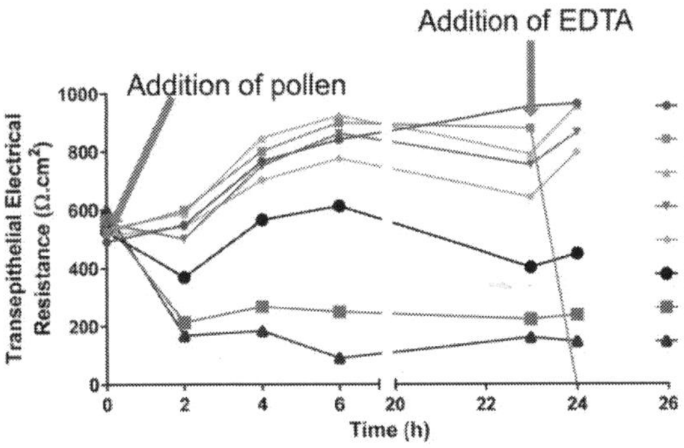


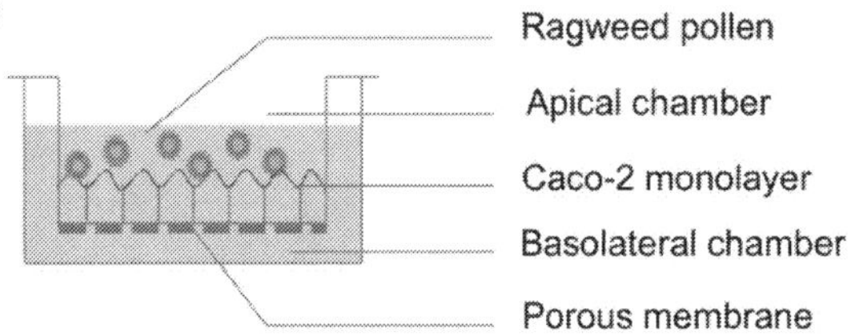


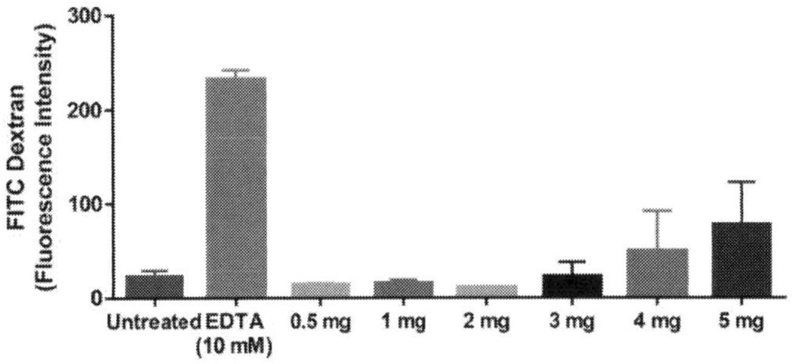


13

# Q2: Cytotoxicity of Pollen

LDH assay used to determine cellular damage caused by pollen

In vivo vs. In vitro:

1. Exposed surface area
2. Concentration
3. Exposure time

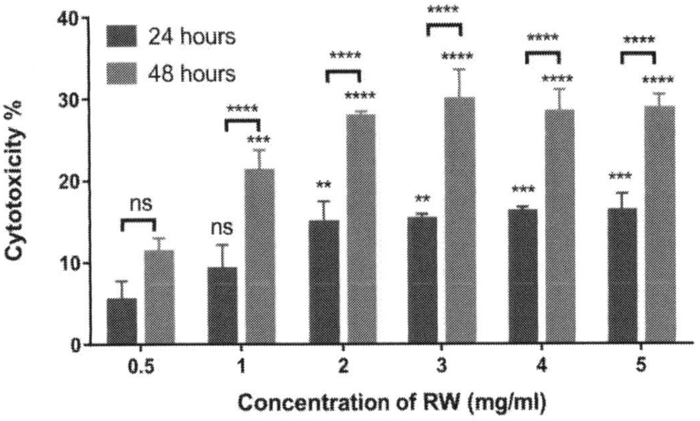


| | |
|---|---|
| ns | $P > 0.05$ |
| * | $P \leq 0.05$ |
| ** | $P \leq 0.01$ |
| *** | $P \leq 0.001$ |
| **** | $P \leq 0.0001$ |

14

## Q3: Pollen Grain Interaction with Macrophage

Lambs Quarter
(22 µm)
+
Macrophage cells

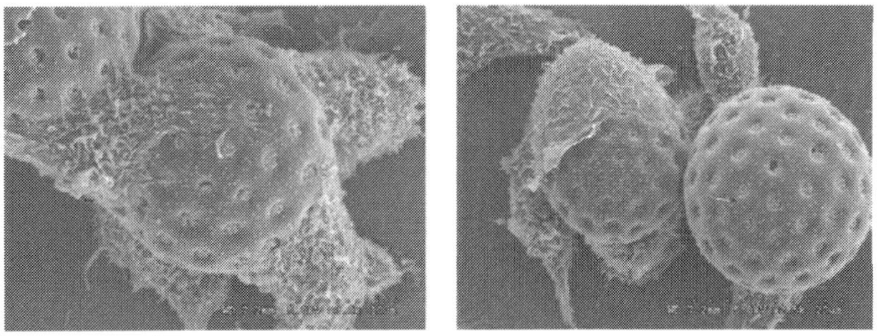

Alder Black
(20 µm)
+
Macrophage cells

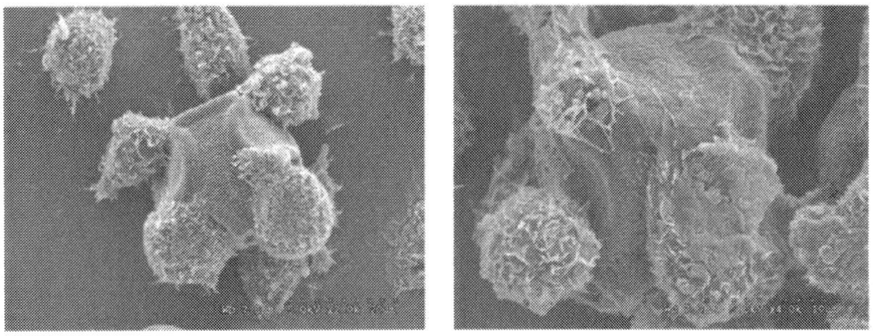

15

## Q3: Pollen Grain Interaction with Macrophage

Sunflower
(30 µm)
+
Macrophage cells

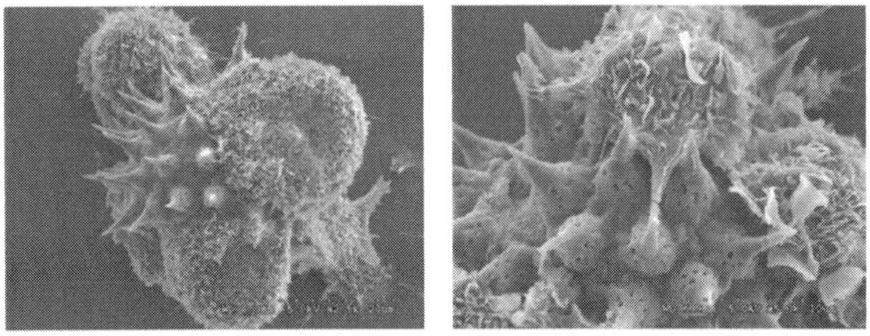

Lycopodium
(25 µm)
+
Macrophage cells

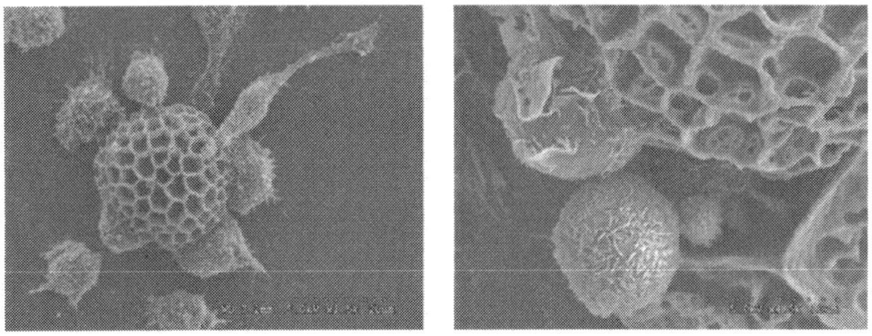

16

## Q3: Pollen Grain Interaction with Macrophage

Ragweed
(15 μm)
+
Macrophage cells

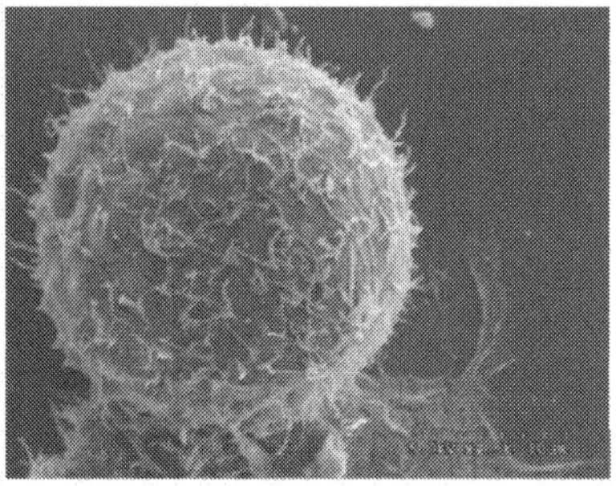

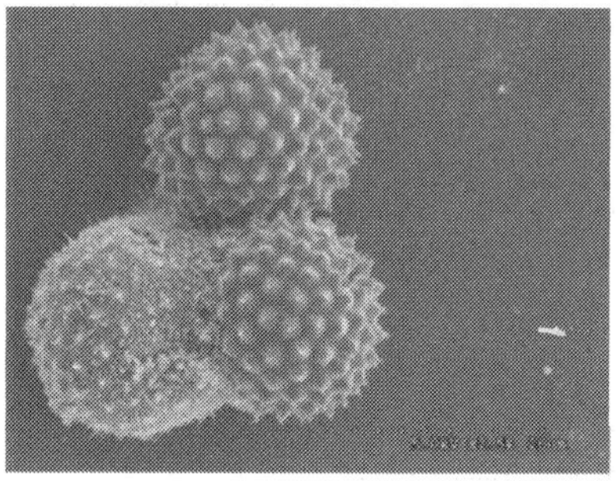

17

## Q3: Pollen Grain Interaction with Macrophage

Ragweed
(15 μm)
+
Macrophage cells

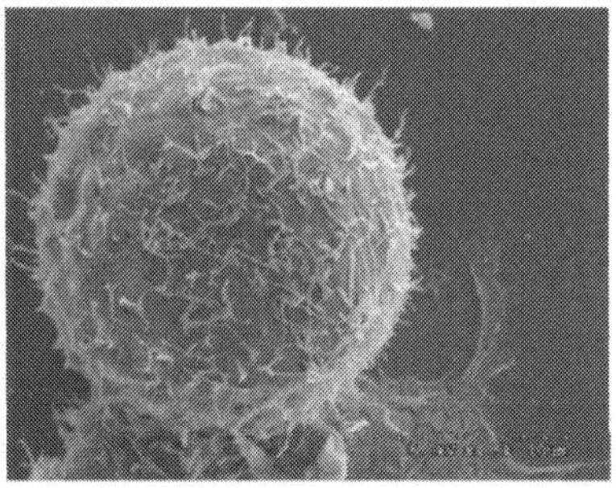

**1.**

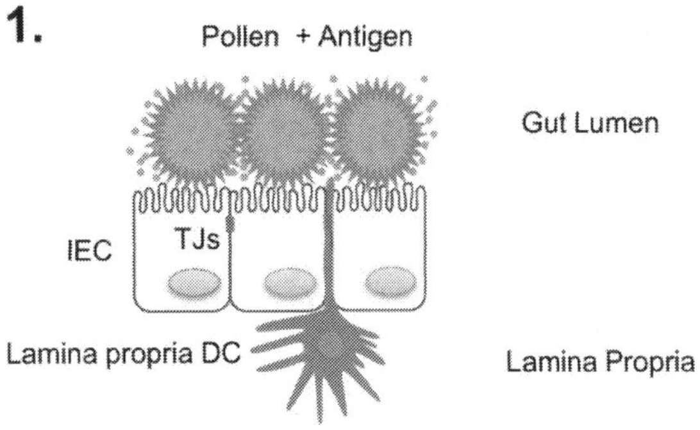


18

# Q3: Pollen Grain Interaction with Macrophage

Ragweed
(15 µm)
+
Macrophage cells

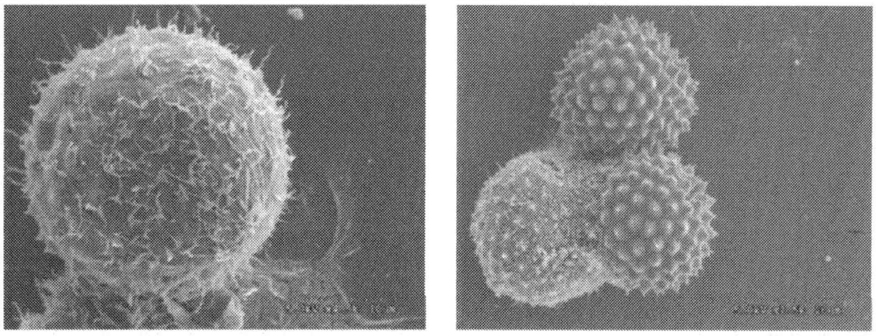

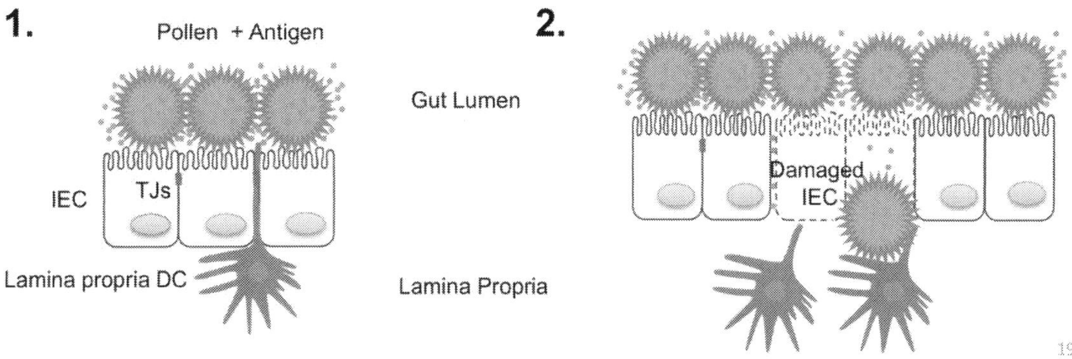


19

# Q3: Cytokines

## BMDC Cytokines

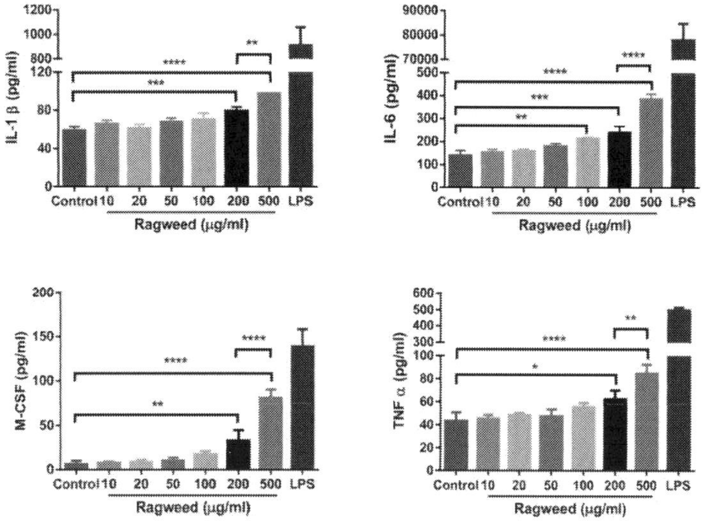


20

# Q3: Cytokines

BMM Cytokines

BMDC Cytokines

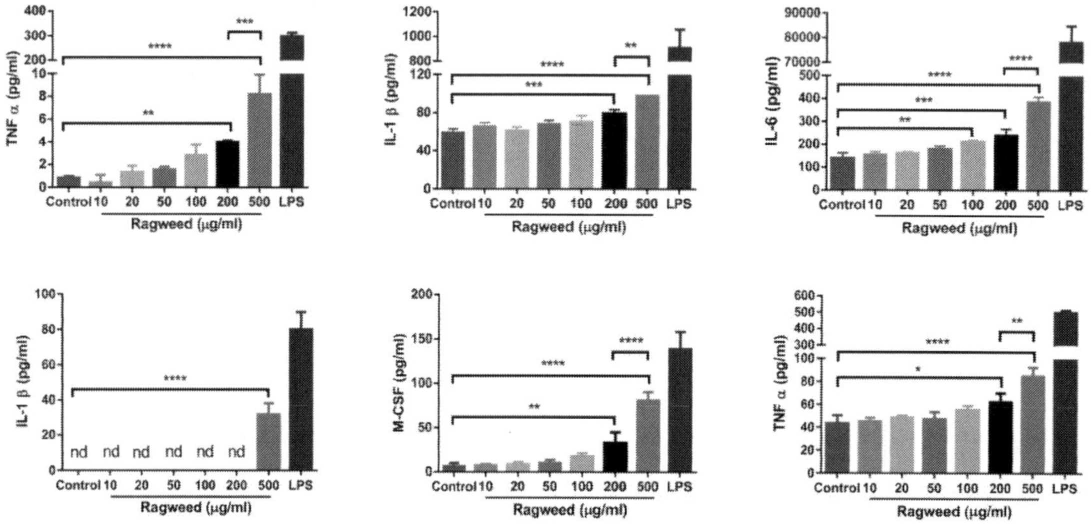


21

# Q4: BMDC Activation by Pollen

CD 40 CD 80 CD 86 MHC Class II

Untreated BMDCs

LPS treated BMDCs

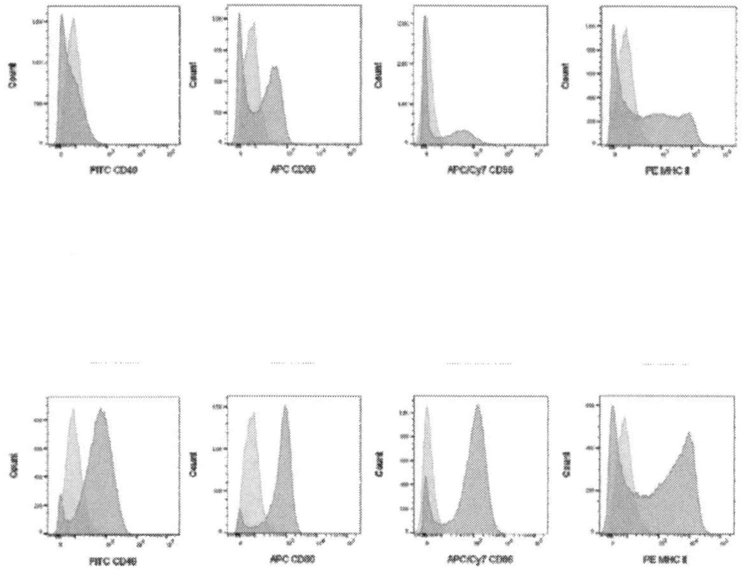

Antibody stained cells

Unstained cells

22

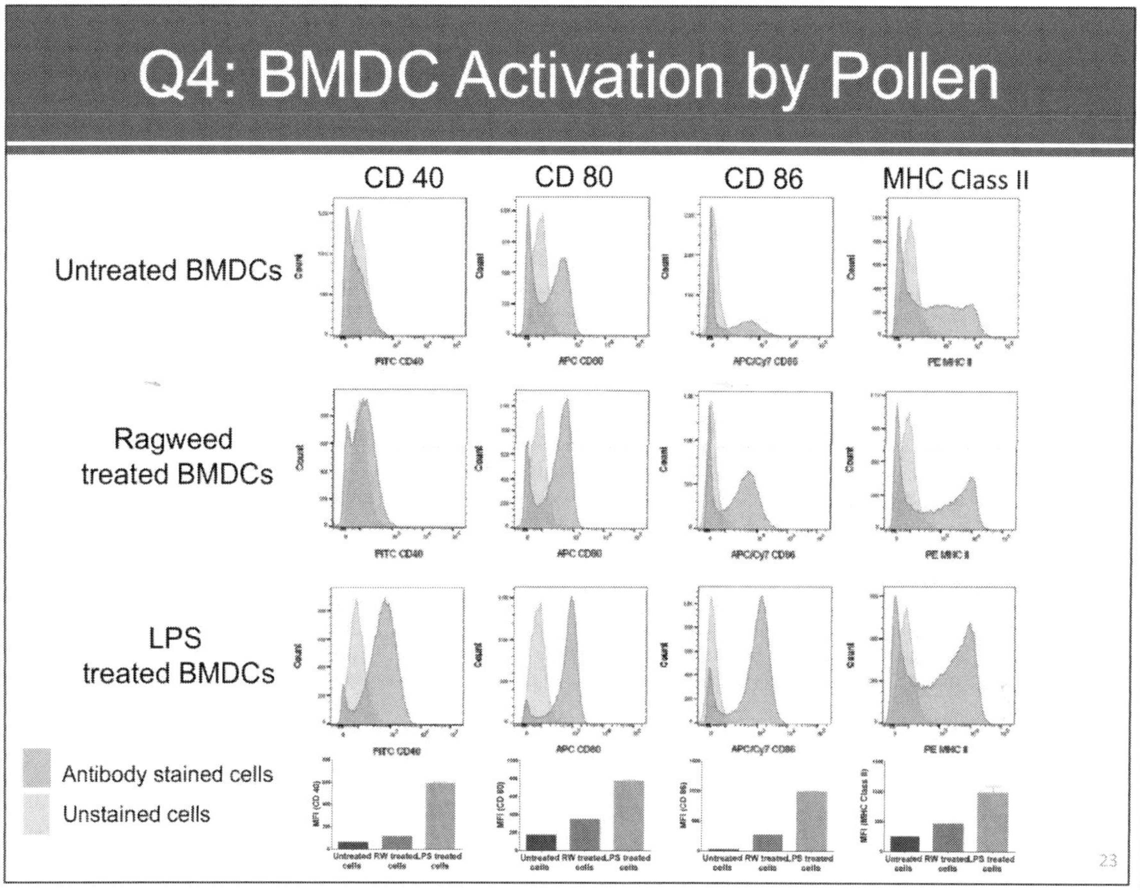
Q4: BMDC Activation by Pollen
CD 40
CD 80
CD 86
MHC Class II
Untreated BMDCs
Ragweed treated BMDCs
LPS treated BMDCs
Antibody stained cells
Unstained cells
23

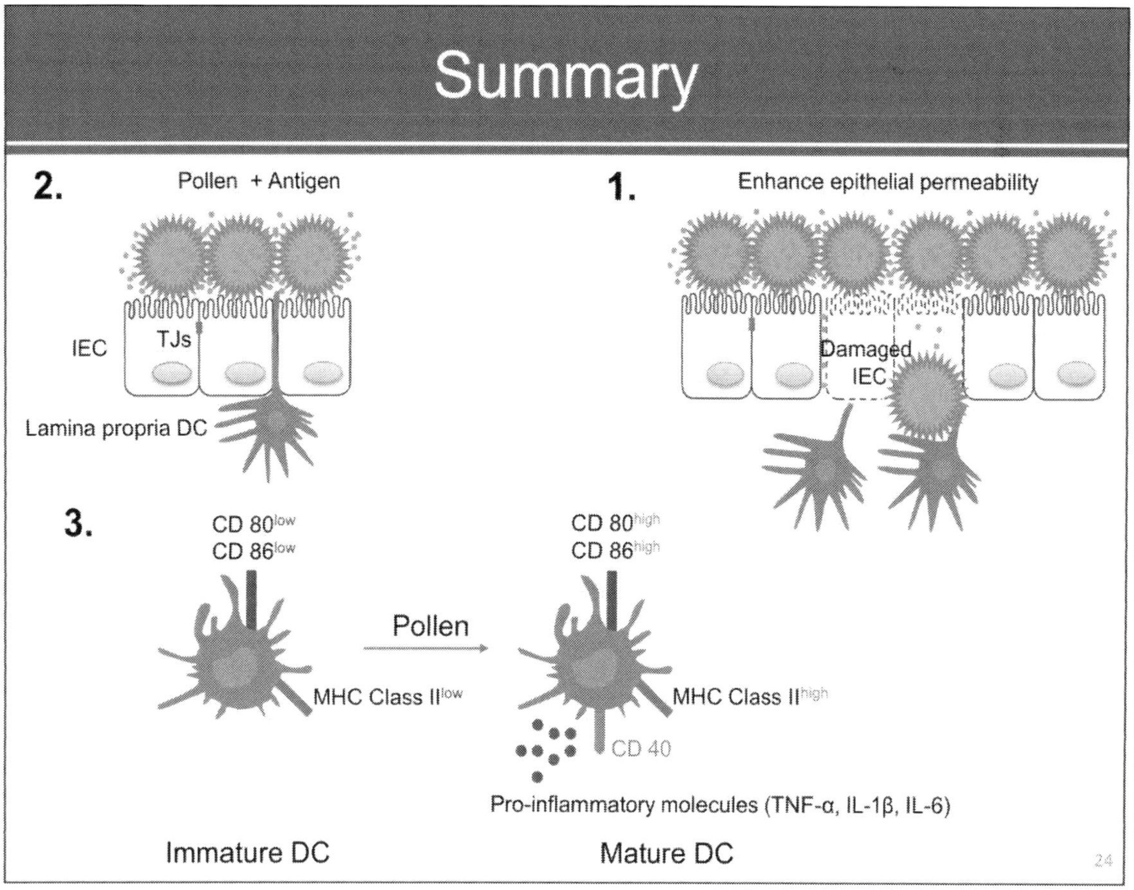
Summary
2.
Pollen + Antigen
IEC
TJs
Lamina propria DC
1.
Enhance epithelial permeability
Damaged IEC
3.
CD 80low
CD 86low
Pollen
MHC Class IIlow
CD 80high
CD 86high
MHC Class IIhigh
CD 40
Pro-inflammatory molecules (TNF-α, IL-1β, IL-6)
Immature DC
Mature DC
24

# Acknowledgement


**GRADUATE STUDENTS**

1. Shashwati Atwe
2. Seth Boese
3. Rohan Ingrole
4. Pedro Gonzalez

**POSTDOCS**

1. Akhilesh Kumar
2. Chang-Hyun Lee
3. Amit Jain
4. Shantanu Lale
5. Mohammed Chowdhury

**COLLABORATORS**

Dr. Silke Paust
Pediatrics and Center for Human Immunobiology, Texas Children's Hospital and Baylor College of Medicine, Houston, TX

Dr. Nitin Nitin
Food Science and Technology & Biomedical Eng, UC-Davis, CA

Dr. Nancy Haigwood
(Director: Oregon National Primate Research Center)

Dr. Jatindra Tripathy
Center for Biotechnology and Genomics, Texas Tech University

**FUNDING SOURCES**

NIH DIRECTOR'S NEW INNOVATOR AWARD

DARPA Young Faculty Award

Eunice Kennedy Shriver NICHD
National Institute of Child Health & Human Development

National Institute of Dental and Craniofacial Research

National Institute of Allergy and Infectious Diseases

CPRIT
CANCER PREVENTION & RESEARCH INSTITUTE OF TEXAS


# Thank You
# Questions?

26

# Effect of Pollen on TJs of Epithelial Cells

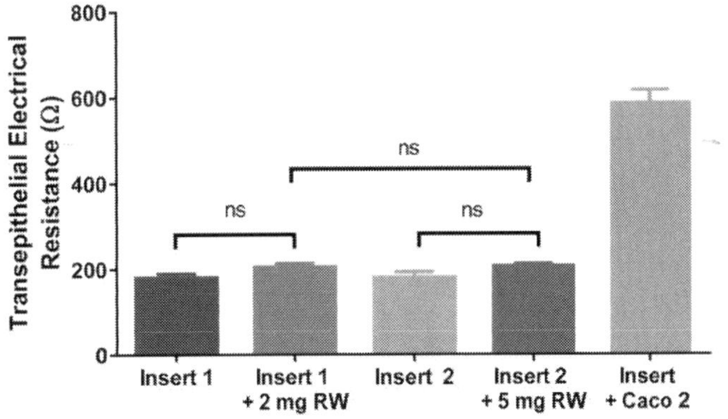


ns P > 0.05

Different concentrations of pollen has no significant effect on insert resistance

27

# Cytotoxicity of Pollen

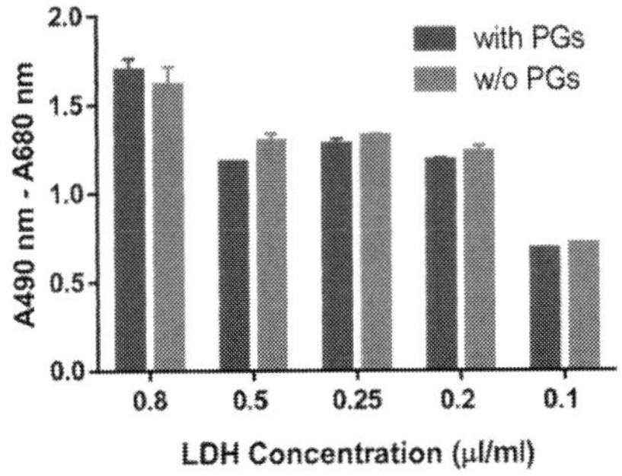


LDH Assay

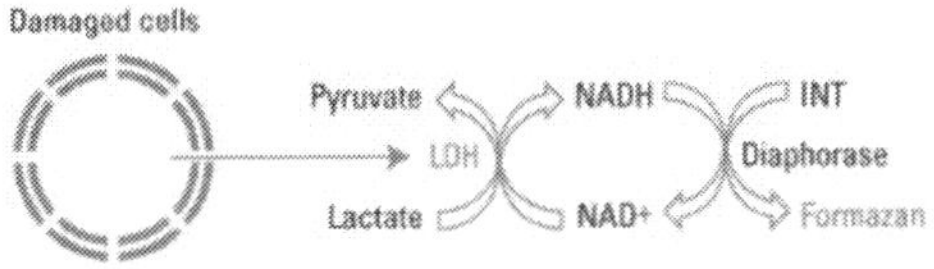


$$\text{Cytotoxicity (\%)} = \frac{LDH_{\text{PG treated cells}} - LDH_{\text{cells only}}}{LDH_{\text{lysed cells}} - LDH_{\text{cells only}}} \times 100$$

Surface area of one well in 96 well plate: 0.32 $cm^2$

The mean surface area of the intestinal tract: 1.41 $m^2$

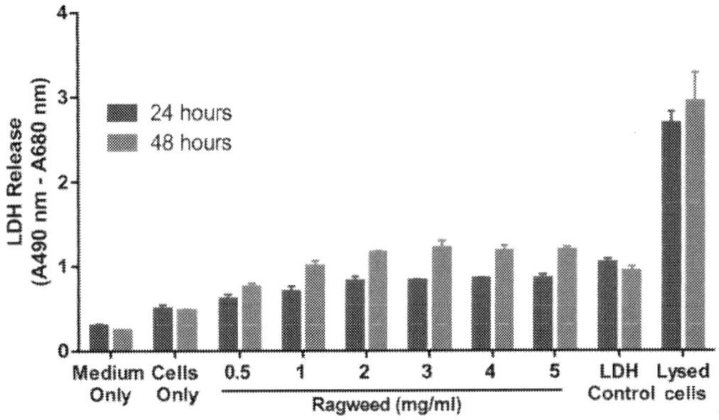


28

# BMDC Activation by Pollen

➢ Pollens can be differentiate from fluorescent cells by their distinct side scattering properties

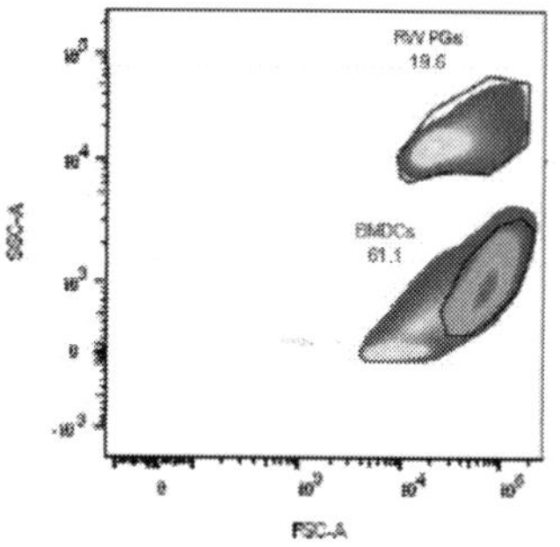


➢ 93% of the cultured cells are DCs (CD11c$^{+}$)

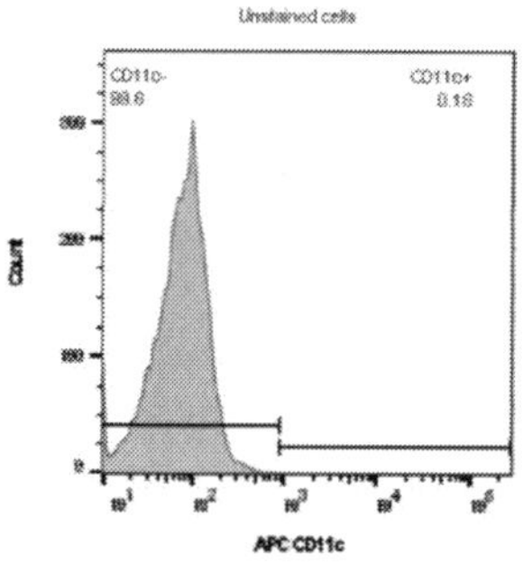


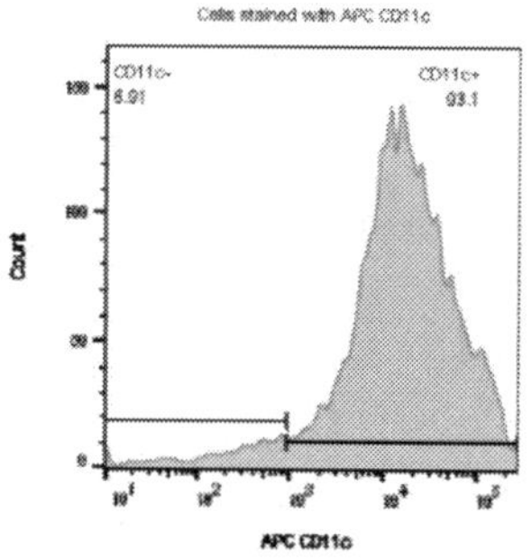


29

# Pollen Grain Interaction with Macrophage

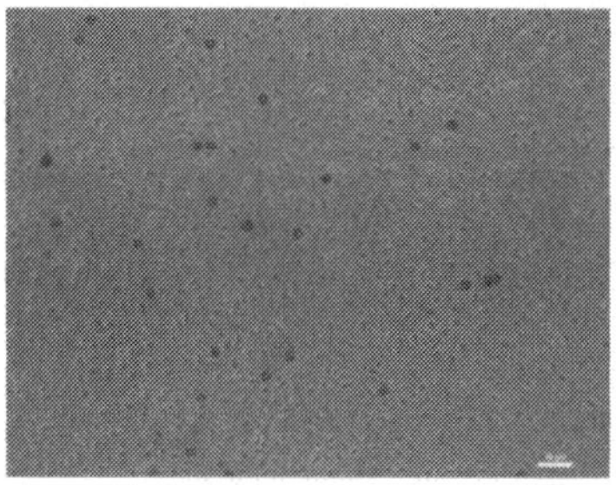

10 μg/ml

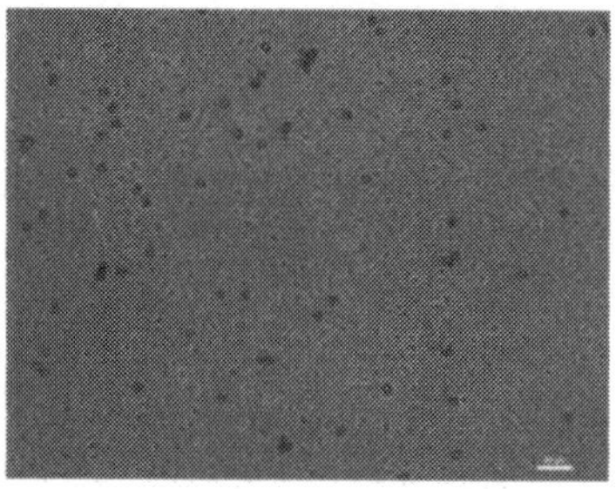

20 μg/ml

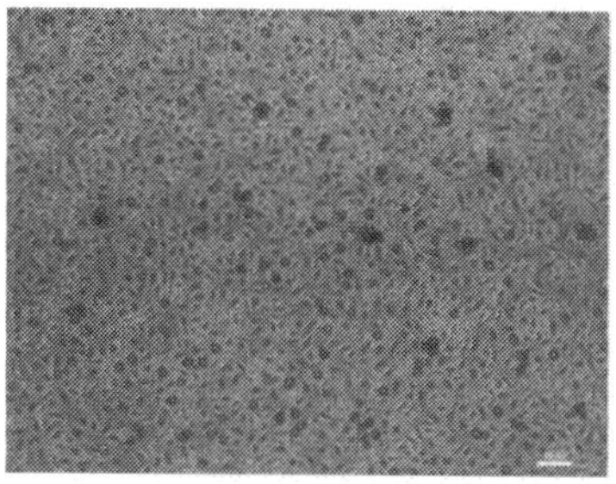

50 μg/ml

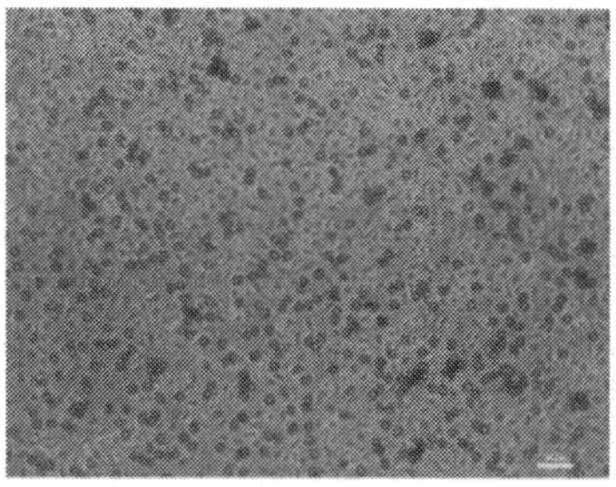

100 μg/ml

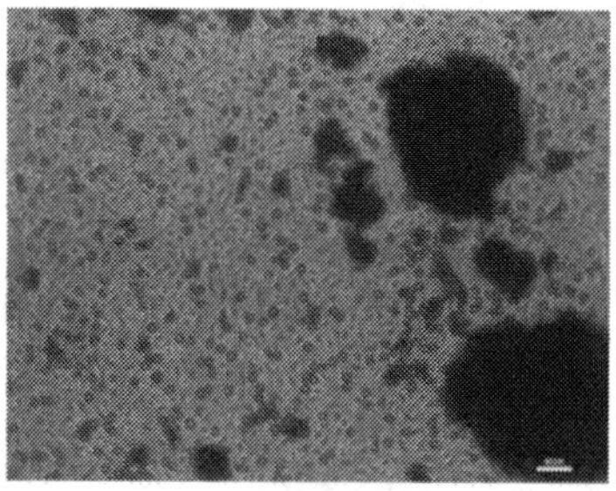

200 μg/ml

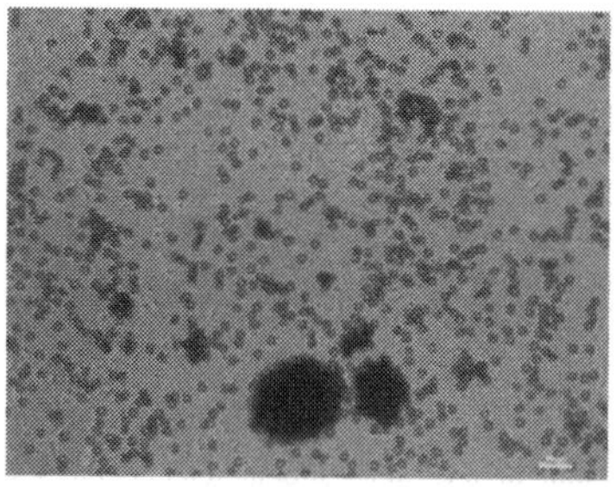

500 μg/ml

30

# Pollen Grain Interaction with Macrophage

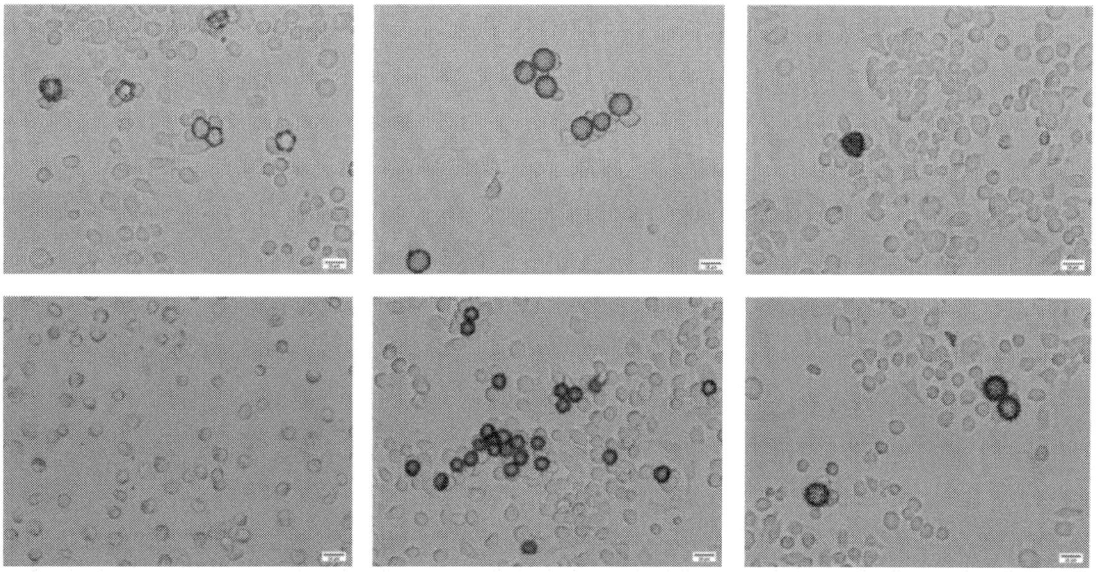

31

# BMDC Activation by Pollen

Aseptically isolate bone marrow (BM) from both hind leg tibia and femur → Culture BM progenitor cells with GM CSF and IL-4 for 7 days → BMDCs culture with RW PGs for cytokine and flow cytometry study

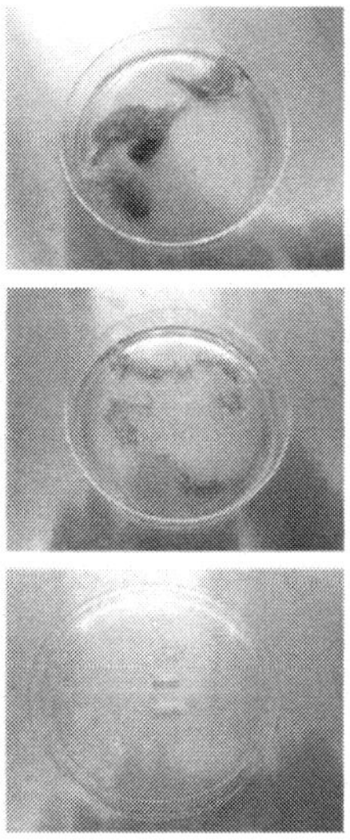

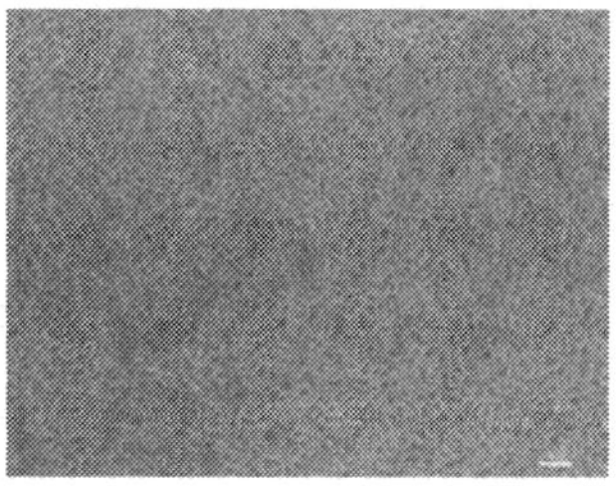

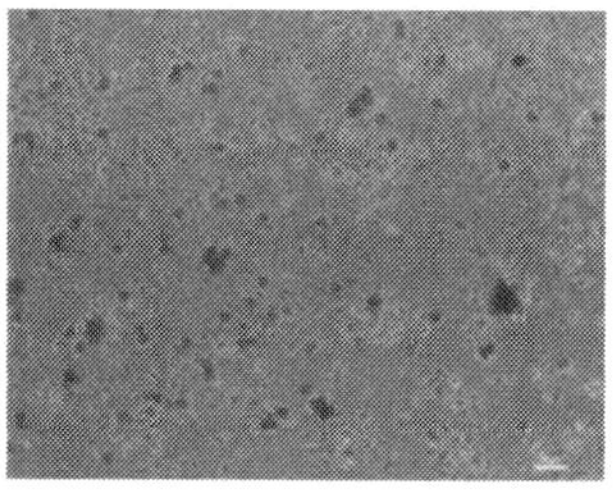

50 ng/ml LPS (*E coli* 0111:B4) used as a positive control

32

# SEM Images of Pollen Grain

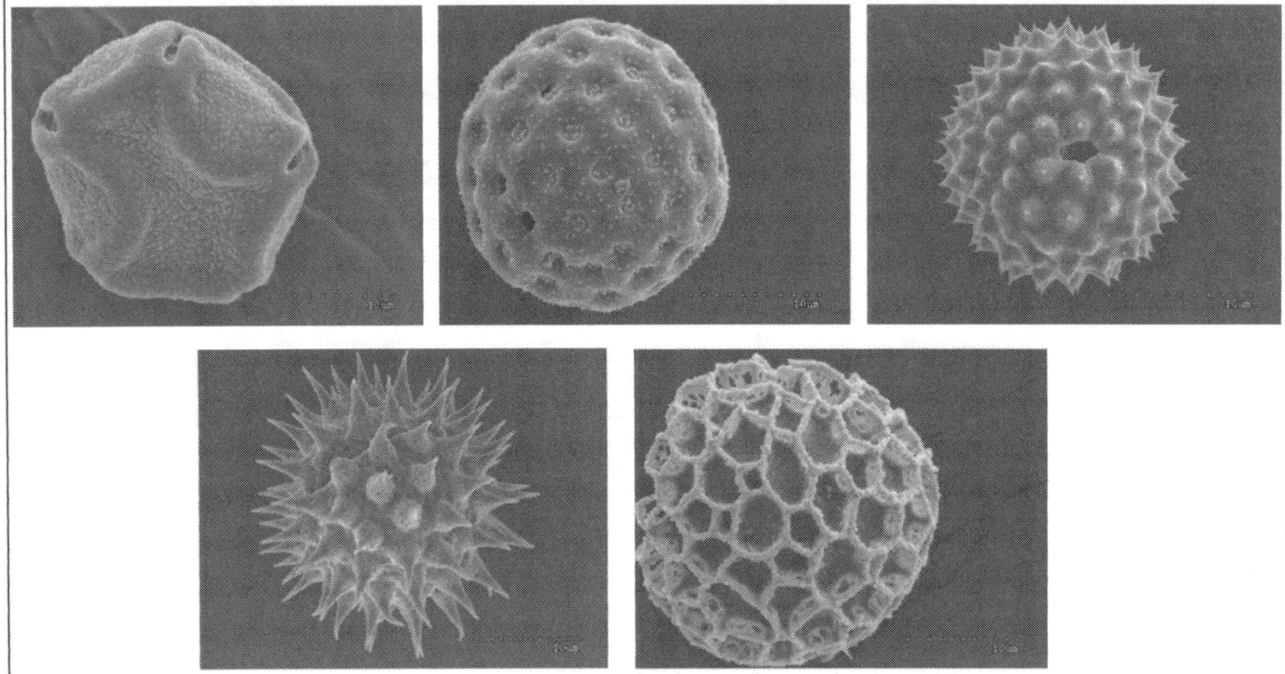

33

# 435256 Transport of Antigen from Administration Site to the Draining Lymph Node in Nanoparticle Formulations

**Abstract:**

Nanoparticles have been shown to enhance the immune response to subunit protein antigens for efficacious vaccinations. The role of particle size in transport from the site of administration to the draining lymph node has been widely investigated. Studies have consistently shown that smaller sized particles, ≤25 nm, are rapidly transported to the draining lymph node, while larger sized particles, ≥100 nm, are transported more slowly. However, the impact of the polymer chemistry and the initial particle size on the transport of these formulations to the lymph node and on the corresponding immune response has not been studied in detail.

Our group has investigated the use of subunit antigens encapsulated into polyanhydride nanoparticles as a vehicle for vaccine delivery and as an adjuvant for the immune response. This has been demonstrated through the effectiveness of a single intranasal administration of a nanoparticle formulation containing soluble and encapsulated *Yersinia pestis* fusion proteins, F1-V, which was protective against a lethal challenge.

In this work we investigate the role of polyanhydride nanoparticle chemistry on the transport of antigen to the lymph node distal from the administration site. The kinetics of soluble and encapsulated antigen delivery to the lymph node and the cellular context were investigated. The role of polymer chemistry on antigen presenting cell internalization and subsequent transport to and localization within the distal lymph node was investigated using a combination of flow cytometry and confocal microscopy. Our results will provide insights into the design of antigen carrier devices needed to impart an efficacious vaccine formulation, and the potential to direct the immune response toward either antibody-mediated or cell-mediated immunity.

# 431298 Simulating Lymphocytes Dynamics within the Lymph Node Towards Complex Cellular Distribution Patterns

**Abstract:**

The mechanisms that guide the migration of lymphocytes and dendritic cells towards specific locations within the lymph node are not well understood. Experimental evidence support both the cellular haptotaxis, induced by chemokine gradients, and the random-walk-like behavior of trafficking cells. A multi-scale simulation approach, including random-walk and haptotactic cellular migration of lymphocytes and dendritic cells, is proposed. A Galerkin finite element simulation method is used to evolve the continuum dynamics, whereas Brownian dynamics are used to study cell activation of discrete coarse-grain cell models. Cell migration and activation within the lymph node are studied by comparing the relevant diffusive mechanisms. In the continuum scale, non-uniform mobility coefficients and chemo-attractive diffusion are included in the species balance. On the other hand, first order kinetics are included in Brownian dynamics to study cell activation. The proposed approach serves to understand complex dynamics of immune cell activation, including the inflammatory response and complement activation.

# 431041 Cadherin-11 Regulates Mechanical Properties of Tissues and Collagen and Elastin Synthesis in-Vivo and in-Vitro

**Abstract:**

Recently our laboratory identified cell-cell adhesion as an important factor determining mesenchymal stem cell (MSC) differentiation. More specifically, we discovered that Cadherin-11 (CDH11) was necessary for MSC differentiation into smooth muscle cells (SMC) through activation of key myogenic transcription factors such as serum response factor (SRF) via the ROCK pathway. Others, showed that differentiation of fibroblast to myofibroblasts during wound healing and production of extracellular matrix (ECM) synthesis that was associated with a shift in cadherin profile from cadherin-2 (CDH2) to CDH11. However, the role of CDH11 in ECM production and deposition is unknown. This is the focus of the present study.

Tissues collected from CDH11 null mice (Cdh11-/-) and wide type (WT) such as aorta, bladder and skin were tested and compared for their mechanical properties. Human dermal fibroblasts and mice dermal fibroblasts were isolated and subjected to experiments. Western blot and real time RT-PCR were employed to measure the level of protein and gene expression, including the main ECM components as collagen and elastin and some relevant transcriptional factors. An engineered surface with immobilized the fusion protein cadherin-11 with the Fc-domain (Cad-Fc) was used as a tool of CDH11 engagement without the complications of high cell densities.

Surprisingly, we observed a significant reduction in mechanical properties of aorta, bladder and skin of Cdh11-/- as compared to WT tissues. The change in mechanical properties was accompanied by reduced collagen and elastin content. At a cellular level, RT-PCR and western blot analyses both confirmed an attenuated expression of collagens and elastin in Cdh11-/- dermal fibroblasts compared to WT. Further, knocking down CDH11 in human dermal fibroblasts significantly reduced the ability of cells to generate force, as evidenced by the extent of hydrogel compaction. In contrast, engagement of CDH11 on CDH11-Fc surface significantly promoted collagen and elastin synthesis, which was accompanied by significant upregulation of key transcriptional factors such as MYOCD, SRF, MRTF-A/B. Surprisingly, CDH11 engagement promoted nuclear localization of MRTF-A, a key event for subsequent transcription. Interestingly, knocking down anyone of these transcriptional factors down-regulated the ECM synthesis. Further experiments demonstrated that both the TGF-β and Rho/ROCK pathways mediated the effects of CDH11 engagement on collagen and elastin synthesis.

Our study identified CDH11 as a critical player in mediating collagen and elastin synthesis both in vivo and in vitro. Engagement of CDH11 enhanced expression of transcriptional factors MRTFA/B, SRF and MYOCD through activation of TGF-β and ROCK pathways. Nuclear localization of MRTF-A was a key event following CDH11 contact. Taken together, our results demonstrate an important role of CDH11-mediated adherens junctions in ECM synthesis with implications in mechanical properties of tissues and wound healing.

# 428551 Intercellular Forces Regulate Mitosis and Migration within Epithelial Tissues

**Abstract:**

**Introduction:** Cell-cell and cell-matrix interactions are crucial for proper embryonic development and tissue homeostasis. The forces generated by these adhesive interactions often transduce mechanical stimuli into changes in intracellular signaling and gene expression. In addition, the spatial distribution of mechanical stress, dictated by the geometry of the tissue and requiring intercellular adhesion, has been shown to define patterns of proliferation within multicellular tissues. However, it is unclear how forces generated locally by individual cells influence this collective behavior. Here, we examined how local cell-cell mechanical communication influences proliferation within an epithelial tissue.

**Materials and Methods:** To increase intracellular tension within individual cells, a photoactivatable RhoA construct was expressed in mouse mammary epithelial tissues cultured on microcontact printed islands of fibronectin. Blue light activates this optogenetic form of RhoA, which leads to a concomitant increase in actomyosin contractility. A control construct lacking the RhoA coding sequence was used for comparison. Timelapse imaging enabled real-time visualization of cellular responses to local activation of RhoA in individual cells within multicellular tissues.

**Results and Discussion:** Optogenetic stimulation of RhoA within epithelial cells led to increased actin polymerization and decreased projected cell area, consistent with activation of the downstream effectors of RhoA. Two-dimensional traction force microscopy confirmed that optogenetic stimulation increased contractility and force generation, compared to control cells. Interestingly, we found that activating RhoA in a subset of cells within the epithelial tissues controlled proliferation of neighboring cells, as determined by timelapse imaging and quantification of cell division frequency. This pattern of proliferation was distinct from the pattern of cell motility within the epithelial tissue, which was also affected by intercellular force transmission. These data suggest that distinct signaling pathways downstream of adhesion forces lead to different emergent phenotypes.

# 428528 Role of the Fibronectin Synergy Site in Modulating Integrin Recruitment and Molecular-Scale Force Generation

**Abstract:**

Integrins are heterodimeric transmembrane adhesion receptors that connect the cell to the extracellular matrix, and that play central roles in transducing mechanical cues into intracellular signaling cascades that control cell migration, differentiation, and survival. Mammalian cells possess at least 26 integrin heterodimers. In most cases the molecular mechanisms by which different integrins fulfill their distinct physiological roles remain poorly understood. In particular, relatively little is known about how different integrin heterodimers may function in the context of mechanotransduction.

Here, we use single-molecule fluorescent molecular tension sensors (MTSs) to measure the forces experienced by $\alpha_5\beta_1$ and $\alpha_v$ -class integrins, two classes of fibronectin-binding integrins that are thought to play distinct roles in cell adhesion and mechanotransduction. We developed MTSs that incorporate the full 9th and 10th type III fibronectin domains, here termed $MTS_{FN9-10}$. Importantly, these probes include a secondary binding site, termed the synergy site, that is specifically required for maximal binding strength and activation of $\alpha_5\beta_1$ but not $\alpha_v$-containing integrins. We observe that fibroblasts seeded on $MTS_{FN9-10}$ form large, plaque-like adhesions rich in $\alpha_5\beta_1$ integrins, a phenotype that is lost in the absence of a functional synergy site. At the single-molecule level, preliminary data indicate that the force per integrin lies between 1 to 5 pN for cells adhering to $MTS_{FN9-10}$-coated surfaces, a range similar to that observed for cells adhering to MTSs containing only a linear RGD binding epitope. In ongoing work, we use $MTS_{FN9-10}$ and related probes to parse out the contributions of the synergy site to traction force generation and force sensing at both $\alpha_5\beta_1$ and $\alpha_v$-containing integrin complexes.

# 423005 Uncovering Mechanisms That Result in Diverse Contact Guidance Behavior Amongst Cancer Cells

**Abstract:**

Cancer cell invasion is driven by migration. Often this migration is directed by aligned collagen fibers, a process called contact guidance. Because of the strong link between directed migration and aligned collagen fiber orientation, extracellular matrix alignment has been suggested as a measurable to determine invasion potential. We have recently shown that different breast cancer cells respond to contact guidance cues differently. MDA-MB-231 cells, which migrate with a mesenchymal mode, sense contact guidance cues with high fidelity, but MTLn3 cells, which migrate with an amoeboid mode, do not sense contact guidance cues. Several studies have shown that modulating contractility and adhesion through Rho-family GTPases and integrin activation allows cells to switch between migrational modes. We were interested if this plasticity in migrational mode affects contact guidance.

In this study, we have used several different approaches for generating contact guidance cues including microcontact printing, soft lithography and epitaxial collagen growth to generate lines, ridges and aligned fibrils of collagen. In addition, we have used pharmacological inhibitors and genetic approaches to examine the role of Rac and Rho GTPases as well as the activation of b1 integrins in modulating contact guidance through focal adhesions and the cytoskeleton. Decreasing contractility by inhibiting Rho-kinase (Y-27632) or partially blocking b1 integrin activation (P5D2 antibody) in MDA-MB-231 cells diminished contact guidance fidelity on epitaxially growth collagen fibrils. One the other hand, increasing contractility (calyculin A (CA)) or activating integrins ($MnCl_2$) increased contact guidance fidelity in MTLn3 on epitaxially growth collagen fibrils. Interestingly, speed was not altered. These changes in motility were explained by changes in F-actin stress fiber density and orientation, myosin light chain phosphorylation and focal adhesion dynamics and maturation. This suggests that the assessment of the cell's ability to generate and transmit force in combination with the organization of the extracellular matrix are important in determining whether contact guidance occurs leading to cancer cell invasion.

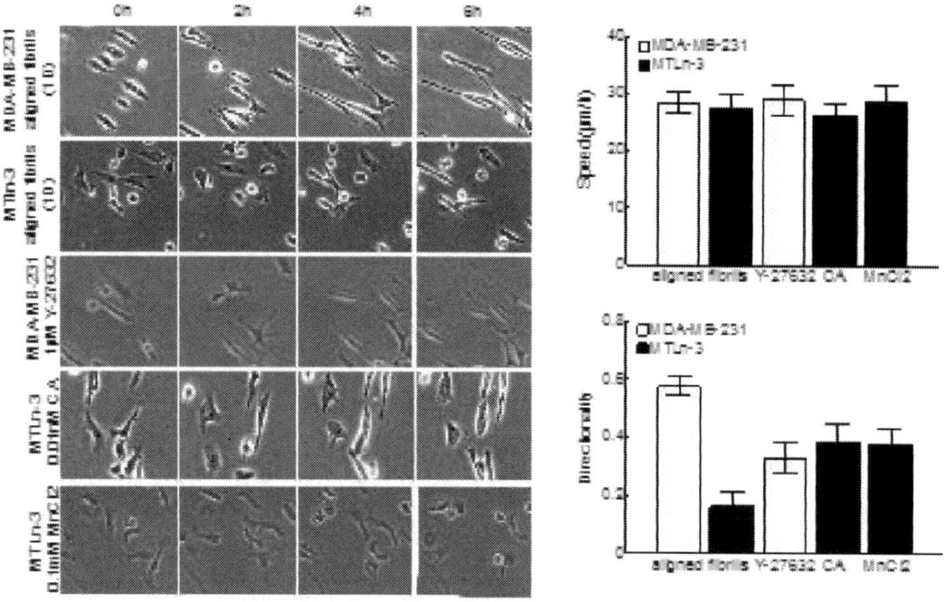


**Figure1: Contact guidance efficiency is altered through contractility and adhesion.** Cells were imaged over time using phase contrast microscopy in the presence of contractility inhibitors (Y-27632), contractility enhancers (calyculin A (CA)) and integrin activators ($MnCl_2$).

# 431989 Extracellular Matrix Control of Metastasis and Dormancy

**Abstract:**

Breast cancer specifically spreads to the bone, brain, liver, and lung, and this phenomenon, tissue tropism, is not well understood. Paget's classical seed and soil hypothesis states that specific interactions between the metastatic cell (the seed) and the recipient tissue microenvironment (the soil) mediate these non-random patterns of spread in breast cancer. We hypothesize that the ability of metastatic cells to survive dormancy, exit quiescence, and colonize a specific tissue depends both upon the ability of the ***seed*** to survive long-term growth arrest, and the ability of the ***soil*** to first sustain survival, and subsequently trigger outgrowth. To this end, we created a biomaterial platform that varies extracellular matrix (ECM) density and composition, and used cell adhesion and migration assays to study how highly metastatic cells differentiate between different tissues. In our recently published paper, we used this biomaterial system to screen the adhesion and motility of cell lines on ECM surfaces. When we compiled this data using a systems biology-like approach, we could predict *in vivo* metastasis with 88% accuracy using a quick screen of adhesion and motility *in vitro*[1].

With this relationship in hand, we are now using this same platform to predict long-term relapse, i.e. dormancy, in very slow growing cell lines. We have found that both integrin binding to the ECM and growth factor supplementation mediate dormancy in cell lines known to display long periods of metastatic latency *in vivo*. We first induce a dormant-like phenotype by culturing cell lines in serum-free medium, and surviving quiescent cells can then be stimulated to grow with serum. We uniformly observe that a laminin-coupled ECM is unable to permit survival of dormant cells, whereas binding to collagen I maintains dormancy. Increased cell density and growth factor supplementation also increase long-term dormancy, but this is a cell type-specific effect that is surprisingly uncorrelated with receptor expression. Specifically, the HCC 1428 cell line, with high epidermal growth factor receptor (EGFR) expression, was unable to maintain a dormant state beyond 7 days, even with epidermal growth factor (EGF) supplementation. In contrast, EGF maintains dormancy of a greater number of ZR-75-1 cells, despite their low EGFR expression. Upon recovery, cells that have been cultured for 1-3 months in serum-free medium proliferate more quickly, are more drug resistant, and are more homogeneously responsive to growth factors than the parental population, suggesting this selective pressure creates an aggressive subpopulation. We are currently investigating the downstream signaling responsible for long-term survival on collagen, and whether growth factor-enhanced survival results from the physical association of growth factor receptors and integrins or changes in receptor expression over time. In sum, we have used a simple biomaterial to understand the roles of integrin binding to the ECM and growth factor stimulation in mediating long-term dormancy, providing insight into combinatorial targets to prevent late breast cancer relapse.

**References:** 1. Barney et al., *Integrative Biology*, 2015.

# 431873 Dynamic Biochemical Tissue Analysis Detects Functional Selectin Ligands on Colon Cancer Tissues

**Abstract:**

A growing body of evidence suggests that the selectin family of cell adhesion molecules, comprised of E-, P-, and L-selectin, engage ligands presented on colon cancer cells to mediate the adhesion of cancer cells to blood vessel walls at the secondary site during metastasis. This adhesion is a rolling adhesion in which selectin/selectin ligand bonds repeatedly form and dissociate under forces generated by hemodynamic blood flow. Currently, standard tissue analyses, such as immunohistochemistry (IHC), do not functionally assay for those molecules that mediate rolling. Recently, our lab developed a novel tissue analysis assay, termed dynamic biochemical tissue analysis (DBTA), which detects functional selectin ligands by probing tissues with selectin-coated microspheres. The selectin microspheres are delivered to tissues under hydrodynamic flow conditions, and if selectin ligands are present, the selectin microspheres will roll on tissues presenting ligands that form functional bonds under the applied force conditions of the assay. When colon cancer tissues of distinct histological classifications were assayed using DBTA, significantly greater quantities of selectin microspheres specifically adhered to cancer tissues relative to noncancerous tissues, and the adhesion patterns of all three types of selectin microspheres demonstrated a heterogeneous presentation of functional selectin ligands on cancer tissues. Furthermore, the rolling velocities of selectin microspheres were a function of the hydrodynamic shear stress and demonstrated the force-dependent nature of selectin/selectin ligand interactions, which are governed by catch-slip bonds. Specifically, microsphere rolling velocities were stabilized at 0.5-0.75 $dyn/cm^2$, yet outside this range the microsphere rolling velocities became erratic as the rate of bond dissociation exceeded the rate of bond formation. Remarkably, when colon cancer tissues were probed with selectin molecules using IHC and DBTA, the two assays disparately detected selectin ligands such that, in certain regions of tissue, selectin ligands were detected using DBTA but not IHC. These results demonstrate that DBTA yields a tissue characterization that is distinct from standard methods of tissue analysis, thereby illustrating the relevance of assays that functionally detect target molecules.

# 436224 Bacterial Adhesion and Early-Stage Biofilm Formation on (poly) Dimethylsiloxane Surfaces with Microtopographic Line Patterns

**Abstract:**

Biofilms are surface-attached aggregates of microbes embedded in extracellular polymeric substances (EPS) produced by attached cells. Found ubiquitously in aqueous environments, biofilms are up to 1,000 times more resistant to antimicrobial agents than planktonic (free-swimming) bacteria of the same genotype. These multicellular structures are involved in 80% of bacterial infections in humans and play an important role in the spread of antimicrobial resistance due to biofilm-associated horizontal gene transfer. Initial attachment is the first step of biofilm formation and plays a critical role in biofilm development. This process can be significantly affected by the properties of the substratum material. For example, surfaces with certain roughness on micrometer, submicrometer, and nanometer length scale (comparable with or smaller than bacterial cells) have been found to affect bacterial adhesion. However, how bacteria respond to these topographic features is not well understood.

To better understand the effect of micron scale surface topography on the early-stage biofilm formation and the underlying mechanism, polydimethylsiloxane (PDMS) surfaces with 5 μm tall line patterns and systematically varied pattern width (5, 10, and 20 μm) and inter-pattern distance (3, 5, 10, and 20 μm) were prepared via soft lithography. By following the adhesion of *Escherichia coli* RP437/pRSH103 on these patterned surfaces, we obtained results revealing that surface topography can affect the orientation and cluster formation of bacterial cells during the early-stage biofilm formation. Interestingly, the attached *E. coli* cells were found to preferentially align perpendicularly to the line patterns when the pattern width got narrower (5 μm); and the mutants of flagella, flagellar stator, and fimbriae genes exhibited defects in such adjustment. In addition, cell attachment and cluster formation on the protruding line patterns were found to increase with pattern width. For example, compared to the smooth PDMS, surfaces with 5 μm wide patterns and 3 μm inter-pattern distance reduced biofilm surface coverage by 5-fold and cell cluster formation by 10-fold. Collectively, these findings provided new information about the effects of surface topography on bacterial adhesion and following formation of multicellular biofilms.

# 411409 Scaling of Dorsal/Ventral Patterning By BMPs in Developing Zebrafish Embryos

**Abstract:**

**Scaling of Dorsal/Ventral patterning by BMPs in developing zebrafish embryos.**

*Yan Huang[1], Francesca Tuazon[2], Joseph M Zinski[2], Mary C Mullins[2], David M Umulis[1]. 1) Purdue University, West Lafayette, IN; 2) University of Pennsylvania, Philadelphia, PA.*

Scale invariance in pattern formation is a phenomenon that occurs when the patterns and organization of tissues, organs or overall organisms remain proportionate with the overall size of the domain or organism. Scale invariance has been observed in numerous contexts including homologous patterning processes that are observed between: 1) individuals within the same species (intraspecies), 2) between different species (interspecies), or 3) between different temporal stages during the development of a domain changing over time (e.g. growth) in an individual (dynamic). Herein we focus on interspecies and intraspecies scale invariance of the dorsal/ventral (DV) patterning during early embryonic development of zebrafish. The DV axis is patterned by a gradient of Bone Morphogenetic Protein (BMP) signaling with a maximum ventrally and a minimum dorsally. Our preliminary data show that embryos with margin circumference reduced surgically by up to 20% are able to grow into the wild-type phenotype and the corresponding BMP gradient scales with normal-size embryos. To elucidate mechanisms explaining these results, we dissected the full extracellular BMP regulatory network into 3 sub-motifs (blastula, Sizzled and ADMP) and developed partial differential equation (PDE) models for each motif. Parameters that are not well-defined by literature (production, binding, degradation) were screened first with the shape of the BMP gradient obtained from previous studies and then with scaling. At steady state, simulation results show that the Sizzled motif is able to scale while the blastula and ADMP motifs fail to scale, contrary to what has been investigated with respect to ADMP's role with the scaling mechanism recently reported in Xenopus. Based on these, we hypothesize that Sizzled and feedback control are key motifs for scale invariance. To validate our hypothesis, we are going to make mutants of extracellular BMP regulators and check scaling among different sized embryos. We are also going to test interspecies scaling of BMP gradient among zebrafish and other species such as *giant danio*. The results from this research will enable us to identify the mechanism of intraspecies and interspecies scaling, to shed light on dynamic scale invariance, and to better understand the earliest stages of life in quantitative detail, specifically, the regulation of BMP signaling.

## (682g) Robust Patterning of the Dorsal-Ventral Axis in the Drosophila Melanogaster Embryo

Carrell, S., North Carolina State University
Thomas, A. S., North Carolina State University
Friedman, J.,
Reeves, G. T., North Carolina State University
Originally presented on: 11/20/2014 14:18:00 - 14:36:00

During the first three hours of *Drosophila* embryogenesis, two conserved signaling pathways act to pattern the dorsal-ventral (DV) axis: the Dorsal pathway and the Dpp pathway. On the ventral side of the embryo, the transcription factor Dorsal, homologous to NF-κB, triggers expression of the genes that initiate the DV pattern. Dorsal is present in a nuclear gradient, with the highest concentration at the ventral midline and a steady decay to about 40% of the embryo?s circumference. The Dorsal gradient is initialized via Toll signaling on the ventral side of the embryo; Toll signaling phosphorylates the inhibitor protein Cactus (homologous to I-κB), marking it for degradation. In the absence of Cactus, Dorsal can enter the nuclei and activate expression of target genes in a concentration-dependent manner. Dorsal also acts as a transcriptional repressor, limiting some genes to expression on the dorsal half of the embryo. One of those dorsally-expressed genes is *decapentaplegic* (*dpp*), which encodes Dpp, a bone morphogenic protein (BMP) ligand. Dpp is present in a concentration gradient, establishing DV gene patterns beyond Dorsal?s reach. BMP signaling occurs when Dpp binds to the receptor Thickveins (Tkv), phosphorylating the receptor Smad protein MAD. Two phospho-MAD molecules bind to the co-Smad Medea; this heterotrimer then enters the nuclei to facilitate expression of target genes. See (A) for a schematic of BMP signaling. Together, the Dorsal gradient and BMP signaling pathway pattern the DV axis of developing *Drosophila* embryos.

Because robustness of tissue patterning is essential for proper embryonic development, regulatory loops must exist to ensure correct placement of target genes in the face of perturbed conditions. As an example, embryos with a half dose of Dorsal protein survive to adulthood. We have analyzed the Dorsal gradient in these heterozygotes by modeling the nuclear gradient of Dorsal as a Guassian, using the equation:

$$C(x) = A exp[-x^2/2\text{\σ}^2] + B$$

where C is the concentration of nuclear Dorsal, x is the normalized DV coordinate (0 being the ventral midline), A represents the amplitude, σ, the width, and B is the basal level of Dorsal present in the nuclei at the dorsal midline. As shown in (B), we found that the Dorsal gradients in these embryos are not only shorter in amplitude as compared to wild type, as expected, but statistically wider and fundamentally different in shape, having a broader domain of peak nuclear concentration. These differences point to the existence of Dorsal gradient regulation in circumstances in which protein levels are compromised. We will present evidence that the Dpp signaling network can alter the Dorsal gradient itself, not only placement of target genes. When levels of MAD (the BMP signal transducer) were overexpressed, we found that the width of the gradient (as measured by σ) expanded significantly. Furthermore, overexpression of Medea (a BMP signaling cofactor) led to the same result: an expansion of the Dorsal gradient. Therefore, feedback through the Dpp signaling network is a prime candidate for enhancing the robustness of patterning of DV gene expression, as shown in (C). These interactions between the NF-κB and BMP signaling pathways may be necessary to ensure robust gene expression in the developing *Drosophila* embryo.

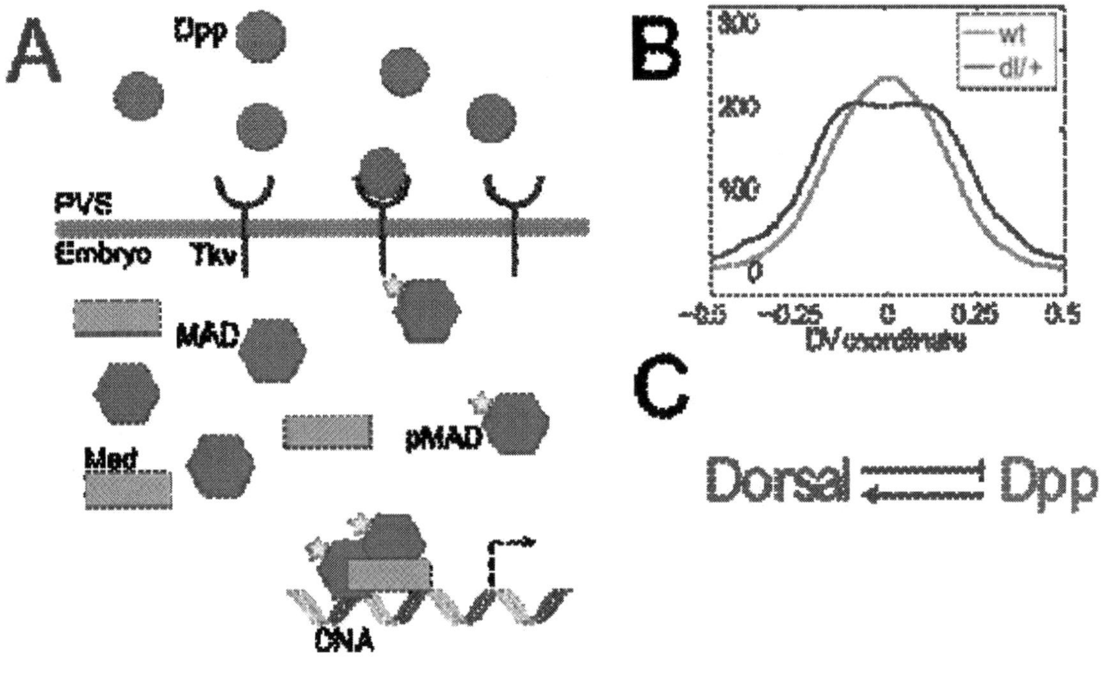

# 426413 Connecting Exogenous Stimuli to Patterned Calcium Signalling in a Regulated Epithelial Microenvironment

**Abstract:**

Intercellular propagation of calcium ions ($Ca^{2+}$) has been observed in a variety of tissues and is mediated by many different stimuli. Calcium signaling plays multiple roles in most cellular processes including cell cycle regulation. To better understand the role of $Ca^{2+}$ signalling in development and homeostasis, we have developed a microfluidic chip which allows for controlled culture and stimulation of *Drosophila* imaginal wing discs under observation on a confocal microscope, a Regulated Epithelial Microenvironment microfluidic chip (REMChip). Key stimulation modalities include chemical perfusion, mechanical compression, mechanical tension, and electrical gradation. Our initial results demonstrate that the wing disc can be cultured for up to 10 hours and that $Ca^{2+}$ signalling is quiescent during mechanical compression, but once compression is removed a $Ca^{2+}$ wave cascades throughout the disc. Through local wounding and growth factor stimulation, we have also quantified and analyzed the patterning of intercellular $Ca^{2+}$ transients, which reflect the underlying mechanics and differentiate state of cells within the tissue.

# 435253 Automatic Long-Term Individual Tracking Microfluidic Platform for Healthspan and Longevity in Caenorhabditis Elegans

**Abstract:**

Over the last two decades, aging studies have found that the evolutionarily conserved genes and signaling pathways could extend lifespans of many organisms. However, it is still unknown whether organisms are extending the healthy lifetime or simply prolonging a period of frailty with increased incidence of age-related diseases. The nematode *Caenorhabditis elegans* is an important model organism for the investigation of the aging process: first, the longevity of the worm population subject to different genetic or environmental regulations can be easily assessed over relatively short lifespans; second, *C. elegans* is capable of exhibiting complex behaviors that are relevant to functional metrics of human healthspan. The current experimental techniques for aging assays in the worm, however, have limitations, such as the inability to exert fine control over environmental conditions, longitudinal tracking of individuals, high experimental temporal resolution, and large-scale experimentation.

To address these limitations, we developed a scalable high-throughput automatic long-term individual tracking platform, integrated with microfluidics and hardware and software technologies; this platform offers the advantages of physical isolation of individual animals, high temporal resolution of behavior tracking, and highly controlled environmental conditions. This platform is composed of three major modules. The first module is the liquid worm culture system, which is comprised of both the microfluidic chip and the off-chip support system for food control and delivery. The second module is to control environmental temperature on chip. This platform can perform the simultaneous investigation of up to 32 independent microfluidic chips, each housing up to 60 worms. These chips can each be operated via independent food delivery paths that permit the investigation of different food conditions. Furthermore, up to 8 different temperature conditions can be used in this system. The third module enables automated experimental control by interfacing with the first two modules and automated data acquisition via camera control and state movement. We can automatically monitor both individual longevity and behavior information with sub-hourly experimental time resolution. Here, we show that this platform allows high-throughput high-content, long-term analysis of health outcomes in large number of animals, with individual resolution and extremely well controlled environmental conditions. We envision this tool to enable large-scale studies of how genes and environment affect aging and age-related diseases.

# 432818 The Role of rDNA in the Nucleolus Formation: A Seeding Process?

**Abstract:**

The nucleolus is a membrane-less organelle, which plays a pivotal role in different cellular functions including ribosomal biogenesis. During embryogenesis, this organelle forms *de novo* through a process that requires the activation of rDNA transcription and the recruitment of nucleolar components to the nucleolus organizer regions. However, the interdependency of these two events is not clear. We developed quantitative approaches that allow us to study this question in wild-type and mutant *Drosophila melanogaster* embryos. Our results show that the assembly of the nucleolar proteins is not necessary for the transcription of rDNA. Moreover, in the absence of rDNA, the nucleolar proteins studied are able to form high concentration assemblies, but as a stochastic event. This shows that rDNA controls the timing and position of these assemblies probably through a nucleation mechanism.

# 428857 Dynamic Tensile Forces Drive Collective Migration through Three-Dimensional Extracellular Matrices

**Abstract:**

**Introduction:** Collective migration drives a variety of dynamic biological and developmental processes, including wound healing, angiogenesis, branching morphogenesis, and cancer invasion. These processes are fundamentally physical, though our understanding of the physical mechanisms by which groups of cells move cohesively through dense three-dimensional extracellular matrices remains incomplete. For this reason, this work sought to investigate the means by which mechanical interactions between a migrating cohort and the surrounding extracellular matrix (ECM) regulates collective migration.

**Approach:** We used arrays of microfabricated tissues to elucidate the physical mechanisms by which multicellular cohorts move through a three-dimensional (3D) ECM. Cells underwent collective invasion predictably from specific locations within the microfabricated tissues, allowing for high-throughput analysis with high statistical confidence. Matrix deformations and the corresponding mechanical forces which accompanied migration were evaluated by tracking the movements of fluorescent beads embedded in the surrounding ECM.

**Results and Discussion:** Using 3D traction force microscopy, we found that cells exert tensile forces on the surrounding ECM while undergoing collective migration. These forces both propelled cells forward and stimulated mechanotransduction signaling at the leading edge of the migrating cohort. Confocal reflectance microscopy revealed that the matrix ahead of the migrating cohort was remodeled into highly aligned fiber bundles along which the cohort preferentially migrated. Moreover, timelapse imaging showed that cohort extension was highly dynamic, and that variations in cohort length were highly correlated and in phase with deformations and tensile forces in the ECM. These results suggest both mechanical and signaling roles for dynamic tensile forces during collective migration.

# 416769 Cell Biomechanics Effects in Limb Development

**Abstract:**

Half century ago, Alan Turing proposed a mathematical theory about morphogenesis based on the formation of spatial patterns. Yet, its relevance in the field of Developmental Biology remained elusive for many years. Remarkably, recently it has been shown that Turing's mechanism underlies a paramount developmental process, i.e. limb formation. This poses the intriguing question of how patterning stability is achieved in a real tissue since, as it grows, cell division events compromise the patterning and, as a consequence, the cell segregation that is critical in the developmental process. Herein we address this question using computational and theoretical means. In particular, we show how cell division dynamics interferes with the patterning process in different ways (from the topological induced disorder to the random binomial distribution of proteins) and we reveal that the interplay of the cell biomechanics and gene regulation is key for understanding tissue homeostais in the context of limb development.

# 428797 Improved Butanol Production from Clostridium Pasteurianum from Crude Glycerol Feedstock

**Abstract:**

Biodiesel production results in crude glycerol waste from the transesterification of fatty acids (10% w/w). The solventogenic *Clostridium pasteurianum*, an anaerobic Firmicute, can produce butanol from glycerol as the sole carbon source. Coupling butanol fermentation with biodiesel production can improve the overall economic viability of biofuels. However, crude glycerol contains growth inhibiting byproducts which reduce feedstock consumption and solvent production. To obtain a strain with improved characteristics, a random mutagenesis and directed evolution selection technique was used. A wild type *C. pasteurianum* (ATCC 6013) culture was chemically mutagenized and the resulting population underwent 10 days of selection in increasing concentrations of crude glycerol (80 to 150 g/L). The best performing mutant (M150B) showed a 72% increase in butanol production in 100 g/L crude glycerol compared to the wild type strain, as well as increased growth rate, a higher final optical density, and less production of the side product PDO (1,3-propanediol). Wild type and M150B strains were sequenced via Single Molecule Real-Time (SMRT) sequencing. Mutations introduced to the M150B genome were identified by sequence comparison to the wild type and published closed sequences. A major mutation in key transcriptional regulator Spo0A was identified. Complementation studies were done with both wild type and mutated *spo0A* genes in a constructed Spo0A-deficient strain. RNAseq identifies differences to the transcriptomes of the strains during growth on crude glycerol. Methylation patterns on genomic DNA identified with SMRT sequencing were used to aid in optimization of transformation efficiency of plasmid DNA to elude *C. pasteurianum* endogenous restriction endonucleases.

# 435138 Exploring Mechanisms for Xylose Transport in the Oleaginous Yeast, Yarrowia Lipolytica

**Abstract:**

The biological production of fuels and chemicals by microorganisms has garnered interest as an alternative to petroleum-derived products. Using lignocellulosic feedstocks enables renewable production without competing for food supplies. However, many industrial microorganisms do not efficiently metabolize xylose and other pentose sugars resulting from lignocellulose hydrolysis. As a result, much effort has focused on engineering microorganisms for efficient xylose utilization. The oleaginous yeast, *Yarrowia lipolytica,* is capable of accumulating lipids >90% of cell biomass, making it ideal for production of fatty acid-based products. While there are conflicting reports in the literature, our experiments show that *Y. lipolytica* is unable to utilize xylose as a sole carbon source. It remains to be understood whether the significant bottleneck in the process lies in the transport of xylose or its metabolism. Here, we explore sugar transporter homologs in *Y. lipolytica* that share strong sequence similarity to well-characterized xylose transporters in Scheffersomyces stipitis and Candida intermedia. We used confocal microscopy to determine whether the identified homologs localize to membrane and determine preferential sugar uptake through these transporters via a series of knockout studies and substrate assays. Furthermore, recombinant expression studies with xylose transporters from *S. stipitis* and *C. intermedia* will be done, engineering the *Y. lipolytica* to have a preferential uptake towards xylose. The work presented here complements the xylose metabolism work currently performed by our lab and will advance the current limited knowledge of *Y. lipolytica* 5-carbon sugar transport and metabolism.

## Engineering *Yarrowia lipolytica* for Production of Medium-Chain Fatty Acids and Alcohols

Charles D. Rutter, Shuyan Zhang, and Christopher V. Rao

Department of Chemical and Biomolecular Engineering
University of Illinois-Urbana Champaign, Urbana, Illinois 61801

### ABSTRACT

Lipids are naturally derived products that offer an attractive, renewable alternative to petroleum-based hydrocarbons. While naturally produced long-chain fatty acids can replace some petroleum analogs, medium-chain fatty acid would more closely match the desired physical and chemical properties of currently employed petroleum products. In this study we engineered *Yarrowia lipolytica*, an oleaginous yeast that naturally produces lipids at high titers, to produce medium-chain fatty acids. Five different acyl-ACP thioesterases with specificity for medium chain acyl-ACP molecules were expressed in *Y. lipolytica*, resulting in formation of either decanoic or octanoic acid. These novel fatty acid products were found to comprise up to 40% of the total cell lipids. Furthermore, the reduction in chain length resulted in a two-fold increase in specific lipid productivity in these engineered strains. The medium-chain fatty acids were found to be incorporated into all lipid classes.

### INTRODUCTION

Renewable fuels and chemicals offer solutions to problems ranging from global warming to supply stability (Fortman et al. 2008; Liu et al. 2014; Lynd et al. 2005). Using yeast as microbial cell factories to produce valuable commodity chemicals offers a promising and sustainable alternative to the continued use of petroleum-based products (Runguphan and Keasling 2014). Fatty acids are molecules with a long alkyl chain, very similar to the chemicals found in petroleum, with a carboxylic acid attached to the end. They accumulate within the cell as energy and carbon storage molecules. Fatty acids can be further converted to other derivatives including (1) methyl and ethyl esters which can be used as biodiesel (Moser 2011), (2) fatty waxes to be used in cosmetics, food, and paper industries (Santos et al. 2014), (3) fatty alcohols with applications as surfactants and industrial solvents (Zheng et al. 2012), and (4) alkanes that can directly be used as fuel (Choi and Lee 2013).

Oleaginous yeasts naturally produce elevated amounts of lipids compared to traditional yeast strains, making them a natural choice as a microbial platform for industrial production of fatty acid derived products (Ageitos et al. 2011; Papanikolaou and Aggelis 2011). *Yarrowia lipolytica* is the most-well studied of these yeasts, and researchers have previously engineered it for increased lipid production by a variety of rational and evolutionary approaches. (Tai and Stephanopoulos 2013; Wang et al. 2014; Liu et al. 2015; Qiao et al. 2015). In addition, many researchers have sucessfully produced novel fatty acid products in *Y. lipolytica* including omega-3-fatty acids, linoleic acid derived unsaturated and polyunsaturated fatty acids as well as alkanes (Xue et al. 2013; Beopoulos et al. 2014; Zhang et al. 2013; Blazeck 2013; Madzak 2015). Although these yeast strains can be engineered to produce large quantities of lipids and related products, the associated alkyl chains are much longer than those found in typical petroleum-derived products, making their physical properties less attractive for industrial and commercial applications (Dwivedi and Sharma 2014; Ramos et al. 2009). Engineering oleaginous yeast for production of medium-chain fatty acids would lead to production of renewable molecules more suitable as replacements for petroleum-based products.

The mechanism of fatty acid synthesis by the multi-enzyme fungal type I fatty acid synthase (FAS) is illustrated in Figure 1. Initiation of fatty acid synthesis occurs when acetyl-CoA is condensed with carbon dioxide by acetyl-CoA carboxylase to form malonyl-CoA. Malonyl-CoA and acetyl-CoA are then incorporated in the FAS complex by transacylase activity to produce malonyl-acyl carrier protein (ACP) or acetyl-ACP, respectively. The ACP will have been previously activated with phosphopantetheine at a conserved serine residue by the phosphopantetheinyl transferase subunit (Lomakin et al. 2007). Malonyl-ACP is then decarboxylated and keto-synthase catalyzes its condensation with acetyl-ACP. Keto-reductase, dehydratase, and enoyl reductase subunits act in series to catalyze removal of the remaining oxygen atom and carbon-carbon double bond to form a saturated acyl-ACP (Landriscina et al. 1972; von Wettstein-Knowles et al. 2006). The acyl-ACP then becomes the substrate for further addition of decarboxylated malonyl-ACP. This cycle of malonyl-CoA addition to the acyl chain continues until elongation is terminated by malonyl/palmitoyl transacylase which catalyzes the transacylation of the acyl-ACP with CoA to produce a fatty acyl-CoA (Leibundgut et al. 2008; Lynen 1969; Lomakin et al. 2007). This is slightly different from fatty acid synthesis in other domains such as bacteria and plants where elongation is terminated by an acyl-ACP thioesterase which catalyzes a transesterification of fatty acyl-ACP with water to produce free fatty acid (Jing et al. 2011; Moreno-Pérez et al. 2012).

In addition to a different mechanism of termination, bacterial, plant, and mammalian fatty acid synthases have widely different structures than their fungal counterpart. In bacterial and plant systems, the different catalytic subunits required for fatty acid synthesis are dissociated (White et al. 2005; Zhang et al. 2006; Wakil 1989). In many eukaryotes including mammals and fungi, however, multi-subunit complexes form that contain all the necessary catalytic subunits in a single enzyme. Furthermore, the mammalian FAS consist of a hexamer of α-subunits while the fungal FAS consists of a hexamer of α-subunits that form a central core capped on either side by 3 β-subunits (Leibundgut et al. 2008; Lynen 1969; Lomakin et al. 2007). The dissociated nature of synthases in bacterial and plant systems make them well suited for modification of fatty acid length by expression of heterologous proteins. Many acyl-ACP thioesterases have been identified that use medium chain acyl-ACP molecules as their substrate (McMahon and Prather 2014; Voelker et al. 1997; Pollard et al. 1991; Dörmann et al. 1993). Numerous studies have shown that expression of a medium chain acyl-ACP specific acyl-ACP thioesterase leads to production of shorter fatty acids in *E. coli* as well as several plant species (Eccleston et al. 1996; Sherkhanov et al. 2014; Lui et al. 2010; Lu et al. 2008; Steen et al. 2010; Voelker et al. 1992).

Most organisms produce only sixteen and eighteen carbon fatty acids, indicating a majority of acyl-ACP-thioesterases and transacylases have specificity toward longer acyl-ACP chains (Cronan Jr and Rock 1996; Harwood 1988). While attempts at engineering *E. coli* for medium chain fatty acid production have been successful, the lack of variety in transacylase specificity combined with the bulwark-like quaternary structure of fungal FAS makes engineering yeast for medium chain fatty acid production challenging. In fact, the first attempt at producing medium chain fatty acids was done by induction of the β-oxidation pathway, which breaks lipids down into acetyl-CoA, along with expression of carnitine O-octanoyltransferase to terminate β-oxidation early to allow accumulation of medium chain fatty acids. This approach was dependent on supplementation of oleic acid in the medium (Chen et al. 2014). One successful strategy for de-novo short chain fatty acid production in *S. cerevisiae* required concurrent expression of a more accessible mammalian FAS with its C-terminal thioesterase domain replaced with an acyl-ACP thioesterase known to produce short chain fatty acids (Leber and Da Silva, 2014).

In this work, we tested whether a heterologous thioesterase would be able to associate with the native FAS complex in *Y. lipolytica* and prematurely terminate elongation to produce medium chain fatty acids. Additionally, we attempted further conversion of medium-chain fatty acids to alcohols by expression of fatty-acyl-CoA reductase. We demonstrate that the expression of different plant and bacterial acyl-ACP-thioesterases reduces the chain length of fatty acids in *Y. lipolytica*. Additionally, the final length of the fatty acids produced depends on the identity of the enzyme expressed and each enzyme produces a single-length fatty acid. An increase in specific lipid yield along with a reduction in final cell mass was also observed during production of medium chain fatty acids. Novel products were detected in all lipid classes. Subsequently, we co-expressed a bacterial fatty-acyl-CoA reductase in an attempt to further convert the medium-chain fatty acids to alcohols. Expression of this reductase in medium-chain fatty acid producing strains results in production of decanol coupled with a reduction in the levels of medium-chain fatty acids.

## RESULTS

### Production of non-native fatty acid species in *Y. lipolytica* by expression of heterologous acyl-ACP thioesterases.

Most microorganisms produce fatty acids predominantly with acyl chains 16 or 18 carbons in length. In an attempt to produce shorter fatty acids, we expressed heterologous acyl-ACP thioesterase enzymes from five different species previously shown to have specificity toward acyl-ACP chains eight carbons in length (McMahon and Prather 2014; Sherkhanov et al. 2014). We evaluated *Y. lipolytica* PO1f strains with acyl-ACP-thioesterases from *Anaerococcus tetradius, Cuphea hookeriana, Cuphea palustris, Clostridium perfringens,* and *Umbellularia californica* randomly integrated into the genome (referred to as PO1f-AtACPT, PO1f-AChACPT, PO1f-CpaACPT, PO1f-CpeACPT, and PO1f-UcACPT, respectively) for their ability to produce medium chain fatty acids. We found that the expression of each enzyme tested was capable terminating the elongation of fatty acid early to produce either decanoic acid or octanoic acid. PO1f-AtACPT, PO1f-ChACPT, PO1f-CpaACPT, and PO1f-CpeACPT produced 36 ± 8%, 44 ± 1%, 57 ± 8%, and 46 ± 8% of their lipids as ten-carbon fatty acid species on a mass basis respectively. PO1f-UcACPT produced 14 ± 5% of their lipids as eight-carbon fatty acid species on a mass basis. The *Y. lipolytica* PO1f wild-type strain showed no production of either eight or ten-carbon fatty acid species when grown under the same conditions. In all cases, production of shorter acids was coupled with reduction in the abundance of eighteen carbon acid species. In the PO1f-UcACPT strain, sixteen-carbon fatty acid abundance increased to 75 ± 5% on a mass basis compared to 34 ± 2% in the wild-type strain.

### Lipid yields from engineered strains.

Total lipid titers and specific lipid production were determined for strains expressing heterologous enzyme during growth under lipid accumulation conditions. Total lipid titers from engineered strains were comparable to that of the wild-type strain with PO1f-AtACPT, PO1f-ChACPT, PO1f-CpaACPT, PO1f-CpeACPT, and PO1f-UcACPT producing 2.9 ± 0.7, 2.3 ± 0.1, 1.7 ± 0.1, 1.8 ± 0.1, and 1.5 ± 0.6 grams of lipid per liter, respectively. A lipid titer of 1.7 ± 0.6 g/L was produced by the wild-type strain. The above strains were found to grow to a final dry cell weight of 4.5 ± 0.4, 4.1 ± 0.2, 2.4 ± 0.2, 3.9 ± 0.1, and 9.8 ± 0.9 grams dry cell weight per liter. The wild-type strain grew to 10.8 ± 1.3 grams dry cell weight per liter. The above strains produced 0.66 ± 0.23, 0.57 ± 0.32, 0.66 ± 0.06, 0.46 ± 0.02, and 0.13 ± 0.05 grams lipid per gram dry cell weight, respectively. The wild-type strain produced 0.15 ± 0.1 grams lipid per gram dry cell weight, indicating that the improved specific lipid productivity was due to reduced growth of engineered strains capable of producing decanoic acid.

### Incorporation of medium chain fatty acids into different lipid classes.

To evaluate the capacity for *Y. lipolytica* to incorporate non-native fatty acids into acylglycerides (mono, di, and tri), sterols, and steryl esters, total lipids were separated into different classes by chromatography and the composition with respect to carbon chain length of each fraction was determined. Lipids harvested from wild type cells were found to be 23% triacylglyceride (TAG) while 17%, 19%, 17%, 21%, 19% and 15% of total lipids were found in the TAG fraction from PO1f-AtACPT, PO1f-ChACPT, PO1f-CpaACPT, PO1f-CpeACPT, and PO1f-UcACPT, respectively. This reduction in TAG abundance coincided with an increase in free fatty acid (FFA) abundance with 17%, 17%, 21%, 25%, and 31% of PO1f-AtACPT, PO1f-

ChACPT, PO1f-CpaACPT, PO1f-CpeACPT, and PO1f-UcACPT lipids present as FFA, respectively, compared to 15% of wild-type lipids present as FFA species.

Medium chain fatty acids produced were found in all lipid classes. In the PO1f-AtACPT strain, 15% and 22% of decanoic acid was found in the TAG and FFA fractions respectively. In the PO1f-ChACPT strain, 20% and 19% of decanoic acid was found in the TAG and FFA fractions respectively. In PO1f-CpaACPT 16% and 21% of decanoic acid was found in the TAG and FFA fractions respectively. In PO1f-CpeACPT 32% and 23% of decanoic acid was found in the TAG and FFA fractions respectively. In PO1f-UcACPT 20% and 28% of octanoic acid was found in the TAG and FFA fractions respectively.

Expression of heterologous enzymes changed the localization of native fatty acids with respect to lipid classes. 19% of the sixteen-carbon fatty acids produced by the wild type were found to be free fatty acids while PO1f-AtACPT, PO1f-ChACPT, PO1f-CpaACPT, PO1f-CpeACPT, and PO1f-UcACPT strains showed 44%, 25%, 46%, 52%, and 60% of their sixteen carbon fatty acids to be present as free fatty acids, respectively. Eighteen-carbon fatty acids from PO1f-AtACPT, PO1f-ChACPT, PO1f-CpaACPT, PO1f-CpeACPT, and PO1f-UcACPT were found to be 36%, 17%, 43%, 56%, and 51% free fatty acids, respectively. Wild-type strains produced 12% of their eighteen carbon fatty acids as free fatty acids. An increase in the FFA abundance of native lipids corresponds to the increase in FFA found in total cellular lipids.

**Conversion of medium-chain fatty acids into alcohols**

Fatty alcohols are one of many classes of valuable lipid based molecules with applications and lubricants, plasticizers, and surfactants (Nakagawa and Tori 1960; Li et al. 2015). In order to convert the medium-chain fatty acids being produced into alcohols, we expressed a fatty-acyl-CoA reducatase from *Marinobacter aquaeolei* concurrently with each of the above acyl-ACP-thioesterase enzymes (AtFAR, ChFAR, CpaFAR, CpeFAR,UcFAR, respectively). Due to the fact that *Y. lipolytica* is naturally capable of catabolizing fatty alcohols the production of medium-chain fatty alcohols was evaluated in the wild-type W29 strain as well as a Pex10 deletion mutant of W29. Alcohol catabolic reactions occur majorly in the peroxisome and Pex10 deletion is known to prevent peroxisome formation (Xue et al. 2013). We found that concurrent expression of acyl-ACP thioesterases and fatty-acyl-CoA reductase results in decanol formation. Decanol concentrations from cultures of *pex10Δ::NAT*-AtFAR, *pex10Δ::NAT*-ChFAR, *pex10Δ::NAT*-CpaFAR, *pex10Δ::NAT*-CpeFAR, *pex10Δ::NAT*-UcFAR were found to be 560, 40, 530, 680, and 570 mg/L, respectively. W29 stains expressing the same enzymes all accumulated < 5 mg/L decanol. In both backgrounds however, decanoic acid concentrations measured were < 10mg/L for all engineered strains.

**DISCUSSION**

Oleaginous yeasts are attractive because they produce lipids at high titers from simple carbohydrates. Numerous researchers have engineered *Y. lipolytica* for increased lipid titers and productivity. Like most other microbes *Y. lipolytica* produces lipids sixteen carbons in length or longer. Medium and short chain lipids have more desirable physical properties compared to the naturally produced long chain lipids for the production of many industrially-relevant chemicals and fuels. In this study, natural lipids were shortened to medium chain lengths of eight or ten carbons by the expression of heterologous acyl-ACP thioesterases. In addition, lipid productivity doubled in some cases, suggesting that this approach can be used to increase flux through the lipid metabolism pathway as well as introduce novel products into an oleaginous yeast strain.

Expression of four acyl-ACP thioesterases in a *Y. lipolytica* host resulted in production of decanoic acid as 40% or more of the total lipids produced. These decanoic acid producing strains did so without drastically changing the abundance of sixteen-carbon lipids, and seemingly the medium chain lipids were produced at the expense of eighteen-carbon lipids. The acyl-ACP thioesterase from *U. californica*, when expressed in *Y. lipolytica,* led to production of a modest amount of octanoic acid. Additionally, the abundance of sixteen carbon lipids nearly doubled in this octanoic-acid producing strain. Again, the increase in octanoic and hexadecanoic acids was coupled to a reduction in eighteen-carbon lipids. These results demonstrates that an acyl-ACP thioesterase with specificity for shorter acyl-ACP molecules is capable of associating with the fatty acid synthase complex and catalyzing the termination of fatty acid elongation to produce medium chain fatty acids. It was previously shown that all enzymes tested catalyzed formation of octanoic acid in an *E. coli* system (McMahon and Prather 2014; Sherkhanov et al. 2014), yet decanoic acid was the product in most cases when they were expressed in *Y. lipolytica*.

It is unknown how a plant or bacterial acyl-ACP thioesterase is capable of associating with the fungal type I FAS, whose structure is known to orient most of the pertinent catalytic sites within a capsule-like structure (Lomakin et al. 2007; Leibungut et al. 2008). One possibility is that the heterologous thioesterase is somehow disrupting the natural structure of the FAS in a way that allows the thioesterase catalytic access to the growing acyl-ACP molecule. This disruption may also explain why cells expressing the thioesterases exhibit reduced growth due to the FAS functioning incorrectly. We note that the change in growth also manifests itself as a change in the color of the culture from pale white to green, possibly due to the production of the shorter-chain fatty acids or altered physiology of the cell. Furthermore, in most cases, the fatty acid products are longer in *Y. lipolytica* than when the same enzyme is expressed in *E. coli* suggesting impeded access of the thioesterase to acyl-ACP substrate in the fungal FAS compared to a bacterial one, indicating that the association of the thioesterase with the FAS is not optimal yet still functional.

We observed that production of decanoic acid resulted in more than double the specific yield of lipids on a dry cell weight basis. An increase in lipid titer during the fermentation, however, was not observed for any engineered strains. This increase in specific yield was not observed in the strain expressing the *U. californica* acyl-ACP thioesterase. Feedback inhibition of several enzymes in the fatty acid metabolism pathway by long chain fatty acyl-CoA is known to occur (Jiang and Cronan 1994; Ohlrogge and Jaworski 1997). Decanoic acid producing strains show drastically reduced levels of long-chain fatty acids compared to the wild type, potentially removing feedback inhibition and increasing productivity in these engineered strains. This effect is not observed in the octanoic acid producing strain, because it produces high amounts of long-chain fatty acids. As mentioned above, the disruption of the native FAS along with reduction in native fatty acids (sixteen and eighteen) and production of potentially toxic medium-chain fatty acids retards cell growth (Viegas et al. 1989). Despite the slower growth of these engineered strains, final lipid titers remain unchanged relative to the wild type and indicate that fermentation by engineered strains results in a lower final density of more productive cells.

Fatty acids naturally segregate into several classes of lipids. This segregation requires the activity of a variety of enzymes that may or may not have activity towards these newly created medium chain fatty acids. We found non-native fatty acids in all the different lipid classes including tri-, di-, and monoacylglycerides, sterol, steryl esters, and free fatty acids. This indicates that the native machinery in *Y. lipolytica* is capable of recognizing and modifying fatty acids as short as octanoic acid. Some engineered strains also showed altered abundances of different lipid classes as compared to the wild type. Typically, the abundance of free fatty acids was increased in the engineered strains with a corresponding decrease in sterol and triacylglyceride abundance. A corresponding change was not observed in the medium-chain fatty acid lipid class abundances but rather in the modification of the native long-chain fatty acids. An increase in the amount of hexadecanoic and octadecanoic acids present as free fatty acids was increased in all engineered strains.

Because these novel fatty acid products are recognized by the native machinery of *Y. lipolytica* they were able to be further converted to alcohols by *M. aquaeolei* FAR, which uses acyl-CoA as a substrate. In this case, significant accumulation of fatty alcohols required abolition of the native alcohol catabolic pathway which was accomplished by abolition of peroxisome formation by deletion of Pex10. In both wild-type and peroxisome null backgrounds, however, low levels of medium-chain fatty acids were detected suggesting that in both cases medium-chain fatty acids are efficiently converted to alcohols but catabolized by native enzymes in the wild-type background.

In this study, we demonstrated that expression of a single acyl-ACP thioesterase with activity toward short acyl-ACP substrates in *Y. lipolytica* results in alteration of the length of fatty acids produced. Enzymes previously shown to produce fatty acids as short as octanoic acid in *E. coli* (McMahon and Prather 2014; Sherkhanov et al. 2014) were able to produce either octanoic or decanoic acid in *Y. lipolytica*. The production of novel medium chain fatty acids accompanied reduction in native long chain fatty acid production, which resulted in a reduction of cell growth and an increase in lipid yields. We also found that the medium-chain fatty acids were found to incorporate in all different lipid classes. Concurrent expression of a fatty-acyl-CoA reductase resulted in further conversion to medium-chain fatty alcohols, which accumulate when the native alcohol catabolic pathway is abolished. Taken together, this data shows that oleaginous yeasts can be engineered to produce medium-chain fatty acids which in turn, results in improved flux through the fatty acid metabolism pathway. Furthermore, these novel fatty acids can be converted to other products through simultaneous action of native and heterologous enzymes making demonstrating *Y. lipolytica* as a promising host for production of molecules derived from fatty acid precursors.

**AKNOWLEDGEMENTS**

This work was funded by the Energy Biosciences Institute. We thank Jeffrey Skerker for his help during the early stages of this project and Chris Somerville for his help with lipid analysis.

## REFERENCES


Ageitos JM, Vallejo JA, Veiga-Crespo P, Villa TG (2011) Oily yeasts as oleaginous cell factories. Appl Microbiol Biotechnol 90(4):1219-1227

Beopoulos A, Chardot T, Nicaud JM (2009) *Yarrowia lipolytica*: a model and a tool to understand the mechanisms implicated in lipid accumulation. Biochimie 91:692–696

Beopoulos A, Verbeke J, Bordes F, Guicherd M, Bressy M, Marty A, Nicaud, JM (2014) Metabolic engineering for ricinoleic acid production in the oleaginous yeast *Yarrowia lipolytica*. Applied Microbiol Biotechnol 98(1): 251-262

Blazeck J, Liu L, Knight R, Alper HS (2013) Heterologous production of pentane in the oleaginous yeast *Yarrowia lipolytica*. J Biotechnol 165(3): 184-194.

Bligh EG, Dyer WJ (1959) A rapid method of total lipid extraction and purification. Can J Biochem Physiol 37(8):911-917

Carroll KK (1961) Separation of lipid classes by chromatography on florisil. J Lipid Res 2(2):135-141

Chandler DS, Qi J, Mattox W (2003) Direct Repression of Splicing by transformer-2. Mol Cell Biol 23(15):5174-5185

Chen L, Zhang J, Chen WN (2014) Engineering the *Saccharomyces cerevisiae* β-oxidation pathway to increase medium chain fatty acid production as potential biofuel. PloS one 9(1)

Choi YJ, Lee SY (2013) Microbial production of short-chain alkanes. Nature 502(7472):571-574

Cronan Jr JE, Rock CO (1996) Biosynthesis of membrane lipids *Escherichia coli* and Salmonella: cellular and molecular biology, 2nd ed. ASM Press, Washington, DC:612-636

Dörmann P, Spener F, Ohlrogge JB (1993) Characterization of two acyl-acyl carrier protein thioesterases from developing *Cuphea* seeds specific for medium-chain-and oleoyl-acyl carrier protein. Planta 189(3): 425-432

Dwivedi G, Sharma M (2014) Impact of cold flow properties of biodiesel on engine performance. Renew Sustain Energy Rev 31:650-656

Eccleston VS, Cranmer AM, Voelker TA, Ohlrogge JB (1996) Medium-chain fatty acid biosynthesis and utilization in *Brassica napus* plants expressing lauroyl-acyl carrier protein thioesterase. Planta 198(1):46-53

Fortman J, Chhabra S, Mukhopadhyay A, Chou H, Lee TS, Steen E, Keasling JD (2008) Biofuel alternatives to ethanol: pumping the microbial well. Trends Biotechnol 26(7):375-381

Freeman Ct, West D (1966) Complete separation of lipid classes on a single thin-layer plate. J Lipid Res 7(2):324-327

Gietz RD, Woods RA (2002) Transformation of yeast by lithium acetate/single-stranded carrier DNA/polyethylene glycol method. Methods Enzymol 350:87-96

Harwood JL (1988) Fatty acid metabolism. Annu Rev Plant Physiol Plant Mol Biol 39(1):101-138

Jiang P, Cronan J (1994) Inhibition of fatty acid synthesis in *Escherichia coli* in the absence of phospholipid synthesis and release of inhibition by thioesterase action. J Bacteriol 176(10):2814-2821

Jing F, Cantu DC, Tvaruzkova J, Chipman JP, Nikolau BJ, Yandeau-Nelson MD, Reilly PJ (2011) Phylogenetic and experimental characterization of an acyl-ACP thioesterase family reveals significant diversity in enzymatic specificity and activity. BMC Biochem 12(1):44

Landriscina C, Gnoni GV, Quagliariello E (1972) Fatty-Acid Biosynthesis. Eur J Biochem 29(1):188-196

Leber C, Da Silva N (2014) Engineering of *Saccharomyces cerevisiae* for the synthesis of short chain fatty acids. Biotechnol Bioeng 111(2) 347-358

Leibundgut M, Maier T, Jenni S, Ban N (2008) The multienzyme architecture of eukaryotic fatty acid synthases. Curr Opin Struc Biol 18:714-725

Lepage G, Roy CC (1984) Improved recovery of fatty acid through direct transesterification without prior extraction or purification. J Lipid Res 25(12):1391-1396
Lomakin IB, Xiong Y, Steitz TA (2007). The crystal structure of yeast fatty acid synthase, a cellular machine with eight active sites working together. Cell 129(2): 319-332.
Li J, Zhang C, Deng M, Luo J. (2015) Investigation of the difference in liquid superlubricity between water-and oil-based lubricants. RSC Advances 5(78): 63827-63833
Lu X, Vora H, Khosla C (2008) Overproduction of free fatty acids in *E. coli*: implications for biodiesel production. Metab Eng 10(6): 333-339
Liu T, Vora H, Khosla C (2010) Quantitative analysis and engineering of fatty acid biosynthesis in *E. coli*. Metab Eng 12(4): 378-386
Liu H, Cheng T, Xian M, Cao Y, Fang F, Zou H (2014) Fatty acid from the renewable sources: a promising feedstock for the production of biofuels and biobased chemicals. Biotechnol Adv 32(2):382-389
Liu L, Pan A, Spofford C, Zhou N, Alper HS (2015) An evolutionary metabolic engineering approach for enhancing lipogenesis in *Yarrowia lipolytica*. Metab Eng 29: 36-45
Lynd LR, van Zyl WH, McBride JE, Laser M (2005) Consolidated bioprocessing of cellulosic biomass: an update. Curr Opin Biotechnol 16(5):577-583
Lynen F (1969) Yeast fatty acid synthase. Methods Enzymol 14: 17-33
Madzak C, Tréton B, Blanchin-Roland S (2000) Strong hybrid promoters and integrative expression/secretion vectors for quasi-constitutive expression of heterologous proteins in the yeast *Yarrowia lipolytica*. J Mol Microbiol Biotechnol 2(2):207-216
Madzak C (2015) *Yarrowia lipolytica*: recent achievements in heterologous protein expression and pathway engineering. Appl Microbiol Biotechnol 99(11): 4559-4577
McMahon MD, Prather KL (2014) Functional Screening and In Vitro Analysis Reveal Thioesterases with Enhanced Substrate Specificity Profiles That Improve Short-Chain Fatty Acid Production in *Escherichia coli*. Appl Environ Microbiol 80:1042-1050
Moreno-Pérez AJ, Venegas-Caleron M, Vaistij FE, Salas JJ, Larson TJ, Garces R, Graham IA, Martinez-Force E (2012) Reduced expression of FatA thioesterases in *Arabidopsis* affects the oil content and fatty acid composition of the seeds. Planta 235(3):629-639
Moser BR (2011) Biodiesel production, properties, and feedstocks. In: Biofuels. Springer, pp 285-347
Nakagawa T and Tori K (1960) Solubilization of long chain alkyl compounds by a non-ionic surfactant, and clouding formation observed in such systems. Kolloid-Zeitschrift 168.2:132-139
Ohlrogge JB, Jaworski JG (1997) Regulation of fatty acid synthesis. Annu Rev Plant Biol 48:109-136
Papanikolaou S, Aggelis G (2011) Lipids of oleaginous yeasts. Part II: Technology and potential applications. Eur J Lipid Sci Technol 113(8):1052-1073
Pollard MR, Anderson L, Fan C, Hawkins DJ, Davies HM (1991) A specific acyl-ACP thioesterase implicated in medium-chain fatty acid production in immature cotyledons of *Umbellularia californica*. Arch Biochem Biophys 284(2): 306-312
Qiao K, Abidi SHI, Liu H, Zhang H, Chakraborty S, Watson N, Stephanopoulos G (2015) Engineering lipid overproduction in the oleaginous yeast *Yarrowia lipolytica*. Metab Eng 29: 56-65
Ramos MJ, Fernández CM, Casas A, Rodríguez L, Pérez Á (2009) Influence of fatty acid composition of raw materials on biodiesel properties. Bioresource Technol 100(1):261-268
Runguphan W, Keasling JD (2014) Metabolic engineering of *Saccharomyces cerevisiae* for production of fatty acid-derived biofuels and chemicals. Metab Eng 21:103-113

Santos TM, Pinto A, Oliveira AV, Ribeiro HL, Caceres CA, Ito EN, Azeredo H (2014) Physical properties of cassava starch–carnauba wax emulsion films as affected by component proportions. Int J Food Sci Technol 49(9): 2045-2051

Sherkhanov S, Korman TP, Bowie JU (2014) Improving the tolerance of *Escherichia coli* to medium-chain fatty acid production. Metab Eng 25:1-7

Steen EJ, Kang Y, Bokinsky G, Hu Z, Schirmer A, McClure A, Keasling JD (2010) Microbial production of fatty-acid-derived fuels and chemicals from plant biomass. Nature 463(7280): 559-562

Tai M, Stephanopoulos G (2013) Engineering the push and pull of lipid biosynthesis in oleaginous yeas *Yarrowia lipolytica* for biofuel production. Metab Eng 15:1-9

Tariq M, Ali S, Ahmad F, Ahmad M, Zafar M, Khalid N, Khan MA (2011) Identification, FT-IR, NMR (1 H and 13 C) and GC/MS studies of fatty acid methyl esters in biodiesel from rocket seed oil. Fuel Process Technol 92(3):336-341

Viegas CA, Rosa MF, Sá-Correia I, Novais JM (1989) Inhibition of yeast growth by octanoic and decanoic acids produced during ethanolic fermentation. Appl Environ Microbiol *55*(1): 21-28

Voelker TA, Worrell AC, Anderson L, Bleibaum J, Fan C, Hawkins DJ, Davies HM (1992) Fatty acid biosynthesis redirected to medium chains in transgenic oilseed plants. Science 257(5066): 72-74

Voelker TA, Jones A, Cranmer AM, Davies HM, Knutzon DS (1997) Broad-range and binary-range acyl-acyl-carrier-protein thioesterases suggest an alternative mechanism for medium-chain production in seeds. Plant Physiol 114(2):669-677

von Wettstein-Knowles P, Olsen JG, McGuire KA, Henriksen A (2006) Fatty acid synthesis. FEBS J 273(4):695-710

Wakil SJ (1989) Fatty acid synthase, a proficient multifunctional enzyme. Biochem 28(11): 4523-4530

Wang G-Y, Zhang Y, Chi Z, Liu G-L, Wang Z-P, Chi Z-M (2014) Role of pyruvate carboxylase in accumulation of intracellular lipid of the oleaginous yeast *Yarrowia lipolytica* ACA-DC 50109. Appl Microbiol Biotechnol:1-9

White SW, Zheng J, Zhang YM, Rock CO (2005). The structural biology of type II fatty acid biosynthesis. Annu Rev Biochem 74: 791-831.

Xue Z, Sharpe PL, Hong SP, Yadav NS, Xie D, Short DR, Damude HG, Ruper RA, Seip JE, Wang J, Pollak DW, Bostick MW, Bosak MD, Macool DJ, Hollerbach DH, Zhang H, Arcilla DM, Bledsoe SA, Croker K, McCord EF, Tyreus BD, Jackson EN, Zhu Q (2013) Production of omega-3 eicosapentaenoic acid by metabolic engineering of *Yarrowia lipolytica.* Nat Biotechnol 31(8):734-740

Zhang YM, White SW, Rock CO (2006) Inhibiting bacterial fatty acid synthesis. J Biol Chem 281(26): 17541-17544

Zhang B, Chen H, Li M, Gu Z, Song Y, Ratledge C, Chen W (2013) Genetic engineering of *Yarrowia lipolytica* for enhanced production of trans-10, cis-12 conjugated linoleic acid. Microb Cell Fact 12(1): 70.

Zhang H, Zhang L, Chen H, Chen YQ, Chen W, Song Y, Ratledge C (2014) Enhanced lipid accumulation in the yeast *Yarrowia lipolytica* by over-expression of ATP:citrate lyase from *Mus musculus.* J Biotechnol 192:78–84

Zheng Y-N, Li L-L, Liu Q, Yang J-M, Wang Z-W, Liu W, Xu X, Liu H, Zhao G, Xian M (2012) Optimization of fatty alcohol biosynthesis pathway for selectively enhanced production of C12/14 and C16/18 fatty alcohols in engineered *Escherichia coli.* Microb Cell Fact 11: 65

# 430012 Evolutionary Engineering of *Geobacillus Thermoglucosidans* for Improved Ethanol Production

**Abstract:**

Thermophiles are organisms that grow at high temperatures (>50°C). Many researchers have proposed that thermophilic organisms are better suited for the production of cellulosic biofuels than mesophilic organisms as the overall processes are less susceptible to contamination and potentially more energy efficient. In this study, we explored the thermophile *Geobacillus thermoglucosidans* for the production of bio-ethanol. This bacterium is a facultative anaerobe, grows at an optimal temperature 60°C, and can ferment diverse carbohydrates including glucose, cellobiose, xylose, and arabinose. However, it natively performs mixed acid fermentation. To improve ethanol productivity in this thermophile, we engineered and evolved two strains of *G. thermoglucosidasius* BGSC 95A1 (95A1) and C56-YS93 (C56). We first deleted lactate dehydrogenase (Ldh) and pyruvate formate lyase (Pfl) in order to eliminate lactate and formation production and then heterologously expressed pyruvate decarboxylase (Pdc) from *Sarcina ventriculi* to directly produce ethanol from pyruvate. Elimination of Pfl, however, slowed growth significantly because the strain was unable to generate acetyl-CoA under anaerobic conditions. To restore growth, we added acetic acid to media to provide a source for acetyl-CoA. The strains were then evolved on glucose and cellobiose. We found that the evolved strains of 95A1 were able to efficiently produce ethanol during growth on glucose or cellobiose whereas the evolved C56 strains performed poorly. Genome sequencing of evolved 95A1 strains identified common loss-of-function mutations in adenine phosphoribosyltransferase (Aprt) and the stage III sporulation protein AA (SpoIIIAA). We disrupted the two genes encoding them and found that deletion of both enhanced the ethanol production, especially Aprt. In conclusion, we were able to engineer a strain of *G. thermoglucosidans* to efficiently produce ethanol from glucose and cellobiose using a combination of metabolic engineering and evolutionary strategies. This work helps establish this thermophile as a platform organism for biofuel and biochemical production.

# 435058 Rewiring Photosynthesis for Terpene Production

**Abstract:**

One of the most fundamental challenges in agriculture and bioenergy is the efficient partition of photosynthates toward a target carbon sink. We used squalene as a model carbon sink to evaluate how photorespiratory glycolate could be redirected to pyruvate and subsequently terpene biosynthesis. Both computational and experimental studies suggested that a functional C2 redirection could increase terpene yield without significantly impacting photosynthetic carbon fixation. The squalene yield in transgenic tobacco implemented with various C2 redirection designs increased up-to four-fold as compared to the lines engineered for terpene biosynthesis only. The C2 redirection has enabled a record level production of squalene at 2.7 mg/g fresh weight (FW) of tobacco leave. Further analysis of carbon flux by feeding of radioactive labelled glycolate indicated that C2 redirection led to increased carbon flux from glycolate to malate and squalene. The C2 redirection plants provided optimal models to study how photosynthesis might be rewired to increase a target compound yield. Metabolomics analysis revealed a significant carbon repartition, as a decrease in the intermediates of sucrose and starch biosynthesis correlated with an increase in malate and pyruvate in the C2 redirection lines. Correlating with the computational study, we have established that redirecting phtorespiratory C2 products to pyruvate could result in significant carbon repartition from storage carbohydrates to terpene without compromising photosynthesis. In addition, recent advances indicated that the integration of carbon repartition by C2 redirection with improved carbon fixation can further increase squalene production to more than 7mg/g FW in tobacco, which represented a more than 10 fold increase of the current state-of-the-art. Together, the results indicated that photosynthetic organisms have the potential to be engineered for efficient production of high value terpene compounds, when proper strategies for carbon repartition and photosynthesis improvement were integrated. The study has provided a new platform for the production of nutraceutical and pharmaceutical compounds.

# 413123 Oxygen-Responsive Genetic Circuits Constructed in Synechocystis Sp. PCC 6803

**Abstract:**

Cyanobacteria exhibit promise as sustainable biotechnology platforms. However, few synthetic regulatory circuits have been developed for these non-model organisms, and the simple circuits that have been created rely on the addition of expensive chemical inducers. Environmental stimuli are often less costly, and can be physiologically relevant to production. For instance, oxygen irreversibly inactivates molybdenum-dependent nitrogenase, the enzyme used by diverse prokaryotes to fix nitrogen. Employing the fumarate and nitrate reduction system from *Escherichia coli*, we developed an oxygen-responsive circuit that can be used to control heterologous expression of nitrogenase in *Synechocystis* sp. PCC 6803. As designed, transcripts for the genes encoding nitrogenase's structural proteins, *nifHDK*, were produced only in the absence of oxygen. In addition, we utilized the oxygen-responsive transcriptional regulator in a two-input AND gate, which allows *Synechocystis* to control heterologous gene expression in response to two signals, both low oxygen and high anhydrotetracycline. To our knowledge, this two-input AND gate is the first complex genetic circuit built in a cyanobacterial strain, enabling more sophisticated regulatory control. This work expands the synthetic biology toolbox for specific gene expression in cyanobacteria, furthering their utility in bio-manufacturing.

# 431226 Elucidating the Metabolism of Rhodococcus Opacus PD630 Via Parallel 13C-Metabolite Fingerprinting

**Abstract:**

*Rhodococcus opacus* PD630 is a promising microbial host for lignocellulosic biofuel production, due to its high tolerance to aromatic compounds and accumulation of triacylglycerol, a biodiesel precursor. To quickly elucidate the metabolism of this non-model organism, several parallel $^{13}C$-fingerprinting experiments were designed to probe multiple pathways in *Rhodococcus opacus* PD630. Our $^{13}C$-fingerprinting approach resulted in several findings: 1) PD630 strongly prefers to utilize the Entner-Doudoroff (ED) pathway over Emden-Meyerhof (EMP) and the pentose phosphate pathway, 2) highly active malic enzyme and anaplerotic pathways, 3) uptake & utilization of exogenous amino acids, and 4) no carbon catabolite repression detected on growth with glucose and phenol. In addition, our results indicate that PD630 uses the ED pathway and gluconeogenesis concurrently for co-utilization of glucose and phenol. Gluconeogenesis is typically inhibited when glucose is catabolized through the EMP pathway, thus PD630's preference for the ED pathway could be a key strategy to avoid this inhibition under growth with multiple carbon sources. These findings have revealed several interesting features of PD630 metabolism as well as demonstrated the potential of parallel $^{13}C$-metabolite fingerprinting. This method is a qualitative alternative to $^{13}C$-Metabolic Flux Analysis that significantly reduces the time and expense for analyzing cell metabolisms. For labs that need quick and inexpensive characterization of non-model species, $^{13}C$-fingerprinting can be the most ideal approach. To promote this technique, we have developed a prototype $^{13}C$-fingerprinting toolkit that consist of $^{13}C$-tracers, an 8-well block that can be used as to cultivate small volumes (~4mL) of culture, a gassing manifold and instructions to enable other microbiology labs to easily adapt this tool into their characterization process of cell metabolisms.

# 437397 Biomaterials Control Cellular Behavior through Materials Engineering

**Abstract:**

< style="padding-left: 47pt; text-align: center;">Biomaterials Control Cellular Behavior through Materials Engineering

Ali Khademhosseini, Ph.D.

Department of Medicine, Brigham and Women's Hospital, Harvard Medical School, Cambridge, MA

Harvard-MIT Division of Health Sciences and Technology, MIT, Cambridge, MA Wyss Institute for Biologically Inspired Engineering, Harvard University, Boston, MA alik@rics.bwh

Dr. Ali Khademhosseini, 65 Landsdowne Street, PRB 252, Cambridge, Massachusetts 02139, United States

< style="padding-left: 5pt; text-align: left;">Keywords: Biomaterials, Hydrogels, Regenerative Engineering, Tissue Engineering, Nanotechnology, Polymer Chemistry, Photocrosslinkable, Bioprinting, Photolithography

< style="padding-left: 5pt; text-align: left;">Abstract: Engineered materials that integrate advances in polymer chemistry, nanotechnology, and biological sciences have the potential to create powerful medical therapies. Our group aims to engineer tissue regenerative therapies using water-containing polymer networks called hydrogels that can regulate cell behavior. Specifically, we have developed photocrosslinkable hybrid hydrogels that combine natural biomolecules with nanoparticles to regulate the chemical, biological, mechanical and electrical properties of gels. These functional scaffolds induce the differentiation of stem cells to desired cell types and direct the formation of vascularized heart or bone tissues. Since tissue function is highly dependent on architecture, we have also used microfabrication methods, such as microfluidics, photolithography, bioprinting, and molding, to regulate the architecture of these materials. We have employed these strategies to generate miniaturized tissues. To create tissue complexity, we have also developed directed assembly techniques to compile small tissue modules into larger constructs. It is anticipated that such approaches will lead to the development of next-generation regenerative therapeutics and biomedical devices.

JENIKE & JOHANSON
SCIENCE ENGINEERING DESIGN

# Powder Characterization and Trouble-Shooting Methods for Successful Process-Scale up

November 9, 2015
2015 AIChE Annual Meeting
Thomas Baxter, Sr. Consultant, tbaxter@jenike.com
James Prescott, Director/Sr. Consultant, jkprescott@jenike.com

## Outline

- Common powder handling challenges during scale-up
  - Obtain reliable flow from bins, hoppers and chutes
  - Achieve target transfer and production rates
  - Reduce segregation and content uniformity (CU) variation
- Powder characterization tests
  - Characterize powder flow (cohesive strength, wall friction)
  - Assess feed rate stability (permeability, compressibility)
  - Analyze segregation potential
- Trouble-shooting methods during process scale-up
  - Address powder flow problems
  - Trouble-shoot feed rate concerns
  - Reduce the risk of segregation and CU concerns

JENIKE & JOHANSON SCIENCE | ENGINEERING | DESIGN 2

# Powder characterization during development

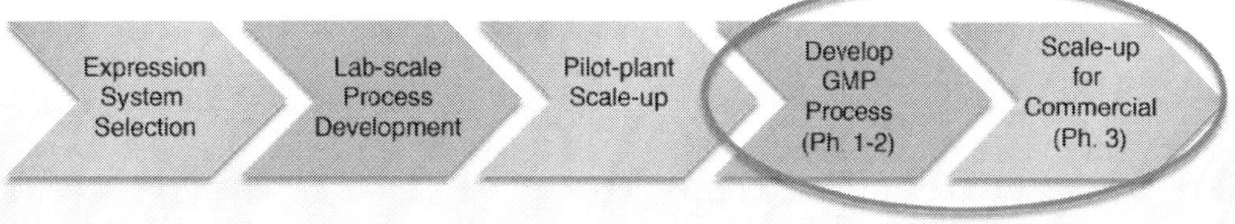


Powder characterization and scale-up concerns during development

- *Early development (Phase I and II)*
  - Characterize API, excipient and blend properties that affect flow and segregation; E.g., cohesive strength, particle size and shape, segregation potential of candidate formulations
  - Optimize formulation: Evaluate/optimize formulas and quantities (API, lubricant, flow aids), assessment of effect of different component suppliers/grades
  - Evaluate manufacturing process alternatives (e.g., granulation vs. direct blend)
- *Late stage development, Registration batches (Phase III), Commercial scale-up*
  - Design equipment to prevent flow and segregation/CU problems
  - Evaluate equipment and/or site transfer
  - Trouble-shoot and address flow and CU problems



# Powder handling challenges: Obtain reliable flow

Cohesive arch

Rathole

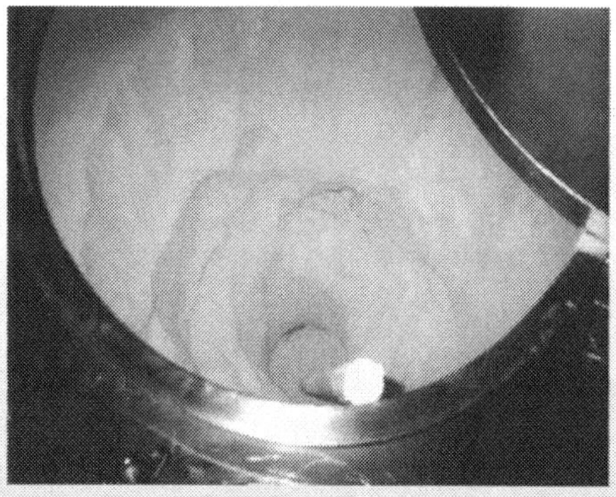

Flow problems can result in:

- Production delays, quality issues, increased operator intervention and exposure

Flow problems can be predicted based upon:

- Flow properties: Cohesive strength, wall friction, compressibility, permeability
- First-principle predictive models
- Key equipment and process parameters: IBC/hopper geometry, wall surface finish, outlet size, storage at rest time, % fill, temperature and % relative humidity

## Powder handling challenges: Achieve target production rates

*E.g., Increased compression variation at higher press speeds*

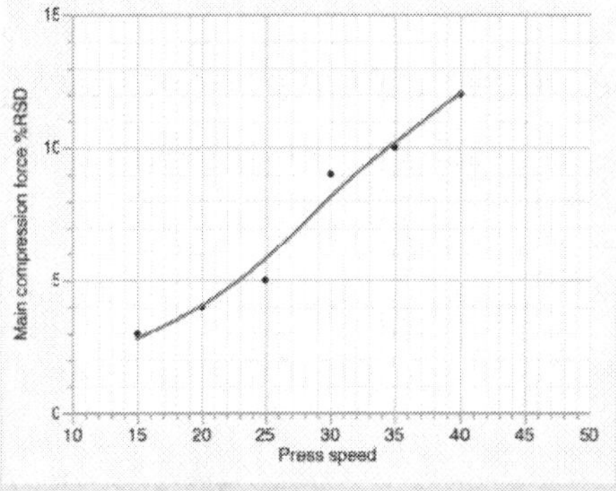


Feed rate limitations can result in:

- Below-target production rates, delays (press shut-downs), quality issues (tablet weight and hardness variation) and increased operator intervention

Flow problems can be predicted based upon:

- Flow properties: *Permeability*, PSD, compressibility, cohesive strength
- First-principle predictive models: "two-phase" (powder:air) flow modeling (*later slides*)
- Key equipment and process parameters: IBC/hopper geometry, outlet size, transfer and feed rates, venting and pressure differential in feed system

JENIKE & JOHANSON SCIENCE | ENGINEERING | DESIGN 5

## Powder handling challenges: Reduce segregation/CU variation

*E.g., Segregation potential in blender to IBC transfer step*

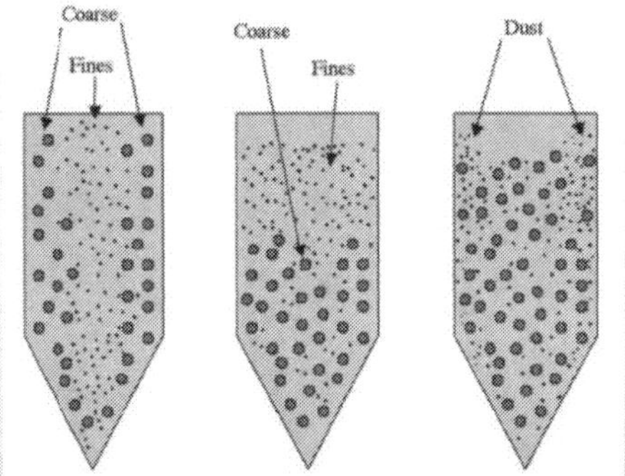


Segregation can result in:

- Quality concerns (CU variation), rejected batches, product validation and launch delays, increased operator intervention

Segregation problems can be predicted and diagnosed based upon:

- Segregation tests and flow properties (flow pattern)
- Stratified sampling data
- Key equipment and process parameters: IBC/hopper geometry (flow pattern), transfer steps, transfer rates, venting and de-acceleration methods

JENIKE & JOHANSON SCIENCE | ENGINEERING | DESIGN 6

## Batch process: Potential powder handling concerns for direct blend process (*italics = higher risk areas*)

- Process steps upstream of blending (weigh, screen, charging)
  - Active loss during transfer after weigh-out
  - Consistent feed through screening step
  - Pneumatic or mechanical transfer of components to blender
- Blending (pre-blend, final and lubrication blend)
  - *Obtain required blend uniformity (BU)*
  - *Successful scale-up of blend processes*
  - *Reliable discharge from blender*
- Transfer of blend to intermediate bulk container (IBC) or press
  - *Potential for segregation during transfer steps (blender-to-IBC and/or IBC-to-press)*
  - *Reliable discharge from IBC*
  - Reliable transfer while minimizing dusting, build-up and active loss
  - Consistent feed into press feed frame at target compression rate
  - *Obtain required content uniformity (CU)*

JENIKE & JOHANSON SCIENCE | ENGINEERING | DESIGN 7

## Continuous process: Potential powder handling concerns for direct blend process (*italics = higher risk areas*)

- Process steps upstream of blending (weigh, screen, *store and meter*)
  - *Reliable storage and discharge from IBCs, esp. for more difficult-to-handle blend components (e.g., API, lubricant)*
  - *Reliable transfer from IBCs to gravimetric feeders (fast/controlled refills)*
  - *Obtain required accuracy from gravimetric feeder at target feed rate*
- Blending (pre-blend, final and lubrication blend)
  - Obtain required BU (at pre-blend and/or final blend stage)
  - Successful scale-up of blend processes (i.e., scale-up may be more dependent on throughput rate than physical scale-up of blender)
  - Two blenders/steps may be required (pre-blend and final/lube blend)
- Transfer of blend to press
  - Potential for segregation during transfer (often minimized vs. batch)
  - Reliable transfer of blend, *esp. potential risks of vacuum conveying* (API loss/buildup, segregation, particle/granule attrition)
  - Consistent feed into press feed frame at target compression rate (*potential concern if running at high compression rates*)
  - Obtain required CU (potential benefit of increased on-line PAT and real time release)

JENIKE & JOHANSON SCIENCE | ENGINEERING | DESIGN 8

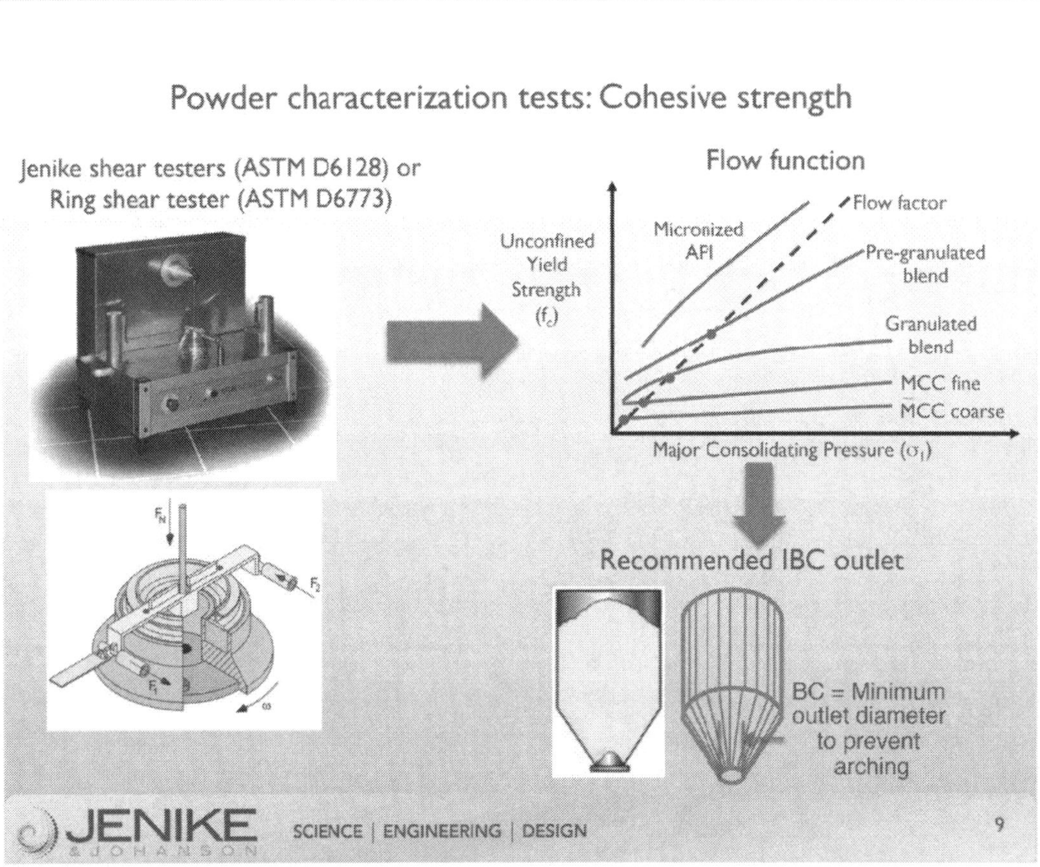
Powder characterization tests: Cohesive strength
Jenike shear testers (ASTM D6128) or
Ring shear tester (ASTM D6773)
Flow function
Unconfined Yield Strength ($f_c$)
Micronized API
Flow factor
Pre-granulated blend
Granulated blend
MCC fine
MCC coarse
Major Consolidating Pressure ($\sigma_1$)
Recommended IBC outlet
BC = Minimum outlet diameter to prevent arching
JENIKE & JOHANSON
SCIENCE | ENGINEERING | DESIGN
9

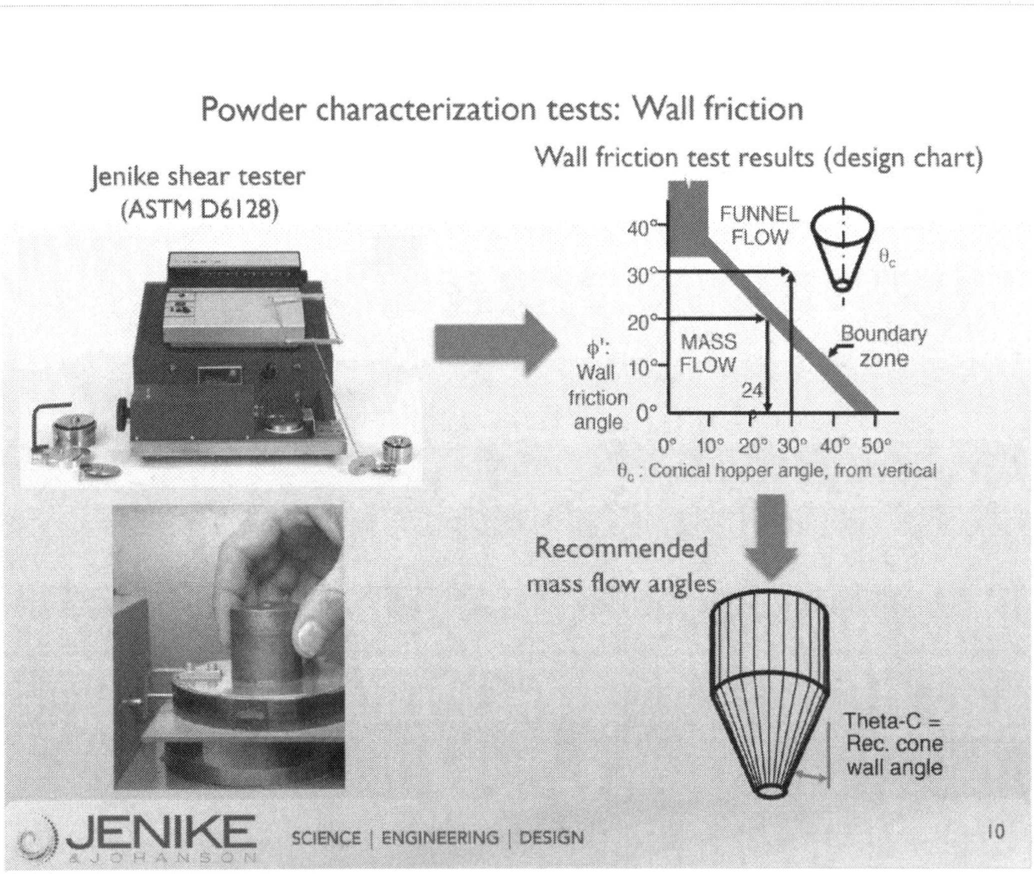
Powder characterization tests: Wall friction
Jenike shear tester
(ASTM D6128)
Wall friction test results (design chart)
FUNNEL FLOW
MASS FLOW
$\theta_c$
Boundary zone
$\phi'$: Wall friction angle
40°
30°
20°
10°
0°
24
0° 10° 20° 30° 40° 50°
$\theta_c$ : Conical hopper angle, from vertical
Recommended mass flow angles
Theta-C = Rec. cone wall angle
JENIKE & JOHANSON
SCIENCE | ENGINEERING | DESIGN
10

## Powder characterization tests: Permeability and compressibility

Permeability tester

Compressibility tester(ASTM 6683)

Two-phase flow analysis and Critical feed rates

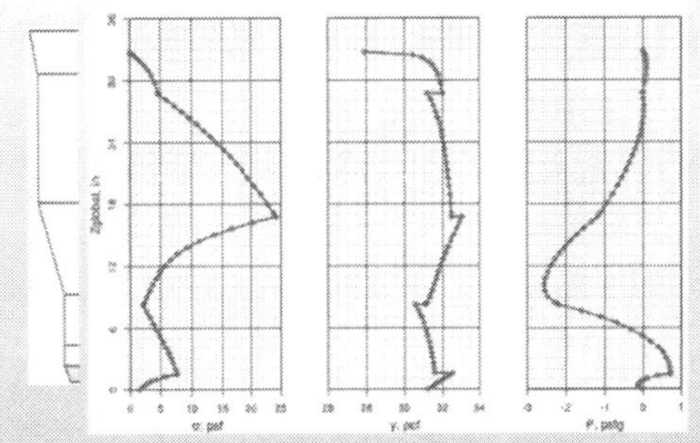




## Powder characterization tests: Fluidization segregation tests

Fluidization segregation tester

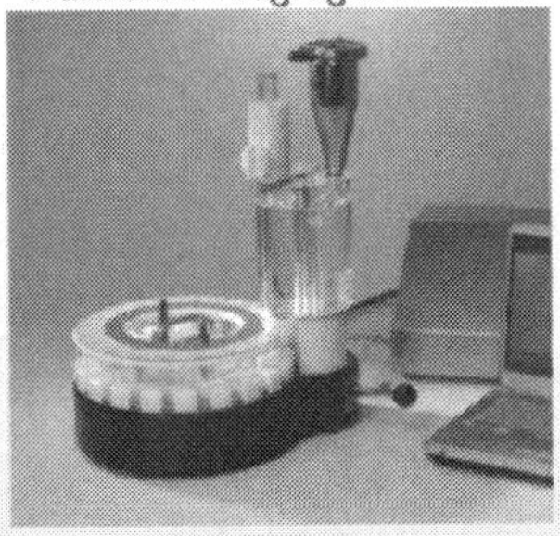

Fluidization segregation test method

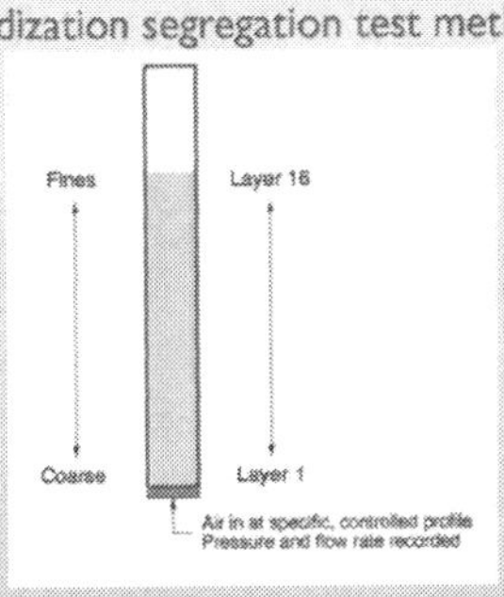


Fluidization segregation test results

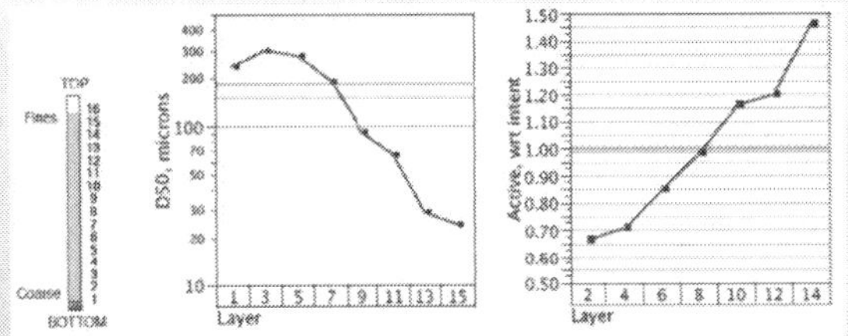

## Powder characterization tests: Sifting segregation tests

Sifting segregation tester

Sifting segregation test method

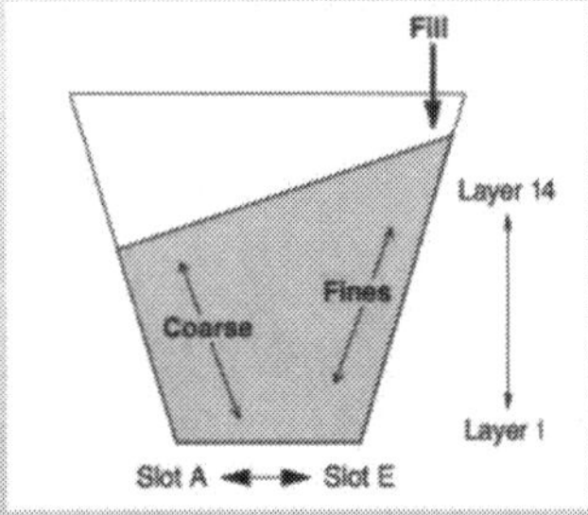


Sifting segregation test results

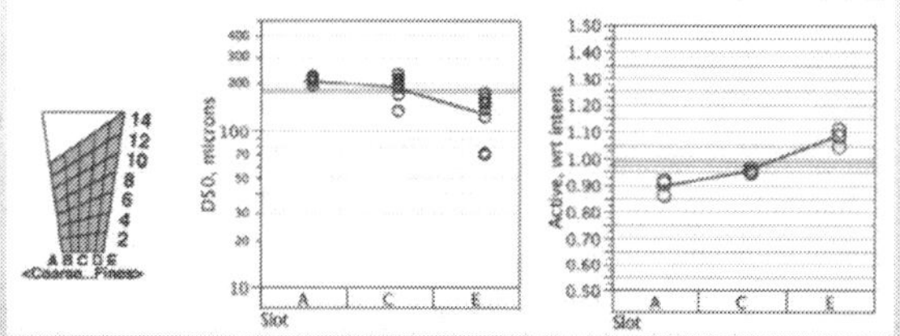


JENIKE & JOHANSON SCIENCE | ENGINEERING | DESIGN 13

## Trouble-shooting: "No Flow" and Erratic Flow Problems

Trouble-shooting methodology:

1) Conduct flow properties tests
2) Identify the flow pattern (mass flow vs. funnel flow)
3) Identify location of arch and/or rathole
4) Identify key equipment parameters
   - Bin and press parameters: capacity, wall angles, interior wall surface, outlet shape/size
   - Transfer chute: height, cross sectional area, interior surface finish, chute angles

Measure flow properties

Identify flow patterns

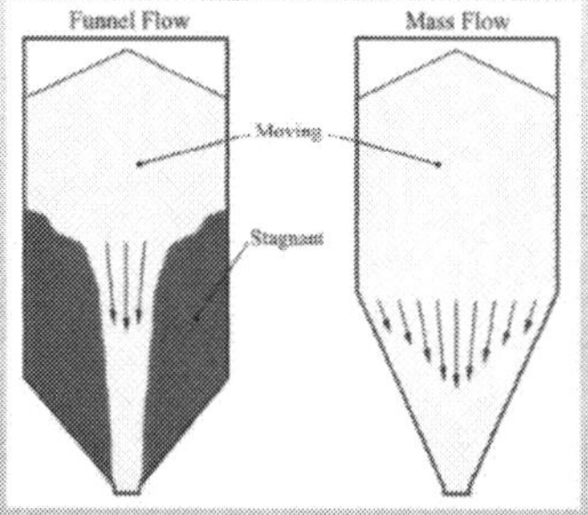


JENIKE & JOHANSON SCIENCE | ENGINEERING | DESIGN 14

## Trouble-shooting: "No Flow" and Erratic Flow Problems

- *Potential* corrective actions to prevent ratholes
  1) Design equipment for *mass flow* (eliminate stagnant material)
  2) Reduce capacity/bin level (reduce effective head)
  3) Size the outlet larger than critical rathole diameter ("Df")
  4) Use an agitator (e.g. Matcon bin, Vibroflow™)
- *Potential* corrective actions to prevent arches (mass flow)
  1) Size the outlet larger than "Bc/Bp"
  2) Use an agitator (e.g. Matcon bin, external vibrator)
  3) Change the storage conditions (Temp., RH, storage at rest)
  4) Change the material (particle size/shape, add flow aid)

Ratholing dimensions (funnel flow)

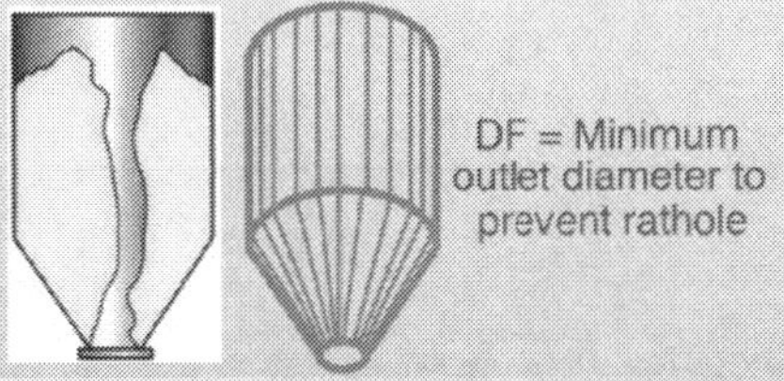


Arching or bridging (mass flow)

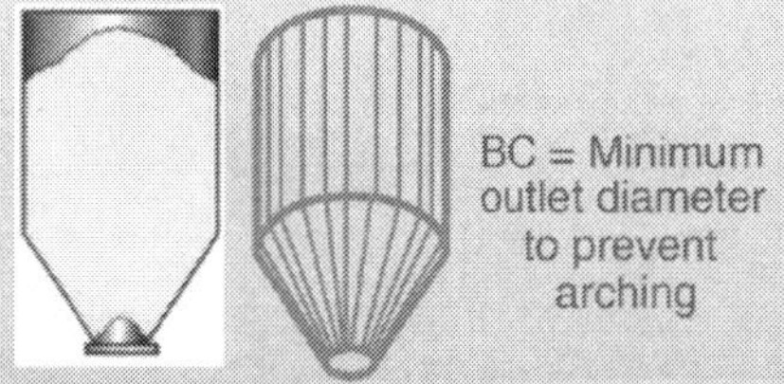




## Trouble-shooting: Feed Rate Problems

- Trouble-shooting methodology
  1) Identify the flow pattern
  2) Conduct flow properties tests (permeability, density)
  3) Identify key process parameters: rate, material level, transfer rate to equipment, desecration time
  4) Identify key equipment parameters
     - Bin and press parameters: capacity, wall angles, interior wall surface, outlet shape/size
     - Transfer chute: height, cross sectional area, interior surface finish, chute angles
     - *Venting/vacuums in the feed system*
  5) Two-phase flow modeling of potential corrective actions

Permeability tester

## Trouble-shooting: Feed Rate Problems

- *Potential* corrective actions for feed rate limitations
  1) Design equipment for mass flow
  2) Increase outlet size (if possible)
  3) Reduce capacity/bin level (reduce effective head/consolidation pressure)
  4) Account for pressure differentials and vacuums (e.g., press vacuum)
  5) Reduce production rate (tableting rate)
  6) Add air permeation (passive or active air permeation, *see analysis below*)
  7) Change the material to increase permeability (increase particle size)

Two-phase flow analysis of press feed system without venting (red) and with venting (green)

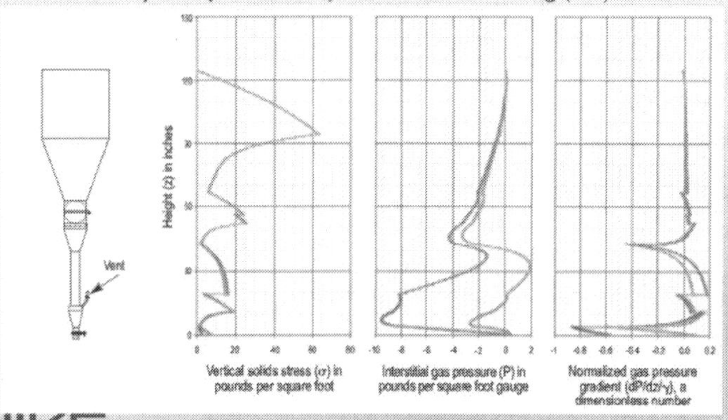




## Trouble-Shooting: Segregation/CU Concerns

Trouble-shooting methods

1) Identify the flow pattern
2) Conduct flow properties and segregation tests
3) Analysis of stratified (systematic) blend and content uniformity data
   - Conduct ANOVA and assess stray values, trending, "hot spots", assay shift
4) Identify key equipment and process parameters
   - Bin and press parameters
   - Transfer chute parameters: dimensions, drop heights
   - Venting and valves; fill and transfer rates

**Conduct segregation tests**

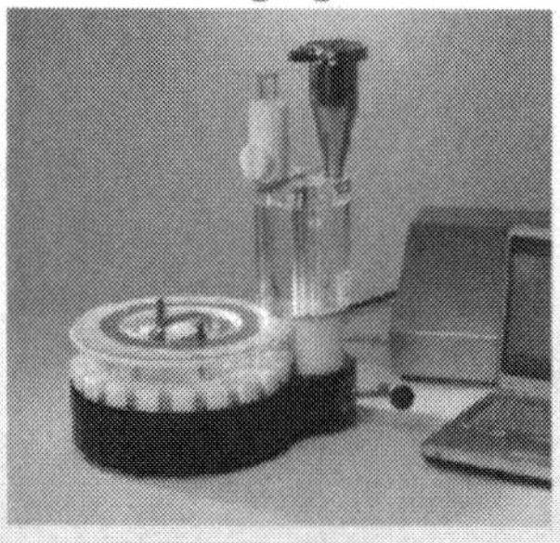

**Analyze stratified CU data**

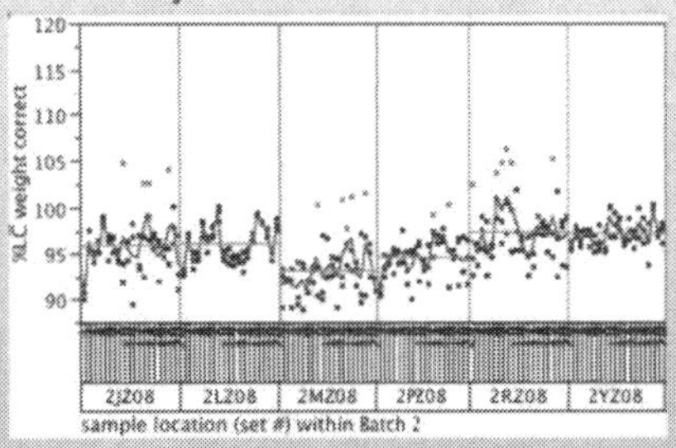


**Characterize CU variation: E.g., "Trending"**

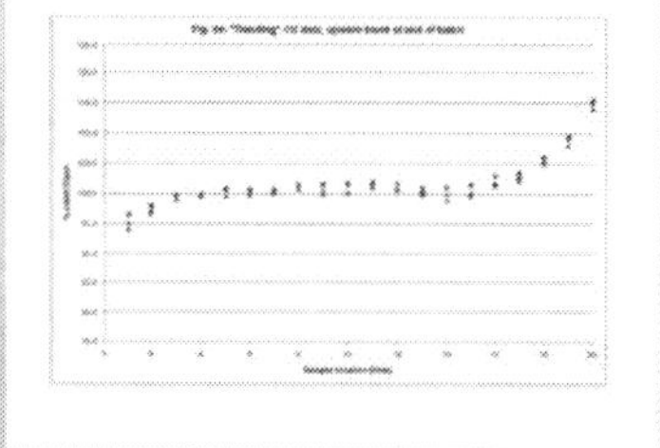

## Trouble-Shooting: Segregation/CU Concerns:

*Characterizing the type of CU variation*

"Trending"

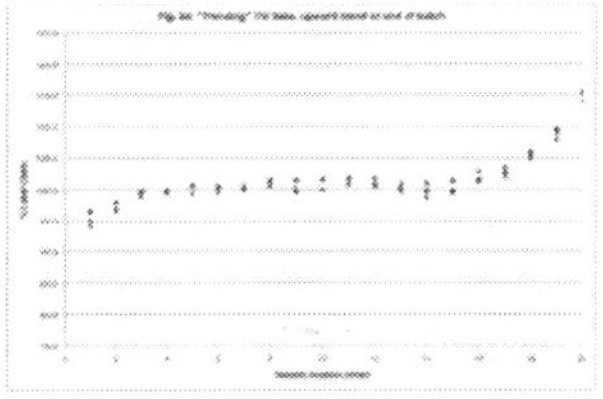

"Wandering"

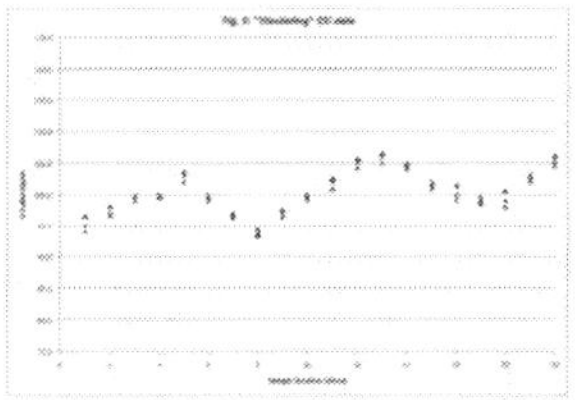

"Scatter"

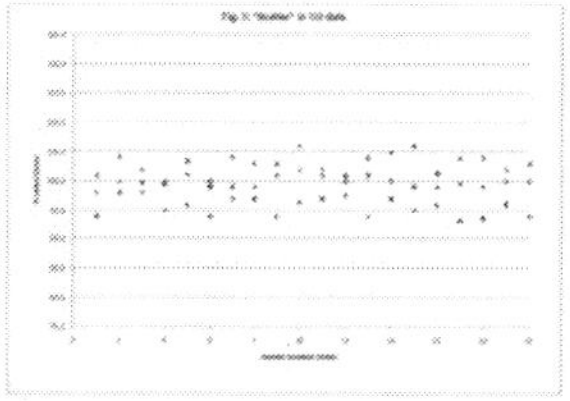

"Hot Spot"

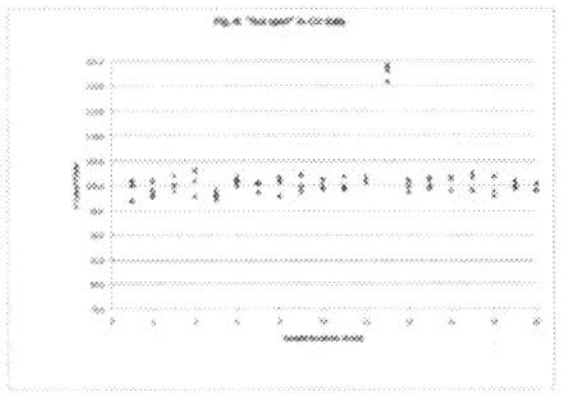

"Stray Value"

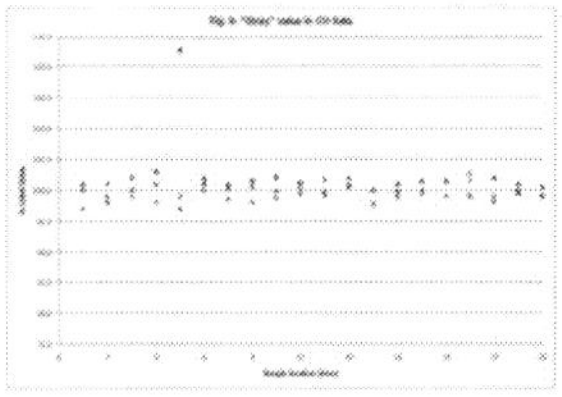

*Note that different investigative methods and corrective actions may be required for the different CU issues*



## Trouble-Shooting : Segregation/CU Concerns:

*CU Trending vs. Flow Pattern*

Note that trending may not always occur at the beginning or end of a batch

Mass flow

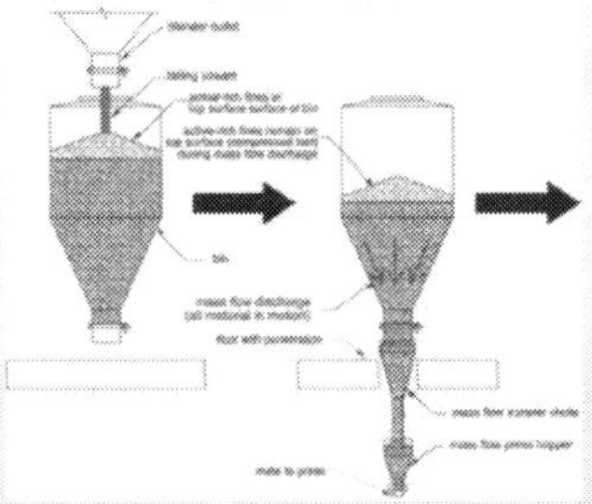

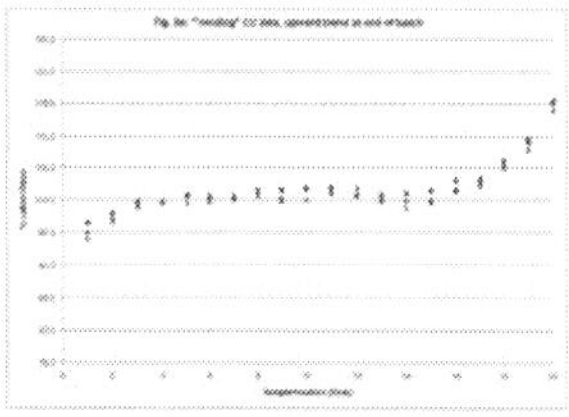

Funnel flow

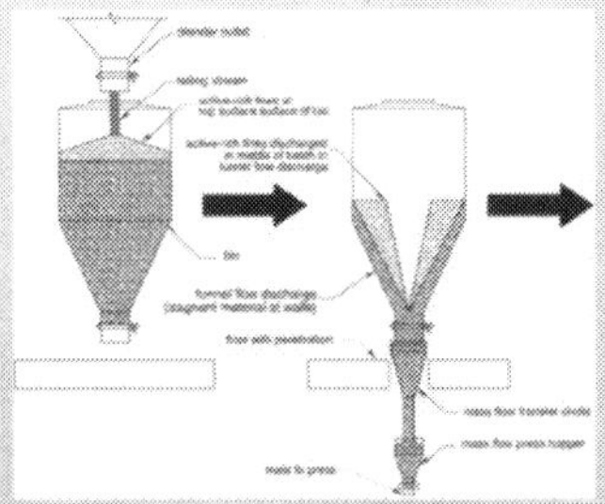

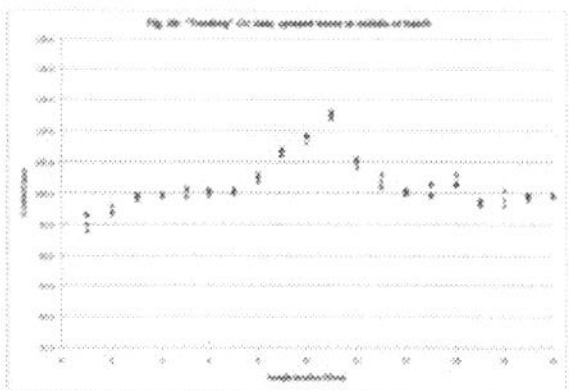

# Trouble-Shooting Segregation/CU Concerns:

*Potential Corrective Actions*

- *Potential* corrective actions to address CU concerns
  1) Change the material (particle size distribution, increase cohesion, granulate, ordered vs. random blend)
  2) Reduce transfer steps (e.g., in-bin blending)
  3) Design for mass flow (may not be effective in addressing fluidization)
  4) Control velocity gradients
  5) Control gravity transfers (blend-to-bin, bin-to-press) with additional valves and venting (*see following slide*)

Mass flow velocity gradients

E.g., Anti-segregation BINSERT®

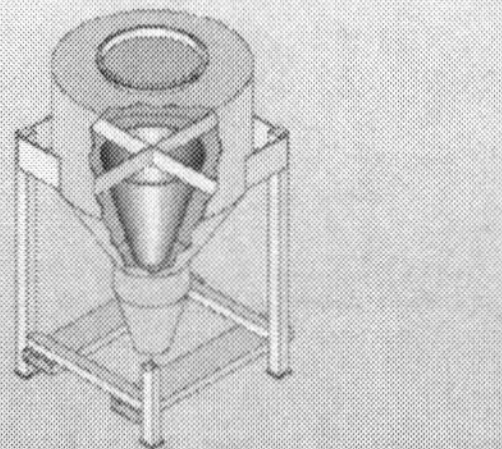



# Trouble-Shooting: Segregation/CU Concerns:

*Example of reducing CU variation*

Transfer chute modification

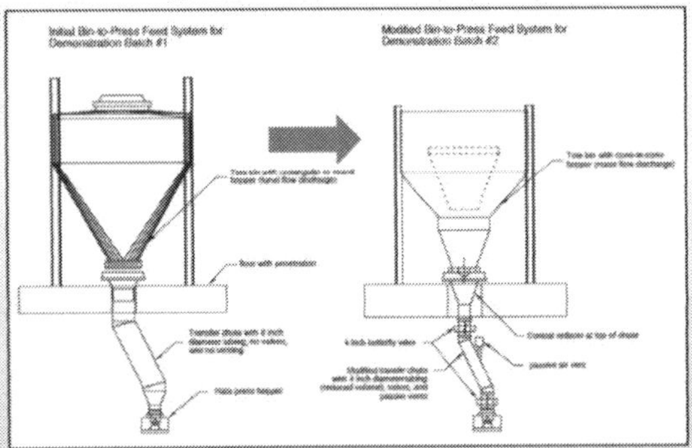

Reduction in CU trending

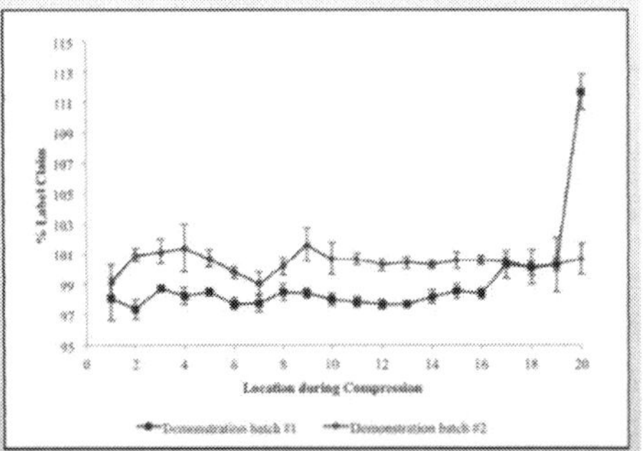


Ref.: "Addressing Segregation of a Low Dosage Direct Blend during Commercial Scale-Up"; Nipun Davar, Thomas Baxter, Pauly Kavalakatt, Sangita Ghosh, Herbert Schock, Pharmaceutical Technology, February, 2011

# Conclusions

- Assess the key powder handling concerns prior to scale-up
  - Design equipment (bins, hoppers and chutes) for reliable flow at target production rates and reduce risks of segregation and CU variation
  - Key batch vs. continuous process concerns may be different
- Conduct powder characterization tests to assess risk
- Use powder test outputs as inputs to process & equipment design
- Utilize trouble-shooting methods during scale-up to address
  - Powder flow problems
  - Feed rate concerns
  - Segregation risks and CU concerns

*The authors wish to acknowledge all their colleagues at Jenike & Johanson for their contributions*

JENIKE & JOHANSON SCIENCE | ENGINEERING | DESIGN 23

# 429404 Establishing Content Uniformity Criteria throughout 3-Stage Validation: Estimate, Confirm, and Control

**Abstract:**

Two recent events have prompted the development of establishing content uniformity throughout a 3-stage process validation. The first was the withdrawal of the "Powder Blends and Finished Dosage Units - Stratified In-Process Dosage Unit Sampling and Assessment". The second was the re-emphasis that the Uniformity of Dosage Units (USU) / USP <905> is intended to be applied to product on the market, not as a release test. The goal is to guard against %RSD of 7.5%. During process validation Stage 1 and 2, an increased sample size is recommended to more accurately determine the process mean and standard deviation. The Stage 3 process validation will use the appropriate sample size based on the true estimate of the standard deviation from the larger sample size in Stage 2. Statistically based sampling plans and acceptance criteria is then developed for release to assure quality product on the market meets the USP <905>. This 3-Stage approach provides a path to estimate, confirm and control the product quality.

# 422206 Fda's Office of Process and Facilities: Facilitating Pharmaceutical Manufacturing Quality and Innovation

**Abstract:**

Christine Moore, Acting Director of the Office of Process and Facilities (OPF), will discuss how FDA CDER has re-organized its quality function into one super office—Office of Pharmaceutical Quality (OPQ) in mid-January 2015. One of the landmarks is that the newly established Office of Process and Facilities (OPF) is dedicated to and fully responsible for pharmaceutical manufacturing process review and facility inspection. The Office of Process and Facilities (OPF) ensures that quality is built into manufacturing processes and facilities over the product lifecycle. It's vision is to be the premiere organization for integrated manufacturing process review and inspection of pharmaceutical regulation, globally. Given the primary functions of newly established OPF and its potential huge impact to the pharmaceutical industry and pharmaceutical engineering practice, PD2M Forum is an appropriate avenue to share these changes.

# 440211 Systems Based Pharmaceutics Alliance: Research Opportunities for Academics

**Abstract:**

Officially launched in early 2014, the Systems Based Pharmaceutics industrial alliance seeks to close the gap between theoretical developments in process modeling and simulation; and the practical implementation of these technologies into the development and manufacture of new medicines.

" The SbP Alliance aims to develop a single model-based framework for drug substance manufacturing, drug product manufacturing, oral absorption and pharmacokinetics. The Alliance will seek to establish best practices in the application of model-based techniques which will be incorporated in validated commercial-quality computer software for the solution of industrially-important problems." (PSE Press Release, 20-Mar-2014)

During its two years of operation, the subject matter experts from the participant companies have implemented models that represent useful and practical decision making tools for the development of new drugs. The focus of the alliance for its Phase I (years 1 and 2), was the integration of models spanning the crystallization, filtration, wet and dry milling of API; and the resulting performance of the crystal in the drug product in models for dissolution, the ASD apparatus, content uniformity and oral absorption.

As the teams were implementing existing published models and approaches, several gaps were found to answer key dependencies to properly mimic dependencies and behaviors of interest in these system. This talk will present (primarily to the academic community) those areas of research that were identified as key areas of academic research where advances would make immediate impact to industrial needs.

# 424463 Advanced Platform Strain Engineering

**Abstract:**

Microbial platform strains are robust strains with a high capacity for production of several related biological products. The effort that goes into designing and building a platform strain for an initial product should ideally reduce the time and cost it takes to get subsequent related products to market. However, the challenges faced by scientists and engineers in taking a highly engineered strain from one product to another are different than if they were to start from scratch. For example, replacing the pathway genes for the first product with those for the new product is often not trivial both from a molecular biology perspective (i.e. multiple copies of the old genes may be present in the genome), and from the pathway perspective (i.e. upstream pathway flux must be balanced for the new enzymes to avoid intermediate toxicity). Additionally, a highly engineered strain may have developed product-specific mutations related to product toxicity and transport that are not beneficial for tolerance of the new product. This presentation will discuss some of these challenges and the solutions that have been successfully applied to Amyris' terpene-producing platform yeast strains.

# 431064 Expanding the Engineered Modular Ester Fermentative Pathways in Escherichia coli

**Abstract:**

Acid production is a major (by)product in chemical production as well as fermentative metabolism. These products are sometimes desired, but are often times undesired. These metabolically undesired products are considered waste carbon the cells do not use for directed formation of the desired product. Here, we establish a systematic approach to utilize waste acids to be anaerobically converted into useful ester molecules via our previously established *Escherichia coli* modular cell platform. The fermentative pathways are designed to convert acids to their respective ac(et)yl-CoAs and produce ethanol as precursor molecules. The ac(et)yl-CoAs and ethanol are then condensed by a library of alcohol ac(et)yl transferases (AATs) to i) demonstrate expansion of the unique fermentative ester library and ii) facilitate a comparative analysis of sequence/structure to function of AATs. We demonstrate the anaerobic fermentation from glucose and acids for directed production of various unique ethyl esters. This work expands anaerobic fermentative ester production for the use in flavorings, fragrances, and solvents, as well as provides comparative insights into a sample library of AATs.

# 418776 Direct Utilization of Sorghum Extract Sugars As a Feedstock for Fatty Acid Production Using Metabolically Engineered Escherichia coli Strains

**Abstract:**

Depleting petroleum resources, increasing energy demand and environmental concerns have attracted significant attention to develop novel processes for production of renewable hydrocarbon fuels. Production of microbial high-density fuels using 'consolidated bioprocessing' is of enormous interest. Current study reports on direct utilization of sorghum extract sugars as a renewable carbon source for free fatty acid production with the help of metabolically engineered *E. coli*. In order to reduce the production cost of fatty acids, direct sucrose utilizing *E. coli* strains were developed by incorporating pUR400 plasmid into ML103 pXZ18 and ML103 pXZ18Z to generate ML103 pXZ18 pUR400 and ML103 pXZ18Z pUR400. Newly created strains showed to utilize mixture of pure sugars and sorghum juice efficiently. The 24h pH adjusted culture of ML103 pXZ18Z pUR400 produced maximum of 6.18 ± 0.52 g/L with 30 g/L of pure sugar mixture and 2.95 ± 0.04 with sorghum juice (15 g/L equivalent sugar concentration.).

**(54e) High Production of Fatty Alcohols By Escherichia coli Modified with Tandem Repeats Assisted Genome Editing (TRAGE) Method**

Xing, J., Institute of Process Engineering, Chinese Academy of Sciences
Wang, Q., Key Laboratory of Systems Microbial Biotechnology, Tianjin Institute of Industrial Biotechnology, Chinese Academy of Sciences
Liu, Y., Institute of Process Engineering, CAS
Originally presented on: 11/9/2015 10:00:00 - 10:20:00

Genome editing can introduce predetermined sequence changes to the target genome. In this work, a new genome editing method of tandem repeat assisted genome editing (TRAGE) was developed. This method can realize seamlessly deleting, substituting or inserting targeted genes in genome by one round of genetic manipulation with high fidelity.

We knocked out *tesC*, *tesB* and *tesA* sequentially for inducing fatty acid starvation to enhance fatty alcohol production. The deletion of *tesC* increased the fatty alcohol titer from 480 to 710 mg/L. Meanwhile, the fatty acid production dropped from 30 mg/L to 15 mg/L. The deletions of all three acyl-ACP thioesterases improved fatty alcohol tilter to 760 mg/L. At the same point, the fatty acid concentration was 8 mg/L.

We performed whole-genome transcriptional analysis. The expression levels of most genes from glycolysis module and fatty acid synthesis module were upregulated. Most genes from TCA cycle and fatty acid degradation module were downregulated.

Deletions of *ldhA*, *pta* and *ackA* from KLCB coupled with over expression of *FAR* were performed. The fatty alcohol tilter was increased from 756 mg/L to 2024 mg/L. Fed-batch fermentations were performed in a 3-L Bioflo 110 fermentor. Total fatty alcohol accumulation reached a maximum of 6.33 g/L after 50 h. Two saturated fatty alcohols (C14:0 and C16:0) and two unsaturated ones (C16:1 and C18:1) are the major components. C14:0 (2.42 g/L) and C16:1 (1.81 g/L) are the two most abundant fatty alcohols, constituting 38.2% and 28.6% of total fatty alcohols, respectively.

# 422811 Engineering Recombinant Escherichia coli for Improved Salicylic Acid Production By Optimizing Synthetic Pathway


**Abstract:**

Salicylic acid (SA) is an important precursor to a wide variety of high-value chemicals and drugs, such as cis,cis-muconic acid (a platform chemical), aspirin and lamivudine (an anti-HIV drug)[1]. Recently, a novel approach for microbial production of SA has been developed via extending the shikimate pathway in *E. coli*[1]. In previous work by our lab, a series of endogenous small molecule sensor-reporter systems were developed by engineering effector specificity of the transcriptional regulator protein AraC[2,3,4]. The AraC variants activate reporter gene expression in response to target small molecules such as mevalonate, triacetic acid lactone, vanillin and SA.

In the present study, we implement our custom-made SA biosensor (AraC-SA) to improve SA production by recombinant *E. coli.* The ribosomal binding site calculator[5,6] was used to design a spectrum of predicted translation rates for six genes involved in the heterologous SA biosynthesis pathway, and the resulting combinatorial expression library of all genes was subjected to AraC-SA-based high-throughput screening and selection. The degradation pathway of the intermediate chorismate was also attenuated to increase flux toward SA. Improved SA production and altered gene expression profile results will be presented. Finally, the influence of increasing expression of a multidrug efflux pump to increase SA tolerance and yield will be discussed.



1. Lin, Y., Sun, X., Yuan, Q. & Yan, Y. Extending shikimate pathway for the production of muconic acid and its precursor salicylic acid in Escherichia coli. *Metab. Eng.* **23**,62–9 (2014).

2. Tang, S. Y. & Cirino, P. C. Design and application of a mevalonate-responsive regulatory protein. *Angew. Chemie, Int. Ed.* **50**,1084–1086 (2011).

3. Tang, S. Y. *et al.* Screening for enhanced triacetic acid lactone production by recombinant Escherichia coli expressing a designed triacetic acid lactone reporter. *J. Am. Chem. Soc.* **135**,10099–103 (2013).

4. Frei, C. S., Qian, S., Gredell, J. A. & Cirino, P. C. The many hats of AraC: a rapidly customizable molecular biosensor. *Manuscr. Prep.*

5. Salis, H. M., Mirsky, E. A. & Voigt, C. A. Automated design of synthetic ribosome binding sites to control protein expression. *Nat. Biotechnol.* **27**,946–50 (2009).

6. Espah Borujeni, A., Channarasappa, A. S. & Salis, H. M. Translation rate is controlled by coupled trade-offs between site accessibility, selective RNA unfolding and sliding at upstream standby sites. *Nucleic Acids Res.* **42**, 2646–59 (2014).

# 436425 Engineering a Synthetic Consortium for Bioconversion of Algal Carbohydrate and Protein to Sesquiterpenes As Advanced Biofuel

**Abstract:**

The existing strategies for algae-based biofuels have primarily focused on biodiesel production through exploiting high algal lipid yields under nutrient stress conditions. However, under conditions supporting robust algal biomass accumulation, carbohydrate and proteins typically comprise up to ~80 of the ash-free dry weight of microalgae biomass. Therefore, the comprehensive utilization of algal biomass to add high energy density fuel compounds with "fit for purpose" properties into the fuel production process will not only diminish the process cost but also improve overall fuel yield. Terpenes are hydrocarbon and hydrocarbon-like (C:O>10:1) compounds with high energy density, and potentially promising candidates for the next generation of biofuel as "drop-in" replacements for petroleum-based fuels. In this study, we demonstrated the feasibility of bioconversion of protein into sesquiterpene compounds as well as comprehensive bioconversion of algal carbohydrate and protein into biofuels, simultaneously. To achieve this, the mevalonate pathway was reconstructed into an *E. coli* chassis with six different terpene synthases (TSs). Strains containing each TS produced a wide spectrum of sesquiterpene compounds under minimal medium containing an amino acid mixture as the sole carbon source. The sesquiterpene production was optimized through three different regulation strategies using chamigrene synthase as an example. The highest total terpene titer reached 166 mg/L, by applying a strategy to minimize mevalonate accumulation *in vivo*. The strain produced the highest yields of total terpene under reduced IPTG induction levels (0.25 mM), lower induction temperature (25·C), and a higher substrate concentration (20 g/L amino acid mixture). A synthetic bioconversion consortium consisting of two engineering strains (DH1-TS and YH40-TS) with reconstructed terpene biosynthetic pathways was designed for comprehensive single-pot conversion of algal carbohydrate and protein to sesquiterpene. At an inoculum ratio 2:1 YH40-TS:DH1-TS, the consortium yielded the highest total terpene concentration up to 187 mg/L, corresponding to was 31 mg fuel/g algae biomass ash free dry weight. This study demonstrated a feasible process for algal biofuel production.

# 429261 A Physical, Chemical and Biological Approach to Targeting Nanoparticles to Lung Endothelium: Leveraging Shape Effects, Cellular Hitchhiking and Antibody Targeting

**Abstract:**

Targeted delivery of nanoparticles to specific tissues remains a challenge due to the rapid removal of nanoparticles from circulation by the reticuloendothelial (RES) system. The liver and spleen, the primary RES organs which are home to phagocytotic macrophages, efficiently remove foreign entities from circulation, including nanoparticles. The majority of past research has addressed this issue via chemical modification of nanoparticles in the form of hydrophilic coatings which shields nanoparticles from these organs by reducing nanoparticle adsorption of opsonins. Recently, numerous other approaches have been developed which exploit the natural stealth abilities of our own circulatory cells use to avoid immune system clearance. One such method, called 'cellular-hitchhiking', facilitates this by non-covalent adsorption of nanoparticles to the surface of red blood cells, the longest circulating cell in the body. Alongside this, approaches using rod-shaped nanoparticles, which avoid RES clearance to a higher extent as compared to their spherical counterparts, have also been developed. In addition to enhanced circulation, both these methods have also been shown to confer enhanced targeting of nanoparticles, specifically to the lungs. Here, we systematically investigate three approaches and establish their roles, both independently and in combination, to improve lung targeting while avoiding RES clearance. Our results show that an approach combining targeting antibodies, rod-shaped particles and cellular hitchhiking offer the best targeting to lungs and RES avoidance as compared to any individual or dual-combination approach studied here. The strategy described here offers a novel means that combine chemical, physical and biological approaches to maximize tissue targeting.

# 435103 Development of Novel Gold-Lipidic Nanocomposites for Improved Imaging and Drug Delivery to Prostate Cancer

**Abstract:**

**Introduction:** Chemotherapy efficacy is limited by toxicity and non-uniform tumor exposure. Liposomal drug carriers are nano-scale, spherical particles with a phospholipid bilayer surrounding an aqueous core. With increased circulation and improved tumor deposition due to the enhanced permeability and retention effect, liposomes increase treatment efficacy and reduce toxicity. However, a lack of understanding regarding tumor growth and metastasis is a challenge for developing treatments. Much current research is devoted to understanding what processes cause tumor behavior. Theranostics, or carriers that combine diagnostic and therapeutic capabilities in one unit, are used to study tumor operation in treatment.

We hypothesized that composite systems, encapsulating gold nanoparticles within liposomes, may be used to improve drug delivery and permit non-invasive imaging. We previously designed a novel gold-lipidic nanocomposite to capitalize on the drug delivery capabilities of liposomes and the imaging contrast power of gold nanoparticles. Here we present the formulation of nanocomposites comprised of 2 nm gold clusters encapsulated within pegylated, long-circulating "stealth" liposomes (Au-LNC).

**Materials and Methods:** Gold clusters 2 nm in diameter were synthesized in the lab through a modified Brust-Schiffrin method. Two different types of clusters were studied, one capped with glutathione (G-AuNP) and one capped with mercaptosuccinic acid (MSA-AuNP), and both types of clusters were water-soluble. Liposomes were prepared using a dry film hydration method with 9:5:1 molar ratio of 1,2-distearoyl-*sn*-glycero-3-phosphocholine (DSPC), cholesterol, and ,2-distearoyl-*sn*-glycero-3-phosphoethanolamine-N-[methoxy(poly(ethylene glycol))-2000 (DSPE-PEG 2000) respectively. The lipids were originally dissolved in chloroform, and the chloroform was removed and a thin film was formed through rotary vacuum evaporation. To form Au-LNCs, the thin lipid film was hydrated with aqueous solution containing gold clusters at the desired concentration. The liposomes were sized through extrusion through a 0.8 μm membrane, and unencapsulated gold particles were removed via dialysis.

Physical and pharmacodynamic studies were completed using atomic absorption spectroscopy (gold quantification), dynamic light scattering (size distribution), zeta potential (particle stability), and MTT and SRB assays (cytotoxicity), and *in vivo* fluorescence/bioluminescence/X-ray images were taken using In Vivo Imaging System (IVIS).

**Results and Discussion:** The nanocomposites have a high gold encapsulation efficiency, with the final composites showing an average encapsulation efficiency of 66 percent and 54 percent for G-AuNPs and MSA-AuNPs, respectively. The composites show narrow size distribution centered at 100 nm, exhibit minimal cytotoxicity at biologically relevant concentrations, and are readily taken up by PC-3 human prostate cancer cells in vitro.

**Conclusions:** These data support our hypothesis that gold-lipidic nanocomposites can be prepared and justify preclinical in vivo studies in murine models of human tumors to improve cancer detection and treatment. On-going research with these systems includes determining the effect of the gold nanoparticles on doxorubicin loading and release. Future aims are to continue examining Au-LNCs in treated animals by multispectral optoacoustic tomography (MSOT).

# 435115 Targeted Nanocarriers for the Delivery of Novel Metastatic Melanoma Chemotherapeutics

**Abstract:**

There are currently two major problems with treating metastatic melanoma. First, it is difficult to deliver a therapeutic concentration of traditional chemotherapeutics without lethal side effects on the patient. The second problem is that there is a large void in the knowledge of genetic mutations that contribute to the disease state which has led to a paucity of targeted chemotherapeutics for metastatic melanoma. This work proposes a potential therapeutic solution that acknowledges both of these issues.

The first task was to identify a target that was specific to metastatic melanoma cells. A significant fraction of metastatic melanomas express gain-of-function mutations in the ErbB4 receptor tyrosine kinase. In previous work, our lab has shown that mutant versions of naturally-occurring ligands for ErbB4 function as partial agonists at ErbB4. The partial agonists stimulate ErbB4 tyrosine phosphorylation, fail to stimulate ErbB4 coupling to cellular proliferation, and inhibit agonist induced ErbB4 coupling to cellular proliferation. The discovery of these partial agonists led to the development of a series of high throughput screening assays to screen a library of small molecules that would potentially function as partial agonists at ErbB4. A sandwich-ELISA (Enzyme Linked Immunosorbent Assay) was developed to identify molecules that stimulate ErbB4 tyrosine phosphorylation. An MTT proliferation assay was used in the development of assays to identify molecules that fail to stimulate ErbB4 coupling and to identify molecules that inhibit agonist induced ErbB4 coupling. The top candidate molecules identified from the screening assays were used to coat the surface of poly(lactic-co-glycolic acid) (PLGA) nanoparticles that contain traditional metastatic melanoma chemotherapeutics. *In vitro* studies were performed to determine the targeting abilities of these partial agonist coated nanoparticles, as well as, whether the partial agonists retained their functionality once bound to the surface of the PLGA nanoparticles.

# 416189 The Impact of Anti-Coagulant in Modulating Vascular-Targeted Drug Carrier (VTC) Adhesion Efficiency in Human Blood Flow

**Abstract:**

**The Impact of Anti-Coagulant in Modulating Vascular-Targeted Drug Carrier (VTC) Adhesion Efficiency in Human Blood Flow**

Daniel J. Sobczynski and Omolola Eniola-Adefeso

Anti-coagulants are commonly used to preserve whole blood or plasma for storage, *in vitro* diagnostics, or performing *in vitro* assays involving blood. The choice of anticoagulant or the absence of it (as in preparation of human serum) is known to influence the proteomic profile of the plasma. In addition, it is known that plasma protein adsorption can have negative effects on vascular-targeted drug carrier (VTC) adhesion. In this work, this phenomenon is explored deeper by considering the impact of the choice of anti-coagulant or absence of it during the preparation of human blood for VTC vascular wall adhesion assays. In particular, a parallel-plate flow chamber assay is used to simulate human blood flow over activated endothelial cells and directly measure VTC binding efficiency of PLGA, silica, and polystyrene nanoparticles. It is observed that for all VTCs tested, adhesion levels in RBC-containing serum flow are drastically reduced relative to citrated plasma, and for polystyrene spheres, adhesion in heparinized whole blood is significantly reduced compared to PLGA and silica spheres. Furthermore, the absence of anticoagulant in non-citrated whole blood re-creates the low adhesion effects seen in RBC-containing serum flow. Overall, the choice of anti-coagulant can modulate VTC adhesion and potentially overestimate VTC adhesion efficiency *in vivo* where anticoagulant is not present. Instead, our results suggest suspension of human RBCs in serum may be a better model for evaluating VTC functionality for human use.

# 420613 Novel Nanoparticle-Mediated Treatment for Fatal Neurodegenerative Diseases

**Abstract:**

Delivery of therapeutics to the brain through noninvasive administration is an impossible task due to the presence of the blood-brain barrier (BBB). The BBB serves as the first level of defense for the brain, selectively permitting the passage of necessary ions and molecules while preventing the transport of 98% of small molecule therapeutics. Patients with lysosomal storage disorders, which occur when a patient is not producing one or more necessary lysosomal enzymes, require central nervous system treatment. Lack of enzymatic activity causes buildup of storage materials in neural cells, leading to loss of body control and premature death. Specifically, in GM1 gangliosidosis, patients are missing β-galactosidase. GM1 gangliosidosis is fatal in infancy, with no clinically available treatment and no hope for parents of diagnosed children. Currently explored treatment methods for GM1 gangliosidosis involve cranial injections of adeno-associated viral vectors in various neural locations in a feline model. Although success has been demonstrated, intravenous (IV) delivery of the missing enzyme would be ideal. We are designing and characterizing the first nanoparticle-mediated treatment of GM1 gangliosidosis through the use of self-assembled polymersomes with high physiological stability and tunable release for IV delivery. When coupled with an endothelial cell receptor, delivery through the BBB and into the lysosome of neural cells will occur, effectively treating patients without invasive surgery.

Polyethylene glycol-b-poly(lactic acid) (PEGPLA) copolymer has been proven to self-assemble, forming vesicle structures using both dimethyl sulfoxide (DMSO) and water. Transmission electron microscopy (TEM) images confirmed vesicle formation. Molecules to be encapsulated were dissolved in stirring water prior to injection of DMSO and PEGPLA. Particle size distributions (PSDs) were determined using dynamic light scattering (DLS). Self-assembly was stopped by liquid nitrogen prior to lyophilization, with mannitol and inulin explored as novel lyoprotectants. Studies indicated that incorporating 2 wt%/v mannitol during formation led to PSDs 54 ± 0.8% smaller after lyophilization. Because of this and mannitol's ability to disrupt the BBB, 2 wt%/v mannitol was used to aid in the formation of brain-deliverable polymersomes encapsulated with Alexa Fluor 488.

Alexa Fluor 488 was loaded into lyophilized polymersomes, which were placed in buffers of pH 7.4, 4.8, and 1 to monitor release. Amine-reactive homobifunctional PEG was introduced during formation to mitigate the attachment of CF 350 Amine, a blue fluorescent ligand, to the polymersome surface. Attachment was confirmed with fluorescent microscopy, XPS, and DLS techniques and monitored with absorption spectroscopy of NHS leaving groups. DLS data indicates that polymersomes loaded with Alexa Fluor 488 in water had diameters of 191 ± 64.7 nm, with no statistical difference in size compared to control. Solvent, block copolymer concentration, injection speed, and needle gauge were varied. At PEGPLA concentrations of 2, 20, and 50 μmol/mL, polymersomes formed at diameters of 139 ± 27.2 nm, 382 ± 21.5 nm, and 1282 ± 663 nm respectively. 2 μmol/mL PEGPLA led to 74% of polymersomes with a diameter less than 200 nm, the goal size for potential delivery through the BBB using receptor mediated transcytosis. Because of this, all loading, release, and attachment studies were done using polymersomes formed with a concentration of 2 umol/mL PEGPLA. 44% of Alexa Fluor 488 added to lyophilized polymersomes was loaded. When placed in pH 7.4 buffer, 37% of the loaded mass of Alexa Fluor 488 was released in 12 hours, compared to 69% in pH 4.8 buffer, indicating increased drug release expected in the lysosome.

Attachment of CF 350 was shown using fluorescence microscopy and increasing concentration of NHS leaving groups after the formation of an amide bond in only 5 minutes, indicating attachment. DLS data also indicated a shift in PSD upon ligand attachment using a mass of 1 mg NHS-PEG(2000)-NHS during polymersome formation. Upon ligand attachment, the overall average polymersome hydrodynamic diameter increased from 138.9 ± 1.58 nm to 170.3 ± 1.50 nm. The amount of polymersomes formed with sizes 100 nm or less decreased from 15.4% to 4.04%, while the amount of polymersomes formed with sizes between 100 and 200 nm increased from 58.8% to 72.4%. XPS data confirmed the presence of nitrogen groups on the surface of self-assembled polymersomes prior to washing.

This work highlights tremendous control over size formed, release location, and ligand attachment possible when utilizing PEGPLA polymersomes as an IV drug delivery vehicle. Varying PEGPLA concentration had the largest impact on polymersome size, with control of this parameter limiting expensive and necessary separation techniques currently used with drug delivery nanotechnologies. Attachment of a ligand with an available amine group proves the capability of polymersomes with amine-functionalized surface groups to bind targeting ligands appropriately. Results are promising towards the goal of creating the first clinical treatment for GM1 gangliosidosis, a disorder impacting the central nervous system, using a combination of enzyme replacement therapy and nanotechnology methods to cross the BBB.

# 435226 Nanoparticle-Based Delivery of Drugs Dramatically Enhances Killing of Filarial Worms

**Abstract:**

Lymphatic filarial diseases represent a significant global cost and social burden, affecting up to 120 million individuals annually. Current antifilarial therapies include the use of mass dosing of multiple drugs, such as ivermectin, diethylcarbamazine, and albendazole. Extended use of these types of drugs have been shown to produce unwanted toxic side effects, which significantly impacts patient compliance. The obligatory endosymbiont bacterium, *Wolbachia*, helps confer enhanced pathogenicity. Therefore many recent mass drug administration efforts have included the antibiotic doxycycline to inhibit the microfilarial burden. Several delivery issues affect treatment with these drugs. Firstly, in order to reach the lymphatic tissues where macrofilariae reside, high doses are required to compensate for the short half-life of the drugs. Additionally, penetration of the soluble drugs through the outer cuticle of the worms is limited. We hypothesize that the use of amphiphilic polyanhydride nanoparticles could significantly improve the current standard of therapy by providing dose sparing, shortened killing times, and improved penetration.

Amphiphilic polyanhydride nanoparticles represent a versatile drug delivery platform. Their surface erosion mechanism results in a sustained and predictable release profile of co-encapsulated drugs that can be tailored for a quick burst release or a slower zero-order release by modifying the copolymer chemistry. This serves to lengthen the amount of time the localized drug concentration is within therapeutic limits, providing potential for a single-dose formulation and improved patient compliance. The nanoparticles stably encapsulate their payloads, conferring thermal protection to the drugs or proteins at ambient temperature. This could potentially reduce the cost of drug administration by eliminating the need for refrigeration during transportation to endemic regions. Our work has shown that polyanhydride nanoparticles are efficiently internalized by macrofilarial worms, allowing lethal co-delivery of antifilarial and antibacterial payloads at suboptimal doses. The nanoparticles can be functionalized to target specific surface receptors, allowing targeted delivery. This potential for specific targeting could provide further dose sparing and enhanced worm killing.

Current *in vivo* studies show that co-encapsulation of ivermectin and doxycycline provided sixteen-fold reduction in total drug dosage relative to comparative macrofilaricidal drugs. This dose sparing is further enhanced by a 50% decrease in the number of required treatments. Previous *in vitro* confocal laser microscopy studies showed extensive tissue penetration by the nanoparticles, compared to negligible penetration for the soluble drugs. These studies are corroborated by recent *in vivo* electron microscopy studies, which demonstrate extensive fragmentation of internal tissues. This observation is hypothesized to be a result of channeling of the nanoparticles, resulting in a subsequent release of concentrated drug. Together, all of these studies provide foundational information for the rational design of a nano-carrier platform that can have a transformative impact on the treatment of filarial diseases.

# 435239 Modulating Vibrio Cholerae Quorum-Sensing-Controlled Communication Using Autoinducer-Loaded Nanoparticles


**Abstract:**

The rise of bacterial antibiotic resistance has created a demand for alternatives to traditional antibiotics. Attractive possibilities include pro- and anti-quorum sensing therapies that function by modulating bacterial chemical communication circuits. We report the use of Flash NanoPrecipitation to deliver the *Vibrio cholerae* quorum-sensing signal CAI-1 ((S)-3-hydroxytridecan-4-one) in a water dispersible form as nanoparticles. The particles activate *V. cholerae* quorum-sensing responses five orders of magnitude higher than does the identically administered free CAI-1, and are diffusive across *in vivo* delivery barriers such as intestinal mucus. This work highlights the promise of combining quorum-sensing strategies with drug delivery approaches for the development of next-generation medicines.

# 422663 The Dionesus Algorithm Provides Scalable and Accurate Reconstruction of Biological Networks to Reveal New Drug Targets

**Abstract:**

**The DIONESUS algorithm provides scalable and accurate reconstruction of biological networks to reveal new drug targets.**

Mark F. Ciaccio and Neda Bagheri. Department of Chemical & Biological Engineering, McCormick School of Engineering, Northwestern University.

**Introduction:** Advances in genomic and proteomic technology brings with it an unprecedented breadth of data on the molecular markers driving cellular function. With this new age in biomedical engineering comes the pressing need to find efficient strategies to glean deeper mechanistic understanding of biological processes. Scalable and rapid algorithms are critical in pushing forward innovation at the interface of medicine, mathematics, and biology. To this end, we created a scalable algorithm, termed DIONESUS, based on partial least squares regression with variable selection to reconstruct networks from large matrices of cue-signal-response relationships. Our scalable algorithm performs with higher specificity and sensitivity than several top-performing network inference algorithms in a fraction of the computational time. We applied DIONESUS to reconstruct the signaling network of a model epidermoid cancer cell line and identified STAT1 as a central modulator of apoptosis specifically in cells that have hyperactivity of the EGF receptor. **Background:** Deriving correlations between biomolecules, such as RNA expression or protein abundance, and cell phenotype can suggest factors that are associated with disease. Deriving the underlying network structure can identify disease drivers by elucidating control structures such as feedback loops and redundant pathways. Several algorithms can reconstruct directed biological networks using information theory, decision trees, or Bayesian networking; algorithms with high accuracy, however, are often limited to small networks [1]. Ideally, network reconstruction can be expanded to genomic or proteomic proportions in order to broaden and deepen understanding of the dysregulation of biological systems in pathological states. **Materials and Methods:** After being validated in three *in silico* networks, DIONESUS was applied to a high confidence dataset characterizing 60 phospho-proteins at four time points as a function of 26 perturbations. We perturbed epidermoid carcinoma cells with diverse growth factors and/or small molecules chosen to activate or inhibit specific subsets of receptor tyrosine kinases. The abundance of 60 phosphosites was quantified using a modified microwestern array, a high-confidence assay of protein abundance and modification [2]. **Results and Discussion:** From the reconstructed network, we identified enhancement of STAT1 activity as a potential strategy to treat EGFR-hyperactive cancers. Quantification of the relationship between drug dosage and cell viability in a panel of triple-negative breast cancer cell lines validated this newly proposed therapy. **Conclusions:** The accurate and scalable DIONESUS algorithm allows for quick iteration between experimental design and computational modeling by incorporating systems-level behavior in order to accelerate drug target discovery.

**References:**

[1] Marbach, Daniel, et al. "Wisdom of crowds for robust gene network inference." *Nat meth* 9.8 (2012): 796-804.

[2] Ciaccio, Mark F., et al. "Systems analysis of EGF receptor signaling dynamics with microwestern arrays." *Nat meth* 7.2 (2010): 148-155.

# 429129 An Integrated Engineering Control System Model of Primary and Secondary Hemostasis

**Abstract:**

Hemostasis, the physiological process that arrests bleeding and keeps blood within an injured blood vessel, consists of a set of reactions and sub-processes that can be divided into various functional components. Normal functioning of the components is necessary to achieve a delicate balance between effective hemostasis and pathological conditions such as thrombosis, one of the major causes of death in the world, or hemophilia. Because the components of the hemostatic process interact in complex ways, how the various components work together to implement fast, effective and stable responses to vascular injury cannot be fully understood using traditional experimental methods alone—quantitative models are required. However, developing a faithful mathematical representation of known mechanistic details of the entire hemostatic process using standard modeling approaches will produce a model containing an inordinately large number of variables and an even larger number of unknown parameters. An alternative conceptual paradigm is required if the important complex details are to be represented adequately, with sufficient fidelity, and in mathematically tractable form.

From a process engineering perspective, we observe that hemostasis is mediated by an automatic biological control system with clearly identifiable control system components such as sensors, controllers, and actuators. Our modeling approach uses such an engineering control system representation as the basis for organizing information efficiently into a holistic model, with the control system components as "functional modules," thereby facilitating development, validation and systematic analysis of the complete hemostatic process as a combination of the component modules.

In this presentation, we will discuss the development of a comprehensive hemostasis model, and its validation against experimental data from the literature. We will first present a high-level engineering control system block diagram for hemostasis that shows the two major, distinct sub-processes: primary hemostasis (responsible for platelet aggregation) and secondary hemostasis (responsible for coagulation) and their interconnections. We will then present mathematical models developed and validated for each sub-process separately and subsequently connected via three known interaction mechanisms: (1) Platelet activation and aggregation (via primary hemostasis) provides biologically active surfaces for the enzyme reactions of the coagulation cascade (in secondary hemostasis); (2) thrombin (produced by secondary hemostasis) is a potent agonist for platelet activation (in primary hemostasis); and (3) platelet activation (in primary hemostasis) triggers the release of coagulation factor V stored within platelets, which in turn promotes thrombin production (in secondary hemostasis). We will then show results of the dynamic characteristics of hemostatic response predicted by our comprehensive model, and demonstrate how the interactions between the components in hemostasis contribute to the overall effectiveness of the hemostatic process.

The validated model will subsequently be used for systematic analysis of the overall integrated process of hemostasis and to generate hypotheses for rational diagnosis and treatment of pathological hemostasis conditions.

# 435404 Development of Kinetic Models of Metabolism for Microbial Production

**Abstract:**

**Development of kinetic models of metabolism for microbial production**


Ali Khodayari[1] (auk241@psu.edu), Satyakam Dash[1] (sud25@psu.edu), Matthew K. Theisen[3], M. Ahsanul Islam[2], Yuting Zheng[2], James C. Liao[3], Gregory Stephanopoulos[2], and **Costas D. Maranas**[1]

[1]The Pennsylvania State University, University Park; [2]Massachusetts Institute of Technology; [3]University of California, Los Angeles.

Improving the accuracy of *in silico* metabolic models is the chief challenge for accurately identifying the metabolic drivers that underpin the production of biofuels and chemicals of interest. Despite several advances, stoichiometry-based approaches are limited to steady-state conditions, and generally are unable to describe the metabolic phenotype in terms of the underlying pool of metabolite concentrations and enzyme abundances. While kinetic models have a potential to address these shortcomings, their constructions are still plagued by a number of challenges, including the paucity of kinetic parameters, diversity of rate laws, and implementation of regulatory events. The Ensemble Modeling (EM) paradigm [1] has addressed some of these challenges by decomposing reactions into their elementary steps thus enabling systematic integration of substrate-level regulatory interactions. Recently, we have constructed a kinetic model of *E. coli*core metabolism including 138 reactions, 93 metabolites, and 63 regulatory interactions under aerobic growth condition with glucose as the sole carbon source [2].

Here, we expand the core kinetic model towards genome-scale by integration of all relevant reactions from the *i*AF1260 model. Model parameterization is performed using experimental data for multiple carbon substrates (glucose, acetate, and pyruvate) under aerobic respiration and fermentative conditions. In addition, we developed a second generation genome scale model (GSM) for *C. thermocellum*(iCth446) which includes 446 genes, 598 metabolites, and 637 reactions. The GSM was subsequently used to construct a core metabolic model of the organism's central metabolism containing 90 reactions and 76 metabolites, with cellobiose as the carbon source under anaerobic growth condition. This core metabolic network spans all major biomass precursors, as well as other chemicals of interest encompassing the cellobiose degradation pathway, glycolysis/gluconeogenesis, the pentose phosphate (PP) pathway, the TCA cycle, major pyruvate metabolism and anaplerotic reactions, alternative carbon metabolism, nucleotide salvage pathway, and 22 substrate level regulatory interactions. The constructed core model serves as the scaffold for building a kinetic model parameterized by using experimentally measured flux data to improve isobutanol production.

1. Tan Y, Liao JC: **Metabolic ensemble modeling for strain engineers**. *Biotechnology journal* 2012, **7**(3):343-353.

2. Khodayari A, Zomorrodi AR, Liao JC, Maranas CD: **A kinetic model of Escherichia coli core metabolism satisfying multiple sets of mutant flux data**. *Metab Eng* 2014, **25**:50-62.

# 433386 GEM-Path: A Novel Pathway Prediction Algorithm and Commodity Chemical Production Atlas

**Abstract:**

The production of 75% of the current drug molecules and 35% of all chemicals could be achieved through bioprocessing. To accelerate the transition from a petroleum-based chemical industry to a sustainable bio-based industry, systems metabolic engineering has emerged to computationally design metabolic pathways for chemical production. Although algorithms able to provide specific metabolic interventions and heterologous production pathways are available, a systematic analysis for all possible production routes to commodity chemicals in *Escherichia coli* is lacking. Furthermore, a pathway prediction algorithm that combines direct integration of genome-scale models at each step of the search to reduce the search space does not exist. Previous work performed a model-driven evaluation of the growth-coupled production potential for *Escherichia coli* to produce multiple native compounds from different feedstocks. In this study, we extended this analysis for non-native compounds by using an integrated approach through heterologous pathway integration and growth-coupled metabolite production design. In addition to integration with genome-scale model integration, the GEM-Path algorithm developed in this work also contains a novel approach to address reaction promiscuity. In total, 245 unique synthetic pathways for 20 large volume compounds were predicted. Host metabolism with these synthetic pathways was then analyzed for feasible growth-coupled production and designs could be identified for 1271 of the 6615 conditions evaluated. This study characterizes the potential for *E. coli* to produce commodity chemicals, and outlines a generic strain design workflow to design production strains.

# 431865 Prediction of Gene Expression Using Goal-Directed Metabolic Models

**Abstract:**

The salient feature of cybernetic models, their goal-oriented control policies, have been used to predict metabolic phenomena ranging from complex substrate uptake patterns, and dynamic metabolic flux distributions to the behavior of gene knockout strains. The key element of this approach is the perspective that metabolic regulation is organized towards achieving goals relevant to an organism's survival. For example, in multi-substrate cultures of *E. coli*, cells invest resources into enzymes for pathways which maximize their growth rate. While goal-oriented control has successfully yielded the prediction of numerous metabolic phenomena, the control policy itself has not been compared with data for metabolic regulation. In this talk, cybernetic control variables are weighed against gene expression data, an indicator of cellular regulation, for two sets of systems.

The first set of systems analyzed consists of bacterial cultures that exhibit diauxic growth in two-substrate environments. For these bacterial models, enzymes are synthesized to maximize the growth rate for a given set of metabolic options at each point in time. Both microarray and qRT-PCR data for pathways pertaining to each substrate are compared with the cybernetic variable for enzyme formation. By demonstrating how cybernetic variables for induced enzyme synthesis mimic transcriptional data, a strong argument for using cybernetic models is made.

The second system modeled describes the behavior of a network of lipids in the mammalian macrophage cell line, RAW 264.7. More specifically, arachidonic acid is converted into various eicosanoids associated with inflammation in response to the detection of a key marker of infection, lipopolysaccharide, by Toll-like receptor 4 on macrophages. The existing framework for modeling single-celled organisms has been modified to describe more complex mammalian metabolic networks. Our approach essentially mimics, in an approximate sense, the maximization of the rate of TNF-alpha production, a crucial factor during inflammation, as an objective function. This is implemented by assuming that the upstream generation of eicosanoids concomitantly results in the downstream transcription of TNF-alpha. To be precise, enzyme synthesis for pathways that yield TNF-alpha production signals is regulated in such a way to ultimately maximize the signal for TNF-alpha formation. This is captured in the model using a linear equation that connects time-dependent changes in TNF-alpha with temporal changes in eicosanoid levels. Fitting this model to metabolite data and comparing cybernetic variables with dynamic gene expression data at the major branch point in this network shows a significant level of agreement. This result serves as an important validation of cybernetic variables using real data from cells. Moreover, it shows that cybernetic variables can be used to infer trends in gene expression data only using information taken from the metabolite level and a description of the metabolic network's organizing principle. This idea has not only been tested for a control condition and LPS treatment but now has also been tested on treatment with compactin, a statin drug that influences arachidonic acid metabolism.

# 419744 Network Motifs and Mechanisms for Oscillatory Dynamics of Biochemical Networks

**Abstract:**

We formulate and examine detailed mechanisms for a number of enzyme and biochemical oscillatory/signalling networks with the aim of uncovering critical subnetworks that generate the excitable/bistable/oscillatory dynamics. Such core models of the specified nonlinear dynamics are often referred to as network motifs. Our method of finding the motifs is based on a theory of stability of chemical networks known as the stoichiometric network analysis. Given that the mechanism is formulated in terms of mass action kinetics, the instabilities of steady states are linked to topological features of the (sub)network diagrams. This approach enables a classification of biochemical networks. Based on our experimental observations we examine in vitro enzyme reactions including the glucose–glucose oxidase– catalase–hydrogen peroxide reaction, the urea–urease reaction and glucose–glucose oxidase–ferricyanide reactions. Topology of the core oscillatory subnetworks are compared against one another as well as against other enzyme reactions known to oscillate, such as oxidase–peroxidase reaction. We also examine some in vivo enzyme systems of biological importance including the mitogen-activated protease kinase (MAPK) and systems determining circa- and ultradian cycles in photosythesizing microbes.

# 428089 Asymmetry in Signal Oscillations Contributes to Efficiency of Periodic Systems

**Abstract:**

Oscillations are an important feature of cell signaling that often result from complicated combinations of positive and negative feedback loops. The encoding and decoding mechanism of oscillations based on amplitude and frequency have been extensively discussed in literature in the context of inter-cellular and intra-cellular signaling [1]. To name a few, the oscillation patterns of cytosolic calcium [2], p53 tumor suppressor protein [3], NF-κB transcription factor [4] and many more have been studied in their respective networks. Although most signaling mechanisms in real physiological systems likely have an extremely complicated mode of modulation, in some cases it is possible to identify the key molecules in a given pathway to determine how one influences another. Among these cases, amplitude-to-amplitude modulation is relatively straight forward, and consequently this idea was applied early in the study of receptor-mediated signaling under sub-saturation regions [1, 3, 5]. In contrast, frequency-modulated mode of cellular signaling, especially decoding, is more convoluted. This mode of signaling has been most extensively studied in the context of calcium signaling, and while it is generally agreed that frequency is the main drive that leads to differentiated downstream events, there is also an argument that cumulative peak duration, not frequency, defines signaling specificity [6]. All of the aforementioned efforts attempt to understand in depth the oscillatory behavior in cellular and intercellular signaling. However, the fundamental question of why cellular signals appear in periodic manner as opposed to a non-oscillatory manner has not been fully answered.

In an attempt to answer this question, we designed a study to analyze oscillatory characteristics of both signaling molecule and system output. To do this, we modified two classic feedback models, the Goodwin oscillator and PER oscillator, and compared periodic signals to steady signals of identical time-averaged values. The two oscillators were selected as experimental systems because: 1) they are both self-sustained oscillators driven by negative feedback; 2) the signaling molecules of these two models exhibit opposite skewness, Goodwin oscillator having fast ascension and slow descension, and PER oscillator having slow ascension and fast descension. Furthermore, it is easy to generate a range of different intensities of skewness in both models by modifying the degree of inhibition, represented by the Hill coefficient or the cooperativity coefficient.

Currently our research into these periodic systems lends the following observations. As the inhibition of both models is increased, the resource generation develops more drastic skewness, which is correlated with an increased efficiency of the oscillatory system as compared to the output of no-feedback counterpart systems. As the cooperativity coefficient is increased in the PER model, the resource tends to skew negatively (toward the right side of the center), indicating slower ascension and faster descension of the signal. This correlates with an increased frequency and output of the system. As the Hill coefficient increases for the Goodwin model, the resource exhibits the opposite effect – it skews positively and is correlated with increased frequency and system output. Combining these results, we hypothesize the existence of regions of degree of inhibition and signaling asymmetry that will improve system output and efficiency. Furthermore, we observe a strong positive correlation between output amplitude and system output (and efficiency), and hypothesize that strong sustained oscillations give rise to a more efficient system. Our hypotheses address the fundamental question of why cellular signals and responses are periodic. From the results, we also search for the underlying mechanism that drives system efficiency.

[1] M. Behar and A. Hoffmann, "Understanding the temporal codes of intra-cellular signals.," *Curr. Opin. Genet. Dev.*, vol. 20, no. 6, pp. 684–93, Dec. 2010.

[2] E. Smedler and P. Uhlén, "Frequency decoding of calcium oscillations," *Biochim. Biophys. Acta - Gen. Subj.*, vol. 1840, no. 3, pp. 964–969, 2014.

[3] M. Lev Bar-Or, R., Maya, R., Segel, L.A., Alon, U., Levine, A.J. and Oren, "Generation of oscillations by the p53-Mdm2 feedback loop: A theoretical and experimental study," *Proc. Natl. Acad. Sci.*, vol. 97, pp. 11250–11255, 2000.

[4] R. Williams, J. Timmis, and E. Qwarnstrom, "Computational Models of the NF-KB Signalling Pathway," *Computation*, vol. 2, no. 4, pp. 131–158, 2014.

[5] J.W. Black, "Operational models of pharmacological agonism," *Proc R Soc L. B Biol Sci*, vol. 220, pp. 141–162, 1983.

[6] L. Zhu, S. Song, Y. Pi, Y. Yu, W. She, H. Ye, Y. Su, and Q. Hu, "Cumulated $Ca2^{+}$ spike duration underlies $Ca2^{+}$ oscillation frequency-regulated NFκB transcriptional activity.," *J. Cell Sci.*, vol. 124, no. Pt 15, pp. 2591–601, Aug. 2011.

# 430552 Dynamic Modeling of Molecular Motor Association with Axonal Amyloid Precursor Protein Vesicles

**Abstract:**

Mano Ram Maurya[1], Lukasz Szpankowski[1,2], Lawrence S.B. Goldstein[3], Shankar Subramaniam[4]

## ABSTRACT

Targeted transport of vesicles, organelles, and other types of intracellular cargo along microtubule tracks is powered by kinesin and cytoplasmic dynein molecular motors. Both classes of motors can attach to the same cargo and thus their binding kinetics decides the distribution of the cargo with these motors. In particular, we have been studying the binding of these motors to amyloid precursor protein (APP) vesicles in mouse hippocampal axons. It has been shown experimentally that several units of each of kinesin and dynein proteins can be bound to APP vesicles. In the present work, we have developed a 16-state biochemical reaction network-based model to capture the dynamics of binding of kinesin and dynein to APP vesicles. Each state represents an APP vesicle with a specified number of associated kinesin and dynein motors, up to a maximum of three units for each. A set of nonlinear coupled ordinary differential equations were generated describing the dynamic mass-balance based on the reaction rates for association/disassociation of each motor species. The system of differential equations was solved to compute the steady-state contribution of each axonal APP vesicular cargo associated with a defined arrangement of molecular motors. Since the rate parameters were not known *a priori*, the endogenous association/dissociation rate of both kinesin and dynein on APP vesicles was estimated through a nonlinear optimization (data-fitting) approach. The resulting model predicted the dynamics in under- and over-expression studies, has been validated against experimental data in heterozygous knockout kinesin mutants, and facilitates prediction of transport dynamics in systems not amenable to experimental manipulation. Furthermore, the systems biology approach used here provides a framework to build detailed predictive kinetic models to characterize the molecular states using macro-level data, which can then be used to generate and test hypotheses leading to the design of novel experiments and further refinement of such models.

**Acknowledgements:** We would like to acknowledge the National Science Foundation (NSF) collaborative grant STC-0939370.

**Key words:** mechanistic dynamic modeling, parameter estimation, systems biology, vesicle transport, kinesin, dynein.

---

1 Equal effort.

2 Current address: Fluidigm Corporation, 7000 Shoreline Court, Suite 100, South San Francisco, CA 94080.

3 Corresponding author. E-mail: lgoldstein@ucsd.edu, Phone: (858) 534-9700, Fax: (858) 246-0162.

4 Corresponding author. E-mail: shankar@ucsd.edu, Phone: (858) 822-0986, Fax: (858) 822-3752.

# 412299 Dynamics and Coexistence in Biochemical Engineering for Applications

**Abstract:**

Speaker: Brian H. DAVISON, Chief Scientist for Systems Biology and Biotechnology and Science Coordinator for Bioenergy Science Center, Oak Ridge National Laboratory,

Dynamics and Coexistence in BioChemical Engineering for applications.

Early work in the Stephanopoulos lab examined mixed microbial cultures and how to maintain diverse competing populations to maintain productivity. We showed through modeling and experiment that productive stability could be maintained in chemostats via allowing fluctuations in time, space and retention.

I will explore how these lessons influenced my work in multiphase bioprocessing systems (such as immobilized cell reactors and extractive bioreactors), the linkage of modeling and experiment in biomass conversion systems into biofuels and bioproducts.

I will also touch on "lessons" from mixed cultures on diversity in science, community service, and leading team science.

# 412301 Effects of Endocrine Disrupters on Adipose Metabolism and Inflammation

**Abstract:**

Contamination of the environment with organic pollutants has emerged as a significant public health concern due to the pervasive nature of these contaminants. Epidemiological studies have linked chronic endocrine disrupting chemicals (EDC) exposure to adverse effects on reproduction, development, and more recently, metabolic diseases [1]. Since Baillie-Hamilton first suggested a link between obesity and synthetic chemical production in the US [2], a growing number of studies have reported that perinatal exposure to certain EDCs, termed obesogens, could predispose the body to weight gain through an adipogenic effect that leads to increased fat mass. To date, studies have mainly focused on the impact of suspected obesogens on stem cell fate and tissue development. Less attention has been paid to clarifying whether these chemicals play a causal role in metabolic disease by disrupting the metabolic regulation of mature adipocytes. For example, EDCs could impair the ability of adipocytes to sequester fatty acids or activate pro-inflammatory pathways, both hallmarks of obesity-related metabolic diseases. Studies have shown that EDCs can bind to nuclear receptors (NRs) in a variety of tissues to disrupt the normal function of the endocrine system [3]. One such NR, peroxisome proliferator activated receptor gamma (PPARg), is a particularly promising link between EDCs and their metabolic effects. This NR recognizes structurally diverse ligands, and regulates the transcription of genes involved in lipid storage homeostasis. Importantly, this NR interacts with nutrient sensors such as AMP-activated protein kinase (AMPK) through feedback mechanisms, thus coupling control over metabolism and signaling in adipocytes [5]. We therefore investigated the effects of several representative EDCs in the context of the regulatory subnetwork anchored by PPARg and AMPK, under the hypothesis that EDCs act via both metabolic and signaling pathways to cause inflammation in adipocytes.

To establish that EDCs elicit an inflammatory response in mature adipocytes, we screened for the effects of tributyltin (TBT), bisphenol A (BPA), and monoethylhexyl phthalate (MEHP) using 3T3-L1 cells. Nanomolar doses of BPA,

MEHP, and TBT did not affect cell morphology or TG accumulation, but significantly increased expression of inflammation markers such as monocyte chemo-attractant protein 1 (MCP-1). To broadly assess the metabolic effects, we profiled the levels of ~50 intracellular metabolites, and performed a partial least squares discriminant analysis (PLS-DA) on the metabolite data. Latent variable score projections from PLS-DA showed that there were clear differences in the metabolite profiles across the treatment groups. The differences were most pronounced for MEHP treated cells. The corresponding loadings showed that multiple metabolites contributed to the differences, with the majority comprising fatty acid species.

Focusing on MEHP as a potent EDC, we performed a dose response experiment to obtain a more detailed analysis of the metabolic and inflammatory responses. In addition to MCP-1, MEHP treatment increased the expression of CXCL1, IL-6, and TNFα, similar to the effects of the prototypical pro-inflammatory cytokine TNFα. Exposure to MEHP also altered the intracellular FFA profile, raising the levels of both saturated and unsaturated fatty acids. Interestingly, MEHP treatment also increased the intracellular AMP/ATP ratio, suggesting activation of AMPK. We next performed untargeted and targeted proteomics experiments to profile the impact of MEHP exposure on the levels of cellular proteins. The proteomic data indicated that the chemical exposure broadly and significantly altered the levels of enzymes and other proteins involved in regulating the lipid balance in adipocytes. These data were consistent with our hypothesis that MEHP impacts transcriptional/translational regulation of metabolic pathways. Several of the up-regulated proteins were downstream targets of PPARg (e.g. GPDH, FABP4), which supports the involvement of this key transcriptional regulator.

Taken together, our results support the hypothesis that the metabolic and inflammatory responses are coupled, and that AMPK and PPARg act as upstream mediators. Our results also support a mechanistic link between EDC exposure and adipocyte inflammation, which underpins many of the metabolic dysfunctions of obesity.

[1] Heindel (2003) Toxicol Sci 76: 247-9. [2] Baillie-Hamilton (2002). J Altern Complement Med 8: 185-92. [3] Grun et a. (2007) Rev Endocr Metab Disord 8: 161-71. [4] Manteiga et al. (2013) Wiley Interdiscip Rev Syst Biol Med 5: 425-47.

# 412302 Cell and Tissue-Based Therapies for Insulin-Dependent Diabetes

**Abstract:**

A living biological substitute for treatment of insulin-dependent diabetes has significant potential in providing a less invasive, more physiologic regulation of blood glucose levels than insulin injections. The critical technologies needed for such a substitute depend strongly on the type of cells used. With cells from another individual (allogeneic) or another species (xenogeneic), encapsulation in semipermeable barriers improves immune acceptance, as it inhibits passage of antibodies and excludes cytotoxic cells of the host. However, immune protection is not complete, and immune suppression may still be needed to prolong survival of the graft. Non-pancreatic cells from the same patient (autologous), targeted by gene transfer vectors or retrieved surgically and genetically engineered *ex vivo*before being returned to the patient, may relax the immune acceptance problems but pose challenges regarding the amount and kinetics of insulin secretion in response to physiologic stimuli. Stem and progenitor cells constitute another promising cell source, however, their reproducible differentiation into pancreatic cells presents significant challenges.

In our laboratory, we focus on encapsulated allo- and xenogeneic pancreatic cells and on non-pancreatic cells genetically engineered to secrete insulin in response to physiologic stimuli. With encapsulated cells, we are developing methods to improve immunoprotection by combining the semipermeable barrier with the local presentation and delivery of pro-survival and insulinotropic factors. Furthermore, we develop technologies for cryopreservation of encapsulated cells and for monitoring grafts in minimally invasive or non-invasive ways. Of particular importance is the use of 19F nuclear magnetic resonance spectroscopy to measure dissolved oxygen levels in vivo. One of our major findings is that the murine peritoneal environment, where encapsulated cells are commonly transplanted, is hypoxic and it becomes more hypoxic upon transplantation of islets or other insulin-secreting cells. With non-pancreatic cells, we genetically engineer hepatic and intestinal endocrine L cells for insulin secretion. These cells are potentially autologous, however, they pose challenges regarding the amount and kinetics of secreted insulin. Results from experiments involving transplantation of murine, insulin-secreting L cells in diabetic mice demonstrated that the graft had a significant positive effect on the glycemic regulation of the animals, however, it did not cure diabetes. The potential and challenges for developing clinical therapies based on these approaches will be discussed.

# 433709 Confessions of a Cell Culture Engineer: How Process Can Impact Protein Products

**Abstract:**

In producing therapeutic proteins from cell culture processes, it is necessary to develop processes capable of delivering consistent protein characteristics, particularly those that could affect the structure or function of the biomolecule and potentially impact the product safety or efficacy. Over the past thirty years, cell culture scientists and engineers have navigated the waters of process understanding in tackling protein product characterization challenges. In many cases, the root cause of protein product attribute variation is related to cell culture process conditions.

The impacts of process conditions on protein attributes may be due to biosynthetic (biological origin) or degradative (chemical or physical origin) effects, and these effects may be driven by intracellular or extracellular factors. This presentation will review industrially relevant experiences, and discuss approaches toward better understanding and control of recombinant protein product characteristics through cell culture manipulations, including medium composition and process design. Specific protein attribute variation addressed will include amino acid sequence, glycosylation (micro and macroheterogeneity), and charge variant distributions, and their potential for impact on various aspects of protein quality.

# 437503 Mapping Gastrointestinal Disease

**Abstract:**

The convergence of engineering and medicine presents a compelling opportunity to apply traditional engineering principles to diagnose and treat challenging diseases. For example, the disease eosinophilic esophagitis (EoE) causes frequent ER visits when the esophagus narrows or strictures, blocking flow of saliva and swallowed solids as small as a pea. Patients with EoE may choke without medical help. Evaluation of esophageal tissues in EoE patients finds patchy invasion of eosinophils, a type of white blood cell normally absent. Although endoscopy and biopsy collection is currently recommended for diagnosis, this procedure is invasive, and up to 20% of patients may be misdiagnosed due to the patchiness of eosinophil invasion. Our team is developing a new class of solid ultrasound contrast agents—a slurry of insulin particles—to provide a less invasive way to diagnose EoE and other gastrointestinal (GI) diseases. Patients simply drink the slurry, the particles bind to eosinophil-enriched tissues, and an external ultrasound probe detects the bound particle distribution as a signal of eosinophil concentrations throughout the esophagus. However, significant engineering and clinical questions remain. Can significant surface binding occur even though the bolus transport time is only 6 to 15 s? What liquid rheology most enhances particle binding? What insight can be gained from the recent treatment successes of viscous budesonide, a suspension of glucocordicoid (steroid) particles? How does stricturing affect esophageal fluid mechanics and agent delivery? Can the adult esophagus be imaged through the rib cage? Can these contrast agents be shaped and distinguished in the multiphase milieu of the GI tract? This talk addresses each of these questions essential to the development of this new class of contrast agents to diagnose a broad range of gastrointestinal diseases.

# 437508 Flow-Induced Segregation in Multicomponent Medical Suspensions

**Abstract:**

Blood is a suspension of particles of various shapes, sizes and mechanical properties and the distribution of these particles during blood flow is important in many contexts. Red blood cells (RBCs) tend to migrate toward the center of a blood vessel, leaving a so-called cell-free layer at the vessel wall, while white blood cells (WBCs) and platelets are preferentially found near the walls, a phenomenon called margination that is critical for the physiological responses of inflammation and hemostasis. Additionally, drug delivery particles in the bloodstream will also undergo segregation phenomena – the influence of these phenomena on the efficacy of such particles is unknown.

This talk describes efforts to gain a systematic understanding of flow-induced segregation phenomena in blood and other complex mixtures, using a combination of theory and direct simulations of flowing suspensions. A kinetic theory model based on pair collisions and wall-induced hydrodynamic migration can capture the key effects observed in direct simulations, including a "drainage transition" in which one component is completely depleted from the bulk of the flow. Experiments performed in the laboratory of Wilbur Lam indicate the physiological and clinical importance of these observations. More generally, having in hand an understanding of the mechanisms underlying these phenomena now allows more rational approaches to development of quantitative models of them and processes that exploit them. This knowledge will also lead to a better understanding of the consequences of these phenomena in physiology and medicine.

# 437559 Nanolipoprotein Particles: Encapsulated or Surface-Bound for Biomedical Applications

**Abstract:**

Nanolipoprotein Particles (NLPs) are disc-shaped nanometer-sized lipid bilayer patches stabilized by a belt of scaffold proteins. NLPs have an average thickness of 5 nm, with a diameter ranging from 10-25 nm depending on the stoichiometric ratios and types of lipids and scaffold proteins being used. This allows NLPs to be compatible with the pore size (5-50 nm) of mesoporous silica. Therefore, we perform entrapment of NLPs using a quick, simple sol-gel processing technique for TMOS that includes evaporation of the majority of the methanol after the hydrolysis reactions. To ensure proper functioning of silica sol-gel entrapped NLPs, we have investigated the phase behavior of the lipids in addition to the secondary structure, localization, and environmental polarity of the scaffold proteins. Our results indicate that silica gel-entrapped NLPs remain intact, with only slightly altered lipid and scaffold protein structure and dynamics. We will briefly discuss the potential to entrap NLPs containing embedded integral membrane proteins (IMPs) for various applications such as biosensing, affinity chromatography, high-throughput drug screening, and bio-reaction engineering.

Further, scaffold proteins can bear polyhistadine tags, which are capable of chelating to $Cu^{2+}$ metal ions. Lipid-phase specific, iminodiacetic acid (IDA) functionalized lipids are also capable of chelating $Cu^{2+}$, providing a mechanism for phase-targeted binding of NLPs. We investigate this binding via fluorescence microscopy and characterize interaction with phase-separated supported bilayers and giant unilamellar vesicles (GUVs). The thermodynamics (enthalpy of lipid mixing and steric pressure of protein crowding) and morphology of binding are also examined. Targeted binding of NLPs bearing functional IMPs and/or other biomolecules to supported lipid bilayers has a variety of applications, including development of nano-array technologies and biosensors.

# 437506 Drug Discovery and Development in a Network of Innovation

**Abstract:**

Earlier this decade, a respected and widely cited study from Tufts University's Center for the Study of Drug Development came up with the astounding figure of $1.2 Billion as the average cost of bringing novel medicinal drugs to commercialization from the earliest R&D stages of discovery biology. Understandably, this large figure was met with shock and disbelief. Shortly thereafter, several business journals embarked on the investigative task of verification of the calculation and data. What they uncovered was even more surprising. Actual R&D spending over a 15-year period (to smooth out annual fluctuations) divided by the number of new drug approvals indicated company-specific estimates in the range of $3.7 Billion (best) to $11.8 Billion (worst) per drug for a cohort consisting of a dozen major companies representing the pharmaceutical industry. The journalists concluded that even after taking allowance that these figures are crude estimates, the pharmaceutical industry had an unsustainable R&D model.

In a dynamic, free-market based society, inefficiencies on such a large scale attract innovators with disruptive models and the pharmaceutical R&D ecosystem is no exception. An analysis of the pharma industry's R&D workflow suggests that the stage between discovery biology and clinical development, known as (preclinical) lead optimization chemistry, is a major bottleneck. There is now ample evidence that R&D innovation in lead-optimization chemistry can reduce significantly the above cited development cost per drug and this presentation will analyze (via information in the public domain) the modified R&D workflows that incorporate a network of innovation. This development is important to chemical engineers as well as the AICHE by association because discovery research in big pharma is the domain of those trained in the "standard" fields of the biomedical sciences whereas the entities in the network of innovation are embracing the integrative and multiscale talents of chemical engineering graduates.

# 437512 Mucus Penetrating Nanoparticles: From Concept to Success in the Clinic

**Abstract:**

The controlled delivery of bioactive molecules to target tissues can significantly improve drug effectiveness while reducing side effects by concentrating medicine at selected sites in the body. Mucus layers coat and protect nearly all entry points into the body that are not coated by skin. Until recently, human mucus was thought to be nearly impenetrable to drug delivery particles even as small as 59 nm in diameter. Particles that become trapped in mucus are typically rapidly cleared from the organ of interest, usually within minutes to a few hours. Thus, while the barrier properties of mucus provide outstanding protection against infection and other potentially toxic substances, they have also thwarted efforts to achieve uniform and sustained drug and gene delivery to mucosal surfaces. This talk will focus on our work to understand the length-scale dependent and adhesion-mediated barrier properties of mucosal fluids, and how this knowledge has guided the development of polymeric nanoparticulate carriers capable of improved drug and gene delivery to the respiratory tract, female reproductive tract, gastrointestinal tract, surface of the eye, and other mucosal tissues.

# 420461 Differential Effects of Acellular Embryonic Matrices on Pluripotent Stem Cell Expansion and Neural Differentiation

**Abstract:**

Extracellular matrices (ECM) derived from pluripotent stem cells (PSCs) provide unique niche that can direct cellular differentiation and tissue regeneration as well as rejuvenation of aged progenitor cells. Decellularized ECMs have also been used as bioink for bioprinting of cell-laden constructs. The unlimited growth capacity of PSCs allows the scalable generation of the associated ECMs which can be used for biomanufacturing. Therefore, derivation and characterization of PSC-derived ECMs are critical for the applications in drug screening, disease modeling and regenerative medicine. In this study, 3-D ECMs of undifferentiated embryonic stem cell (ESC) aggregates (AGG), spontaneously differentiated embryoid bodies (EB), and ESC-derived neural progenitor cell (NPC) aggregates were decellularized and their capacities to direct proliferation and neural differentiation of the reseeded mouse ESCs and human induced pluripotent stem cells (iPSCs) were characterized. Proteomic analysis by liquid chromatography–tandem mass spectrometry (LC-MS/MS) revealed protein expression profiles that reflected distinct niche properties for each tested ECM group. The reseeded mouse ESCs and human iPSCs responded to different types of ECMs with high expression of pluripotent markers (Oct-4 and Nanog) for the cells grown on the AGG-ECM and the accelerated neural differentiation (expressions of β-tubulin III and Nestin) for the cells grown on the NPC-ECM. The role of Wnt/β-catenin signaling in the cell-matrix interactions was also studied and the expression of β-catenin was reduced in the cells grown on the NPC-ECM. This study demonstrates that PSC-derived ECMs are able to influence stem cell proliferation and differentiation by providing a spectrum of stem cell niche microenvironments during embryonic tissue development.

# 426194 A Modular GAG/Chitosan/MSC Approach to Generating Tissue Engineered Bone

**Abstract:**

Modular tissue engineering involves the construction of large (macro-scale) tissue structures from smaller (micro-scale) tissue units or modules. The field has the potential to solve various cell-density related challenges associated with traditional tissue engineering. Previous studies have demonstrated the value of vascularization enhancing strategies as a necessary component in bone defect regeneration. In this study, we investigated the fabrication of macro-scale, bone-like constructs by fusion of micro-scale modular units composed of natural polyelectrolyte polymers, hydroxyapatite (HAP) granules, and mesenchymal stem cells (MSCs). The micro-scale modules were formed by complex coacervation whereby an ionic reaction between chondroitin 4-sulfate (C4S) and chitosan produced an insoluble polyelectrolyte membrane. Briefly, a C4S solution with suspended MSCs and HAP granules was extruded as 400 micron diameter droplets and collected into stirred chitosan solution. The insoluble polyelectrolyte complex was formed at the droplet-chitosan interface, encapsulating MSCs and HAP granules suspended in the C4S solution. The resulting microcapsule modules were transferred to dish culture, and maintained in a standard osteogenic culture medium. Analysis of the cultured modules over time indicated that C4S/chitosan/HAP microcapsules supported MSC proliferation and differentiation to an osteoblastic lineage. Osteoblastic activity was assessed by quantification of: bone-specific protein secretion (osteocalcin, osteopontin, collagen), calcified matrix deposition, and increases in alkaline phosphatase (ALP) activity. Moreover, the micro-scale modules could be fused post-culture to create larger (1.5 $cm^3$) tissue constructs which possessed inherent pore spaces between the spherical, mineralized units. Fusion was accomplished by generation of additional polyelectrolyte complex between packed modules using a sequential polymer perfusion procedure. The compressive strength and elastic modulus of fused tissue constructs increased significantly during induced MSC osteogenesis, compared to values for constructs contain non-differentiating MSCs: fused constructs fabricated from microcapsules containing MSCs undergoing osteoinduction for 4 weeks exhibited a compressive strength of 6.2 MPa (+ 3.1 MPa), significantly higher than control fused constructs without differentiating MSCs (0.0059 + 0.0011 MPa). Moreover, microCT data shows that differentiating MSCs are able to extensively mineralize the walls and interior of C4S/Chitosan microcapsules: microcapsules containing MSCs osteoinduced for 4 weeks, but no initial HAP granules, exhibited a bone volume fraction of 182.5 mg HAP/$cm^3$, compared to a bone volume fraction of zero for capsules without differentiating MSCs or HAP granules included initially in capsule fomration. Inter-module pores were found to facilitate prevascularization of constructs by culturing accessory cells, such as endothelial cells, on the exterior of the modules prior to fusion. Current studies are evaluating the effects of alternate fusion chemistries on construct mechanics and the physiology of encapsulated and accessory cells. Results to date indicate that the C4S/chitosan/HAP based module system can form the basis of either a 3D printing based or an injectable matrix based strategy for generation of vascularized bone.

# 421975 Agent-Based Modeling of Porous Scaffold Degradation and Vascularization: Optimal Scaffold Design Based on Architecture and Degradation Dynamics

**Abstract:**

Agent-based modeling of porous scaffold degradation and vascularization: Optimal scaffold design based on architecture and degradation dynamics

Hamidreza Mehdizadeh [a], Elif S. Bayrak [a], Chenlin Lu [a], Sami Somo [a], Banu Akar [a], Eric M. Brey [b,c], Ali Cinar [a,*]

[a] *Department of Chemical and Biological Engineering, Illinois Institute of Technology, 10 W 33rd St, Suite 127, ChBE desk, Chicago, IL 60616, USA*

[b] *Department of Biomedical Engineering, Illinois Institute of Technology, Suite 314, 3255 S Dearborn St, Chicago, IL 60616, USA*

[c] *Research Service, Hines Veterans Administration Hospital, Hines, IL, USA*

Biomaterial scaffolds serve as substrates for angiogenesis and tissue growth by providing essential mechanical and structural support required for appropriate cell function and morphogenesis. The degradation behavior of scaffolds affects angiogenesis significantly since it changes the porous structure and mechanical properties of the scaffold dynamically over time. It is important for the scaffold to degrade at a balanced rate and maintain the scaffold integrity during tissue formation. One problem that may arise during the tissue regeneration process is the mechanical failure of the tissue engineering construct as a result of fast scaffold degradation before production of new tissue.

An agent-based model (ABM) of biomaterial scaffold vascularization [1] is extended to consider the effects of scaffold degradation on the dynamics of angiogenesis. A statistical model developed by Metters et al. to describe the bulk degradation of PLA-*b*-PEG-*b*-PLA hydrogel is applied to incorporate scaffold degradation factors into the model [2]. The modified ABM is focused on the combined effects of scaffold porous architecture and degradation on vascularization process. The degradation rate of scaffolds can be manipulated and altered based on the rate degree of crosslinking, hydrolysis reaction rate constant, polymer composition, and/or polymer concentration. A multi-layer scaffold model having regions with different degradation rates within the scaffold structure was designed. The ABM of the multi-layer scaffold, was used to explore the optimal conditions for scaffold vascularization.

The simulation results indicate that scaffold degradation plays an important role in angiogenesis. Specifically, it improves vascularization more when degradation rate is higher and scaffold porosity is lower. Simulation with multi-layer scaffold model shows that vascularization was successfully continued after the previous scaffold layer was fully disintegrated because a considerable portion of the present layer was still remaining and providing sufficient mechanical strength and cell-attachment sites for angiogenesis. In general, it is observed that degradation can be utilized as a means of enhancing vascularization and the multi-layer scaffold model can further optimize the scaffold vascularization without losing mechanical support of the biomaterial scaffolds. Simulations show the flexibility of the developed ABM and exemplify the types of investigation it enables. These results can be used in combination with experimental research to design optimal scaffold structures.

Reference:

[1] H. Mehdizadeh, S. Somo, E.S. Bayrak, E.M. Brey, A. Cinar, "Three-dimensional modeling of angiogenesis in porous biomaterial." *Biomaterials, 34(12),* pp.2875-2887, January 2013.

[2] A.T. Metters, C.N. Bowman, K.S. Anseth, "A Statistical Kinetic Model for the Bulk Degradation of PLA-*b*-PEG-*b*-PLA Hydrogel Networks." *J. Phys. Chem. B, 104(30),* pp.7043-7049, July 2000.

# 432595 Engineering Vasculogenesis to Determine the Vascular Anatomy Dependence of Circulating Tumor Cell Extravasation

**Abstract:**

Circulating tumor cells are known to preferentially metastasize to certain organ sites. A large number of studies have outlined the metastasis mechanism, however there still remain questions regarding tissue specific tropism. We hypothesize that vascular anatomy, ensuing hemodynamics, and near tissue microenvironment are critical to the attraction of metastasis cells. To address this, we have engineered implantable tissue microenvironments with tunable hydrogel scaffolds that reproducibly formed vascularized tissue analogues in vivo. Specifically we subdermally implanted 4 different hydrogel scaffolds with varying pore diameters (53-75, 150-180, 250-300 and 425-500 μm) in a single host mouse. After the scaffolds were implanted for 5 weeks, histological characterization revealed that scaffolds accommodated distinct vasculatures as a function of pore diameter. We next simulated circulating tumor cells by intravenously injecting 1-2 million cancer cells. Six hours later we retrieved the implanted scaffolds and determined the correlation between vasculature anatomy and systemic migration of cancer cells. Our results indicate that tissue dependent vascular anatomy is critical in recruiting specific types of circulating tumor cells. Engineered tissue microenvironments are expected to become a valuable platform for dissecting the metastasis mechanism and important parameters of the pro-metastatic niche.

NANO LETTERS

Letter

pubs.acs.org/NanoLett

Open Access on 06/10/2015

# Self-Folding Single Cell Grippers

Kate Malachowski,[†] Mustapha Jamal,[†] Qianru Jin,[†] Beril Polat,[†] Christopher J. Morris,[‡] and David H. Gracias*,[†]

[†]Department of Chemical and Biomolecular Engineering, The Johns Hopkins University, 3400 N. Charles St., Baltimore, Maryland 21218, United States

[‡]United States Army Research Laboratory, Sensors and Electron Devices Directorate, 2800 Powder Mill Rd., Adelphi, Maryland 20783, United States

*Supporting Information*

ABSTRACT: Given the heterogeneous nature of cultures, tumors, and tissues, the ability to capture, contain, and analyze single cells is important for genomics, proteomics, diagnostics, therapeutics, and surgery. Moreover, for surgical applications in small conduits in the body such as in the cardiovascular system, there is a need for tiny tools that approach the size of the single red blood cells that traverse the blood vessels and capillaries. We describe the fabrication of arrayed or untethered single cell grippers composed of biocompatible and bioresorbable silicon monoxide and silicon dioxide. The energy required to actuate these grippers is derived from the release of residual stress in 3–27 nm thick films, did not require any wires, tethers, or batteries, and resulted in folding angles over 100° with folding radii as small as 765 nm. We developed and applied a finite element model to predict these folding angles. Finally, we demonstrated the capture of live mouse fibroblast cells in an array of grippers and individual red blood cells in untethered grippers which could be released from the substrate to illustrate the potential utility for in vivo operations.

KEYWORDS: *Self-assembly, origami, lab-on-a-chip, surgery, nanomedicine, nanomechanics, robotics*

Because of the large size of tools that are typically utilized for surgical diagnostics and biological analyses, cellular samples are often large in size. Consequently, the data collected from tissue biopsied samples and related assays average over a multitude of cells. However, that average may not accurately represent the behavior of individual cells, particularly if the cells of interest are a small fraction of the population. Further, it can be challenging to draw conclusions about dynamic or transient behaviors of single cells by looking at large populations.[1−3] Tumors have long been known to be heterogeneous populations of cells with varying phenotypes and genotypes, proliferation rate, potential for metastasis, and drug responsiveness, yet we are only beginning to understand how these heterogeneities affect their progression.[4−7] Single cell analyses may be necessary to differentiate the behavior of a cell subpopulation from the bulk sample, particularly in the fields of cancer biology, genomics, proteomics, stem cell biology, and hematology.[3] This work is especially important as treatments for cancer, immune diseases, and tissue regeneration move toward personalized medicine.[8]

A wide range of techniques are available for in vitro single cell analysis, and each has advantages and disadvantages in terms of efficiency, cell manipulation, imaging capability, sensitivity, and ability to mimic or actually perform in vivo.[9,10] These methods include flow cytometry,[9] optical traps,[11−16] microfluidic traps and devices,[2,17−26] microwells,[27,28] microtubes,[29] and 2D surface patterns.[30−34] Several miniaturized robotic devices have been created to trap and manipulate particles and cells with precise control.[35−37] For example, Chronis et al. demonstrate the manipulation of a 10 μm cell using a wired, electrothermally actuated SU-8 gripper.[35] This device can manipulate cells with high precision, but the electrical wires that control its actuation and its large back-end design limit throughput and in vivo utility. Another SU-8 device, by Sakar et al., provides untethered manipulation of single cells via magnetic forces with minimal fluid disturbance due to its micrometer size and biocompatibility.[36] However, these devices are passive, trapping cells in a recess, and thus they may lose their grip on a cell if moved in the wrong direction or in all three dimensions. A significant hurdle in the

Received: May 19, 2014
Revised: June 5, 2014
Published: June 10, 2014

creation of active single cell devices with moving parts is the challenge in harnessing energy at small size scales and in a highly parallel and untethered manner.

An ideal in vitro device would combine the high throughput efficiency of flow cytometry, the incorporation of patterned microfeatures for biomolecular analyses, and the 3D manipulation precision of optical tweezers. An ideal in vivo device should be composed of biocompatible and possibly bioresorbable materials while facilitating tissue excision or targeted capture, robust gripping, and retrieval in an autonomous manner.[38] Here, we describe an important step toward achieving tools that combine both of these in vitro and in vivo capabilities for single cell studies and to potentially access tiny conduits in the body. Our approach is inspired by previous studies on the stress-based roll-up and self-folding of thin films and the energy required to enable gripper motion is derived from the differential residual stress in nanoscale bilayers.[39−64] The approach utilizes photolithography, which is a high throughput technique capable of fabricating 500 000 to 10 million single grippers on a 3 in. wafer or potentially over 100 million on a 12 in. wafer which is the size of wafers used currently in CMOS fabrication facilities. Additionally, the grippers can be actuated to close around single cells en masse, creating devices with patterns in all three dimensions. The thickness of the films can be varied to control the fold angle, while the sharpness of the tips could aid with the capture and containment of cells. As compared to previously described stimuli responsive "$\mu$-grippers" that were used to biopsy cell samples and porcine organs under in vitro, ex vivo, and in vivo conditions,[54,65−67] these grippers are 30 times smaller, requiring significantly thinner hinges and different materials to achieve a tight radius of curvature. While we previously utilized the larger $\mu$-grippers only in the gastrointestinal (GI) tract, we envision that these single cell sized grippers could be used in tighter spaces such as within the circulatory, urinogenital, or central nervous system. In these regions, however, there are more stringent requirements on biocompatibility and biodegradability for these applications.[38]

In light of the more stringent biodegradability requirements and the need for small devices, considerable thought was given to the materials chosen for these single cell grippers. Silicon (Si) and silicon dioxide ($SiO_2$) react with water via hydrolysis to form $Si(OH)_4$[68,69] and thus dissolve into DI water and various biofluids[68,70] as previously reported by Hwang et al. in their work on transient electronics.[70] Additionally, electronic devices made from Si, $SiO_2$, and other metals were implanted subdermally into mice with no significant inflammatory reactions and almost complete dissolution in 3 weeks. Silicon monoxide (SiO) is a two-phase, nonhomogenous mixture of amorphous Si and $SiO_2$ and has been previously paired with $SiO_2$ to form tightly rolled tubes with microscale radii of curvature when deposited by electron beam (e-beam) evaporation in nanometer-scale thicknesses.[29,52,59,60,76,77] Thus, we selected these two silicon-based oxides for our single cell grippers for their biocompatibility, bioresorbability, and self-curling properties. We independently verified the dissolution of SiO and $SiO_2$ in phosphate-buffered saline solution (PBS) over a 20 day period at 37 °C. Thirty nanometer films of $SiO_2$ dissolved at a rate of 2 nm/day at 37 °C, while SiO dissolved more slowly at less than 1 nm/day (Figure S1).

Grippers were fabricated with flexible, prestressed bilayer hinges, connected to rigid segments (Figure 1a). The prestressed bilayer was constructed from e-beam evaporated

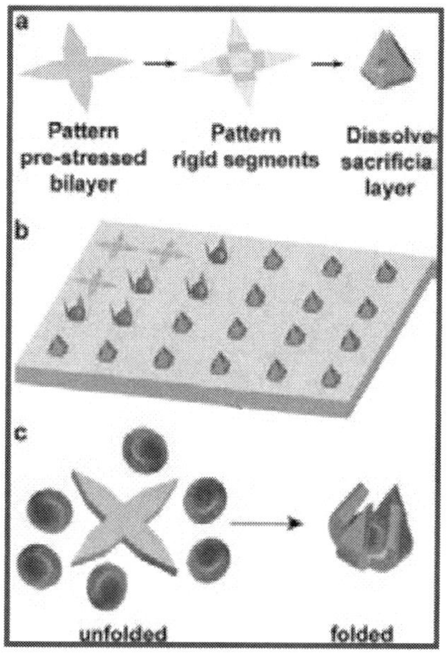


**Figure 1.** Illustration of single cell gripper fabrication and use on substrates or as untethered tools. (a) Fabrication scheme for creating single cell grippers. The prestressed actuator hinge is a $SiO/SiO_2$ bilayer, while the rigid segments are made of SiO. Upon dissolution of the sacrificial layer, the arms are released and self-actuate to close around cells. An optional thermoresponsive trigger layer can be molded atop the grippers. (b) Illustration of cells captured by single cell microgrippers arrays. (c) Illustration of untethered single cell grippers and red blood cell capture.

thin films of SiO and $SiO_2$. The rigid segments were formed from thicker films of e-beam evaporated SiO. A stimuli-responsive hinge trigger could be patterned or molded atop the grippers to control actuation.[54,65−67,71] These devices can either be arrayed on a substrate for use as a single cell in vitro analytical assay device or completely released to be used as free-floating or untethered tools (Figure 1b−c). Detailed thicknesses and evaporation conditions are in the Suppporting Information (Table S1).

We developed several gripper variants with three or four arms, varying in size from 10 to 70 $\mu$m in length (tip-to-tip when open), which is an appropriate size range to grasp a variety of individual cells (Figure 2a). The alternating rigid frames and flexible hinges are evident in the open grippers (Figure 2b, d). Grippers folded at angles ranging from 90° to 115° depending on the bilayer film thickness, corresponding to folding radii ranging from to 765 nm to 5 $\mu$m. The film thickness could be adjusted to assemble tightly folded grippers in a range of sizes. For example, 9 nm of SiO and 27 nm of $SiO_2$ were deposited to assemble the 50 $\mu$m grippers shown in Figure 2b and c, while a bilayer of 3 nm of SiO and 3 nm of $SiO_2$ was used to assemble the 10 $\mu$m grippers in Figure 2d and e. Despite their small sizes, these grippers were fabricated using photolithography on a projection mask aligner with 500 nm resolution. Photolithography and registry of multiple film layers became increasingly difficult as the size of the grippers decreased for the specific tools used, leading to a lower limit for an open tip to tip gripper size of about 10 $\mu$m. In principle, smaller grippers could be fabricated using serial e-beam lithography, but quantities would be limited due to the serial nature of that technique, and the sizes would be smaller than those of single cells thereby limiting use.

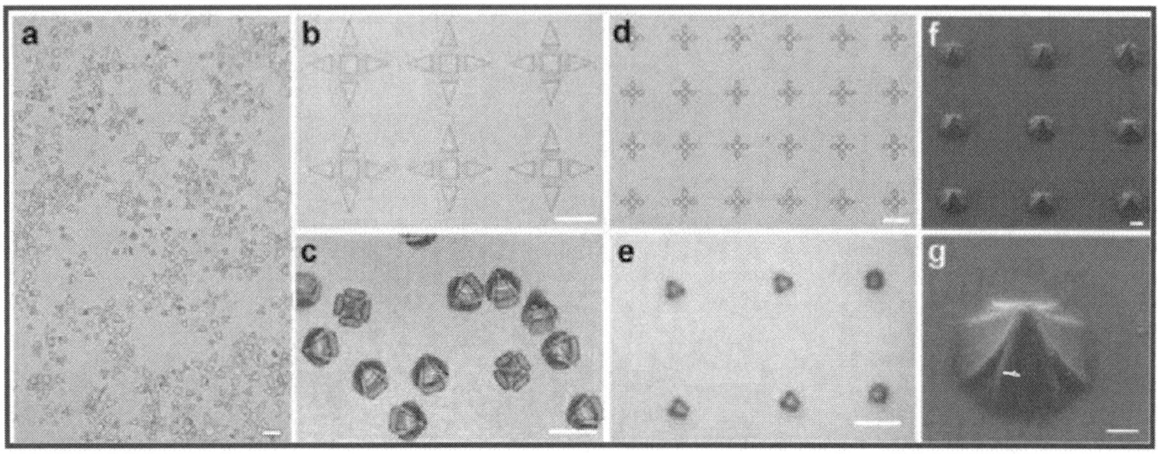


**Figure 2.** Optical images of single cell grippers before and after closing. (a) Optical image of grippers released from the substrate with open arms prior to closing, in sizes ranging from 10 to 50 $\mu$m. (b–c) Zoomed optical images of 50 $\mu$m grippers (b) prior to release from the substrate and (c) closed tightly after release. (d–e) Optical images of 10 $\mu$m grippers (d) open and (e) closed. Scale bars are (a, b, c) 25 $\mu$m and (d, e) 10 $\mu$m. (f–g) SEM images at different magnifications of closed single cell grippers attached to the substrate. Scale bars are (f) 10 $\mu$m and (g) 5 $\mu$m.

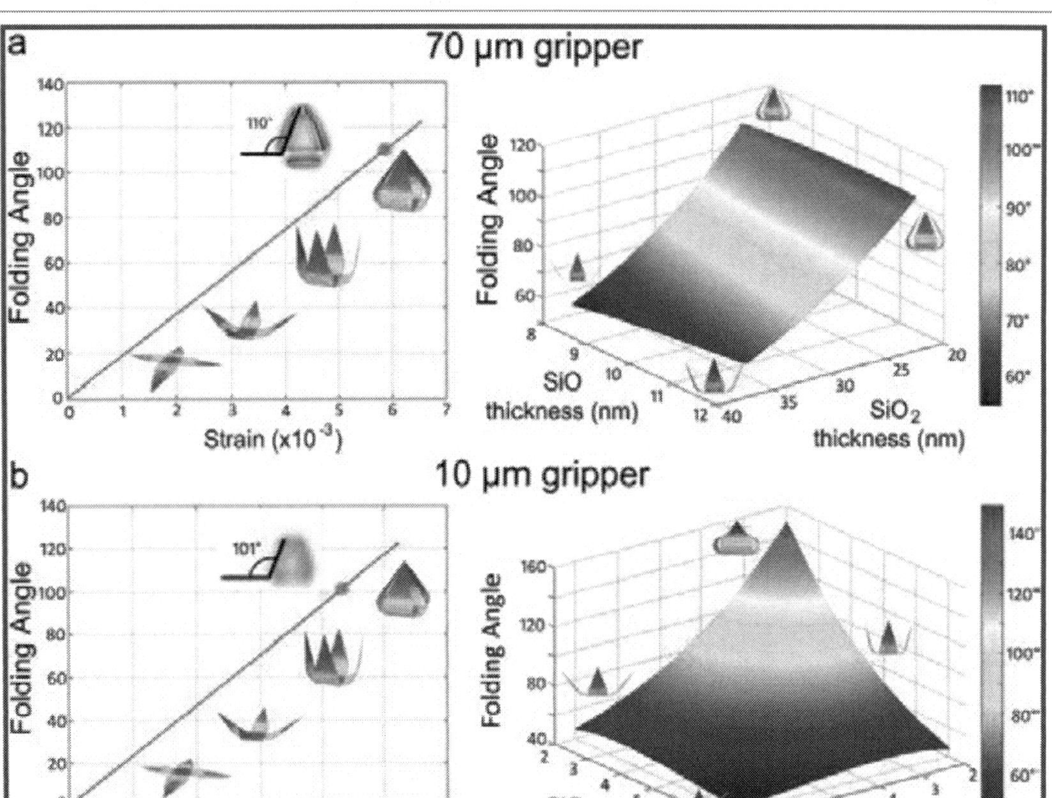


**Figure 3.** Characterization of thin film stress and gripper folding angle. (a) Graphs depicting the effect of mismatch strain (left panel) and $SiO/SiO_2$ thickness (right panel) on folding angle for the 70 $\mu$m gripper. (b) Graphs depicting the effect of mismatch strain (left panel) and $SiO/SiO_2$ thickness (right panel) on folding angle for the 10 $\mu$m gripper. The inset images of a folded gripper with angle measurement in the left panels are optical microscopy images of actual folded grippers for comparison to the modeled gripper folding. The red star on these graphs indicates the experimentally observed folding angle for these experimentally observed grippers. More details of the parameters used to generate the models and graphs are in the Supporting Information.

The radius of curvature of each gripper hinge is related to the film thickness, mechanical properties of the materials, and residual stress of each layer within the prestressed bilayer. It is noteworthy that previous designs of $\mu$-grippers were made from either a chromium/copper (Cr/Cu) bilayer or a chromium/gold (Cr/Au) bilayer.[54,66,67] In those designs, the Cr layer had significant tensile stress (~1 GPa), while the Cu or Au layer was relatively neutral in stress. This stress differential caused the grippers to fold due to the shared boundary between the two layers. These previously utilized metallic combinations, however, were unable to curl with a radius of curvature less than about 30 μm, limiting a multisegmented gripper device to larger than 200 μm in tip to tip size when open. However, as we describe in this work, the $SiO/SiO_2$ combination provided a sufficiently small radius of curvature for single cell grippers, with radii as small as 765 nm.

We characterized the residual stress within the SiO and $SiO_2$ films. We found the stress in both SiO and $SiO_2$ to be compressive, with less compressive stress as thicknesses increased from 10 to 100 nm (Figure S2a). These stress values varied significantly with thicknesses below 100 nm but were consistent with the expected range of compressive stress for the deposition conditions used.[72,73] However, a more important parameter for $SiO_2$ film stress was the time of exposure to room air following deposition in an evaporation chamber. We observed an increasing tensile component in $SiO_2$ films over time (Figure S2b), while the stress in SiO films remained generally constant. We attribute the change in $SiO_2$ stress to the absorption of water by $SiO_2$, and the subsequent reaction between water and dangling Si bonds that leads to a tensile stress component that grows linearly with the logarithm of aging time.[74] By using $SiO_2$ films to form the inside of each concave folded hinge, we ensured that the growth of this tensile stress component helped each hinge fold with a sufficiently small radius of curvature.

We examined the effect of preload strain and film thickness on the folding angle using an analytical curvature model[75] and a computational finite element analysis (FEA) simulation (Figure 3 and Figures S3−S5). Details of the models are given in the Supporting Information. We modeled the effect of strain on folding angle for a 70 μm and a 10 μm gripper (Figure 3) and found that as the strain increased the folding angle also increased which is expected. Larger mismatch strain allows the gripper to fold more tightly. One can tune the strain by choosing different materials and adjusting thin film deposition condition. Film thickness is also a tunable property in gripper fabrication and greatly affects the folding angle. We performed a thickness sensitivity analysis by modeling the folding angle versus SiO and $SiO_2$ film thickness for a 70 μm and a 10 μm gripper. In general, due to larger bending stiffness, the folding angle decreases as film thickness increases; thus for the smallest grippers, films on the order of 3−5 nm were required. However, grippers with films that are too thin may be too fragile for use in real biological applications. These plots, and other detailed plots given in the Supporting Information, serve as design guides for determining the necessary thicknesses for each layer within the prestressed bilayer to achieve a desired folding angle.

One application we envision for these devices is an in vitro arrayed analytical device that could be used to entrap many single cells for biological assays. As these devices can be patterned in all three dimensions and grip en masse, many individual cells can be trapped, assayed, and imaged with a high yield. To demonstrate this application, we fabricated 50 μm grippers that remain attached to the substrate upon release of the arms. The arms were patterned on a Cu sacrificial layer and were thus able to fold. The center or "base" of the gripper was patterned directly onto the Si wafer so that it remained attached during the release and folding process. We pipetted live L-929 mouse fibroblasts in media on top of the open grippers. The grippers closed around individual cells after 2−6 h in warm culture media due to the slow etching action of the ions in the media (Figure 4). Thus, in this application, no hinge trigger is required as cells are captured during the release of the gripper arms from the substrate. The gripper arms closed around each cell, and the cells remained viable as evidenced by the green fluorescence of the cell bodies from the calcein stain of the live/dead assay. Some grippers were empty, but when occupied, each of the grippers contained only one cell. Our best observed yield for successfully filled grippers was 48% for an area of approximately 75 grippers. The concentration of cells used in the suspension significantly impacted the yield of capture. The optimal concentration created a single layer of well-distributed cells, wherein each gripper was touching at least one cell. This optimal concentration varied with the size of the grippers. The $SiO/SiO_2$ grippers also are optically transparent and thus are ideal for imaging the entrapped cells using optical microscopy techniques. These grippers have slit openings at the intersection of the arms, and consequently, nutrients, waste, and other biochemicals can flow easily to and from the cells, yet

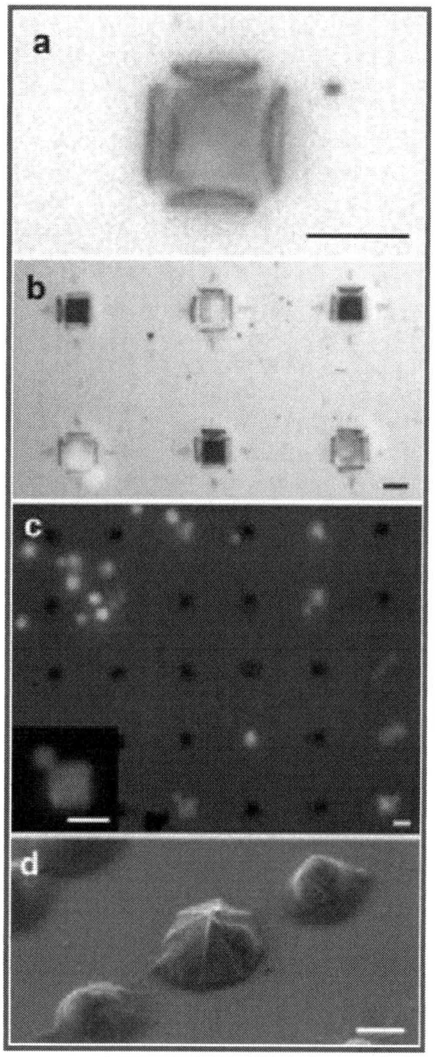


**Figure 4.** Single cell microgripper arrays. (a−c) Individual cells captured within the arms of grippers. Since the films are optically transparent, cells captured by the grippers can be readily visualized using optical microscopy. (c, inset) The cell shown is entrapped by a gripper, as evidenced by the square appearance of the cell when viewed from the top, which matches the square shape of the base of the gripper. (d) SEM image of a cell trapped within the arms of a gripper, surrounded by untrapped cells. Scale bars are 10 μm.

we observed that the grippers held all cells in place during staining and imaging. We performed a live/dead assay by staining with calcein AM and ethidium homodimer after the cells were captured (Figure 4a–c). The cells fluoresced green, demonstrating that they are alive, and that the assay chemicals successfully diffused between the grippers arms. Thus, the grippers did not kill the cells, and they allow the cells access to any chemicals within the media.

To verify that the cells were in fact contained inside the grippers, we fixed the cells and performed scanning electron microscopy (SEM) on an array of grippers with isolated fixed cells (Figure 4d). This image confirms that the cell was contained within the arms of the gripper, as opposed to floating on top of the gripper. It is noteworthy that the cells conformed to the shape of the gripper, as evidenced both in the SEM and by the square shape of the green fluorescing cells (Figure 4c inset, 4d), highlighting potential interactions with the faces of the gripper.

We also investigated applicability of grippers to capture red blood cells from a beagle blood sample (Figure 5). While the previous experiment captured L-929 fibroblasts and remained anchored to the substrate for further analysis, here we release the grippers from the substrate to show the potential for cells to be captured and moved in solution. We chose a 35 μm open gripper size which is about 4–5 times the size of the red blood cells (6–8 μm in size). The central palm of these grippers is matched to the size of the red blood cells ensuring that the cells could be entrapped within the gripper in a snug fit. Initially, red blood cells were pipetted onto partially released grippers. Many grippers were able to trap single blood cells within their arms.

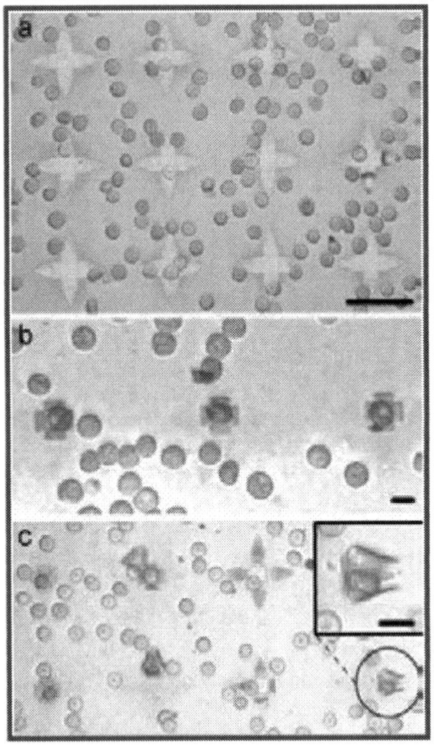


**Figure 5.** Capture of single red blood cells and untethered single cell grippers. (a–c) Optical images of red blood cells trapped in 35 μm $SiO/SiO_2$ grippers. (a) Grippers with red blood cells prior to folding and release from the substrate. Scale bar is 35 μm. (b–c) Red blood cells captured by the grippers. Scale bars are 10 μm.

Optical profilometry and microscopy of the grippers confirmed that the cells were trapped within the grippers. This experiment highlights the potential for these devices as in vivo cell capture tools. This application would require a biocompatible, thermo- or chemo-responsive hinge trigger, as previously demonstrated on larger grippers.[54,65−67,71] Whereas these larger grippers were able to capture large clumps of tissue in vivo, the single cell grippers could potentially enable the capture of individual cells.

We noted previously that both SiO and $SiO_2$ dissolve over time in biological fluids; therefore, grippers composed of such materials have significant potential for both in vitro and in vivo applications. Since the dissolution rate of these materials in bodily fluids is known to be dependent on a number of parameters such as method of thin film deposition, thickness, pH, and volume of fluid, further studies will be needed to design and optimize bioresorbability and safety for specific surgical applications in different biological environments in vivo.[78,79] For example, given the faster rate of dissolution of $SiO_2$ in PBS, grippers designed for in vivo use could be fabricated with $SiO_2$ rigid segments for rapid elimination from the body. Grippers fabricated with SiO rigid segments, like those shown in these figures, are best suited to in vitro applications for biocompatibility and slow degradation. If needed, such tools could also be created with magnetic elements using highly stressed bilayers of nickel (Ni) with rigid Ni panels for guidance through narrow conduits using magnetic fields (Figure S6). As an additional form of motion control, patterned biomarkers on the grippers could enable targeting of specific diseased cells in vivo.

In summary, we have designed and fabricated grippers capable of capturing and isolating single cells. These single cell grippers, made from biocompatible, optically transparent materials, can be arrayed for high throughput in vitro assays and imaging or released for use as untethered tools. We employed varying sizes of these grippers to capture individual fibroblasts and red blood cells. These cells were alive and could be assayed or fixed for imaging. Because these devices are fabricated in 2D and subsequently folded into 3D, future studies could explore patterned topography such as spikes, holes, nanoscale roughness, and biochemical surface functionalizations in specific designs onto one or more device walls. This approach could enable multiple assays to be run at one time on a single cell. Additionally, our group has previously demonstrated the fabrication of many differently shaped polyhedra.[51,80] Future studies could utilize pyramidal grippers and other regular polyhedra to study the effect of 3D confinement on cell growth and morphology. Our process is amenable to other lithographic approaches such as e-beam or nanoimprint lithography for subcellular gripping capabilities. Finally, this work highlights the potential for the creation of untethered active single cell capture devices. When doped with magnetic elements or made from magnetic materials such as nickel, these devices could be guided using magnetic fields[36,54,65−67,81] and consequently used as in vitro single cell capture devices as an alternative to laser microdissection of tissue samples. Due to their small size, they could also be utilized as bioresorbable surgical tools and in vivo single cell capture devices that are able to traverse conduits within the circulatory, central nervous, and urogenital systems.

## ■ ASSOCIATED CONTENT

### Supporting Information

Thin film stress and folding characterization. This material is available free of charge via the Internet at http://pubs.acs.org.

## ■ AUTHOR INFORMATION


### Corresponding Author

*E-mail: dgracias@jhu.edu.


### Author Contributions

M.J. and Q.J. contributed equally. D.H.G. and K.M. conceived of these devices and designed the experiments. K.M. conducted all experiments. M.J. performed SEM microscopy and optical microscopy for Figures 2 and 4 and consulted on fabrication strategies and cell experiments. Q.J. conducted finite element analysis modeling and performed the SiO/$SiO_2$ dissolution study. B.P. aided with red blood cell experiments and assembly of supplemental figures. C.J.M. performed stress analyses and consulted on modeling. D.H.G., K.M., and Q.J. wrote the paper. All authors commented on and/or edited the manuscript and figures.

### Notes

The authors declare no competing financial interest.

## ■ ACKNOWLEDGMENTS

This research was supported by NIH 1DP2OD004346 and the National Science Foundation grants CBET-1066898 and CMMI-1200241, the Army Research Laboratory, and Northrop Grumman. We thank Professor Benjamin Schaffer for assistance with the modeling software.

## ■ REFERENCES


(1) Altschuler, S. J.; Wu, L. F. *Cell* **2010**, *141*, 559−563.
(2) Ma, C.; Fan, R.; Admad, H.; Shi, Q.; Comin-Anduix, B.; Chodon, T.; Koya, R. C.; Liu, C.-C.; Kwong, G. A.; Radu, C. G.; Ribas, A.; Heath, J. R. *Nat. Med.* **2011**, *17*, 738−743.
(3) Di Carlo, D.; Tse, H. T. K.; Gossett, D. R. Introduction: Why Analyze Single Cells? In *Single-Cell Analysis: Methods and Protocols*; Lindstrom, S., Andersson-Svahn, H., Eds.; Springer Science + Business Media: New York, 2012; Vol. 853.
(4) Navin, N.; Krasnitz, A.; Rodgers, L.; Cook, K.; Meth, J.; Kendall, J.; Riggs, M.; Eberling, Y.; Troge, J.; Grubor, V.; Levy, D.; Lundin, P.; Maner, S.; Zetterberg, A.; Hicks, J.; Wigler, M. *Genome Res.* **2010**, 68−80.
(5) Calbo, J.; van Montfort, E.; Proost, N.; van Drunen, E.; Beverloo, H. B.; Meuwissen, R.; Berns, A. *Cancer Cell* **2011**, *19*, 244−256.
(6) Pietras, A. *Adv. Cancer Res.* **2011**, *112*, 255−281.
(7) Mannello, F. *BMC Med.* **2013**, *11*, 169−173.
(8) Beckman, R. A.; Schemmann, G. S.; Yeang, C.-H. *Proc. Natl. Acad. Sci. U.S.A.* **2012**, *109*, 14586−14591.
(9) Lindström, S.; Andersson-Svahn, H. *Lab Chip* **2010**, *10*, 3363−3372.
(10) Lindström, S.; Andersson-Svahn, H. *Single Cell Analysis: Methods and Protocols*; Springer Science + Business Media: New York, 2012; Vol. 853.
(11) Ashkin, A.; Dziedzic, J. M.; Bjorkholm, J. E.; Chu, S. *Opt. Lett.* **1986**, *11*, 288−290.
(12) Townes-Anderson, E.; St. Jules, R. S.; Sherry, D.; Lichtenberger, J.; Hassanain, M. *Mol. Vision* **1998**, *4*, 12.
(13) Berns, M. W.; Tadir, Y.; Liang, H.; Tromberg, B. *Methods Cell Biol.* **1998**, *55*, 71−98.
(14) Xie, C.; Dinno, M. A.; Li, Y.-q. *Opt. Lett.* **2002**, *27*, 249−251.
(15) Lúcio, A.; Santos, R.; Mesquita, O. *Phys. Rev. E* **2003**, *68*, 041906.
(16) Zhang, H.; Liu, K.-K. *J. R. Soc. Interface* **2008**, *5*, 671−690.
(17) Wheeler, A. R.; Throndset, W. R.; Whelan, R. J.; Leach, A. M.; Zare, R. N.; Liao, Y. H.; Farrell, K.; Manger, I. D.; Daridon, A. *Anal. Chem.* **2003**, *75*, 3581−3586.
(18) Peng, X.; Li, P. *Anal. Chem.* **2004**, *76*, 5273−5281.
(19) Peng, X.; Li, P. *Anal. Chem.* **2004**, *76*, 5282−5292.
(20) Valero, A.; Merino, F.; Wolbers, F.; Luttge, R.; Vermes, I.; Andersson, H.; van den Berg, A. *Lab Chip* **2005**, *5*, 49−55.
(21) Di Carlo, D.; Wu, L. Y.; Lee, L. *Lab Chip* **2006**, *6*, 1445−1449.
(22) Di Carlo, D.; Aghdam, N.; Lee, L. *Anal. Chem.* **2006**, *78*, 4925−4930.
(23) Roman, G. T.; Chen, Y.; Viberg, P.; Culbertson, A. H.; Culbertson, C. T. *Anal. Bioanal. Chem.* **2007**, *387*, 9−12.
(24) Nilsson, J.; Evander, M.; Hammarström, B.; Laurell, T. *Anal. Chim. Acta* **2009**, *649*, 141−157.
(25) Park, S.; Zhang, Y.; Wang, T. H.; Yang, S. *Lab Chip* **2011**, *11*, 2893−2900.
(26) Ding, X.; Lin, S.-C. S.; Kiraly, B.; Yue, H.; Li, S.; Chiang, I.-K.; Shi, J.; Benkovic, S. J.; Huang, T. J. *Proc. Natl. Acad. Sci. U.S.A.* **2012**, *109*, 11105−11109.
(27) Charnley, M.; Textor, M.; Khademhosseini, A.; Lutolf, M. *Integr. Biol.* **2009**, *1*, 625−634.
(28) Lindström, S.; Andersson-Svahn, H. *Biochim. Biophys. Acta* **2011**, *1810*, 308−316.
(29) Smith, E. J.; Xi, W.; Makarov, D.; Monch, I.; Harazim, S. M.; Bolanos Quinones, V. A.; Schmidt, C. K.; Mei, Y.; Sanchez, S.; Schmidt, O. G. *Lab Chip* **2012**, *12*, 1917−1931.
(30) Kane, R. S.; Takayama, S.; Ostuni, E.; Ingber, D. E.; Whitesides, G. M. *Biomaterials* **1999**, *20*, 2363−2376.
(31) Azioune, A.; Storch, M.; Bornens, M.; Thery, M.; Piel, M. *Lab Chip* **2009**, *9*, 1640−1642.
(32) Gautrot, J. E.; Trappmann, B.; Oceguera-Yanez, F.; Connelly, J.; He, X.; Watt, F. M.; Huck, W. T. S. *Biomaterials* **2010**, *31*, 5030−5041.
(33) Leclair, A. M.; Ferguson, S. S. G.; Lagugne-Labathet, F. *Biomaterials* **2011**, *32*, 1351−1360.
(34) Mandal, K.; Balland, M.; Bureau, L. *PLoS One* **2012**, *7*, e37548.
(35) Chronis, N.; Lee, L. P. *J. Microelectromech. Syst.* **2005**, *14*, 857−863.
(36) Sakar, M. S.; Steager, E. B.; Kim, D. H.; Kim, M. J.; Pappas, G. J.; Kumar, V. *Appl. Phys. Lett.* **2010**, *96*, 043705.
(37) Kim, S.; Qiu, F.; Kim, S.; Ghanbari, A.; Moon, C.; Zhang, L.; Nelson, B.; Choi, H. *Adv. Mater.* **2013**, *25*, 5863−5868.
(38) Fernandes, R.; Gracias, D. H. *Mater. Today* **2009**, *12*, 14−20.
(39) Prinz, V. Y.; Seleznev, V. A.; Gutakovsky, A. K.; Chehovskiy, A. V.; Preobrazhenskii, V. V.; Putyato, M. A.; Gavrilova, T. A. *Phys. E (Amsterdam, Neth.)* **2000**, *6*, 828−831.
(40) Schmidt, O. G.; Eberl, K. *Nature* **2001**, *410*, 168.
(41) Kazuyoshi, K.; Fleischmann, T.; Saravanan, S.; Vaccaro, P. O.; Aida, T. *Jpn. J. Appl. Phys.* **2003**, *42*, 4079−4083.
(42) Chua, C. L.; Fork, D. K.; van Schuylenbergh, K.; Lu, J.-P. *J. Microelectromech. Syst.* **2003**, *12*, 989−995.
(43) Prinz, V. Y. *Russ. Phys. J.* **2003**, *46*, 568−576.
(44) Zhanga, L.; Goloda, S. V.; Deckardta, E.; Prinz, V. Y.; Grützmachera, D. *Phys. E (Amsterdam, Neth.)* **2004**, *23*, 280−284.
(45) Sasaki, M.; Briand, D.; Noell, W.; de Rooij, N. F. *IEEE J. Sel. Top. Quantum Electron.* **2004**, *10*, 455−461.
(46) Huang, M.; Boone, C.; Roberts, M.; Savage, D. E.; Lagally, M. G.; Shaji, N.; Qin, H.; Naim, J. A.; Liu, F. *Adv. Mater.* **2005**, *17*, 2860−2864.
(47) Arora, W. J.; Nichol, A. J.; Smith, H. I.; Barbastathis, G. *Appl. Phys. Lett.* **2006**, *88*, 053108.
(48) Moiseeva, E.; Senousy, Y. M.; Harnett, C. K. *J. Micromech. Microeng.* **2007**, *17*, N63−N68.
(49) Stellman, P.; Buchner, T.; Arora, W. J.; Barbastathis, G. *J. Microelectromech. Syst.* **2007**, *16*, 932−949.
(50) Bassik, N.; Stern, G. M.; Jamal, M.; Gracias, D. H. *Adv. Mater.* **2008**, *20*, 4760−4764.
(51) Leong, T. G.; Benson, B. R.; Call, E. K.; Gracias, D. H. *Small* **2008**, *4*, 1605−1609.

(52) Mei, Y.; Huang, G.; Solovev, A. A.; Bermudez Urena, E.; Monch, I.; Ding, F.; Reindl, T.; Fu, R. K. Y.; Chu, P. K.; Schmidt, O. G. *Adv. Mater.* **2008**, *20*, 4085–4090.
(53) Solovev, A. A.; Mei, Y.; Ureña, E. B.; Huang, G.; Schmidt, O. G. *Small* **2009**, *5*, 1688–1692.
(54) Leong, T. G.; Randall, C. L.; Benson, B. R.; Bassik, N.; Stern, G. M.; Gracias, D. H. *Proc. Natl. Acad. Sci. U.S.A.* **2009**, *106*, 703–708.
(55) Huang, G.; Mei, Y.; Thurmer, D. J.; Coric, E.; Schmidt, O. G. *Lab Chip* **2009**, *9*, 263–268.
(56) Bassik, N.; Stern, G. M.; Gracias, D. H. *Appl. Phys. Lett.* **2009**, 95, 091901.
(57) Mei, Y.; Thurmer, D. J.; Deneke, C.; Kiravittaya, S.; Chen, Y.-F.; Dadgar, A.; Bertram, F.; Bastek, B.; Krost, A.; Christen, J.; Reindl, T.; Stoffel, M.; Coric, E.; Schmidt, O. G. *ACS Nano* **2009**, *3*, 1663–1668.
(58) Jamal, M.; Bassik, N.; Cho, J.-H.; Randall, C. L.; Gracias, D. H. *Biomaterials* **2010**, *31*, 1683–1690.
(59) Shenoy, V. B.; Gracias, D. H. *MRS Bull.* **2012**, *37*, 847–854.
(60) Harazim, S. M.; Xi, W.; Schmidt, C. K.; Sanchez, S.; Schmidt, O. G. *J. Mater. Chem.* **2012**, *22*, 2878–2884.
(61) Xi, W.; Solovev, A. A.; Ananth, A. N.; Gracias, D. H.; Sanchez, S.; Schmidt, O. G. *Nanoscale* **2012**, *5*, 1294–1297.
(62) Smith, E. J.; Xi, W.; Makarov, D.; Mönch, I.; Harazim, S.; Quiñones, V. A. B.; Schmidt, C. K.; Mei, Y.; Sanchez, S.; Schmidt, O. G. *Lab Chip* **2012**, *12*, 1917–1931.
(63) Soler, L.; Magdanz, V.; Fomin, V. M.; Sanchez, S.; Schmidt, O. G. *ACS Nano* **2013**, *7*, 9611–9620.
(64) Chalapat, K.; Chekurov, N.; Jiang, H.; Li, J.; Parviz, B.; Paraoanu, G. S. *Adv. Mater.* **2013**, *25*, 1.
(65) Bassik, N.; Brafman, A.; Zarafshar, A. M.; Jamal, M.; Luvsanjav, D.; Selaru, F. M.; Gracias, D. H. *J. Am. Chem. Soc.* **2010**, *132*, 16314–16317.
(66) Gultepe, E.; Randhawa, J. S.; Kadam, S.; Yamanaka, S.; Selaru, F. M.; Shin, E. J.; Kalloo, A. N.; Gracias, D. H. *Adv. Mater.* **2013**, *25*, 514–519.
(67) Gultepe, E.; Yamanaka, S.; Laflin, K. E.; Kadam, S.; Shim, Y.; Olaru, A. V.; Limketkai, B.; Khashab, M. A.; Kalloo, A. N.; Gracias, D. H.; Selaru, F. M. *Gastroenterology* **2013**, *144*, 691–693.
(68) Rimstidt, J. D.; Barnes, H. L. *Geochim. Cosmochim. Acta* **1980**, *44*, 1683–1699.
(69) Iler, R. K. *J. Colloid Interface Sci.* **1973**, *43*, 399–408.
(70) Hwang, S.-W.; Tao, H.; Kim, D.-H.; Cheng, H.; Song, J.-K.; Rill, E.; Brenckle, M. A.; Panilaitis, B.; Won, S. M.; Kim, Y.-S.; Song, Y. M.; Yu, K. J.; Ameen, A.; Li, R.; Su, Y.; Yang, M.; Kaplan, D. L.; Zakin, M. R.; Slepian, M. J.; Huang, Y.; Omenetto, F. G.; Rogers, J. A. *Science* **2012**, *337*, 1640–1644.
(71) Randhawa, J. S.; Leong, T. G.; Bassik, N.; Benson, B. R.; Jochmans, M. T.; Gracias, D. H. *J. Am. Chem. Soc.* **2008**, *130*, 17238–17239.
(72) Hill, A. E.; Hoffman, G. R. *Br. J. Appl. Phys.* **1967**, *18*, 13.
(73) Fang, M.; Hu, D.; Shao, J. *Chin. Opt. Lett.* **2010**, *8*, 119–122.
(74) Leplan, H.; Robic, J. Y.; Pauleau, Y. *J. Appl. Phys.* **1996**, *79*, 6926–6931.
(75) Nikishkov, G. P. *J. Appl. Phys.* **2003**, *94*, 5333–5336.
(76) Friede, B.; Jansen, M. *J. Non-Cryst. Solids* **1996**, *204*, 202–203.
(77) Schulmeister, K.; Mader, W. *J. Non-Cryst. Solids* **2003**, *320*, 143–150.
(78) Rui, L.; Cheng, H.; Su, Y.; Hwang, S.-W.; Yin, L.; Tao, H.; Brenckle, M. A.; Kim, D.-H.; Omenetto, F. G.; Rogers, J. A.; Huang, Y. *Adv. Funct. Mater.* **2013**, *23*, 3106–3114.
(79) Kang, S.-K.; Hwang, S.-W.; Cheng, H.; Yu, S.; Kim, B. H.; Kim, J.-H.; Huang, Y.; Rogers, J. A. *Adv. Funct. Mater.* **2014**, DOI: 10.1002/adfm.201304293.
(80) Pandey, S.; Ewing, M.; Kunas, A.; Nguyen, N.; Gracias, D. H.; Menon, G. *Proc. Natl. Acad. Sci. U.S.A.* **2011**, *108*, 19885–19890.
(81) Kummer, M. P.; Abbott, J. J.; Kratochvil, B. E.; Borer, R.; Sengul, A.; Nelson, B. J. *IEEE Trans. Robotics* **2010**, *26*, 1006–1017.

# 427739 Reconstruction of Human Respiratory Platform for in Vitro Drug Development

**Abstract:**

Recapitulating the complex cellular and functional environment of a lung in vitro would provide critical advantages over conventional pulmonary toxicity and assessment models. However, the human lung is a complex organ system that contains numerous cells types and performs multiple mechanical/physiological functions. To capture multiple critical features of the human lung system in vitro, we have undertaken a stepwise approach to engineer a complex microfluidic platform. Our objective was twofold - develop a platform with two key lung components bronchiole and alveolar compartments, and simulate the unique structural complexity of the lung. To reconstitute the complex lung physiology and the microenvironment that maintains cell differentiation, we fabricated our human lung organ platform by integrating both bronchiolar and alveolar lung compartments. We stacking laser patterned substrates integrated with silicone based adhesive transfer tapes. The bronchiolar component contains a porous polyester membrane to create the air liquid interface required for the growth of bronchiole tissue. Artificial human bronchiole tissue was reconstructed by culturing human bronchial epithelial cells (HBTECs) on the membrane. Human alveoli environment was mimicked by growing human alveolar epithelial cells (A549) on inflatable polydimethylsiloxane (PDMS) membranes. HBTECs can be seeded on porous membrane and differentiated with air liquid interface after 2 weeks. A549 cells were viable before and after with cyclic stretching (25% of surface area changed) of the PDMS membrane for 2 days. The bronchiolar and alveolar cells can be cultured and differentiated in bronchiolar and alveolar tissues culture system. These systems mimic complex lung organ physiology suitable for systemic absorption, distribution, metabolism, excretion, and toxicology (ADMET) studies and the recapitulation of human pulmonary diseases. Validation of the lung construct was performed using the well-established drug amiodarone.

# 434612 Stiffness Mediated Regulation of Hepatocyte Functions

**Abstract:**

Liver damage as a consequence of injury or disease can lead to fibrosis causing restructuring of the liver microenvironment (LME). The increase in the liver stiffness is a strong indication of the progression of liver diseases. However, there is a lack of understanding in the effect of increased stiffness on the liver cells *in vivo*. There is a critical need to engineer *in vitro* models that will mimic various stages of the diseased liver to serve as accurate models for studying disease mechanism of individual cell types. Such models need to incorporate the dynamic changes in LME including the change in liver stiffness.

In our study, we investigated the effect of substrate stiffness on the functions of cultured primary hepatocytes utilizing a polydimethyl siloxane (PDMS) based substrate with tunable stiffness. We employed a **soft** substrate to represent the healthy liver tissue stiffness and a **stiff** substrate to represent diseased liver tissue and compared the cellular properties with the cells cultured on tissue culture polystyrene surfaces (TCPS). We observed that hepatocytes cultured on the soft substrate displayed a consistently more differentiated phenotype for a prolonged duration as compared to the hepatocytes that were cultured on the stiff substrate and TCPS. We further demonstrated that hepatocyte cultured on soft substrate showed better maintenance of hepatocyte-specific functions of urea and albumin synthesis. Similarly, cytochrome P450 (CYP) activity on soft substrate displayed a multi-fold increase in CYP activity, as compared to TCPS. Together, these results indicate that the substrate stiffness plays a significant role in modulating hepatocyte behavior. The intricate LME-hepatocytes signaling pathways through which healthy and diseased liver cells communicate within the LME will be further investigated in our future studies.

# Advancements in the Prediction and Control of Batch High Shear Granulation and Continuous Twin Screw Granulation


Author: Don Clancy
Title: Investigator - Modeling Lead
Department: Modeling Center of Excellence,
Division: R&D Product Development,
Company: GlaxoSmithKline
Contact E-mail: Donald.J.Clancy@gsk.com

## 1.1 Introduction to Modeling Batch High Shear Granulation

Useful predictive models for batch and continuous granulation utilize many of the same concepts, and it's constructive to understand batch high shear granulation prior to continuous twin screw granulation. Batch high shear granulation has been explored more extensively, and in some ways is less complex than continuous. The models that have been found to be useful to date for process control and design space exploration are a combination of first principles mechanistic linked with linear regression to predict an outcome such as granule particle size distribution, milled granule density, or tablet dissolution. The prediction in each case is going to be product specific, but the first principle nature of the mechanistic terms utilized in the linear regression models is shown to hold true across multiple drug products.

Most pharmaceutical high shear granulation processes granulate for a specified amount of time, or specify a maximum motor impeller power or torque before stopping the granulation process. Since most products aren't particularly sensitive to granulation endpoint, this simple control methodology is often adequate. Sometimes though, products can be particularly sensitive to changes in granulation extent. As an example, Drug Product "P" is a micronized hydrophobic API with 80% drug content in the granule. Relatively small changes in granule porosity and density have significant impacts on resulting tablet dissolution, and tablet breaking force. For this reason, the "Work" control model shown as Equation 1 that integrates both power and time was first developed and applied to drug product "P" and has controlled it's commercial manufacture.(Clancy, Method and Apparatus for Predicting Properties of Granulated Materials and Dosage Forms made Therefrom, 2010)

From a first principles perspective, "work done" is the instantaneous power applied to the system multiplied by the time the power was applied. Since power varies over time in granulation, the integral of power with respect to time needs to be considered. Typically the power measured in granulation is the power flow to the motor, and some power is lost due to the motor and power train inefficiencies. It's useful to subtract a baseline power ($P_0$) which is associated with the power usage needed by the motor and drive mechanism from the overall power used by the impeller, as shown in Equation 1.

$$\text{Work} = \int_0^{\text{Time}} \left(\text{Power}_{\text{Impeller}} - P_0\right) dt \tag{1}$$

Where "$\text{Power}_{\text{Impeller}}$" is the measured power of the main impeller during granulation, and $P_0$ is the steady state baseline power measured prior to start of the granulation process with nothing in the bowl.

Graphically what Equation 1 is calculating is the net work done, which is the cumulative amount of power added above baseline, as the shaded in area shown in Figure 1.

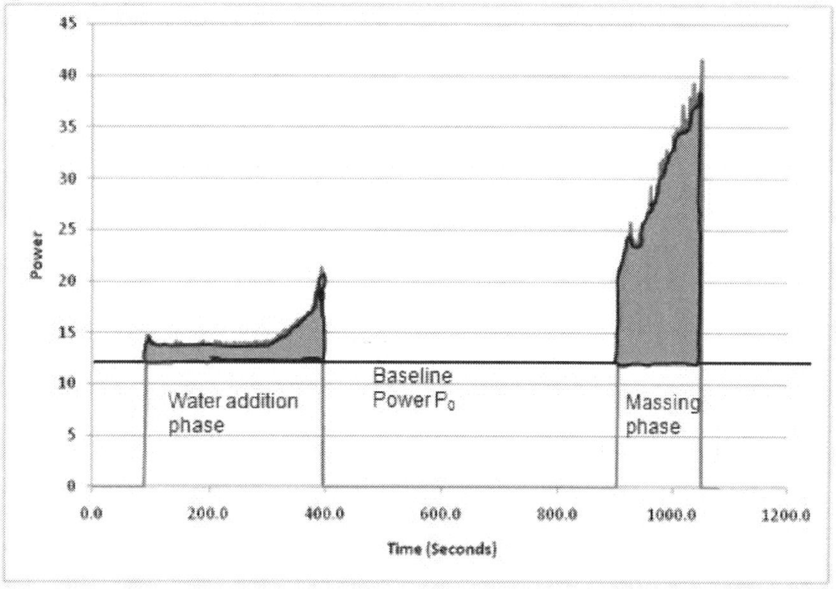


**Figure 1: Calculation of Granulation Work Corresponds to Shaded Area**

Calculation of a definite integral isn't generally possible or convenient, thus equation 1 is approximated using numeric integration and rapid time slicing. Equation 2 is the actual numeric calculation utilized to calculate the integral shown in Equation 1.

$$W_i = \left(P_i - P_0\right) \cdot \Delta t + W_{i-1} \qquad (2)$$

$W_i$ is the work calculated at the $i^{th}$ time slice, $P_i$ is the impeller power at the $i^{th}$ time slice, $\Delta t$ is the time difference measured in seconds between "i" and "i-1" time slices. The output from above equation is integrated over time of the granulation once water input has started.

Work done per unit dry powder mass, water fraction added, and tablet thickness at a constant tablet mass can be used as parameters for empirical linear regression models to predict tablet dissolution. Examples of this are shown for three different drug products in Figure 2, Figure 3, and Figure 4.

The dissolution of tablets from drug product "P" can be modeled as a 4-dimensional surface of tablet dissolution as a function of work, water Fraction, and tablet thickness, as shown in Figure 2. The emperical model for dissolution in this case is a 10-term linear regression model as a function of work, water, tablet thickness and their cross-products with an R2 value of 92%. The data to form this model comes from a 10-batch response surface DoE as function of Work and Water, and split to 5 levels of compressed tablet thickness to make 50 sub-batches in a split plot experimental design. The raw data for this design is available in Appendix A, Table 1.

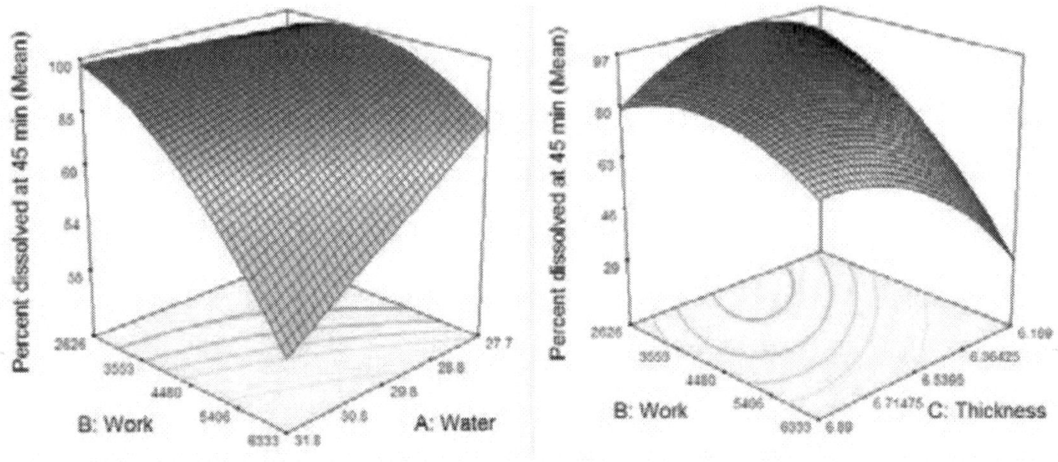


**Figure 2: Percent Tablet Dissolved at 45-minutes for Drug Product "P" at Constant Target Tablet Thickness (Left) or Constant Target Granulation Water Fraction (Right)**

If tablet thickness is held constant, granule dissolution and tablet hardness can be predicted as a 3-dimensional surface as function of granulation work and water fraction. An example surface for Drug Product "D" from a 9-batch response surface DoE is shown in Figure 3. In this case the tablet dissolution can be predicted from a simple 4-term empirical linear regression model considering work, work-squared, and water with an R2 value of 99%. The DoE data from this experiment is available in Appendix A, Table 2.

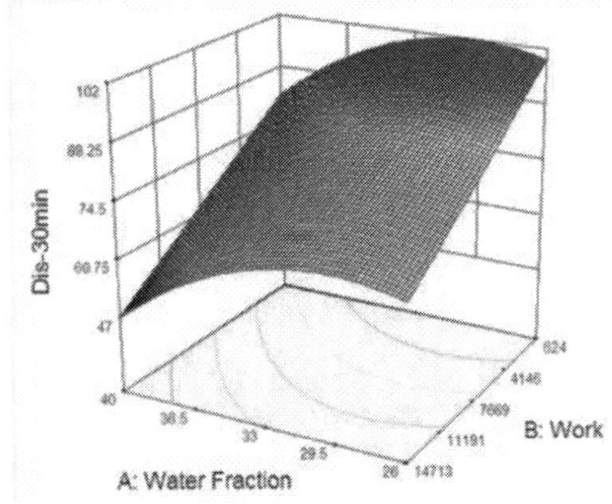


ANOVA for Response Surface Reduced Quadratic Model

Analysis of variance table [Partial sum of squares - Type III]

| Source | Sum of Squares | df | Mean Square | F Value | p-value Prob > F | |
|---|---|---|---|---|---|---|
| Model | 3180.72 | 3 | 1060.57 | 136.42 | < 0.0001 | significant |
| A-Water Fraction | 177.32 | 1 | 177.32 | 22.99 | 0.0049 | |
| B-Work | 328.35 | 1 | 328.35 | 42.52 | 0.0013 | |
| A² | 130.80 | 1 | 130.80 | 16.94 | 0.0092 | |
| Residual | 38.61 | 5 | 7.72 | | | |
| Cor Total | 3199.33 | 8 | | | | |

| | | | |
|---|---|---|---|
| Std. Dev. | 2.78 | R-Squared | 0.9879 |
| Mean | 78.56 | Adj R-Squared | 0.9807 |
| C.V. % | 3.54 | Pred R-Squared | 0.9664 |
| PRESS | 107.37 | Adeq Precision | 28.065 |

**Figure 3: Percent Tablet Dissolved at 30-minutes for Drug Product "D" from 9-batch Response Surface Experiment at 65-L Scale**

## 1.2 Modeling Batch High Shear Granulation by Sampling during Granulation

The concept has been explored of sampling from high shear granulation at varying time points and using the results from sampled batches to predict the behavior of whole batches on basis of Work done per unit mass. To accomplish this, the power input per unit mass needs to be integrated with respect to time to calculate a Work done per unit mass for each sample. The equation used for calculation is shown as Equation 3.

$$W_i = \left(\frac{P_i - P0}{mass_i}\right) \cdot \Delta t + W_{i-1} \tag{3}$$

Where "$mass_i$" is the mass of powder remaining in the bowl at the ith time point on a dry weight basis, and $W_i$ is now scaled as the work done per unit of mass.

Use of sampling can have the useful outcome of reducing the material requirements to produce a granulation model by a factor of three to five. Sampling should be used with the caution that there is some power needed to restart the motor and impeller. For smaller or newer granulators this restart power appears to be less significant than with larger or older electric motors. In a typical sampling experiment 20% to 25% of the batch is removed at each of four or five time points, and work done per unit of dry powder mass is calculated at each time point.

An example of sampling three batches at constant water level at 300-L scale for drug product "P", compared to dissolution of a non-sampled batch is shown in Figure 4. Each of three batches was sampled at varying time points to make four samples per batch with each sample 25% of the batch. The results of the 12 sub-sampled batches at varying work/mass was used to calculate an optimum granulation endpoint that minimized the dissolution RSD while providing close to optimum mean dissolution. The results of the complete batch were within the standard error of prediction for dissolution mean and RSD for models built on the sampled batches.

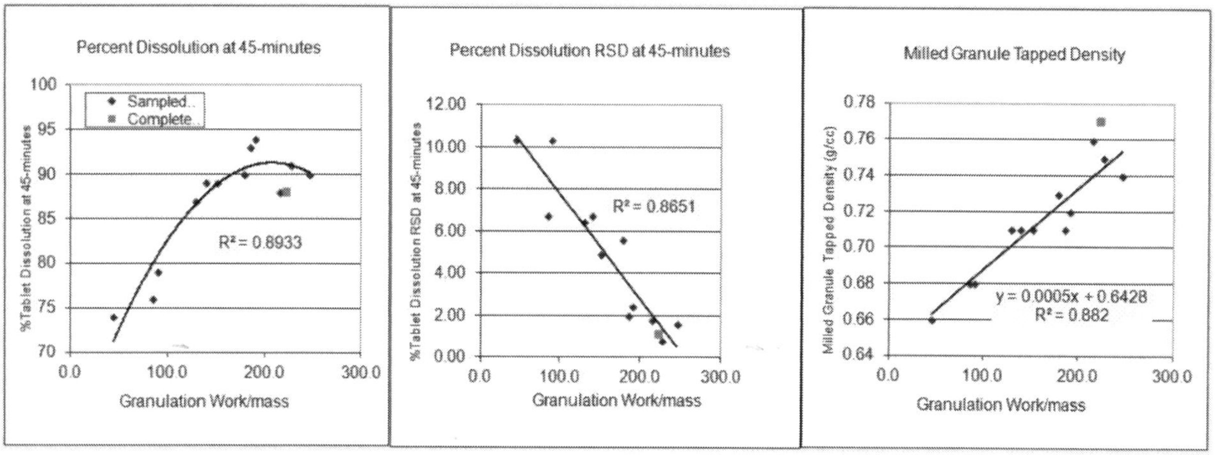


**Figure 4: Granulation Work/Mass Models from Sampling Three Batches Four Times Each Compared to a Non-Sampled Batch**

There is some evidence that models built from sampling will slightly overestimate the optimum granulation work/mass, for example the milled granule tapped density of the non-sampled batch shown in Figure 4 is slightly higher than anticipated from a model built from the sampled batches. This has been attributed to the power needed for startup and shutdown. For this particular 300-L granulator there is an initial power spike as the power is turned on, and some mathematical exploration was done that if this initial power spike were removed, the models improved. The initial power spike varies from equipment scale and motor, with some older motors having too large and variable of an initial power spike to allow meaningful sampling experimentation to occur. The reproducibility and size of the initial power spike and overall stability of the baseline power ($P_0$) should be looked at carefully if a sampled batch granulation model is desired. In some cases the initial startup power spikes can be reduced or eliminated by warming the motor up for 20 to 30 minutes prior to adding the granulation mixture to the bowl, and ensuring all moving components are lubricated.

Given the caveat about ensuring use of a granulator motor without a large initial power spike, sampling during granulation can be a useful strategy in R&D to allow modeling and assessment of how robust a formulation is to variation in granulation. Figure 5 shows an example of a 20-point experiment at 25-L scale for product "S" composed of four batches of varying water level each sampled five times. This sampled 20-point experiment was useful for the drug product team in defining the optimum work and water levels needed for subsequent production of clinical batches at 25-L scale. The data table from this experiment is available in Appendix A, Table 3.

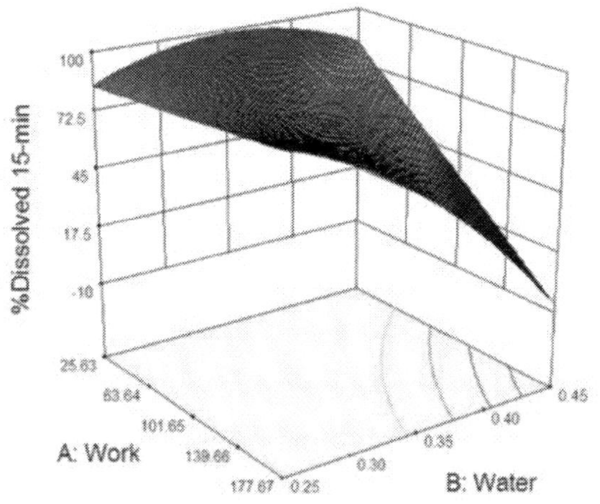


**ANOVA for Response Surface Reduced Quadratic Model**

**Analysis of variance table [Partial sum of squares - Type III]**

| Source | Sum of Squares | df | Mean Square | F Value | p-value Prob > F |
|---|---|---|---|---|---|
| Model | 3408.42 | 4 | 852.10 | 26.44 | < 0.0001 |
| A-Work | 1244.67 | 1 | 1244.67 | 38.62 | < 0.0001 |
| B-Water | 1467.69 | 1 | 1467.69 | 45.54 | < 0.0001 |
| AB | 626.50 | 1 | 626.50 | 19.44 | 0.0010 |
| B² | 528.56 | 1 | 528.56 | 16.40 | 0.0019 |
| Residual | 354.52 | 11 | 32.23 | | |
| Cor Total | 3762.94 | 15 | | | |

The Model F-value of 26.44 implies the model is significant. There is only a 0.01% chance that a "Model F-Value" this large could occur due to noise.

| | | | |
|---|---|---|---|
| Std. Dev. | 5.68 | R-Squared | 0.9058 |
| Mean | 79.08 | Adj R-Squared | 0.8715 |
| C.V. % | 7.18 | Pred R-Squared | 0.7680 |
| PRESS | 872.92 | Adeq Precision | 18.778 |

**Figure 5: Response Surface of Drug Product "S" Dissolution at 15-minutes to Granulation Work/Mass and Water Fraction in Granulation from Sampled Experimentation at 25-L Scale**

## 1.3 Impact of Raw Material Particle Size and Surface Area Changes on High Shear Granulation Modeling

As material is milled into finer particle size it requires more granulation work to achieve the same endpoint characteristics. A good example experiment was done at 65-L scale with drug product "P" that illustrates this relationship.

As material is milled to a finer size a variety of physical properties change the particle size distribution (PSD) becomes finer, specific surface area (SSA) increases, and bulk density decreases. A challenge is getting a good representative number describing the entire batch as there are typically significant sampling and replication variances for PSD and SSA. A novel technique was developed on a Freeman Powder Rheometer that provided the best signal to noise ratio of these techniques when comparing the RSD of replicate measurements and samples within batches.

A powder Rheometer is simply a very sensitive torque instrument mounted on a mixing blade. The mixing blade can be set to a specified RPM, then travels vertically with the rotating blade tracing a helix into and then out of a powder sample. The methodology established placed 30 grams of dry powder (corresponding to approximately 100 mL of volume) into the 200-mL sample apparatus. Tip speed was set to 300 mm/sec with a downward and upward helix angle of traverse of 5º. This rotation should fluidize the powder bed upwards as the blade traverses downwards through the bed. Ten traverses were done to "condition" the powder to remove affects of sample history. Twenty more replications were done, and the instantaneous work done by the impeller at 1-cm height from the bottom of the sample was then measured and called "Rheometer Work" measured in milliJoules (or milliNewton*meters) per 30 grams.

Rheometer work of the dry API powder was found to correlate to the particle size distribution X90 and the specific surface area of the API for drug product "P" as shown in Figure 6.

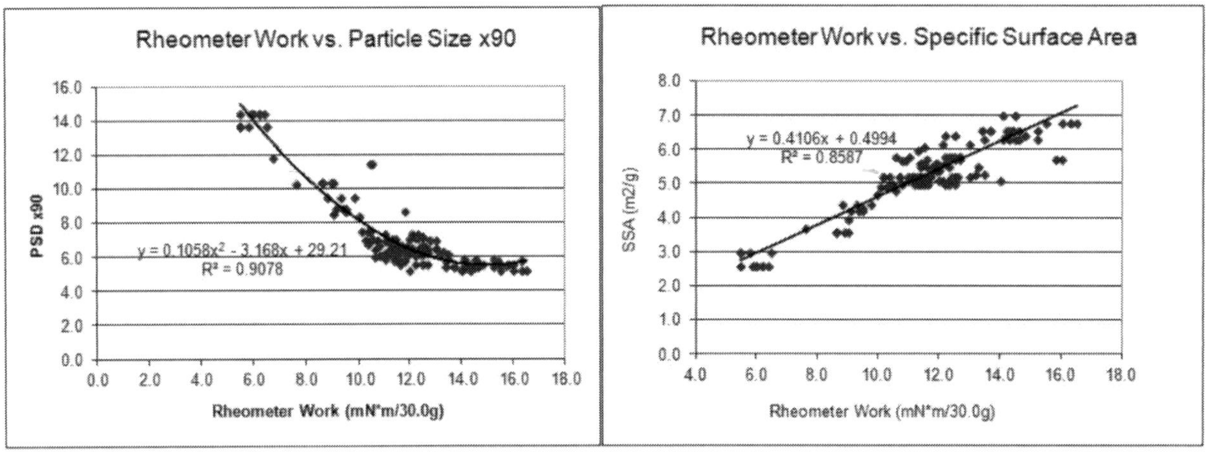


**Figure 6: Relationship of API Surface area and Particle size X90 to Powder Rheometer Work**

Specific surface area had an RSD of replicate measurements on a sample of 6%, PSD X90 had an RSD of replicate measurement of 4%, and the measurement of "Rheometer Work" had replicate measurement RSD of 2%. Sampling variance was also a significant concern, and the powder Rheometer uses 60 times as much sample as SSA and PSD testing, significantly reducing the impact of sampling variation.

To evaluate the impact of variation in API milling, a batch of drug product "P" API was micronized to three differing micronization work done of 60, 180, and 280 KJ/KG, resulting in 3 lots of API with particle size X90 of 13.2, 6.5, and 5.8 microns. These three lots of API were then tested on the Powder Rheometer, resulting in Rheometer Work values of 6.8, 11.0, and 12.2 mJ/30 grams.

The three lots of API were then granulated at 65-L scale with a constant water amount, and sampled at 4 different time points during granulation. For this drug product, the API is 80% of the granulation mixture, with remainder consisting of 5.9% PVP binder, 12.4% microcrystalline cellulose, and 1.7% sodium starch glycolate. The granulation Work per unit mass was calculated for each sample as described in the previous section. The DoE data from this experiment is available in Appendix A, Table 5.

If granulation work per unit mass is then used to correlate the resulting milled granule density, tablet dissolution, and tablet dissolution RSD, poor correlations are observed (Figure 7). Clearly there was a significant impact of the changes in API particle size distribution.

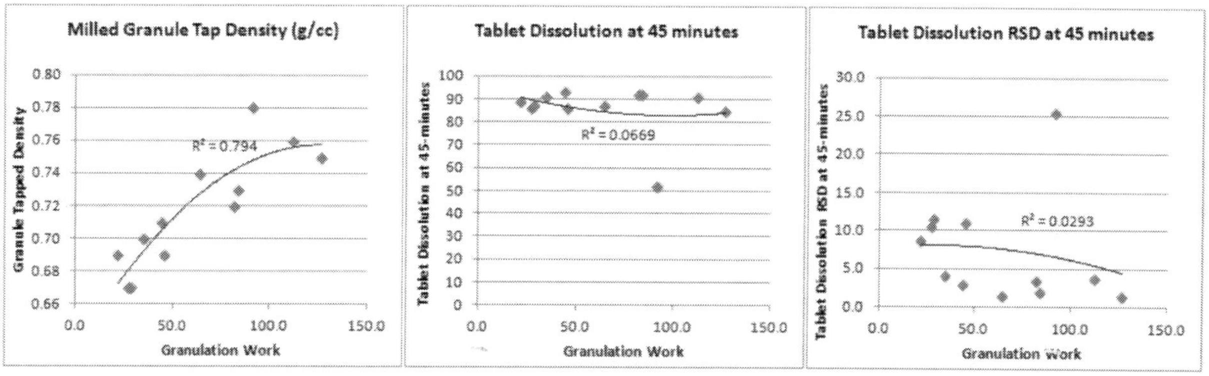


**Figure 7: Poor Correlation of Granulation Work/mass to Output Granule and Tablet Characteristics**

If the granulation work per unit mass is then normalized by dividing by the Powder Rheometer work done on 30 grams of dry sample, then the characteristic models for drug product "P" can be resolved. Milled granule tapped density, tablet dissolution, and tablet dissolution RSD can then be predicted by polynomial fits with goodness of fit ranging from 94% to 83% (Figure 8).

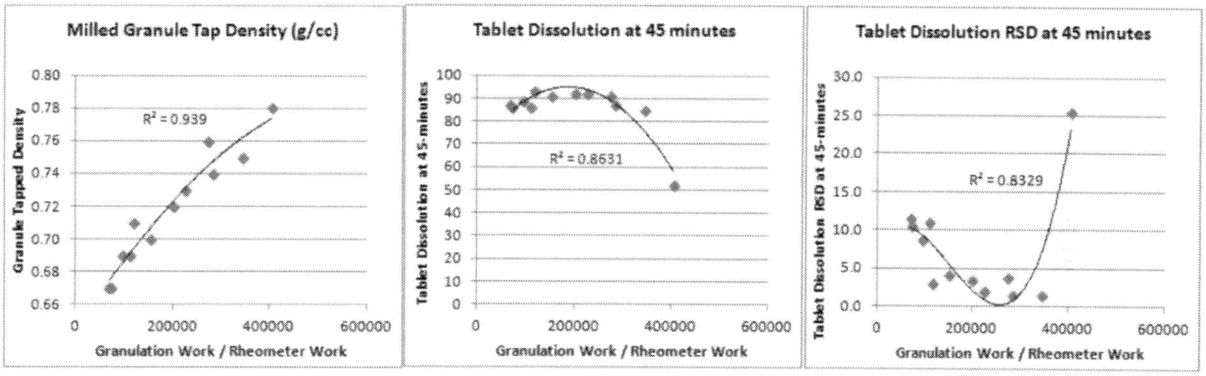


**Figure 8: Correlation of (Granulation Work/mass)/Rheometer Work to Output Granule and Tablet Characteristics**

There was a linear relationship noted between Rheometer work and the API specific surface area, thus the (Granulation Work / Rheometer Work) term could be expressed as a "Granulation Work per unit surface area" basis with similar results. From perspective of modeling the most precise measurement was chosen to be utilized, thus this experiment prompted work to qualify a Freeman Powder Rheometer for GMP purposes.

It was also possible to model this experiment as a response surface of granulation work per unit mass and API D90 value. This model had significantly lower goodness of fit for prediction of dissolution mean (74%), and problems with variance inflation in the estimation of model parameters. This response surface model is shown in Figure 9.

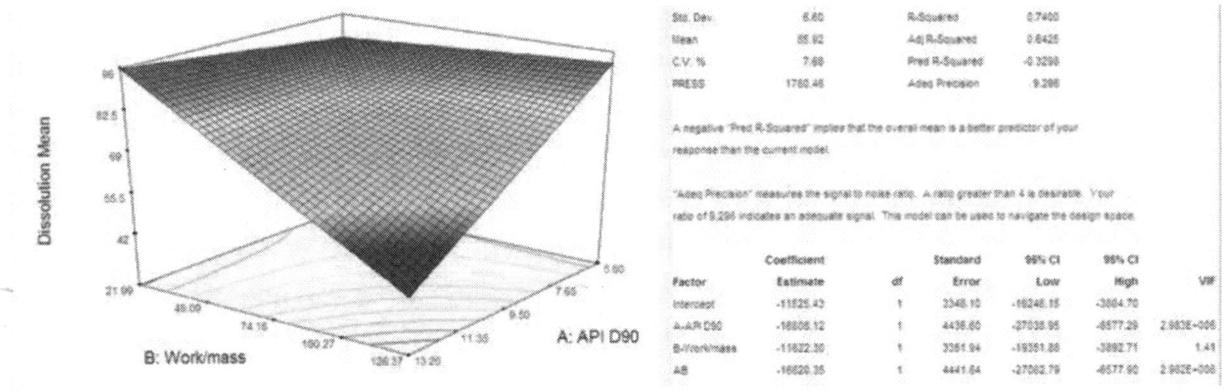


**Figure 9: Correlation of Granulation Work/mass and Particle Size D90 to Output Tablet Dissolution**

Evidence suggests that major changes in raw material particle size and surface area have a significant impact on the granulation process. Finer powder of higher surface area is likely to require more granulation Work to achieve a similar endpoint. Since measurements of PSD and SSA have challenges of reproducibility and sampling variance, a more precise model can potentially be developed by examining how much work it takes for a Rheometer blade to traverse a dry sample while fluidizing the powder. This work found that dividing the granulation work by the Rheometer Work resulted in a dimensionless parameter that was predictive of endpoint variation.

## 1.4 Models Describing Scale-up and Equipment Transfer of Batch High Shear Granulation

Research has been done to understand how to transfer the granulation process between different granulators of varying impeller geometry or scale when "Work/mass" is being utilized as the control model. The main impeller blade angle and rotations per minute (RPM) have significant impacts on the amount of water needed to start the granulation process, and therefore affect the total amount of water and Work needed. Correlations were developed relating the Work and water requirements to main impeller Froude number or RPM, diameter, and blade angle. [8] With changes in impeller angle, a GSK designed 6-L granulator was made by FluidAir that mimics the performance of a 300-L granulator used in production.(Clancy, Campbell, Zhang, Gupta, & Oh, 2011)

Scale-up of high shear granulation is generally accomplished by ensuring similar geometry at each scale, and ensuring the main impeller tip speed is held constant across scale, typically at 5.5

meters/second. The main impeller tip speed has significant impacts on granule consolidation and breakage and needs to be fixed at a constant value. Low tip speed for example can result in runaway granule growth resulting in low density granules the size of golf balls. High tip speed can effectively mill the granules to point where little granule growth occurs though the granules that do form and survive will be small and highly dense.

It's impossible to fix all the important factors to consider when moving across scale. Fixing the impeller tip speed results in changes of g-forces as RPM and blade diameter are changed. Thus, the key to understanding scale-up of high shear granulation is to realize that the amount of water needed at each scale is not necessarily constant. One possible explanation is that the spread of water among surfaces of the mixture is affected by the g-forces being applied by the main impeller blade. Use of absorbent materials such as microcrystalline cellulose and super-disintegrants within the granulation mixture may amplify this affect as some of the water can be absorbed and will be released as these ingredients are subjected to g-forces.

An observation of impeller load behavior as water is added in granulation for drug product "P" is shown in Figure 10 for four different equipment scales. As the equipment scale was increased more water was need to invoke a power response where the impeller power started rising.

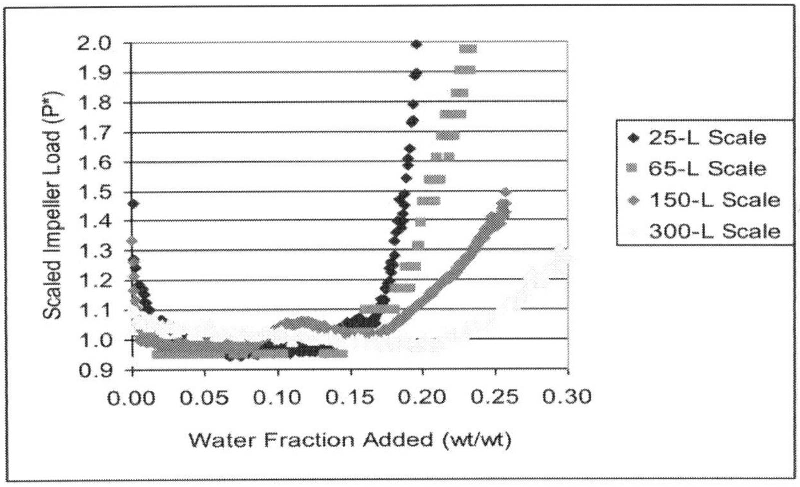


**Figure 10: Change in Impeller Power During Water Addition as Granulation Scale is Changed**

Initially, the impeller power decreased due to hysteresis effects and the time for the impeller motor to warm up then rises at a later point due to commencement of the granulation process. The amount of water added at the point of inflection in the impeller power indicates the beginning of densification in the granulation process and is called "Xsat" in this analysis.

$X_{sat}$ can be found using one of two mathematical techniques. The first technique is to find the point where the impeller load mean value for 10 seconds is 3-sigma higher than the previous 30 seconds worth of data. A second way is to find the amount of water added at a point of minima of the following function.

$$\text{Objective\_Function} = (Power_{15sec} - P_0)/Mass_{H2O} \quad (4)$$

$Power_{15sec}$ is the 15-second backwards data averaging of the impeller power. 15-second backwards data averaging is used to smooth the impeller load data to prevent extraneous noise from providing a false minimum. The amount of backwards averaging needed will depend on the noise in the impeller power data. At 300-L scale, only 5-seconds of backwards averaging was used as the signal observed had a higher signal to noise ratio. The variable $Mass_{H2O}$ is the amount of water added. The two methodologies for Xsat determination agree with an R^2 of 90% for the data presented in this paper. The minima approach is chosen and utilized in this paper because it is simpler and more robust to calculate in a Programmable Logic Controller (PLC) than the statistical technique, and is not affected as strongly by signal noise, thus provides better models. An example of what taking the minimum of the objective function looks like during water addition, is shown in Figure 11.

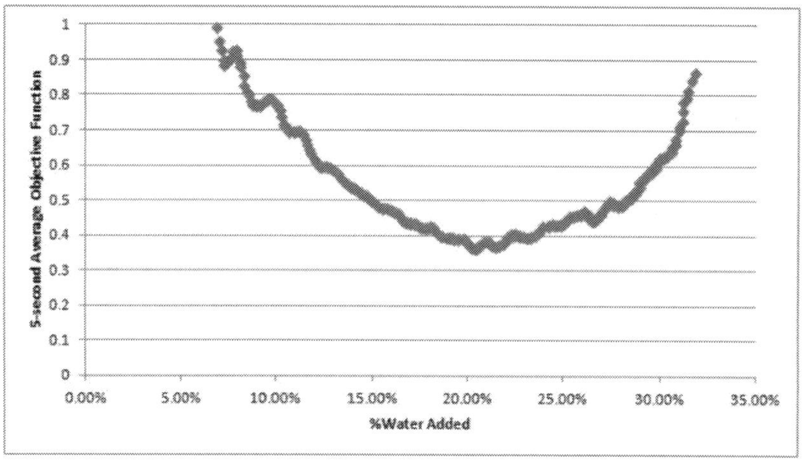


**Figure 11: Calculation of an Xsat of 20% Water as the Minimum of Equation (4)**

Observation was that an incremental approximately 2% more water was needed for each scale from 25-L, to 65-L, to 150-L, and 300-L in order to get an impeller power increase. Optimal granulation was consistently found using water amount of 8% to 10% in excess ($X_s$) of the point where impeller power starts to increase (Xsat), see Figure 12. The optimal water added at 300-L scale is 30%, or 9% in excess.

| Granulator Scale (L) | Average Xsat Values at Tip Speed 5.5 m/sec | Optimal Water Requirement | Water Required in Excess of Xsat |
|---|---|---|---|
| 25 | 14.7% | 23% | 8.3% |
| 65 | 16.0% | 26% | 10.0% |
| 150 | 19.4% | 28% | 8.6% |
| 300 | 21.1% | 30% | 8.9% |

**Figure 12: Optimal water requirement at the various granulator scales to achieved Xat and suitable granules for Drug Product "P"**

Centrifugal force is necessary for accelerating the powder bed into toroidal (roping) flow, which is the predominant flow pattern at the later stages in granulation. The centrifugal force also creates a compressive force pressing the powder bed toward the inner walls of the bowl. Thus Froude Number (*Fr*) (Equation 5) is a hypothesized scaling criteria for granulation, and can be thought of as a measure of the centrifugal forces being applied to the power.

$$Fr = \frac{4\pi^2 RN^2}{60^2 g} \quad (5)$$

where *N* is the impeller rotational speed (rpm), *R* is the impeller radius (m), and g is the gravitational constant (9.81 m $s^{-2}$).

Xsat was noted to be linearly related to impeller Froude Number when granulators of dimensionally similar bowl (blade height/bowl height) are utilized at the same tip speed. In Figure 13, a plot of the average Xsat determined at each scale against Froude Number is presented. The linearity between Xsat and Froude Number is measured by the high correlation coefficient ($R^2$ = 0.97), which suggests that for dimensionally similar bowl at constant tip speed, possibly the Froude Number may be a reasonable predictor of Water or Work requirements. This may explain why Froude Number has been postulated as potential scaling factor; however, subsequent experimentation utilizing impellers of differing design and tip speeds suggests this relationship is limited to dimensionally similar blades at same tip speed.

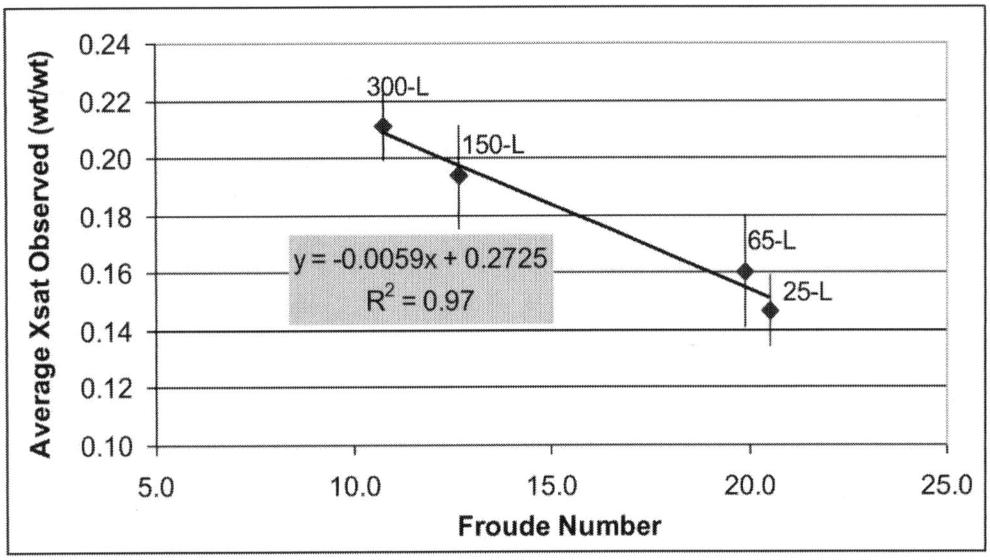


**Figure 13: A plot of average Xsat values versus Froude Number for the various Fielder granulator scales at constant tip speed.**

Subsequent experimentation found that Xsat can be manipulated by RPM at fixed tip speed (scale), and also by the blade angle. A lower angle blade at smaller scale and higher RPM can replicate the results of a higher angle blade at larger scale and lower RPM. Studies across multiple scales from 6-L, 25-L, 65-L, 150-L, and 300-L were then pursued, with details of the studies documented elsewhere.(Clancy, Campbell, Zhang, Gupta, & Oh, 2011)

If the desire is to mimic the behavior of a large scale granulator using small scale equipment it's useful to predict the correct blade angle. When a linear regression analysis was carried out an all the data generated from the studies conducted during the scale up of Drug Product "P" process impeller speed and blade angle gave the best fit model of Xsat with a 94.4% prediction coefficient. The 3-term model is as follows:

$$\text{Xsat} = 0.797 - 3.25\text{x}10^{-4}\,\text{xN} - 1.85\text{xsin}(\theta) + 1.53\text{xsin}(\theta)^2 \qquad (6)$$

Figure 14 shows the plot of actual Xsat against the predicted Xsat for the different granulator scales. No correlation to Froude number was observed once dissimilar blade angles and differing tip speeds were utilized in the study. Also, when the data is limited to geometrically similar blades to Fielder granulators (eliminating a flared tip 45-degee blade standard to FluidAir), the above correlation fits the data with a 99% coefficient. None of the other possible squared terms or cross products was significant at the 95%

confidence interval, suggestive that the inflection point changes linearly with impeller speed and nonlinearly with sine of the blade angle.

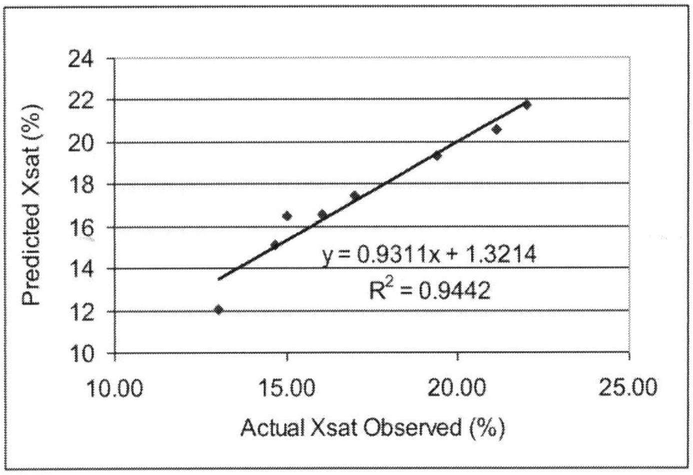


**Figure 14**: Actual Xsat Observed versus Predicted Xsat from impeller speed and blade angle model (Equation 6)

Another potential physical explanation for the trend observed in Figure 13 and Figure 14 comes from the observation that it appears to be important to maintain the same distribution of coverage of water spraying from the nozzle onto the powder surface passing below. If the geometry of the blade is held constant the powder bed at small scale passes below the spray nozzle many more times during water addition, thus achieving better spray distribution at small scale relative to large scale. The number of passes the powder takes below the spray nozzle for a given time period should be proportional to RPM x $Sin^2$(Theta). A plot of 1/(RPM x $Sin^2$(Theta)) creates a linear line of with nearly the same goodness of fit as the above fitted statistical trend, and can also be used to estimate the appropriate angle of the blade (Figure 15).

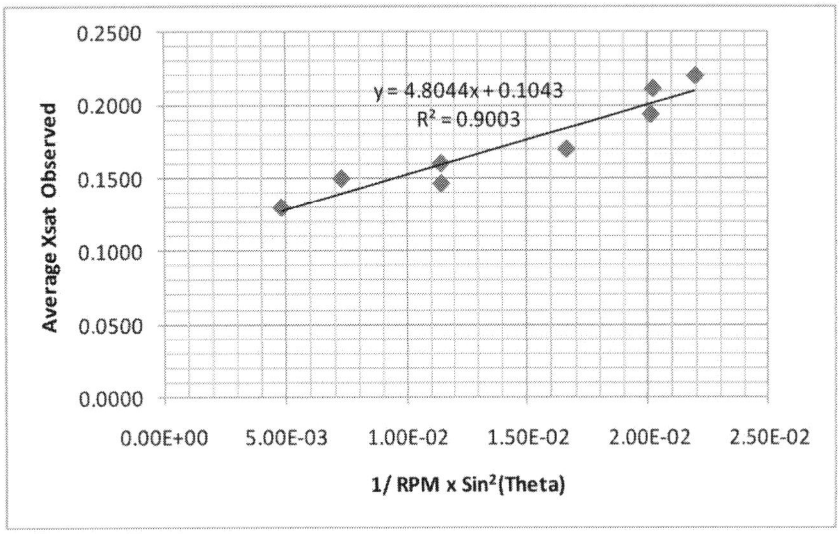


**Figure 15**: Actual Average Xsat Observed versus Scaling Factor of 1/( RPM x $Sin^2$(Theta))

The scaling factor of RPM x $Sin^2$(Theta) may already be designed into some Fielder granulators. The 25-L and 65-L granulator used in this study was a fixed-speed unit with two bowls built into the same piece of equipment. It is noted that the blade angle was varied and the impeller speed was fixed with result that RPM x $Sin^2$(Theta) was held constant between the two bowls. Similarly the 150-L and 300-L granulators had slight variation in blade angle such that this potential scaling factor was also held constant between those two machines; however, at a different value compared to the 25/65-L granulator. RPM x $Sin^2$(Theta) was the criteria utilized to design an experimental custom blade for the 6-L granulator utilized in this study. RPM x $Sin^2$(Theta) appears to be a less-perfect approximation compared to Equation 6, but may provide a close enough approximation for use with engineering the blade angle for use in scaling up of most products.

## 1.5 Evaluating the Significance of Work, Xsat and the Amount of Water Added Within Scale

Observation of the impeller power inflection point "Xsat" proved to be useful when explaining scaling differences, and also shows significant use in monitoring drug products controlled by the Work model at constant scale. As an example, for drug product "P" manufactured on a 300-L Fielder granulator, a total of 40 granulation batches were manufactured and presented as data in Table 4. The first ten batches were part of the experimental DoE described with the response surface shown in Figure 3. The next 30 batches were produced using the optimum granulation work and water amount found from the DoE. Dissolution profiles of the film coated tablets were used as an assessment of end product quality. The 40-batch data set with work endpoints and film coated tablets dissolution are tabulated in Appendix A, Table 4.

For this drug product held at constant tablet thickness and tablet weight, granulation work and granulation water amount are good predictors of dissolution performance. It is possible to use a quadratic model to define a region of proven acceptable quality such that if the Work, Water, and Thickness are controlled, acceptable dissolution performance will be achieved. Thus for this drug compound, Work, Water, and Thickness are utilized as effective control parameters to ensure passing dissolution.

Monitoring of 30 batches following the DoE showed there was variation in impeller load inflection point during water addition, and dissolution below 90% was noted to occur at water excess (Xs) values of less than 8% (Figure 16). This drug product has an undergranulation failure mode resulting in reduced dissolution mean and increased dissolution RSD at lower extents of granulation.

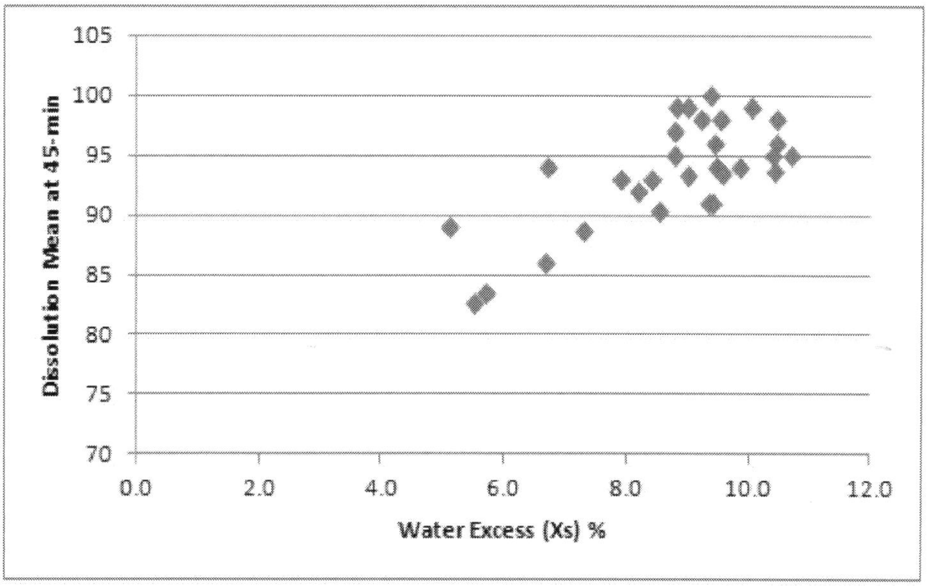


**Figure 16: Variation in Water Excess and Tablet Dissolution for Drug Product "P" in 30 runs at Optimum Work and Water Setting at 300-L Scale**

The general conclusion was that even at constant scale, there may be variation in the amount of water needed to start the granulation process ($X_{sat}$) which causes variation in the impeller load during water addition. If the drug product is sensitive to granule variation, the variation in excess water (Xs) can result in small to moderate changes in granule tapped density and drug dissolution. This is a risk factor that will need to be considered when establishing models and process control based on a granulation Work model, and might prompt need for a more complex model such as the SaWW model described in the next section. However it should be noted that even for Drug Product "P", the Work control model has been used for 700+ granulations to date of this writing with no failures below specification of 75% dissolution. Need for a more complex control model beyond "granulation work" such as the "SaWW" model discussed in the next section is only for cases where a finer degree of control is needed.

## 1.6 A Single Equation to Model Granulation – SaWW Model

The scale up development work and also monitoring granulation batches at 300-L scale suggested it's not the absolute amount of water added, but the amount of water added above the critical point where an increase is observed in the main impeller power. This led to development of the "Saturated Weighted Work" (SaWW) equation which can be used to predict granulation endpoint across equipment blade angle changes, water amount changes, Xs, or Work changes. The SaWW equation describes the granulation process using the idea that water in excess of a certain critical amount and work input can be integrated together to create a single predictive equation.

$$SaWW = \int_0^t \left(Power_{Impeller} - P_0\right) \cdot X_S \, dt \tag{7}$$

Where $X_s$ is the amount of water above the saturation point, which is defined as follows:

$$X_S = (X_{H2O} - X_{SAT}) \tag{8}$$

$X_{H2O}$ is the fraction of water added at that point of time. $X_{sat}$ is defined as the fraction of water needed at the point the impeller power begins to increase, indicating onset of granulation. $X_{sat}$ changes with scale, and also with raw material physical property changes, and can be found using Equation 4 described in the previous section discussing scale-up.

The SaWW model, or a model that calculates the appropriate Work for a given water excess ($X_S$) could potentially be used as an improved control methodology when programmed into a PLC, but recommendation at this time is to use the Work model by first intent as most drug products are not as sensitive to granulation as drug product "P" and online measurement of Xsat may unnecessarily increase the complexity of the control code and control model.

SaWW, or Work and $X_S$ were found to provide better model fits compared to Work and Water as predictive model terms for production and DoE data evaluated, and offers the advantage of reducing the number of factors needed to estimate to fit a linear regression model. As an example, for Drug Product "P", the 50-point DoE data available in Table 1 illustrated in Figure 2 as a complex 4 dimensional surface from a 10-factor linear regression model can be reduced to a 3-dimensional surface and a 6-term linear regression model with a similar goodness of fit (R^2=90%). The response surface of SaWW and compressed tablet thickness is shown in Figure 17. For the production monitoring data set in Table 4, the resulting tablet dissolution can be correlated to a single polynomial line as shown in Figure 18.

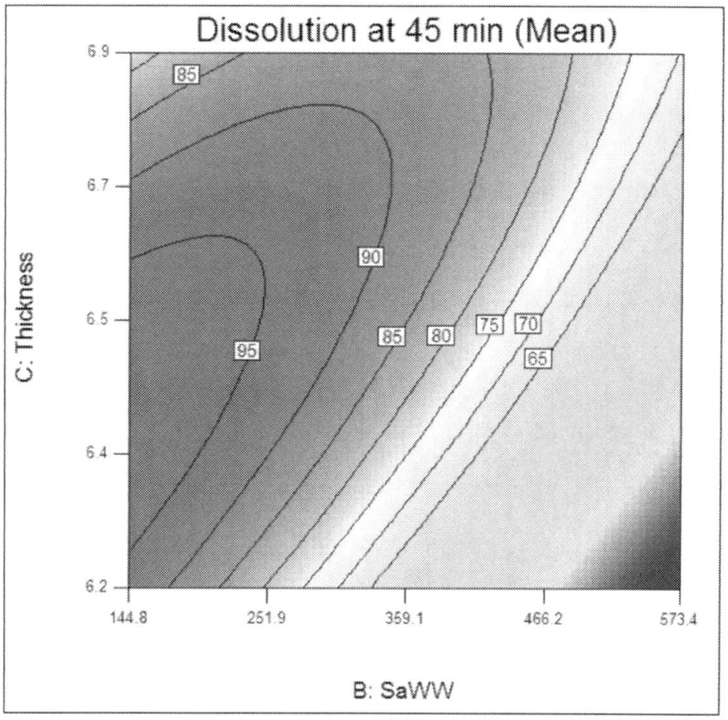


**Figure 17: Drug Product "P" Data from 10-batch DoE with 5 compression levels to 50 observations, Dissolution Modeled as a single 3-dimensional surface using SaWW Model with R^2=90%**

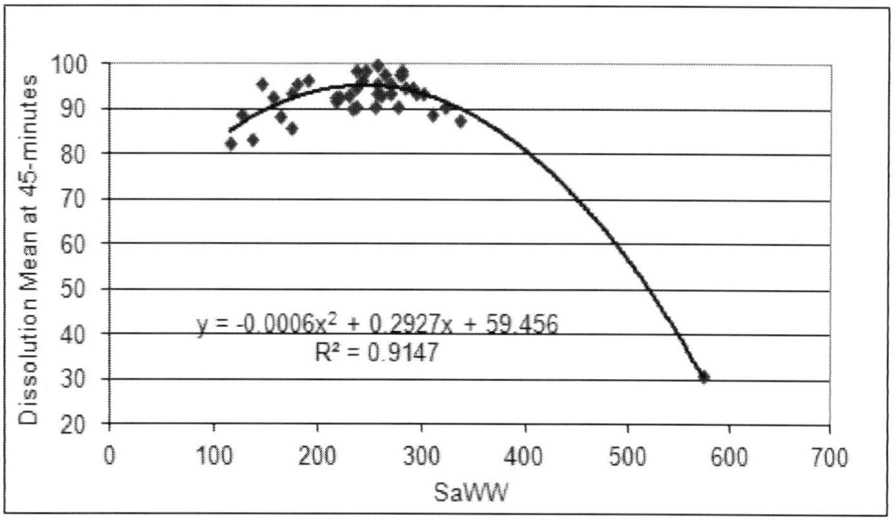


**Figure 18: Drug Product "P" Data from 10-batch DoE at constant target tablet thickness, plus 30 subsequent batches, Dissolution Modeled as a single 2-dimensional plot using SaWW Model**

The observation of the SaWW model describing the water and work interactions into a single parameter is not limited to just drug product "P", but has applied to all of them discussed in this chapter. For instance the SaWW model applied to Drug Products "D" (Figure 19 and Figure 20), finds the same observation of the prediction being reduced to a single characteristic line. This single line is different for each drug product, but the underlying mechanistic equation used for prediction is the same.

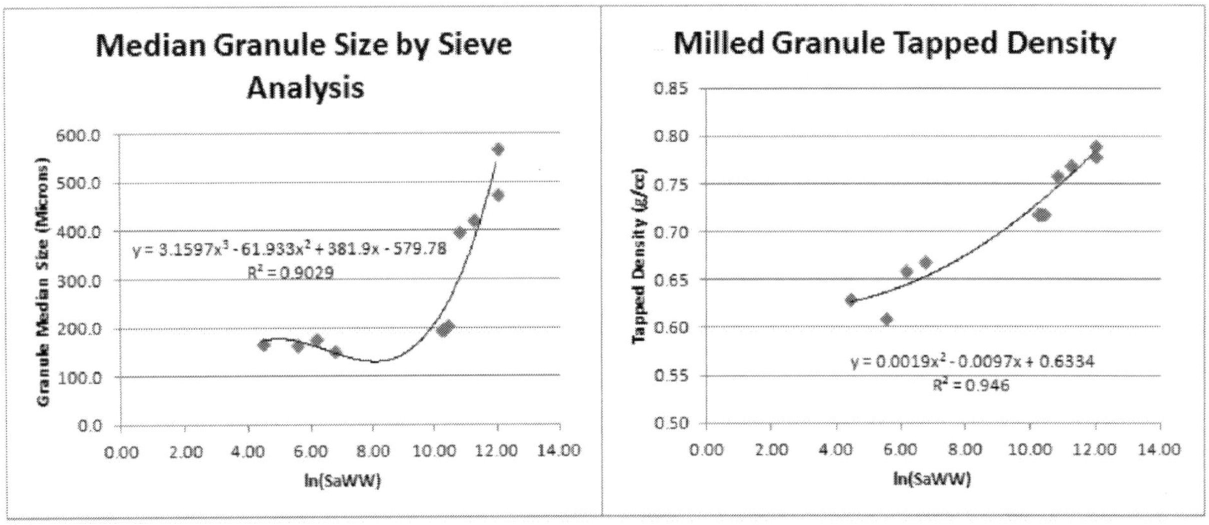


**Figure 19: Drug Product "D", Response of Tapped Density, Granule Size to the natural log of SaWW**

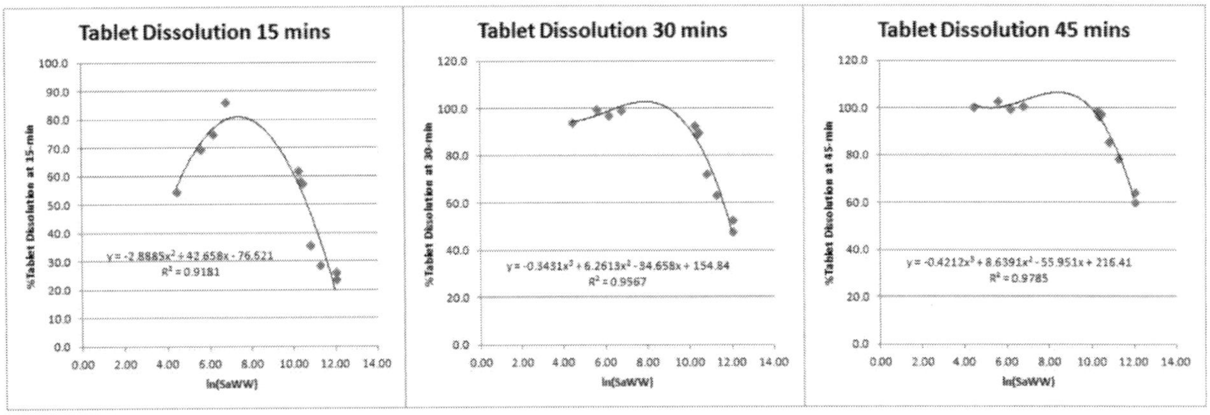


**Figure 20: Drug Product "D", Response of Tablet Dissolution to the natural log of SaWW**

Drug Product "S" (data Table 3) is modeled versus the SaWW model in Figure 21. This particular DoE examined water addition over a much larger range than typical, from 25% to 40% water. This allows an examination if the impact of water excess is truly linear over a wide range. The model equation used was equation 9 to test the significance of an exponent "N" on the water excess term. The model results with N=1 or N=3 is shown in Figure 21.

$$SaWW = \int_0^t \left(Power_{Impeller} - P_0\right) \cdot \left(X_S\right)^N dt \qquad (9)$$

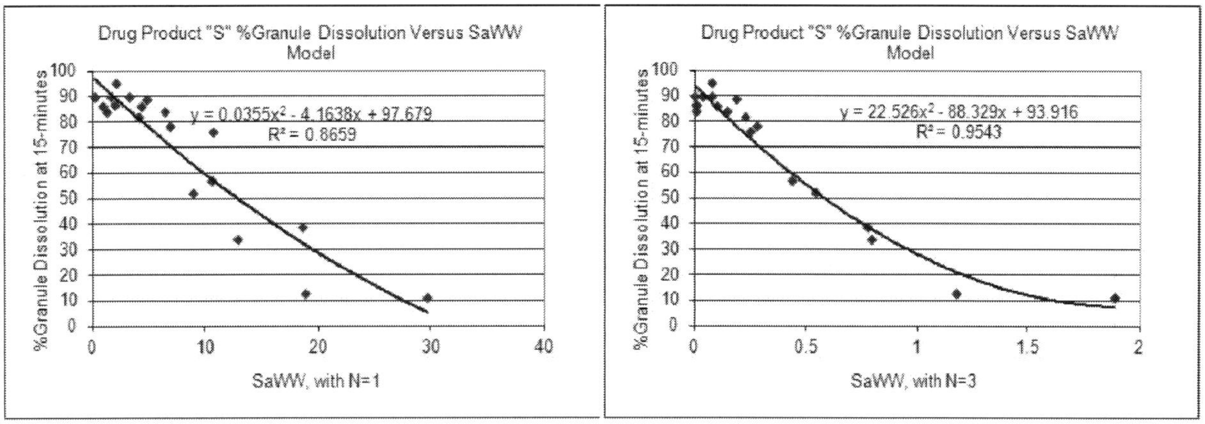


**Figure 21: Drug Product "S" Data from 20-batch Sampled DoE at constant target tablet thickness, Dissolution Modeled as a single 2-dimensional plot using SaWW Model, N=1 on left, and N=3 on right.**

The fits to granule dissolution data were notably improved using N=3 rather than N=1, indicating changes in water amount over a very broad range, may show a nonlinear impact of water excess, and Equation 9 is recommended.

The power of the SaWW model is the characteristic lines for a given product can be found with very few experiments. It only takes three to four observations to define a nonlinear line. Testing this line at multiple Work and Water levels then allows observation of reproducibility of the model prediction. Fewer linear regression terms are needed to fit the models, thus giving much higher capability to estimate the error distribution.

The strength and weakness of the SaWW model will lie in the capability to measure Xsat, the point of impeller inflection, as precisely as possible. If Xsat is significant and measured precisely, then the SaWW model will provide better model fits compared to models based on Work and Water amounts alone. As long as Xsat can be measured in a robust fashion, and is significant, then use of SaWW to control the process directly is likely to provide improved levels of process control compared to use of the Work equation.

## 1.7 Modeling Twin Screw Continuous Wet Granulation

Twin screw extrusion has been extensively studied and developed in the polymer industry since the 1950's. The first application to large scale OTC pharmaceutical granulation began in the 1970's and 1980's, primarily by GSK for antacid production. More recently, starting in 2002 twin screw extrusion came under more intensive study as an efficient and continuous means of replacing high shear wet granulation.(Keleb, Vermeire, Vervaet, & Remon, 2002) Use of "kneading" mixing elements (Figure 22) in the screw conveyer improves the granule flow, and is the primary source of granule consolidation.(Djuric, 2008)

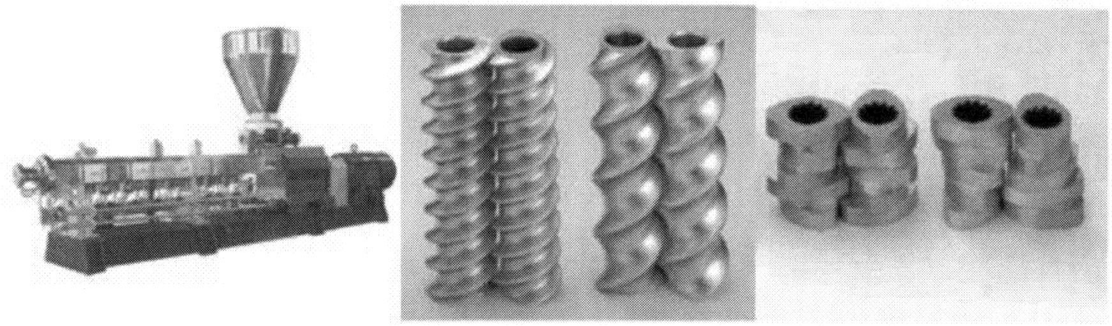

**Figure 22: Twin Screw Extruder (left) with Conveying Elements (Center) and Kneading Elements (Right)**

An excellent publically available data set that illustrates many of the principles for fitting a model to twin screw granulation data is from work done at Ghent University.(Vercruysse, et al., 2012) The experiment granulated a theophylline mixture in a 6-parameter DoE. The relevant data analyzed utilizing the first-principle equations described in this section is shown in Appendix A, Table 6.

The signal that can be best correlated from the granulation data is the observed granule yield, defined as material between 150 micron and 1400 microns in size. Too little granulation extent reduces yield due to increased fines, but too much granulation results in yield loss due to granules larger than 1400 microns in size. Regression analysis of the DoE parameters to predict granule yield finds the most significant terms are the number of kneading elements, and the barrel temperature. There's a possible cross product of throughput and screw speed that also appears to be significant. A regression model constructed from these parameters is shown in Figure 23 with impacts of factors shown in Figure 24.

## Response Yield (Percent Granules between 150-1400 micron)

### Actual by Predicted Plot

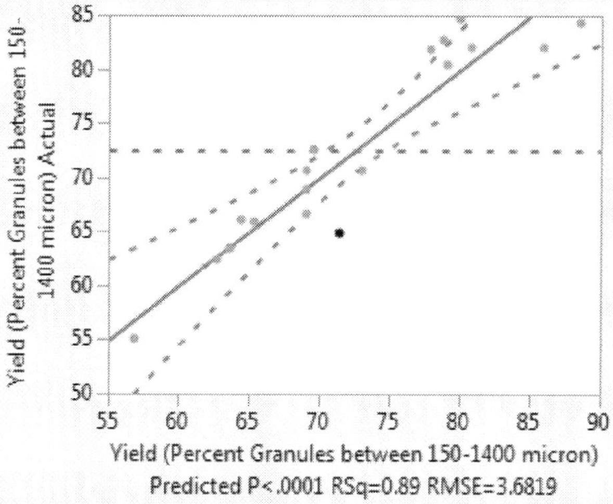


### Summary of Fit

| | |
|---|---|
| RSquare | 0.885191 |
| RSquare Adj | 0.837354 |
| Root Mean Square Error | 3.681898 |
| Mean of Response | 72.52778 |
| Observations (or Sum Wgts) | 18 |

### Parameter Estimates

| Term | Estimate | Std Error | t Ratio | Prob>\|t\| |
|---|---|---|---|---|
| Intercept | 44.971531 | 7.06944 | 6.36 | <.0001* |
| Throughput (kg/hr) | -0.209037 | 0.127923 | -1.63 | 0.1282 |
| Screw Speed (RPM) | 0.0080413 | 0.005482 | 1.47 | 0.1682 |
| Kneading Elements | 1.6275576 | 0.20184 | 8.06 | <.0001* |
| Barrel Temp (°C) | 0.4616674 | 0.128307 | 3.60 | 0.0037* |
| (Throughput (kg/hr)-17.9167)*(Screw Speed (RPM)-784.722) | 0.0019671 | 0.000736 | 2.67 | 0.0203* |

**Figure 23: Regression Model of Twin Screw Theophelline DoE**

### Prediction Profiler

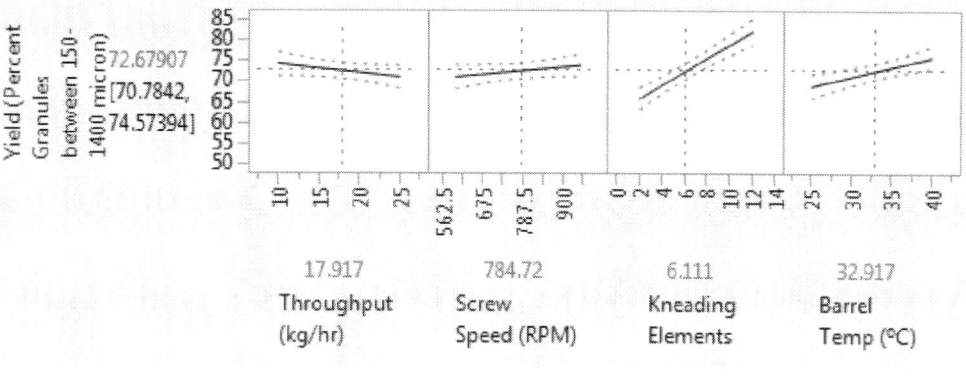


**Figure 24: Factor Prediction Profiler for DoE Parameters for fit Granule Yield**

Alternatively, it's possible to construct a model for twin screw granulation using first principles and the knowledge developed for high shear granulation, discussed earlier in this chapter. From first principles, the power input into the powder mixture is a function of torque and RPM as shown in equation 10.

$$Power = Torque \cdot \frac{RPM \cdot 2 \cdot \pi}{60} \quad (10)$$

High shear granulation was best modeled as power per unit dry powder mass, thus the above power term needs to be divided by the mass throughput measured in grams per second.

$$Power\_per\_mass = Torque \cdot \frac{RPM \cdot 2 \cdot \pi}{60 \cdot throughput} \quad (11)$$

There's an effect of temperature where higher temperature is noted to drive the granulation further. This effect is hypothesized to be due to changes in rheology. A mixture that requires more energy to overcome viscous forces or surface area interactions will require more power to reach an equivalent endpoint. Thus the impact of temperature is hypothesized to reduce the viscosity of the binding liquid thus allowing easier deformation and consolidation of the wet granules. The power per unit mass in the above equation is divided by the liquid viscosity at the granulation temperature resulting in Equation 12.

$$Power\_per\_Mass\_Viscosity = Torque \cdot \frac{RPM \cdot 2 \cdot \pi}{60 \cdot throughput \cdot \mu} \quad (12)$$

Plots of power per unit mass, and then power per unit mass divided by viscosity are shown in Figure 25. Use of viscosity to describe the temperature effect is noted to improve the correlation coefficient. The resulting model provides a better correlation coefficient ($R^2$=90%) than that achieved with linear regression, and only needs to estimate two fitted parameters.

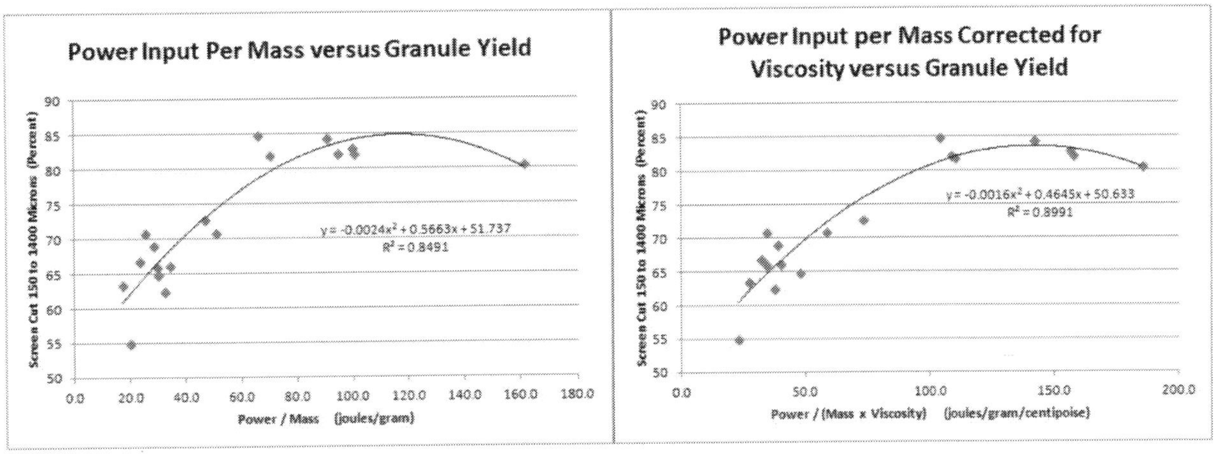


**Figure 25: Prediction of Granule Yield from First Principle Equations**

The same relationship to power is noted to be valid for both high shear granulation and twin screw granulation. If the twin screw granulation has motor power measured instead of torque then power per unit mass can be calculated as:

$$Power\_per\_mass = \frac{(P - P_0)}{throughput} \tag{13}$$

Where P is the measured power, and $P_0$ is the baseline power needed to turn the impeller without solids present. The resulting units from equation 13 are joules/gram, which is equivalent to the Work/Mass models described in the description of batch high shear granulation modeling.

## 1.8 The Impact of Feeder Variability on Twin Screw Wet Granulation

Continuous operations make the assumption that the flow into the equipment is continuous, and unfortunately powders that need to be granulated do not tend to flow in a continuous fashion. It is useful to consider the impact of granulation variation induced by variation in inlet powder flow to the twin screw granulator. If drug product tablet dissolution is sensitive to changes in granulation extent, the impact of feed variation is in-batch variation of granulation extent, resulting in increased in-batch variation in granule properties, and potentially reduced dissolution mean with higher RSD. In cases of very poorly flowing powders, the observation is the tablet dissolution properties achieved from twin screw granulation is not as good as those observed in high shear granulation. The situation can be improved by improving the powder flow of the mixture being granulated, through use of flow aids like fumed silica, or by limiting the region of operation of the powder hopper feeding the twin screw granulator.

Example hoppers used for feeding dry powder into twin screw granulation are manufactured by K-Tron, or Brahbender (Figure 26), and utilize either dual or single helical impellers (Figure 27).

**Figure 26: K-Tron Hopper (Left) and Brahbender Hopper (Right)**

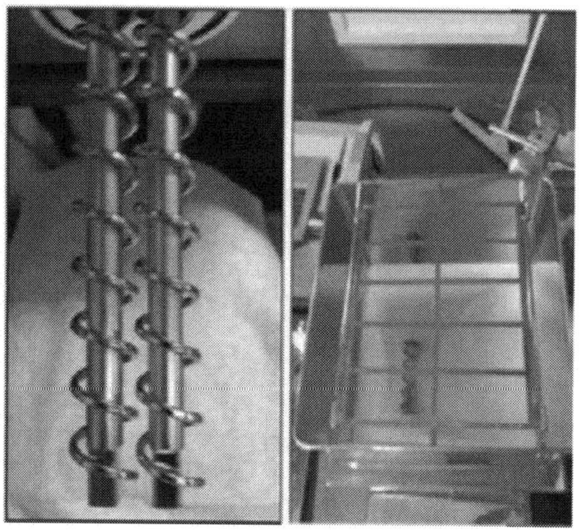

**Figure 27: Dual Helical Impeller of K-Tron Hopper (Left) and Single Helix of the Brahbender Hopper (Right)**

Figure 28 shows two different powders being fed from a K-Tron dual screw hopper. The freely flowing powder on left comes out as nearly a continuous stream, whereas the poorly flowing powder on right exits the hopper in cohesive packets.

**Figure 28: Freely flowing powder from K-Tron Hopper (left), Very cohesive powder (right)**

To examine the variability of powder flow from hoppers, ten different powders were analyzed using a 6-inch Brookfield PFT shear cell, with resulting density and flow index analytical data shown in Figure 29.

| Rank # | Material Name | Flow Function Slope | Flow Function Intercept | Fill Bulk Density (kg/m^3) | Flow Index at 0.5 KPA | Flow Index at 1 KPA | Flow Index at 2.4 KPA | Flow Index at 5 KPA | Flow Index at 10 KPA | Type of Flow at 1 KPA |
|---|---|---|---|---|---|---|---|---|---|---|
| 1 | Manitol Pearlitol 200SD | 0.03 | 0.002 | 484.4 | 0.034 | 0.032 | 0.031 | 0.0304 | 0.0302 | Free Flowing |
| 2 | Avicel PH102 | 0.14 | 0.04 | 367.3 | 0.22 | 0.18 | 0.157 | 0.148 | 0.144 | Easy Flowing |
| 3 | Avicel PH101 | 0.22 | 0.023 | 303 | 0.266 | 0.243 | 0.230 | 0.2246 | 0.2223 | Easy Flowing |
| 4 | HPMC Pharmacoat 603 | 0.22 | 0.116 | 401.7 | 0.452 | 0.336 | 0.268 | 0.2432 | 0.2316 | Cohesive |
| 5 | Improved 6% API Formulation | 0.06 | 0.277 | 378.8 | 0.614 | 0.337 | 0.175 | 0.1154 | 0.0877 | Cohesive |
| 6 | Improved 12% API Formulation | 0.14 | 0.338 | 332.9 | 0.816 | 0.478 | 0.281 | 0.2076 | 0.1738 | Cohesive |
| 7 | Manitol Pearlitol 160C | 0.16 | 0.366 | 585.3 | 0.892 | 0.526 | 0.313 | 0.2332 | 0.1966 | Very Cohesive |
| 8 | 6% API Formulation | 0.19 | 0.555 | 378.8 | 1.3 | 0.745 | 0.421 | 0.301 | 0.2455 | Very Cohesive |
| 9 | 12% API Formulation | 0.23 | 0.561 | 332.9 | 1.352 | 0.791 | 0.464 | 0.3422 | 0.2861 | Very Cohesive |
| 10 | 100% API lot 76 | 0.16 | 0.835 | 134.9 | 1.83 | 0.995 | 0.508 | 0.327 | 0.2435 | Very Cohesive |

**Figure 29: Shear Cell Analysis of 10 Powders Utilized in Hopper Study**

Each powder was then loaded into the K-Tron hopper, and the feedrate setting was varied over values of 10%, 15%, 25%, 50%, 75%, and 99% of maximum RPM. After powder flow was established, data recording was enabled on a Mettler Toledo scale, collecting dynamic weight measurement at 8 measurements per second, with data averaging for 3 seconds per weight observation. The average feedrate for the trial was recorded, as well as the 3-second relative standard deviation (RSD) of the powder flowrate. Example data for Manitol 200SD is shown in Figure 30, and this process was repeated for the 9 other powders resulting in a total of 126 observations.

| Name | Hopper Mass | Feed setting | Feed Rate (grams/sec) | Flow RSD% for 3 second time interval |
|---|---|---|---|---|
| Manitol 200SD | 2.8 | 99% | 5.472 | 1.35% |
| Manitol 200SD | 2.55 | 75% | 4.113 | 1.59% |
| Manitol 200SD | 2.99 | 50% | 2.587 | 2.00% |
| Manitol 200SD | 3.2 | 25% | 1.092 | 2.57% |
| Manitol 200SD | 1.92 | 99% | 5.404 | 2.89% |
| Manitol 200SD | 1.53 | 75% | 4.133 | 0.70% |
| Manitol 200SD | 1.73 | 50% | 2.623 | 2.54% |
| Manitol 200SD | 1.26 | 25% | 1.136 | 2.25% |
| Manitol 200SD | 0.97 | 15% | 0.541 | 5.59% |
| Manitol 200SD | 2.147 | 15% | 0.527 | 5.53% |
| Manitol 200SD | 2.33 | 10% | 0.215 | 14.93% |

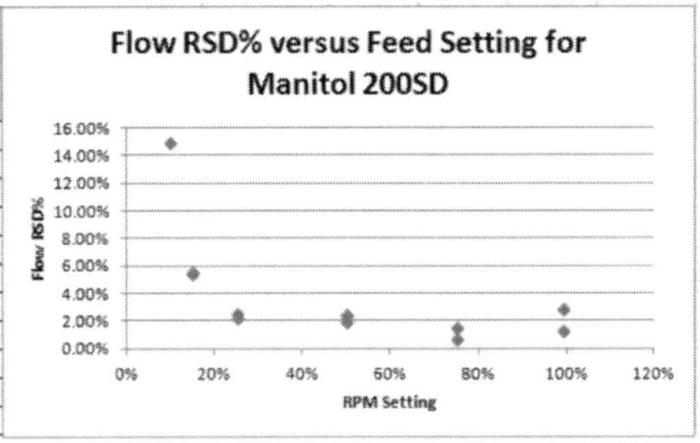


**Figure 30: K-Tron Flow Data for Manitol 200 SD (left), Observed Relationship between RPM Setting and Flow RSD% (right)**

At a fixed RPM, 25% for example, a correlation is observed between the shear cell flow index at 1 KPa, and the 3-second hopper flow RSD (Figure 31). Correlations were attempted at shear cell consolidation stresses ranging from 0.5 KPa to 10 KPa, with conclusion that the RSD data is best fit to shear cell data at 1 KPa consolidation pressure.

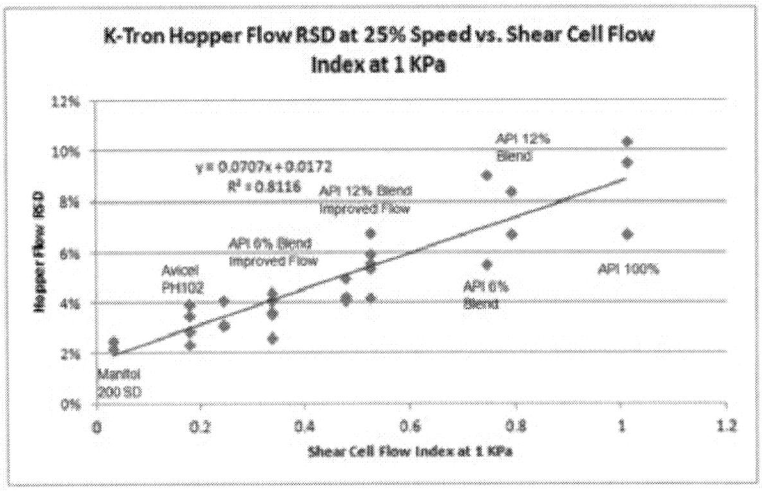


**Figure 31: K-Tron Hopper Flow RSD at 25% Feed Setting versus Shear Cell Flow Index at 1-KPa**

Modeling all the data collected on the K-Tron Hopper at all the RPM settings results in in a 5-term linear regression model built from 126 observations. The response surface is best fit to the reciprocal of feeder RPM, measured as seconds per feed screw revolution. The model for the K-Tron hopper, parameter estimates, and the resulting response surface predicting RSD fraction is shown in Figure 32.

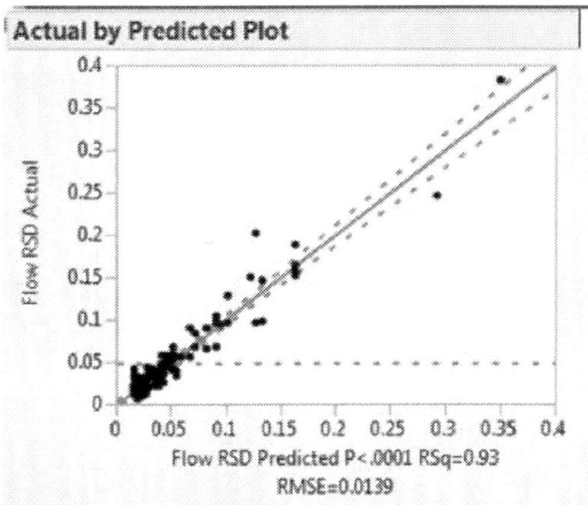


**Summary of Fit**

| | |
|---|---|
| RSquare | 0.928332 |
| RSquare Adj | 0.925963 |
| Root Mean Square Error | 0.013944 |
| Mean of Response | 0.048511 |
| Observations (or Sum Wgts) | 126 |

**Analysis of Variance**

| Source | DF | Sum of Squares | Mean Square | F Ratio |
|---|---|---|---|---|
| Model | 4 | 0.30473011 | 0.076183 | 391.8343 |
| Error | 121 | 0.02352547 | 0.000194 | Prob > F |
| C. Total | 125 | 0.32825558 | | <.0001* |

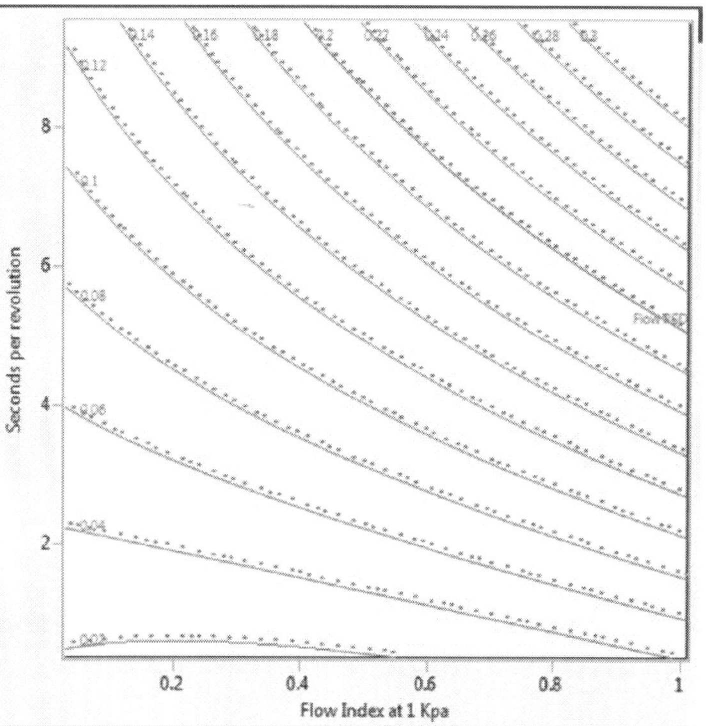


**Parameter Estimates**

| Term | Estimate | Std Error | t Ratio | Prob>\|t\| |
|---|---|---|---|---|
| Intercept | -0.011143 | 0.002509 | -4.44 | <.0001* |
| Flow Index at 1 Kpa | 0.0449248 | 0.004573 | 9.82 | <.0001* |
| Seconds per revolution | 0.0217191 | 0.000776 | 27.99 | <.0001* |
| (Flow Index at 1 Kpa-0.47865)*(Flow Index at 1 Kpa-0.47865) | 0.0449677 | 0.014787 | 3.04 | 0.0029* |
| (Flow Index at 1 Kpa-0.47865)*(Seconds per revolution-1.4832) | 0.0228292 | 0.00197 | 11.59 | <.0001* |

**Figure 32: K-Tron Twin-Screw Hopper Powder Flow RSD Fraction Model, Response Surface to Powder Flow Index at 1-KPa, and Number of Seconds per Feed Screw Rotation**

The K-Tron hopper is remarkably robust, and was able to successfully feed powder up to a flow index of 0.995 at 1-KPa, albeit with increasing levels of variation measured every 3-seconds. The disadvantage of the K-Tron in this particular configuration is that even for very good flowing powders, such as Manitol 200SD, there was still a significant relationship between hopper speed setting and the observed RSD of flow. It should be noted there is also an impact of screw pitch, a finer pitch impeller is able to have more discharge events per second thus lower RSD, but cannot handle as poorly flowing of powders.

The same experimental procedure was also applied to the Brahbender hopper, though fewer powders could be fed through this hopper. Powders above a flow index of 0.5 to 0.6 at 1-KPa have a probability of forming cohesive arching resulting in starving the impeller screw of powder. The Brahbender hopper was able to poorly feed a powder with a flow index of 0.75 at 1-KPa, but with numerous observations of cohesive arching and powder flow entirely stopping for seconds. Only six test powders could be fed through the Brahbender, resulting in 72 observations for the experiment. The resulting model to predict powder flow RSD fraction and response surface diagram is shown in Figure 33.

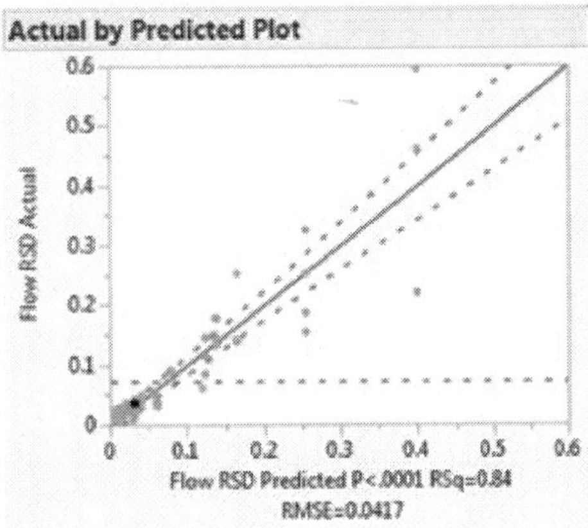


**Summary of Fit**

| | |
|---|---|
| RSquare | 0.842022 |
| RSquare Adj | 0.83259 |
| Root Mean Square Error | 0.041742 |
| Mean of Response | 0.072676 |
| Observations (or Sum Wgts) | 72 |

**Analysis of Variance**

| Source | DF | Sum of Squares | Mean Square | F Ratio |
|---|---|---|---|---|
| Model | 4 | 0.62222723 | 0.155557 | 89.2771 |
| Error | 67 | 0.11674108 | 0.001742 | Prob > F |
| C. Total | 71 | 0.73896832 | | <.0001* |

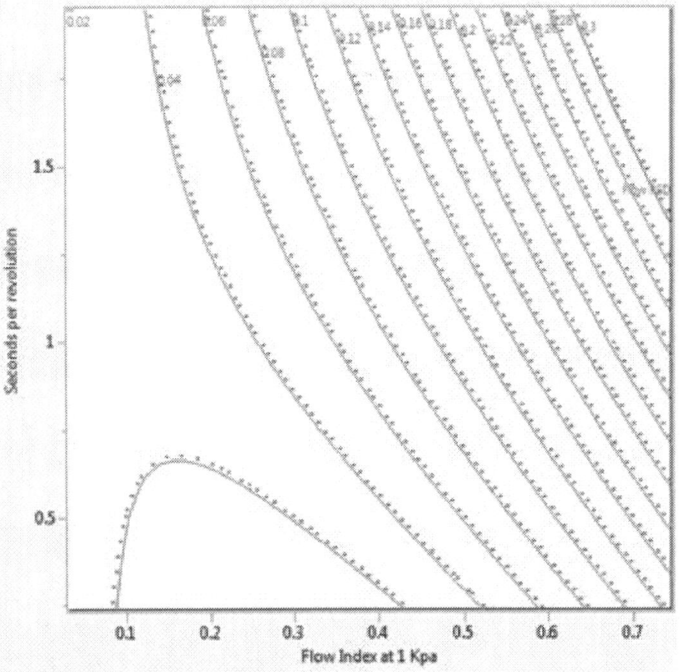


**Parameter Estimates**

| Term | Estimate | Std Error | t Ratio | Prob>\|t\| |
|---|---|---|---|---|
| Intercept | -0.064341 | 0.010601 | -6.07 | <.0001* |
| Flow Index at 1 Kpa | 0.1889273 | 0.027612 | 6.84 | <.0001* |
| Seconds per revolution | 0.0564204 | 0.008212 | 6.87 | <.0001* |
| (Flow Index at 1 Kpa-0.30083)*(Flow Index at 1 Kpa-0.30083) | 0.5057237 | 0.094609 | 5.35 | <.0001* |
| (Flow Index at 1 Kpa-0.30083)*(Seconds per revolution-0.86368) | 0.233865 | 0.030841 | 7.58 | <.0001* |

**Figure 33: Brahbender Hopper Powder Flow RSD Fraction Model, Response Surface to Powder Flow Index at 1-KPa, and Number of Seconds per Feed Screw Rotation**

The significant advantage the Brahbender Hopper has is that for powders that flow well, up to a flow index of approximately 0.4, there is comparatively less impact of feed rate (RPM) variation compared to the K-Tron hopper, as illustrated by the response surface in Figure 33. This makes the Brahbender Hopper the preferred choice for feeding materials that flow well, such as excipient mixtures. The K-Tron should be chosen if the flow index is greater than 0.4 to 0.5 at 1-KPa consolidation stress.

To this point, the response surfaces shown are the RSD relative to a 3-second measurement of mass discharge. Three seconds is an arbitrary number which was decided upon based on the settling time of the scale. As the time period of measurement is reduced, the RSD will increase, so it's important to consider what the residence time is of the powder underneath the powder feed point and the water input point. The input ports where the water and powder enter the granulator require 3.66 revolutions of the impeller to traverse, resulting in a residence time of 0.5 and 1.5 seconds at 450 and 150 RPM respectively (Figure 34). As this residence time is decreased, more variable packets of powder then encounter the water addition nozzle, resulting in variable extents of moisture fraction, and thus more variable granulation.

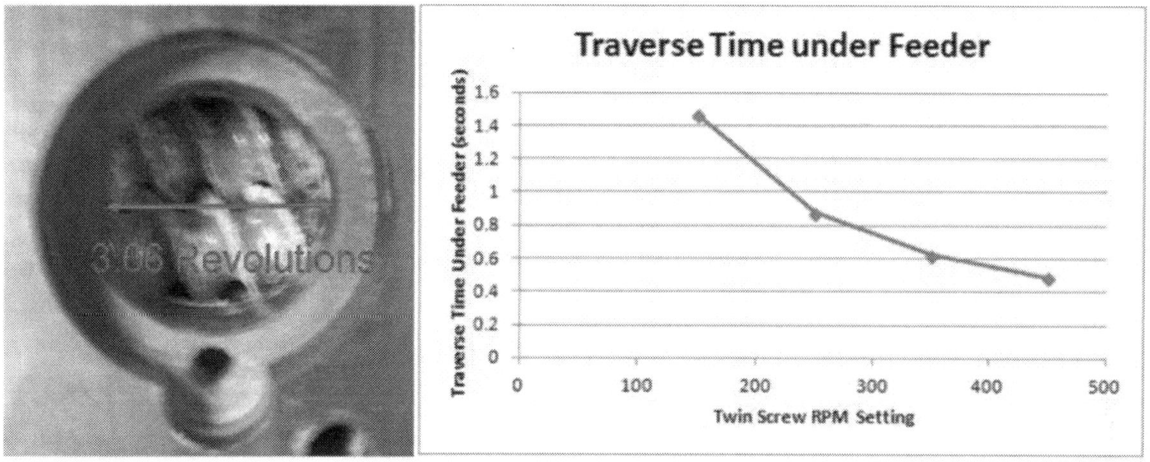


**Figure 34: Impact of Twin Screw Granulator RPM on Residence Time Under Powder and Liquid Feed**

The RSD should decrease the more powder discharge events can be averaged together from the hopper feeding the twin screw. If the measurement time is extended from 3 seconds to 12 seconds, a nonlinear trend is observed in the relative RSD that is measured. Theoretically this RSD should scale based upon the reciprocal of the square root of time, based on theory of how standard deviation of a population mean tends to vary with increasing number of observations. Figure 35 shows the impact of measurement time period and also the near-linear response if plotted versus the reciprocal of square root of time. This near-linear relationship can then be extrapolated to faster time periods, larger values of 1/(time^0.5), to get an estimate of the increase in RSD expected as the twin screw RPM setting is changed (Figure 36).

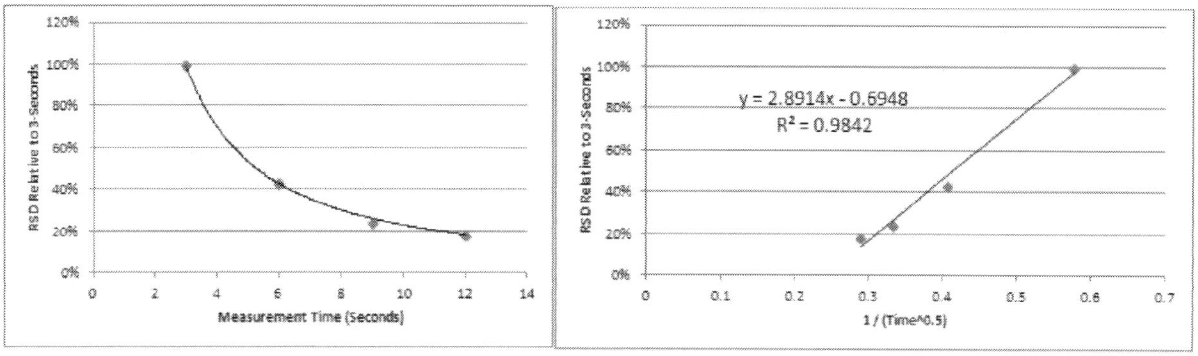


**Figure 35: Impact of Change in Measurement Time on Calculated RSD Relative to 3-Seconds**

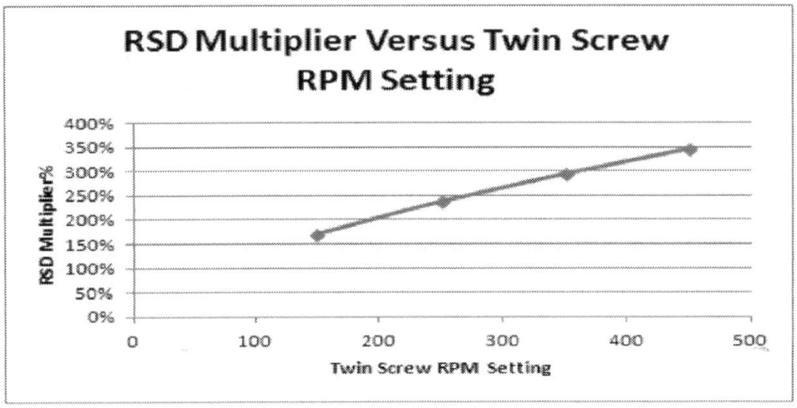


**Figure 36: Impact of Twin Screw Granulator RPM Setting and RSD Multiplier Compared to 3-Second Data**

The result of input powder variability is then going to be water wetting variability, and then variability in granulation extent in the resulting product granule. It's important to develop and include back-mixing zones that increase the powder residence time to seconds before and after the water addition zone to reduce the impact of this phenomenon for poorly flowing powders.

Figure 37 shows two case study predictions assuming impeller configuration is set up with no back mixing elements, for a powder with a flow index of 0.8 at 1-KPa. The first case, shown on left sets the K-Tron feed setting to 50% (1.08 revolutions per second) and a twin screw speed of 150 rpm. For a water target of 20%, 99% of the material would be expected to have wetting between 15% and 25%. Case two on the right shows it's possible to get much more variability by reducing the K-Tron feed setting to 20% while increasing the twin screw speed to 450 RPM. In this second scenario the 99% limits of wetting is between 3% and 37%, and significant variation in granule properties is expected.

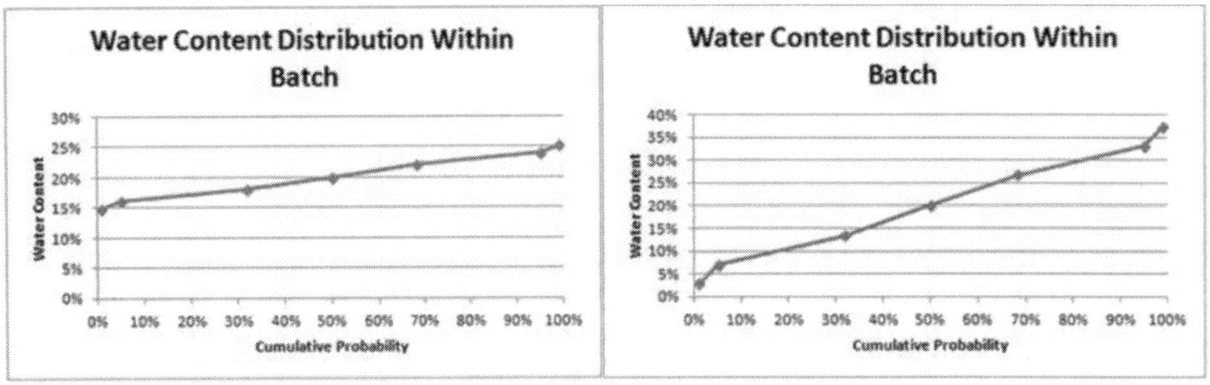


**Figure 37: Model Predictions of Water Distribution for Case 1 (left) and Case 2 (right)**

In reality the twin screw granulator would not be set up with no back mixing elements, and operated in such an extreme case of slow K-Tron feeder speed and rapid screw RPM. The situation is unlikely to be as extreme as this, and the impact of powder feed variation will be more subtle, more similar to the case study shown on the left in Figure 37. For drug products that are sensitive to granulation, it does illustrate the need to consider powder flow properties, it's impact on feed variability, the residence time distributions, the impact of back mixing, and the resulting granulation consistency. Models using the equations provided in this section can be used to estimate the worse-case water wetting consistency, before the addition of back mixing elements.

## 1.9 Conclusion

Batch high shear granulation and twin screw granulation can both be successfully modeled using equations relating to the granulation work done per unit mass of dry powder. Models based on work per unit mass can and have been used to successfully control granulation processes in real time, even for drug products fairly sensitive to granulation extent. If an even finer degree of control or potentially more precise modeling is desired or there is desire to predict water and work requirements across scale, then it's useful to measure the minimum amount of water needed to cause a significant change in impeller power during water addition, referred to as "Xsat" in this discussion. For high shear granulation, observation of Xsat as it varied with scale allow the generation of models that can predict how the water requirements will change with scale, and allows construction of granulation impeller blades designed to ensure similar water requirements across scale.

For high shear granulation, an equation (named SaWW) that integrates both granulation work and the amount of water in excess of Xsat is found to be useful in providing better models and reducing the number of dimensions that need to be considered when creating a linear regression response surface model. It appears likely the SaWW model shown as Equation 7 is a good fundamental equation for modeling most granulation processes over moderate changes in water excess, and will reduce the resulting predictions to a single characteristic line that requires fewer experiments to define. If large changes in water amount are explored (e.g. moving from 25% to 40% water addition), it's possible there may be significant nonlinearity in the impact of water excess, in which case Equation 9 should be explored.

Continuous twin screw granulation adds the complexity that the powder feeding the granulator does not necessarily feed in a continuous fashion, but rather in a variable fashion than can be predicted using the response surface models for each hopper type. A measure of powder flow using the Brookfield PFT shear cell flow index at 1-KPa was found to correlate to the resulting variability. Models for worst case water-wetting variation, assuming minimal back mixing are discussed. For drug products where dissolution is sensitive to variation in granulation extent, it is useful to consider optimizing the flow properties through API physical properties or through use of flow aids such as fumed silica, to reduce the variability in feed rate to the twin screw granulator. It's also useful to consider use of back mixing elements to reduce the severity of the feed variability.

## 1.10 Works Cited

Clancy, D. (2010). Method and Apparatus for Predicting Properties of Granulated Materials and Dosage Forms made Therefrom. *US Patent* , 20100318228.

Clancy, D., Campbell, G., Zhang, J., Gupta, M., & Oh, C. (2011). Closing the gap in series scale up of high shear wet granulation process using impeller power and blade design. *Powder Technology* , 184-192.

Djuric, D. (2008). *CONTINUOUS GRANULATION WITH A TWIN SCREW EXTRUDER.* Düsseldorf: Heinrich-Heine-Universität Düsseldorf.

Johannes von Orelli, H. L. (Volume 287, Issues 1–2, 9 December 2004). Search for technological reasons to develop a capsule or a tablet formulation with respect to wettability and dissolution. *International Journal of Pharmaceutics* , Pages 135–145.

Keleb, E., Vermeire, A., Vervaet, C., & Remon, J. (2002). Continuous twin screw extrusion for the wet granulation of lactose. *International Journal of Pharmaceutics* , 69-80.

Matthew J. Mollan Jr., P. a. (2004). Continuous Processing in Pharmaceutical Manufacturing. *Pharmaceutical Manufacturing* .

Trafton, A. (2012, March 12). Continuous drug manufacturing offers speed, lower costs. *MIT News* .

Vercruysse, J., Dias, C., Peeters, E., Fonteyne, M., Delaet, U., Assche, U., et al. (2012). Continuous twin screw granulation: Influence of process variables on granule and tablet quality. *European Journal of Pharmaceutics and Biopharmaceutics* , 205-211.

von Orelli, J., & Leuenberger, H. (2004). Search for technological reasons to develop a capsule or a tablet formulation with respect to wettability and dissolution. *International Journal of Pharmaceutics* , 135-145.

Wallack, T. (2007, September 28). Novartis to give MIT $65m to find new way to produce drugs. *The Boston Globe* .

Wilburn, K. (2010). *The Business Case for Continuous Manufacturing of Pharmaceuticals.* Massachusetts: Massachusetts Institute of Technology.

# Appendix A: DoE and Repeat Run Data Tables

| Granule Batch | Tablet Batch | Granulation Elapsed Time (seconds) | Power for 10-seconds at end of batch (%Motor Load) | Water Delivered (%mass/ mass) | Equation (2) Work (%Load* seconds) | Xsat | Equation (8) Xs(%) | Equation (7) SaWW (%Load*sec) | Tablet thickness (mm) | Milled Granule Tapped Density at 1250 taps (g/cc) | Percent Dissolved at 45-min | Dissolution RSD 45-min |
|---|---|---|---|---|---|---|---|---|---|---|---|---|
| 1 | A | 481 | 41.9 | 29.9% | 4268.1 | 20.2% | 9.7% | 322.730 | 6.21 | 0.73 | 65 | 22.7 |
| 1 | B | 481 | 41.9 | 29.9% | 4268.1 | 20.2% | 9.7% | 322.730 | 6.34 | 0.73 | 88 | 7.2 |
| 1 | C | 481 | 41.9 | 29.9% | 4268.1 | 20.2% | 9.7% | 322.730 | 6.51 | 0.73 | 91 | 1.1 |
| 1 | D | 481 | 41.9 | 29.9% | 4268.1 | 20.2% | 9.7% | 322.730 | 6.67 | 0.73 | 89 | 1.7 |
| 1 | E | 481 | 41.9 | 29.9% | 4268.1 | 20.2% | 9.7% | 322.730 | 6.86 | 0.73 | 87 | 0.7 |
| 2 | A | 484 | 43.1 | 31.8% | 6364.2 | 20.2% | 11.6% | 573.387 | 6.21 | 0.81 | 5 | 41.4 |
| 2 | B | 484 | 43.1 | 31.8% | 6364.2 | 20.2% | 11.6% | 573.387 | 6.36 | 0.81 | 13 | 86.4 |
| 2 | C | 484 | 43.1 | 31.8% | 6364.2 | 20.2% | 11.6% | 573.387 | 6.51 | 0.81 | 31 | 49.6 |
| 2 | D | 484 | 43.1 | 31.8% | 6364.2 | 20.2% | 11.6% | 573.387 | 6.68 | 0.81 | 69 | 24.6 |
| 2 | E | 484 | 43.1 | 31.8% | 6364.2 | 20.2% | 11.6% | 573.387 | 6.85 | 0.81 | 77 | 2.4 |
| 3 | A | 442 | 47.3 | 31.8% | 4125.6 | 21.9% | 9.9% | 309.117 | 6.23 | 0.77 | 70 | 22.8 |
| 3 | B | 442 | 47.3 | 31.8% | 4125.6 | 21.9% | 9.9% | 309.117 | 6.35 | 0.77 | 89 | 1.3 |
| 3 | C | 442 | 47.3 | 31.8% | 4125.6 | 21.9% | 9.9% | 309.117 | 6.53 | 0.77 | 89 | 2.2 |
| 3 | D | 442 | 47.3 | 31.8% | 4125.6 | 21.9% | 9.9% | 309.117 | 6.70 | 0.77 | 89 | 2.4 |
| 3 | E | 442 | 47.3 | 31.8% | 4125.6 | 21.9% | 9.9% | 309.117 | 6.86 | 0.77 | 82 | 2.6 |
| 4 | A | 501 | 30.0 | 27.7% | 2796.6 | 20.5% | 7.2% | 144.777 | 6.21 | 0.73 | 88 | 5.7 |
| 4 | B | 501 | 30.0 | 27.7% | 2796.6 | 20.5% | 7.2% | 144.777 | 6.36 | 0.73 | 96 | 1.2 |
| 4 | C | 501 | 30.0 | 27.7% | 2796.6 | 20.5% | 7.2% | 144.777 | 6.55 | 0.73 | 96 | 1.3 |
| 4 | D | 501 | 30.0 | 27.7% | 2796.6 | 20.5% | 7.2% | 144.777 | 6.71 | 0.73 | 94 | 1.5 |
| 4 | E | 501 | 30.0 | 27.7% | 2796.6 | 20.5% | 7.2% | 144.777 | 6.89 | 0.73 | 90 | 5.3 |
| 5 | A | 664 | 33.9 | 27.7% | 5587.9 | 20.1% | 7.6% | 301.534 | 6.20 | 0.72 | 69 | 36.3 |
| 5 | B | 664 | 33.9 | 27.7% | 5587.9 | 20.1% | 7.6% | 301.534 | 6.36 | 0.72 | 94 | 1.6 |
| 5 | C | 664 | 33.9 | 27.7% | 5587.9 | 20.1% | 7.6% | 301.534 | 6.50 | 0.72 | 94 | 1.5 |
| 5 | D | 664 | 33.9 | 27.7% | 5587.9 | 20.1% | 7.6% | 301.534 | 6.69 | 0.72 | 93 | 2.8 |
| 5 | E | 664 | 33.9 | 27.7% | 5587.9 | 20.1% | 7.6% | 301.534 | 6.87 | 0.72 | 80 | 8.7 |
| 6 | A | 413 | 33.5 | 31.8% | 2647.0 | 22.1% | 9.7% | 178.592 | 6.19 | 0.73 | 92 | 3.1 |
| 6 | B | 413 | 33.5 | 31.8% | 2647.0 | 22.1% | 9.7% | 178.592 | 6.34 | 0.73 | 96 | 0.8 |
| 6 | C | 413 | 33.5 | 31.8% | 2647.0 | 22.1% | 9.7% | 178.592 | 6.50 | 0.73 | 96 | 0.6 |
| 6 | D | 413 | 33.5 | 31.8% | 2647.0 | 22.1% | 9.7% | 178.592 | 6.69 | 0.73 | 93 | 2.4 |
| 6 | E | 413 | 33.5 | 31.8% | 2647.0 | 22.1% | 9.7% | 178.592 | 6.87 | 0.73 | 84 | 9.7 |
| 7 | A | 533 | 41.0 | 29.9% | 4934.7 | 20.7% | 9.2% | 337.302 | 6.20 | 0.73 | 60 | 41.5 |
| 7 | B | 533 | 41.0 | 29.9% | 4934.7 | 20.7% | 9.2% | 337.302 | 6.34 | 0.73 | 90 | 0.9 |
| 7 | C | 533 | 41.0 | 29.9% | 4934.7 | 20.7% | 9.2% | 337.302 | 6.52 | 0.73 | 88 | 1.5 |
| 7 | D | 533 | 41.0 | 29.9% | 4934.7 | 20.7% | 9.2% | 337.302 | 6.68 | 0.73 | 89 | 2.2 |
| 7 | E | 533 | 41.0 | 29.9% | 4934.7 | 20.7% | 9.2% | 337.302 | 6.87 | 0.73 | 84 | 3.3 |
| 8 | A | 413 | 30.2 | 30.1% | 2626.3 | 21.0% | 9.1% | 156.056 | 6.20 | 0.70 | 89 | 7.2 |
| 8 | B | 413 | 30.2 | 30.1% | 2626.3 | 21.0% | 9.1% | 156.056 | 6.37 | 0.70 | 94 | 2.5 |
| 8 | C | 413 | 30.2 | 30.1% | 2626.3 | 21.0% | 9.1% | 156.056 | 6.52 | 0.70 | 93 | 1.4 |
| 8 | D | 413 | 30.2 | 30.1% | 2626.3 | 21.0% | 9.1% | 156.056 | 6.71 | 0.70 | 87 | 7.7 |
| 8 | E | 413 | 30.2 | 30.1% | 2626.3 | 21.0% | 9.1% | 156.056 | 6.89 | 0.70 | 86 | 10.5 |
| 9 | A | 544 | 29.5 | 28.0% | 3612.5 | 20.5% | 7.5% | 190.096 | 6.21 | 0.70 | 90 | 10.9 |
| 9 | B | 544 | 29.5 | 28.0% | 3612.5 | 20.5% | 7.5% | 190.096 | 6.36 | 0.70 | 97 | 0.9 |
| 9 | C | 544 | 29.5 | 28.0% | 3612.5 | 20.5% | 7.5% | 190.096 | 6.53 | 0.70 | 97 | 0.8 |
| 9 | D | 544 | 29.5 | 28.0% | 3612.5 | 20.5% | 7.5% | 190.096 | 6.70 | 0.70 | 93 | 4.8 |
| 9 | E | 544 | 29.5 | 28.0% | 3612.5 | 20.5% | 7.5% | 190.096 | 6.88 | 0.70 | 83 | 6.2 |
| 10 | A | 456 | 39.3 | 30.1% | 3712.7 | 20.9% | 9.3% | 275.916 | 6.20 | 0.74 | 80 | 8 |
| 10 | B | 456 | 39.3 | 30.1% | 3712.7 | 20.9% | 9.3% | 275.916 | 6.36 | 0.74 | 91 | 3.8 |
| 10 | C | 456 | 39.3 | 30.1% | 3712.7 | 20.9% | 9.3% | 275.916 | 6.52 | 0.74 | 91 | 0.6 |
| 10 | D | 456 | 39.3 | 30.1% | 3712.7 | 20.9% | 9.3% | 275.916 | 6.68 | 0.74 | 89 | 1.9 |
| 10 | E | 456 | 39.3 | 30.1% | 3712.7 | 20.9% | 9.3% | 275.916 | 6.86 | 0.74 | 81 | 10 |

**Table 1: Drug Product "P" Granulation DoE at 300-L Scale, 10 granulation batch response surface design for Granulation Work and Water, Split Plot Design to 50 compression batches**

| | DoE Factors | | | Calculated DoE Factors | | | | | Dried Granule Properties | | | Tablet Properties | | |
|---|---|---|---|---|---|---|---|---|---|---|---|---|---|---|
| Run ID # | Water Addition (% w/w) | Wet Massing time (min) | Water Addition Rate ( g/min kg) | Work | Xsat | Xs | SaWW | ln(SaWW) | Tapped Bulk Density (g/ml) | Median Particle Size | Compaction Compressibility (%) | Tablet Dissolution 15 mins | Tablet Dissolution 30 mins | Tablet Dissolution 45 mins |
| 1 | 40 | 0 | 29 | 5032.1 | 24.25 | 15.75 | 48207 | 10.78 | 0.76 | 396.6 | 20 | 35.7 | 72.0 | 86.0 |
| 2 | 33 | 2 | 24 | 5113.3 | 25.50 | 7.50 | 27550 | 10.22 | 0.72 | 194.0 | 18 | 62.0 | 93.0 | 99.0 |
| 3 | 40 | 0 | 19 | 7874.0 | 24.45 | 15.55 | 75101 | 11.23 | 0.77 | 421.3 | 21 | 28.7 | 63.3 | 79.0 |
| 4 | 26 | 0 | 29 | 624.2 | 23.21 | 2.79 | 257 | 5.56 | 0.61 | 165.1 | 18 | 69.7 | 99.3 | 103.0 |
| 5 | 33 | 2 | 24 | 4582.4 | 23.67 | 9.33 | 30046 | 10.31 | 0.72 | 199.2 | 18 | 58.0 | 89.0 | 97.0 |
| 6 | 26 | 0 | 19 | 1196.5 | 24.80 | 1.20 | 84 | 4.43 | 0.63 | 168.3 | 17 | 54.7 | 94.0 | 100.7 |
| 7 | 40 | 4 | 29 | 11870.8 | 22.99 | 17.01 | 158878 | 11.98 | 0.78 | 472.8 | 14 | 26.3 | 52.7 | 64.7 |
| 8 | 26 | 4 | 19 | 2424.8 | 25.26 | 0.74 | 844 | 6.74 | 0.67 | 152.0 | 17 | 86.0 | 99.0 | 101.0 |
| 9 | 26 | 4 | 29 | 1447.3 | 25.54 | 0.46 | 478 | 6.17 | 0.66 | 176.6 | 17 | 75.0 | 97.0 | 100.0 |
| 10 | 33 | 2 | 24 | 5170.4 | 24.08 | 8.92 | 32919 | 10.40 | 0.72 | 202.7 | 16 | 57.3 | 89.7 | 97.7 |
| 11 | 40 | 4 | 19 | 14713.2 | 25.26 | 14.74 | 159812 | 11.98 | 0.79 | 566.9 | 15 | 23.7 | 48.0 | 60.3 |

**Table 2: Drug Product "D" 11 granulation batch Factorial DoE at 65-L Scale**

| Granulation Batch | Sample | Description | Water Content Added | Water Addition Time (seconds) | Granulation Time (seconds) | Impeller Load (kW) | Work (kj/kg) | SaWW (N=1) (kj/kg) | SaWW (N=3) (kj/kg) | %Granules Dissolved in 15-min |
|---|---|---|---|---|---|---|---|---|---|---|
| 1 | 1 | 25% water sample 1 | 0.2540 | 320 | 360 | 0.808174 | 25.63106 | 0.219555 | 0.000556 | 90 |
| 1 | 2 | 25% water sample 2 | 0.2540 | 320 | 540 | 0.941444 | 39.00426 | 0.552105 | 0.001525 | |
| 1 | 3 | 25% water sample 3 | 0.2540 | 320 | 660 | 0.885915 | 49.55154 | 0.805422 | 0.002264 | 86 |
| 1 | 4 | 25% water sample 4 | 0.2540 | 320 | 960 | 0.808174 | 65.86184 | 1.174112 | 0.003339 | 84 |
| 1 | 5 | 25% water sample 5 | 0.2540 | 320 | 1240 | 0.808174 | 99.23494 | 1.869198 | 0.005366 | 87 |
| 2 | 1 | 35% water sample 1 | 0.3540 | 300 | 380 | 1.178368 | 31.57435 | 1.592449 | 0.033889 | 90 |
| 2 | 2 | 35% water sample 2 | 0.3540 | 300 | 540 | 1.178368 | 48.88172 | 3.153912 | 0.070921 | 90 |
| 2 | 3 | 35% water sample 3 | 0.3540 | 300 | 660 | 1.230196 | 60.52855 | 4.215445 | 0.096096 | 86 |
| 2 | 4 | 35% water sample 4 | 0.3540 | 300 | 960 | 1.256109 | 83.33616 | 6.310204 | 0.145776 | 84 |
| 2 | 5 | 35% water sample 5 | 0.3540 | 300 | 1200 | 1.282023 | 130.5264 | 10.6802 | 0.249414 | 76 |
| 3 | 1 | 40% water sample 1 | 0.4106 | 320 | 380 | 1.337552 | 31.02979 | 2.069927 | 0.074342 | 95 |
| 3 | 2 | 40% water sample 2 | 0.4106 | 320 | 540 | 1.415293 | 50.81866 | 4.694684 | 0.186448 | 89 |
| 3 | 3 | 40% water sample 3 | 0.4106 | 320 | 660 | 1.415293 | 67.07937 | 6.882321 | 0.279885 | 78 |
| 3 | 4 | 40% water sample 4 | 0.4106 | 320 | 960 | 1.522649 | 93.87543 | 10.53338 | 0.435825 | 57 |
| 3 | 5 | 40% water sample 5 | 0.4106 | 320 | 1200 | 1.548562 | 151.172 | 18.53245 | 0.777474 | 39 |
| 4 | 1 | 45% water sample 1 | 0.4546 | 300 | 380 | 1.759573 | 36.7678 | 4.000549 | 0.226612 | 82 |
| 4 | 2 | 45% water sample 2 | 0.4546 | 300 | 540 | 1.8114 | 63.06521 | 8.894825 | 0.543823 | 52 |
| 4 | 3 | 45% water sample 3 | 0.4546 | 300 | 660 | 1.785487 | 83.94737 | 12.77302 | 0.795179 | 34 |
| 4 | 4 | 45% water sample 4 | 0.4546 | 300 | 960 | 1.759573 | 116.4073 | 18.73007 | 1.181271 | 13 |
| 4 | 5 | 45% water sample 5 | 0.4546 | 300 | 1200 | 1.574476 | 177.6654 | 29.6435 | 1.888599 | 11 |

**Table 3: Drug Product "S" Experiment at 65-L Scale, 4 granulation batches each sampled 5 times at Designated Time Points, Resulting in 20-sample DoE**

| Batch ID | Water Fraction | Granulation Work | Xsat | Xs | SaWW | Dissolution Mean at 45-min |
|---|---|---|---|---|---|---|
| 1 | 29.93 | 4268 | 20.2% | 9.7 | 322.7 | 91 |
| 2 | 31.81 | 6364 | 20.2% | 11.6 | 573.4 | 31 |
| 3 | 31.81 | 4126 | 21.9% | 9.9 | 309.1 | 89 |
| 4 | 27.74 | 2797 | 20.5% | 7.2 | 144.8 | 96 |
| 5 | 27.74 | 5588 | 20.1% | 7.6 | 301.5 | 94 |
| 6 | 31.80 | 2647 | 22.1% | 9.7 | 178.6 | 96 |
| 7 | 29.93 | 4935 | 20.7% | 9.2 | 337.3 | 88 |
| 8 | 30.14 | 2626 | 21.0% | 9.1 | 156.1 | 93 |
| 9 | 27.95 | 3612 | 20.5% | 7.5 | 190.1 | 97 |
| 10 | 30.14 | 3713 | 20.9% | 9.3 | 275.9 | 91 |
| 11 | 30.15 | 3545 | 20.8% | 9.4 | 236.9 | 91 |
| 12 | 30.15 | 3569 | 19.8% | 10.3 | 290.7 | 95 |
| 13 | 30.15 | 3587 | 21.4% | 8.7 | 236.3 | 95 |
| 14 | 30.14 | 3581 | 20.3% | 9.8 | 268.4 | 94 |
| 15 | 30.15 | 3573 | 22.0% | 8.1 | 215.7 | 92 |
| 16 | 30.15 | 3555 | 19.5% | 10.7 | 282.4 | 95 |
| 17 | 30.15 | 3545 | 21.8% | 8.4 | 216.6 | 93 |
| 18 | 30.15 | 3563 | 20.7% | 9.4 | 255.3 | 94 |
| 19 | 30.15 | 3572 | 20.8% | 9.4 | 255.4 | 96 |
| 20 | 30.15 | 3421 | 19.7% | 10.4 | 269.2 | 96 |
| 21 | 30.15 | 3611 | 20.6% | 9.5 | 278.6 | 98 |
| 22 | 30.15 | 3552 | 19.7% | 10.4 | 277.1 | 98 |
| 23 | 30.15 | 3637 | 20.1% | 10.0 | 279.0 | 99 |
| 24 | 30.15 | 3606 | 21.4% | 8.8 | 235.6 | 99 |
| 25 | 30.15 | 3567 | 21.4% | 8.8 | 241.5 | 97 |
| 26 | 30.15 | 3601 | 21.2% | 9.0 | 245.0 | 99 |
| 27 | 30.15 | 3592 | 21.0% | 9.2 | 262.6 | 98 |
| 28 | 30.14 | 3591 | 20.8% | 9.3 | 256.3 | 100 |
| 29 | 30.14 | 3553 | 19.8% | 10.4 | 292.7 | 94 |
| 30 | 30.14 | 3543 | 25.1% | 5.1 | 124.5 | 89 |
| 31 | 30.14 | 3486 | 24.7% | 5.5 | 114.3 | 83 |
| 32 | 30.15 | 3521 | 24.5% | 5.7 | 135.7 | 84 |
| 33 | 30.14 | 3554 | 22.9% | 7.3 | 162.2 | 89 |
| 34 | 30.14 | 3599 | 20.6% | 9.5 | 228.6 | 94 |
| 35 | 30.16 | 3645 | 22.3% | 7.9 | 219.3 | 93 |
| 36 | 30.15 | 3614 | 23.5% | 6.7 | 173.9 | 94 |
| 37 | 30.15 | 3618 | 21.2% | 9.0 | 259.7 | 93 |
| 38 | 30.15 | 3623 | 21.7% | 8.5 | 231.7 | 90 |
| 39 | 30.16 | 3613 | 20.9% | 9.3 | 254.0 | 91 |
| 40 | 30.15 | 3628 | 23.5% | 6.6 | 173.4 | 86 |

**Table 4: Data from Ten Experimental Granulation Batches for drug product "P" followed by 30 runs at Optimum Work and Water Setting at 300-L Scale**

| API Batch | Micronization Work (kj/kg) | API D90 (microns) | Rheometer Work (mJ/ 30grams) | Granulation Sample Number | Granulation Work (kj/kg) | Granulation Work/ Rheometer Work | Milled Granule Tap Density (g/cc) | Tablet Dissolution at 45-minutes | Tablet Dissolution RSD at 45-minutes |
|---|---|---|---|---|---|---|---|---|---|
| 1 | 180 | 6.5 | 11.0 | 1 | 27.2 | 74117 | 0.67 | 86 | 10.6 |
| 1 | 180 | 6.5 | 11.0 | 2 | 44.0 | 120030 | 0.71 | 93 | 3.0 |
| 1 | 180 | 6.5 | 11.0 | 3 | 83.5 | 227782 | 0.73 | 92 | 1.9 |
| 1 | 180 | 6.5 | 11.0 | 4 | 126.4 | 344642 | 0.75 | 85 | 1.4 |
| 2 | 60 | 13.2 | 6.8 | 1 | 22.0 | 97018 | 0.69 | 89 | 8.7 |
| 2 | 60 | 13.2 | 6.8 | 2 | 34.8 | 153436 | 0.70 | 91 | 4.1 |
| 2 | 60 | 13.2 | 6.8 | 3 | 64.6 | 284842 | 0.74 | 87 | 1.4 |
| 2 | 60 | 13.2 | 6.8 | 4 | 91.5 | 403851 | 0.78 | 52 | 25.4 |
| 3 | 280 | 5.8 | 12.2 | 1 | 28.6 | 70446 | 0.67 | 87 | 11.5 |
| 3 | 280 | 5.8 | 12.2 | 2 | 45.2 | 111125 | 0.69 | 86 | 11.0 |
| 3 | 280 | 5.8 | 12.2 | 3 | 81.9 | 201398 | 0.72 | 92 | 3.4 |
| 3 | 280 | 5.8 | 12.2 | 4 | 112.1 | 275592 | 0.76 | 91 | 3.8 |

**Table 5: Drug Product "P" Micronization and Granulation DoE at 65-L Scale Illustrating the Impacts of API raw material variation on Granulation**

| | DoE Parameters | | | | | | Observed | Calculated Parameters | | | Granule Characteristics | |
|---|---|---|---|---|---|---|---|---|---|---|---|---|
| Run Number | Throughput (kg/hr) | Screw Speed (RPM) | Kneading Elements | Angle of Kneading | Barrel Temp (ºC) | Type Binder Addition | Screw Torque (N m) | Power/Mass (joule/gram) | Water Dynamic Viscosity (kg/m s) | Power/ (Mass* Viscosity) | Yield (Percent Granules between 150-1400 micron) | Milled Granule D50 (microns) |
| 1 | 17.5 | 775 | 4 | 60 | 32.5 | Wet | 1.5 | 25.0 | 7.37E-04 | 34.0 | 70.7 | 442 |
| 2 | 25 | 600 | 2 | 90 | 25 | Wet | 2.2 | 19.9 | 8.68E-04 | 22.9 | 55 | 352 |
| 3 | 25 | 950 | 2 | 30 | 40 | Dry | 2.1 | 30.1 | 6.35E-04 | 47.3 | 64.8 | 399 |
| 4 | 10 | 600 | 12 | 30 | 40 | Dry | 4.0 | 90.5 | 6.35E-04 | 142.4 | 84.3 | 567 |
| 5 | 25 | 600 | 12 | 30 | 40 | Wet | 7.3 | 66.0 | 6.35E-04 | 103.9 | 84.7 | 583 |
| 6 | 25 | 950 | 2 | 30 | 25 | Wet | 2.4 | 34.4 | 8.68E-04 | 39.6 | 66.1 | 435 |
| 7 | 25 | 600 | 12 | 60 | 25 | Dry | 5.6 | 50.7 | 8.68E-04 | 58.3 | 70.7 | 445 |
| 8 | 25 | 950 | 12 | 30 | 25 | Dry | 6.6 | 94.5 | 8.68E-04 | 108.9 | 82.1 | 542 |
| 9 | 25 | 600 | 2 | 90 | 40 | Dry | 1.9 | 17.2 | 6.35E-04 | 27.1 | 63.4 | 405 |
| 10 | 10 | 600 | 2 | 30 | 25 | Dry | 1.3 | 29.4 | 8.68E-04 | 33.9 | 65.9 | 423 |
| 11 | 10 | 950 | 2 | 90 | 25 | Dry | 0.9 | 32.2 | 8.68E-04 | 37.1 | 62.4 | 394 |
| 12 | 25 | 950 | 6 | 90 | 40 | Wet | 4.9 | 70.2 | 6.35E-04 | 110.5 | 81.8 | 544 |
| 13 | 10 | 950 | 12 | 30 | 25 | Wet | 4.5 | 161.2 | 8.68E-04 | 185.6 | 80.5 | 536 |
| 14 | 17.5 | 775 | 4 | 60 | 32.5 | Wet | 1.7 | 28.4 | 7.37E-04 | 38.5 | 68.9 | 430 |
| 15 | 17.5 | 775 | 4 | 60 | 32.5 | Wet | 1.4 | 23.4 | 7.37E-04 | 31.7 | 66.7 | 418 |
| 16 | 10 | 950 | 2 | 30 | 40 | Wet | 1.3 | 46.6 | 6.35E-04 | 73.3 | 72.6 | 458 |
| 17 | 10 | 600 | 6 | 90 | 40 | Wet | 4.4 | 99.5 | 6.35E-04 | 156.6 | 82.8 | 542 |
| 18 | 10 | 950 | 12 | 60 | 40 | Dry | 2.8 | 100.3 | 6.35E-04 | 157.8 | 82.1 | 533 |

**Table 6: Theophylline DoE on Twin Screw Granulator at Ghent University Recalculated Using First Principle Equation**

# Model based approach to control pinacol precipitation risk in telescoped Miyaura Borylation-Suzuki coupling process


Marimuthu Andiappan, Jeremy Merritt, Shankar Vaidyaraman

Small Molecule Design and Development (SMDD)

Eli Lilly and Company, Indianapolis, USA


AICHE Annual Meeting 2015

1

## Outline

- Telescoped Miyaura Borylation-Suzuki coupling process
- Problem: Pinacol precipitation risk
- Distillation model, Empirical pinacol solubility model
- Use of Model
  - Predict & Avoid the risk of failure (Pinacol precipitation)
  - Design a robust process

2

## Miyaura borylation/Suzuki cross-coupling reaction (C-C bond formation)

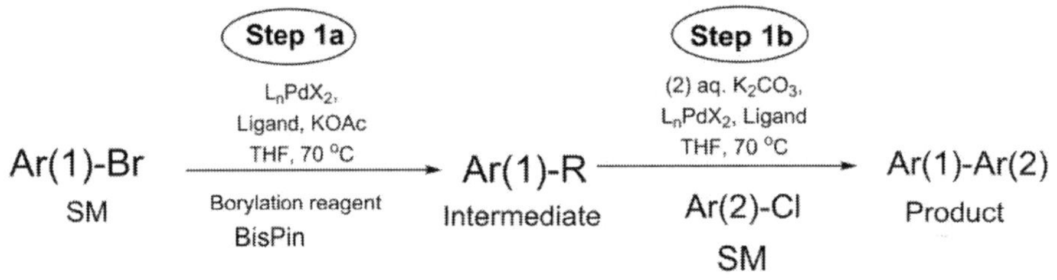


- Miyaura Borylation/Suzuki Cross-coupling reaction
  - 25% of all reactions performed in pharmaceutical industry consist of cross-coupling reactions
- Step 1a – Borylation
  - 1 equiv. Aryl bromide (LR), 1.4 equiv. borylation reagent - BisPin, 0.03 equiv. $L_nPd^{II}X_2$ catalyst and 0.03 equiv. ligand
- Step 1b – Suzuki reaction
  - 1.5 equiv. Aryl chloride and 0.02 equiv. $L_nPd^{II}X_2$ catalyst

3

## Telescoped Miyaura borylation/Suzuki cross-coupling reaction: Big picture

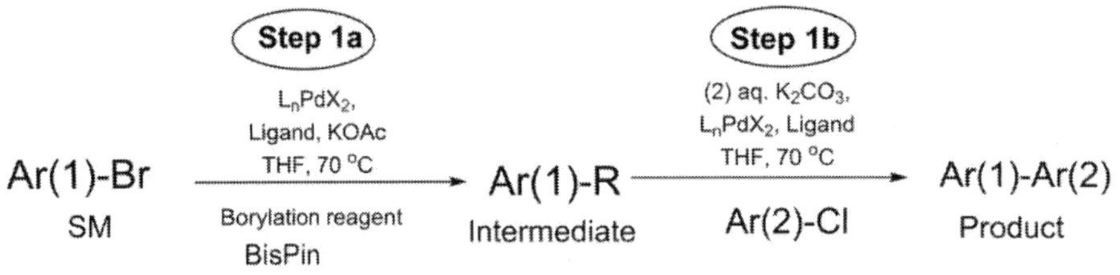


- Telescoped process (One pot, two step process): Minimize extra isolations associated with intermediate
  - Energy Efficient
  - High overall yield
  - Significant waste reduction

- E factor = Kg waste/Kg product
  - Oil refining : 0.1
  - Bulk Chemicals: 1-5
  - Fine chemicals: 5-50
  - Pharmaceuticals: 25-100

4

# Problem: Byproduct precipitation in the final isolated wetcake

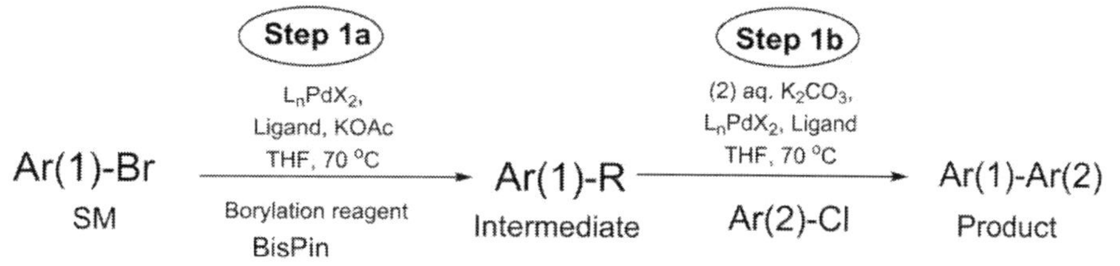


- THF process: 70 % Yield (Reactions 1a & 1b, Aqueous phase separation, Crystallization to 0°C, Filtration & Washing at 0°C, Vacuum drying at 50°C)
- Modified THF process: ~ 80% Yield (Organic contents: distilled from ~15 Vols. to ~6 Vols.)
  - Byproduct precipitation in wetcake

Vols. = Volume (litre) of mixture per Kg of limiting reagent (L/Kg)

5

# Characterization of the wetcake – NMR, XRD, DSC

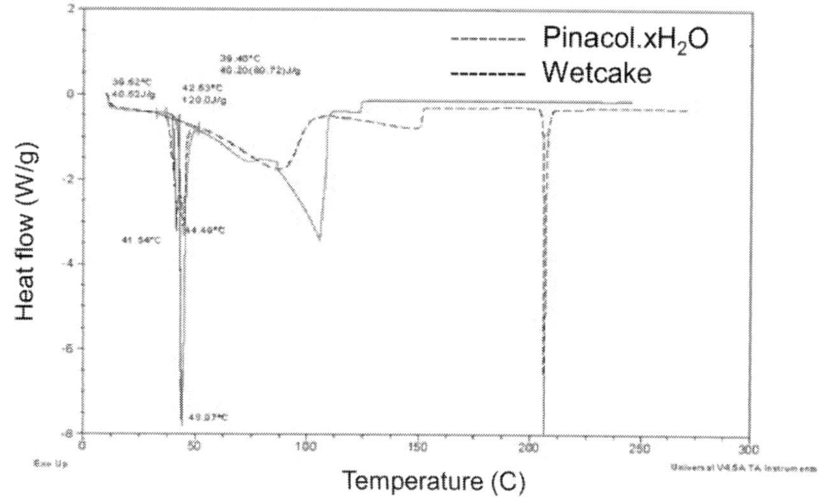


DSC

- Pinacol properties
  - Melting point: 40-43 °C
  - Different forms*: Hexahydrate, monohydrate, anhydrous

*Ref: X. Hao, S. Parkin and C. P. Brock *Acta Cryst.* (2005). **B61**, 689-699

6

# Source of Pinacol: BisPin (Borylation reagent)

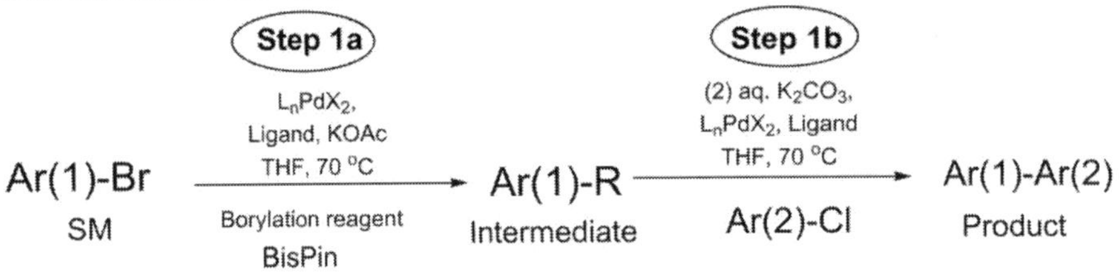


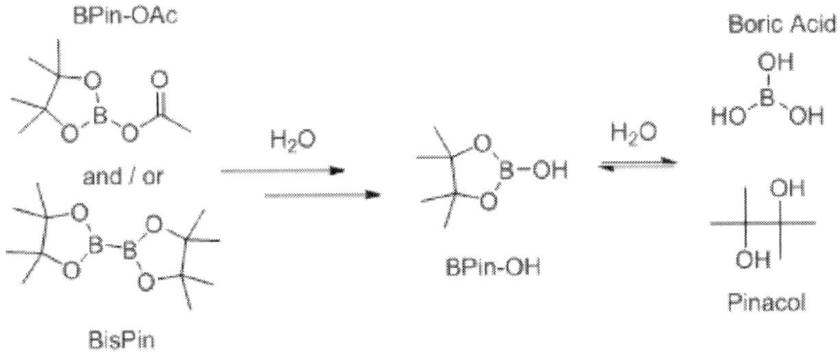


7

# BisPin to Pinacol Mass Balance: 10 g batch

| Component | Expected Pinacol Quantity (g) | Aqueous (% of mass balance) | Filtrate (% of mass balance) | Washes (% of mass balance) | Total Mass Balance |
|---|---|---|---|---|---|
| Pinacol | 12.2g (100%) | 1.78 g (14.57 %) | 9.6 g (78.63 %) | 0.39 g (3.18 %) | 11.77 g 96.38% |

- ~100% mass balance – BisPin to Pinacol

# Pinacol precipitation problem: Operational risk

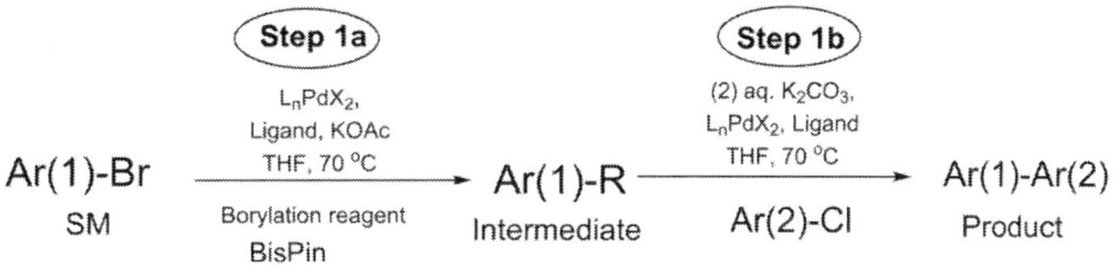


- THF process: ~ 80% Yield (with distillation, ~15 Vols. to ~6 Vols.)
  - Pinacol precipitation in wetcake (Pinacol M.pt: ~40°C, B.pt: ~170°C, Isolation temp: 0°C)
  - No pinacol in dried product: Evaporates at 50°C (vacuum)
  - Pinacol could precipitate in large scale vacuum drier or may not be able to dry – **Operational risk**

9

# Pinacol Solubility Model

- Pinacol Solubility data in THF rich organic mixture: GC, XRD, water wt% by KF measurement

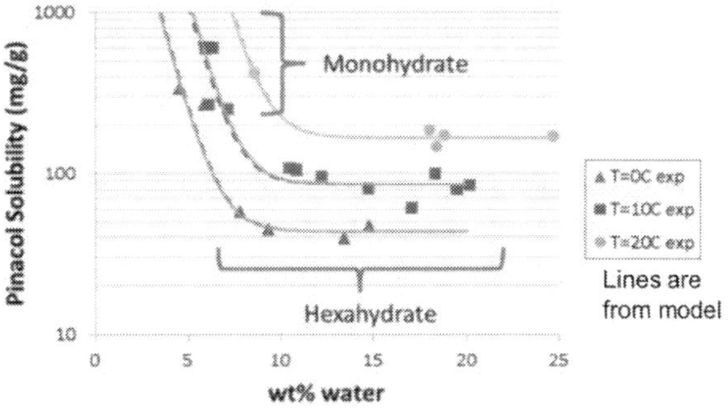


Pinacol (empirical) solubility model

$$\mathrm{Sol}\left(\frac{\mathrm{mg}}{\mathrm{g}}\right)[\mathrm{KF(wt\%)},\mathrm{T(C)}] = \mathrm{A(T)} * \exp(\mathrm{KF_c(T)} - \mathrm{KF}) + \mathrm{S_0(T)}$$

*Ref:for pinacol forms: X. Hao, S. Parkin and C. P. Brock *Acta Cryst.* (2005). B**61**, 689-699:.

10

# Distillation Model

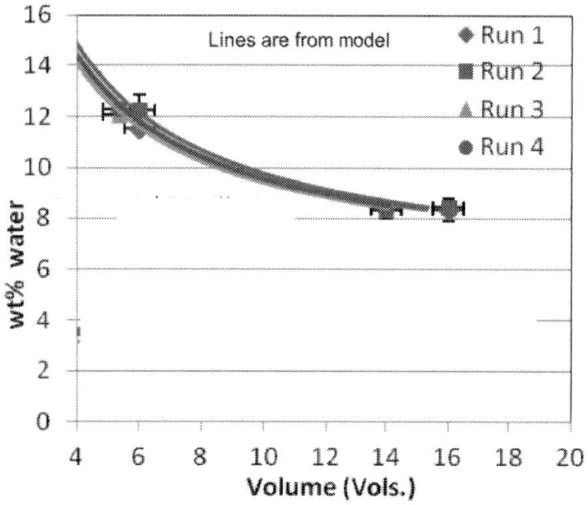


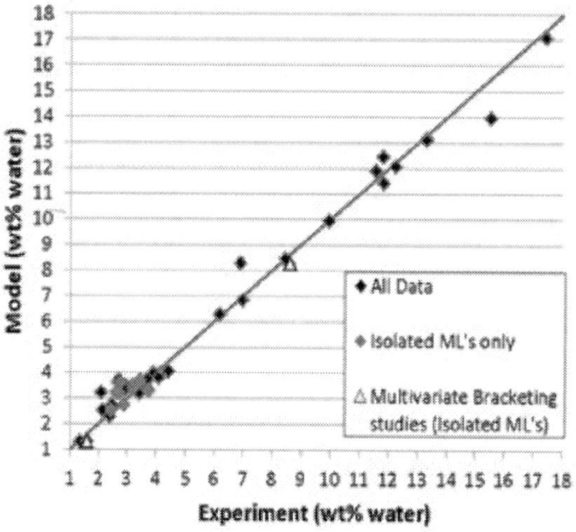


Overall fit for different set of data

- Distillation model: DynoChem tool, NRTL model

11

# Use of (solubility+distillation) model

- Understand the risk of failure (Pinacol precipitation)
- Avoid the risk of failure (Pinacol precipitation)
- Identify PAR ranges for key process parameters

12

## Why did pinacol precipitation happen when distilled? (~80% yield)

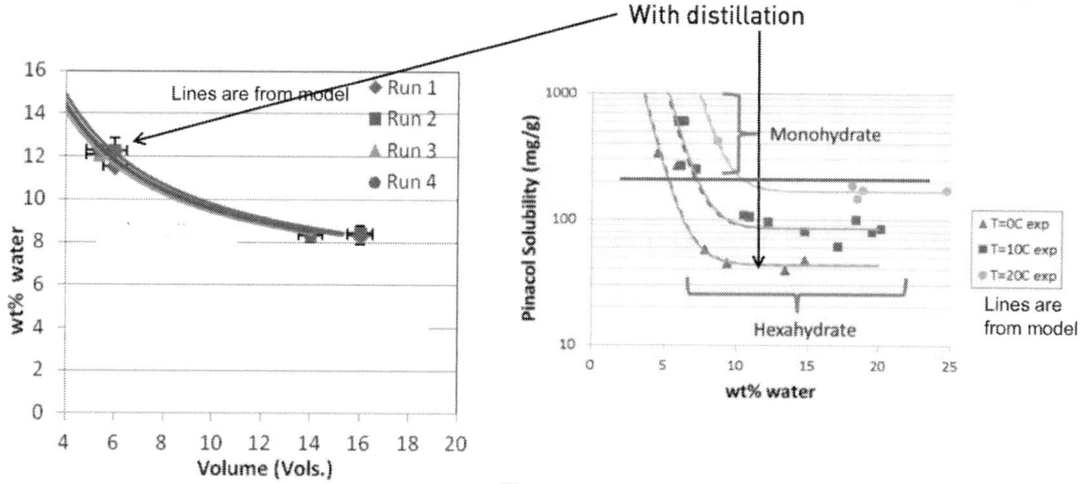


Distillation model ✚ Pinacol solubility model

- Pinacol concentration in the concentrated liquor (~6 vol., 1.4 equiv. BisPin) is ~200 mg/g (solubility is ~40 mg/g)

13

## How to achieve KF below 5% in ~6 vols. of concentrated slurry?

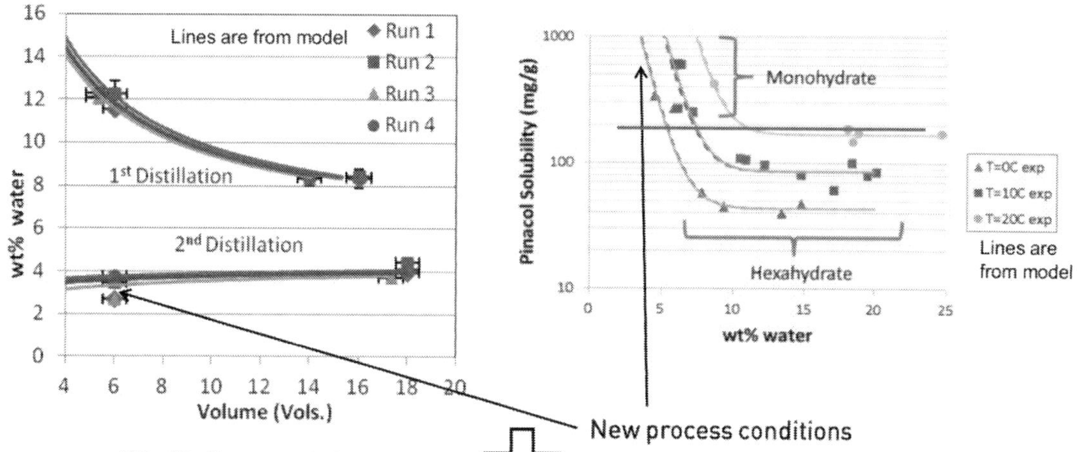


Distillation model ✚ Pinacol solubility model

- Water-THF azeotrope is ~4.9% water
- THF dilution and 2nd distillation to bring water wt% below 5 wt% in the concentrated liquor & Pinacol solubility higher than 200 mg/g
- **New process : ~80% yield and no pinacol precipitation in lab & plant scales**

14

## Use of (solubility+distillation) model

- Understand the risk of failure (Pinacol precipitation)

- Avoid the risk of failure (Pinacol precipitation)

- Identify PAR (Proven Acceptable Ranges) for key process parameters

15

## What are PAR (proven acceptable ranges) for key process parameters ?

- BisPin equiv.
  - Target= 1.2 equiv.
- 1$^{st}$ Distillation end volume
  - Target= 6 Vol.
- Fresh THF volume
  - Target= 12 Vol.
- 2$^{nd}$ distillation end volume
  - Target= 6 Vol.
- Isolation temperature
  - Target= 0 °C

16

# Variability in key process parameters

- Routine operational variability ($\sigma$) for Lilly manufacturing equipments

| Parameter | Routine variability ($\pm \sigma$) | Range Assessed for Criticality ($\pm 6\sigma$) |
|---|---|---|
| Charge Quantity (by scale); Solids | ± 1% | ± 6% |
| Charge Quantity (by mass flow meter); Solvents | ± 2% | ± 12% |
| Batch Quantity (by Radar Level Detector) | ± 3% | ± 18% |
| Temperature | ± 2 °C | ± 12 °C |

Ref: Seibert, K. et al.; *The use of routine process capability for the determination of process parameter criticality in small-molecule API*. J. Pharm Innov. 2008; 3: 105

17

# What are PAR (proven acceptable ranges) for key process parameters ?

- BisPin equiv.
  - Target= 1.2 equiv.
- 1st Distillation end volume
  - Target= 6 Vol.
- Fresh THF volume
  - Target= 12 Vol.
- 2nd distillation end volume
  - Target= 6 Vol.
- Isolation temperature
  - Target= 0 °C

18

## Pinacol (Sol.-Conc./Conc.) in % at 0°C

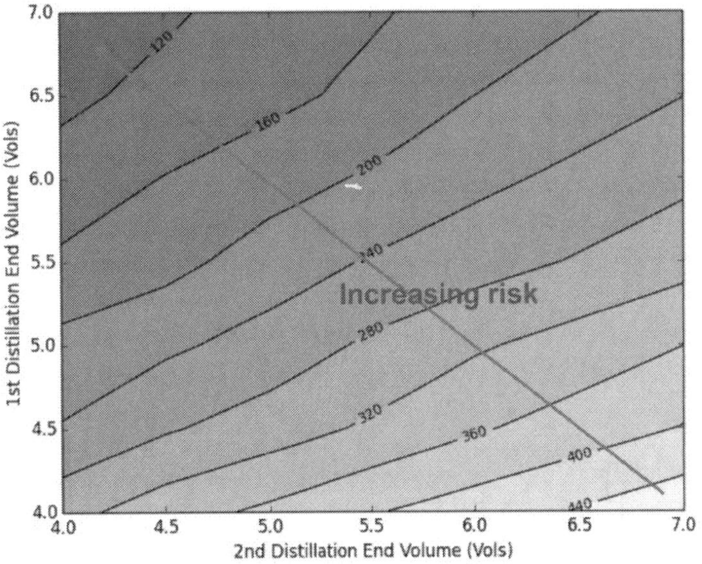


- Target
  - 1st dist. = 6 Vol.
  - 2nd dist. = 6 Vol.
- No pinacol precipitation risk at 0°C (targeted isolation temperature)

19

## Pinacol (Sol.-Conc./Conc.) in % at -6°C

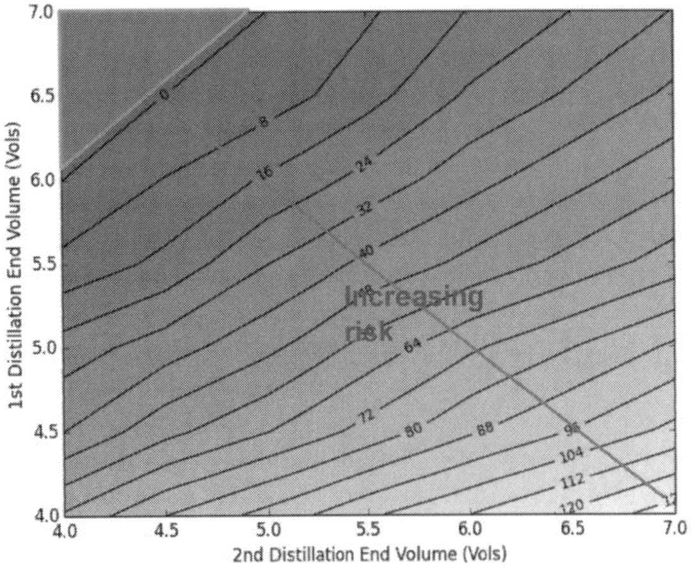


- Target
  - 1st dist. = 6 Vol.
  - 2nd dist. = 6 Vol.
- Pinacol precipitation risk at -6°C when under-distilled in 1st distillation and over-distilled in 2nd distillation

20

# PAR ranges – Univariate

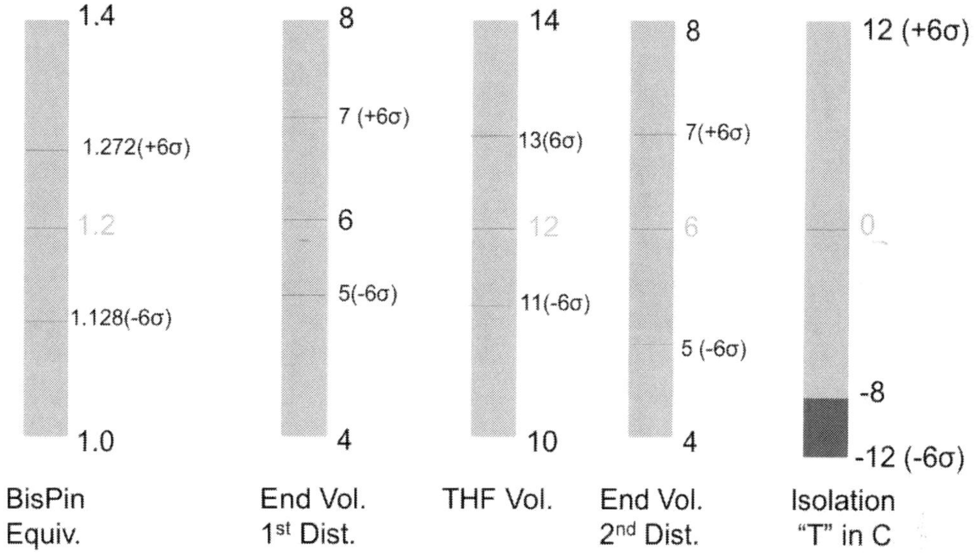


Yellow - The univariate range for a variable that results in no pinacol precipitation given that other variables are at their target

Red region - The univariate range for a variable that results in pinacol precipitation given that other variables are at their target

Ref. for visualization: Catagnoli, C. et al. Organic Process Research & Development, Vol. 14, No. 6, page 1407-1419, 2010.

21

# PAR ranges: Multivariate (Green - All combinations)

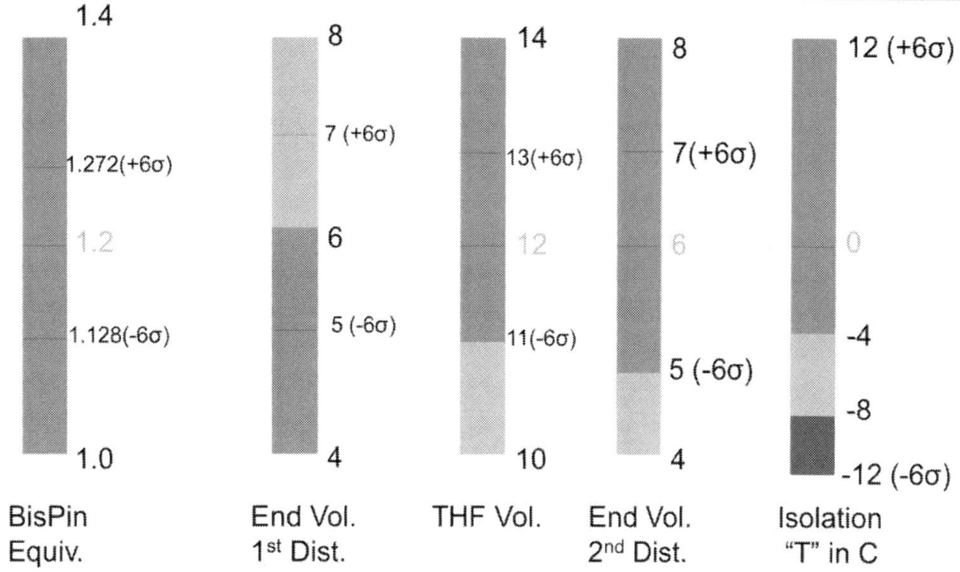


- Yellow+Green region - The univariate range for a variable that results in no pinacol precipitattion given that other variables are at their target
- Green region - Any combination of variable values in green region will result in no pinacol precipitation

22

# PAR ranges: Multivariate (Green - All combinations)

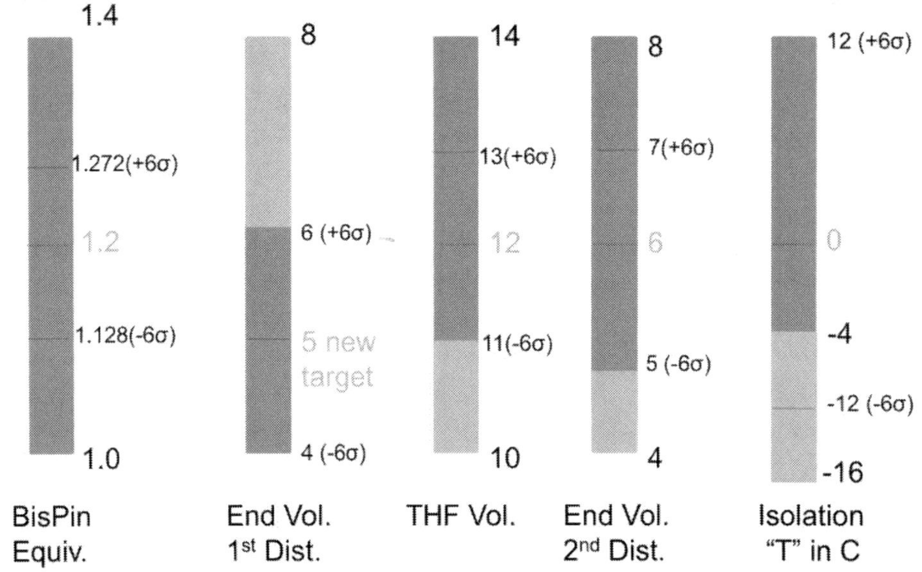


- Yellow+Green region - The univariate range for a variable that results in no pinacol precipitation given that other variables are at their target
- Green region - Any combination of variable values in green region will result in no pinacol precipitation

23

# PAR ranges – Univariate

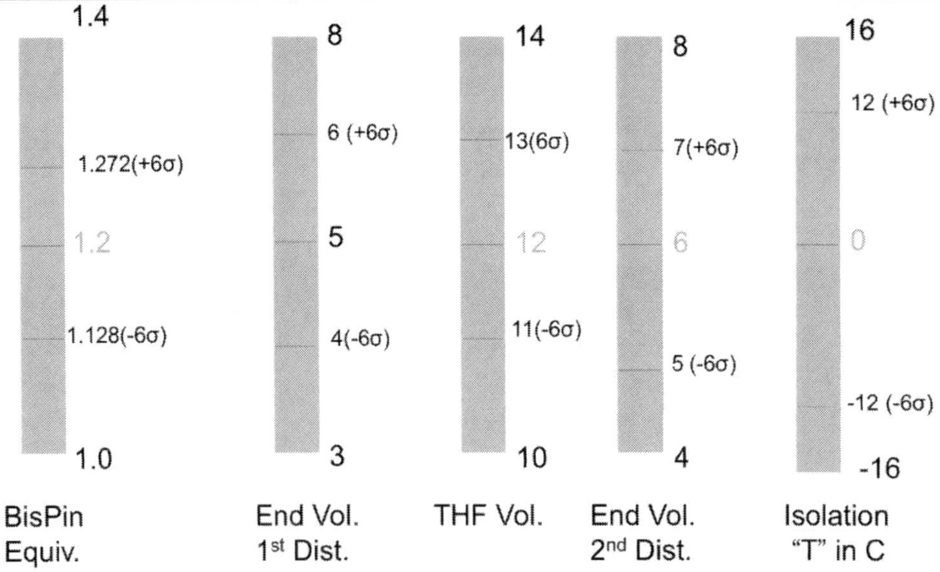


Yellow - The univariate range for a variable that results in no pinacol precipitation given that other variables are at their target

Red region - The univariate range for a variable that results in pinacol precipitation given that other variables are at their target

Ref. for visualization: Catagnoli, C. et al. Organic Process Research & Development, Vol. 14, No. 6, page 1407-1419, 2010.

24

# Design of experiment to verify model prediction

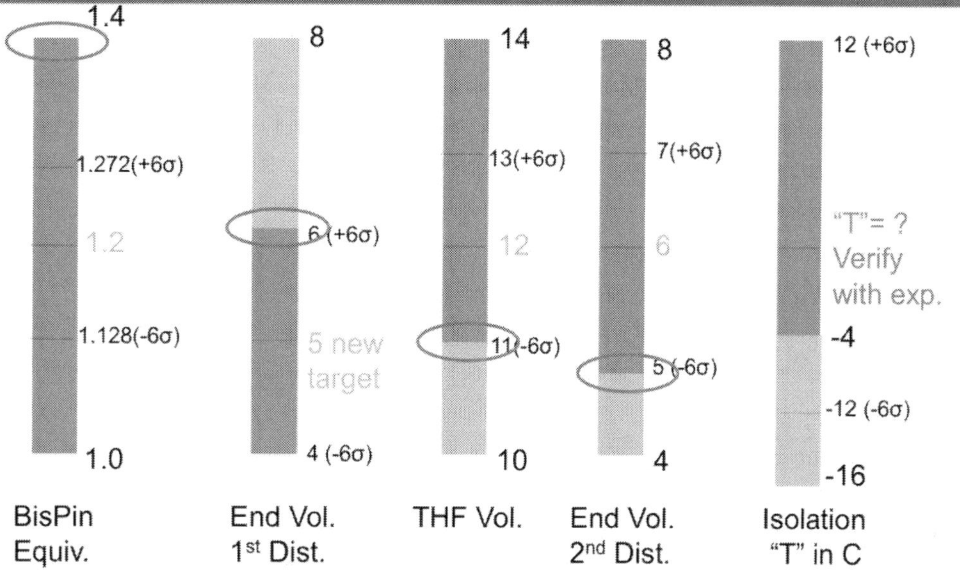


- Yellow+Green region - The univariate range for a variable that results in no pinacol precipitation given that other variables are at their target
- Green region - Any combination of variable values in green region will result in no pinacol precipitation

25

# Model vs Experiment verification

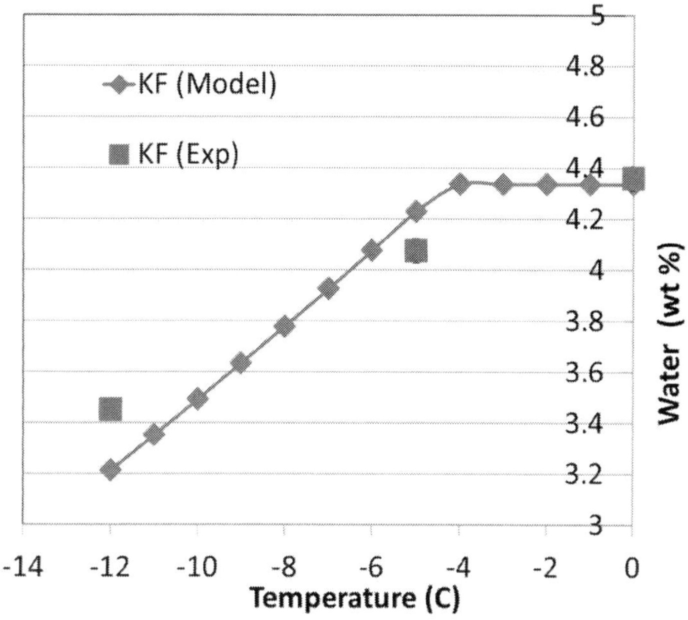


- No pinacol precipitation at 0°C
- Pinacol precipitation at -5°C and below

26

## Summary

- Distillation+Solubility models in combination with experimental data to
  - Understand pinacol precipitation in the lab scale
  - Design new process with high yield and no pinacol precipitation
  - Predict the design space (PAR ranges)
  - Predict the operational flexibility
- This approach should be applicable to other Miyaura Borylation-Suzuki couplings to predict & avoid pinacol precipitation risk

27

## Acknowledgements


- Team - Kevin Sullivan, Rachel Richey, Mark Pietz, Brad Held, Nathan Furr, Jen Kuehne-Willmore, Nick Magnus, Doug Kjell
- Stan Kolis
- Kevin Seibert
- Shekhar Viswanath
- Dave Varie


28

## Thank You

## Questions?

29

## Option 2 – Make all Parameter ranges green

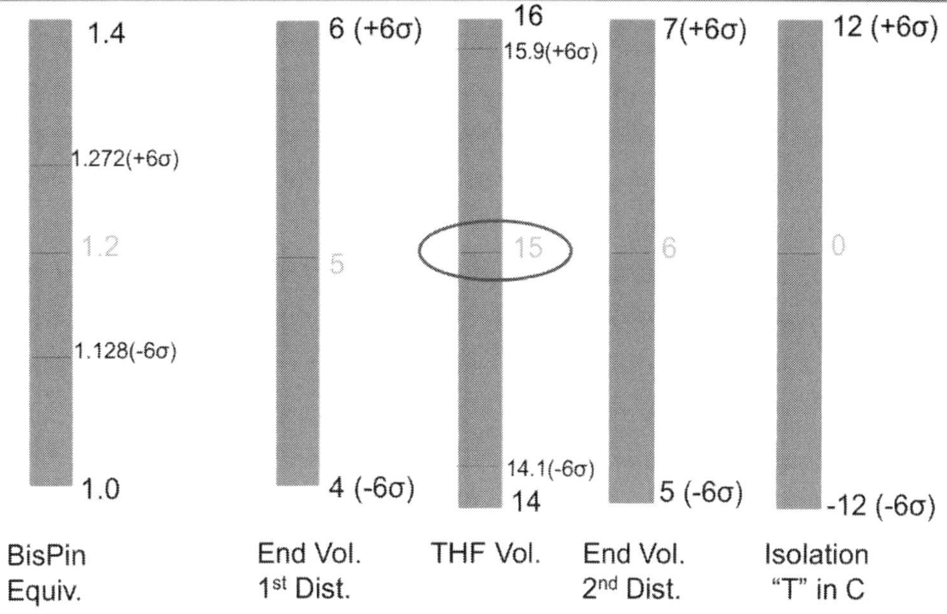


- The improvements (THF target from 12 to 15 Vol) help the yield and quality (no pinacol precipitation risk)
- There is a balance: Very low level of water leads to potency drop due to inorganics

# Corrective action in the plant

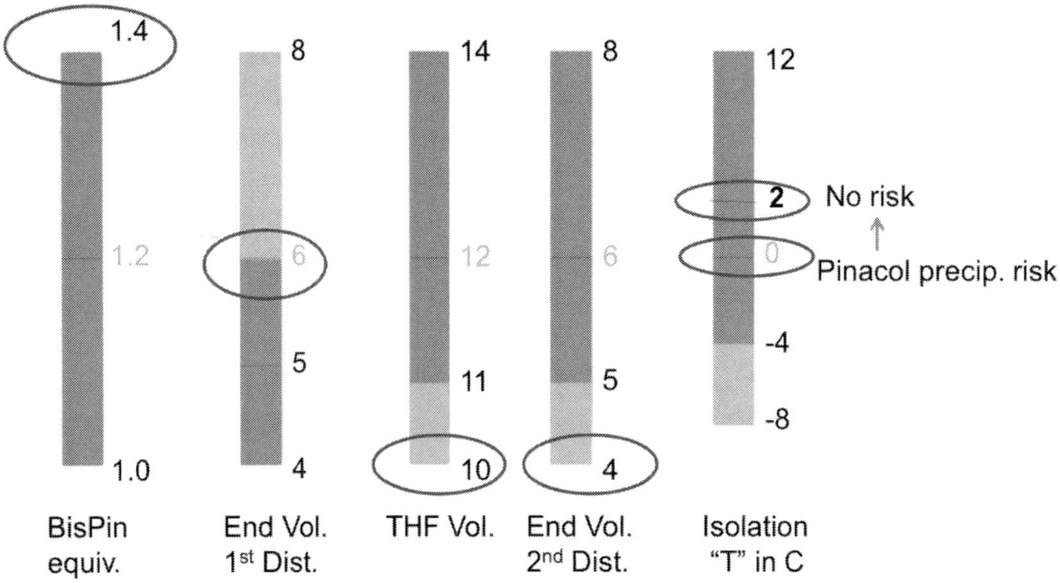


- Adjust the downstream set points when there are deviations in the upstream to eliminate pinacol precipitation in the wetcake.

31

# PAR ranges (six sigma, actual numbers)

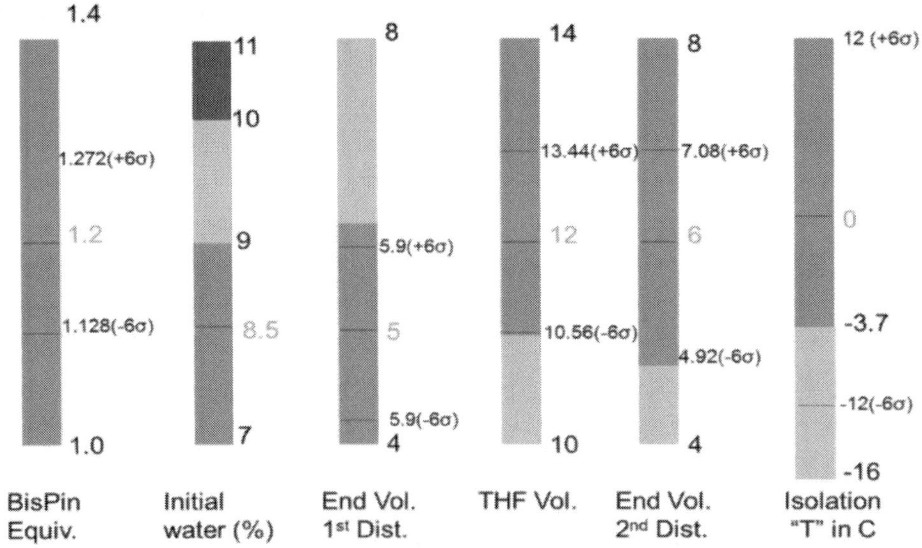


- Yellow+Green region - The univariate range for a variable that results in passing specs given that other variables are at their target
- Green region - Any combination of variable values in green region will result in passing specs

32

# 421537 Continuous Solid-Liquid Separation in Pharmaceutical Manufacturing

**Abstract:**

Continuous manufacturing processes offer many advantages for pharmaceutical production. Yet, continuous production is only slowly being adopted in the industry, main issue being a lack of ready-to-use continuous manufacturing equipment and missing process knowledge. In contrast to classical chemical production, small scale equipment would often be required for production of modern low-volume-high-value patient centred medication. In upstream production large efforts are already undertaken to close the existing technology gap [1]–[3]. Also, in secondary manufacturing many groups have formed, working on topics such as continuous wet granulation, -blending, -direct compaction or -thermal drying.[4], [5] However, at the interface between primary and secondary manufacturing, solid-liquid removal is a crucial step during pharmaceutical production.[6], [7] Nevertheless, little work has been reported regarding the development of continuous equipment for solid-liquid separation of pharmaceutical products, particularly to allow treatment of small process streams.

Commercially available continuous equipment for filtration and thermal drying was selected and thoroughly analyzed. Representative model APIs were chosen to evaluate the equipments aptitude to handle small process streams in a continuous pharmaceutical manufacturing environment.

The equipment's suitability for pharmaceutical production could be proven. Operation strategies for further enhancement of filtration performance were developed and successfully implemented. The achieved increase of permeate rates significantly sustains economic efficiency of the developed cross-flow filtration setup. To allow solvent removal below regulatory limits a spin-flash dryer was used, enabling treatment of highly viscous slurries under extremely robust process conditions within a given design space.

References:

[1] H. Zhang, R. Lakerveld, P. L. Heider, M. Tao, M. Su, C. J. Testa, A. N. Dantonio, P. I. Barton, R. D. Braatz, B. L. Trout, A. S. Myerson, K. F. Jensen, and J. M. B. Evans, "Application of continuous crystallization in an integrated continuous pharmaceutical pilot plant," *Cryst. Growth Des.*, vol. 14, pp. 2148–2157, 2014.

[2] M. O. Besenhard, A. Thurnberger, R. Hohl, E. Faulhammer, J. Rattenberger, and J. G. Khinast, "Continuous API-crystal coating via coacervation in a tubular reactor.," *Int. J. Pharm.*, vol. 475, no. 1–2, pp. 198–207, Nov. 2014.

[3] M. Sen, A. Rogers, R. Singh, A. Chaudhury, J. John, M. G. Ierapetritou, and R. Ramachandran, "Flowsheet optimization of an integrated continuous purification-processing pharmaceutical manufacturing operation," *Chem. Eng. Sci.*, vol. 102, pp. 56–66, Oct. 2013.

[4] J. Vercruysse, U. Delaet, I. Van Assche, P. Cappuyns, F. Arata, G. Caporicci, T. De Beer, J. P. Remon, and C. Vervaet, "Stability and repeatability of a continuous twin screw granulation and drying system.," *Eur. J. Pharm. Biopharm.*, vol. 85, no. 3 Pt B, pp. 1031–8, Nov. 2013.

[5] R. Singh, M. Ierapetritou, and R. Ramachandran, "An engineering study on the enhanced control and operation of continuous manufacturing of pharmaceutical tablets via roller compaction.," *Int. J. Pharm.*, vol. 438, no. 1–2, pp. 307–26, Nov. 2012.

[6] H. Zimmer and M. Drudel, "Optimized drying. Thermal processes. Mass spectrometric online gas analysis for improved process understanding, simplified maintenance, and shorter drying time," *CIT Plus*, vol. 10, no. 10, pp. 92–93, 2007.

[7] J. Burgbacher and J. Wiss, "Industrial Applications of Online Monitoring of Drying Processes of Drug Substances Using NIR," *Org. Process Res. Dev.*, vol. 12, no. 2, pp. 235–242, Feb. 2008.

# 421691 Model-Based Risk Mitigation Strategy for Late Stage Development of a Drug Substance Manufacturing Process

**Abstract:**

This work presents a strategy to identify options for the development of drug substance manufacturing processes through late stage while maximizing the use integration of computer based modeling and simulation technology and targeted laboratory experimentation. Our proposal considers the molecule specific product-associated risks and the process operational-associated risks as they systematically relate to and interact with each other across the entire GMP sequence. An input to this strategy is a new *unit operation model classification system* designed to provide estimates of modeler effort when a risk is studied and addressed. The classification system is employed to complete a process-wide risk , and a model-based development strategy is designed to mitigate (or eliminate) the multiple process risks. The output of this strategy is a set of proposed options for executing late stage development and serves as planning tool which provides quantitative predictions of the effort required for each proposed option. and the impact (measured as process risk reduction achieved) for each of the development options is predicted The various model-based development options are ultimately designed to produce the commercial-scale operating conditions and ranges as well as the necessary information package to write the CTD and successfully transfer the process to commercial manufacturing.

# 415050 Prediction of Feasible Operating Region for an Integrated Set of Unit Operations

**Abstract:**

**Title: Prediction of Feasible Operating Region for an Integrated Set of Unit Operations**

**Authors: Carla Luciani, Salvador Garcia, Kevin Seibert, Thomas Wilson, Eric Crick**

**In this work, the prediction of feasible operating regions for a set of integrated unit operations will be discussed. Particularly, three unit operations will be investigated: a Suzuki coupling reaction, a liquid-liquid extraction, and metal removal in a series of continuous adsorption columns. The prediction of the feasible operating region is aided by a mechanistic model built.**

# 425275 Technoeconomic Optimization for Green Continuous Pharmaceutical Manufacturing of Analgesics

**Abstract:**
Continuous Pharmaceutical Manufacturing (CPM) has recently emerged as a promising alternative to the currently dominant batch production methods which are expensive to develop and are characterised by low operational asset efficiency. Advances in continuous organic synthesis routes have induced demonstrations of full (end-to-end) continuous pharmaceutical production and comparative economic analyses explicitly illustrate CPM advantages (Schaber et al., 2011), thus attracting the strong and sustained interest of global pharmaceutical corporations (Jiménez-González et al., 2011) and authoritative regulatory bodies (Lee et al., 2015).

The foundation of every truly novel CPM process is the continuous synthesis chemistry, which has been demonstrated for numerous Active Pharmaceutical Ingredients (APIs), from common painkillers and blockbuster anti-cancer medications, to key therapeutic molecules against debilitating tropical diseases (Bogdan et al., 2009; Hopkin et al., 2013; Kopetzki et al., 2013). Landmark experimental studies have demonstrated viable end-to-end pilot-scale processes from raw feedstocks to final product formulations (Heider et al., 2014). Nevertheless, fully continuous separations must be fully elucidated and are critical toward maximising CPM potential at production scale (Baxendale et al., 2015). The importance of systematic process systems engineering methodologies is evident toward evaluating and demonstrating CPM benefits, both in primary/upstream (Jolliffe and Gerogiorgis, 2015) as well as in secondary/downstream (Sen et al., 2013) pharmaceutical processing.

The present paper addresses the problem of technical and economic optimisation of a conceptual upstream flowsheet for API production, with explicit consideration of reaction and separation unit operations. A nonlinear optimisation model is formulated to maximise the Net Present Value (NPV), subject to explicit mass and molar balance constraints for a reaction cascade comprising three plug flow reactors (PFRs), with explicit inclusion of revenue, fixed capital (CapEx) and operating expenditures (OpEx), plant operation lifetime and inflation rate. Chemical reaction kinetics have been determined from experimental data analysis of a previously published continuous organic synthesis pathway (Bogdan et al., 2009). Furthermore, feedstock price variation and heat transfer requirements are explicitly included in the model, to achieve a realistic estimation of operating costs.

As API recovery from the final separation affects revenue significantly, continuous final separation schemes and the impact of solvent selection on product recovery are explicitly addressed. The range of available options for continuous separation is systematically analysed via the inclusion of UNIFAC equations which correlate API solubility and product recovery: several possible green solvents (to ensure low toxicity and minimal environmental impact) and process operating conditions (temperature, solvent feed ratio) have been considered. Constraints include product purity and design sustainability (via waste limitation). Optimisation results indicate that the CPM plant design achieves profitability (positive NPV) within an acceptably short time. Product yields are in good agreement with published results, and an acceptably low environmental impact is envisaged on the basis of E-factor comparisons to current pharmaceutical practice metrics.

LITERATURE REFERENCES

1. Baxendale, I.R., Braatz, R.D., Hodnett, B.K. et al., 2015. Achieving continuous manufacturing: technologies and approaches for synthesis, workup, and isolation of drug substance. Continuous Manufacturing Symposium (May 20–21, 2014). *J. Pharm. Sci.* **104**: 781–791.
2. Bogdan, A.R., Poe, S.L., Kubis, D.C. et al., 2009. The continuous-flow synthesis of ibuprofen. *Angew. Chem. Int. Ed.* **48**: 8547–8550.
3. Heider, P.L., Born, S.C., Basak, S. et al., 2014. Development of a multi-step synthesis and workup sequence for an integrated, continuous manufacturing process of a pharmaceutical. *Org. Process Res. Dev.* **18**: 402–409.
4. Hopkin, M.D., Baxendale, I.R., Ley, S.V., 2013. An expeditious synthesis of imatinib and analogues utilising flow chemistry methods. *Org. Biomol. Chem.* **11**: 1822–1839.
5. Jiménez-González, C., Poechlauer, P., Broxterman, Q.B. et al., 2011. Key green engineering research areas for sustainable manufacturing: a perspective from pharmaceutical and fine chemicals manufacturers. *Org. Process Res. Dev.* **15**: 900–911.
6. Jolliffe, H.G., Gerogiorgis, D.I., 2015. Process modelling and simulation for continuous pharmaceutical manufacturing of ibuprofen. *Chem. Eng. Res. Des.* **97**: 175–191.
7. Kopetzki, D., Lévesque, F., Seeberger, P.H., 2013. A continuous-flow process for the synthesis of artemisinin. *Chem.-Eur. J.* **19**: 5450–5456.
8. Lee, S.L., O'Connor, T.F., Yang, X. et al., 2015. Modernizing pharmaceutical manufacturing: from batch to continuous production. *J. Pharm. Innov.* **10**: 1–9.
9. Schaber, S.D., Gerogiorgis, D.I., Ramachandran, R. et al., 2011. Economic analysis of integrated continuous and batch pharmaceutical manufacturing: a case study. *Ind. Eng. Chem. Res.* **50**: 10083–10092.
10. Sen, M., Rogers, A., Singh, R. et al., 2013. Flowsheet optimization of an integrated continuous purification-processing pharmaceutical manufacturing operation. *Chem. Eng. Sci.* **102**: 56–66.

# 436336 Enabling Simultaneous Saccharification and Fermentation in Ionic Liquid for High-Yield Conversion of Cellulose into a-Ketoglutaric Acid

**Abstract:**

Ionic liquids (ILs) are benign solvents that are highly effective for biomass pretreatment. However, their application for scale-up biorefinery is hampered by toxicity of ILs on enzymes and microorganisms. Multiple washing steps for IL separation and recovery are typically required before saccharification and fermentation. To overcome this limitation, it is very critical to develop IL-compatible enzymatic and microbial biocatalyst systems to perform the simultaneous saccharification and fermentation in IL, named SSF-IL. While enzyme mixtures were developed to remain reasonably active in up to 10% (v/v) 1-ethyl-3-methylimidazolium acetate [EMIM][OAc] for biomass saccharification, most microbes reported to date could not thrive and perform efficient fermentation at high IL concentrations, >5% ILs. Here, we harnessed the robust metabolism of a *Yarrowia sp.* that can thrive in at least 10% IL. We demonstrated that this *Yarrowia sp.* could be used a microbial platform to produce high-value chemicals, such as alpha-ketoglutaric acid, up to ~92% of the maximum theoretical yield from cellulose in SSF-IL containing 10% [EMIM][OAc].

# 415922 Conversion of Aromatic Compounds to Lipids By Engineered *Rhodococcus* Strains

**Abstract:**

Lignocellulosic biomass is a renewable feedstock that can be converted to biofuels or high-value products using a biorefinery. However, pretreatment of lignocellulosic biomass often releases toxic inhibitors for the downstream fermentation. In addition, a major component of lignocellulosic biomass, lignin, remains an untapped carbon resource due to its recalcitrance and toxic monomeric units (phenolics). Our work focuses on developing *Rhodococcus opacus* as a new chassis for conversion of phenolics to triacylglycerol (TAG), a biodiesel precursor. *R. opacus* is a promising host for bioconversion of phenolics due to its high tolerance to aromatics and its ability to grow on phenolics as a sole carbon source. In addition, *R. opacus* can accumulate TAGs up to ~80% of cell dry weight (CDW) under nitrogen-limiting conditions. To enhance its innate phenolic-degrading capacity, we applied adaptive evolution, a growth-based strain selection method, by sequentially sub-culturing cells in aromatic compounds as sole carbon sources. After >200 generations, our top adapted strains demonstrated a >50% increase in final optical density ($OD_{600}$) and >30% increase in growth rate as well as high lipid accumulation compared to the wild type strain. Whole genome sequencing, RNA-seq, and $^{13}C$-fingerprinting analysis identified possible phenolic tolerance and utilization mechanisms such as phenolic degradation pathway upregulation. We will present progress towards development of *R. opacus* as a microbial cell factory.

# 416849 Discovery of Design Strategies for Enabling Pyrolytic Sugars Tolerance and Utilization By Escherichia coli

**Abstract:**

Lignocellulosic biomass has been extensively studied and utilized to produce a variety of compounds since it has been considered as an alternative source of carbon and energy for the production of biofuels and biochemicals to meet the increasing demands of petroleum-based products. There have been a variety of methods developed to pretreatment biomass for extracting fermentable substrates, however, plenty of organic coproducts are also produced during this process and inhibit the biological utilization. Fast pyrolysis, in our project, was employed to produce sugar syrup from red oak along with aldehydes, ketones, phenolic compounds, and so on, which inhibited the growth and sugar utilization of ethanologenic *E. coli* KO11. Thus, we developed three strategies to solve this problem. The first is to characterize the impacts of the pyrolytic sugars on cell membrane in terms of membrane fluidity, membrane leakage and membrane composition to study the response of *E. coli* to the sugars. The results of membrane analysis showed that the pyrolytic sugars did cause membrane leakage and affect the membrane fluidity of *E. coli*, which demonstrate that the cell membrane damage may be one of the factors resulting in the inhibition of pyrolytic sugars. The second is to reduce the toxicity of pyrolytic sugars on *E. coli* by either removing some of the inhibitors in the pyrolytic sugars by alkali treatment or protecting cells from the sugars by the way of encapsulation using Ca alginate. Here we used sodium hydroxide and calcium hydroxide to treat the pyrolytic sugars at high temperature and pH values. The alkali-treated sugars showed less toxicity in terms of both reduced damage on cell membrane and better fermentability. Besides, the cells entrapped into the Ca alginate were still able to produce ethanol at high titer while no growth of the free cells was observed at such concentrations. And the last one is to improve the resistance of *E. coli* to the pyrolytic sugars by the use of directed evolution. After long-term directed evolution, the tolerance of *E. coli* KO11 to the pyrolytic sugars was improved and reverse engineering was followed to study the impacts of mutant genes of the tolerant strain obtained from genomic sequencing on the phenotypes of increased tolerance and reproduce it on the KO11 or other microbes.

# 433422 Acetone-Butanol-Ethanol (ABE) Production from Fermentation of Hot-Water Extracted Hemicellulose Hydrolysate of Woody Biomass

**Abstract:**

In Kraft pulping process, a large fraction of hemicellulose of wood is discharged into black liquor recovering only its heating value, thus underutilizing the carbohydrate resource. In this study, the hemicellulose portion of pulp feed was investigated as feedstock for acetone-butanol-ethanol (ABE) fermentation using Clostridium *acetobutylicum*. For this purpose, hot-water treatment was applied to Poplar (hardwood) and Southern pine (softwood) to extract the hemicellulose portion prior to the pulping process without degrading the quality of the cellulose fiber. In ABE fermentation by Clostridium *acetobutylicum*, both glucose and xylose are utilized as carbon source giving similar yields in solvent production. In this work, the culture was shown to consume up to 57 g/L of hexose or xylose, producing ABE solvents at the levels of 17-18 g/L with a total solvent yield of 30-31% (g-solvent/g-sugar). During the hemicellulose extraction process, various toxins are generated including furfural, HMF and lignin-degraded phenolic compounds. The hemicellulose hydrolysate was therefore detoxified prior to fermentation. The effectiveness of different detoxification methods; overliming, ion-exchange and activated carbon adsorption, is assessed. Another issue in the process was that hot-water treatment produces hemicellulose sugars in oligomers. It needs to be hydrolyzed to monomers before it can be metabolized by the microorganism. Thus a SSF scheme was employed incorporating Clostridium *acetobutylicum* and commercial enzymes having xylanase and mannanase activities. The process details of hot-water extraction, performance of detoxification as well as the bioconversion profiles are presented.

# 435036 The Role of Endogenous Xylose Metabolism in Yarrowia Lipolytica

**Abstract:**

The biological production of fuels and chemicals by microorganisms has garnered interest as an alternative to petroleum-derived products. Using lignocellulosic feedstocks enables renewable production without competing for food supplies. However, many industrial microorganisms do not efficiently metabolize xylose and other pentose sugars resulting from lignocellulose hydrolysis. As a result, much effort has focused on engineering microorganisms for efficient xylose utilization. The oleaginous yeast, *Yarrowia lipolytica,* is capable of accumulating lipids >90% of cell biomass, making it ideal for production of fatty acid-based products. While there are conflicting reports in the literature, our experiments show that *Y. lipolytica* is unable to utilize xylose as a sole carbon source. Here, we systematically investigated the xylose pathway in *Y. lipolytica* using synthetic biology and metabolic engineering approaches. Through our efforts, we determined that *Y. lipolytica* contains a few key enzymes required for xylose metabolism. The elucidation of these endogenous enzymes was confirmed both *in vivo* and *in vitro* via utilization of intermediate substrate sources in xylose degradation, enzymatic assays and relative qPCR. The elucidation of functional and non-functional enzymes of the xylose pathway in *Y. lipolytica* will enable the engineering of robust xylose utilization in this oleaginous yeast. Finally, we will present data toward engineering efficient xylose metabolism in *Y. lipolytica*. As an oleaginous yeast with facile genetic tools, and well defined fermentation characteristics, *Y. lipolytica* is poised to be important for processing of lignocellulosic feedstocks for sustainable fuels and chemicals.

# 436594 Inhibitory Effect of Cellobionic Acid on Cellobiohydrolase and Cellobiose Dehydrogenase

**Abstract:**

The development of microbial fermentation platforms for the production of organic acids has gained interest in the last decade due to the reliability and cost-effectiveness of such processes. In recent years, carboxylic acids have emerged as specialty acids due to their unique physiochemical properties. Cellulosic biomass, which is available at low cost and in widespread abundance, is a potential substrate for the bio-production of carboxylic acids. We have engineered*Neurospora crassa* for the production of cellobionic acid (CBA) directly from cellulose with exogenous laccase and redox mediator addition. In this system, cellulases convert cellulose to cellobiose, cellobiose dehydrogenase (CDH) oxidizes cellobiose to CBA, and CDH is continuously regenerated by laccase and a redox mediator. While inhibition of cellulases by cellobiose has been extensively studied, cellobionic acid inhibition of cellulases has not received much attention, despite the discovery that oxidative enzymes produced by cellulolytic organisms generate aldonic acids from cellulosic substrates. In this study, we investigate the inhibition of cellobiohydrolase I by cellobiose and CBA using p-nitrophenyl-β-D-lactopyranoside as a model cellulosic substrate. Additionally, CBA inhibition of CDH is evaluated. The kinetic properties and mechanism of CBA inhibition are studied and compared to that of cellobiose inhibition.

# 424720 Engineering Catalytic Biofilms for Hydrolysis of Mannan Substrates

**Abstract:**

In biotechnology industry, enzymatic hydrolysis of polysaccharides is often facilitated by isolated enzymes or whole cells harbouring specific enzymes. Isolated enzymes are often sensitive to operational conditions and the whole cell-based polysaccharide hydrolysis are confined only to batch and fed batch systems. Biofilms offer attractive catalytic systems for continuous and robust biotransformations. Biofilms consist of surface attached microorganisms that are embedded within a matrix of self-produced extracellular polymeric substances and they are the most ubiquitous and resilient form of microbial life on earth. However, because of the presence of a polymeric biofilm matrix, diffusion limitations often pose a great challenge to biofilm-mediated whole-cell catalysis. In this study, we engineered a biofilm through the expression of a mannanase Man5C on cell surface to achieve an efficient biofilm-enabled biocatalytic activity to hydrolyze the amorphous mannan substrates. Using *Shewanella oneidensis* as a model organism, in which a biofilm matrix-associated surface protein BpfA has been identified in our previous study, we displayed Man5C on the cell surface of *S. oneidensis* through genetically fusing Man5C (~36 kDa) onto the C-terminus of BpfA (~285 kDa). We demonstrated that the cells with BpfA-Man5C on their surface exhibited the characteristic Man5C activity by hydrolyzing glucomannan in both growth and resting cell conditions. Intriguingly, the Man5C activity of the BpfA-Man5C could be well retained (~80%) after seven 24-h cycles, suggesting a high recyclability of the engineered biocatalyst. We further used this engineered strain to develop biofilms in a fluidized bed biofilm reactor and investigated the factors influencing the extracellular Man5C activity exhibited by the catalytic biofilms. Our results showed that the catalytic biofilms of the BpfA-Man5C cells exhibited a significantly higher hydrolysis activity than the planktonic cultures. Further, the effect of biofilm thickness on the overall biocatalytic activity in the fluidized bed bioreactor was elucidated. In our experimental conditions, an optimal biofilm thickness was determined as 20-30 μm, exhibiting an Man5C activity of ~1.4 g mannose/(h•mg biofilm).

# 441936 Trans-Tympanic Delivery of Antibiotics to Treat Otitis Media

**Abstract:**

**Trans-tympanic delivery of antibiotics to treat otitis media**

Rong Yang, Daniel Kohane

Harvard Medical School, Children's Hospital Boston

Otitis media (OM) is a major child health burden in the United States. Twelve to 16 million physician visits per year in the United States are attributed to OM, making it the most common specifically treated childhood disease.[1, 2] Acute otitis media (AOM) has a prevalence of 90% within the first 5 years of life[3] and is the most common reason for antimicrobial prescribing in US children. Frequent recurrences of this highly prevalent disease (1/3 of all children in the U.S. having 6 or more episodes of AOM by age 7)[4, 5] are believed to be partially responsible for the ongoing increase in antibiotic resistance among pathogenic bacteria as current therapy for AOM typically consists of a 10-day course of broad spectrum oral antibiotics [6, 7]. The annual impact of antibiotic-resistant infections on the U.S. economy have ranged as high as $20 billion in excess direct health care costs, and as much as $35 billion in lost productivity from hospitalizations and sick days.[8]

Local, sustained delivery of active therapeutics directly to the middle ear for the treatment of OM could potentially minimize systemic exposure and allow for much higher middle ear drug concentrations than from systemic administration. However, the tympanic membrane (TM), while only about 100 μm thick, presents a barrier that is largely impermeable to all but the smallest lipophilic molecules due to its keratin- and lipid-rich stratum corneum[9]. We successfully designed and synthesized a non-invasive trans-tympanic OM treatment platform that produces sustained and sufficient drug flux across the TM to treat acute OM in chinchillas. A multicomponent hydrogel with reverse thermal gelation was applied in situ. Design of the molecular structure of the hydrogel was directed by two principles: 1) enhance the gelation process, particularly in the presence of small molecules that enhance the drug flux across the TM; and 2) improve adherence of the gel to TM. One simple chemical modification leads to faster gelation rate and considerably improved mechanical properties of the gel formed in situ. We chose ciprofloxacin as the antibiotic because a) it is small (331 da), moderately hydrophobic (log P = 0.28), can be dissolved at relatively high concentration in aqueous solution at acidic pH (pKa = 6.16), and has a broad antibacterial spectrum, b) it is currently used successfully clinically to treat acute otorrhea in children with tympanostomy tubes. The effects of small molecule additives on the mechanical properties of the hydrogel were investigated systematically and can be used as important reference for the design of future multicomponent polymeric systems. As a proof-of-concept, as designed multicomponent formulation was applied in single application to chinchillas infected with non-typable *Haemophilus influenzae* (NTHi) and was sufficient to provide high concentrations of antibiotics localized to the middle ear, resulting in eradication of bacterial OM without the systemic exposure of oral therapy. Additionally, application of the formulation can alleviate the formation of biofilm, which has been implicated in the pathogenesis and persistence of OM.

**Reference**

(1) Berman, S. Otitis media in children. *N Engl J Med* **1995**, *332*, 1560-5.

(2) Fried, V. M.; Makuc, D. M. Rooks, R. N. *Ambulatory health care visits by children: principal diagnosis and place of visit.*; 137; Washington, D.C.: Government Printing Office, 1998.: 1998.

(3) Teele, D. W.; Klein, J. O. Rosner, B. Epidemiology of otitis media during the first seven years of life in children in greater Boston: a prospective, cohort study. *J Infect Dis* **1989**, *160*, 83-94.

(4) Faden, H.; Duffy, L. Boeve, M. Otitis media: back to basics. *Pediatr Infect Dis J* **1998**, *17*, 1105-12; quiz 1112-3.

(5) Lanphear, B. P.; Byrd, R. S.; Auinger, P. Hall, C. B. Increasing prevalence of recurrent otitis media among children in the United States. *Pediatrics* **1997**, *99*, E1.

(6) Bluestone, C. D. Klein, J. O., in *Otitis media in infants and children*, BC Decker, Hamilton, Ontario, Canada **2006**.

(7) Bluestone, C. D. Klein, J. O., in *Otitis media in infants and children*, BC Decker, Hamilton, ON **2007**.

(8) Antibiotic / Antimicrobial Resistance. http://www.cdc.gov/drugresistance/

(9) Doyle, W. J.; Alper, C. M.; Seroky, J. T. Karnavas, W. J. Exchange rates of gases across the tympanic membrane in rhesus monkeys. *Acta Otolaryngol* **1998**, *118*, 567-73.

# 431159 Controlled Trigger and Drug Release from a Liposomal Nanocarrier Modulated with NIR Light

**Abstract:**

Liposomes have been widely assessed as drug nanocarriers, however they face the inherent contradiction of minimizing non-specific drug release and initiating fast release at the targeted site. Encapsulated or tethered hollow gold nanoparticles (HGN) photo-activated with near infrared (NIR) light can be used as a novel mechanism to trigger and control the nanocarrier content release. NIR light is an appropriate triggering source as it can penetrate several centimeters with minimal thermal injury to normal tissues. HGN strongly absorb NIR light in the range of 700-900 nm, depending on their diameter-to-shell thickness ratio. Nanoparticles that exhibit strong absorption of NIR light, offer the opportunity to convert optical energy into thermal energy. The photo-thermal heating of the HGN, generated by short laser pulses, results in large temperature gradients that lead to the formation of transient vapor nanobubbles in aqueous solutions. The collapse of the nanobubbles induces the rupture of cell, endosome, or liposome membranes and the subsequent drug release with minimal damage to the surroundings. We systematically synthesized different sizes and shapes of HGN to absorb NIR light, these nanoparticles were irradiated with a pico-second pulsed laser. We show that we can selectively photoactivate HGN at specific wavelengths in the NIR region to remotely trigger the content release of a DPPC liposome loaded with an anticancer drug, cisplatin, in order to deliver the drug to an in vitro system of prostate cancer cells. Near-complete cell killing was observed after triggering the release. This study builds a better understanding of the novel mechanisms of controlled drug release with potential applications for in vivo models.

# 433835 Dendrimer Based Systemic Therapies for the Treatment of Glioblastoma

**Abstract:**

Malignant glioma is the most common and most aggressive primary brain tumor. Despite the advances in treatment, the median survival remains at 16.4 months. Recent advances in nanotechnology have offered multiple novel platforms for targeted, sustained and controlled delivery of therapeutics in order to overcome the limitations of traditional small molecule drugs. However, the small 'cut off size' of the brain tumor microvasculature pores, the relatively large and heterogeneous intervascular distances in combination with the tortuous and dense extracellular matrix, impose serious limitations in the ability of nanoparticles to reach target cells. Hydroxyl-terminated poly(amidoamine) (PAMAM) dendrimers, with their small size, near-neutral surface charge, and safety profile may offer new opportunities to address these challenges. In addition, dendrimers have demonstrated promising results in targeting neuroinflammation, promising their potential to be taken-up by tumor associated macrophage (TAM). In this study we show systemically delivered hydroxyl terminated PAMAM dendrimer uniformly and selectively distributed through the entire solid tumor and peritumoral area 15 min after injection with subsequent co-localization and retention in TAM even at 48 hours post injection. The rapid clearance of systemically administered dendrimers from major organs promises minimal off-target adverse effects of conjugated drugs. This comprehensive study of the pharmacokinetics and biodistribution of dendrimers in a rodent gliosarcoma model provides crucial information for the design and engineering of dendrimer-drug conjugates for effective treatment of glioblastoma.

# 434789 Investigating the Effects of Polyethylene Glycol (PEG), and Poly-L-Lysine (PLL) Molecular Weights and Grafting Ratios on the Physicochemical Properties and the Transduction Efficiency of Peg-PLL/Fiberless Adenovirus Vectors Targeting Prostate Cancer Cells


**Abstract:**

Viruses are considered highly efficient gene delivery vectors. Adenovirus, in particular, is a relatively safe virus with neither severe pathogenicity nor oncogenecity. Adenovirus has been extensively used as a gene delivery vector. Usefulness of the viral vector, however, is limited by drawbacks such as immunogenicity, promiscuous tropism, and inefficient infectivity of certain types of cells. Many efforts have been made to develop non-viral gene delivery vectors that are safer, less immunogenic, and targeted, but these vectors typically suffer from poor efficiency.

Our goal was to develop a hybrid vector that provides advantages over both viral and non-viral vectors. For this purpose, we developed a vector formed of a fiberless adenovirus (Ad-FL), polyethylene glycol (PEG), poly-L-lysine (PLL), and an RNA aptamer that served as a targeting ligand for prostate cancer cells expressing prostate-specific membrane antigen (PSMA). The virus fiber, which is known to play a role in both virus tropism and initiating an inflammatory response and virus clearance, was replaced with the PEG-PLL grafted copolymer. This approach reduced the drawbacks associated with the fiber and provided advantages often associated with PEG, such as reduced immunogenicity and improved biodistribution. In addition, incorporating the PSMA targeting aptamer allowed for targeting prostate cancer cells expressing PSMA.

The PEG-PLL/Ad-FL hybrid vector was prepared by conjugating different molecular weights (1, 2, 5 and 10 kDa) of heterobifunctional PEG with a thiol-reactive maleimide (MAL) group and an amine-reactive N-hydroxyl succinimidyl ester (NHS) group (MAL-PEG-NHS), with different molecular weights of PLL (15-30 and 30-70 kDa) at 10% and 25% degrees of PEGylation. The MAL-PEG-PLL grafted copolymer was then complexed with Ad-FL before finally conjugating a thio-modified RNA targeting aptamer. The targeting efficiency and overall performance of the hybrid vector was evaluated by targeting the PSMA-positive LNCaP prostate cancer cell line and the PSMA-negative PC3 prostate cancer cell line that served as a negative control. The effects of PEG and PLL molecular weights and degree of PEGylation on hybrid vector performance were evaluated and compared to the native adenovirus. The nanoparticle size and surface charge were characterized using dynamic light scattering, and the transduction efficiency was determined using a beta-galactosidase reporter protein measured using a chemiluminescence-based assay.

The results showed that the PEG-PLL/Ad-FL hybrid gene delivery vector was able to overcome some of the significant drawbacks of adenoviral vectors and simultaneously retain high transduction efficiency and impart flexible targeting.

# 434868 Protein Encapsulation for Delivery to Erythrocytes

**Abstract:**

Therapeutic proteins are promising candidates for the treatment of a wide variety of diseases. Although proteins have great potential as therapeutics, they are often limited by short circulation half-lives. One approach to overcome this challenge is to encapsulate and target a protein to long-circulating erythrocytes. Toward this goal, a grafted copolymer, composed of poly-L-lysine (PLL) and polyethylene glycol (PEG), was used to encapsulate a model protein, bovine serum albumin (BSA), which was then targeted to erythrocytes using an ERY 1 peptide ligand.

The aim of the present study was to identify conditions for best encapsulating the protein and to target the resulting nanoparticles to erythrocytes. A library of nine grafted copolymers was synthesized using 4-15 kDa, 15-30 kDa, or 30-70 kDa PLL and PEG:PLL grafting ratios of 2, 10, or 20. BSA was encapsulated with the copolymers through electrostatic self-assembly of the protein and the grafted copolymer. The resulting nanoparticles were stabilized by cross-linking PLL-g-PEG with glutaraldehyde, and the nanoparticles were functionalized with the ERY 1 peptide for targeting to erythrocytes.

SDS-PAGE gel retardation assays confirmed BSA could be efficiently encapsulated. For several copolymers, nearly complete encapsulation of the protein was achieved, and encapsulation of BSA into nanoparticles was found to be strongly dependent on the molecular weight of the PLL backbone and the copolymer to protein mass ratio. The formation of nanoparticles was observed using SEM, TEM, and dynamic light scattering. The nanoparticles had a core/shell structure and their size ranged from approximately 40-100 nm. The BSA retained catalytic activity toward p-nitrophenyl ester, despite encapsulation and exposure to glutaraldehyde. Additionally, the nanoparticles were found to be stable against polyanions and protease degradation. Confocal microscopy and flow cytometry demonstrated that the ERY 1 peptide enabled enhanced binding of the nanoparticles to erythrocytes under both static conditions and within a physiological flow loop circulating whole blood. Under conditions meant to simulate the circulatory system, approximately 15% of erythrocytes had detectable levels of nanoparticles after 4 hours in circulation. The results of this study demonstrated that therapeutic proteins can be encapsulated within polymer nanoparticles in an active state and delivered to red blood cells circulating in a physiological flow loop.

# 426004 Adsorption By Encapsulation and Controlled Release of 5-Fluorouracil on Mesoporous F300 MOF

**Abstract:**

Metal-Organic Frameworks (MOFs) are promising functional materials for encapsulation of medicinal drugs by adsorption and the subsequent controlled drug delivery. It is important to learn about molecular-level detail of chemical bonding between biologically active adsorbates and major structural units in the MOF encapsulation sorbents. 5-fluorouracil (5-FU) is a major anti-cancer drug that features a quick metabolism and hence needs to be encapsulated for a controlled drug delivery. We report mechanistic studies of encapsulation of 5-FU by adsorption on non-toxic F300 MOF, chemical bonding between 5-FU and adsorption sites in F300, and kinetics of controlled drug release to bodily fluid. Adsorption of 5-FU from non-toxic solvent DMSO yields an adsorption complex with F300, with encapsulated amount of 5-FU up to 45 wt. %. To learn about adsorbed state of 5-FU, we have used, for the first time, the front face 3-dimensional photoluminescence spectra at 77 K, the FF-Cryo-3D PL spectra. The FF-Cryo-3D PL spectra of F300 MOF at $\lambda_{exc}$>390 nm show vibronically resolved emission from the BTC linker connected to the Fe(III) coordinatively unsaturated site (CUS). Upon encapsulation of 5-FU, vibronically resolved FF-Cryo-3D PL spectra of F300 undergo a drastic change with quenching of vibrations of the BTC linker, due to the formation of molecular complex with 5-FU. These findings were complemented by the near-UV-Visible and Near InfraRed Diffuse Reflectance Spectra (UV-Vis-NIR DRS), ATR-FTIR, Raman spectra and DFT calculations. Release of encapsulated 5-FU to simulated intestinal fluid (SIF) at 37 °C was studied by the *in-situ* FF-PL spectroscopy in suspension; controlled drug release occurs on time scale of >24 h. Non-toxic mesoporous MOF F300 is a promising novel encapsulation platform for the synthesis of the new biologically active composite functional materials and for targeted drug delivery of anti-cancer drugs and other small molecule heterocyclic drugs. Acknowledgments: A.S. thanks Rutgers University for his Research Council Award.

# 433878 Free Surface Electrospinning of Microemulsions Containing Vitamin E

**Abstract:**

An estimated 90% of active pharmaceutical ingredients (APIs) in research and development are poorly soluble or insoluble in water. These APIs exhibit poor bioavailability in solid dosage forms as a result of their poor solubility. In order to increase the release rate of APIs, we consider free surface electrospinning of a microemulsion as a means to produce submicron size domains of API dispersed in an amorphous excipient. Microemulsions containing vitamin E, a poorly soluble API, Kolliphor EL, a surfactant, and Polyvinylpyrrolidone, an excipient, are electrospun to produce highly porous nanofibers with high specific surface area, which promotes rapid drug release. Turbidity measurements were taken as a means to standardize the preparation process and to evaluate the stability of produced microemulsions. In free surface electrospinning, multiple jets are formed from a free liquid surface, resulting in higher productivities over conventional needle-base electrospinning. This method gives the opportunity to alter the quality of fibers. The working distance, applied voltage and spindle rotation were examined. The produced electrospun nanofibers were characterized by scanning electron microscopy. High performance liquid chromatography (HPLC) was utilized to evaluate the dissolution rate of the final material. The fibers contained significant amounts of Vitamin E encapsulated within the excipient, and release rates were significantly improved over compared thin films. This technique may be utilized to improve the downstream production of pharmaceuticals, which would result in lower operating costs and improved uniformity over current batch manufacturing processes.

# 432090 Novel Hydrogel Mediated Generation of Endothelialized Islet Organoids from Hpsc Derived Pancreatic Progenitor Cells

**Abstract:**

Pluripotent stem cells are unique in their abilities for self renewal and capacity to differentiate into organ-specific mature cell types, rendering them attractive for the treatment of degenerative diseases. Currently, human pluripotent stem cell (hPSC) based regenerative therapies are transitioning from two dimensional (2D) adherent differentiation towards three dimensional (3D) platforms which more adequately resemble native cellular environments. Such 3D aggregation has the potential to create multicellular 'organoids' relevant for studying advanced organ development, disease physiology, and drug testing. Specifically, over the last few years, hPSCs have been used to generate gut, retinal, and liver organoids[1]. A majority of these current methods employed for cellular aggregation rely on external pressures or the confinement of cells within three dimensional artificial scaffolds to enforce aggregation. These approaches however differ from spontaneously organizing structures formed by the inherent forces and interactions of only the integrated cell population. In this regard, we have developed a novel hydrogel platform which not only promotes spontaneous aggregation of hESC derived pancreatic progenitor cells (hESC-PPs) into robust, mechanically stable, controlled spheroids but also render them free from embedding or forced cellular interactions.

Pre-differentiated hESC-PP cells seeded on the pre-formed hydrogel spontaneously aggregated within 14 hours producing one aggregate per well of a 96 well plate. These hESC-PP aggregates were tunable for size and cellular composition, robust enough for mechanical manipulation, and viable over extended culture in situ. Additionally, this system allowed for the efficient application of functional assays on individual aggregates, enabling high throughput analysis in 3D. Self-organized hESC-PP aggregates exhibited significant enhancement in relevant PP gene expression, along with two-fold increase in overall cell population positive for Pdx-1, a key PP protein marker. Furthermore, a significant portion of the Pdx-1 population co-expressed NKX6.1 protein marker, an early marker for potential endocrine maturation. Moreover, self-organization also improved the efficiency of chemically driven hESC-PP maturation towards beta-like cells compared to conventional cultures. An endothelial presence during in vivo islet development is crucial for extra cellular matrix and microvascular development[2]. Therefore co-seeding hESC-PPs with endothelial cells on the hydrogel resulted in self-organized multicellular pancreatic 'organoids' to support pancreatic development. Interestingly, the endothelialization of the resulting organoid also induced spontaneous maturation into beta-like cells in the absence of specific chemical induction. In conclusion, our hydrogel platform has direct application not only to improve pancreatic tissue engineering, but also has the flexibility to be further adapted to multiple cell types and other tissue engineering applications.

1. Lancaster, M.A. and J.A. Knoblich, *Organogenesis in a dish: modeling development and disease using organoid technologies*. Science, 2014. **345**(6194): p. 1247125.

2. Stendahl, J.C., D.B. Kaufman, and S.I. Stupp, *Extracellular matrix in pancreatic islets: relevance to scaffold design and transplantation*. Cell Transplant, 2009. **18**(1): p. 1-12.

# NANOG Restores the Impaired Extracellular Matrix Synthesis and Mechanical Strength in Senescent Stem Cells.

Na Rong, Panagiotis Mistriotis, Xiaoyan Wang, Georgios Tseropoulos, Stelios T. Andreadis

**Department of Chemical & Biological Engineering, University at Buffalo**

University at Buffalo
The State University of New York

## The importance of Collagens in human body

Collagens:

- 1/3 of the total protein in humans
- Exhibit a broad range of functions
  - ➢ Tissue scaffolding
  - ➢ Cell adhesion
  - ➢ Cell migration
  - ➢ Angiogenesis
  - ➢ Tissue morphogenesis
  - ➢ Tissue repair

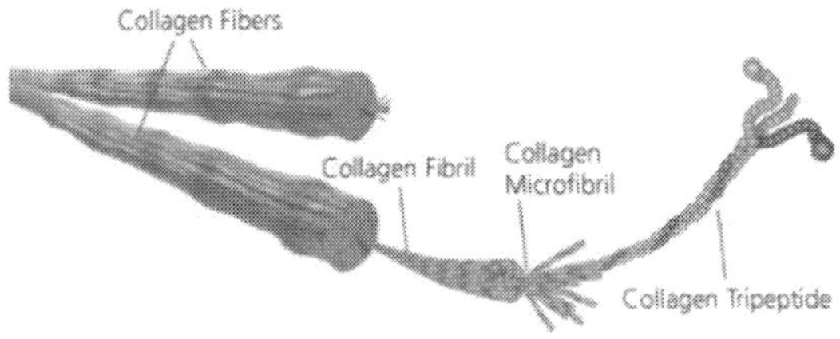


Image taken from :sigmaaldrich.com

# Types of Collagen

- 28 collagens that contain at least one triple-helical domain.
- Collagen type Ⅰ, Ⅱ, Ⅲ constitute about 90% of total collagen.

| Type | Constituent | Distribution | Pathology |
|---|---|---|---|
| Ⅰ | α1 ( Ⅰ ), α2 ( Ⅰ ) | Most connective tissues, especially in dermis, bone, tendon, ligament | OI, EDS(Ehlers-Danlos Syndrome), Osteoporosis |
| Ⅱ | α1 ( Ⅱ ) | Cartilage, vitreous humour, cornea | Osteoarthrosis, Chondrodysplasias |
| Ⅲ | α1 (Ⅲ) | As collagen Ⅰ expect absent in bone and tendon | EDS (vascular type), Arterial aneurysms |

Shoulders and Raines, 2009, Annu. Rev. Biochem

# Aging decreases Collagen expression

**Skin**

Type Ⅰ, Ⅲ collagen content in normal skin

| Age Group | Type Ⅰ collagen (ug/g) | Type Ⅲ collagen (ug/g) | Type Ⅰ/Ⅲ |
|---|---|---|---|
| Fetus | 264.71±5.88 | 278.87±6.18 | 0.95±0.03 |
| Adolescent | 279.12±7.65 | 123.27±5.30 | 2.27±0.13 |
| Adult | 241.79±8.23 | 98.41±5.58 | 2.46±0.15 |
| Elderly | 209.50±14.31 | 71.30±7.41 | 2.97±0.40 |

Cheng et al., 2011, African Journal of Biotecnology
Image taken from wisegeek

Young Skin Elderly Skin

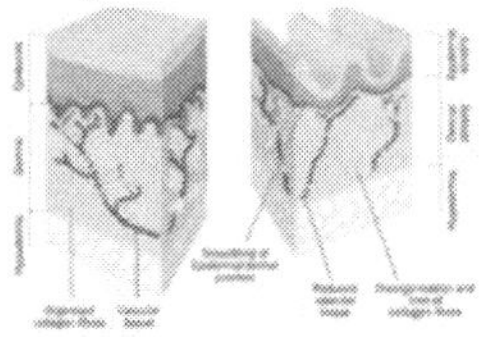

# Aging affects the mechanical properties of engineered tissues

**Isolate Mesenchymal Stem Cells:**
- **3 neonatal sheep** (<3 days old)
- **3 adult sheep** (4–4.5 years old)

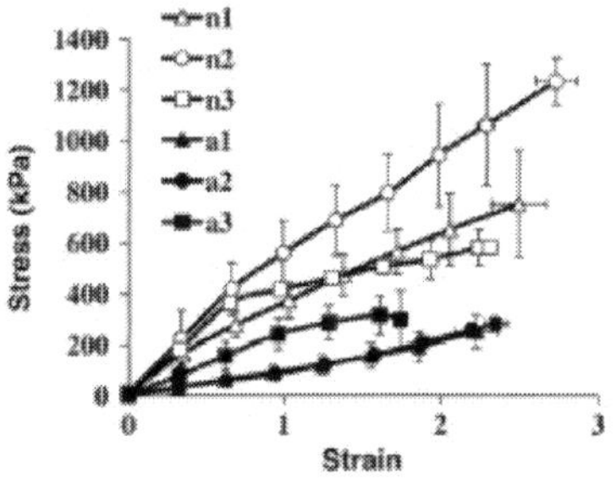


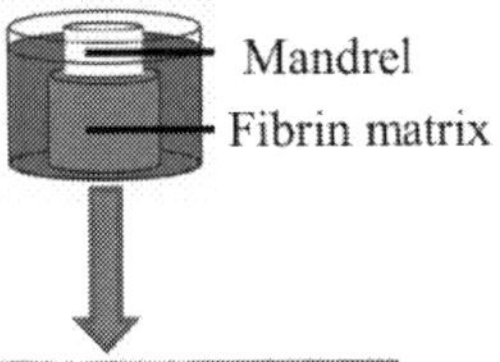


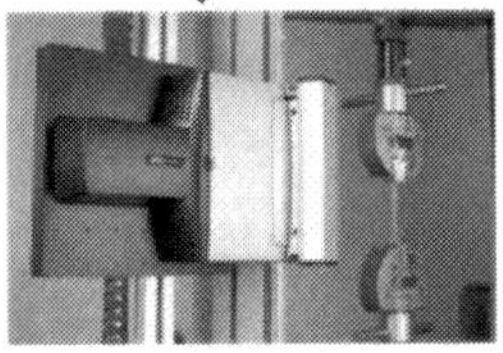

Mechanical testing
- **UTS**
- **Young's Modulus**

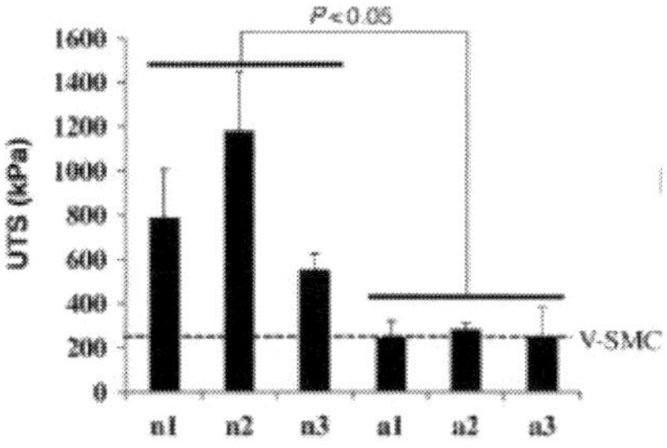


Han et al., 2010, Cadiovasc. Res.

# Goal of the work

## CAN WE RESTORE COLLAGEN EXPRESSION IN AGED CELLS???

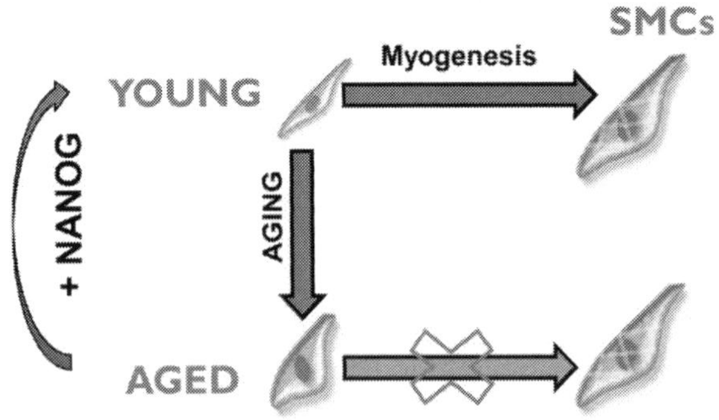

# Systems to study the effect of NANOG on aging cells

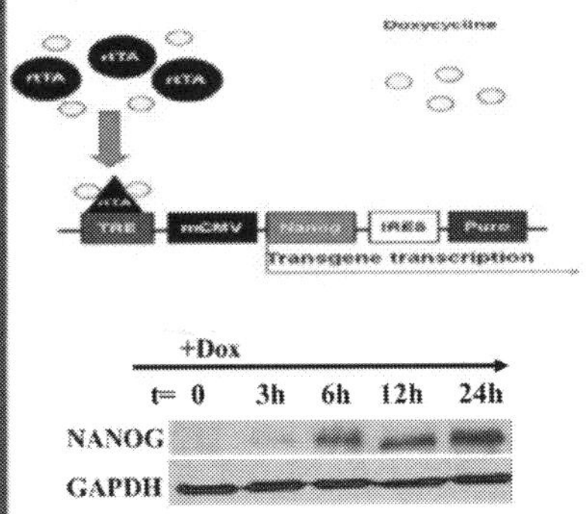


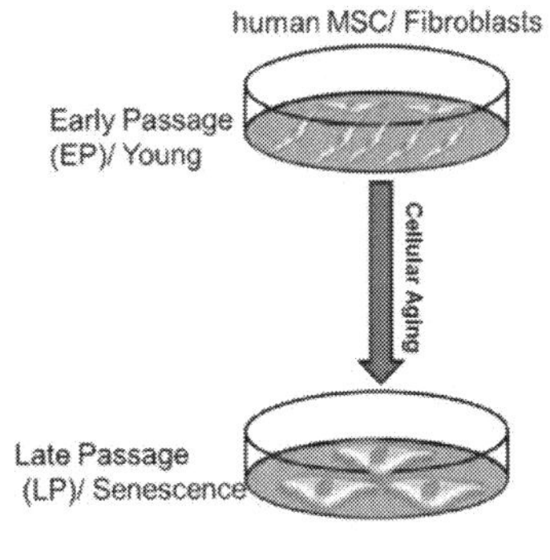


HGPS derived myofibroblasts

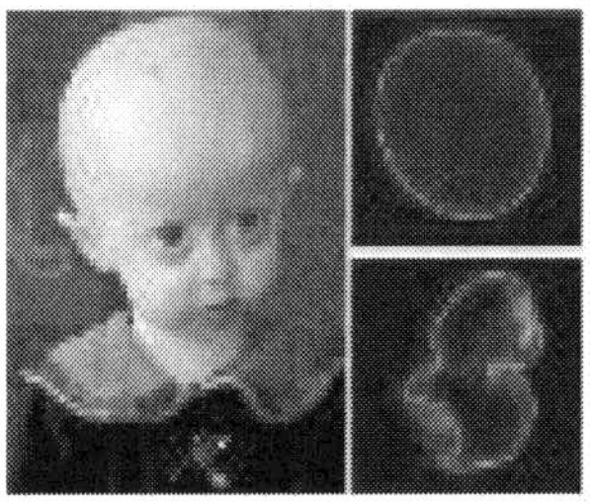

HGPS: Hutchinson-Gilford Progeria Syndrome

Panagiotis Mistriotis et al.,2015.

# NANOG restores myogenic gene expression in senescent cells

Senescent hHF-MSC

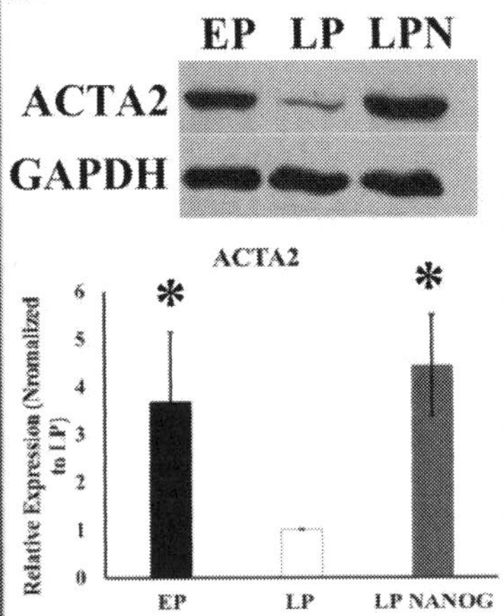


**EP**: Early Passage hMSCs
**LP**: Late Passage hMSCs
**LP N**: LP expressing NANOG
*** p< 0.05 compared to LP**

Senescent myoFibroblasts

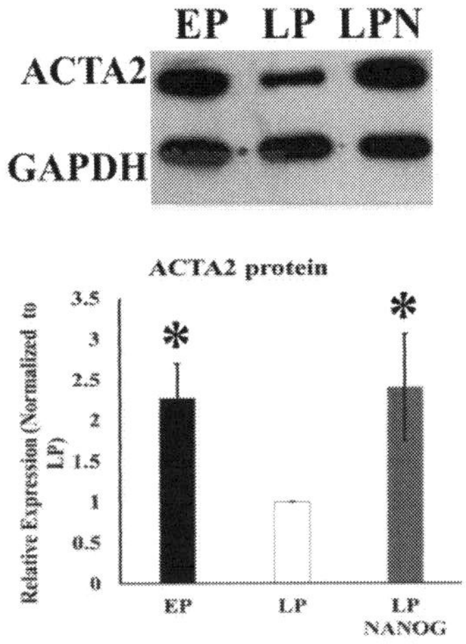


HGPS myoFibroblasts

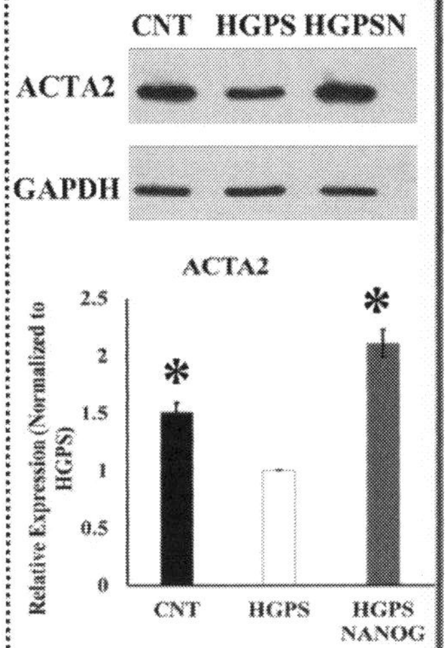


Panagiotis Mistriotis et al.,2015.

# NANOG restores Actin polymerization of senescent cells

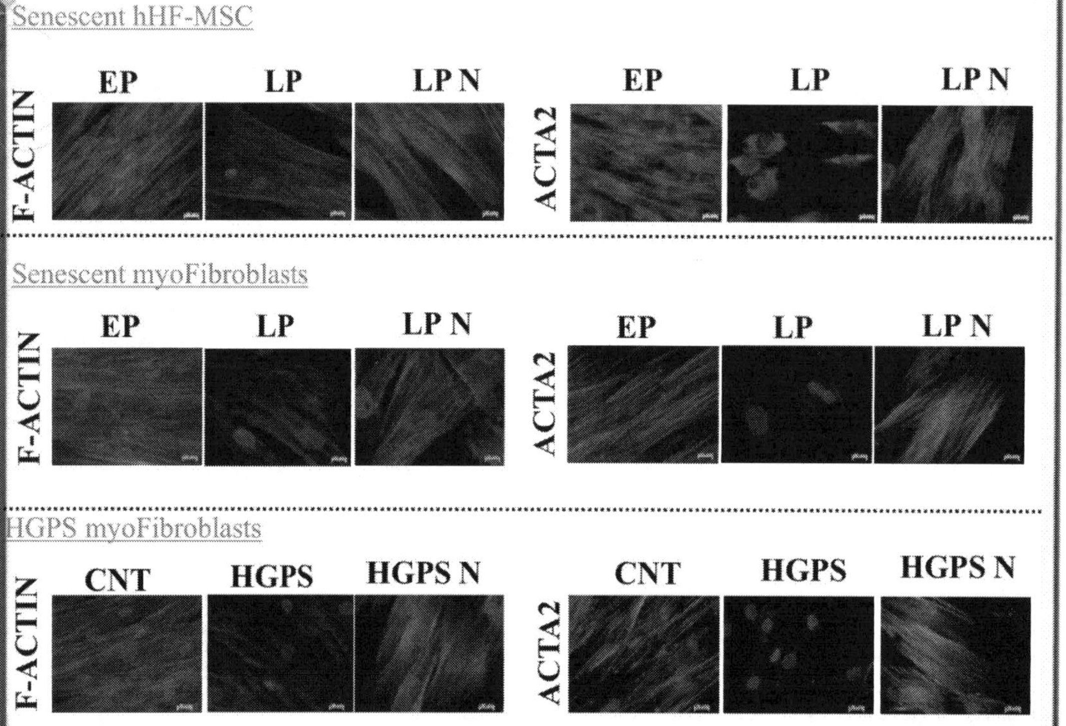


Panagiotis Mistriotis et al.,2015.

# NANOG restores contractile capacity of senescent cells

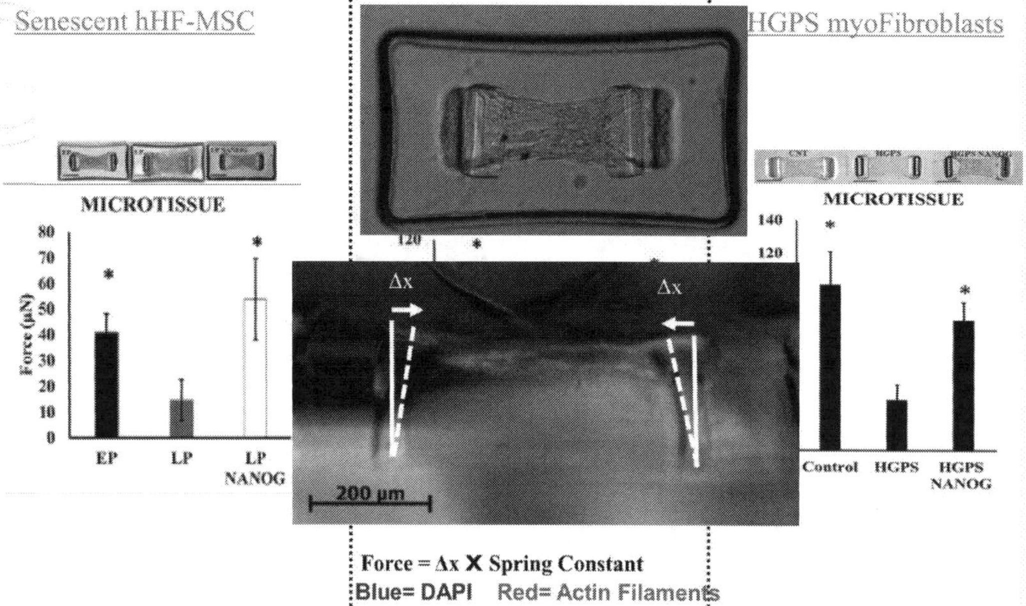


Force = Δx X Spring Constant

Blue= DAPI Red= Actin Filaments

**EP**: Early Passage hMSCs
**LP**: Late Passage hMSCs
**LP N**: LP expressing NANOG
*** p< 0.05 compared to LP**

Panagiotis Mistriotis et al.,2015.

## HYPOTHESIS

**ECTOPIC EXPRESSION OF NANOG RESTORES THE LOSS OF COLLAGEN EXPRESSION DUE TO CELLULAR AGING**

## NANOG restores the mechanical properties of engineered tissues

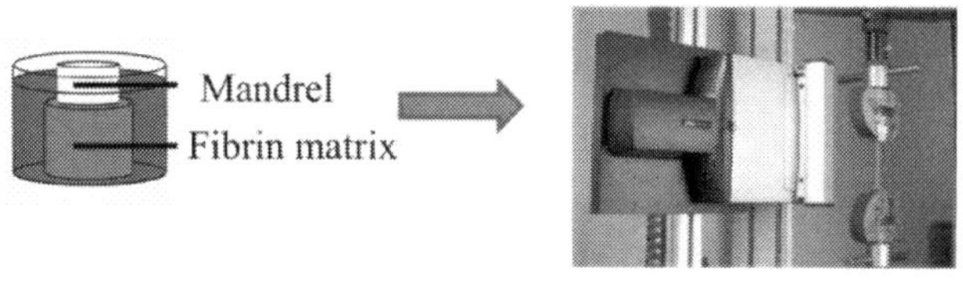


Mechanical testing
- **UTS**
- **Young's Modulus**

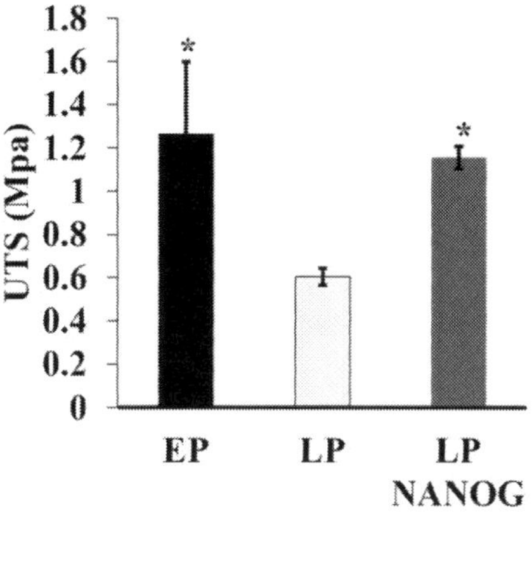


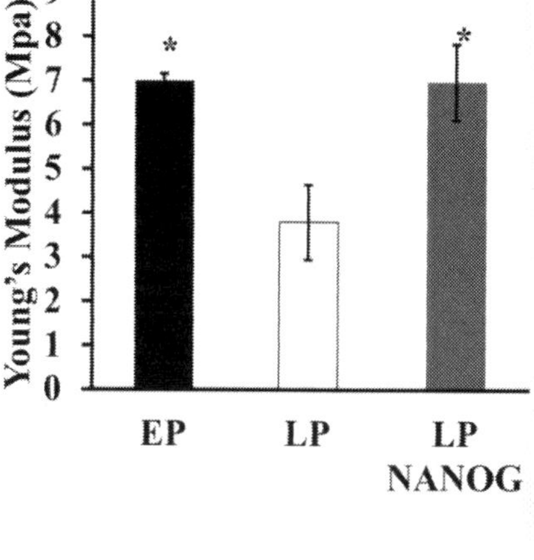

# NANOG increases Collagen 1 & 3 expression

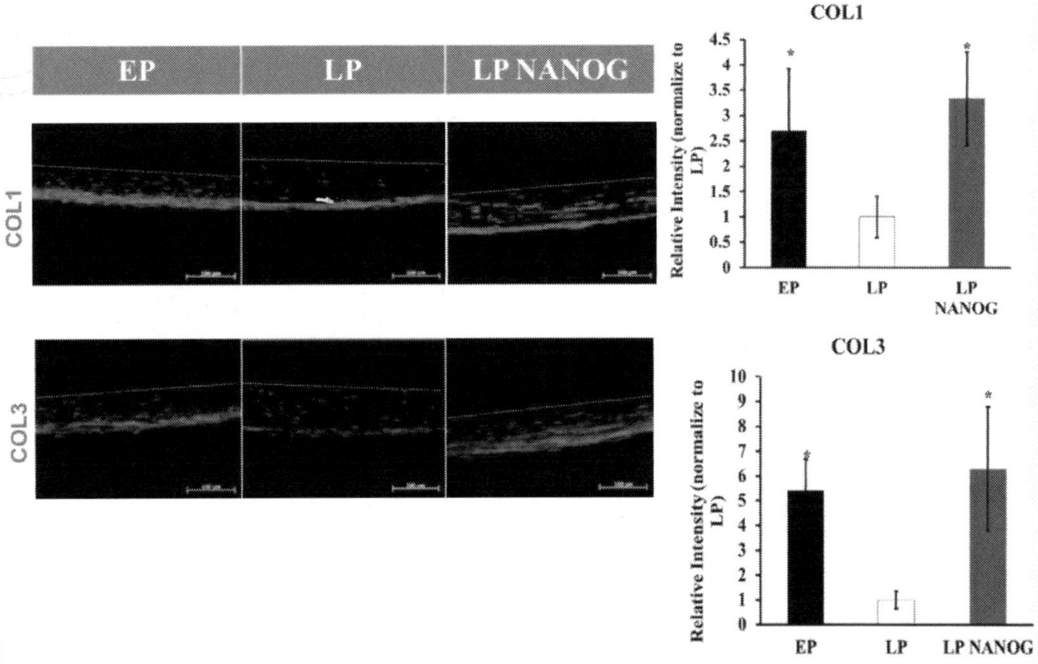


# NANOG increases Collagen 3 expression

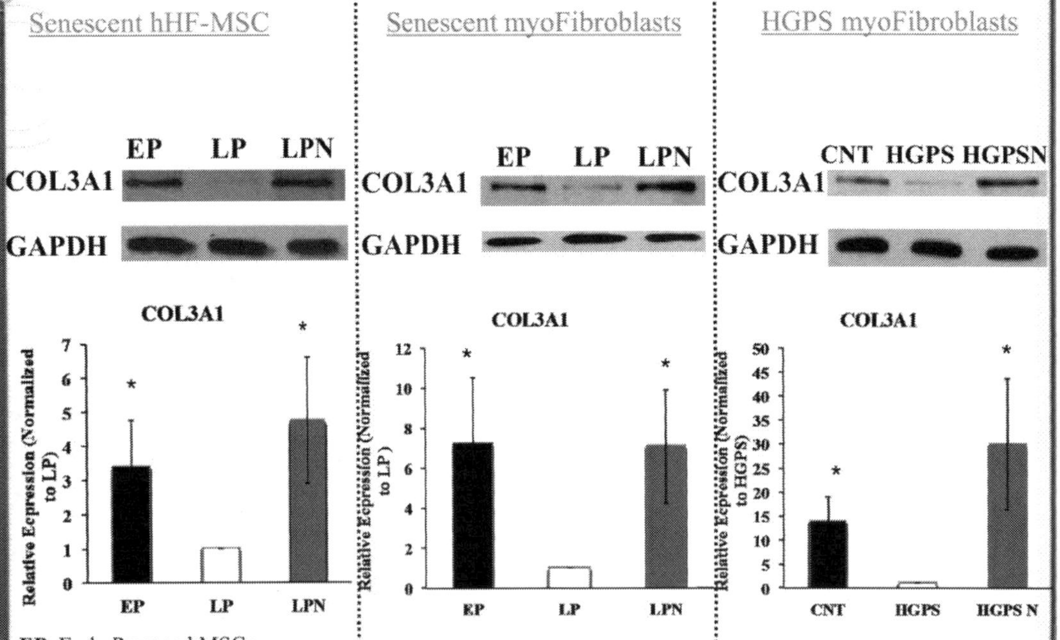


**EP**: Early Passage hMSCs
**LP**: Late Passage hMSCs
**LP N**: LP expressing NANOG
*** p< 0.05 compared to LP**

# RNA-seq analysis

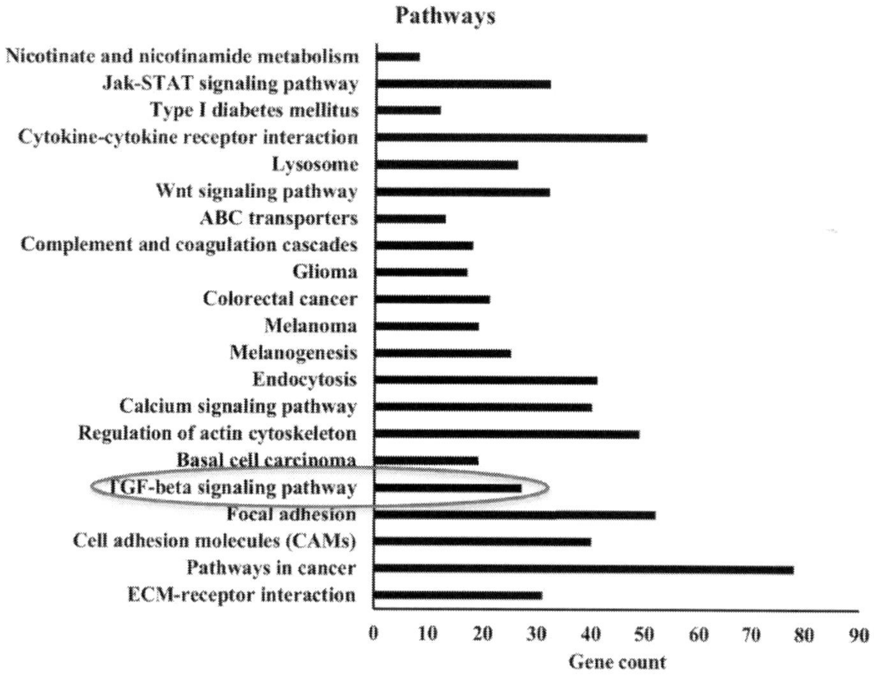


# TGF-beta pathway

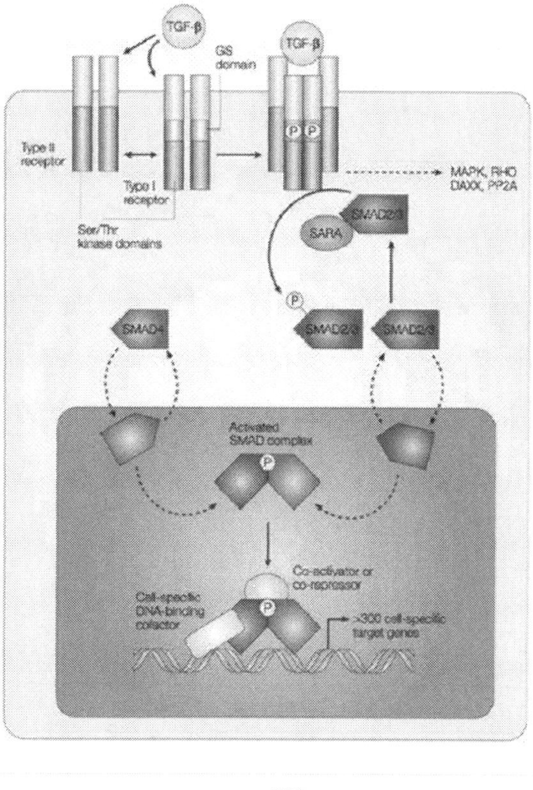


Peter M. Siegel and Joan Massagué 2003, Nature reviews

# RNA-seq analysis

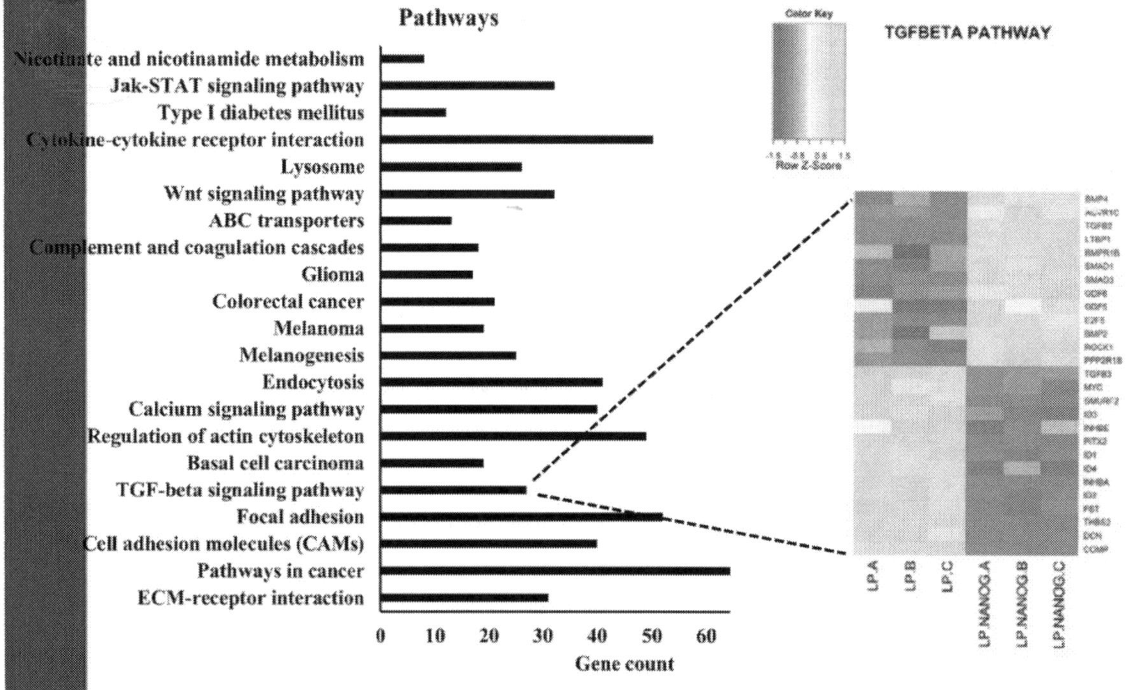


# NANOG increases phosphorylation of SMAD2

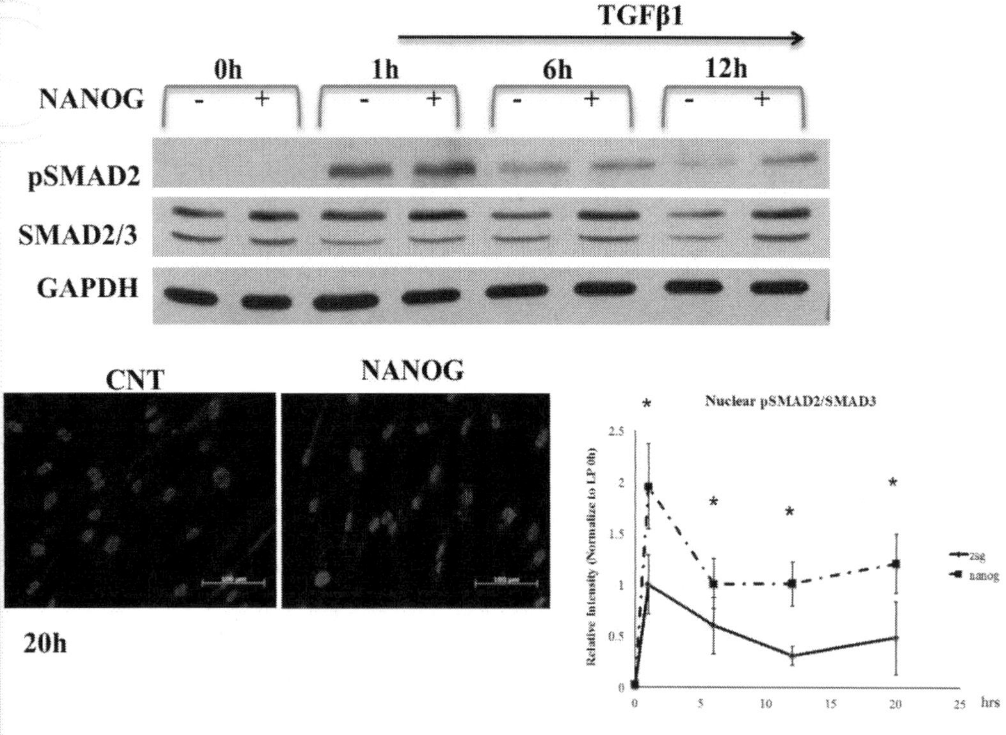

# NANOG increases SMAD2&3 levels

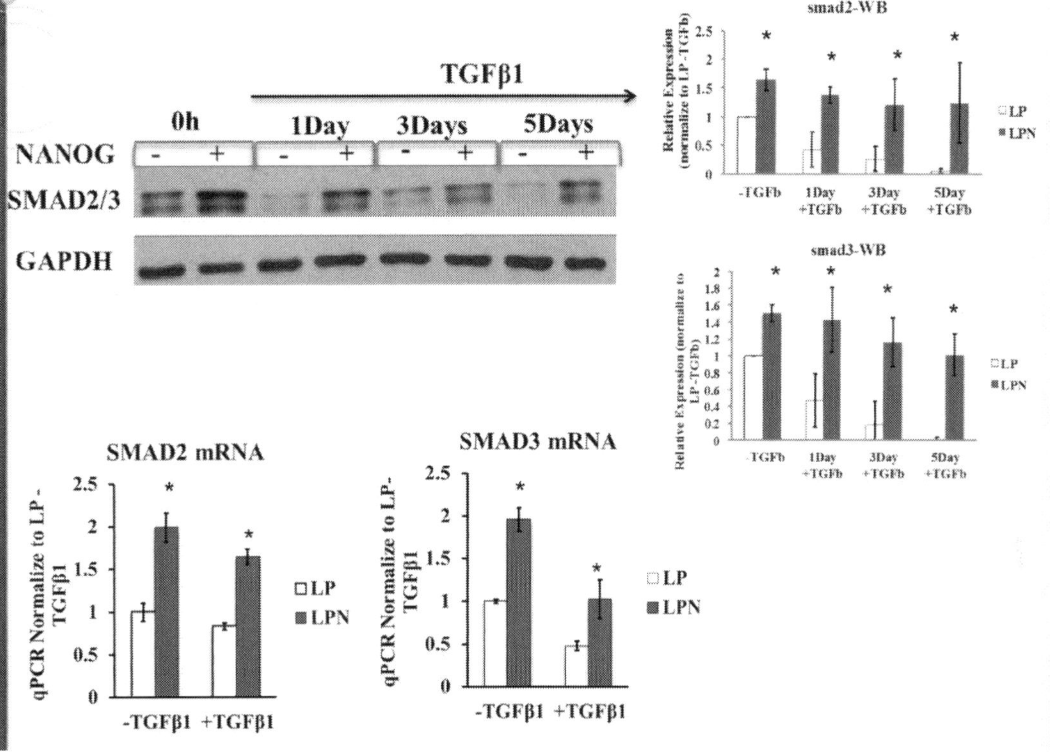


# Inhibition of SMAD pathway abolished NANOG mediated Collagen 3 expression

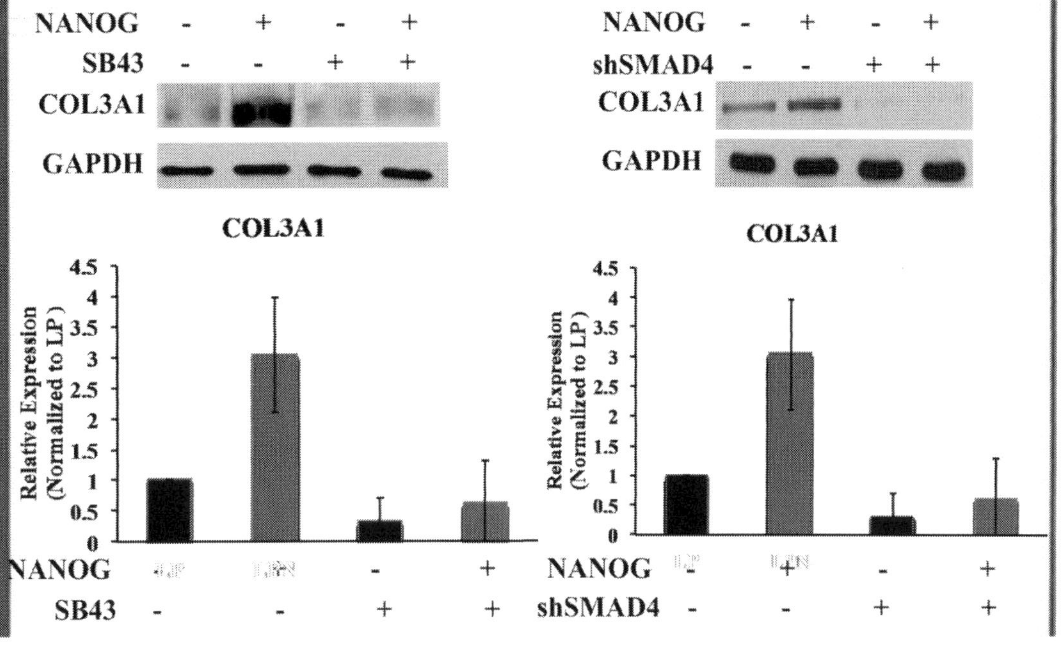

## Nanog increases Collagen 3 through SMAD3

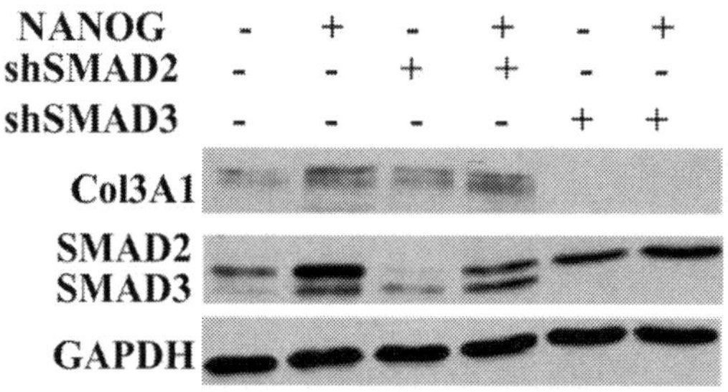


## Conclusion

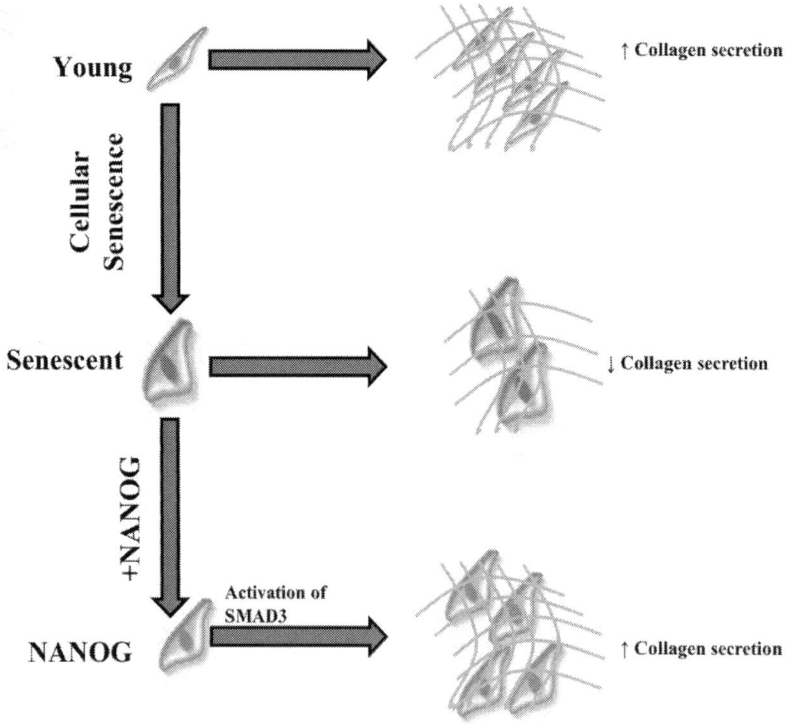

# Thank You!!!!!

**Funding sources**
- NSF
- NIH

Panagiotis Mistriotis et al., ***ECTOPIC EXPRESSION OF NANOG RESTORES THE MYOGENIC AND CONTRACTILE CAPACITY OF SENESCENT STEM CELLS*** , 2015.

University at Buffalo
*The State University of New York*

# Contraction mechanism

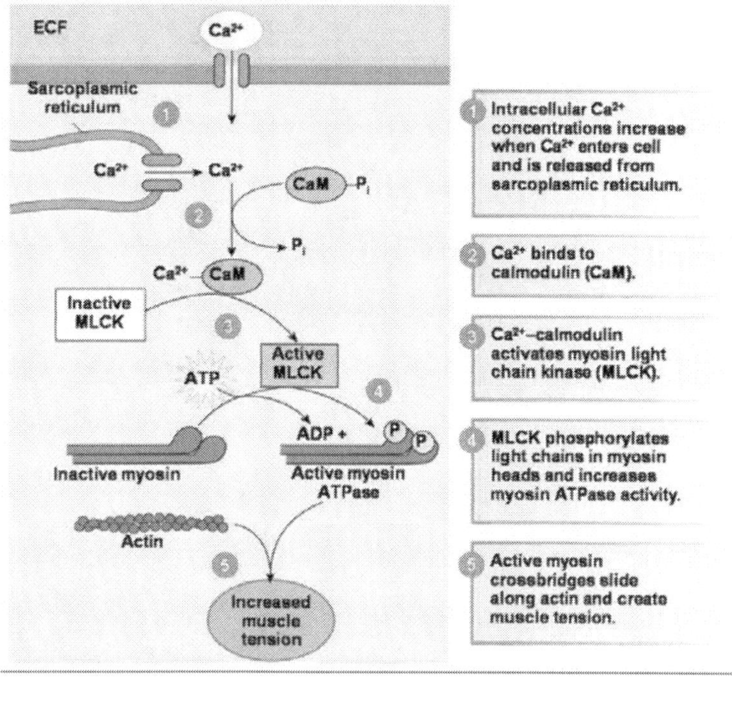

# Conclusion

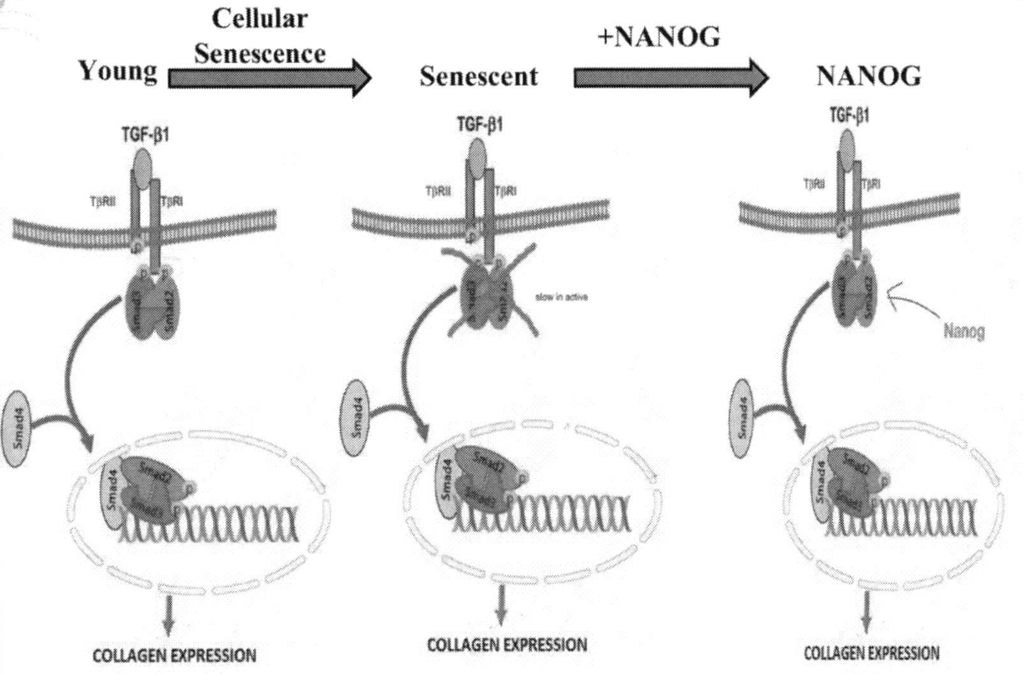


# Effect of Nanog on proliferation

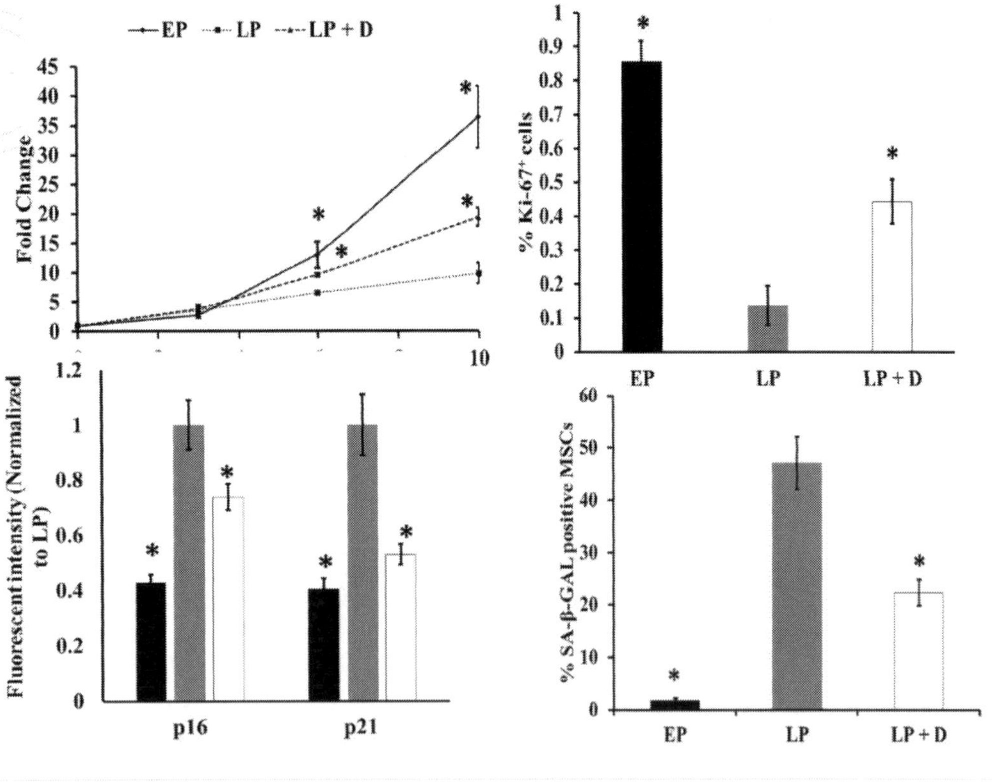

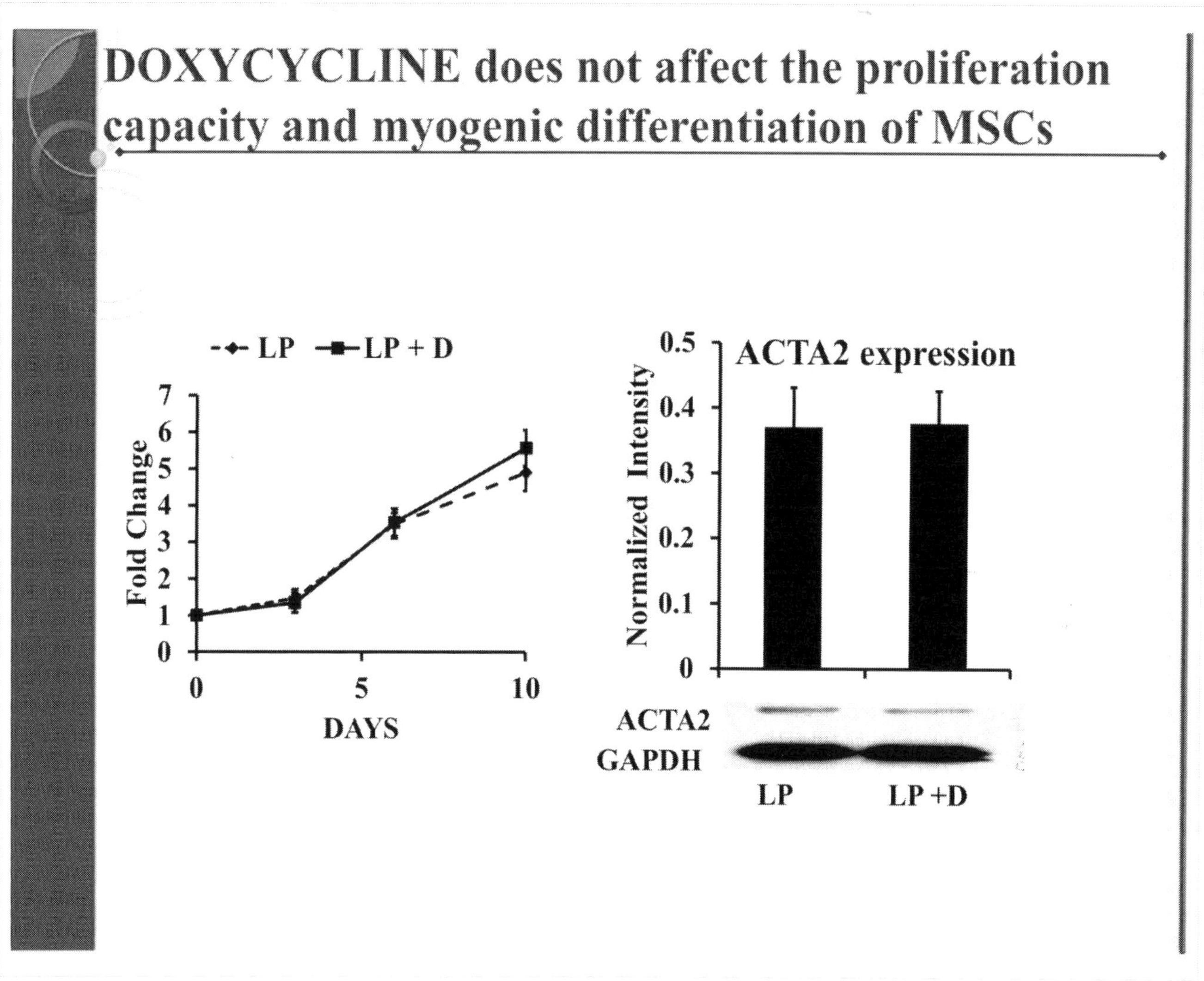
DOXYCYCLINE does not affect the proliferation capacity and myogenic differentiation of MSCs
LP
LP + D
Fold Change
7
6
5
4
3
2
1
0
0
5
10
DAYS
0.5
0.4
0.3
0.2
0.1
0
Normalized Intensity
ACTA2 expression
ACTA2
GAPDH
LP
LP +D

# 413077 Generation of Complex in Vitro Systems for Improved Evaluation of Nanoparticle-Induced Bioresponses

**Abstract:**

Cellular systems are required to test the safety of newly developed materials, chemicals, and biological-based products to ensure their safety. Both *in vitro* and *in vivo* methodologies are being utilized for evaluation, however, limited correlation exists between these models. This discrepancy is due to the drastic, fundamental differences between basic cell culture conditions and fully functioning organisms. One way to overcome this limitation is through the development of enhanced *in vitro* environments that retain the advantages of cellular systems but more accurately mimic physiology. Studies are beginning to emerge that attempt to bridge the gap between these models by improving upon standard *in vitro* techniques through the addition of physiologically relevant variables. Alterations currently being explored include, the formation of cellular co-cultures, the use of physiological fluid instead of cell culture media, generation of a 3-dimensional growth substrate, and the introduction of shear stress through dynamic flow. However, to date, the simultaneous combination of more than one of these advancements has yet to be executed.

In the work presented here, we successfully developed and implemented two enhanced *in vitro* microenvironment that more closely mimicked an *in vivo* system, while retaining the advantages of cell-based techniques. These systems, which consisted of multiple modulations to standard *in vitro* techniques, were employed for the evaluation of nanoparticle (NP) bioeffects and the nano-bio interface. NPs were specifically chosen owing to their unique physicochemical properties that distinguish them from chemical species, such as their insoluble nature, ability to undergo ionic dissolution, agglomeration tendencies, and NP sedimentation effects. In particular, silver and gold NPs (AgNPs and AuNPs) were selected, as they are the most prevalent and highly utilized composition in the consumer market. As inhalation is a primary route of exposure for NPs, that was our target cellular system.

In the first microenvironment, alveolar epithelial cells (A549) were exposed to AgNPs and AuNPs in artificial alveolar fluid (AAF) and under dynamic flow. Dynamic flow was established through the use of a peristaltic pump and was set to a volumetric flowrate that produce physiologically accurate linear velocities. The NPs demonstrated significant behavioral alterations in the microenvironment, specifically targeting NP agglomeration, spectral profiles, and kinetic rates of dissolution. For example, following dispersion in AAF, aggregate sizes of AuNPs increased over 2-fold from a traditional media environment. Beyond NP effects, the cellular response was greatly varied following the addition of AAF or shear stress. In agreement with greater agglomeration and decreased ionic dissolution, the stress and cytotoxic responses to AgNPs were significantly reduced in AAF. While the influence of dynamic flow on the NPs was negligible, it greatly modified both A549 morphology and the nano-bio interface. In a flowing system, NP deposition was reduced over 20%, with TEM imaging confirming modified endocytosis mechanisms.

The goal of the second constructed microenvironment was to allow for augmented evaluation of biological responses following NP exposure; therefore the system was expanded to include a co-culture consisting of A549 alveolar epithelial cells and U937 macrophages. The effectiveness of this co-culture model was first evaluated by confirming that the U937 cells were preferentially engulfing the sub-toxic dose of AgNPs. Next, dynamic flow was introduced and again, we observed a dramatic drop in NP deposition: approximately 40% less than the static environment. Interesting, for the co-culture model, the shear stress induced by dynamic flow did translate to greater biological stress, as evaluated through LDH release and ROS production. However, the inflammatory response, assessed through secretion of target cytokines, following AgNP exposure was decreased under flow, correlating with the lower degree of NP deposition. Taken together, these results support the emerging idea that numerous environmental variables can influence NP properties and behaviors, with the potential to affect downstream bioeffects and overall safety evaluations. To that end, the utilization of a complex and physiologically relevant models is necessary to accurately and effectively evaluate the safety and performance of NPs in nano-based applications.

# 415076 Regenerative Engineering of the Endothelial Cell Glycocalyx Using Nanomedicine

**Abstract:**

The long term goal of this project is to apply regenerative engineering to rebuild the endothelial cell (EC) surface sugar coat, the glycocalyx (GCX). The GCX lines and protects the blood vessel wall, interfaces with flowing blood and mediates mechanotransduction, plays an important role in vascular homeostasis, and in its absence causes EC de-differentiation and dysfunction [Ebong et al., Integr Biol (Camb) 6, 3, 2014] that promote atherosclerosis leading to cardiovascular disease. Our regenerative engineering goal is to reconstitute the GCX to treat atherosclerosis and cardiovascular disease. In a present study, we are using a nanoparticle-based approach to replace GCX components and to deliver drugs that stimulate their synthesis. Here, we report on foundational experiments in which we assess how EC-specific nanoparticle delivery is impacted by baseline GCX conditions. We exposed rat fat pad EC (RFPEC) with intact GCX to ultra-small PEGylated gold nanoparticles [Kumar et al., Transl Cancer Res 2, 4, 2013]. RFPEC with intact GCX did not exhibit any nanoparticle uptake. In contrast, some nanoparticles were retained by RFPEC with protein deficient and collapsed GCX, characteristic of certain disease conditions. A more substantial number of nanoparticles were retained by RFPEC with enzymatically degraded GCX heparan sulfate, the most abundant component of the GCX. In another case, after enzymatic heparan sulfate degradation, we induced GCX regeneration by adding heparan sulfate to the culture media for its incorporation into the GCX. HS regeneration resulted in restoring blockage of nanoparticle entry into RFPEC. This work indicates that the GCX integrity and composition influence nanoparticle uptake by EC. The observed passive targeting of nanoparticles to GCX-deficient EC is promising for future work on this project. We look forward to further developing our PEGylated gold and other nanoparticles to delivery replacement GCX components or drugs to induce EC to regenerate GCX. Funding: Northeastern University and IGERT Nanomedicine Science & Technology Program at Northeastern University (NSF/DGE-096843).

# 427942 Serotonin Control of the Activated and Osteoblastic Phenotypes of Valvular Interstitial Cells

**Abstract:**

Degenerative valvular heart diseases can be regarded as resultant of the mechanochemical environment where cells reside. This study investigates how serotonin, a small molecule involved in a number of biological processes including valvular degeneration, regulates valvular interstitial cell acquisition of activated (aVIC) or osteoblastic (obVIC) phenotypes. Valvular interstitial cells (VIC) are fibroblast-like cells, which have the ability to transform into an activated phenotype (aVIC) marked by increased alpha-smooth muscle actin or alternatively, into osteoblast-like cells (obVIC) marked by bone proteins. Each of these phenotypes is characteristic of specific degenerative valvular diseases. Primary valvular interstitial cell cultures were established on soft substrates in osteogenic media and markers of each specific phenotype were monitored over 8 days. We observed that osteogenic culture conditions drive the acquisition of osteoblastic phenotypes strongly in aortic VIC when compared to mitral VIC. The transformation to activated phenotype is also observed under osteogenic conditions. This finding is in agreement with naturally-occurring valvular degeneration modes (calcific versus myxomatous). In both mitral and aortic VICs, phenotype transformation to obVIC was observed as an early change followed by increased number of aVICs. In a different set of experiments, inhibitors of serotonin signaling (serotonin type 2 receptors and tryptophan hydroxylase 1 inhibitors) were used to assess the effects of serotonin signals on the distinct phenotypes. We observed that inhibition of serotonin signals reversed phenotype transformation, which was assessed using markers of expression of activated and osteogenic markers, calcium and collagen deposits. These results lead to the conclusion that aortic interstitial cells have higher potential for calcification than mitral interstitial cells and that such potential is regulated at least partly by serotonin signals. These findings provided a better understanding of signaling pathway regulating valvular cell phenotype transformation and can contribute to the development of effective therapeutic strategies of degenerative valvular diseases.

# 432309 Cancer Cell Migration Decisions in Engineered Environments

**Abstract:**

In many pathological and physiological settings, cells must integrate numerous signals from their environment. Often times, these signals are spatially inhomogeneous and are presented in the form of gradients or asymmetric structures like aligned fibers. For instance, cancer cells must interpret both chemotactic and contact guidance cues in the tumor microenvironment. While thousands of studies have examined how individual directional cues influence migration, relatively few have examined how cells respond to simultaneously presented directional cues. Furthermore, while there is an appreciation that 3D environments more closely match the *in vivo* tumor microenvironment, it is not clear which aspect of 3D environments (topology, confinement or mechanical properties) regulates directed migration, particularly in the context of multi-cue sensing.

In this talk, I will discuss the competition and cooperation of chemotactic and contact guidance cues in directing cancer cell migration. We employ several different techniques to generate gradients of growth factors and aligned extracellular matrix structures, inducing chemotaxis and contact guidance, respectively. We have also developed hybrid environments that allow us to present contact guidance cues while tuning topology, confinement and mechanical properties. These hybrid environments show dramatic differences in the ability of contact guidance cues to direct migration. For instance, environments where MDA-MB-231 breast cancer cells contact aligned collagen fibrils with both ventral and dorsal surfaces are more directional than cells migrating on 2D surfaces. In 2D multi-cue environments, we observe that gradients of epidermal growth factor (EGF), a known chemoattractant, can enhance the directionality of cells migrating in response to contact guidance cues when presented parallel to the contact guidance cue. However, when presented perpendicular to the contact guidance cue, directionality is unchanged. Understanding how different cues compete or cooperate to direct migration as well as how topology, confinement and mechanical properties alter competition and cooperation is critical in understanding how cancer cells sense and respond to the tumor microenvironment. In addition, these multi-cue environments could be designed to be used to separate subpopulations of highly invasive cancer cells or to screen chemotherapies that block invasion.

# 429552 Matrix Stiffness and Interstitial Flow Regulate Breast Cancer Malignancy in a Tumor-Microenvironment-on-a-Chip (TMOC) System

**Abstract:**

In the past 50 years, extensive studies have explored the different mechanical and chemical cues that regulate tumor malignancy. Based on these studies, numerous attempts have been made to replicate the tumor microenvironment in a high-throughput lab-scale system. Such a "tumor-microenvironment-on-a-chip (TMOC)" would greatly reduce the time and costs associated with cancer drug screening. Many TMOC-like systems have been prepared with mechanical and chemical properties similar to those in the *in vivo* tumor environment. To date, however, no TMOC system has been prepared with both tunable interstitial flow and matrix stiffness—two key parameters known to regulate cancer malignancy. To this end, we embedded breast cancer cells in a collagen matrix with varying degrees of stiffness, ranging from that of healthy tissue to that of cancerous tissue. These breast cancer cells were then cultured for 6 days with or without a continuous interstitial flow, and cancer malignancy upregulation was determined by measuring the downregulation of estrogen receptor-alpha (ER-alpha). More specifically, the downregulation of ER-alpha is commonly associated with a luminal-to-basal transition in breast cancer cells, which is a key biomarker for aggressive breast cancer. Cancer cells in the cancer tissue-like gel had significantly lower ER-alpha expression than cells grown in the healthy tissue-like gels, independent of interstitial flow. Interestingly enough, for cells in the healthy tissue-like gel, the presence of flow led to a higher level of ER-alpha, suggesting that some amount of mechanical flow will prevent a luminal-to-basal transition. For future studies, we will incorporate additional growth factors and cells to complete the tumor microenvironment, in turn creating a precision TMOC system.

# 412294 Antibiotic Biosynthesis

**Abstract:**

In this presentation, I will review major advances that have occurred in the field of antibiotic biosynthesis over the past 25 years, and then discuss some of the challenges that lie ahead of the field as it looks out into the future.

# 412296 Engineering Static and Dynamic Control of Malonyl-CoA Derived Chemicals in E.coli

**Abstract:**

Our research goal it to utilize the richness, versatility but also simplicity of microbial organisms in order to make them ideally suited to convert cheap, renewable resources into either high-value or commodity chemicals.

For the purpose of reprogramming the cellular network in order to achieve high-yield production, we have developed *in silico* model of the genome-wide metabolism *Escherichia coli.* Through the application of Metabolic Flux Analysis, we predicted genetic modifications such as deletions and gene expression attenuations that lead to dramatic increases in production levels. Such Systems Biology approaches have resulted in robust production levels leading to potentially commercially viable processes for the synthesis of natural products, in particular ones that derive from malonyl-CoA. We also report the engineering of both positive and negative feedback controls for dynamic tuning of metabolic flux in *E. coli.* Specifically, we have identified a dual transcriptional regulator that can act either as an activator or a repressor for two different promoters. The level of activation or repression is dependent on the level of intracellular malonyl-CoA. As a proof of concept, we demonstrated that the expression of two reporter proteins can be exclusively switched between the on and off state. By engineering this synthetic malonyl-CoA controller, both a malonyl-CoA source pathway and a malonyl-CoA sink pathway were dynamically modulated so that carbon flux could be efficiently redirected to synthesize our target compounds, fatty acids.

# 412289 Measuring Metabolism of Individual Cell Populations in Mixed Microbial Cultures


**Abstract:**

Microbial communities play an important role in biochemical engineering, medicine, food production, and waste water treatment. The capabilities of multi-microorganism systems are often enhanced by synergistic interactions at different levels. For example, it was recently shown that co-culture systems have unique advantages over mono-culture systems in optimizing product yield resulting from synergistic interactions and optimal division of labor.

Here, we present a novel framework for measuring metabolism of individual cell populations in mixed microbial cultures using $^{13}C$ metabolic flux analysis ($^{13}C$-MFA). We demonstrate for the first time that it is possible to measure metabolic fluxes in multiple species simultaneously in a mixed microbial culture without the need for physical separation of cells or proteins, or overexpression and detection of species-specific products. Instead, metabolic fluxes of each species are estimated directly from isotopic labeling of total biomass obtained using conventional mass spectrometry approaches (GC-MS). In addition to determining metabolic fluxes, this approach estimates relative population size of each species in the mixed culture, and inter-species metabolite exchange. As such, this approach enables detailed studies of microbial communities including species dynamics and interactions between community members. We have applied this methodology to study two model co-culture systems: 1) a co-culture of two *E. coli* knockout strains, D*zwf* (knockout of the first step in the pentose phosphate pathway) and D*pgi* (knockout of the first step in glycolysis pathway); and 2) a co-culture of *E. coli* and yeast. For the first time, we have elucidated how these two model microbes grow and interact with each other when grown simultaneously in a co-culture.

In summary, we have developed a new fluxomic methodology that adds a new dimension to metabolic engineering and significantly extends the scope of $^{13}C$-MFA studies to a large number of multi-cellular systems that are of significant importance in biotechnology and medicine.

# 434755 Tools for Engineering Non-Growth Metabolism


**Abstract:**

Stationary phase metabolism that is not carbon-limited is ideal for metabolic engineering. Without cell growth, feedstocks can be efficiently converted to product and toxicity that leads to growth-inhibition is irrelevant. However, to date, most metabolic engineering tools have focused on engineering growing cells, with only modest effort being applied to non-growing cells. While there is great promise, non-growth metabolic engineering presents new challenges, as post-translational (e.g., allosteric) enzyme regulation dominates control and is not amenable to analysis by transcriptomics/proteomics or engineering by altering promoters.

Our lab is developing a suite of tools to engineer non-growth metabolism in *E. coli*. I will discuss computational and experimental efforts to understand, predict, and control non-growth metabolic regulation for optimal biosynthesis. We are developing non-growth metabolic models that accurately reproduce different non-growth conditions and predict metabolic perturbations. These models are complemented by experimental strategies to uncover and overcome condition-specific metabolic regulation. Finally, we are developing next-gen synthetic biology circuits to manage the transition to non-growth and optimize product synthesis. This work should be applicable to a broad range of products and substantially improve process economics by improving titers, yields, and productivity.

Our strategy of model-driven metabolic engineering is deeply rooted in the training experience and research approach from my time in the Stephanopoulos lab. It is an honor to be mentored by Greg, and a pleasure to honor his achievements.

# 412297 From Metabolic Engineering to Metabolomics and Back

**Abstract:**

Metabolic engineering is a field where the level of a specific metabolite is quite obviously a very important thing, because that metabolite is the valuable product that one is trying to produce. If knowing the level of one metabolite is useful, it stands to reason that knowing the levels of many metabolites --- immediate precursors and products of the target metabolite, as well as other related molecules --- should be even more useful. It is perhaps surprising, then, that more metabolic engineers are not immersed in the field of metabolomics. At the very least, there is quite a natural research segue from metabolic engineering to metabolomics; beyond that, though, there is true synergy between the fields that surprisingly few have harnessed. While there are certainly instrument limitations and shortcomings of analyses in the field compared to some other more established systems-scale experimental approaches, metabolomics provides a downstream, phenotypic snapshot of the cell that can be critical to understanding metabolism and metabolic dynamics.

Here, we will discuss the intersection of metabolomics and metabolic engineering in a broad sense. While the step from metabolic engineering to metabolomics is obvious, the path back to metabolic engineering can go in many directions. We will discuss the potential for metabolomics to drive computational, semi-rational metabolic engineering (even in an age of directed evolution) by automating and expanding an intuitive process. We will consider the broader impacts of this modeling approach as we consider similar applications to understanding mammalian cell metabolism. We will also discuss the potential of metabolites as regulators, and what this means for potential future directions of metabolic engineering. Expanding on this, we will also discuss potential roles of metabolites as therapeutics, an idea with prior history and a firm grounding in biology. While past applications of metabolomics in metabolic engineering have been useful, the future potential is more broad and vast than most people realize.

# 412293 Making Escherichia coli Grow on Methanol

**Abstract:**

The quest for using methanol as a substrate to produce useful metabolites such as commodity chemicals, chemical precursors, biopolymers, neutraceuticals and, more recently, biofuels, goes back to more than 40 years. The concept originated from the development of processes to produce single-cell protein (SCP) from methanol or methane using native methylotrophs. While technologically successful, SCP did not succeed in the marketplace, but led to many technological innovations in bioreactor design. The more recent increasing availability and lower prices of natural gas is resulting in increased methanol supplies, and has thus renewed the interest in using methanol as a substrate or co-substrate with sugars. The latter derives from the fact that methanol is considerably more reduced than most sugars and thus has to potential to enable the biological production of metabolites with significantly increased yields. The biology of natural methylotrophs has generated a large body of knowledge, but little of that has translated into biotechnological applications despite the sustained research enterprise. The last 10-15 years has also seen efforts to engineer other organisms to use methanol for growth, but without significant success yet. *E. coli* is an obvious platform for synthetic methylotrophy. We will discuss the native biology of methylotrophy and how it can be engaged to develop superior methylotrophic strains using synthetic biology and metabolic engineering. In principle, *E. coli* and most other organisms need only 3 genes/enzymes to be able to use methanol as a substrate, and tried though it may have been, an *E. coli* or other non-methylotrophic organism that can effectively grow on methanol as a single substrate has not been reported. There are several reasons for that, and many barriers to be overcome: the choice of desirable enzyme co-factors, thermodynamic roadblocks and, most importantly, regulatory constraints that make the quest for synthetic methylotrophy an exciting challenge. We will discuss how we systematically dissected the issues of synthetic methylotrophy to arrive to problem resolution. SUPPORTED by the US DOE ARPA-E agency through contract no. DE-AR0000432.

# 432303 Phenotypic and Gene Expression Variability Underlies Adaptive Resistance in Heterogeneous Bacterial Populations

**Abstract:**

The rapid rise of drug-resistant superbugs and the declining antibiotic pipeline are serious challenges to global health. The root cause of the antibiotic resistance crisis is the ability of bacteria to evolve tolerance to antibiotics and other environmental toxins. Bacteria exhibit extensive intra- and inter- population heterogeneity, which provides an advantage in adaptation to diverse environmental pressures, but renders identification of underlying mechanisms of adaption challenging. Here, we used RNA-Sequencing to locate key regulators of adaptive resistance by investigating transcriptomes of bacterial populations upon prolonged exposure to diverse chemical toxins and comparing to gene expression profiles in unadapted populations. Our lab-evolution experiments show that initially antibiotic-sensitive isogenic bacteria adapted to the same toxin exhibit rapid population-level phenotypic divergence. To investigate the mechanisms underlying heterogeneity across populations, we evolved clinically derived multi-drug pathogens and lab strains of *Escherichia coli* to a range of dissimilar toxins, including antibiotics and biofuels. We find a set of genes that are differentially expressed across multiple adaptation conditions. Further, we observe the presence of significant shifts in gene expression variability across unadapted and adapted populations. A more focused quantification of mRNA expression of stress-response genes corresponding to the multiple antibiotic resistance (*mar*) regulon, the general stress response, and the SOS response, all known to regulate adaptation, confirms the underlying gene expression heterogeneity. Hierarchical clustering and principal component analysis reveal that adapted populations exhibit variability in transcript levels that is distinct from unadapted populations; certain stress response genes are observed to increase or decrease expression variability with respect to wild type, suggesting the presence of dissimilar regulation of gene expression during adaptation. By propagating strains with corresponding stress response genes either knocked out or perturbed via CRISPR-interference mechanism, we identify unique relationships between gene expression variability and impact on adaptation that differentiate adapted or unadapted populations. This study provides insight into the dynamic nature of the bacterial adaptive transcriptome and evidence that gene expression variability can be used to identify bacterial populations undergoing adaptive resistance. The identification of genes contributing to adaptive resistance may be an avenue towards the development of antimicrobial agents that hinder the onset of drug-resistance.

# 432675 Understanding the Poplar Microbiome Structure in Response to Host Stress

**Abstract:**

Plant roots in the soil provide a rich carbon source through root exudates and organ turnover that attracts soil bacteria. The bacteria associated with plants are diverse, with thousands of unique bacteria measured as operational taxonomic units (OTUs) identified in natural samples. The associated bacteria can have beneficial effects on the host plant through nutrient acquisition, signaling and defense, and buffering of extreme environmental conditions. The plant microbiome has been shown to be dependent on host plant genotype, soil type, and environmental conditions. Given the importance of plants as feedstocks for food and energy, it is important to understand how the microbiome changes in response to natural stresses and how these changes may supplement the host response.

Using *Populus* trees as host plants, we investigated how abiotic stresses affect the host plant and belowground microbiome. Drought stress was implemented by watering plants at complete wilting to identify potential microbes that are beneficial in buffering the plant against drought. Drought plants were significantly reduced in overall growth, with extreme decreases in photosynthesis rate due to closed stomata especially at the lowest water conditions. The effects of drought on the host plant and environment significantly changed the endophytic microbiome relative to control, with increases in alpha and gamma proteobacteria. Only 15 of the top 20 OTUs were shared between control and drought plants, representing 73% and 65% of the total identified OTUs in the system, suggesting that drought microbiome is more diverse relative to controls. In parallel, a shade stress was induced by limiting light input to near the $CO_2$ compensation point, affecting carbon allocation and the chemical environment of the host, ultimately leading to changes in the microbiome. 16 OTUs were shared with the control top 20, representing 65% of the total diversity. 4 of the differential OTUs were different than what was observed in drought stress. The microbiome of shade and drought plants clustered together by principal coordinate analysis, suggesting a core microbiome that is enriched under host stress conditions. With this information we can begin to identify bacteria that are conserved in all *Populus* host growth conditions, as well as bacteria that enriched under specific environmental stresses. This research will lead to holistic understand of how the microbiome community functions to supplement the host phenotype.

# 430660 Understanding Candida Albicans and Candida Glabrata Mixed Species Biofilm Growth and Co-Culture

**Abstract:**

Candidiasis, fungal infection caused by *Candida* species, affects patients of all ages and ranges from thrush in the oral cavity to disseminated deep tissue infections. *Candida* infections may be difficult to treat due to inherent or acquired antifungal resistance, and limited number of available therapeutic agents; with immunocompromised patients being the largest at-risk population. Biofilms, which are often associated with the pathogenesis of *Candida*, are generally polymicrobial in nature, and are held together by an extracellular matrix (ECM) that is attached to a surface. In the human host, this surface may by a catheter, mucosal surface of the gastrointestinal tract, or lining of the stomach. Biofilms are difficult to treat largely due to the protective ECM, which reduces the penetration and efficacy of antifungal agents. *C. albicans* and *C. glabrata* are the two most commonly isolated species from *Candida* infections, and have been isolated together, which makes polymicrobial populations an interesting and clinically relevant area of study. In this work, we characterize the impacts of co-cultures in biofilm formation and population dynamics using mixed cultures of *C. albicans* and *C. glabrata*. When grown as biofilms, we find that biofilm formation of the polymicrobial population depends on the relative ratio of starting cell concentrations of *C. albicans* and *C. glabrata* in an unexpected manner. The specific ratio we find to yield the highest biomass is when the ratio of *C. albicans* to *C. glabrata* is low. *C. glabrata*, in comparison to *C. albicans*, is not very successful at forming biofilms when grown as a single species. Interestingly, we find that the higher ratios of *C. albicans* to *C. glabrata* in fact result in less biofilm formation than the low *C. albicans* to *C. glabrata* counterpart. When co-cultured in liquid medium in suspension, one of the species consistently exhibits a growth advantage over the other. Our results suggest interspecific interactions between these two fungal pathogens, where a polymicrobial population causes unexpected changes in biofilm formation. The underlying mechanisms of these inter-specific interactions are being investigated.

# 437407 Engineering Multicellular Motility-Based Biosensors Systems

**Abstract:**

The ability to robustly control cellular motility in response to an external stimulus can facilitate the development of biosensors or the construction of micron-sized robots that use microbes to enable movement. By separating the sensing and actuating cells, and coordinating their behavior via cell-cell communication, we can improve system design through modularity, signal amplification and logic gate-type responses to multiple inputs. Our first goal was to use a synthetic biology approach to build a system where the motility of *Escherichia coli* is controlled by a cell-cell communication signal. Towards this goal, we have designed and constructed a system where *E. coli* motility is controlled by an acyl homoserine lactone (AHL) quorum sensing (QS) signal molecule. Unlike previous efforts that used transcriptional activation-based QS systems to control motility, we used the repressor-based esa QS system to enable tight regulation of a key motility gene. We have demonstrated that we can restore motility in an *E. coli* strain with a *motA* deletion by expressing *motA* under the control of the esa promoter, which is tightly repressed by the QS transcriptional regulator (EsaR) in the absence of AHL. Expression of *motA* is triggered by AHL, which caused EsaR to dissociate from the *esa* promoter. To quantify motility, we measured the swimming of cells through semi-solid agar plates. We observed that cells do not swim in the absence of AHL, and swimming is induced in the presence of micromolar concentrations of AHL. Overall, our engineered communication-dependent motility (CoMot) cells show AHL dependent motility. We replaced wild type EsaR with a mutant that responds to nanomolar AHL concentrations to increase the AHL sensitivity of the cells (CoMot+). Interestingly, directional swimming of the CoMot and CoMot+ cells was observed up an AHL gradient even though chemotaxis was not directly controlled. Further, we have demonstrated that the cells respond to AHL produced by a second strain of *E. coli*. When the AHL production is inducible, the combination of AHL-producers and the CoMot cells enable a novel biosensing framework. The CoMot strains engineered here can be used to build synthetic networks and biosensor systems where a motility output is tightly regulated and controlled by cell-cell communication.

# 429325 Improving the Residence Time of Probiotics in the Gut through Functional Metagenomics

**Abstract:**

The human large intestine houses trillions of microorganisms which collectively form the highly diverse microbial community known as the gut microbiota. The gut microbiota is becoming increasingly recognized as an organ, performing many functions critical to the maintenance of health, including extraction of nutrients, production of vitamins, and defense against pathogens. Like an organ, pathologies in the structure of the gut microbiota (termed "dysbioses") can have significant negative impacts to the human host, including obesity, malnutrition, cancer, and death. However, the gut microbiota is currently unique among the organs in that it is highly engineerable, enabling improvement of function through the substitution of beneficial microbes for less desirable ones. Towards this goal, probiotics are orally ingested microorganisms which improve gut health through provision of additional metabolic functions or through competition for niche space with pathogens. Unfortunately, current probiotic strains have short residence times in the gut, necessitating continual administration for treatment of chronic conditions. In order to improve the gut residence time of the important probiotic *E. coli* Nissle, we extracted tens of gigabases of metagenomic DNA from several highly distinct gut microbial communities and selected colonization-enhancing gene fragments via passage in both gnotobiotic and conventional mice. Bioinformatic and biochemical analyses of these fragments indicated mechanisms of improved colonization, and integration of these colonization factors into probiotic *E. coli* Nissle significantly improved overall gut residence time. Subsequently, these fragments were tested in combination to investigate synergistic and antagonistic effects between colonization factors. Taken together, this work has the potential to significantly improve our understanding and use of probiotics while laying an important foundation for engineering the human gut microbiota.

# 420101 Agent-Based Models of Heterogeneous Tumor Populations Predict Multi-Scale Dynamics and Inform Therapeutic Perturbations

**Abstract:**

Cancer accounted for 8.2 million deaths in 2012, a number that continues to grow despite the efforts of researchers and medical professionals alike (World Health Organization, 2014). The challenge in understanding cancer, and thereby developing successful therapies, stems from its inherent heterogeneity and the complexity of its microenvironment. It is therefore essential to study, and account for, the manner in which tumors interact with the environment. To address this need, we developed a tunable, multi-scale *in silico* model that predicts the emergent behavior of a tumor in response to a variety of environmental and therapeutic perturbations.

Continuous and deterministic models cannot account for the discrete, stochastic, and heterogeneous nature of tumor cells, prompting researchers to consider a different paradigm: bottom-up rather than top-down modeling. Bottom-up approaches characterize the low-level building blocks of a system and couple them together to predict dynamic responses. Here, these building blocks, or agents, are individual healthy or cancerous cells. These cells are coupled in a 3D microenvironment that accounts for crowding, nutrient resources, and cytokine signaling (both secretion and uptake). Unlike cellular automata (CA) models, the agent-based modeling (ABM) framework allows both spatial and temporal resolution so these cells can divide, invade, and apoptose across space and time.

Our multi-scale ABM incorporates inter- and intra-cellular interactions, nutrient diffusion, and paracrine signaling. Cell agents, both healthy and cancerous, are equipped with a simplified, 14-component EGFR network that interacts with the environment (Zhang et al., 2009). Each cell agent can exist in one of five phenotypes—quiescent, proliferative, invading, apoptotic, or necrotic—and follows rules defined by its phenotype, local environment, and neighboring cells. More specifically, the intracellular network is most responsible for proliferation and invasion, and—through the corresponding dynamics—couples the cell to the physical environment. Glucose availability and crowding is most responsible for a cell's entrance to, and exit from, quiescence. Cancerous agents have a modified rule set based on the "hallmarks of cancer" described by Hanahan & Weinberg (2000). Examples of distinctions include increased tolerance for crowding and an accelerated cell cycle.

The flexibility of agent-based modeling is demonstrated by the ease with which new agents and perturbations can be introduced. A sensitivity analysis of the model demonstrated significant sensitivity of parameters affecting glucose conditions, highlighting glucose concentration as a potential control input. Variable glucose conditions, in combination with existing treatment strategies—excision, radiation, and chemotherapy—will be simulated to predict therapeutic efficacy. These tunable perturbations can be introduced at any time point and in any combination. Preliminary results quantify the effects of increasing excision radius under different glucose conditions, revealing important tradeoffs that should be explored in future experiments as well as strategies to be validated in the clinic. The development and analysis of tumor growth under different parameters and therapies demonstrate the versatility of ABMs for guiding translational cancer therapy and informing more personal medicine.

Hanahan, D., & Weinberg, R. A. (2000). The hallmarks of cancer. *Cell*, **100**(1), 57-70.

World Health Organization. Global status report on noncommunicable diseases. 2014.

Zhang, L., Want, Z., Sagotsky, J. A., & Deisboeck, T. S. (2009). Multiscale agent-based cancer modeling. *J Math Biol*, **58**, 545-559.

# 433845 Mathematical Modeling of Podocytes in Diabetic Kidney Disease

**Abstract:**

Diabetic kidney disease (DKD) is the primary cause for end-stage renal failure, is among the severe complications of diabetes, and is a significant factor leading to morbidity and mortality in both type I and type II diabetic patients. Hyperglycemia is known to initiate and exacerbate the pathophysiology of DKD. In the early stages of DKD, hyperglycemia induces increases in the rate of blood filtration through capillary beds (glomeruli) within the filtration units (nephrons) of the kidney [1]. The increase in glomerular filtration rate (GFR) contributes to glomerular injury, eventual damage to the tubular sections of nephrons downstream of the glomeruli, and loss of renal function. Damage to the epithelial cells called podocytes surrounding the glomerular capillaries is a critical factor in glomerular injury but is not immediately detectable with non-invasive clinical measurements of urine albumin concentration until after proteinuria develops due to substantial leakage of proteins through damaged glomeruli [1 – 2]. However, it is important to slow the rate of progression before irreversible loss of podocytes and nephrons occurs. In this work, we use mathematical modeling to investigate factors that contribute to the onset of nephropathy in DKD and seek to predict nephropathy in advance of detection of proteinuria by describing and simulating the mechanism of podocyte damage and loss. While mathematical models have been proposed for various renal physiological and pathophysiological processes, a quantitative description of the dynamics of podocyte injury at the onset and during the progression of DKD is lacking in the published literature. We have formulated a mathematical description of the mechanism of podocyte injury in DKD by including the contributions from glucose metabolism in the kidney during hyperglycemia, nephron hemodynamics, podocyte cell-signaling, and the dynamics of biomechanical forces that actively and passively alleviate or exacerbate effects of damage. These processes interact across multiple length and time scales. The mathematical model consists of a system of ordinary and partial differential equations and algebraic equations that are solved numerically. The intensity and duration of elevated GFR during periodic hyperglycemia is varied in simulations to investigate the influence on the transition to irreversible podocyte loss. The simulation results from the mathematical model are compared to published experimental data to assess the model ability to accurately and quantitatively predict how podocyte loss depends on filtration rate and the onset of proteinuria after hyperglycemic events.

[1] Lopez-Novoa JM, Martinez-Salgado C, Rodriguez-Pena AB, Lopez-Hernandez FJ. Common Pathophysiological Mechanisms of Chronic Kidney Disease: Therapeutic Perspectives. Pharmacol Ther 2010;128:61-81.

[2] Garg P, Holzman LB. Podocytes: Gaining a Foothold. Exp Cell Res 2012;318:955-63.

## *Growing bone through microsphere scaffolds*

Ornella Tempo
McNair Scholar
Rowe Scholar

November 9th, 2015

UCONN

## Who cares?

- Every year, more than 500,000 to 600,000 bone graft procedures are performed in the United States to address orthopedic injuries [1,2]
- In 2001 more than 3 million bone graft procedures were performed globally, with estimates of $3.7 billion through 2017 [3,2]

UCONN | SCHOOL OF ENGINEERING

More than one in four Americans has a musculoskeletal impairment

AAOS

American Academy of Orthopaedic Surgeons

UCONN | SCHOOL OF ENGINEERING

## Current Treatments

**Autograft**

- Require harvesting the bone from the iliac crest which involves an additional procedure

**Allografts**

- Allografts involve transplanting donor bone tissue; this increases the risk of immunoreactions and transmission of infections[1]

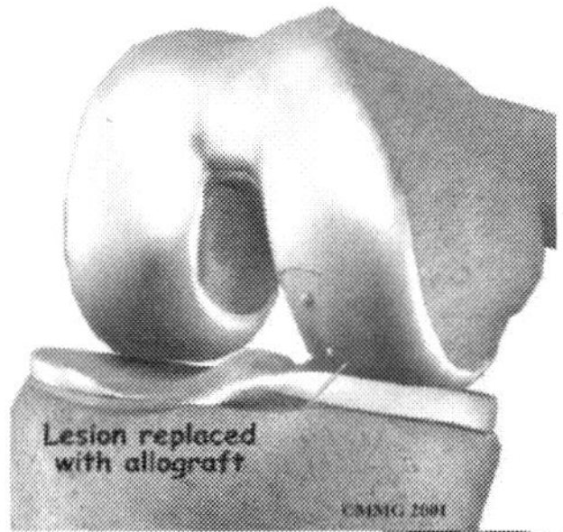


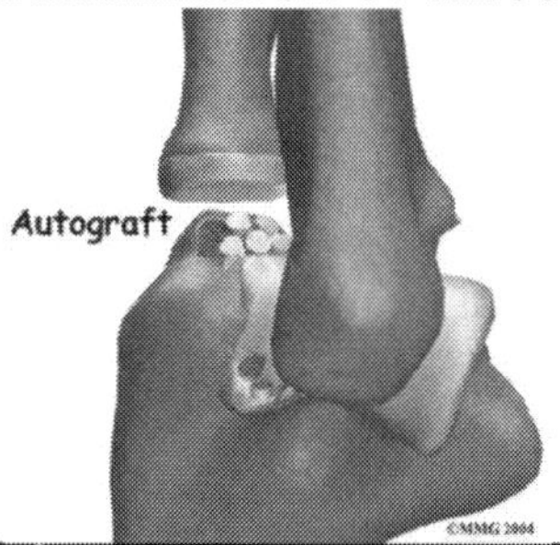


UCONN | SCHOOL OF ENGINEERING

# Regeneration of lizards

**6-9 weeks for complete regeneration**

**Have you ever seen the Spiderman movie?**

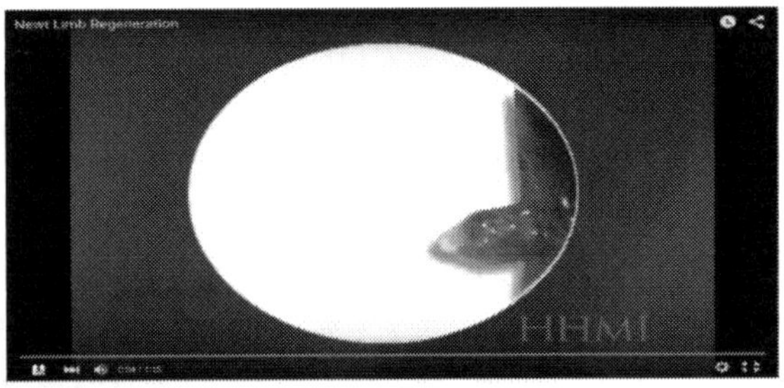


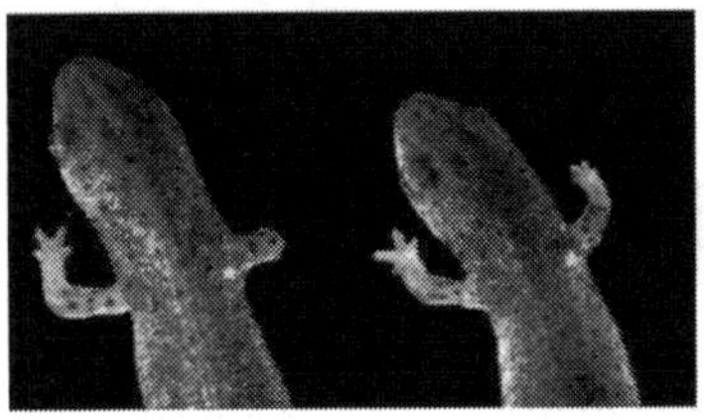

UCONN | SCHOOL OF ENGINEERING

## Orthopedic Biomaterials Market 2011-2021

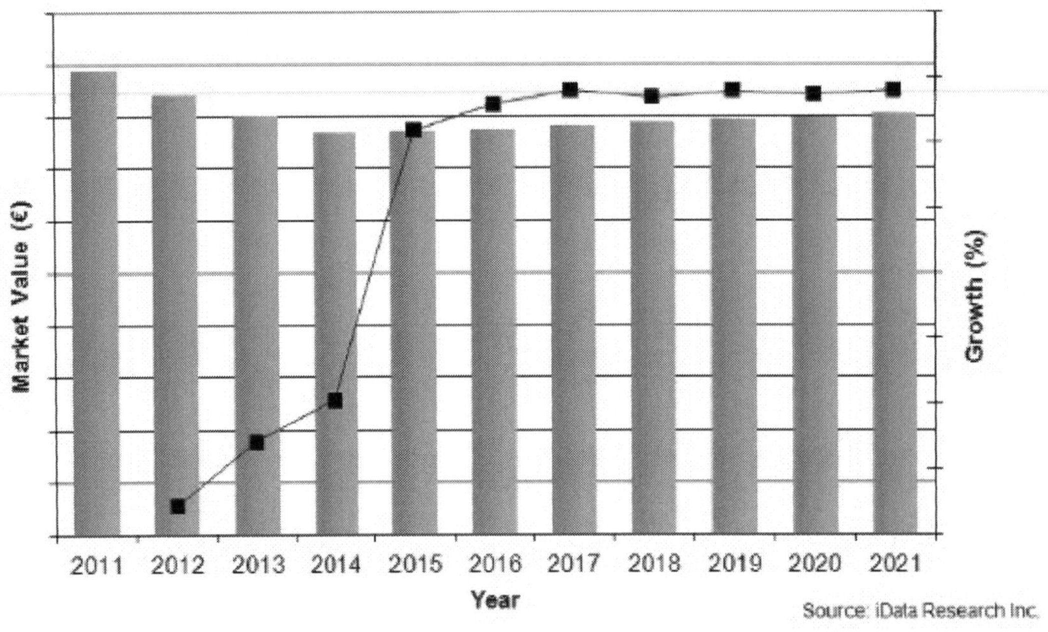


UCONN | SCHOOL OF ENGINEERING

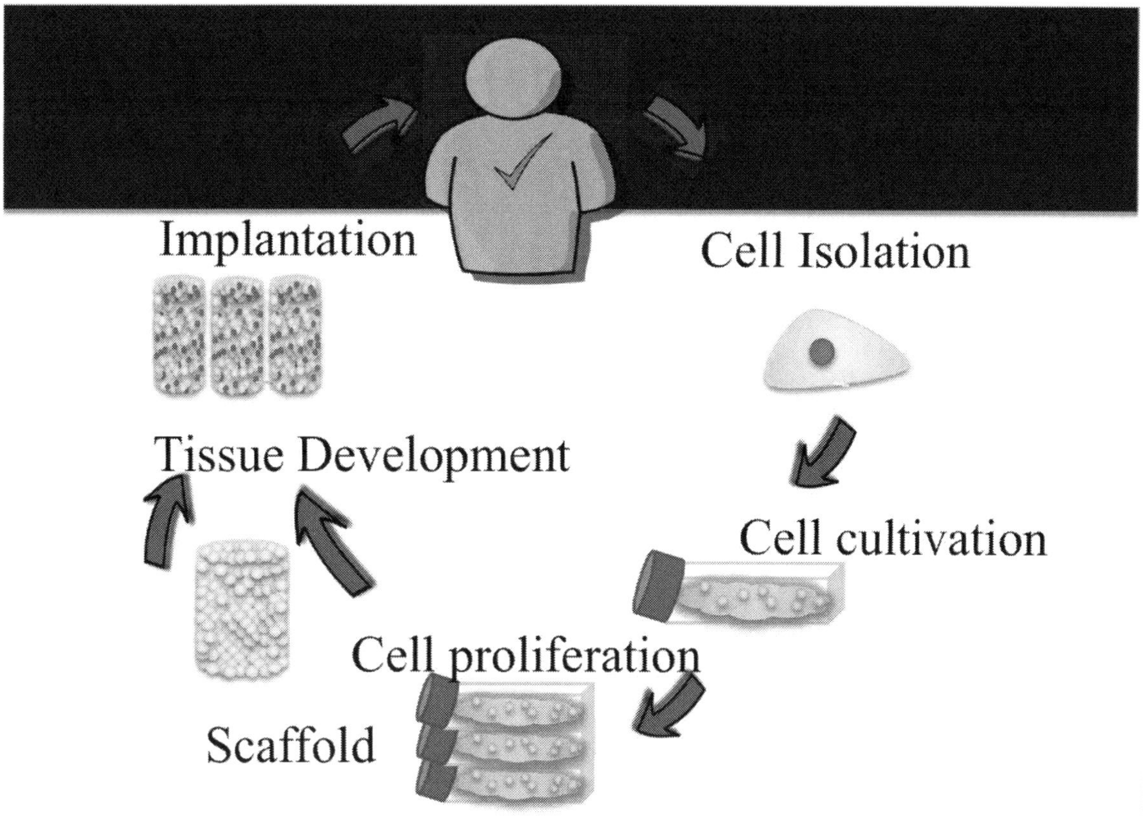

**Objective**

- To fabricate composite PLGA/$CaO_2$ sintered microsphere scaffolds as an alternative to current bone repair strategies
- To perform a degradation study to analyze the kinetics and bioactivity of released $CaO_2$

**Hypothesis**

- The hydrolytic degradation of composite PLGA/$CaO_2$ 3-dimensional (3D) scaffolds would result in a measurable *in vitro* release of the degradation products such as calcium and hydrogen peroxide over 21 days

UCONN | SCHOOL OF ENGINEERING

## Hydrolytic degradation of calcium peroxide

•PLGA scaffolds loaded with calcium peroxide produces $H_2O$ as a degradation product

- **$CaO_2 + H_2O \rightarrow CaO + H_2O_2$**
- **$2H_2O_2 \rightarrow O_2 + 2H_2O$**
- **$CaO + H_2O \rightarrow Ca(OH)_2$**
- **$Ca(OH)_2 \rightarrow Ca^{+2} + 2(OH)$**

UCONN | SCHOOL OF ENGINEERING

## Microsphere Scaffolds

**Why microspheres?**

- Simple Fabrication
- Control over morphology
- Physicochemical Characteristics
- Versatility of controlling release kinetics of encapsulated factors
- Polymer offers flexibility in degradation kinetics

**0.5% $CaO_2$**
**25x SEM magnification**

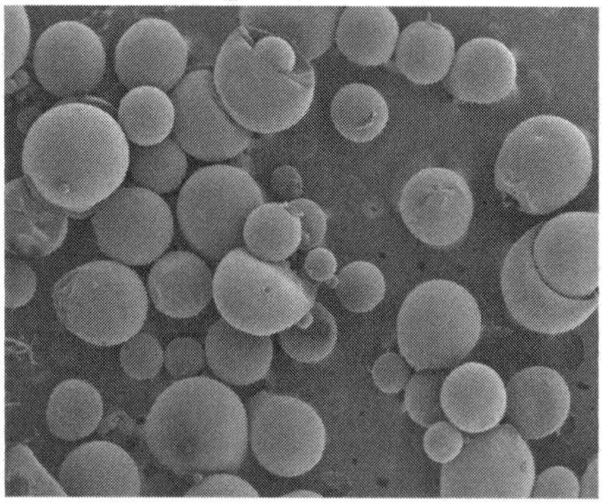

UCONN | SCHOOL OF ENGINEERING

# Sintered microsphere scaffold fabrication

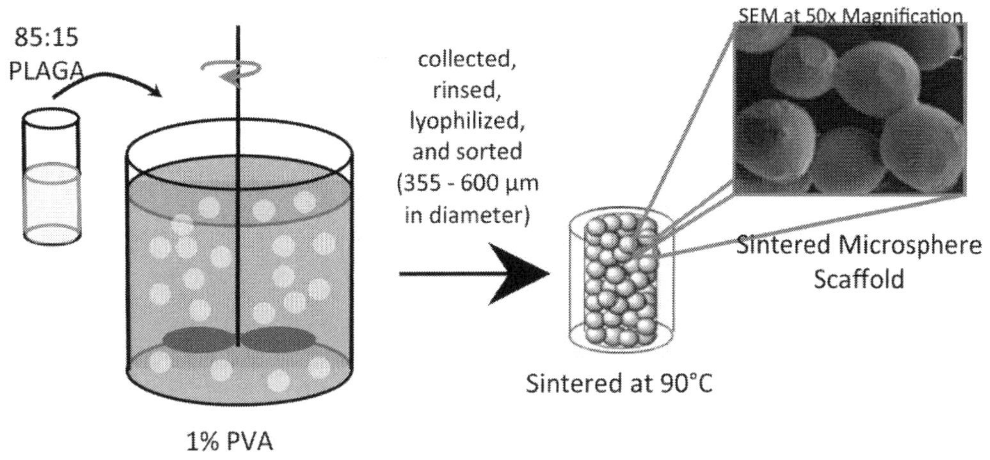


UCONN | SCHOOL OF ENGINEERING

# Degradation study

- Calcium Release: The release of $Ca^{2+}$ ions was determined by submerging scaffolds in 1 ml of calcium-free phosphate buffered solution with 500 μL samples taken and replaced at 1, 2, 4, 8 and 12 hours, and 1, 2, 3, 4, 5, 6, 7, 9, 11, 13, 15, 17, 19, and 21 days.

- Quantities of $Ca^{2+}$ were measured using the Calcium (CPC) Liquicolor kit and the hydrogen peroxide was measured with a PeroXOquant Quantitative Peroxide kit

UCONN | SCHOOL OF ENGINEERING

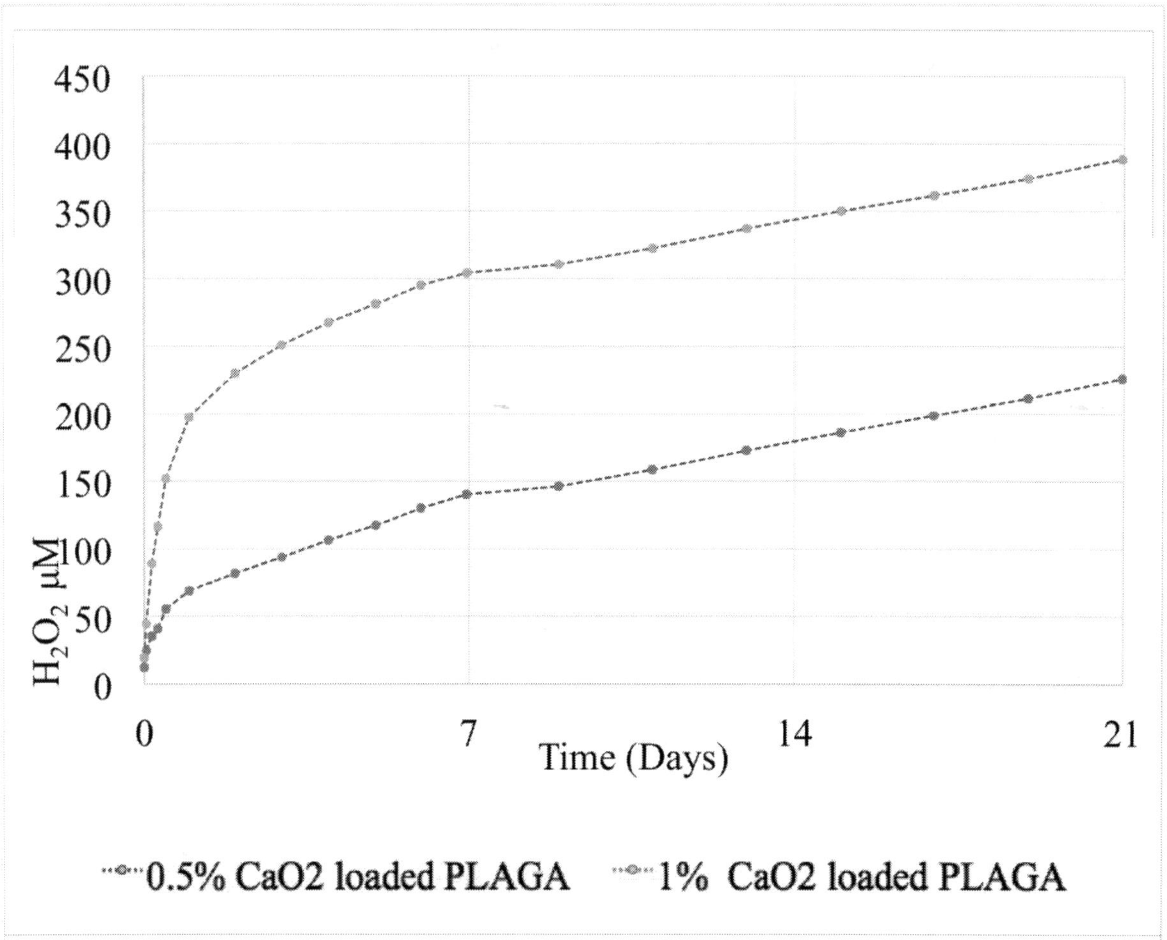
450
400
350
300
250
200
150
100
50
0
$H_2O_2$ μM
0
7
14
21
Time (Days)
0.5% CaO2 loaded PLAGA
1% CaO2 loaded PLAGA

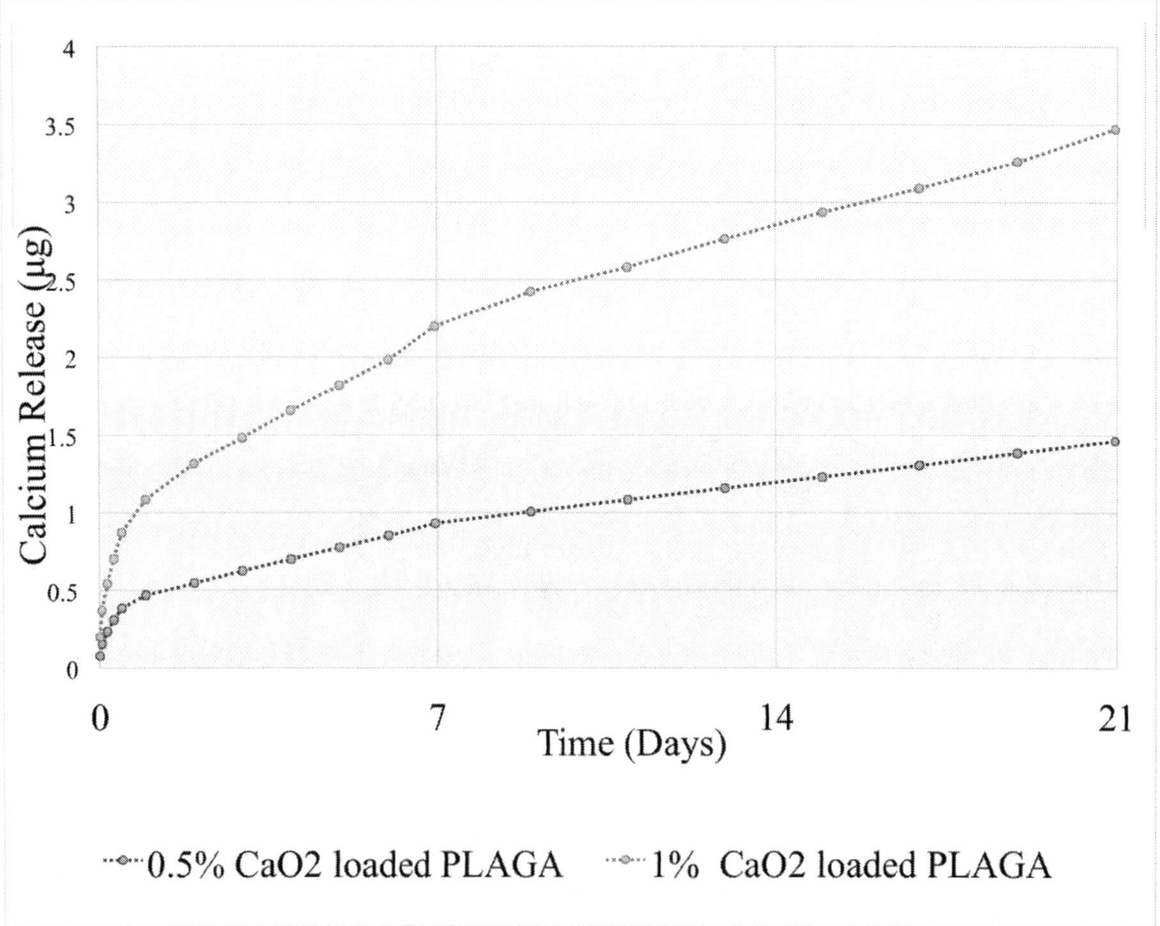
4
3.5
3
2.5
2
1.5
1
0.5
0
Calcium Release (μg)
0
7
14
21
Time (Days)
0.5% CaO2 loaded PLAGA
1% CaO2 loaded PLAGA

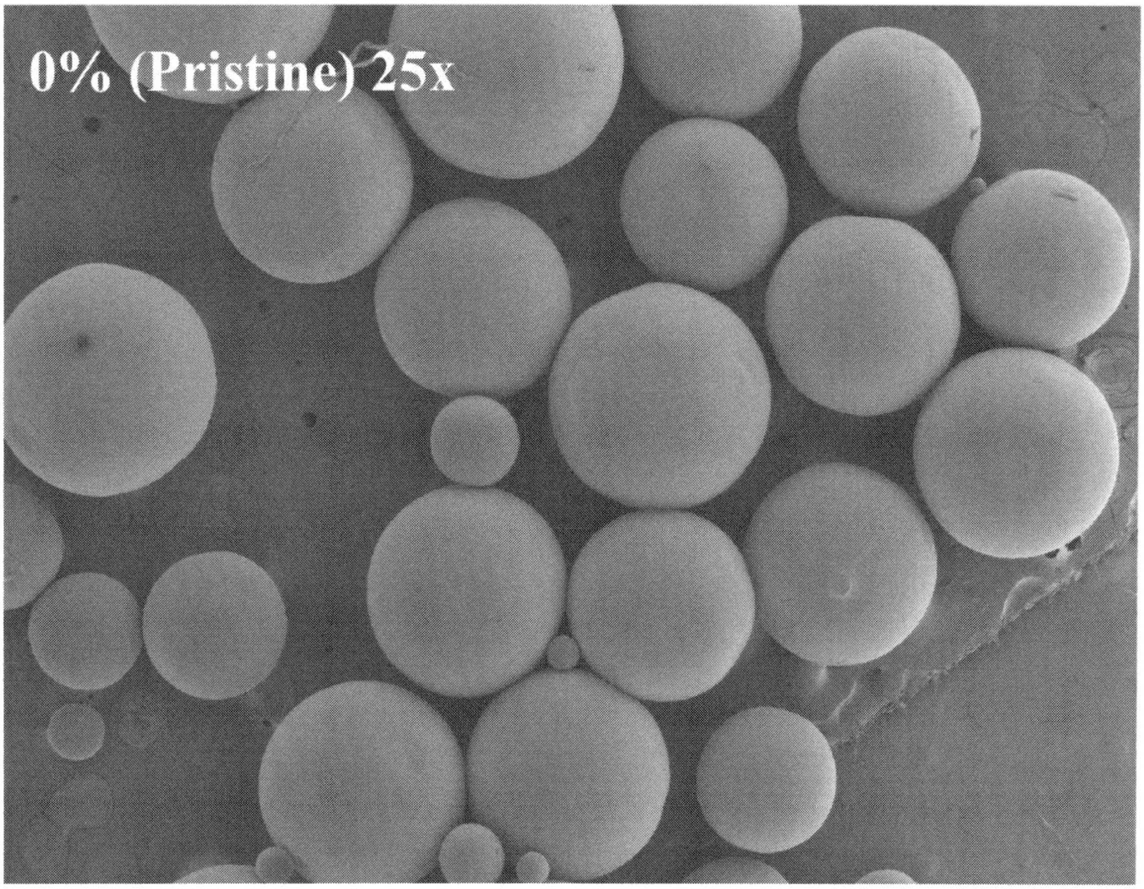
0% (Pristine) 25x

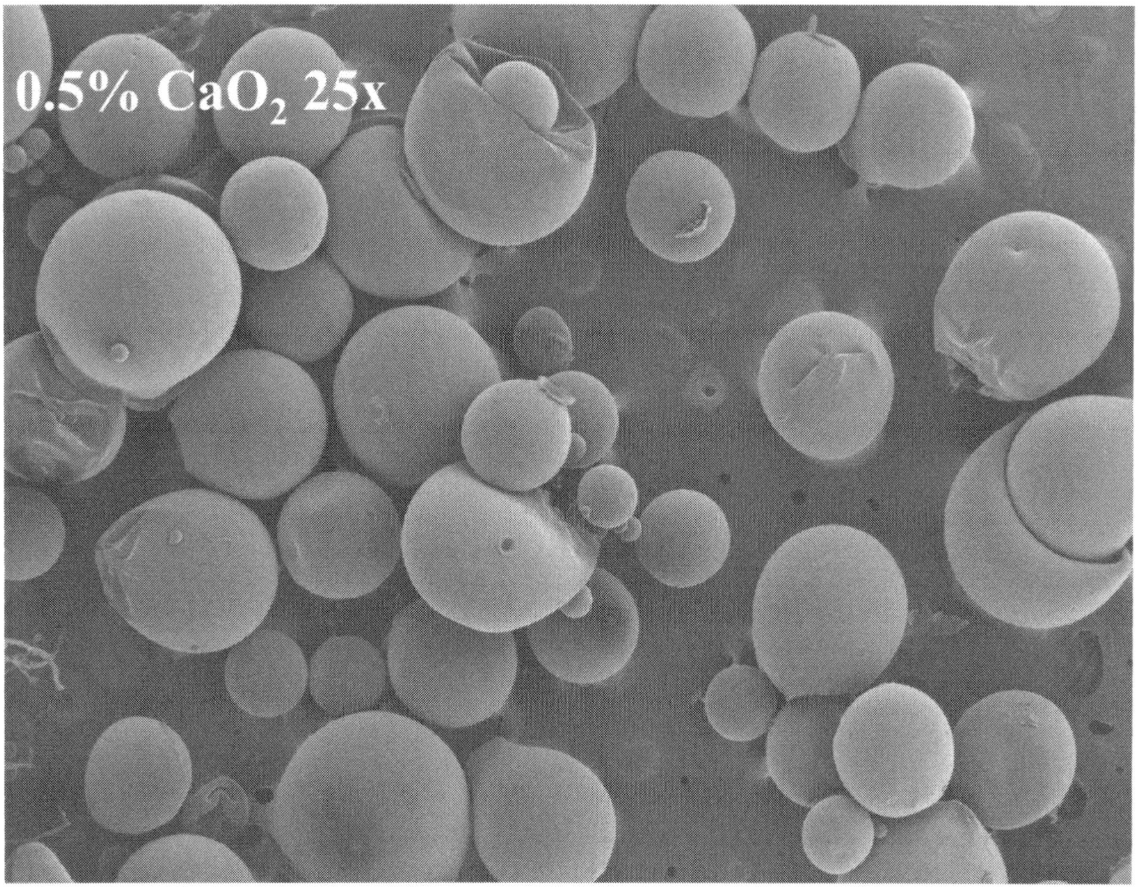
0.5% CaO2 25x

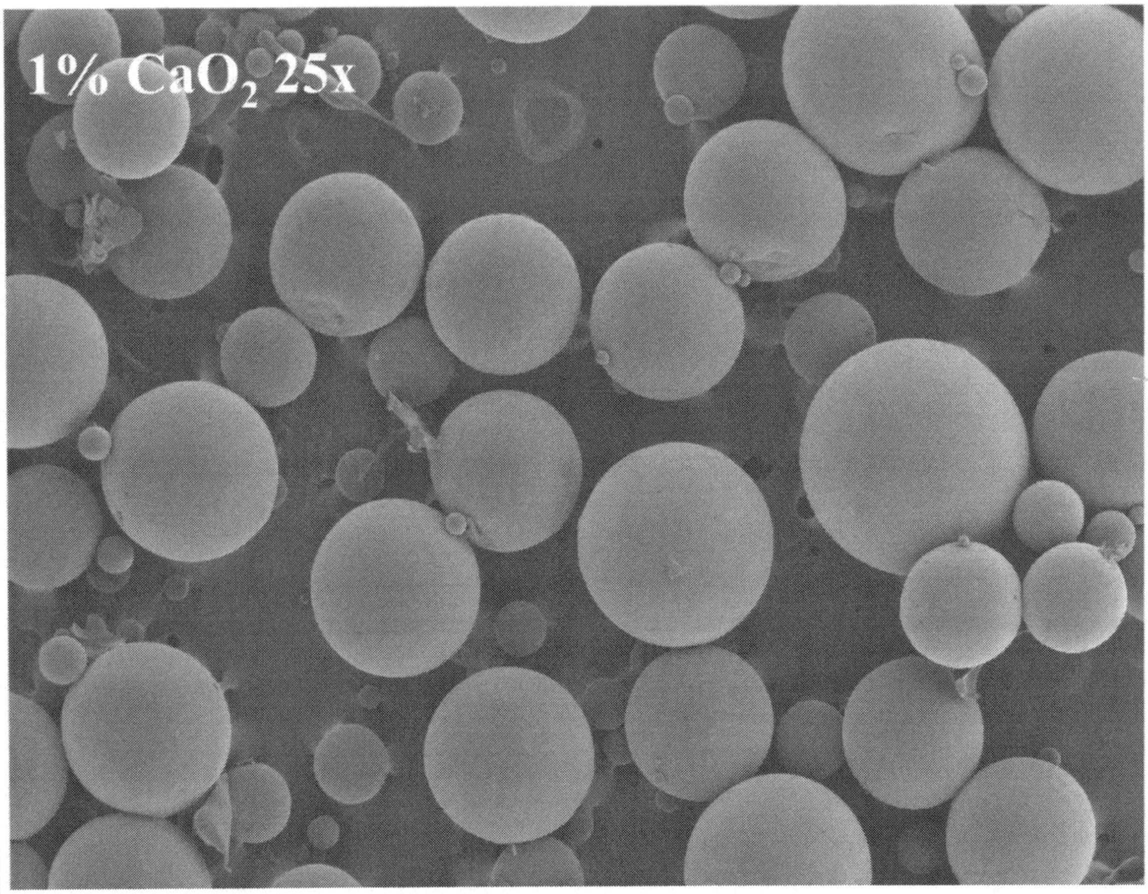


## Scaffolds 5mm

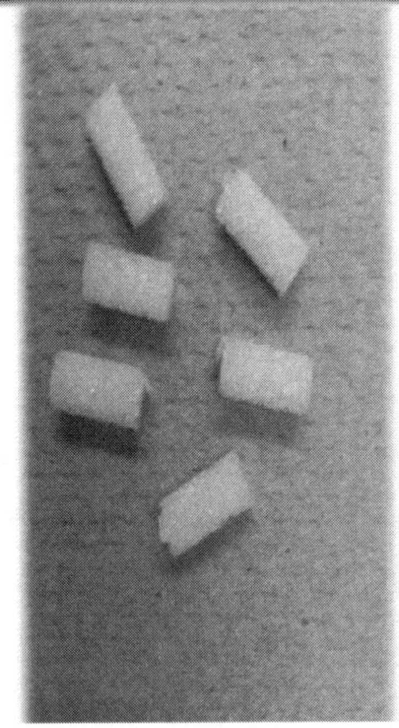

# Summary

- The *in vitro* release kinetics revealed the structure-function relationship of the material to develop a correlation between *in vitro* and *in vivo* release.
- Calcium levels for low concentration of $CaO_2$ on the scaffolds increased as time increased. Hydrogen peroxide released levels increases as the concentration of calcium peroxide increases.
- Loading scaffolds with calcium peroxide does not affect its morphology, this is important for cells mobilization, adhesion, proliferation, differentiation and signaling.

- future studies:
  - the observed calcium release could influence the osteogenic differentiation of stem cells without other supplemental factors
  - as noted in recent studies, This new treatment could increase patient access to orthopedic treatments as it is designed to require fewer surgical procedures and no additional technological assistance.

# Thank you!

- McNair Scholars Program
- Rowe Scholars Program
- Yusuf Khan, Ph.D.
- Ph.D. Candidate, Keisha Ashe
- Renée Gilberti, Ph.D.

UCONN | CENTER FOR ACADEMIC PROGRAMS

RONALD E. MCNAIR
POST-BACCALAUREATE
ACHIEVEMENT PROGRAM

UCONN | SCHOOL OF ENGINEERING

# 419875 Local Delivery of rhBMP-2 from a Compression-Resistant Graft in a Canine Lateral Ridge Augmentation Model

**Abstract:**

**Statement of Purpose:** Large mandibular defect reconstruction presents a continual challenge in oral and maxillofacial surgery. Growth factors such as recombinant human bone morphogenetic protein-2 (rhBMP-2) incorporated in scaffolds for tissue engineering promote cellular infiltration, induce osteoblast differentiation, and enhance new bone formation. RhBMP-2 is often delivered via an absorbable collagen sponge (ACS), but this may require a secondary space maintenance system, such as titanium mesh or polymer membrane[1]. Biodegradable polyurethane (PUR) biocomposites are reported to be effective carriers for rhBMP-2 and support new bone growth[2]. Mastergraft ceramic (CM) is an osteoconductive, biphasic ceramic with mineral content similar to bone. PUR/CM composites are osteoconductive and have sufficient stiffness to resist soft tissue prolapse. In the present study, we investigated space maintenance and new bone formation with an injectable PUR/CM composite with and without rhBMP-2 in a canine lateral ridge augmentation model.

**Methods:** The biodegradable PUR was synthesized from a lysine triisocyanate (LTI) and polyethylene glycol (PEG) prepolymer, a polyester triol (450 g/mol), and triethylene diamine catalyst. Treatment groups included the composite containing 45 wt% CM with no rhBMP-2, a low dose of 200 µg/mL rhBMP-2, or a high dose of 400 µg/mL rhBMP-2 (n=4/group). An ACS with a titanium mesh incorporating 400 µg/mL rhBMP-2 was used as a clinical control. For the PUR composites, lyophilized rhBMP-2 was hand-mixed with the PUR and injected into a lateral ridge defect (2/animal) measuring approximately 13-14 mm mesiodistally, 8-9 mm apico-coronally, and 3-4 mm bucco-lingually. The volume of the injected graft allowed for the expansion to 45-55% porosity to fill the defect space, which occurred within a set time of 7-9 min. After this time, the soft tissue was closed over the defect. For the ACS samples, the rhBMP-2 was resuspended in sterile water and absorbed on the collagen sponge for 15 min prior to placement in the defect. A piece of titanium mesh was fastened over the sponge to prevent soft tissue prolapse. Animals were sacrificed at 16 weeks and new bone formation evaluated by radiographs, µCT, histology, and histomorphometry.

**Results:** In a previous study we compared PUR/CM and PUR/bioactive glass (BG) composites with a low (100 µg/mL) or high (400 µg/mL) dose of rhBMP-2 in a canine saddle defect model. For both PUR/CM and PUR/BG groups, a high dose of rhBMP-2 led to better preservation of the host ridge width as measured by µCT. PUR/CM composites (a slow-resorbing ceramic) led to improved space maintenance and more new bone than PUR/BG composites (a fast-resorbing ceramic). In the canine lateral ridge augmentation study, the compression-resistant PUR/CM composites are compared to an ACS (clinical control), which requires titanium mesh for space maintenance. To investigate dose dependence on new bone formation, the PUR/CM composite was augmented with 0, 200, or 400 µg/ml rhBMP-2. The canine lateral ridge augmentation study is designed to answer the following questions: (1) Will the biocomposite bone grafts maintain space and prevent prolapse in lateral ridge defects without protective membranes? (2) What is the optimal dose of rhBMP-2 from the PUR/CM composites? (3) How does defect healing compare between PUR/CM composites and the ACS control? In initial surgeries, clinical placement of the PUR/CM composites revealed the ease of conforming to the osseous defects. Once set, the composites maintained their shape during suturing and all sites healed satisfactorily. Preliminary µCT results at 16 weeks showed more new bone formation in high dose PUR/CM composites when compared to composites without rhBMP-2. Work is ongoing to test how the PUR/CM high dose composites compare to the ACS control.

**Conclusions:** PUR/CM composites with a high dose of rhBMP-2 support new bone formation and remodeling in a canine saddle defect model. We have applied this biocomposite to a canine lateral ridge augmentation model to test the hypothesis that the compression-resistant PUR/CM composite without protective membranes is an effective alternative for rhBMP-2 delivery compared to an ACS control.

Acknowledgements: This work was supported by the Armed Forces Institute of Regenerative Medicine (W81XWH-08-2-0034) and Medtronic, Inc.

**References:**

1. von Arx T. Int. J. Oral Maxillofac. Surg 2002;31:190-199.

2. Li B. Biomaterials 2009;30:6768-6779.

# Nerve Regeneration Using Nerve Growth Factor and Lysophosphatidylcholine

Ryan Wood, Matthew Landeen, Tyler Brown, Austin Thompson, Jonathan Jacobs, Trent Taylor, Stephen Wirthlin, Stephen Wong, Keaton Karlinsey, Mark Rigby, Alonzo Cook

# Wallerian Degeneration

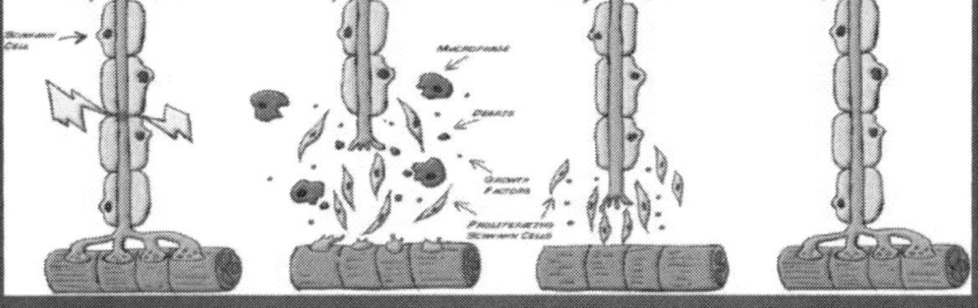


Following injury, Schwann cells detach from the axons, start proliferating and help the recruited macrophages to clear the cellular and myelin debris. At the same time, expression of stimulating factors by Schwann cells create a favorable environment for nerve growth.

Faroni, A. et al. **Peripheral nerve regeneration: Experimental strategies and future perspectives** Advanced Drug Delivery Reviews, Volumes 82–83, 2015, 160–167 http://dx.doi.org/10.1016/j.addr.2014.11.010

# Nerve Growth Environment

- During development, neuronal growth is directed by proteins to the target tissue. These proteins include surface-bound matrix molecules, neurotrophins, and growth factors.
- After injury, neurotrophin receptors are upregulated on the distal portion of the nerve, on denervated Schwann cells and on the growth cones of regenerating axons.
- Research has found that Schwann cells near the target tissue could prevent nerve re-innervation after long periods of denervation.

# Current Solutions

- Reattachment
- Autograft (Gold Standard)
- Allograft (Not yet widely used)
- Nerve Conduits (Surgical alternative)

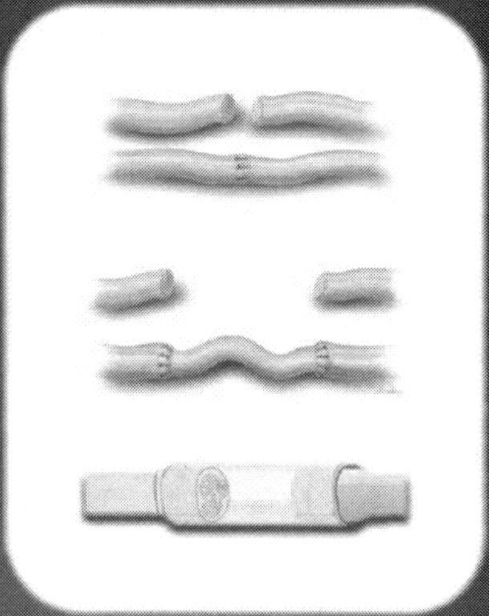

## Chemical Treatments

- Some Examples:
  - Neurotrophic Factors
  - Support Cells
  - ECM Proteins

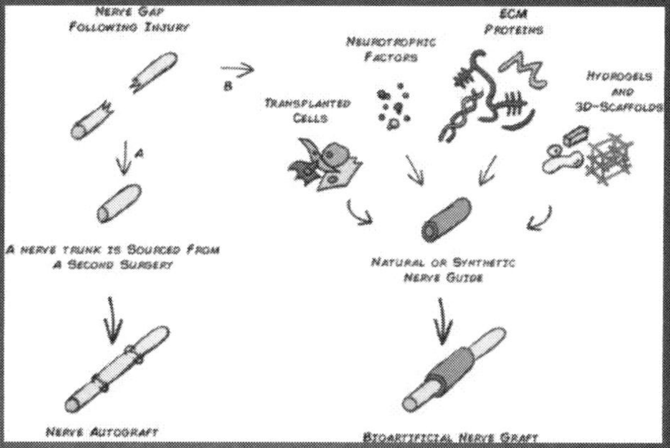


## Nerve Growth Factor (NGF)

- NGF is the most characterized neurotrophin and is known to be generated by the cells of target tissues, such as skin, muscle, etc.
- NGF supports survival, growth, and differentiation of neurons in the peripheral nervous system.
- NGF acts through two-types of receptors, the tropomyosin-related kinase A (TrkA) and the p75 receptor for neurotrophins.

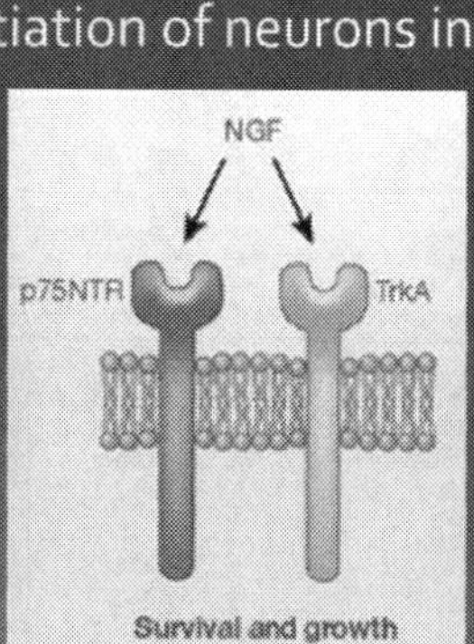

# Lysophosphatidylcholine (LPC)

- LPC is a bioactive lipid that gets released from the plasma membrane and the hydrolytic degradation of phosphatidylcholine.
- LPC acts on both the nerves and the on the Schwann cells surrounding the nerves.
- With the nerves, LPC acts through phosphorylation of the Trk-A receptor in the presence of NGF which helps prevent neurite death.
- With Schwann cells, LPC induces prompt demyelination through phospholipase activity and promotes NGF receptor expression.

# Proposed Solution

- Lysophosphitydalcholine injections will increase expression of the Nerve Growth Factor receptors which, when coupled with Nerve Growth Factor administration, will result in an increased rate of regeneration.

- 3 main groups: Crush, Crush + NGF, Crush + NGF + LPC

# Results

- Electrophysiology

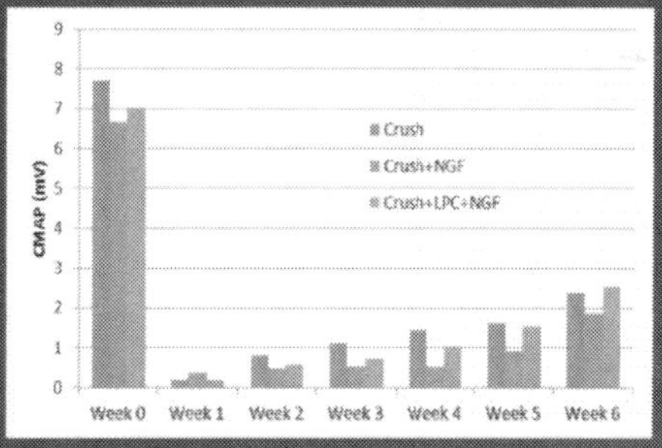


- Histology

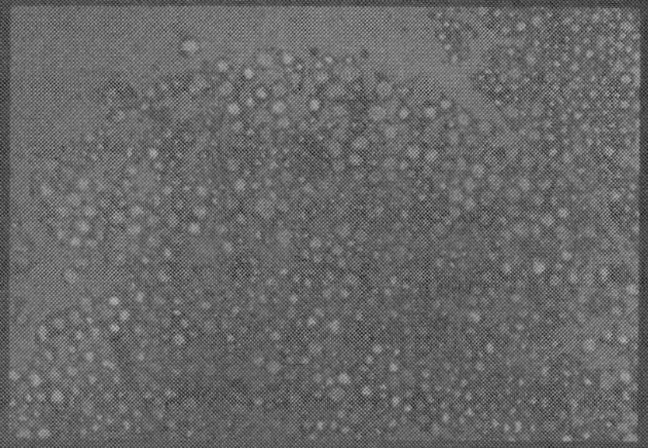

# Future Work

- Optimizing NGF and LPC concentrations
- LPC with Multiple NGF injections
- Test healing capacity of just electrical stimulation
- Compare the early stages of demyelination between a crush and a crush receiving LPC

# The Lab

- Dr. Alonzo Cook
- Dr. Scott Steffensen
- And all the undergrads!!

- Funding from Brigham Young University

# 433822 Rational and Combinatorial Analysis of 3D Biomaterials for Optimized Responses of Neural and Non-Neural Cells for Neural Tissue Engineering

**Abstract:**

Peripheral nervous system (PNS) regeneration after an injury is limited by presence of scar tissue, inhibitory myelin, cell death and lack of growth promotive substrate. Rational design of biomaterials supporting neurons and non-neural cells (Schwann cells and endothelial cells) should lead to an ideal growth promotive nerve guidance channel. We hypothesize rational selection of scaffold, matrix proteins and growth factors and screening of cell responses to combinatorial library will lead to discovery of optimal tissue-specific biomaterial supporting all cell types. To address this goal, we have designed and developed a novel high-throughput screening platform for generation of composite biomaterials and evaluation of cell responses. Composite libraries are generated using a high-throughput printer which combines scaffold and matrix protein at relevant concentrations. Cell encapsulated 3D biomaterials are cultured in microarrays and cell responses are evaluated through high throughput imaging. A confocal flatbed scanner and confocal high content array scanner are used for image acquisition and cell proliferation, morphology and metabolic activity are assessed. We evaluated the cell responses of HUVEC, brain endothelial cells, schwann cells and neurons. Collagen type I was used as base scaffold and matrigel, laminin, fibronectin and collagen type IV were used as matrix proteins. Relevant growth factors to cell type VEGF and FGF (endothelial cells), NGF and BDNF (neurons and schwann cells) are screened. High throughput screening results indicate cell response is dependent upon both collagen and matrix protein concentration and each cell type exhibited unique behavior. The high-throughput screening system provides an efficient way to screen combinatorial biomaterials in a cost-effective, resourceful manner and identify biomaterial 'leads'. The results cell-specific screens will provide useful information for designing optimized tissue-specific biomaterials.

# 422555 Recellularization of Whole Decellularized Porcine Hearts with Human Cardiac Fibroblasts and Endothelial Cells

**Abstract:**

**Recellularization of Whole Decellularized Porcine Hearts**
**with Human Cardiac Fibroblasts and Endothelial Cells**

Nima Momtahan, Beverly L Roeder, Alonzo D Cook

Cardiovascular diseases in the United States account for approximately 30% of all deaths and 60% on a global scale [1]. During myocardial infarction, a portion of the heart does not receive a supply of nutrients and oxygen which leads to irreversible cell damage. For patients who progress to end stage heart disease, the only option is to receive a heart transplant; however, the demand for heart transplants is higher than the supply of suitable hearts. Advances in technology (e.g. total artificial heart, ventricular assist devices, improved diagnostic techniques, etc.) have slowed the accelerating demand for end-stage organ transplantation, yet are not sufficient to satisfy the need. Clinical efforts to restore lost function after an infarction through cell delivery to the injured areas of the heart have not been completely effective [1]. Tissue engineered hearts that are produced by decellularization and then recellularization with the patient's own cells hold promising advantages such as creating an unlimited source of donor hearts [2]. Also tissue engineered hearts are non-immunogenic, meaning they will not be rejected by the recipient's body because the natural scaffold is non-immunogenic, and the cells are sourced from the same patient. Advances made in research of stem cells such as the discovery of iPS (induced pluripotent stem) cells as well as improved decellularization techniques that produce naturally vascularized scaffolds with preserved components, have brought a lot of hope to create the ideal tissue engineered heart.

DNA and some membrane proteins of the cells can initiate an immune response leading to the rejection of the tissue. Ideal decellularization will be achieved by completely removing all cells and cell debris from the heart, while keeping the heart scaffold's vasculature intact and also preserving the important growth factors, glycosaminoglycans (GAGs), and proteins in the extracellular matrix (ECM) [3]. In order to replicate the normal heart's function, the ECM should be reseeded with the necessary cell types and be able to provide support for the cells from preserved structural and signaling molecules. The heart consists of many cell types, including cardiomyocytes, endothelial cells, vascular smooth muscle cells, pericytes, and cardiac fibroblasts in which fibroblasts outnumber all other cell types.

In this work, we improved the decellularization process of porcine hearts in terms of time and cost by using a custom-designed perfusion apparatus. Automation systems were implemented in the system to control pressure and temperature during the process, as well as programmed gradual pressure increase. A combination of type-1 distilled water, phosphate buffered saline, sodium dodecyl sulfate and Triton X-100 with minimum detergent exposure to the ECM were used for the decellularization process. Decellularized samples were characterized by histology and immunohistochemistry staining and tested for residual DNA, GAGs and collagen. Biocompatibility of the decellularized ECM was demonstrated with a hemolysis assay which measured the rupture of erythrocytes in contact with the scaffolds. An immunogenicity test was performed by measuring the stimulation of macrophage cells when they were in contact with the decellularized ECM compared to a fresh heart.

After decellularization, the right ventricles of the hearts were dissected, cannulated through the main right coronary artery and any large open blood vessels were closed using a cyanoacrylate tissue adhesive (Vetbond®). Sections of the hearts were placed in sealed, agitated and oxygenated bioreactors and sterilized with a protocol optimized for being less destructive and enhancing cell attachment to the ECM. Sections were then perfused with growth medium for 24 hours and reseeded with $1\times10^8$ fluorescence-labeled human umbilical vein endothelial cells (HUVECs) and human cardiac fibroblasts (HCFs). The HUVECs were introduced through vasculature perfusion and HCFs by needle injection into the ECM at multiple time points. After at least 7 days of growth, histological images showed HUVECs dispersed throughout the right ventricle, covering the vasculature and HCFs present in the myocardium.

This study introduced and validated a decellularization method that is optimized for porcine hearts which are similar to human hearts in terms of size and anatomy. This method may be very beneficial in aiding the development of tissue engineered human hearts. We showed that the decellularized ECM has minimal DNA residue and stimulation of macrophages which means that the ECM should not be immunogenic when transplanted into the body. Also the hemolysis assay that was developed showed minimal erythrocyte cell rupture which demonstrates cell compatibility of the decellularized ECM. Furthermore, perfusion recellularization of the right ventricles with HUVECs and HCFs demonstrated the feasibility of recellularizing whole decellularized porcine heart tissues. The next steps will be towards repopulating the ECM with cardiomyocytes in addition to supportive cells, in order to obtain a functional heart tissue that can be transplanted into an infarcted heart to help restore its lost function.

1. Lanza, R., R. Langer, and J.P. Vacanti, *Principles of tissue engineering*. 4th ed. 2014: Academic press. p. 771-789.
2. Robertson, M.J., et al., *Optimizing recellularization of whole decellularized heart extracellular matrix*. PLoS One, 2014. **9**(2): p. e90406.
3. Momtahan, N., et al., *Strategies and Processes to Decellularize and Recellularize Hearts to Generate Functional Organs and Reduce the Risk of Thrombosis*. Tissue Eng Part B Rev, 2015. **21**(1): p. 115-132.

# 420095 Recellularization of Whole Porcine Kidneys with Human Epithelial and Endothelial Cells

**Abstract:**

Recellularization of whole porcine kidneys with human epithelial and endothelial cells

Nafiseh Poornejad, Evan M. Buckmiller, Jaron J. Lundwall, Ho Hin Ma, Beverly L. Roeder, Alonzo D. Cook

Brigham Young University, Provo, Utah

Whole organ engineering is a promising alternative for organ transplantation. Murine, porcine and human kidneys have been successfully decellularized and rat kidneys have previously been recellularized with human umbilical vein endothelial cells (HUVECs), rat neonatal kidney cells, and murine stem cells [1, 2]. In this study previous technology was scaled up to porcine kidneys as a model of human-sized kidneys. Decellularization of porcine kidneys was optimized using pressure control and tonic cycles and the effects of freezing/thawing conditions on decellularized extracellular matrix were investigated [3]. Sterilization methods for whole kidneys were also optimized to improve cell adhesion and proliferation. Human renal cortical tubular epithelium (RCTEs) cells and murine endothelial cells (MS-1) were used to repopulate the epithelial and endothelial niches of renal tissue, respectively. More than $6\times10^8$ endothelial or epithelial cells were perfused through the whole decellularized porcine kidneys for 30 min, either through the ureter for epithelial cells or through the artery for endothelial cells, and then media perfusion was continued for at least 7 days. Immunohistological staining demonstrated that tubules were repopulated with epithelial cells and endothelial cells were located in the complicated vasculature of the kidney. Metabolic activity of reseeded RCTEs was determined by resazurin reduction assay and glucose assay kit. The epithelial cells activity throughout the kidney will be checked by creatinine clearance measurement. This study demonstrated the feasibility of regeneration of whole human-sized kidneys using decellularized porcine kidneys and human cells.

[1] Ross EA et al. Journal of the American Society of Nephrology 20, 2009.

[2] Song JJ et al. Nature Medicine 19,2013.

[3] Poornejad N et al. Organogenesis 11, 2015.

# 431265 Three-Dimensional Rapid Prototyping of Vascular Substitutes for Medical Applications

**Abstract:**

## Three-Dimensional Rapid Prototyping of Vascular Substitutes for Medical Applications

Connor Dodge, Alex Bischoff, Jason Griggs, Kyle Larsen, Sarah Livingston, Sterling Rosqvist, Tomonori Baba, Tyler Hendricks, Alonzo D. Cook PhD.

Brigham Young University, Provo, Utah, USA

Autologous transplant of a patient's living veins is the most common clinical practice for atherosclerotic disease of the peripheral arteries. Often, it is non-ideal to transect a sick patient's veins to substitute for an artery. The search for an ideal arterial substitute has taken exciting new turns. The newest research focuses on recreating and then implanting arteries from a patient's own cells in vitro, with no immunological effects and having every functional property of a living artery. Our blood vessel research team has entered the tissue engineering field in its most exciting effort: the scalable rendering of cell-seeded vascular constructs with rapid prototyping machines or 3D printers. Gabor Forgacs Ph.D. and other researchers have pioneered the organ printing field. To date, these scientists have managed to build 3-dimensional scaffolds of organs and vasculature while simultaneously seeding living cells using the same machines.

We have built and are modifying a 3D printer that will allow us to test several different avenues of 3D organ genesis. We are currently testing the viability of lyophilized alginate gel, fibronectin coated alginate gel, and methacrylated alginate gel as bio-compatible organ scaffolds. Using these three materials as scaffolds, we hope to learn more about the ideal conditions for cell growth and replication on a polymer scaffold.

We plan to construct blood vessels by 3D printing a cylindrical shape, then preparing it in the correct manner; by either lyophilization, coating with fibronectin, or methacrylation, and attaching the structure to a pump and circulating live endothelial cells through the lumen of the polymer scaffold. We currently perform tests with MS1 endothelial cells, and plan to include smooth muscle cells in our upcoming tests to observe the interaction between endothelial and smooth muscle cells on a polymer scaffold. We expect that with the right pump speed and environment, endothelial cells will attach to the lumen of the polymer scaffold and align themselves as they replicate to cover the inside surface, creating a non-thrombogenic section of blood vessel.

To test functionality we will compare tensile strength with Young's modulus models, test for blood contacting with thrombogenicity measurements in partnership with Thrombodyne Inc., ensure synchronization of smooth muscle contractions, and eventually perform autologous animal tests to identify signs of immunogenic rejection.

### Acknowledgements

Funding was provided by Brigham Young University and the Office of Research and Creative Activities (ORCA).

# 433871 Kinetic Characterization of Proliferation and Dedifferentiation of Müller Cells

**Abstract:**

**Kinetic Characterization of Proliferation and Dedifferentiation of Müller Cells**

**Alonzo D. Cook, Chris Burns, Ryan Gillis, Steven Christiansen**

Approximately 11 million Americans suffer from dry age-related macular degeneration (dry AMD), an eye disease that occurs in the macula of the eye when retinal pigment epithelial (RPE) cells experience oxidative stress and die. In several animals (including rats, zebra fish, and chicks), photoreceptor regeneration is possible with the appropriate stimulus. Müller cells have been shown to be the critical cell in photoreceptor regeneration. In zebra fish and chicks, Müller cells dedifferentiate to a proliferating progenitor cell when photoreceptors are damaged. These progenitor cells are then transported to their final area in retinal tissue where they differentiate to photoreceptors and integrate into the retinal tissue. Activation of cellular signaling pathways (Notch, MAPK, and others) is integral to this process. On each level of regeneration (dedifferentiation, proliferation, differentiation, migration and integration), these signaling pathways guide cell behavior, producing new photoreceptors that replace dead or damaged photoreceptors. A therapy guided by this knowledge could replace more intrusive methods to potentially cure dry AMD.

Dedifferentiation and differentiation of Müller cells is a potential option for curing dry AMD. The therapy would involve inducing Müller cells to transform into the needed photoreceptors. This would likely involve initially dedifferentiating the Müller cell into a progenitor state and then inducing the progenitor cells to become photoreceptors. Dedifferentiation is the process by which a cell that has already developed into a fully functioning adult cell returns back to a previous state. The dedifferentiation step has been qualitatively observed in vitro. Determination of the proliferation kinetics was performed in-house by isolating Müller cells and obtaining time-resolved cell counts. A Wistar albino rat eye was dissected and the retina was removed, dissociated and cultured. The Müller cells were isolated and divided into several different 25 mL flasks. An initial cell concentration was measured with a microfluidic cell counter. At 12 hour time intervals, a flask was removed and cell counts were taken.

Assuming a 1$^{st}$ order model of proliferation dependent on cell concentration, the rate constant was found to be 0.09/hr for cells cultured with fetal bovine serum. This rate constant dropped to 0.03/hr when fetal bovine serum was removed from the growth media. This value matches data reported by Limb[1] and is similar to the value derived from data reported by Zhao[2]. The shape of the cells was observed to shift to a dendritic neuronal shape. The dedifferentiated nature of the cells was confirmed by immunocytochemistry, observed qualitatively in the cells used in the proliferation experiments, and quantitatively described by protein ratios according to the method of Goldman, et al[3]. The concentration of proteins typical of photoreceptor progenitor cells was assumed to be proportional to the number of cells, and the rate constant for dedifferentiation was determined to be 0.35/hr. This is significant because when combined with proliferation kinetic data, it is clear that dedifferentiation occurs much faster than proliferation.

This project will continue by repeating the proliferation kinetic experiments and attempting to determine the rate constants for differentiation into photoreceptors. Also the model will be modified to handle non-first order rates. Eventually, once the system is defined, the model will allow the optimal design of therapies for retinal degeneration.

References

1. Limb, Astrid; et al. Differences between the neurogenic and proliferative abilities of Müller Glia with stem cell characteristics and the ciliary epithelium from the adult human eye. *Exp Eye Res.* 93(6): 852-61. Dec 2011.

2. Zhao. Induction of Retinal Progenitors and Neurons from Mammalian Müller Glia under Defined Conditions. *J. Bio. Chem.* Feb 2014.

3. Goldman, Daniel; et. al. Ascl1a regulates Müller Glia dedifferentiation and retina regeneration via a Lin-28-dependent, let-7 miRNA signaling pathway. *Nat Cell Bio.* 12(11): 1101-7. Nov 2010.

# 438527 Controlled Release from Polyelectrolyte Complex Drug Carriers

**Abstract:**

Comparison studies of release of a model drug, Rhodamine B, from two different types of polyelectrolyte complex beads (PECs) made of poly(styrenesulfonate, PSS)-poly(allylamine, PAA) and chitosan-alginate were investigated. The release was carried out by using a modified Franz diffusion cell into deionized water at 37°C. The long term sustained release (> 7 days) of Rhodamine B was obtained with PSS-PAA PECs due to their tighter networks. By tuning the compositions of PSS-PAA, the time needed to completely release the drug could be varied from two days to 14 days. For the case of chitosan-alginate PECs, 100% accumulated release occurred within 24 hours. With the incorporation of a thermo-responsive polymer, e.g., poly(N-isopropylacrylamine), the release could be extended to two or more days for the chitosan-alginate PECs. By drying the beads, the release of Rhodamine B was found to be further extended.

# 438450 3D Printed Microfluidic Device for Dynamic Investigation of the Blood Brain Barrier

**Abstract:**

The "Blood Brain Barrier" (BBB), surrounding the capillaries in the Central Nervous System, is a tissue composed of endothelial cells joined to each other by tight junctions. It serves to impede the passage of any impurities from the blood to the brain and is essential to brain homeostasis, disruption of which can result in a myriad of brain disorders like Alzheimer's, Multiple Sclerosis, and Parkinson's disease. Current in-vitro models of the BBB include a seeded Transwell ® Membrane under static conditions, where the apical and basal side fluids do not exhibit dynamic flow and apply shear stress on the membrane. Being a simplified model of the tissue, it does not adequately model the tissue as it is in-vivo, where there is dynamic flow of both basal (cerebrospinal fluid) and apical (blood) fluids. Consequently, we have designed a microfluidics device, the μTRANS chip, to model the BBB under dynamic conditions, allowing us to study not only the tissue structure of the BBB but also the cellular level dynamics of it under different environmental conditions. Our initial results, revolving around the construction of the device, its characterization, and the biochemical/mechanical modulations of the tissue will be presented at the conference. In terms of construction, the device was built using a seeded Transwell ® membrane (consisting of a layer of human brain endothelial cells on the apical side and a layer of astrocytes on the basal side) inserted between two glass slides. Each of the glass slides served as a distinct chamber, containing two 3D printed micro-channels, interconnected by smaller micro-channels. The smaller channels were designed to allow for fluid exchange between the larger micro-channels. Gold electrodes were also built into the device to analyze the permeability of the BBB using electrochemical impedance spectroscopy (EIS). The channels, glass substrate layers, and Transwell ® Membrane were all placed into a 3D printed vacuum chamber as the final step in construction, in order to maintain contact between the different layers of the device during experimental modulations. The permeability characteristics of the BBB were studied using fluorescence and EIS, specifically in the form of transendothelial electrical resistance tests (TEER), under different biochemical and mechanical conditions, including varying levels of pH, shear stress, and nutrients. We believe that our ability to study the BBB in a dynamic environment allows us to not only accurately model the tissue under physiological conditions but to control its characteristics using biochemical and mechanical means.

# 438448 Self-Assembly Simulations of Polymer Functionalized Virus Capsids

**Abstract:**

Viral capsids have been the subject of a significant amount of research in a variety of fields. Their potential relevance to biomedical science arises from their ability to encapsulate therapeutic agents and deliver them to targeted cellular sites. The capsids can be decorated with polymers to target specific sites or promote controlled aggregation for creating novel materials. We used coarse-grained molecular dynamics to simulate the aggregation dynamics of amphiphilic functionalized icosahedral capsids, representing the Cowpea Mosaic Virus capsid, when grafted with diblock copolymers of polyethylene glycol and polylactic acid. Additional variations of the system includes changing the polymer grafting density and capsid volume fraction to show how these changes affect the self-assembled structures. Our observations suggest that (a) the excluded volume of the icosahedral and (b) the polymer grafting density play the largest role on aggregation dynamics and morphologies when compared to the other variants.

# 438410 Investigation of the Interaction Between a Novel Drug Delivery System and an Epithelial Cell Layer

**Abstract:**

Surgery and other invasive processes are the current treatment options for patients with diabetic retinopathy, glaucoma, and age-related macular degeneration, which are eye diseases that affect at least nine million Americans. We are developing a targeted drug-delivery system, which would provide a non-invasive option for treatment. The system features a collagen membrane with drug-loaded poly(lactic-co-glycolic acid) (PLGA) nanoparticles incorporated into a thin membrane. This membrane can be worn by a patient in conjunction with a contact lens, and provides for continuous drug release. The purpose of this project is to study the interactions of lidocaine-loaded nanoparticles in a collagen membrane with a layer of corneal epithelial cells. Lidocaine-loaded PLGA nanoparticles were made and characterized by measuring size and drug content. Nanoparticles were mixed into a collagen solution and then gelled to form a thin membrane. Corneal epithelial cells were cultured on semipermeable well inserts to form a confluent layer of cells. The cell layer was exposed to the collagen membrane containing the lidocaine-loaded nanoparticles. Measurements were taken to determine cell viability and permeability, as well as the amount of lidocaine that crosses the cell layer. Results showed that lidocaine was released from the system and delivered across the cell layer. The release profile shows an initial burst release, followed by steady drug release over the testing period. Cell viability decreased with the application of the membranes, compared to a control sample of cells only. Further work is being done to determine the cause of cell death and to test the system for longer periods of time

# 438407 Quantitative Analysis of Fundus Images for Grading of Vitreous Haze

**Abstract:**

**Purpose:** Intraocular inflammatory processes lead to progressive accumulation of cells and protein exudate in the vitreous, which can be detected by ophthalmoloscopic inspection of the eye as decreased visibility of the retinal vasculature, optic nerve head, and other fundus details. The objective of this work was to develop a quantitative method for grading the blurriness of ocular fundus images in order to provide automated grading of the severity of uveitis.

**Methods:** A computer algorithm was developed to quantify image blurriness through application of entropy filtering and Fourier spatial-frequency analysis. The algorithm was refined using a set of 8 reference images that were created by optically filtering a single standard fundus picture to varying degrees that simulate increasing vitreous haze. It was then applied without modification to a dataset of 120 digital fundus images collected from patients with different grades of uveitis. The algorithm performance was evaluated against the acutance method, which is commonly used to quantify the overall sharpness of edges in an image. Computed scores of vitreous haze were also compared for both methods in a masked fashion with the subjective readings of three expert clinicians.

**Results:** The Pearson correlation coefficient between computed automated grades of the algorithm and clinician grades of blurriness was 0.81-0.88 ($R^2$ = 0.66-0.78). Acutance measurements calculated from the raw images and from images with background fluctuations subtracted out did not perform as well. The Pearson correlation coefficient between acutance and clinician grades was found to be -0.05 (R = 0.003) and 0.33 (R=0.109), respectively, which is quite weak.

**Conclusions:** We have developed a computer algorithm for grading vitreous haze in an unbiased and quantitative manner that correlates strongly with the subjective readings of expert clinicians. The algorithm performance with fundus images is much better than that of standard acutance measures.

McCormick

Northwestern Engineering

Department of Chemical and Biological Engineering

# Developing a Strategy for Constructing Modular Biosensors

Andrew K.D. Younger[1], **Neil C. Dalvie[2]**, Austin G. Rottinghaus[3], and Joshua N. Leonard[2,4,5,*]

[1] Interdisciplinary Biological Sciences Graduate Program, Northwestern University, Evanston, IL, 60208, USA
[2] Department of Chemical and Biological Engineering, Northwestern University, Evanston, IL, 60208, USA
[3] Department of Chemical and Biological Engineering, University of Wisconsin-Madison, Madison, WI, 53706, USA
[4] Chemistry of Life Processes Institute, Northwestern University, Evanston, IL, 60208, USA
[5] Member, Robert H. Lurie Comprehensive Cancer Center, Northwestern University, Evanston, IL, 60208, USA

1

McCormick

Northwestern Engineering

Department of Chemical and Biological Engineering

# Motivation: Quantifying metabolism

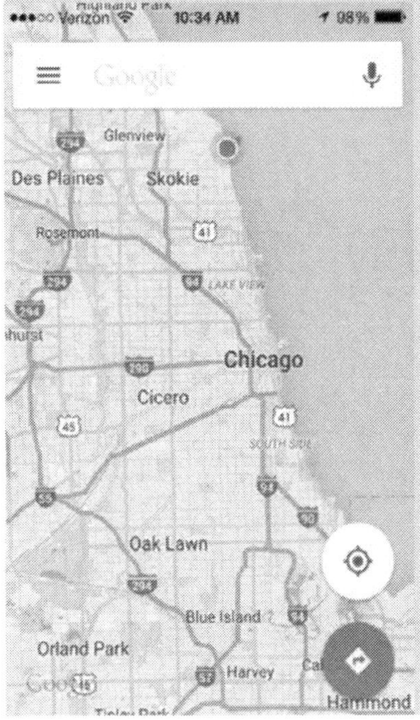


2

McCormick

Northwestern Engineering

Department of Chemical and Biological Engineering

# Motivation: Quantifying metabolism

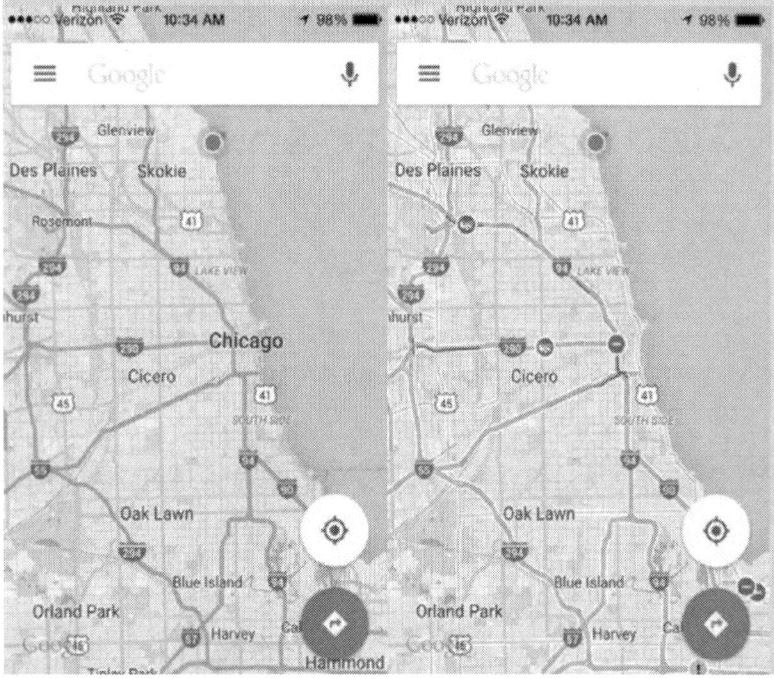


3

McCormick

Northwestern Engineering

Department of Chemical and Biological Engineering

# Motivation: Quantifying metabolism

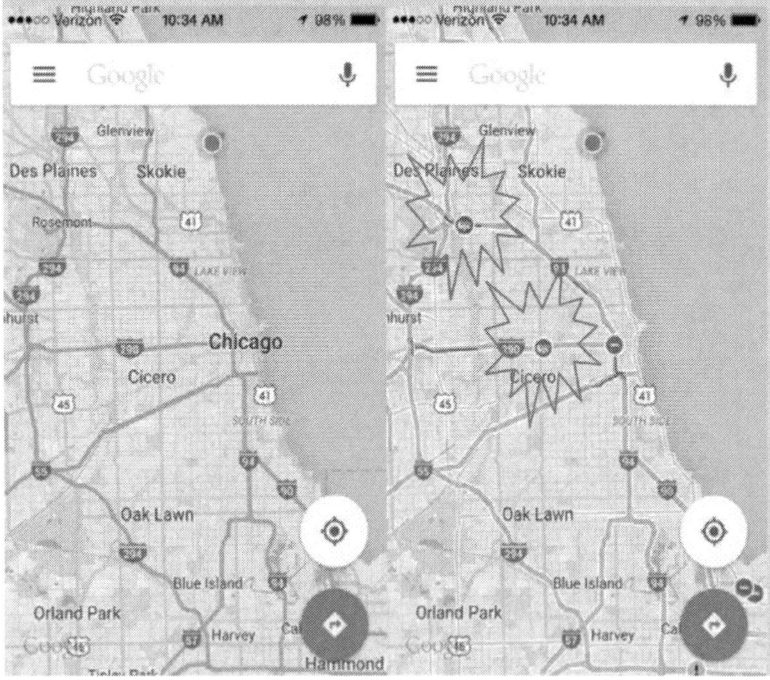


4

**McCormick**

Northwestern Engineering

**Department of Chemical and Biological Engineering**

# Motivation: Quantifying metabolism

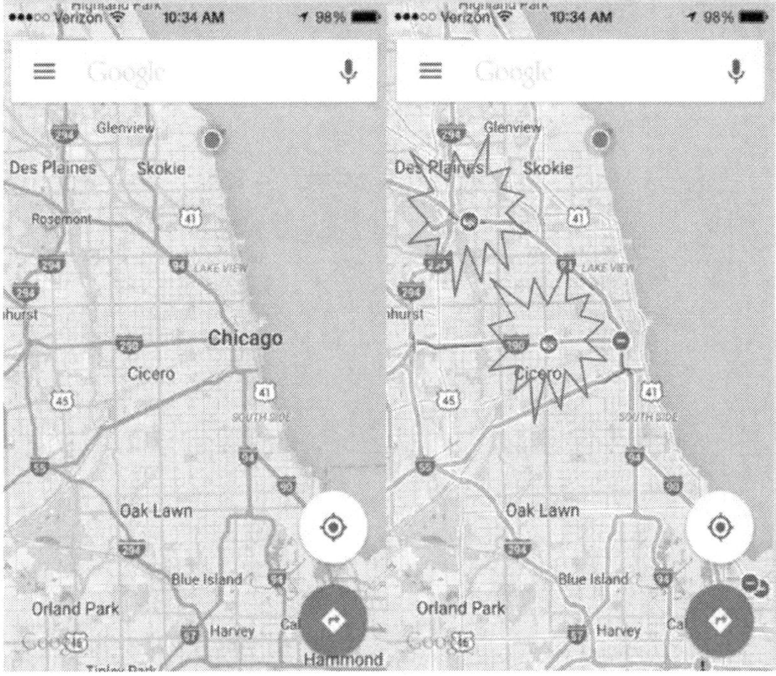


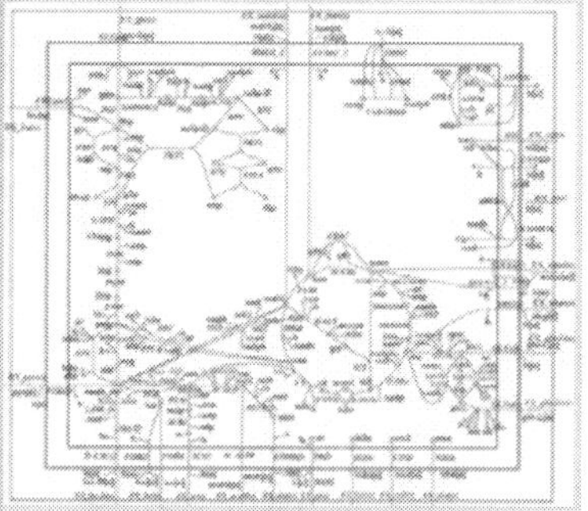

5

Orth *et al.* 2011

**McCormick**

Northwestern Engineering

**Department of Chemical and Biological Engineering**

# Motivation: Quantifying metabolism

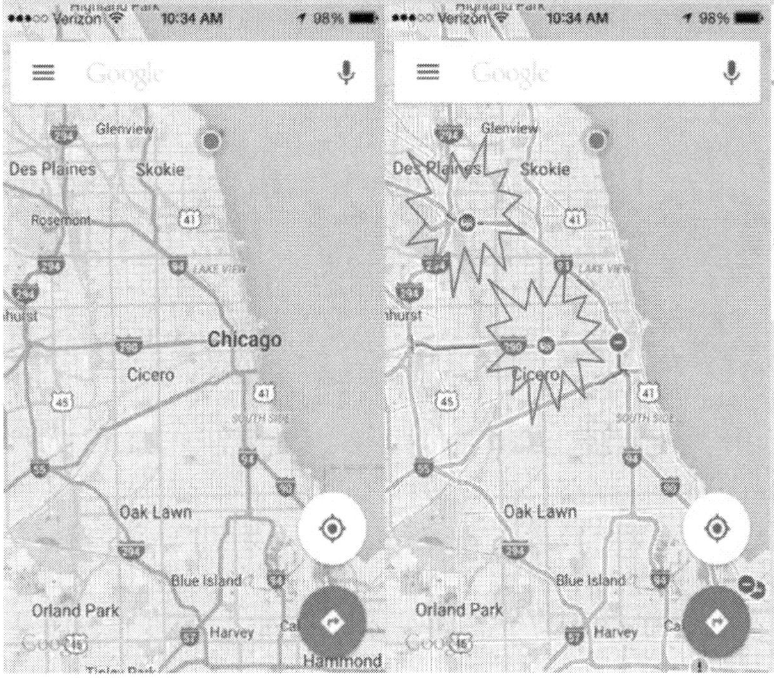


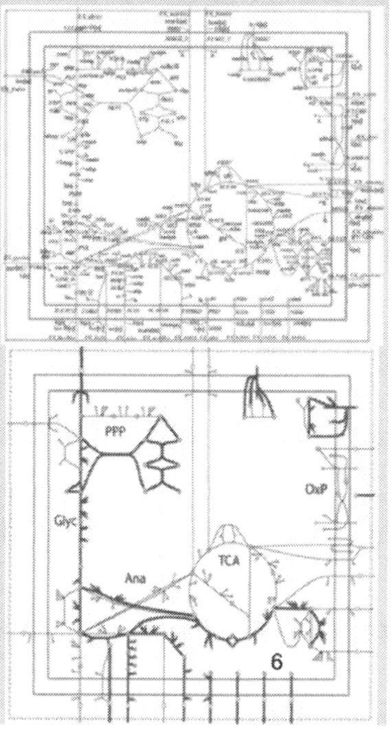


6

McCormick

Northwestern Engineering

Department of Chemical and Biological Engineering

# Useful functions of metabolite biosensors

## High Throughput Library Screening

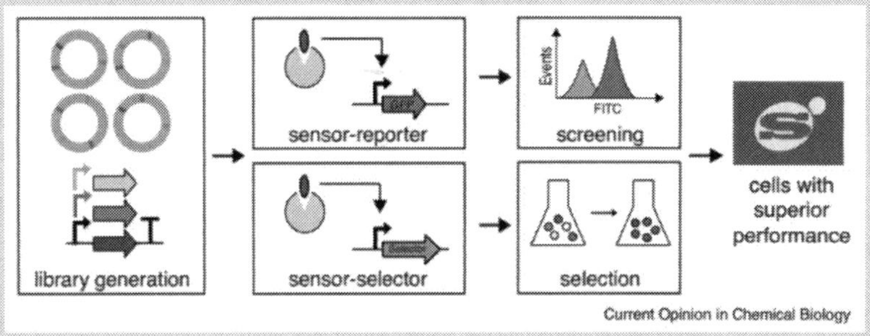


7

Zhang *et al.* 2015

McCormick

Northwestern Engineering

Department of Chemical and Biological Engineering

# Useful functions of metabolite biosensors

## High Throughput Library Screening

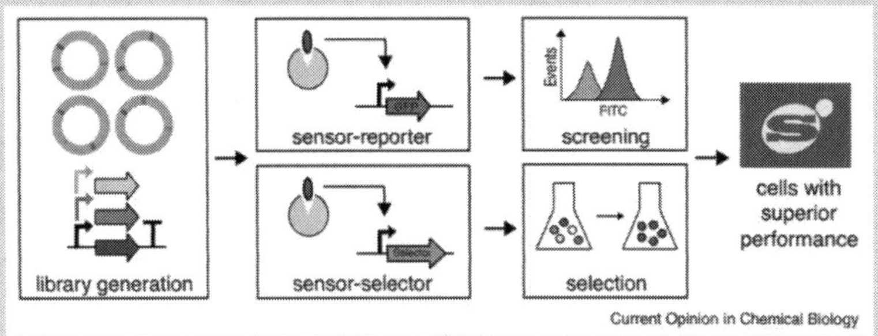


## Feedback control

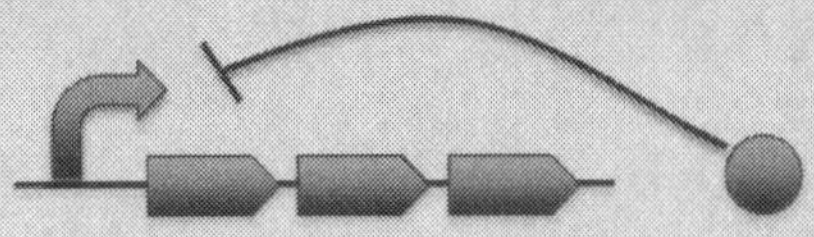

8

Zhang *et al.* 2015

**McCormick**

Northwestern Engineering

**Department of Chemical and Biological Engineering**

# Useful functions of metabolite biosensors

High Throughput Library Screening

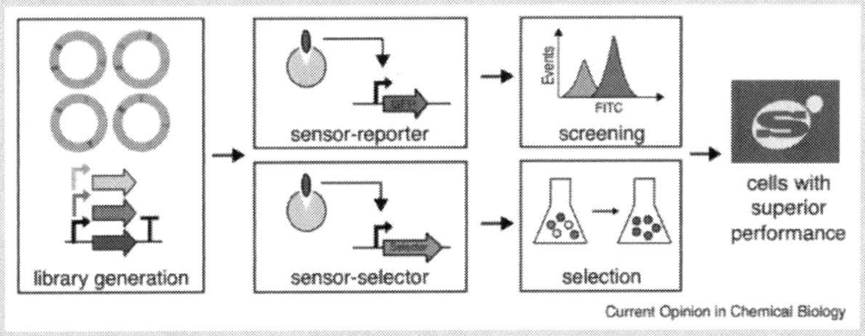


Feedback control

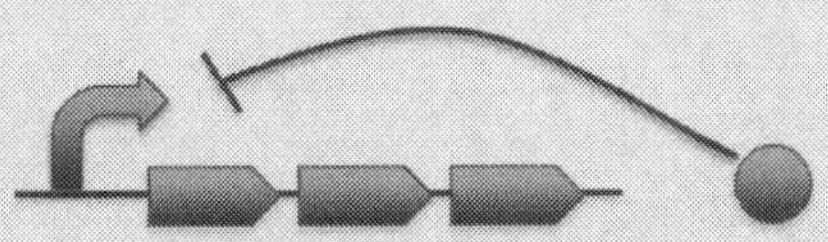

Limited natural options

9

Zhang *et al.* 2015

**McCormick**

Northwestern Engineering

**Department of Chemical and Biological Engineering**

# Metabolite responsive transcriptional regulators

<u>Goal:</u>

Ligand responsive transcriptional regulator

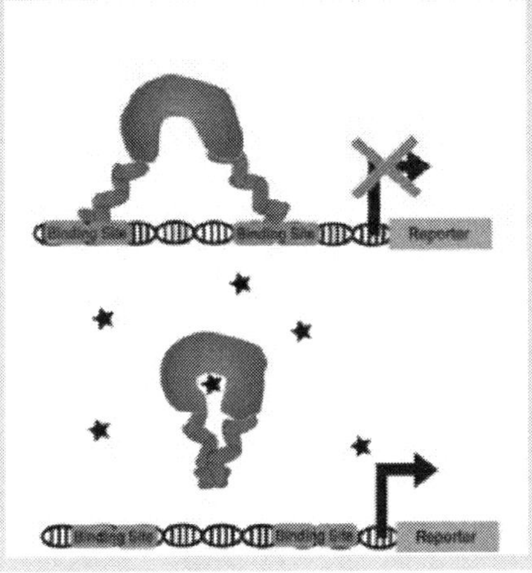


<u>Approach:</u>

Modular design

-Convert ligand binding protein to biosensor

-DNA binding domain

10

# Metabolite responsive transcriptional regulators

<u>Goal:</u>

Ligand responsive transcriptional regulator

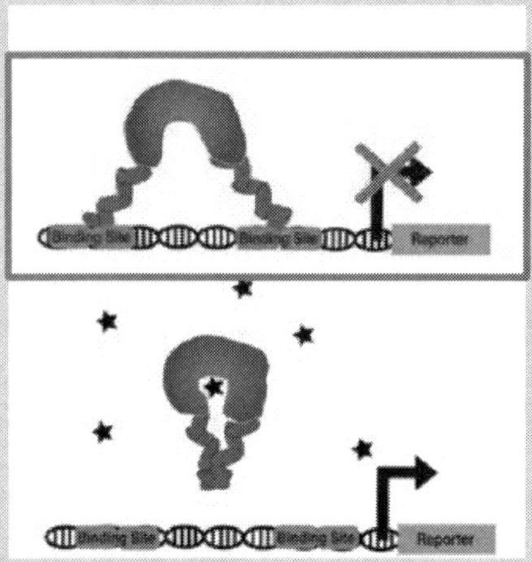


<u>Approach:</u>

Modular design

-Convert ligand binding protein to biosensor

-DNA binding domain

11



# DNA binding domain: BCR1 zinc finger

12

PDB #1A1L Thomas Splettstoesser

McCormick

Northwestern Engineering

Department of Chemical and Biological Engineering

# DNA binding domain: BCR1 zinc finger

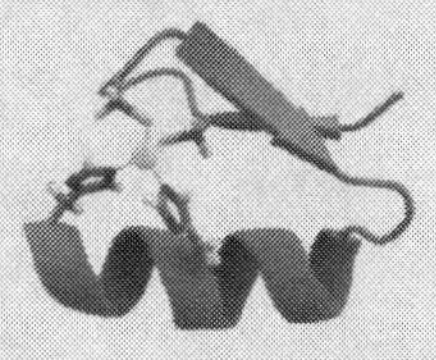

9 base pair binding site

13

PDB #1A1L Thomas Splettstoesser

McCormick

Northwestern Engineering

Department of Chemical and Biological Engineering

# DNA binding domain: BCR1 zinc finger

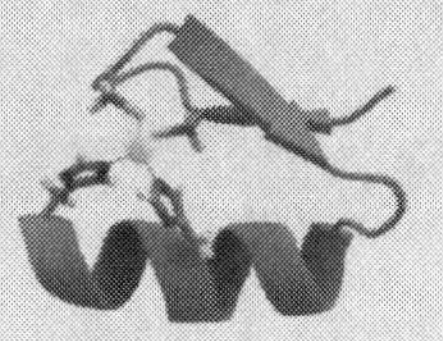

9 base pair binding site

Constitutive GFP reporter

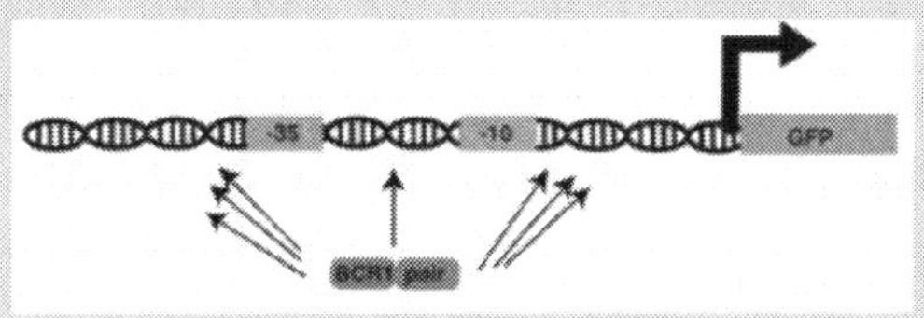


14

PDB #1A1L Thomas Splettstoesser

McCormick
Northwestern Engineering
Department of Chemical and Biological Engineering

# DNA binding domain: BCR1 zinc finger

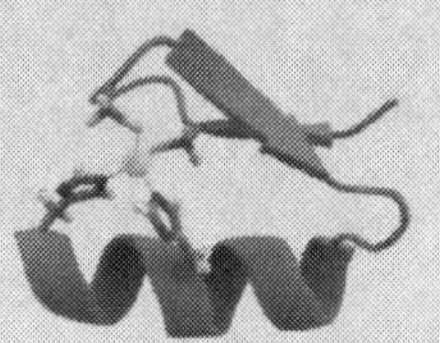

9 base pair binding site

Constitutive GFP reporter

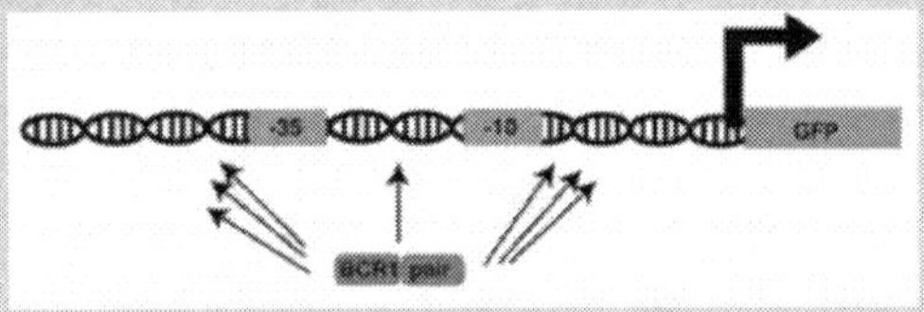


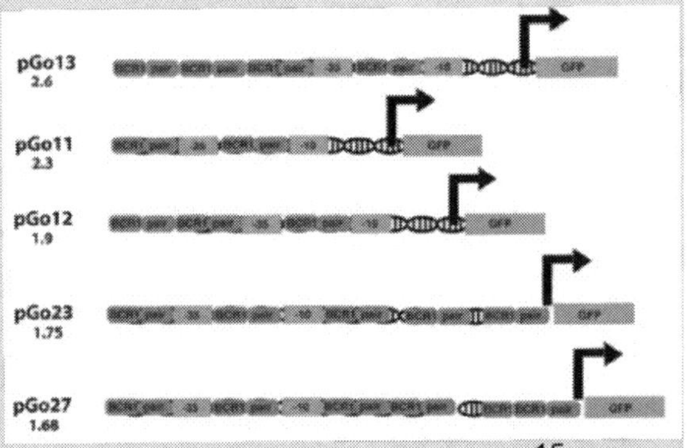


15

PDB #1A1L Thomas Splettstoesser

McCormick
Northwestern Engineering
Department of Chemical and Biological Engineering

# Investigate expression of promoter library

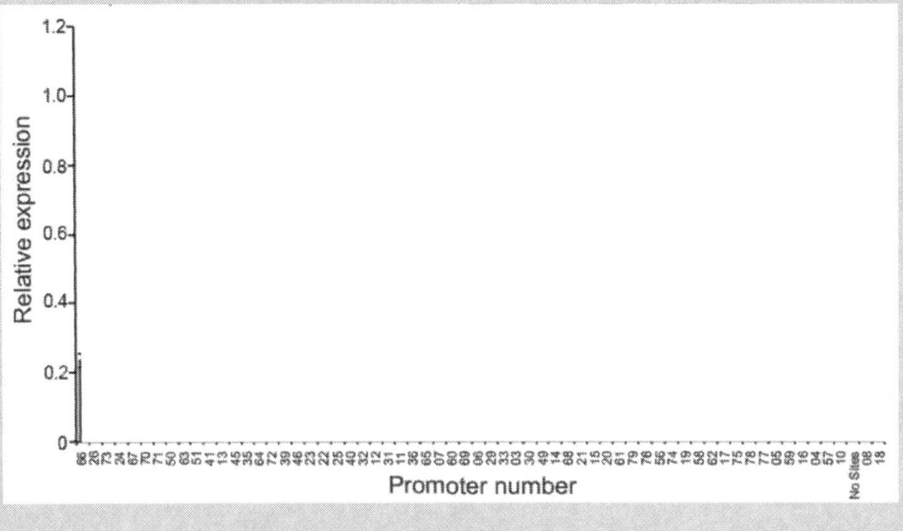

McCormick

Northwestern Engineering

Department of Chemical and Biological Engineering

# Investigate expression of promoter library

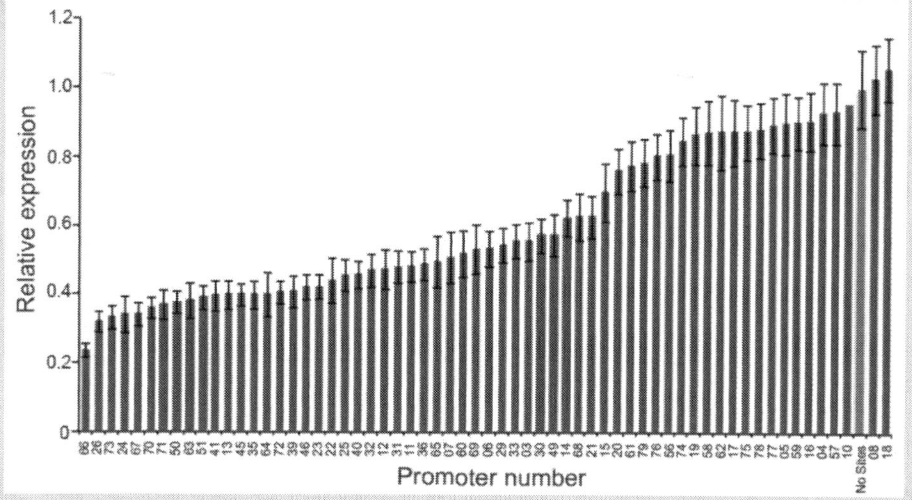


McCormick

Northwestern Engineering

Department of Chemical and Biological Engineering

# Investigate expression of promoter library

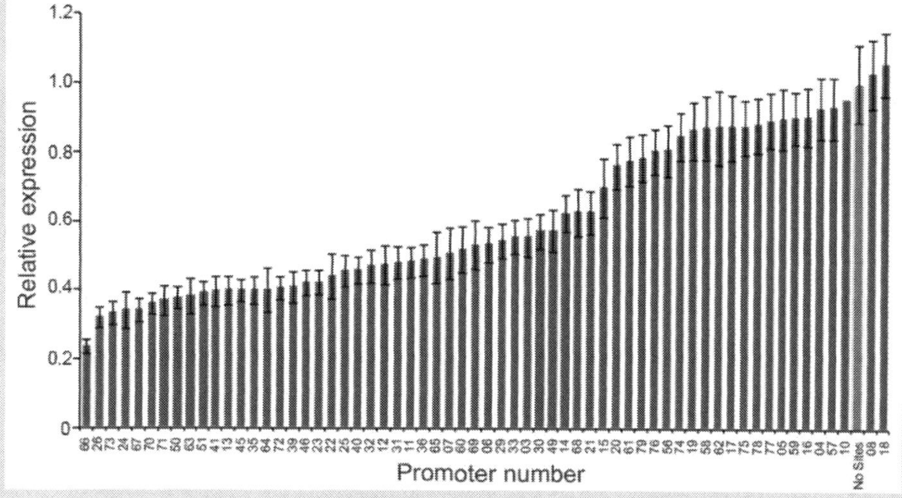


Wide range of repression – Why?

# Method for quantitatively describing the library

Goal: Select a set of descriptors that describe promoters.

Risk: Too many descriptors will over fit the system.

19

McCormick
Northwestern Engineering
Department of Chemical and Biological Engineering

Method for quantitatively describing the library

Goal: Select a set of descriptors that describe promoters.

Risk: Too many descriptors will over fit the system.

*Assumption:* The only thing that affects reporter expression is the binding of the biosensor

Features only describe the location of zinc finger binding sites

20

McCormick

Northwestern Engineering

Department of Chemical and Biological Engineering

| Number | Shorthand | Description of Feature |
|---|---|---|
| 1 ★ | *BS in core* | Number of binding sites in the core |
| 2 ★ | *pairs in core* | Number of pairs in the core |
| 3 | *BS upstream* | Number of binding sites upstream of the -35 region |
| 4 | *pairs upstream* | Number of pairs upstream of the -35 region |
| 5 | *BS downstream* | Number of binding sites downstream of the -10 box |
| 6 | *pairs downstream* | Number of pairs downstream of the -10 box |
| 7 | *length* | Length of the promoter in base pairs |
| 8 ★ | *spacer -10* | Base pairs downstream of the -10 box, before the first site |
| 9 ★ | *spacer -35* | Base pairs upstream of the -35 region, before the first site |
| 10 ★ | *spacer inside -10* | Base pairs upstream of the -10 box, before the first site |
| 11 ★ | *spacer insider -35* | Base pairs downstream of the -35 region, before the first site |
| 12 | *core middle space* | Non-binding site base pairs in the core |
| 13 | *ave downstream spacing* | Average number of base pairs between sites downstream |
| 14 | *ave upstream spacing* | Average number of base pairs between sites upstream |
| 15 | *down length* | Base pairs downstream of the -10 box |
| 16 | *up length* | base pairs upstream of the -35 box |
| 17 | *lopsidedness* | Absolute value of (sites upstream) - (sites downstream) |

21

McCormick

Northwestern Engineering

Department of Chemical and Biological Engineering

# Determining important features

Define output:

"repressibility" = (-1)*expression

*Questions:*

What is the relative importance of features in determining repressibility?

How many of these features are actually significant?

22

McCormick

Northwestern Engineering

Department of Chemical and Biological Engineering

# Approach: Three independent feature selection methods

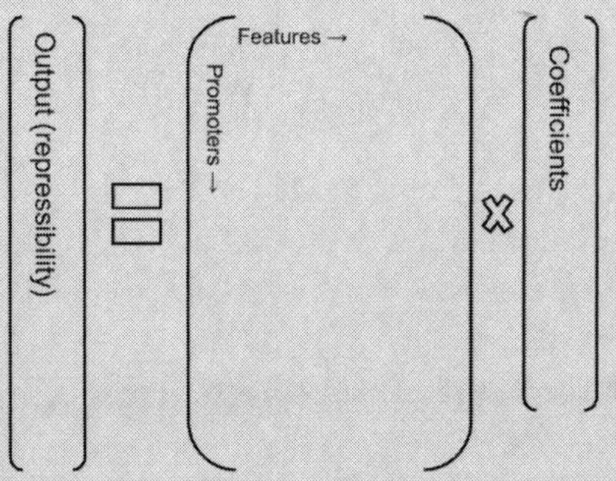


- Partial least squares regression
- Random forest
- Lasso regression

Iterative

Regularized

23

McCormick

Northwestern Engineering

Department of Chemical and Biological Engineering

# How many features are significant?

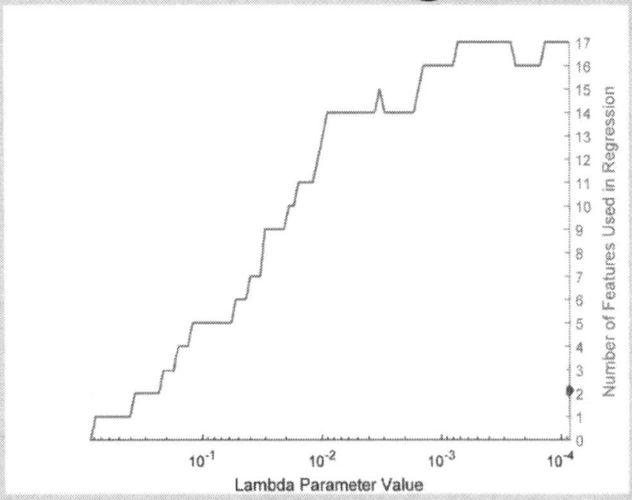


Lasso regression shrinks the coefficients of unimportant features to 0

After a certain number of features, the fits stops improving

Decreasing penalty on coefficients

24

McCormick

Northwestern Engineering

Department of Chemical and Biological Engineering

# How many features are significant?

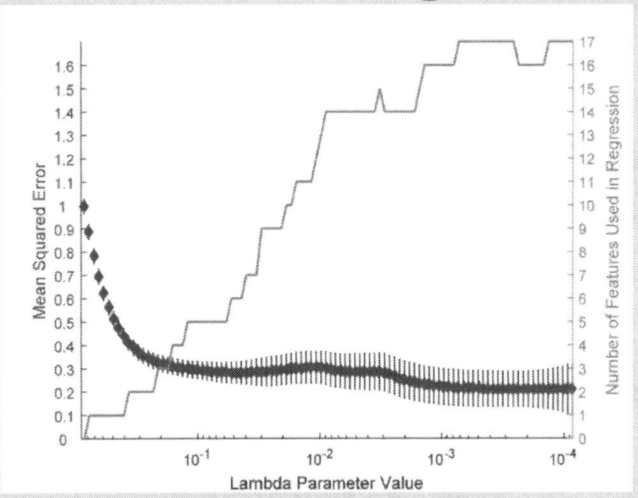


Lasso regression shrinks the coefficients of unimportant features to 0

After a certain number of features, the fits stops improving

Decreasing penalty on coefficients

25


McCormick

Northwestern Engineering

Department of Chemical and Biological Engineering

# How many features are significant?


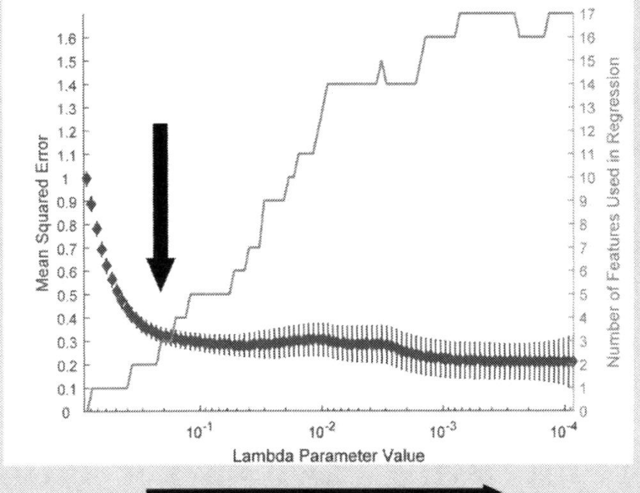



Lasso regression shrinks the coefficients of unimportant features to 0

After a certain number of features, the fits stops improving

Decreasing penalty on coefficients


26

McCormick
Northwestern Engineering
Department of Chemical and Biological Engineering

# How many features are significant?

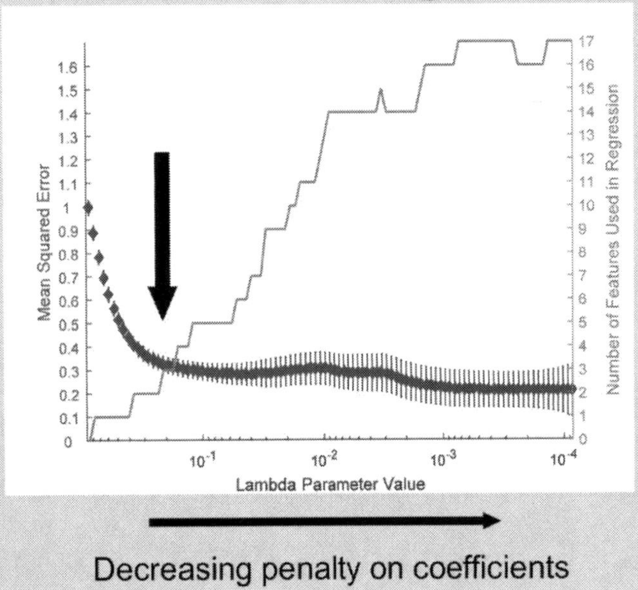


Lasso regression shrinks the coefficients of unimportant features to 0

After a certain number of features, the fits stops improving

**~3 important features**

27

McCormick
Northwestern Engineering
Department of Chemical and Biological Engineering

# Comparison of all three methods

Green = positive regression coefficient (positively correlated with repressibility)

Pink = negative regression coefficient (negatively correlated with repressibility)

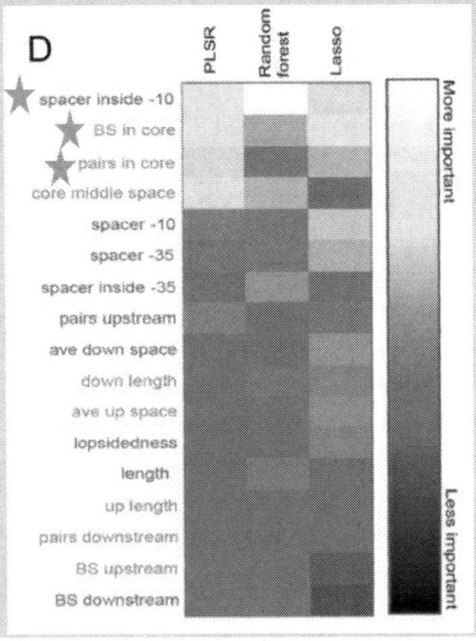


28

McCormick

Northwestern Engineering

Department of Chemical and Biological Engineering

## Rules for zinc finger repression

Crowding promoter boxes, especially -10, tends to create strong repression

Binding site quantity does not matter
- No titration effect

Regions away from the boxes are not important (upstream, downstream)

*Proposed mechanism:* The zinc finger blocks the sigma factor when near the -10 box

29

McCormick

Northwestern Engineering

Department of Chemical and Biological Engineering

## Application to biosensors

Promoter library analysis revealed elegant design rules for synthetic promoters

Useful feature selection methods for describing transcription factor binding

Application of this DNA binding domain in biosensors

McCormick
Northwestern Engineering
Department of Chemical and Biological Engineering

# Application to biosensors

Promoter library analysis revealed elegant design rules for synthetic promoters

Useful feature selection methods for describing transcription factor binding

Application of this DNA binding domain in biosensors

***Development of a small molecule responsive transcriptional regulator platform.***
Andrew Younger

Session 734: Emerging Tools for Synthetic Biology

Thursday 3:15, Salt Palace Convention Center 150G

McCormick
Northwestern Engineering
Department of Chemical and Biological Engineering

# Acknowledgements

**Josh Leonard**
**Andrew Younger**
**Austin Rottinghaus**
Rachel Dudek
Andy Scarpelli
Michelle Hung
Kelly Schwarz
Taylor Dolberg
Alia Zander
Danny Wells

**Collaborators**
Declan Bates (U. Warrick)
Paul Freemont (Imperial College London)
Jay Keasling ( UC Berkeley)
Anne Osbourn (John Innes Center)
Susan Rosser (U. Edinburgh)

Chemistry of Life Processes Institute

Office of Undergraduate Research UR@NU

NSF

EPSRC Pioneering research and skills

32

McCormick

Northwestern Engineering

Department of Chemical and Biological Engineering

McCormick

Northwestern Engineering

Department of Chemical and Biological Engineering

# Decision tree regression: Random forest

Uses decision trees on subsets of features

Quantify how much accuracy is lost when feature is not present

Creates a ranked list of importance

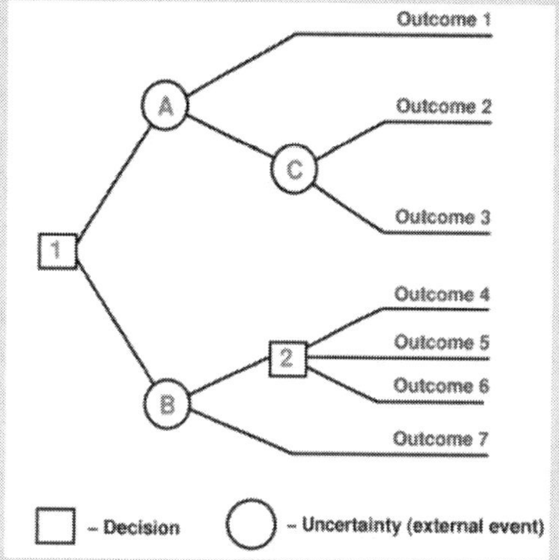


34

McCormick
Northwestern Engineering
Department of Chemical and Biological Engineering

# Partial Least Squares Regression: Permutation Test

Shuffle output randomly

Obtain a variance for each feature coefficient

Compare the real output to the random variance

Again, creates a ranked list

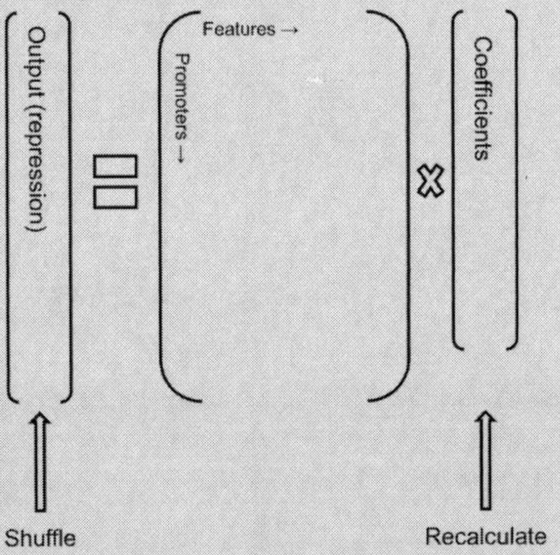


35

McCormick
Northwestern Engineering
Department of Chemical and Biological Engineering

# Regularized regression: Lasso

Constrains $\sum|\beta|$
(sum of absolute coefficients)

penalty

$$\sum_{i=1}^{N}(y_i - \sum_j \beta_j x_{ij})^2 + \lambda \sum_j |\beta_j|$$

Eliminates features by shrinking unimportant coefficients to 0

Creates a ranked list of importance (again)

36

# 438528 Immunomodulatory Amphiphilic Polyanhydride Microparticles for Peripheral Nerve Regeneration

**Abstract:**

: Peripheral nerve repair using implanted biomaterial conduits seeded with cells face many challenges, including the host immune response. Tissue engineering approaches often employ the use of drug delivery vehicles to supply sustained release of trophic factors which support the differentiation and continued growth of the implanted cells. However, host immune cells responding to inflammatory cytokines released by the damaged nerve may inhibit this growth. This project focuses on the fabrication of amphiphilic degradable polyanhydride micro particles that can release surface adsorbed Interleukin-4 and encapsulated nerve growth factor simultaneously. Interleukin-4, an anti-inflammatory cytokine, serves to promote an M2-like phenotype in invading macrophages as a method to support wound healing. These M2-like macrophages may reduce the inflammatory response, thereby generating an environment which supports cell growth and tissue regeneration, thus allowing implanted cells to benefit from the sustained release of neurotrophic factors. The release of nerve growth factor helps stimulate the peripheral nerve regeneration. We have focused on the fabrication and release kinetics of the two macromolecules from the polyanhydride microparticles.

# 409106 A Case Study of Data Analytics for the Manufacturing of a Monoclonal Antibody

**Abstract:**

The U.S. biotechnology sector has had double-digit growth rates in recent years (Imarc Group, 2012). In 2012, sales of biologics were approximately $63.6 billion, with monoclonal antibodies (mAbs) representing the largest fraction of this market with approximately 39% of sales (Aggarwal, 2014). Mathematical modeling of the manufacturing process is one possible way to both support the growing biologics market as well as decrease costs via improved control and understanding of process operations. Modeling can play an important role in understanding, controlling, and optimizing the process steps used in these processes (Tziampazis & Sambanis, 1994).

In this case study, analysis techniques are demonstrated in both bench-scale as well as process-scale datasets from the production of a monoclonal antibody product. Bench-scale data are analyzed using response surface methodology as well as regularization techniques. Process-scale data are analyzed using correlation analysis, latent variable methods, and regularization techniques. In all cases, this work shows that careful attention must be paid to model calibration and validation, particularly in biopharmaceutical applications where the amount of data is often small. This work also demonstrates that, where available, unit operations should not be isolated, but instead processed in sequence. Most importantly, this work demonstrates that there can be predictive value in modeling critical quality attributes using only a small, heterogeneous dataset. References

Aggarwal, S., 2014. What's fueling the biotech engine - 2012 to 2013. *Nature Biotechnology,* 32(1), pp. 32-39.

Imarc Group, 2012. *Global Biopharmaceutical Market Report and Forecast (2012-2017),* s.l.: International Market Analysis Research & Consulting Group.

Tziampazis, E. & Sambanis, A., 1994. Modeling of cell culture processes. *Cytotechnology,* Volume 14, pp. 191-204.

# 420144 Application of a Closed-Loop Analysis Approach to Achieve Process Robustness of a 25,000L Mammalian Cell Culture Process

**Abstract:**

Rapid advances in process data acquisition has enabled a large number of parameters to be monitored at any given time throughout the cell culture process. Modern large-scale cell culture facilities are fit with digital data-acquisition and aggregation systems that allow to scientists and engineers to analyze data. Although enormous amounts of data are collected through these systems in real-time, substantial effort is placed to extract meaningful information from the data. Recently, strategies have been developed to provide a more holistic view of the process, allowing analysts to extract useful process information that may not be captured via analysis of individual parameters. By unearthing these interactions, process improvement opportunities may arise that were not apparent prior to interrogating the data. Recently, multivariate analysis (MVA) has been used to observe and compare complex datasets for troubleshooting or characterization.

In this case study, a closed-loop process was utilized to implement process changes for increasing process yield and robustness. First, principles of MVA were applied to online and offline data from a 25,000L fed-batch CHO process to identify parameters that contribute to overall process variability. From this analysis, several parameters were identified as key contributors to variability in culture titer and glycosylation levels of the recombinant fusion protein produced from the culture. Results from this MVA were confirmed with appropriate experiments using a representative 5-L scale down model of the 25,000L bioreactor. Results from the scale-down model and MVA were used to implement at-scale process changes to reduce culture performance variability and increase overall process yield. Thus, the combined experimental and modeling approach enabled actionable changes to large-scale process, yielding increases in overall process yield.

# 428472 Clone and Media Selection in Mab Biosimilar Process Development

**Abstract:**

In the biopharmaceutical sector the topic of biosimilars remains a highly controversial issue. Until 2014, eleven different biosimilar active ingredients have been approved by the European Medicines Agency (EMA), while issuing a European marketing authorization to the first biosimilar monoclonal antibody (mAb) in 2013. Experience indicates that 7 to 8 years are required to launch a biosimilar on the market, at a cost of $100 – 250 million, while antibodies are likely to cost more. From the bioengineering perspective, this requires a target-directed process development approach.

This work presents a novel approach for the screening stage of the process development of a biosimilar. The objective of this stage is to evaluate and select both, cell line and media candidates, as well as media supplements and their levels. The procedure relies on the use of high-throughput experiments performed in deep well-plates. The results are analyzed by advanced multivariate data analysis tools, with the aim of maximizing the probability of reaching the bio-target in the following development stages. Regarding the manifold quality characteristics of the target molecule including its glycosylation and charge profiles, as well as aggregated and high and low molecular weight species, an approach combining principal component analysis (PCA) and decision trees is applied to provide selection criteria. The robustness of the approach is then tested on an external validation data set. Finally, the scalability of the derived process design space is discussed by a multivariate comparison to micro-bioreactor and lab scale experiments.

# 425441 Efficient Experiment Design and Model Generation to Assist Process Decision Making

**Abstract:**

Statistically designed experiments and empirical modeling techniques are commonly used in the pharmaceutical industry to define process ranges and enhance process knowledge. Parallel and automated experimental techniques are frequently used to execute the designs, as the experimental plan can be large. In some cases, parallel experimental techniques cannot be used due to mixing considerations or other setup specific issues.

This work presents a model-driven experimental strategy that was used to characterize a two reaction telescoped synthesis to product an active pharmaceutical intermediate. Additional complexity in this synthetic step includes the heterogeneity in the reactions and analytical stability issues. This process step required labor intensive experiments which had to be executed one at a time This work describes a process characterization strategy that started with an efficient D-optimal experimental design that was subsequently augmented based on findings from exploratory data analysis. The team used a model-driven approach, where analysis of the data was partially accomplished by the studying of the resulting statistical models and fitting characteristics. The resulting models were used to guide pilot plant parameter variation and were verified by the resulting on-scale data collection. The models were also used to perform robustness analysis of the system, where the expected process outcome variability was estimated by simulating expected process parameter variability. This experimental and modeling strategy provided actionable information with a modest number of experiments that allowed the project team to make decisions in a timely manner.

# 428862 Determination of Reaction Profiles Using Online FTIR Spectroscopy Coupled with Kinetics Model-Based Multivariate Curve Resolution

**Abstract:**

Fourier Transform Infrared (FTIR) spectroscopy is an important tool for the monitoring and controlling Active Pharmaceutical Ingredients (API) manufacturing processes. High spectral resolution and information content are particularly suitable for monitoring chemical reactions where differentiation of structurally similar molecules is required to maintain the quality of products. With a brief introduction to the setups of online FTIR in pharmaceutical reaction monitoring, this presentation will focus on a novel approach that uses the kinetics model-based multivariate curve resolution algorithm as a "calibration-free" tool for real-time analysis of batch reaction processes.

In this application, online FTIR was used to collect data from a hydrogenation process that makes an API (drug substance). The FTIR spectra are fed into the kinetics model-based multivariate curve resolution algorithm to generate the concentrations of product and starting material in real-time without using traditional calibration models. Experiments in the lab have shown that the results from kinetics model-based multivariate curve resolution were similar to those generated by Partial Least Square (PLS) prediction models in terms of analytical accuracy. This offers a "calibration-free" alternative to traditional multivariate (chemometrics) calibration for overcoming spectral overlapping in quantitative analysis. Because the process parameters are explicitly expressed in the kinetics model, their values can be changed easily to reflect the changes in the process condition. Therefore, unlike statistical calibration models that often require adjustment or complete rebuild once process conditions are changed, the kinetics model-based multivariate curve resolution algorithm is able to adjust to the new conditions and perform normally as before. This has been demonstrated in the scale up of the hydrogenation process to pilot plant, in which the batch surface area, hydrogen pressure and agitation were different from those in the lab experiments. The kinetics model-based multivariate curve resolution algorithm worked well without any adjustment. This study has demonstrated an alternative approach for meeting the challenges in supporting fast changing process development and scale up without using multivariate (chemometrics) calibration models that require lengthy adjustment or rebuild once process conditions are changed.

# 421708 Application of Multivariate Latent Variable Methods to Drug Product Development

**Abstract:**

This work presents three case studies of the use of Multivariate Latent Variable Methods (LVM) in pharmaceutical development: i) a root-cause investigation into the compression behavior of powder blend for oral dosage form manufacture, and further implications for commercial scale-up; ii) the study of the feasibility to transfer a roller compaction process from one roller compaction unit to another of a different brand and design; and iii) the use of LVM to develop a solvent replacement system based on solvatochromic properties of solvents and the further coupling with optimization technology for a D-Optimal selection of solvents for experimental screening.

# 429799 Improving Yield of Drug Product Manufacturing with Hybrid Approach

**Abstract:**

For remaining competitive in times of a growing public pressure on drug prices and the shrinking research productivity in the industry, drug product manufacturing must be performed increasingly efficient. The presented work is an expansion of Eberle et al. (2014), which comprises of a four-level approach for yield enhancements and is applied in an industrial case study to the manufacturing of injectable drug products, also referred to as Parenterals.

The four levels of the presented hybrid approach combine heuristic and statistical, data-based elements to take advantage of process specific information both from expert knowledge as well as process data. On the first level, the set of loss causing production processes at a unit under review are listed before identifying the processes with most improvement potential. Then, on the second level, expert knowledge is consulted for generating loss scenarios and assigning measurable parameters. Descriptive statistical modelling is performed on the third level before deriving enhancements or escalating process investigations on the fourth level.

As a case study, the presented approach was applied on an annual production output of a manufacturing facility for injectable drug products of Roche in Switzerland. The sampled output comprises of six products from liquid vials and pre-filled batches. Identified costs for lost product units on the first level revealed, that about 20 % of costs for non-commercialized products result from sampling for quality control activities. This loss cause was followed in magnitude by two different glass container defects, i.e. imperfectly formed and scratched vials. Scratched vials and their possible generation during filling processes were modelled with a variety of explanatory variables. However, the statistical models on these variables did not lead to the derivation of implementable information and, as a consequence of that, to an escalation of process investigations. An investigation of scratches with a vial replica that includes a pressure-sensitive outer surface was initiated; enabling the quantification of physical exposures on the glass body. Based on the resulting exposure information, processing steps were prioritised by their impact on glass scratches, hot spots were diagnosed and the statistical model was refined. In summary, the presented data-based approach structured the use of production data and supported efficiency enhancements of production processes.

References: Eberle L., Sugiyama H., Papadokonstantakis S., Graser A., Schmidt R., Hungerbühler K., 2014, Data-Based Tiered Approach for Improving Pharmaceutical Batch Production Processes, Proceedings of the 24th European Symposium on Computer Aided Process Engineering, Elsevier, Amsterdam, the Netherlands.

# 431431 Benefits of Quality By Design (QbD) for Consistent Product Quality

**Abstract:**

The pharmaceutical industry continues to embrace the principles outlined in ICH Q8(R2), Q9, Q10 and Q11 and to move from a Quality by Testing mentality to a Quality by Design paradigm. This talk will provide a high level overview of how the pharmaceutical industry can benefit by using QbD principles to develop robust manufacturing processes for small molecule drug products in order to consistently deliver quality drug products to the patient. Through examples, the talk will illustrate the QbD tools utilized by applicants during pharmaceutical development. It will also touch upon Integrated Quality Assessment (IQA) in the newly formed Office of Process and Facilities within the Office of Pharmaceutical Quality.

# 430873 Predicting Roller Compaction Process Capability

**Abstract:**

Roller compaction is a common processing option in the tableting process of small molecule pharmaceutical compounds. Because roller compaction can be described using a physical model of the process, it is possible to predict manufacturing behavior from laboratory experiments. However, it is desired to estimate the process robustness incorporating uncertainty at the defined manufacturing conditions. This presentation will describe an application of probabilistic modeling to leverage process models, lab data, manufacturing data, and uncertainty to best incorporate potential sources of process variability.

Using Markov Chain Monte Carlo, the process model can be regressed to laboratory data or pilot manufacturing data, resulting in an estimate of the possible outcomes of the measured process quality attributes, i.e. ribbon density. These outcomes are generated from a Bayesian network that integrates all the sources of variability (measurement error, model error, and input variability) associated with the process, predicting process capability (Cpk) and identifying any hidden deficiencies in the input data.

The framework quantifies the expected Cpk at the time of process qualification, increases the fundamental scientific knowledge of the process and through a sensitivity analysis, provides a rational mechanism to prioritize resources for future measurements.

# 415192 Impact of API Properties and Excipient Source on the Process and Performance of a Typical High Shear Wet Granulation Formulation

**Abstract:**

In a pharmaceutical formulation, some excipients, such as binder and disintegrant are used for their particular functionalities. While present only in small quantities, changes in these excipients can sometimes have significant impact on the processiblity and performance of a drug product. When a formulation contains high drug loading, properties of the API can also interplay with the process parameters to impact the properties of drug intermediate and drug product. In this study, we evaluated the effect of differences in binder, disintegrant, as well as API properties on the process and performance of a typical high drug loading high shear wet granulation formulation. The factors evaluated include:

1. Types of disintegrant (Sodium Starch Glycolate vs. Croscarmellose Sodium)
2. Source of the binder Hydroxylpropyl cellulose (Ashland Klucel vs Nisso)
3. API solubility and wettability
4. API particle size and surface area

The responses evaluated include:

1. Amount of granulation water needed for desirable granule properties
2. Granule growth ratio
3. Granule compactability
4. Tablet disintegration

Salicylic acid, aspirin, citric acid as well as some BMS compounds were used in this study, which covers a broad range of solubility and contact angle (0.002-590 mg/ml, 0-146°). It was found that the amount of granulation water needed for optimal granule properties is highly sensitive to the solubility and contact angle of the API. For highly soluble API, water amount needed for granulation is significantly reduced. Granule growth ratio has a significant dependence on the particle size/surface area of the API. Using the same granulation parameters, smaller API particle size results in higher granule growth ratio. For the low solubility API that was selected for the disintegrant type and binder source study, no significant difference was observed in the processibililty and disintegration of the drug product.

# 422303 Implementation of Advanced Multilayer Plant-Wide Control Architecture into a Direct Compaction Continuous Pharmaceutical Manufacturing Process

**Abstract:**

Continuous pharmaceutical manufacturing provides an appropriate platform where feed-forward and feedback control strategies, including necessary alarms, can be implemented to improve the product quality, minimizing product rejection and operating expenses. The precise control of the quality of the pharmaceutical product requires corrective actions on the process/raw material variability before they can influence the product quality. It is therefore highly desirable to implement a combined feed-forward/feed-back control system into a pharmaceutical manufacturing plant [1] where multivariate measurements of raw and intermediate materials, finished products, and process parameters are used to control quality. The feed-forward controller takes into account the effect of process disturbances proactively while the feedback control system ensures the end product quality consistently. The integrated feed-forward/feedback control system ensures minimum variability in the final product quality irrespective of process and raw material variations.

In this work, efficient multilayer plant-wide control architecture has been designed and implemented into a direct compaction continuous tablet manufacturing pilot-plant. The proposed control architecture consists of four layers. In the first layer, necessary sensors including PAT tools have been implemented for real time process monitoring. In second layer, feedback control loops have been implemented to ensure the critical quality attributes. In third layer, feed-forward control loops have been integrated with feed-back control loops to proactively mitigate the effects of process and raw materials variability and thereby to achieve more precise control of end product quality with minimum rejections. In final layer of the control architecture, the necessary alarms and signals have been implemented to control the plant operation more efficiently and reject the out of specs powder blends and tablets automatically.

The critical process parameters and quality attributes are monitored at each unit operation via a centralized control platform using advanced monitoring tools. For example, feeder flow rate and feeder screw rotational speed have been monitored at feeding operation. Powder blend composition, uniformity (RSD) and bulk density are monitored at the blending operation. Tablet weight, thickness, hardness, compression forces, ejection force, and operating parameters (e.g. fill depth, feed frame speed) are monitored at the tableting operation. NIR is applied first time to measure the powder bulk density in real time. As part of second layer of control architecture, supervisory feedback loops to control powder blend composition [2], main compression force, tablet weight and hardness are implemented. The tablet weight and hardness control loops have been decoupled so that both critical quality attributes can be controlled simultaneously. Main compression force and tablet weight are controlled through a cascade control arrangement. The necessary, feed-forward control loops are implemented so that the upstream measurements can be used to take downstream mitigation actions in real time. For example, measured powder bulk density was feed-forwarded in real time to the tablet press to proactively mitigate the effect of powder bulk density on tablet weight and hardness via changing the fill depth. Finally, the necessary alarms are implemented for unexpected events management including sending the warning/alarms signals to the operator, stopping the production if needed and rejecting powder blends and tablets out of specification. The four layers of control architecture have been implemented using industry standard control platform (e.g. DeltaV), PAT data management tool (synTQ), OPC communication protocol, fieldbus devises, chemometric tools and real time monitoring tools. The proposed control strategy is found to be better than implementations using only feedback control or without control strategy and therefore demonstrates potential to further improve pharmaceutical tablet manufacturing operations.

The objective of this presentation is two-fold. First to highlight the combined feed-forward/feedback based four layers of control architecture, and second to demonstrate its performance to control the direct compaction continuous tablet manufacturing process.

**References**

1. Singh, R., Muzzio, F., Ierapetritou, M., Ramachandran, R. **(2015)**. A combined feed-forward/feed-back control system for a QbD based continuous tablet manufacturing process. PROCESSES Journal, 3, 339-356.
2. Singh, R., Sahay, A., Karry, K. M., Muzzio, F., Ierapetritou, M., Ramachandran, R. **(2014)**. Implementation of a hybrid MPC-PID control strategy using PAT tools into a direct compaction continuous pharmaceutical tablet manufacturing pilot-plant, International Journal of Pharmaceutics, 473, 38–54.

# 435447 Systematic Framework for the Design and Analysis of Control Systems for Continuous Pharmaceutical Manufacturing


**Abstract:**

In the pharmaceutical industry, fulfilling the Food and Drug Administration (FDA) requirements for product quality is one of the main concerns [1]. In order to assure product quality, batch operation is the most used manufacturing mode [2]. During batch production, the quality of the batch subject to operator action and quality is often controlled by statistical sampling and rejection of an entire batch if sampling indicates deviations from specifications. This leads to significant amounts of wasted material and significant overall cost [3]. The cost for manufacturing goods represent a large part of the products sales. For instance, the manufactured goods cost represent twenty-seven percent of the patented products sales, while for generic products the percentage is fifty percent [4]. This cost can be minimized by implementing improvements in the manufacturing process [5]. One of the possible improvements is to evolve from batch operations to continuous processing. Continuous manufacturing and on-line process monitoring have the potential to improve product quality and reduce manufacturing cost; the potential annual savings across the industry are projected to be as high as fifty billion dollars [3]. In 2002 the FDA initiated Process Analytical Technology (PAT). The goal of PAT is to augment the understanding and control the manufacturing process [6,7]. Through the implementation of QbD and PAT, the companies have to demonstrate understanding of how the operating conditions, process design and raw material variability affect the product quality [4]. Control design plays a significant role for continuous manufacturing implementation. It is by implementing appropriate process control strategies that poor quality product manufacturing can be mitigated or prevented [1]. By implementing Quality by Control (QbC), it is possible to reduce the operation cost, reduce the off-spec or segregated material and improve the safety environment.

In this contribution a generic framework is proposed for design and analysis of a hierarchical feedback control scheme for continuous manufacturing. The framework proposes three control levels that have to be considered in the control design. Level 0 includes the built-in control systems of the individual unit operations, in Level 1 PAT is used in simple feedback control schemes (property feedback control – PFC) to control critical quality attributes (CQAs), whereas level three uses dynamic models for prediction and advanced control schemes, such as model predictive control (MPC) for enhanced control performance. Control performance indicators suitable for the comparative analysis of various control schemes are also proposed. The potential impacts of using level 1 property feedback control (PFC) and level 2 model predictive control on product quality and process performance are analyzed using the proposed performance indicators. The proposed generic framework is exemplified using as a model system the feeder and blender unit operations. The control schemes are evaluated under realistic disturbance and measurement scenarios, at high and low active pharmaceutical ingredient (API) content.


**References**


[1] Ramachandran, R.; Arjunan, J.; Chaudhury, A.; Ierapetritou, M.G. Model-Based Control-Loop Performance of a Continuous Direct Compaction Process. *J. Pharm. Innov.* **2011**, 6, 249-263.

[2] Singh, R.; Boukouvala, F.; Jayjock, E.; Ramachandran, R.; Ierapetritou, M.; Muzzio, F. Flexible Multipurpose Continuous Processing of Pharmaceutical Tablet Manufacturing Process. *PharmPro Magazine*, **2012**, 27, 22–25.

[3] Reklaitis G. The Pharmaceutical Manufacturing Institute. **2013**, (White Paper) Purdue University, West Lafayette, IN.

[4] Rogers, A. J.; Hashemi, A.; Ierapetritou, M. Modeling of Particulate Processes for the Continuous Manufacture of Solid-Based Pharmaceutical Dosage Forms. *Processes*. **2013**, 1, 67-127.

[5] Rogers, A. J.; Inamdar, C.; Ierapetritou, M.G. An Integrated Approach to Simulation of Pharmaceutical Processes for Solid Drug Manufacture. *Industrial & Engineering Chemistry Research*. **2013**, 53, 5128-5147.

[6] US Food and Drug Administration. *Guidance for Industry, Q8 Pharmaceutical Development*; Food and Drug Administration, 2006

[7] Lee S. L., O'Connor T. F., Yang X., Cruz C. N., Chatterjee S., Madurawee R. D., Moore C. M. V., Yu L. X. Modernizing Pharmaceutical Manufacturing: from Batch to Continuous *Production. J Pharm Innov.* **2015**

# 413370 An Optimization Framework to Combine Design Space Maximization with Design of Experiments

**Abstract:**

While pharmaceutical quality by design enables the specification of a design space in terms of process inputs, the space of process outputs may be more relevant to safety and product quality. Thus, while design of experiments techniques are often used to explore the effect of process parameters, it is often desirable during development, and for the purpose of process understanding, to maximize the explored space in terms of process outputs or quality attributes. However, timeline and resource constraints limit the number of experimental trials available to properly characterize both spaces, prompting the need for systematic design of experiments. An optimization framework based on the sometimes conflicting objectives of maximizing the size of the output or quality attribute design space and maximizing the D-optimality of the input parameters is proposed and applied to a spray-coating process. This example shows that optimizing only one of the two objectives yields unacceptable results in the neglected objective. Two strategies are explored. First, a multi-objective optimization approach is presented, enabling visualization of possible tradeoffs as a Pareto-optimal front—a set of solutions where no objective can be further improved without degrading another objective value. This allows the decision maker to choose among a set of efficient designs. It also provides insights into the potential benefits of increasing (or decreasing) the number of experiments. Second, a bilevel formulation is also presented which enables process engineers to quickly identify good compromise designs without the need to map out multiple points along the Pareto-optimal front.

# 421707 Control Strategy for Manufacturing of Dry Powder Inhalation Capsules: A Regulatory Perspective

**Abstract:**

Dry Powder Inhalation (DPI) is a leading drug delivery technology for the treatment of chronic pulmonary diseases, such as asthma and chronic obstructive pulmonary disease (COPD). DPI capsules provide a portable, simple-to-use, and cost-effective way to deliver inhalation medication. However, manufacturing of low-dose DPI capsules with consistent delivered dose and fine particle dose is not an easy task due to the complex interactions between drug formulation, delivery device design, and patient usage. Quality by design (QbD) provides an effective tool to systematically evaluate, understand, and refine the formulation and manufacturing process to ensure predefined product quality to meet the needs of the patient. In this presentation, the manufacturing process of low-dose DPI capsules will be used as an example to illustrate how to implement QbD principles to enhance product and process understanding and to implement effective process controls to ensure reproducible commercial manufacturing for consistent product quality. The purpose of this talk is to provide a regulatory perspective with regard to a QbD-based control strategy for the manufacturing process aspects of low-dose DPI capsules.

# 428945 Remote Manipulation with Microlablets and Control of Enzymatic Reaction

**Abstract:**

Progress in microfabrication technologies makes it possible to design and fabricate autonomous self-assembling microparticles called lablets (under 100 mm in size), which can implement programmable microscale electronic chemistry, forming a bridge between electronic and chemical computing. Due to limited energy storage capacity of the lablets and high material density, their movement in aqueous environment over larger distances and attachment to a predefined substrate (docking station) remains a challenge. The aim is precise positioning of the lablet above the docking station in aqueous environment, followed by controlled attachment of the lablet on pre-defined part of the docking station. The position of the lablet with respect to the docking station has to be within the range of 10's of micrometers.

The lablets are mainly composed of silicon (density 2.57 $g.cm^{-3}$), hence they are denser than water and thus remote manipulation with the lablets above the docking station in aqueous environment is not possible due to their sedimentation. 150 - 200 mm small surfactant stabilized oil droplet (kerosene, density 0.80 $g.cm^{-3}$) were be attached to the lablets in order to decrease their overall density. The oil microdroplets with attached lablets were observed to be floating on the water-air interface. Since one side of the lablets surface is formed from silicon (Si) while the other one contains also gold electrodes (for communication with the dock station), it is required to attach the lablets to the oil droplets in such a way that the gold electrodes will be immersed only in the water phase and thus be able to communicate with the dock station. The immersion of the lablets in the oil phase can be changed by hydrophilic/hydrophobic modification of the lablet's surface.

In order to make the remote manipulation possible, magnetic (iron-oxide) nanoparticles with a hydrophobic modification (oleic acid stabilized) were added to the oil (kerosene) phase to create a ferrofluid. The ferrofluid will consequently move in a magnetic field and carry the lablet with it. The magnetic field is created by four electromagnets (the distance between two opposite electromagnets will be 25 mm), allowing us to precisely manipulate with the lablets in the X,Y (planar) direction. Depending on the voltage applied and the concentration of magnetic particles in the ferrofluid, lablets can be moved with a velocity of 10'-100's mm/s. The attachment of the lablet to the docking station will be achieved by placing another electromagnet bellow the docking station, hence we will be able to move/elongate the ferrofluid microdroplets in the Z (vertical) direction that will consequently lead to the connection of the lablet with the docking station.

The lablets can also act as carriers of active substances, i.e. liposomes with encapsulated substrate attached on the lablets surface. The release of the encapsulated content can be triggered by i) radiofrequency heating which causes a reversible phase transition of the lipid bilayer; or by ii) electroporation of the lipid bilayer. By controlled release of the substrate, we are able to trigger an enzymatic reaction on-demand. This principle makes it possible to deliver unstable or reactive active ingredients that cannot be formulated into traditional dosage forms.

# 435312 Enzyme Immobilization By Polydimethylsiloxane Stamping for Glucose Sensor Fabrication

**Abstract:**

## Abstract


Enzyme Immobilization by Polydimethylsiloxane Stamping for Glucose Sensor Fabrication



Bo Wang, Bonhye Koo, Harold G. Monbouquette*

*Chemical and Biomolecular Engineering Department, University of California, Los Angeles, CA 90095, USA*



Most glucose sensors are based on electrochemical detection of hydrogen peroxide ($H_2O_2$) generated from the glucose oxidase (GOx)-catalyzed oxidation of glucose, and immobilization of the enzyme is a critical step in biosensor fabrication. Here, a new and effective method for GOx immobilization is presented. The enzyme is successfully transferred from a polydimethylsiloxane (PDMS) stamp surface onto a chitosan-coated electrode. Compared to manual enzyme immobilization methods, the enzyme layer generated by PDMS stamping is thinner, which leads to fast sensor response and high sensitivity. The sensor construction procedure begins with the coating of polypyrrole and Nafion on a platinum (Pt) electrode surface sequentially as permselective polymers to exclude anionic and cationic interferents (*e.g.*, ascorbic acid and dopamine).[1] A chitosan film is electrochemically deposited on top of the Nafion to serve as a suitable adherent layer for enzyme immobilization.[2] GOx and bovine serum albumin (BSA) are transferred together onto the chitosan surface using a PDMS stamp. Then, the stamped layer is exposed to glutaraldehyde vapor for cross-linking. The layering process of biosensor fabrication is monitored by electrochemical impedance spectroscopy (EIS) and the chitosan surface before and after enzyme stamping is characterized by scanning electron microscopy (SEM). In testing with varying concentrations of glucose and interferents, the sensors show high sensitivity, low detection limit, good selectivity, and fast response time. This enzyme immobilization method can be used in principle to immobilize various enzymes for creation of biosensors for applications both *in vitro* and *in vivo*.


## Reference


1. Tseng, T.T.C. and H.G. Monbouquette, *Implantable microprobe with arrayed microsensors for combined amperometric monitoring of the neurotransmitters, glutamate and dopamine.* Journal of Electroanalytical Chemistry, 2012. **682**: p. 141-146.

2. Tseng, T.T.C., J. Yao, and W.C. Chan, *Selective enzyme immobilization on arrayed microelectrodes for the application of sensing neurotransmitters.* Biochemical Engineering Journal, 2013. **78**: p. 146-153.

# 428540 Using the Personal Glucose Meter to Quantify a Wide Range of Targets

Monday, November 9, 2015: 3:53 PM
Canyon C (Hilton Salt Lake City Center)

**Tian Lan**, Research Park, University of Illinois at Urbana-Champaign, Champaign, IL, Yu Xiang, Chemistry, TsingHua University, Beijing, China and Yi Lu, Chemistry, University of Illinois at Urbana-Champaign, Urbana, IL

**Using the personal glucose meter for convenient quantification of wide range of targets**

**Tian Lan, Yu Xiang and Yi LU**

**Abstract:**

Convenient, low-cost and quantitative detection of broad range of targets has the great potential to revolutionize today's practice for medical and environmental monitoring. Despite years of research, very few such devices have been commercialized at a large scale. Today's personal glucose meter (PGM) is the culmination of decades of R&D and commercialization efforts aimed to satisfy the daily need of millions of diabetes patients worldwide to monitor their blood glucose levels. The PGM is the almost the ideal sensing platform, which offers convenience, low-cost and quantitative, except it only quantifies one target, glucose.

Instead of designing a new platform that can be time consuming and costly, we have developed an innovative method that adapts the widely available PGM for quantifying a wide range of targets, including metal ions, small molecules, proteins and nucleic acid sequences. An enzyme invertase, which hydrolyzes sucrose into glucose, is conjugated to recognition molecules, such as aptamers, DNAzymes, antibodies and DNA, and relays the non-glucose target binding events to the presence and concentration of glucose, which can be measured using PGM. Using this approach, we have demonstrated quantification of metal ions using DNAzymes, small molecules using aptamers and antibodies, various protein markers using antibodies, and viral DNA fragments using complementary DNA as shown in Table-1.

**Table 1**. New targets detected and quantified by BGM using our novel technology

| Type | Example | Detection limit |
|---|---|---|
| Diabetic biomarkers | HbA1C | 3 mg/L |
| Disease biomarkers | INF-γ (a TB marker) | 2.8 nM |
| Cancers | Prostate cancer antigen (PSA) | 0.4 ng/mL |
| Small organic molecules | Cocaine adenosine | 5.3 μM |
| Organic toxins | Ochratoxin A | 6.8 ng/mL |
| Metal ions | Pb and U | 9.8 nM |
| Viral DNA | Hepatitis B | 40 pM |

In conclusion, we have demonstrated an innovative way of leveraging the existing PGM hardware for quantification of non-glucose target using different recognition molecules. The technology can be used to develop new sensing platform using the technologies developed for the PGM over the last decades.

# 430671 Sequence-Specific Nucleic Acid Detection Based on Blockade of a Nanopore in a Thin Glass Membrane

**Abstract:**

A new platform for sequence-specific nucleic acid (NA) detection with binary response has been demonstrated. The detection of nucleic acid sequences is useful for diagnosing the presence of bacteria, especially those proven to be harmful to public health. A platform for low cost, reliable detection of the presence of pathogenic bacteria has a wide variety of applications, ranging from the diagnosis of bacterial infections to identification of contaminated food or water. Most existing methods are complex and require special reagents, as they rely on polymerase chain reaction (PCR) for target sequence amplification and fluorescence or enzyme labels. In our previous work, we described our first-generation, PCR-free, label-free system based on polystyrene beads conjugated with uncharged peptide nucleic acid (PNA) as sequence-specific probes, and demonstrated a detection limit of 10 fM.[1,2] A drawn glass pipette tip served as the micropore and was placed between two buffer-filled chambers, one of which contained the bead-PNA conjugates. Since the bead-PNA conjugates carry little to no charge, they do not respond to an electric field applied across the pore. However, in the presence of target NA sequence, the bead-PNA conjugates acquire enough negative charge to become mobile by hybridizing the NA sequence of interest. Upon pore blockade by the beads, a change in the pore conductance occurs, which is reflected in a step-decrease in current that signals the presence of the target nucleic acid.[1,2]

At lower concentrations of NA where fewer target NA molecules hybridize per bead, smaller beads are expected to have a higher mobility. Accordingly, by decreasing both the size of the pore and the beads, detection limits below 10 fM may be achieved. Motivated by this analysis, we have fabricated thin (1 μm or thinner) glass membranes in which a nanopore is milled with a focused ion beam (FIB). This device has two benefits: 1) By milling a pore via FIB, the pore size can be reduced significantly compared to drawn glass pipette tips, which lie in the micrometer range; 2) Since the nanopore is milled in a planar substrate, it can be mass-manufactured and can be assembled into devices more easily. To date, the nanomachined devices have proven capable of distinguishing between complementary and non-complementary sequences of ssDNA at various bead sizes. In addition, a promising trend has been observed showing that with decreasing bead size, higher mobility is attained at lower NA concentrations. These results support the expectations that detection limits less than 10 fM may be achieved with the nanomachined device and that this work could give rise to a diagnostic platform for the rapid, sensitive, and cost-effective detection of NA of specific sequence.

[1] Esfandiari L, Monbouquette HG, Schmidt JJ. Sequence-specific Nucleic Acid Detection from Binary Pore Conductance measurement. J. Am. Chem. Soc. 2012; 134: 15880-15886.

[2] Esfandiari L, Lorenzini M, Kocharyan G, Monbouquette HG, Schmidt JJ. Sequence-Specific DNA Detection at 10fM by Electromechanical Signal Transduction. Anal. Chem. 2014; 86: 9638-9643.

# 429352 Quantitative Analysis of Multivalent Protein Binding Using a Novel Combinatorial Lipid Array

**Abstract:**

Nolan Worstell[1], Joshua Weatherston[1], Pratik Krishnan[1], Ishan Bajaj[1], M.M. Faruque Hasan[1], Hung-Jen Wu[1]

[1]Dept. Chemical Engineering, Texas A&M University, USA

Most biochemical processes in the intracellular environment begin by membrane recruitment, protein transport from 3-D space to a 2-D membrane surface. This is achieved by multivalent binding interactions that synergistically combine weak affinity monovalent lipid-protein binding events to increase binding affinity and specificity. These multivalent binding events are essential for protein recognition, but the mechanism is poorly understood. To better understand the multivalent binding mechanism, quantitative analysis of multivalent membrane recruitment onto the cellular membrane is essential. We have developed a nanocube sensor coupled with complex reaction analysis to quantitatively explore the multivalent binding mechanism. The nanocube sensor is surrounded by a lipid bilayer that possesses the same physical and chemical properties as cell membranes. This biomimetic surface then enables the label-free detection of multivalent binding interactions by observation of the absorption spectra shift of the localized surface plasmon resonance (LSPR) peak. This biosensor works with a standard laboratory plate reader for high-throughput binding kinetic analysis. The simple protocol ("mix-and-then-detect") allows any end users to perform the analysis in their own laboratories. We have successfully explored many essential multivalent binding events, including Cholera toxin-glycolipid and Shiga toxin-ganglioside interactions. Moreover, we have introduced a complex reaction analysis technique to analyze the rich kinetic data acquired by the high-throughput lipid array and model the binding cooperativity among multivalent binding proteins, ligands, and accessory molecules. This technique can also be used to study other biomolecule adhesion, including nucleic acid and pathogen adhesions.

# 410635 Real-Time Detection of Lipid Bilayer Assembly and Detergent-Initiated Solubilization

**Abstract:**

The cellular membrane governs numerous fundamental biological processes, and therefore, developing a comprehensive understanding of its structure and function is critical. However, because of its inherent complexity, this challenge is as yet unsolved. In an attempt to develop a model, two different experimental approaches are being pursued in parallel: performing single cell experiments (top down) and using biomimetic structures (bottom up), such as lipid bilayers. One challenge in many of these experiments is the reliance on fluorescent probes for detection. In the present work, we have used a label-free detection method based on an evanescent optical sensor known as an optical resonant cavity. In this approach, we are able to detect the self-assembly and solubilization of lipid bilayers in real-time. Specifically, using these silica devices, there are two independent detection mechanisms which are able to confirm the formation and detergent assisted solubilization of the lipid bilayers: 1) a refractive index change and 2) a material loss change. Both mechanisms can be monitored in parallel, on the same device, thus allowing for cross-confirmation of the results. To verify the proposed method, we have detected the formation of self-assembled phosphatidylcholine lipid bilayers from small unilamellar vesicles (SUVs) on the device surface in real-time. Subsequently, we exposed the bilayers to two different detergents (non-ionic Triton X-100 (TX-100) and anionic sodium dodecyl sulfate (SDS)) to initiate solubilization, and this process was also detected in real-time. After solubilization, the device returned to its initial state, exhibiting minimal hysteresis. The experimental wash-off was also collected and analyzed using dynamic light scattering (DLS).

# 416775 Biosensing Mechanism of Floating Gate Transistors Utilizing Electrolyte Dielectrics


**Abstract:**

## BIOSENSING MECHANISM OF FLOATING GATE TRANSISTORS UTILIZING ELECTROLYTE DIELECTRICS



Scott P. White, Kevin D. Dorfman, C. Daniel Frısbie,

Department of Chemical Engineering and Materials Science

University of Minnesota, Minneapolis, MN 55455, USA



In a recent publication,[1] we reported a new label-free biosensing strategy based on floating gate transistors (FGTs) that uses only solution processed, electrolyte dielectrics rather than vacuum deposited solid dielectrics. The device uses the benchmark organic polymer poly(3-hexylthiophene) or P3HT, in conjunction with an ion-gel that permits low-voltage operation. Our previous experiments demonstrated high sensitivity for DNA hybridization, which we hypothesized was due to changes in the surface potential upon binding of the analyte. In this presentation, we show that our device is sensitive to changes in both the interfacial capacitance and the surface potential of the active sensing area. This new insight allows the electrolyte dielectric FGT strategy to be readily applied to non-charged analytes (in contrast to previous FGT biosensors) while also simplifying the design of capture molecules.

To demonstrate the ability to sense interfacial capacitance, we designed a set of experiments where we systematically changed the interfacial area between the gold electrode (the sensing area) and the ion-gel. The total interfacial capacitance can thus be tuned and correlated to: (1) the current through the semiconductor and (2) the voltage when the semiconductor spikes in conductivity, $V_T$. When the interfacial area adjacent to the semiconductor is altered, we show that the device response can be predicted by a simple model of two capacitors in series. However, when the interfacial area of the sensing surface (decoupled from the semiconductor) is altered it results in large changes in output due to charge storage in the floating gate. These experiments and the corresponding scaling laws that we have developed highlight differences between operating with electrolyte over solid dielectrics and outline the capacitive component of the sensing mechanism.

To demonstrate the different sensing responses for capacitance and surface potential, we adjusted the work function of the floating gate electrode by selectively immobilizing self-assembled monolayers (SAMs) onto a single electrode/ion-gel interface through microfluidic channels. The SAMs were formed through thiol/gold chemistry but with a varied functional group chemistry that changes the work function from -250 to +320 meV when measured in an inert environment. The effective $V_T$ of the semiconductor changes significantly when SAMs are formed with (1) a direction highly dependent on the orientation of the SAM but (2) largely independent on the functional group chemistry. These results are explained using a simple thermodynamic model with the high ionic strength of the ion-gel (~3.8 M) screening the electric field of the SAM functional group dipole.



1. White, S.P.; Dorfman, K.D.; Frisbie, C.D. Label-Free DNA Sensing Platform with Low-Voltage Electrolyte-Gated Transistors. *Anal. Chem.* **2015,** *87* (3), 1861-1866

# 418082 Electrochemical Piezoelectric Cantilever Sensors: Vibration-Induced Release from Biosensor Surface

**Abstract:**

**Objectives.** The objective of this study is to develop a sensor that uses the principles of cantilever mechanics and electrochemical impedance spectroscopy (EIS) for detecting a biological analyte, and to subsequently release it through cantilever resonance vibration. The 'de-binding' technique can be used as a confirmation step, or to regenerate the sensing surface. The model analytes used in this study to demonstrate the principles are: Protein G and *E. coli* O157:H7.

**Methods.** Piezoelectrically-excited cantilevers were designed with a sputtered gold electrode to facilitate impedance spectroscopy measurements at the cantilever surface. Impedance spectroscopy was used in tandem with classic dynamic cantilever resonance for detecting target analyte; the measured resistance to charge transfer from the electrode to ferricyanide in solution, and cantilever resonant frequency served as the measured signals of the two techniques, respectively. All resonant frequency and EIS measurements were conducted within a specially-designed flow cell, which incorporated a reference electrode, platinum counter electrode, and the sensor gold surface as a working electrode. After Protein G or *E. coli* O157:H7 were attached to the sensor surface, the applied excitation potential which induces vibration due to the bonded piezoelectric material to the glass cantilever, was increased until the surface acceleration was sufficient to release the target analytes. Finite element modeling was employed to evaluate the relative surface accelerations for various dynamic modes present in liquid.

**Results.** The immobilized Protein G or *E. coli* cells were removed from the sensor electrode when the excitation potential was increased from 100 mV to 10 V, with the assistance of a negative DC potential applied to the electrode. The resistance to charge transfer decreased by ~ 40% when the 10V excitation potential was applied to a sensor saturated with Protein G with the assistance of a -700 mV DC potential applied to the electrode. Little to no removal of analyte was observed with the DC potential alone. The signal returned to its original value when the surface was exposed to fresh Protein G, indicating regeneration of the sensing surface.

**Conclusion.** . The vibration of the piezoelectric sensor triggered release of the physisorbed Protein G and antibody-bound *E. coli* with the assistance of electrostatic repulsion. The vibration-induced release approach is useful for repeated sensing requirement applications, and can be used to separate dilute analytes from complex matrix. We show in this study that sensor surface can be regenerated.

# 426334 New Ligands Designed for Improving Antibody Purification with Mixed-Mode Chromatography

**Abstract:**

Mixed-mode chromatography (MMC) is new technology for bioproduct separation, especially for antibody purification. The mixed-mode ligands typically combine multiple binding modes, including hydrophobic interaction, electrostatic forces and hydrogen bonding, resulting in a combination of protein-ligand interactions. High capacity, salt-tolerance, good stability and relatively low cost are the major advantages of MMC for direct capture processes. However, compared with Protein A capture, the binding selectivity of mixed-mode ligands to antibodies still needed to be improved with a rational design strategy.

Based on the evaluation of the binding modes between some natural Fc-specific ligands and Fc fragment of IgG with molecular simulation, novel MMC ligands was designed for antibody purification, including amino-benzimidazole (ABI), a combining of amino acid and amino-benzimidazole (AA-ABI) and some tetrapeptide ligands. The hydrophobic groups could provide the orientated binding to target protein and the charged groups would induce the electrostatic repulsion for the efficient elution. New MMC ligand was coupled to the agarose gel, and the adsorption behaviors were investigated. High affinity to IgG was found with high static adsorption capacity at the neutral pH, and the flow rate had little impact on the dynamic binding capacity of IgG. In addition, the elution could be achieved at mild acidic pH with high recovery. After the optimization of operation pH and salt addition, IgG could be separated from the albumin containing feedstock and monoclonal antibodies (mAbs) could be purified from CHO cell culture supernatant with high purity and recovery. The results indicated that the resins developed were better than commercial MMC resins and comparable to Protein A resin.

For biopharmaceuticals, downstream processes require highly selective and robust technologies to achieve extremely high purity. The complexity of the interactions between the mixed-mode ligand and the desired protein can improve the binding selectivity. MMC with the specially-designed ligands would be a potential alternative to Protein A-based capture for the cost-effective purification of antibody.

This work was supported by National Natural Science Foundation of China and the Zhejiang Provincial Natural Science Foundation of China.

# 431648 Antibody Purification Via an Affinity Membrane Chromatography Method Utilizing Nucleotide Binding Site Targeting Small Molecule

**Abstract:**

Antibodies have extraordinary specificity and affinity to antigens, which in turn makes them important candidates to be used in numerous applications including detection, diagnosis, and therapy. Over last two decades, therapeutic antibodies have continued to be evaluated extensively for treatment of many diseases including cancer and autoimmune diseases. Even though antibody therapies are very efficacious for patients, monoclonal antibody-based treatments are expensive; therefore many patients cannot afford these treatments. A major contributor to the cost of antibody treatments are mostly linked to the downstream production process, or more specifically the usage of Protein A (or G) affinity columns for purification of antibodies. These columns are expensive and have short life cycles with several obstacles that preventing them to be used repeatedly. Here, we design a novel affinity membrane chromatography technique for purification of monoclonal and polyclonal antibodies from cell culture media and ascites fluids. This method utilizes the nucleotide-binding site (NBS) that is located on variable domain of Fab region nearly all antibodies to facilitate capturing antibodies on a membrane affinity column. Regenerated cellulose membranes has been choose as a matrix in this method due to their major advantages over traditionally used resin-based affinity systems. Rituximab was used for proof of concept experiments. A small molecule was attached to the membrane in order to target the NBS for capturing antibodies on the column. Antibody capture was accomplished by injection of samples while running equilibration buffer (50 mM sodium phosphate pH 7.0) and elution of the antibody was achieved by running a gradient of mild elution buffer (3M NaCl in 50 mM phosphate pH 7.0). The results indicate that the tryptamine column efficiency for selected antibodies was >98%, with a purity level of >98%. This small molecule affinity purification method is further tested for a number of necessary parameters such as: injection concentrations, volumes, wash/bind time, elution gradient, antibody/protein-contaminant combinations, effects of injection buffer, post-purification antigen binding activity of antibodies, and column reusability and stability. This method provides a superior alternative to the protein-A affinity purification method that is widely used for purification of humanized and chimeric antibodies. Results from these findings will help to increase the availability of antibody based treatment and diagnostic systems to patients.

# 426926 Integrated Processes for High Yield Protein Pegylation

**Abstract:**

Considering the production cost of therapeutic proteins, post translational modifications, as PEGylation, have to be highly efficient and optimized, not to waste the valuable and highly purified protein to be modified. Because of the multiple conjugation sites often present at the protein surface, PEGylation suffers from limited yield when it comes to selectively target one single degree of PEGylation. Regarding this limitation, the integration of a chromatographic step in the process can bring highly competitive advantages. Ion-exchange chromatography is used as fractionation technique in order to isolate and recycle the unreacted protein from the PEGylated products. This allows operating the reactor with short reaction times so as to minimize the production of multi-PEGylated proteins. That is, the coupling reaction is stopped before the desired product (i.e. the mono-PEGylated protein) can further react, thus leading to limited conversion but high yield. The recycling of the unreacted protein was then considered to drive the protein overall conversion to completion.

In a second part of this work, the integration is pushed at is paroxysm by investigating an on-column PEGylation process. Integrated processes, efficiently combining reaction and separation in a single step are very attractive to perform protein PEGylation at high yield and high conversion. By adsorbing the protein on an ion-exchange media before contacting it with the functionalized PEG solution, this approach has the potential to inhibit the formation of multi-PEGylated species by restricting the coupling reaction to the few accessible sites. After developing the batch on-column PEGylation process, a multicolumn counter-current chromatography set-up was considered to turn the batch on-column PEGylation process to a continuous one. Using the parameters previously estimated in batch mode, the reactive multicolumn countercurrent solvent gradient purification (rMCSGP) process was simulated and the performance of the process assessed. The rMCSGP can potentially improve the throughput, purity and yield for the production of mono-PEGylated protein.

# 423719 A Split Intein-Based Self-Cleaving Tag for Recombinant Protein Purification in a Mammalian Cell Expression System

**Abstract:**

A Split Intein-based Self-cleaving Tag for Recombinant Protein Purification

In a Mammalian Cell Expression system

Tzu-Chiang Han

Dr. David W. Wood

Inteins are genetic elements that can self-excise from various host proteins in a post-translational protein splicing process. With a simple mutation of the intein, one can yield isolated, controllable intein cleaving activity for tag removal purposes. By combining self-cleaving inteins with conventional affinity tags, self-cleaving affinity tags have been developed. The self-cleaving tags benefit the tag removal process by a simple, inducible cleaving reaction without any addition of protease. The phenomenal tag removal strategy works successfully in prokaryotic cell systems in our previous studies; however, the *in vivo* premature cleaving of the precursor protein remains the bottleneck of its application in eukaryotic cell systems. Here we present our most recent work in developing a split intein-based recombinant protein purification strategy in a mammalian system. Split intein consists of two complementary fragments of intein that can undergo reconstruction to form a fully-active intein by *in vitro* assembling, while no splicing activity shown individually. The affinity tag-fused intein (N-fragment) was first expressed in *E.coli* and further conjugated with the beads in a chromatography column. The target protein that fused with the C-fragment of intein was expressed in a mammalian system and then mixed thoroughly with the tagged N-fragments of the intein. After an affinity-based purification, the target protein can be eluted by inducing the self-cleaving reaction of the re-assembled inteins, resulting in a tagless, purified protein. With the split-intein system, we provide a feasible scheme to overcome the unwanted *in vivo* premature cleaving and generate a potential platform technology for the manufacture of non-antibody glycoprotein therapeutics at large scale.

# 413490 Protein Recovery from Potato Processing Wastewater: Pre-Treatment and Fouling Minimization

**Abstract:**

Waste treatment has been a challenge for the potato processing industry for many years. The wastewater effluent with high concentrations of potassium and Chemical Oxygen Demand (COD) caused by the presence of starch, proteins, amino acids and sugars, imposes expensive treatment processes for the companies. However, this waste effluent contains high amounts of valuable by-products. Starch content of this waste stream that range from 15 to 20%, is being recovered using simple separation systems such as hydro-cyclones in some manufacturing sites but the commercially valuable protein ranging from 1-1.5 g/L is still being discharged into the waste stream. Potato proteins have high functionality and nutritional value and are considered as one of the best plant proteins and can be utilized in food applications if recovered in native form with high quality and be isolated by the methods that do not result in denaturation. Different techniques such as traditional concentration, heat coagulation and precipitation, ion-exchange using Expanded Bed Adsorption (EBA) and UF have been used to recover the potato fruit juice (PFJ) protein (potato fruit juice is the juice generated from potato starch production that mainly contains water, salts and protein). The methods all resulted in relatively high yield of recovered protein, but they either have failed to recover completely un-denatured high quality protein due to the usage of heat and acids or are complex and expensive to be used commercially. Ultrafiltraion membrane (UF) has been reported as an appropriate technology to recover relatively high yield of high quality potato protein. However, the optimization of membrane process to mitigate fouling problems, enhance removal of toxins and impurities such as microfiber and glycoalkaloids and achieve high quality protein products has not been systematically examined. In this study, fouling and protein recovery from potato process water was investigated utilizing a number of potential pre-treatments and filtration conditions to target high quality protein production. Sedimentation, centrifugation, prefiltration with paper filter and microfiltration were assessed as pre-treatments to remove particulate and fibrous matter prior to a final concentration with 10 kDa molecular weight cutoff ultrafiltration with the aim of removing low molecular weight components.

barrywehmiller bw betterworld

**194 Monday Nov. 9, 2015 3:15 pm**
**Bioseparations and downstream processing**
150D/F (Salt Palace Convention Center)
Caryn L Heldt, Chair
Seongkyu Yoon, Co-Chair

## Single Use Centrifuge " UniFuge® in Bioseparations"

*David R. Richardson, Single Use systems, Pneumatic Scale Angelus. Division of Barry Wehmiller. Clearwater, FL*

The large scale economic collection of proteins is increasingly an important for the Biotechnology industry. Generally, proteins are produced by cell culture, engineered to produce the proteins of interest by insertion of a recombinant plasmid containing the gene for that protein. Since the cell lines used are living organisms, they must be fed with a complex growth medium. Separation of the desired protein from the fermentation broth poses challenges.

Advances in cell culture techniques have greatly increased the cell populations and titers of the target proteins in culture fluid. This increase in upstream efficiency has led to challenges in downstream processing at the cell-harvest stage. When the product is extracellular cell harvesting is used to separate the product from the cells and cellular debris. Other applications collect viable cells. Most Biopharm companies today are seeking ways to integrate single use technology (including harvest) into a complete SU process path. The UniFuge®, Single Use Centrifuge fits very well into this SU initiative. This presentation will Demonstrate several UniFuge processes.

Pneumatic Scale Angelus

barrywehmiller bw betterworld

# Continuous Flow
## Centrifuge Platforms

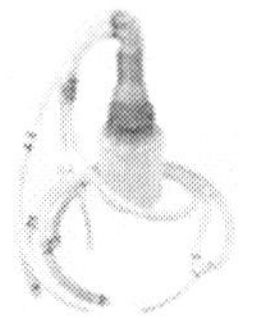

**Disposable Single Use Systems**

**UniFuge®**
- Low shear viable cell recovery to 3,000 x g
- Batch and fed batch harvesting
- Scalable: 250ml/min → 6 L/min

**Centritech®**
- Low shear cell harvest, perfusion, media exchange
- Scalable: 250ml/min → 150 L/hr

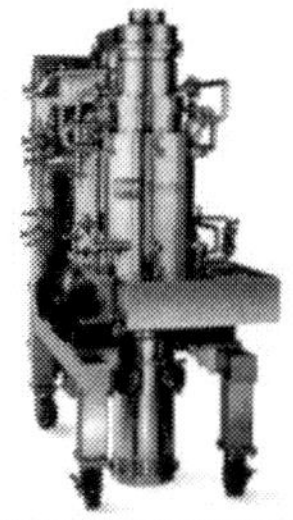

**Cleanable Systems**

**CARR Powerfuge®** - up to 20,000 x g
- Bacteria, fungi, proteins, sub-micron particles
- Scalable: 250ml/min → 1,700 L/hr

**CARR Viafuge®** - Low shear cell harvest
- Mammalian, insect, plant cells (>99% viability)
- Scalable: 250ml/min → 2,400 L/hr

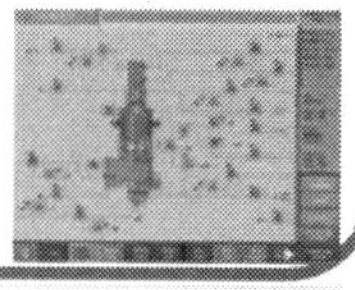

Pneumatic Scale Angelus

barrywehmiller betterworld

*Single Use*

# UniFuge® Single Use Centrifuge

Pneumatic Scale Angelus

barrywehmiller betterworld

## UniFuge ®
## Introduction

### UniFuge ® -Single Use Centrifuge

- UniFuge is a Single-use tube bowl centrifuge based on proven platforms. Installed in vaccine manufacturing 2010.
- **Mammalian** Applications: CHO, MD, HAL2, MRC5, SF9,HEK, PERC6, stem cells-MSC, vaccines.
  - No shear
  - Clarification, separation, Mab collection, cell banking
  - Sterile cell harvest, stem cell concentration and washing
  - Vaccines- Human and animal
- 10-1000 liter bioreactor clarification: 200-3000 g
  - Process Feed rates 0.25 – 6 L/min (to 360 L/hr)

Pneumatic Scale Angelus

## UniFuge®
## Process Buffer Options

- **Media displacement**
- **Re-suspension-freezing buffer**
- **Cell washing, concentration**
- **All-Process steps are automated**

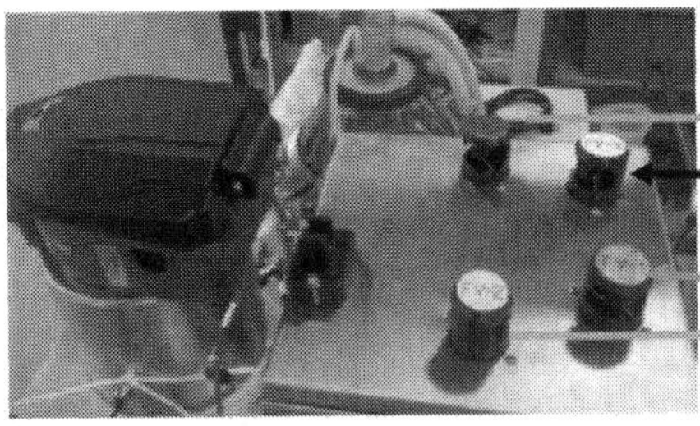


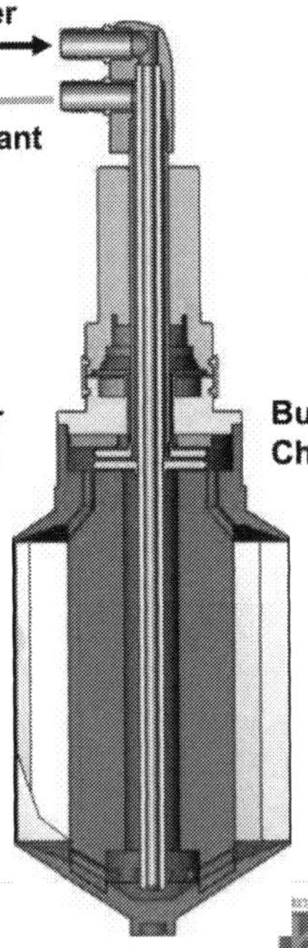


Pneumatic Scale Angelus

# UniFuge®

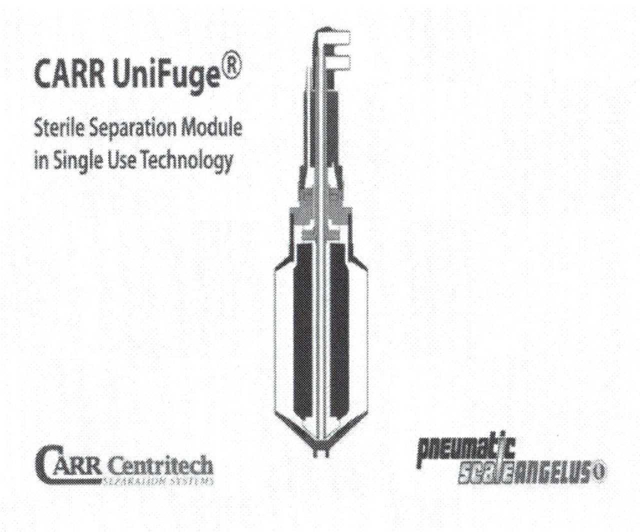


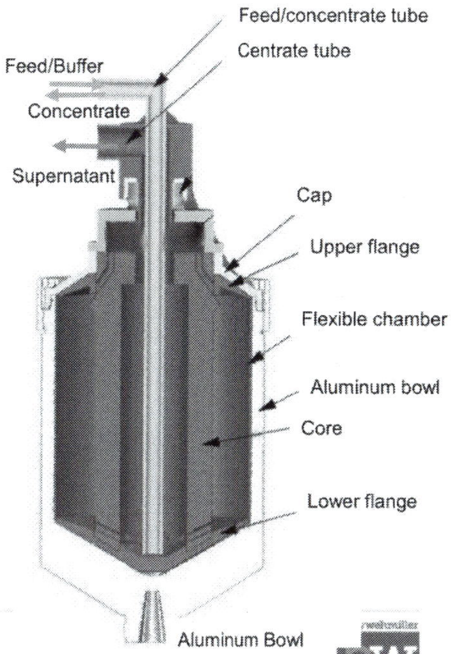


Pneumatic Scale Angelus

# UniFuge® Pilot Layout

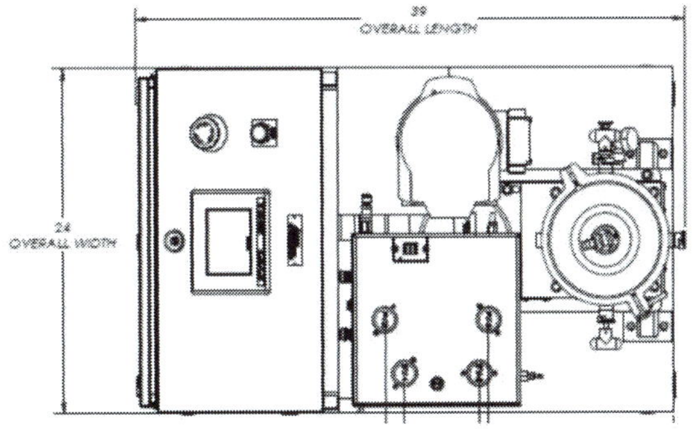


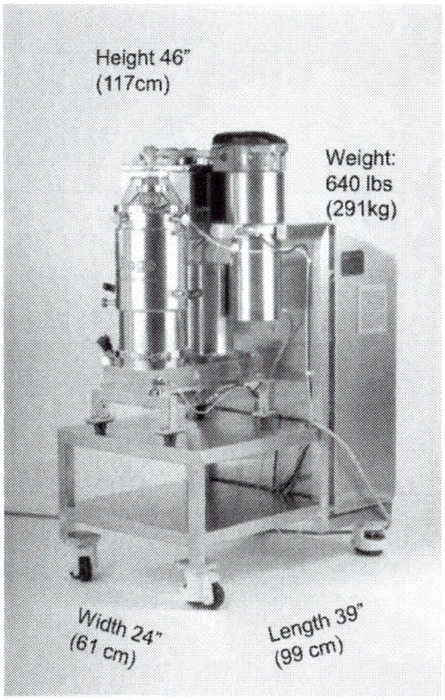


- Electrical: 120/230VAC, 1 phase, 50/60Hz, 15/7.5A
- Pneumatic: 80 PSI, 10 SCFH
- Feed: 200 ml/min to 6L/min, ½" tubing
- Cooling jacket-1L/min; max pressure: 90 psi

Pneumatic Scale Angelus

## Module Production Facility- ISO Class 7 (Class 10,000). WHK is ISO 13485: 2003 Certified and FDA Registered as a Medical Device Manufacturer.

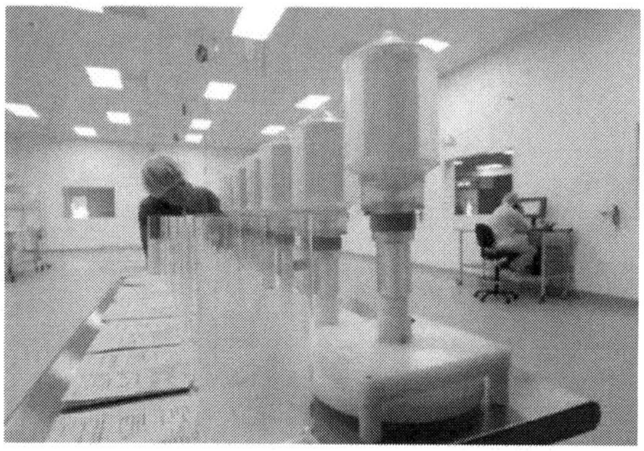

Pneumatic Scale Angelus

barrywehmiller bw betterworld

## UniFuge®

*Single Use Component*

### Single use-Module Assembly

- Module with 4-tube set, disposable pressure sensor & 0.2 micron air filter
- Terminally sterilized (gamma)
- USP Class VI materials
- COC
- 1.7 liter sediment holding space

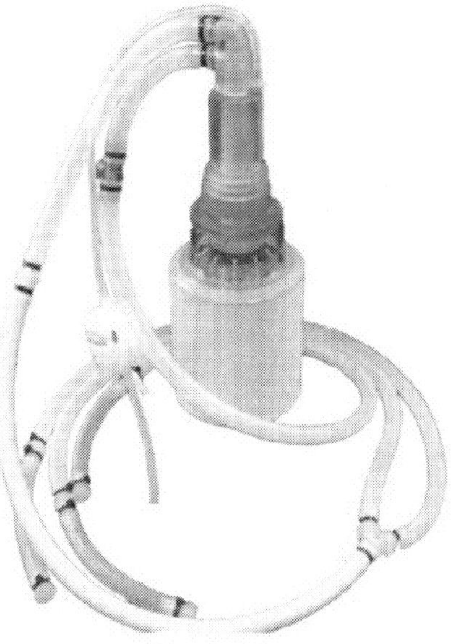

Pneumatic Scale Angelus

barrywehmiller bw betterworld

# HMI Screenshot

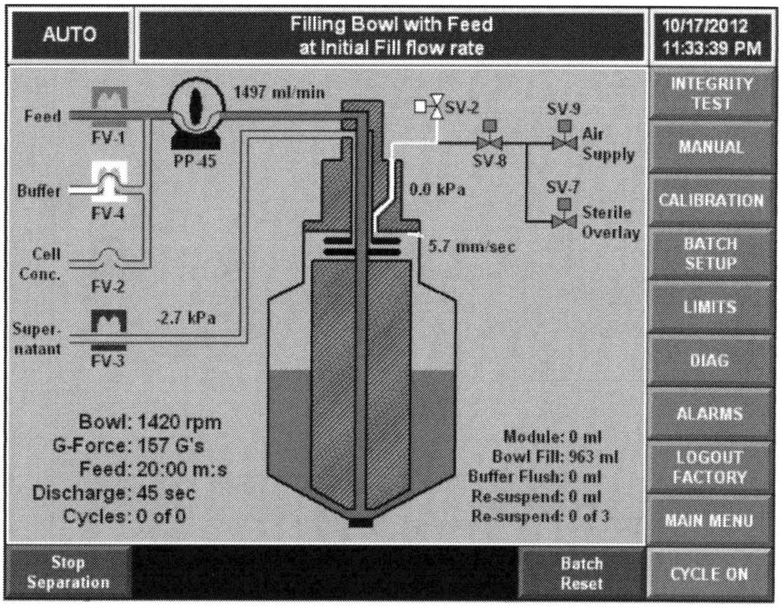


Pneumatic Scale Angelus

# UniFuge® Pilot

*Field Trial Data*

## Viability Table- Feed Centrate- UniFuge *-measured after freeze thaw.

| Cell Line | Volume | Feed Flow rate | G-force | Feed Viability | Harvest Viability | % recvovered | centrate |
|---|---|---|---|---|---|---|---|
| CHO 1.96E6 Suspension | 23 liters | .75 LPM | 500 | 93% | 90% | 90 | 0 |
| Myloid progenitor 4.93E6 | 5.6 | 1.5 LPM | 700 | 95% | 93* | 93 | 0 |
| Hek Suspension | 8 | 2 | 3000 | 90 | 89 | 89 | 0 |
| MPC -10 Micron | 50 | .75 lpm | 750 | 95 | 93* | 100 | 0 |
| Msc-stem | 200 | 1.25 | 700 | 97 | 95 | 101 | 0 |

Pneumatic Scale Angelus

# CHO 1-6 LPM NTU/Vmax MA CMO

| G force | Feed rate lpm | Feed volume liters | Ntu | Vmax at 20 psi |
|---|---|---|---|---|
| 3000 | 2 | 3 | 87 | 356 Vs 156 d.s. |
| 3000 | 3 | 3 | 94 | 217 |
| 3000 | 4 | 3 | 111 | |
| 3000 | 5 | 3 | 120 | |
| 3000 | 6 | 3 | 144 | |
| 3000 | 1 | 3 | 77 | |
| 4000 | 2 | 3 | 80 | 322 |
| 4000 | 4 | 3 | 102 | |
| 4000 | 6 | 3 | 118 | 80 |
| 4000 | 1 | 3 | 72 | |
| 4000 | 5 | 3 | 115 | |
| 4000 | 4 | 22 | 117 | |
| Control 3500 rpm | | | 61 | |
| Feed | | | 1260 | |

15E6 cells/ml 70% viability

Pneumatic Scale Angelus

barrywehmiller bw betterworld

# UniFuge® Loading and Pressure

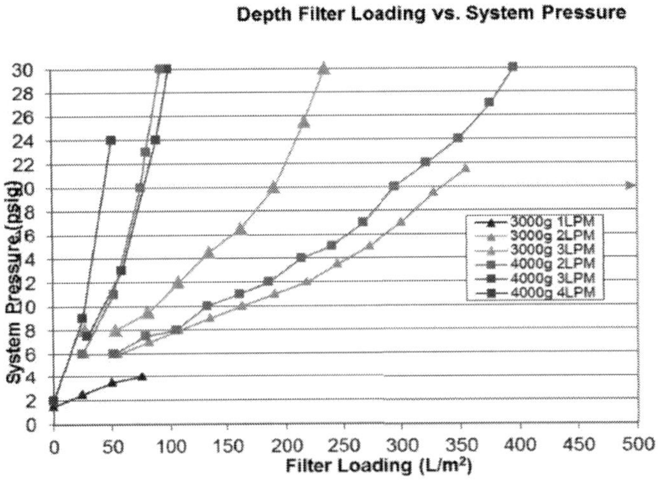


Millipore X0HC grade depth filter followed in series by a Sartorius Sartopore 2 0.45/0.2um filter

Pneumatic Scale Angelus

barrywehmiller bw betterworld

# 27 % viability CHO 12E6 JP

| Parameter set<br>Result Table 1: | Feed rate lpm | Feed volume liters | Ntu | Discharge volume | G force | Time at NTU sample | Avg. NTU pooled (2 cycles) | Pmax rating |
|---|---|---|---|---|---|---|---|---|
| 1 | 1 | 2 | 39 | | 4000 | 0 | | #1 |
| 1 | 1 | 7 | 44 | | 4000 | 5 | | |
| 1 | 1 | 12 | 45 | | 4000 | 10 | | |
| 1 | 1 | 17 | 44 | | 4000 | 15 | | |
| 1 | 1 | 22 | 51 | | 4000 | 20 | | |
| 1 | 1 | 27 | 48 | | 4000 | 25 | | |
| 1 | 1 | 33 | 50 | 1.4/1.8 | 4000 | 33 | 51 | |
| | | | | | | | | |
| 2 | 3.4 | 2 | 159 | | 4000 | 0 | | #3 |
| 2 | 3.4 | 8.8 | 56 | | 4000 | 2 | | |
| 2 | 3.4 | 15.6 | 74 | | 4000 | 4 | | |
| 2 | 3.4 | 22.4 | 74 | | 4000 | 6 | | |
| 2 | 3.4 | 29.2 | 72 | | 4000 | 8 | | |
| 2 | 3.4 | 33 | 72 | 1.4/1.7 | 4000 | 10 | 86 | |
| | | | | | | | | |

Pneumatic Scale Angelus

barrywehmiller bw betterworld

# 27% Viability CHO 12E6 JP

| Parameter set | Feed rate lpm | Feed volume liters | Ntu | Discharge volume | G force | Time at ntu sample | Avg ntu of pooled 2 cycles | Pmax rating |
|---|---|---|---|---|---|---|---|---|
| 3 | 3.4 | 2 | 192 | | 3000 | 0 | | #4 |
| 3 | 3.4 | 8.8 | 63 | | 3000 | 2 | | |
| 3 | 3.4 | 15.6 | 66 | | 3000 | 4 | | |
| 3 | 3.4 | 22.4 | 65 | | 3000 | 6 | | |
| 3 | 3.4 | 29.2 | 63 | | 3000 | 8 | | |
| 3 | 3.4 | 33 | 64 | 1.3/1.6 | 3000 | 10 | 85 | |
| | | | | | | | | |
| | | | | | | | | |
| 4 | 6.0 | 2 | 124 | | 4000 | 0 | | #2 |
| 4 | 6.0 | 8 | 90 | | 4000 | 1 | | |
| 4 | 6.0 | 14 | 83 | | 4000 | 2 | | |
| 4 | 6.0 | 20 | 79 | | 4000 | 3 | | |
| 4 | 6.0 | 26 | 80 | | 4000 | 4 | | |
| 4 | 6.0 | 30 | 81 | | 4000 | 5 | 88 one cycle | |
| | | | | | | | | |

Pneumatic Scale Angelus

barrywehmiller bw betterworld

## UniFuge® vs Disk stack

**UniFuge® vs. Westfalia SC6 disk stack clarifier**

- Pmax 356 vs. 156 at 2 lpm.
- No CIP or set up
- Quiet operation
- Low maintenance
- No dis-assembly

Pneumatic Scale Angelus

barrywehmiller bw betterworld

U2k- Scale up. Alpha 2 test unit being built

- U2K- 5 x UniFuge Scale
- Single use
- Continuous discharge
- For 2000 bioreactors
- High density/acid precipitates

Pneumatic Scale Angelus

barrywehmiller bw betterworld

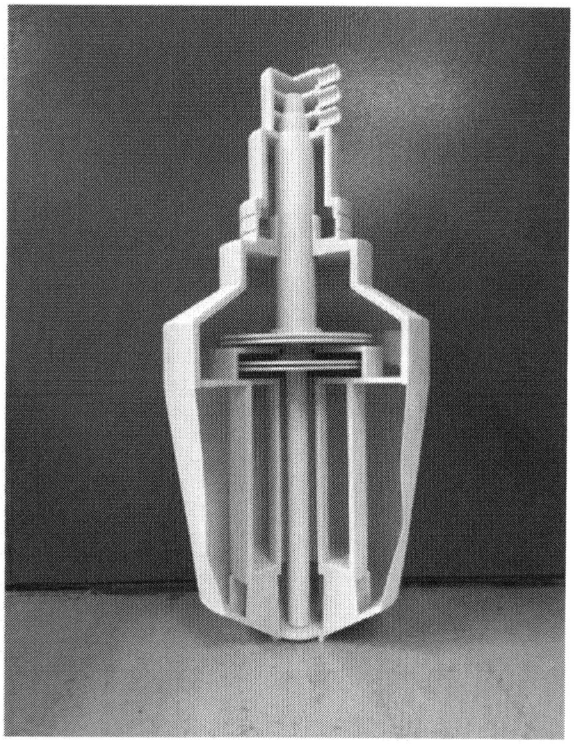

Pneumatic Scale Angelus

barrywehmiller bw betterworld

# U2k specs

Pneumatic Scale Angelus

CARR Centritech Separation Systems

U2K®

We are excited to introduce the UniFuge®'s big brother, U2K®! Combining state-of-the-art technology and engineered research, the U2K® will bring disposable centrifugation to full production scale.

- High cell density CHO up to 10E6 to 50E6 cells/ml
- Pcv up to 25% at 50E6 cells/ml
- Acid precipitate application available
- 5 x UniFuge® flow rates; up to 30 LPM
- Continuous operation
- Continuous feed flow
- Continuous discharge flow
- AB or Siemens PLC
- Data acquisition available
- 50" L x 34" W x 78" H (approx.) Cart Size
- 400-600kg (approx.) Cart Weight
- 400/460 V, 50/60 Hz, 3 Phase, 20 Amp
- 90 PSI - 1 SLPM
- 4,000 G Force (approx.)

# UniFuge® Single Use Centrifuge

UniFuge

Pneumatic Scale Angelus

barrywehmiller
bw
betterworld

# 414046 Effect of Ionic Conditions on the Purification of Plasmid DNA Using Ultrafiltration Membranes

**Abstract:**

It is well known that the structure of plasmid DNA varies significantly with solution ionic strength due to intramolecular electrostatic interactions between the charged phosphate groups. The objective of this work was to examine the effect of ionic conditions on the transmission of different plasmid isoforms through small pore size ultrafiltration membranes, with the goal of identifying conditions that can be used to enhance the purification of the supercoiled plasmid DNA which is of interest in gene therapy applications and as DNA-based vaccines.

Experimental were performed with 3.0 and 16 kbp plasmids using composite regenerated cellulose and polyethersulfone ultrafiltration membranes. The different isoforms were generated by enzymatic digestion of the supercoiled plasmid, with the identity of the isoforms confirmed by agarose gel electrophoresis. The sieving coefficient of the linear isoform was nearly independent of solution ionic strength, but increased significantly with increasing filtrate flux due to the elongation of the highly flexible plasmid in the converging flow field into the membrane pores. In contrast, the transmission of the open-circular and supercoiled plasmids both increased with increasing NaCl concentration. Similar trends were observed using solutions of $MgCl_2$. The dependence on the ionic condition was more dramatic for supercoiled plasmid, mainly due to changes in the superhelix structure in varying salt environments. The separation between the supercoiled and linear isoforms was enhanced at low ionic strength due to the increase in retention of the supercoiled plasmid. In contrast, the separation between the supercoiled and open-circular isoforms was enhanced at high ionic strength, with the supercoiled plasmid collected in the permeate. These results clearly demonstrate the potential for enhancing the performance of membrane systems for plasmid DNA separations by controlling the ionic conditions to modify the transmission of the different plasmid DNA isoforms.

**Keywords:** DNA, ultrafiltration, ionic strength, isoforms, DNA deformation

# 412287 The State of Flux: Stability As a Criterion for Model Building and Prediction

**Abstract:**

Stability in a metabolic system may be unattainable if incorrect amounts of enzymes are used. Upon loss of stability, some metabolites may accumulate or deplete leading to the irreversible loss of the desired operating point. Changes in enzyme amount which cause loss of stability in continuous systems can lead to lower production even in batch quasi-steady systems by the inability to maintain productivity. Here we use Ensemble Modelling for Robustness Analysis (EMRA) to analyze stability to enzyme changes in three cell-free enzymatic systems. In these systems, variations in the level of certain enzymes or substrate feed rate have been predicted to cause instability in steady state simulations. Instability can be caused by a kinetic trap in which two reactions both require the same metabolite. The predictions of instability by EMRA are supported by the lower productivity in batch experimental tests reported in the literature. The EMRA method uses no *a priori* knowledge of the kinetic parameter values of the enzymes, and is generalizable to in vivo systems.

# 412292 Integration of Metabolic Pathways to Develop Optimal Yeast Strains for Producing Biofuels and Chemicals

**Abstract:**

Plant cell wall hydrolysates contain fermentable sugars (e.g., glucose and xylose) and toxic fermentation inhibitors. Therefore, a fermenting microorganism will need to utilize both glucose and xylose under the presence of toxic levels of fermentation inhibitors. We engineered *Saccharomyces cerevisiae* for efficient and rapid fermentation of the cellulosic sugars. As *S. cerevisiae* cannot ferment xylose, a xylose-fermenting pathway from *Scheffersomyces stipitis* was introduced into *S. cerevisiae* and the resulting engineered strain was evolved to produce ethanol from xylose with a high yield and productivity. In addition to ethanol, we demonstrated that the xylose-fermenting yeast can be modified to produce polymer precursors, such as lactic acid and 2,3-butanediol. We further engineered the xylose-fermenting *S. cerevisiae* for assimilating cellobiose which is a disaccharide that can be produced from cellulose. When a cellobiose-fermenting pathway from *Neurospora crassa* was introduced into the xylose-fermenting yeast, the resulting strain was able to co-consume cellobiose and xylose simultaneously. While the engineered yeast can ferment cellobiose and xylose, the presence of acetic acid in lignocellulosic hydrolysates hindered efficient utilization of the cellulosic sugars. In order to alleviate the toxicity of acetate towards the xylose and cellobiose co-fermenting strain, we introduced an acetate reduction pathway which can enhance xylose fermentation through redox coupling. As a result, we constructed an optimal yeast strain capable of producing ethanol from a mixture of cellobiose, xylose, and acetate. Our results not only demonstrate the synergistic effects from integration of multiple metabolic pathways, but also exemplify the Design-Build-Test and Learn (DBTL) cycle of synthetic biology.

# 412286 Metabolic Engineering through Evolution

**Abstract:**

The field of Metabolic Engineering has recently undergone a transformation that has led to a rapid expansion of the chemical palate of cells. The incorporation of tools from synthetic biology and directed evolution has enabled a newfound control of cells. Now, it is conceivable to produce nearly any organic molecule of interest using a cellular host—from biofuels to biopolymers to pharmaceuticals. However, these feats require the ability to "hijack" native cellular machinery and metabolism and navigate the complexity inherent in cellular regulation. In this regard, there is a growing use of directed evolution within the design cycle. The combination of evolved pathways and synthetic parts with "traditional" metabolic engineering established a blended approach of "metabolic pathway protein engineering". This talk highlights how the evolution of small parts, enzymes, pathways, and whole cells can rapidly accomplish metabolic engineering goals.

# 412291 Application of 13C Flux Analysis to Identify Molecular Disease Mechanisms: From Cells to Whole Animals


**Abstract:**

$^{13}C$ metabolic flux analysis (MFA) is the gold standard approach for quantifying rates of biochemical reactions inside living cells. It has been widely applied to debottleneck the metabolism of industrial host organisms, but it is now being increasingly used to investigate metabolic phenotypes of human disease models. As an example of the latter, my group has applied $^{13}C$ MFA to quantitatively map the metabolic alterations that occur in liver cells exposed to toxic levels of free fatty acids (FFAs). In humans, abnormal hepatic lipid accumulation caused by FFA overload leads to a condition known as nonalcoholic fatty liver disease (NAFLD), which is closely associated with obesity and type-2 diabetes and is estimated to affect up to 30% of the U.S. population. Unfortunately, the factors that control disease severity are poorly understood, and therapeutic strategies for preventing or reversing NAFLD are limited.

Based on $^{13}C$ MFA studies of cultured hepatocytes and hepatic cell lines, we have uncovered a novel lipotoxicity mechanism by which saturated FFAs promote abnormal cell metabolism and oxidative stress in liver cells. This process involves dysregulation of intracellular calcium homeostasis that subsequently promotes increased mitochondrial metabolism and accumulation of toxic reactive oxygen species. Surprisingly, $^{13}C$ MFA revealed that these metabolic alterations were not driven by increased FFA oxidation, but instead were fueled by increased entry of glutamate carbon into the citric acid cycle (CAC). I will discuss these findings as well as my lab's ongoing work to develop an *in vivo* $^{2}H/^{13}C$ MFA approach to simultaneously assess multiple gluconeogenic, CAC, and anaplerotic fluxes in the livers of conscious, unstressed mice. Our goal is to establish a scalable *in vivo* MFA platform that can be used to examine liver metabolic phenotypes in mouse models of NAFLD and to test specific hypotheses suggested by our *in vitro* studies.

# 412290 Metabolic Engineering of Complex Plant Processes in Bacteria

**Abstract:**

Specialized (secondary) metabolites are products of complex biochemistries found in nature and include a vast number of chemical candidates (>200,000) with a myriad of applications in drugs, food additives, consumer products and industrial chemicals. However, sourcing of these valuable materials remains a significant barrier to more widespread utilization. The low accumulation of these molecules in nature (ppm levels) translates to inefficient plant extraction processes which remain economically unviable and present an undue burden on our land resources. In addition, the structural complexity of these natural products, which are typically heavily functionalized molecules with multiple chiral centers, precludes the development of economical chemical synthesis routes to these molecules. Therefore, developing tools and technologies for the rapid and efficient construction of microbial strains capable of producing specialized natural products is a high priority research area for the metabolic engineering community.

We have developed a new metabolic engineering approach, multivariate modular metabolic engineering (MMME), for systematically engineering such complex multi-step biosynthetic pathways. Recently, we also expanded the scope of our approach to merge quick construction and optimization tools with a thorough characterization of strains (using transcriptomics, proteomics and metabolomics) to expedite strain development. These enabled not only the rapid construction of microbial strains for synthesizing complex biochemicals in useful accessible quantities, but also provided several key insights on natural product biosynthesis and the origins of biosynthetic diversity of these specialized products. Here, I focus on the MMME for optimizing the multi-step pathways with complex chemistries, including multi-step P450 chemistry engineered in bacterial cells for scalable synthesis of complex molecules.

# 412288 Integrated Regulation of Amino Acid and Fatty Acid Metabolism

**Abstract:**

Metabolism is central to virtually all cellular functions and contributes to a range of diseases. A detailed understanding of how biochemical pathways are regulated is necessary to understand and control cell function. Quantitative information describing the flow of metabolites through biochemical networks provides critical insights into how different nutrients contribute to energy metabolism and biosynthesis. To this end we apply stable isotope tracers, mass spectrometry, and metabolic flux analysis (MFA) to study central metabolism in mammalian cells. Using these approaches we have characterized how proliferating and differentiated cells regulate flux of glucose and amino acids into mitochondria for biosynthesis. Through this analysis we have characterized how these critical metabolic pathways are influenced by nutrient availability and the cellular microenvironment. Ultimately, the application of MFA to these cellular models has improved our ability to characterize intracellular metabolic processes, providing a mechanistic understanding of cellular physiology and metabolic function.

# 430845 Genome-Scale Modeling of the Core Proteome of Metabolism and Expression, and Its Transcriptional Regulation

**Abstract:**

The cell 'chassis' is of fundamental importance for synthetic biology applications. To predictively engineer this chassis, a systems biology understanding of it is critical. In this work, we broaden our understanding of the chassis using integrated genome-scale models of metabolism and expression. We first define a core proteome consisting of 356 gene products, accounting for 44% of the *Escherichia coli* proteome by mass based on proteomics data. This core proteome includes 212 genes not found in previous comparative genomics-based core proteome definitions, accounts for 65% of known essential genes in *E. coli*, and has functional overlap with minimal genomes (*Buchnera aphidicola* and *Mycoplasma genitalium*) (AUC = 0.78). Based on transcriptomics data across growth conditions and genetic backgrounds, the systems biology core proteome is significantly enriched in non-differentially expressed genes, and depleted in differentially expressed genes. Compared to the non-core, core gene expression levels are also similar across genetic backgrounds. Furthermore, core genes exhibit significantly more complex transcriptional and post-transcriptional regulatory features (40% more transcription start sites per gene, >20% longer 5'UTR). Thus, genome-scale systems biology approaches rigorously identify a functional core proteome needed to support growth. This framework, discerned and validated using high-throughput datasets, facilitates a mechanistic understanding of systems-level core proteome function through in silico models. We then investigate how the core proteome is regulated using a genome-scale, transcriptional regulatory network of *E. coli*. Taken together, the tightly regulated core proteome of *E. coli* comprises a fundamental repertoire of genes and biological functions that are of importance to a broad range of microbial synthetic biology applications.

# 423632 Investigation of the Antibiotics-Triggered Perturbation in the Metabolism of Pseudomonas Aeruginosa Via a Metabolic Modeling Approach

**Abstract:**

Although antibiotics are known to eliminate pathogens through a variety of mechanisms [1, 2], it is reported that antibiotics may induce the transition in the microbial phenotypes and promote the formation of persisters, which are dormant variants of regular cells that are highly tolerant to antibiotics [3]. For example, Dr. Wood's group at Pennsylvania State University reported that *E. coli* pretreated with antibiotics such as rifampicin, tetracycline, and carbonyl cyanide m-chlorophenyl hydrazine increased the microbial persistence dramatically. They demonstrated that the bacterial persistence resulted from halted protein synthesis and a variety of environmental cues [4]. The presence of persister-specific resistance is suggested to account for the recalcitrance of infectious diseases [5, 6]. It is thus important to study the mechanisms via which antibiotics cause pathogens to change their metabolism to form persister cells. Experimental approaches such as transposon insertion have been used to study the phenotype transition mechanisms during antibiotic treatment. Screening knockout libraries has not produced mutants that lack persisters, indicating that dormancy mechanisms are not regulated by single genes or enzymes [3, 7]. This understanding motivated us to develop a systems level approach that can incorporate the interaction of multiple genes/enzymes/metabolites to investigate potential persister-forming mechanisms triggered by antibiotics. Since genome-scale models contain the genes/enzymes/metabolites for metabolic pathways that determine the microbial growth rate, they can be used to pinpoint the genes/metabolites involved in the phenotypic transition. In particular, some metabolites in genome-scale models behave like signaling molecules in triggering persister formation. One example is acyl-homoserine lactone, which promotes the persister formation in *Pseudomonas aeruginosa* through quorum sensing [8]. Studying the metabolic alternation between different phenotypes may provide cues to potential signaling metabolites and pathways to manipulate the formation of persister cells. Although research has been conducted on the reaction-flux distribution upon the antibiotic treatment [9], no genome-scale modeling approach has been published to quantify the effect of antibiotics on bacterial phenotype transition.

In this work, we present the first genome-scale modeling approach to quantify the effect of antibiotics on the metabolic flux redistribution, and thus investigate potential mechanisms for the persister formation that is triggered by antibiotics. Specifically, we first developed an approach to integrate gene expression data with metabolic models to identify the reactions that positively reflect the metabolic activities of microorganisms. On the basis of these metabolic-activity indicating reactions, we then developed the first genome-scale approach to predict microbial metabolic perturbation of *P. aeruginosa* upon antibiotic treatment. *P. aeruginosa* is selected as the example microorganism in this work, as it is one of the leading causes of nosocomial infections in hospitalized patients and it displays resistance to a wide array of antibiotics by forming a biofilm in chronic infectious processes [10, 11]. The results showed that 1110 metabolic reactions within the network, 171 reactions increased flux, 372 reactions decreased fluxes and 10 reversible reactions changed flux direction. Parts of our results were further validated by experiment data related to persister cell formation. The flux profile of cell with antibiotic treatment provides clues on the detection of novel signaling pathways that trigger cell persistence. The exchange rates of several metabolites with significant alternation were also proposed. Specifically, 14 metabolites including acetate, fumarate, gluconate, succinate, citrate, ethanol, glycerol, and pyruvate increased their exchange rates, while 6 metabolites including 4-hydroxybenzoate, orotate and allantoin decreased their exchange rates. These metabolites may be used in further controlling persister formation in nutrition supplement. Given the metabolic network information, this approach can be applied to study the metabolism adjustment in phenotype transition of other microorganisms.

# 433487 A Modular Computational Framework for Multi-Scale Simulation of Chemotaxis in Complex Environments

**Abstract:**

Chemotaxis has long served as a model system for signal transduction in both bacteria and eukaryotes. Numerous mathematical models have been developed over the years to explore different aspect of these pathways. These models have tended to focus either on the intracellular mechanisms governing pathway dynamics or the coarse-grained, population-level behavior of swimming/migrating cells in chemical gradients. Only a few models have attempted to link the two. In this work, we developed a modular multi-scale computational framework for simulating cell behavior in complex chemical environments using a hybrid-continuum approach. This framework enables us to directly employ intracellular pathway models for individual cells to determine population-level, macroscale cell behavior. Our strategy is based on the particle-in-cell approach. Here, the aggregate effect of many cells in a Lagrangian frame of reference is interpolated onto Eulerian (stationary) mesh points, where the moments of the continuum distribution (e.g. concentration) are also computed and updated. The typical cycle involves the following: (i) integration of the pathway equations of motion for individual cells; (ii) interpolation of the cell effects (e.g. metabolism) onto the stationary mesh; (iii) computation of the new field values using this information; and (iv) interpolation of the fields from the mesh back to the particle locations. To validate this framework, we investigated bacterial chemotaxis in complex chemical and spatial environments. As we demonstrate, our simulation results are quantitatively in agreement with the corresponding experimental results. While this framework was developed to study chemotaxis, we will also illustrate how it can be used to simulate at multiple scales any biological system involving cell populations interacting via chemical signaling.

# 434178 Spatial Patterns of Growth and Inhibition in Spreading Infections of Vesicular Stomatitis Virus

**Abstract:**

Viruses infect their multi-cellular hosts through processes that play out over multiple scales: a virus particle infects a susceptible host cell, producing virus progeny that then spread to other cells within the host and initiate new rounds of infection. Infected cells make not only virus progeny, but also virus-like defective interfering particles (DIPs), which lack essential genes needed for replication. DIPs are found in natural isolates of influenza A, dengue and West Nile viruses, but their roles in infection spread and disease pathogenesis are not known. DIPs alone are non-infectious. However, in co-infections with intact virus, DIP genomes can interfere with virus growth by competing for viral polymerase, replicating at the expense of viral genomes, or by binding packaging proteins, assembling DIP progeny at the expense of intact virus particles. To elucidate how DIP-virus interactions play out in spreading infections we engineered wild-type and DIP strains of vesicular stomatitis virus to encode and express red and green fluorescent proteins, respectively. Moreover, the reporter-DIP was further designed to carry a deletion of the gene for the viral surface glycoprotein G (VSV-ΔG-eGFP). With this system we were able to monitor the dynamics of wild-type and defective virus gene expression through their respective reporter proteins in single co-infected cells. In co-infections of wild-type (RFP) the reporter-DIP inhibited production of infectious virus to the same extent as standard non-reporter DIPs, but reporter-DIPs were less potent in their ability to inhibit wild-type (RFP) reporter protein production. Heterogeneity in protein expression patterns was apparent among cells co-infected with the same dose of wild-type and reporter-DIPs, but wild-type protein expression (RFP) overall decreased as DIP reporter protein (GFP) expression increased. When infections were permitted to spread spatially across a population of susceptible cells, the heterogeneity in reporter protein increased. Unlike spreading infections in the presence of natural DIPs, which can completely arrest infection spread, co-infection with reporter-DIP produced many local areas of wild-type virus escape, and complete arrest of spread was not observed. These results suggest that competition for packaging protein G can reduce wild-type virus production at the cell level to the same extent as natural DIPs, but this reduction is not sufficient to fully inhibit wild-type virus spread in host cell populations. More broadly, these results highlight how similar patterns of virus growth and inhibition at the level of individual cells can exhibit distinctly different properties of infection spread or containment when virus and DIPs propagating spatially across a population of susceptible cells.

# 432782 In silico Dynamic Gene Expression Models Quantify the Impact of Regulatory Logic Gates on Network Inference

**Abstract:**

Network inference algorithms are widely used to infer biological networks, but their capacity to resolve regulatory logic gate relationships remains largely unknown. In this study, we built an *in silico* model of dynamic gene expression to quantify the influence of logic gate regulation on the accuracy and confidence of network inference algorithms in order to elucidate the strengths and limitations of inferring gene regulatory networks.

Network graphs are a simple and intuitive means to represent the underlying topology and information transfer inherent to gene regulatory systems. Accurate network models provide valuable insight into functional modules and information flow, highlighting targets for therapeutic intervention. With high throughput technologies, network inference has become a conventional method to uncover regulation among genes and identify strategies to modulate system behavior. However, every method has its limitations: the DREAM consortium recently demonstrated specific network motifs, and method of inference, impact accuracy of the resulting inferred network model[1]. A significant source of this error is likely attributed to regulatory logic gates and its associated kinetic parameters, neither of which are considered in common practice.

An *in silico* model was built to simulate a gene regulatory network. Six different networks–each containing five nodes —define the activation and inhibition among genes. Genes with multiple inputs are assigned an AND, OR, NAND, or NOR logic gate relationship. A thermodynamic ODE model generates dynamic gene expression data based on the network interactions and logic gate relationships. GENIE3, a well-established inference algorithm that can deal with combinatorial and non-linear interactions, was selected to infer the putative gene regulatory network[2]. Each *in silico* data set was inferred 100 times using GENIE3 and scored against a permuted data set in order to calculate each edge's statistical significance. The network inference and scoring objective were performed over a range of 441 logic gate kinetic parameters

The GENIE3 algorithm infers AND-gates with a higher confidence than OR-gates for 94% of the tested kinetic parameter space for the bi-fan network structure. In the feed-forward network structure, the AND-gate relationship is inferred with a higher confidence for 90% of the kinetic parameter space. In the bi-fan network, GENIE3 infers OR-gates with higher confidence when both inputs into a single gene have low kinetic parameters. However, for the feed-forward loop, the OR-gate is inferred with a higher confidence when only one edge within the network motif has a low kinetic parameter value. Lastly, GENIE3 shows little inference bias between NAND and NOR gates. Further simulations will evaluate Bayesian, regression, and correlation based inference algorithms, as well as the SUM logic gate relationship.

GENIE3, one of the top performing inference algorithms in the DREAM challenge, demonstrates bias when inferring specific logic gate relationships, limiting its predictive insight. For instance, in drug design, it is critical to consider biases in an inferred mechanism because AND-gates would be targeted and perturbed differently than OR-gates. These biases would also be critical in designing combinatorial drug targets and predicting the impact of upstream perturbations on the mechanism of interest. By considering the limitations and biases of network inference algorithms, we will be able to build more informed network models for the future design of drug targets or perturbation studies from inferred network models.

**References:**

(1). Marbach, Daniel *et al*. "Wisdom of Crowds for Robust Gene Network Inference." Nature Methods 9,8 (2012): 796-804

(2). Huynh-thu, Van Anh *et al*. "Inferring Regulatory Networks from Expression Data Using Tree-Based Methods." PLoS ONE, 5(9): e12776, 2010.

# 432630 Coarse-Graining of Heterogeneous Neural Dynamics

**Abstract:**

In biology, the structure of networks plays an important role in the resulting dynamics both intracellularly and in the interactions between cells, especially in the nervous system. The realism of simulations of large networks of neurons can be enhanced by introducing cell heterogeneity: neurons are characterized by different values of physiological parameters, as well as by different connectivities in the network. In this case, we find that the (evolving) system state often quickly becomes a smooth function of these parameters. We make use of this phenomenon to express the system state through a significantly reduced number of variables, allowing for acceleration of several computational tasks.

The problem of coarse-graining a large heterogeneous network of neurons thus becomes that of finding the appropriate basis functions (the appropriate "observables") given the (multidimensional) distribution of the parameters. We use nonparameteric techniques to generate such basis function sets which are orthogonal with respect to arbitrary multidimensional parameter spaces. For high-dimensional spaces of heterogeneous parameters, we use analysis of variance (ANOVA) as an alternative coarsening technique.

With a coarse description, we show how to accelerate tasks such as finding coherent steady states and synchronized limit cycles. We demonstrate applications of these techniques both for networks of simple coupled oscillators, and for a much more complicated simulation of circadian dynamics in the suprachiasmatic nucleus of the mammal hypothalamus.

# 415023 Effects of Nonlinear Responses of the Huang-Ferrell Model of MAPK Cascades on the Character of Autocrine Communication

**Abstract:**

Autocrine communication in cellular systems is mediated by the interaction between extracellular ligands and membrane-bound receptors. In our previous works, we have studied the effects of ligand-receptor kinetics, transport parameters, and interstitial flow on the propagation velocity of autocrine signaling. However, the process of signal transmission strongly depends on the dynamics of intracellular processes. Signal coming from a ligand-receptor complex into the intracellular space is typically amplified in MAPK phosphorylation/dephosphorylation cascade. It was recently shown that the widely accepted Huang-Ferrell model of MAPK signaling offers bistable (switch-like) and oscillatory behavior. These nonlinearities can strongly alter the release of new ligand molecules into intracellular matrix as the MAPK output modifies various transcription factors or activates protease precursors. Here we developed a mathematical model of autocrine communication that considers important reaction-transport processes in the intracellular and extracellular domains together. By the method of stoichiometric network analysis, we have found the minimal subnetwork of MAPK cascade that offers oscillatory behavior and steady state multiplicities. This subnetwork was then incorporated into the model of autocrine signaling, where it plays a role of positive or negative feedback for the process of ligand release. By numerical simulation of the model equation, we study transient behavior of the developed model at the interface between cells with and without high MAPK activity. We identify and classify possible responses of cellular systems to specific regimes of MAPK cascades in parametric space.

# 422070 Transient Dynamics of a Four State Actomyosin Model of Muscle Contraction

**Abstract:**

A model of muscle action has been used to compute the steady state response of mechanical and chemical variables in muscle action. the merits of a modelof this action, however, depends on both itssteady state and transient resonses. This model depends on a posted high strain and low strain state for the actomyosin *ADP complex (AMADP). Combining these two strain states with a simple averaging rule and mass action kinetic models for the biochemical cycleallows for computation of both steady state and trasient responses.The overall theory and the steady state responses of this model were published (Long and Phoonsiri, 2006). The transient responses are considered here.Compliance corrections have not yet been included in the model.The model simply links the nono-scale to the macro scale using the half sarcomere as the basic system for analysis.

A comparison of computed results is made with well-known mechanical tests widely published in the literature. In addition to mechanical tests, a number of species transients are also presented. These pose additonal possible results that could be compared against experiment once the experiments are done.

# 415633 Development of a Novel Method for Investigation of Polymer/siRNA Intracellular Trafficking

**Abstract:**

Despite substantial investment in RNA interference (RNAi)-based therapies, few clinical trials have shown promising results. This is due largely to a lack of methods and materials for the efficient delivery of small-interfering RNA (siRNA). Many new materials have been designed specifically for this purpose. Investigators commonly design such materials based on reasonable assumptions about the limiting barriers to efficient delivery and structure-activity relationships. Such an approach is often unsuccessful, however, because the intracellular mechanisms of siRNA delivery (e.g., cell surface binding, cellular uptake, endocytic escape, unpackaging of the nucleic acid, etc.) are unclear. Increased understanding of intracellular trafficking mechanisms and kinetics, and how these processes are affected by material properties, is required for rational design of siRNA delivery agents.

Here we report investigation of the kinetics of cellular internalization and endocytic processing of siRNA complexed with 25-kDa polyethylenimine (PEI) in HeLa cells. Conventional techniques including microscopy and cellular fractionation did not provide sufficient precision, prompting us to develop new methods for quantifying intracellular trafficking. GFP fusions with endocytic markers of early endosomes (Rab5), late endosomes (Rab7) and lysosomes (LAMP-1) were expressed individually in HeLa cells. Confocal fluorescence microscopy was used to quantify colocalization of fluorescently labeled siRNA with these markers. We observed high colocalization of siRNA with Rab5-GFP and Rab7-GFP at short times, which decreased linearly over the next 4 h. In contrast, little colocalization with LAMP-1-GFP was initially observed, but the siRNA was strongly localized in lysosomes from 30 minutes to 4 h post-transfection. However, significant overlap of the marker proteins in different organelles limited the precision of conventional microscopy to assess intracellular localization. In addition, laser-scanning confocal microscopy inherently favors observation of large particles (i.e., aggregates) over single polyplexes. Thus, we developed a unique technique for resolving the location of siRNA within the endocytic pathway. The method relies on the known endocytic trafficking kinetics of horseradish peroxidase (HRP) and the enzyme's ability to catalyze polymerization of 3,3'-diaminobenzidine (DAB). In the presence of H2O2 and DAB, HRP-containing vesicles are cross-linked, making them easy to isolate for analysis of siRNA content. Using this new method, we observed ~60% of siRNA within early endosomes within 5 min, a rapid decrease to 15-20% by 30 min, and a slower decrease to <5% at 3 h post-transfection. siRNA localized in lysosomes was first detected (<5%) 30 min post-transfection, increased to ~20% by 30 min (due to apparent transfer of siRNA from the early endosomes) and further increased to only 16% by 3 h. Finally, this assay allowed us to observe a steady increase of siRNA in cytosol, presumably due to release from endocytic vesicles.

This method allows for more precise determination of intracellular localization and analysis of intracellular trafficking kinetics of siRNA/PEI polyplexes (or any nanoparticle delivery system) compared to more commonly employed methods. We are employing this method to compare intracellular trafficking of various siRNA delivery vectors and construct a quantitative model of the transfection process.

# 434426 Engineering Extracellular Vesicles As Therapeutic Delivery Vehicles

**Abstract:**

Extracellular vesicles (EVs) are biological nanoparticles that transfer proteins and nucleic acids between cells and have great potential as tunable therapeutic delivery vehicles; they are relatively easy to engineer, well-tolerated *in vivo*, and naturally capable of intracellular delivery of functional biomolecules. In order for this promise to be realized, robust and general methods for delivering therapeutic cargo molecules via EVs and targeting therapeutic EVs to specific destinations *in vivo* are required.

Although EVs are known to load native RNAs and deliver them to recipient cells, functional delivery of exogenous cargo molecules has proven challenging. Many questions remain, including how the cargo of endocytosed EVs is released into the cytoplasm, and how this process can be enhanced. The degree to which RNAs loaded into EVs are intact and functional is also not fully understood. To answer these questions, we investigated functional RNA delivery by EVs with or without expression of the membrane fusion-inducing viral protein, VSVG. We found that expression of VSVG enhanced RNA delivery by EVs. We also investigated the relationship between RNA levels (intact or fragmented) in EVs and functional delivery to recipient cells and determined that increased intact RNA loading does not necessarily lead to increased expression in recipient cells. We determined the time course over which EV-delivered RNAs are expressed in recipient cells and found significant expression of EV-delivered RNA over 48 hours. Altogether, we have developed a system for robust EV-mediated delivery of functional RNA to recipient cells and investigated the details of this system allowing for optimal RNA delivery by EVs.

To enhance delivery of therapeutic EVs to target recipient cells, we have also developed a technology for enhancing the display of targeting ligands on the exterior of EVs. Previous attempts to target EVs to specific recipient cells have been hindered by inconsistent targeting using different ligands or different EV-producing cell types. We determined that degradation of targeting peptides by endosomal proteases occurs during EV biogenesis. To overcome this challenge, we have developed a glycosylation strategy that protects peptides from degradation while maintaining their ability to bind target proteins. We applied this peptide targeting strategy to achieve enhanced delivery of EVs to neuroblastoma cells. The tools we have developed enable programming EVs for targeted delivery of RNA cargos in a robust and straightforward manner that can be applied to a wide variety of therapeutic targets.

# 433269 AAV Ancestral Reconstruction Library Enables Selection of Broadly Infectious Viral Variants

**Abstract:**

Adeno-associated virus (AAV) vectors have achieved clinical efficacy in treating several diseases. Enhanced vectors are required to extend these landmark successes to other indications, however, and protein engineering approaches may provide the necessary vector improvements to address such unmet medical needs. To generate new capsid variants with potentially enhanced infectious properties, and to gain insights into AAV's evolutionary history, we computationally designed and experimentally constructed a putative ancestral AAV library. Combinatorial variations at 32 amino acid sites were introduced to account for uncertainty in their identities. We then analyzed the evolutionary flexibility of these residues, the majority of which have not been previously studied, by subjecting the library to iterative selection on a representative cell line panel. The resulting variants exhibited transduction efficiencies comparable to the most efficient extant serotypes, and in general ancestral libraries were broadly infectious across the cell line panel, indicating that they favored promiscuity over specificity. Interestingly, ancestral AAVs were more thermostable than modern serotypes and did not utilize sialic acids, galactose, or heparan sulfate proteoglycans for cellular entry. Finally, ancestral variants mediated 19-31 fold higher gene expression in muscle compared to AAV1, a clinically utilized serotype for muscle delivery, highlighting their promise for gene therapy. The putative mutational tolerance and evolvability of this library could be further harnessed in directed evolution studies to overcome gene therapy challenges such as targeted gene delivery and immune evasion.

# 429143 A Modular, Antibody Conjugate Dual-Delivery Strategy for Enhancing Endosomal Escape of Delivered siRNA


**Abstract:**

Title: A Modular, Antibody Conjugate Dual-Delivery Strategy for Enhancing Endosomal Escape of Delivered siRNA

RNA interference (RNAi) has emerged over the last 15 years as a specific and potent means of targeted gene silencing at the mRNA level. Since its initial discovery in *C. elegans*, researchers have sought to harness this mechanism, by way of short interfering RNAs (siRNA), as an efficient route to silencing disease-causing genes. However, inefficient cytosolic localization of delivered siRNA continues to be a key challenge in RNAi therapeutic development. The most widely used delivery strategy currently involves cationic polymers or lipid nanoparticles to encapsulate the siRNA cargo, deliver it to the cell, and facilitate endosomal release to the cytoplasm. However, the design of a single polymer or lipid formulation that can overcome the multiple biological barriers to delivery has proven difficult.

In this talk, we introduce a modular dual-delivery strategy that decouples cellular internalization from endosomal escape through the use of antibody conjugates. Cellular internalization of these conjugates will be dictated solely by the antibody while endosomal escape will be attained with an endosomal escape agent (EAA). Decoupling internalization from escape will enable us to screen EEAs that disrupt membranes for cytosolic translocation of siRNA independently of its endocytic pathway. Two bioconjugates, an antibody-siRNA and antibody-EEA, will be prepared using the same antibody to ensure compartmental co-localization within the cell, then co-delivered to initiate RNAi.

Toward this aim, I will introduce a novel method for the assembly of both RNA-transporting and EEA-transporting antibody conjugates. This method enables efficient and reagent free bioconjugate assembly at sub-micromolar concentrations, typical of commercial antibodies, without significant disruption of antibody binding affinity. The modularity of this system allows us to then explore the effect of entry pathway on gene silencing by substituting the targeting antibody. Furthermore, I will present *in vitro* data on novel biomaterials with lytic properties for use as EEAs. This novel dual-delivery approach will be used as a platform to elucidate structure-activity relationships of potent EEAs and effects of entry pathway on the efficient delivery of siRNAs.

# 409568 Hybrid Liposomal Polymeric Nanoparticle Delivery of Combination Chemo-, Immuno- and RNAi Therapy for Triple Negative Breast Cancer

**Abstract:**

Breast cancer continues to remain a major health care concern in the U.S. The triple-negative breast cancer (TNBC) subtype tends to be particularly aggressive with poor prognosis. Triple combination therapy by encapsulating a chemotherapeutic agent, immunotherapeutic agent and interfering RNA (RNAi) may enhance treatment response in TNBC. First-line treatment of TNBC currently includes docetaxel, or pegylated liposomal doxorubicin followed by docetaxel. These drugs are toxic to healthy normal cells in addition to cancer tissues. Furthermore, metastatic tumor treatment can fail due to multi-drug resistance (MDR) to standard therapies. We aim to mitigate MDR through combination treatment and reduce peripheral toxicity by drug encapsulation and active targetting.

Co-delivery of cytotoxic agent docetaxel, tyrosine kinase inhibitor sunitinib, and anti-VEGF RNAi is a promising alternative that may mitigate the pro-survival tumor response to chemotherapy. Solid lipid nanoparticles will be synthesized by encapsulating docetaxel within PLGA nanoparticles (NPs). The NPs will be surrounded with a lipid bilayer while embedding sunitinib and anti-VEGF oligonucleotides at the interface of the PLGA polymers and the lipid bilayer to create liposomal chemo- immunotherapeutic nanocarriers (LICHINs, see Figure 1). Furthermore, we will coat LICHINs with folic acid, actively targeting folate receptor alpha, commonly overexpressed in TNBC cells.

NPs will be characterized based on particle size, polydispersity, and zeta potential. Carrier production will be optimized for uniformity based on these characteristics, as well as encapsulation efficiency and surface morphology. Treatment efficacy will be evaluated for cytotoxicity in MDA-MB-231 cells. The impact of RNAi delivery on down-regulation of VEGF will also be evaluated by assay. If successful, LICHINs will represent a novel delivery platform for targeting tumor cell proteins and mRNA via chemo, immuno and gene therapy while minimizing treatment toxicity.

# 430921 Using Nucleic Acid Aptamers to Modulate the Specificity, Binding, and Quantity of Drug to Create a Personalized Treatment Approach Towards Small Cell Lung Carcinoma

**Abstract:**

**Using Nucleic Acid Aptamers to Modulate the Specificity, Binding, and Quantity of Drug to Create a Personalized Treatment Approach Towards Small Cell Lung Carcinoma**

Ricky J. Whitener[1, 2, 3], Katherine A. Windham[1,2], Jacek Wower[1, 3], Mark E. Byrne[2, 3, 4]

[1] RNA Biochemistry Laboratories, Department of Animal and Dairy Sciences, College of Agriculture

[2] Biomimetic & Biohybrid Materials, Biomedical Devices, and Drug Delivery Laboratories, Department of Chemical Engineering, Auburn University

[3] US Dept of Education GAANN Graduate Fellowship Program in Biological & Pharmaceutical Engineering

[4] Biomimetic & Biohybrid Materials, Biomedical Devices, and Drug Delivery Laboratories, Department of Biomedical Engineering, Rowan University

Lung cancer remains the most frequently diagnosed and deadliest cancer with 1.8 million new cases and 1/5$^{th}$ of all cancer deaths annually. Small cell lung carcinoma (SCLC), one of two subtypes of lung cancer, accounts for 15% - 20% of all lung cancer diagnoses. The survival rates for SCLC patients have not significantly improved over the last 40 years. Untreated SCLC is rapidly fatal within two to four months. In our studies, we use doxorubicin, one of the best drugs for treating SCLC. We constructed a novel high payload nanocarrier for doxorubicin. The release of the drug can be modulated to account for the genomic characteristics and health status of the patient. At present, we evaluate the effectiveness of our nanocarrier in vitro using cultured SCLC cells.

Our nanocarrier consists of a 15 nm gold nanoparticle (AuNp) detectable by x-rays, single-stranded DNA anchors covalently bound to the AuNp surface, and DNA aptamers. We used aptamers that are known to bind to the cultured SCLC cell line NCI-H69 and modified them by adding a drug-binding site that can accommodate 5-7 molecules of doxorubicin. Each aptamer is equipped with a single-stranded segment that is complementary to the single-stranded DNA anchor. Such design produces a drug nanocarrier that can be readily programmed with either one or several cancer-specific aptamers. By increasing the size of the AuNps, we can produce nanocarriers that carry more aptamers and deliver higher payloads of drug to the surface of the cancer cell. Alternatively, we can produce smaller nanocarriers that can be more readily taken up by cancer cells.

Presently, we are studying the cellular uptake of 15-nm nanocarriers using transmission electron microscopy (TEM), hyperspectral microscopy, and other techniques that allow for visualization of particles inside cells. In the cellular uptake tests, we are using the cultured SCLC cell line NCI-H69 that was used for the selection of SCLC-specific aptamers. For control experiments we are using cultured non-SCLC cells NCI-H661 that are not recognized by the aptamers attached to our nanocarrier. The TEM studies are complemented by cell viability and carrier/drug effectiveness tests (XTT Assay). Preliminary results indicate that by manipulating the size of the nanocarrier, the structure of the DNA aptamers, and its drug-binding segment, we have created a an effective yet flexible, versatile, highly tunable platform that is capable of recognizing a specific SCLC tumor, and is able to deliver controlled amounts of doxorubicin to individualized patients as dictated by their health status.

# 426047 Zein As a Natural Material for the Controlled Delivery of Non-Viral Gene Delivery Vectors: Zein Films and Nanoparticle Coatings

**Abstract:**

**Zein as a Natural Material for the Controlled Delivery of Non-Viral Gene Delivery Vectors: Zein Films and Nanoparticle Coatings**

Eric Farris[1], Jessica Taylor[1], Christopher Davidson[1], and Angela K. Pannier[1]

[1]Department of Biological Systems Engineering, University of Nebraska-Lincoln, Lincoln NE, 68588

Nonviral strategies for the delivery of nucleic acids are considered a promising alternative to viral vectors due to their increased safety and lower immunogenicity. However, nonviral vectors still lack the efficiency needed to reach clinical relevance. The inefficiency, which can be attributed to many factors, such as degradation of the DNA by nucleases and other enzymes, and instability of the vector in physiological conditions, can potentially be overcome through use of particulates that physically incorporate plasmid DNA. The use of particulates incorporating DNA are promising vehicles for gene delivery as they have the ability to protect DNA from degradation and allow for sustained and controlled release of DNA for increased transfection.

Another strategy at increasing the efficiency of nonviral systems involves immobilizing delivery vectors to a substrate prior to cell seeding. This strategy, termed substrate-mediated gene delivery, allows for an increased concentration of DNA, which is localized at the cell surface for increased uptake and transfection. Zein, the major prolamine, or storage protein, in corn, is biocompatible and biodegradable and has been shown to have unique properties that allow it to form particulates, films, and coatings through coacervation or evaporation induced self-assembly. Moreover, zein has very low aqueous solubility and has been shown to be stable in acidic environments. Due to these properties zein has the ability to be tailored for applications in substrate-mediated gene delivery as well as in nanoparticulate coatings for oral DNA delivery applications.

For this work, our objective was to investigate zein's ability to form films for substrate-mediated gene delivery, as well as investigate zein as a material for particulate DNA delivery, in particular as a protective coating material for non-viral gene delivery complexes. For substrate-mediated gene delivery applications, zein was found to be able to form uniform films on a variety of substrates, including glass, titanium, and gold. Observation of the films by scanning electron microscopy indicated the films were composed of nanospherical particles, the size of which could be modulated by altering the pH of the initial zein solutions to alter the surface roughness and resulting hydrophobicity of the films. Zein films were shown to adsorb both poly(ethylene imine)/DNA (PEI/DNA) complexes and Lipofectamine 2000/DNA complexes and once cells were seeded on these immobilized complexes, transfection in HEK 293T cells was demonstrated. Furthermore, zein films were shown to increase cellular proliferation, which is often critical to gene delivery efficacy.

Due to the enzymatic and pH resistant properties of zein, we next investigated the use of zein as a protective coating for DNA delivery systems for use in oral gene delivery and DNA vaccination applications. Zein was used to encapsulate chitosan/DNA (CS/DNA) complexes via a water-in-oil emulsion method to generate Zein/CS/DNA micro/nanoparticles. The resulting microparticles ranged in size from 5 to 10 μm and DNA loading in the particles was found to increase from 1.68 to 7.72 mg DNA/g of zein by increasing the concentration of the zein solution from 3 to 7% (w/v). Zein particles are capable of protecting DNA cargo within the stomach for up to two hours, and after incubation in simulated intestinal fluid (SIF), zein degrades rapidly, releasing CS/DNA particles that were shown to mediate transfection in HEK 293T cells. A significant increase in transgene expression mediated by the ZN/CS/DNA particles after incubation in SIF was observed with increasing ZN concentration, demonstrating the ability of this system to release intact and transfection competent CS/DNA NPs in a site specific manner for targeted gene delivery in the intestinal environment. Overall, these studies demonstrate the versatility of zein as a natural material for controlled delivery systems for non-viral gene delivery.

# 413094 IQ Consortium Enabling Technologies for Pharma

**Abstract:**

The International Consortium for Innovation and Quality in Pharmaceutical Development (IQ) is a technically-focused organization of pharmaceutical and biotechnology companies with a mission of advancing science-based and scientifically-driven standards and regulations for pharmaceutical products worldwide. Since 2013, it has established an 'Enabling Technologies Working Group' to identify opportunities among member companies in the pre-competitive space to enable cost-effective and relevant solutions to improve efficiency and quality in the R&D process with respect to API process development. The working group has identified projects of mutual interest, and subject-matter experts and champions to lead technical area sub-teams with the goal of creating proposals to progress on these projects. This presentation provides an update on the progress of the overall 'Enabling Technologies' effort with focus on modeling, crystallization, drying, solubility, and automation.

# 419660 Precompetitive Collaboration for Pharmaceutical Crystallization Processing

**Abstract:**

In 2014, the IQ Enabling Technologies Working Group conducted a survey of participating pharmaceutical companies to identify areas of common interest. Among these, crystallization development was the topic that generated the most interest, with 11 of 12 members of the Working Group expressing a desire to participate in this sub-team which kicked off at the beginning of 2015.

One of the first activities the team performed was to brainstorm a list of potential areas for collaboration, and then each company ranked their relative interest in each of the topics. Topics ranged from best practices for salt/polymorph selection to lab equipment for process development to modeling to sharing experiences related to product filings.

The highest scoring areas were related to sharing of best practices. The team agreed to focus initially on knowledge sharing, with the recognition that team members might be able to help each other close certain gaps and improve each company's specific practices. The knowledge sharing would lead into performing a collective gap analysis in order to identify prioritized projects that will require funding and external collaboration. The intent is for these prioritized projects to kick off in 2016, likely under the banner of a new construct being proposed as a spin-off from the IQ consortium.

This presentation will focus on both the outcomes of the collaboration thus far and, perhaps more importantly, on the approaches the team has taken to advance its objectives in the face of numerous challenges.

Allotrope Foundation

Building an Open Framework for Laboratory Data

# Collaborating to Deliver an Innovative Framework for Laboratory Data Based on Open Standards

James M. Vergis

Pre-Competive Collaborations in Pharma Industry

AIChE Annual Meeting

November 9, 2015



*"Every decision we make about our products is based on data. To deliver safe and efficacious medicines to the patient, we must be confident in the quality and integrity of our data. Our integration project with Allotrope Foundation will deliver standard metadata and data stored in a standard format, addressing the root cause of our data challenges and revolutionizing the way we generate, access and use data to make decisions and deliver our pipeline of medicines."*

Dr. Annette Doherty
GSK Senior Vice President
Product Development & Supply

## ALLOTROPE PRESS RELEASE SEPTEMBER 30, 2015

# Why are we here today?

- Provide an update to our peers in the industry and relay the message "the Allotrope Framework is a reality"
- Demonstrate how vendors and Allotrope members are collaborating to leverage the Framework through integration projects
- Invite you to participate in 2016



# Allotrope Foundation

| | |
|---|---|
| Member Companies | • Subject Matter Experts<br>• Project Funding |
| Secretariat<br>DrinkerBiddle | • Project Management<br>• Legal & Logistical Support |
| Professional Software Firm<br>OSTHUS success with R&D | • Framework Development<br>• Technical Leadership |
| Partner Network | • Requirements & Specifications<br>• Contributions, PoC Applications |

# Allotrope Foundation

**Member Companies**

| | | |
|---|---|---|
| AbbVie | Biogen | Genentech/Roche |
| Amgen | Boehringer Ingelheim | GlaxoSmithKline |
| Baxter | Bristol-Myers Squibb | Merck & Co. |
| Bayer | Eli Lilly | Pfizer |

**Secretariat**

DrinkerBiddle

- Project Management
- Legal & Logistical Support

**Professional Software Firm**

OSTHUS success with R&D

- Framework Development
- Technical Leadership

**Partner Network**

| | | |
|---|---|---|
| ACD/Labs | LEAP Technologies | Thermo Scientific |
| Agilent | Mettler Toledo | Waters |
| Biovia | Persistent Systems | Erasmus Univ. Med Center |
| BSSN | Riffyn | University of Southampton |
| IDBS | Sartorius | |
| Mestrelab Research | Shimadzu | |



# What is the problem we are trying to solve?

## Comparison of Music vs. Scientific Data

| | DATA SOURCE | DATA | VALUE FROM DATA |
|---|---|---|---|
| MUSIC | PANDORA |  |  |
| SCIENTIFIC | • Electronic Lab Notebook<br>• Chromatography Data System<br>• Data Archive | 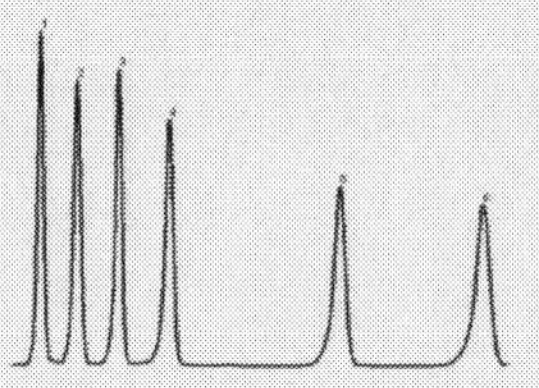 |  |

# What is the problem we are trying to solve?

## Comparison of Music vs. Scientific Data

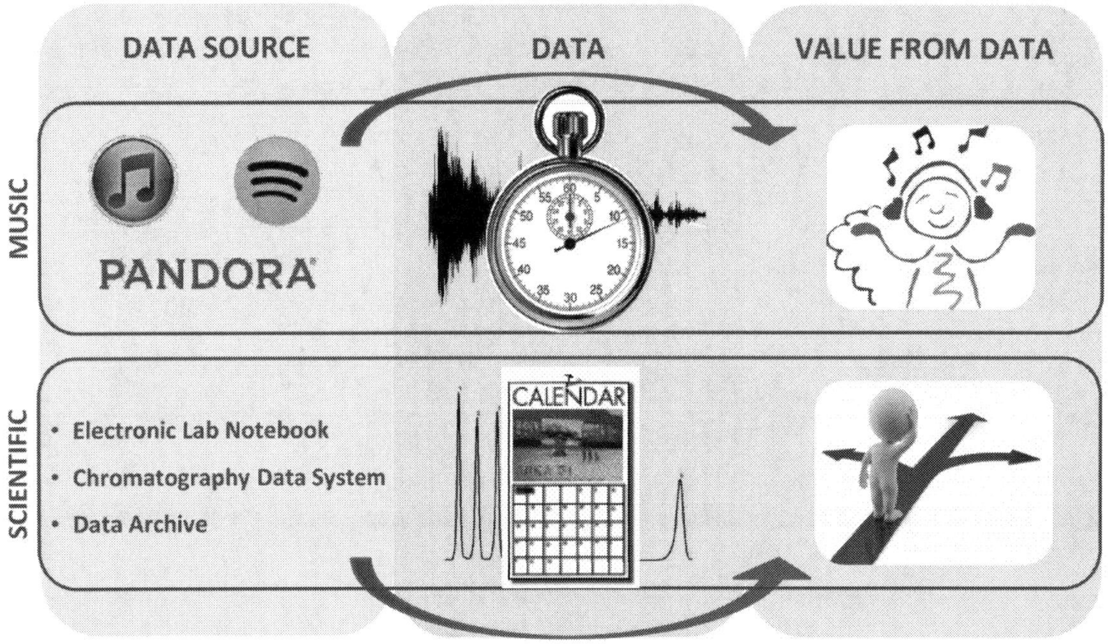


 2015 AIChE Annual Meeting 7

# Why is Data for Music so Much Better?

### Think about music...

Music is typically stored in a small number of *standard, non-proprietary* formats...

...with contextual *metadata* that are *complete, consistent & correct*

| | |
|---|---|
| Artist | Genre |
| Album | Date |
| Song | Artwork |

...enabling the user to *find, share* and *enjoy* it years later from any device *easily*!

### Think about scientific data...

Scientific data is typically stored in a wide variety of *non-standard, proprietary* formats...

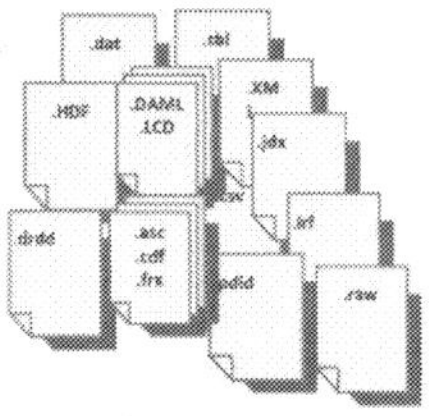

...with contextual *metadata* that are *missing, inconsistent & incorrect*

| | |
|---|---|
| Material | Process |
| Equipment | Result |

...making it *costly, difficult or practically impossible* to find and get value from it.

 2015 AIChE Annual Meeting 8

# Allotrope Data Format and Taxonomies will Simplify our World

Vendor-Specific Formats

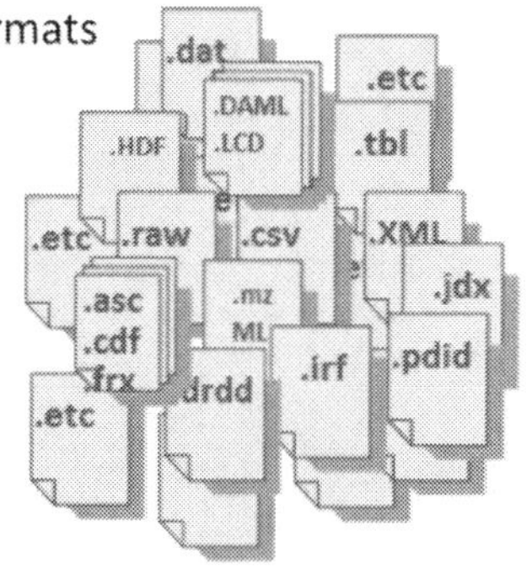


Data in Standard Format

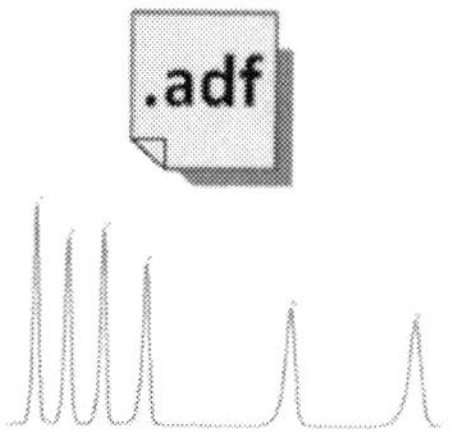


Regulatory Guidance
Methods
Recipes
SOPs
...

Standard Metadata from Taxonomies

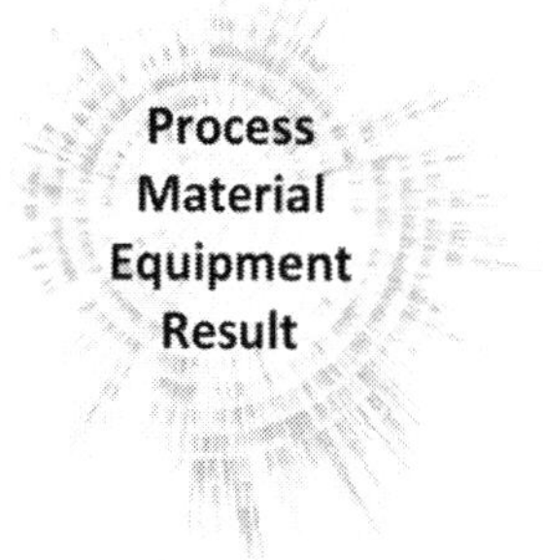





# Allotrope Data Format and Taxonomies will Simplify our World

Vendor-Specific Formats

Data in Standard Format

Regulator
Methods
Recipes
SOPs
...

data from

# Scope: Not Just Analytical Data

**Holistic solution across multiple data generation scenarios**

- Measurement process
  *offline, online, PAT*
- Research through Manufacturing process
  *chemistry, formulation, bioprocessing*
- Records management
  *record retention, regulatory submissions, reporting*

**Requirements from range of perspectives & roles**

- Regulators, bench scientist, data analysts, modelers, manufacturing, archivists, IT




# Allotrope Data Format (ADF):
## Standardizes our Data

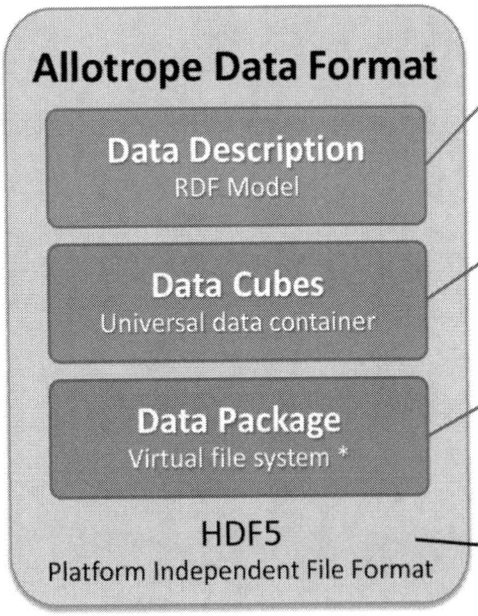


Contains:
- Method, instrument, sample, process, result, etc.
- Data cube metadata
- Data package metadata
- ...

Data represented by one- or multidimensional arrays.

Data represented by arbitrary formats, incl. native instrument formats, images, pdf, video, etc.

Specifically designed to store and organize large amounts of numerical data.

** Use is optional*

# Allotrope Foundation Taxonomies (AFT):
## Standardizes our Metadata

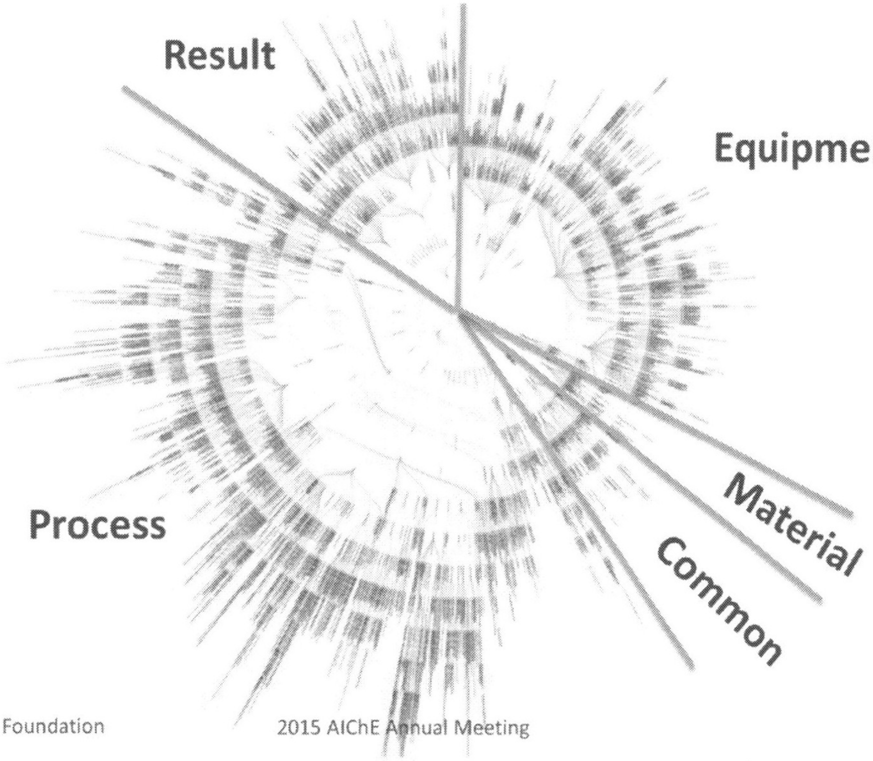




# Allotrope Foundation Taxonomies (AFT):
## Standardizes our Metadata

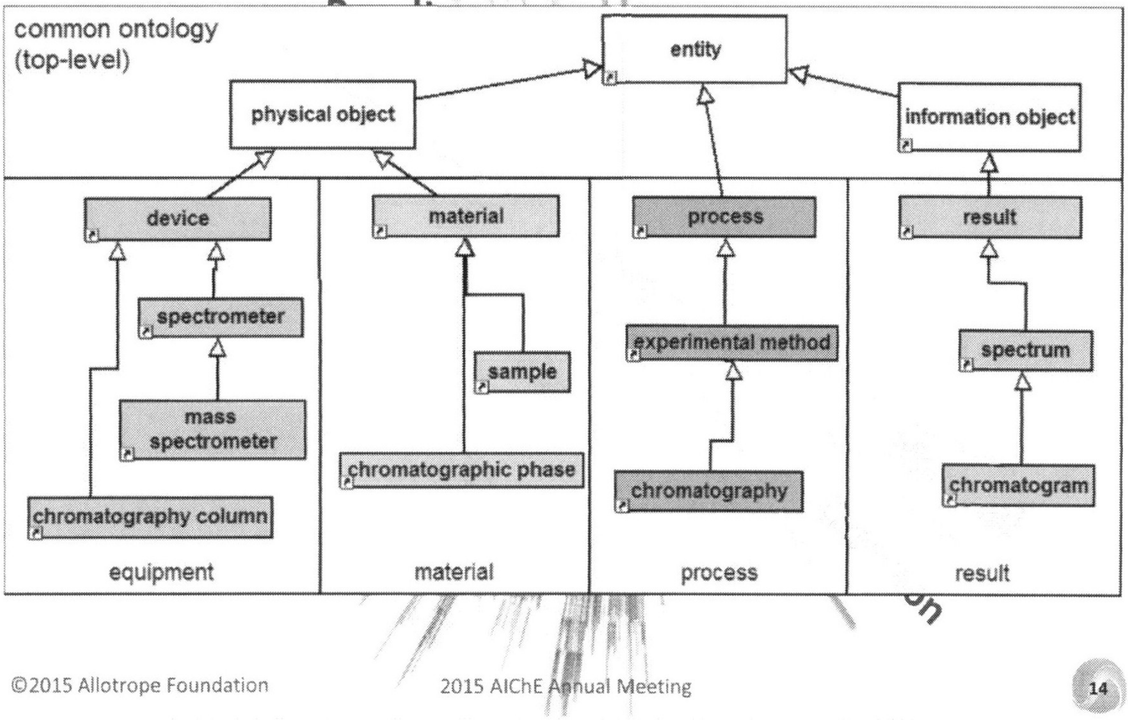

# Example: ADF for NMR

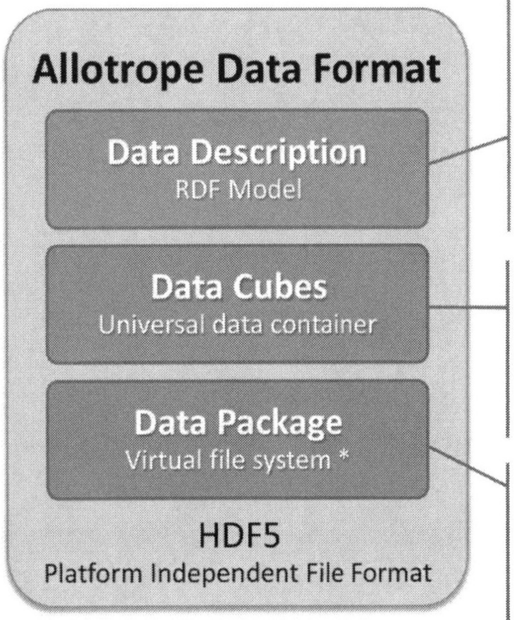


Contains semantic descriptions of:

- Context sensitive metadata – ex. sample ID
- Pulse programs, gradient programs
- Instrument parameters - power & pulse lengths, delays, other settings incl. selectable DF settings, LP settings, NUS parameters as needed for expt(s).
- Processing macros & parameters: apodization functions, ZF settings, etc.
- Peak lists, integrals, assignments, etc.

Contains binary representation of:

- FID(s) (1D/nD)
- Transformed data (maybe multiple complementary freq domain representations – ex. signal & resolution enhanced 1D's)

Contains intact files in virtual FS:

- Structure as MOL file
- PDF Reports of peak lists, integrals, assignments
- Report templates.
- PDF plots of region(s) of interest



# Allotrope Capabilities Today

- **September 2015 internal release package**
  - Allotrope Data Format (ADF)
  - Allotrope Taxonomies (AFT)
  - Application Programming Interfaces and Class Libraries
  - Documentation

- **Supports the following techniques**
  - gas chromatography
  - Karl Fischer
  - liquid chromatography
  - mass spectrometry
  - NMR
  - thermogravimetric analysis
  - ultra violet spectroscopy
  - capillary electrophoresis
  - cell counter
  - cell culture analyzer
  - blood gas analysis
  - balance
  - pH

# Integration Projects: Reduce it to Practice

- Divide and conquer: members share the costs and benefits
- Early adoption and acceleration of AF Framework development through delivery of Framework components
- Accelerate vendor adoption via collaboration on delivering specific capabilities through the integration projects
- Ensure validity of the Framework through real-world requirements and implementation
- Benefit the sponsoring companies in terms of shared development of new functionality & shared experience
- Drive integration of the Framework into the IT/IS Roadmap

 2015 AIChE Annual Meeting 17

# Integration Projects: Reduce it to Practice

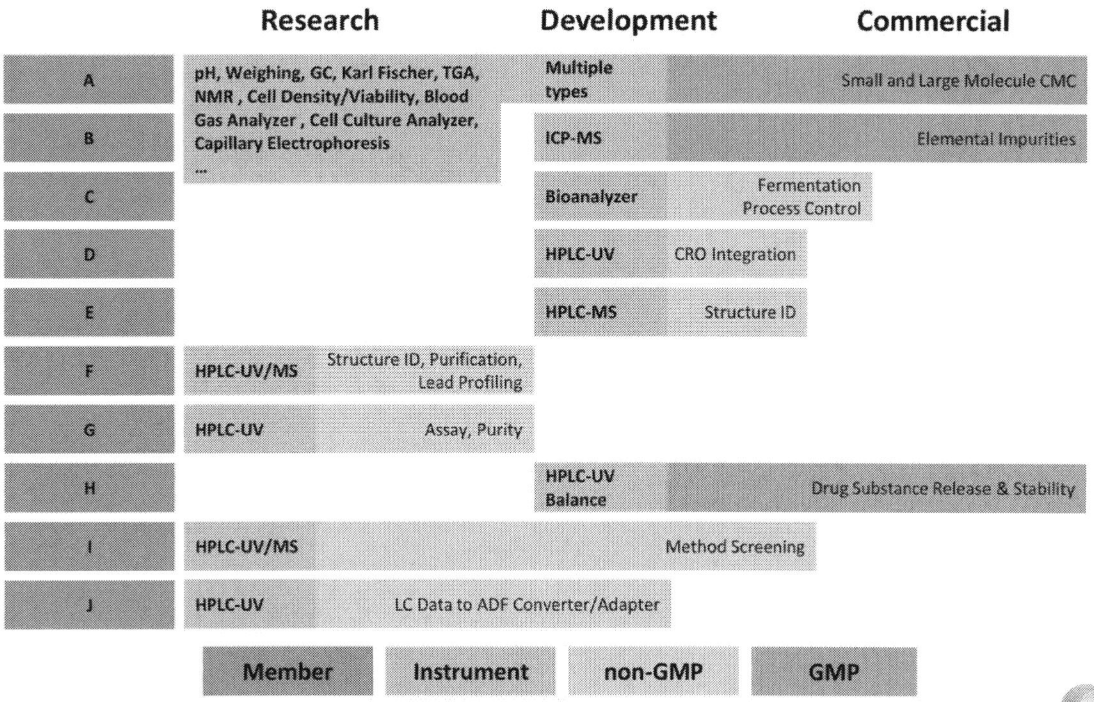


 2015 AIChE Annual Meeting 18

# 2016 AND BEYOND




# GMP Compliance

- **Compliance (21 CFR Part 11)**
  - Digital and Electronic Signatures
  - Validation of the Framework
- **Security**
  - Encryption
- **Data Integrity**
  - Checksums
  - Audit Trail
  - Thread Safety
  - Veto Listener to Triple Store
  - Data Validation
  - Entailment Regime

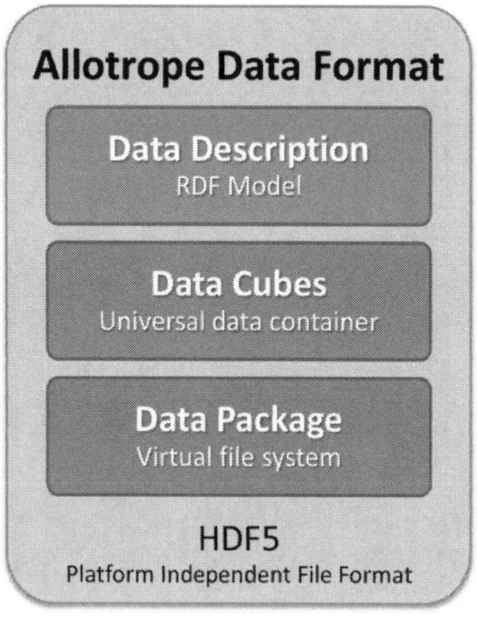

# More Taxonomies Available in 1Q 2016

- **Instrument Taxonomies**
  - Protein Characterization System
  - Microplate Reader
  - Flow Cytometer
  - Protein Purification System
  - Millipore Filtration System
  - Differential Scanning Calorimeter (DSC)
  - X-Ray Powder Diffractometer (XRPD)
  - Particle Sizer
  - Roller Press
  - Tablet Press
  - Tablet Coater
  - Dynamic Vapor Sorption (DVS)
  - solid-state NMR (ssNMR)
  - Infrared Spectroscopy (IR)
- **Other Taxonomies**
  - IDMP-Identification of Medicinal Products
  - Extend Process (S88)
  - Extend Materials
  - Extend Equipment
  - Investigation/Study/Analysis
  - Phase of Development
  - Route of Administration
  - Product Quality Attributes
  - Container Categories
  - Chemical Reactions

**Funded by Individual Allotrope Member**

**Shared with Allotrope Foundation**



# Growing an Ecosystem

- **Framework is applicable beyond pharmaceutical industry**
  - The data problems are universal, the instruments and software are the same, only the experiments and context are different
- **The Framework is as good as we make it**
  - New capabilities are realized only through new integration projects
- **Framework has been funded by Founding Members**
  - New members still welcome to join in 2016
- **Business Model will promote adoption and growth by vendor and end-user communities**
  - Licensing model will be simple
- **Public Commercial release in December 2016**
  - Integration of the Framework into your infrastructure will take time

## Opportunities across R&D and Manufacturing

- **Allotrope Framework is designed to support *any* Process, *any* Material, *any* Equipment and *any* Result**

- **Members have an opportunity to address other big data challenges**
  - Offline analytical chemistry
  - PAT / Online
  - Process Chemistry Campaigns
  - Drug Product Manufacturing Campaigns
  - Safety Assessment
  - Bioanalysis – *e.g.*, DMPK, biologicals
  - ???




## Next steps

- **Think about Allotrope & discuss it internally at your company**
  - What pain points do you want to fix?
  - Why would you want to join now?
  - Why not collaborate with your peers to create a sustainable solution for the industry?

- **Encourage your colleagues to attend one of the upcoming Allotrope Cross-Industry Workshops**
  - January 20, 2016: San Francisco, CA @ Genentech
  - April 2016: Ingelheim, Germany @ Boehringer Ingelheim
  - September 2016: Indianapolis, ID @ Eli Lilly and Co.

# Thank you!

- Any questions, please contact the Secretariat at more.info@allotrope.org or james.vergis@dbr.com

- **2016 Workshops**
  - January 20, 2016: San Francisco, CA @ Genentech
  - April 2016: Ingelheim, Germany @ Boehringer Ingelheim
  - September 2016: Indianapolis, ID @ Eli Lilly and Co.

http://www.allotrope.org for more information and to register

## What Will be Discussed Today

- ❑ Motivation for the Design Space Verification (DSV) Lab
- ❑ Scope of the DSV Pilot Project
- ❑ DSV Project implementation
- ❑ Data Standards and Data Flow
- ❑ Examples of how we plan to deploy the DSV Lab
- ❑ Key Accomplishments and Learnings

## Motivation - Scale-up Verification

- ❑ *Avoid surprises during scale-up that may lead to API supply issues, increased costs and longer timeline*
- ❑ *Understand the capability of the commercial process and reduce failure rates*
- ❑ *Implement a strategy that addresses concerns from regulators*

*Would you trust data from this experiment?*

***Typical Regulatory Questions from recent GSK Filings***

- ❑ *"If the large scale batches were manufactured within a narrow sub-range of the process parameters, then the full [ranges] cannot be approved on large scale and thus should be tightened accordingly. Alternatively, a [verification] strategy should be presented for future variation outside the set points already validated at commercial scale.*
- ❑ *"If the design space is not established with commercial scale batches, a discussion on the scale up aspects and a verification protocol for the scale dependant CQAs should be submitted. The design space verification protocol should allow to demonstrate that for any new operating ranges of the DS, product quality attributes are maintained."*

## Motivation – The Informatics Challenge

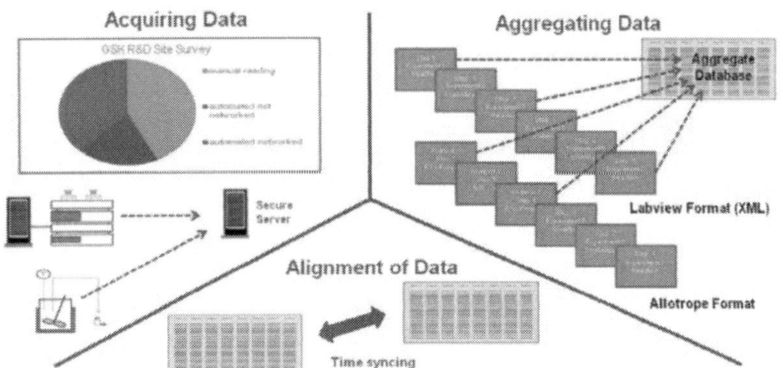

## DSV Lab Project Scope

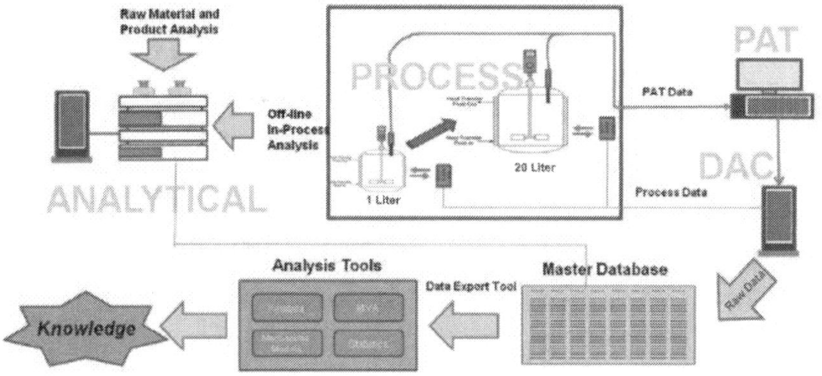




## Equipment/Control System Interface

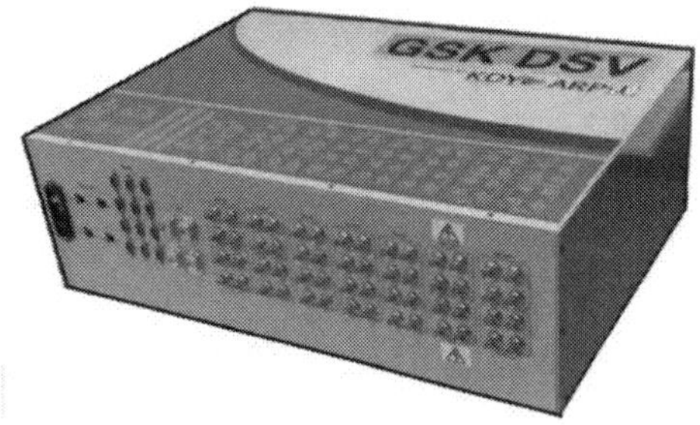


| Connection Type | Number | Purpose |
|---|---|---|
| Ethernet | 4 | Connections to Labview control system and database |
| Serial Ports | 12 | Agitation control, dosing (e.g. Pumps, gravimetric, syringe, etc.) |
| USB Ports | 4 | Temperature control, PAT Systems |
| 24V DC, 120V, RTD Power | Many | Turbidity, pH, conductivity, liquid flow, gas flow, level, etc. |

## Advanced Manufacturing Technology (AMT)

*Project Support Model*

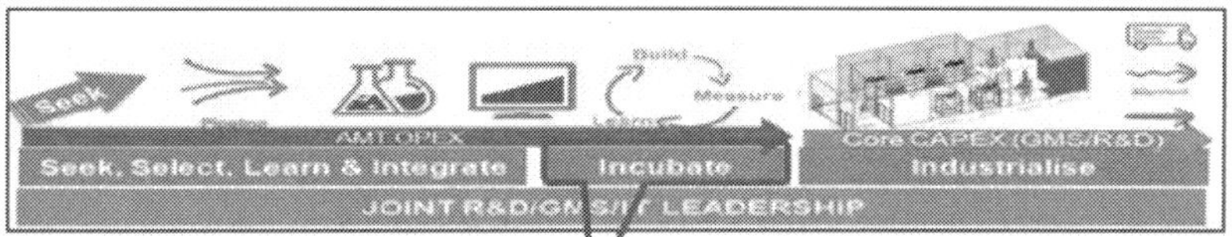


DSV Lab Project set-up in the ***Incubate*** framework:

- ❑ Pilot of ready-now technology in three phases
- ❑ Funded by AMT to demonstrate PoC
- ❑ Share resources AMT and other Development lines
- ❑ Read-out: of Phase 1 end of 4QR15
- ❑ Roll-out of Industrialization solution to be funded by R&D Product Development (PD)

Resource requirement:

- ❑ AMT provides Project Leader to drive activities
- ❑ Cost $190K
- ❑ PD provides access to Scale-up Lab and 0.8 FTE support until completion
- ❑ Product Quality & Risk Management provides single point of contact for regulator interactions - < 0.1FTE



## DSV Lab Data Flowchart

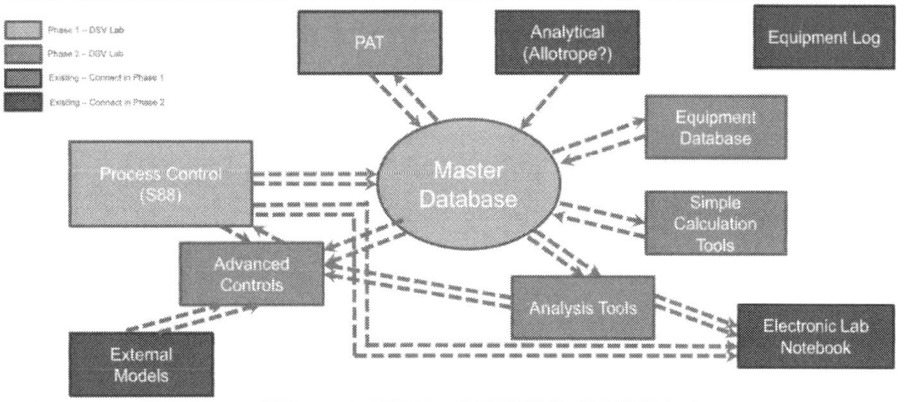

## S88 Alignment – Process Data Standards

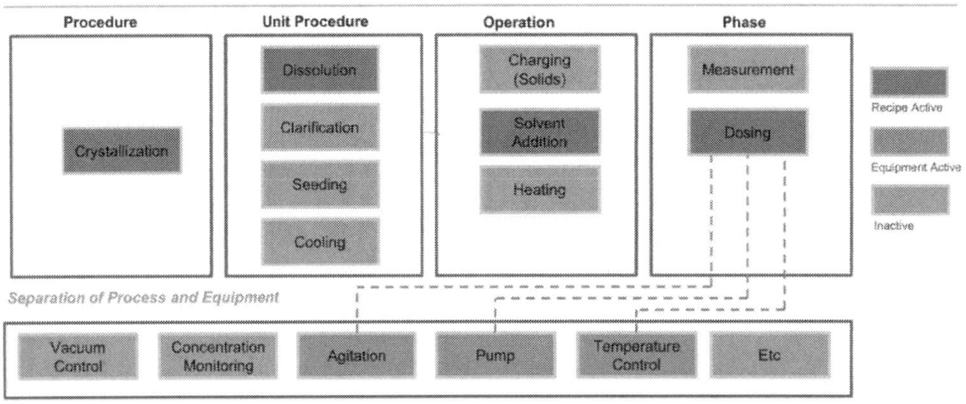


Note: Some Metadata entered upon experiment initiation including; Project Number, Step, User ID and Notebook/Experiment No.

## DSV Lab Master Database

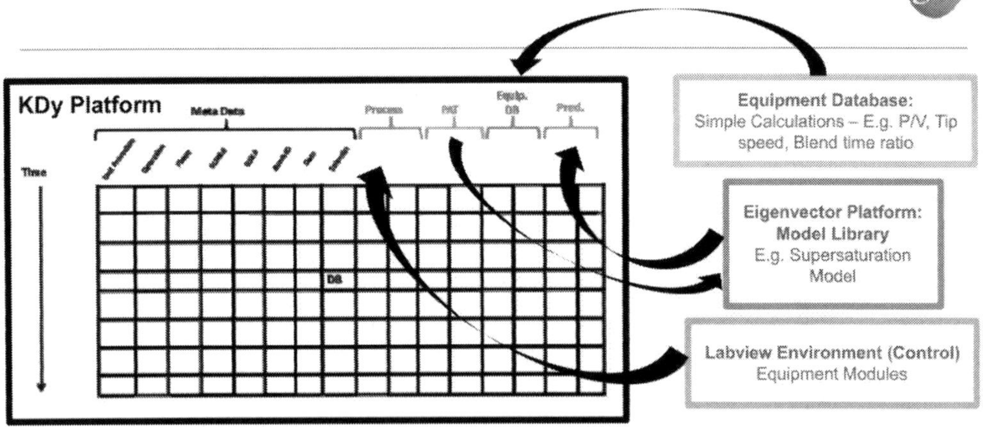

# Scale-up Approach

*Understanding and verifying scale-up*

1. Define design space at small scale
   - Mechanistic model
   - DoE
2. Define design space at intermediate scale
   - Mechanistic model
   - DoE
3. Scale (dis)similarity (MVA)
   - Size and directions of factor effects
   - Reaction/ Cryst profile over time
   - Predictions at target scale
4. Explore other sources of variability (MVA)
   - Design factors sufficient?
   - Scale dependent factors
   - Natural variability of other process parameters/attributes
5. Combine information at different scales (MVA of all historical experiments)
   - Amount of CQA variability described
   - All experiments/scales described equally well
   - Size and directions of design effects
   - Natural variability effects
   - Factors changing with scale effects
6. Target attribute value at large scale (MVA)
   - Use interpretation of factor effects to obtain product with targeted attribute value
   - Simulations to identify process conditions with high likelihood to obtain product with targeted attribute value
7. Root cause analysis (MVA) scale effects
   - Link between process variation and reaction/crystallization profile
   - Model or control scale dependence
8. Define design space based on multiple scales
   - Bayesian approach incorporating between scale variability



# 3. Scale similarity

*Relative size and direction of effects similar?*

gsk

**1L Crystallization**

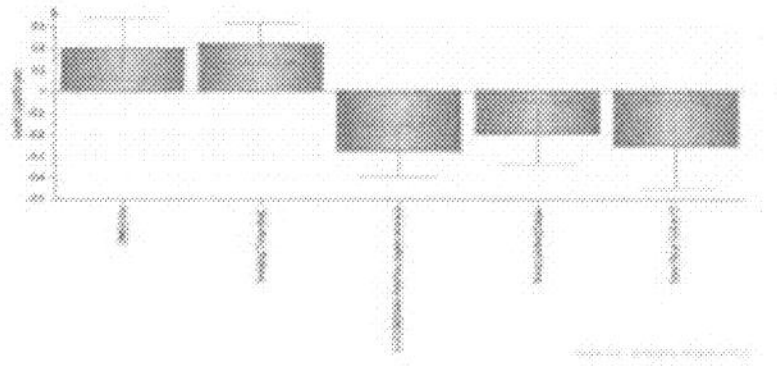

**20L Crystallization**

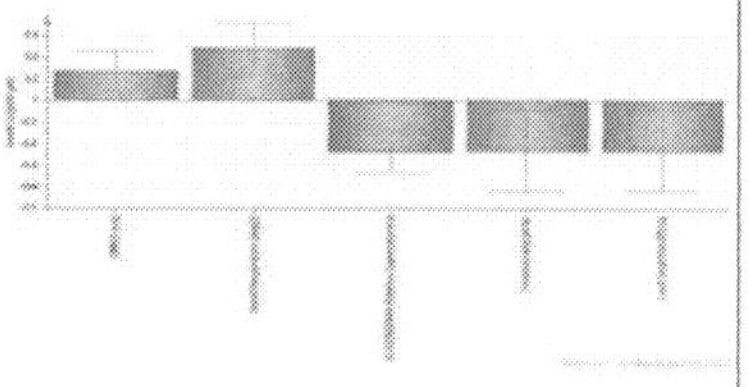

**Example: effect of crystallization variation (5 factors) on x50 API**

The direction of the effects is the same at both scales for all factors.

Only the relative size of the DMSO volume is different for the 20L scale caused by decreased stretch at that scale.

## 3. Scale similarity
*Compare process profile over scales*

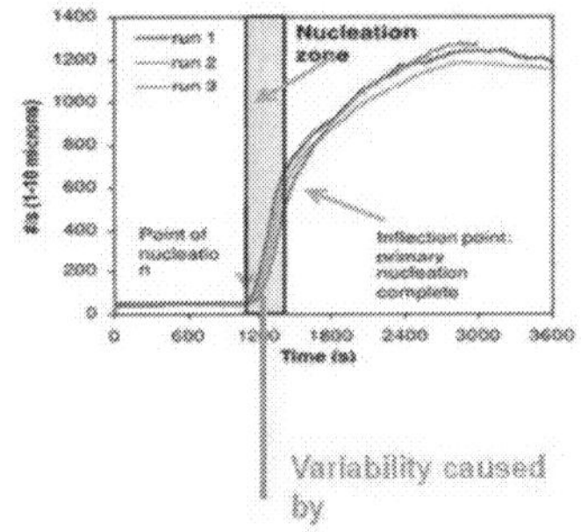


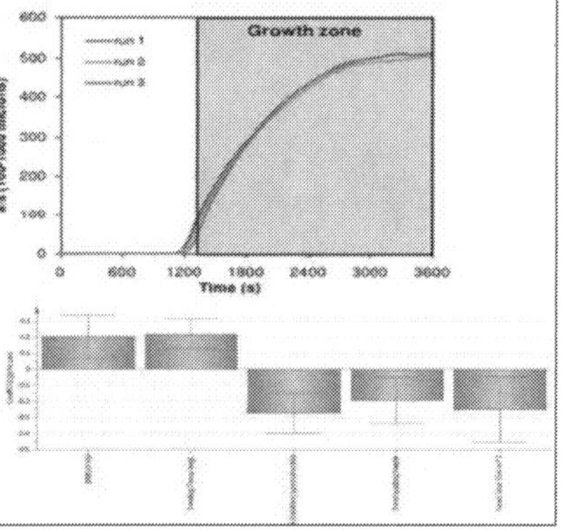


Illustrative example as no data available yet.

## 8. Define design space based on multiple scales

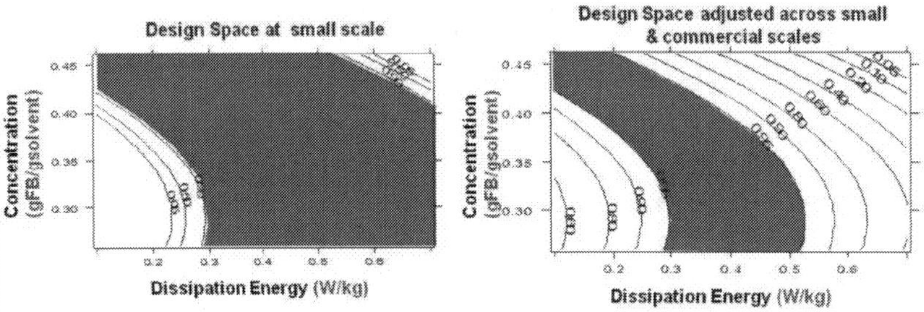


API crystallisation example demonstrated that design space was smaller (for the same level of assurance – 95%) when adjusting for commercial batches

## Key Accomplishments and Learnings

- ❑ ***Informatics improvements*** – The pilot project will provide data needed to make intelligent investments in informatics improvements
- ❑ ***S88 implemented*** – We expect to be able to retrieve critical data more easily which will increase confidence in and understanding of scale-up
- ❑ ***Managing change*** – The project has triggered internal discussions about various informatics options (e.g. Process, eLNB or analytical centric systems)
- ❑ ***Implementation challenges*** – Improvements will require new ways of working by platform users. Good implementation and support plans are required to achieve success.
- ❑ ***Additional opportunities*** – We have identified many new improvement opportunities not apparent at project inception.

## Acknowledgements

*Sponsors;*

- Vance Novack *(Global Head of API Chemistry)* and Phil Dell Orco (*Global Head of Particle and Process Engineering)*

*Contributors*

- Jeff Sterbenz and Rick Atwell - *Upper Merion Site Scale-up Lab*
- Allan Salant, Doug McVey, Justin Palpant - *KDY Automation*
- Melanie Dumarey and Mohammed Yahyah - *Manufacturing Simulation and Modeling*
- Jim Rydzak – *Analytical Sciences*
- Brian Lewis- *Research & Development Information Technology*
- API Project Team –*Huan Wang, Dan Patience et al*

# 416559 Challenges for Academic and Industrial Partners in Major Research Collaborations - a Case Study

**Abstract:**

A major challenge in the management and delivery of academic-industrial collaborations is the alignment of expectations of both the academic and industrial partners. As the number of partners increases, the scale and diversity of the collaboration can present hurdles to successful implementation of the collaboration.

This presentation will discuss some of these challenges with reference to the Synthesis and Solid State Pharmaceutical Center (SSPC) which has 9 funded academic research organisations and 22 industrial partners. Some strategies to successfully address these challenges will be discussed, from both an academic and an industrial perspective.

# 422671 Erc-Sops - a Pre-Competitive Model for Strategically-Driven Research and Technology Development and Commercialization

**Abstract:**

In this talk, we share the experience of the last 9 years of C-SOPS, an NSF-ERC focused on Pharmaceutical products and manufacturing processes.

Creation of the academic/industrial partnership for C-SOP started well ahead of NSF funding. Our first meetings with industry took place approximately one year before the pre-proposal was submitted, and about 2 years before our center was funded. In the initial discussion with pharmaceutical (and other) companies, we focused on primarily two questions - the scope of the research program, and the collaborative model. A strategy emerged from these consultations, indicating that we should focus on pre-competitive research topics of maximum interest to pharmaceutical manufacturers. Once these companies joined the center, we anticipated that many f their technology providers would also be motivated to participate.

This approach succeeded as intended. In nine years, C-SOPS has executed over 60 membership agreements, currently has 40 members, including not only pharmaceutical companies, but also suppliers of manufacturing equipment, control equipment, sensors and process analyzers, modeling software, and others. These members have contributed cash memberships in excess of $7 million, in kind equipment and software iciness in excess of $10 million, and have funded collaborative projects in excess of $20 million.

The pre-competitive model makes IP developed at the center available to all members on a royalty-free basis, provided that is used in-house for non-commercial purposes. Commercial use (defined narrowly as using center IP to define and enable a company's commercial product) requires a commercial license, which is offered simultaneously to all center members.

Overlaying the pre-competitive research program is a large and rapidly growing set of projects that are funded directly by member companies. These projects also benefit from center IP and know-how, but because they are funded directly by members, IP deriving from these projects is negotiated on a case by case basis. Whenever possible, these projects are funded via service agreements, and negotiations regarding commercial disposition of IP rights is delayed until after the IP is developed.

In our experience, this model has been an extremely useful component of our center engagement model and has helped attract substantial levels of support over almost a decade of uninterrupted growth.

# The Role of NCATS, SBIR to Support Precompetitive Collaboration

Asaf Alimardanov, PhD

November 9, 2015

NCATS

NIH National Center for Advancing Translational Sciences

# National Center for Advancing Translational Sciences (NCATS)

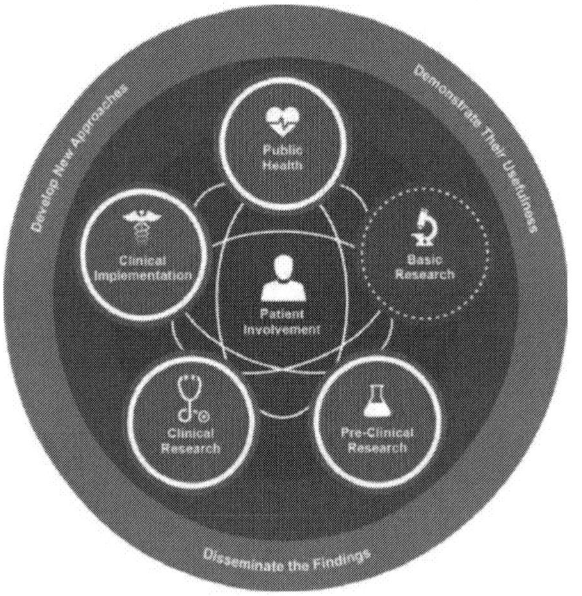


NIH National Center for Advancing Translational Sciences

NCATS

# NCATS SBIR & STTR

## *FOCUS*

To transform the translational science process so new treatments and cures for disease can be delivered to patients faster

## *GOAL*

Seeks to increase small business participation in federally supported R&D and private-sector commercialization of technology

## *FUNDING*

$18 million annually, allocated by NCATS for small business innovation research and technology transfer

NIH National Center for Advancing Translational Sciences

NCATS 3

# What are the SBIR & STTR Programs?

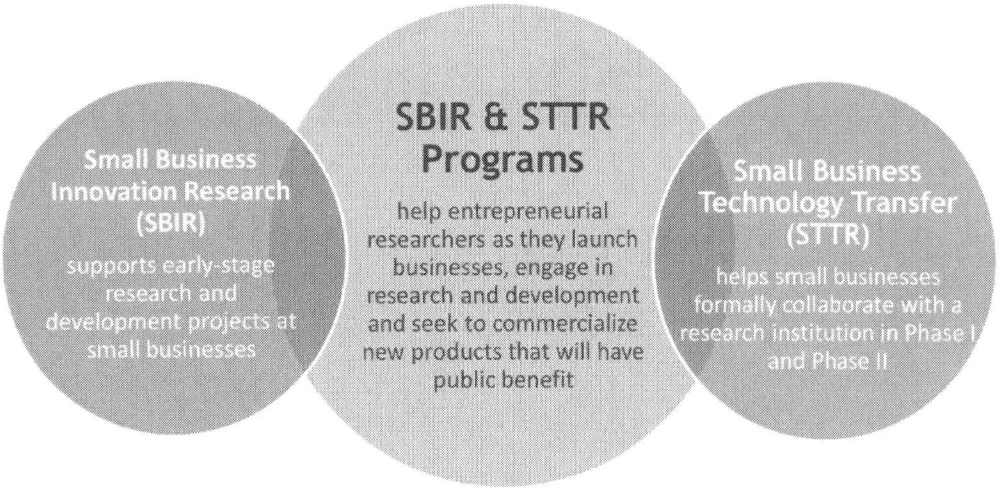


NIH National Center for Advancing Translational Sciences

NCATS 4

# Benefits of NIH SBIR & STTR Funding

- **One of the largest funding sources** of early stage life sciences in the country **(stable & predictable)**
- **IP rights** are **retained by the small business**
- **Not a loan, non-dilutive** capital
- Projects undergo **NIH's rigorous scientific peer review** process, which awardees leverage **to attract other funding and collaborations**

NIH National Center for Advancing Translational Sciences NCATS 5

# NIH SBIR/STTR FY2016 Funding

Set aside requirements for Agencies will increase:

| FY | Set-aside SBIR | Set-aside STTR |
|---|---|---|
| **2013** | **2.7%** | **0.35%** |
| **2014** | **2.8%** | **0.40%** |
| **2015** | **2.9%** | **0.40%** |
| 2016 (current) | 3.0% | 0.45% |
| **2017** | **3.2%** | **0.45%** |

NIH National Center for Advancing Translational Sciences NCATS 6

# SBIR and STTR: One of the Largest Sources of Early- Stage Financing

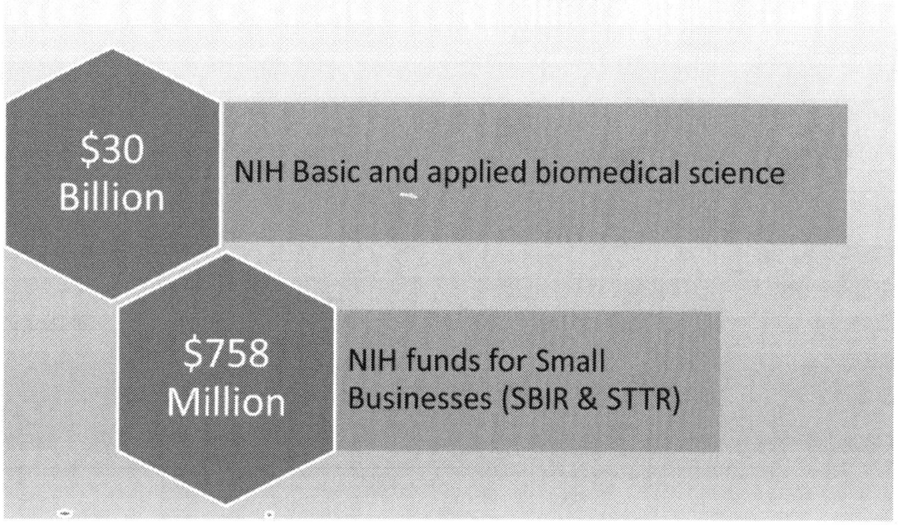


NIH National Center for Advancing Translational Sciences

# Budget Hard Caps

### Award Guidelines

| Program | Phase I | Phase II |
|---|---|---|
| SBIR | $150,000 | $1,000,000 |
| STTR | $150,000 | $750,000 |

### Award Hard Caps

| Program | Phase I | Phase II |
|---|---|---|
| SBIR | $225,000 | $1,500,000 |
| STTR | $225,000 | $1,500,000 |

See NOT-OD-15-101 for details

NIH National Center for Advancing Translational Sciences NCATS 8

# Budget Hard Cap Waiver

- Agencies (not applicants) must apply for a waiver from the SBA for budgets over the hard cap
- SBA has approved a topics list for NIH for 2016
- NIH ICs may have their own budget guidelines
- Appendix A in "Program Description and Research Topics": https://sbir.nih.gov/funding#omni-sbir

**Contact NIH Program Staff to discuss!**



# SBIR/STTR Program Structure Overview & Commercialization Process

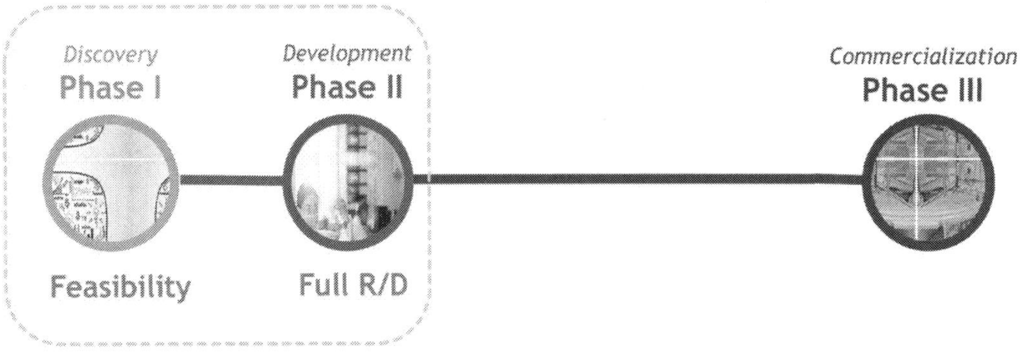


**SBIR/STTR** | **Non-SBIR/STTR**

**Phase I**: $150,000 total costs for 6 months (SBIR) or 1 year (STTR)

**Phase II:** Normally do not exceed $1,000,000 total costs for 2 years ($750,000 for STTR)

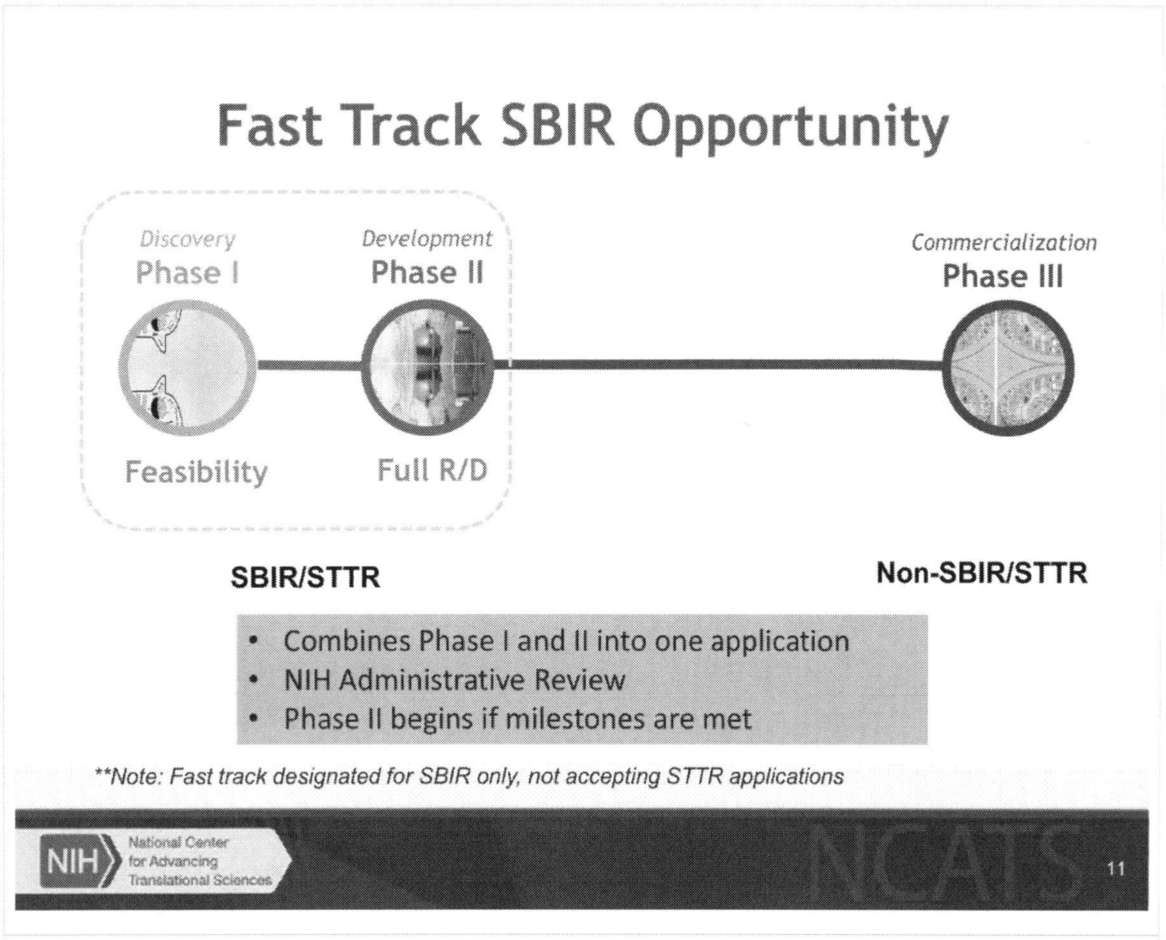

Fast Track SBIR Opportunity
Discovery
Phase I
Feasibility
Development
Phase II
Full R/D
Commercialization
Phase III
SBIR/STTR
Non-SBIR/STTR
• Combines Phase I and II into one application
• NIH Administrative Review
• Phase II begins if milestones are met
**Note: Fast track designated for SBIR only, not accepting STTR applications
NIH National Center for Advancing Translational Sciences
NCATS
11


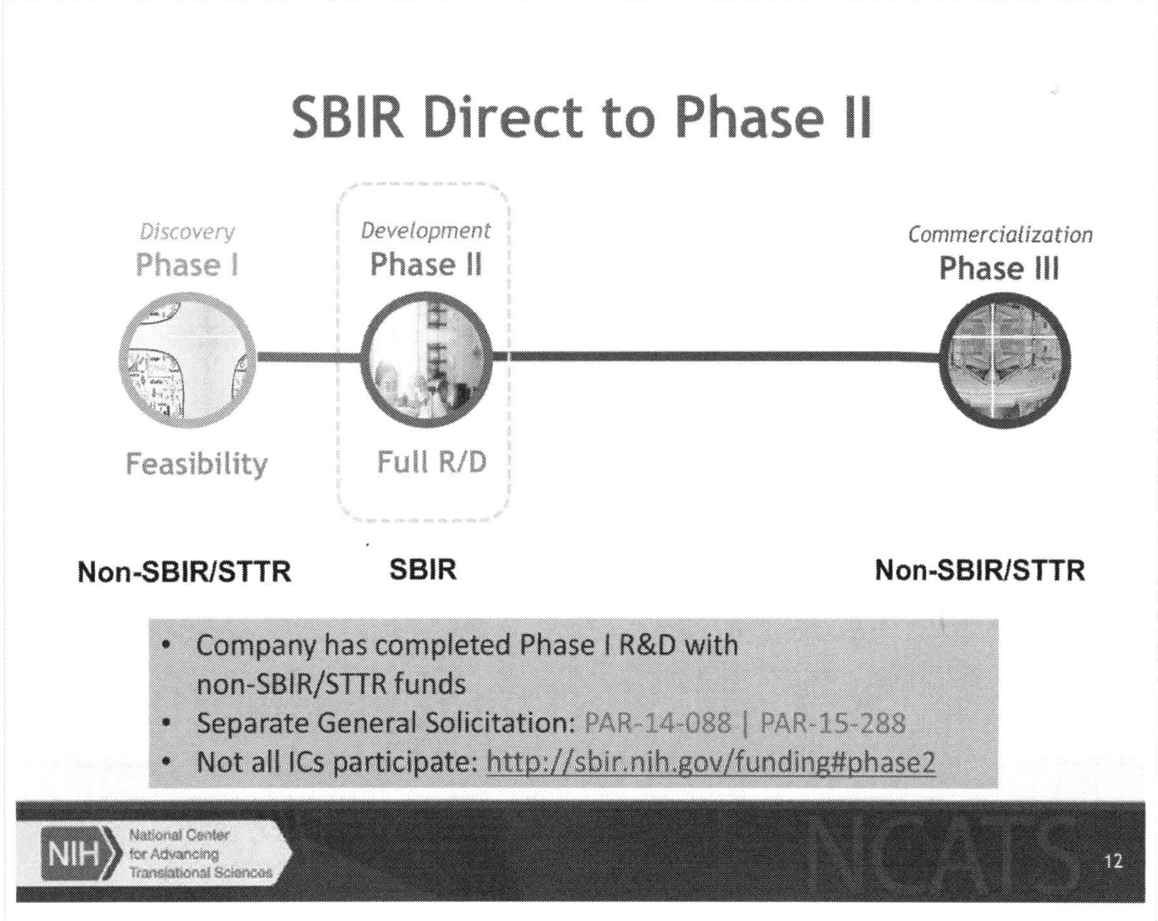

SBIR Direct to Phase II
Discovery
Phase I
Feasibility
Development
Phase II
Full R/D
Commercialization
Phase III
Non-SBIR/STTR
SBIR
Non-SBIR/STTR
• Company has completed Phase I R&D with non-SBIR/STTR funds
• Separate General Solicitation: PAR-14-088 | PAR-15-288
• Not all ICs participate: http://sbir.nih.gov/funding#phase2
NIH National Center for Advancing Translational Sciences
NCATS
12

# NCATS Funding Overview

The NCATS SBIR and STTR programs support small businesses and research organizations advancing technologies that can be commercialized to improve translational science and health care

| | |
|---|---|
| **Omnibus Solicitation** | • Investigator-initiated grant funding<br>• Deadlines: April 5, September 5, January 5 |
| **Direct to Phase II Grant** | • Designed to enhance the pace of technology development<br>• Deadlines: April 5, September 5, January 5 |
| **Grant Solicitations in Targeted Areas** | • Grant to advance a particular technology/ research area<br>• Due dates may vary |
| **Contract Solicitation** | • Contract opportunity to advance areas of high research interest<br>• Typically due in October or November |



# Eligible for SBIR Funding?

**Criteria for applying to SBIR:**

✓ U.S. businesses with **500 or fewer employees**

✓ **PI Primary employment with small business** at the time of the award and duration of the project

✓ **More than 50% U.S.-owned** by individuals and independently operated

OR

✓ More than 50% owned and controlled by other business concern/s that is/are > 50% owned and controlled by one or more individuals

OR

✓ More than 50% owned by multiple venture capital operating companies, hedge funds, private equity firms or any combination of these

## Eligible for STTR Funding?

**Criteria for applying to STTR:**

✓ **An established cooperative research and development effort** delineated as:

- Minimum 40% by small business concern; minimum 30% by U.S. college or university, non-profit research organization or Federally-Funded R&D Center (FFRDC)

✓ **Formalized intellectual property agreement**

- Should provide the necessary IP rights in order to carry out follow-on R&D and commercialization

✓ **Primary employment of the principle investigator with either the small business or research institution**

NIH National Center for Advancing Translational Sciences NCATS 15

## Key Differences Between SBIR and STTR Funding

| | |
|---|---|
| SBIR | • Permits research institution partners (e.g., universities)<br>• Small business concern may outsource ~33% of Phase I activities and 50% of Phase II activities |
| STTR | • Requires research institution partners (e.g., universities)<br>• Minimum 40% of the work should be conducted by the small business concern (for profit), and minimum of 30% by a U.S. research institution (non-profit) |

**Awards always made to small business**

NIH National Center for Advancing Translational Sciences NCATS 16

## FEATURED SBIR EXAMPLE

NCATS

NIH National Center for Advancing Translational Sciences

17

# Collaboration with IQ Consortium

- Emerging Technologies group: survey of IQ member companies
- Discussion at Emerging Technologies group meeting
- Two potential topics for SBIR identified
  - Online Real Time Metals Analysis at Low ppm Level
  - Simple and Robust Reaction Progress Analyzer

NIH National Center for Advancing Translational Sciences

NCATS

18

## Online Real Time Metals Analysis at Low ppm Level

- Allow rapid measurement of trace metal levels (at least <5 ppm)
- Inexpensive, Small footprint, Co-localized
- Robust and have an ease of use advantage, i. e. not require any specialized training
- Ideally, able to be used for continuous monitoring of metal content in continuous processing operations
- Ideally, detect more than one transition metal typically used for catalysis (Pd, Rh, Cu, Zn, Fe, Ir, etc.)

NIH National Center for Advancing Translational Sciences NCATS 19

## Metals Analysis at Low ppm Level

- Strong response
- Unique approach in each proposal
- Thorough peer review
- Three applications funded for Phase I
- Feasibility work ongoing
- Potential pharma end users agreed to act as consultants

NIH National Center for Advancing Translational Sciences NCATS 20

## CURRENT NCATS FUNDING OPPORTUNITIES

NCATS

NIH National Center for Advancing Translational Sciences

21

# Investigator Initiated Grants

## Omnibus "Parent" SBIR/STTR Grant Solicitation

SBIR: PA-15-269 (R43/R44) STTR: PA-15-270 (R41/R42)

**We encourage applications for topics within the NCATS mission**

**Read the "Program Descriptions and Research Topics" Section in the Solicitation for more details**

Due January 5, April 5, September 5

NIH National Center for Advancing Translational Sciences

NCATS

22

# Addressing Unmet Needs

*NCATS SBIR/STTR Research Interests*

Drug Discovery & Development → Diagnostics & Devices → Research & Bioinformatics → Clinical Research

NIH National Center for Advancing Translational Sciences NCATS 23

# NCATS SBIR/STTR Topics of Interest

**Drug Discovery & Development**

- Small molecule and biologics analytical characterization
- Accelerate bioengineering approaches to the development and clinical application of biomedical materials, devices, therapeutics and/or diagnostics
- Technologies to determine alternative uses for existing therapeutic interventions
- Protein-protein interaction assays for high-throughput screening of rare disease related projects
- Tools and technologies to enable assaying of compound activity on currently "non-druggable" targets
- Innovative platforms for identification and prioritization of targets for therapeutic intervention with clear clinical impact
- Fluorescence probes to replace antibodies for determination of cellular protein translocation
- Co-crystallization high-throughput screening techniques
- Tools and technologies that increase the predictivity/efficiency of medicinal chemistry, biologic or other intervention optimization
- Use of continuous flow manufacturing technology to address therapeutics shortages
- Methodologies and technologies to substantially lower the cost of manufacturing biologics
- Development of novel technologies for enzyme replacement therapies to solve a major bottleneck in rare disease research
- Development of non-AB biologics, cell-based therapies and gene therapy discovery amid technology development

NIH National Center for Advancing Translational Sciences NCATS 24

# NCATS SBIR/STTR Topics of Interest

**Specialized Topics**

- Platform delivery technologies for nucleic acid therapeutics
- Bioreactors for reparative medicine
- Development of appropriate pediatric formulations and drug delivery systems
- Onsite tools and technologies for heart, lung and blood clinical research point-of-care

NIH National Center for Advancing Translational Sciences NCATS 25

# ADDITIONAL SMALL BUSINESS RESOURCES

NCATS

NIH National Center for Advancing Translational Sciences 26

## SBIR/ STTR Technical Assistance Opportunities

NIH offers distinct technical assistance programs to SBIR Phase I and Phase II awardees

- Niche Assessment Program
- Commercialization Assistance Program
- I-Corps at NIH offers a nine-week boot camp to help awardees explore potential markets for their federally funded innovations

Applicants can utilize their own technical assistance vendor (must be included in the budget) in lieu of these programs

NIH National Center for Advancing Translational Sciences NCATS 27

## Dedicated NCATS Small Business Resources Site

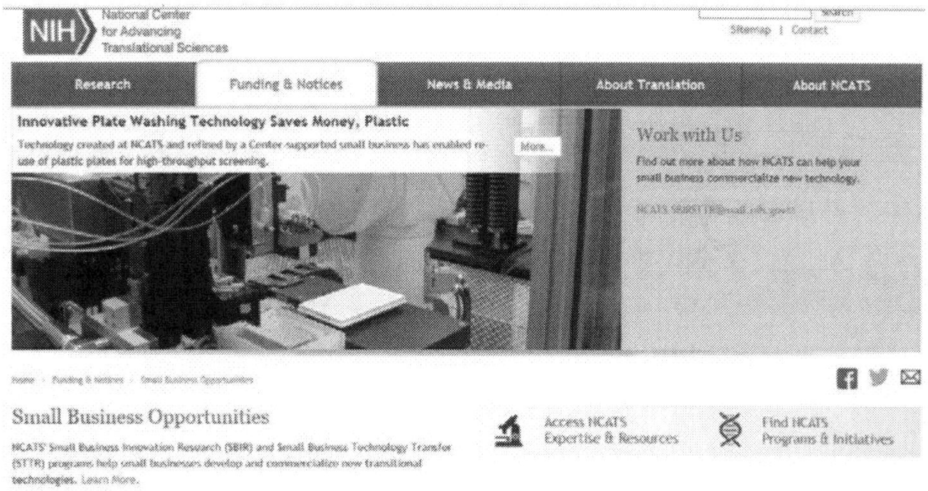


http://ncats.nih.gov/smallbusiness

NIH National Center for Advancing Translational Sciences NCATS 28

# Key NCATS Contacts and Resources

| | |
|---|---|
| **Email** | **Lili Portilla, M.P.A.**<br>NCATS-SBIRSTTR@mail.nih.gov |
| **Online** | http://ncats.nih.gov/smallbusiness |
| **Stay Connected:** | @ncats_nih_gov<br>Join the conversation #NCATSsbir |
| **Searching for funding opportunities?** | http://grants.nih.gov/grants/guide/index.html<br>**Use activity codes:** R41/R42 & R43/R44 |

NIH National Center for Advancing Translational Sciences

NCATS 29

# 429500 Alginate Capsule Composition Influences the Pancreatic Differentiation of Human Embryonic Stem Cells

**Abstract:**

**Introduction**: Type 1 diabetes (T1D) is an autoimmune disease affecting millions of people worldwide wherein the beta cells of the pancreas are destroyed, resulting in insulin dependence. Shortage of donor islets combined with immune rejection limits islet transplantation from becoming a viable therapy. We have shown previously that calcium alginate encapsulation for differentiation of human embryonic stem cells (hESCs) results in efficient differentiation to insulin-producing cells. The unlimited supply of hESCs and the immunoisolation capability of alginate capsules, suggest this approach as an alternative treatment for T1D. Islet encapsulation studies demonstrate that capsule composition has a significant effect on immunoisolation. However, capsule composition can modify the differentiation of encapsulated cells. Hence in this study we are evaluating the effect of material properties on pancreatic differentiation of encapsulated hESCs.

**Materials and Methods**: A single cell suspension of ROCK inhibitor treated hESCs were suspended in 1.1% alginate and 0.2% gelatin followed by encapsulation by drop wise addition into a divalent cation bath. hESCs were encapsulated in barium alginate, crosslinked with 10, 15, 20, 50 or 100 mM $BaCl_2$. The hESCs were differentiated entirely under encapsulation and characterized after completion of pancreatic differentiation. Encapsulated hESCs were induced to the pancreatic islet like state following a stagewise differentiation protocol. Definitive endoderm (DE) was induced with ActivinA and Wnt3Afor 4 days followed by pancreatic progenitor induction with KAAD-cyclopamine and KAAD-cyclopamine with retinoic acid for 2 days each, respectively. DE stage differentiation was characterized by gene and protein expression of SOX17, FOXA2, EOMES, and CER. Pancreatic commitment was analyzed by PDX1 gene and protein expression.

**Results:** Increasing the concentration of $BaCl_2$ from 10 to 100 mM $BaCl_2$ resulted in an increase in substrate stiffness from 4 to 75 kPA, as determined by AFM analysis. Viability and proliferation of encapsulated hESCs were very sensitive to capsules stiffness, with only 4-7 kPa capsules supporting cell growth and colony formation. Diffusional differences were confirmed to be minimal under the tested conditions. DE stage differentiation was found to be enhanced as capsule stiffness was increased, as shown by both gene and protein analysis of SOX17, FOXA2, EOMES, and CER. Pancreatic commitment was analyzed by PDX1 gene and protein expression. *PDX1* gene expression was highest in the 4 kPa stiffness gel (10 mM $BaCl_2$), and was significantly reduced as stiffness increased. The same trend was observed with PDX1 protein expression, which showed the highest number of cells positive for PDX1 in the 10 mM $BaCl_2$ condition.

**Discussion:** Our study demonstrates that capsule materials properties can significantly modulate the efficiency of chemically induced pancreatic differentiation. Specific encapsulation conditions can significantly enhance the efficiency of chemical induction. In general, higher crosslinking was found to be detrimental to the viability and differentiation of hESCs. Research is currently underway for encapsulating hESCs in a high throughput alginate array platform. This will allow for a more thorough investigation of the effect of substrates stiffness on differentiation, as well as delineate the signaling pathways which govern pancreatic differentiation. Results of these studies will be presented and discussed.

# 425787 3D Biomaterial Platforms for the Generation of hESC Derived Dopaminergic Neurons

**Abstract:**

Cell replacement therapies are becoming increasingly viable treatments for neurodegenerative diseases. As one key example, a major strategy to address Parkinson's disease (PD) is to replenish dopaminergic (DA) neurons in the midbrain. To this end, human embryonic stem cells (hESCs) or human induced pluripotent stem cells (hiPSCs) are differentiated into DA neurons in vitro, followed by surgical implantation into the brain. Unfortunately, experiments using animal models and in clinical trials indicate that on average 99% of the cells die following surgical implantation. Considerable resources are consumed to generate enough cells for implantation, given that only 1% survive to become DA neurons that can potentially alleviate PD symptoms. On average 100,000 DA neurons are needed to effectively treat PD in humans, requiring 10 million cells to be implanted per patient. There is therefore a crucial need to devise strategies to efficiently generate a large number of DA neurons. We are using biomaterial cell production platforms to address these challenges.

In previous work, we generated 10-20 fold higher numbers of viable, pluripotent, hESCs in a fully defined, xeno-free 3D gel, compared to equivalent culture on a 2D surface. Here, we have used a 3D differentiation culture to generate large numbers of DA neurons using dual-SMAD inhibition, followed by Wnt and SHH activation and neurogenesis. Production of DA neurons in such a next-generation 3D format calls for extensive characterization to monitor neurogenesis and verify the midbrain DA fate of the resulting neurons. The fate of the hESCs was monitored using immunocytochemistry and qPCR at regular intervals during the patterning and differentiation stages, providing dynamic information on the expression of various markers of interest. Subsequently, the final dopaminergic fate of the neurons was verified with immunocytochemistry, qPCR and electrophysiology.

Differentiation in 3D cultures resulted in equivalent or better levels of relevant marker expression compared to neurons differentiated in a traditional 2D format. FOXA2/LMX1A co-expression demonstrated their midbrain lineage, and TH expression confirmed their dopaminergic identity. Using an equivalent level of resources as for a 2D culture, we were able to generate 20 fold higher numbers of DA neurons. Marker analysis with immunocytochemistry and qPCR provided dynamic information on the expression of various markers of interest. Interestingly, we discovered that some markers, including FOXA2, were expressed at a higher level from an earlier stage in 3D culture compared to differentiation on a 2D surface. Moreover, immunocytochemistry and qPCR showed robust expression of mature dopaminergic markers including DAT and Nurr1. Furthermore, the DA neurons were shown to be electrophysiologically active and able to generate action potentials.

In summary, 3D culture in functionalized biomaterials improves the yield of DA neurons generated from hESCs in vitro. Promising results from current studies warrant further optimization of differentiation in 3D hydrogels to generate enriched populations of DA neurons. Generating large numbers of quality DA neurons could prove useful for a variety of applications including surgical implantation in PD, understanding development of DA neurons, and evaluating PD disease models.

# 432041 Linking Human Pluripotent Stem Cell Cycle Regulation and Culture Density during Expansion and Differentation

**Abstract:**

Cell cycle progression is a major regulator orchestrating human pluripotent stem cell (hPSC) self-renewal and differentiation. Self-renewing hPSCs exhibit a very short $G_1$ phase and proliferate faster than their differentiating counterparts. Among the stem cell environment variables, which affect the regulation of cell cycle, we focused in this study on the culture density. The association between hPSC cycle and culture density has been largely empirical.

The cycle of hPSCs was examined under various culture densities and media [1]. The fraction of cells residing in the $G_1$ phase increased from 25% to over 60% in high-density culture (>40 × $10^4$ cells/cm$^2$) while hPSCs still maintained their expression of pluripotency markers such as NANOG, TRA-1-60 and TRA-1-81. Cells cultured at higher densities also upregulated their expression of the cycle inhibitor CDKN1A (p21) but not of p27 and p53. A population balance equation (PBE) model was constructed based on experimental data to rationalize in quantitative terms the link between culture density, cell cycle progression, and the states of self-renewing and differentiating hPSCs. The proliferation kinetics of each subpopulation were determined. Media commonly used for maintenance of self-renewing hPSCs exhibited the slowest kinetics of induction of differentiation ($k_{diff}$), but supplementation with BMP4 led to 14-fold higher $k_{diff}$ values.

To that end, a cell-cycle fluorescence reporter system (FUCCI) [2] has been employed to further investigate the progression of the cell cycle in pluripotent and differentiating hPSCs. In conjunction with aforementioned framework, we have estimated the physiological state functions of hPSCs under different conditions. Our findings will contribute to the deepening of our understanding of cell cycle regulation during self-renewal and differentiation thereby leading to improved strategies for generating therapeutically useful progeny from hPSCs.

References:

[1] J. Wu, Y. Fan, E.S. Tzanakakis. Increased culture density is linked to decelerated proliferation, prolonged $G_1$ phase, and enhanced propensity for differentiation of self-renewing human pluripotent stem cells. *Stem Cells Dev.* 24(7):892-903, 2015.

[2] A. Sakaue-Sawano, H. Kurokawa, T. Morimura et al. Visualizing spatiotemporal dynamics of multicellular cell-cycle progression. *Cell.* 132(3):487-98, 2008.

# 433918 Exploring the Myelinogenic Potential of Human Keratinocytes Derived Neural Crest Cells: Implications for Demyelinating Diseases

Monday, November 9, 2015: 4:15 PM
151A/B (Salt Palace Convention Center)

**Vivek K. Bajpai**, Chemical & Biological Engineering, University at Buffalo, Buffalo, NY, Xiaoyan Wang, Chemical & Biological Engineering, University at Buffalo, BUFFALO, NY, Ryan Zeiger, Chemical & Biological Engineering, University at Buffalo, Amherst, NY and Stelios Andreadis, Chemical & Biological Engineering, University at Buffalo, Amherst

**Introduction:** During embryonic development, neural crest stem cells (NCSC) migrate laterally along the length of developing notochord and give rise to diverse cell types (e.g. peripheral neurons, Schwann cells, melanocytes and skeletal and connective tissue). In this work, we have shown, for the first time, that human epidermal stem cells (also called keratinocytes (KC)) derived from neonatal and adult skin can be coaxed to acquire functional NCSC fate under defined culture conditions without any transgene overexpression. We further demonstrated using an in-vivo chicken embryo model that human KC derived neural crest cells (KC-NCSC) can migrate and differentiate into functional neural crest derivatives in the developing embryo. Finally, we differentiated KC-NCSC towards myelin producing Schwann cells, which were successfully implanted in the mouse model of dysmyelination to examine their potential to myelinate neurons.

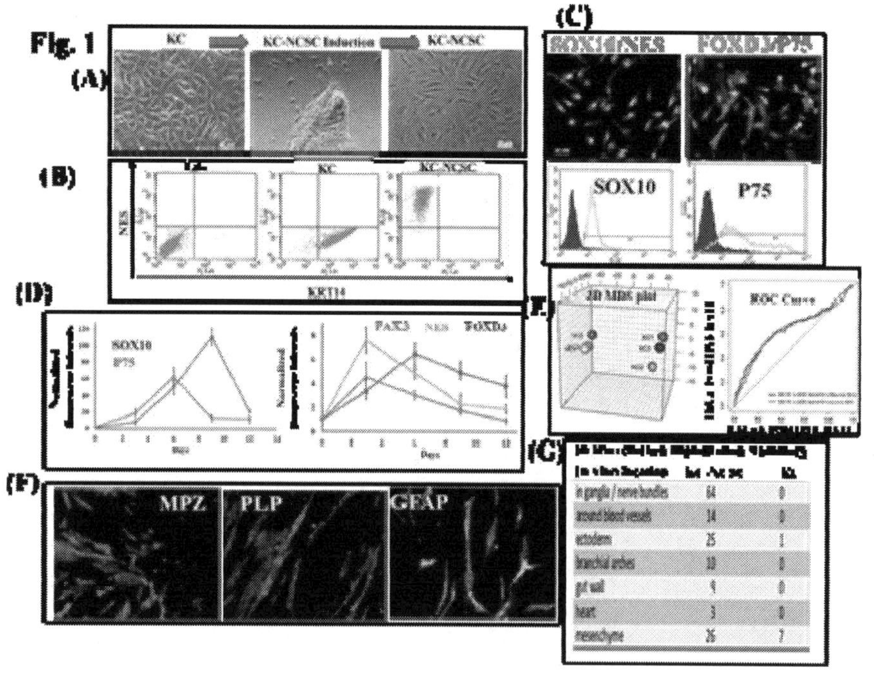


**Materials & Methods:** Human neonatal foreskin KC and adult skin KC were isolated and KRT14+ KC were induced to undergo neural crest fate conversion under an NC induction medium (NCIM) treatment for 8-12 days. KC derived neural crest

stem cells (KC-NCSC) were analysed transcriptionally (through global RNA sequencing (RNASeq) and qRT-PCR) as well as translationally (through flow cytometry, immunocytochemistry, and immunoblotting). To demonstrate that KC possess the ability to undergo NC transformation at single cell level, clonal cultures of KC were established and induced to acquire NC fate. For chicken embryo experiments, we implanted KC-NCSC into cranial portion of 10-13 somite stage chicken embryos and examined embryos 2 and 4 days afterwards. To coax KC-NCSC to undergo Schwann cell differentiation, KC-NCSC were cultured in differentiation media containing FGF2, CNTF, GDNF, and NRG1 for 15-20 days. Further, we injected KC-NCSC derived Schwann cells into a Shiverer/Rag2 (-/-) mouse model after modifying them with lentivirus carrying mCherry reporter to allow us monitor mCherry+ cells after explantation. Mice were sacrificed at day 7 day and day 30 intervals and were studied for myelination using immunohistochemistry.

**Results and Discussion:** After isolation, KC grew as typical epithelial colonies and after 8-12 days of induction treatment small spindle shaped cells appeared to be delaminating from tighter KC colonies (Fig1A). These were termed as KC-NCSC as they were uniformly positive for NC genes i.e. SOX10, FOXD3, PAX3, KIT, NGFR, NES and lacked KRT14/KRT5 (KC markers) as evidenced by transcriptional and translational analyses (Fig1B, C and D). KC-NCSC induction upregulated key epithelial to mesenchymal transition (EMT) genes like SNAI1, SNAI2, TWIST, FOXC2, VIM, CDH2 and downregulated CDH1. Illumina based RNA sequencing analysis showed that in contrast to KC, KC-NCSC possess global transcriptional profile similar to native NC as demonstrated by ROC analysis and 3D MDS plot (Fig1E). Clones of KC-NCSC expressed characteristic NC genes confirming clonal NC potential of KC. Under appropriate differentiation conditions, KC-NCSC differentiated and matured along functional neuronal, Schwann cell, melanocyte and mesenchymal linages as confirmed by mRNA and protein analysis as well as functional tests for each lineage. Upon implantation in chicken embryos, KC-NCSC migrated extensively and differentiated into multiple NC derivatives including neurons, glia, myocytes and melanocytes as confirmed by immunohistochemistry (results summary in Table1G). KC-NCSC efficiently differentiated into Schwann cell phenotype as confirmed by immunocytochemistry for key schwann cell genes like MPZ (P0), Myelin basic protein (MBP) and PLP as well as upregualtion of these genes at the transcriptional level (Fig1F).

**Conclusions:** In summary, this work establishes KC as novel source of bonafide NCSC and KC-NCSC derived Schwann cells, which were successfully transplanted into murine dysmyelinating disease model for studying their myelinogenic potential. This study has significant implications for the use of cellular therapies for treatment of devastating demyelinating diseases like multiple sclerosis.

# 426226 Implantable Microenvironments for Stem Cell and Cancer Research

**Abstract:**

Bone marrow provides unique microenvironments for hematopoietic stem cells. It is also known as a major reservoir for many circulating tumor cells and may represent a constant source of metastatic cells that lay dormant and disseminate from the marrow. However, inaccessibility of the bone marrow space has limited systemic investigation of these vital cellular processes in normal and pathological settings. In this presentation, we will introduce implantable humanized stem cell niche that can support hematopoietic progenitor cells in an accessible, ectopic site under the skin. The implants consist of human bone marrow stroma cells seeded in a 3D microfabricated biomaterial that mimics the physical and anatomical features of the bone marrow. When implanted subcutaneously in immunocompromised mice, these stromal cell-laden biomaterials formed a highly vascularized and organized marrow-like tissue over a period of time. These constructs represent 2% of the native marrow volume and recruited host Lin-Sca-1+c-kit+ hematopoietic progenitor cells to the site at 20% of normalized marrow levels. We demonstrated that intravenously injected human CD34+ progenitor cells and human leukemic cells were spontaneously recruited to the implanted microenvironments. Microenvironment dependent tumor metastasis was also modeled generating an orthotropic tumor that shed circulating tumor cells into the bloodstream. Finally we employed genetically engineered stromal cells that synthesis specific human cytokines to determine cytokine dependent human CD34+ cell behaviors. We envision that implantable and tunable marrow-mimicking tissue microenvironments could enable various aspects basic and translational stem cell and cancer research.

# 432545 Metabolic Reprogramming during Three-Dimensional Culture of Human Mesenchymal Stem Cell Enables Stem Cell Phenotypic Enhancement

**Abstract:**

## Introduction

Human mesenchymal stem cells (hMSCs) are primary cell source in cell therapy for a wide range of diseases. However, immediately after isolation and upon culture expansion, MSCs acquire and accumulate genetic and phenotypic changes in culture and that many proliferating cells progressively enlarge, exit the cell cycle, and become senescent. Emerging studies have shown that in vitro self-assembled 3D MSCs have intriguing biological properties, including enhanced multi-lineage differentiation potential and more primitive stem cell phenotype. It has been shown that Wnt signaling and Notch signaling play significant roles in hMSCs aggregate functional activity, including anti-inflammation, anti-apoptosis, and endothelia cell-supporting function. The mechanism of hMSCs stemness enhancement during cellular aggregation is however remains uncertain. Moreover, as 3D culture of MSC been proposed as preconditioning strategy to enhance MSCs function, understanding the adaptive changes of MSCs metabolism during self-assembly has important implication in optimizing aggregate properties and function activation. The objective of present study is to investigate energy metabolism in regulating MSCs phenotype changes in 3D aggregates of hMSCs cells with aim to develop approachable strategy for MSCs expansion with enhanced stem cell phenotype.

## Materials and Methods

hMSC aggregates were formed by seeding cells into ultra-low adherent surface plates. To assess the role of specific metabolic pathway in regulating MSCs phenotype enhancement under aggregation, several metabolic modulators, including glycolysis inhibitor (2-DG), pentose phosphate pathway (PPP) inhibitor (DHEA), were used to treat hMCs in aggregate and adherent conditions, and their metabolic and cellular impacts were discussed.

## Results

The results revealed that hMSCs in self-assembled 3D cell aggregates have improved "stemness" compared to hMSCs grown in adherent condition, indicated by up-regulation of three pluripotent stem cell genes, including Nanog, Oct4, and Sox2. This phenotypic change of hMSCs in cell aggregates is associated with metabolic reprogramming towards increased glycolytic and decreased pyruvate oxidation metabolism pattern. mRNA level of glycolysis enzymes, including HK2, PKM2, and LDHA are up-regulated in hMSCs aggregates, with mRNA level of PDK1 down-regulated, an enzyme involved in pyruvate oxidation within TCA cycle. Among the reprogrammed metabolic pathways involved in hMSCs aggregate, pentose phosphate pathway (PPP) was identified playing an significant role as inhibiting the rate limiting enzyme in PPP, G6PDH, by DHEA dramatically decreased mRNA level of Nanog, Oct4, Sox2, HK2, PKM2, and LDHA, thus abolished the phenotypic enhancement and metabolic reprogramming of hMSCs aggregates. The regulatory role of PPP during stemness enhancement of hMSCs is attributed to its stress sensing ability during cell aggregation. The redox balancing substrate NADPH and stress-response genes, including HIF-1a, NFkB and Notch were all increased in hMSCs aggregates, and were decreased dramatically upon DHEA treatment. Together, the results revealed the mechanistic connection between metabolic reconfiguration and hMSCs potency under 3D aggregation, and identified the novel role of PPP as stress sensor during hMSCs phenotypic enhancement.

## Conclusions

The results of present study demonstrate that energy metabolic transition plays an important role in hMSC stemness improvement under 3D culture. Understanding the metabolism transition associated with hMSCs phenotypic changes has important implication for adapting 3D aggregation culture into as a non-genetic preconditioning method to potentiate hMSC properties for cell therapy.

**References:**

Riganti, C., Gazzano, E., Polimeni, M., Aldieri, E., Ghigo, D., The pentose phosphate pathway: An antioxidant defense and a crossroad in tumor cell fate. Free Radical Biology and Medicine, 2012.53 (2012) 421–436

Stincone, A., Prigione, A., Cramer, T., et al., The return of metabolism: biochemistry and physiology of the pentose phosphate pathway. Biol. Rev., 2014. Sep 22. doi: 10.1111/brv.12140

Kruger, A., Gruning, N.M., Wamelink, M.M., et al., The pentose phosphate pathway is a metabolic redox sensor and regulates transcription during the antioxidant response. Antioxidants & Redox Signaling, 2011.15(2):311-324

# 425247 Adult Human Mesenchymal Stem Cell Differentiation at the Population and Single-Cell Levels Under Alternating Electric Current

**Abstract:**

Progress in developing clinically-relevant biomedical applications (such as regeneration and repair of damaged bone tissues) using stem cells has been slow because the conditions to generate either the required numbers of undifferentiated cells or lineage-specific cells remain, at best, partially understood. Cells involved in the process of new bone formation include (i) osteoblasts (differentiated, bone-forming cells) and (ii) bone-marrow derived mesenchymal stem cells (undifferentiated, multipotent, with the potential to differentiate at least into osteoblasts, chondrocytes, and adipocytes upon appropriate stimulation). To date, biochemical compounds (such as bone morphogenetic proteins) have been utilized to promote the differentiation of mesenchymal stem cells into osteoblasts. In contrast, the effects of biophysical stimuli on stem cell functions remain unknown.

For these reasons, the present *in vitro*study used interdisciplinary approaches and novel laboratory setups to examine and optimize the effects of alternating electric current alone (i. e., in the absence of supplemented exogenous growth factors) to promote the osteodifferentiation of adult human mesenchymal stem cells. Motivation for the study was provided by the physiological milieu: bone and its constituent cells exist, and function, in an environment composed of biochemical as well as by biophysical (specifically, mechanical and electrical) stimuli.

For this purpose, adult human mesenchymal cells were cultured on flat, indium-tin-oxide-coated glass (pre-coated with fibronectin) in the absence of exogenous growth factors. A custom-made laboratory set-up was used to expose adult human mesenchymal stem cells (passage 3-5) under standard cell culture conditions to alternating electric current (sinusoidal waveform, in the ranges of 5-40 μA and 5-10 Hz frequency) for various durations (in the range of 1-24 hours) daily, for up to 21 consecutive days. The cellular/molecular analyses utilized two approaches: (i) at the cell population level; and, for the first time, at the (ii) single-cell level. Quantitative real-time polymerase chain reaction (PCR) was used to determine gene expression in mesenchymal stem cells at the population and single-cell levels. Genes indicative of the lineage specific osteoblastic pathway (specifically, *TAZ*, *RUNX2*, *SP7*, *SPP1*, and *BGLAP*), as well of the chondrogenic (*COL2A*) and adipogenic (*FABP4*) lineages were monitored.

Optimal osteodifferentiation of adult human mesenchymal stem cells was obtained at the cell population level when cells were exposed to a sinusoidal, 10 μA, 10 Hz, alternating electric current for 6 hours daily for up to 21 consecutive days. Under these conditions the mesenchymal stem cells expressed early (*TAZ*, *RUNX2*, and *SP7*), and late (*SPP1* and *BGLAP*) genes indicative of exclusive osteodifferentiation because genes indicative of the chondrocyte and adipocyte lineages were not expressed at all time points tested. The single-cell analysis provided evidence of the heterogeneity of the adult human mesenchymal stem cells and confirmed expression of genes indicative of the early stages of osteodifferentiation.

In addition to providing fundamental information pertinent to stem cell physiology, the unique biophysical stimulus examined provides a still untapped alternative approach, (which does not require supplemented exogenous growth factors) in order to obtain critically-needed lineage specific differentiated cells (e.g., osteoblasts) for tissue engineering and tissue regeneration applications. In this respect, the present study could have major impact in bioengineering, biotechnology, and in specific therapeutic modalities in the clinical milieu.

# 431932 Development and Scale up of a Continuous Flow Reaction for API Intermediate Manufacture

**Abstract:**

Continuous processing is a growing interest in pharmaceutical industry for the development and the manufacture of active pharmaceutical ingredient (API) and upstream intermediates. Enhanced heat and mass transfer rates inherent to continuous flow reactors are enabling chemical researchers to explore reactions that are traditionally difficult to investigate in common batch reactors. By selecting reactors with small internal volume and operating at the appropriate flow rates, it is possible to explore reactions at short reaction times. These features render flow reactors powerful tools for studying highly energetic reactions with unstable intermediates. Furthermore, it may be possible to gather critical reaction data more quickly with these reactors as the continuous flow operations enable rapid, sequential experimentation.

We present a lab-based flow reactor system used to investigate the Aldol reaction used in the production of an API intermediate. This system was utilized to understand the impact of various reaction parameters on conversion and selectivity. To better understand and predict the reaction performance during scale up, the reaction was studied in the lab under various mixing conditions. The lab-scale results are in agreement with the process performance at pilot scale.

# 433910 Development and Implementation of Flow Chemistry for a Curtius Rearrangement

**Abstract:**

The Curtius rearrangement is the conversion of a carboxylic acid moiety into an amine with retention of stereochemistry via a high energy acyl azide intermediate. An investigation of the Curtius rearrangement to generate an asymmetric isocyanate (Curtius product) via the acyl azide and reaction of the isocyanate with BnOH will be described. A flow procedure to produce the Curtius product was developed. Attempts to conduct the entire sequence from trans acid to amino ester using flow chemistry will be described. Multikilograms of the amino ester have been produced safely through the Curtius reaction using batch or flow processing in high overall yield.

# 410930 Development and Scale-up of Acrylamide Formation Using an Impinging Jet

**Abstract:**

In a synthetic step of a pre-clinical drug candidate, a problematic dimer impurity was observed. Early work identified the origin of this dimer impurity as reaction between the starting material and product, indicating that efficient mixing was critical to achieving low levels of this impurity. A fit-for-purpose flow chemistry solution using an impinging jet was developed and implemented for this process. Kinetics work performed on this reaction also led to mechanistic understanding for the formation of additional impurities, including a chloro impurity that was identified as a GTI (genotoxic impurity). Reaction kinetics and response surface simulations were used to reduce the chloro impurity to acceptable levels. Using these insights, this process was successfully scaled to 12 kg in a cGMP manufacturing environment.

*The work presented here was sponsored by AbbVie. AbbVie contributed to the design, research, and interpretation of data, writing, reviewing, and approving the presentation. Daniel Caspi, Benoit Cardinal-David and Alex Huters are employees of AbbVie.*

# 416213 Robust Parallelization of Triphasic Micro/Millireactors for Sustainable Pharmaceutical Manufacturing

**Abstract:**

**Robust Parallelization of Triphasic Micro/Millireactors for Sustainable Pharmaceutical Manufacturing**

Swee Kun Yap, Wai Kuan Wong, Nicholas Xiang Yang Ng, Ning Yan, and Saif A. Khan

Department of Chemical and Biomolecular Engineering

National University of Singapore, SINGAPORE

Metal-catalyzed gas-liquid reactions, such as hydrogenation, carbonylation and halogenations are ubiquitous in pharmaceutical and fine chemical manufacturing. Such reactions are typically carried out in stirred batch reactors, and face several challenges in the form of mass and heat transport limitations, which often require energy intensive solutions such as high stirring rates and pressurization. There is considerable interest in continuous-flow micro/millireactor technology for gas-liquid reactions due to the tremendous acceleration in heat and mass transport possible in such systems, in addition to the prospect of facile phase separation.[1] We have previously demonstrated the design and application of a triphasic segmented flow millireactor for rapid nanoparticle-catalyzed reactions with facile catalyst recovery.[2,3] Here, we present the first demonstration of robust eight-fold parallelization of triphasic millireactors for sustainable pharmaceutical manufacturing. Specifically we demonstrate a novel parallelized reactor system for rapid hydrogenation reactions with facile catalyst recovery and recycle. Parallel operation of multiphase microfluidic circuits presents several unique challenges over their single-phase counterparts; in particular, nonlinear pressure-flow characteristics lead to severe problems with robust and equal routing of the same multiphase flow in the various arms of a parallelized flow network. Biphasic circuits have been studied extensively in recent years, and parallelized biphasic circuits have been demonstrated recently.[4] The dynamics of triphasic circuits, comprising a gas and two liquid phases, are considerably more complex. In our demonstration of numbering-up to eight parallel *triphasic* flow millireactors, the segmented flow in each millireactor consists of aqueous NPs, organic substrate and hydrogen gas at near atmospheric pressure. Feed solutions (aqueous NPs and organic substrate) are drawn from their respective reservoirs by means of peristaltic pumps, the latter being preferred over syringe pumps due to limited syringe capacities. However, flow rate fluctuations are typical of peristaltic pumps, and these fluctuations dramatically influence the stability of the three-phase flows in the parallelized reactor system. We also demonstrate the design of inline hydraulic damper systems for fluid delivery, which are able to damp out pressure fluctuations from the pumps, thereby allowing for the formation of a smooth triphasic flow in all the eight millireactors. We model, from the principles of fluid mechanics, the flow profile of fluid in the hydraulic dampers to better understand the influence of tube material in regulating pressure fluctuations in a parallelized multiphase flow system.[5] Our system is then applied in the hydrogenation of nitrobenzene, a model pharmaceutical substrate. The parallelized system yields a consistent conversion of $(80 \pm 4)\%$ across all eight reactors over the course of several hours, with a throughput of ~10 mL/h. Finally, we demonstrate continuous catalyst recovery and recycle, where the aqueous PtNPs are subsequently continuously recycled. In summary, we present a robust and general scheme for parallelization of triphasic flows, and demonstrate an industrially relevant application in sustainable pharmaceutical manufacturing.

**REFERENCES:**

[1] E.R. Murphy, J.R. Martinelli, N. Zaborenko, S.L. Buchwald, K.F. Jensen, Angew. Chem. Int. Ed. 46, pp. 1734-1737, 2007.

[2] S.K. Yap, Y. Yuan, L. Zheng, W.K. Wong, J. Zhang, N. Yan, S.A. Khan, Green Chem. 16, pp. 4654-4658, 2014.

[3] S.K. Yap, Y. Yuan, L. Zheng, W.K. Wong, N. Yan, S.A. Khan, J. Flow Chem. 4, pp. 200-205, 2014.

[4] M. Al-Rawashdeh, L.J.M. Fluitsma, T.A. Nijhuis, E.V. Rebrov, V. Hessel and J.C. Schouten, Chem. Eng J., 181-182, 549 (2012).

[5] "The Physics of Pulsatile Flow," M. Zamir, Biological physics series, Springer, 2000

# 416384 Multi-Step Continuous GMP Production of a Highly Potent Compound

**Abstract:**

The development of a GMP continuous process to synthesize a highly potent compound is disclosed. A dedicated, inexpensive equipment set run in laboratory fume hoods was utilized on manufacturing scale with in-process monitoring. This process included four unit operations including a continuous crystallization and filtration. Two chemical reactions were run plug flow tube reactors. One of these PFRs was designed to strip off a gas byproduct continuously during reaction and decouple liquid residence time from gas residence time. Continuous anti-solvent driven crystallization was done using two MSMPRs in series. Continuous filtration, washing, drying, and re-dissolving eliminated the isolation of a potent intermediate. Throughput was 4 kg per day in-situ penultimate from the GMP manufacturing plant. Finally, the robustness of the process and a comparison of lab and commercial scales highlighting the challenges that were encountered will be discussed.

# 416697 Continuous Processing: Application to Process Design and Material Supply

**Abstract:**

Continuous processing represents a powerful methodology that can be applied both to process design and material supply from reactions to isolations. Within process design, augmenting typical approaches with continuous operation enhances the mapping of the design space. Specific points within the design space, represented by each steady state, can be probed and characterized generating deeper understanding and allowing for the extraction of kinetic information not otherwise readily accessible. Continuous operation also increases the options with regard to material supply offering advantages with regard to impurity profile, control of solid state attributes such as PSD and form, capacity and safety. However, given that continuous processing is still within its infancy with regard to adaption by the pharma industry, there are significant technical and regulatory challenges to be overcome to successfully supply material. A number of case studies applying continuous processing to process design and material supply will be presented.

# 431745 Engineering of a Peptide Probe for Beta-Amyloid Aggregates

**Abstract:**

Aggregation of β-amyloid (Aβ) is central to the pathogenesis of Alzheimer's disease (AD). Aβ aggregation produces amyloid assemblies, such as oligomers and fibrils. In contrast to non-toxic Aβ monomers, Aβ oligomers and fibrils can act directly as major toxic agents and indirectly as pools of the toxic entities, respectively. Thus, the detection of Aβ aggregates is of diagnostic interest and should benefit enhanced molecular understanding of AD. Among many molecular platforms, peptide-based ligands hold promise as Aβ probes due to their relative simplicity, ease of optimization and facile conjugation to other molecular contexts. In this regard, Aβ hydrophobic segments (critical in Aβ self-assembly) or variants thereof can serve as lead molecules for Aβ probe development. Unfortunately, the resulting peptides are either highly self-aggregation-prone or their probe potential has not been thoroughly examined. In the present study, we characterized a novel peptide ligand, KLVFWAK, which was created by simple point mutations of an Aβ hydrophobic segment ($^{16}$KLVFFAE$^{22}$). We found that KLVFWAK displayed low self-aggregation propensity and was preferentially bound to Aβ oligomers and fibrils relative to Aβ monomers. Interestingly, binding of KLVFWAK to Aβ aggregates occurred at a non-homologous Aβ segment (e.g., Aβ C-terminal domain) rather than the homologous $^{16}$KLVFFAE$^{22}$. We also show that detection of Aβ aggregates during incubation of fresh Aβ was possible with KLVFWAK, further supporting KLVFWAK's high probe potential for Aβ aggregates. In short, this study presents creation of a non-self-aggregating peptide ligand for Aβ aggregates through simple point mutation of an Aβ-derived segment.

# 433129 A Versatile Solid Binding Peptide to Control Protein-Inorganic Interactions

**Abstract:**

Car9 is a dodecaptide that was originally identified for its ability to bind graphitic materials through electrostatic and ∏-∏ interactions and later found to adhere to silica. Beyond carbon alloform discrimination and the affinity purification of proteins, we show here that the Car9 tag is a remarkably versatile tool that supports: (i) one-step, chemically addressable micro-contact printing of functional proteins; (ii) controlled release of proteins from sol-gel materials; (iii) manipulation of carbon nanostructures; (iv) and synthesis of hybrid materials with unique optical properties.

# 431546 Photo-Induced Formation of Vesicles in-Situ

**Abstract:**

The production of synthetic vesicles has seen a large increase in interest due to their possible applications ranging from artificial cell membranes and microreactors to delivery mechanisms for drugs and cosmetics. While formation techniques have been developed they lack an enhanced level of spatial and temporal control, which can be established through the use of photo-initiated reactions. To achieve this the Copper-catalyzed Azide-Alkyne Cycloaddition (CuAAC) reaction, which has been used previously in the formation of vesicles through the use of an alkyne functionalized lysolipid and azide functionalized aliphatic tail, was initiated by substituting the reducing agent with a water soluble, visible light photo-initiation system. Fluorescent microscopy in conjunction with a lipid bound dye gave proof to the spatio-temporal control over formation as well as the greater control over vesicle size and increased formation density. To expand beyond the use of radical generating photo-initiators to spur vesicle formation, newly functionalized lysolipids containing thiol functionalities have been developed. These new lipids allow for vesicle formation by the thiol-Michael reaction, which can be initiated by photo-bases. Together, this combination of light controlled formation systems allows for a large expansion on the control over synthetic vesicles.

# 415425 An Integrated Fabrication-Conjugation Approach for Biopolymeric-Synthetic Hybrid Polymeric Microspheres and Improved Protein Conjugation Capacity and Kinetics Via Robust Micromolding and Bioorthogonal Click Reactions

**Abstract:**

Hydrogel microparticles with controlled structures have gained increasing attention in various biomedical applications as biosensing platforms, drug carriers, and building blocks for tissue engineering. However, it still remains challenging to fabricate such microparticles in a simple and reliable manner. We present facile fabrication of chemically functional and monodisperse biopolymeric–synthetic hydrogel microspheres with controlled macroporous structures that can be utilized as conjugation platforms for large biomolecules. Specifically, we exploited a simple micromolding-based approach utilizing surface tension-induced droplet formation followed by photo-induced radical polymerization in order to fabricate highly uniform poly(ethylene glycol) (PEG)-based hydrogel microspheres, in which a potent aminopolysaccharide chitosan providing chemically reactive and abundant amine groups is incorporated. The as-prepared microspheres showed macroporous and/or intriguing core-shell structures stemming from spontaneous polymerization-induced phase separation. Fluorescent labeling studies utilizing an amidation reaction on the chitosan's primary amines confirmed stable incorporation of the chitosan moieties with retained chemical reactivity, showing potential of such amines as conjugation handles. The utility of our microspheres as biomolecular conjugation platforms was thoroughly examined in terms of conjugation capacity and kinetics by conjugating model biomolecules (fluorescein-labeled single-stranded DNA and red fluorescent protein R-phycoerythrin) via two high-yield bioorthogonal reactions (strain-promoted alkyne–azide cycloaddition (SPAAC) and tetrazine–*trans*-cyclooctene (Tz–TCO) cycloaddition reactions). The conjugation results showed well-defined and macroporous network structures of the microspheres, leading to enhanced protein conjugation capacity and kinetics and allowing programmable protein conjugation. In this presentation, we will also highlight simple methods to further fine-tune the macroporous strucutres via inert PEG porogens.

# 429109 Gating Protocell Permeability to RNA

**Abstract:**

We are interested in designing simple models of primitive cells in order to study how cellular life may have emerged from prebiotic chemistry. An integral feature of the artificial cell is the coordination of both membrane biophysical processes and genetic replication; the coupling of which would enable RNA replication alongside membrane division and advance the design of a synthetic cell. Short polynucleotide transport through vesicle membranes is expected to be a key step in facilitating nonenzymatic RNA replication inside vesicle membranes. While nucleotide monophosphates have been shown to permeate certain fatty acid membranes, the permeability of larger polynucleotides, like dinucleotide and trinucleotide oligomers, has not been extensively explored. Using monoacyl fatty acids to construct model membranes and short RNA polynucleotides, we studied how membrane composition, polynucleotide size, and divalent cations affect polynucleotide permeability through vesicle membranes. We find that RNA as large as a trinucleotide oligomer, can penetrate fatty acid vesicle membranes. We observe that membrane permeability can be further enhanced by exposing vesicles to short durations of heat, all the while maintaining membrane stability. Finally, using the conditions we found optimized membrane permeability, we demonstrate that chemically activated RNA trinucleotides transport across vesicle membranes to nonenzymatically copy template strands encapsulated in the interior of fatty acid vesicles. Our results showing that membrane permeability to short RNA oligomers can be gated with divalent cations and heat exposure provides insight into the environmental conditions that may have promoted genetic replication in a primitive cell. In addition, the construction of a temperature responsive membrane that enables polynucleotide passage can advance the design of vesicle bioreactors or sensors that utilize short oligonucleotides as substrates.

# 433902 Astaxanthin Encapsulation By Mpeg-PCL Self-Assembled Polymeric Micelle

**Abstract:**

Astaxanthin is a lipid soluble compound widely used as a food colorant, nutritional supplement, antioxidant, and therapeutic agent. The poor water solubility has limited the bioavailability and application of astaxanthin. The purpose of this study is to increase the aqueous solubility of astaxanthin by encapsulation into self-assembled methoxypolyethylene glycol-poly( -caprolactone) (MPEG-PCL) polymeric micelle. Astaxanthin loaded MPEG-PCL diblock copolymer was directly dissolved in water assisted by ultrasound. The obtained astaxanthin encapsulated micelle solution was characterized by UV-Vis, FT-IR, dynamic light scattering, and gel permeation chromatography. The release profile of astaxanthin-loaded MPEG-PCL was also evaluated. As a result, astaxanthin was successfully encapsulated into biocompatible MPEG-PCL micelles. It can be concluded that encapsulation into MPEG-PCL micelle can be used as an effective strategy to enhance the bioavailability and delivery of astaxanthin.

# 433516 A Ratiometric Sensor Using Single Chirality Near-Infrared Fluorescent Carbon Nanotubes: Application to in Vivo Monitoring

**Abstract:**

The development of ratiometric fluorescent SWCNT sensors has been limited by a lack in techniques to obtain preparative scales of chirality-separated single walled carbon nanotubes (SWCNT). Recent advances in SWCNT separation and functionalization [1] have enabled the creation of a near-infrared ratiometric optical sensor. Herein, we use single chirality-separated SWCNT, independently functionalized to recognize either nitric oxide (NO), hydrogen peroxide ($H_2O_2$), or no analyte (remaining invariant), to create sensor responses from the ratio of distinct near-infrared emission peaks. This ratiometric approach provides a measure of analyte concentration which is invariant to the absolute intensity emitted from the sensors, and hence, more stable to external noise and detection geometry. We specifically demonstrate two distinct ratiometric sensors functionalized for specificity based on Corona Phase Molecular Recogntition [2]. Each sensor is responsive either to $H_2O_2$, or to NO, via selective quenching of 1100 nm emission, with an invariant 1000 nm emission wavelength as the optical reference. To synthesize these ratiometric sensors from chirality-separated SWCNT, we introduce a method for rapid and efficient coating exchange of single chirality sodium dodecyl sulfate (SDS)-SWCNT. As a proof of concept, spatial and temporal patterns of the ratiometric sensor response to $H_2O_2$, and separately NO, were monitored in leaves of living plants in real time. This ratiometric optical sensing platform can enable the detection of trace analytes in complex environments such as strongly scattering media and biological tissues [3].

1. Jain, R.M., et al., *Quantitative Theory of Adsorptive Separation for the Electronic Sorting of Single-Walled Carbon Nanotubes.* Acs Nano, 2014. **8**(4): p. 3367-3379.

2. Landry, M.P.*, Zhang, J.Q.*, et al., *Molecular recognition using corona phase complexes made of synthetic polymers adsorbed on carbon nanotubes.* Nature Nanotechnology, 2013. **8**(12): p. 959-968.

3. Giraldo, J.P., et al., *Plant nanobionics approach to augment photosynthesis and biochemical sensing (vol 13, pg 400, 2014).* Nature Materials, 2014. **13**(5): p. 530-530.

# 425944 Discovery of Novel Strong Yeast Promoters for Refactoring Fungal Natural Product Gene Clusters

**Abstract:**

Refactoring fungal natural product gene clusters in *Saccharomyces cerevisiae* represents an attractive strategy for discovery of novel natural products. Robust function of these gene clusters involving large proteins like polyketide synthases, fatty acid synthases, and non-ribosomal peptide synthetases especially depends on strong expression levels to drive production forward. Previous advances in yeast promoter engineering have focused on improving a limited set of well-characterized promoters, but few efforts have been applied to expand the diversity and range of promoters available. We have taken a two-pronged approach to identify new promoters for use in *S. cerevisiae* that extend the range of promoter strength and expression timing. Utilizing a yeast-optimized destabilized GFP reporter system, we leveraged preexisting expression data to identify several potent, previously unutilized native yeast promoters. Among them, the promoter for phosphoenolpyruvate carboxykinase (PCK1) displays expression 4-fold higher than the strongest reported constitutive native yeast promoter, TEF1p, and also 1.5-fold higher expression than the most commonly used inducible promoter, ADH2p. Hypothesizing that the vital role of PCK1 in catabolism being conserved among yeasts would lead to similar promoter strengths in related species, we cloned 17 promoters controlling the orthologous genes from various yeast species. Seven of them were found to be active in *S. cerevisiae* and maintained relatively similar expression profiles with varying strengths, resulting in a library of sequence-varied promoters for concerted expression of genes. We have also demonstrated the utility of these promoters by overexpressing multiple heterologous pathways to increase the titer of the corresponding natural products.

# 433216 Gene-Centric Discovery of a Novel Coumarin Biosynthetic Pathway in Higher Plants

**Abstract:**

As sessile organisms, plants have evolved a remarkable array of specialized biochemical pathways in order to interact with and manipulate their environment. The products of specialized metabolism have been a rich source of bioactive molecules used for therapeutic and agrochemical applications, among others. However, the discovery and utilization of novel plant biochemistries has traditionally been a slow process.

Here, we present one instance of the application of a "gene-centric" approach for pathway discovery in plants. Leveraging large amounts of freely available genomic and transcriptomic data enables us to make predictions about the function of biosynthetic genes of interest, while metabolomic analysis of plant tissue allows for the direct testing of such predictions, leading to marked acceleration of the pathway discovery process. In particular, we report the elucidation of the biosynthesis of sideretin, a novel, highly oxidized coumarin with unusual redox properties involved in soil iron uptake. Starting with a single candidate gene from the cytochrome P450 family in the model plant *Arabidopsis thaliana*, we have systematically uncovered the biosynthesis of sideretin from phenylpropanoid precursors, and have managed to reconstitute the pathway in a heterologous host as well as *in vitro*. We have further shown that this pathway is widespread among evolutionarily distant clades of flowering plants, and therefore represents a conserved strategy for iron uptake, with implications for engineering crops with enhanced nutritional uptake capability. Finally, we present a chemical synthesis of sideretin from commercially available precursors and its detailed physical and chemical characterization. Overall, our work demonstrates the power of a gene-centric approach for understanding and harnessing plant specialized metabolism.

# 415836 Enabling Selection in Directed Evolution of Enzymes Via Cellular Engineering

**Abstract:**

Directed evolution has been used extensive to engineer enzymatic activity for the synthesis of fuels, chemicals, and bioactive compounds. The process of directed evolution consists of an iterative process of creating mutant libraries and choosing desired phenotypes through screening or selection until the enzymatic activity reaches the desired goal. Screening is accomplished through linking the desired property to some visual output using colorimetric or fluorometric assays, while selection links the desired property to survival of host cell. Between the two, selection is more desirable because it offers far higher throughput capabilities with lesser effort. Furthermore, in screening, every mutant in the library needs to be examined individually for the desired property, whereas using a selection enables elimination of the majority of undesirable candidates. However, selection is harder to implement since it requires identification of a unique selection mechanism for every given directed evolution undertaking. That is, given an enzymatic reaction, the challenge is to automatically identify a consumption pathway from the desired product to a metabolite native to the host cell. In our prior work, presented at International Workshop on Bio Design Automation, June 2014, we developed an algorithm, ASF (Automated Selection Finder), for constructing a selection pathway utilizing metabolites and reactions catalogued in a database (e,g., KEGG). Here, we improve on our earlier work by providing a systemic way to engineer the host cell to maximize the yield through the selection pathway.

Given a selection pathway, we describe in this work a computational method for coupling the pathway with cell survival. Our method aims to maximize the yield of the selection pathway. To reach this goal, our method restricts carbon uptakes, making the selection pathway the only possible carbon source within the host. The added pathway thus becomes essential for producing cellular biomass. Additionally, our method identifies possible knockout targets to improve the yield through the pathway. To assess the effectiveness of our approach, we applied our method to construct selection pathways for several beneficial metabolites including xylitol, aniline, methanol, and D-ribulose-1,5-bisphosphate. We used *Escherichia coli* as the host organism. For all compounds except aniline, high-consumption pathways were identified. Further, for xylitol and D-ribulose-1,5-bisphosphate, our method identified high-consumption pathways that were experimentally validated and confirmed in the literature.

# 428958 Cheminformatic Elucidation of Enzymatic Substrate Promiscuity

**Abstract:**

Enzymatic substrate promiscuity—the ability of an enzyme to perform a singular chemical reaction on several different substrates—presents an opportunity to expand the scope of existing biosynthetic pathways to include more efficient, direct routes to chemicals in common laboratory and industrial organisms and processes. To date, promiscuous enzymes are currently characterized for substrate specificity against a narrow distribution of chemicals that may not be representative of the enzyme's versatility. This limitation exists largely because activity assays are slow and expensive, in spite of the sheer number of chemicals available for purchase (~$10^7$ according to the ZINC database). While testing large numbers of chemicals may be impractical, computational approaches may allow for existing data to be used more effectively and new data to be collected more strategically. In this instance, machine learning, and the support vector classifier (SVC) in particular may be a useful tool to predict whether a given compound will have observable activity with a given enzyme. In our current study, we develop SVCs for predicting substrate activity in addition to an SVC-based active learning algorithm that can be used to select batches of optimal compounds to test against enzymes of interest. Cross-validation of the constructed SVCs shows that existing data can predict the activity of untested compounds closely related to the tested compounds, but cannot accurately predict the activity of compounds that are more distantly related to the existing training set. The active learning algorithm addresses this shortcoming by selecting samples that can improve global accuracy with fewer experiments. The application of these algorithms to wide arrays of metabolic enzymes would result in a library of SVCs that can predict high-probability promiscuous enzymatic reactions and could prove a valuable resource for the design of novel metabolic pathways.

# 432397 High Throughput Screening Assay for Alcohol Acetyl/Acyltransferase

**Abstract:**

The biological synthesis of esters has recently been the focus of a number of metabolic engineering efforts. Long chain waxy esters have potential as diesel fuel substitutes and volatile esters are valuable as natural food additives (e.g. isoamyl acetate and phenyl ethyl acetate) and industrial solvents (e.g. ethyl acetate). The biosynthetic pathways that produce these compounds share a common terminal reaction step, the condensation of an alcohol with an acetyl- or acyl-CoA by alchohol-O-acetyltransferase (AATase, E.C. 2.3.1.84 and 2.3.1.74). With respect to enzyme engineering, rational design and directed evolution approaches to improve AATase activity have been limited due to the low throughput nature of gas chromatography (GC) and high performance liquid chromatography (HPLC). To address this bottleneck, we developed a high throughput screening assay to evaluate AATase activity. The spectrophotometric assay can be performed in 96-well plate format and is applicable to high throughput screening ($Z \geq 0.5$). AATases orthologs from Saccharomyces and non-Saccharomyces yeast and also fruit species were screened with various short- and medium-chain acyl-coA and ethanol as substrates. A key finding from these screening experiments was the identifications of a high activity AATase from tomato fruit. The developed assay is robust and easy to use, thus providing new tool for rational design and directed evolution studies of AATases.

# 430687 Crystallograhic Investigation of Imidazolium Ionic Liquid Effects on Enzyme Structure and Inactivation

**Abstract:**

Owing to their unique and diverse molecular features, ionic liquids (ILs), which are liquid salts, have been the subject of immense focus as solvents for a broad spectrum of biocatalytic reactions. Despite this immense focus, there remains a critical need to develop rational approaches to engineer enzymes for improved activity and stability in the presence of ILs. We have recently engineered a quadruple mutant of lipase A (lipA) from *Bacillus subtilis* that has enhanced stability towards the IL 1-butyl-3-methylimidazolium chloride ([BMIM][Cl]) via mutation of residues that alter the surface charge of the enzyme. To understand the molecular basis for the improved stability, as well as, more generally, how ILs interact with enzymes on the structural level, we have for the first time determined the x-ray crystal structure of an enzyme in the presence of ILs. Specifically, we have solved the x-ray structure of wild-type and mutant lipA crystals soaked in aqueous solutions with different concentrations (0, 5, 10, 20 % v/v) of [BMIM][Cl] at resolutions between 1.2-1.9 Å. Notably, the concentration of [BMIM][Cl] in the soaking solution had little effect on the tertiary structure of either wild-type or mutant lipA. However, significant differences were observed in the regions of Y49 and G158, which were mutated in the IL-tolerant lipA, including differences in the occupancy of [BMIM] ions. Additionally, the binding modes of [BMIM] suggested the interaction of [BMIM] with the enzyme was driven by hydrophobic interactions with the alky tail of the cation. Interestingly, binding of the cation via cation-pi stacking, which may be synergistic with interactions with the hydrophobic alkyl tail, were also observed. While painting a detailed picture of the interaction of ILs with enzymes, these results ultimately provide unprecedented insight into the mechanism of IL-induced denaturation of enzymes.

# 425815 Engineering Bacterial Microcompartments for Use As Biosynthetic Nanoreactors

**Abstract:**

Bacterial microcompartments are subcellular bacterial organelles bound by a porous protein membrane. Enteric and other bacteria use these organelles to sequester metabolic pathways to enhance growth and pathogenicity. We aim to develop a tunable nanobioreactor system based on these protein structures to spatially organize heterologous biosynthetic pathways in bacterial hosts. To this end, we develop tools to control both the formation of the microcompartments and their enzymatic content.

By engineering the transcriptional regulation of the microcompartments, we can orchestrate their formation and exercise control over the loading of heterologous enzymes to the nanobioreactor, tuning the catalytic activity of the microcompartments. We also engineer the N-terminal signal sequences responsible for enzyme targeting to the compartments, creating new signal sequences based on a common hydrophobic motif and allowing fine control over the loading phenotype of the reactors. Our results have implications for the controlled encapsulation of heterologous synthetic pathways to control enzyme stoichiometry and small molecule transport.

# 433233 Sequence-Specific Antisense Inhibitors of Antibiotic Resistance and Mechanism of Adaptive Resistance

**Abstract:**

The recent surge of drug-resistant superbugs and shrinking antibiotic pipeline are serious challenges to global health. In particular, the emergence of β-lactamases has caused extensive resistance against the most frequently prescribed class of β-lactam antibiotics. Furthermore, resistance has also developed to the β-lactam/β-lactamase inhibitor combinations due to extended spectrum β-lactamases such as carbapenemases and New Delhi metallo-β-lactamase 1, providing another avenue for widespread antibiotic resistance. Here, we develop novel synthetic peptide nucleic acid based antisense inhibitors that target the start codon and ribosomal binding site of the TEM-1 β-lactamase transcript and act via translation inhibition mechanism. We show that these antisense inhibitors are capable of re-sensitizing drug-resistant *Escherichia coli* to β-lactam antibiotics exhibiting 10 fold reduction in the minimum inhibitory concentration (MIC). To study the mechanism of resistance, we adapted *E. coli* at MIC levels of the β-lactam/antisense inhibitor combination and observed a non-mutational, bet-hedging based adaptive antibiotic resistance response as evidenced by phenotypic heterogeneity as well as heterogeneous expression of key stress response genes. Mutants that developed tolerance to the RNA inhibitor/β-lactam combination exhibited heterogeneous expression of representative stress response genes corresponding to the multiple antibiotic resistance (mar) regulon, general stress response, and SOS response; implying role of gene expression heterogeneity in development of resistance. Our data shows that both the development of new antimicrobials and understanding the cellular response during development of tolerance to could aid in mitigating the impending antibiotic crisis.

# 433589 Rapid, High-Throughput Screening of Proteins with Non-Canonical Residues Using Cell-Free Protein Synthesis

**Abstract:**

The site-specific incorporation of non-canonical amino acids (ncAA) can enhance protein characteristics towards improved therapeutics, medical diagnostics and biocatalysts. However, predicting the characteristics of proteins with such novel residues is still difficult. Simulation methods, which often depend heavily on homology, are ill-suited to deal with proteins containing novel residues. Conventional *in vivo* screening methods require a plasmid DNA library, whose cloning-based techniques are time- and labor-intensive. Therefore, there is an unmet need for a rapid, high-throughput screening system to probe the characteristics of proteins with ncAA residues.

Cell-free protein synthesis is an attractive alternative platform for ncAA incorporation due to the open and accessible nature of the reaction environment. Further, cell-free systems can express proteins from PCR-produced linear templates, and thus can avoid cumbersome cloning steps. Here, we engineer a screening technique that coupling PCR-based library generation and cell-free protein synthesis. Unique ncAA incorporation sites in a model biocatalyst were screened in parallel to optimize enzymatic activity and functional stability, while significantly reducing labor time compared to popular *in vivo* methods.

# 434703 Methods and Application of Ligand Charge Engineering Towards Improved Tumor Detection

**Abstract:**


**Methods and Application of Ligand Charge Engineering towards Improved Tumor Detection**

Brett A. Case, Nicole Olson, Benjamin J. Hackel

Small, engineerable protein scaffolds – such as affibodies and fibronectin domains – have been demonstrated as effective targeting agents for molecular imaging and therapy. These affinity ligands, modifiable through directed evolution towards numerous targets, have favorable pharmacokinetic properties for select applications. Small size drives rapid clearance from the blood and healthy tissue making them suitable for imaging with short-lived positron emitting isotopes including $^{68}$Ga ($t_{1/2}$ = 68 min.) and $^{18}$F ($t_{1/2}$= 110 min.). However off-target uptake and retention, particularly in the clearance organs, hinders the effectiveness of many scaffolds. Such signal prevents comprehensive imaging proximal to these organs and presents toxicity concerns for therapeutic and diagnostic applications. Anecdotal data suggests that there may be a correlation between excretory retention and protein charge, necessitating comparative analysis across ligands of differing charge. However, significant complications can arise when modulating protein charge. As charged residues are exchanged for neutral amino acids, or vice-versa, the protein's native fold may be altered leading to the formation of insoluble aggregates. Additionally, loss of intramolecular structure reduces thermal stability and may be detrimental to target binding dependent upon paratope positioning.

This study aimed to create and analyze protein ligands with reduced charge using an affibody targeted to epidermal growth factor receptor (EGFR) as a basis. The affibody scaffold is a 58 amino acid, three helical bundle based on the B domain of *staphylococcal* protein A, typically diversified for binding at 13 positions on the first and second helices. The EGFR affibody (EA68), a hybrid of an evolved paratope and an evolved framework, exhibited good affinity ($K_d$ = 5.3±1.7 nM) and stability ($T_m$ = 73°C). Analysis of charge mutants emphasized maintaining or increasing thermal stability, recombinant production in the soluble fraction of *E. coli*, and target affinity. Three experimental models were used create mutants based on EA68 and with 3 positive and 3 negative residues replaced by neutral amino acids. Natural homolog sequence frequency guided the design of charge-reduced ligands covering narrow sequence space (11 clones purified from bacterial productions) as well as broad (>650,000 clones via a yeast surface display library). In addition, a consensus design approach was used based on sequence tolerance observed within the functional members of the library.

Charge reduction was generally well tolerated among all models. Soluble recovery yield equaled or surpassed that of parental EA68 in all cases; thermal stability ranged between 48°C and 71°C with a mean of 62°C; and EGFR affinity remained below 20 nM for all but two clones with 8 mutants retaining single nanomolar dissociation constants. Narrow and broad models each provided a preeminent ligand with yields in excess of 5 mg/L, affinities of 6.9±1.4 nM and 1.3±0.3 nM, and stabilities of 72°C and 69°C respectively.

Methods for modifying charge while retaining or improving desirable ligand characteristics including bacterial production yield, thermal stability, and target affinity, particularly in the context of an EGFR-targeting affibody will be discussed. Moreover, application of these results to additional scaffolds will be shared. In addition to providing a thorough data set elucidating the ability to modulate charge within this hybrid affibody, the functional mutants provide a diverse family of charge distributions with which to evaluate biodistribution. Comparison of physiological distribution, plasma clearance, and tumor targeting -- via PET/CT imaging and excised tissue gamma counting -- in a xenograft mouse model will be covered.

# 415215 Thiazoline Ring Formation in an Engineered Intein Serves As a Splicing Switch

**Abstract:**

We have engineered an intein which autocatalytically forms a thiazoline ring at the N-terminal splicing junction and provided the first crystallographic evidence of this chemistry in an intein. The mutation that causes the aberrant thiazoline ring chemistry suggests that the highly conserved B-block Thr plays a pivotal role in orienting the resulting tetrahedral intermediate for resolution into a thioester, giving insight into the splicing mechanism. We demonstrate the stability of the thiazoline ring at mild conditions as well as sensitivity to hydrolytic opening of the ring under acidic conditions and high temperature. The reversibility of this thiazoline "trap" is of significant interest to biotechnological applications requiring a splicing activity switch.

# 423754 Constructing, Evaluating, and Screening Bioconjugates on the Yeast Surface

**Abstract:**

The introduction of non-canonical amino acids (ncAAs) into proteins offers attractive opportunities for engineering new classes of reagents, diagnostics, and therapeutics. We present here a ncAA-compatible yeast surface display platform that enables the construction, evaluation, and screening of bioconjugates on the yeast surface. This work utilizes azide- and ketone-containing antibody constant regions (Fcs) and antigen-binding Fcs (Fcabs) to demonstrate several important aspects of the platform. Binding experiments indicate that ncAA-substituted constructs remain well-folded and, in the case of Fcabs, able to bind to target antigens. Moreover, azide-containing constructs can be covalently modified on the yeast surface under mild conditions while retaining their thermal stability. Importantly, model sorting experiments lead to the efficient enrichment of constructs based on either chemical modification status or antigen binding. Cells displaying azide-containing Fcs can be selectively modified and enriched from an excess of cells displaying Fcs containing only canonical amino acids. Mixtures of successfully constructed conjugates in which only a small portion recognize a target antigen can be enriched for the conjugates exhibiting antigen recognition. The approaches presented here will enable high throughput characterization and screening of libraries of protein-small molecule hybrids for the discovery of therapeutic leads and numerous other biological applications.

# 413337 Rapid Uptake of Fluorescent Peptides into Intact Mammalian Cells Using a b-Hairpin Sequence Motif

**Abstract:**

Fluorescent, peptide-based reporters are an attractive method to quantify intracellular enzyme activity in intact cells due to their relative ease of synthesis and the fact that they do not require complex genetic engineering. While unstructured peptides are often rapidly degraded in the cytosol, structured sequences, such as a β-hairpin 'protectides', resist intracellular degradation which makes them ideal candidates for enzyme reporters. Recent work has verified the utility of incorporating a protectide on the N-terminus of a short peptide substrate to measure enzyme activity.[1] One challenge in the use of peptide-based reporters is the efficient delivery of the peptide into intact cells. While this can be accomplished by external manipulation of the cells, a preferred method is to utilize a cell penetrating peptide (CPP) sequence to facilitate transport across the plasma membrane. This presentation highlights the identification and characterization of a short peptide sequence that incorporates a β-turn motif to act, not only as a protectide, but also as a CPP. The protectide, named OWRWR [Ac-OWVRVpGO(FAM)WIRQ-NH
permeate HeLa cells using both fluorometry and microscopy techniques at temperatures ranging from 4°C-37°C. The stability of OWRWR under conditions similar to the cytosolic environment was verified using a degradation assay and reverse phase high performance liquid chromatography (RP-HPLC). The β-turn sequence was confirmed using circular dichroism (CD). The internalization efficiency of OWRWR was compared to a commercially available CPPs and non-permeating control peptides. Interestingly, it was discovered that the identity of the N-terminal amino acid residue played an essential role in the ability of this protectide to function as a CPP. An ornithine (O) to arginine (R) substitution at the N-terminus completely eliminated the permeability of peptide into intact cells. As such, subsequent characterization of a library of protectide variants was performed to identify the impact of the identity of the positively charged residues in the β-turn sequence. Finally, the protectide was incorporated into the N-terminus of an enzyme substrate to generate a long-lived, cell permeable fluorescent probe for detecting intracellular enzyme activity.

1. Yang, S.; Proctor, A.; Cline, L.; Houston, K.; Waters, M.; Allbritton, N., beta-Turn sequences promote stability of peptide substrates for kinases within the cytosolic environment. *Analyst* **2013,** *138* (15), 4305-4311.

# 415054 Quantum Molecular-Sequencing (QM-Seq): Single Molecule DNA and RNA Sequencing

**Abstract:**

The sequences and structure of nucleic acids in bio-macromolecules like DNA and RNA can define their biological function by controlling the downstream expression of genes, proteins, and other cell-regulatory functions. Small variations in this genetic coding in individual cells can lead to mutations, which can play a key role in physiology. For example, recent data on genetic variation in Ebola virus through its origin and transmission provides valuable information about mutations that altered protein sequences, and new targets for diagnosis, vaccines and therapy. While nanoelectronic sequencing techniques can in-principle be used to distinguish biochemically distinct nucleobases, it relies on identifying unique "*electronic fingerprints*" for different nucleotides. So far only partial deoxynucleotides have been successfully identified for single molecule, in part due to the large conformational entropy of molecules which causes dispersion of electronic spectroscopic signatures. Furthermore, attempts at using magnitude of electronic current, current decay properties, and its derivative ($dI/dV$) have been unable to provide deterministically unique fingerprints for biochemical states of nucleic acids. Here, we present a novel Quantum Molecular-Sequencing (QM-Seq) method which provides on fast nanoelectronic spectroscopy and unique electronic fingerprints of all DNA (A, G, T, C) and RNA nucleotides (A, G, C, U). We will show that QM-Seq spectra allows us to detect the biochemical state of all individual molecules, using facile identification of isomers and tautomer species in nucleic acids. We also demonstrate simultaneous identification of single nucleotide modifications (methylation here), which highlights the potential of QM-Seq as a versatile technique for epigenomics and personalized medicine.

References:

Nagpal, Chatterjee, Ribot, Patent Pending

Ribot, Chatterjee, Nagpal, J. Phys. Chem. B, 119, 4968 (2015)

PRAAN Biosciences Inc.

# 430090 Spatiotemporal Modeling of Laser Tissue Welding of Plasmonic Nanoparticles

**Abstract:**

Occurrences of surgical site infection can be decreased by suture-free techniques that create a rapid hydroseal. Plasmonic nanocomposites composed of a collagen protein matrix embedded with gold nanorods convert near-infrared radiation to fuse apposing tissue in colorectal anastamoses as a suture-free technique known as laser tissue welding. In brief, the exogenous chromophore gold nanorods convert the near-infrared light to heat sufficient to denature and then interdigitate native tissue proteins. The results of a spatiotemporal model of heat transfer in laser tissue welding is presented, supplemented by *ex vivo* experimental results. As high as 68% of native tissue ultimate tensile strength and 48% of burst pressure are retained through laser tissue welding at clinically relevant times and at the lowest laser power densities yet recorded (2.5 $W/cm^2$ and lower). These experimental results are accompanied by a spatiotemporal model of the heat generated by the nanocomposites during laser exposure and how that heat is transferred to the peripheral tissue. The model results suggest that this laser welding technique using collagen-gold nanorod composites can be tuned to minimize thermal damage to peripheral tissue and apposing tissue while still providing a rapidly formed, fluid-tight seal to friable tissue.

# 425761 Nano-Endoscope for Light Manipulation in Single Cells

**Abstract:**

Fundamental understanding of cell biology requires development of new instrument for single cell analysis [1,2]. Nanoscale endoscopes are a new class of instruments that can be inserted into cell for minimally invasive intra-cellular measurements [3]. However, most of such devices are too complex, very expensive to fabricate and operate. Here we report a single-cell endoscope reproducibly fabricated with nanoscale precision. It is automatically controlled with nano-manipulator inserting it in precise regions in individual cells. The endoscope can simultaneously deliver excitation light and collect emitted fluorescent signal from labelled cell organelles. Here we present FDTD simulations for optimization for forward and back coupling of light through the endoscope. Then, we experimentally demonstrate optical signal manipulation using endoscope inserted into cells.

ADDIN EN.REFLIST 1. Walling, M.A. and J.R.E. Shepard, "Cellular heterogeneity and live cell arrays". *Chemical Society Reviews*, 2011. **40**(7): p. 4049-4076.

2. Zheng, X.T. and C.M. Li, "Single cell analysis at the nanoscale". *Chemical Society Reviews*, 2012. **41**(6): p. 2061-2071.

3. Ladanov, M., S. Cheemalapati, and A. Pyayt, "Optimization of light delivery by a nanowire-based single cell optical endoscope". *Optics Express*, 2013. **21**(23): p. 28001-28009.

# 426563 Nano-Photosensitizers for Photochemical Sensitization of Pancreatic Cancer to Chemotherapy

**Abstract:**

It is becoming increasingly clear that the complexity of cancer biology requires well-thought out, mechanistically sound combinations to address the problem of multiple and often compensatory pathways involved in cancer development and progression. Toward this end, we leverage nanotechnology to developed mechanism-guided, photodynamic therapy-based combination treatments. The goal of the current study is to demonstrate such a combination in orthotopic models of pancreatic cancer with photodynamic therapy and a recent clinically successful chemotherapeutic formulation (liposomal irinotecan). In orthotopic pancreatic cancer models, we demonstrate that low-dose photodynamic therapy and nanoliposomal irinotecan, working *via* three-way complementary mechanisms (reduced hypoxia, survivin and ABCG2 transporter) to control cancer compensatory pathways, dramatically inhibit tumor growth and prolong animal survival, offering immediate opportunities for clinical translation of pancreatic cancer treatment. Results from initial studies and mechanistic basis of the observed synergism along with clinical translation potential and a forward looking multi-agent nanoconstruct will be discussed.

# 434221 Photodynamic Therapy of Melanoma By Engineered Titanium Dioxide Nanoparticles

**Abstract:**

Metastatic melanoma has a very poor prognosis, and conventional chemotherapy treatment causes severe toxic response to healthy cells due to lack of selectivity in targeting the tumor cells. The advancement of nanotechnology has opened up avenues for the use of engineered nanoparticles for diagnosis, imaging and targeted treatment of cancer. In this work, we hypothesize that engineered hybrid titanium dioxide ($TiO_2$) nanoparticles (NPs) that possesses tumor targeting peptide ligands and tunable photoactivity will result in a superior cancer therapeutic strategy. $TiO_2$ NPs make them attractive materials for photodynamic therapy of cancer due to their a) tunable size and photoemissive properties b) efficient photoactivity to induce cytotoxicity c) conjugation with targeting ligands and ability to carry drug payloads. Peptides are excellent targeting agents due to a) high potency and selectivity in targeting cancer cells b) tunable physiochemical properties due to high chemical diversity c) enhanced tumor penetration, higher stability and lower antigenicity. We have investigated the toxicity of two size variants of $TiO_2$ NPs (5 and 25 nm) on normal human cells and cancer cells. Our results, indicated that NPs were not cytotoxic to cells in concentration range tested (0.01 - 100 μg/mL). A dose - dependent phototoxicity of UV irradiation was observed and cell death was predominant when UV intensity was greater than 20 $J/cm^2$. UV photoactivated 25 nm particles show enhanced cytotoxic effect compared with 5 nm particles in normal and cancerous cells. Photoactivity measurements indicated 25 nm particles had a higher photocatalytic activity than 5 nm particles. Transferrin receptor mediated pathway was selected for targeting cancer cells. Flow cytometry analysis indicated transferrin receptor was overexpressed in melanoma cells by 2 – 3 fold compared to the normal dermal fibroblasts and keratinocytes. We have designed peptide ligands to target the transferrin receptor overexpressed on cancer cells. These results provide us the opportunity to employ transferrin receptor mediated targeting of melanoma cancer cells using photoactivated titanium dioxide nanoparticles.

# 428576 Characterization of New Classes of Bright and Photostable Fluorescent Nanoconjugates for Biological Imaging

**Abstract:**

Recent advances in optical microscopy have enabled tremendous improvements in the spatial and temporal resolution of fluorescence-based imaging techniques. However, there is a strong need for the development of advanced fluorescent probes to study biological events in living cells, which will enable a molecular-scale understanding of biological processes. To this end, we have developed a new class of fluorescent probes based on dye-conjugated dendrimers called fluorescent dendrimer nanoconjugates (FDNs). In particular, we use polyamidoamine (PAMAM) dendrimers as molecular scaffolds for the attachment of multiple organic dyes and "helper" molecules. We then characterize these probes using single molecule fluorescence imaging and total internal reflection fluorescence microscopy (TIRF-M). We observe that FDNs are much brighter compared to single fluorophores, which allows for high-resolution localization and detection. In addition, we link photoprotectant molecules such as triplet state quenchers (TSQs) directly onto FDNs, thereby generating "self-healing" probes which have enhanced photostability when compared with native fluorescent dyes. In this work, we characterized two separate synthetic strategies for creating self-healing dye conjugates, by calculating important photophysical parameters such as total accumulated photons and stability of single molecule intensity over time. In the first generation of probes, we attached the fluorescent dye Cy5 and photoprotective Trolox molecules to the PAMAM dendrimer, which allowed us to vary the Trolox:dye ratio. This led to an increased stabilization with the higher synthesized Trolox:dye ratios. However, while this strategy allows for control over the Trolox:dye ratio, it does not allow for precise control over the distance between the molecules which has been shown to be an important parameter. Thus, in a second generation of self-healing FDNs, we first covalently attached the Cy5 and Trolox into one dye conjugate, which we then labeled onto the dendrimer in varying amounts. This keeps the Trolox:dye ratio at 1:1, and it also ensures constant close contact between the two species and results in large increases in brightness, total photon output and stabilization of fluorescence. We tested these probes *in vitro* under a variety of different solution conditions to gain insights into the mechanisms of stabilization, along with testing in a cell immunofluorescence experiment as a proof-of-principle demonstration of biological utility. Overall, we anticipate that these new fluorescent probes will enable new insight and mechanistic understanding in molecular and cellular biology.

# 421711 Dispersion of Anti-TB Drugs with Magnetite Nanoparticles to Overcome Mycobacterial Drug Resistance

**Abstract:**

Despite progress in tuberculosis (TB) treatment, emergence of multi-drug resistant (MDR) and extensive-drug resistant (XDR) strains of *Mycobacterium tuberculosis* still remains a global health challenge. Among the various mechanisms responsible for drug resistance, decreased intracellular drug levels due to combination of lowered influx of drugs into the cells (decreased membrane permeability) and increased efflux of drugs outside the cells (over-expression of efflux pumps) plays a very important role in increasing the tolerance of *Mycobacterium* towards the anti-TB drugs. Thus, development of novel strategies to enhance this effective intracellular drug concentration is crucial. In our earlier work, we demonstrated that nanoparticles can be used to enhance the efficacy of anti-TB drugs to target intrinsic resistance in *Mycobacterium smegmatis*, a model mycobacteria. In the present work, we demonstrate for the first time that nanoparticles can be used to overcome the acquired drug resistance in mycobacteria, such as in clinical isolates. To mimic the situation of clinical drug resistance, we have generated *in vitro* hyper-resistant mutant strains of *M. smegmatis*, which confers a very high level of resistance to anti-TB drugs compared to the wild type strain.

Polyacrylic acid coated magnetite nanoparticles (PAA-MNP) were used in combination with anti-TB drugs to enhance the drug efficacy in hyper-resistant mutants of *M. smegmatis*. PAA-MNP's synthesized by thermal decomposition route formed a stable dispersion with an average size of 9 nm. These nanoparticles were found to interact with the mutant strains of *M. smegmatis* via surface attachment and internalization as revealed by Transmission Electron Microscopy (TEM) and Inductively Coupled Plasma–Atomic Emission Spectroscopy (ICP-AES), respectively. Next, we have investigated the effect of these coated nanoparticles alone, drug alone and a combination of coated nanoparticles with drugs on growth profile of the mutants. These nanoparticles were non-toxic to mutants when used alone, thus making them a superior candidate for therapeutic use. A synergistic enhancement in growth inhibition of mutants, upto 16-folds was observed using anti-TB drugs (rifampicin or isoniazid or norfloxacin) in combination with PAA-MNP in comparison to free drug alone.

In order to mechanistically understand the role of coated magnetite nanoparticles in overcoming drug resistance, we have performed extensive biochemical analysis and drug transport studies in the hyper-resistant mutants. As the alteration of drug transport due to permeability barrier and active efflux of drugs plays a major role in development of drug resistance, we have measured the intracellular drug levels in mutants in presence of drug alone and drug in combination with nanoparticles. PAA-MNP's were able to enhance the intracellular levels of various anti-TB drugs in mutants by atleast 4-folds when compared to free drug. Furthermore, the real-time transport kinetics of fluorescent tracers revealed that PAA-MNP enhances the drug uptake by increasing the influx of drugs into the cells while simultaneously inhibiting efflux of the drug out of cells, thereby enhancing the intracellular drug concentration.

Altogether, these results suggest that coated magnetite nanoparticles can act as a new class of permeability enhancers and efflux inhibitors, which in combination with anti-TB drugs, can effectively target drug resistant mycobacteria.

**(275a) Nanosensors-Encapsulated Microcapsule for Glucose Sensing**

Xie, X., Massachusetts Institute of Technology
Zhang, W., Harvard University
Abbaspourrad, A., Harvard University
Weitz, D. A., Harvard University
Anderson, D., Hope College

Originally presented on: 11/10/2015 8:30:00 - 8:52:00

**Non-invasive implantable systems that are able to continuously track blood glucose levels are needed for diabetes diagnostics and monitoring. Although many nanomaterial colloidal solutions have been developed as optical sensors to detect glucose in aqueous samples, the liquid nature of these sensor solution limits their applications as implantable devices for glucose sensing in vivo. One potential solution is embedding the nanosensors in biocompatible hydrogel matrix, yet the direct contact between the sensor surface and the polymer chains of the hydrogel tends to disrupt the performance and sensitivity of the sensors. In this work, glucose responsive nanosenors were encapsulated in liquid-core microcapsules. Glucose is allowed to diffuse through the capsule shell freely, while the nanosensors are trapped inside the capsules without leaking away. The interaction between glucose and the nanosensors produces detectable optical signals. The microcapsules were fabricated by forming water-oil-water template double emulsion using microfluidic techniques. The capsules' stability and permeability to glucose were demonstrated. Three types of glucose-responsive nanosensors, glucose oxidase (GOx)-functionalized quantum dots, GOx-functionalized gold nanorods, and boronic acid-functionalized carbon nanotubes, were encapsulated in the microcapsules individually, and they responsively exhibited optical signals in the present of glucose. Furthermore, implantable devices were fabricated by embedding these capsules in calcium alginate hydrogel, and their responses to glucose were demonstrated.**

# 430221 Protein Efflux Mapping in Single Living Cells with Synthetic Optical Nanosensors

**Abstract:**

Recent advances in metabolic engineering of *Escherichia coli* have provided an abundance of platforms for natural product biosynthesis and protein export [1]. However, real-time optical detection of *E. coli* metabolites remains a challenge. Previous work has shown great promise for the use of near-infrared fluorescent nanomaterials for optical detection of biomolecules [2, 3]. We present a platform for near-infrared optical detection of proteins exported from *E. coli*. By electrostatic conjugation of an aptamer-anchor polynucleotide sequence to a single-wall carbon nanotube, we are able to detect both RAP1 GTPase and HIV-1 integrase proteins, which each produce increases in near-infrared fluorescence of their respective aptamer-SWNT conjugates. Herein, we engineer the RAP1 gene, which is native to yeast, into *E. coli*. Subsequently, we monitor the productivity of RAP1 protein export by individual engineered *E. coli* cells over the course of 1 hour. We show that the process of induction, protein synthesis, and protein export is a highly stochastic process yielding variability in protein secretion on a single-cell level. We further show that our platform can be extended to the detection of protein from *E. coli* infected with T7 bacteriophages carrying the RAP1 gene. Our platform shows promise of extension for single-cell optical detection of a broad range of metabolic products with high spatial and temporal resolution.

1. Keasling, J.D., *Synthetic biology and the development of tools for metabolic engineering*. Metabolic Engineering, 2012. **14**(3): p. 189-195.
2. Landry, M.P.*, Kruss, S.*, et al., *Neurotransmitter Detection Using Corona Phase Molecular Recognition on Fluorescent Single-Walled Carbon Nanotube Sensors*. Journal of the American Chemical Society, 2014. **136**(2): p. 713-724.
3. Landry, M.P.*, Zhang, J.Q.*, et al., *Molecular recognition using corona phase complexes made of synthetic polymers adsorbed on carbon nanotubes*. Nature Nanotechnology, 2013. **8**(12): p. 959-968.

# 433673 Advances in SERS Based Point of Care Diagnostic Assays

**Abstract:**

In the development of next generation bioassays designed for detecting the emergence and spread of diseases in point-of-care (POC) and remote settings, more rapid and portable analytical methods are required. Colloidal based SERS bioassays have distinct advantages for use in portable diagnostic systems over surface sensors. Nanoparticles in solution allow for both a freely diffusing sensor surface and a high surface area to volume ratio. Therefore colloidal systems are not impacted by mass transfer limitations, allowing for the rapid analysis of small sample volumes. In addition, paramagnetic capture assay schemes provide both separation and concentration mechanisms, minimizing signals from non-specific interactions. Recent developments in our laboratory in the fabrication of gold-coated magnetic nanoparticles provide an additional alternative single nanoparticle assay scheme. The nanoparticles function as SERS substrates while simultaneously allowing for manipulation by a magnetic field. Detection strategies have been demonstrated for the sensitive detection of Rift Valley Fever virus and West Nile virus nucleic acids and proteins in multiplexed formats. In addition, we have demonstrated the advantages of using antibody fragments over IgGs as a more compact recognition element in our immunoassay schemes.

# 434940 Functionalization of Nanostructured ZnO Surfaces with Selective Linker Molecules Towards Biomolecule Detection

**Abstract:**

In this study, the influence of surface functionalization on charge transfer in nanostructured zinc oxide (ZnO) surface was investigated. Sensing sites with tailored surfaces offer novel immobilization strategies for selective functionalization and specific interaction of protein biomolecules, which can lead to enhancement in specific output signal response and hence selectivity. Functionalization of nanostructured ZnO surfaces were evaluated using two organic molecules with thiol and phosphonic acid functional groups. Dithiobis(succinimidyl propionate) (DSP) and 11-Aminoundecylphosphonic acid (AUPA) covalently link to ZnO providing NHS binding sites for controlled immobilization of antibody. Fluorescence intensity measurements with varying concentrations of DSP and AUPA labeled with Rhodamine B were performed, to identify the amount of functionalization. We observed saturation in fluorescence intensity with increasing concentration of linker molecules. Such modified ZnO surfaces were characterized using contact angle measurements, X-ray photoelectron spectroscopy (XPS) and electrically using electrochemical impedance spectroscopy (EIS) and Mott-Schottky (MS) analysis. XPS spectra revealed selective binding of thiol to zinc interstitials and phosphonic to oxygen vacancies on nanostructured ZnO. To demonstrate the potential of functionalized surfaces for biomolecule detection, cardiac troponin-T, a protein biomarker specific for acute myocardial infarction was chosen. The influence of selective functionalization on charge transfer and perturbations at ZnO electrode-electrolyte interfaces due to biomolecular binding was studied using EIS and MS analysis. Our results demonstrate that leveraging zinc interstitials increases the sensitivity of biomolecule detection over larger dynamic range and detection limit at 10 fg/mL.

# 433015 Using Phage and Phage Genomes for Generating Highly Amplified Signals for Protein Sensing

**Abstract:**

For early detection of many diseases, it is critical to be able to diagnose small amounts of biomarkers in blood or serum. One of the most widely used sensing assays is the enzyme-linked immunosorbent assay (ELISA), which typically uses detection monoclonal antibodies conjugated to enzymes to produce colorimetric signals. In order to increase the overall sensitivities of these sensors, in the first half of the talk, I will describe our use of a dually-modified version of filamentous bacteriophage that produces significantly higher colorimetric signals in ELISAs than what can be achieved using antibodies alone. Since only a few proteins at the tip of the micron-long bacteriophage are involved in antigen binding, the ~4000 other coat proteins can be augmented - either by chemical functionalization or genetic engineering - with hundreds to thousands of functional groups. In this work, we demonstrate the use of bacteriophage that bear a large genomic fusion which allows it to bind specific antibodies on coat protein 3 (p3) and multiple biotin groups on coat protein 8 (p8) to bind to avidin-conjugated enzymes. In direct ELISAs, the anti-rTNFa conjugated bacteriophage show about 3 to 4-fold gains in signal over that of anti-rTNFa, demonstrating their use as platforms for highly sensitive protein detection.

In the second half of the talk, I will highlight our recent efforts of using the genetic information of the M13 bacteriophage to generate highly amplifiable signals in a single solution isothermally. While one method for producing amplified signals in the presence of antigens has been to use oligonucleotide-conjugated antibodies followed by real-time or isothermal PCR to amplify the DNA strands, the DNA conjugation methods used can affect an antibody's ability to bind its protein target, and many rounds of linear amplification are often required for sufficient signal generation. In contrast, because the viruses naturally contain a large, 9.5 kilobase genome, combining isothermal rolling circle amplification (RCA) with multiple priming sites can yield statistically significant positive signals at sub-femtomolar antigen concentrations. The phage therefore provides a convenient container to carry the detection genome to the analyte. I will show that since the genome can be released without or minimal heat denaturation of the phage, and that the number of DNA amplification steps (primer annealing, extension, and dehybridization) are limited, massive amounts of DNA can be detected in real time by luminescence in a much shorter time, in a single reaction tube, at room temperature, and without a specialized PCR instrument. I will also show that the virus genome can be easily modified with restriction sites at different locations so different analytes can be identified through digestion with a single restriction enzyme and gel electrophoresis.

# 413030 Multiplexed Biomarker Detection Using Phase Change Nanoparticles

**Abstract:**

Many biomarkers are specific for early stage cancer detection, but no individual biomarker is ideal to distinguish lethal cancer from indolent ones due to lack of tumor specificity. Parallel to efforts of finding specific biomarkers, a feasible way of providing better predictive value is to detect multiple biomarkers from a sample and collectively assess response pattern of more biomarkers. But, existing techniques such as microarrays and nanoparticles are not suitable to detect multiple low concentration biomarkers in a tiny amount of sample due to low sensitivity, low multiplicity and extensive sample preparation. We have detected multiple DNA and protein biomarkers with a panel of phase change nanoparticles of metals and alloys. Nanoparticles attached on a substrate through ligand-biomarker interactions are readout using differential scanning calorimetry, where the peak position and area reflect presence and concentration of biomarkers. This research is innovative for several reasons. (1) It uses an unexplored phenomenon of solid materials, i.e., the temperature of a solid will not rise above its melting point until the entire solid is molten. (2) Sharp melting peak, large thermal scan range and wide choices of materials enhance multiplicity this detection without adding system complexity. (3) The surface grafting density of ligands on nanoparticles is controlled to detect multiple biomarkers with concentrations differing several orders of magnitude.

We have identified ten different metals that form binary eutectic alloys among any two of them are identified from phase diagram database. These metals (i.e., aluminum, bismuth, cadmium, copper, gadillium, indium, lead, magnesium, palladium, silver) can form 45 binary eutectic alloys, 120 ternary eutectic alloys, 210 quaternary eutectic alloys, and so on. The total number of metals and eutectic alloys are 1,023. Pandat 8 software (CompuTherm) has been used to derive compositions of ternary and higher order eutectic alloys formed by these ten metals. Two thermophysical properties (melting temperature and latent heat of fusion at eutectic composition) had been extracted from data. After obtaining eutectic compositions, nanoparticles had been made by thermally decomposing of organometallic precursor in a high boiling point solvent (ethylene glycol) by using polyvinyl alcohol (PVA) as surfactant. We have used these nanoparticles to detect proteins and single strand DNA (ssDNA).

# 424157 Selective Vaporization of Superheated Nanodroplets for Rapid, Sensitive Acoustic Biosensing

**Abstract:**

This work describes a simple, scalable ultrasound imaging detection method in which aggregation of superheated droplets allows detection of small concentrations of biomolecules without the need to wash out excess reagent. While many contrast agents can be functionalized with ligands that can bind membrane-bound targets, fewer agents are able to sense small quantities of soluble, nonreactive biomarkers and relay their location using a modality capable of deep-tissue imaging. The mechanism of biosensing in this work derives from the discovery that biomolecule-specific aggregation of superheated nanodroplets greatly changes their ability to be vaporized into microbubbles. While in their liquid form the nanodroplets possess poor acoustic contrast properties owing to their incompressibility, these superheated nanodroplets can be vaporized using High Intensity Focused Ultrasound (HIFU), resulting in the generation of high-contrast microbubbles in situ. We show that even small-order aggregates of droplets joined via biotin-streptavidin interactions can specifically be vaporized using HIFU, thus allowing sensing of analytes with a dynamic range of 100 fM to 100 nM. Because of the bulk droplet aggregation mechanism of detection, full activation of signal can be achieved within 10 min. Finally, this work provides evidence of a mechanism for aggregation-based droplet vaporization that will aid the design of future ultrasound contrast agents and biosensors for in vivo imaging.

# 416924 Preparation of Enteric Capsules Using a Fugitive Salt

**Abstract:**

Preparation of Enteric Capsules Using a Fugitive Salt

Jin Zhao[1], Roland Adden[2], Robert Schmitt[1], Wes Spaulding[1], Uma Shrestha[1], Debbie Holbrook[1], Josh Imperial[1], Shrikant Khot[1]

(1) Pharmaceutical Excipients R&D, Dow Pharma & Food Solutions, Midland, MI 48674

(2) Food & Nutrition R&D, Dow Pharma & Food Solutions, Bomlitz, ND 29699, Germany

To protect acid sensitive active pharmaceutical ingredients (API), current technology typically involves coating multi-particulates, capsules or tablets with a known enteric polymer. Such coating processes usually involve organic solvents that are flammable, toxic, and explosive, or involve aqueous dispersions that are hard to handle, tend to block coating nozzles and have poor stability(*1-3*). To overcome these drawbacks, films and capsules were directly prepared using aqueous enteric solutions based on a fugitive salt.

Films and capsule shells were prepared using aqueous hypromellose acetate succinate (HPMCAS) solutions based on a model fugitive salt, ammonium bicarbonate. The rheology of such aqueous solutions was adjusted to fit manufacturing requirements by varying the polymer concentration and the addition of triethyl citrate. The resulting films demonstrated comparable mechanical strength to films made from organic solutions. The API diffusion across the film also demonstrated acid resistance that satisfied USP enteric requirement.

HPMCAS capsules were produced by a dipping process where the aqueous solution with model fugitive salt enhanced the dipping and drying process (Figure 1a). The residual by-product of the neutralization reaction such as $H_2CO_3$ can act as plasticizer via association with the succinic groups – C(O) – R – COOH. When the aqueous composition is dried on the molding pin, the residual ammonium bicarbonate, decomposes at the drying temperature. Accordingly, the evaporation of $NH_3$, $CO_2$ and $H_2O$ facilitates gelation of the polymer on the mold pin resulting in a homogeneous film. The resulting dried capsule does not comprise substantial amounts of residual alkaline material. Even though the gel strength of the aqueous HPMCAS solution is significantly lower than that of HPMC solutions, the low gelation temperature and quick formation of the HPMCAS film leads to a straightforward capsule manufacture technique advantaged over the current state of the art (*4*). The dissolution of the enteric capsules and subsequent API release also demonstrated the necessary acid resistance to satisfy USP enteric requirements (Figure 1b).

The findings from this research can help pharmaceutical scientists utilize truly enteric capsules as opposed to enteric coated conventional capsules. This can protect APIs from degradation in the acidic gastric environment, prevent irritation of the gastric mucosa that often leads to ulceration, and achieve delayed and targeted release in the small intestine.

(a)

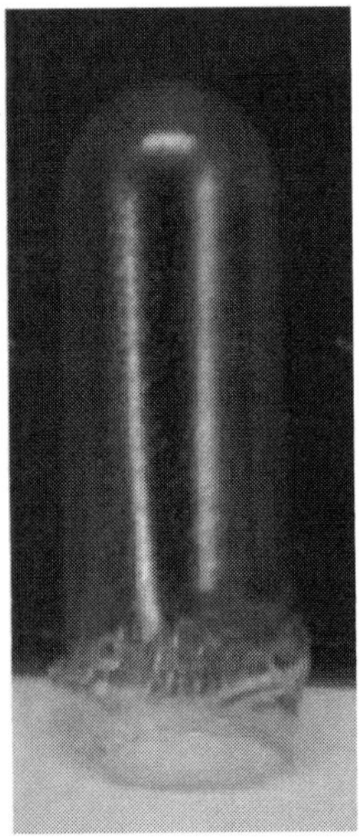

(b)

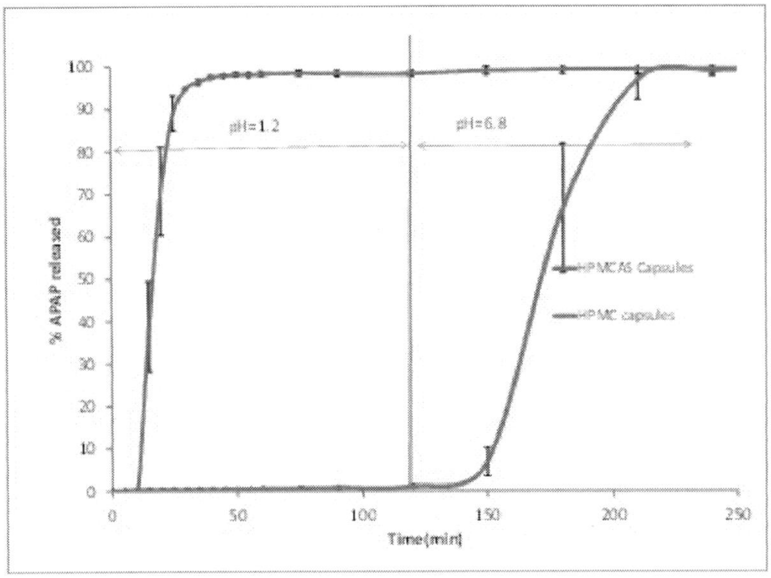


Figure 1. Enteric capsules made from HPMCAS aqueous solutions (a) and demonstrated enteric performance acetaminophen loaded capsules (b).

**Acknowledgements**

The authors wish to thank Shrikant Khot for his insights and suggestions throughout this research.

**References**

1. M. A. Osman, R. B. Patel, D. S. Irwin, P. G. Welling, Absorption of theophylline from enteric coated and sustained release formulations in fasted and non-fasted subjects. *Biopharmaceutics & drug disposition* **4**, 63-72 (1983).

2. M. A. Osman, R. B. Patel, D. S. Irwin, P. G. Welling, Absorption of theophylline from enteric coated and sustained release formulations in fasted and non-fasted subjects. *Biopharmaceutics & drug disposition* **4**, 63-72 (1983)

3. P. Mura, F. Maestrelli, M. Cirri, M. L. Gonzalez Rodriguez, A. M. Rabasco Alvarez, Development of enteric-coated pectin-based matrix tablets for colonic delivery of theophylline. *Journal of drug targeting* **11**, 365-371 (2003).

4. D. Cade, H. straub. (Capsugel France SAS, WO2013/164121).

# 419844 High-Throughput Polymer Screening for Oral Drug Delivery: From Discovery to Design

**Abstract:**

Spray-dried dispersions are fascinating polymer–drug mixtures that exploit the amorphous state of a hydrophobic drug to dramatically elevate its aqueous solubility above equilibrium for oral administration. However, these materials have limited fundamental understanding of what polymeric attributes prolong drug supersaturation and combat crystallization in solution. We present an automated protocol to combinatorially explore a vast architectural parameter space for five polymer platforms and elucidate how specific microstructures dictate drug dissolution. Over 60 reversible addition-fragmentation chain transfer (RAFT) polymerizations were conducted in three high-throughput experiments, varying constituent properties such as polymer amphilicity, thermoresponsive phase behavior, and hydrogen bonding capability. The intermolecular interactions of well–defined lead compounds with model drug phenytoin were screened and studied in vitro. This study reports the first results of a polymer–focused screening assay that successfully solubilized an otherwise-intractable drug for oral delivery.

# 429258 Improving Drug Delivery By Employing Terahertz Technologies to Monitor Manufacturing Processes


**Abstract:**

Terahertz spectroscopy and imaging are powerful new techniques for monitoring of pharmaceutical manufacturing processes. A number of parameters that critically affect drug delivery are able to be measured and continuously evaluated by these methods. These include bulk physical properties such as density, porosity, solid fraction and hardness as well as chemical properties such as crystalline/amorphous state of an API or blend uniformity. Furthermore, the unique ability of terahertz light to penetrate deeply into opaque materials allows one to determine the thickness and uniformity of functional coatings on tablets.

Such methods are easily integrated into the pharmaceutical manufacturing process. Their rapid and non-destructive nature enables them to be employed in-line to assess and monitor key Critical Quality Attributes (CQAs) and maintain product quality.

This talk gives an overview of a number of applications where CQAs critical to effective drug delivery have been measured using these techniques. These include:

- Continuous, in-line monitoring of roller compaction ribbon density
- Measurement of tablet porosity and hardness and their correlation to dissolution properties
- Rapid and precise quantification of percent crystallinity of API in a mixture
- Determination of coating thickness and uniformity in functionally coated tablets

Due to these diverse applications and simplicity of use, Terahertz Process Analytical Technology (PAT) is rapidly becoming a powerful and effective tool for obtaining optimal and consistent drug delivery characteristics.

# 423774 Hot Melt Coating with Lipid Excipients for User-Friendly Solid Oral Dosage Forms

**Abstract:**

**Hot melt coating with lipid excipients for user-friendly solid oral dosage forms**

[1]Diogo Gomes Lopes, [2]Karin becker, [3]Detlev Haack, [4]Michael Stehr, [2]Andreas Zimmer, [1]Sharareh Salar-Behzadi*

1- Research Center Pharmaceutical Engineering (RCPE) GmbH, Graz, Austria

2- Institute of Pharmaceutical Science, Department of Pharmaceutical Technology, University of Graz, 8010 Graz, Austria

3- HERMES ARZNEIMITTEL GmbH, Großhesselohe, Germany

4- CREMER OLEO GmbH & Co. KG, Witten, Germany

The simple and effective orally delivery of an active pharmaceutical ingredient (API) is an important factor influencing treatment efficacy, tolerability and patient compliance. To further improve the user-friendliness and acceptance of solid oral dosage forms, multiparticulate systems are becoming more common. As they have proven especially popular with patients requiring regular doses or those that find it difficult to swallow solid tablets, such as children, the elderly or individuals suffering from esophagitis. However, for this method to prove effective, most APIs must be effectively coated. Traditional methods to coat solid oral dosage forms mostly use solvents, an approach that suffers from a range of drawbacks including environmental concerns, lengthy processing times and the risk of dissolving drug. Hot melt coating (HMC) is an innovative and effective alternative that is reliable, cheap and fast, while simultaneously offering an opportunity to better manipulate and control characteristics such as taste, stability and release rate.

In this work, multiparticulate systems of different APIs are coated with lipid formulations via HMC. Drug release profiles and the solid state of triacylglycerides (TAGs) based excipients are monitored over storage. Small angle X-ray scattering, scanning electron microscopy and differential scanning calorimetry are used to characterize polymorphism, surface disruption (blooming) and complex networks of crystallites.

Our results show that surface blooming does not play a critical role in the drug release from TAGs coatings, but that the apparent diffusion coefficient of drugs can be dramatically reduced by increasing TAGs crystallite size and resulting tortuosity. This work brings new insights on the drug release from solid lipid dosage forms being an important step towards avoiding conventional complex formulations and preventing the overuse of excipients with unknown toxicity.

# 428444 Evaluation of Static or Dynamic Curing Process for Improved Controlled Release, Dry Powder Coatings of Ethylcellulose Using a Rotor Coater

Tuesday, November 10, 2015: 10:10 AM
Ballroom D (Salt Palace Convention Center)

**Shawn Engels**[1], Cody A. Schoener[2], Uma Shrestha[3], Brian Jensen[1] and Kody Bellach[4], (1)Freund-Vector Corporation, Marion, IA, (2)Dow Pharma & Food Solutions R&D, The Dow Chemical Company, Midland, MI, (3)Dow Pharma and Food Solutions, The Dow Chemical Co., Midland, MI, (4)Freund-Vector Corporation, marion, IA

**Evaluation of Static or Dynamic Curing Process for Improved Controlled Release, Dry Powder Coatings of Ethylcellulose using a Rotor Coater**

*Shawn Engels[‡], Cody Schoener[†], Uma Shrestha[†], Kody Bellach[‡], and Brian Jensen[‡]*

*[‡]Freund-Vector Corporation, Marion, IA 52302, USA*

*[†]The Dow Chemical Company, Midland, MI 48674 USA*

***Purpose:*** To assess how different curing conditions influence controlled drug release from ethylcellulose, high productivity coatings using a rotor process technology. The rotor process technology enables the user to achieve high levels of weight gain in a short period of time while still demonstrating sustained drug release with micronized ethylcellulose. For this work, static and dynamic curing was assessed at different temperatures and lengths of time to improve film coalescence to provide more stable dosage forms. Static curing was completed in an oven while dynamic curing was completed in the rotor coater. Optimized curing conditions and times will enable formulators to achieve the most robust product.

***Methods:*** Acetominophen (APAP) was dry powder coated onto sugar spheres (#20 – 25; Colorcon Inc., USA) using polyvinyl pyrrolidone (K2932, ISP Inc., USA) as binder on a Granurex GXR-35 rotor process (Freund-Vector Corp., USA). A drug to binder ratio of 97:3 w/w was maintained. Experimental ethylcellulose samples (The Dow Chemical Company, USA) were dry powder coated onto the APAP beads with triethyl citrate (TEC) as a plasticizer using the same rotor process as above. An ethylcellulose to TEC ratio of 3:1 w/w was maintained throughout the coating. Once the APAP sugar spheres were coated to 20% weight gain of ethylcellulose, samples were either removed and static cured or kept in the rotor coater and cured. For static curing, samples were placed on trays and placed inside an oven for 2 hours at 60°C. For dynamic curing, samples continued to rotate at 250 rpm while forced, hot air was applied from above at 40 or 60°C between 15 and 60 minutes. Dissolution and SEM were evaluated for curing influence on drug release. USP Apparatus I (baskets) was utilized for in-vitro dissolution studies and drug release data was compared using similarity factor ($f_2$) analysis.

***Results:*** Initial experiments demonstrate that dynamic curing provided more controlled drug release compared to static curing (Figure 1A). Dynamic curing for at 40°C for 15 minutes was significantly slower (by $f_2$ analysis) in drug release compared to static curing at 60°C for 2 hours. This indicated the combination of the force hot air plus the frictional movement caused by the rotor bed movement enables improved film coalescence versus a static tray of sample placed in the oven. Dynamic curing at 40°C for 15 minutes versus dynamic curing at 60°C for 60 minutes was similar based on $f_2$ analysis indicating a short cure time of at least 15 minutes might provide full film coalescence and a stable coating. Further curing assessment on drug release will be presented on samples which were placed on stability for 1, 3, and 6 months in storage conditions of either 30°C and 65% relative humidity or 40°C and 75% relative humidity.

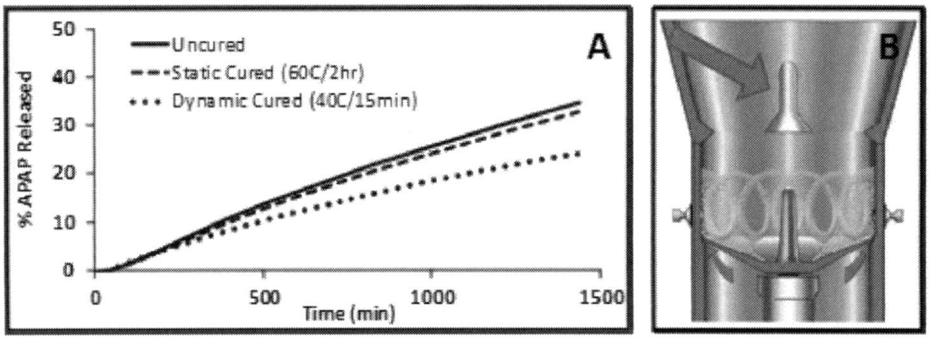


Figure 1. (A) Influence of static vs dynamic curing on APAP drug release and (B) cross-sectional view of rotor coater. Hot air is forced upon the product bed from above (large arrow at top).

***Conclusion:*** Using the rotor process and force hot air, formulators can achieve sustained release and fully coalesced films by dynamically curing. Dynamic curing as short as 15 minutes at 40°C can significantly reduce curing times compared to static curing. The combination of dynamic curing and dry powder coating using the rotor process and micronized ethylcellulose also provides high productivity opportunities for pharmaceutical companies when compared to traditional organic or aqueous based ethylcellulose wurster spray coatings.

# 423668 Gluing Pills: A Novel Approach for Processing of Multi-Layer Tablets

**Abstract:**

**Gluing Pills: A Novel Approach for Processing of Multi-Layer Tablets**


**[1]Sharareh Salar-Behzadi, [1]Sandra Stranzinger, [2]Josef Lingitz, [1]Stephan Sacher, [1,3]Johannes Khinast**

**1- Research Center Pharmaceutical Engineering (RCPE) GmbH, Graz, Austria**

**2- M & R Automation GmbH, Graz, Austria**

**3- Institute for Process and Particle Engineering, Graz University of Technology, Inffeldgasse 13/3, 8010 Graz Austria**


Multi-layer tablet have several advantages over conventional tablets. For example, multi-layer tablet dosage forms can be designed as osmotically-controlled tablets, fixed-dose combinations for delivering incompatible APIs, controlling the releases of multiple APIs at different rates and releasing actives at different sites.

In the production of multi-layer tablets it is critical to control the weight of each layer and to assure a good adhesion between layers. It is also desirable to achieve a clear demarcation plane between the layers, since this is not only appealing but also assures the absence of cross-contamination.

The conventional technology to produce multi-layer tablets is a special rotary press, where each layer has a special compression step. The weight control of this technology is rather complex, making use of either a force control system or tablet thickness. The adhesion between layers is dependent on the material properties and requires a certain surface roughness of the initial layer for particle interlocking and adhesion with the second layer. Furthermore, fines below an average size of 150 microns should be avoided to provide a sharp resolution between the two layers. Problems are often encountered. Inadequate individual layer weight control, disproportionate layers-weight ratio coupled with low drug loads, cross contamination between the layers, insufficient multi-layer tablet hardness and a tendency of delamination of the layers are some of the disadvantages of rotary tablet multi-layer compaction.

In this study we present an innovative approach for the processing of multi-layer tablets. Each layer will be compressed separately, in the form of thin tablet bodies. Then, these tablet bodies will be glued to each other in an automated process. The adhesion between layers depends mainly on the properties of the adhesion gel, physical properties of tablets and the homogeneity of the adhesive layer.

To investigate the impact of these properties, tablets were produced with different excipients and different compaction forces. The impact of potentially critical material attributes (CMA), such as the degree of the plasticity of excipients, particle size distribution, compaction force and punch velocity on critical quality attributes (CQA) of tablets (rawness, specific surface area, hardness and friability), were investigated. Afterwards, these CQAs together with the type of gel binder, its concentration and viscosity were considered as CMAs for glued tablets and their impact on the quality of the adhesion between layers was investigated. Moreover, the penetration and cross-contamination between layers have been studied using a worst case scenario, i.e., tablets with high concentrations of water-soluble API, glued with water soluble gel binder.

This innovative and fast gluing method offers significant advantages, e.g., the ability to control the weight and compaction force in each tableting process, thus avoiding the variability in tablet hardness, disintegration time and dissolution profile of each layer. Moreover, this approach enabled us to design multilayer-tablets with advanced performance and low risk of cross contamination between layers.

# 409093 An Engineering Approach to Discover Novel Antivirulence Strategies

**Abstract:**

Antibiotic resistance is a growing public health threat that is worsened by a declining antibiotic pipeline. Strains resistant to "last line of defense" antibiotics have begun to appear and the danger of our antibiotic arsenal becoming obsolete is quite real. Since current antibiotic discovery has failed to keep pace with the rise of resistance, novel methodologies are required to address this looming crisis. Antivirulence therapies have been receiving increased attention as a novel class of anti-infectives. Rather than targeting essential bacterial functions, as current antibiotics do, antivirulence therapies target essential host-pathogen interactions required for infection such as adhesion, quorum sensing, and susceptibility to immune attack. These therapies are less prone to resistance development than are antibiotics, and have the potential to fill the widening therapeutic gap caused by the rise of multi-drug antibiotic resistance.

Reactive oxygen and nitrogen species (RNS, ROS) are antimicrobials generated by immune cells to kill bacteria. The importance of RNS and ROS to immune function is evidenced by the many pathogens, including *Mycobacterium tuberculosis*, *Neisseria meningitides*, *Vibrio cholerae*, *Salmonella enterica*, and enterohemorrhagic *Escherichia coli* (EHEC), that depend on ROS and/or RNS detoxification to establish or sustain an infection. Inhibitors of these defense systems promise to be potent antivirulence therapies; however, known agents are either toxic to humans or poorly transported into bacterial cells. In this talk, I will discuss how we use methodologies from systems biology and metabolic engineering to quantitatively study RNS and ROS stress in bacteria, as well as how we have used the resulting knowledge to identify novel ways to sabotage the RNS and ROS defense systems bacteria require for infection.

# 432001 Communication and Collaboration in Synthetic Microbial Communities

**Abstract:**

The biosynthesis of compounds of medical and industrial importance often requires engineering and optimization of complex metabolic pathways. Traditionally, these processes have employed a clonal population of recombinant microbes. There are many limitations of using a single population of microorganism that could be alleviated or addressed by using a mixed community, including metabolic load and the number of exogenous elements that can be cloned and optimized in a single cell. In order to control interactions required for these cells to work together, it is important to generate robust communication pathways between biotechnologically relevant species. To this end, we have developed new regulatory components and systems that can be used to turn gene expression on or off in response to a cell-cell communication signal, and enable interspecies communication between representative Gram-negative (*Escherichia coli*) and Gram-positive (*Bacillus megaterium*) organisms. We are also employing new strategies to control the growth and coexistence of species in co-cultures by manipulating ecological interactions; a key challenge for applications with microbial consortia. Finally, I will describe our recent efforts to distribute functions between these microbes in a model bioprocess. This work has yielded a set of key technologies enabling the division of labor between specialized species and strains for bioprocessing and a broad range of other applications.

# 409204 Systematic Mining of Noncoding RNAs for Regulating Complex Phenotypes

**Abstract:**

An important question in strain engineering is: *how to search the genome for targets that regulate a phenotype of interest?* Given the natural role that noncoding RNAs play in coordinating complex stress-responses, there is increasing interest in exploiting these regulators to construct phenotypes of interest that can further expand the current potential for microbial cells. An emerging need in this frontier is the proper identification and selection of functional RNAs that can be engineered for optimal regulation of desired genes and relevant metabolic pathways. Thus far, the number of mechanisms and RNA scaffolds that are being exploited for these applications are highly constrained. Although a large number of "omics" approaches are being attempted to expand the number of RNAs that globally regulate phenotypes, it is unclear how to interpret this *massive* amount of data in the context of biological pathways and utilize it to uncover underlying biological mechanisms that can be relevant to strain engineering.

This talk will address characterization tools that we are developing to automate the identification of global regulators that could be exploited in constructing synthetic schemes to regulate complex phenotypes. Specifically, we will present a general pipeline for mechanistically understanding and engineering desirable complex phenotypes based on a *rational and predictive* systems analysis of global regulation and on the intracellular assaying of relevant targets. Our approach has been applied to microbial organisms with extremophilic traits and has combined bioinformatics, genetics, biophysical models and synthetic parts for prediction of pathways.

# 427263 The Vision of a Sustainable Bioeconomy in an Era of $50 Oil

**Abstract:**

Biotechnology is rapidly coming of age as enabling technology for the production of biobased chemicals and biofuels. This is a process of creative destruction whereby one type of manufacturing technology based on chemistry is gradually being replaced by another based on biology. Sustainability is a major driving force, but, biobased processes have additional unique advantages, such as superb specificity and ability to convert with high efficiency renewable feed stocks. These advantages tend to counterbalance the high space times of competing chemical processes. A critical question is how biotechnology will fare in a period of declining energy and fossil feedstock prices. I argue that, despite temporary excursions, the fundamental relationship between real costs of fossil and renewable feedstocks will be maintained and the outcome of the competition between biotechnology and chemistry will be decided on the technical merits of the corresponding processes. In this regard, metabolic engineering becomes of critical importance in advancing the competitiveness of biotechnology and realizing the vision of sustainable bioeconomy.

# 437507 Challenge and Promise of Extracellular Nanovesicles in Early Detection and Treatment of Cancer and Other Systemic Diseases

**Abstract:**

Exosomes and other nano-size extracellular vesicles (EVs) are stable membrane-bound nanoparticles secreted into the circulation by diverse cell types, including epithelial cells, mesenchymal cells, lymphocytes, and tumor cells. They are 20-120 nm in diameter and contain membrane and cargo, which includes nucleic acids and proteins, originating from the parent cell. They are found in astronomical numbers in extracellular spaces and body fluids, including blood, urine, saliva, breast milk, semen, lymphatic, and spinal fluids. In blood alone, there are over $10^{16}$ circulating exosomes, the physiological functions of which are not completely understood. It has already been shown that exosomes are important in modulating immune response and are thought to play a role in short and long-range intercellular signaling by fusing with recipient cells and releasing their cargo. Their population is heterogeneous, reflecting the diversity of cells that secrete them which, in patients, include abnormal (e.g. malignant) cells. The stability of these vesicles allows their isolation from body fluids, characterization by molecular content and biophysical properties, which together can potentially be used for medical diagnostics. However, the fundamental difficulty in using molecular content of exosomes in early and routine diagnostics is the large background of nanovesicles originating from normal cells found in the circulation that interferes with the analysis even when the most sensitive techniques (such as digital droplet PCR) are used. Our laboratory has focused on identifying biophysical properties of cancer exosomes that most distinguish them from other nanovesicles in the blood and use these properties to fractionate the sample and enrich in tumor exosomes. In this approach, we aim to decrease the influence of the background in order to improve the accuracy of downstream molecular measurements, ultimately improving the sensitivity to allow for an early detection of cancer and other systemic diseases.

# 437537 Contributions Chemical Engineers Can Make to Therapeutic Delivery to the Lungs

**Abstract:**

As Chemical Engineers we sometimes wonder what we can contribute to medical science. I will give examples from our work, and suggest that the first step is collaborating with medical researchers who understand the physiology and disease pathology that needs to be addressed. With that starting point our understanding of fluid mechanics, material science, and self-assembly enables us to make important contributions which the medical community is not equipped to provide. We will give two examples of nanoparticle delivery to the lungs that address problems in TB and non-small cell lung cancer. The first example is the creation of inhalable aerosol particles comprising therapeutic nanoparticles and a matrix phase to create low-density particles that enable delivery to the deep lung. The second example is the creation of gel microparticles that deliver nanoparticles through venous filtration. The gel microparticles are created by emulsion processing and polymer gelation so that the IV injected particles are trapped in the vasculature in the lungs where they deliver anti-cancer drugs over extended periods.

# 430000 Separation of Bacteria from Blood Components


**Abstract:**

Blood bacteremia, or a bacteria infection in the blood stream, is of major concern in terms of mortality and morbidity, particularly when antibiotic resistant bacteria are involved. Persons presenting bacteremia symptoms are often given standard doses of general antibiotics and a blood sample is taken to culture the bacteria for identification. The problem is that culturing and identifying the pathogen in order to administer the correct antibiotic response usually requires 24 to 48 hours, during which time the patient could become severely sick if the particular pathogen is resistant to the general antibiotic already administered.

Our overall goal is to develop methods that are much faster, even as fast as one hour, that can identify antibiotic resistant bacteria from a person's blood. This implies that there is no time to culture the bacteria in the blood to obtain sufficient numbers of bacteria to identify resistance by current conventional means. We are developing methods to quickly identify the bacteria in the blood using molecular mechanisms such as direct DNA or RNA identification without PCR amplification of certain gene sequences and without culture amplification of the bacteria. The first step in this process is to separate the bacteria from the blood components that would interfere with the subsequent assays. The means separating the bacteria from the red blood cells (RBCs), white blood cells (WBCs) and platelets (PLTs). The target time is 10 minutes or less for separation, after which the recovered bacteria will be lysed and their DNA analyzed for identification within 50 minutes.

This submission presents our successes and failures in using porous hollow fibers, flat membrane filters and centrifugal processes for separation.

While the use of porous hollow fibers with 3-5 um pores is good in theory, the application is difficult because hollow fibers with uniform cylindrical pores are not commercially available. Porous hollow fibers with elliptical pores of 5 um size are available, but these trap bacteria (not pass bacteria) because of the pore shape. Pores with tortuous pathways also trap bacteria that would otherwise pass through a cylindrical pore.

Flat membranes with uniform pores are commercially available, and have been used in this work with bovine blood. However, the bacteria recovery is low, and the area needed to prevent filter cake build-up is so large that a small disposable device that could be used in a hospital becomes impractical in terms of size and expense.

Centrifugal separation can easily separate RBCs (~8 um) and WBCs (~15 um) from platelets (~3 um) and bacteria (0.5 to 2 um) because of the large size of the cells. The dynamic sedimentation velocity goes as the diameter squared, and red cells are relatively dense because of their hemoglobin content. In these preliminary studies we have used bovine blood spiked with E. coli. We have had success in centrifugal separation with high bacterial recovery as only a small fraction of bacteria gets trapped in packed cells. We have developed and will present some unique centrifugation procedures that increase the efficiency or separation of cells from bacteria.

# 437510 Engineering of Articular Cartilage: Challenges and Prospects

**Abstract:**

**Engineering of articular cartilage: Challenges and prospects**

Nehal Abu-Lail, Chrystal R. Quisenberry, Arshan Nazempour, and Bernard Van Wie

Voiland School of Chemical Engineering & Bioengineering, Washington State University, Pullman, WA 99164-6515.

**Introduction:** Articular cartilage (AC) is the smooth white connective tissue of diarthrodial joints. It acts as a lubricant to the moving joins as well as a distributor of loads applied on them. However, AC tissues are subjected to extreme mechanical loads, harsh chemical conditions and worse of all, they are devoid of blood and nerve vessels. Mainly due to the latter, AC tissues have limited regenerative capabilities when injured or damaged. AC failure results in osteoarthritis (OA) or can cause musculoskeletal morbidity. Currently, no successful long-term treatments of OA are present. Patients with OA usually have to replace joints. In the US alone, a joint is replaced ~ every 90 seconds and it is even worse in Europe. With no treatments, AC presents an ideal candidate for tissue engineering. Tissue engineering is an interdisciplinary field in which biological concepts are combined with engineering tools and quantitative methods to model environments capable of producing functional tissues that can be used to replace native diseased tissues. Here, we are utilizing a unique centrifugal bioreactor, an array of mechanical and chemical cues and a variety of cell types to engineer AC *in vitro*. We hypothesize that biomechanical and chemical stimulation will be sensed by the cell *via* mechanotransduction through β1-integrins and N-cadherins and this will prompt cells to secrete extracellular matrix. We further hypothesize that this stimulation will improve the Young's modulus of cartilage tissues by increasing it to near native cartilage properties.

**Materials and Methods**

Chondrocytes, the primary cell type in AC, remodel the tissue's extracellular matrix (ECM) in response to mechanochemical cues from their environment. Unfortunately, during growth, they dedifferentiate. With that limit, human adipose-derived stem cells (hASCs) or bone marrow stem cells can be used as a cell source. Cells were seeded in our novel centrifugal bioreactor (CBR) in a variety of ways. To mimic native AC environment, oscillating hydrostatic pressure and transforming growth factor-β3 (TGF-β3) were also applied. These cues lead to differentiation of stem cells into chondrocytes to form a cartilage tissue with desirable properties. Cells were generally grown for 21 days in the reactor as well as in static controls of micromass and pellet. The functionality of the engineered AC tissues was tested via an array of techniques. First, atomic force microscopy (AFM) was used to measure the mechanical properties of the cell and cartilage tissue as well as to map the location and density of β1-integrins on the cell surface as a measure of mechanotransduction. mRNA sequencing was used to probe for 10 markers of chondrogenesis. Finally, biochemical analyses were carried to quantify DNA and glycosaminoglycans (GAG) contents of the engineered cartilage.

**Results and Discussion**: CBR tissues grown with only oscillating hydrostatic pressure had Young's moduli 44-fold higher than the micromass or pellet static controls. With the addition of TGF-β3 to the CBR tissues, a 1.8-fold increase was observed, bringing the Young's modulus nearer to that of native cartilage. It was found that the average adhesion force for a single antibody-β1-integrin interaction was 80 pN, independent of treatment group. The micromass and pellet cultures had two to five-fold larger β1-integrin counts compared to the cultures engineered in the CBR. Because β1-integrins facilitate cell-ECM attachment, a deficient ECM may require more β1-integrins to attach to the cell. Analysis of N-cadherins on the surfaces of cells in response to stresses is ongoing.

**Conclusions:** Cartilage tissues grown from hASCs had higher Young's moduli, nearing that of native cartilage, when oscillating hydrostatic pressure and TGF-β3 were applied synergistically. The surface expression of β1-integrin was dependent on the tissue culture method. In the future, adding a scaffold with gradients in its properties to the CBR may bring us a step close to engineering AC with properties close to those of native cartilage.

**Acknowledgements:** This work was supported by an NSF EAGER grant 1212573, the NIH Protein Biotechnology Training Program 24280305, a NASA Space Grant, a WSU DRADS fellowship, and a Harold P. Curtis Scholarship.

# 437505 15 Years of Chemical Engineering Advances in Medicine

**Abstract:**

There is an acute shortage of organs due to disease, trauma, congenital defects, and most importantly, age related maladies. The synthetic materials used in tissue engineering applications today are typically composed of millimeter or micron sized particles and/or fiber dimensions. Although human cells are on the micron scale, their individual components, e.g. proteins, are composed of nanometer features. By modifying only the nanofeatures on material surfaces without changing surface chemistry, it is possible to increase tissue growth of any human tissue by controlling the endogenous adsorption of adhesive proteins onto the material surface. In addition, our group has shown that these same nanofeatures and nano-modifications can reduce bacterial growth without using antibiotics, which may further accelerate the growth of antibiotic resistant microbes. Inflammation can also be decreased through the use of nanomaterials. Finally, nanomedicine has been shown to stimulate the growth and differentiation of stem cells, which may someday be used to treat incurable disorders, such as neural damage. This strategy also accelerates FDA approval and commercialization efforts since new chemistries are not proposed, rather chemistries already approved by the FDA with altered nanoscale features. This invited talk will highlight some of the advancements and emphasize current nanomaterials approved by the FDA for human implantation. It will also provide a perspective on the role chemical engineers have played in the field of nanomedicine, in terms of pioneering the field, contributing state-of-the-art knowledge, commercialization, and education.

# 437364 Detection of Mycobacterium Tuberculosis Volatile Biomarkers Using a Titanium Dioxide Nanotube Based Sensing Platform

**Abstract:**

This research presents a metal functionalized titanium dioxide ($TiO_2$) nanotube based sensor developed for detection of volatile biomarkers (methyl *p*-anisate and methyl nicotinate) associated with tuberculosis (TB). In this work self-aligning $TiO_2$ nanotubes were fabricated, and functionalized with cobalt for specific binding to the volatile biomarkers of interest. Mimics of the volatile biomarkers were dissolved in water at variable concentrations, and nitrogen gas was flowed through the biomarker solutions and delivered to the sensor. A potentiostatic scan was used to measure current across the device under constant potentials ranging from -1.0 volts to 1-volts methyl nicotinate and methyl *p*-anisate. The sensor response showed a $3 \times 10^5$ change in current for methyl nicotinate, and $7.61 \times 10^6$ change in current for methyl *p*-anisate when compared to baseline. The sensor also showed little response to five other VOCs (acetone, ethanol, methanol, benzene, and phenol) which suggest the sensor is functionalized to be specific for the target VOCs associated with TB. Results showed detection of the volatile biomarkers yield orders of magnitude change in current which is detected easily. The sensor is simple to operate, responds in minutes, and has potential applications in non-invasive diagnosis of TB and other diseases that have distinct volatile biomarkers.

# 434371 Continuous Plug-Flow Crystallization with Anti-Fouling Control

**Abstract:**

While plug-flow crystallization (PFC) has been touted as a promising continuous pharmaceutical unit operation, it is prone to fouling or "encrustation", a phenomena by which uncontrolled crystallization occurs on the crystallizer's surface. This results in operational issues such as flow blockage, increased thermal resistance, and reduced supersaturation, which in turn lead to limited continuous operation and reduced crystal quality and yield. Here, we introduce a model-based and model-free anti-fouling control (AFC) via spatial and temporal heating and cooling cycle. The two methods are compared with optimized PFC operation without AFC under two conditions: (1) maximization of crystal growth in the presence of encrustation and (2) minimization of encrustation while maintaining sufficient crystal growth. Specifically, the study contrasts an open-loop control approach, analogous to quality-by-design (QbD), with a feedback control approach, referred to herein as quality-by-control (QbC) concept. The computational and experimental results highlight the pros and cons of the different AFC designs as well as demonstrates the effect of PFC designs in terms of number of temperature segmentations, the segment length and the cycling frequency between the heating and cooling cycles, to the control performance.

# 416773 Mechanistic Details and Process Implications of Azide Mediated Conversion Route of Carboxylic Acid to Isocyanate

**Abstract:**

**Mechanistic Details and Process Implications of Azide Mediated Conversion Route of Carboxylic Acid to Isocyanate**

Anuj A. Verma, Herbert Ogutu, Yuanyuan Li, Ann M. Czyzewski, Su Yu, & Xiu C. Wang

Process Research & Development, AbbVie Inc., North Chicago, IL

The conversion of a carboxylic acid to an isocyanate via the Curtius rearrangement is a commercially attractive route in the pharmaceutical and chemical industries. Traditionally, the carboxylic acid is converted to acyl azide at 20°C by using diphenyl phosphoryl azide (DPPA) as the azide source. The acyl azide is subsequently heated to enable Curtius rearrangement to form isocyanate. A safety concern is identified in handling a pot of acyl azide across scales of operation and it is desired to minimize the inventory of that azide. Herein, we describe a process which minimizes that inventory. The process synthesis was accomplished by developing a generalized azide chemistry/reaction engineering platform (attached to Curtius reactions) to critically evaluate and identify optimal processes which are inherently safe at different scales of operation. The platform involves a detailed mechanistic understanding of product formation along-with impurity generation, and represents a step above traditional kinetic modeling by *elucidating optimal reagent selection rules* in addition to traditional reaction engineering solutions to reduce impurity generation and increase the product formation rate.

*AbbVie contributed to the design, research, and interpretation of data, writing, reviewing, and approving the publication. Anuj A. Verma, Herbert Ogutu, Yuanyuan Li, Ann M. Czyzewski, Su Yu, and Xiu C. Wang are AbbVie employees who worked on this project.*

# Validation of a New Product Introduction Involving Continuous Unit Operations in a Batch/Continuous Hybrid Process

*Ashley Humenik, Process Engineer*
*Eli Lilly S.A. – Irish Branch*

*AIChE Annual Meeting, Salt Lake City*
*10th November 2015*

## New Product Introduction

- Technical Transfer from Development to Manufacturing
- Validated at Kinsale in 2015
- 2 Continuous Unit Operations:
  - Flow Palladium Adsorption Removal in Packed Columns
  - Thermal Cleavage of the BOC protection group in Plug Flow Reactor

# Continuous Process Flow Drawing

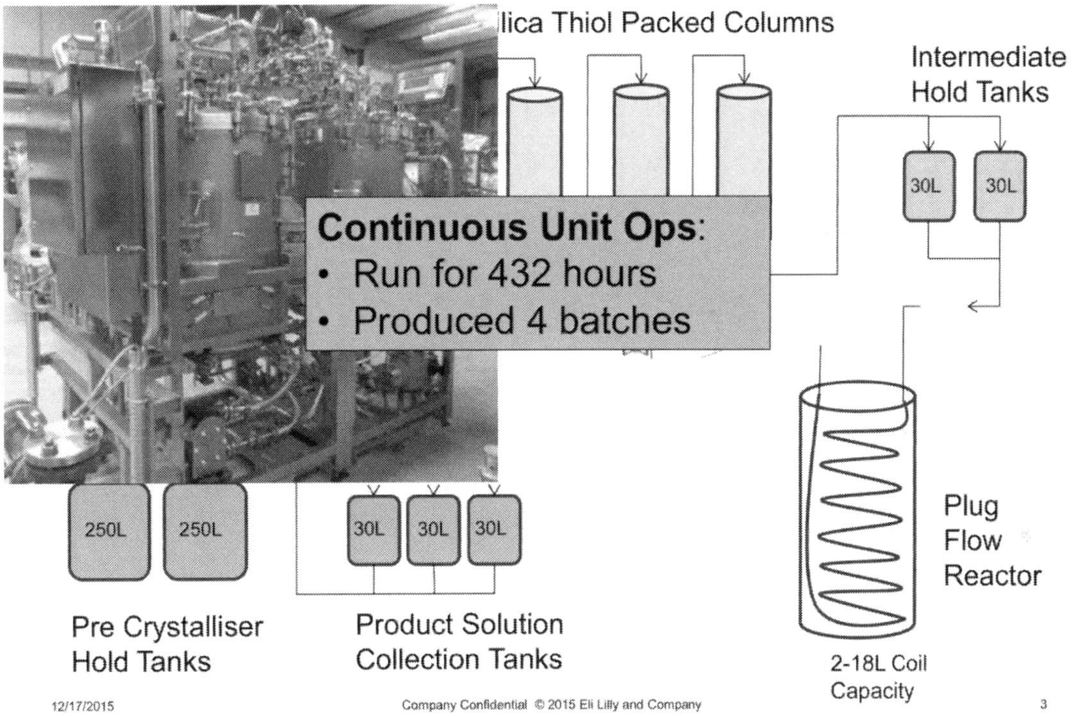


**Continuous Unit Ops**:
- Run for 432 hours
- Produced 4 batches





# Palladium Removal

- Mechanistic model determined proven acceptable ranges (PAR)
- Two process parameters have multivariate interaction: Adsorption temperature & Superficial flow velocity (v)
- Assumptions: Campaign size & Equipment setup

$$v = \frac{L_C}{\tau_C}$$

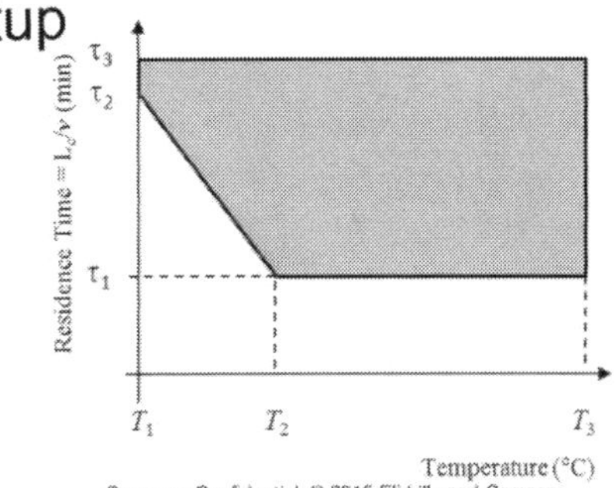

≥12.13kg silica thiol

## Column Setup & Considerations

- Silica Thiol Particle Size Distribution (PSD)
- Silica Thiol Resin Stability
- Height of packing
- Flowrate
- Isothermal operation – Heat tracing installed
- Column flushing & Gas displacement
- Negligible axial dispersion (approaching plug flow, no bypassing)

12/17/2015 Company Confidential © 2015 Eli Lilly and Company 5

## Thermal BOC Cleavage

- Mechanistic model determined proven acceptable ranges (PAR)
- Two process parameters have multivariate interaction: Reaction temperature & Residence time
- Assumptions: Single liquid phase, Plug flow, Steady State

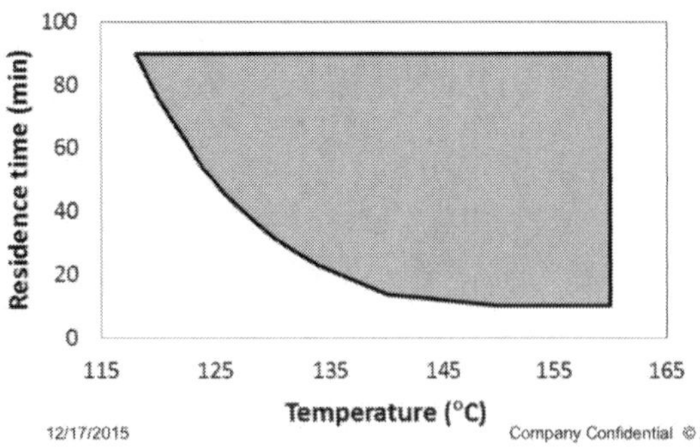

6L volume coil
3/8" Hastelloy tube

12/17/2015 Company Confidential © 2015 Eli L 6

## Characterisation of System Residence Time Distribution (RTD)

- Flushed system with pure THF until baseline was achieved
- Solvent Switch to 90:10 THF:$H_2O$
- FTIR used to determine residence time ($\tau$) at 50% of the water response and system volume ($V = \tau \times q$)

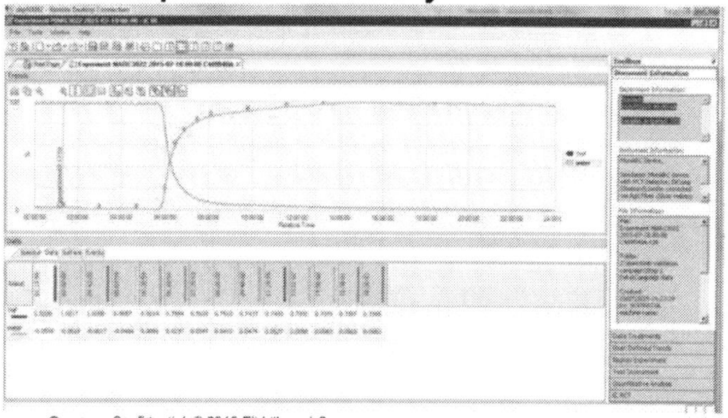





## Batch Definition

- Offline samples analysed using Karl Fischer (KF) Method
- Time to achieve KF sample on system exit ≥99% of the original KF value ($T_1$)
- Calculated residence time multiplier (N) to achieve above KF:
  - $N = \frac{T_1\ (min)}{\tau\ (min)}$
- Calculated solvent push-out time (min):
  - $\frac{System\ Volume\ (ml)}{Process\ Flowrate\ (\frac{ml}{min})}$ X N

## Safety Considerations

- Solution transfer at high pressure (300psig) and high temperature (140°C)
  - Potential spill leading to fire or personnel injury
  - Potential overpressurisation of tubular reactor
  - Potential leak leading to vapour cloud explosion (VCE)



## Unit Operations Setup & Considerations

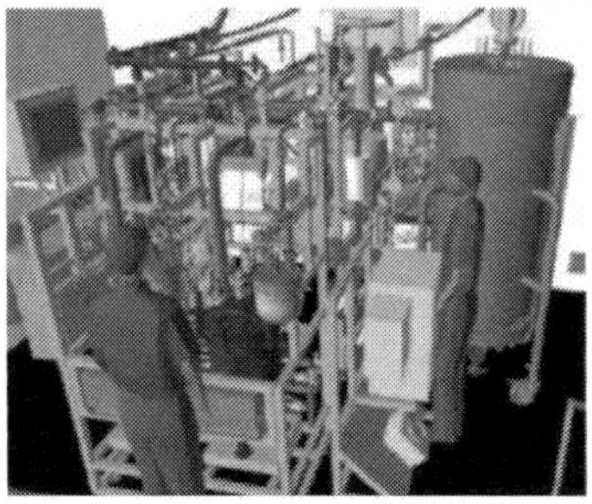

- Same mass flowrate
- Intermediate hold tanks
  - Not plug flow
  - Increased dispersion
  - Cascade control
  - Reactor flow range (not setpoint)
  - Minimum weight setpoint (2kg)
    - Reduced Batch to Batch Carryover
- 300 psig back pressure
  - Maintain evolved $CO_2$ in solution

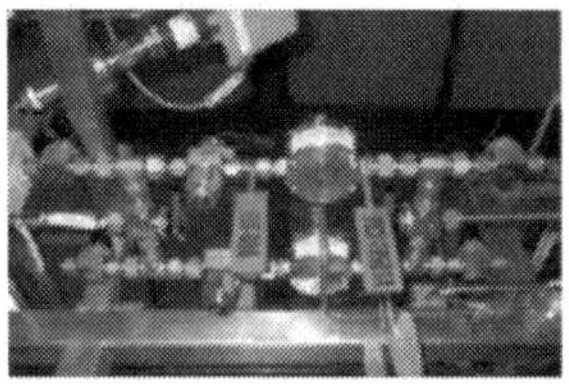

# Plant Operations & Process Monitoring

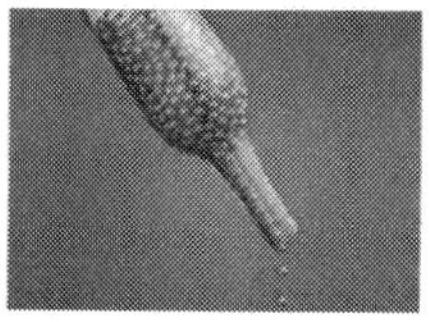

- Continuous = 'Intentional' Bottleneck

- Simultaneous activities:
  - Front end batch reactions
  - Continuous
  - Back end batch isolation

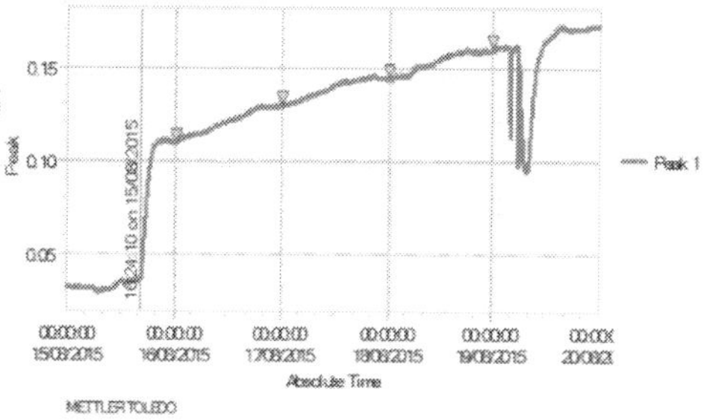


- 'DECISION' phase
  - Process monitoring of continuous unit ops

# Unplanned Stoppage Considerations

- Stoppages will occur
  - Little time to decide best response
- Option to restart or push-out
  - Possible segregation of material
- Agree a decision tree to handle various cases

- 'Short stop' of <u>less than 30 minutes</u> duration
  - Testing completed during Solvent Run showed all process parameters remained within acceptable range.

# Conclusion

- Successfully validated 2 manufacturing steps
  - 2 continuous unit operations
  - 4 batches
  - Quality material meeting specification
  - No safety issues

# Acknowledgements

- E. Crick
- K. Seibert
- C. Luciana
- T. Wilson
- G. Flanagan
- C. O'Mahony
- S. Moursy
- R. Spencer
- R. Moylan
- A. Corrigan
- R. O'Sullivan
- M. Collins
- E. Cotter
- J. Cunningham
- C. Duffy
- B. Healy
- C. Healy
- B. Murphy
- P. Desmond
- S. Jeffery
- I. Mossop
- L. Maguire
- B. Walsh
- J. Cashman
- H. Moynihan
- V. Stallard
- K. Lydon
- P. Harrington
- P. Harrigan
- L. McCarthy
- J. Grace
- D. Smith

# Questions?

12/17/2015 Company Confidential © 2015 Eli Lilly and Company 15

# The Development of an Integrated Control Strategy for a Continuous Primary Pharmaceutical Process

Hannah Robinson
GlaxoSmithKine

Applications of Continuous Processing in the manufacture of Pharmaceuticals: Drug Substance Part II
AIChE Annual Meeting Nov 2015

## Agenda

- Introduction
- Case Study
  - Development to Industrialisation of a Flow Process
- Summary

## Introduction

## Introducing Continuous to Pharma

**Batch**

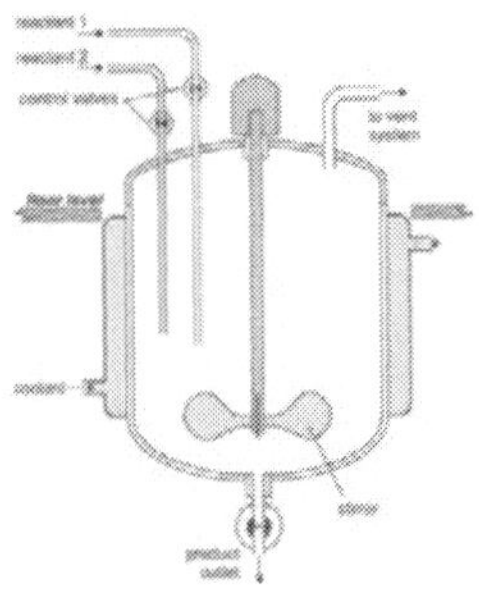

- Historical workhorse of Pharma
- Control  is scale dependent
- Quality mindset batch dominated

**Continuous**

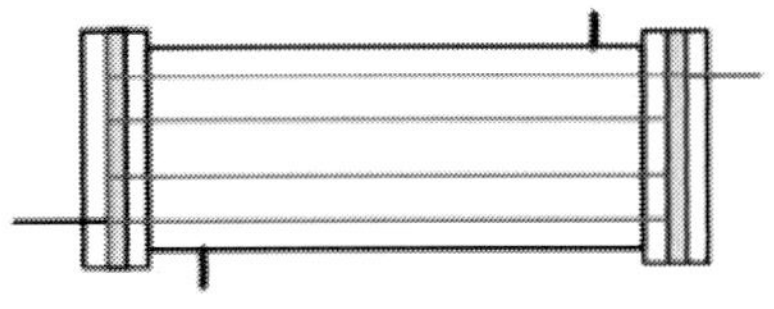

- Workhorse of other industries
- Uniform control easier to achieve
- Constant production provides challenge to traditional quality approaches

# Case study – Development to Industrialisation of a Flow Process

## Process Overview

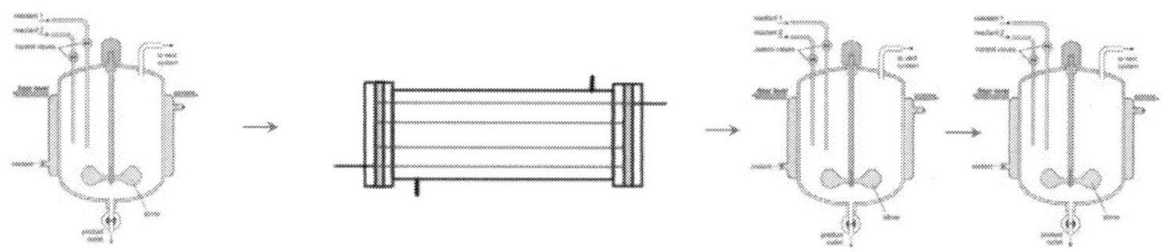

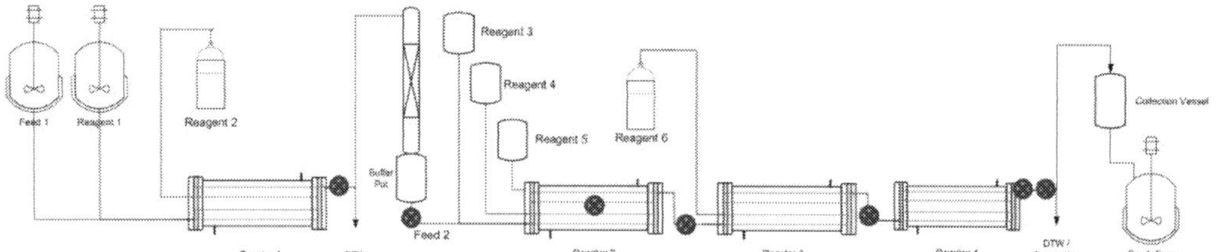

## Process Understanding

- Kinetics
- Calorimetry
- Physical property behaviour of streams
- Mixing sensitivity

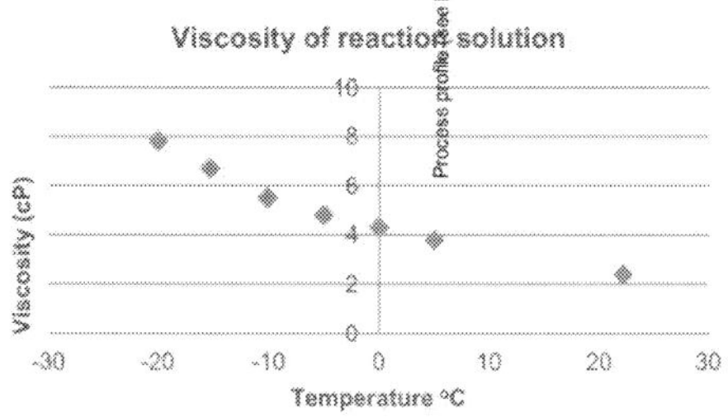


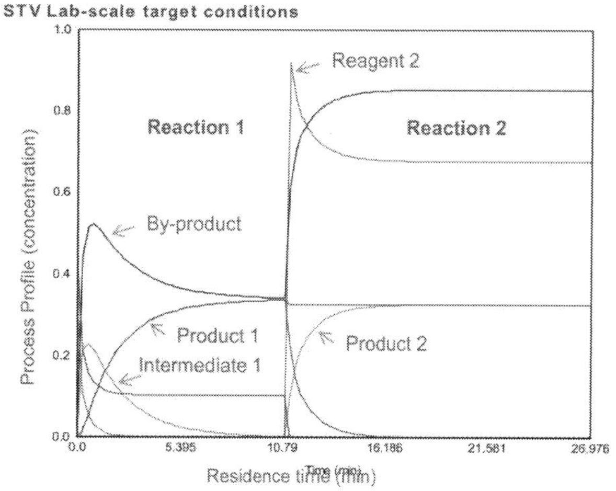


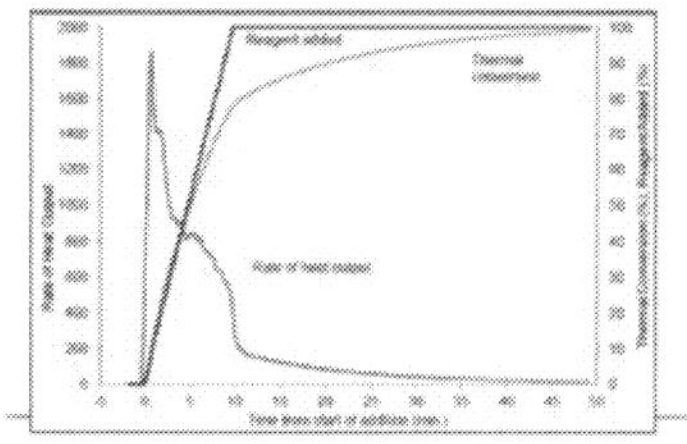

## Equipment Characterisation

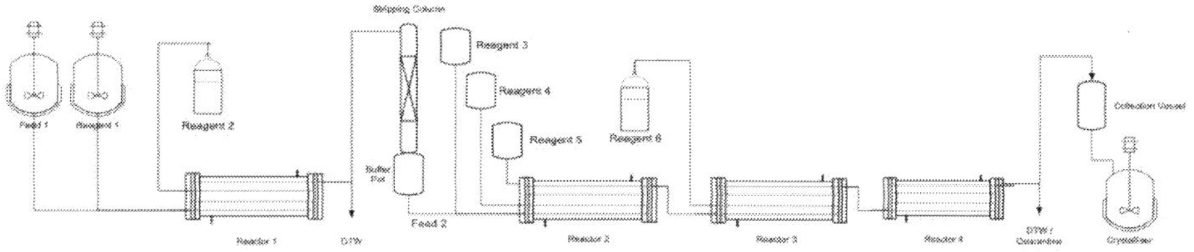

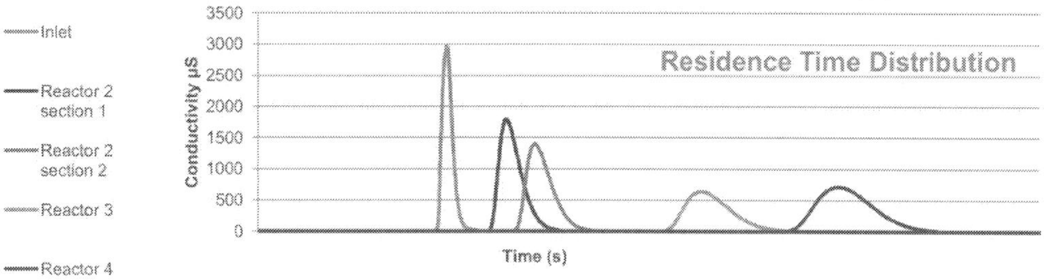


- Mixing intensity and Re
- Reactor volumes and surface area
- Heat transfer capability
- Residence time distribution

## Lab processing data

Online UPLC

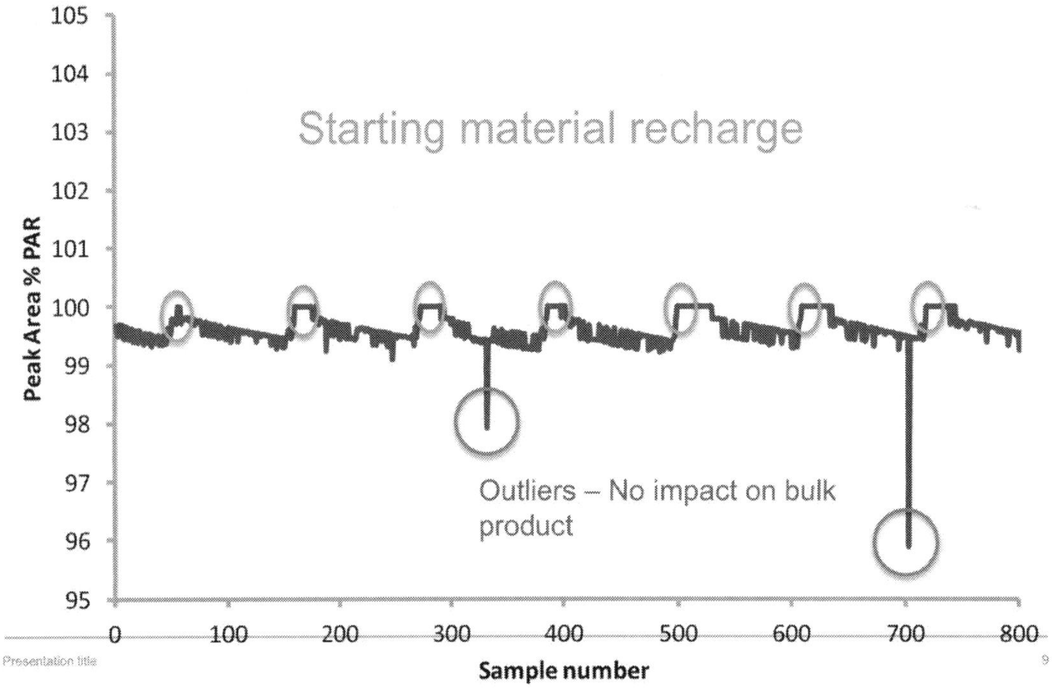


Presentation title 9

## Transfer to Manufacturing

# Operational controls

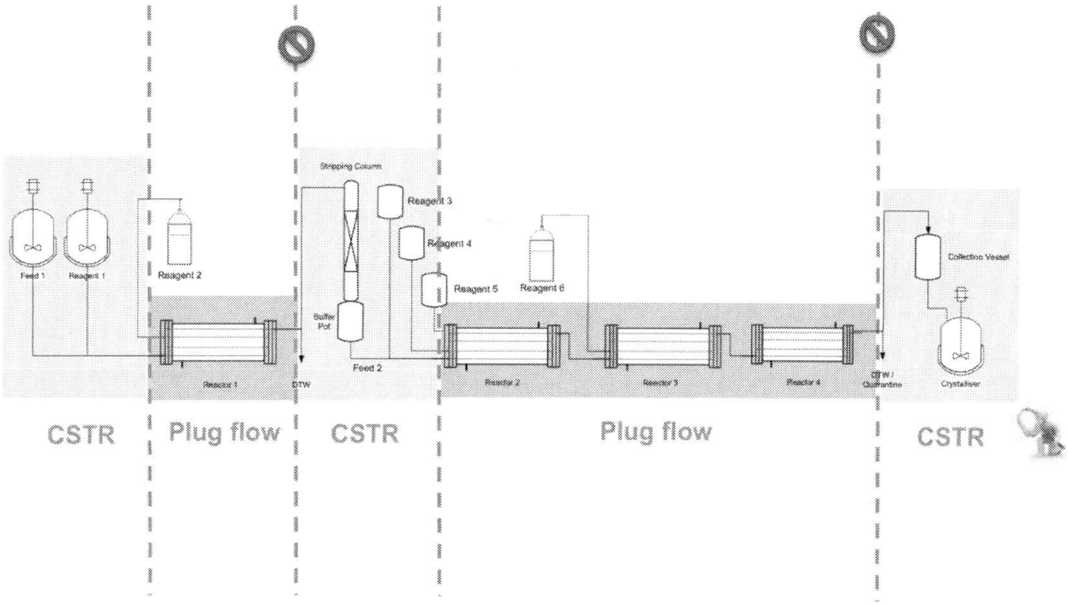




# Operational controls

## Axial Dispersion model

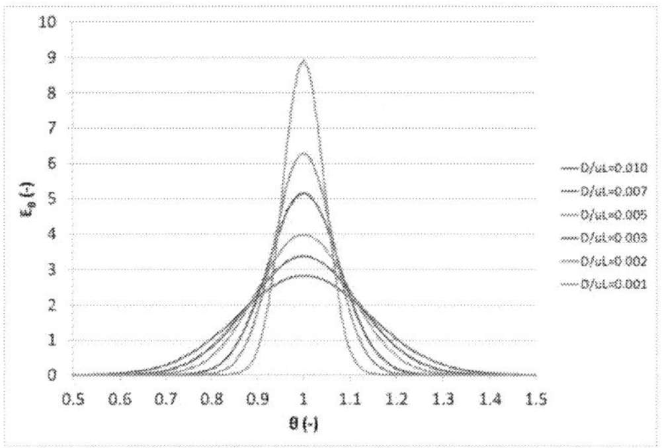


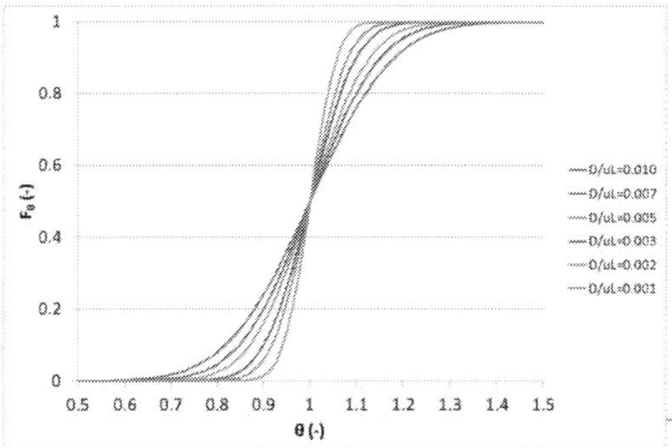


- Quality control based on combination of dispersion and time to meet a pre-defined acceptable quality
- Acceptable quality defined by process requirements of design space
- Characterisation of equipment system → D/uL < 0.01 for equations to hold

**Totalised residual flow post residence time θ**

| D/uL | 0.01 | 0.004 | 0.001 |
|---|---|---|---|
| θ=1.0 | 135g | 81g | 47g |
| θ=1.1 | 48g | 12g | 0.9g |
| θ=1.2 | 12g | 0.6g | <0.001g |
| θ=1.3 | 2g | 0.01g | Trace |
| θ=1.4 | 0.2g | <0.001g | Trace |

## Control phases

Disturbances

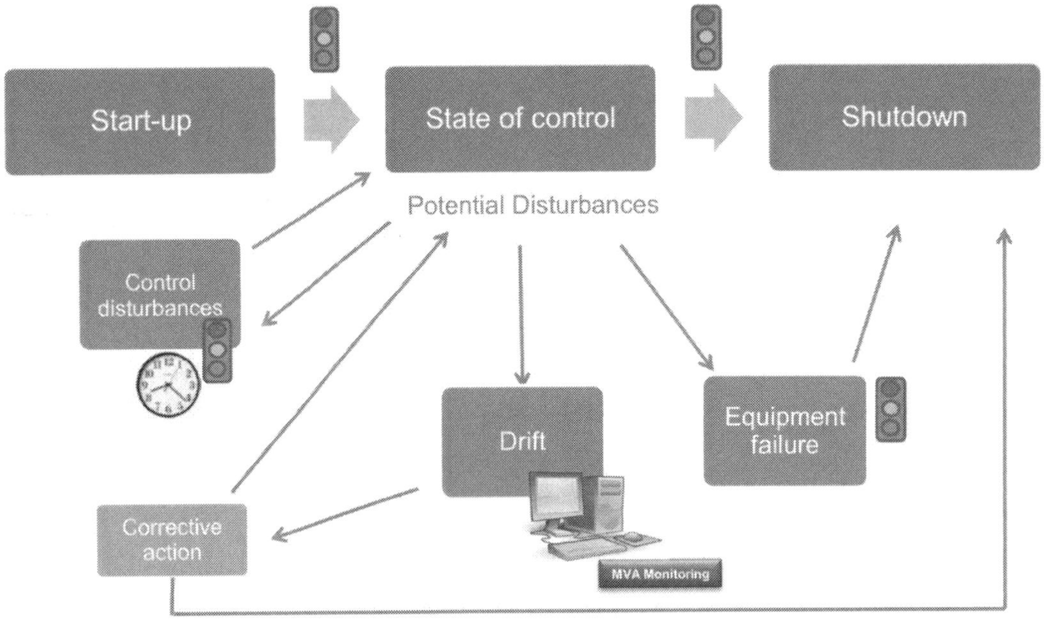


## MVA monitoring

3 hours post model

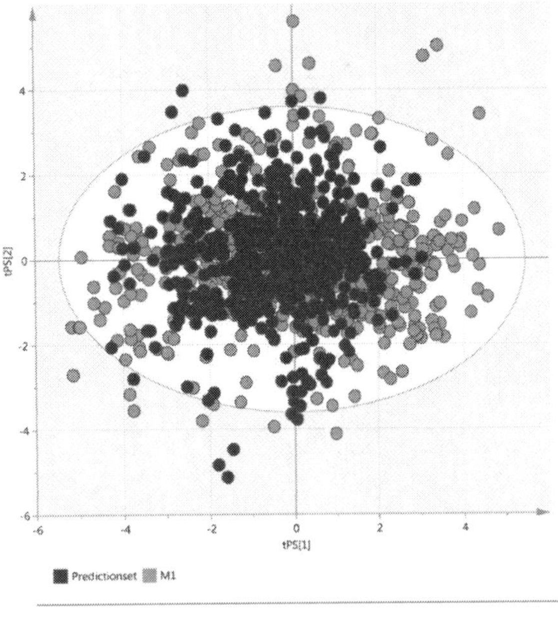

## MVA monitoring

3 hours post model

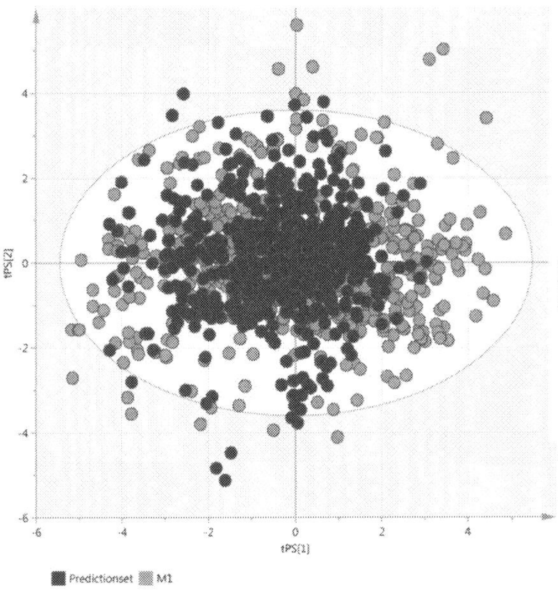


14 hours post model

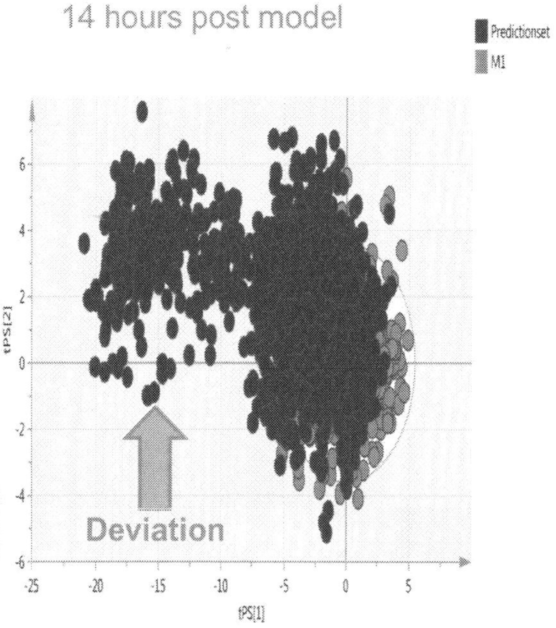


## MVA monitoring

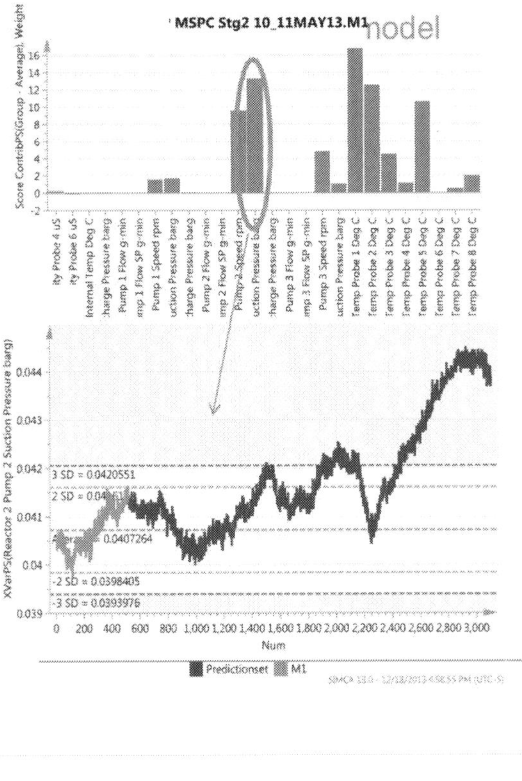


14 hours post model

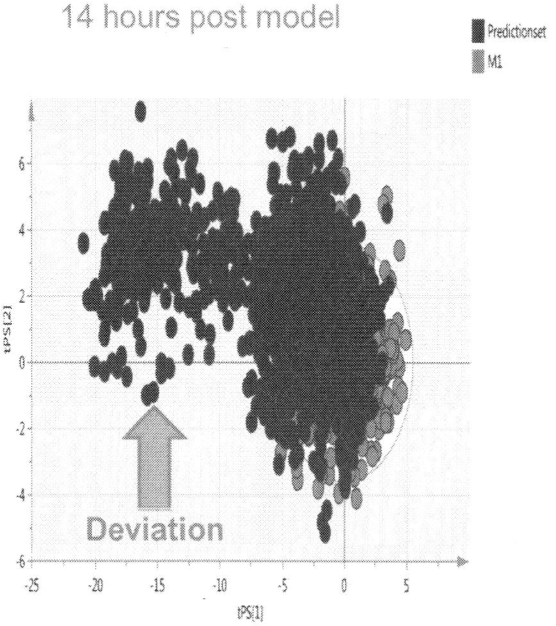

## Control system infrastructure

| | |
|---|---|
| Quality Traffic Light | Quality alarms activated depending on sequence phase<br>Quality met? Yes / No → Collect / DTW |
| Ratio control | Active during controlled state,<br>To maintain stoichiometry,<br>To allow co-ordinated step up step down of each sectionof the process |
| Alarms, trips and interlocks / Sequence Phases | Safety system active throughout<br>Sequence phases drive allowable activities |
| CM / EM | Individual control loops – maintain SP |
| MVA process monitoring | Trending of tags, to detect system drift → operational control<br>Prompt for operations to look at control system to establish need for manual action |
| System Characterisation RTD | Providing timings for start-up, shutdown, alarm settings and DTW actions |

## Control trends

Start-up Sequence

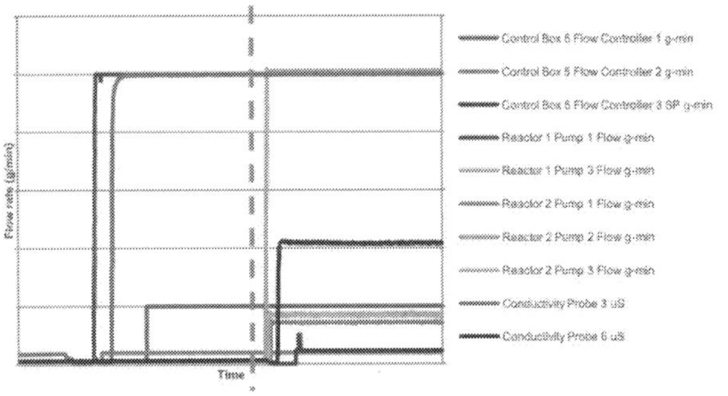


Controlled state

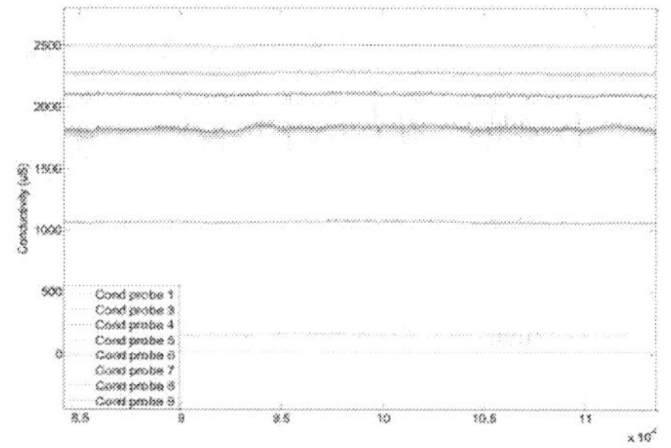


Control Failure

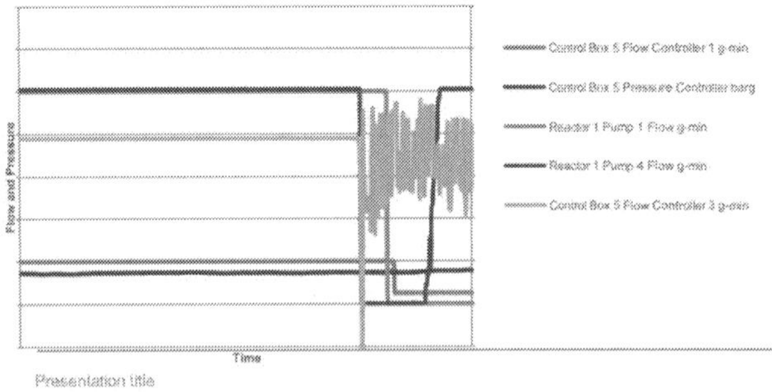


Control Blip

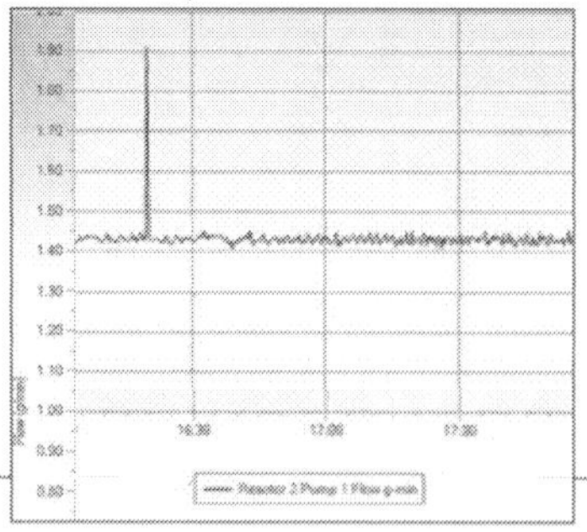

## Summary

## Summary

- Use of continuous technology processes within the Pharma industry requires a change in batch operation mindset
  - Some additional quality levers
  - Some additional challenges
- Continuous processes offer many opportunities for utilisation of combined process and equipment understanding
  - Predictive modelling
  - Scaling capability
  - Operational controls
- Continuous processes offer a wealth of data which can be used pro-actively to manage the process and equipment
  - Data display
  - Use of MVA

## Manufacturing Installation

## Acknowledgements


- Malcolm Berry
- Peter Shapland
- Mark Hughes
- Matt John
- Gill Turner
- Irene Areri
- Peter Hamilton
- Melanie Dumarey
- Yahyah Yahyah
- Neil Hodnett
- Andrew Rutter
- Gareth Alford
- Dave Johns
- Rob Wadley
- Heather Todd
- Peter Clements
- Ben Hagger
- Mathew Callaghan-Till
- Lesley Wong
- Wei-Feng Tan

# 427812 Development & Application of Continuous Crystallization in an Integrated Continuous Pharmaceutical Pilot Plant

**Abstract:**

Development & Application of Continuous Crystallization in an Integrated Continuous Pharmaceutical Pilot Plant

Haitao Zhang, Richard Lakerveld, Patrick Heider, Justin Quon, James Evans, Allan S. Myerson, Bernhardt Trout

Novartis-MIT Center for Continuous Manufacturing

Department of Chemical Engineering, Massachusetts Institute of Technology, E19-502b, 77 Massachusetts Avenue, Cambridge, Massachusetts 02139, USA.

Abstract

The concept of continuous manufacturing to address key challenges faced by the pharmaceutical industry has gained significant interest. A number of research programs related to this topic have been sponsored by pharmaceutical companies, including Pfizer, Novartis, Merck, and Eli Lilly, etc. The Novartis-MIT Center for Continuous Manufacturing has developed a seamless, end-to-end pilot plant for continuous manufacturing of pharmaceuticals including new pieces of technologies to support integrated continuous manufacturing (ICM). Continuous crystallization plays a critical role within ICM processes in terms of final product quality control. Crystallization processes are designed to obtain active pharmaceutical ingredients (API) crystals with controlled purity, size, shape, and polymorphic form in high yield. Knowledge of the continuous crystallization process conditions required to fabricate crystals with controlled characteristics is critical during process development. Real-time control using process analytical technology (PAT) tools is required for the implementation of continuous crystallization within integrated continuous manufacturing (ICM) of pharmaceuticals. However, appropriate selection of PAT tools is challenging, and the design and operation of automated control loops for continuous crystallization within a continuous pharmaceutical process brings forward important questions.

The aim of this contribution is to investigate the performance of a continuous crystallization step operated for a prolonged period of time within an ICM process. The integration of different unit operations and the steady-state operation of those unit operations are of primary concern. The aim is to keep the critical material attributes (CMAs) for each unit operation including the continuous crystallization units within an accepted range at all times through the control of critical process parameters (CPPs). Consequently, the control strategy should possess the capability to tolerate variations from upstream units and should minimize disturbances that propagate to downstream units. Therefore, the selection and implementation of PAT tools and the design of a control strategy are of vital importance for successful ICM of pharmaceuticals. Such a control strategy would rely on automated feedforward and feedback control loops.

Herein, we discusses the process design and operation of a continuous reactive crystallization of aliskiren hemifumarate as part of an ICM pilot plant. Several PAT tools were used within automated control loops to satisfy specifications on the critical materials attributes of the crystallization process. The operational performance of the process was maintained for periods of time over 100 h. The purity of the targeted product exceeded 99%, and the process yield reached 91.4% with the physical properties in specification.

U.S. Food and Drug Administration
Protecting and Promoting Public Health
www.fda.gov

# Continuous and Discontinuous Anti-Solvent Crystallization:

## An Integrated Real-time Process Monitoring and Modeling Approach


**Huiquan Wu, Ph.D.**

Office of Process and Facilities/Office of Testing and Research
OPQ/CDER/FDA


AIChE Annual Meeting, Salt Lake City, UT, Nov. 10, 2015

1


U.S. Food and Drug Administration
Protecting and Promoting Public Health
www.fda.gov


# Disclaimer

The views and opinions expressed in this presentation are only of author, and should not be construed to represent FDA views or policies

2

# Outline

- Why crystallization matters?
- Continuous and Discontinuous Anti-solvent Crystallization
- Modified classical nucleation theory model
- Conclusions
- Acknowledgement

3



# Why Crystallization Matters?

- The drug substance's quality attributes are directly impacted by drug substance crystallization process
  - Crystallinity
  - Solubility
  - Purity
  - Size and Shape
  - Solid state and polymorphic forms
  - Hygroscopicity
  - Residual solvent level
  - Impurity level
- Link to the critical quality attributes (CQAs) of the final drug product

4

U.S. Food and Drug Administration
Protecting and Promoting Public Health
www.fda.gov

# FDA Research on Crystallization Process

- Focus
  - Provide scientific basis and case studies to support FDA's Pharmaceutical Quality Initiatives
  - Gain Process Understanding and Product Understanding
- Work done so far
  - Model system: Naproxen-Eudragit L100-Alcohol-Water
  - Evaluate discontinuous and continuous crystallization
    - Discontinuous: Stepwise addition in multiple aliquots of a fixed amount of antisolvent
    - Continuous: Addition of the same amount of antisolvent at a fixed flow rate
    - All other conditions the same
  - Establish design space and nucleation mechanisms
  - Model construction and verification
- Work to be done in the future
  - Continuous crystallization with other formulation systems

5

U.S. Food and Drug Administration
Protecting and Promoting Public Health
www.fda.gov

# An Integrated Approach to Pharmaceutical Crystallization

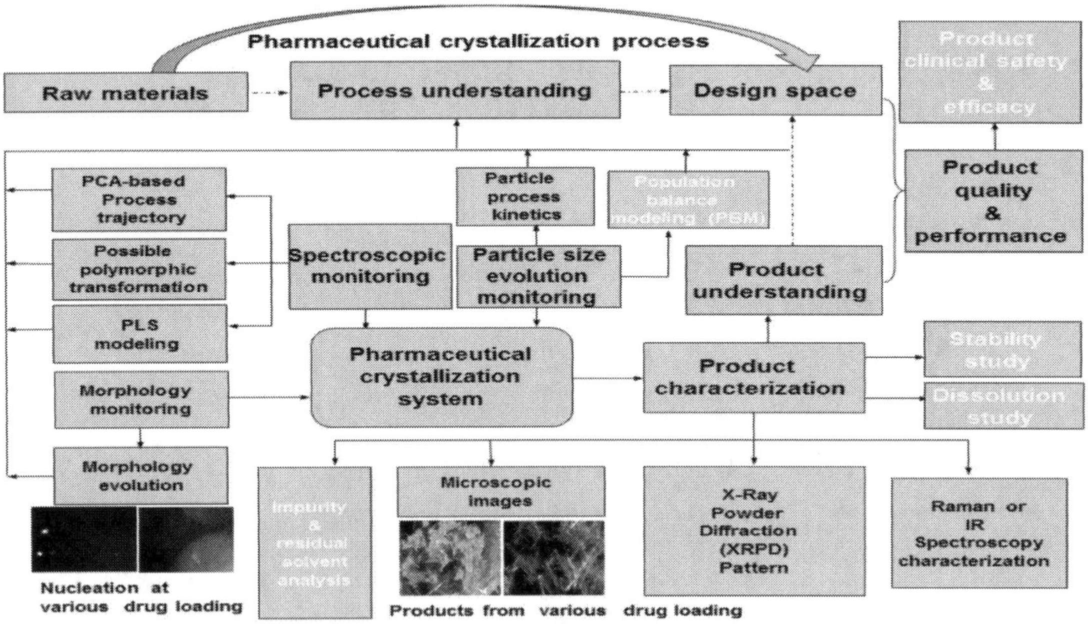


6

# Schematic of Experimental Setup

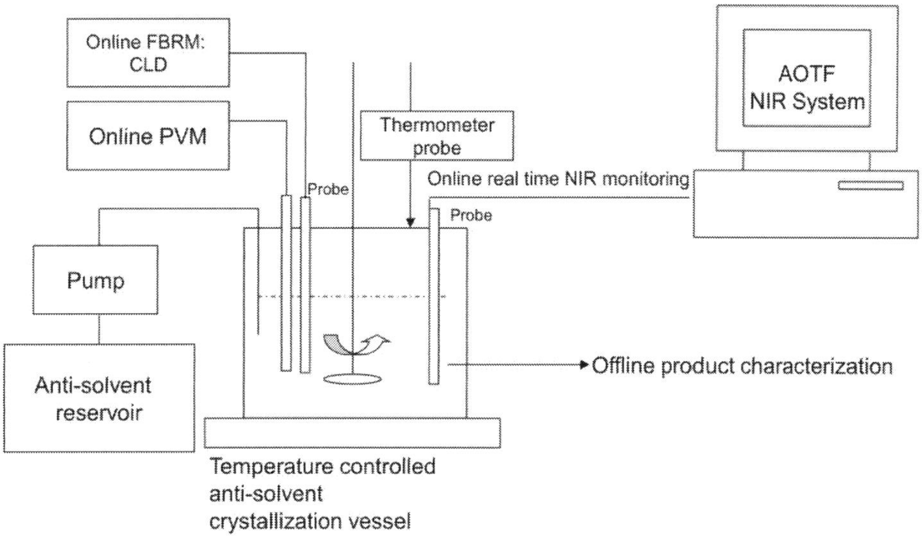


7



# Real Time FBRM Chord-Length-Distribution

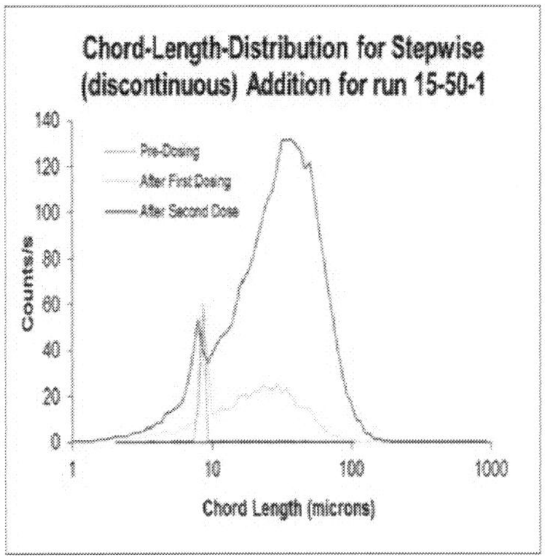


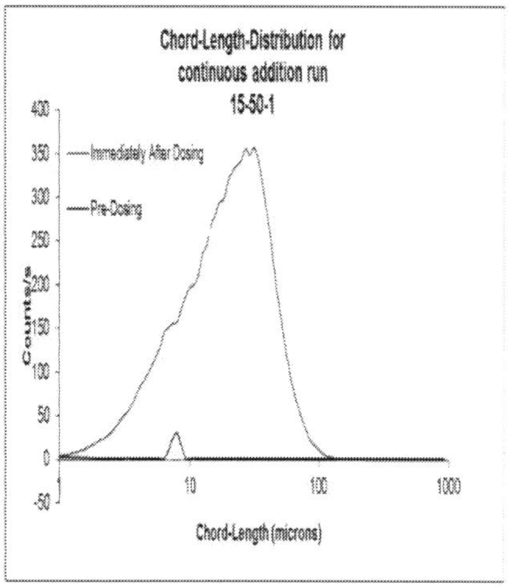


- Continuous addition mode demonstrated a faster generation rate for fine chords (i.e., fine crystals) than discontinuous addition mode

8

U.S. Food and Drug Administration
Protecting and Promoting Public Health
www.fda.gov

# Real Time PVM Images

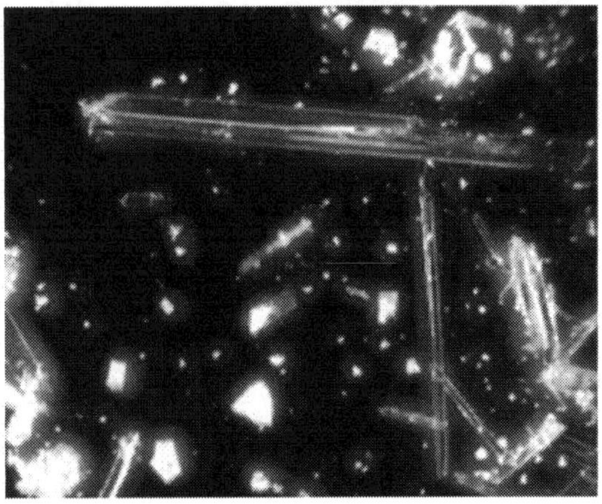

After discontinuous addition of 150 ml anti-solvent at 15°C and 50 rpm

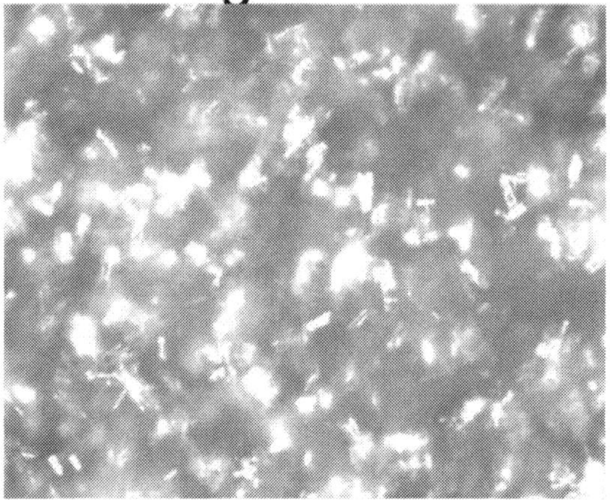

After continuous addition of 150 ml anti-solvent at 15°C and 50 rpm

- Continuous addition mode can generate more uniform CSD than discontinuous addition mode =>a better controllability of the physicochemical properties of the crystalline product

9

U.S. Food and Drug Administration
Protecting and Promoting Public Health
www.fda.gov

# Induction Time vs. Drug Loading (Discontinuous)

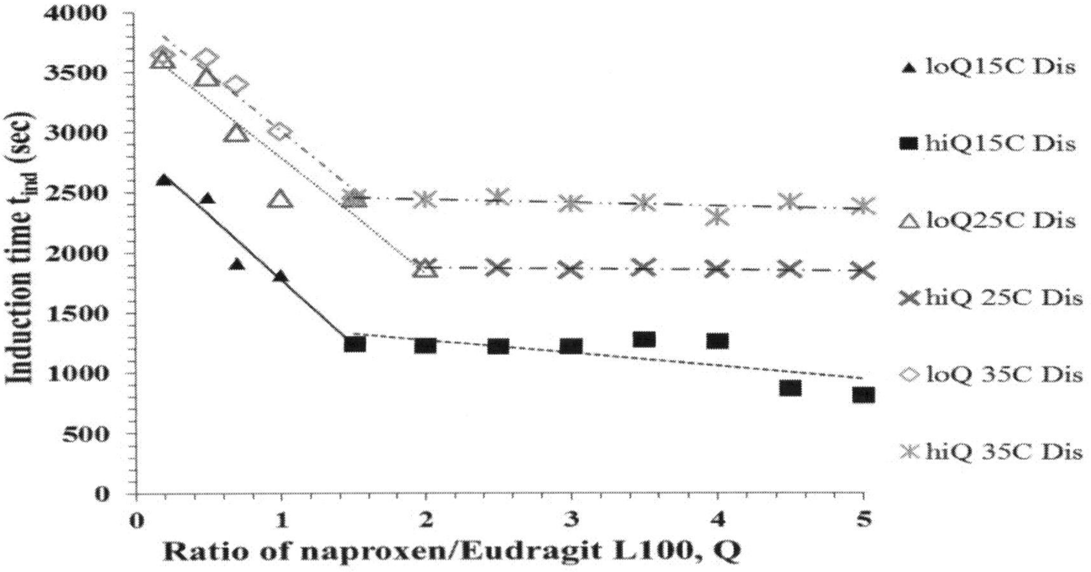


10

U.S. Food and Drug Administration
Protecting and Promoting Public Health
www.fda.gov

# $t_{ind}$ Measurement Verification (Discontinuous)

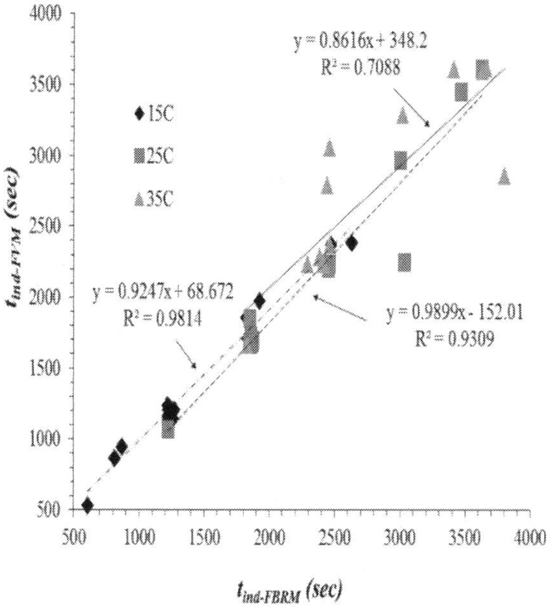


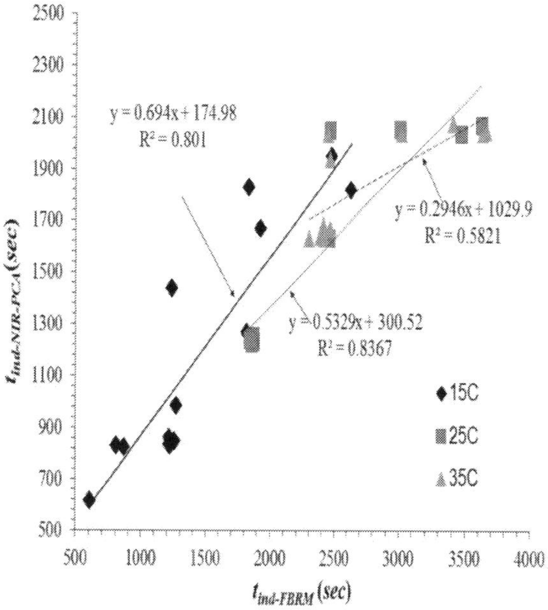


11

U.S. Food and Drug Administration
Protecting and Promoting Public Health
www.fda.gov

# Classical Nucleation Theory (CNT) and Modified CNT (MCNT)

- Solubility calculated by Jouyban-Soltanpour-Acree Equation (2010)
- Super-saturation calculated by mass balance
- For homogeneous nucleation mechanism

$$\ln(t_{ind}V) \approx ln(\frac{N_c}{A_{hom,J}}) + \frac{16\pi\gamma^3 v^2}{3(kT)^3}\frac{1}{(lnS)^2}$$

- For heterogeneous nucleation mechanism

$$\ln(t_{ind}V) \approx ln(\frac{N_c}{A_{het,J}}) + \frac{16\pi\gamma^3 v^2}{3(kT)^3} f(\theta)\frac{1}{(lnS)^2}$$

12

U.S. Food and Drug Administration
Protecting and Promoting Public Health
www.fda.gov

# $\ln(t_{ind\text{-}FBRM}V)$ vs. $(\ln S)^{-2}$ for Discontinuous

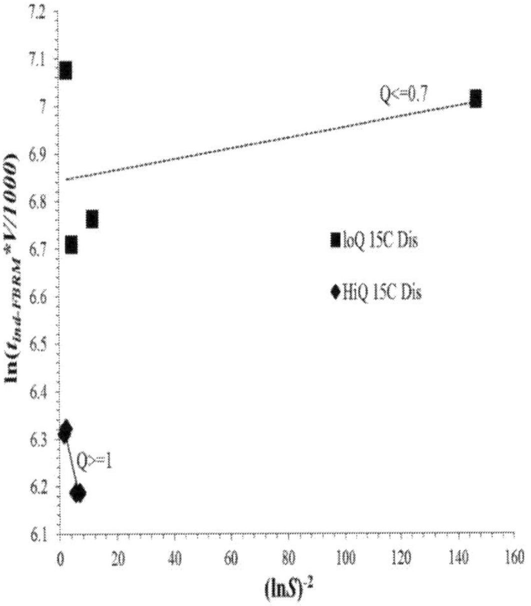


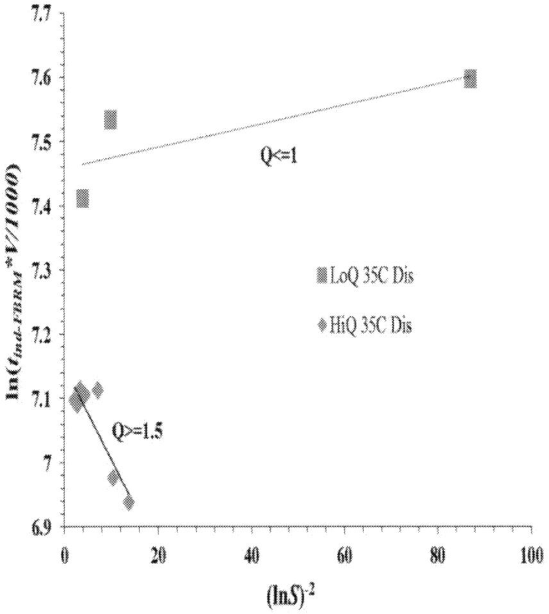




U.S. Food and Drug Administration
Protecting and Promoting Public Health
www.fda.gov

# $\ln(t_{ind\text{-}NIR\text{-}PCA}V)$ vs. $(\ln S)^{-2}$ for Discontinuous

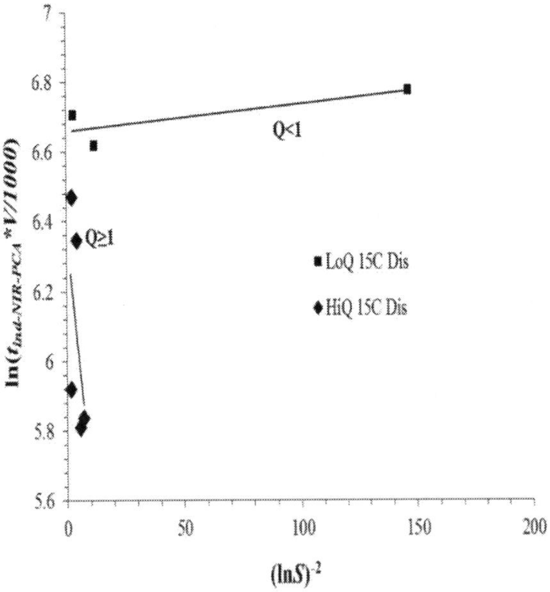


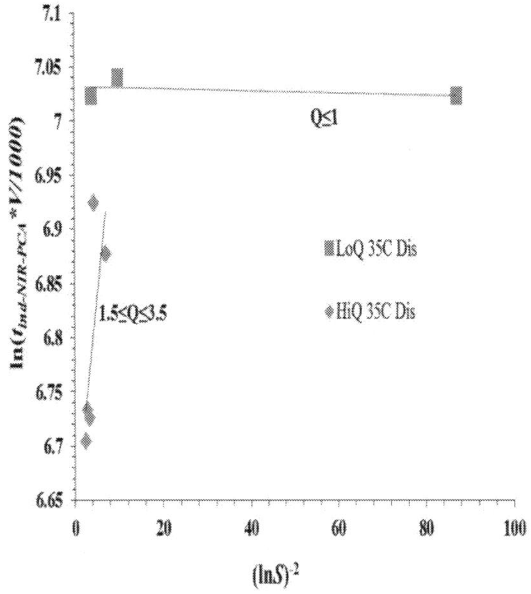

U.S. Food and Drug Administration
Protecting and Promoting Public Health
www.fda.gov

# Continuous Addition of Anti-Solvent

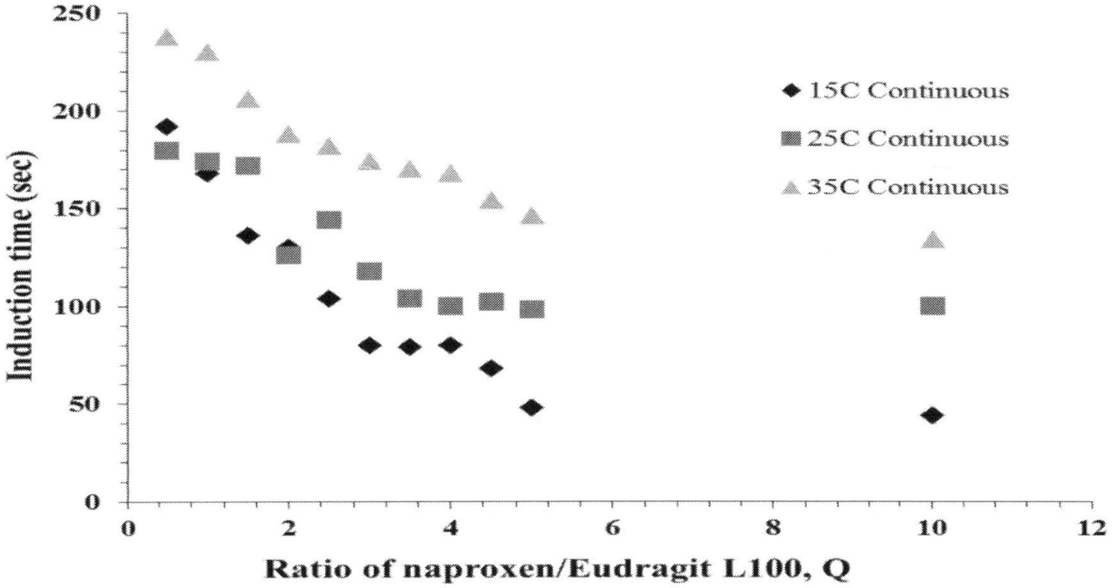


15

U.S. Food and Drug Administration
Protecting and Promoting Public Health
www.fda.gov

# Continuous Addition: CNT

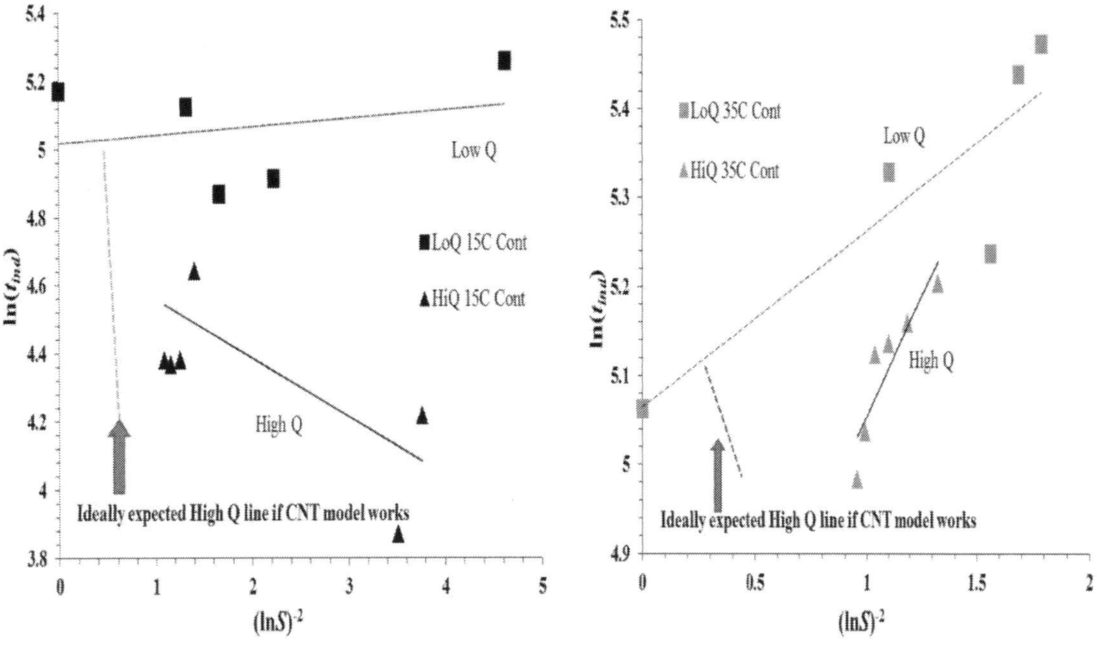


16

U.S. Food and Drug Administration
Protecting and Promoting Public Health
www.fda.gov

## Continuous Addition: MCNT

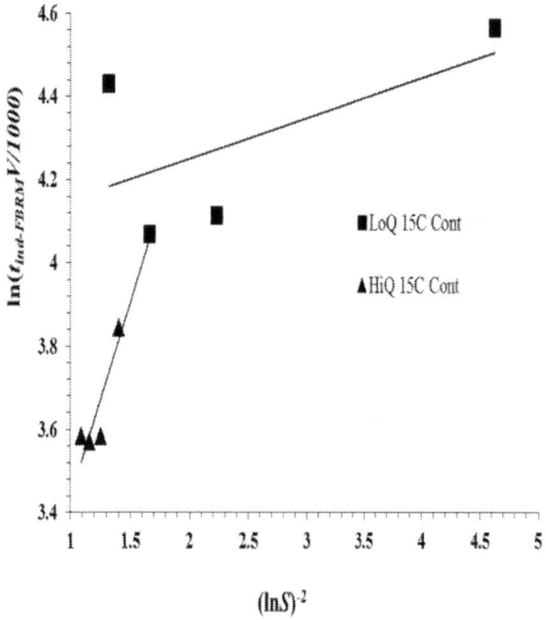


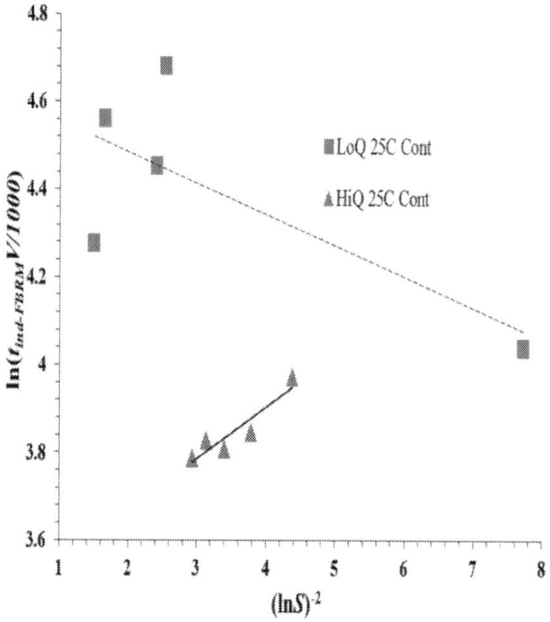


17

U.S. Food and Drug Administration
Protecting and Promoting Public Health
www.fda.gov

## Conclusions

- Integrated PAT measurements (FBRM, PVM and NIR-PCA) together with modified classic nucleation theory can be used to model both continuous and discontinuous antisolvent crystallization process
- Modified classic nucleation theory can capture the dynamic features of continuous and discontinuous anti-solvent addition
- Continuous crystallization can offer advantages over discontinuous crystallization
  - Achieve a desired and more uniform particle size distribution
  - Generate fine crystals faster
- Above approaches offer the opportunity to develop continuous drug substance manufacturing processes

18

U.S. Food and Drug Administration
Protecting and Promoting Public Health
www.fda.gov

# Acknowledgements

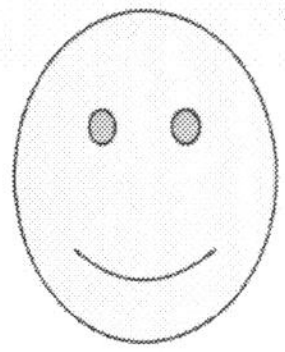


- FDA CDER OPQ Office of Testing and Research
  - ORISE Fellows: Maury White, Ryan Foringer
  - Cindy Buhse, Vincent Vilker (Retired), M.A. Khan(moved)
- FDA CDER OPQ Office of Process and Facilities
  - Rapti Madurawe, Jennifer Maguire, Robert Iser
- FDA CDER RSR 12-42; RSR-13-32; FDA CDER OTR PAT-12


19

U.S. Food and Drug Administration
Protecting and Promoting Public Health
www.fda.gov

# References


- H. Wu and M. Khan, 2010. *J. Pharm. Sci.*, 99(3): 1516-1534.
- H. Wu, M. White, M.A. Khan, 2011. *Int. J. Pharm.*, 405:63-78.
- H. Wu and M.A. Khan, 2011. *J. Pharm. Sci.*, 100 (5): 1969-1986.
- H.Wu, M. White, R. Berendt, R. Foringer, M.A. Khan, 2014. *Ind. Eng. Chem. Res.*, 53 (4) :1688–1701.
- H. Wu, Z. Dong, H. Li, M.A. Khan, 2015. *Org. Process Res. Dev.*, 19(1) : 89-101.
- H. Wu, B. Kurtyka, G. Wu, J. Maguire, R. Madurawe, C. M.V.Moore, R. Iser, M.A. Khan, R.C. Lyon, P. Faustino, L. Buhse, S. Lee, L. Yu. *European Pharmaceutical Review*, 2015, Nov. Issue.


20

# 431784 Process Modeling and Simulation for Continuous Pharmaceutical Manufacturing of Diphenhydramine

**Abstract:**

Continuous Pharmaceutical Manufacturing (CPM) has been recognized as an increasingly promising multidisciplinary research domain, addressing the significant obstacles hampering sustainable profitability (increasing R&D costs, globalized competition, licensing protocols, legal and regulatory landscape changes). Introducing a viable alternative to traditional batch production processes, CPM has an expanding potential capable of achieving lower costs (Schaber et al., 2011), higher yields, more economical heat and solvent use, high mass transfer rates, rapid scalability and elimination of intermediate storage (Calabrese and Pissavini, 2011). The technical viability and process efficiency of CPM prototypes must however be investigated in detail, in order to ensure successful pilot-scale demonstrations (Mascia et al., 2013) which can be used toward reliable production-scale implementations keenly pursued by pharmaceutical corporations (Poechlauer et al., 2013).

Organic flow synthesis literature must thus be (and indeed, has been) extensively surveyed to identify a wide range of Active Pharmaceutical Ingredients (APIs) whose production has already been accomplished in continuous mode at laboratory scale. A wide range of promising API candidates have thus been identified as suitable for CPM implementation (Jolliffe and Gerogiorgis, 2015a), including ibuprofen and artemisinin, which have already been examined in recent publications (Jolliffe and Gerogiorgis, 2015b). Systematic flowsheet synthesis, process modelling and simulation are essential towards rapidly assessment of CPM potential, and are already widely applied for CPM case studies (Gernaey et al., 2012).

The present paper focuses on a plantwide process modeling and simulation study for the production of diphenhydramine, a first-generation antihistamine which has been developed and has anticholinergic, antitussive, antiemetic and sedative properties and is mainly used to treat allergies, drug-induced parkinsonism and other neurological symptoms; moreover, it has a strong hypnotic effect and is and FDA-approved nonprescription sleep aid, especially in the form of diphenhydramine citrate. Our process model is developed on the basis of a published continuous flow synthesis (Snead and Jamison, 2013), which demonstrates that this economically and societally important API can be produced using Plug Flow Reactors (PFRs) with molten ammonium salts, thereby allowing for heating heat well above the boiling point of all reaction components. Moreover, this method also achieves solvent and waste minimization, thereby reducing operating costs and suppressing hazards associated with excess and disposal of chemicals.

Process modelling and steady-state simulations have been conducted toward process design for the respective CPM routes, with novel kinetic expressions developed on the basis of published kinetic data from experimental campaigns and thermodynamic property estimations essential for process design. Economic evaluations of the respective plant construction and operation venture can accordingly be performed on the basis of plantwide mass and molar balances, in order to quantitatively evaluate the viability of a production-scale CPM process.

LITERATURE REFERENCES

1. Calabrese, G.S., Pissavini, S., 2011. From batch to continuous flow processing in chemicals manufacturing. *AIChE J.*, **57**(4): 828-834.

2. Gernaey, K.V., Cervera-Padrell, A.E., Woodley, J.M., 2012. A perspective on PSE in pharmaceutical process development and innovation, *Comput. Chem. Eng.*, **42**(1): 15-29.

3. Jolliffe, H.G., Gerogiorgis, D.I., 2015a. Process modelling and simulation for continuous pharmaceutical manufacturing of ibuprofen, *Chem. Eng. Res. Des.*, **97**: 175-191.

4. Jolliffe, H.G., Gerogiorgis, D.I., 2015b. Plantwide design and economic evaluation of two Continuous Pharmaceutical Manufacturing (CPM) cases: Ibuprofen and artemisinin, *Comp.-Aided Chem. Eng.* (in press).

5. Mascia, S., Heider, P.L., Zhang, H. et al., 2013. End-to-end continuous manufacturing of pharmaceuticals: Integrated synthesis, purification and final dosage formation, *Angew. Chem. Int. Ed.* **52**(47): 12359-12363.

6. Poechlauer, P., Colberg, J., Fisher, E., et al., 2013. Pharmaceutical roundtable study demonstrates the value of continuous manufacturing in the design of greener processes, *Org. Proc. Res. Dev.* **17**(12): 1472-1478.

7. Schaber, S.D., Gerogiorgis, D.I. et al., 2011. Economic comparison of integrated continuous and batch pharmaceutical manufacturing: a case study, *Ind. Eng. Chem. Res.* **50**(17): 10083-10092.

8. Snead, D.R., Jamison, T.F., 2013. End-to-end continuous flow synthesis and purification of diphenhydramine hydrochloride featuring atom economy, in-line separation, and flow of molten ammonium salts, *Chem. Sci.* **4**(7): 2822-2827.

# 422358 Enzymatic Assembly for Increased Methanol Utilization

**Abstract:**

Methanol is an important feedstock in the chemical industry but has had trouble gaining traction as a feedstock in biotechnological processes due to the difficulty of bacteria utilizing it as a carbon and energy source. The first enzyme, methanol dehydrogenase (MDH), which converts methanol to formaldehyde favors the reverse reaction and this is the key in limiting engineered E. coli cells from utilizing methanol at a fast rate. In metabolic engineering there are myriad ways to increase flux through a pathway to drive product formation. To bypass this thermodynamic barrier, we created a "kinetic trap" to prevent formaldehyde from accumulating and enable methanol to be consumed continuously at the highest rate. Several protein-ligand interaction pairs were used to link methanol dehydrogenase (MDH) with hexulose phosphate synthase (HPS), which irreversibly converts formaldehyde and ribulose 5 phosphate to hexulose 6-phosphate. The resulting MDH-HPS complexes enabled a 20-fold improvement in hexulose 6-phosphate production compared to unassociated enzymes confirming that enzymatic assembly improves methanol utilization.

# 432599 New Enzymes for the Hybrid Enzymatic and Organic Electrocatalytic Cascade for the Complete Oxidation of Glycerol


**Abstract:**

**New enzymes for the hybrid enzymatic and organic electrocatalytic cascade for the complete oxidation of glycerol**



Sofiene Abdellaoui[a], David P. Hickey[a], Matthew S. Sigman[a] and Shelley D. Minteer[a]

[a]*Department of Chemistry, University of Utah, 315 S 1400 E Rm 2020, Salt Lake City, Utah, 84112, USA.*

*minteer@chem.utah.edu*


The complete electrochemical oxidation of the biofuel, glycerol, to $CO_2$ using a hybrid enzymatic and organic catalytic system has been demonstrated[1]. This system combines an organic catalyst, 4-amino-(2,2,6,6-Tetramethylpiperidin-1-yl)oxy (TEMPO-$NH_2$) with oxalate oxidase (OxOx), resulting in the complete electrochemical oxidation of glycerol at a carbon electrode (Figure 1). This hybrid approach consists of five initial oxidative steps (by TEMPO), resulting in the oxidation of glycerol to mesoxalic acid (**1** → **6**, Figure 1). A combination of OxOx and TEMPO then facilitates the oxidation of mesoxalic (**6**) acid to glyoxalic acid (**7**), oxalic acid (**8**), and finally, $CO_2$.

Nevertheless, this system is partly limited, by the weak overlap in pH profiles of OxOx and TEMPO, the first working at acidic pH (pH 4.0) and the latter preferring alkaline pH. Moreover, the electrons generated in the cascade reaction by enzymatic oxidative steps are not mediated to the electrode (4 / 16 electrons) and the OxOx substrate range is limited.

In order to improve this system, we are investigating new enzymes to replace OxOx. Like this latter, Oxalate decarboxylase (Oxdc) belongs to the cupin superfamily of proteins and incorporates mononuclear manganese ions coordinated by four amino acids[2,3]. This enzyme is able to transform oxalate to formate and $CO_2$, with $O_2$ acting as a unique cofactor which can be used to catalyze a side oxidation reaction with o-phenylenediamine[4] or 2,2'-azino-bis(3-ethylbenzothiazoline-6-sulphonic acid) (ABTS). This study also aims to investigate other enzymes working at higher pH and the ability to work with redox mediators as dehydrogenases.

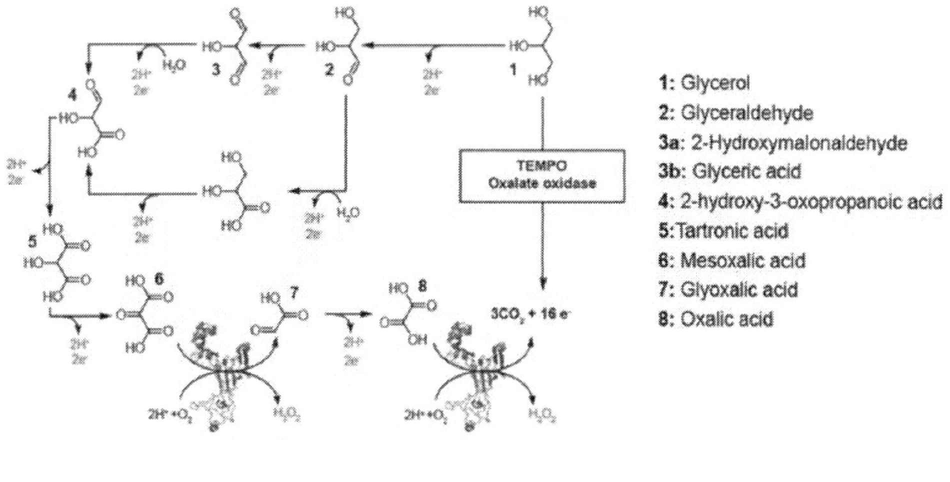


Figure 1: Electrocatalytic Oxidation Cascade of Glycerol by TEMPO and Oxalate Oxidase[1]


(1) Hickey, D. P.; McCammant, M. S.; Giroud, F.; Sigman, M. S.; Minteer, S. D. *Journal of the American Chemical Society* **2014**, *136*, 15917.

(2) Moomaw, E. W.; Angerhofer, A.; Moussatche, P.; Ozarowski, A.; Garcia-Rubio, I.; Richards, N. G. J. *Biochemistry* **2009**, *48*, 6116.

(3) Tanner, A.; Bowater, L.; Fairhurst, S. A.; Bornemann, S. *Journal of Biological Chemistry* **2001**, *276*, 43627.

(4) Emiliani, E.; Bekes, P. *Archives of biochemistry and biophysics* **1964**, *105*, 488.

# 425193 Biocatalytic Carboxylation of Furans for the Production of 2,5-Furandicarboxylic Acid

**Abstract:**

This work seeks to explore the relationship between, and applications of, two co-dependent enzymes, 2,5-furandicarboxylic acid decarboxylase I and II, which were initially identified as isofunctional components of a 5-hydroxymethylfurfural degrading pathway in Cupriavidus basilensis. (1)
Once the nature of the interaction between these enzymes is understood, we apply traditional reaction engineering, aided by thermodynamic modeling using the electrolyte Perturbed-Chain Statistical Associating Fluid Theory equation of state, to drive the natively catalyzed reaction in reverse. (2) In this manner we produce a reaction pathway with novel chemistry enabling the production of 2,5-furandicarboxylic acid, a key substitute for the terephthalic acid in polyesters, from furfural and carbon dioxide which are bio-renewable and readily abundant resources.

1. Koopman, F.; Wierckx, N.; de Winde, J. ; Ruijssenaars, H. J. Identification and characterization of the furfural and 5-(hydroxymethyl)furfural degradation pathways of Cupriavidus basilensis HMF14. Proc. Natl. Acad. Sci. U.S.A. 2010, 107 (11), 4919-4924.
2. Cameretti, L. F.; Sadowski, G.; Mollerup , J. M. Modeling of aqueous electrolyte solution with the perturbed-chain statistical associated fluid theory. Ind. Eng. Chen. Res. 2005, 44 (9), 3355-3362.

# 429828 Innovative Reactor Design and Execution Towards Development of a Biocatalytic Route for a Drug Substance Intermediate

**Abstract:**

This work highlights ongoing efforts in the development of a biocatalytic route to a key advanced intermediate in support of an active clinical development program. At the forefront of this development, the team was tasked with identification and characterization of scale-appropriate reactors to perform the reaction. The biocatalytic transform is a hydroxylation, and as such requires oxygenation of the aqueous medium during the course of the reaction. Complicating the matter, the reaction requires catalytic quantities of iron which is susceptible to background oxidation, creating a need for a careful balance between adequate chemical reaction rate, and over-oxidation of iron, ultimately leading to a stalled reaction.

In this work, we discuss the initial reactor design and characterization as well as preliminary demonstration of the system at scales from 10 to 1000 ml. In particular, this work highlights development and application of novel reactor designs and experimental strategies towards the establishment of a robust manufacturing process. Furthermore, advanced experimentation techniques, including integrated data capture and analysis, have been employed to build process understanding as rapidly as possible, especially considering the complex reactor design requirements this chemistry poses. Additionally, we have repurposed an AMBR15, a parallel micro bioreactor commonly employed for mammalian cell culture, to enable rapid screening of a wide variety of experimental parameters. This experimental strategy has enabled deep insight into the suspected mode of catalyst deactivation, and offers an opportunity for robust operation and parameter optimization moving forward.

# 425265 Drivers of P450 Selectivity in the Taxol Biosynthetic Pathway

**Abstract:**

Isoprenoids are a large and diverse family of natural products with uses ranging from next-generation biofuels (isopentenol and farnesene), flavors and fragrances (menthol and nootkatone), as well as pharmaceuticals (paclitaxel and artemisinin). One of the biggest contributors to the success of isoprenoids is their diversity. This diversity is generated primarily through both the initial cyclization step (in which a linear backbone may undergo a complex rearrangement), and from oxidation steps catalyzed by a family of enzymes known as cytochrome P450 monooxygenases (P450's). Paclitaxel (Taxol®), originally derived from the pacific yew tree, Taxus cuspidata, is a potent chemotherapeutic agent with sales exceeding $3 billion/year. Despite the importance of this molecule and intense study of its synthesis, both biological and chemical, its biosynthesis and pathways to alternative products remain poorly understood and industrial production relies on non-recombinant plant-cell culture. Here, we investigate P450's as a key driver in the natural diversity of the Taxol-related compounds and their importance in high-level production of Taxol precursor molecules. We illustrate the drivers of substrate promiscuity and reaction selectivity on an enzymatic level and then leverage this understanding to improve the heterologous production of Taxol-precursor molecules. This work lays a foundation for reliable, scalable production of paclitaxel both through reconstruction in heterologous hosts and through a deeper understanding, native plant-cell culture.

# 430613 Elucidation of Six New Enzymes That Complete the Biosynthetic Pathway to the Etoposide Aglycone in Mayapple

**Abstract:**

Etoposide is an effective anticancer agent currently utilized in the clinic. The drug is semi-synthetically derived from podophyllotoxin, a lignan secondary metabolite isolated from *Podophyllum* rhizomes. However, the growing demand for the compound has critically endangered the traditional source, *Podophyllum hexandrum*. Though other podophyllotoxin-producing plants have been identified, it is desirable to relieve the severe pressure on natural sources. Current chemical synthetic routes, however, are not economically viable. Alternatively, we suggest metabolic engineering of a tractable host for overproduction, requiring the elucidation of the minimum set of genes sufficient for podophyllotoxin biosynthesis.

Herein, we present a platform for accelerated discovery in non-model plants. This involves the development of RNA-Sequencing conditions for the identification of potential gene candidates, and the screening of these candidates in a *N. benthamiana* system coupled with liquid chromatography-mass spectrometry metabolomics. We describe the usage of this platform for the identification of six new enzymes from *Podophyllum hexandrum* that complete the biosynthetic pathway to 4'-desmethyl-epipodophyllotoxin, a more direct precursor to Etoposide than podophyllotoxin. We also describe the full reconstitution of the pathway in *N. benthamiana* from the early lignan precursor, pinoresinol. The discovered genes enable future metabolic engineering efforts in *Saccharomyces cerevisiae* for industrial production of the anti-cancer agent.

# 430774 Recombinant Expression of Thermostable Endo-Arabinase in Vitro and in Planta for Generating Functional Oligosaccharides from Plant Cell Wall for Colon-Specific Health Benefits

**Abstract:**

Lignin-deficient(LD) plant fibers, e.g., sugar beet pulp and rice bran, are rich sources of functional cell wall polysaccharides. Used primarily as low-value animal feed, these established feedstock can be capitalized on for generation of value-added bioproducts, which will improve the economic viability and competitiveness of sugar beet and rice industry. The overall *goal* of this project is to develop an efficient enzymatic platform for generating functional oligosaccharides, specifically, feruloylated arabino-oligosaccharides (FAOs), from LD plant fibers. FAOs may be used in food and feed applications for healthful colon functioning through prebiotic, anti-inflammatory and mucosal immunomodulatory activities. FAOs of defined structure can be released from LD fibers through selectively cleaving the arabinan chain of cell wall polysaccharides by a key glycohydrolase - arabinase (ABN). In this study, a thermostable endo-1,5-a-L-ABN from *Bacillus thermodenitrifican* was expressed in yeast *(Pichia pastoris)* to produce significant quantity of enzyme for testing its activity of digesting plant fibers. Recombinant enzyme was secreted into culture media at a yield of -70 mg/L and showed a specific enzymatic activity of 350 U/mg. ABN was also expressed *in planta* (in tobacco) to assess its function in *post-harvest* modification of cell wall polysaccharides to facilitate *in situ* FAOs releasing. Transgenic tobacco accumulated -20 (Jg/gFW active ABN, and were not phenotypically different from wild-type plants. Finally, T84 epithelial cell culture was established to assay the anti-inflammatory activities of FAOs. Preliminary study indicated that FAOs isolated from rice bran significantly enhanced the expression of occludin, a biomarker of epithelial cells' immune response to FAOs.

# 425141 Improved Biocompatiblity with Novel Biomaterials

**Abstract:**

Synthetic or natural materials have been commonly used in many biomedical applications, while their biocompatibility with cells or human body is a central concern. Recently, the development of novel polymeric biomaterials with new chemistry and/or structures has significantly improved the biocompatibility in several major biomedical applications. One important advance are zwitterionic polymers, which contain balanced-charged biomimetic moieties similar to the hydrophilic moieties in the cell membranes. Zwitterionic polymers are superhydrophilic, they can strongly bind water molecules, and therefore they are highly resistant to nonspecific protein adsorption. This important property renders their ignorable nonspecific interaction with living cells and thus no negative interaction with the living cell or tissue functions. In this talk, we will present several most recent examples of novel biocompatible materials and their excellent performance in applications such as tissue engineering.

# 429873 Drug Delivery Particles Made from Zwitterionic Polymer Micelles

**Abstract:**

Zwitterionic polymers have found a wide range of applications based on excellent non-fouling properties. It has been demonstrated that their use as surface coatings can effectively resist binding from proteins, cells, and full blood. Nanoparticles coated with zwitterionic polymers exhibited long blood circulation time in vivo. These excellent properties were attributed to the unique superhydrophilicity of these polymers providing strong hydration effects. In this presentation, I will introduce a so-called sharp contrast micelle system. Each micelle molecule composes of a superhydrophilic zwitterionic polymer domain and a superhydrophobic lipid domain. The polarity contrast between the two domains is drastically "sharper" than most conventional micelle molecules. We will discuss the special synthetic route that led to the reaction between the two polarity distinct blocks to form the sharp contrast surfactant. The unique behavior of this sharp contrast micelle and its use for drug delivery purposes will be demonstrated. Particularly, colloid systems made from this micelle achieved unprecedented stability comparing with conventional stabilizing technologies.

# 431837 Influence of Composition and Reaction Conditions on the Static and Dynamic Properties of Biodegradable Poly(curcumin beta amino ester) Networks

**Abstract:**

For the past 30+ years, hydrogels have garnered a lot of attention due to their applications in drug delivery, tissue engineering and regenerative medicine. Hydrogels are typically biocompatible, owing to their high water absorption capacity and the ability to tailor mechanical properties to achieve biomimetic performance. Biodegradable hydrogels are advantageous for biomedical applications as the hydrogel can degrade and be removed from the body once its function is achieved. A one-step hydrogel synthesis method was recently developed enabling preparation of biodegradable poly(antioxidant β amino ester) hydrogels for antioxidant delivery without the use of free radical chemistry [1]. Poly(antioxidant β amino esters) synthesized using this method improve the delivery of antioxidants that display limited aqueous solubility and environmental instability. The drug of interest in this work is curcumin, which has been shown to possess antioxidant, anti-inflammatory, anti-cancer and anti-angiogenesis properties. In this study, we explore a series of poly(curcumin β amino ester) PCbAE hydrogels formed by polymerization of commercial diacrylate (PEGDA) and a primary diamine in combination with acrylate-functionalized curcumin (i.e., curcumin diacrylate or CDA). PCbAE undergoes hydrolytic degradation to release curcumin in its original form. The network properties of these gels were characterized using controlled swelling measurements, aqueous degradation and dynamic mechanical analysis (DMA) as a function of independent parameters such as reaction condition, relative amount of curcumin diacrylate (CDA), amine crosslinker, diacrylate monomer and ratio of total acrylate to amine (RTAA). In order to efficiently design PCβAE for specific biomedical applications, it is essential to understand the impact of synthesis conditions on the thermomechanical and drug release properties of these materials.

An improved film synthesis protocol based on ring casting with *in-situ*polymerization was developed to synthesize PCβAE films with reproducible network properties. Cross-linked polymer networks were formed by pre-polymerization of primary diamine with PEGDA in anhydrous organic solvent followed by addition of CDA. The liquid reaction mixture was deposited in an enclosed casting ring and incubated at 60°C in a convection oven, resulting in the formation of polymer films of thickness 0.1 to 0.3 mm. The relative content of curcumin diacrylate in PCβAE was systematically varied from 0:100 (CDA:PEG400DA) to 90:10 (CDA:PEG400DA) using the same primary diamine (TTD) and an RTAA value of 1.0. Curcumin, by its rigid character, alters the flexibility and architecture of the network, which then affects the degradation and release profile. An increase in CDA content resulted in an increase in glass transition temperature from -39°C to 67°C, as measured by DMA, and overall degradation times ranging from 3 hours to 24 hours were obtained. To assess the effect of length and relative hydrophilic character of the crosslinker on the hydrogel degradation and mechanical properties, three primary diamine crosslinkers viz. 4,7,10-trioxa-1,13-tridecanediamine (TTD), 2,2'(ethylenedioxy) bis ethylamine (EDBE) and hexamethylene diamine (HMD) were employed. These were combined with commercial poly(ethylene glycol) diacrylates of varying molecular weight (diethylene glycol diacrylate (DEGDA), PEG400DA and PEG700DA) to provide a wide range of network structures for the incorporation of curcumin diacrylate with controlled release profiles. A detailed discussion of the spectrum of performance properties accessible with these compositional variants will be presented.

**References:**

1. Wattamwar PP, Biswal D, Cochran DB, Lyvers AC, Eitel RE, Anderson KW, Hilt JZ, and Dziubla TD. Acta Biomater 2012;8(7):2529-2537.

# 414056 Encapsulation of Single ß Cells in Microgels Via Integrated Microfluidic Flow Focusing and Photopolymerization

**Abstract:**

Encapsulation of ß cells in biocompatible hydrogels is a highly promising therapy for Type 1 Diabetes Mellitus, which avoids post-transplantation immunorejection. Many factors affect post-encapsulation cell viability and function, including monomer molecular weight and concentration, solution composition, and cell-cell interactions, among others. The described microfluidic droplet generation technology provides a platform for the encapsulation of single cells and the high throughput screening of cell response to encapsulation. Polyethylene glycol diacrylate (PEGDA), a versatile and widely used hydrogel-forming macromer, is employed as the encapsulating phase and is photopolymerized in situ immediately following droplet formation. However, atmospheric oxygen inhibits the photopolymerization of PEGDA, an issue that is dramatically exacerbated at the microscale, and particularly so when using a fluorinated oil carrier phase within channels fabricated from oxygen permeable poly(dimethylsiloxane) (PDMS). To overcome the challenge, we present a nitrogen micro-jacketed microfluidic device to locally control the oxygen concentration in the system. The reduction of oxygen allowed free radical polymerization to proceed while also mitigating the production of deleterious reactive oxygen species, with which cell laden hydrogel microparticles can be produced. Employing the oxygen-controllable microfluidic device, we were able to encapsulate single cells within photopolymerized microscale (~25 µm) droplets with high post-encapsulation cell viability. This platform was successfully used for -cell encapsulation and matrix parameter screening.

# 421182 Agent-Based Simulation of Vascularized Bone Formation in Degradable Scaffolds

**Abstract:**

The success of engineered bone tissue growth in biomaterial scaffolds is determined by the rich vascular supply and high mineralization. Blood vessel formation through angiogenesis from host tissue is the common mechanism for scaffold vascularization. Bone growth can be induced by either external growth factor supply or ceramic incorporation to stimulate cell differentiation and growth. Calcium phosphates, such as tricalcium phosphate (TCP), and hydroxyapatite (HA) are widely used materials due to their high biocompatibility and osteoinductivity. Interactions between blood vessel formation, bone cell growth and modified biomaterial structure are often difficult to study with experimentation alone. Computational models are powerful tools to test different hypothesis and strategies to explore promising alternatives in an organized manner. Due to the complexity and heterogeneity involved in the biomaterial environment, conventional mechanistic models are sometimes inadequate to describe the underlying biological processes. Agent-based models (ABM) represent the individual decision makers in the system such as cells, organelles, genes, or molecules by software agents while allowing an easy integration with continuous models for more realistic environment description. We have previously developed an ABM to study angiogenesis (Mehdizadeh et al., 2013) and osteogenesis (Bayrak et al., 2014) in porous, growth-factor-loaded biomaterials. The current work extends and modifies the model to investigate bone tissue regeneration and study the effects of HA and TCP incorporation to degradable, porous, biomaterial structure. The model showed bone regeneration can be induced using HA- and TCP-loaded scaffolds and suggested the optimum ratio between these materials.

## References

Bayrak, E. S., Mehdizadeh, H., Akar, B., Somo, S. I., Brey, E. M. & Cinar, A. Agent-Based Modeling Of Osteogenic Differentiation Of Mesenchymal Stem Cells In Porous Biomaterials. Engineering In Medicine And Biology Society (EMBC), 2014 36th Annual International Conference Of The IEEE, 2014. IEEE, 2924-2927.

Mehdizadeh, H., Sumo, S., Bayrak, E. S., Brey, E. M. & Cinar, A. 2013. Three-Dimensional Modeling Of Angiogenesis In Porous Biomaterial Scaffolds. *Biomaterials*, 34, 2875-2887.

# 425271 Substrate Modulus and Pore Size of 3D Scaffolds Fabricated By Templated Fused Deposition Modeling Regulate Osteogenic Differentiation

**Abstract:**

The properties of the extracellular matrix, including elastic modulus, porosity, pore size, and curvature, are known to regulate cell fate in a number of physiological processes. Biomimetic 3D systems are needed for investigating interactions between cell populations and the microenvironment. Three dimensional polyurethane (PUR) scaffolds with tunable rigidity (10-900 MPa) and pore size (423-557 μm) were fabricated by a new template-Fused Deposition Modeling (t-FDM) process to investigate the effects of substrate modulus and pore size on osteogenic differentiation of rat bone marrow-derived mesenchymal stem cells (BMSCs). Total protein assay indicated that modulus and pore size had minimal effects on cell proliferation. In contrast, BMSCs were more metabolically active and migrated faster on rigid substrates. Gene expression of osteogenic markers, including Runx2, Collagen-1, Fibronectin, and Osteopontin was up-regulated on rigid scaffolds. Mineralization of BMSCs on D21 was significantly up-regulated on rigid scaffolds with decreasing pore size, suggested by Alizarin Red S staining and SEM. These findings would guide the rational design of cell-responsive scaffolds that recapitulate the bone microenvironment for restoration and repair of bone damaged by trauma or disease.

# 421523 Preparation of Tissue Plasminogen Activator Loaded Microbubbles for Potential Application in Ischemic Stroke Treatment

**Abstract:**

Microbubbles are well known as ultrasound contrast agents for medical ultrasound imaging due to their ability to scatter and reflect ultrasound. By incorporating drugs into their shell layers, microbubbles could be used as drug carriers. Ultrasound could serve as a trigger which facilitates the release and local delivery of drugs from the microbubbles. The only FDA approved thrombolytic agent for ischemic stroke patients is tissue plasminogen activator (tPA). However, due to the small therapeutic time-window and side effect of intracerebral haemorrhage, the proportion of patients that has received intravenous injection of tPA is fairly low. In this study, we aim to fabricate tPA-loaded microbubbles as combination therapy agent to overcome the small therapeutic time frame and reduce the dose of tPA which could make the thrombolytic agents safer. Coaxial electrohydrodynamic atomization (CEHDA) method was explored to produce the tPA-loaded microbubbles. First, bovine serum albumin (BSA) was used as shell layer material. The effects of operating conditions such as voltage, core and shell flow rate ratio and tPA/BSA volume ratio on the microbubbles size were investigated. Moreover, phospholipid was explored as the shell materials of microbubbles to carry tPA. The results showed coaxial electrohydrodynamic atomization (CEHDA) has the capability to fabricate tPA-loaded microbubbles. BSA microbubbles with mean diameter of ~41 um can be obtained under optimal operating conditions. Above the optimum applied voltage of 12.8kV, no further decrease of mean microbubble diameter can be observed. However, the bubble size can be further decreased to ~38 um and ~28 um by increasing the conductivity and adding 1,3-propanediol into shell layer, respectively. The results also showed that by using lipid as shell material, the tPA-loaded microbubbles size can be further diminished to ~8 um. The acoustic performance was examined and the results showed that the burst of fabricated phospholipid microbubbles can be triggered by ultrasound with 2MHz frequency. Compared with BSA microbubbles, lipid microbubbles with tPA loaded in shell layer showed a higher potential for the ischemic stroke treatment.

# 419285 A Wearable Integrated Sensor Array for Perspiration Analysis

**Abstract:**

Wearable technologies can play a significant role in realizing personalized medicine through continuously monitoring an individual's physical and physiological states. Most currently developed wearable technologies are capable only in tracking the physical activities of an individual and fail to provide insight into the individual's state of health. Human sweat contains the physiologically rich information needed to infer an individual's sate of health and is an excellent candidate for non-invasive monitoring. The wearable sweat sensors can serve as an ideal platform for a wide range of real-time healthcare monitoring such as exercise-induced dehydration and medical diagnosis.

# 428893 Curved and Self-Folded Polymer-Graphene Sensors

**Abstract:**

Graphene has exceptional electronic, mechanical and optical properties of relevance to sensing applications. In order to create three dimensional curved and flexible sensors we describe the use of strain engineering in graphene/SU-8 bilayers to form curved and folded sensors and discuss applications as stimuli responsive materials, electronic and optical sensing.

# 433911 Paper-Based Home Biosensor for Detecting Phenylalanine from a Sample of Whole Blood

**Abstract:**

Biosensors are important tools for measuring concentrations of specific analytes found in biological fluids. Paper-based microfluidic devices can offer many benefits to the field of biosensor development. Paper-based materials are inexpensive, easily disposable, and readily available. Small volumes of fluid can be used in these devices, which decrease the need for costly reagents. And capillary flow in paper materials removes the need for pumping equipment. The topic of this presentation is a paper-based, biosensing device that can be used for therapeutic, home monitoring by people with the genetic disorder Phenylketonuria (PKU.)

People affected by PKU are required to maintain a strict, low Phenylalanine (Phe) diet to avoid irreversible, adverse health affects. Maintaining appropriate levels of Phe can be especially challenging for young children, adolescents and pregnant women. Much like a glucose test kit helps a person with diabetes maintain appropriate glucose levels, a device that can monitor the concentration of Phe in a finger prick of blood could aid a person with PKU maintain appropriate Phe levels and adhere to diet therapy. Current testing for Phe is laboratory-based, requiring expensive equipment or trained technicians.

Our proposed solution is a semi-quantitative, paper-based assay that utilizes a well-known, two-step chemical reaction to produce a colorimetric response that correlates Phe concentration to color intensity. The assay accepts a finger prick volume of blood (40 μL) into an asymmetric polymeric membrane that filters out cellular components and allows plasma to wick into two glass fiber pads where the first reaction takes place for six minutes. These pads have been impregnated with Phenylalanine Dehydrogenase (PheDH) and the oxidized form of Nicotinamide adenine dinucleotide ($NAD^+$.) Phe is catalyzed by PheDH and the reaction reduces $NAD^+$ into NADH. A third glass fiber pad, impregnated with Nitro blue tetrazolium chloride (NBT) and methoxy Phenazine Methyl Sulfate (mPMS,) is then folded to make contact with a pad containing the products of the first reaction to allow for a second reaction to take place for an additional minute. NBT is reduced by NADH from the first reaction (in a process aided by mPMS) to its formazan dye product. This visible color change is recorded with a desktop scanner. Statistically relevant intensity differences have been measured in concentrations of blood spiked with 0 (normal), 3.75 (slightly elevated) and >7.5 (substantially elevated) mg/dL Phe using this device.

The focus of this presentation will be on optimization of the flow path of reagents to achieve maximum signal-to-noise ratio in the detection region by controlling the patterning and drying of reagents in the multi-component assay.

# 429106 Novel Electrical Impedance System for Rapid Detection of Mycobacteria


**Abstract:**

Tuberculosis (TB) is a common infectious disease of the human respiratory system caused by *Mycobacterium tuberculosis*. While it has a high mortality rate, it can be treated with timely detection. Hence the rapid detection of *mycobacteria* is an essential component of efforts to control the spread of TB. The current gold standard for detection of TB (especially in low-resource environments) involves culture-based diagnostic techniques. These techniques are able to infer the presence of *live* microorganisms in a sample by detecting changes to the properties of the suspension (such as O2/CO2 levels, pH, electrical conductivity, temperature etc) brought about by bacterial metabolism. Although these methods are relatively inexpensive and reliable, they are time consuming. For slow growing *mycobacterium*species, which have a doubling time of ~24 hours, this method of detection can take anywhere from 7-56 days. The objective of our research is to reduce the time to detect mycobacterium in a sample and thereby prevent the delay in treatment.

We achieve this using a multi frequency electrical impedance spectroscopy (EIS) to detect the presence of mycobacterium in a sample that has been treated to eliminate all organisms except mycobacteria. Under the influence of an AC field, the membrane of a *living* cell, which encloses a cell at a different potential than that of its surroundings (membrane potential), acts as a capacitor. Thus the "bulk capacitance" of a suspension is proportional to the number of living cells present. Consequently, any increase in the bacterial number is reflected as an increase in the bulk capacitance of the system – which we monitor over time. A "significant" change in the value of the bulk capacitance from its baseline (time t=0) value indicates the presence of growing microorganisms (presumably mycobacteria) in the sample of interest.

This basic approach (multi-frequency EIS) has been previously used by our group to perform blood cultures (wherein blood suspected of being infected with bacteria is added to growth media and the presence of live organisms is inferred) significantly faster than currently methods that rely on bacterial metabolism. In this study, we demonstrate the applicability of this technique towards detecting *mycobacteria*, and obtain estimates of the amount of time that could be saved in detecting*mycobacteria* upon using our method.

For our study, we used *Mycobacterium bovis* (a zootonic analog of *M. tuberculosis). M. bovis* was cultured in modified Middlebrook 7H9 media, and samples were prepared with three different initial loads of M. bovis: "low" (2-5 CFU/ml), "medium" (100-200 CFU/ml) and "high" (~3000 CFU/ml). Each sample (and a control) was assayed using both the established technique (BACTEC MGIT ™ culture) and our method, which involves periodic (once every 12-24 hrs) sampling of the culture fluid, measurement of electrical impedance at frequencies ranging from 10KHz to 100 MHz, an analysis of EIS data to calculate the bulk capacitance of the sample, and tracking the calculated bulk capacitances over time.

We found that for control samples, the bulk capacitance of the suspensions did not change in a statistically significant manner from their baseline values. On the other hand, the bulk capacitance values different samples containing *Mycobacteria* increased over time and become statistically significantly different from baseline when concentrations rose to ~ 10,000 CFU/ml. Thus our times to detection (TTDs) were about 60-70 hrs (< 3 days) for "high" concentration samples, 84-108 hrs (< 5 days) for "medium" concentration samples, and about 132-156 hrs (< 7 days) for "low" concentration samples. In contrast the TTDs for the laboratory standard was ~ 6 days for "high" concentration samples, and > 40 days for "medium" and "low" concentration samples (we are still awaiting results from the University Hospital microbiology lab).

Hence our method promises to deliver results significantly faster than current automated systems used in commercial clinical microbiology laboratories.

ARIZONA STATE UNIVERSITY

# The Development of a New Differential Pressure Flow Meter for Bidirectional Measurement of Human Breath Flow

Devon Bridgeman

ASU

ARIZONA STATE UNIVERSITY

## Applications of Flow Measurement in Human Breath

- Two main types of breath analysis
  - Forced
    - Spirometry, patient is asked to exhale as hard as possible
  - Relaxed
    - Most methods, patient breaths at a natural or close to natural pace

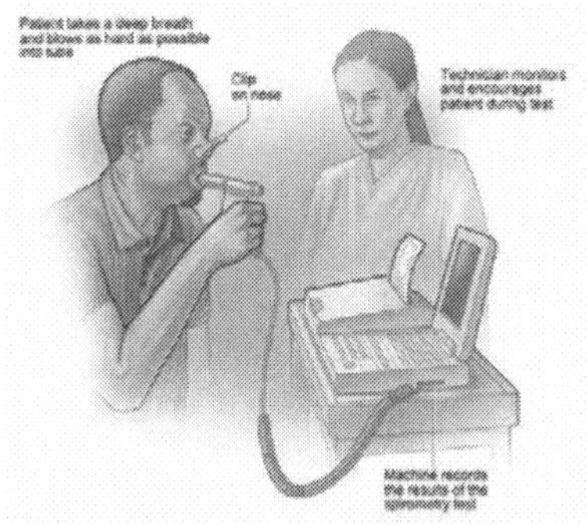


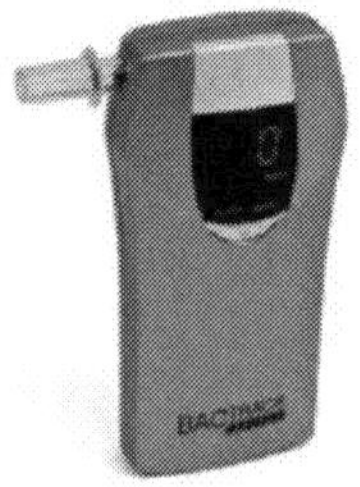

1

ASU

ARIZONA STATE UNIVERSITY

# The Issue

- With any flow measurement, there is a tradeoff between the accuracy, resistance, and cost
  - For any clinical or consumer application, cost is a large factor

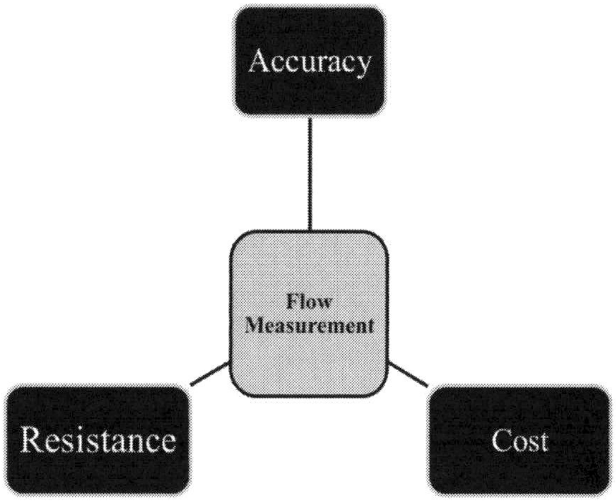


2

ARIZONA STATE UNIVERSITY

# Human Breathing

- Added challenges with breath
  - Very dynamic measurement
  - High sensitivity to flow-resistance
- Even more added challenges with rebreathing
  - Periodic temperature swings
  - Periodic humidity swings
  - Directionality
  - Higher sensitivity to flow-resistance

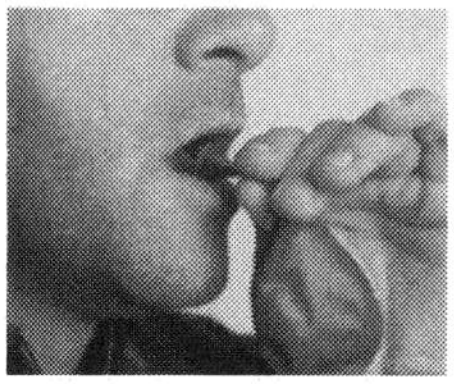

3

ARIZONA STATE UNIVERSITY

# Differential Pressure Flow Designs

- For low-cost monitoring, differential pressure is attractive
  - Often fixed- or variable-orifice meters are used for relaxed flows
    - Rough tradeoff between sensitivity and resistance with fixed-orifice designs
    - Wear issues with variable-orifice designs
  - Fleisch meters often used in clinical settings
- Pitot Tubes not seen in this type of application

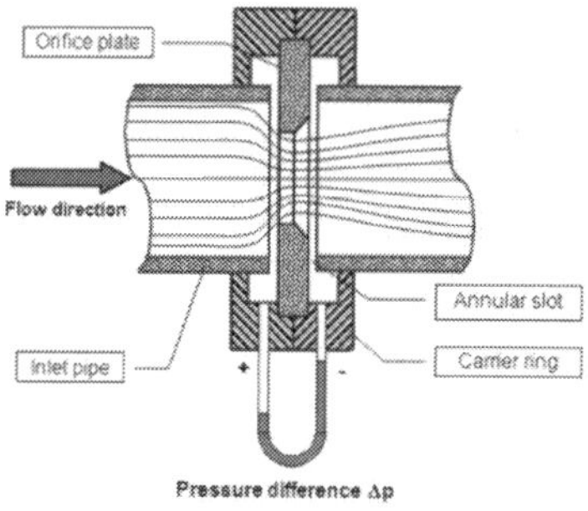


Sagemetering.com

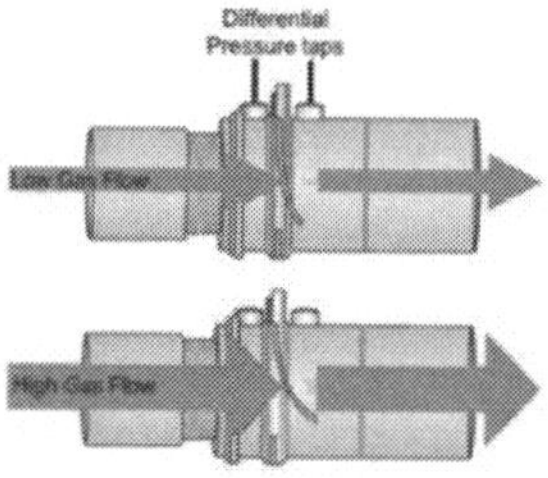


apsf.com

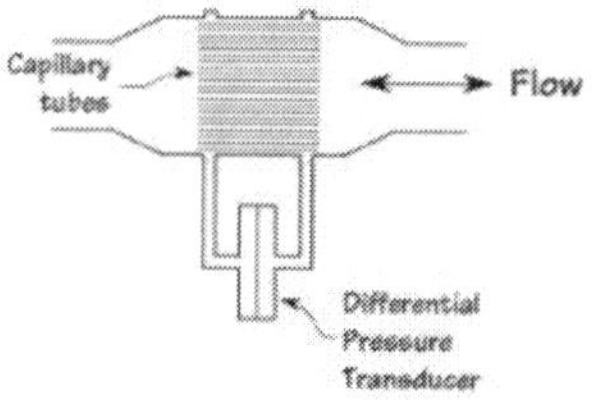


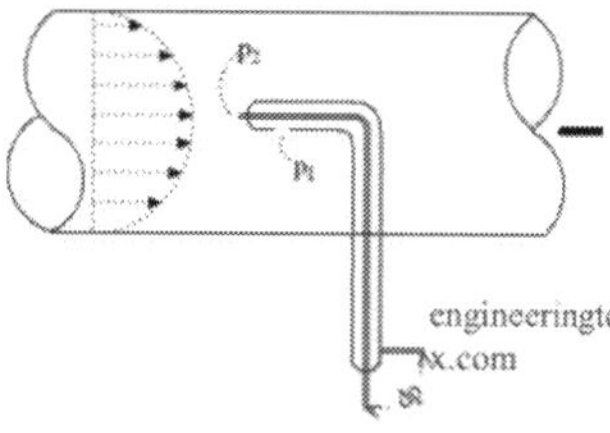

engineeringtoolbox.com

4

ARIZONA STATE UNIVERSITY

# The Confined Pitot Tube

- The Confined Pitot Tube (CPT) flow meter combines multiple effects
  - Stagnant Pressure
  - Confinement
  - Inertial Effects

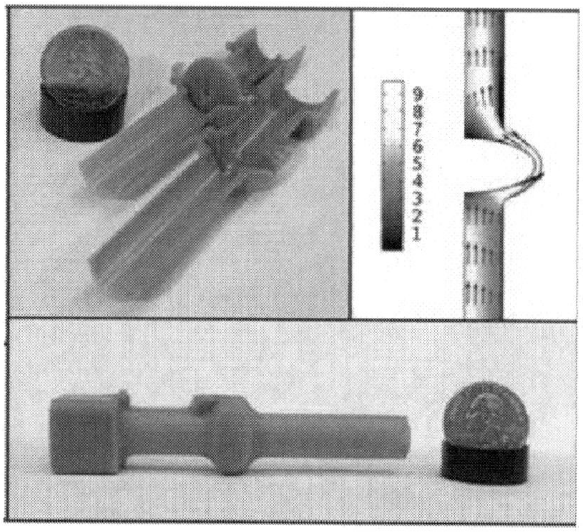


5

ARIZONA STATE UNIVERSITY

# Pressure Distribution

- Pressure profile shows a large stagnant pressure in the center of the disk
- Negative pressure after the bend
- Two options for sampling
  - 1-Point
  - 2-Point

P1 Vs. P2

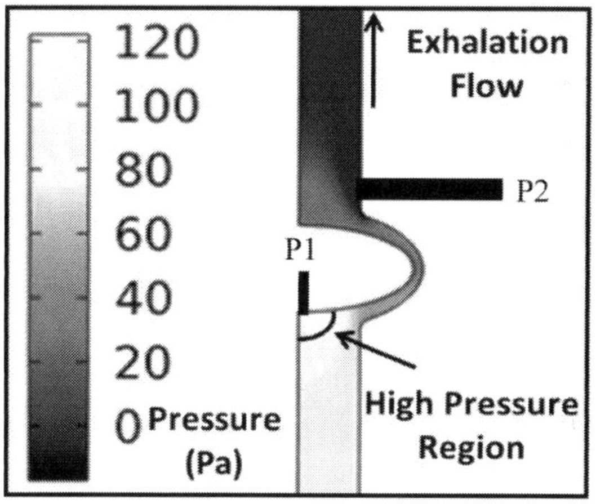


6

ASU

ARIZONA STATE UNIVERSITY

# Flow Meter Comparison

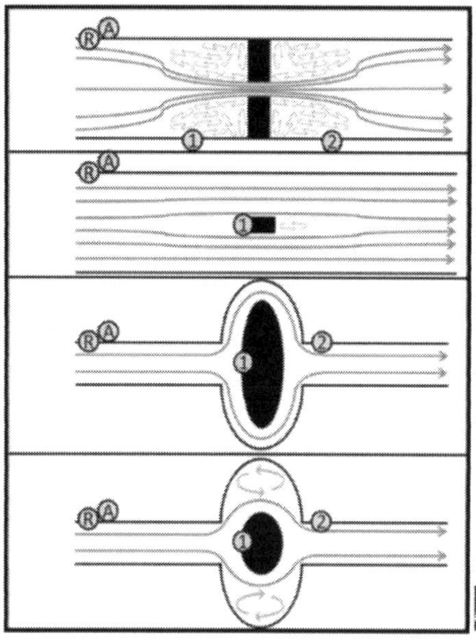


7

ASU

# Comsol® CFD Simulations

- Simulations were run in Comsol Multiphysics to determine an optimal design
  - Non-trivial transitions between laminar and turbulent flow
  - Solution: Run the simulations under both sets of physics and compare experimentally
- A range of inner disc sizes probed

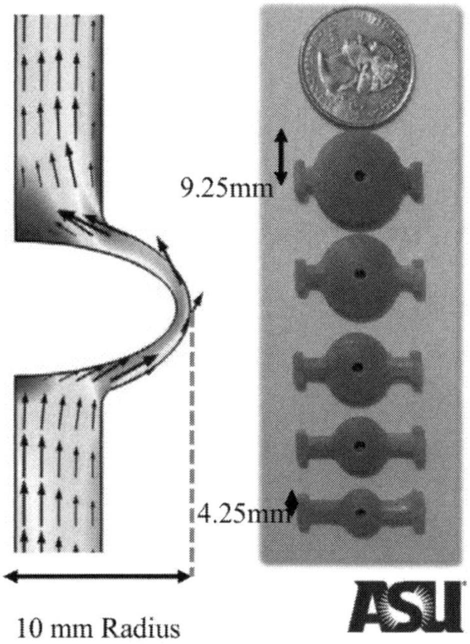


8

ARIZONA STATE UNIVERSITY

# Static Flow Tests

- Tests done at constant flows of dry compressed gas
- Sensiron flow meter used as reference

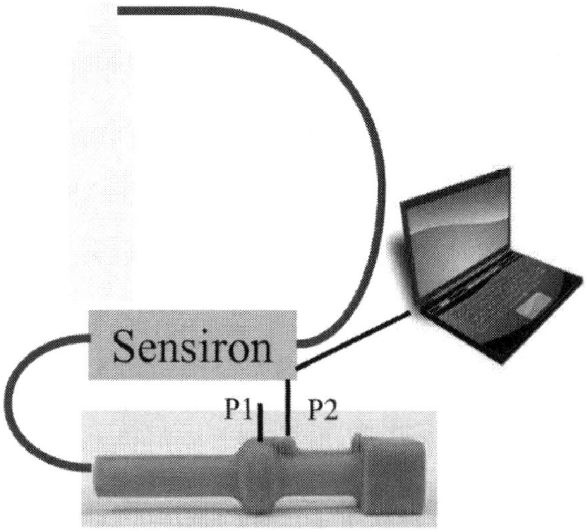


9

ARIZONA STATE UNIVERSITY

# Experiments Vs. Simulations: Small Radii

1-Point Sampling 2-Point Sampling

4.25 mm Disk

5.5 mm Disk

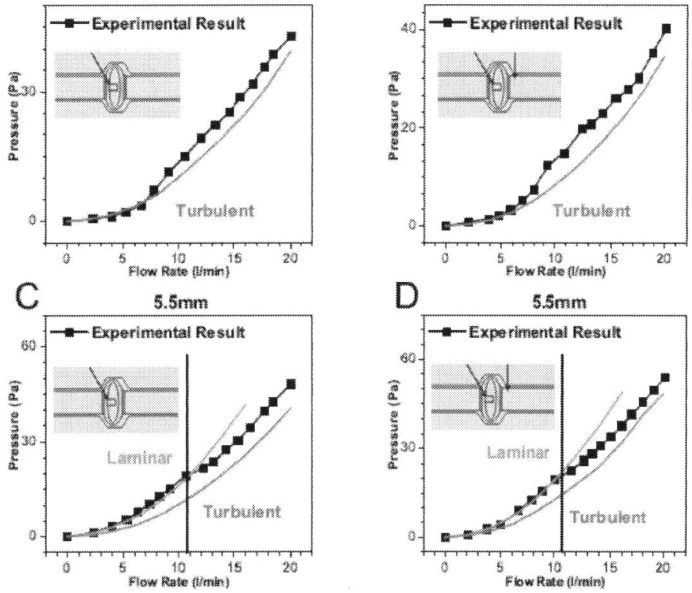


10

ARIZONA STATE UNIVERSITY

# Experiments Vs. Simulations: Large Radii

1-Point Sampling 2-Point Sampling

0.8cm Disk

0.925 cm Disk

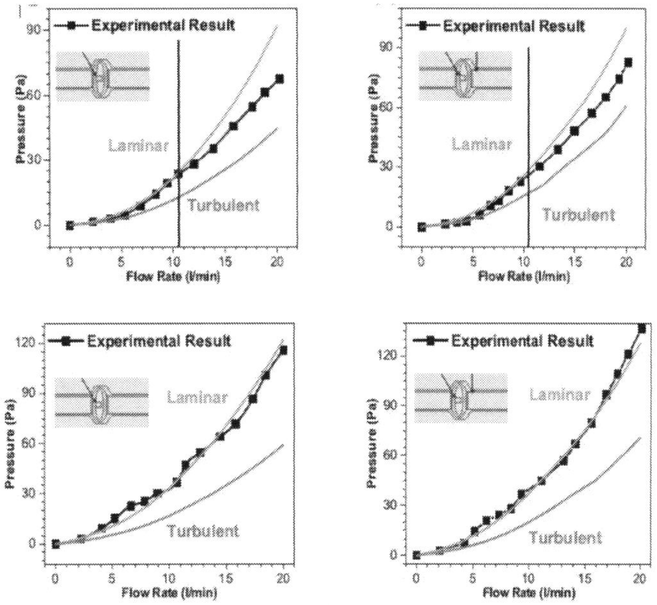


11

ARIZONA STATE UNIVERSITY

# Comparison to a Fixed-Orifice

- If we compare to a fixed-orifice flow meter (2.5mm radius), we see:
  - More rapid scaling
  - Comparable performance at low flows

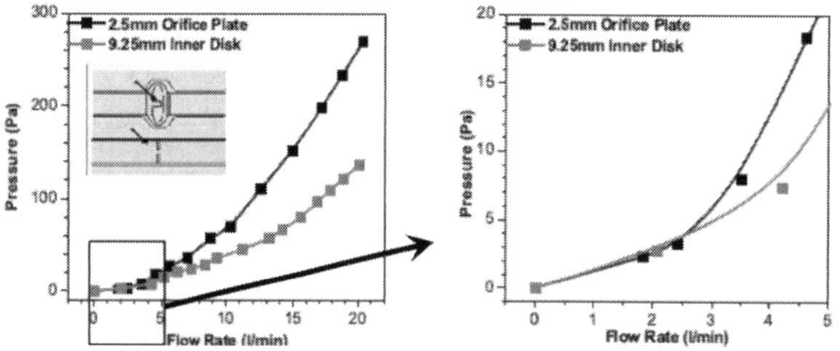


12

ASU

ARIZONA STATE UNIVERSITY

# ...and on the Other Hand

- The flow-resistance scales much slower with the most-confined setup
- As we go to smaller inner disk diameters, the flow-resistance drops off rapidly

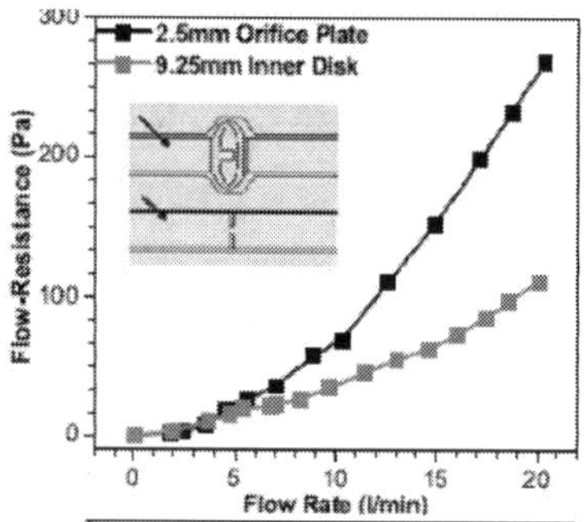


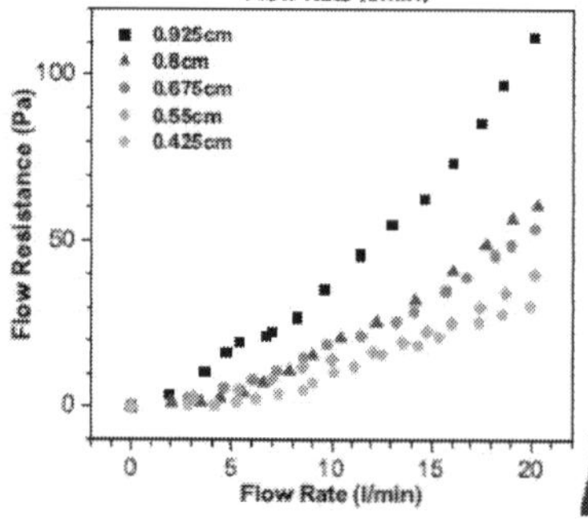


13

ASU

ARIZONA STATE UNIVERSITY

# Artificial Breathing Patterns

- We wanted to probe breathing waveforms on a known platform
- Modified Vacumed® breath simulator

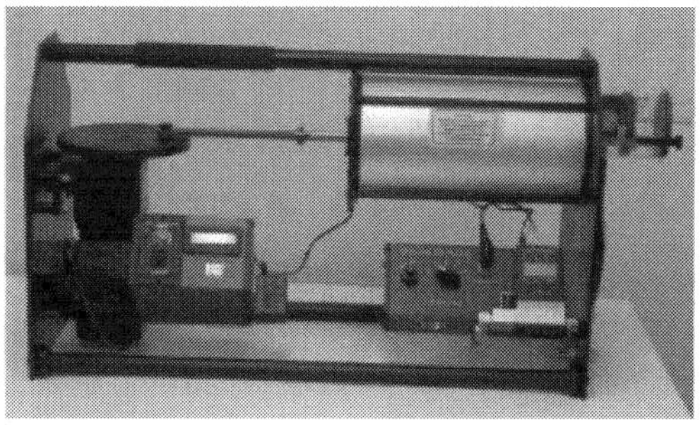

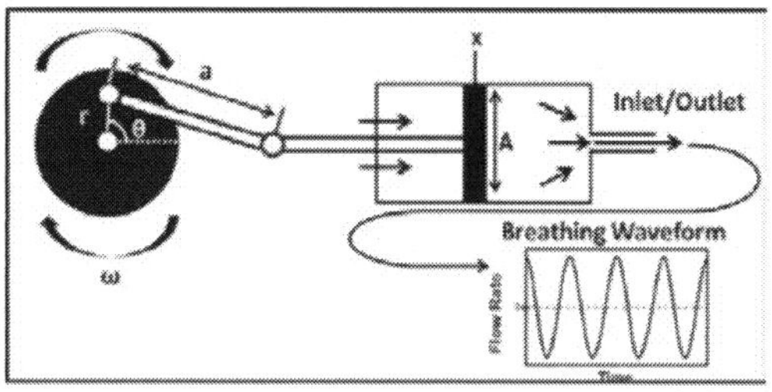


14

ASU

ARIZONA STATE UNIVERSITY

# Artificial Breathing Results

- Using a multiple-equation method, we are able to read something very similar to what we expect

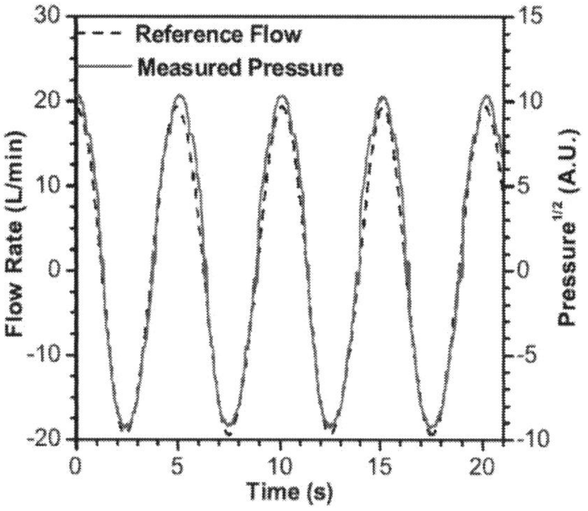


15

ASU

ARIZONA STATE UNIVERSITY

# Actual Breath Tests

- Comparison to an Oxycon Mobile unit
  - Volume comparison
  - Turbine Flow meter
- Strong trend, although off by a linear factor
  - Easily Corrected!

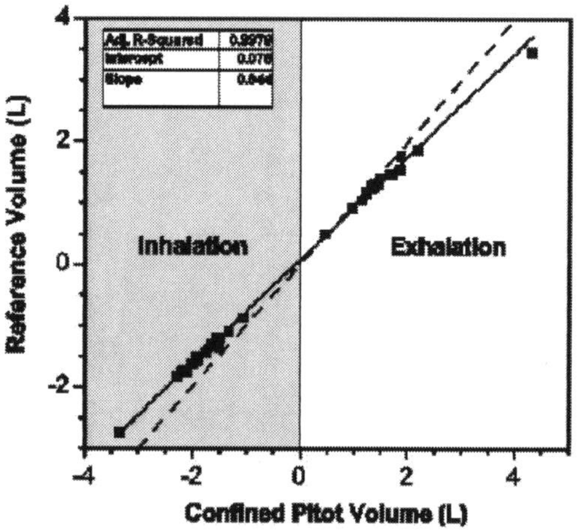


16

ASU

ARIZONA STATE UNIVERSITY

# Actual Breath Test Errors

- Volume error comparison
- Standard deviation of less than 50mL!
  - Not bad when considering the published volume accuracy of the commercial unit!

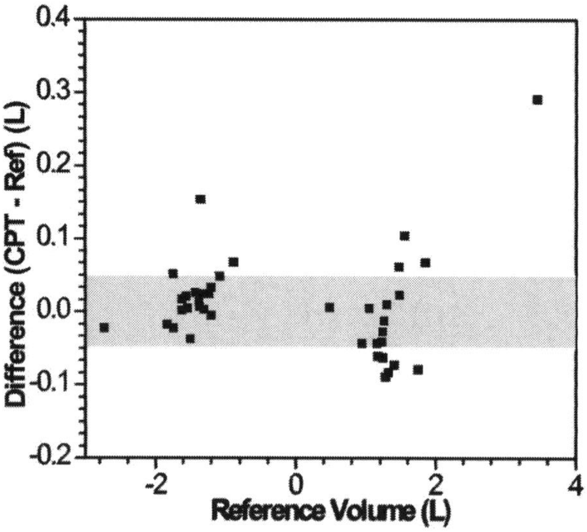


17

ASU

# Conclusions

- The Confined Pitot Tube combines the effects of stagnant pressure, flow confinement, and inertia
- At high confinement, the flow meter displays mostly laminar behavior
- The Confined Pitot Tube provides good performance at a low flow-resistance
- The Confined Pitot Tube is suitable for measuring human breathing patterns

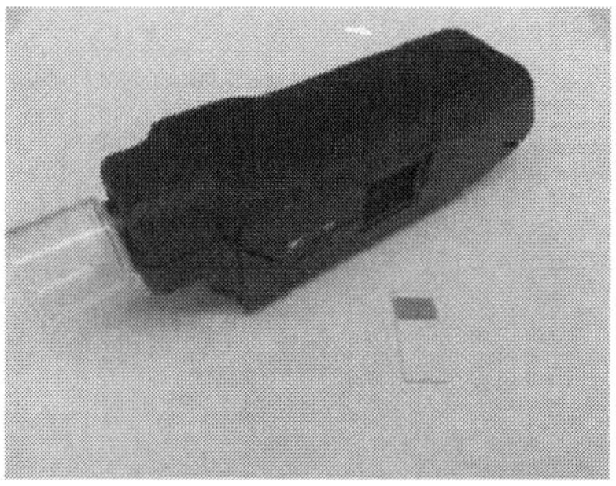

18

ASU



# Thank You!

- I would like to thank:
  - Dr. Erica Forzani
  - Dr. Xiaojun Xian
  - Dr. Francis Tsow
  - Dr. N.J. Tao

our GPSA graduate and professional student association gpsa.asu.edu

NIH

ASU BIODESIGN INSTITUTE ARIZONA STATE UNIVERSITY

19

ASU

ARIZONA STATE UNIVERSITY

# Gas Properties

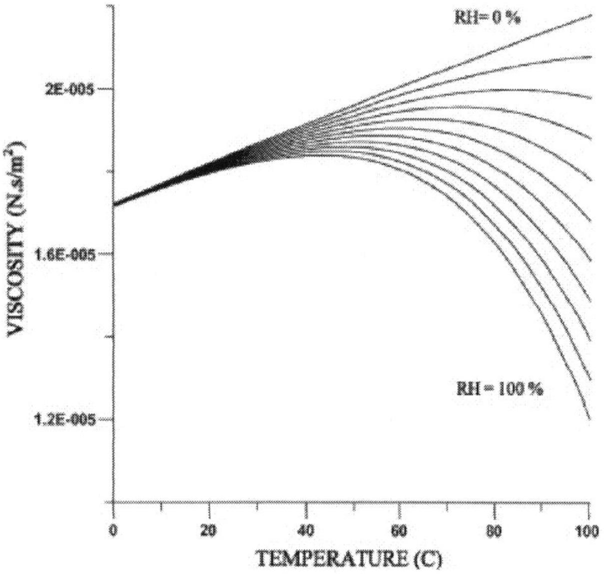


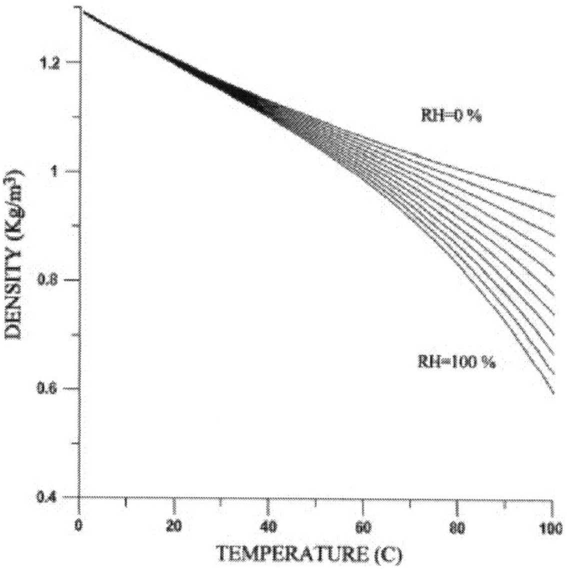


ASU

# 433772 Controlled ZnO Nanostructures Alignment and Electrode Spacing – an Approach Towards Multiplexed Biomarker Detection

**Abstract:**

We demonstrate a methodology to tailor the morphology of zinc oxide (ZnO) nanostructures growth in aqueous chemical bath which in turn dictates the multiplexing capabilities of nanostructured microelectrode sensor platform. A key challenge in multiplexed biomarker quantification is to detect low concentrations of target analytes with high specificity and sensitivity from physiological fluids. Technological advancements provide capabilities to synthesize and integrate nanomaterials into high density arrays leading to development of miniaturized, portable electrical biosensor platforms that meets these requirements. In this work, we demonstrate the multiplexed electrical detection of cardiac biomarkers, troponin-T (cTnT) and troponin-I (cTnI) using one dimensional (1D) ZnO nanostructures as our sensing platform. Electrode characteristics such as material, dimensions and its surface modifications dictate the performance of electrochemical biosensors. In this regard, we demonstrate the methodology to design ultra-small electrodes using 1D ZnO nanostructures via hydrothermal synthesis. We have studied and identified the influence of various aqueous chemical bath parameters on morphology of ZnO nanostructure formation and characterized using SEM and AFM. We have established previously that the density of ZnO nanostructure on microelectrode platform can be effectively controlled by varying precursor concentration, which in turn modulates the electrode behavior (micro/ nanoelectrode). Nanoelectrode behavior enables biomolecule detection under steady state condition due to improved electrochemical reaction rates and increased signal to noise ratio. Here, we discuss the various electrode configurations that enables detection of multiple analytes simultaneously in physiological buffer. Electrode spacing and dimensions that offer minimum interconnect issues and no signal crosstalk were identified through COMSOL simulation and verified experimentally. The charge perturbations due to biomolecular binding events on nanostructured sensing sites were recorded using electrochemical impedance spectroscopy (EIS). Results demonstrate amplified signal response and larger dynamic range for cTnT and cTnI detection with limit at femtogram/mL. Multiplexed detection of these two biomarkers yield about 90% sensitivity and selectivity that individual detection of cTnT and cTnI on separate microelectrode platform. Concurrent detection of these cardiac biomarkers have greater significance in improving quality of life for those suffering from acute myocardial infarction.

# 425292 Developing Low-Cost Nanowire Sensors Based on a Seed-Mediated Solution Process

**Abstract:**

This talk describes the translation of research discovery and patented technology made in our lab into nanosensor devices. Nanowires have been applied to sensing for over 10 years but few nanowire sensors have reached the market. The major barriers are the complexity of manufacturing and difficulty in connecting nanowires in devices. Our technology can potentially overcome these barriers by using a different approach and with materials not previously used as nanowires. We discovered that when mixing organic crystalline compounds with nanoparticles in a solution, nanowires nucleated on nanoparticles upon solvent evaporation. We developed and tested a working hypothesis that links the nanowire formation to the nanoconfinement effect imposed by the nanoparticle seed. The nanoconfinement concept has been successfully applied to the synthesis of nanowires of conductive tetrathiafulvalene charge-transfer and potassium tetracyanoplatinate salts on gold nanoparticle seeds using electrocrystallization. We further synthesized the organic nanowires directly on microelectronic substrates by using features of the substrate pattern as nucleation points for nanowire formation. We tested the nanowire device for electrochemical sensing of different vapors. This research contributes new nanowire materials as signal transducers in nanosensor devices built on commercial electronic components.

# 434528 Dynamic Biochemical Tissue Analysis of P-Selectin Ligands Expressed By Colon Cancer


**Abstract:**

Growing evidence suggests that selectin/selectin-ligand interactions are involved in mediating the adhesion of bloodborne circulating tumor cells at distant sites during the later stages of cancer metastasis. Presently, immunostaining is the standard method for detecting selectin ligands expressed on tissue, but this approach cannot ascertain if the selectin ligands are functional. That is, detection of purported selectin ligands by static (no-force) immunostaining under equilibrium conditions does not necessitate that the reactive antigens are able to mediate cell adhesion with hemodynamic flow (force) under non-equilibrium conditions. As a direct consequence, the relevance of functional selectin ligands as biomarkers is presently unclear because of the lack of adequate detection methods. Therefore, we developed a method, dynamic biochemical tissue analysis (DBTA), to detect and characterize functional selectin ligands *in situ* by perfusing selectin-coated microspheres (probes) over cancer tissues in a microfluidic device. DBTA using P-selectin microspheres was performed on cancer tissue sections in conjunction with static biochemical tissue analysis (SBTA) using P-selectin chimera, antibodies against purported selectin ligand carbohydrate moieties sLeX and sLeA (HECA-452, KM231, and CSLEX), and antibodies against P-selectin ligands (CD24, CD44, and PSGL-1). Using DBTA, calcium-dependent selectin/selectin ligand adhesive interactions in the form of P-selectin microsphere rolling were observed on multiple, distinct cases of colorectal cancer tissue. Examination of serial sections with SBTA for purported selectin ligand carbohydrate structures in the same regions of tissue displaying P-selectin microsphere rolling revealed significant inconsistencies; not all regions in which sLeX and sLeA moieties were detected using SBTA displayed DBTA P-selectin probe adhesion while some regions that lacked sLeX and sLeA SBTA detection displayed DBTA P-selectin probe adhesion. Furthermore, in the regions of tissue that displayed P-selectin microsphere adhesion, CD24 and PSGL-1 epitopes were not detected in SBTA of serial sections. While one case displayed CD44 expression that was in agreement with DBTA, not all regions of this tissue section that displayed specific reactivity with the P-selectin microspheres used in DBTA were recognized with the CD44 antibody in SBTA.

Follow-up CD45 SBTA ruled out the possibility of microsphere interaction with infiltrated leukocytes, in agreement with the lack of PSGL-1 detection. These results imply the presence of less well-recognized, potentially novel P-selectin ligands in the tissue sections.

To further characterize these potentially-novel, functional P-selectin ligands, a more refined method of interaction analysis that discretizes DBTA probe rolling into a sequence of succinct pauses (i.e., brief stationary adhesion) was used. In this pause time analysis, microsphere rolling recorded at 200 frames per second was assessed using a cross correlation tracking algorithm to obtain a robust point estimate that represents the off-rate (inverse lifetime) bond breakage of the ensemble of receptor ligand complexes that mediate microsphere adhesion to the tissue surface. This analysis revealed the inverse lifetime of P-selectin DBTA probe adhesion to functional selectin ligands expressed on colorectal cancer tissue is force-dependent, with a global minimum off-rate observed at 0.75 dyne/cm$^2$. More specifically, increasing levels wall shear stress initially decreased (range: 0.12 to 0.75 dyne/cm$^2$) then increased (range: 0.75 to 1.5 dyne/cm$^2$) P-selectin DBTA probe off-rate to the tissue surface. Appropriately, this method of adhesion inverse lifetime analysis was then subsequently applied to the aforementioned colorectal cancer tissue section that revealed the presence of CD44 (a P-selectin ligand heavily associated with colorectal cancer). Characterization of P-selectin DBTA probe adhesion in a 0.01 mm$^2$ region of tissue that displayed CD44 recognition using SBTA was compared to an adjacent region of equal size that displayed no recognition of CD44 using SBTA. The maximum likelihood off-rate point estimate of 40 separate P-selectin DBTA probes perfused over a CD44-positive region at 0.50 dyne/cm$^2$ was 0.6 inverse seconds while the corresponding CD44-negative region value was 0.4 inverse seconds, resulting in a U-value of 1152.5 (equivalent to $P<0.001$), further supporting the presence of distinguishable P-selectin ligands. Taken together, these results unequivocally demonstrate DBTA's ability to detect functional P-selectin ligands and generate an adhesion parameter that lays the framework for evaluating aberrantly expressed functional selectin ligands as functional cancer biomarkers.

# 434767 Simple and Robust Antibody Fragment Immobilization for Detection of Antigens

**Abstract:**

Arrays of antibody fragments have great potential in diagnostics and screening of molecule libraries, but synthesis of arrays containing a large number of different antibody fragments requires a robust immobilization approach that does not depend on optimization for individual antibody fragments. We have developed a simple and robust immobilization method that takes advantage of the high-affinity and specific interaction of biotin and streptavidin to immobilize single-chain variable fragment (scFv) antibodies while maintaining their ability to bind their target antigens. scFvs are biotinylated *in vivo* during recombinant expression in *Escherichia coli* cells by genetically fusing the scFvs to biotinylation sequences. We efficiently immobilized a variety of biotinylated scFvs onto streptavidin-coated plates directly from cell lysates, with no need for purification prior to immobilization. The immobilized scFvs retain the ability to bind their target proteins and retain the specificity of the soluble scFvs. Antigens can be detected from a purified sample or from cell lysate. Furthermore, the immobilization of the scFvs is stable for at least 40 days after storage in buffer at 4 °C or -20 °C with no loss of binding activity. The simplicity and robustness of our method make it a promising approach for immobilizing a wide variety of antibody fragments and other proteins in an array format for diagnostics, screening, or any other application that necessitates immobilization of large number of functional antibody fragments.

# 432226 Oriented Fab Fragment Immobilization By Site-Specific Biotinylation at the Nucleotide Binding Site for Enhanced Ebola Detection

**Abstract:**

Oriented immobilization of antibodies and antibody fragments has become increasingly important because of the efforts to reduce the size of diagnostic and sensor devices to miniaturized dimensions for improved accessibility to end-user. Reduced dimensions of sensor devices necessitate the immobilized antibodies to conserve their antigen binding activity for proper operation. Fab fragments are becoming more commonly used in small-scaled diagnostic devices due to their small size and ease of manufacture. Fab fragment, with its small size and simple production methods compared to intact antibody, are good candidates for use in miniaturized diagnostic devices and targeted therapeutic applications. However, commonly used modification techniques are not well suited for Fab fragments as they are often more delicate than intact antibodies. Fab fragments are of particular interest for sensor surface functionalization but immobilization results in damage to the antigen binding site and greatly reduced activity due to their truncated size that allows only a small area that can bind to surfaces without impeding antigen binding. The nucleotide-binding site (NBS) is a highly conserved region between the variable light and heavy chains at the Fab domains of all antibodies, and a small molecule that binds specifically to this site. In this study, we describe an NBS-UV photocrosslinking functionalization method (UV-NBS$^{Biotin}$) in which a Fab fragment is site-specifically biotinylated via UV energy exposure (1 J/cm$^2$) without affecting its antigen binding activity. This study demonstrates successful immobilization of biotinylated Ebola detecting Fab fragment (KZ52 Fab fragment) via the UV-NBS$^{Biotin}$ method yielding 1031 fold and 2 fold better antigen detection sensitivity compared to commonly used immobilization methods: direct physical adsorption and NHS-Biotin functionalization, respectively. Utilization of the UV-NBS$^{Biotin}$ method for site-specific conjugation to Fab fragment represents a proof of concept use of Fab fragment for various diagnostic and therapeutic applications with numerous fluorescent probes, affinity molecules and peptides.

# 434380 Automated Screening of *C. Elegans* Lipid Droplet Phenotypes Using Microfluidics and Granulometry

**Abstract:**

Obesity is a serious health issue currently plaguing 72 million Americans. While the genetic origin has yet to be fully understood, the risk has been linked to fat distribution and storage. The nematode, *C. elegans*, is a convenient model to understand key fat storage mechanisms conserved through evolution, which can give insight towards pathways controlling these processes in humans. Fat in *C. elegans* is primarily stored in organelles called lipid droplets, which when labeled can be imaged *in vivo* through the transparent animal. Classification of various lipid droplet phenotypes in worms would allow for discovery of important genes associated with lipid storage and metabolism. However, current limitations in imaging and analysis of lipid droplets in worms, such as low throughput and labor intensive quantification, are some of the factors that have hindered further growth in lipid research. Engineering methods and tools to improve object quantification would dramatically enhance the ability to study lipid droplets in *C. elegans*.

To address these issues, we use a microfluidic device to facilitate imaging and developed a modified granulometry algorithm to rapidly detect and quantify lipid droplets. We integrated technologies to investigated genes that affect lipid droplets and their storage by performing an automated real-time on-chip forward genetic screen of *C. elegans* for phenotypes with varying lipid droplet size distributions. With this platform, we automatically screened over 3,000 animals and successfully sorted and recovered various mutant strains in a much higher throughput manner, approximately two orders of magnitude faster, than previously achievable. The development of this integrated experimental platform has enabled the potential discovery of factors that contribute to fat accumulation that are currently unknown.

# 429802 Quantum Molecular-Sequencing (QM-Seq): Nanoelectronic Single Molecule DNA/RNA Sequencing and Epigenetics

**Abstract:**

The sequences and structure of nucleic acids in bio-macromolecules like DNA and RNA can define their biological function by controlling the downstream expression of genes, proteins, and other cell-regulatory functions. Small variations in this genetic coding in individual cells can lead to mutations, which can play a key role in physiology. For example, recent data on genetic variation in Ebola virus through its origin and transmission provides valuable information about mutations that altered protein sequences, and new targets for diagnosis, vaccines and therapy. While nanoelectronic sequencing techniques can in-principle be used to distinguish biochemically distinct nucleobases, it relies on identifying unique "electronic fingerprints" for different nucleotides. So far only partial deoxynucleotides have been successfully identified for single molecule, in part due to the large conformational entropy of molecules which causes dispersion of electronic spectroscopic signatures. Furthermore, attempts at using magnitude of electronic current, current decay properties, and its derivative (dI/dV) have been unable to provide deterministically unique fingerprints for biochemical states of nucleic acids. Here, we present a novel Quantum Molecular-Sequencing (QM-Seq) method which provides on fast nanoelectronic spectroscopy and unique electronic fingerprints of all DNA (A, G, T, C) and RNA nucleotides (A, G, C, U). We will show that QM-Seq spectra allows us to detect the biochemical state of all individual molecules, using facile identification of isomers and tautomer species in nucleic acids. We also demonstrate simultaneous identification of single nucleotide modifications (methylation here), which highlights the potential of QM-Seq as a versatile technique for epigenomics and personalized medicine.

References:

Nagpal, Chatterjee, Ribot, Patent Pending Ribot, Chatterjee, Nagpal, J. Phys. Chem. B, 119, 4968 (2015)

PRAAN Biosciences Inc.

# 432361 Engineer a Protein-Based Sensor for Quantifying Histone Epigenetic Modifications

**Abstract:**

Epigenetic modifications are heritable changes in chromatin that affect gene expression and contribute to determining cellular phenotype. Post-translational modifications (PTMs) of histone proteins are an important category of epigenetic modifications. Aberrant changes in histone acetylation levels, one of the most common histone PTM, have been associated with early-stage tumor genesis; and are emerging as promising biomarkers for cancer detection and monitoring. Current detection approaches of histone acetylation are either tedious to perform or require expensive instrumentation, which hinders their transition to clinics. To address this challenge, we have developed a sensitive histone acetylation probe by combining protein engineering, nanomaterial and fluorescence spectroscopy. The protein sensor is developed by engineering the bromodomain of human polybromo-1 (PB1). The developed probe can recognize acetylation of H3K14 with high specificity. The acetylation level is quantified by using fluorescence correlation spectroscopy (FCS). The performance of our sensor was characterized using histone proteins with defined levels of acetylation as well as extracts from human cells lines. Our sensor offers a simple and sensitive quantification of total acetylation levels of H3K14.

# 425638 Cytoplasmic Microrheology from Particle Trajectory Analysis in Motile Cells

**Abstract:**

The mechanical properties of the cytoplasm play a crucial role in the transport and interaction of intracellular components as well as the cell's response to its physical environment. Like other complex fluids, cytoplasmic mechanics can be probed with passive microrheology, by tracking the thermal motion of individual particles over time. In cells that are actively moving or deforming, however, large-scale persistent flow of the cell contents tends to obscure the stochastic motion from which material properties of the cytoplasm can be derived.

We present a novel technique for analyzing single particle trajectories confounded by slowly-varying flow, in order to characterize the underlying stochastic motion. Our method leverages the separation of time-scales to rigorously subtract out the persistent component of the motion so as to enable robust extraction of the diffusion coefficient and subdiffusive scaling exponent. This technique allows the characterization of a flowing complex fluid as a viscous or viscoelastic material.

We apply this method to analyze the motion of lysosomes within motile neutrophils crawling in a two-dimensional environment, demonstrating that the cytoplasm of these cells behaves as a viscous fluid. The technique is widely applicable to passive microrheology studies in the presence of spatially heterogeneous slowly-varying drift, such as cytoplasmic microrheology in moving cells or particle motion in bulk flow of a complex medium.

# 426658 High-Throughput Linear Optical Stretcher for Mechanical Characterization of Blood Cells

**Abstract:**

Cell mechanical properties are a label-free biomarker capable of indicating cell health. Differences in individual cell deformability in cases such as cancer, malaria, and sickle cell anemia have been used to differentiate between normal and diseased cell state. Established individual cell measurement techniques are typically low throughput, making acquisition of statistically relevant population data impractical. To overcome this limitation, we have developed a linear optical stretcher as a high-throughput mechanical property cytometer. Custom, inexpensive, scalable optics image a linear diode bar source into a microfluidic channel, where cells are hydrodynamically focused into the optical stretcher. Upon entering the stretching region, antipodal optical forces deform cells in flow. Each cell relaxes as it flows out of the trap and is compared to the stretched state to determine deformation. The deformation response of untreated red blood cells and neutrophils were compared to chemically treated cells. Statistically significant differences were observed between normal, diamide-treated, and glutaraldehyde-treated red blood cells, as well as between normal and cytochalasin D-treated neutrophils. Based on the behavior of the pure, untreated populations of red cells and neutrophils, a mixed population of these cells was tested and the discrete populations were identified by deformability.

## (339d) The Role of a-Catenin in Cadherin-Mediated Mechanical Signaling in Human Embryonic Stem Cells

Huang, E. Y. H., Stanford University
Kee, Y. S., Stanford University
Price, A. J.,
Weis, W. I., Stanford University
Sebastiano, V.,
Dunn, A. R., Stanford University
Originally presented on: 11/10/2015 13:24:00 - 13:42:00

Controlling stem cell renewal and differentiation is crucial to tissue engineering and regenerative medicine as a whole. Substrate stiffness and cell-cell junctional forces have been shown to be critical factors that influence proliferation, survival, and differentiation in human mesenchymal stem cells and other cell types. However, how mechanical cues regulate proliferation and self-renewal in human embryonic stem cells (hESCs) and other pluripotent cell types is poorly understood. Work from our and other laboratories supports an emerging picture in which the protein αE-catenin functions as a central mechanosensor at cell-cell junctions. Genetic and developmental evidence likewise suggest that αE-catenin acts as a potent regulator of the Hippo pathway, a central regulator of cell proliferation, by preventing nuclear translocation of the transcriptional coactivator yorkie-associated protein (YAP).

Using hESCs as a model system, we sought to determine the molecular mechanisms that connect force-dependent changes in αE-catenin conformation to downstream signal transduction that controls cell fate in pluripotent stem cells. To do so, we made use of an R551A αE-catenin mutant which mimics the force-activated conformation of αE-catenin present at cell-cell junctions. We found that in epithelial cells, R551A αE-catenin reduces cell proliferation, sequesters YAP at the junctions, and disrupts apical-basal polarity by recruiting the tight junction protein ZO-1 to the lateral cell-cell junctions. In contrast, there are no observable differences in protein localization or proliferation rates when R551A αE-catenin was expressed in hESCs. Instead, R551A αE-catenin preferentially localized to the apical junctions, indicative of its recruitment to either tight junctions or the polarity complex (**Figure 1**). Despite this difference, hESCs cultured on engineered polyacrylamide gels of varying stiffness still exhibit differential Hippo pathway activation and marked differences in cytoskeletal structures. These results suggests that, while hESCs are critically sensitive to mechanical inputs, they do so via a mechanism distinct from the one previously described in epithelial stem cells[1]. Ongoing work examines the role of αE-catenin and its conformation in controlling germ layer organization during hESC differentiation, and aims to discover the identity of the proteins responsible for mechanosensing specifically in pluripotent stem cells.

[1] Schlegelmilch, K. *et al.* (2011) *Cell.* 144(**5**):782-95

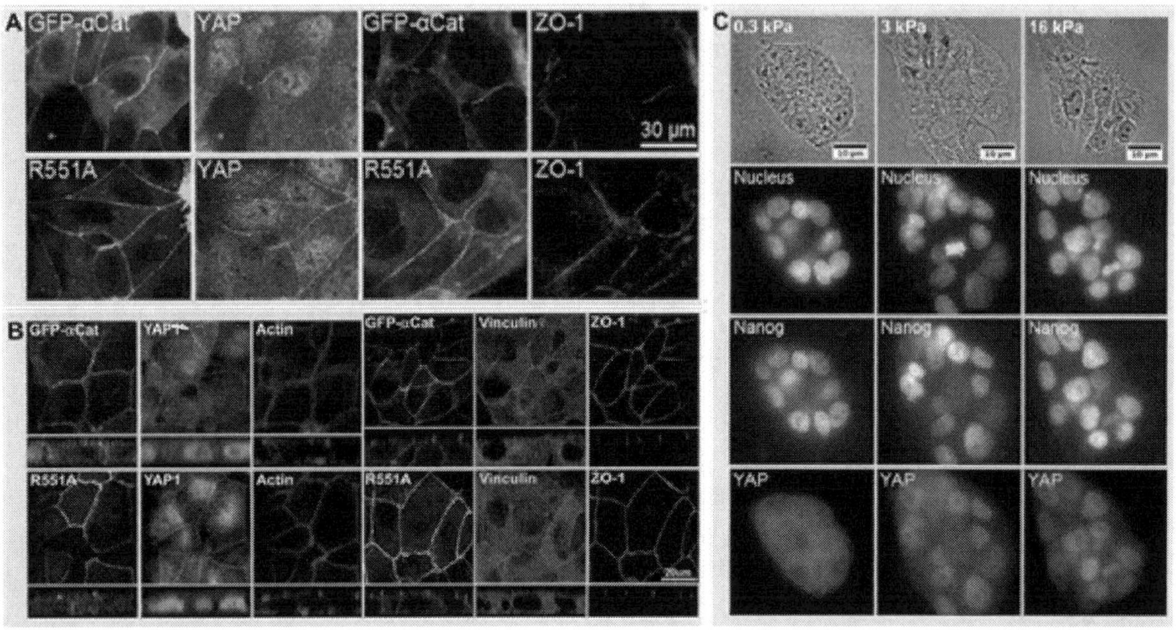


**Figure 1. A** Immunofluorescent staining of R2/7 epithelial cells expressing wild-type and R551A αE-catenin. **B** hESCs expressing wild-type and R551A αE-catenin are stained with YAP, actin, vinculin and ZO-1. Below each image is a y-z projection of a confocal stack. **C** hESCs cultured on polyacrylamide gels with varying stiffness. Cells are stained with YAP, pluripotency marker Nanog and nuclear dye Hoechst.

# 435816 Biophysical Regulation of Epigenetic Reprogramming during TGFbeta1-Induced Epithelial-Mesenchymal Transition

**Abstract:**

Epithelial-mesenchymal transition (EMT) is a key developmental program that is co-opted in the progression of pathological conditions including fibrosis and cancer. During EMT, epithelial cells exhibit drastic morphological changes, alterations in gene expression patterns, and global epigenetic remodeling. We have previously demonstrated that biophysical signals regulate cell morphology and gene expression during EMT; however, the mechanisms governing epigenetic remodeling during EMT are not completely understood. Here, we employed microfabricated cell and tissue arrays to determine how biophysical cues modulate histone modifications during transforming growth factor (TGF)-β1-induced EMT. We find that matrix geometry regulates histone methylation and acetylation and we identify key factors governing the epigenetic reprogramming that occurs as a cell transitions from an epithelial to a mesenchymal phenotype. Results provide insight into how biophysical cues contribute to epigenetic remodeling and may suggest ways to regulate EMT related processes in the context of health and disease.

# 413723 Single Molecule Mechanical Characterization of the Mitotic Kinesin Motor Kif15

**Abstract:**

Proper formation of the mitotic spindle is essential for chromosome separation and cell division. Single molecule studies have been essential in evaluating the mechanistic behavior of molecular motors, including conventional kinesin and myosin. Dimeric Kinesin-12, or Kif15, has been shown to serve as a backup mechanism for cell division, promoting spindle assembly in the absence of tetrameric Kinesin-5, or Eg5. The mechanism by which Kif15 aids in the assembly of the spindle is unknown. Here, single molecule motility assays were performed using optical tweezers to evaluate the mechanistic behavior of Kif15 on microtubules (MTs). Constructs containing various sub-components of the motor were utilized to determine the role of each domain, including the motor heads and stalk (N420); motor heads, stalk, and MT binding domain (N700); and the full length motor (FL) with its inhibitory tail (Coil-2). From these, characteristics such as step size, dwell times, and stall forces were measured. Binding assays were also performed on the MT binding domain (Coil-1) to determine the rupture force and the nature of the Coil-1/MT interaction. Motility and rupture assays were repeated on subtilisin-digested MTs (dMT) in which the negatively-charged carboxyl-terminal tail of MTs, called the E-hook, is removed. Together, these results provide insight into the interaction between Kif15 and MTs and how motor sub-components contribute to spindle assembly in tetrameric-Eg5-independent cells.

# 433670 The Chemical Structure of Fungal Cell Membrane Lipids Affects Their Biophysical Properties


**Abstract:**

*Cryptococcus neoformans* is an environmental fungal pathogen and the most common cause of brain infection in immunocompromised individuals. *Cryptococcus* infections afflict a large population, especially among HIV/AIDS patients, resulting in an estimated 625000 deaths per year. Recent studies from our group have shown that changes in the chemical structure of the membrane-localized lipid, glucosylceramide (GlcCer), are directly linked with cryptococcal virulence. While the virulent *C. neoformans* strain synthesizes GlcCer with a double bond in carbon position 8 and a methyl group in carbon position 9, genetically modified strains that lack the methyl group (strain Δ*smt1*) and the double bond (strain Δ*sld8*) in their GlcCer structure become avirulent in the mouse model. It has been suggested that changes in GlcCer structure might affect membrane biophysical properties and contribute to loss of virulence. In an attempt to investigate this potential mechanism, the present study focuses on the role of GlcCer chemical structure in regulating the biophysical properties of the cell membrane in *C. neoformans*.

Three strains were used in this study: the highly virulent *C. neoformans* wild-type strain H99 (WT), strain Δ*smt1*, and strain Δ*sld8*. GlcCer was purified from all strains using column chromatography and characterized using thin layer chromatography and liquid chromatography-mass spectrometry. Purified lipids were formulated into vesicles either alone or in combination with ergosterol and the partially unsaturated lipid, 1-palmitoyl-2-oleoyl-*sn*-glycero-3-phosphocholine (POPC) to generate a simplistic membrane model. The thermotropic behavior of lipid vesicles was investigated by monitoring the fluorescence anisotropy of the fluorescent probe, diphenylhexatriene, in the temperature range of 16 °C to 60 °C. The tendency of lipids to phase segregate and form ordered domains was examined using Forster resonance energy transfer (FRET). Gas chromatography-mass spectrometry (GC-MS) was used to analyze ergosterol levels in the cell membrane of each strain.

The thermotropic behavior of purified GlcCers was dependent on their structure. The temperature at which purified lipids transitioned from ordered to disordered phase (Tm) was higher in GlcCer purified from the mutant strains compared to the WT ($T_{m\text{-}WT}$= 34.7 ± 0.0 °C, $T_{m\text{-}\Delta smt1}$= 40.7 ± 1.1 °C, $T_{m\text{-}\Delta sld8}$= 43.2 ± 1.1 °C), suggesting an increased ability of GlcCer from mutants to form ordered domains at physiological temperatures. Similar trends were observed in membrane models, in which the more saturated GlcCer structures showed higher lipid order. GlcCer structures purified from all strains formed ordered domains as evidenced by FRET studies. GC-MS analysis revealed that the mutant strains contained up to 4 times higher amount of ergosterol, a sterol known to promote ordered lipid domain formation. Taken together, these data suggest that mutations in GlcCer chemical structure are likely to increase lipid order and domain formation in the cell membrane and disrupt normal cell behavior. These results provide information on the role of lipid chemical structure on cell membrane physical properties and are of importance in the fields of fungal pathogenesis and membrane biophysics.

# 405063 Efficient Detection of Amyloidâusing PMMA-Tag Fused VHH Antibody

**Abstract:**

VHH domains represent the smallest antigen-binding unit with a molecular size of 15 kDa in comparison to scFv fragments (30 kDa), Fab fragments (60 tkDa), and whole antibodies (150 kDa) [1]. For this reason, VHH fragments have been considered to have great potential in various industrial applications. Furthermore, the genetically fused PMMA tag to VHH antibody allow the high rate of refolding and immobilization. Recently, Amyloid beta (Aβ) denotes peptides of 36–43 amino acids that are crucially involved in Alzheimer's disease as the main component of the amyloid plaques found in the brains of Alzheimer patients. In the present study, we examined whether PMMA-tag fused VHH antibody could offer the potential for detecting Amyloid β antigen. To do it, the $(His)_6$-tagged recombinant protein expressed as inclusion bodies in *Escherichia coli* Rosseta (DE3 component cell) were purified by affinity chromatograph on an Histrap $Ni^{2+}$-Chealate column using a linear gradient of imidazole. His-tag-VHHs was purified as a monomer, eluting on the range from 0 to 400 mM imidazole under pH 6.0. As a results, the highly purified VHH-PM antibodies showed different recovery rate and antigen-binding activity, according to the different strains of VHH-PM antibodies. The recovery rate of Anti-Aβ-VHH-PM (L1-3) and Anti-Aβ-VHH-PM (61-3) was lower in pH 8.5. However, Anti-Aβ-VHH-PM (61-3) was highly recovered at pH 6.0. Furthermore Anti-Aβ-VHH-PM (L1-3) refolded in pH 8.5 showed high antigen binding activity, compared to that Anti-Aβ-VHH-PM (61-3). It was demonstrated that the optimal condition of refolding in Anti-Aβ-VHH-PM was changeable with pH and strain.

# 425853 Comparative Proteomic Analysis of Host and High Biopharmaceuticals Producing CHO Cell Lines

**Abstract:**

One important target in biopharmaceutical industry is to achieve a high production of therapeutic protein. However, the effect of different host cells on protein expression has not been fully understood, so it is very challengeable to rationally design host cells and construct production cell lines. In this study, comparative proteomics was performed to investigate the host cell proteins that could correlate with the expression of recombinant therapeutic proteins. The global protein maps and biopharmaceutical production of four cell lines were analyzed and compared, including adapted ATCC CHO/dhfr-, CHO/dhfr-/EPO, Life Tech CHO DG44, and CHO DG44/IgG. All the host cells and production cell lines were cultivated in suspension culture using chemical defined medium. The productivity analysis in a simple fed-bath cell culture with glucose feeding was performed in shaker flask, showing that the ATCC CHO/dhfr-/EPO produced erythropoietin with titer of 421 mg/L and the CHO DG44/IgG produced IgG with titer of 1,385 mg/L. The LC/MS/MS was used to create the profiles of the global intracellular proteins of these four cell lines, and more than 1,000 host proteins were analyzed. The host proteins associated with metabolism of carbon and energy in glycolysis pathway, TCA cycle and amino acids were investigated. Other proteins involved in, such as protein processing (e.g. protein disulfide-isomerase A3 and endoplasmin), cytoskeletal structure (e.g. alpha-actinin-4 and vimentin) and cell cycle control (e.g. heat shock protein HSP 90-alpha and cyclin-dependent kinase 1), were also analyzed. The possible host cell regulators that correlate with different dhfr- host cells and different therapeutic proteins were identified although further investigation is needed. This study provided a deep insight into the regulation mechanism of biopharmaceutical productions by CHO cells. The results could be used to direct the rational cell engineering for new-generation host cells in the future.

# 432085 Proteomic Time Profiling of Mab Producing CHO Fed-Batch Bioreactor Cultures

**Abstract:**

The production of recombinant proteins such as monoclonal antibodies (mAbs) is usually based on complex mammalian cell processes. Rising omics technologies have enabled the characterization of manufacturing processes on a molecular level. Getting a deeper insight into cellular mechanisms is crucial for the understanding and optimization of manufacturing processes, which is also highly promoted by regulatory agencies through the Quality by Design (QbD) initiative. In particular, the investigation of proteomic profiling can be helpful for the identification of important up- and down regulated pathways which have an influence on the process performance and quality of the produced biopharmaceuticals.

The time evolution of the intracellular protein abundance of a monoclonal antibody (mAb) producing Chinese hamster ovary cell line during fed-batch cultivation was investigated. Two different reactor scales were compared distinguishing possible scale-dependent differences. Multivariate data analysis (MVDA) such as principal component analysis (PCA) was used to emphasize highly up- or down expressed proteins over time. A classification of important biological functions was performed by gene ontology annotation using the proteins with highest fold-change. Furthermore, profiling of the produced mAb and the corresponding quality attributes on a proteomic level is discussed targeting possible operational changes. Scale and time dependent proteomic profiling in combination with multivariate statistical analysis will be highlighted for future applications.

Malvern

# Characterizing and Understanding Protein Stability at High Concentration

Wei Qi, Samiul Amin and E.Neil Lewis
Salt Lake City, AICHE
Nov. 2015



## Outline

- Challenges from High Concentration Formulation
- Raman Spectroscopy and Dynamic Light Scattering
- Case studies
    - Model Proteins – Thermal/Reducing conditions
    - Thermal stability of mAb and its modified version
    - Other applications
- Summary

# What A Difference a Few Years Makes

**2011**

| | | | |
|---|---|---|---|
| 1 | Lipitor<br>Pfizer Inc | 7,429,953 | Stats |
| 2 | Plavix<br>Bristol-Myers Squibb Company | 6,580,260 | Stats |
| 3 | Nexium<br>AstraZeneca Pharmaceuticals, LP | 5,963,917 | Stats |
| 4 | Abilify<br>Otsuka Pharmaceutical Co., Ltd | 5,032,032 | Stats |
| 5 | Advair Diskus<br>GlaxoSmithKline | 4,492,496 | Stats |
| 6 | Seroquel<br>AstraZeneca Pharmaceuticals, LP | 4,492,001 | Stats |
| 7 | Singulair<br>Merck & Co., Inc. | 4,449,747 | Stats |
| 8 | Crestor<br>AstraZeneca Pharmaceuticals, LP | 4,266,282 | Stats |
| 9 | Cymbalta<br>Eli Lilly and Company | 3,562,147 | Stats |
| 10 | Humira<br>AbbVie, Inc. | 3,421,540 | Stats |

*www.drugs.com*

**2014**

| Rank in 2013 (in 2012) | Product | Company | To treat | 2013 sales ($US m) | 2012 sales ($US m) | |
|---|---|---|---|---|---|---|
| 1 (1) | Humira | AbbVie | Rheumatoid arthritis / autoimmune diseases | 10,659 | 9,616 | Protein |
| 2 (2) | Enbrel | Pfizer/Amgen | Rheumatoid arthritis / autoimmune diseases | 8,776 | 8,496 | Protein |
| 3 (4) | Remicade | Johnson & Johnson/ Merck & Co | Rheumatoid arthritis / autoimmune diseases | 8,386 | 7,899 | Protein |
| 4 (3) | Seretide/Advair | GlaxoSmithKline | Asthma | 8,251 | 7,834 | |
| 5 (6) | Lantus | Sanofi | Diabetes | 7,592 | 7,155 | Protein |
| 6 (5) | Rituxan | Roche | Rheumatoid arthritis | 7,503 | 6,877 | Protein |
| 7 (9) | Avastin | Roche | Bowel cancer | 6,751 | 6,282 | Protein |
| 8 (7) | Herceptin | Roche | Breast cancer | 6,562 | 6,253 | Protein |
| 9 (8) | Crestor | AstraZeneca | Cholesterol | 5,622 | 6,149 | |
| 10 (10) | Abilify | Otsuka Holdings | Depression / schizophrenia | 5,500 | 5,334 | |

*The Guardian, March 2014*

**Protein Research- Academic and Industry Interests !**



# Commercial Antibody Formulations

| Product name | Dosage form | Concentration | pH | Formulation |
|---|---|---|---|---|
| ILARIS® (canakinumab) | Lyophilized | 150 mg/mL | NA | 92.4 mg/mL sucrose, L-histidine and L-histidine HCl, 0.6 mg/mL polysorbate-80 |
| XOLAIR® (omalizumab) | Lyophilized | 125 mg/mL | NA | 125 mg/mL omalizumab, 90 mg/mL sucrose, 1.7 mg/mL L-histidine hydrochloride monohydrate, 1.1 mg/mL L-histidine, 0.3 mg/mL polysorbate-20 |
| RAPTIVA® (efalizumab) | Lyophilized | 100 mg/mL | ~6.2 | Approximately 82 mg/mL sucrose, 4.5 mg/mL L-histidine hydrochloride monohydrate, 2.9 mg/mL L-histidine, 2 mg/mL polysorbate-20 |
| SYNAGIS® (palivizumab) | Lyophilized | 100 mg/mL | NA | 47 mM histidine, 3 mM Glycine, 5.6% mannitol |
| SIMPONI® (golimumab) | Liquid | 100 mg/mL | ~5.5 | 0.88 mg/mL histidine and histidine–HCl monohydrate, 41 mg/mL sorbitol, 0.16 mg/mL polysorbate-80 |
| STELARA® (ustekinumab) | Liquid | 90 mg/mL | 5.7–6.3 | 1 mg/mL L-histidine and L-histidine–HCl, 76 mg/mL sucrose, 0.04 mg/mL polysorbate-80 |
| HUMIRA® (adalimumumab) | Liquid | 50 mg/mL | ~5.2 | 6.2 mg/mL sodium chloride, 0.86 mg/mL sodium citrate, 1.3 mg/mL citric acid monohydrate, 12 mg/mL mannitol, 1 mg/mL polysorbate-80 |
| CAMPATH® (alemtuzumab) | Liquid | 30 mg/mL | NA | 8 mg/mL sodium chloride, 1.44 mg/mL dibasic sodium phosphate, 0.2 mg potassium chloride, 0.2 mg/mL monobasic potassium phosphate, 0.1 mg/mL polysorbate-80, 0.0187 mg/mL disodium edentate dihydrate |
| AVASTIN® (bevacizumab) | Liquid | 25 mg/mL | 6.2 | 60 mg/mL trehalose, 5.8 mg/mL sodium phosphate monobasic monohydrate, 1.2 mg/mL sodium phosphate dibasic anhydrous, 0.4 mg/mL polysorbate-20 |
| HERCEPTIN® (trastuzumab) | Lyophilized | 21 mg/mL | ~6 | Approximately 20 mg/mL α,α-trehalose dehydrate, 0.5 mg/mL L-histidine HCl, 0.32 mg/mL L-histidine, 0.09 mg/mL polysorbate-20, 1.1% bacteriostatic water |
| VECTIBIX® (panitumumab) | Liquid | 20 mg/mL | 5.6–6.0 | 5.8 mg/mL sodium chloride, 6.8 mg/mL sodium acetate |
| ROACTEMRA® (tocilizumab) | Liquid | 20 mg/mL | 6.5 | 15 mM phosphate, 50 mg/mL sucrose, 0.5 mg/mL polysorbate-80 |
| ARZERRA® (ofatumumab) | Liquid | 20 mg/mL | 6.5 | 8.55 mg/mL sodium citrate, 0.195 citric acid monohydrate, 5.85 mg/mL sodium chloride |
| SOLIRIS® (eculizumab) | Liquid | 10 mg/mL | NA | Chloride, phosphate dibasic, phosphate monobasic, polysorbate-80 |
| LUCENTIS® (ranibizumab) | Liquid | 10 mg/mL | 5.5 | 10 mM histidine–HCl, 10% trehalose, 0.01% polysorbate-80 |
| REMICADE® (infliximab) | Lyophilized | 10 mg/mL | ~7.2 | 50 mg/mL sucrose, 0.05 mg/mL polysorbate-80, 0.22 mg/mL monobasic sodium phosphate monohydrate, 0.61 mg/mL dibasic sodium phosphate dihydrate |
| RITUXAN® (rituximab) | Liquid | 10 mg/mL | ~6.5 | 9 mg/mL sodium chloride, 7.35 mg/mL sodium citrate dihydrate, 0.7 mg/mL polysorbate-80 |
| ZENAPAX® (daclizumab) | Liquid | 5 mg/mL | ~6.9 | 3.6 mg/mL sodium phosphate monobasic monohydrate, 11 mg/mL sodium phosphate dibasic heptahydrate, 4.6 mg/mL sodium chloride, 0.2 mg/mL polysorbate-80 |
| SIMULECT® (basiliximab) | Lyophilized | 4 mg/mL | NA | 1.4 monobasic potassium phosphate, 0.20 mg/mL disodium hydrogen phosphate (anhydrous), 0.32 mg/mL sodium chloride, 4 mg/mL sucrose, 16 mg/mL mannitol, 8 mg/mL glycine |
| ERBITUX® (cetuximab) | Liquid | 2 mg/mL | 7.0–7.4 | 8.48 mg/mL sodium chloride, 1.88 mg/mL sodium phosphate dibasic heptahydrate, 0.42 mg/mL sodium phosphate |
| REOPRO® (abciximab) | Liquid | 2 mg/mL | pH 7.2 | 10 mM sodium phosphate, 150 mM sodium chloride, 0.001% polysorbate-80 |

* All formulations are expressed as mg/mL after reconstitution. Because of fill volume variability, some concentrations are approximate.

European Journal of Pharmaceutics and Biopharmaceutics 78 (2011) 208–212

Malvern © 2015 Malvern Instruments Limited | Company Confidential www.malvern.com

## Analytical Challenges from High Concentration

- Characterization of proteins at the formulated conditions, with excipients.
- Stability profile, either storage conditions or forced degradation.
- Understanding of high concentration formulation, i.e viscosity.
- Conventional spectroscopic technologies suffer from low concentration requirement, excipients effect, and provide limited information.

*Far UV CD --- Secondary Structure*
***Low concentration, interference from excipients.***
*Intrinsic Fluorescence --- Tertiary Structure*
***Low concentration, fluorescence quenching.***
*FTIR --- Secondary Structure*
***High concentration, water subtraction.***



## Protein Characterization --- Scales

Molecular Heterogeneity | Nano/Micro Heterogeneity

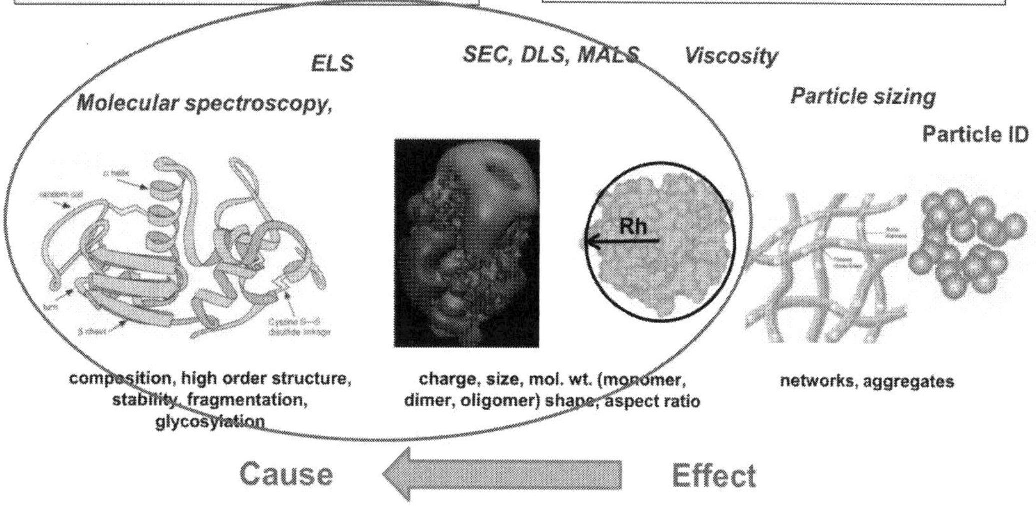


Structure, Function, Dimension, Behavior

**Combine Elastic and Inelastic Scattering**

# Zetasizer ZS Helix – Hybrid DLS-Raman

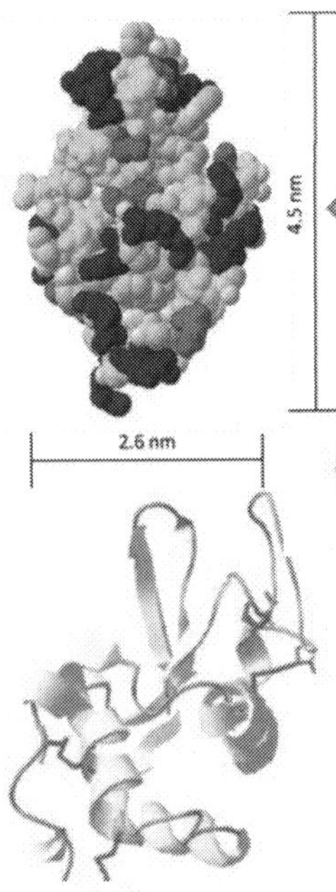


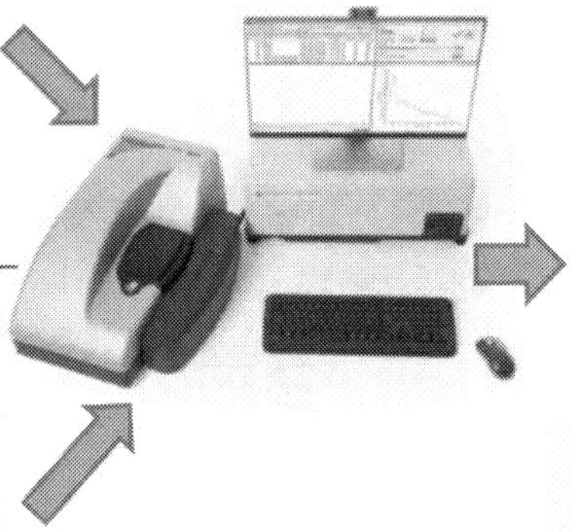

Hydrodynamic Size
Interaction Parameters ($k_D$)
Colloid Stability

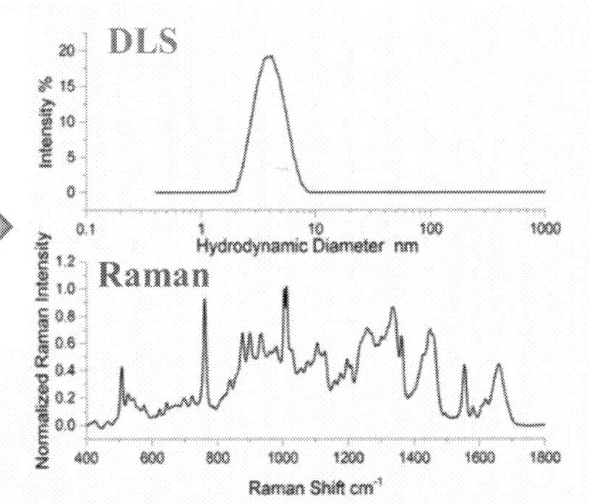


- Same Time, Same Sample
- Two –fold information
- High concentration preferred
- Varied Stresses: pH, thermal, etc.
- Requires as little as 20 µL

Secondary and Tertiary Structure
Aromatic Side Chain
Disulfide Bond
Conformational Stability

*Lewis, US Patent Pending*



# Dynamic Light Scattering (DLS)

Correlation of short time scale (µs) intensity fluctuations leads to the diffusion coefficient (D), hydrodynamic radius ($R_H$), & polydispersity index (PDI).

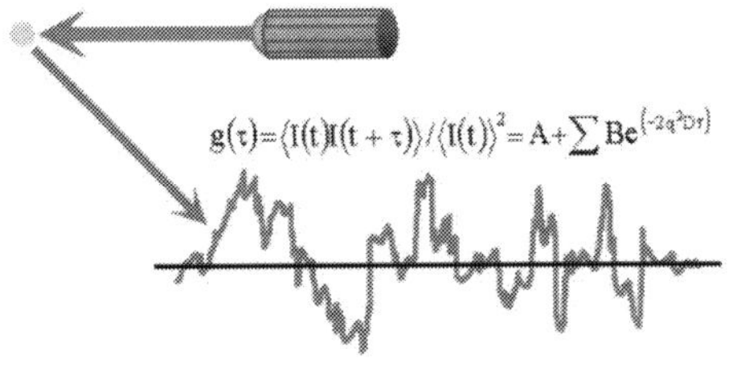


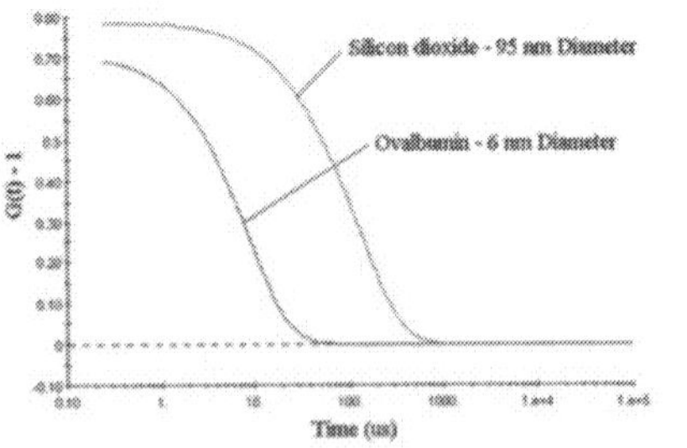


*Stokes-Einstein Equation*

$$R_H = \frac{kT}{6\pi\eta D}$$

q = Scattering vector
D = Diffusion coefficient
$R_H$ = Hydrodynamic radius
k = Boltzmann constant
T = Temperature
η = Solvent viscosity

## Raman Spectroscopy

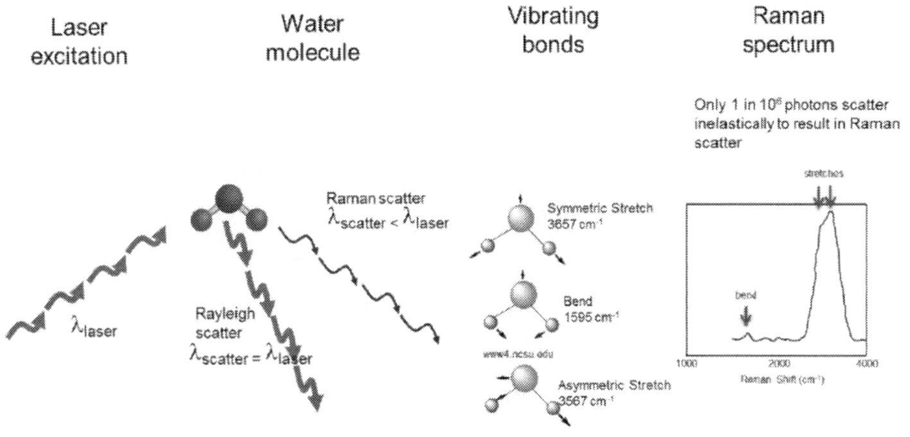


- A Non-Invasive, Non-destructive inelastic light scattering/vibrational spectroscopic technique.
- Different selection rules – polarizability (Raman) vs dipole moment (IR)
- Water weak to Raman spectrum



## Typical Protein Raman Spectrum

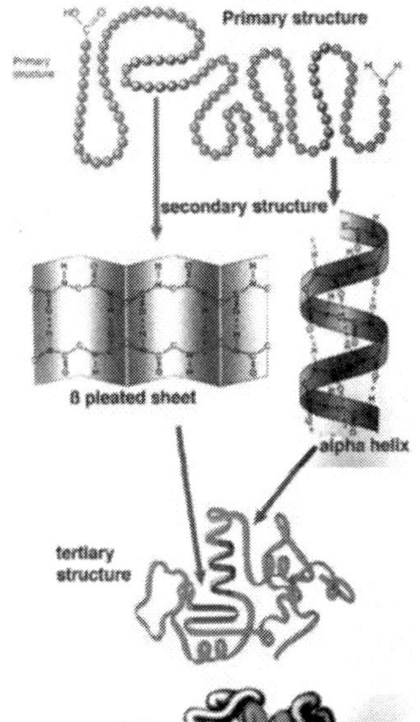


*http://medicinenewbie.blogspot.com/2011/05/professional-i-exam-2-may-2011-part-4.html*

Raman spectrum of proteins provides

- Backbone (Amide I, III) → helical/sheet
- Side chain orientation (Trp)
- Hydrophobic or hydrophilic environment (Trp)
- Hydrogen bonding patterns (Trp, Tyr)
- Disulfide bond (*g-g-g, g-g-t, t-g-t*)

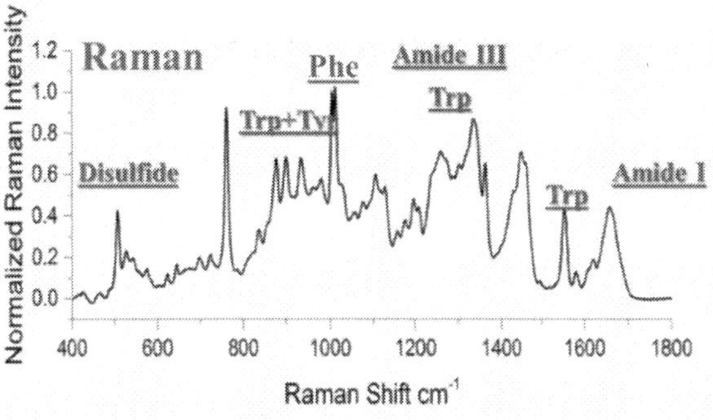

## Lysozyme --- Thermal induced Unfolding

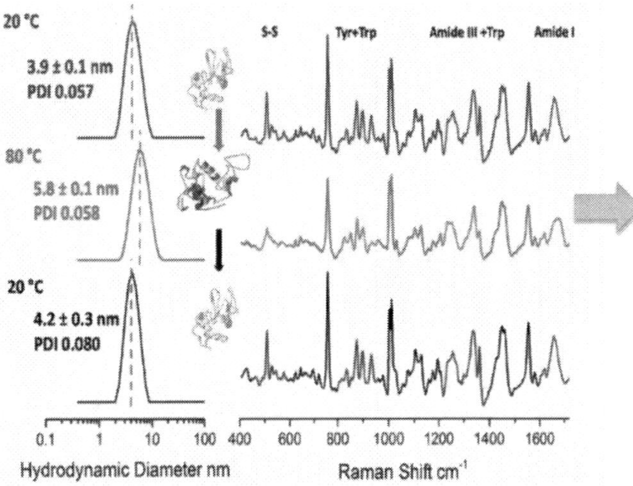


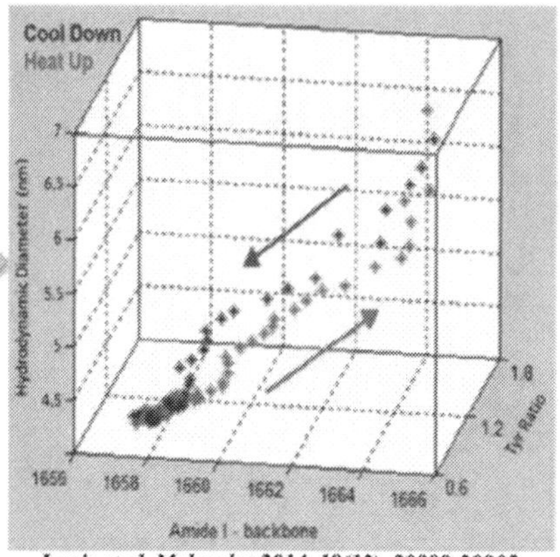


*Lewis et al. Molecules 2014, 19(12), 20888-20905*

- Real time profile --- full reversible thermal event
- Simultaneous characterization --- structure and size
- A thermal induced unfolding event without aggregation associated

Malvern  www.malvern.com

## Lysozyme --- TCEP induced Reducing

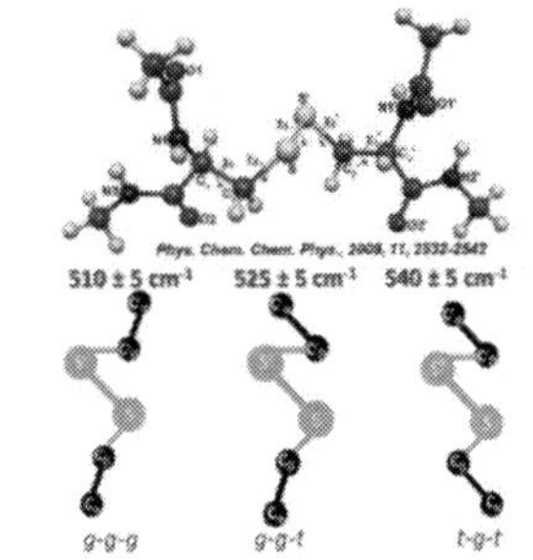


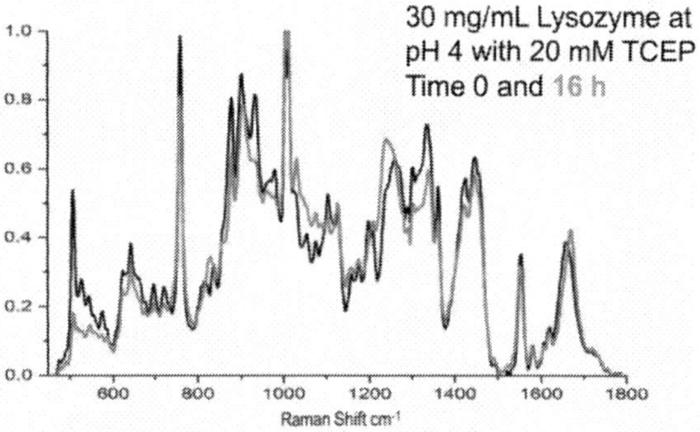


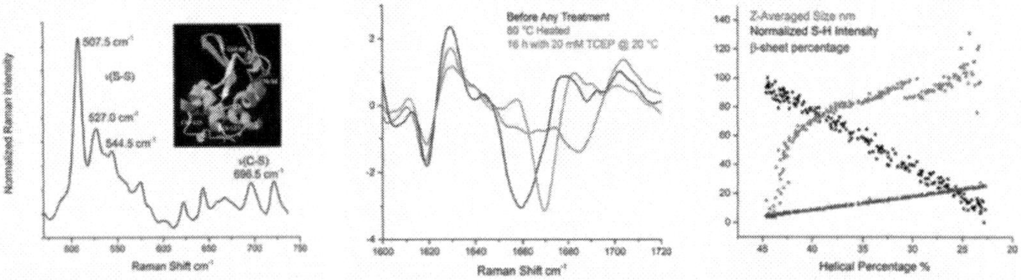


- A conformation change with aggregation/gelation process.
- Different denatured states for reducing and thermal stress.

Malvern  www.malvern.com

## RNase A --- Thermal induced Unfolding

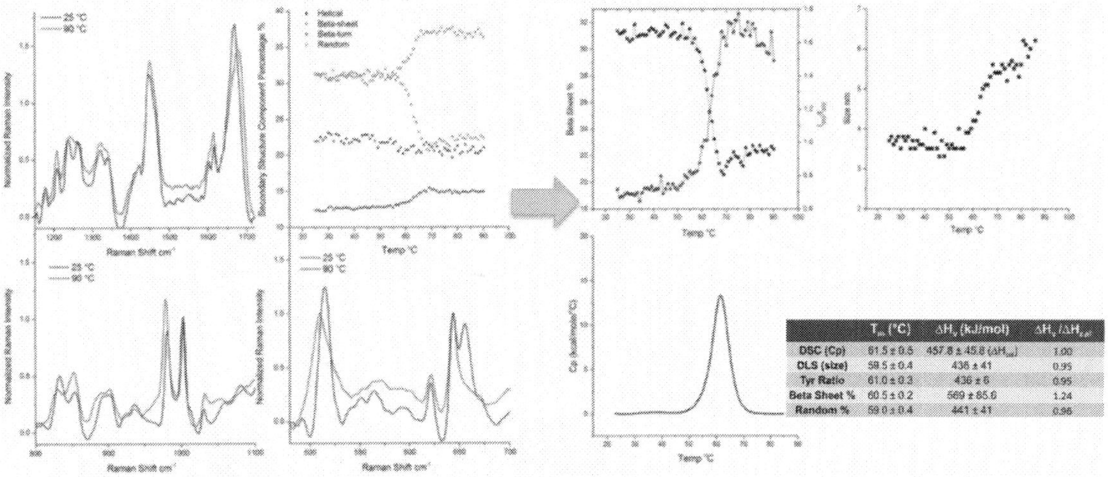


| | $T_m$ (°C) | $\Delta H_v$ (kJ/mol) | $\Delta H_v / \Delta H_{cal}$ |
|---|---|---|---|
| DSC (Cp) | 61.5 ± 0.5 | 457.8 ± 45.8 ($\Delta H_{cal}$) | 1.00 |
| DLS (size) | 59.5 ± 0.4 | 436 ± 41 | 0.95 |
| Tyr Ratio | 61.0 ± 0.3 | 436 ± 6 | 0.95 |
| Beta Sheet % | 60.5 ± 0.2 | 569 ± 85.6 | 1.24 |
| Random % | 59.0 ± 0.4 | 441 ± 41 | 0.96 |

- Clear change of secondary structure (Amide I, III), Tyr and disulfide region.
- Simultaneous/cooperative transitions for tertiary, secondary structure and size.
- The van't Hoff enthalpy and calorimetry enthalpy ratio indicates a highly cooperative two state transition.



## RNase A --- TCEP induced Reducing

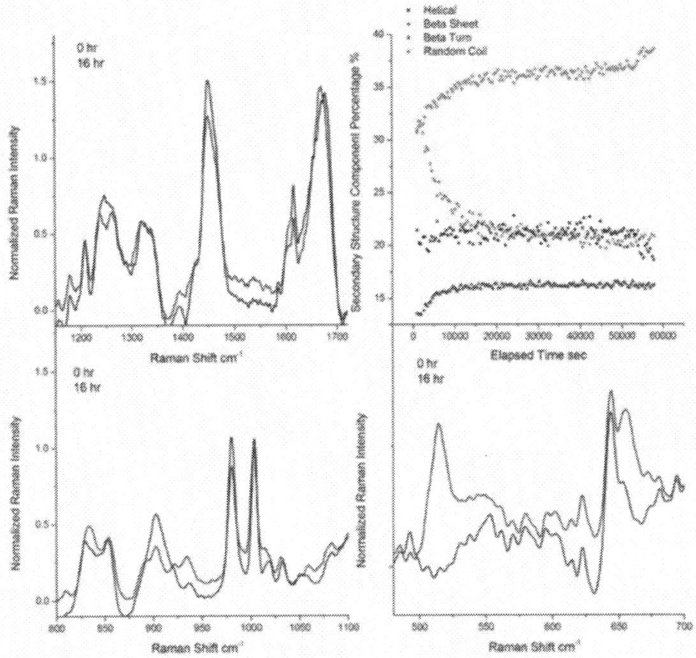


- 30 mg/mL RNase A pH 5.5, 20 mM TCEP
- Clear transition from beta-sheet to random coil/beta turn upon TCEP reduction.
- The Tyr ratio at 850/830 also changed, indicating a modification of the local hydrogen bonding environment, but less change than for thermal stress.
- The disulfide S-S mode at 515 $cm^{-1}$ disappeared and as before (ref. heating) there is a similar decrease for the C-S mode at 655 $cm^{-1}$.

# RNase A --- TCEP induced Reducing Kinetics

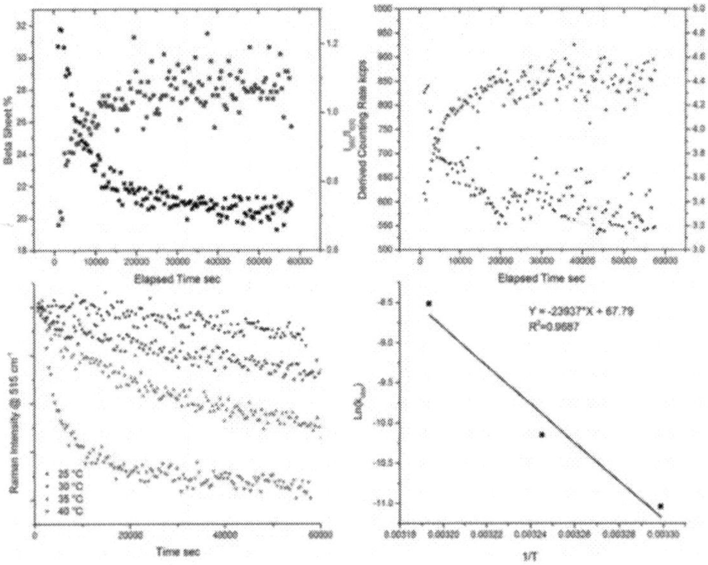


- Upon reduction, there is a decrease in the derived count rate and a small size increase, indicating a gradual expanding structure.
- Reduction experiments at multiple temperatures can be used to investigate the kinetics and activation energy of reduction.
- A reducing induced conformation change, but not associated with aggregation !



# mAb X --- Thermal Induced S-S Region Change

25 °C

63 °C

67 °C

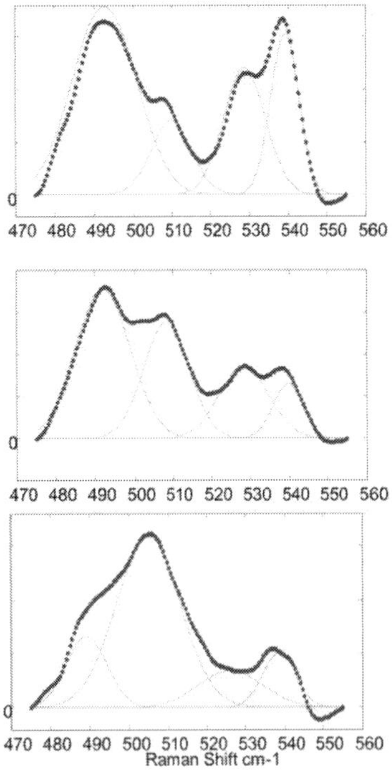


Disulfide bond conformers

Intensity at 539 $cm^{-1}$
trans-gauche-trans

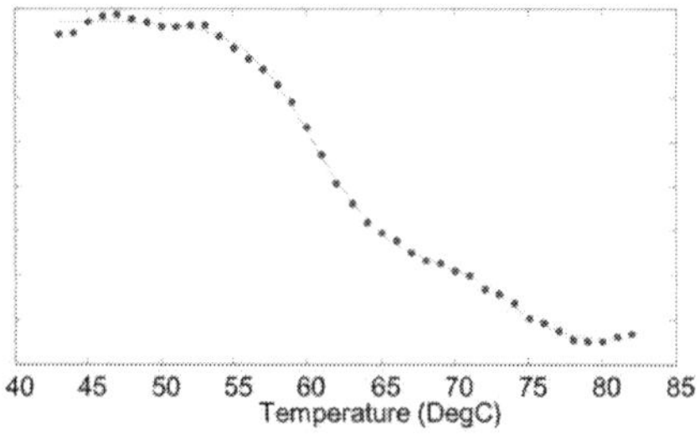


mAb X at 150 mg/mL Temp
Ramp 25-90 °C
Disulfide bond conformers –
double transition @ 65.0 & 60.2 °C

# mAb Y and modified mAb Y --- Thermal Stability

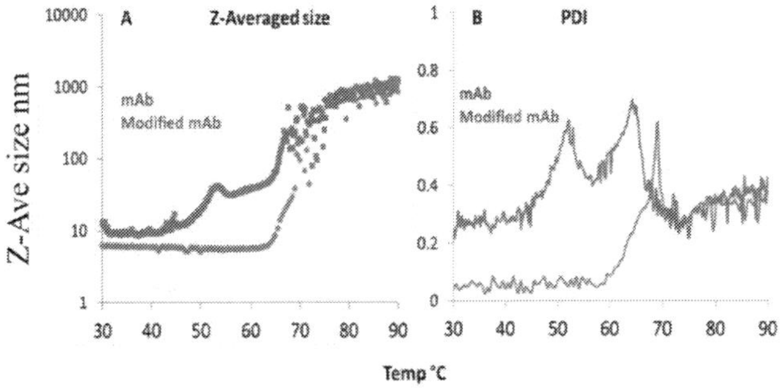


Sizing results

- mmAb double transition: ~53 & 65°C
- mAb single transition: ~ 69 °C
- PDI trend for mmAb – oligomerization followed by aggregation

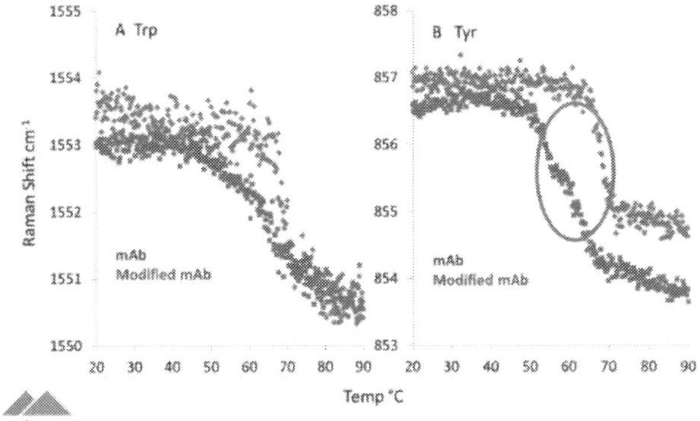


Raman results

- Minimum secondary structure change
- mmAb double transition: ~54 & 63 °C
- mAb single transition: ~ 69 °C

***Tyrosine H-bonding environment most perturbed during low temperature transition***



# Oligomerization of modified mAbY

## *Isothermal Incubation at 46, 53 and 60 °C*

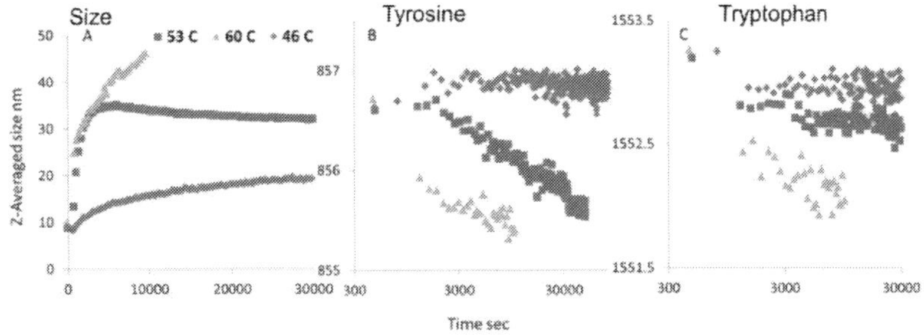


- 46 °C - pre-transition
  - size slowly increases
  - Raman maker bands show no change
- 53 °C – in-transition
  - size quickly increases from 10 to ~35 nm and then stabilizes
  - Tyr peak position drops, while the Trp peak position shifts slightly
  - Tyr peak position shifts from 856.8 to 855.5 $cm^{-1}$, the value seen at the first transition in the T ramps
- 60 °C – post-transition
  - size quickly increases to over 40 nm within 3 hours
  - Trp and Tyr peak positions quickly shift to lower frequencies
  - Tyr peak position shifts to ~855.5 $cm^{-1}$.

## Oligomerization of mmAb

### *Isothermal Incubation at 46, 53 and 60 °C*

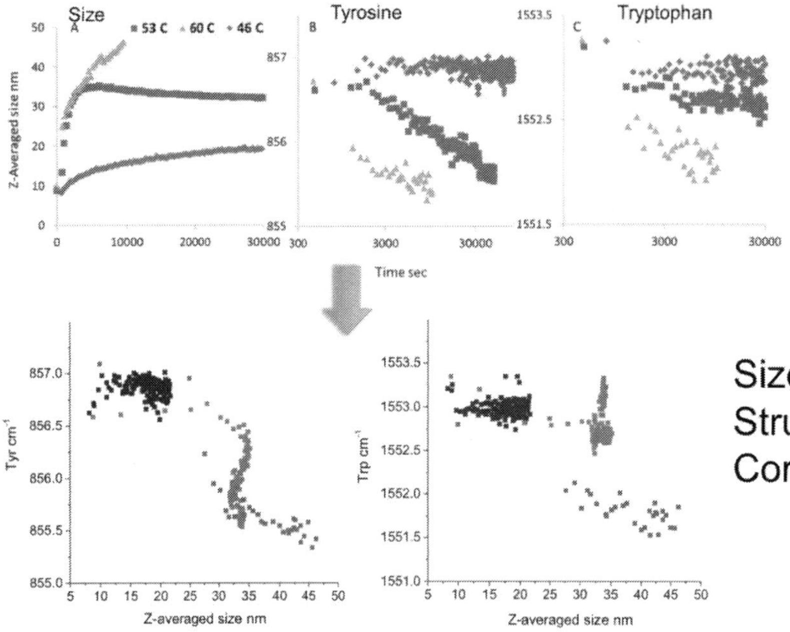


Size and Structure Correlation !





## Protein Aggregation in Solution: Conformational and Colloidal Stability

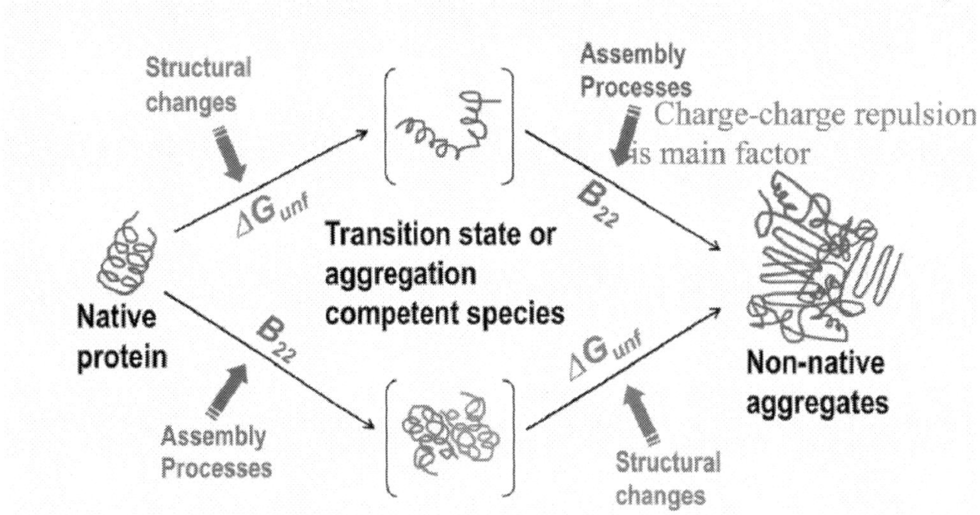


**High protein concentration put additional challenges for formulation development and real time stability characterization.**

***Chi et al., 2003. Pharm. Res. 20:1325***

## Other Applications

- Hydrogen-deuterium exchange.
- Biosimilars Characterizations (Insulins, mAbs, etc.).
- Coupled with Micro-rheology with low frequency marker.
  *Steve Blake's Talk on β-lactoglobulin at 2:20 pm in Rm. 206B*
- Chemical denaturation (GdnHCl/Urea/$H_2O_2$)/Chemical modifications.
- Lyophilization.

## Summary

- High concentration formulation brings new analytical challenges.
- Raman spectroscopy could benefit the user by providing abundant structure information simultaneously. Like all the spectroscopic techniques, it provides an ensemble information.
- Coupled with dynamic light scattering for size distribution, the platform could be versatile to probe the protein at the formulation conditions with selected stresses, in the real time, if preferred.
- Could be widely applied to many potential application areas.
- Samples at high concentration and want a lot of information.... DLS/Raman !

Malvern  www.malvern.com

Malvern

## Acknowledgements

**Malvern Instruments**
**Bioscience Development Initiative, Columbia MD.**

**University of Colorado, Denver**
**Prof. John Carpenter**

## Questions?

wei.qi@mavlern.com
**443-878-1319**

# 408788 Construction of Robust, Multi-Input Genetic Circuits and Memory Devices for Environmental Signal Sensing and Programmed Actuation

**Abstract:**

Signal sensing is essential for bacteria that live in fluctuating environments. However, even a group of isogenic bacteria, subject to identical culture conditions, often exhibit heterogeneity in phenotypes. This multi-stability can be explained by stochastic transitioning among multiple stable states of gene expression which are often the results of positive feedback loops in gene regulatory networks. Although such a "bet-hedging" strategy can increase bacterial fitness in unpredictable environments, such stochastic processes can prevent synthetic biologists from predictably engineering microbial cells that need to face uncontrolled environments. Thus, to fully exploit microbes in synthetic biology applications, complex gene regulatory systems and their behaviors in diverse conditions need to be better understood. To this end, we propose to construct "robust" genetic sensors and circuits that endow bacteria with the ability to sense environmental signals (e.g., temperature, light, pH, and oxygen level), perform computational operations, and implement programmed actuations (e.g., generation of toxins to kill parasite eggs).

By building sensors and circuits from the bottom-up, we have gained design principles for construction of programmable cells. Specifically, we have built robust logic gates that contain modular sensors for temperature or oxygen levels that can be more meaningful signals in the environment than chemical inducers that are used in the lab. In addition, we have constructed bistable switches that can generate noise-tolerant responses to fluctuating environmental inputs. These switches are based on the sequestering interaction between a transcription factor (an activator) and an anti-activator, along with multiple positive feedback loops. Importantly, these switches are functional over a wide range of parameters and endow the cells with a long-lasting memory. We are assembling these modular sensors, logic gates, and memory devices to construct robust, programmable probiotic strains that would reproduce in the intestine where parasite eggs are produced, and would come out of the human body with the parasite eggs and kill them. We will present progress towards development of such engineered microbes, which can be programmed to kill parasite eggs only when user-defined conditions are met.

# 409914 Synthetic Control of Transcription: A Decade's Perspective on Synthetic Promoter Design

**Abstract:**

Synthetic biology provides the means of designing novel parts to aid in the controlled, tunable, and regulated expression of individual genes, circuits, and pathways. Over the past decade, significant progress has been made in the area of promoter engineering. This talk will cover several advances in the area that range from random point-mutation guided approaches to new, model based design of synthetic parts. By utilizing such approaches, it is possible to create libraries of transcriptional control that can span several 100's of fold with respect to mRNA level. An emphasis will be placed on newer approaches to minimize the size of synthetic transcriptional control elements in yeast. Next, this talk will discuss the synthetic design of expression cassettes and operons and the impact of genetic context for synthetic parts—a particularly important aspect for characterization and predicable function. By synthetically altering upstream regions, terminator components, and promoter structure, defined and reproducible function can be obtained. Collectively, these results demonstrate novel approaches to synthetically alter and control gene transcription—a central goal of metabolic engineering and synthetic biology efforts. Once we have mastered these aspects, it is possible to design transcriptional control from scratch. This talk will conclude with progress toward this goal.

# 408893 Getting through the Gatekeepers: Changing the Selectivity of Semi-Permeable Cellular Membranes

**Abstract:**

The phospholipid bilayer has long been described as the cellular gatekeeper, preventing the passive transport of proteins, small molecules, and ions across cellular and organelle boundaries. It was recently discovered that some bacteria use organelle-like compartmental systems as well, contrary to the long-held belief that bacteria lack such organization, and these bacterial compartment membranes are made entirely of proteins. We explore the gatekeeping functions of each of these distinct types of membranes, studying protein and small molecule transport and sequestration in each case. Using various synthetic biology approaches, we engineer these systems in order to increase the yield of biochemical products such as butanol. Specifically, I will discuss how we can 1) sequester metabolic pathway enzymes and intermediates in the protein compartments and 2) gain control and alter the specificity of lipid membrane transporters for small molecules and proteins.

# 437446 Engineering Designer Biocatalysts and Pathways for Microbial Glycobiology

**Abstract:**

Carbohydrates add a level of diversity across all forms of life that is unparalleled by the information content of nucleic acids and proteins. The lack of a simple template to translate a glycan code into defined sugar structures contributes to this complexity and impedes efforts aimed at the production of biologically important glycans and glycoconjugates. With the discovery of glycoprotein synthesis in bacteria and functional transfer of glycosylation pathways between species, *Escherichia coli* cells have become a tractable host for understanding glycosylation and the underlying glycan code of living cells. Moreover, efforts to manipulate the pathways from sugar nucleotides to glycolipids to glycoproteins have transformed *E. coli* into a living factory for scalable, bottom-up production of complex glycoconjugates by design. Here, I will discuss our efforts to develop *E. coli* for the biosynthesis of a diverse array of glycan structures, which can be used to tailor the activity, stability, half-life, and shelf-life of a diverse array of protein targets ranging from biopharmaceuticals to industrial enzymes. I will also discuss our efforts to unify protein glycosylation in *E. coli* with the advanced techniques of protein engineering such as chimeragenesis, protein-fragment complementation, and directed evolution. The result is a powerful new platform for engineering the enzymes, pathways, end-products, and genomes of glycoengineered bacteria for creating the next generation of biotechnology-derived products.

# 440216 Imaging Approaches to Study and Manipulate Neural Behavior


**Abstract:**

### Laboratory for Non-Invasive Biological Interfaces

Many of today's biggest questions in areas such as neuroscience, cancer and immunology concern the function of specific cells and molecules deep inside living organisms. Yet, our ability to study biological function within this critical in vivo context is currently limited by the lack of suitable imaging technologies. While optical imaging using fluorescent reporters such as the green fluorescent protein (GFP) enables exquisite observation of molecular and cellular events in microscopic specimens, it has limited penetration into intact organisms. Meanwhile, non-invasive technologies such as magnetic resonance imaging (MRI) and ultrasound, although capable of imaging deep tissues with high spatial and temporal resolution, currently lack sensitive molecular imaging agents analogous to GFP. If such agents could be developed, they would transform many areas of biomedicine by enabling the observation of dynamic cellular processes in living, breathing animals and, ultimately, patients.

To develop breakthrough technologies for non-invasive biomedical imaging, the Shapiro Laboratory pursues fundamental advances at the interface of molecular and cellular engineering with deeply penetrant forms of energy such as magnetic fields and sound waves. We have shown that molecules ranging from paramagnetic proteins to hollow protein nanostructures can be readily detected with MRI and/or ultrasound, and are now developing these molecules as non-invasive imaging agents of neural activity, brain injury and cancer. In parallel, we are developing technologies that allow specific cellular functions to be activated remotely using sound waves. These technologies are expected to contribute to the study of basic biological function and disease, and to form the basis of future medical approaches to diagnosis and personalized therapy.

More information about the Shapiro Lab at Caltech is online at http://shapirolab.caltech.edu.

# 427902 Design of Dendrimer-Doxorubicin Conjugates for Transport Modulation Across in Vitro Pulmonary Epithelium and Their Solution Formulation in Pressurized-Metered Dose Inhalers

**Abstract:**

Lung cancer is the leading cause of cancer death among both men and women in the United States. Adenocarcinoma accounts for more than 40% of all lung cancer cases. Doxorubicin (DOX) is an FDA-approved chemotherapeutic that has been widely used as primary anticancer drug in the treatment of a variety of cancers including lung cancers. However, rapid elimination, uncontrolled release, and life-threatening cardiotoxicity, has hindered the applicability of DOX and other potent anti-cancer agents. The potential use of DOX in the treatment of lung adenocarcinoma with DOX is also limited by its low accumulation in the lungs, which is generally administered through systemic route. The development of polymeric carriers for the modulation of transport, targeting and controlled release of potent anti-cancer agents and their aerosol formulation for local lung delivery is, therefore, of great relevance in the treatment of lung adenocarcinoma.

In this work, a series of PEGylated poly(amidoamine) dendrimer nanocarriers with acid-labile and acid-nonlabile DOX conjugates were synthesized and characterized. We investigated the effect of PEGylation density (low, medium and high) and number of DOX conjugates (low and medium) on the release of DOX from the dendrimer nanocarrier, *in vitro* cytotoxicity, and transport of dendrimer nanocarriers across a model of the lung epithelium (Calu-3 cells). PEGylation retards the release of DOX in acidic medium. PEGylation density also affects cytotoxicity as seen by a slight increase in $IC_{50}$ for DOX-conjugated dendrimer compared to free DOX, due to the controlled release of the therapeutic. PEGylation degree significantly helped dendrimer nancarriers escape mucus trapping compared to bare dendrimer, aiding their transport to the basolateral side. Reduced transepithelial electrical resistance (TEER) and limited cellular internalization demonstrated that paracellular transport plays a vital role in transporting dendrimer-DOX conjugates across airway epithelium. The PEGylated dendrimer-DOX conjugates were directly formulated in the propellant-based metered-dose inhalers (pMDIs) without the aid of any surfactants and co-solvents, forming a solution aerosol formulation. The solution formulation tends to deliver dendrimer-DOX conjugates with high efficiency to the alveolar region (up to 80%). The results suggest the targeting, transport, release and formulation of the dendrimers can be controlled by careful design of the nanocarriers.

# 430373 Controlling Mechanical Properties and Dissolution Rate of Strip Films Loaded with Griseofulvin Nanoparticles

**Abstract:**

Polymer strip films have been demonstrated to be an ideal delivery form for poorly water-soluble drugs. Specifically for active pharmaceutical ingredient (API) particles that have been reduced in size to enhance their solubility and dissolution rate, polymer strip films provide a means by which these nano/micro-size particles can be delivered relatively quickly while retaining their nano/micro size upon delivery. However, the ability to intelligently manipulate the mechanical properties and dissolution rate of such films for the delivery of poorly water-soluble API nanoparticles has yet to be investigated. To this end, hydroxypropyl methylcellulose (HPMC) films were prepared, loaded with model API, griseofulvin, nanoparticles produced via wet stirred media milling in aqueous suspension. These films consisted of various plasticizers and HPMC molecular weights. Addition of glycerin, triacetin, and polyethylene glycol as plasticizers successfully reduced film strength and increased film elasticity with no significant impact on dissolution rate. Increasing the molecular weight of the polymeric film-former generally led to slower film dissolution rate, accompanied by a small increase in tensile strength. All films exhibited good uniformity of content in terms of film thickness and drug distribution (<6% RSD for 0.7 $cm^2$ samples). Results demonstrate that, when used in tandem, adjustment of plasticizer content and polymer molecular weight can be used to successfully tailor the mechanical properties and dissolution rate of polymer films loaded with poorly water-soluble drug nanoparticles without sacrificing the advantages or integrity of the strip film format, allowing for intelligent film formulation design for targeted applications.

# 406800 Lipid-Polymeric Hybrid Nanoparticle Based Nicotine Vaccine: Size Matters

**Abstract:**

**Lipid-polymeric hybrid nanoparticle based nicotine vaccine: size matters**

Zongmin Zhao , Yun Hu, Reece Hoerle, Meaghan Devine, Chenming Zhang

Department of Biological Systems Engineering, Virginia Tech, Blacksburg, VA 24061

Presenting author

Email addresses:

Zongmin Zhao: zmzhao@vt.edu

Yun Hu: hyun86@vt.edu

Reece Hoerle: rhoerle@vt.edu

Meaghan Devine: meaghd1@vt.edu

Chenming Zhang: chzhang2@vt.edu

**Abstract**

Tobacco use is prevalent worldwide and has consistently been the top preventable cause of many serious diseases, resulting in huge mortality, morbidity, and economic loss in recent decades. Due to the limited success of FDA approved pharmacological therapies, it is necessary to develop more effective treatment for smoking cessation. Currently, nicotine vaccine that can induce nicotine specific antibodies is considered as a promising approach against tobacco addiction. However, traditional protein-hapten conjugate nicotine vaccines have shown less than desired immunological efficacy due to their poor recognition and internalization by immune cells. Previous study in our group showed that negatively charged nanohorn supported liposome could effectively enhance the immunogenicity of nicotine protein conjugate based vaccine. In this study, nanohorn was substituted with Poly (lactic-co-glycolic acid) (PLGA) nanoparticle (NP) and a novel lipid-polymeric hybrid NP based nicotine vaccine was developed. In particular, the effects of particle size on the immunogenicity of hybrid NP based vaccine was investigated. 100 nm hybrid NPs were found to be taken up by dendritic cells more efficiently than 500 nm hybrid NPs. Mice immunization results demonstrated that both 100 nm and 500 nm nanovaccines could induce strong immune response against nicotine. However, a significantly higher titer of antibody against nicotine was achieved by 100 nm nanovaccine than 500 nm nanovaccine. Moreover, no evidence of systemic toxicity was observed on mice immunized with all the hybrid NP based nicotine vaccines. These results from this study suggest that lipid-polymeric NP based nicotine vaccine is a promising candidate for treating nicotine dependence.

# 434194 High Throughput Synthesis of Nanovaccine Formulations

**Abstract:**

Using nanoscale vehicles to deliver protein antigens in the development of vaccines has been a well-studied area of research. High throughput techniques utilizing combinatorial libraries have the potential to be a valuable resource in rapidly synthesizing and screening nanoscale biomaterial-based carriers for drug and vaccine delivery. In this study, we highlight the development of an automated high-throughput robot for synthesizing polyanhydride nanoparticles containing protein antigens. Polyanhydrides are a class of safe and biodegradable polymers that have been widely used as a vaccine adjuvant/carrier. The method we have developed was based on a multiplexed homogenizer and has the capacity to handle parallel streams of monomer or polymer solutions to synthesize polymers and/or nanoparticles. Copolymers were synthesized from sebacic acid (SA), 1,6-bis(*p*-carboxyphenoxy)hexane (CPH), and 1,8-bis(*p*-carboxyphenoxy)-3,6-dioxactane (CPTEG). Nanoparticles loaded with the model antigen, ovalbumin, were synthesized using anti-solvent nanoprecipitation. Concentrations of surfactant were varied to test its effect on protein encapsulation efficiency as well as protein release kinetics. We observed that surfactant concentration did not significantly affect protein release rate, but greatly enhanced protein encapsulation efficiency. Particles synthesized with the high throughput method were compared to particles made with conventional anti-solvent nanoprecipitation to validate the high throughput approach. Protein burst and release rates between the two methods were similar, even though particles synthesized using the high-throughput technique were observed to be smaller. Finally, we demonstrated that the high-throughput method could be adapted to functionalize the surface of the particles to aid in the design of targeted drug and vaccine delivery systems. Our results showed that the developed high-throughput system is a viable alternative to conventional methods for studying vaccine and drug delivery vehicles, with the potential to more rapidly optimize the design of these systems.

# 434315 Preparation and Characterization of Pegylated, Acid-Labile Dendrimer-Doxorubicin Conjugates and Their Aerosol Formulation in Propellant-Based Inhalers

**Abstract:**

Lung cancer is the leading cause of cancer death among both men and women in the United States. Adenocarcinoma accounts for more than 40% of all lung cancer cases. Doxorubicin (DOX) is an FDA-approved chemotherapeutic that has been widely used as primary anticancer drug in the treatment of a variety of cancers including lung cancers. However, rapid elimination, uncontrolled release, and life-threatening cardiotoxicity, has hindered the applicability of DOX and other potent anti-cancer agents. The potential efficacy of DOX in the treatment of lung adenocarcinoma is also limited by low accumulation in the lung tissue, which is usually administered through systemic route. The development of polymeric carriers for the modulation of transport, targeting and controlled release of potent anti-cancer agents and their aerosol formulation for local lung delivery is of great potential relevance in the treatment of lung adenocarcinoma.

In this work, a series of PEGylated poly(amidoamine) dendrimer nanocarriers with acid-labile DOX conjugates were synthesized and characterized. We employed a two-step PEGylation strategy to increase the payload of the hydrophobic DOX. We investigated the impact of pH (neutral and acidic pH), PEGylation density (low, medium and high) and number of DOX conjugates (low and medium) on the release of DOX from the dendrimer nanocarrier, the kinetics of carrier uptake, intracellular release kinetics of DOX from the nanocarrier, and toxicity in an alveolar adenocarcinoma cell line (A549). PEGylation retards the release of DOX in acidic medium, and also intracellularly, as determined by co-localization studies with confocal microscopy. We also observed that the kinetics of cellular entry of the nanocarrier with DOX increased significantly compared to free DOX, as determined by flow cytometry. At the highest PEGylation density, the rate of internalization of the nanocarrier containing DOX was even higher than that of free DOX. PEGylation density also affects cytotoxicity as seen by an increase in $IC_{50}$ for DOX-conjugated dendrimer compared to free DOX, due to the controlled release of the therapeutic. PEGylated dendrimer-DOX conjugates were directly formulated in the propellant-based metered-dose inhalers (pMDIs) with a trace amount of ethanol (ca. 0.2%), forming a solution aerosol formulation. The stability and aerosol characteristics increased significantly as PEG density increased and DOX payload decreased. The formulation of highly PEGylated dendrimer-DOX conjugates resulted in aerosols with high deep lung deposition (up to 80%). These results are rationalized based on the phase behavior of the polymer in propellant HFA.

# 414380 A Drug-Sparing Approach to the Preparation of Drug Nanosuspensions: High-Intensity Vibratory Milling

**Abstract:**

Almost 40% of the new chemical entities in drug pipeline of pharmaceutical companies have low aqueous solubility, which leads to poor bioavailability *in vitro* and *in vivo*. Reduction of drug particle size down to nano-scale has been an effective way of increasing the dissolution rate of poorly water-soluble drugs. In this study, we aim to elucidate the impact of the process parameters on the breakage kinetics and particle size during the preparation of drug nanosuspensions via high-intensity vibratory milling, which can be used as a drug-sparing approach in early development. Griseofulvin (GF) particles were wet-milled in the presence of two stabilizers, hydroxypropyl cellulose (HPC) and sodium dodecyl sulfate (SDS), to prepare stable GF suspensions under various processing conditions. Laser diffraction (LD), scanning electron microscopy (SEM), and X-ray powder diffraction (XRPD) were used to characterize the suspensions. Several milling experiments were performed to determine the optimal bead size at low power density. Then, the process was intensified with the optimal bead size and operated at a much higher power density by increasing the vibration intensity and bead loading step-wise with the objective of increasing the breakage rate. After examining the breakage kinetics upon intensification, additional experiments were performed with various bead sizes to explore if and how the optimal bead size depends on the power density. Finally, a comparative assessment of the performance of vibratory milling vs. wet stirred media milling has been carried out in view of the power density, resultant breakage kinetics, and final drug particle size attained. This study provides a systematic process intensification approach as well as guidance for bead size selection and suggests process/design improvement strategies for fast preparation of drug nanosuspensions via vibratory milling.

# 435203 Metabolomic Modularity Analysis (MMA) to Quantify Human Liver Perfusion Dynamics

Tuesday, November 10, 2015: 3:15 PM
155A (Salt Palace Convention Center)

**Gautham V. Sridharan**[1], Bote Bruinsma[1], Martin L. Yarmush[1] and Korkut Uygun[2], (1)Center for Engineering in Medicine (CEM) at Massachusetts General Hospital – Harvard Medical School - Shriners Hospital for Children, Boston, MA, (2)Center for Engineering in Medicine, Harvard Medical School, Boston, MA

Biological cells are complex dynamic systems, comprising thousands of interacting nucleic acids, proteins, and metabolites acting in concert to perform coordinated and well regulated tasks. The ability to effectively quantify *system-wide* cellular responses to various stimuli is paramount to our understanding of complex disease-states and how pharmacological agents elicit desired outcomes. Investigators are therefore increasingly relying on large scale –omics data (transcriptomics, proteomics, metabolomics) to experimentally capture changes in signaling and metabolic pathway components. While the burden of data acquisition has lightened with improved instrumentation, data interpretation in the proper biological context remains challenging. Metabolomics data, in particular, are difficult to interpret in the context of pathways because a single metabolite may be utilized by several functional modules. Both univariate and multivariate statistical tools can determine which metabolites explain a differential response among experimental treatment groups. However, these data-driven methods do not utilize the vast domain-specific biochemical knowledge of known stoichiometric and regulatory interactions that can be used to systematically uncover which signaling or metabolic pathways are activated by a perturbation. To this end, mapping metabolomics data onto well curated metabolic graph networks can help identify interactions that may not be intuitive from simply observing the metabolic pathway maps offered by the KEGG database.

In this study, we introduce Metabolomic Modularity Analysis (MMA) as a graph-based algorithm to systematically identify modules of reactions enriched with metabolites flagged to be statistically significant based on univariate analysis. Briefly, a metabolic network is abstracted as a reaction-centric graph network where a non-directional edge is drawn between two reactions if a metabolite produced by the first reaction is consumed by the second. The length of the edge between reactions is weighted such that if the two reactions are involved in metabolites that are statistically significant, their edge distance is shorter. When applying Newman's hierarchical partition algorithm based on relative reaction-pair shortest paths, the resulting modules feature reactions from the surrounding local topological neighborhood, which favors adjacent reaction pairs that involve statistically significant metabolites. A defining feature of determining reaction-centric modularity is that interactions between reactions mediated by the production and consumption of cofactors and other hub metabolites are also accounted for.

We apply MMA on time-course metabolomics data collected from biopsies during subnormothermic machine perfusion (SNMP) of nine discarded human livers that were rejected for transplantation. These livers had endured various degrees of warm ischemia time (WIT) prior to organ procurement and three of them exhibited over 30% macrovesicular steatosis. Of the 155 primary metabolites measured, the pre-perfusion levels (t=0 hours) of 33 of them were found to be significantly correlated with WIT, suggesting that they could be putative biomarkers for ischemic injury. MMA was performed on the human ReconX metabolic network (7439 reactions and 2626 metabolites) after flagging these 33 metabolites in the cytosolic compartment as being significant. For a network this large, computing the adjacency distance matrix requires $_{7439}C_2$, or $2.77 \cdot 10^7$ computations for the initial network, which causes total run times on the order of days with a laptop, rendering the method impractical for repeated analysis. Fortunately, Mathworks® now offers Matlab's Distributed Computing Server (MDCS) on Amazon's Elastic Compute Cloud (EC2). For this study, one cluster node (c3.8xlarge) with 16 workers was rented at a rate of $1.68/hour for the cluster and $1.12/hour for MDCS. Several key parts of the MMA algorithm are parallelizable and was therefore able to complete a full run in 2.7 hours, with a total cost of $8.40 for the 3 hours of server time. The MMA partitioning resulted in 4755 hierarchical modules, of which 223 contain at least one significant metabolite. To highlight an example, one such module contained four metabolites significantly correlated with WIT; arachidic acid (arach[c]), cholesterol (chsterol[c]), stearic acid (ocdca[c]), and palmitic acid (hdca[c]). The stoichiometry of reactions in the module suggests that ischemia may have an impact on coenzyme A (coA[c]) pools as a cause for why these four metabolite levels are affected (Figure 1). Ongoing work involves applying MMA to identify modules activated by perfusion itself by flagging metabolites significantly different between various time points and determine how ischemia or fat content affects the modules uncovered.

In this study, we demonstrate that graph network-based analytics to quantify metabolomics data is made practical and feasible using parallel computing on Amazon's cloud. Prospectively, similar graph-based algorithms can be employed to analyze proteomics and transcriptomics data as well and ultimately devise ways to incorporate all information to best characterize cellular dynamics.

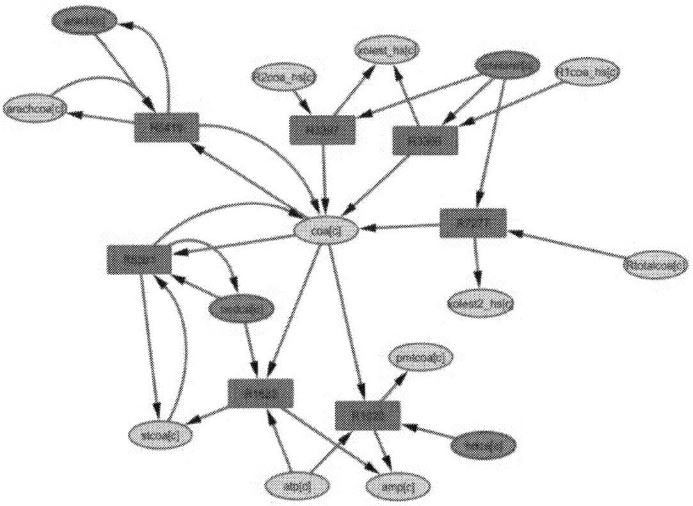


**Figure 1** Example of module identified using MMA. Metabolite nodes are represented as ellipses, while reaction nodes are represented as rectangles. Metabolite nodes colored in red are those whose pre-perfusion levels are correlated to WIT (p<0.10).

# 419144 OM-FBA: Integrate Multi-Omics Data to Flux Balance Analysis to Better Understand Cell Metabolism

**Abstract:**

omFBA: Integrate Multi-omics Data to Flux Balance Analysis to Better Understand Cell Metabolism

Weihua Guo, Xueyang Feng*

Department of Biological Systems Engineering, Virginia Polytechnic Institute and State University, Blacksburg, VA 24060

Constraint-based metabolic modeling such as flux balance analysis (FBA) has been widely used to simulate cell metabolism. Thanks to its simplicity and flexibility, numerous algorithms have been developed based on FBA and successfully predicted the phenotypes of various biological systems. However, their phenotype predictions may not always be accurate in FBA because of using the objective function that is assumed for cell metabolism. To overcome this challenge, we have developed a novel algorithm, namely omFBA, to integrate multi-omics data (e.g. transcriptomics, proteomics and metabolomics) into FBA to obtain omics-guided objective functions with high accuracy. In general, we first collected multi-omics data and phenotype data from published database (e.g. GEO database) for different microorganisms such as *Saccharomyces cerevisiae*. We then developed a "Phenotype Match" algorithm to derive an objective function for FBA that can lead to the most accurate estimation of the known phenotype (e.g. biomass yield). The derived objective function was next correlated with multi-omics data via regression analysis to generate the omics-guided objective functions, which will be further used to accurately simulate cell metabolism. We have applied omFBA in both *E. coli* and *S. cerevisiae* and found that the ethanol yield and biomass yield can be accurately predicted in most of the cases tested (~70%) by using transcriptomics data. In addition, omFBA could be incorporated with other existing algorithms easily by using the omics-guided objective functions. Compared to current approaches that attempt to integrate omics data to FBA (e.g. ROOM), omFBA can be more advantages in extendibility and comparability.

# 422695 A Hidden Markov Model Reconstruction of a Cellular Signaling Network in Embryonic Fibroblasts from Time-Course Gene Expression Profiles Reveals the Mechanism of the SPRY2 Tumor-Suppressor

**Abstract:**

**A hidden Markov model reconstruction of a cellular signaling network in embryonic fibroblasts from time-course gene expression profiles reveals the mechanism of the SPRY2 tumor-suppressor.**

Mark F. Ciaccio and Neda Bagheri. Department of Chemical & Biological Engineering, McCormick School of Engineering, Northwestern University.

**Introduction:** Cellular phenotype is controlled, in part, by the abundance of signaling proteins as well as the dynamic parameters governing their activation. For this reason, time-course gene expression data is invaluable for understanding the molecular regulatory mechanisms that underlie cancer behavior. Sprouty Homolog 2 (SPRY2), a known tumor suppressor, not only attenuates the amplitude of Receptor Tyrosine Kinase (RTKs) signaling but also decreases the duration of activity [1]. However, the mechanism by which SPRY2 inhibits RTK-signaling is poorly understood. We performed RNA-sequencing at six time points on cell lysates after stimulation with Fibroblast Growth Factor (FGF) in Sprouty-null Mouse Embryonic Fibroblasts (MEFs) and wild-type controls. We trained a hidden Markov model on a transcription-factor/DNA binding database [2], and used it to predict the transcription factors responsible for bifurcation points where clusters of time-course transcriptional profiles diverge. The differential activity of transcription factors informed how the signaling network is rewired in the absence of SPRY2.
**Background:** SPRY2 is down-regulated in many types of cancers including those of the colon and breast [1]. Understanding the mechanism of SPRY2 in signal transduction can lead to new therapies for cancer. **Materials and Methods:** We quantified the expression of gene transcripts using RNA-seq at six time points in SPRY-knockout vs. wild-type MEF cells. A refined version of the Dynamic Regulatory Events Miner (DREM) algorithm [2] was used with a protein-DNA binding database to infer the most likely active transcription factors at each time point. Differences in edge confidences between signaling networks inferred from the changes in activity of transcription factors were used as the basis to infer the location of SPRY2 within the signaling network. **Results and Discussion:** We found SPRY2 increased binding of the ubiquitin ligase, CBL-B, to several RTKs including FGFR. This suggests that the Sprouty-proteins locate ubiquitin ligases to the cellular membrane, thereby marking the receptors for degradation. **Conclusions:** Hidden Markov models are an efficient means to infer the mechanism of proteins from time-course gene expression profiles. This study suggests the importance of the regulation of RTK-degradation in the etiology of cancer and highlights the importance of incorporating regulation of protein stability into computational models of cancer.

**References:**

[1] Edwin, Francis, et al. "Intermolecular interactions of Sprouty proteins and their implications in development and disease." *Mol pharm* 76.4 (2009): 679-691.

[2] Bar-Joseph, Ziv, et al.. "Studying and modelling dynamic biological processes using time-series gene expression data." *Nat Reviews Genet* 13.8 (2012): 552-564.

# 435867 Systematic Analysis of Adaptive Resistance and Fractional Responses of Melanoma Cancer Cells to RAF/MEK Inhibition

**Abstract:**

Activation of BRAF via a V600E mutation is the most prevalent genetic change in human melanoma (the deadliest form of skin cencaer), found in >50% of tumors, and induces constitutive activation of pro-mitogenic RAF/MEK/ERK signaling. Treatment of $BRAF^{V600E}$ melanomas with drugs, such as vemurafenib, that inhibit RAF/MEK signaling is effective in the short term, but remission is not durable. Drug resistance appears to involve short-term adaptive responses that compensate for RAF/MEK inhibition via up-regulation of other pro-growth mechanisms. Understanding and ultimately preventing adaptive responses is key to durable therapy. Systematic data comparing $BRAF^{V600E}$ tumor cells is generally lacking and it is not known whether adaptation is fundamentally similar across cell types or among individual cells within a cell population.

We apply a systematic approach to studying the responses of human melanoma cell lines to five RAF and MEK inhibitors, with the overall goal of (i) characterizing variability in adaptation with time, dose, cell type and across individual cells, (ii) discovering new or poorly characterized adaptive mechanisms, and (iii) demonstrating the effectiveness of a high-throughput approach involving multiplex measurement, single-cell analysis and computational modeling. The data involves time-course measurement of total level and activity of signaling proteins and cell state markers using array-based methods and single-cell immunofluorescence assays as well as measurement of apoptosis and cell viability under the same conditions. Statistical modeling using partial least squares regression (PLSR) revealed which of the changes in the ~200,000 point dataset were phenotypically consequential.

We found that responses to RAF inhibitors are remarkably diverse and involve multiple pathways that can be up or down-regulated over time, with significant variability across cell types and individual cells. We identified a role for JNK/c-Jun signaling in altering the cell-cycle distribution of melanoma cells, causing apoptosis-resistant cells to accumulate and drug maximal effect (Emax) to fall; co-drugging with RAF and JNK inhibitors or JUN knockdown reverse this effect. Our study shows that a systems-level approach (combining high density time-dependent measurements, quantitative modeling and single-cell analysis) may provide a general framework for evaluating new drugs with adaptive and paradoxical response, and identifying potentially useful combination therapies.

# 434054 Synechocystis Sp. PCC 6803: Light Plus Endogenous Regulation Governing Gene Expression Patterns in the Diurnal Cycle

**Abstract:**

Cyanobacteria are the only photosynthetic prokaryotes with circadian behavior. Based on global transcriptomic data, several cyanobacteria including the strain *Synechocystis* sp. PCC 6803 have been analyzed thus revealing their oscillatory temporal expression. *Synechocystis* cells tune the transcriptional level of genes in response to the oscillation of daily diurnal cycle. However, it is still unclear how changes in cellular physiology are coordinated by this cyclic light exposure. In this study, we measured and subsequently analyzed expression patterns of nearly 3,500 genes in *Synechocystis* 6803 over two consecutive diurnal periods. Based on this analysis, at least 39% of genes with many from major cellular processes including central carbon metabolism, transport, and regulation were found to exhibit oscillating expression profiles. The pattern of gene expression also led to the development of two distinct transcriptional networks of co-regulated oscillatory genes. These networks described how *Synechocystis* regulated its cellular processes at the end of the dark period so as to make efficient use of light at the early light period. Interestingly, the light peaking genes showed damping patterns even in the middle of the light period which led to the inference not only the presence/absence of light but also endogenous regulation controlled the oscillatory gene expression patterns. We then measured important physiological parameters including glycogen concentration, ATP level, and $NADPH/NADP^+$ ratio in two consecutive light/dark cycles and, thereby, analyzed the physiological changes and their correlations with temporal expression patterns. This global transcriptional analysis of *Synechocystis* sheds light on the global coordination of cellular processes as evidenced in the transcriptional levels of genes.

# 429940 Applying Adaptive Laboratory Evolution to Engineer Platform Strains

**Abstract:**

Adaptive Laboratory Evolution (ALE) is a powerful technology for developing both compound-specific and platform bioprocessing strains. To leverage the utility of high-throughput sequencing after ALE, sophisticated experimentation is necessary to perform controlled high-throughput ALE experiments with the goal of understanding the range of phenotypes available to cellular production systems and how they arise. The presented work will describe an automated technology built for performing ALE experiments and its application towards highlighted use cases. These ALE use cases include adapting strains to higher tolerance of a desired compound, adaptation to utilize non-natural substrates, and adaptation to various bioprocessing conditions. The genomic changes which enable such adaptation will also be discussed along with interpretation of optimal phenotypes using genome-scale modeling and omics data characterization. ALE use is increasing in industrial biotechnology and sophisticated platforms for conducting experiments and analyzing outcomes will be necessary to make it a viable technology in the strain generation pipeline.

# 412226 Genome Editing in *Yarrowia Lipolytica* Using a Crispr-Cas9 System with RNA Polymerase II Transcribed Single Guide RNA


**Abstract:**

*Yarrowia lipolytica* has been the focus of a wide range of studies due to its ability to make and accumulate lipids to a high level. Here, we engineer a single plasmid CRISPR-Cas9 system for *Y. lipolytica* that enables markerless gene knockouts and insertions. A functional system was created using a codon optimized Cas9 from *Streptococcus pyogenes* and single guide RNA (sgRNA) expressed from the well characterized constitutive TEF promoter. RNA cleavage by flanking ribozymes generated the mature sgRNA, ensuring proper folding and preventing export to the cytosol by mRNA transport processes. Gene knockout efficiency of ≥75% was achieved when targeting the *ku70* gene without a cotransformed homologous recombination (HR) donor. The *ku70* gene was targeted as it has been shown that *Y. lipolytica* strongly favors non-homologous end-joining (NHEJ) to repair double strand breaks. Knockout of *ku70* suppresses NHEJ and allows HR to serve as the primary double strand break repair mechanism. Using the generated *Δku70* strain and the CRISPR-Cas9 system, markerless gene tagging with fluorescent proteins is being done to investigate protein localization to lipid droplets and to the endoplasmic reticulum.

# 437872 Food, Pharmaceutical & Bioengineering Division Award: Molecular Bioprocessing As an Emerging Paradigm for Advancing Medicine

**Abstract:**

Nature is unparalleled in its structural and functional diversity. In many cases, nature has provided us with a blueprint to design and assemble both natural and synthetic building blocks to create a new generation of functional, organized, and responsive materials. We have taken cues from nature to design materials with unique structural and functional properties, along with new process technologies with the ability to produce a wide range of biomimetic structures. In this talk I will highlight our recent efforts to exploit the interface of biology with materials science to address clinical applications. In particular, we have identified and engineered cell-lytic enzymes, and generated hybrid enzyme-containing surfaces with tailored activity against hospital-acquired infections (e.g., MRSA), food-borne illnesses (e.g., Listeria), and bacillus spores. Such activity provides a safe and potentially broadly applicable route to eliminating toxic compounds and pathogenic microorganisms from common surfaces. We have also developed a genetically encoded protein-based nanoparticle-generating system for remote regulation of gene expression by low-frequency radio waves (RFs) or a magnetic field. In mice with stem cell or viral expression of these genetically encoded components, remote stimulation of insulin transgene expression with RF or a magnet lowers blood glucose. We have also demonstrated a similar RF/magnetic field platform that can inhibit hypothalamic glucose-sensing neurons to regulate metabolism and behavior in mice. This robust, repeatable method for remote regulation *in vivo* may ultimately have applications in basic science, technology and therapeutics.

# 437878 Area 15A - Food - Plenary - Beyond Cell's Constraints: In Vitro Synthetic (Enzymatic) Biosystems for Next-Generation Biomanufacturing

**Abstract:**

The largest production challenge of biocommodities (e.g., biofuels, renewable chemicals, and even food) from carbohydrates, other energy sources (e.g., solar energy and methane) or other carbon sources (e.g., $CO_2$ and coal) is economics with satisfying three manufacturing criteria: high product yield, fast productivity and easy product separation. Living natural or engineered cells have their inherent constraints, such as self-duplication, bioenergetics, cell membrane, and so on, resulting in their limited success in high-energy efficiency biotransformation, especially in the energy area. In vitro synthetic biosystems (IVSBs) via the in vitro assembly of numerous standardized and exchangeable enzymes or their complexes and/or (biomimetic) cofactors can implement complicated biological reactions that microbes cannot do.

In this talk, I will introduce basic concept of in vitro synthetic biosystems for biomanufacturing, and present our recent efforts: (1) sweet hydrogen production from hexose and pentose sugars, three times the highest yields of hydrogen-producing microorganisms; (2) complete oxidation of sugars to electricity, better than microbial fuel cells in terms of electron yield and productivity; and (3) low-cost biotransformation of cellulose to starch without the barrier of cellular membrane. In a word, in vitro synthetic biosystems based on stabilized enzyme mixtures would lead to a biomanufacturing paradigm shift and carbohydrate would be the best energy storage carrier, replacing crude oil.

# 437880 Area 15B/PD2M - Pharmaceuticals - Plenary: Holistic Pharmaceutical Creation

**Abstract:**

Pharmaceutical creation is plagued with inefficiencies. The technologies used are decades old, if not older, and the current approach to process development is to make incremental changes as opposed to quantum leaps. This is starting to change with the incorporation of new approaches, such as continuous manufacturing and the incorporation of developability and manufacturability in the discovery process. In this presentation, we discuss new technologies to change the efficiency of pharmaceutical creation by integrating discovery, development, and manufacturing.

# 437881 Area 15C - Bioengineering - Plenary: Antibiotic Probiotics Reduce Pathogen Counts in Host GI Tracts

**Abstract:**

Foodborne gastrointestinal infections are significant causes of morbidity and mortality worldwide. Alarmingly, often because of the extensive, non-therapeutic use of antibiotics in agriculture, foodborne bacteria are emerging that are resistant to our most potent drugs.

We will discuss a novel approach to reduce the use of antibiotics in food-producing animals and to treat gastrointestinal infections. We engineer probiotic, lactic acid bacteria (LAB) that express and release antimicrobial peptides (AMPs). LAB are part of the gastrointestinal microflora and can be safely delivered with known benefits to humans and animals. AMPs are proteins that can be readily produced by LAB. One unique aspect of our approach is the use of synthetic promoters that precisely regulate the delivery of AMP molecules.

At the heart of proposed efforts are multiscale models that guide explanations and predictions of the antagonistic activity of recombinant LAB against pathogenic strains. Models are developed to quantify how AMPs kill bacteria at distinct but tied scales. Using atomistic simulations the various interaction steps between peptides and cell membranes are explored. Mesoscopic models are developed to study ion transport and depolarization of membranes treated with AMPs. [1] Stochastic kinetic models are developed to quantify the strength of synthetic promoters and AMPs expression.[2]

Experimentally, we engineer lactic acid bacteria to inducibly produce antimicrobial peptides. We have developed a library of synthetic biological constructs.[3] We test modified bacteria against pathogenic bacteria. We will present in vitro results against salmonella and enterococcus.[4-6] We will also present proof-of-concept results of *in vivo* studies with turkey poults and mice.

**References**

[1] Kaznessis Y, "Multiscale Models of Antibiotic Probiotics", Curr Opin Chem Eng, **2014**, 1;6:18-24

[2] Smadbeck P, Kaznessis YN. "A closure scheme for chemical master equations. Proc Natl Acad Sci U S A". **2013** Aug 27;110(35):14261-5. doi: 10.1073/pnas.1306481110.

[3] Volzing K, Billiouris K, Kaznessis YN. "proTeOn and proTeOff, New Protein Devices That Inducibly Activate Bacterial Gene Expression", ACS Chem Biol. **2011**, 6:1107-1116.

[4] Volzing K, Borrero J, Sadowsky MJ, Kaznessis YN. "Antimicrobial Peptides Targeting Gram-negative Pathogens, Produced and Delivered by Lactic Acid Bacteria." ACS Synth Biol. **2013**. 15;2(11):643-50. doi: 10.1021/sb4000367 .

[5] Borrero J, Dunny G, Kaznessis YN, " Modified Lactic Acid Bacteria Detect and Inhibit Multiresistant Enterococci." ACS Synth Biol. **2014**, 10.1021/sb500090b

[6] Geldart K, Borrero J, Kaznessis YN. "A Chloride-Inducible Expression Vector for Delivery of Antimicrobial Peptides Against Antibiotic-Resistant Enterococcus faecium", Applied and Environmental Microbiology, **2015** Jun 1;81(11):3889-97. doi: 10.1128/AEM.00227-15.

# 437882 Area 15D/E - Engineering Fundamentals in Life Science - Plenary: IRE1 Signaling, ER Stress and Cancer

**Abstract:**

Recent research in cancer have found that endoplasmic reticulum (ER) stress and the unfolded protein response (UPR) are highly induced in various tumors and are closely associated with cancer cell survival and resistance to anti-cancer treatments. Concomitantly, epidemiological evidences have linked elevated levels of free fatty acids (FFAs) with hepatocellular carcinoma (HCC) tumorigenesis and increased mortality. Of particular interest are saturated long chain-FFAs, notably palmitic acid, which have been implicated in ER stress and numerous diseases, including cancer. We and others have found that palmitic acid activates the IRE1α (inositol-requiring enzyme 1 α) pathway in liver and pancreatic beta-cells. However, the molecular mechanisms of how palmitic acid induces ER stress have not been well established. Through molecular dynamic simulations, mutation studies, and FRET analysis, we uncover that palmitic acid interacts with the cytosolic and transmembrane domains of the IRE1 protein to impact phosphorylation and dimerization of the IRE1 protein and their downstream signaling.

Additionally we found that elevated palmitic acid induces liver cells to undergo epithelial to mesenchymal (EMT), a prerequisite step for cells undergoing metastasis. Analyzing The Cancer Genome Atlas (TCGA) liver cancer dataset, we found the induction of the EMT program is closely associated with the expression of the fatty acid uptake protein CD36, which we confirmed in human HCC tumor samples. Treatment of human liver cancer cells with palmitic acid exacerbated the EMT phenotype, whereas chemical inhibition of CD36 mitigated these effects. Given these results and the literature information linking IRE1α signaling to EMT, we investigated how induction of ER stress by palmitic acid through direct interaction with IRE1α leads to the activation its enzymatic activities and the promotion of EMT.

# The Use of GastroPlus™ Modeling in Enabling Drug Discovery Research - A Case Study

Ching Kim Tye, PhD (Bristol-Myers Squibb)

AIChE ANNUAL CONFERENCE, NOV 10, 2015

Bristol-Myers Squibb

## Co-authors

2

- John Crison
- Zhanbin Wang
- Ching Chiang Su
- Michael J Hageman

Bristol-Myers Squibb

# Outline

3

- Background – GastroPlus™ and its general usage
- How pH affects drug absorption
- Case study
  - The use of GastroPlus™ modeling in pH-dependent absorption risk assessment to enable decision making at preclinical stages

Bristol-Myers Squibb

# GastroPlus™

4

- GastroPlus
  - Based on Advanced Compartmental Absorption and Transit (ACAT) model
  - Originally developed for drug absorption for oral administration
  - Physiological Based Pharmacokinetic (PBPK) Model
    - **Mechanistic-based** by mathematically transcribing anatomical and physiological descriptions of the phenomena involved in ADME processes
  - Significant publications since late 1990 - main usage during early clinical development and formulation optimization
- GastroPlus peer-reviewed publications
  - 159 compiled on Simulation Plus website (as of 10/19/2015)

Bristol-Myers Squibb

## Advanced Compartmental Absorption and Transit Model (ACAT™)

5

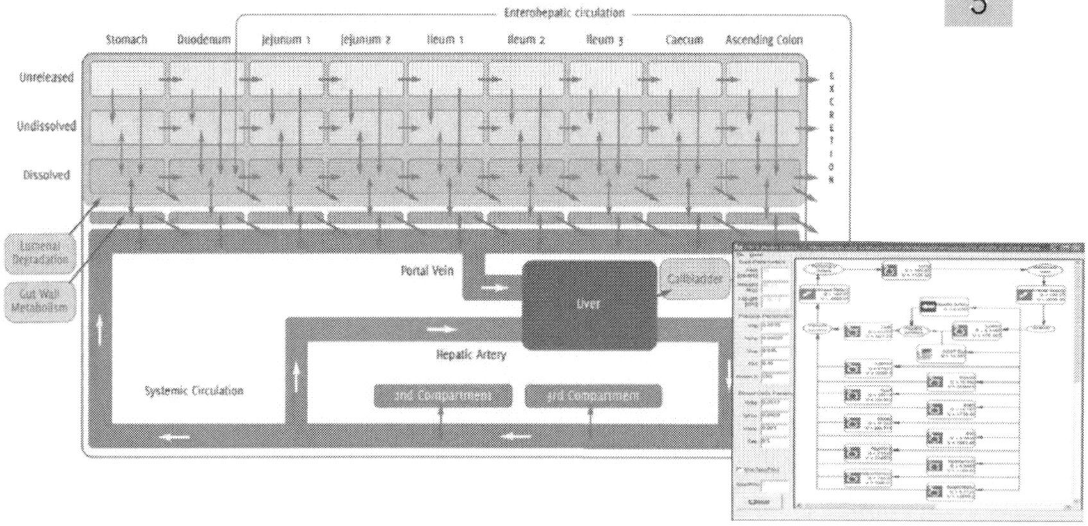


SimulationsPlus

## Applications of Absorption Modeling in Drug Discovery and Development

6

| Drug Discovery | Pre-clinical Development | | Clinical Development |
|---|---|---|---|
| **Lead Optimisation** | **Clinical Lead Selection** | **Entry Into Human** | **Phase1 Phase2 Phase3** |
| Predict absorption from *in vitro* and *in silico* data<br>Explore limitations on absorption via parameter sensitivity analysis | Integrate *in vitro* and *in vivo* data to explore formulation possibilities<br>Assist development of formulations for toxicological studies | Predict oral pharmacokinetics in human<br>Extended biopharmaceutical evaluation to define clinical formulation strategy | Predict food effect<br>Design extended release formulations<br>Develop In Vitro In Vivo Correlations |

Neil Parrott and Thierry Lave, Applications of physiologically based absorption models in drug discovery and development, Molecular Pharmaceutics, Volume 5(5), 2008.

Bristol-Myers Squibb

## How pH Affects Drug Absorption?

7

- Drug absorption is mainly through passive diffusion across cells - passive lipid diffusion
- The Henderson-Hasselbalch equation
  - Acid: pH = pKa + log $[A^-]/[HA]$
  - Base: pH = pKa + log $[B]/[HB^+]$
- Weak acids and weak bases can be ionized or un-ionized as the pH of the media changes. Ionized drugs are less likely to participate in the lipid diffusion across the biological membrane.

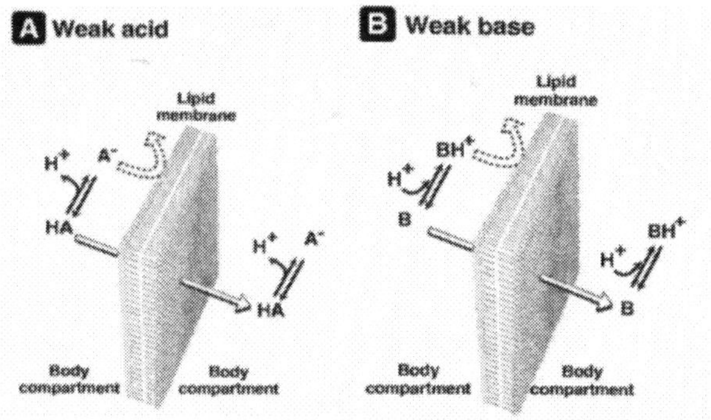


Bristol-Myers Squibb

## The Relationships Between pH, pKa and Ionization

8

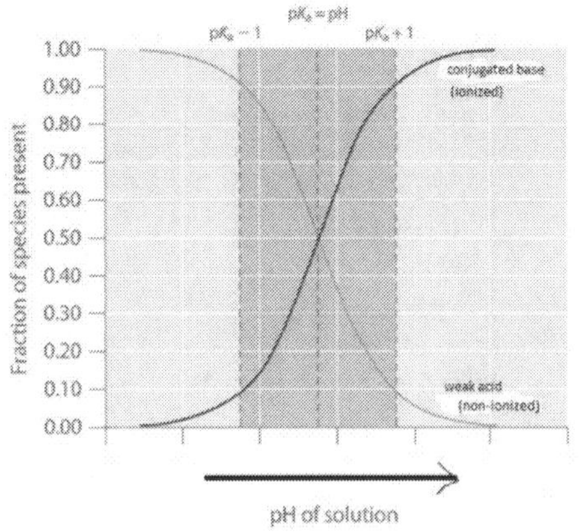

| | Stomach | Jejunum | Ileum | Colon | |
|---|---|---|---|---|---|
| *pH* | 1.4-2.1 | 4.4-6.6 | 6.8-8.6 | 5-8 | fasting |
| | 3-7 | 5.2-6.2 | 6.8-8.0 | 5-8 | fed |

Bhal, S. K. Lipophilicity Descriptors: Understanding When to Use Log*P* & Log*D*., *2012*

- Weak acid ionized at
  - 50% at its pKa
  - 91% at pKa +1
  - 99% at pKa + 2
- Opposite for weak base

Bristol-Myers Squibb

## Advanced Compartmental Absorption and Transit Model (ACAT™)

9

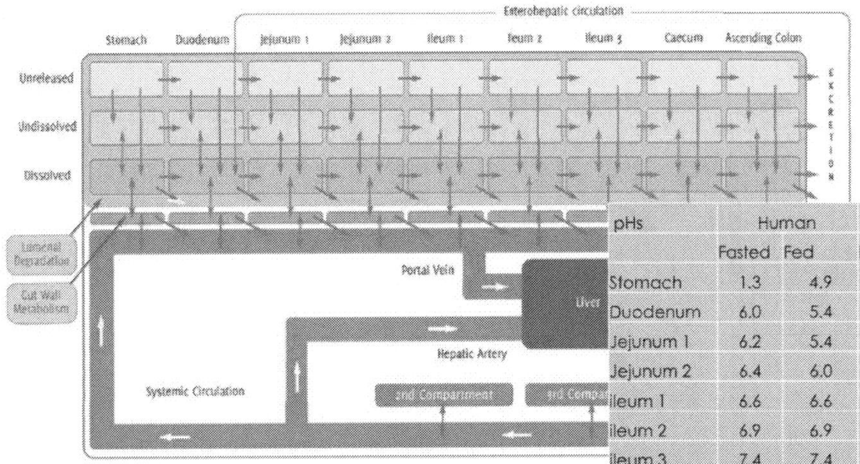


| pHs | Human | | Rat | |
|---|---|---|---|---|
| | Fasted | Fed | Fasted | Fed |
| Stomach | 1.3 | 4.9 | 3.9 | 3.2 |
| Duodenum | 6.0 | 5.4 | 5.9 | 5.0 |
| Jejunum 1 | 6.2 | 5.4 | 6.1 | 5.1 |
| Jejunum 2 | 6.4 | 6.0 | 6.1 | 5.1 |
| ileum 1 | 6.6 | 6.6 | 5.9 | 5.9 |
| ileum 2 | 6.9 | 6.9 | 5.9 | 5.9 |
| ileum 3 | 7.4 | 7.4 | 5.9 | 5.9 |
| Caecum | 6.4 | 6.4 | 6.6 | 5.9 |
| Asc colon | 6.8 | 6.8 | 6.2 | 5.5 |

Opt LogD Model SA/V6.1

SimulationsPlus

## Case Study: The Use of GastroPlus™ to Enable Drug Discovery

10

- Compound BMS-1
- Physicochemical properties
  - Molecular weight ~ 500
  - Log D (at pH 6.5) = 3.4
  - High permeability (Caco-2)
  - Weak base, pKa = 2.5
  - Solubility
    - Low solubility in aqueous (<0.001 mg/ml)
    - pH dependent solubility
    - Good solubility in biorelevant media: FaSSIF and FeSSIF
    - Very soluble in polyethylene glycol (PEG)

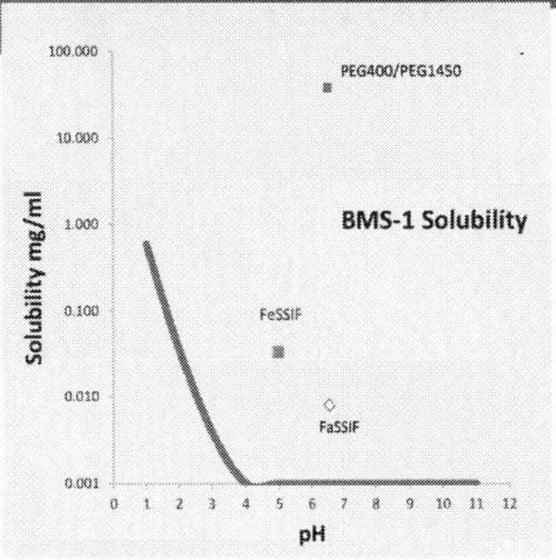


FaSSIF = Fasted State Simulated Intestinal Fluid, pH 6.5
FeSSIF = Fed State Simulated Intestinal Fluid, pH 5

Bristol-Myers Squibb

## Preclinical Bioavailability

11

- Based on a solution formulation
- Variable bioavailability across species
- High fraction absorbed

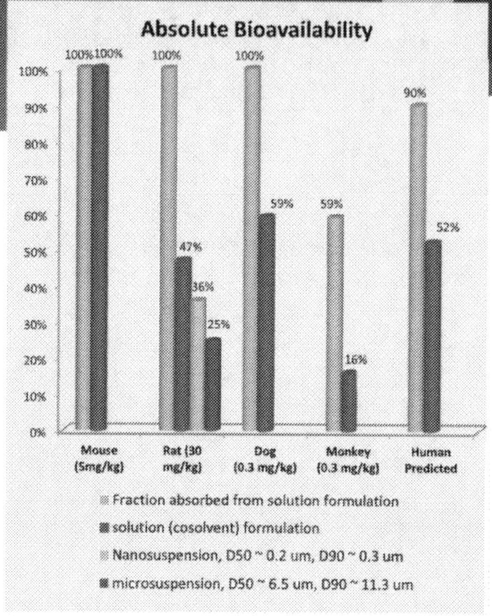


## The Question
## How Much pH-dependent Absorption Risk?

12

BMS-1 properties

- 600x higher solubility at pH 1 than at pH ≥ 4, a potential for pH-dependent absorption
- A highly potent oncologic compound with a low therapeutic index

A significant pH-dependent effect will be problematic for development.

Therefore, here are the questions:

- Is this a viable drug candidate?
- How much is the risk of significant pH-dependent absorption in human?
- If there is a pH-dependent absorption, can this be mitigated?

Bristol-Myers Squibb

## Processes Needed to be Understood for pH-Dependent Absorption

13

- pKa and ionization, acid/base
- Solubility
- Supersaturation
- Precipitation timing and ease of redissolve
- Permeability
- stability
- Particle size
- Formulation factors (dissolution & release kinetics)
- And the interplay of these factors

Bristol-Myers Squibb

## pH-Dependent Absorption Risk Assessment Methods

14

Noted that during discovery research, the dose is uncertain and API availability is very low

| | Method Type | Method Details | Benefits and Limitations |
|---|---|---|---|
| 1 | In vitro | pH shift transfer dissolution tests | ▪ Simple<br>▪ Set up to mimic human conditions<br>▪ High demand of API |
| 2 | In vivo | Pre-dosed rats with pentagastrin or famotidine | ▪ Most resource intensive<br>▪ *In vivo* system can be more representative<br>▪ Not human |
| 3 | In silico | GastroPlus modeling | ▪ No API needed<br>▪ Allow prediction in different doses<br>▪ Require physicochemical & biopharmaceutical property data<br>▪ Based on animal model |

Bristol-Myers Squibb

## *In Vitro* Microdissolution: pH-Shift Transfer Method*

15

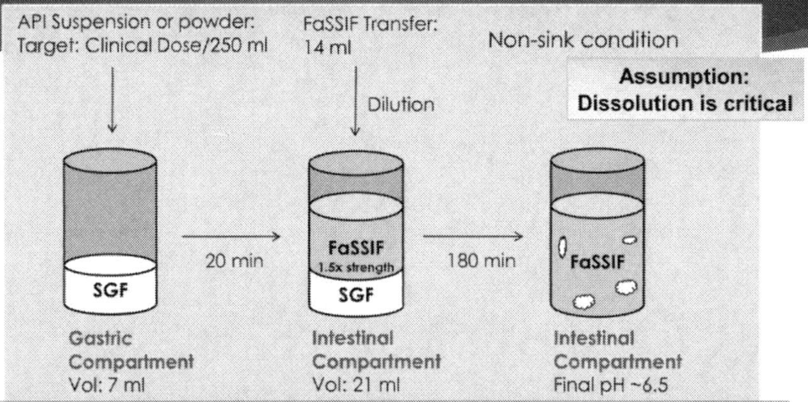


**Test A: $SGF_{pH2} \rightarrow FaSSIF_{pH6.5}$ (equivalent to un-treated or normal)**
**Test B: $SGF_{pH6} \rightarrow FaSSIF_{pH6.5}$ (equivalent to famotidine-treated or achlorhydria)**

* Neil Mathias *et. al.* Assessing the risk of pH-dependent absorption for new molecular entities: a novel in vitro dissolution test, physicochemical analysis, and risk assessment strategy. Molecular Pharmaceutics 2013, 10, 4063-4073

## *In Vitro* MicroDissolution Tests Indicated Moderate pH-Effect (at dose = 175 mg)

16

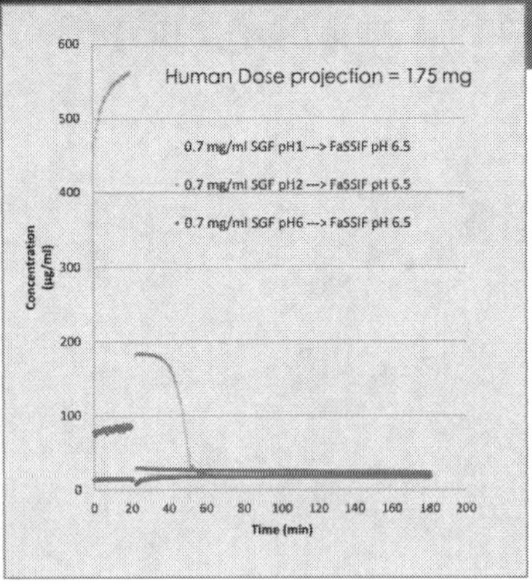


**Table 4. *In Vitro*–*in Vivo* Correlation: pH-Effect Risk Categories***

| risk category | *in vitro* AUC Ratio | clinical AUC or $C_{max}$ ratio |
|---|---|---|
| high | <0.2 | <0.5 |
| moderate | 0.2–0.5 | 0.5–0.8 |
| low | >0.5 | >0.8 |

* Neil Mathias et al. Assessing the risk of pH-Dependent Absorption for New Molecular Entities. Mol. Pharmaceutics 2013, 10, 4071

Test 1: pH effect = $\frac{\text{AUC ( SGF pH 6 to FaSSIF pH 6.5)}}{\text{AUC ( SGF pH 2 to FaSSIF pH 6.5)}} = 0.50$

Test 2: pH effect = $\frac{\text{AUC ( SGF pH 6 to FaSSIF pH 6.5)}}{\text{AUC ( SGF pH 1 to FaSSIF pH 6.5)}} = 0.16$

- Result indicated potential moderate pH-effect

Bristol-Myers Squibb

# *In vivo* (Rat) Evaluation Showed Significant pH-Effect at 20mg/kg

17

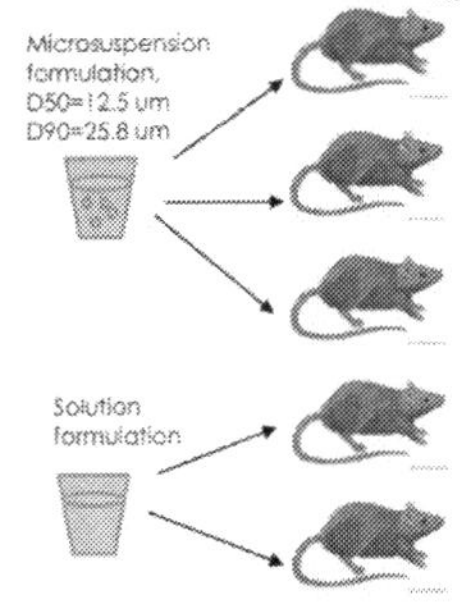


| Treatment | Dose (mg/kg) | Cmax (µM) | Tmax (h) | T half (h) | AUC (µM*h) | Bioavailability (%) | AUC changes (%) |
|---|---|---|---|---|---|---|---|
| control | 20 | 0.61 ± 0.15 | 4 ± 0 | 8.9 ± 4.4 | 4.9 ± 1.3 | 30 | 100% |
| Pretreatment: pentagastrin 0.25 mg/kg, SC 30 min prior | 20 | 0.68 ± 0.21 | 3.3 ± 1.5 | 5.0 ± 1.5 | 4.5 ± 1.0 | 27 | 92% |
| Pretreatment: famotidine 25 mg/kg, PO 60 min prior | 20 | 0.28 ± 0.08 | 4.8 ± 1.5 | 14 ± 11 | 2.7 ± 0.4 | 16 | 55% |
| control | 30 | 2.1 ± 0.1 | 0.7 ± 0.3 | 13 ± 11 | 11.9 ± 1.3 | 48 | 100% |
| Pretreatment: famotidine 25 mg/kg, PO 60 min prior | 20 | 1.53 ± 0.26 | 1.8 ± 1.9 | 5.1 | 9.5 ± 0.4 | 57 | 120% (dose normalized) |

- Microsuspension formulation – pretreatment with famotidine resulted in 55% AUC reduction. This indicates a potential pH-effect at 20 mg/kg.
- Solution formulation – No significant AUC changes with famotidine pretreatment. This indicates that a solution formulation could be a strategy for pH-dependent risk mitigation.

Bristol-Myers Squibb

# GastroPlus Modeling: Parameters Inputs and Model Building

18

- Physicochemical
  - MW, compound structure, partition coefficient, pKa, pH-solubility profile, biorelevant solubility, particle size, precipitation time, permeability*
- Pharmacokinetics
  - Use PKPlus model by fitting experimentally determined mean plasma concentration-time profile following intravenous administration to calculate clearance
- Physiological: Rat - physiological - Fasted
- A model was developed based on good fits across
  - Intravenous administration
  - Oral solution formulations at 5 mg/kg and 30 mg/kg
  - Oral suspension formulations at 30 mg/kg

N = 3, resource constraints in Discovery but data fits fairly well

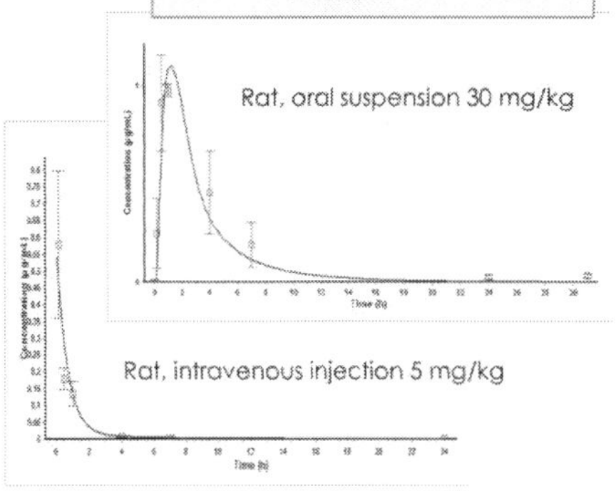


* rat intestine permeability was based on an internal correlation of Caco-2 permeability and intestinal permeability established for literature compounds.

Bristol-Myers Squibb

# Parameter Sensitivity Analysis (PSA)

19

- With the validated rat PK model, based on the microsuspension data, PSA was run for
  - stomach pH between 1 to 8
  - Dose between 0.5 & 100 mg/rat

Bristol-Myers Squibb

# GastroPlus Simulation for Stomach pH Sensitivity Effect in Rats (microsuspension, 30 mg/kg)

20

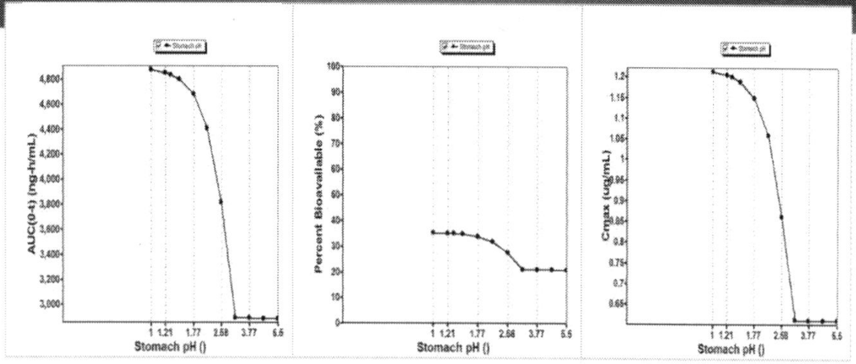


- Based on available rat data, it is predicted that there may be ~ 50% more drug available in a stomach with pH 1.5 in compare to pH > 3

Bristol-Myers Squibb

## GastroPlus Output - The Effect of pH-Effect Absorption is Dose-Dependent

21

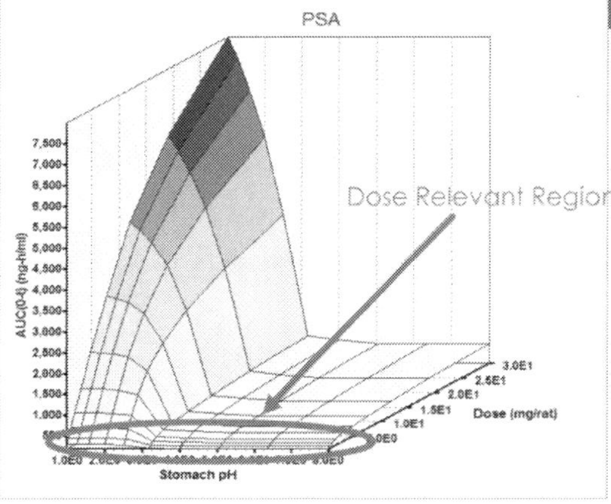


- As the program progressed, it was determined that the dose requirement will be low.
- The model predicted that the risk would be low
- A risk mitigation option available: solubilized formulations
- This allowed the program to move forward into early development quickly

Bristol-Myers Squibb

## The No Effect Prediction was Later Confirmed in a Dog Study

22

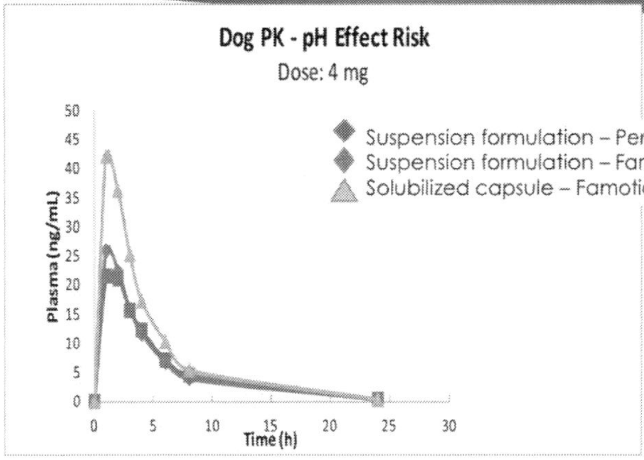


Amy Saari, Meg McCallum, Yan Xu

Bristol-Myers Squibb

# Summary

23

The use mechanistic-based modeling (GastroPlus™) in combination with *in vitro* and *in vivo* data was successfully used to evaluate the risk of pH-dependent absorption across a wide range of doses to enable decision making in Discovery Research.

The benefits of mechanistic-based GastroPlus™ modeling in Discovery

- ACAT physiological model available for commonly used preclinical toxicology and pharmacology species including, mouse, rat, rabbit, cat, dog, etc
- Use animals to assess formulation performance and recommend formulations strategy for human testing - parameter sensitivity analysis
- Dose is a critical factor in the risk equation, however dose is usually not well defined during preclinical stage - modeling & simulation allow information to be gathered and projected for different doses.

Bristol-Myers Squibb

# 430408 Modeling of Drug Delivery Across the Blood-BRAIN Barrier for the Treatment of Autism Spectrum Disorder

**Abstract:**

Modeling of Drug Delivery Across the Blood-Brain BARRIER FOR the Treatment of Autism Spectrum Disorder

*(Submitted to AIChE Session: Recent Challenges in the Biopharmaceutical and Pharmaceutical Fields)*

Jamelle Simmons[1], Luke Achenie[2,3], and Yong W. Lee[1,3]

1. Virginia Tech, School of Biomedical Engineering and Sciences (SBES)
2. Virginia Tech, Department of Chemical Engineering
3. Virginia Tech Center for Autism Research (VTCAR)

**Introduction:** Autism spectrum disorder (ASD) is classified as a developmental disability with deficits in areas of communication and interaction. Signs are noticeable during early childhood and are lifelong. Between 2000 and 2010, ASD prevalence rose from 1 in 150 to 1 in 68 [1]. Previous research in our lab has found elevated levels of xanthine oxidase (XO) in male autism mouse brains along with increased neuronal loss and microglial activation. We hypothesize that inhibition of XO will reduce brain injury in ASD. In this presentation we will discuss the modeling of two mechanisms for transport of drugs across the blood-brain-barrier (BBB). We will also discuss our initial attempts to do multiscale modeling of transport using agent-based modeling under the Netlogo modeling environment.

**Materials and Methods:** Sodium fluorescein salt (Na-Fl, MW=376.27, Sigma-Aldrich) was used as a fluorescent substitute for our xanthine oxidase inhibitor febuxostat (MW=316.374). C6 (rat glioma) cells were cultured in DMEM supplemented with fetal bovine serum (FBS) and penicillin-streptomycin (Penn-Strep) on the basal side of Corning¨ Transwell¨ polyester membrane inserts (12mm with 0.4 micron pores). After C6 cells reached confluence, RBE4 (rat brain capillary) cells were seeded on the apical side of the inserts and cultured in DMEM/Ham's F-12 supplemented with FBS, Penn-Strep, basic fibroblast growth factor (bFGF), and hydrocortisone (Hydro). Four groups were studied, (1) uncoated inserts, (2) collagen coated inserts, (3) RBE4+Hydro mono-culture, and (4) RBE4+C6+Hydro co-culture. Transport experiments were run with apical chambers containing Na-Fl in RBE4 media with basal chambers containing only RBE4 media. Spectrophotometric analysis was conducted on basal chamber samples taken at 0, 3, 6, 9, and 12 hrs. Effective diffusion ($D_{eff}$) constants were calculated for each transport group and used to construct partial differential equation (PDE) diffusion models.

**Results and Discussion:** Effective diffusion constants ($cm^2/hr$) were calculated from the average of 3 and 6hr time points for each well type with (1) $D_{eff} = 1.02 \times 10^{-4}$ for blanks, (2) $D_{eff} = 1.43 \times 10^{-4}$ for collagen coated wells, (3) $D_{eff} = 6.85 \times 10^{-5}$ for RBE4+Hydro wells, and (4) $D_{eff} = 7.54 \times 10^{-4}$ for RBE4+C6+Hydro wells. Diffusion models were plotted for each well type (Figure 1).

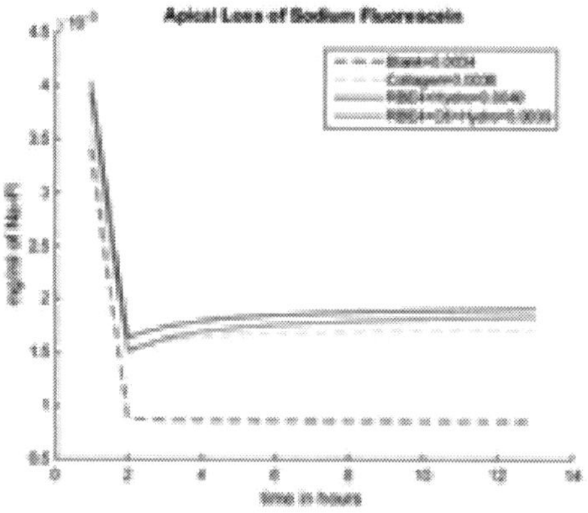


**Figure 1.** Loss of tracer drug from apical layer to basal layer over 12 hours. More sodium fluorescein is loss to the basal layer in the blank inserts compared to other Transwell models. The loss of tracer to the basal layer in the collagen coated inserts occurs slower than the blank insets but faster than the RBE4 models. Loss of tracer between RBE4+Hydrocortisone and RBE4+C6+Hydrocortisone models are similar. Experimental cumulative transport of Na-Fl across *in vitro* models were (1) 51.26% (SD=6.088, N=4) in blank wells, (2) 50.25% (SD=8.595, N=4) in collagen coated wells, (3) 37.304% (SD=2.957, N=5) in RBE4+Hydrocortisone wells, and (4) 37.648% (SD=3.625, N=5) in RBE4+C6+Hydrocortisone wells.

**Conclusions:** Current results indicate that our *in vitro* BBB models are capable of inhibiting the free diffusion of molecules compared to controls resulting in lower effective diffusion constants. Future studies will involve validation of the predictive capabilities of each diffusion model followed by microfluidic and animal studies to study the effects of diffusion across the BBB when flow dynamics are being considered *in vitro* and *in vivo*.

**References:** [1] (2015, March 25). Retrieved from http://www.cdc.gov/ncbddd/autism/index.html

# 422560 From in-Vitro Data to Virtual Patients: Development of an in-Silico Model of Acne Pathogenesis for Prioritizing Targets for Novel Acne Therapies and Prediction of Late Stage Clinical Outcomes

**Abstract:**

Quantitative Systems Pharmacology (QSP) is an integrated systems-level modeling approach that combines the fields of pharmacology and systems biology. It quantitatively links the pathogenesis of the disease to the mechanism of action of drugs and clinical outcomes by utilizing data from a number of different sources like in-vitro target engagement assays, in-vivo studies as well as patient data from clinical trials. It thus allows mechanistic modeling of the effect of different drugs on disease progression and late stage clinical outcomes and has emerged as an important tool to advance the discovery and development of therapeutic drugs.

In this talk, the application of QSP for modeling the pathogenesis of a skin disease, acne, will be presented. Acne is a multifactorial disease with limited understanding of its pathogenesis. Therefore, a QSP based approach was used to understand the interplay between the processes involved in acne such as sebum production, P. acnes growth and inflammation etc. and their role in disease progression. Also, the effect of modulating these acne processes on disease severity has been evaluated to guide the development of new acne treatments. In addition, different virtual patients have been simulated which can help in identifying the patient subsets that can benefit most from new acne treatments under development within GSK. We have also evaluated the efficacy of three GSK compounds in reducing acne severity as well as compared them to currently available treatments using this model. The talk thus presents an example of how a QSP model of a disease, developed by utilizing current literature knowledge and in-silico tools, can help guide decisions in different stages of drug development.

# 422856 Quantitative Systems Pharmacology (QSP) Modeling and Simulation in Drug Discovery to Support Pharmacology: Strategy and Applications

**Abstract:**

Significant challenges are facing the pharmaceutical industry in developing new drugs, and the high rate of attrition demands new approaches in research and development. Advances in modeling and simulations (M&S) have the promise to enable modern drug discovery and development in accelerating decision making. Model based drug development has shown value in increasing probability of success and M&S tools has become integrated in drug discovery and development processes. In this presentation an overview of the applications of M&S to support drug discovery and development projects will be discussed. Examples of modeling drug pharmacokinetics processes including absorption, distribution, metabolism and elimination, as well as modeling systems pharmacology concepts will be presented. Strategies for implementation of M&S tools in drug discovery will be discussed and recommendations for a recipe of success will be proposed.

# 435605 How Modeling Tools Benefit Early to Late Stage Formulation Development and the Design of Clinical Studies

**Abstract:**

Pharmacokinetic absorption modeling and simulation is widely used through many drug development stages as a powerful tool for assessing biopharmaceutical risk to, and in silico data. PK absorption modeling tools leverage in vitro, and in vivo to greatly impact formulation development, clinical study design, and regulatory submission. The aim of this presentation is to demonstrate the applications of modeling tools and how they affect drug development decisions.

Four case studies of absorption modeling and simulations combined with conventional analysis tools will be presented. Commercially available software (GastroPlus, v8.5) was used to build PBPK models for the dog and human physiologies. These models were parameterized based on the compound's physicochemical properties, in vitro biorelevant dissolution in fasted state simulated intestinal fluid (FaSSIF), fed state simulated intestinal fluid (FeSSIF) and in vivo data (clearance, plasma time profile, etc.) for each drug.

Rodent and human physiology models in the GastroPlus program were tested, modified, and optimized to obtain a best reliable model to enable decision making based on certain stage of the project. Parameter sensitivity analysis was used to analyze the impact of formulation factors (dose level, particle size, dissolution rate, etc.) to AUG and Cmax in simulated human clinical studies. The case studies cover:

1. Comparing the exposures and pharmacokinetic profile of crystalline material and amorphous dispersions.
2. The impact of drug dose and administration with food on exposure.
3. The impact of particle size on expsoure
4. The impact of patient variability on single and multiple doses using virtual clinical trial simulation

An exploratory clinical study was conducted with larger particle size material, higher dose levels and a standard to evaluate the food effect, dose linearity and particle size risk, and the results aligned with the GastroPlus predictions. The PBPK model output was used to provide guidance on evaluating the biopharmaceutical risk in humans and served as enable decision on formulation development and influencing the clinical development plan.

# 410121 A Potential Sporolactobacillus Strian and Its Enhanced Production of an Optically Purified D-Lactic Acid for the Stereocomplex Polylactic Acid

**Abstract:**

Plastics that are widely used nowadays are from petroleum based resources. The petroleum based plastics are generally classified as non compostable or partial compostable with the help of some additives added while processing into the plastic products. Due to a large plastic consumption and a long term degradation profile, this eventually leads to the environmental pollution. As a result, research on the compostable plastic has become of interest. Polylactic acid (PLA) is a good example of compostable plastic made from the renewable resources such as plant biomass and agricultural commodities. During the initial phase of development, PLA was made from polymerization of the optically pure L-lactic acid since the L-form can be easily found in nature and has long been extensively used in food and pharmaceutical industries. Nonetheless, poly L-lactic acid (PLLA) still cannot be competitive with the petroleum based plastics due to the poor mechanical and thermal properties causing several problems in compounding and fabrication. The deficiency in such properties can be reduced by the stereocomplex structure of PLA. This structure can be tailored made by polymerizing the optically pure L-lactic acid with the optically pure D-lactic acid which is rarely found in nature due to its non – biocompatible property. Nevertheless, some bacteria can produce this stereoisomer of lactic acid as the protecting agent under stress conditions. We have isolated and screened the D-lactate producers from the natural resources and later identified as *Sporolactobacillus laevolacticus* SK5-2. Extensive fermentation optimization was conducted in a 5 L stirred fermentor. Compared to those commercial strains available, our SK5-2 gave the very high D-lactate yield of 0.92 g per g glucose and productivity of 3.83 g/L×h with the final titer of 115 g/L and the optical purity of 99.0% ee from the simple medium containing the initial glucose of 120 g/L.

# 435051 Recovery and Purification of Recombinant Butyrylcholinesterase from Nicotiana Benthamiana Plants

**Abstract:**

**Recovery and Purification of Recombinant Butyrylcholinesterase from *Nicotiana benthamiana* Plants**

**Salem Alkanaimsh,** Kalimuthu Karuppanan and Karen A. McDonald

Chemical Engineering and Materials Science, University of California, Davis, CA, 95616

Human butyrylcholinesterase (hBuChE EC 3.1.1.8) is a 574 amino acid cholinesterase-hydrolyzing enzyme. Organophosphates (OP) are highly toxic inhibitors of the acetylcholine-hydrolyzing enzymes like acetylcholinesterase. The resulting accumulation of acetylcholine can lead to respiratory collapse and death. Current therapies are based on elevating the serum levels of OP bioscavengers like BuChE. The major limitation of this therapy is high cost, with plasma-derived hBuChE costing more than $10,000/treatment. Limitations like cost and availability necessitate an alternative expression platform capable of large scale, low-cost production of a fully active and efficacious recombinant hBuChE (rhBuChE). The development of an effective rhBuChE is a pressing national security concern in terms of protecting the nation's warfighters and civilian population from the threat of attack with OP agents. Therapeutic proteins (e.g. hBuChE) can be produced in plant leaves such as *Nicotiana benthamiana* using transient agroinfiltration. *Nicotiana benthamiana* represents a valuable biopharmaceutical protein expression system due to its scalability, ability to glycosylate proteins and elimination of risk of viral infection. We have a developed a viral expression system based on *Tobacco mosaic Virus* (TMV) that express a FLAG tagged rhBuChE at a level of 10 μg/g fresh weight. Downstream processing is a major challenge of *Nicotiana benthamiana* based where it represents 80% of the overall production costs ADDIN EN.CITE <EndNote><Cite><Author>Wilken</Author><Year>2012</Year><RecNum>170</RecNum><DisplayText><style R.</author><author>Nikolov, Zivko L.</author></authors></contributors><titles><title>Recovery and purification of plant-made recombinant proteins</title><secondary-title>Biotechnology Advances</secondary-title></titles><periodical><full-title>Biotechnology Advances</full-title></periodical><pages>419-433</pages><volume>30</volume><number>2</number><keywords><keyword>Recombinant protein</keyword><keyword>Purification</keyword><keyword>Downstream processing</keyword><keyword>Transgenic plants</keyword><keyword>Protein extraction</keyword><keyword>Bioseparations</keyword><keyword>Process economics</keyword><keyword>Extraction</keyword><keyword>Chromatography</keyword></keywords><dates><year>2012</year><pub-dates><date>3//</date></pub-dates></dates><isbn>0734-9750</isbn><urls><related-urls><url>http://www.sciencedirect.com/science/article/pii/S0734975011001303</url></related-urls></urls><electronic-resource-num>http://dx.doi.org/10.1016/j.biotechadv.2011.07.020</electronic-resource-num><access-date>2012/4//</access-date></record></Cite></EndNote>[1]. Thus, development of an effective downstream process for purifying rhBuChE is important. Use of an anti-FLAG affinity resin to capture rhBuChE gives only a 20% yield recovery. Procainamide resin which tightly binds to cholinesterases can be used to purify the protein and to enhance the yield percentage. However prior to this chromatography step, different pretreatment strategies were employed such as pH reduction, ammonium sulfate fractionation, polyethylene glycol (PEG) fractionation or two phase aqueous separation of plant extracts to remove plant pigments and native plant protein (e.g. Rubisco).

ADDIN EN.REFLIST 1. Wilken, L.R. & Nikolov, Z.L. Recovery and purification of plant-made recombinant proteins. *Biotechnology Advances* **30**, 419-433 (2012).

# 436035 Bioprocess Platform Design for Kefiran Production in Semi-Industrial Scale

**Abstract:**

Kefiran is a water soluble polysaccharide produced by the lactic acid bacteria *Lactobacillus kefiranofaciens* and composed of equal amount of glucose and galactose. This biopolymer has wide application in food and pharmaceutical industries based on its functional properties as antibacterial, antifungal, anti-inflammatory, immunomodulatory, and many other health benefits. In spite of many research were conducted for the wide applications of kefiran, little information is available on the industrial production of this important compound. However, the industrial production of kefiran is carried in submerged cultivation system using *L. kefiranofaciens.* The production of kefiran in industrial scale faces two main challenges. First, the low growth rate of the producer bacteria; second, the low yield of kefiran when cells cultivated in submerged culture. Thus, this study was focused at first on medium and bioprocess optimization of kefiran production in mono-culture system using different approaches. Furthermore, the production of polysaccahrides was significantly enhanced by using mixed culture composed of *L. kefiranofaciens* and *Saccharomyces cerevisiae* and medium optimization of this mixed culture cultivation system. Further improvement was then achieved when process scaled up to 16-L stirred tank bioreactor and the maximal kefiran production was achieved using fed-batch cultivation strategy under controlled pH and DO in level of 6.0 and 30% saturation, respectively. In the present work, complete industrial platform was designed for kefiran production in semi-industrial scale.

# 413245 Regulating Pyruvate Carboxylase in the Living Culture of *Aspergillus Terreus* NRRL1960 By L-Aspartate for Enhanced Itaconic Acid Production

**Abstract:**

Itaconic acid, also know as methylene succinic -acid ($C_5H_6O_4$), is an unsaturated 5-carbon dicarboxylic acid. It can be derived into many high value added chemicals or materials. To date, it is a potential building block that can be produced by fermentation using *Aspergillus terreus*. However, the low production yield from fermentation process rather limit the supply to the market. The lower production yield was claimed as the result of completing TCA cycle towards biomass synthesis with the sufficient phosphate and nitrogen sources. While under limiting phosphate and nitrogen, the glycolytic flux was somewhat shunted; thus lowering the flux in TCA cycle and consequently that towards itaconic acid production route. In this study, pyruvate carboxylase, one of the key responsible enzymes involved in itaconic acid production, was allosterically inhibited by L-aspartate. The presence of 10 mM L-aspartate in the production medium directly suppressed pyruvate carboxylase expression in the living *A. terreus*. By the repression, oxaloacetate in the cytosol was lowered. Subsequently, malic acid concentration was decrease in the cytosol resulting in the improved transport of malate and cis-aconitate, citrate isomer, towards the the malate/citrate antiporters as the response of the living culture of *A. terreus* to complete TCA cycle. The transport of cis-aconitate via the antiporters induced citrate synthase and 6-phosphofructo-1-kinase activities in response to balance the fluxes of TCA intermediates. This eventually resulted in the improved itaconic acid production yield of 8.33% and final concentration of 60.32% compared to those obtained from the control fermentation with the shortened lag time to produce itaconic acid during the production phase.

# 425846 Analysis of Amino Acids Metabolism in Butanol Fermentation By Clostridium Tyrobutyricum

**Abstract:**

The objective of this study was to understand the metabolism of amino acids in biobutanol production by the engineered *Clostridium tyrobutyricum*, AKOKO-*adhE2*. First, we investigated the expression of the proteins involved in amino acids consumption and metabolism using proteomics analysis. The proteomics study showed that the expression of amino acid ABC transporter substrate binding protein was upregulated while the aminotransferase was downregulated during butanol production. The amino acids associated with the intermediates in the central metabolic pathway of butanol formation were then analyzed. The metabolites analysis indicated that the uptake rate of alanine (associated with pyruvate), leucine and oxoproline (acetyl-CoA) was reduced while the consumption rate of α-ketoglutamate and succinate (butyryl-CoA) was increased with the production of butanol. Butanol production was increased from 15 g/L to 19 g/L by yeast extract supplement. Finally the kinetics profiles of all 20 amino acids were collected to develop a systematic understanding of amino acids metabolism, which could be used to rationally optimize the butanol production medium or regulate the metabolism of the key metabolite in butanol production.

# 427165 Fed-Batch Cultivation for Induction Strategy When Induced with Lactose to Increase Production of Thermostable Phytase in Semi-Industrial Scale 16-L Bioreactor By E. coli BL21(DE3)

**Abstract:**

A recombinant *Escherichia coli* BL21 (DE3) which harbours the thermostable phytase gene from *Bacillus* sp.MD2 is used as the host cell. Lactose was used as an alternative inducer. Lactose can be metabolized as a carbon source and can contribute to increase the metabolic overflow which leads to acetic acid accumulation. The important of this study to observe the growth kinetics of *E.coli*BL21(DE3) when induced with lactose without addition of antibiotic in the semi-industrial scale 16-L bioreactor towards expression of phytase gene which is growth in the glycerol-minimal medium as the production medium.The present work underlines the great importance of production medium and induction strategy for efficient production of recombinant proteins towards excretion as extracellular phytase.Different modes of induction strategies were conducted: one-pulse, intermittent by feeding two pulse of lactose, constant feeding and DO-stat concomitantly feeding with glycerol in the post-induction phase to determine the factors limiting the cell growth and distribution of phytase activity.It can be concluded that, addition of lactose as the only carbon source in the fermentation broth in the post induction phase was not only for the cell growth but also for heterologous protein expression by the host cells. A combination of continuous glycerol feeding rate of 15 g $L^{-1}$ $h^{-1}$ and intermittent feeding of lactose without supplementation of amphicillin can maintain the specific growth rate of 0.11 $h^{-1}$ with the maximum cell mass about 18.3 g $L^{-1}$ due to dissolved oxygen limitation during cultivation. The plasmid was maintained stably in the culture throughout the fed-batch process without the use of an antibiotic. The ratio distribution of phytase activity between extracellular: periplasmic: cytoplasmic was 7.75:1.28:1. The extracellular phytase activity about 111.19 U $mL^{-1}$ of the total phytase activity about 161.0 U $mL^{-1}$ with productivity of almost 32,200.00 U $L^{-1}$ $h^{-1}$ after 5 hours of induction. In conclusion, the total phytase productivity is governed by the lactose feeding rate per unit biomass of host cells in the post-induction phase.

# AUTHOR INDEX

# AUTHOR INDEX

# AUTHOR INDEX

# AUTHOR INDEX

# AUTHOR INDEX

# AUTHOR INDEX

# AUTHOR INDEX

# AUTHOR INDEX

# AUTHOR INDEX

# AUTHOR INDEX

# AUTHOR INDEX

# AUTHOR INDEX

# AUTHOR INDEX

# AUTHOR INDEX

# AUTHOR INDEX

# AUTHOR INDEX

# AUTHOR INDEX

# AUTHOR INDEX

# AUTHOR INDEX

# Food, Pharmaceutical and Bioengineering Division 2015

Core Programming Area at the 2015 AIChE Meeting

Salt Lake City, Utah, USA
8-13 November 2015

Volume 2 of 2

ISBN: 978-1-5108-1860-6

**Printed from e-media with permission by:**

Curran Associates, Inc.
57 Morehouse Lane
Red Hook, NY 12571

CURRAN ASSOCIATES INC. proceedings.com

**Some format issues inherent in the e-media version may also appear in this print version.**



Printed by Curran Associates, Inc. (2016)

For permission requests, please contact AIChE
at the address below.

AIChE
120 Wall Street, FL 23
New York, NY 10005-4020

Phone: (800) 242-4363
Fax: (203) 775-5177

www.aiche.org

**Additional copies of this publication are available from:**

Curran Associates, Inc.
57 Morehouse Lane
Red Hook, NY 12571 USA
Phone: 845-758-0400
Fax: 845-758-2633
Email: curran@proceedings.com
Web: www.proceedings.com

# TABLE OF CONTENTS

## VOLUME 1

## VOLUME 2

# 427792 Effects of Operating Parameters on Tablet Quality in a Continuous Manufacturing Line

**Abstract:**

As continuous manufacturing is gaining increasing importance within the pharmaceutical industry there still remains a knowledge gap in building quality into the final product. In the context of establishing continuous manufacturing technology, characterization of these processes becomes important. As there are many parameters impacting the product quality, it is important to understand the interplay between these process parameters and tablet performance. Hence, the present study evaluates experimentally the effect of variation in critical process parameters (CPP) on the critical quality attributes (CQA) of tablets produced by direct compression continuous manufacturing. The study was performed on a model API, acetaminophen (APAP), using lactose monohydrate and magnesium stearate (MgSt) as excipients. Multivariate effects were analyzed using Quality-by-design principles. A fractional factorial design with 30 experiments, including 3 repeated center points, was used to study the influence of four process variables – drug concentration (%), compaction force (KN), blending speed (rpm), and feed frame speed (rpm) – on the final product. Design of experiments was executed on a continuous tablet manufacturing line consisting of three feeders, a mill, a blender, and a tablet press. Tablets were characterized for hardness, dissolution, content uniformity, and weight variability. Additionally, they were subjected to NIR transmission spectroscopy analysis and ultrasound testing. The processed NIR spectral datasets were analyzed using PCA, and it was determined that the first and second principal components track tablet drug content and compaction force. Analyses of designed experiments also revealed that drug concentration and compaction force significantly affect tablet dissolution. The results were further used to optimize the process and establish boundary limits for critical process parameters that influence the product to achieve desirable tablet quality.

# 416476 Enhanced Spray Drying Throughput with High Productivity Hypromellose Acetate Succinate

**Abstract:**

New drug entities frequently suffer from low solubility and present significant formulation challenges. One of the leading technologies utilized to overcome the low solubility of drugs is the formation of amorphous solid dispersions via spray drying of the API with an excipient. While a number of excipients have been utilized to create amorphous solid dispersions, hypromellose acetate succinate (HPMCAS) has proven to be highly effective and versatile in creating dispersions that demonstrate enhanced free drug levels in solution and inhibit recrystallization. To date, there are no excipients that have been specifically developed for spray drying applications and the excipients used in this application space were all developed for other applications such as enteric coatings and are not optimized for the spray drying process.

The present study compares AFFINISOL™ High Productivity HPMCAS, a new grade of HPMCAS designed specifically for spray drying, to standard HPMCAS. The High Productivity HPMCAS has reduced viscosity compared to standard commercial HPMCAS, allowing for increased throughput and reduced use of organic solvent in the spray drying process. Spray drying performance of the two HPMCAS materials was compared by measuring solution mass flow rate as a function of applied pressure during nozzle atomization testing. A head pressure was applied to the solution container and the resulting mass flow through the nozzle was measured using a Micromotion Coriolis mass flow meter. The size of the spray dried particles were measured using a Malvern Mastersizer 2000 Scirocco particle size analyzer with a variable dispersive air pressure. It was found that to match the atomization performance of standard HPMCAS, the concentration of High Productivity HPMCAS in solution had to be increased by 1.7x relative to standard HPMCAS. In addition, viscosity vs. shear rate at constant temperature of the HP HMPCAS was measured and compared to that of standard HPMCAS. Furthermore, solutions at low concentration show a Newtonian viscosity behavior as a function of the shear rate, without any onset of pseudoplasticity in the detectable shear rate region. With an increase in concentration a small onset of pseudoplasticity becomes apparent and is more pronounced for the conventional HPMCAS than for the High Productivity HPMCAS. Finally, this study will discuss the in vitro drug release rate and crystallization inhibition of amorphous spray dried dispersions formed with itraconazole and phenytoin.

# 425187 Measuring Blend Hydrophobicity By Single Droplet Penetration Technique

**Abstract:**

Hydrophobicity is recognized as a critical property for pharmaceutical intermediate blends. For lubricated blends, the hydrophobicity is especially important as it affects the drug release in later *in vitro* dissolution tests. Hydrophobicity also affects the characteristic wetting time during wet granulation processes. In this work, a method was developed to monitor the penetration process of a droplet on powder surface. Using a high-speed camera and video processing technique, the droplet penetration profile was obtained. As a case study, blends with varying hydrophobicity were prepared using a modified Couette shear cell. Comparison of droplet penetration profiles showed that blends with higher hydrophobicity had longer penetration time. Interestingly, a "waiting phase" was observed before the droplet initiates the penetration for less hydrophobic blends. A non-wetting liquid was used in addition to DI water. The penetration model was used to describe the droplet penetration process with the assumption of constant drawing area. The method to compare blend hydrophobicity using penetration profiles from two liquids was demonstrated. This study presented a potential technique to measure blend hydrophobicity to indicated process changes.

# 427913 Amorphous Solid Dispersion of Poorly Soluble API By Hot Melt Extrusion (HME) and Spray Drying (SD): A Rational Screening Approach

**Abstract:**

**Amorphous solid dispersion of poorly soluble API by hot melt extrusion (HME) and spray drying (SD): A rational screening approach**

*A.Paudel[1], A. Mercuri[1], S. Mohr[1], M. Bresciani[1], J.G. Khinast[1]*

*[1]Research Center Pharmaceutical Engineering, Graz, Austria*

Polymeric amorphous solid dispersion (ASD) of poorly water soluble active pharmaceutical ingredients (APIs) are one of the promising strategies of solubility and dissolution rate enhancement, and hence, tackling the growing challenges in oral bioavailability (1). Hot melt extrusion (HME) and spray drying (SD) are two industrial technologies for manufacturing ASD (2). Major hurdle associated to the ASD is the possible risk of crystallization of amorphous form during storage and/or administration, thus negating the claimed solubility advantage. Since solid-state API-carrier molecular miscibility and carrier-induced supersaturation play a critical role towards stabilization of ASD and in vivo performance, selection of suitable carrier (s), optimal drug loading and appropriate manufacturing process/parameters is crucial for the success of this formulation strategy (3). HME and SD both exhibit different pros and cons associated with processability of the range of carriers, available solvent range, thermal degradation and molecular, particulate and bulk-level properties of resulting amorphous intermediates. Thus, there is a different degree of downstream processability for finished dosage development.

This work aims towards a multi-methodological approach, using a model drug candidate and a wide diversity of ASD carriers, such as Soluplus, HPMC, HPMC-AS, PVPVA 64 and their blends with Eudragit E or HPC: (i) solution-state excipient-induced biorelevant supersaturation, (ii) film-based mini-formulation preparation via solution casting (SC) and melt-quench-cooling (MC) using selected carriers, (iii) solid-state characterization and mini-dissolution studies, (iv) formulating ASD formulation via HME and SD and (v) solid-state characterization, in vitro biorelevant dynamic dissolution and accelerated short-term physical stability of formulations prepared by HME and SD.

The first set of carriers was selected based on the extent of API supersaturation generation and its maintenance in fasted-state simulated intestinal fluid (FaSSIF). These were Soluplus, HPMC and HPMC-AS/HPC blends. SC and MC films were prepared of API with the varying ratios of latter polymers using HME relevant temperatures and SD relevant solvents, respectively. Drug-carrier miscibility and interaction, glass transition, crystallinity and wettability of film formulations were evaluated using several methods, such as DSC, pXRD, ATR-FTIR spectroscopy, contact angle etc. Based on the outcome, supersaturating formulations with a homogenous amorphous phase were generated by HME and/or SD using different carriers (Soluplus, HPMC and HPMC-AS/HPC). Solid-state characterization, *in vitro* biorelevant dynamic dissolution and accelerated short term stability study were performed on formulations prepared by HME and SD.

It was found that supersaturation screening and small-scale dissolution tests can partially predict the *in vitro* behaviour of ASDs. Film based screening (SC vs MC) aids in predicting the performance rank order of ASD formulations. The results showed that strong functional interactions like API-carrier H-bonding is not always a prerequisite for stabilizing miscible API-polymer systems. Comparable *in vitro* dissolution behaviour of amorphous and partially crystalline Soluplus based dispersion implied that complete amorphicity of API in dispersion is not essential for enhancing *in vitro* performance of solid dispersion. ASD derived from HME and SD can behave completely different when exposed to elevated T/%RH, even though identical carrier with identical composition and physical states are used.

**References**

1. Van den Mooter, G. (2012) The use of amorphous solid dispersions: A formulation strategy to overcome poor solubility and dissolution rate. *Drug Discovery Today: Technologies 9*, e79–e85.

2. Paudel, A., Worku, Z.A., Meeus, J., Guns, S., Van den Mooter, G. (2013) Manufacturing of solid dispersions of poorly water soluble drugs by spray drying: formulation and process considerations, *International journal of pharmaceutics 453, 253-284.*

3. Paudel, A., Van Humbeeck, J., Van den Mooter, G. (2010) Theoretical and experimental investigation on the solid solubility and miscibility of naproxen in poly (vinylpyrrolidone), *Molecular pharmaceutics* 7, 1133-1148.

GlaxoSmithKline

# Pharmaceutical Processes Selection from Shear Cell Analysis and Compaction Simulation

**Don Clancy**
**Modeling Center of Excellence**

Presented to AIChE
Wednesday, November 11, 2015: 10:10 AM
Ballroom D (Salt Palace Convention Center)

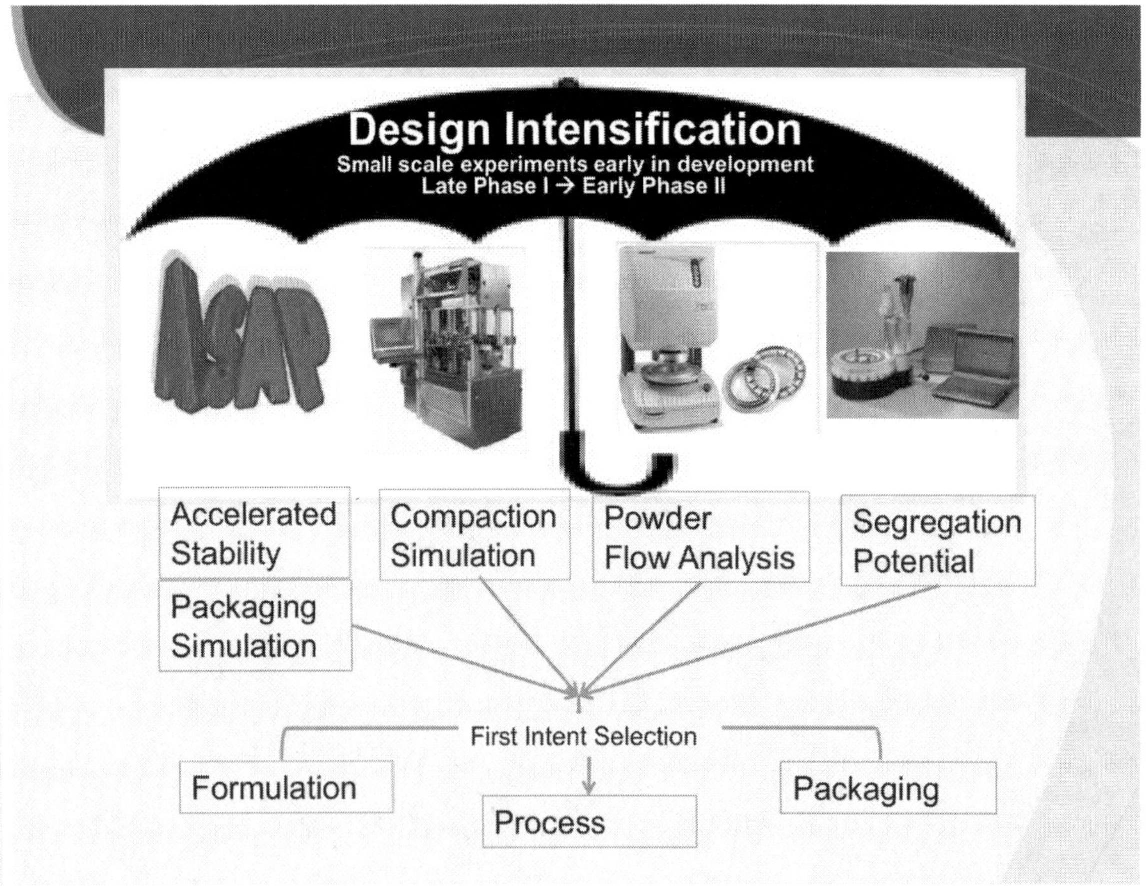

## First Intent Selection of Secondary Processing Routes

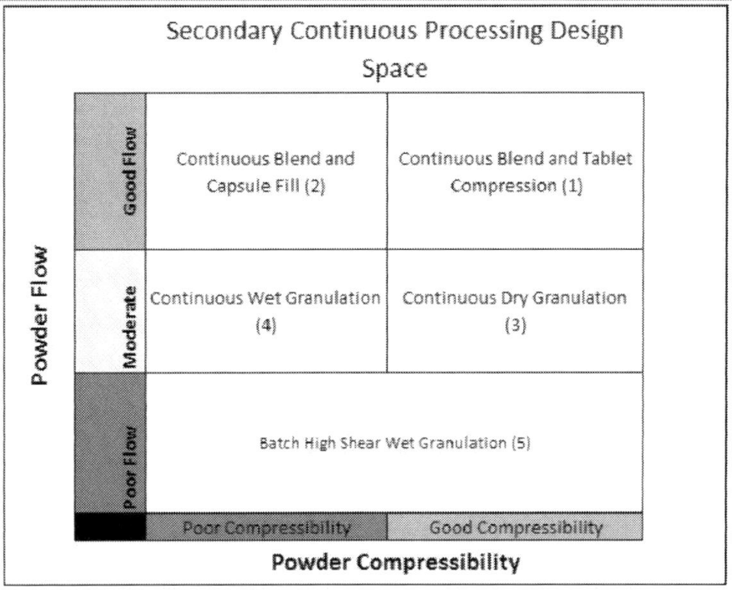

Shear Cell

Compaction Simulation

- This research focus started with the concept that small scale testing in a shear cell and a compaction simulator can provide the information necessary for first intent selection of secondary process routes.

gsk GlaxoSmithKline

# About Shear Cell Analysis

# Method: Annular Shear Cell - Brookfield Powder Flow Tester

Principle: Tester Cell Lid

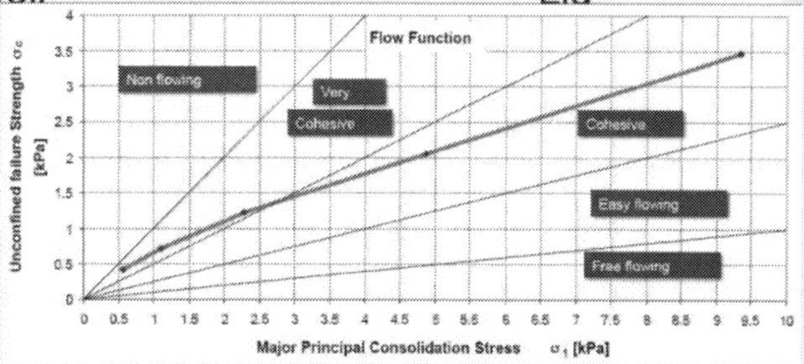


- Cell is filled with the powder being poured loosely and excess material is scraped off
- Powder is compressed by the lid and sheared rotationally
- Shear stress values are recorded by the tester indicating the force required to get the particles to flow against each other
  - This step is repeated for various consolidation loads allowing to build a flow function

# Simplified Explanation of How "Flow Index" is Calculated

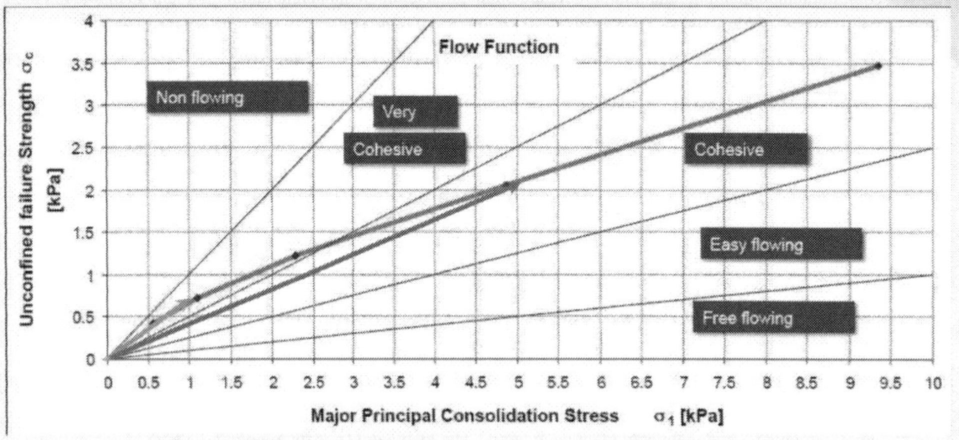


- Two shear cell terms that are used extensively are the "Flow Function", and the "Flow Index."
- "Flow Index" (FI) is the slope of the line that goes from the origin through the measured failure strength at a given major principal consolidation stress.
  - Flow Index at 1-KPa is slope of green line or approximately 0.65.
  - Flow Index at 5-Kpa is slope of blue line, or approximately 0.42.
- The "Flow Function" (FF) is the inverse of the "Flow Index." FF = 1/FI
- A lower Flow Index is indicative of better flow.

GlaxoSmithKline

# Tablet Compression

Don Clancy and Laura Wehmeyer

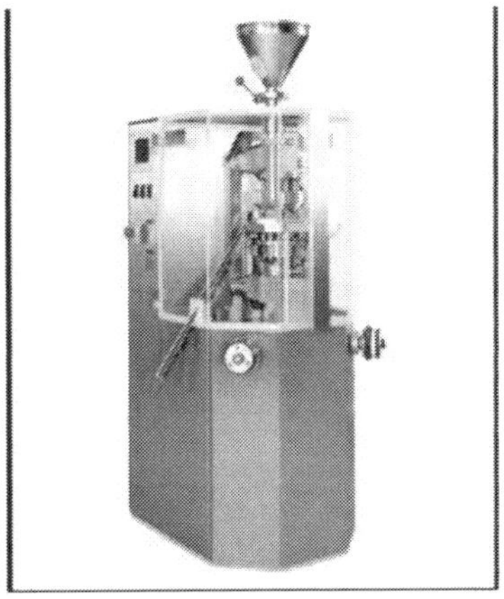

## Rational for Defining a Powder Flow Limit for Tablet Compression

- Powder flow increasingly affects the tablet weight variability as the time given for die fill decreases.
- Tablet weight variability affects tablet content uniformity, which has a typical limit of 6% RSD.
- Conservatively, the tablet weight uniformity should not exceed 3% RSD, as both weight variability and blending API content uniformity affect overall tablet content variability.

# Typical Experimental Design for evaluating Compression

| Blend # | Formulation | Flow Index at 1 KPA | RPM | Tablet Weight (mg) | RSD% tablet weight |
|---|---|---|---|---|---|
| 1 | 79.2% Fastflo Lactose, 19.8% Avicel PH102, 1% MgSt | 0.10 | 13 | 202.28 | 1.10% |
| 2 | 19.8% Fastflow Lactose 79.2% Avicel PH102, 1% MgSt | 0.17 | 13 | 200.82 | 1.51% |
| 3 | 99% Avicel PH102, 1% MgSt | 0.21 | 13 | 186.98 | 1.70% |
| 4 | 99% Avicel PH101, 1% MgSt | 0.28 | 13 | 169.3 | 2.05% |
| 1 | 79.2% Fastflo Lactose, 19.8% Avicel PH102, 1% MgSt | 0.10 | 66 | 195.6 | 1.26% |
| 2 | 19.8% Fastflow Lactose 79.2% Avicel PH102, 1% MgSt | 0.17 | 66 | 155.9 | 2.66% |
| 3 | 99% Avicel PH102, 1% MgSt | 0.21 | 66 | 162.94 | 2.37% |
| 4 | 99% Avicel PH101, 1% MgSt | 0.28 | 66 | 138.98 | 5.96% |
| 1 | 79.2% Fastflo Lactose, 19.8% Avicel PH102, 1% MgSt | 0.10 | 120 | 169.4 | 1.31% |
| 2 | 19.8% Fastflow Lactose 79.2% Avicel PH102, 1% MgSt | 0.17 | 120 | 133.62 | 2.61% |
| 3 | 99% Avicel PH102, 1% MgSt | 0.21 | 120 | 148.72 | 3.10% |
| 4 | 99% Avicel PH101, 1% MgSt | 0.28 | 120 | 146.2 | 4.32% |

- This is an example experiment design for a single size die cavity on the Korsch XL300.

# Tablet Compression – Gravity Fed Analysis

**Actual by Predicted Plot**

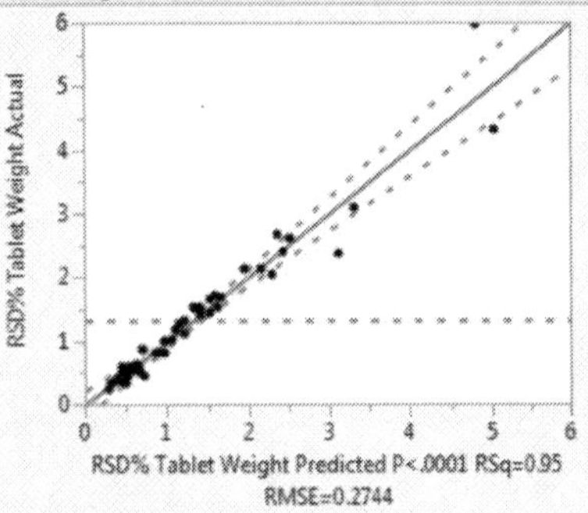


**Summary of Fit**

| | |
|---|---|
| RSquare | 0.945261 |
| RSquare Adj | 0.935916 |
| Root Mean Square Error | 0.274434 |
| Mean of Response | 1.321794 |
| Observations (or Sum Wgts) | 49 |

**Analysis of Variance**

| Source | DF | Sum of Squares | Mean Square | F Ratio |
|---|---|---|---|---|
| Model | 7 | 53.323252 | 7.61761 | 101.1446 |
| Error | 41 | 3.087875 | 0.07531 | Prob > F |
| C. Total | 48 | 56.411127 | | <.0001* |

**Analysis of Variance**

| Source | DF | Sum of Squares | Mean Square | F Ratio |
|---|---|---|---|---|
| Model | 7 | 53.323252 | 7.61761 | 101.1446 |
| Error | 41 | 3.087875 | 0.07531 | Prob > F |
| C. Total | 48 | 56.411127 | | <.0001* |

**Parameter Estimates**

| Term | Estimate | Std Error | t Ratio | Prob>\|t\| |
|---|---|---|---|---|
| Intercept | 0.46523 | 0.168395 | 2.76 | 0.0085* |
| Flow Index at 1 KPA | 11.078386 | 0.799136 | 13.86 | <.0001* |
| Tablet Weight (mg) | -0.001521 | 0.000163 | -9.32 | <.0001* |
| 1/RPM | -9.125136 | 2.848993 | -3.20 | 0.0026* |
| (Flow Index at 1 KPA-0.14642)*(Flow Index at 1 KPA-0.14642) | 31.193683 | 12.82952 | 2.43 | 0.0195* |
| (Flow Index at 1 KPA-0.14642)*(Tablet Weight (mg)-538.497) | -0.010197 | 0.00299 | -3.41 | 0.0015* |
| (Tablet Weight (mg)-538.497)*(Tablet Weight (mg)-538.497) | 1.6606e-6 | 7.439e-7 | 2.23 | 0.0311* |
| (Flow Index at 1 KPA-0.14642)*(1/RPM-0.02267) | -218.5304 | 37.43267 | -5.84 | <.0001* |

# Contour Plot for RSD% versus Tablet Weight and Flow Index at 1 KPA – Gravity Fed Presses

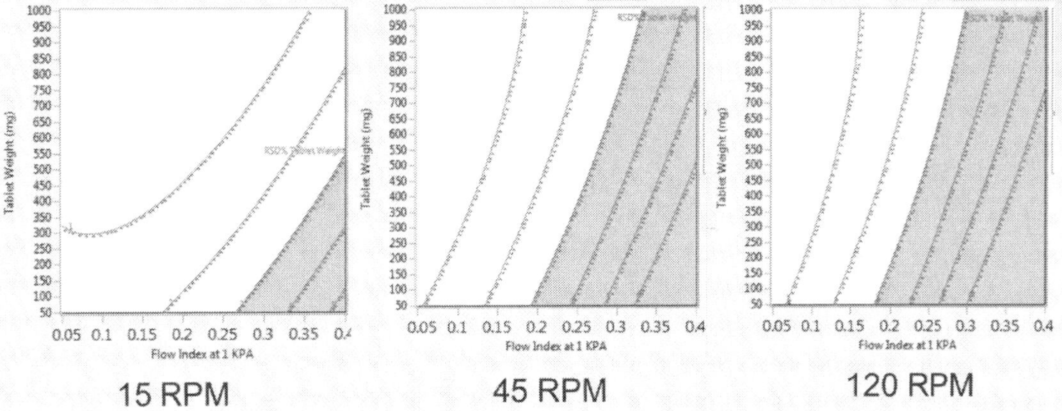


- 200 mg tablets require a flow index of 0.2 at 1 KPA for 3% RSD at 120 RPM, which is where the design space upper limit is set.
- The limit is dependent on tablet weight, as 3% RSD upper limit, shows possible flow index for larger tablets could reach 0.3 for a 500 mg tablet.
- Model is extrapolated above 0.28 flow index due to problems with arching in the feed hopper preventing further material with worse flow properties than 0.28 from being processed. Larger presses with larger hopper feed outlets will have less problems with arching in the feed hopper.

# High Speed Compression - Fette 102i with Paddle Feeder

**Whole Model**

**Actual by Predicted Plot**

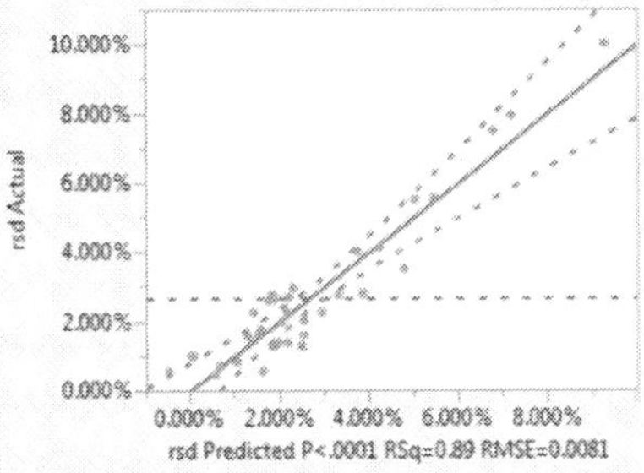

**Summary of Fit**

| | |
|---|---|
| RSquare | 0.894036 |
| RSquare Adj | 0.862639 |
| Root Mean Square Error | 0.008119 |
| Mean of Response | 0.026702 |
| Observations (or Sum Wgts) | 36 |

**Analysis of Variance**

| Source | DF | Sum of Squares | Mean Square | F Ratio |
|---|---|---|---|---|
| Model | 8 | 0.01501471 | 0.001877 | 28.4755 |
| Error | 27 | 0.00177959 | 0.000066 | Prob > F |
| C. Total | 35 | 0.01679430 | | <.0001* |

**Parameter Estimates**

| Term | Estimate | Std Error | t Ratio | Prob>\|t\| |
|---|---|---|---|---|
| Intercept | -0.027731 | 0.004961 | -5.59 | <.0001* |
| Flow Index | 0.0873655 | 0.011679 | 7.48 | <.0001* |
| RPM | 0.000275 | 3.36e-5 | 8.18 | <.0001* |
| Tablet wt. | 5.1275e-5 | 6.561e-6 | 7.82 | <.0001* |
| (Flow Index-0.235)*(Flow Index-0.235) | -0.289955 | 0.115668 | -2.51 | 0.0185* |
| (Flow Index-0.235)*(RPM-65) | 0.0013241 | 0.000284 | 4.66 | <.0001* |
| (RPM-65)*(RPM-65) | 2.0141e-6 | 1.011e-6 | 1.99 | 0.0566 |
| (Flow Index-0.235)*(Tablet wt.-356.161) | 0.0001756 | 4.878e-5 | 3.60 | 0.0013* |
| (RPM-65)*(Tablet wt.-356.161) | 1.0893e-6 | 1.579e-7 | 6.90 | <.0001* |

## Fette 102i – 24 Station Paddle Fed up to 172,800 tablets per hour

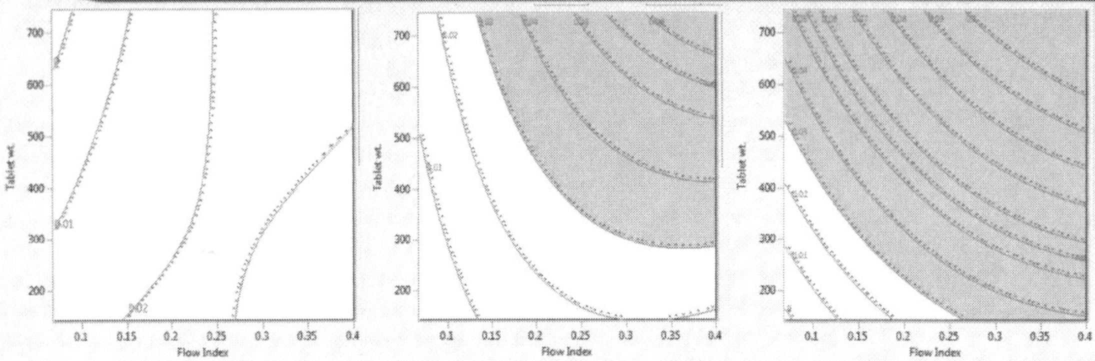


10 RPM    65 RPM    120 RPM

- The Fette 102i with a paddle feeder is far more capable of handling poorly flowing materials at lower speeds. Material with flow index as high as 0.44 was successfully processed even at 65 rpm for 200 mg tablets.
- The same approximate limitation of a flow index of 0.23 for a 200 mg tablet weight is observed at 120 rpm, which is not much different than the observation on the Korsch.
- The extreme limitation for larger tablets >400 mg might be due to limitations of the feed rate through this particular paddle feeder.

gsk GlaxoSmithKline

## Part II:

## Powder Flow Requirements for Encapsulation

**With thanks to:**
Fernando Muzzio, et.al. - Rutgers

## Predicting Capsule Weight Variation

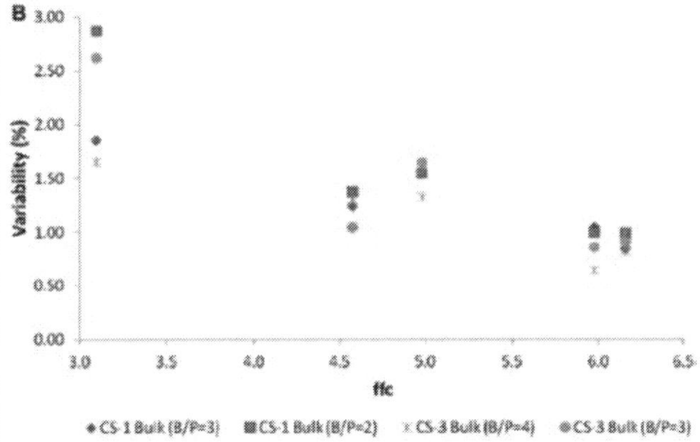


- Capsule weight variation can be reasonably predicted from the Powder Flow Function (ff) (left graph) measured by shear cell at 6 KPa.
- For capsule RSD<3%, a Flow Function greater than 3 is needed. This corresponds to a Flow Index of 0.33 at 6 Kpa. The approximate corresponding flow index at 1 Kpa is 0.4.

Reference Link

## First Intent Selection of Secondary Processing Routes Using Powder Flow and Compressibility

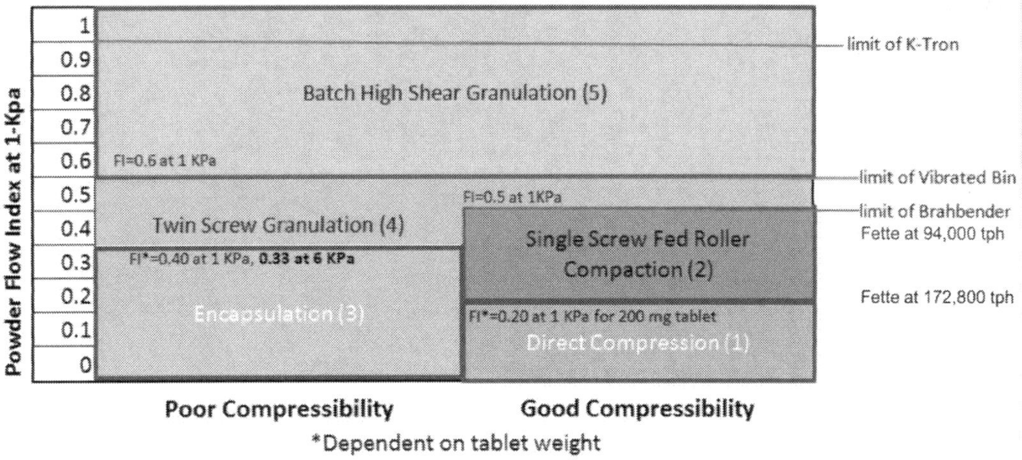


- The resulting "Design Space" confirmed by experimentation is shown in the above figure.
- Further information on segregation potential and melting point can also affect this process selection.

# Part III: Powder Flow Requirements for Twin Screw Granulation

Don Clancy

**Significant Contributions:**
Vicky He, Matt Bahr, Scott Croy, Sayantan Chattoraj

## "K-Tron" Twin Screw Hopper

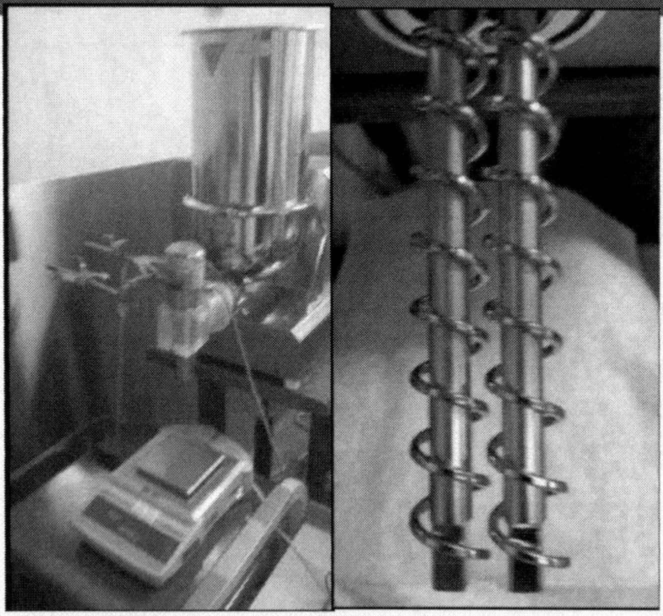

- K-Tron is the main hopper used for Twin Screw granulation.
- The hopper features twin helical screw conveyors.
- Outlet of the twin screw conveyors leads to a vertical "Cramming" unit to push powder into the twin screw granulator.

# Good Versus Poor Powder Flow From Screw Conveyed Hoppers

Good powder flow (movie link) Poor flowing powder (movie link)

- Powders cannot flow continuously like water. When a pressure is put on a powder, it will not move until it reaching point exceeding failure pressure. Then it will move as a sudden avalanche.
- Good flowing material will appear to flow near continuously as shown on left, but poor powder flow will discharge as packets of material as shown on right.
- Variation can be observed by weighing the powder output at rapid time intervals. In this study, the fastest our scale can weigh dynamically is once every 3 seconds.

# K-Tron Hopper 3-Second Flow Variability Related to Shear Cell Flow Index at 1 KPa

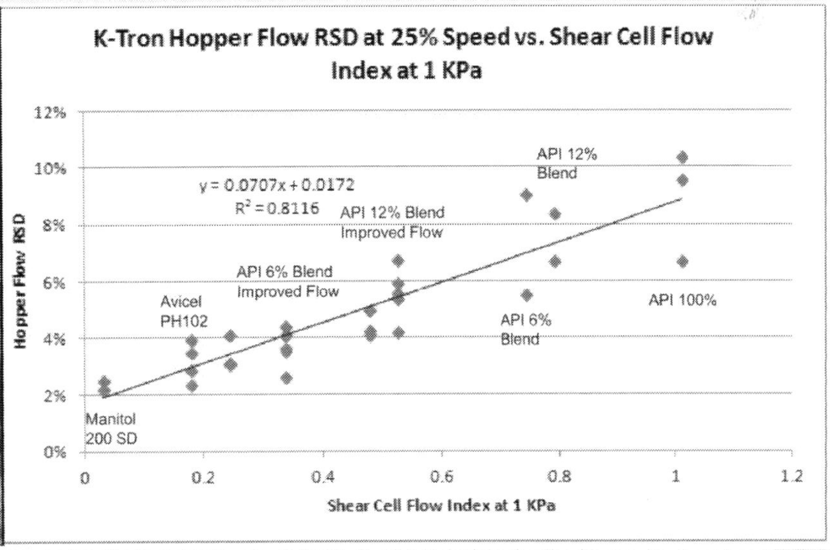


- The K-Tron hopper feeding twin screw granulation is remarkably robust, and has successfully fed pure PHI API with flow index up to 1.0, though with increasing levels of instantaneous variation in feed rate.

## Dual Helix Hoppers (K-Tron) –Discharge Variability

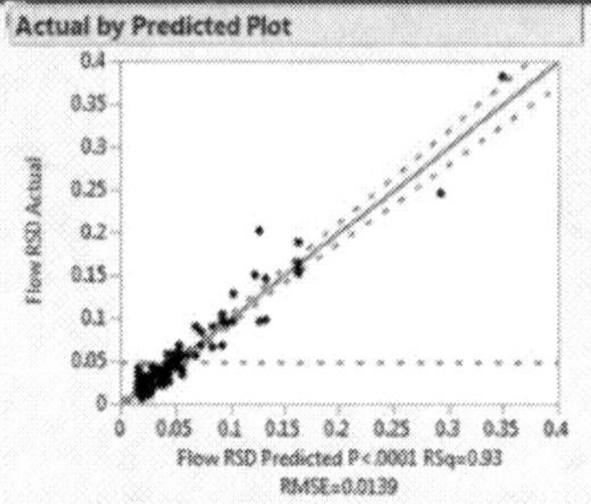


**Summary of Fit**

| | |
|---|---|
| RSquare | 0.928332 |
| RSquare Adj | 0.925963 |
| Root Mean Square Error | 0.013944 |
| Mean of Response | 0.048511 |
| Observations (or Sum Wgts) | 126 |

**Analysis of Variance**

| Source | DF | Sum of Squares | Mean Square | F Ratio |
|---|---|---|---|---|
| Model | 4 | 0.30473011 | 0.076183 | 391.8343 |
| Error | 121 | 0.02352547 | 0.000194 | Prob > F |
| C. Total | 125 | 0.32825558 | | <.0001* |

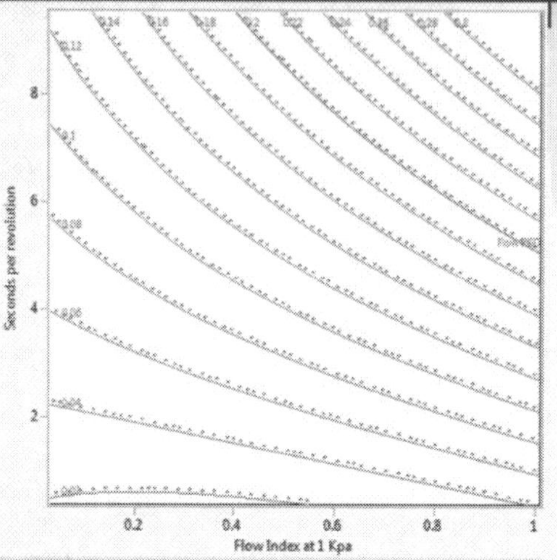


**Parameter Estimates**

| Term | Estimate | Std Error | t Ratio | Prob>\|t\| |
|---|---|---|---|---|
| Intercept | -0.011143 | 0.002509 | -4.44 | <.0001* |
| Flow Index at 1 Kpa | 0.0449248 | 0.004573 | 9.82 | <.0001* |
| Seconds per revolution | 0.0217191 | 0.000776 | 27.99 | <.0001* |
| (Flow Index at 1 Kpa-0.47865)*(Flow Index at 1 Kpa-0.47865) | 0.0449677 | 0.014787 | 3.04 | 0.0029* |
| (Flow Index at 1 Kpa-0.47865)*(Seconds per revolution-1.4832) | 0.0228292 | 0.00197 | 11.59 | <.0001* |

- This is a remarkably resilient hopper, and will not be the real limitation.
- Discharge powder flow RSD is mostly impacted by time needed for discharge screw rotation (RPM), with some impact of flow properties at slow discharge rates.

## Conclusions Regarding K-Tron Dual Helix Hopper Feed to Continuous Twin Screw Granulation

- The K-Tron hopper is a remarkable piece of equipment and unlikely to create a hard limit of flow index that can be fed. Material with a flow index as high as 0.94 at 1 Kpa consolidation pressure was successfully fed with no incidence. This is the worst flowing API, GSK has with a rod-like crystalline structure.

- Instead, there's a softer limit imposed by the variability in flow that's correlated to the flow index and the feeder RPM. A more variable inlet flow will lead to a more variable continuous granulation process, and a narrower design space for the granulation.

GlaxoSmithKline

# Part IV: Requirements for Single Screw Hopper fed Unit Operations (Roller Compaction)

Don Clancy

June 2014

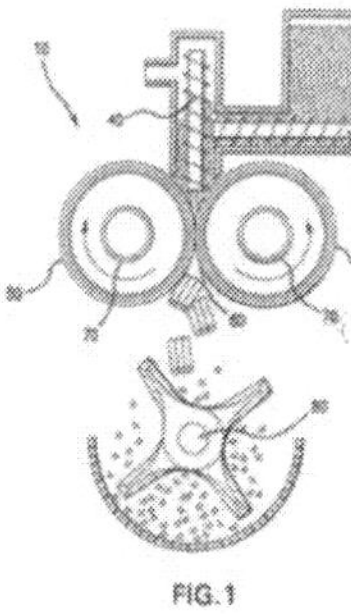

FIG. 1

## Part IV - Flow Requirements for Single Screw Hoppers - Brahbender

View from Top

- This Brahbender hopper features a single screw, and has a useful advantage of being able to work with very little powder. It's able to deliver powder with only 30 to 60 grams of powder retention left in the hopper.
- Single screw hoppers such as this are not able to feed as wide a variety of materials as the dual screw K-Tron.

# Single Helix Hopper (Brahbender) –Discharge Variability

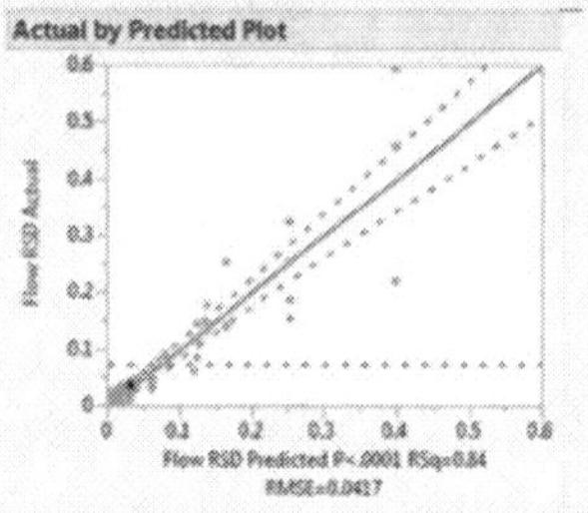


**Summary of Fit**

| | |
|---|---|
| RSquare | 0.842022 |
| RSquare Adj | 0.83259 |
| Root Mean Square Error | 0.041742 |
| Mean of Response | 0.072676 |
| Observations (or Sum Wgts) | 72 |

**Analysis of Variance**

| Source | DF | Sum of Squares | Mean Square | F Ratio |
|---|---|---|---|---|
| Model | 4 | 0.62222723 | 0.155557 | 89.2771 |
| Error | 67 | 0.11674106 | 0.001742 | Prob > F |
| C. Total | 71 | 0.73896832 | | <.0001* |

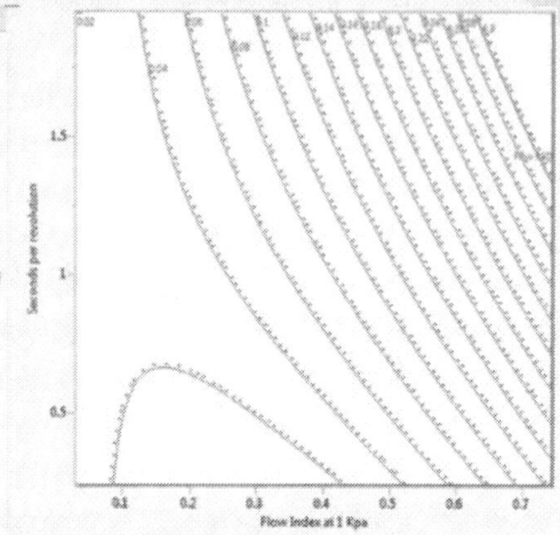


**Parameter Estimates**

| Term | Estimate | Std Error | t Ratio | Prob>\|t\| |
|---|---|---|---|---|
| Intercept | -0.064341 | 0.010601 | -6.07 | <.0001* |
| Flow Index at 1 Kpa | 0.1889273 | 0.027612 | 6.84 | <.0001* |
| Seconds per revolution | 0.0564204 | 0.008212 | 6.87 | <.0001* |
| (Flow Index at 1 Kpa-0.30083)*(Flow Index at 1 Kpa-0.30083) | 0.5057237 | 0.094609 | 5.35 | <.0001* |
| (Flow Index at 1 Kpa-0.30083)*(Seconds per revolution-0.86368) | 0.233865 | 0.030841 | 7.58 | <.0001* |

- Heavily impacted by powder flow properties, only suitable for flow index <0.5 at 1 Kpa.
- Problems with arching and unpredictability at >0.5 flow index at 1 Kpa.
- Same helix impeller design used for feeding roller compaction and twin screw

# Beware of Bins Upstream of Continuous Feeders!

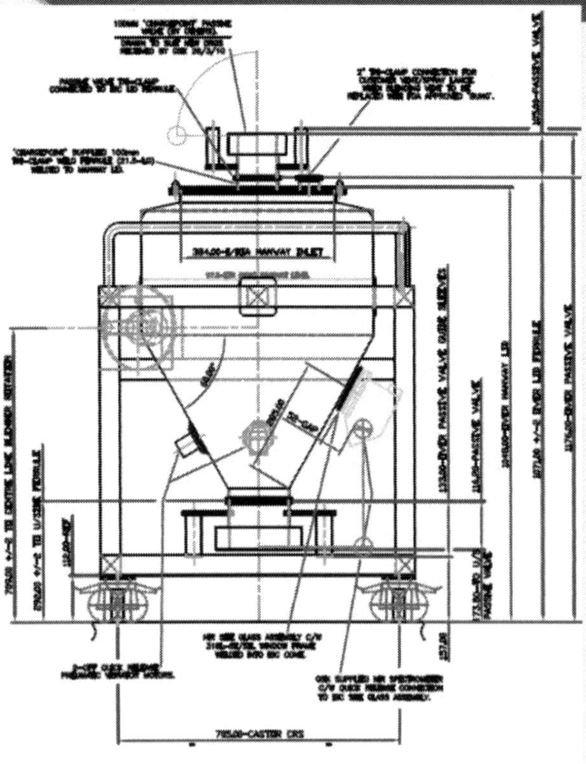

- This bin is used to blend excipients and API together prior to twin screw granulation or roller compaction
- The limitation is the narrow outlet, which is 15 cm in diameter.
- Powders arch over the outlet and powder flow can be intermittent. The arching dimension needs to be less than 15-cm, which corresponds to a flow index of approximately 0.6 at 1 KPa for most powders.

## First Intent Selection of Secondary Processing Routes Using Powder Flow and Compressibility

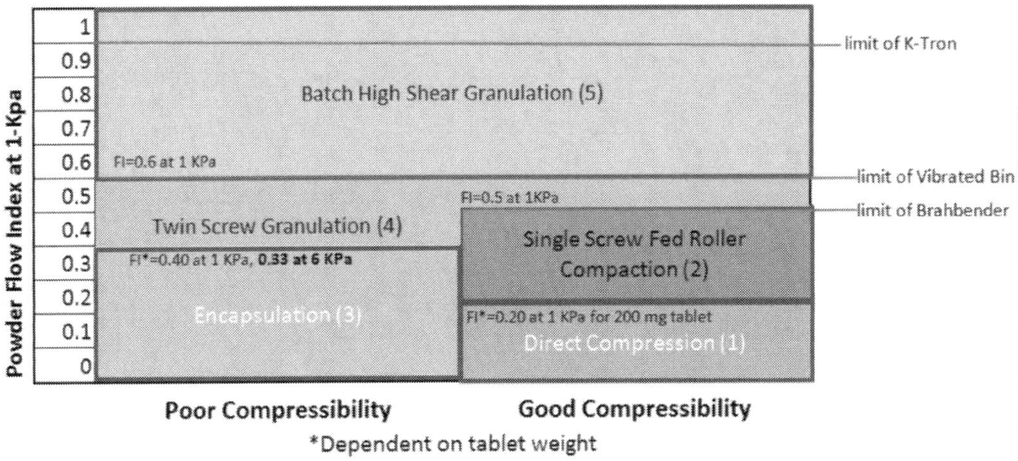


Final mapping of flow requirements for processing.

GlaxoSmithKline

# Example Case Study:

# Formulation Development of An Expensive Drug Product for Roller Compaction


Don Clancy, Maria Barnett and Francis Ignatious


Nov 2014

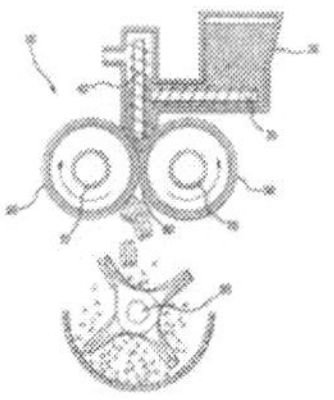

FIG. 1

## Background to Case Study

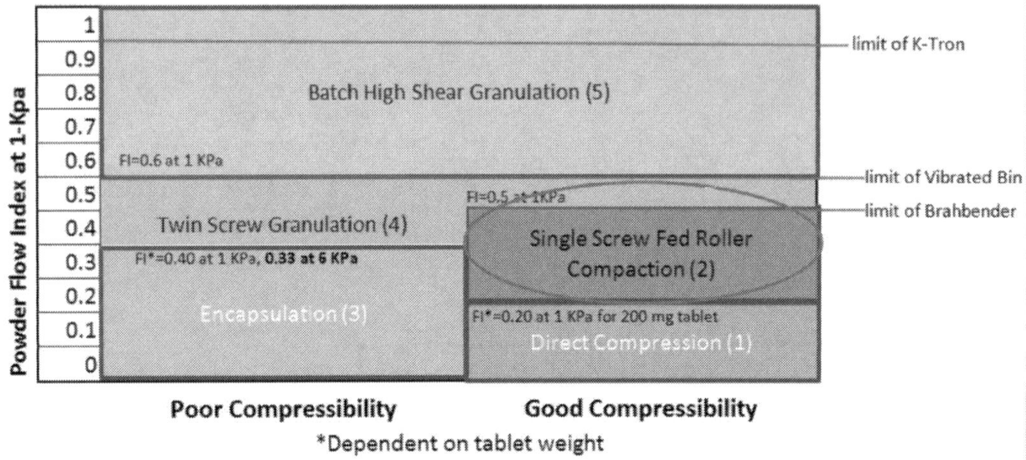


- Formulation exhibits good compressibility characteristics.
- Desire is to minimize tablet size by increasing drug content to the limit of what can be robustly processed by roller compaction.

## API Content in Blends Versus Flow Index at 1 KPa

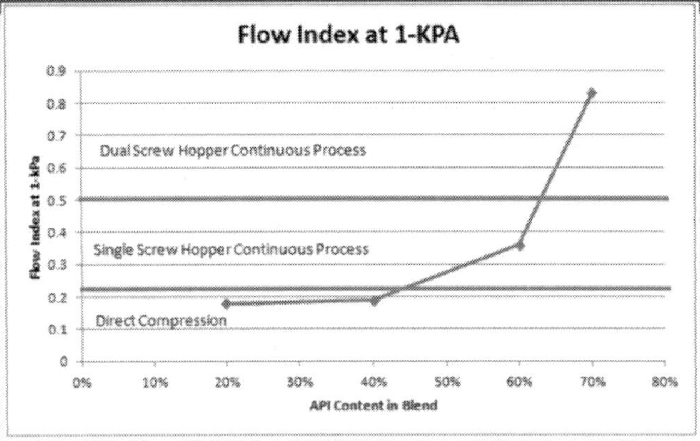


Flow Function Results:

| Data Set # | Flow Function Slope | Flow Function Intercept [kPa] | Flow Index | Batch Number |
|---|---|---|---|---|
| 1 | 0.06 | 0.259 | 0.18 | EE771611D_20%Blend |
| 2 | 0.08 | 0.521 | 0.19 | EE771611C_40%Blend |
| 3 | 0.09 | 0.839 | 0.36 | EE771611B_60%Blend |
| 4 | 0.08 | 0.767 | 0.83 | EE771611A_70%Blend |
| 5 | 0.10 | 0.102 | 0.22 | Avicel PH102_142385515 |

- To qualify for direct compression, the flow index needs to be better (smaller) than Avicel PH102 at 1 KPa pressure. In this case, the 20% and 40% blends could be directly compressed without significant concerns of tablet weight variability problems.

- The 60%, and 70% blends have significantly poorer flow compared to Avicel PH102, and would need to be granulated prior to compression.

# IBC Bin Requirements for formulation with flow aid

☑ Rat-hole Diameter: Bin Diameter 1 .000 m Height 1 .000 m ☑ Arching Dimension:

Results:

| Data Set # | Arching Dimension (m) | Rat-hole Diameter (m) | Batch Number |
|---|---|---|---|
| 1 | 0.030 | 0.216 | EE7716110_20%Blend |
| 2 | free flow | 0.296 | EE771611C_40%Blend |
| 3 | 0.070 | 0.528 | EE771611B_60%Blend |
| 4 | 0.301 | 1.000 | EE771611A_70%Blend |
| 5 | 0.041 | 0.213 | Avicel PH102_142385515 |

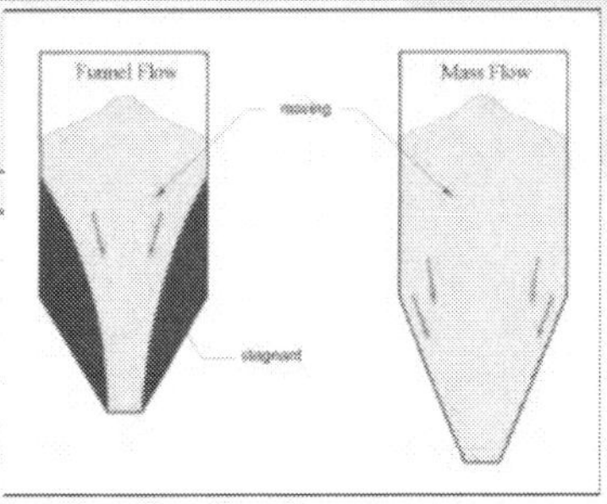


- Of interest is the 60% blend (data set#3), and whether the 60% and 70% formulations can be fed reliably out of an rectangular IBC bin to feeder into roller compaction.
- The 60% blend needs a bin opening outlet greater than 7-cm in diameter to prevent stable arch formation. Flow type will be funnel flow which will result in a rat hole that forms approximately 0.5 meter in diameter.
- The 70% blend will not feed reliably from an IBC bin. It will be able to arch over a 30-cm outlet. Assuming it's agitated to break the arch, the rat hole diameter equals the bin diameter. A rat hole will form and then no further flow would happen after that point. Approximately half the batch in the bin would be within the stagnant region.

# Impact of 0.5% Flow Aids at 55% API Content, High Shear Blending

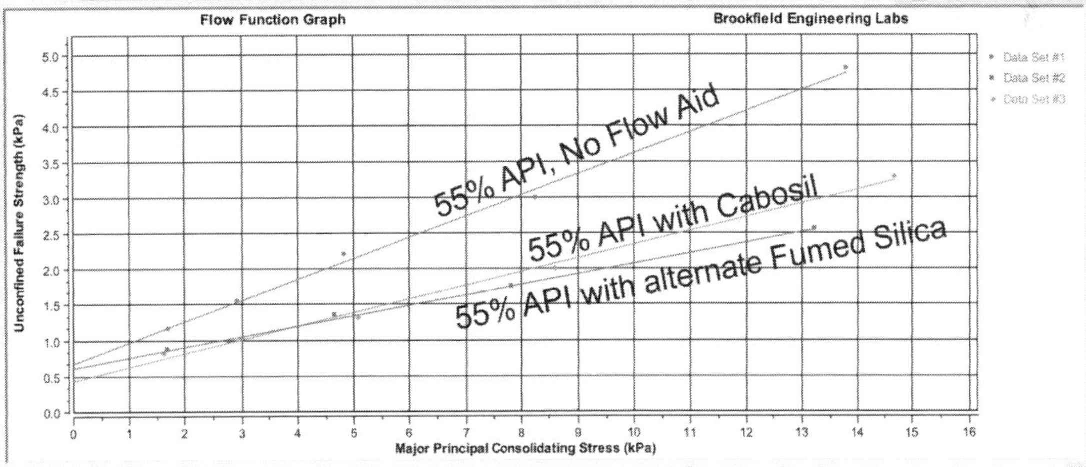


- 0.5% Fumed silica has a substantial impact of improving flow properties of this blend, and making it possible to feed into roller compaction.
- There's very little difference between Cabosil and the other type of fumed silica, though Cabosil is showing slightly improved flow at low pressure.

# Conclusions from Case Study

- The 20% formulation has good flow properties, is classified as "easy flowing close to free flowing" and would be a candidate for direct compression with little tablet weight variation.
- The 40% formulation has good flow properties at low pressure that would be classified as "easy flowing." This formulation could be directly compressed with tablet weight variation similar to Avicel PH102.
- The 60% formulation would need to go through a granulation process to improve the flow properties before going on to compression. This formulation can be hopper fed from a typical single screw conveyed hopper without expecting problems. It can be fed from a bin with at least a 7-cm outlet.
- The 70% formulation can not be fed from an IBC bin, or a single screwed hopper. It can be fed on a double screw conveyed hopper, like the "K-Tron" hopper for the twin screw.
- A fumed silica flow aid is needed. There's little difference between the fumed silica suppliers on the resulting improvement in flow.

# Conclusion

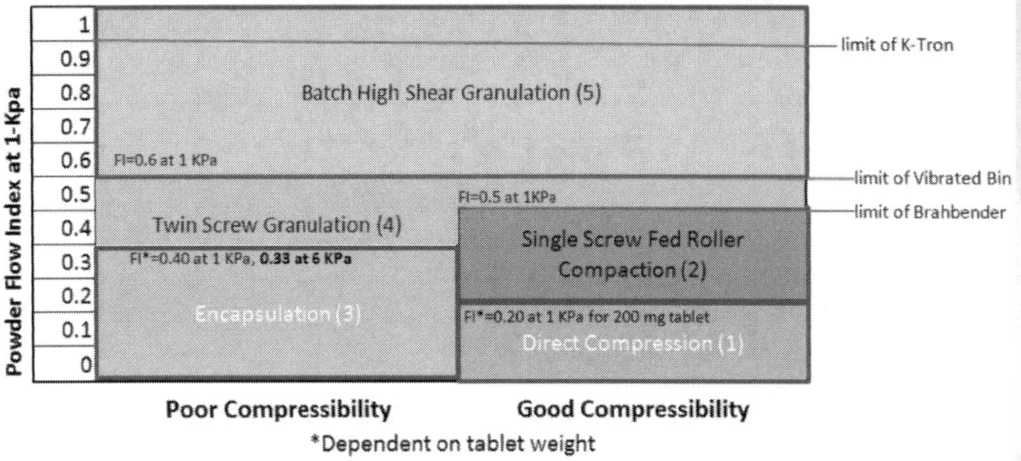


The use of a shear cell, and a compaction simulator, can provide product teams with useful information to designate the process needed, or alternatively provide feedback on formulation options to optimize the product for a given process.

## Acknowledgements


- Robert Yule - GSK
- Sayantan Chattoraj - GSK
- Laura Wehmeyer - GSK
- Fran Muller - GSK
- Vicky He - GSK

# 433552 High-Throughput Biomimetic Assay Designed to Quantify Antimalarial Efficacy

**Abstract:**

**High-throughput Biomimetic Assay Designed to Quantify Antimalarial Efficacy**

*Megan A. Ketchum, Jeffrey D. Rimer, Peter G. Vekilov*

*Department of Chemical and Biomolecular Engineering, University of Houston, 4800 Calhoun Rd., Houston, TX 77204*

Crystallization is associated with the pathogenesis of several pathological and infectious diseases in humans. In malaria, parasites utilize crystallization to prevent eradication by sequestering toxic hematin (released by the catabolization of hemoglobin) into innocuous crystals termed *hemozoin*. The sequestration of hematin into hemozoin is the target of numerous antimalarial drugs that inhibit crystallization, thereby increasing hematin supersaturation and eliminating the parasite from the body due to the cytotoxicity of hematin.[1] Parasite resistance to current antimalarials is spreading, fueling the demand for the development of new antimalarial drugs. Assays play an integral role in the discovery of new pharmaceutical drugs for diseases. High throughput assays are used by companies to test large libraries of drugs containing thousands of compounds to identify lead candidates. Here, we discuss a new method to screen for effective growth inhibitors of hematin crystallization. A relevant assay must mimic the parasite digestive vacuole (DV), which is a complex environment that is comprised of membrane interfaces, an acidic aqueous solution with pH 4.8 – 5.5, and lipids. While there is still debate regarding the medium (aqueous or organic) from which soluble hematin reaches the crystals and associates to them, we have provided convincing evidence that the lipid environment is crucial for crystal growth.[2, 3] To this end, we use a two-phase system of citric buffer saturated octanol (CBSO) to emulate physiological conditions in the DV. We designed an assay for quantifying hematin crystallization in the presence of putative growth modifiers to screen molecules as potential antimalarial drugs. This assay successfully determines IC
new drugs. The use of biomimetic solutions also provides a method of analyzing the fundamental mechanism(s) of drug-crystal interactions, which may prove to be an integral component of drug design that is capable of circumventing combinatorial approaches used to select and screen potential drugs.

References

1. Olafson, K.N., M.A. Ketchum, J.D. Rimer, and P.G. Vekilov, *Mechanisms of Hematin Crystallization and Inhibition by the Antimalarial Drug Chloroquine*. Proceedings of the National Academy of Sciences U S A, 2015. **112**: p. 4946-4951.

2. Ketchum, M.A., K.N. Olafson, E.V. Petrova, J.D. Rimer, and P.G. Vekilov, *Hematin crystallization from aqueous and organic solvents*. Journal of Chemical Physics, 2013. **139**(12).

3. Olafson, K.N., J.D. Rimer, and P.G. Vekilov, *Growth of Large Hematin Crystals in Biomimetic Solutions*. Crystal Growth & Design, 2014. **14**(5): p. 2123-2127.

# 424089 Utility of DNA Capture Resins in the Development of Minimally Invasive Tools for Genetic Monitoring of Pancreatic Health

**Abstract:**

The diagnosis of pancreatic cancer carries an especially grim prognosis, with a median patient survival of less than 6 months, and an average 5-year survival of <5%. One reason for this poor outlook is the fact that pancreatic cancer is often diagnosed very late in the cancer development, resulting in the existence of distant metastases in more than one-half of patients at the time of diagnosis. There is a significant need for effective ways of detecting this cancer, early in its progression, which could lead to the evaluation and identification of better treatments for this deadly disease. Because ~95% of pancreatic cancers are exocrine in nature, pancreatic fluid represents a rich source of potential biomarkers that could lead to earlier diagnoses, and better clinical outcomes. However, the only method that is currently available for sampling pancreatic fluids is endoscopy, which is both time-consuming and highly invasive.

We are engineering a tool that would allow for the sampling and genetic screening of pancreatic fluids in a minimally invasive manner. We have identified materials that efficiently capture DNA from raw, untreated pancreatic fluid, thereby allowing it to be purified for downstream analysis. We have also performed time-course studies to evaluate fluctuations in pancreatic fluid DNA content upon secretin-stimulated secretion. These samples are now being utilized, in conjunction with next-generation sequencing, to identify resident genetic mutations. Here, I discuss these results and also provide data on our ability to detect rare mutant alleles within these fluid samples.

# 422060 Fluorophore-Gold Nanoparticle Contrast Agent for Specific and Sensitive MMP-14 Detection

**Abstract:**

Matrix metalloproteinases (MMPs) are a family of endo-peptidases that contain zinc ions in their structures and are highly expressed in cancer. MMPs play crucial roles in the invasion, metastasis, proliferation, and development of cancer cells. MMP-14, one of the MMPs, is a membrane bound enzyme and is expressed at a high level in several aggressive cancer types, particularly in aggressive breast cancer (e.g. MDA-MB 231). It is known to promote the cancer cell detachment from the primary site, degrades and rebuilds the extracellular matrix (ECM), and stimulates tumor angiogenesis. MMP-14 also activates other MMPs, particularly proMMP-2, and is involved in the activation of extracellular signal-regulated kinase (ERK). Thus, specifically and sensitively detecting MMP-14 is important for early diagnosis of aggressive breast cancer, and specially designed, fluorophore mediated molecular sensing may be a good way to do so.

One way to increase the specificity of the fluorophore mediated molecular sensing is designing the contrast agent to emit its fluorescence only when this agent encounters the biomarker of interest. Increasing sensitivity of the contrast agent can be done by enhancing the fluorescent emission level per fluorophore.

Locally modulating fluorescent emission level of a fluorophore can be achieved by properly utilizing the electromagnetic (EM, plasmon) field generated by the gold nanoparticle (GNP) because GNP is small and can form a strong EM field around it. The EM field strength is highest on the GNP surface and exponentially decreases with the distance from the surface. Therefore, the fluorescent emission level can be controlled by appropriately varying the distance between the fluorophore and GNP. When the fluorophore is placed on the surface or very close to the GNP, its fluorescence is quenched significantly. When it is placed at an appropriate distance from the GNP, its fluorescence can be enhanced extensively. With this relationship between the EM field strength generated by a GNP and fluorescent emission level of a fluorophore, we have been developing a fluorophore contrast agent that is highly specific and sensitive for MMP-14 detection.

The fluorophore-GNP contrast agent for MMP-14 consists of a GNP, a near-infrared red (NIR) fluorophore Cypate (Cy), and spacers that link the GNP and Cy. An NIR fluorophore is chosen for a deep penetration to the tissue with a minimal background noise caused by the tissue auto-fluorescence, most of which are in the UV or visible wavelength. The GNP and Cy are linked by two different spacer types: The short spacer (SS) contains a substrate motif for MMP-14 and it is sufficiently short to place Cy close to GNP for the fluorescence quenching. When the agent encounters MMP-14, the SS is cleaved and Cy is released from the GNP, restoring its fluorescence. The long spacer (LS) is a stable, molecular sequence with a length for the Cy to be at the distance from a GNP for its maximum fluorescence. Ultimately, these two spacers will be conjugated to a single GNP together (also with an MMP-targeting molecule) so that the resulting contrast agent generates the Cy maximum fluorescence only when it encounters MMP-14.

In our process of optimizing the SS integrity, first, the length of the peptide sequence of SS was estimated by molecular simulation. Then, candidate peptide sequences possessing the MMP-14 substrate motif were conjugated to Cy (SS-Cy). Then the resulting SS-Cys were conjugated to the GNP (GNP-SS-Cy) and tested for their fluorescence levels and the cleavability of their SSs when they meet MMP-14. The fluorescence of the optimized GNP-SS-Cy was quenched up to 85% of its free form. When adding MMP-14, the fluorescence was restored up to 90%. We are currently testing the efficacy of the selected substrate motif in various cell lines. For the LS, we chose to use an amino acid sequence. The length of the optimal LS was first theoretically estimated and commercially available peptide candidates with potential length were selected. At a proper distance the fluorescence is estimated to be enhance up to 2.5 times of that of the free form. The GNPs conjugated with the Cy linked LS (GNP-LS-Cys) are currently tested for their fluorescence level.

The resulting products of the optimized Cy-SS/LS-GNPs with targeting moieties will be able not only to detect the cancer but also to diagnose cancer type. Lately, GNPs have used for various biomedical purposes for both diagnostic and therapeutic. In our study, the plasmon generated by the GNP was appropriately utilized to develop an optical contrast providing both enhanced specificity and sensitivity. Although this particular product is developed for MMP-14, the same concept can be used for the detection of any biomarkers that changes the spacer length, including enzymes, DNA, RNA biomarkers. In addition, GNPs are natural contrast agent for X-ray and/or CT, allowing contrast enhancement for two different imaging modalities.

# Design and Simulation of an Automated Rare Blood Cell Detector

*Zhixi Qian[a], Eugene D. Boland[2], Paul W. Todd[2], and Thomas R. Hanley[1]*
*[1]Department of Chemical Engineering, Auburn University, AL 36849*
*[2]Techshot, Inc., 7200 Highway 150, Greenville, IN 47124*

## Abstract

Circulating tumor cells (CTC) plays an essential role in cancer's early detection and treatment. Meanwhile, the CTC's detection devices is limited in use due to its cost and complicity. To detect tumors at an early stage and improve patient's treatment quality and life expectancy, a point-of care cell detector can help in detecting CTC at clinics, nursing home or patient's home. Our goal is to design such a point-of-care CTC detector. The process of cell labeling and magnetic separation is simulated with the help of computational fluid dynamics (CFD) software. Models for fluid mixture as well as magnetic particles' track is set up and tested. We checked the relationship between magnetic ponderomotive force and magnet pole's size, and modified the magnetic particle velocity formula and a better description of magnetic particle's motion inside the magnetic trap. Consider a balance between and diffusion and sedimentation, we find an optimized value for the mixing process. Finally, a combine of theory calculation and laboratory experiments, the CTC detector is proved able to label and capture the minor CTCs in blood sample.

Key words: circulating tumor cells; CFD simulation; magnetic separation; cell labeling; static mixer

**Introduction**

In recent years there's an increasing demand of using circulating tumor cell (CTC) counts in cancer management. The primary objective of this research is to design, simulate and test an accurate, cost-effective, user-friendly point-of-care CTC detection device. The preliminary design includes three syringes for blood sample, buffer and reagents, a static mixer, a magnetic filter, a microfluidic "chip" and a waste receptor. The device is smaller and cheaper than traditional detection devices. Also, device operation is easier with only one operator required. The device can be used in clinics, nursing homes or residences thus allowing earlier detection of most fatal cancers, including pancreatic cancer, ovarian cancer and melanoma. Our recent research focused on the analysis and simulation of static mixer and the flow cytometer and a further analysis of factors affecting the fluid mixing.

The fluid dynamic analysis of the staged mixer was accomplished using the FLUENTTM analysis package, which performs modeling of fluid flow and tracking particles' movement in complex geometries. The simulation showed that inside the magnetic trap, the fluid performs a well-developed laminar flow. And it's estimated most magnetic particles will be captured at first pair of magnets.

An experiment was set up to test the magnetic trap's separation ability. Various conditions were tested, including flow rate, magnetic size and type. In addition, we tested the blood cell's sedimentation inside the magnetic trap. Then it's possible to optimize the device's flow rate. Hemacytometer and Hyperflux was used in analyzing particle patterns of the mixture.

**Devices and methods**

During the research task CFD software simulation and mathematic methods were applied. Computational fluid dynamic analysis of the channel design was accomplished using the FLUENTTM analysis package, which performs modeling of fluid flow and tracking particles' movement in complex geometries. With the help of ICEM CFDTM, a mesh file of the channel geometry was developed, and a laminar flow model was employed for calculation. For the simulation of magnetic particles, the channel was simplified to a 2-D rectangle shape. A discrete model was applied to simulate the particle movement. Tracking of particles movements was performed to show the particle trapping efficiency.

The trajectories of magnetic particles were simulated by choosing a set of coordinates within the channel and adding the flow and magnetophoretic velocity vectors as a function of position in the mesh. At each step in time the velocity is recalculated to account for simultaneous changes in streamline velocity, dy/dt, and magnetophoretic velocity, dx/dt.

To test the magnetic trap's capture efficiency, a hemacytometer is used in counting and calculation of magnetic particle captured. The red blood cell (RBC) and magnetic beads are put in sample tubes. After counting the origin concentration, the sample containing tumor cells is diluted and changes its concentration to a value closer to actual condition. Then the tumor cells will get labeled by magnet beads inside the static mixer. The mixture will then flow through the magnet trap, where tumor cells will stayed inside with free-moving magnet beads. The origin sample, the captured cell and escaped fluid will be analyzed in Hyperflux to check for its magnetic mobility and sizing information as well.

Magnet beads used in the testing are Dynabeads® Biotin Binder (InVitrogen/Dynal), the producer InVitrogen in Oslo, Norway and the catalog number is 11047. The magnet bead's magnetophoretic mobility range about 1.3~2.0*10-11m3/TAS and the concentration of beads is 4*108 beads/mL in the original product.

The Hyperfluxtm Velocimeter system is a particle analyzer and velocimeter that is used to measure the speed of movement of cells based on magnetophoretic mobility. It is driven by a calibrated magnetic field within the device. In addition to measured speed, the Hyperfluxtm also measures and generates 22 other characteristics about each identified particle including: size, sedimentation rate, shape, and intensity. The image sets created by the high-definition video camera are stored in memory can be used for processing and data evaluation. These image sets were captured on using the IKOVision program with specifically selected particles and operation conditions.

The device's magnetic separation ability was tested using known labeled target cells. The magnetic beads and cells were mixed in centrifuge tubes on a shaking bed for 20 minutes at 23 C. Thus all target cells are considered as labeled.

First we use a high concentration cell sample that will provide an indication of the magnetic trap's performance. It's expected that the device is able to keep a relative high and steady capture efficiency at a high tumor cell concentration. Then we dilute the cell suspension to the lowest practical concentration and check the separation patterns using hemacytometer counts and with the help of Hyperflux system. The system can check the diameter and mobility of particles, thus the difference of magnetically labeled cells and unlabeled cells can be understood.

Then we attach the static mixer upstream of the magnetic trap to test whether the system can successfully perform a test from labeling cells to separating tumor cells. After proving performance, whole blood is used in the final test, thus leading to a conclusion concerning potential clinical performance.

**Results and discussion**

At low cell concentration, the capture ratio cannot have a precise value owing to the low number of cells available in the hemacytometer, which was the rationale for the research at high cell counts described in the previous section. The starting and separated samples from these tests were subjected to detailed mobility analysis using the Hyperflux Velocimeter.

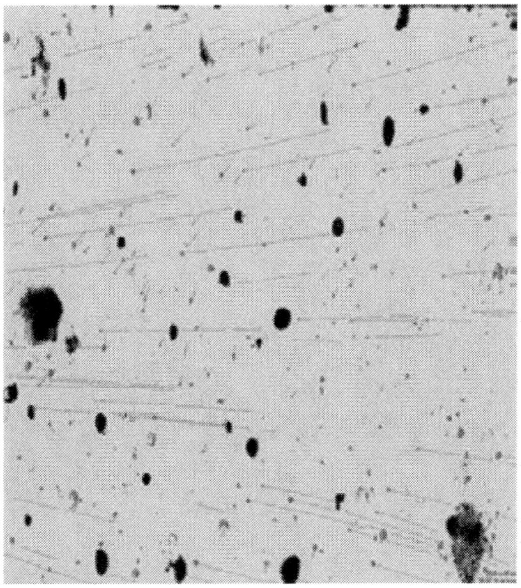

Figure above is the track screenshot for original sample. The red dots with long tracks are particles with high mobility, including cells labeled by magnet beads, and the red dots with short blue lines are particles with low mobility, which are not affected by the magnetic force.

n the starting sample, the ratio of underflow particles (those with zero or less velocity) is 29.1% of the analyzed tracks. In the 2-dimensional graph of diameter vs. mobility in Figure 5-9 there are points gathered around the y-axis (with a mobility less than 0.002), and at the right side of the xy-plot, the scattered dots represent labeled cells and magnetic beads which haven't labeled cells.

The track images for cells that were collected by the magnetic trap are shown below:

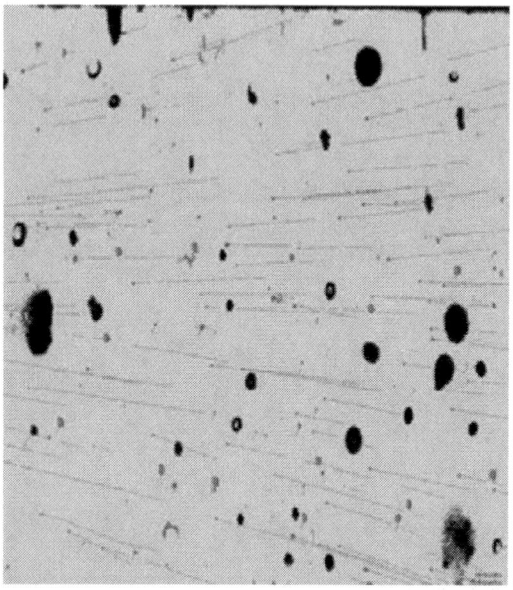

Compare with origin sample, less low mobility particles are tracked and more high mobility cells are observed.

And for the waste collected, the track is mainly from the particle with low mobility which is shown in fig below:

Fig 5-14: track images for waste of sample1

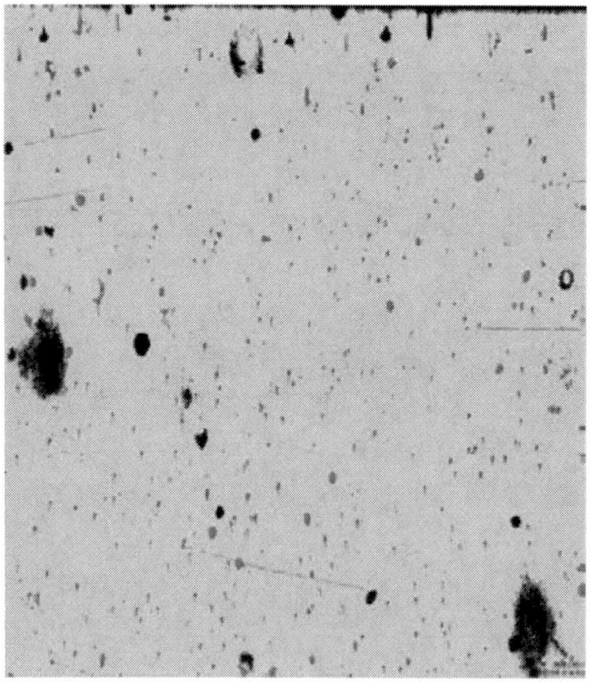

And the Hyperflux peak also showed a scattered peak in fig 5-15. Which comes mainly from the first peak in origin sample.

From the chart, there's a plenty of unlabeled particles in the part of waste, a combine of waste fluid and collected cell.

All these indicated the magnet trap's separation process. Three different types of particles are involved: particles with low mobility and varied diameter, which is unlabeled cells, they are mostly gathering in the waste after separation; particles with high mobility and low diameter, they are usually unattached magnet beads and will gather at collected sample. The labelled cells have a large diameter and sometimes even higher mobility as there're multiple magnet beads attached on the cells.

The result of Hyperflux tracking showed the magnet bead's mobility is between $1.3\sim2.0*10^{-11}m^3/TAS$, which is a lower value than in chapter 3, so the simulation result is different as well:

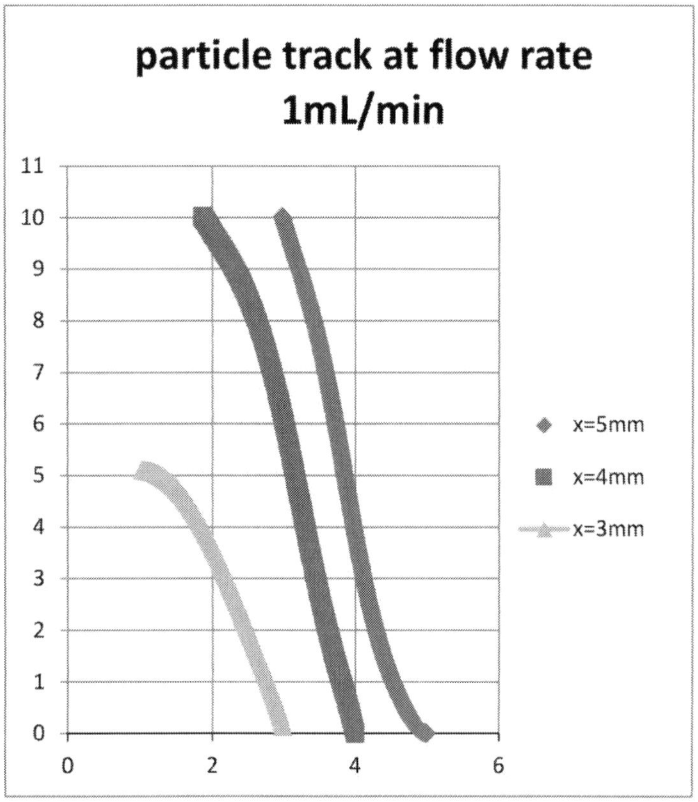

The first stage of beads can't capture all beads inside the trap, however, the beads which have a distance from magnet less than 2mm can be captured. After 1-stage magnet capture, those beads at x=5mm has moved to position of x=3mm, which will able to be captured at the second stage. So the 3-stage magnet ensures the capture ability.

**Conclusion**

The result of high concentration capture showed the device's ability in handling large amount of tumor cells. In blood sample, the tumor cell's concentration is less than 1000 cells/ml, so the test condition with $10^6$~$10^7$ cells/ml is greatly exceeding the upper limit of the magnet trap. Still, the device can stably capture over 60% of labeled tumor cells. This proves the number of labeled cell device can capture is over $10^6$ cells, which is greatly larger than the amount of cells actually in blood sample. And this showed the system is adequate in performing the task of cell separation.

The analysis result from the Hyperflux showed difference particles and their properties in the mixer. Unlabeled cells are varied in size and have a mobility around 0, free-moving magnet beads have a small size and high mobility, and the labelled cell have a larger size than free-moving magnet beads and the highest mobility. In the analyze report of magnet beads, we may see the origin sample contains all three types of particles. After separation. Most particles with low mobility is flowing through the separator and into the waste flow, the captured cells and free-moving magnet beads stay inside the device. The Hyperflux result proves the device can separate the labelled cells from unlabeled one with a good selectivity.

The result of magnetic bead's track showed difference type of magnetic beads can determine the performance of separation. A higher mobility results in a larger drag force performed on the cell. The multistage design proves necessary in supporting a wider range of magnet bead's choice. In this test, the labelled cells are collected within the first two sets of magnetic poles, indicating labelled cells labeled with lower magnet mobility can still be captured by the magnetic trap.

## References

Afshar, R., Moser,Y, Lehnert, T and Gijs, M. A. 2011. Three-dimensional magnetic focusing of superparamagnetic beads for on-chip agglutination assays, Annul. Chem., 83(3) : 1022-1029.

Allard, W. J., Matera, J., Miller,M.C., Repollet, M., Connelly, M. C., Rao, C., Tibbe, A. G. J., Uhr J. W and Terstappen, L. W. M., 2004. Tumor cells circulate in the peripheral blood of all major carcinomas but not in healthy subjects or patients with nonmalignant diseases, Clin. Cancer Res.10: 6897-6904.

Cancer Data and Statistics, http://www.cdc.gov/cancer/dcpc/data/

Etzioni, R., Urban N. and Ramsey S., 2003. The case for early detection, Nature Reviews (Cancer), 3: 1-10.

Fan, T., Q. Zhao, J. J. Chen, W-T. Chen and M. L. Pearl, 2009. Clinical significance of circulating tumor cells detected by an invasion assay in peripheral blood of patients with ovarian cancer, *Gynecologic Oncology*, **112**, 185-191

Haun, J. B., Yoon, T-J. , Lee H. and Weissleder, R. 2010. Magnetic nanoparticle biosensors, Nanomedicine and Nanobiotechnology, 2: 291-304.

Hsieh, H. B., Marrinucci, D. , Bethel, K. , Curry, D. N. , Humphrey, M. , Krivacic, R. T. , Kroener, J. , Kroener, L., Ladanyi, A. , Lazarus, N, Kuhn, P. , Bruce R. H. and Nieva, J. 2006. High speed detection of circulating tumor cells, Biosensors and Bioelectronics, 21(10): 1893–1899.

# 434587 Biofilm Mitigation on Implanted Devices

**Abstract:**

Each year in the United States thousands of people have medical implants removed due to bacterial biofilm infections obtained during implantation. These biofilms are notoriously difficult to treat for a variety of reasons, including transport resistance from the extracellular polymeric substance, changes in gene regulation, and decreased metabolism from nutrient limitations. Despite decades of work on antifouling surfaces and antimicrobial agents, these infections typically require invasive explantation and re-implantation surgeries coupled with long-term antibiotic treatments. One promising new approach to address this problem is *in situ* thermal mitigation. In this study, *Pseudomonas aeruginosa* biofilms were subjected to temperature shocks at 50°C, 60°C, 70°C, and 80°C with controls run at 37°C for an exposure times of 1, 2, 5, 10, 15, 20, and 30 minutes. To guarantee instant, uniform thermal exposure these shocks were performed by immersion and removal from pre-heated wells, though our lab is currently developing magnetic/polymer composite coatings which can supply the thermal shock using an externally applied magnetic field. Following thermal shock, the remaining viable colony forming units (CFU) were quantified by serial dilution and plating of the homogenized biofilm. The resulting bacterial load in $CFU/cm^2$ fit the following equation:

$$\log(CFU/cm^2) = \log(CFU/cm^2)_0 - [0.079 + 0.044 \log(t)] * (T - 37)$$

where $\log(CFU/cm^2)_0$ is the initial bacterial load, t is the exposure time, and T is the thermal shock temperature. However, biofilms cultured under different conditions showed a difference in susceptibility to the heating, indicating that different areas of the body could lead to different needs for successful treatment. This work is the first step in a new approach to the goal of eliminating a major liability of implanted medical devices and the billions of dollars spent on explantation, recovery, and re-implantation.

# 434042 Second Window Near Infrared Fluorescent Imaging for Deep Tissue Animal Imaging

**Abstract:**

Biomedical imaging is an integral part of clinical decision-making during screening, diagnosis, staging, therapy planning and guidance, treatment and real-time monitoring of patient response, because of its ability to provide morphological, structural, metabolic and functional information at various spatio-temporal scales of interest, while being a minimally invasive and highly targeted source of physiological evidence. The main clinical challenge, however, is to be able to image biological features with sufficient sensitivity for detection at the cellular level (in a complex tissue environment at deep penetration), which could then be used to identify and treat small tumor masses before the angiogenic switch growth phase, which is below the threshold of detection of most current imaging technologies. There are few main imaging modalities available clinically, classified according to the image contrast mechanism: X-ray (2D film imaging and computed tomography, CT), positron emission tomography (PET), single photon emission computed tomography (SPECT), magnetic resonance imaging (MRI), ultrasound (US), intravital microscopy (confocal, multiphoton), optical imaging, or some combination of these. Despite the significant advances in both the imaging instrumentation and algorithms for image processing, most of the aforementioned techniques suffer from poor sensitivity and/or resolution and/or depth of focus in the three spatial dimensions (3D). The most promising technique for high-resolution deep-tissue whole body imaging, using relatively safe molecular probes and excitation sources, at a reasonably low cost, appears to be optical imaging.

In recent years, there has been tremendous interest in the exploration of optical imaging *in vivo*, due to the development of near-infrared fluorescent probes, effective targeting agents, and custom-built imagers. Particularly, NIR light has been theoretically predicted to penetrate through ~ 10 cm of human tissue, with simulations suggesting a signal-to-noise (*S*/*N*) improvement of over 100-fold by imaging in the second near-infrared window (NIR-II: 900 - 1,400 nm) compared to the first window (NIR-I: 650 - 900 nm). Despite the recent development of various NIR-II fluorescent probes and custom built imagers based on InGaAs detectors, there are three significant challenges to imaging in the NIR-II wavelength domain with InGaAs detectors. The first challenge is the limited selection of fluorescent probes which emit in the NIR-II range favorable for medical imaging. Similarly to NIR-I or visible fluorescent imaging, only part of the fluorescent signal is utilized for contrast imaging to avoid spectral overlap with autofluorescence. The second challenge is to maximize the *S*/*N* ratio for high-sensitivity deep-tissue imaging. Due to the intrinsic low SNR of InGaAs detectors compared to Si detectors, strategies are required to depress SNR of InGaAs detectors by either imaging methodologies or data processing algorithms. The third challenge is diffuse light scattering by heterogeneous turbid biological media, imposing trade-off between depth and resolution.

In this work, we designed an optical system in NIR-II window that combines both hyperspectral and hyperdiffuse measuring. HSI resolves the first two challenges of SNR by providing information in frequency domain, and allows novel type of investigation (including pixel-wise spectral analysis and 3-D reconstruction) as well as improves result confidence. HDI resolves the third challenge, by not only excluding the emission scattering in the processed results, also presenting pixel-wise diffuse scattering information for contrast imaging and 3-D reconstruction. The hyperspectral and diffuse imaging system (900-1700 nm) has the capability to distinguish the optical signatures of the primary pump laser, background, various types of tissue autofluorescence, reporter fluorescence, as well as the diffuse scattering effect of the fluorescence signal upon transport through heterogeneous turbid optical media. Further by applying our imaging system to various tissue and whole-animal, we demonstrated that it is capable of (0) no background or autofluorescence information is required to be collected beforehand, and instead, these background and aufluorescence information is proved to be useful for both diagnostic and localization reasons; (1) detecting 1 mm sized particles up to 9 cm depth through a breast-tissue mimic phantom; (2) identifying spectral and diffuse signature of various tissues; (3) reconstructing 3-D fluorescent images based on newly interpreted information from HSC and HDC; (4) identifying important organs using label-free reflection mode hyperspectral imaging for feature location and registration. This study opens up exciting new possibilities for clinical translation of NIR-II imaging as a viable platform theranostic technology; for early diagnostics, as a real-time surgical assistance tool, and for monitoring patient response to therapies.

# 433393 Specifically Tuned Light Activated Nano-Therapeutics for Selective Killing of Multi Drug Resistant Bacterial Strains

**Abstract:**

The recent surge of drug-resistant superbugs and shrinking antibiotic pipeline are serious challenges to global health. Here we present a novel nanoparticle-based therapy where the nanoparticles specifically target selective cell phenotypes based on their biochemical potentials. Using a specific light induced targeting mechanism, these light-activated reactive species (LARS) have the capability to target and kill specific-pathogens on illumination with visible light, without affecting the growth of other cells. While the growth of selected cells and the targeted pathogens is unaffected in dark, we demonstrate that LARS can eliminate a broad range of multi-drug resistant bacterial strains including extended spectrum beta lactamase *Escherichia coli*, carbapenem resistant *Klebsiella pneumoniae*, and Methicillin resistant *Staphylococcus aureus*. Furthermore, we demonstrate a potential therapeutic intervention with co-culture of *E. coli* and HEK293 cells, providing evidence that LARS is capable of removing an infection while leaving mammalian tissue intact. Finally, we designed different LARS species which are capable of promoting proliferation of targeted cells when induced with light. The ability to tune the biochemical potential of nanoparticles allows for selective cell phenotypes over a range of organisms, and has potential applications in antimicrobials and targeted organism-specific therapeutics.

# 429152 Increase the Feeding Value of Corn Ethanol Coproducts Via Genetic Engineering of Yeast

**Abstract:**

With fast growth in recent several years, the U.S. corn ethanol industry provides enormous amount of co-products, for instance, distiller dried grains with solubles (DDGS), to the feed market, making the ethanol industry as one of the largest contributors to the US feed supply. Although DDGS is a good source for protein, the low content and low digestibility of several essential amino acids (lysine, methionine, threonine, arginine and tryptophan) is a limiting factor when considering the inclusion of DDGS in animal diets and its impact on animal performance. High inclusion of DDGS in animal diets causes amino acid imbalances with subsequent consequences on animal performance and excess excretion of nitrogen into the environment. Addition of synthetic sources of amino acids can alleviate amino acid imbalances and result in significant decreases in the nitrogen discharge and manure odor. Commercially available synthetic amino acids for feed supplementation are only available for some amino acids while costs can fluctuate if supplies are short. Many amino acids remain too costly for commercial feeding purposes. Regardless of the availability of amino acid supplements, the value of an ingredient such as DDGS is improved if the content and digestibility of limiting amino acids such as lysine, tryptophan and arginine are increased. *Saccharomyces cerevisiae* has been considered as a natural food product. Non-living dry yeast powder can often be used as a food supplement for human nutritional and pharmaceutical purposes. In corn ethanol co-products, yeast biomass is an important composition and is expected to provide more of these limiting amino acids. Although genetic modification of yeast has been intensively carried out to improve the baker's yeast for ethanol production and some of the engineered strains have been implemented and commercialized for industrial ethanol production, no genetic engineering research has been specifically focused on the animal feed improvement for the corn ethanol coproducts. Using genetic manipulation to engineer yeast strains is considerably easier than engineering feed grains while engineered yeast is also more likely to be adopted by the industry. The goal of this study is to significantly increase the feeding value of corn ethanol coproducts through genetically engineering the yeast strain to enrich several key amino acids (first limiting lysine, commercial costly tryptophan and arginine) during the fermentation so that the yeast biomass can significantly improve the nutritional value of DDGS. Specifically, the metabolic pathway of lysine, arginine and tryptophan in *S. cerevisiae* is genetic engineered to generate the biomass with more of these amino acids. Quite different from the current mainstream emphasis on the ethanol production, this research focuses on the improvements of animal feeds generated from corn ethanol industry and will potentially benefit all sides involved in this industry, including the animal feeds and the corn farmers, and potentially decrease the environmental impacts of the whole corn ethanol industry.

# 432905 Towards Biofuel Production in Synechocystis Sp. PCC 6803: Expanding the Molecular Biology Toolbox for Pathway Engineering

**Abstract:**

Biofuel production in microorganisms has great potential to fulfill this nation's need for fungible "green" transportation fuels. Because of the simplicity of bacterial cells, photosynthetic cyanobacteria make good candidates as a platform for biofuel production. *Synechocystis* sp. PCC 6803 is an attractive production platform due to rapid doubling times, a sequenced genome and an established molecular biology toolbox. This molecular biology toolbox was established to probe the biology of photosynthesis and has not been optimized for pathway engineering which is necessary for biofuels production. This presentation will discuss our efforts to develop an alternative counter selection marker and to develop and characterize promoters for gene expression under varying conditions. The use of these tools to study and optimize biofuel production in Synechocystis will also be discussed.

## (466c) Effect of Different Feeding Strategies of Glucose and Short-Chain Carboxylic Acids on the Production of Natamycin By Streptomyces Natalensis

Elsayed, E. A., National Research Centre
Farid, M., National Research Centre
Wadaan, M., King Saud University
EL Enshasy, H., Universiti Teknologi Malaysia
Originally presented on: 11/11/2015 9:06:00 - 9:24:00

Natamycin is an industrially important tetraene antibiotic produced in submerged culture by different strains of *Streptomyces* sp. Natamycin has a wide spectrum activity against molds and yeasts with low toxicity against mammalian cells. It is used in the treatment of bronchopulmonary aspergillosis, mycotic keratitis, yeast keratomycosis and candidiasis. Natamycin is used in many food industries for food preservation and has been granted the GRAS status (Generally Regarded As a Safe) by FDA. This present work was carried out to investigate the optimization of natamycin production through different feeding strategies of glucose as a limiting substrate as well as short-chain carboxylic acids (acetic, propionic and butyric) and their effect on cell growth and natamycin biosynthesis. The optimization was performed in shake-flask level as well as 2-L stirred-tank bioreactor. Cultivations have been evaluated under different feeding strategies of glucose as a limiting substrate, as well as different short-chain carboxylic acids (acetate,propionate, and butyrate) . The obtained results showed that the highest natamycin level was obtained at 20 $g.L^{-1}$ glucose. Moreover, increasing glucose concentration resulted in a significant decrease in natamycin level. In addition, natamycin decomposition was observed at lower glucose concentrations (10-20 $g.L^{-1}$) after glucose depletion. The addition of acetic and propionic acids at concentrations below 2 $g.L^{-1}$ at cultivation start showed stimulatory effects on natamycin production. Furthermore, the addition of acetic and propionic acids in a mixture (7:1) at a total concentration of 2 $g.L^{-1}$, increased antibiotic production significantly. Conclusively, the addition of short chain carboxylic acids increased antibiotic yield and decreased the production time.

# 433138 Systematic Characterization of a Panel of Strong Promoter/Terminator Pairs to Facilitate Scheffersomyces Stipitis Engineering

**Abstract:**

Systematic Characterization of a Panel of Strong Promoter/Terminator Pairs to Facilitate *Scheffersomyces stipitis* Engineering

*Scheffersomyces stipitis* is an unconventional yeast that has the highest native capability for xylose fermentation, therefore holding a lot of potentials in lignocellulosic biomass conversion. However, its broad application as a popular host in biorenewables field is limited by the lack of genetic manipulation tools. Here, we report the identification and characterization of a panel of strong constitutive promoters and terminators. Based on RNA-seq analysis of *S. stipitis* growing in both mixed sugars and single sugar conditions, nine promoters were selected for further comprehensive characterization under conditions with various sugar and oxygen levels. The strengths of majority of these promoters were more than 10-fold higher than that of the commonly used TEF1 promoter, representing a collection of strongest Scheffersomyces promoters reported in literature. In parallel, the transcriptional termination efficiency of the selected terminators was studied and compared to the tight TEF1 terminator. Such a systematic characterization of potent promoters and terminators not only paves the way for manipulating pathways in *S. stipitis* to produce industry-relevant high-value biochemicals from mixed sugars, but also provides a foundation for future mechanism study of xylose uptake and conversion in the native host, which will benefit the entire biorenewables field.

# 424924 Isotopically Nonstationary 13C Flux Analysis of Isobutyraldehyde Production in Synechococcus Elongatus

**Abstract:**

Recent studies have demonstrated the feasibility of converting energy from sunlight and carbon from $CO_2$ directly into biofuels using photosynthetic microorganisms. Despite the advances made in cyanobacterial biofuels production, the productivities achieved are currently too low for industrial feasibility and few tools are available that specifically address the challenges of determining and redirecting metabolic flux in photosynthetic microbes. Our group is developing novel approaches that use $^{13}C$ metabolic flux analysis (MFA) to quantitatively assess in vivo metabolic phenotypes of photoautotrophs. Previously, we mapped carbon fluxes in the cyanobacterium *Synechocystis* sp. PCC 6803 by applying isotopically nonstationary MFA (INST-MFA), which involves model-based regression of transient $^{13}C$-labeling patterns of intracellular metabolites. The flux analysis revealed unanticipated photosynthetic inefficiencies tied to oxidative metabolic pathways, despite minimal photorespiration.

We have now applied a similar modeling approach to mapping photoautotrophic metabolism in another strain of cyanobacteria, *Synechocococcus elongatus* PCC 7942. Metabolism was quantified in the wild-type strain of *S. elongatus,* as well as a strain engineered to produce isobutyraldehyde (IBA, a direct precursor of isobutanol). GC-MS profiling of isotopically labeled intracellular metabolites, growth, and IBA production were used to create flux maps of central carbon metabolism in the WT and IBA-producing strains. The flux analysis results quantitatively describe increased flux through phosphoenolpyruvate carboxylase (PEPC) rather than pyruvate kinase (PK) in both studied strains, with an increased overall PEPC flux in the IBA-producing strain. This pointed to an alternative carbon route to pyruvate formation, a major precursor to IBA (phosphoenolpyurvate → oxaloacetate → malate → pyruvate → → → IBA).

Based on these results, we generated three cyanobacterial single gene overexpression strains in the IBA-producing background for reactions involved in the pyruvate formation bypass: IBA/PEPC$^{ox}$, IBA/MDH$^{ox}$, and IBA/ME$^{ox}$. We found that the IBA/ME$^{ox}$ strain showed a significant improvement in IBA production, while maintaining a growth rate comparable to its IBA-producing parental strain. We also overexpressed pyruvate kinase in the IBA-producing background strain, but found that IBA production did not increase and the growth rate actually decreased.

# 434294 2D and 3D Biomaterial Platforms for Cancer Drug Screening

**Abstract:**

The *in vivo* tumor microenvironment is composed of both various cell types and the extracellular matrix (ECM), which is known to provide the mechanical support and chemical cues to regulate tumor cell growth, motility, and their responses to chemotherapeutics. Biomaterials have been widely used to recapitulate the ECM for studying of cell-matrix interactions in different types of cancers, and different 2D or 3D high-throughput biomaterial platforms have been developed as a tool to predict tumor cells' response to chemotherapeutics. Although the 3D platforms are considered to be ideal for mimicking the complex 3D tissue architecture, they are usually difficult to fabricate and scale up for high-throughput drug testing as opposed to 2D models. We have taken a holistic approach and evaluated drug response across various drugs with different cell lines on plastic surfaces, on 2D gel surfaces and within 3D gels (either as single cells or spheroids) with different stiffnesses to identify the appropriate model for each type of drug mechanism.

We used a high-throughput PEG-PC hydrogel biomaterial platform as a 2D biomaterial drug screening system. Briefly, we co-polymerized poly(ethylene glycol) dimethacrylate (PEG) and the zwitterion 2-Methacryloyloxyethyl phosphorylcholine (PC) to form an optically clear hydrogel via radical polymerization in a silane-treated glass 96-well plate, and the hydrogel modulus can be controlled by varying the PEG crosslinker content. For the 3D biomaterial, we used 3D poly(ethylene glycol)-Maleimide (PEG-Mal) gels. PEG-Mal prepolymer solution, which contains either single cells or spheroids, was crosslinked by PEG-dithiol, which was dissolved in triethanolamine, and the PEG-Mal concentration was varied to control the modulus. For spheroid formation, single cells were encapsulated and cultured in poly(N-isopropylacrylamide) (pNIPAAm) hydrogels for 14 days to form multicellular spheroid structure. We have used these systems to test breast carcinoma cells SkBr3 and MDA-MB-231's response to a common chemotherapeutic drug (doxorubicin) and three different targeted drugs (sorafenib, lapatinib, and temsirolimus), which inhibits receptor tyrosine kinases and their downstream signaling pathways, and we use a proliferation assay to quantify the cells' inhibitory concentration at 50% or IC50.

Thus far, for most of tested drugs, we have observed that both cell lines' drug responses are more sensitive to the change in stiffness on 2D PEG-PC gels, with higher IC50s on stiffer gels, compared to 3D PEG-Mal gel. For both cell lines, their IC50s are more sensitive to the change in testing platforms when treated with sorafenib and lapatinib as opposed to treating with temsirolimus and doxorubicin, and SkBr3 cells' drug response is more sensitive than MDA-MB-231 cells. We also observed that encapsulating SkBr3 spheroids resulted in a reduction in phosphorylation level of HER3 while there was not a significant change in HER2 phosphorylation. We are currently investigating the mechanisms behind these differences across these platforms. We propose that this approach will identify appropriate models for different type of drugs, therefore reducing the unnecessary use of complex models, saving cost, and facilitating the drug development.

# 419954 Three-Dimensional Cell Culture Platform for High-Throughput Toxicity and Differentiation Screening of Human Neural Stem Cells

**Abstract:**

Drug discovery and toxicology use animal models for many purposes such as identifying therapeutic targets, assessing the efficacy of drug candidates and testing the safety of potential therapeutics and commercially available chemicals. While animal studies are valuable for these purposes, they are associated with high costs, low-throughput and results that do not translate well to humans. There is, therefore, significant interest in developing *in vitro* systems that use human cells to reduce costs, increase throughput, and improve reliability and predictability in drug and toxicity screening processes. Importantly, differences in toxicity between different types of human cells, such as stem cells compared to terminally differentiated cells, is largely unstudied. To this end, we have developed a microarray chip (microchip) platform for three-dimensional cell culture and high-throughput screening. Human neural stem cells (hNSC) were grown and differentiated on the microchip for periods up to two weeks and used to screen for chemical toxicants or modulators of differentiation. The cytotoxic dose responses and $IC_{50}$ values for 12 chemicals was measured for hNSC and their differentiated progeny on the microchip platform. Lower $IC_{50}$ values were observed for all-*trans* retinoic acid, 5-fluorouracil and cytosine arabinoside against hNSCs in comparison to their differentiated counterparts, which were predominantly astrocytic. This work demonstrates progress in developing *in vitro* screening systems for predicting human adult and developmental neurotoxicity. In addition, this work begins to address how different cell types have differential sensitivities to chemicals, which should be considered when performing screens.

# 433581 High-Throughput Amikagel Platform for Identification of Synergistic Treatments Against Tumor Dormancy

**Abstract:**

United States statistics show prostate cancer (PCa) as one of the leading cancers in male population in 2014 and causing an estimated 30,000 deaths [1]. A significant proportion of patients experience cancer metastasis to distant secondary disease sites such as bone. PCa shows unique tropism towards bone [2]; almost 66% of PCa patients experience bone metastases in some form or other, making it a crucial prognostic factor [3]. Often in these cases of prostate metastasis, tumor cells initially undergo prolonged periods of dormancy, which are followed by relapse [4]. Increasing evidence points towards a definitive phase of prolonged G0/G1 arrest rather than balanced proliferation to support prolonged viability before relapse [5]. Ablation of these cancer cell systems in their dormant state can not only eliminate the disease, but also reduce the risk of tumor relapse and death.

We have previously developed a high-throughput hydrogel system ('Amikagel' system) that captured prostate tumor dormancy in high-throughput. Dormant prostate cancer cells showed very high resistance to conventional cancer chemotherapeutics such as mitoxantrone and docetaxel. Here, we report that significant cell death in dormant prostate cancer cells was achieved by targeting the protein folding pathway in the endoplasmic reticulum of the cells. Targeting the endoplasmic reticulum in concert with cellular adaptations against unfolded protein response, allowed for identification of drugs that can synergize to induce dormant prostate cancer cell death. Our results suggest a new paradigm of drugs that can synergistically induce death in the dormant prostate cancer cells by targeting the endoplasmic reticulum and unfolded protein response. The talk will discuss further amplification to the synergy by modulating cellular ion levels. We show that high-throughput Amikagel platform captures tumor dormancy for large scale phenotype specific drug screens for lead identification and formulation.

[1] Siegel R, Ma J, Zou Z, Jemal A. Cancer statistics, 2014. CA: a cancer journal for clinicians. 2014;64:9-29.

[2] Tombal B, Lecouvet F. Modern detection of prostate cancer's bone metastasis: is the bone scan era over? Advances in urology. 2011;2012.

[3] Rigaud J, Tiguert R, Le Normand L, Karam G, Glemain P, Buzelin J-M, et al. Prognostic value of bone scan in patients with metastatic prostate cancer treated initially with androgen deprivation therapy. The Journal of urology. 2002;168:1423-6.

[4] Páez D, Labonte MJ, Bohanes P, Zhang W, Benhanim L, Ning Y, et al. Cancer dormancy: a model of early dissemination and late cancer recurrence. Clinical Cancer Research. 2012;18:645-53.

[5] Wells A, Griffith L, Wells JZ, Taylor DP. The dormancy dilemma: quiescence versus balanced proliferation. Cancer research. 2013;73:3811-6.

# 431632 A Pipette-Integrated Nanoliter-Volume Well Plate for Cell-Based Drug Assays

**Abstract:**

**A pipette-integrated nanoliter-volume well plate for cell-based drug assays**

Swastika S. Bithi and Siva A. Vanapalli

Department of Chemical Engineering, Texas Tech University, Lubbock, TX 79409

There is a growing interest in conducting drug screens with primary cells derived from human tissues and biofluids to predict patient outcomes. In contrast to immortalized cell lines, primary cells are a scarce resource and yet preclinical studies demand diverse assays probing specific targets, off-targets and cytotoxicity. Therefore there is a need for a technology that can screen drugs with minimal requirements on cell sample and compound volumes. Currently, multiwell plates and pipetting systems are the established methods for drug assays. However, this technology is difficult to scale down to nanoliter volumes due to fluid evaporation and pipetting errors. In recent years, drop-based microfluidics has emerged as a powerful technology to compartmentalize cells in volumes down to picoliters. Despite its great potential, drop-based microfluidics has not been configured to conduct cell based drug assays with the same ease and parallelized fluid handling capability as well plates and pipetting systems. In this study, we develop a new method to integrate pipettes into a uniquely designed microfluidic well plate for storing arrays of nanoliter droplets and demonstrate cell-based drug assays. The method involves a microchannel network that contains traps to store nanoliter-scale fluid volumes in a grid format similar to that of a multiwell plate. Both the carrier fluids as well as the assay fluids are delivered into the device using multi-channel pipettes. We find that the nanoliter volumes containing either soluble reagents or cells can be stored in a reproducible manner despite pipetting errors as the trapped droplet volume is preset by the geometry of the trap. We investigate potential loss of reagent fluid due to evaporation and find that there is negligible change in droplet volumes during 24 hours. We instrument our microfluidic well plate to an automated microscopy system and pursue cell-based assays using human leukemia cancer cell lines. Each nanoliter-scale droplet is designed to store 1 – 50 cells. Tests of cell viability show no adverse effects of the carrier fluid and system components. Cytotoxicity assays conducted with this method and the results compared to a conventional well plate assay. In summary, our method has significant potential for conducting drug screens using only as little as 200 cells and a microliter of compound volumes without any need for sophisticated flow-control interface. Such a technology is expected to find applications in testing of drugs with primary cells and patient-derived samples.

# 435314 Microfluidic Systems for High-Throughput Functional Imaging of Mechanosensing Neurons in Caenorhabditis Elegans

**Abstract:**

Mechanosensation is an important modality to sense the environment in most animals, but the exact mechanism of how information is sensed and transduced is not well understood. *Caenorhabditis elegans*, a soil nematode and an important model organism for neuroscience, is a good model for studying mechanosensory biology. *C. elegans* has gentle and harsh touch neurons, which mediate behavior responses. Conventional methods for studying mechanosensation uses painstaking laser ablation and behavior assay by eyelash touches or glued worm mechanically stimulated by stylus. The limitations of these methods include relatively low throughput and difficulty to control the exact stimulus repeated and robustly on large number of worms, so studying functions of genes or screening drugs is impractical.

To address these technical limitations, we designed a microfluidic device for trapping and delivering mechanical stimuli to single worms while monitoring their neural activity. We used this mechanosensing device to monitor genetically encoded calcium indicator GCaMP6 changes in touch receptor neurons. Reponses to anterior touch require ALML, ALMR, and AVM, while responses to posterior touch require PLML and PLMR. We show the ability of the device to deliver mechanical stimulus in a graded manner. The higher pressure generated reliable increases in the fluorescence ratio averaged over the cell body. It is indicating our controlled stimuli correspond to gentle touch typically delivered in behavioral tests by eyelash stroke. We also show that the controlled mechanical stimuli can be delivered to ALM, AVM, and PVM neurons, and calcium changes of these neurons can be recorded. To assess the effect of mechanical stimulation other than cell body, we also record along ALM processes. Calcium transients show smaller peak and faster decays. Upon repeated stimulation, many sensory systems show habituation. Further, we demonstrate how the *C. elegans* gentle touch system habituates upon repeated stimulation.

We have shown in this work that the microfluidic device is flexible in stimulating different parts of the *C. elegans* and allowing for function calcium imaging. The assay is semi-quantitative in that the intensity and duration of the stimuli can be easily controlled. We envision this device as a general mechanosensing manipulation platform for neuronal imaging that will contribute to the neuronal functional map of the *C. elegans* nervous system, as well as serving as a vehicle for discovering genes affecting the function of neuronal functions and therapeutics for neuronal diseases.

# 422738 Systems-Level Assay of Phosphoproteins Using the Microwestern Array Elucidates the Mechanism of Action for Bioactive Natural Products

**Abstract:**

**Systems-level assay of phosphoproteins using the microwestern array elucidates the mechanism of action for bioactive natural products.**

Mark F. Ciaccio, Neda Bagheri. Department of Chemical & Biological Engineering, McCormick School of Engineering, Northwestern University.

**Introduction:** Technological advances in the collection and isolation of bioactive natural products have provided a rich treasure-trove of potential drug candidates. Compounds derived from natural sources such as aspirin, morphine, and penicillin have brought invaluable improvement to longevity and quality of life. While the untapped potential of natural products seems limitless, the number of compounds that successfully pass clinical trials remains prohibitively low. An efficient pipeline to identify the mechanism of action of drug candidates can (i) identify new purposes, (ii) mitigate potential side effects, and (iii) suggest chemical combinations to improve therapeutic potential. Here we take the interdisciplinary approach of combining a systems-level biological assay, termed the microwestern array, coupled with sparse regression, to discover the mechanism of action of naturally-derived small molecules in a cancer signaling network. **Background:** New technology that allows for direct quantification of protein abundance and phosphorylation states provides a particularly useful source of data as proteins are often the functional entities of cellular decision-making processes. Because the microwestern array incorporates the separation of proteins using electrophoresis, the sizes of proteins can be cross-referenced against molecular standards, eliminating much of the uncertainty that convolutes the quantification of proteins due to non-specific antibody-antigen binding common in antibody-based assays such as ELISA or reverse phase lysate arrays. **Materials and Methods:** A non-contact arrayer was used to spot 2 nL of cell lysate from drug-treated and control stimulations on to an acrylamide gel. Each sample was printed 96 times in the configuration of a 96-well plate. After electrophoresis, the samples were transferred on to a nitrocellose membrane, clamped in a multiwell gasket, and probed with 96 distinct antibodies. Fluorescence of the signals, corresponding to phosphorylation abundance, was quantified using a 700-nm infrared scanner. From the cue-signal response matrix, directed relationships between bioactive compounds and phosphorylation sites were inferred using partial least squares regression. **Results and Discussion:** We identified PTEN as a target of the common antioxidant, n-acetylcysteine. This suggests that this commonly-mutated protein may function as a sensor for the redox state of the cell and may be used to modulate cellular response to oxidative stress. In addition we identified 4EBP1 as a target negatively regulated by wortmannin suggesting that this natural product has pleiotropic effects beyond its use as a canonical PI3K inhibitor. **Conclusions:** Use of the microwestern array technology to assay changes in phosphorylation levels after small-molecule perturbation proves to be a powerful new method to harvest the potential of bioactive natural products in drug discovery.

**References:**

[1] Harvey, Alan L., et al. "The re-emergence of natural products for drug discovery in the genomics era." *Nat. Rev. Drug Discov.* 14. (2015): 111–129.

[2] Ciaccio, Mark F., et al. "Systems analysis of EGF receptor signaling dynamics with microwestern arrays." *Nat meth* 7.2 (2010): 148-155.

# 418270 High Throughput Identification of Clinically Approved Pharmacological Agents Which Prime Cells for Efficient Nonviral Gene Delivery

**Abstract:**

Human mesenchymal stem cells (hMSCs) are multipotent stem cells that can be derived from multiple sources within the adult body. Because of their unique roles in immune regulation and wound healing, hMSCs are under much investigation for therapeutic applications involving tissue engineering, treating protein deficiencies, and cancer treatment. All of these applications would be greatly aided by the introduction of exogenous DNA to genetically guide stem cell differentiation, induce production of therapeutic proteins, or to secrete anti-tumorigenic proteins. Non-viral gene delivery, although considered much safer than viral genetic modification, remains especially inefficient in hMSCs, with efficiencies as low as 1% of cells expressing transgene. Previous studies within our lab have shown up to 13-fold increases in transgene expression by hMSCs when cells were primed by treatment with a glucocorticoid (GC) prior to transfection with DNA complexes. Using BSA-Cortisol, a protein-conjugated GC incapable of crossing the cell membrane, we have been able to confirm that this effect is mediated by binding of the GC to the cytoslic GC receptor, which through unknown mechanisms inreases total nuclear plasmid translocation.

The promising priming effects of the ubiquitously prescribed GCs on transfection efficiency of hMSCs has also led us to search for other clinically approved pharmacological agents that might have similar enhancements to gene delivery. We began this search by performing a high-throughput, high-content screen of the 727 compounds of the NIH Clinical Collection for priming effects on the transfection efficiency of HEK293T cells by both polyethyleneimine /DNA and chitosan/DNA complexes, identifying several compounds for both formulations whose priming significantly increased or decreased transfection relative to vehicle controls, including Diazepam, which was shown to increase transfection, and Epigallocatechin gallate, which was shown to decrease transfection. The screen was performed in a 96 well plate format, with each well of the compounds under test assessed for effects on transfection efficiency, cell proliferation, and cytotoxicity, utilizing high-content fluorescence microscopy imaging. With these data, in combination with annotations of each compound's effects from literature and databases, we are able to hypothesize the mechanisms by which observed increases or decreases to transfection are being achieved, including the induction of mitosis, reduction of cytotoxicity, or by exclusion, a compound specific interaction with the mechanisms of transfection. In addition, we employ bio/chem-informatics data mining and clustering approaches to correlate the data from this screen to our previous microarray studies, which compiled the gene expression profiles of successful and unsuccessful transfection, as well as existing databases and models of drug, protein, and DNA interactions, to improve our general understanding of how genomic and proteomic targets can be potentially modulated to improve non-viral gene delivery. This combinatorial screening, informatics, and drug re-purposing approach, which began with the non-obvious identification of glucocorticoid transfection priming effects, has been fruitful in identifying compounds that: illuminate pharmacologically tunable mechanisms in transfection, may enable the efficient nonviral transfection of various stem cell types, and could potentially be instrumental as adjuvants that boost transfection efficiency of nonviral gene delivery therapies in vivo to clinically relevant levels.

# 437432 Continuous Chromatography: Breaking the "Bottleneck" in Downstream Processing

**Abstract:**

Multi-column continuous chromatography has recently become an enabling technology that will "break the bottleneck" in downstream processing of bio-pharmaceuticals. This development is not only driven by the tremendous increases in expression levels in the upstream fermentations, especially for monoclonal antibodies, but also by the introduction of biosimilars/biobetters the market. Continuous chromatography has shown promises in reduction of manufacturing cost. However, up to today there has not been a reported case at the production scale. What are the remaining technical barriers when implementing the technology in the GMP environment? This presentation highlights how to overcome the major barriers when developing this technology platform.

# 418599 Design and Simulation of a Continuous Bioreactor

**Abstract:**

A novel continuous bioreactor consists of an outer shell cylinder and inner core cylinder, each with adjustable rotational velocities to control the mixing in the reactor and provide various shear environments. Inlet flows are controlled to provide additional medium and oxygen to the reactor volume. A spiroid is attached to the inside of the bioreactor to increase the gas-liquid contact area and improve oxygen transfer. This bioreactor has dramatic advantages in cell cultivation because of the high cell survival rate, enhanced growth rate, and overall high productivity.

Computational fluid dynamics (CFD) simulations were successfully performed to describe the range of flow patterns possible in this bioreactor. Simulations are conducted for multiphase fluids with various operating conditions to simulate the actual reaction environment in the bioreactor. Velocity profiles, shear distributions and turbulent energy dissipation were shown to predict the high performance of the bioreactor. In addition, various experiments along with mathematical models were used to calculate the volumetric mass transfer coefficients at different operating conditions for bioreactors with and without the spiroid. The comparison of the results supported the improved performance of the bioreactor with the spiroid. Cell lines were selected for cultivation in the bioreactor.

**Key words**: computational fluid dynamics, continuous bioreactor, oxygen transfer, cell cultivation

# 422745 New Synthetic Adsorptive Depth Filtration Media for the Primary and Secondary Clarification of Mab from CHO-S Harvests

**Abstract:**

We report the design and performance characteristics of a synthetic depth filtration media that demonstrates a greatly improved filtration capacity and decreased effluent turbidity for the clarification of mAb from CHO-S harvests. This new depth filtration media is constructed entirely from synthetic components, which include a silica-based filter aid, acrylic fibers, and a charged polymeric binder resin. When compared to commercial depth filtration media comprising mixtures of diatomaceous earth and cellulose, the synthetic depth filter media is shown to have greatly decreased organic extractibles and reduced WFI flushing requirements by TOC analysis. The synthetic depth filter media also presents a significantly increased filtration capacity versus the commercial benchmark due to the incorporation of a pre-filtration layer. As an added benefit, the silica filter aid is shown to impart a modest HCP clearance capability through an adsorption mechanism. The impact of HCP clearance on downstream process purification steps is briefly explored.

# 418751 Strategies for Implementation of a New CHO Platform Process


**Abstract:**

Platform processes are integral for the rapid development of new therapeutic molecules within the biopharmaceutical space. Since platform processes are inherently limited in their capacity for change, there are serious strategic questions related to timing and implementation of modifications to a platform. Under discussion are the implications of changing a platform process with a focus on mammalian cell culture development. This includes necessity and timing for the change, the method of exploration of new platform setpoints and conditions, qualification of the platform, and characterization of platform robustness.

A case study is presented regarding application and implementation of a new platform process. Exploratory studies were used to identify incremental improvements to an existing platform. Changes included culture feed and feed timing, gassing conditions, and seed train. When evidence was generated to support use of the new platform, a series of qualification studies on multiple established molecules were conducted to support the use of the new platform. Finally, after the platform was accepted a collection of process robustness studies were performed based on a platform risk analysis. These studies will provide the process knowledge to support a quality by design approach through both early and late stage development.

# 431560 The Impact of Cell Culture Media Variability on the Process Development of Biosimilars

**Abstract:**

Sourcing limitations and the efforts to minimize introduction of adventitious viral or prion agents in the production of biologics have over the years accelerated the adoption of serum-free cell culture media. The subsequent use of animal component free media containing plant derived hydrolysates in cell culture processes has come with observed variability in cell growth, productivity and product quality of biologics. In its purest and supposedly most consistent form, cell culture media has evolved to chemically defined media. As part of a media vendor qualification study, four proprietary media formulations manufactured by five vendors were tested for composition, particle size distribution and solubility characteristics. Media from two vendors were then used in functional cell culture testing. There was a direct correlation between the media powder particle volume mean diameter (VMD), time to solution and final media turbidity. Differences in media turbidities before and after filtration indicated particulate removal during filtration step, suggesting that some components were not going into solution and therefore unavailable to cells in the bioreactor. Differences between expected and measured levels of media components such as trace metals suggested that there were other extraneous sources of trace elements. Functional cell culture testing results showed that these variations in concentrations of media components had some product quality impact that can affect comparability. Media quality, among other process variables, needs to be controlled for consistency in cell culture process performance.

# 410968 Advances in Electron Diffraction of Three-Dimensional Microcrystals

**Abstract:**

Crystallography has been an invaluable tool for studying the function of biomolecules on the atomic scale. Because creating large well-ordered crystals necessary for X-ray crystallography is still a major hurdle, there has been a great deal of interest in developing new methods that can provide diffraction data from microcrystals. Previously, we reported on a new technique, called MicroED, which uses electron diffraction in a cryo-transmission electron microscope to obtain structural data from extremely small three-dimensional microcrystals. In this presentation I will present on the major advances since the initial work was presented. These advances include improved sample preparation methods and new protein structures, some which could only be solved by MicroED.

# 429585 Regulation of Multispanning Membrane Protein Topology Via Kinetic Annealing of the Stop-Translation Ensemble

**Abstract:**

The topology of a multispanning membrane protein is typically expected to be established during cotranslational membrane integration such that the orientation of the N-terminal transmembrane domain dictates the overall topology of the protein with respect to the cell membrane. However, this mechanism is inconsistent with the behavior of dual-topology proteins for which the overall topology of the protein can be biased via mutation of charged residues on the soluble loops. In particular, EmrE is a dual-topology protein for which the inclusion of positive loop charges, even close to the C-teminus, leads to dramatic shifts in its overall topology. In this work, we use a recently developed coarse-grained simulation model of Sec-facilitated membrane integration to investigate this behavior in EmrE and its mutants. Excellent agreement is obtained between the membrane protein topologies predicted from the simulations and those observed experimentally, suggesting that the simulations capture the underlying molecular processes that govern topological shifts. We find that, in contrast to some single-topology multispanning proteins, a large fraction of the transmembrane domains of EmrE and its mutants are misintegrated at the end of ribosomal translation. These misintegrated configurations subsequently anneal to fully integrated multispanning configurations via the flipping of soluble loops across the membrane. The energetic barriers associated with flipping the soluble loops enforce kinetic pathways that dictate the final topology of the fully integrated protein. This proposed mechanism, in which kinetic annealing of the stop-translation ensemble dictates the topology of multispanning proteins, is consistent with a range of experimentally observed features of the topogenesis of EmrE and its mutants; furthermore, the mechanism offers new experimentally-testable predictions regarding the effect of translocon mutations on topogenesis. This work provides insight into the topogenesis of dual-topology proteins, with additional relevance to the topogenesis and folding of broader classes of multispanning membrane proteins.

# 431191 GDH to Adh: Complete Redesign of Glucose Dehydrogenase to Alcohol Dehydrogenase

**Abstract:**

Biocatalysts are increasingly used in industry to create enantiomerically pure compounds; routes employing dehydrogenases have been especially successful. Glucose dehydrogenase belongs to the extensive superfamily of SDRs (short chain dehydrogenase/reductase) and is currently a favored enzyme for cofactor regeneration. Glucose is its only substrate at a reasonable rate and the reaction is favored towards oxidation of glucose.

A thermostable glucose dehydrogenase, previously developed at GeorgiaTech, was submitted to several rounds of site-directed mutagenesis, supported by database analysis and structure-guided design, to create an alcohol dehydrogenase based on a glucose dehydrogenase scaffold with a markedly changed substrate specificity. Key to the identification of the necessary amino acid hot spots within the sequence was a bioinformatics approach that involved an extensive data base analysis of existing SDRs

using specific algorithms. The resulting 3-4 amino acid exchanges led to a complete change of substrate specificity away from glucose but respectable activity towards substrates such as cyclopentanol and 1,3-cyclohexandiol. By achieving the goal of de novo design of a glucose dehydrogenase, we demonstrate the successful implementation of bioinformatic tools and modeling that can accelerate development of novel substrate specificity in an otherwise limited enzyme system.

# 428885 Conformational Heterogeneity of Protein Binding Interfaces

**Abstract:**

Recent successes in simulating protein structure and conformational dynamics have demonstrated the power of molecular dynamics to predict the long timescale behavior of proteins. However, the structural origins of a protein's ability to bind diverse signaling molecules including hormones, downstream effectors etc. are not well understood. In this study, we investigate two key signaling proteins Calmodulin (a ubiquitous calcium sensor and a crucial signaling hub in many pathways aberrantly activated in disease) and nitrogen regulatory protein C (NtrC, a key bacterial two component response regulator).[1,2] By employing unbiased Markov state model-based distributed molecular dynamics simulations (total sampling of over 2 milliseconds), we construct a dynamic picture of the activation pathways of these proteins that is consistent with experimental observations and predicts new mutants that could be used for validation of the mechanism. Moreover, these results suggest a novel mechanistic paradigm for conformational switching and predict new binding interfaces that could be candidates for pharmacological intervention.

References:

1. A network of molecular switches controls the activation of the two-component response regulator NtrC. D. Vanatta, D. Shukla, M. Lawrenz and V. S. Pande, Nature Communications, In press, 2015.
2. Conformational Heterogeneity of the Calmodulin Binding Interface. D. Shukla, A. Peck and V. S. Pande, Nature Communications, under review, 2015.

# 426094 Single-Molecule Investigation of Surface Heterogeneity and Crowding on Protein Conformation at the Solid-Liquid Interface

**Abstract:**

Protein-surface interactions are a concern for a number of fields, including biocatalysis, biosensing, biomaterials, and chromatography. Typically these interactions have been studied through ensemble measurements, averaging the heterogeneous behaviors of individual molecules. Here we apply single-molecule fluorescence microscopy and protein engineering techniques in order to study large numbers ($10^4$-$10^5$) of individual T4 lysozyme proteins upon adsorption at the silica-water interface. Site-specific labeling of this model protein with pairs of fluorophores provides simultaneous structural information via Forster Resonance Energy Transfer (FRET). Analysis of individual molecular trajectories with FRET provides quantitative information about protein adsorption, diffusion, and desorption as a function of protein folded state. Super-resolution mapping of the surface reveals nanoscale heterogeneities on the fused silica surface, despite careful sample preparation. At low protein concentration, the diffusive behavior of proteins is dominated by adsorption and immobilization of unfolded protein to a small number of surface sites. As the concentration of protein on the surface increases, these strong adsorption sites become passivated and the surface appears more homogeneous. Further increase in protein concentration increases the rate of protein unfolding, suggesting crowding effects on the surface may be denaturing for T4 lysozyme on fused silica.

# 425260 The Relationship of Glycoside Hydrolase Ligand Binding Free Energy to Processive Ability

**Abstract:**

In nature, organisms secrete synergistic glycoside hydrolase (GH) enzyme cocktails to deconstruct polysaccharides to sugars. The cocktails consist of multiple classes of processive, non-processive, and accessory enzymes that aid in substrate accessibility and reduce product inhibition. Non-processive enzymes attack the amorphous regions of polymer crystals and cleave glycosidic linkage once or twice to create accessible free chain ends; whereas, processive enzymes usually attach to chain ends and hydrolyze many glycosidic linkages in sequence to produce disaccharide units prior to dissociation. Processive enzymes are responsible for the majority of hydrolytic bond cleavages, and hence, are the most logical targets for activity improvements towards efficient and economical biomass conversion. However, the molecular-level mechanism and factors responsible for GH processivity still elude description. From structural data and analysis of reaction products it has been suggested that the presence of a long tunnel or deep enzyme active site cleft in GH indicates processivity and lack thereof indicates non-processivity. However, this strict delineation is not always accurate. We hypothesize that the processive GHs active site architecture contributes to relatively high ligand binding affinities and low substrate dissociation rates, while non-processive enzymes will demonstrate comparatively weak binding allowing the enzyme to more freely dissociate. To investigate our hypothesis, we have chosen the model GH system, *Serratia marcescens* Family 18 chitinases, which includes two processive chitinases, ChiA and ChiB, and a non-processive endo-chitinase, ChiC. In line with our hypothesis, we expect that ChiA and ChiB will exhibit stronger ligand binding than non-processive ChiC. Free energies of binding a chito-oligosaccharide to each chitinase calculated using free energy perturbation with Hamiltonian replica exchange molecular dynamics support our hypothesis. Further, we find that interaction of the protein with pyranose rings in the substrate side of the binding cleft contribute the most to affinity. Comparison to experimental values obtained with Isothermal Titration Calorimetry reveals the formation of a catalytically active conformation significantly affects heat evolution, and thus affinity. Molecular dynamics simulations were used to characterize protein dynamics contributing to ligand binding affinity. Overall, our characterization of ligand binding affinity in *S. marcescens* Family 18 chitinases suggests that binding in the pre-hydrolytic substrate binding sites significantly contributes to affinity and likely processivity.

# 431918 Effect of Reynolds and Viscous Stress, and Kolmogorov Length Scale on Hemolysis

**Abstract:**

Turbulent blood flow in artificial hearts and ventricular assist devices (VAD) [1, 2] cause red blood cell (RBC) damage, which is a major concern when designing prosthetic heart devices. Although the effects of specific turbulent flow characteristics on the blood cells are uncertain,[3-5] it is generally known that turbulence affects RBCs causing trauma and hemolysis – an effect that increases when cells are exposed to turbulent stresses.[5]

Blood damage is mostly represented using empirical power law models that assume that hemolysis would be a function of the magnitude of the shear stresses and exposure time to high stresses. While turbulent stresses, the well-known Reynolds stresses, might be responsible for hemolysis, having the same effect as viscous stresses in laminar flows, other researchers have examined the size of flow eddies relative to the size of the RBCs in order to identify the mechanism responsible for RBC trauma. If the Reynolds stresses are similar in effect as viscous stresses are for laminar flow, then one expects to see a large increase in the hemolysis at some threshold value of the Reynolds stress, as a result of the exponential feature of the power law relationship. However, the value of a threshold turbulent stresses for hemolysis has been controversial.[6] The significance of Reynolds stresses and viscous stresses on hemolysis might be identified by performing a threshold analysis. In this study, we investigate the effect of time averaged, area averaged Reynolds stresses and viscous stresses to hemolysis by doing a threshold analysis for 5 Couette viscometer experiments and 4 capillary tube experiments. Moreover, we have also searched for relationships with hemolysis using extensive quantities, such as the total surface area from subpopulation distributions of eddies with different Kolmogorov Length Scales (KLS) for 13 jet experiments as an addition to Couette and capillary tube experiments. The relation between the hemolysis seen experimentally to the eddy surface area and the Reynolds and viscous stresses in the flow were investigated by using Reynolds-Averaged Navier-Stokes models of turbulence (k-ε and k-ω SST turbulence models, and finite volume techniques) to simulate different experimental conditions in a Couette viscometer, a capillary tube, and jet. Simulation results showed that for both capillary flow and flow in a Couette viscometer, there was no evidence of a threshold for hemolysis in terms of Reynolds and viscous stresses. Therefore, Reynolds and viscous stresses are not good predictors of hemolysis. Eddy analysis results of Couette, capillary, and jet experiments showed that hemolysis is related directly to the occurrence of eddies with diameters of up to about 10 μm. Thus, while it is easier to predict Reynolds stresses or total shear stresses in medical devices than to predict or measure KLS, the eddy size analysis presented can be applied in conjunction with various turbulence simulations, possibly across a wide range of conditions and devices. The prediction of the Kolmogorov eddy size distribution might then lead to an evaluation of whether a particular design of a medical device is more or less susceptible to hemolysis, and what changes need to be done in the design to increase the size of the Kolmogorov scales.

References

1. Antiga L, Steinman DA. Rethinking Turbulence in Blood. Biorheology. 2009;46:77-81.

2. Hund SJ, Antaki JF, Massoudi M. On the Representation of Turbulent Stresses for Computing Blood Damage. International Journal of Engineering Science. 2010;48:1325-31.

3. Aziz A, Werner BC, Epting KL, Agosti CD, Curtis WR. The cumulative and sublethal effetcs of turbulence on erythrocytes in a stirred-tank model. Ann Biomed Eng. 2007;35:2108-20.

4. Bludszuweit C. Three-dimensional numerical prediction of stress loading of blood particles in a centrifugal pump. Artif Organs. 1995;19:590-6.

5. Kameneva MV, Burgreen GW, Kono K, Repko B, Antaki JF, Umezu M. Effects of Turbulent Stresses upon Mechanical Hemolysis: Experimental and Computational Analysis. American Society for Artificial Internal Organs. 2004;50:418-23.

6. Goubergrits L. Numerical modeling of blood damage: current status, challenges and future prospects. Expert Review of Medical Devices. 2006;3:527-31.

# 428153 Inhibition of Triglyceride Synthesis Does Not Sensitize Hepatic Cells to Palmitate Lipotoxicity or Interfere with Oleate-Mediated Rescue

**Abstract:**

Obesity has reached epidemic proportions both in the U.S. and abroad, creating rapidly increasing healthcare costs. Metabolic diseases associated with obesity, including type 2 diabetes, non-alcoholic fatty liver disease (NAFLD) and cardiovascular disease, have also seen concomitant increases in population incidence rates. NAFLD, the hepatic manifestation of metabolic syndrome, is characterized by chronic ectopic lipid accumulation in the liver and is estimated to affect over one-third of the U.S. general population and up to 80% of obese and diabetic individuals. An estimated 10-25% of NAFLD patients progress to a more severe condition known as non-alcoholic steatohepatitis (NASH), characterized by chronic inflammation and hepatic cell death that can eventually lead to cirrhosis. Despite its high prevalence, the molecular mechanisms underlying the progression from NAFLD to NASH are not well understood and represent a critical area of investigation in order to develop more targeted and effective therapeutics for fatty liver disease.

*In vitro* experiments in a wide variety of cell types including Chinese hamster ovary (CHO) cells, pancreatic beta cells, breast cancer cells, HeLa cells and hepatic cells have consistently demonstrated the acute lipotoxic effects of saturated fatty acids, such as palmitate (PA), but not unsaturated fatty acids. Palmitate-mediated lipotoxicity is characterized by endoplasmic reticulum (ER) stress, ER calcium release, reactive oxygen species (ROS) accumulation, caspase activation and cell death. Cells treated with monunsaturated fatty acids (MUFAs), such as oleate (OA), exhibit none of these features but instead demonstrate significant increases in triglyceride (TG) synthesis and accumulation. Adding OA to PA-treated cells can fully reverse PA lipotoxicity though mechanisms that are not well defined but which have been previously attributed to increased TG accumulation and sequestration of PA into neutral lipid stores. However, this hypothesis has never been directly tested by experimentally modulating the relative partitioning of PA/OA between TGs and other lipid fates in hepatocytes. In this study, we test the alternative hypothesis that the rescue effect of PA/OA co-treatment (relative to treatment with PA alone) is due to its ability to normalize the composition of phospholipid membranes, rather than its ability to divert PA into neutral TG species. This hypothesis was suggested by our prior lipidomic profiling study that showed dramatic accumulation of saturated phospholipid species in response to PA treatment of liver hepatocytes, which was completely reversed by PA/OA co-treatment.

The goal of this study was to better understand the mechanism by which oleate rescues palmitate-induced lipotoxicity in hepatic cells and the role that triglyceride and phospholipid metabolism plays in this phenomenon. We show that addition of OA to PA-treated hepatocytes decreases overall phospholipid saturation while rescuing PA-induced apoptotic cell death. As shown previously, this rescue effect is accompanied by enhanced TG synthesis and, specifically, increased incorporation of PA into TG stores. However, simultaneous knockdown of both liver isoforms of diacylglycerol acyltransferase (DGAT), the rate-limiting step in TG synthesis, significantly reduced TG accumulation but without altering the ability of OA to prevent PA lipotoxicity. In both wild-type and DGAT knockdown cells, co-treatment with OA significantly reduced overall phospholipid saturation in comparison to cells treated with PA alone. These data indicate that OA's protective effects are not due to its ability to divert PA into inert TG stores, but instead may be due to OA's ability to normalize the composition of cellular phospholipids in the presence of a lipotoxic PA load.

# 425767 The Heterogeneous Response of Bone Marrow Metastases to EGFR Targeting Drugs and Chemotherapy

**Abstract:**

**Introduction:** Metastasis, the spread of cancer from a primary tumor to a secondary tissue site, is the leading cause of cancer-associated deaths. Breast cancer most often metastasizes to bone, brain, and lung tissues and becomes challenging to treat once it has become metastatic, and most often patients with metastases are treated with hormone therapy or chemotherapy. Hormone therapy targets specific cell receptors and chemotherapy is toxic to all cells and patients commonly develop resistance to these treatments making them ineffective in the long term. Triple negative breast cancer (TNBC) is both difficult to treat, because of its lack of receptor expression, and it aggressively metastasizes. We focus here on bone marrow because metastatic tumors disrupt critical bone functions. Within the bone marrow microenvironment, breast cancer cells could potentially become drug resistant because of limited drug diffusion, as well as survival signals from cells and the extracellular matrix[1]. Of particular interest to us, mesenchymal stem cells (MSCs) derived from bone marrow have been shown to facilitate both drug resistance[2] and metastasis[3]. In this work we are interested in studying the drug response of epidermal growth factor receptor (EGFR)-overexpressing TNBC cell lines when exposed to soluble factors from the bone marrow microenvironment.

**Materials and Methods:** EGFR targeting drugs (Afatinib, Erlotinib, Gefitinib, & Lapatinib) and chemotherapy (Doxorubicin) at 0-50 μM were tested on EGFR-overexpressing TNBC cell lines (MDA-MB-231, BT-549, MDA-MB-468, & Hs578T) for 48 hours in 96 well plates. These tests were done with and without MSC conditioned medium. MSC conditioned medium was made by growing MSCs for 72 hours and collecting the medium at this time point. This medium contained the soluble factors released by the MSCs. The drug response was mesured by using the CellTiter-Glo ATP assay and the concentration of drug required to kill 50% of the cells ($IC_{50}$) was calculated. The growth rate of the cell lines in the two media conditions without drug was measured at 1, 2, 4, and 7 days using the CellTiter 96 proliferation assay. The cell morphology at selected conditions was analyzed with manual cell image tracing and ImageJ. In separate experiments, hTERT MSCs were treated with a very low dose of drug to create drug-treated conditioned medium, and this was used to study how pretreating the MSCs with a drug affects cancer cell response to drugs. The levels of 15 soluble factors in the conditoined medium were measured with a high-throughput bead based protein quantification system (MAGPIX).

**Results and Discussion:** We used a simple drug screening platform that incorporates factors from the microenvironment to show that bone marrow-derived MSC soluble factors affect cancer cell drug response. The Hs578T cell line is resistant to Doxorubcin in the presence of MSC released factors and the surviving cells have a pronounced nuclear shape change. We are working to identify the specific factors that are responsible for the observed results. The MAGPIX results show that SDF-1α may be one of the soluble factors causing this effect. We have also observed that MDA-MB-231 cells die in the MSC conditioned medium after a few days without the presence of any drug. Cancer cell drug resistance to Lapatinib has been observed with Lapatinib-treated MSC conditioned medium. This suggests that treating the MSCs with drug changes the profile of the soluble factors that they release.

**Conclusions:** Soluble factors released by MSCs provide protection to Hs578T cells treated with Doxorubicin. Interestingly, this same conditioned medium kills the MDA-MB-231 cells even in the absence of drug. Ongoing studies include using conditioned medium from MSCs that were treated with drug to more accurately capture what happens *in vivo* when a patient is dosed with a drug.

**References:** [1] Meads, MB. Clinical Cancer Research. 2008. 14(9):2519-26. [2] Roodhart, JM. Cancer Cell. 2011. 20(3):370-83. [3] Karnoub, AE. Nature. 2007. 449(7162):557-63.

# 434851 The Dependence of Gamma-Secretase Substrate Cleavage on Lipid Microdomains in the Context of Spatial Diffusion-Reaction Models

**Abstract:**

In these studies, the influence of cholesterol-rich lipid domain structures on the enzymology of the intramembrane protease γ−secretase is probed using microsphere-supported biomembranes. This important protein is a particularly complex enzymatic system composed of four subunits that catalyzes the cleavage of over 80 known transmembrane protein substrates. Despite extensive studies and decisive roles in both the Notch signaling pathway and in amyloid precursor protein (APP) degradation and processing in Alzheimer's disease, deciphering γ−secretase regulation and catalysis has been hampered by membrane-associated enzymology. Multiple lines of evidence in cell-based systems has implicated cholesterol and cholesterol-rich membrane microdomains, termed lipid rafts, in the regulation of substrate cleavage of γ-secretase relating to Alzheimer's disease pathology. In this work we have incorporated controlled cholesterol-rich domain formation into our recently developed method to investigate and directly visualize active γ-secretase and its substrates in microsphere-supported biomembranes. This enables us to directly probe enzyme and substrate mobility, localization and cleavage in phase-separated domains to investigate the cholesterol dependence of γ-secretase.

In this work we have characterized the structures, phase colocalization and compositions of gamma-secretase proteolipobead systems by employing correlative optical and cryoelectron microscopy from the micro- to the nanoscale. We have probed the formation of $L_0$ domains in situ, monitored using 3D FRET phase detection. We have probed the phase partitioning of enzyme, substrates and cleavage products in PLB systems. A combination of biomembrane mobility methods are being used to probe the diffusivities and mobile fractions of lipid, substrates and gamma-secretase. The results of phase partioning and diffusivity measurements are being integrated into bottom-up spatial diffusion-reaction models to correlate microenvironment with enzyme function.

# 433400 Chronic Hypertension Increases Aortic Endothelial Hydraulic Conductivity By Upregulating Endothelial Aquaporin-1 Expression

**Abstract:**

Blood pressure drives the advective transport of large solutes such as LDL cholesterol from the blood through rare widened inter-endothelial junctions into the vessel wall where they can bind to extracellular matrix in the subendothelial intima (SI) layer; this is believed to be the earliest event in atherogenesis. Water and small molecules can also cross the far more numerous normal endothelial junctions and, we have previously shown, transcellularly via aquaporin-1 membrane proteins that the endothelial cells avidly express. This flow can potentially dilute LDL concentration that has built up in the SI and slow its kinetics of binding there; so increasing this water flow may be beneficial in slowing early atherogenesis. It is first worth asking why the endothelial cells express this protein if water can traverse the endothelium paracellularly.

In addition to other functions, e.g., NO transport, of aquaporin, it may be the case that regulating aquaporin expression allows the vessel/cells to actively control both the force/area that the endothelial cells feel and the water flux that they allow across them in response to changes in environmental conditions, e.g., chronic hypertension. We have developed a theory that relates aquaporin expression to both endothelial and vessel wall hydraulic conductivity. It predicts how SI thickness and vessel hydraulic conductivity should vary with both aquaporin expression and transmural pressure. We test this theory on two high-renin rat models of hypertensive rat models, one genetic (spontaneously hypertensive rats and their normotensive cousins, the Wystar Kyoto rats) and Sprague-Dawley rats, some of which have been made hypertensive by renovascular surgery. In both cases aquaporin-1 expression (measured by quantitative immunohistochemistry) of the hypertensives is 2-3 times that of the normotensives and, at lower pressures, the hydraulic conductivity of the hypertensives is 2-3 times that of the normotensives. Our theory explains this pressure dependence. In addition, acute chemical upregulation of endothelial aquaporin-1 expression (by two hour forskolin treatment) has a similar and consistent effect on both aquaporin-1 expression and endothelial hydraulic conductivity, despite the fact that medial hydraulic conductivity in chronic hypertensives is far lower than in acutely treated normotensives. Upregulated aquaporin-1 expression and function may thus be a response to hypertension that critically determines conduit artery vessel wall viability and long-term susceptibility to atherosclerosis.

# 432554 A Mechanistic Model of Stress Hyperglycemia Evolution in the Intensive Care Unit

**Abstract:**

Many patients in the intensive care unit (ICU) experience elevated blood glucose levels, so-called stress hyperglycemia. It has been shown that controlling glucose levels using intensive insulin therapy (IIT) can reduce mortality [1, 2]. However, a prospective multi-center study [3] of over 6,000 patients showed increased mortality for those receiving IIT. Retrospective analysis [4] shows that IIT caused increased occurrence of hypoglycemia, which consequently reduced survival within the IIT cohort. The conclusion drawn from the aforementioned studies is the existence of a glucose range that, when patients are controlled therein, yields lower mortality, approximately 80 to 120 mg/dL. A model-based decision support system (DSS) can provide patient-specific treatment and improve blood glucose control to range without increasing risk of hypoglycemia. The critical component of a patient-tailored DSS is a mathematical model that can resolve the dynamic changes resulting in a unique individuals metabolic state.

The underlying mechanism of stress hyperglycemia is a complex network of biological signaling pathways that decrease sensitivity to insulin [5] and increase endogenous glucose production (EGP). In this work, mathematical modeling is used to identify and characterize the complex biological pathways leading to stress hyperglycemia. The two modeled metabolic regulatory processes involved in dynamic modification of metabolism in the ICU are: (i) the hypoglycemia counterregulatory response and (ii) the acute inflammatory response. The counterregulatory response occurs following a hypoglycemic event, common in the ICU [4], and increases blood glucose via increased EGP and insulin resistance leading to reduced insulin-mediated glucose uptake (IMGU). Similarly, the acute inflammatory response includes cytokines such as TNF-α and hormones such as cortisol, which also decrease IMGU in humans [7, 8]. These two metabolic regulatory pathways are coupled with an model of glucose and insulin homeostasis [6] from literature to resolve patient specific variations in metabolic state.

Patient glucose measurements are fit by adjusting a single, time-varying insulin sensitivity parameter. The time-varying parameter profile used to match model glucose values with recorded glucose values for each patient is subsequently used to predict explanatory biomarker concentrations belonging to the two metabolic regulatory pathways previously described. Specifically, TNF-α, cortisol, epinephrine, and glucagon dynamics are simulated to match the insulin sensitivity profile for each patient. Mathematical models of acute inflammation and counterregulation are used as explanatory mechanisms driving insulin sensitivity, which enforce pathway specific dynamics on overall glucose dynamics. Both pathways are evaluated to determine whether they act individually or concurrently to produce observed insulin resistance and stress hyperglycemia in ICU patients.

This model of glucose and insulin, combined with mechanistic counterregulatory and inflammatory dynamics, serves as a simulation platform to generate clinically-relevant critically ill patient metabolic profiles. A library of time-varying metabolic parameter profiles is fit to a subset of the 215 trauma victims from two hospital centers in the Cologne area participating in the German Trauma Registry effort, to provide a virtual patient cohort matching the dynamics of clinical response. Such a virtual patient platform is useful for developing DSS control strategies, as well as to better understand possible patient differentiation metrics for separating treatment cohorts (*e.g.*, counterregulation driven, inflammation driven, or both) for which treatment strategies may differ as a result of their metabolic upset.

References

1. Van den Berghe G, Wouters P, Weekers F, Verwaest C, Bruyninckx F, et al. (2001) Intensive insulin therapy in critically ill patients. New England journal of medicine 345: 1359-1367.

2. Krinsley JS (2004) Effect of an intensive glucose management protocol on the mortality of critically ill adult patients. Mayo Clinic Proceedings. Elsevier, volume 79, pp. 992-1000.

3. Finfer S (2009) Intensive versus conventional glucose control in critically ill patients. New England Journal of Medicine 360: 1283-1297.

4. Finfer S (2012) Hypoglycemia and risk of death in critically ill patients. New England Journal of Medicine 367: 1108-1118.

5. Brealey D, Singer M (2009) Hyperglycemia in critical illness: A review. Journal of Diabetes Science and Technology 3: 1250-1260.

6. Lin J, Razak NN, Pretty CG, Le Compte A, Docherty P, et al. (2011) A physiological intensive control insulin-nutrition-glucose (ICING) model validated in critically ill patients. Computer Methods and Programs in Biomedicine 102: 192-205.

7. Plomgaard P, Bouzakri K, Krogh-Madsen R, Mittendorfer B, Zierath JR, et al. (2005) Tumor necrosis factor-α induces skeletal muscle insulin resistance in healthy human subjects via inhibition of akt substrate 160 phosphorylation. Diabetes 54: 2939-2945.

8. Rizza RA, Mandarino LJ, Gerich JE (1982) Cortisol-induced insulin resistance in man: Impaired suppression of glucose production and stimulation of glucose utilization due to a postreceptor defect of insulin action. The Journal of Clinical Endocrinology & Metabolism 54: 131-138.

# 432471 Whole Animal to Single Cell Near Infrared Fluorescent Imaging and Prediction of Therapeutic Protein Distribution to Improve Efficacy of Antibody Drug Conjugates


**Abstract:**

Abstract:

Biologics are the fastest growing area in the pharmaceutical industry, particularly in cancer therapy because they bind tumor specific antigens with limited off-target effects. Antibody Drug Conjugates (ADC's) are one of the newest classes of biologics to be approved by the FDA for solid tumors. Development of these agents is exceedingly complex given the inter-related pharmacokinetics and pharmacodynamics of the biologic and small molecule. Optimizing the dose and the drug antibody ratio (DAR) to achieve a therapeutic microdistribution within the tumor while maintaining safe systemic concentrations are important determinants of ADC efficacy. Here we have developed experimental and computational tools for understanding ADC distribution across several length scales.

In this study, we developed a dual near infrared fluorescent dye labeling technique to follow biologics at high temporal resolution from whole animal to organ, tissue, cellular, and subcellular spatial resolution to understand metabolism and distribution of ADC's. For proof-of-principle, we quantified the metabolism of two antibodies, cetuximab and anti-A33, and one protein, EGF, *in vitro* and *in vivo* using a non-residualizing fluorophore to track intact protein and a residualizing fluorophore to measure cumulative uptake (Figure 1A and 1B). Fluorescent microscopy images qualitatively show tumor distribution, heterogeneity, and variability at the cellular, tissue, and organ scales (Figure 1C), while biodistribution studies quantitatively show cumulative uptake into different organs. Quantitative localization in tumors showed high %ID/g (percent injected dose per gram) for both antibodies as expected from their slow clearance. High renal uptake and metabolism was seen for EGF due to rapid filtration in the kidneys. Histological examination of the tumors showed heterogeneous distribution at low doses with increasing penetration distance and homogeneous distribution as dose increased.

Supplementing the experimental work, a computational model incorporating tumor distribution (reaction-diffusion partial differential equations) into a whole animal physiologically based pharmacokinetic (PBPK ordinary differential equation) model allowed for the simultaneous simulation of tissue and systemic distribution. The ability to track biomolecules across these spatial resolutions *in vitro*, *in vivo*, and *in silico* will both aid in understanding the multi-scale distribution of novel therapeutics and facilitate the development of predictive computer models for understanding therapeutic antibody distribution. In particular, these techniques are being applied to improve the distribution of the clinically relevant antibody-drug conjugate (ADC) Kadcyla, where the local heterogeneity can be quantified at the cellular level to determine the impact of tumoral distribution on efficacy.

In conclusion, a technique for simultaneously imaging residualizing and non-residualizing near-infrared fluorescently labeled biologics was developed to track the distribution and metabolism of biologics from the whole animal to tissue and cellular resolution. These data were paired with a predictive reaction-diffusion equation model to understand the impact of dose, local metabolism, and expression on tumoral distribution. This approach can be used to follow antibodies and other protein therapeutics *in vivo* to understand the interplay between the pharmacokinetics and pharmacodynamics of this class of agents.


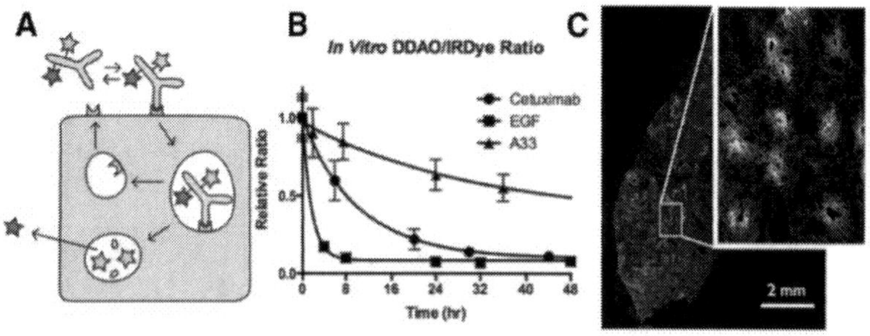


Figure 1: Metabolism and distribution using a dual NIR imaging approach. (A) An antibody dually labeled with a non-residualizing dye (red) and a residualizing dye (green) binds, internalizes and subsequently degrades. Once degraded the non-residualizing dye leaks out, while the residualizing dye is trapped. Non-residualizing dye approximates intact antibody while residualizing shows cumulative uptake. (B) In vitro ratio imaging of non-residualizing dye (DDAO) to residualizing (IRDye800CW) dye shows different rates of protein degradation. EGF is rapidly degraded, cetuximab moderately, and A33 slowly. (C) 30μg (~1.5mg/kg) IV injection of cetuximab in A431 tumor xenograft nude mice showing perivascular distribution. Images of frozen sections taken using confocal microscope. Scale bar 2 mm.

# 434832 Regulatory Science Perspective on Continuous Manufacturing

**Abstract:**

The continuous manufacturing of pharmaceutical drug products is gaining momentum within the industry. The successful implementation of continuous manufacturing represents the culmination of innovative quality and regulatory approaches from the previous decade, including PAT, increased product and process understanding, and advanced control strategies. The FDA's Office of Pharmaceutical Quality supports the development and implementation of control strategies that lead to increased quality assurance and availability of medicines to the public. This presentation will address the control strategy considerations for implementation of a continuous manufacturing process, focusing on assurance of quality aspects. The discussion will include concepts such as science-based justification for material traceability, sampling frequency and throughput/run time. The importance of understanding failure modes and implementing of controls at appropriate levels will be highlighted with emphasis on increased process understanding and demonstration of a state of control. In conclusion, design and operation of a continuous manufacturing process is based on science and engineering principles, but a control strategy that ensures product quality and consistency is the key to addressing the regulatory challenges for this technology.

# 430211 Key Considerations to Achieve Manufacture Excellence in Continuous Manufacturing Using Hot Melt Extrusion

**Abstract:**

***Key Considerations to Achieve Manufacture Excellence in Continuous Manufacturing Using Hot Melt Extrusion***

Hot melt extrusion is one of the most important enabling processing technology for manufacturing amorphous solid dispersions (ASDs) to enhance oral drug absorption. The extrusion process is inherently continuous. Back-to-back extrusion runs without shutdown have been demonstrated successfully in commercial manufacturing over the past decade with improved manufacture efficiency and flexible batch size to meet dynamic market supply needs. Manufacturing excellence is ensured through fundamental understanding of formulation degradation mechanism and impact of process/formulation variability on product quality as well as implementation of a robust quality control strategy including but not limited to PAT tools. Best practices to achieve manufacture excellence in continuous manufacturing using hot melt extrusion will be shared through a case study. Techniques and control strategies to enable fully continuous operation and real-time release will be reviewed and discussed.

# 414679 Considerations in Choosing Between the Use of Process Data and Analytical Data to Monitor and Control Cqa's in Continuous Drug Product Processes

**Abstract:**

Continuous drug product manufacturing introduces new risks and opportunities in providing assurance of quality. Many of these involve the ability to measure or predict product quality and reject/accept/adjust the process in real-time. The primary approach discussed in the literature for this is through the intensive use of on-line and at-line analytical testing (often referred to as PAT). A clear definition of the objectives and a first-principles approach to addressing them can result in a greater reliance on achieving routine quality though process monitoring and control and the focusing of analytical testing on those areas not being addressed with process data. An example is provided for the continuous direct compression platform where potency, blend uniformity and content uniformity can all be assured by the use of process data. This requires the process meet the criteria of being in a state of control and having good to excellent capability and also having acceptable risk of special cause upsets. When used in conjunction with process data this approach might also change validation requirements for on-line/at-line analytical methodologies to a less demanding level of rigor depending on its intended purpose.

# 430155 Nonlinear Steady-State Data Reconciliation for a Continuous Tableting Line

**Abstract:**

The robust operation of a continuous process requires a sufficient level and quality of online process measurements and an appropriate level of process control. As the pharmaceutical industry transitions from batch to continuous operation the questions of how many measurement points and what quality of online measurements are necessary or sufficient and the level of sophistication of the process control systems required are very much under discussion. Under the aegis of the NSF ERC on Structured Organic Particulate Systems, we have been investigating these questions in the context of continuous production of tablets. Tablets are the most common dosage form and thus the conversion from batch to continuous production of this drug product form has thus received quite a bit of attention [1].

In this paper we specifically focus on answering questions concerning the selection of and impact of online measurements. To address these questions, we first develop a nonlinear steady state data reconciliation model for a continuous dry granulation line. The data reconciliation model of such a process network consists of the relationship between all of the measured and relevant unmeasured variables of the process together with a description of the error distributions of these measurements [2, 3]. It is assumed that the number of measurements is larger than the degree of freedom of the network model, hence that there is redundancy. Given a set of measurements, an optimization problem is posed which has as its objective function a likelihood type function. This function is the minimization of the difference between the measured values and their best estimates, weighted by the standard deviation of the measurement error of the associated sensor model, with the above mentioned reconciliation model forming the constraint set [3, 4].

The solution of the maximum likelihood problem constitutes the best real time estimate of the state of a processing system taking into account all of the measurements and their error characteristics. Such an estimate can serve as basis for checking whether all process variables are within acceptable bounds, for identify gross measurement errors, for control action or for process optimization steps. Additionally the reconciliation formulation provides a framework for investigating the impact of adding or eliminating measurements, of replacing instruments with given error characteristics with those with tighter variances or of introducing measurement redundancy [5].

While the data reconciliation problem has received much attention in the process systems engineering literature, most of the attention has been directed at linear models, which consist of the component mass balances of the process [3]. Nonlinearities have principally been introduced through the addition of energy balances, which in their most common form involve equations that are bilinear in flow and composition or flow and temperature [3]. Additionally, it is generally assumed that measurement errors are normally distributed. In this paper we consider nonlinearities which arise because of the inclusion of online measurements of properties such as density and variables such as screw speed, roll speed, etc. The model nonlinearities have implications on computation performance, which is important for real time application. Finally, given the nature of some of the spectroscopic measurements used in the tableting application [6, 7], the assumption of normality of the measurement error comes into question [5]. We illustrate the approach using case studies carried out on a continuous tableting line.

**References**

[1] Reklaitis G. The Pharmaceutical Manufacturing Institute. **2013**, (White Paper) Purdue University, West Lafayette, IN.

[2] F. Carl Knopf. Modeling, Analysis and Optimization of Process and Energy Systems. First Edition. *John Wiley & Sons, Inc*. **2002.**

[3] Narasimhan S. & Jordache C. Data reconciliation & Gross Error detection. Houston Texas. *Gulf Publishing Company*. **2000.**

[4] Narasimhan S., Bhatt N. Deconstructing principal component analysis using a data reconciliation perspective. *Computers and Chemical Engineering*. **2015**, 77, 74–84

[5] Cencica O. and Frühwirth R. A general framework for data reconciliation. Part I: Linear constraints. *Computers and Chemical Engineering*. **2015**, 75, 196–208

[6] Austin, J.; Gupta, A.; McDonnell, R.; Reklaitis, G.V.; Harris, M.T. The use of Near-Infrared and Microwave Resonance Sensing to Monitor a Continuous Roller Compaction Process. *Wiley Online Library*. **2013**.

[7] Austin, J.; Gupta, A.; McDonnell, R.; Reklaitis, G.V.; Harris, M.T. A novel microwave sensor to determine particulate blend composition on-line. *Analytica Chimica Acta*. **2014,** 819, 82-93.

# 415000 Novel Use of Cognex Vision System for on-Line Level Monitoring/Control in a Continuous Direct Compression Manufacturing Process

**Abstract:**

The continuous direct compression manufacturing process consists of several loss-in-weight (LIW) feeders, a horizontal mixer, a surge hopper, and tablet press equipped with a near infrared (NIR) probe at the die-filling station of the tablet feed frame. The surge hopper atop the tablet feed frame helps both dampen upstream flow fluctuations and maintain a consistent head of powder above the feed frame so as not to starve the tablet press of powder. Use of a Cognex Vision system allowed for a novel means of measuring level in the surge hopper to gauge the mass balance of material flowing into and out of the surge hopper. With the Cognex Vision system (or comparable level sensor) measuring fluctuations in surge level, the tablet press rotor/turret speed can be adjusted and has been shown to have minimal impact to tablet properties through monitoring of the force control on the main compression stage. Additionally, measurement of the surge level enables application of a reduced system model which, in part, would forecast when an upstream disturbance in API concentration would reach the feed frame and when that disturbance would clear the feed frame.

# 433765 A Novel Tablet Coating Technology to Support a Continuous Tableting Line

**Abstract:**

Continuous Manufacturing is rapidly being developed and implemented by the Pharmaceutical Industry, driven mainly by the reduced Research and Development costs and improved manufacturing quality. For example, manufacturing lines for 10 kg/hr to 100 kg/hr have been implemented for Continuous Direct Compression, Continuous Wet Granulation and Drying, and Continuous Dry Granulation in a tablet manufacturing line. However, to support this throughput range of tablet manufacturing, continuous tablet coating technologies have been very limited and not attractive in supporting the plug flow manufacturing concept.

A novel tablet coating technology was developed to support the 10-100 kg/hr continuous tableting line which is able to coat small sub-batches (3-6 kg) very rapidly (7-15 minutes) with a high degree of accuracy. Using this method, each sub-batch is able to be tracked from the upstream tableting process and critical process parameters monitored during coating to ensure product quality of each sub-batch. Employing a small-footprint twin coating system, the coating process can be linked to the tablet manufacturing line uninterrupted and while maintaining the plug flow from the upstream processes. Additionally, this coating method enables in-line coating thickness and build rate measurement.

# 410319 PAT Data Management for Continuous Manufacturing – the Next Generation


**Abstract:**

The mind shift from batch production to continuous manufacturing has been made in the pharmaceutical industry for a couple of years, but the IT infrastructure to cover the analyzer integration, data management and advanced process control was put under pressure by this new paradigm. The current generations of PAT data management platforms are bridging the gap when it comes to full robust and reliable integration of analyzers, high throughput dataflows, and a further and deeper integration in the process automation layer.

Also optimized time alignment is critical to combine the right pieces of information. In this presentation we will dig a bit deeper in time alignment applied on hot melt extrusion lines with dynamic line speeds, and how this can be translated into optimized time alignment. Automated selection of PAT data collection methodology with specific time alignment settings will be discussed based on actual process parameters and decision trees. This will show the importance of a very close integration between the automation layer and the PAT data management on top. And we will go even a step further…

Also the user interface is an important aspect in smoothening the integration of PAT data management into continuous manufacturing. UX or user experience is the new trend to optimize the user interfaces for your operators into the IT platforms. Logic user-focused workflows are guiding the operator though the different necessary steps of operation. Some examples from the new SIPAT 4.1 operator dashboard will be shown after it was revamped by a team of UX specialists.

A last item to be discussed is data security. The IT landscape is more and more the backbone of all the decisions taken on the quality of your products. This makes your network a vulnerable source for hacking, sabotage and potential terrorism. In how far are you aware that data from your analyzers could be captured on the fly and slightly changed, so the wrong decisions are made in the end? Man-in-the-middle attacks have caused serious issues in the past and should be avoided at all times. We have gone that additional step by offering data solutions, fully ensuring and securing your data security.

# 426743 Hofmeister Effects on Aggregation and Prion Strain Generation By Amyloid Beta 1-42

**Abstract:**

Aggregation of amyloid-beta peptides to form plaques is a key event in the pathogenesis of Alzheimer's disease (AD). Amyloid beta$_{1-42}$ is one of the most abundant species in amyloid plaques and is considered to play a central role in pathogenesis of AD. It has been shown to aggregate *in vitro* like many other amyloid proteins, however, the effect of environmental factors on aggregation of Amyloid beta$_{1-42}$ has not been investigated systematically. In this work we have studied the effect of ions on the aggregation of Amyloid beta$_{1-42}$. Previous work from our group on amyloid formation by Sup35NM protein containing the prion domain of yeast prion protein Sup35p has shown that strongly hydrated anions (kosmotropes) initiate nucleation quickly and cause rapid fiber elongation, whereas poorly hydrated anions (chaotropes) delay nucleation and reduce the elongation rate. The Hofmeister salts also determined the amyloid strain that was preferentially formed with strong strains being formed in kosmotropes and weak strains in chaotropes. Here, we have investigated how Hofmeister ions affect the aggregation kinetics of Amyloid beta$_{1-42}$ peptide. Preliminary results indicate that aggregation behavior of Amyloid beta$_{1-42}$ in the presence of Hofmeister salts, particularly the highly chaotropic perchlorate, is widely different from Sup35NM which suggests that these proteins may aggregate by different mechanisms. Further, the biophysical properties of the amyloid strains formed by Amyloid beta$_{1-42}$ in the presence of Hofmeister salts will also be studied to gain insights into the aggregation mechanism of Amyloid beta$_{1-42}$.

[1] Jonathan Rubin, Hasan Khosravi, Kathryn L. Bruce, Megan E. Lydon, Sven H. Behrens, Yury O. Chernoff, Andreas S. Bommarius.(2013), Ion-Specific Effects on Prion Nucleation and Strain Formation. *J. Biol. Chem.* 18; 288(42):30300-8 doi: 10.1074/jbc.M113.467829

[2] Yeh, V., Broering, J. M., Romanyuk, A., Chen, B., Chernoff, Y. O. and Bommarius, A. S. (2010), The Hofmeister effect on amyloid formation using yeast prion protein. Protein Science, 19: 47–56. doi: 10.1002/pro.281

[3] Aditi Sharma, Kathryn L. Bruce, Buxin Chen, Stefka Gyoneva, Sven H. Behrens, Andreas S. Bommarius, Yury O. Chernoff. Role of the environment of cross-seeding reaction in the prion species barrier. *J. Biol. Chem*. To be submitted.

# 427822 Quercetin and Metabolites Reduce Amyloidβ-Induced Apoptosis Associated with Alzheimer's Disease

**Abstract:**

Alzheimer's disease (AD), the most common form of dementia, is characterized by extracellular plaques in the brain created when monomeric amyloid-β (Aβ) protein aggregates into fibrillar structures. Soluble Aβ aggregates, including oligomers that form along the reaction pathway, are believed to be the primary toxic species and have been shown to increase the production of reactive oxygen species (ROS). Polyphenols have been suggested as a complimentary AD therapeutic based on epidemiological evidence that polyphenol-rich diets correlate with a reduced incidence of AD. In particular, many flavonoids, a subclass of polyphenols, have demonstrated the ability to inhibit Aβ aggregation thereby neutralizing the protein's damaging effects. Alternatively, polyphenols may counteract Aβ-induced cellular responses by neutralizing ROS through their antioxidant properties. This study sought to identify polyphenols that can reduce Aβ-induced apoptosis by inhibiting Aβ oligomer formation and/or reducing ROS.

Polyphenols investigated include quercetin (QUE) and some of its most common metabolites, rhamnetin (RHA), isorhamnetin (IRHA), and tamarixetin (TAM). To identify polyphenols capable of inhibiting the formation of oligomers, polyphenols were added in 10-fold excess to monomeric $A\beta_{1\text{-}42}$prior to induction of oligomerization. The resulting quantity of oligomers formed was analyzed via SDS-PAGE and Western blot. To identify polyphenols capable of neutralizing ROS, antioxidant capacity was evaluated using the OxiSelect™ Oxygen Radical Antioxidant Capacity Activity (ORAC) Assay. To verify that high antioxidant capability led to a reduction in Aβ-induced intracellular ROS, SH-SY5Y human neuroblastoma cells were treated with either oligomers alone or simultaneously with oligomers and polyphenol. Addition of DCFH-DA, which is converted into fluorescent DCF in the presence of intracellular ROS, was used to determine the resulting ROS. To assess the effect of oligomer inhibition and antioxidant capacity on Aβ-induced apoptosis, SH-SY5Y cells were treated with either oligomers formed in the presence of polyphenols or with oligomers formed in the absence of polyphenols followed by addition of polyphenols at concentrations capable of reducing ROS. Cells were stained using TUNEL, which identifies breaks in the DNA strand characteristic of apoptosis. Apoptosis was visualized using a fluorescence microscope and quantified using MATLAB.

A volumetric analysis was used to quantify the amount of oligomers formed in the absence or presence of polyphenols. Only IRHA was unable to reduce Aβ oligomers 250-100 kDa in size, while QUE reduced these oligomers by 88.3 ± 2.4%. QUE, RHA, and TAM all reduced Aβ oligomers <100 kDa in size by approximately 50%, while IRHA remained the weakest inhibitor, only reducing oligomers by 26.2 ± 3.8%. Antioxidant capabilities were quantified relative to Trolox, a vitamin E analog. All compounds exhibited antioxidant capability similar to Trolox, although IRHA and TAM had decreased antioxidant capacity compared to QUE and RHA. This trend was also observed in the reduction of intracellular ROS, with only QUE and RHA significantly reducing Aβ-induced intracellular ROS. When present during oligomerization, all compounds were able to reduce Aβ-induced apoptosis by at least 30%. Polyphenol reduction of ROS also protected against apoptosis.

Characterization of polyphenols for oligomer inhibition and antioxidant capabilities led to the identification of polyphenols with varying degrees of both activities. This panel of compounds demonstrated that both inhibiting Aβ aggregation and reducing ROS can attenuate Aβ-induced apoptosis associated with AD. Future studies will examine the capabilities of these polyphenols to synergistically attenuate apoptosis to determine their potential as dual action drugs for AD.

WISCONSIN
UNIVERSITY OF WISCONSIN-MADISON

# Cystatin C

## A Cerebrospinal Fluid Protein with Both Amyloid and Anti-Amyloid Activity

Tyler J. Perlenfein and Regina M. Murphy
Department of Chemical and Biological Engineering
University of Wisconsin – Madison

# What makes cystatin C (CysC) so interesting?

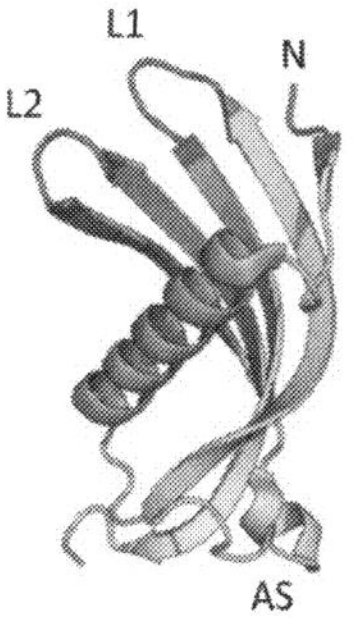


- Important protease inhibitor (e.g. cathepsin B)
- Concentration in CSF is **enriched 5-fold** compared to plasma
- Known to be upregulated in Alzheimer's brain tissue
  - Clustered in neurons throughout temporal lobe only in AD patients
- **Also found in Aβ plaques** in AD-affected tissue
- Is **amyloidogenic**!

CysC-Aβ interaction **prevents fibril formation**:
- Selenica & Grubb et al, Scand. J. Clin. Lab Inv., 2007
- Sastre & Levy, Neurobiology of Aging, 2004

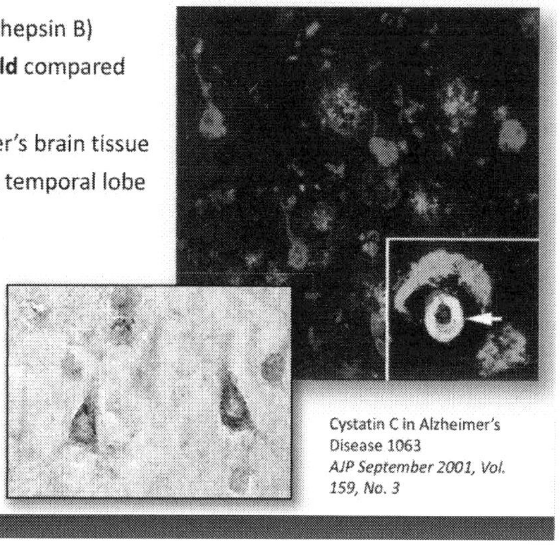

Cystatin C in Alzheimer's Disease 1063
*AJP September 2001, Vol. 159, No. 3*

# Amyloid behavior of CysC

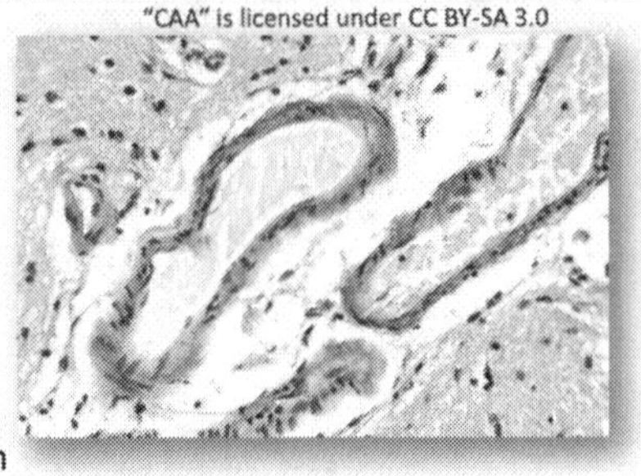


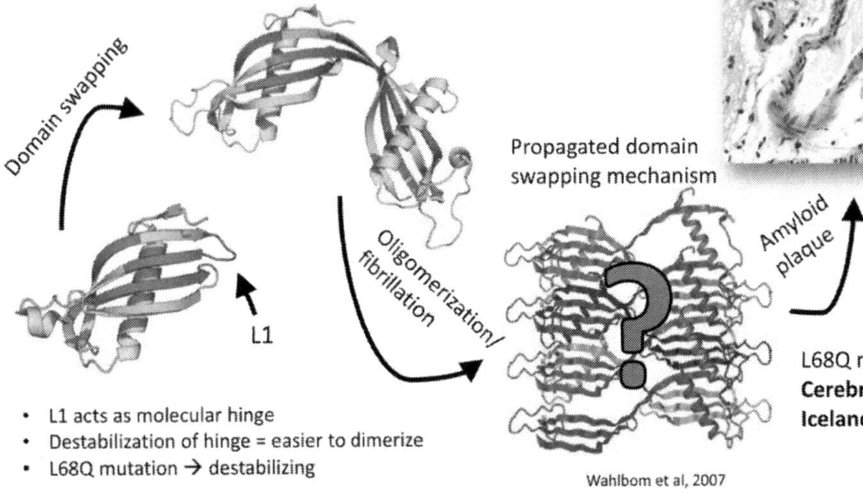


- L1 acts as molecular hinge
- Destabilization of hinge = easier to dimerize
- L68Q mutation → destabilizing

L68Q mutation responsible for:
**Cerebral amyloid angiopathy – Icelandic type**

# CysC production & purification scheme

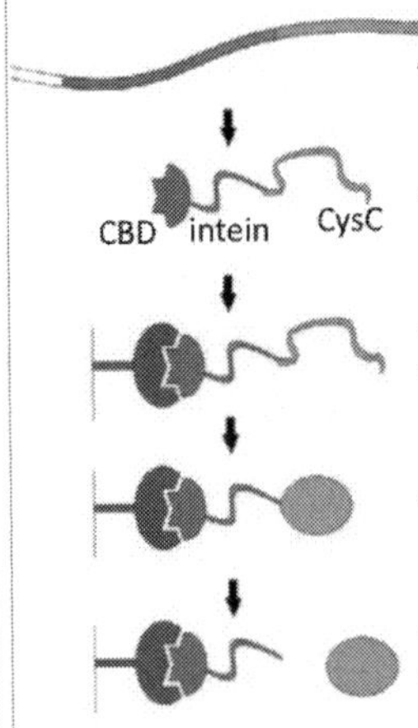


- CysC gene inserted into pTWIN1 plasmid (Impact TWIN system)
- *E. coli* expression yields denatured fusion protein:
  - Chitin binding domain – intein linker – CysC
- Purify by binding to chitin column
- Wash away cell debris
- Refold CysC while on the column
- Elute CysC by cleaving intein at C-terminus
- Results in purified CysC yield of 15 mg/L of culture

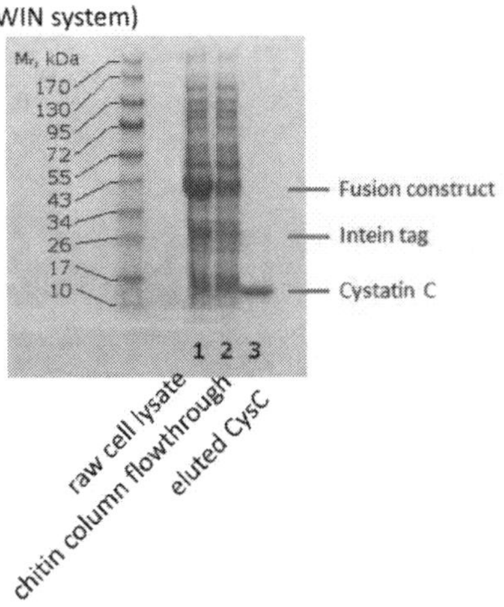


T.J. Perlenfein and R.M. Murphy,
*Protein Expression and Purification*, 2015

# Our CysC has native structure and activity...

CD spectroscopy

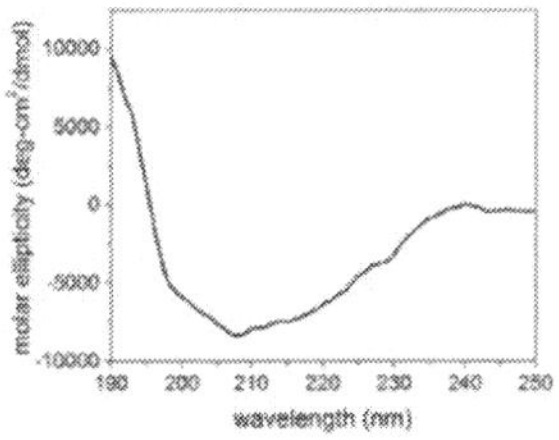


Endopeptidase inhibitory activity

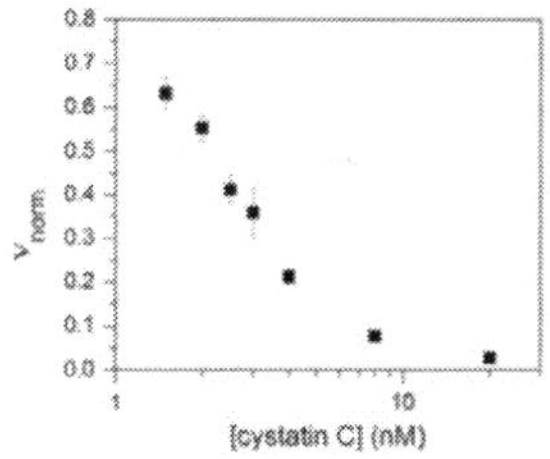


CD spectrum agrees with literature

Cathepsin B activity as function of CysC

- $K_I$ = 60 pM, also consistent with literature

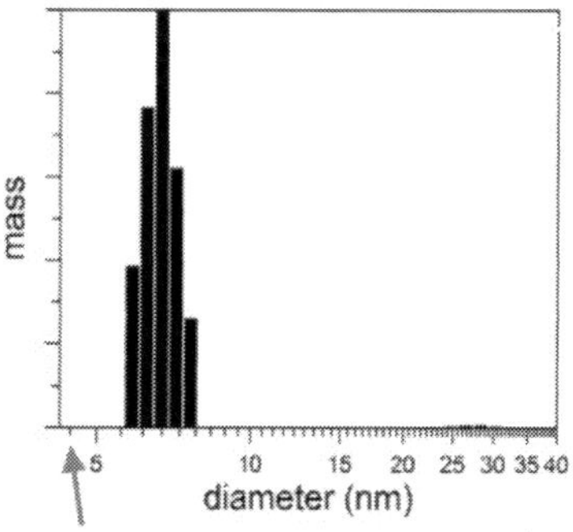


...but size is too big by light scattering!

Monomer should be here

T.J. Perlenfein and R.M. Murphy,
*Protein Expression and Purification*, 2015

# CysC oligomers formed during purification

Crosslink & SDS-PAGE

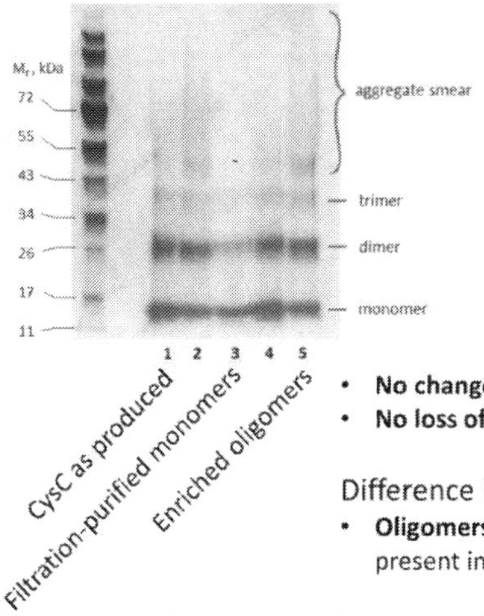


CD spectroscopy

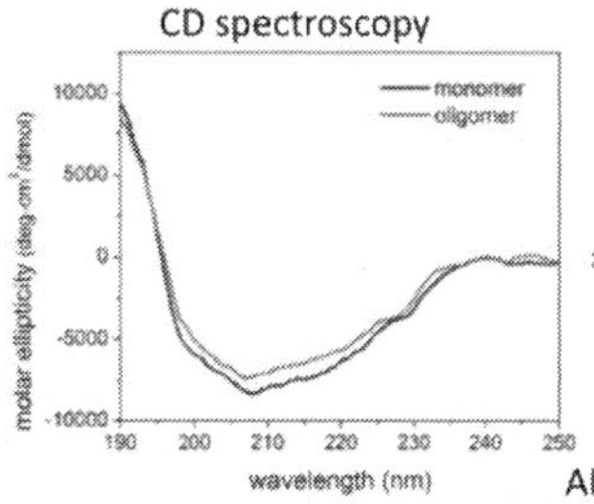


Inhibitory activity

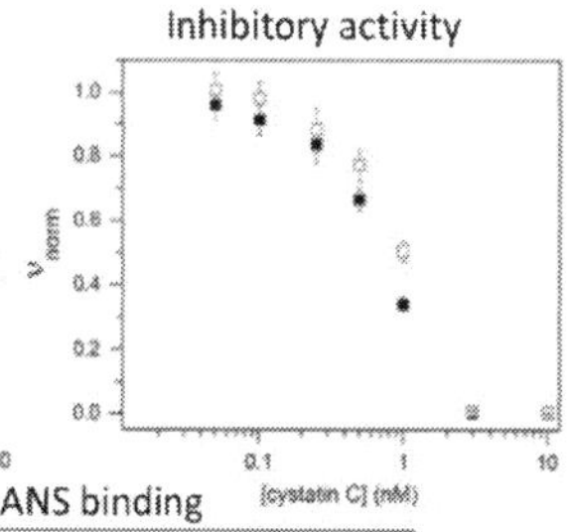


ANS binding

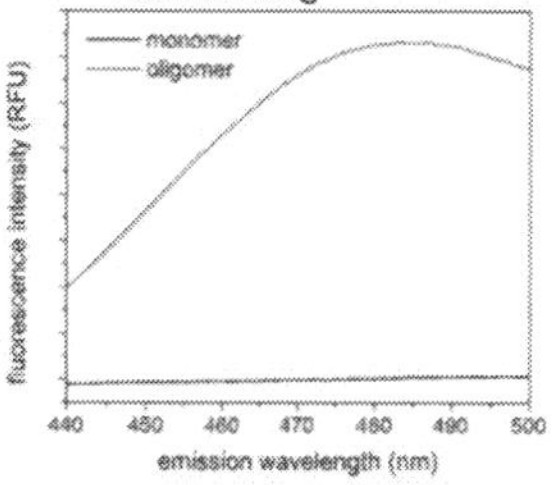


- **No change to 2' structure** upon oligomerization
- **No loss of activity** upon oligomerization

Difference in ANS binding

- **Oligomers form hydrophobic pockets** – not present in monomers

# Oligomers are not amyloid...

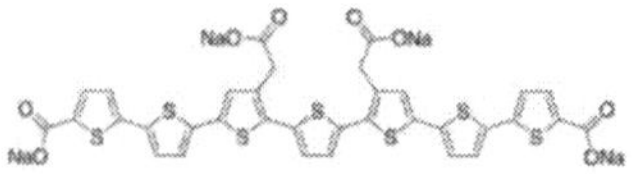


ThT binding

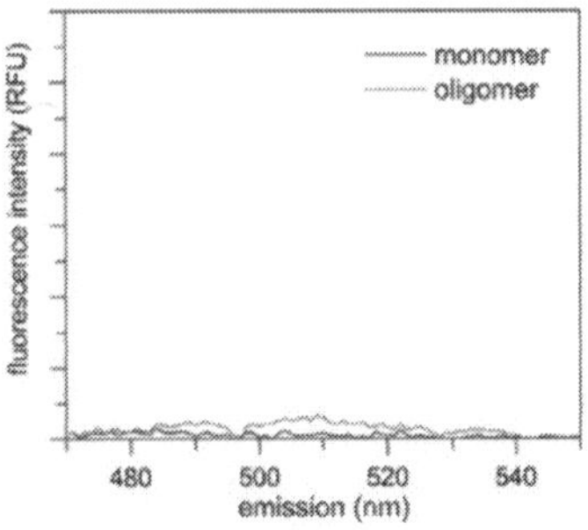


No ThT binding

TEM imaging

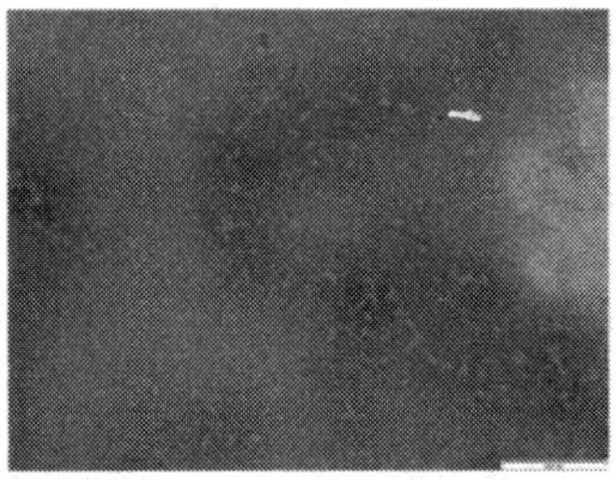

Mostly amorphous morphology

h-FTAA binding

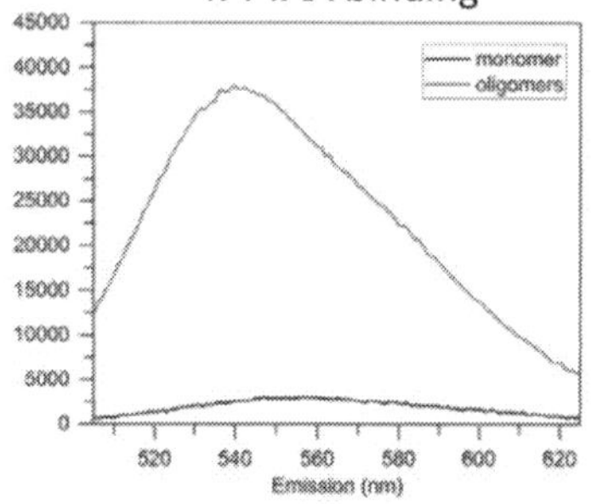


Strong binding to sensitive pre-amyloid dye

...but pre-fibrillar assembly

For more info on h-FTAA:
K.P.R. Nilsson et al, *Org. Biomol. Chem.*, 2011

# Oligomer formation concept

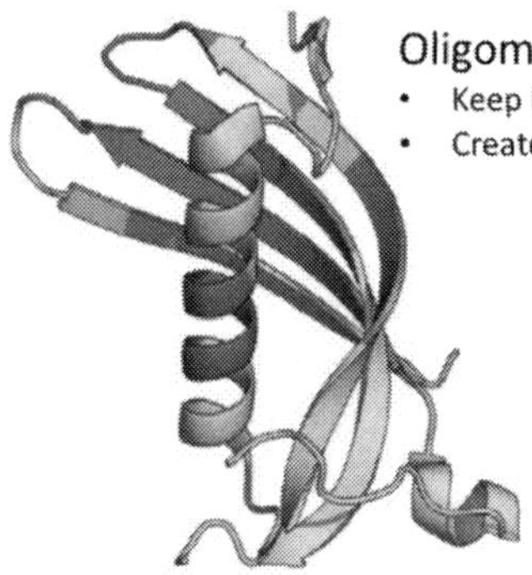

Oligomer model must:

- Keep loop regions free (for cathepsin B inhibition)
- Create hydrophobic pockets (for ANS binding)

Possible mechanism of formation:

Helical bundling

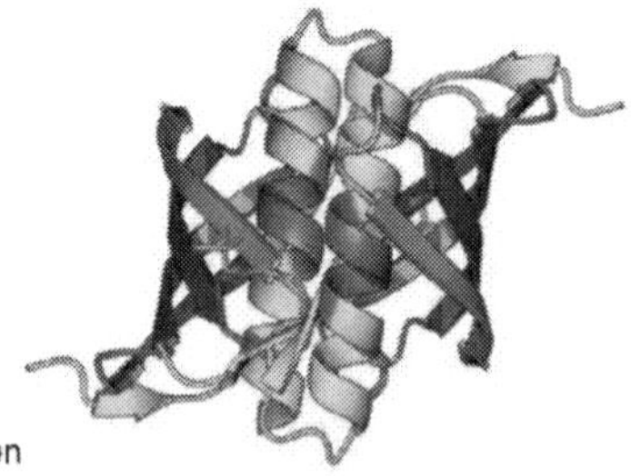

Beta sheet stacking

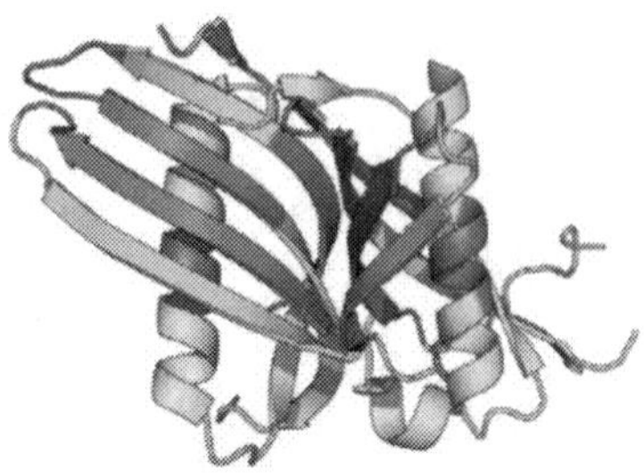

Red = hydrophobic, aggregation-prone region
Pink = hydrophobic stretch

# Production of dimers from pure monomers

Monomer denaturation by W106 fluorescence

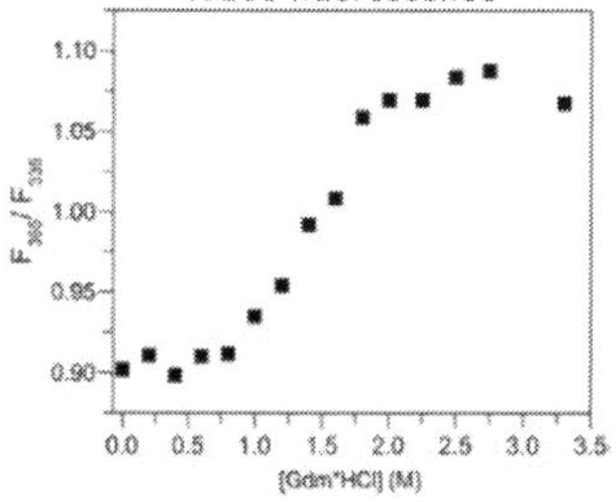


Dimer formation by native PAGE

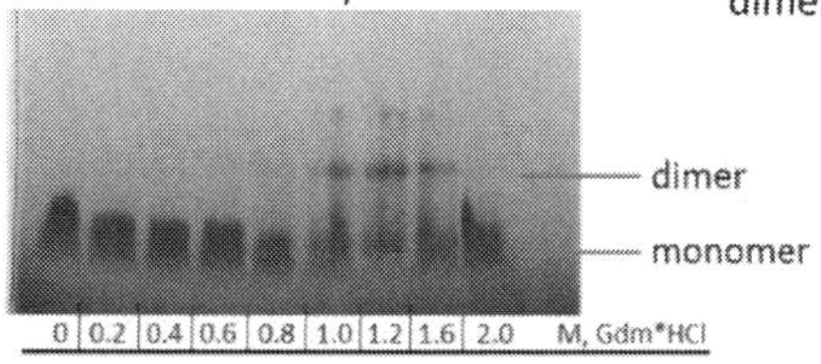


- Dimers formed in mildly denaturing condition
- Dimers were stable upon fast dilution into native buffer
- Loss of activity upon dimerization

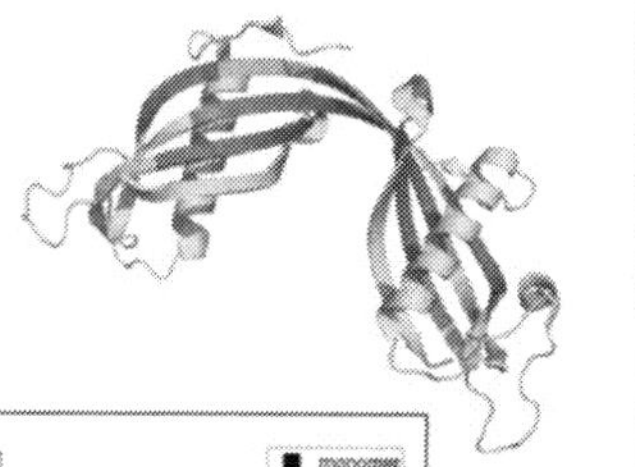

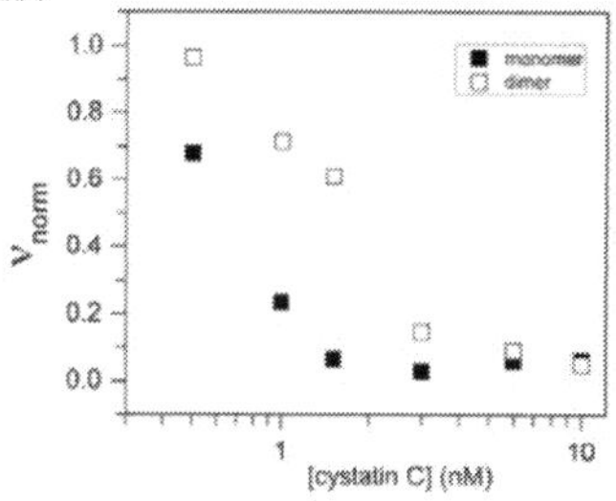


# Structural comparison of CysC multimer states

ANS

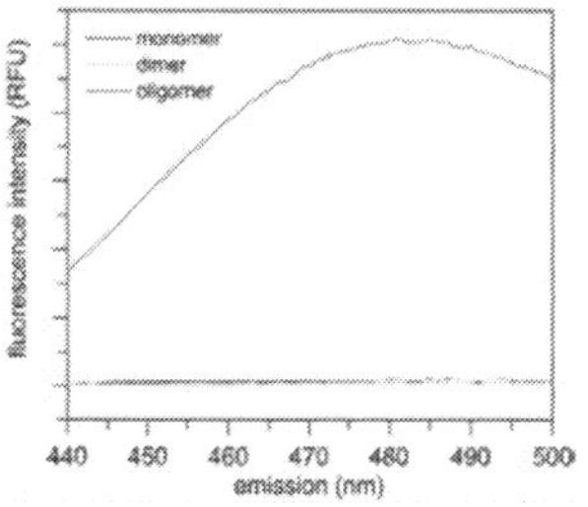


h-FTAA

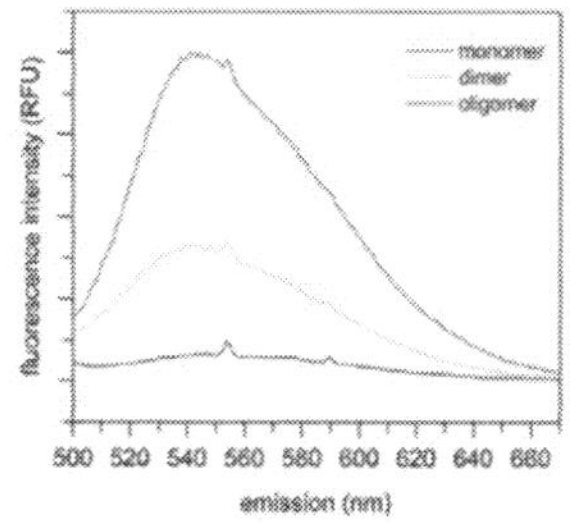


ThT

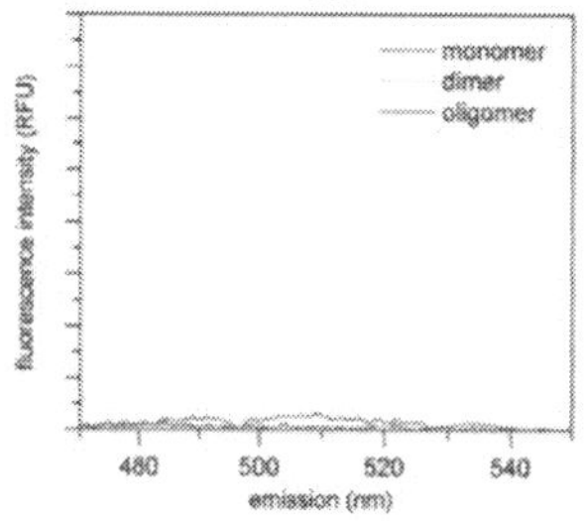


| | Monomer | Dimer | Oligomer |
|---|---|---|---|
| Structure | Native fold | Non-native (extended β-sheet interface) | Native secondary structure |
| ANS binding | - | - | Very strong |
| ThT binding | - | - | - |
| h-FTAA binding | - | Small | Medium |
| Inhibit Cathepsin B? | Yes | No | Yes |

# What about CysC – Aβ interaction?

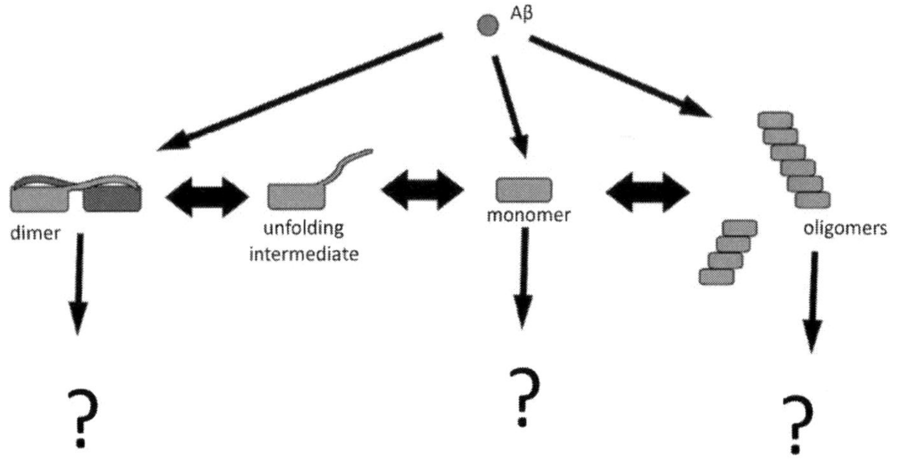


# CysC monomer does not stop Aβ aggregation…

Dynamic Light Scattering (DLS)

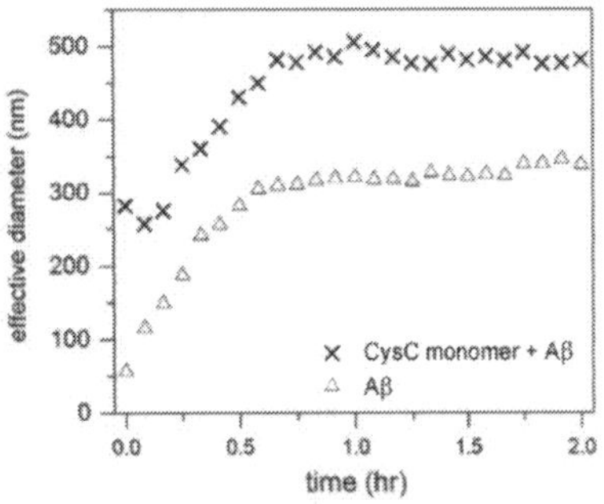


With CysC monomer: size grows larger & faster – very surprising!

ThT time course study

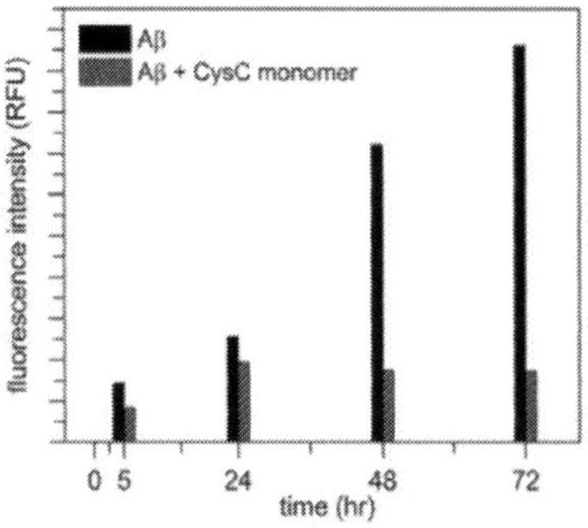


…But does suppress fibril formation

# Dimer inhibits fibrils...

Dynamic Light Scattering (DLS)

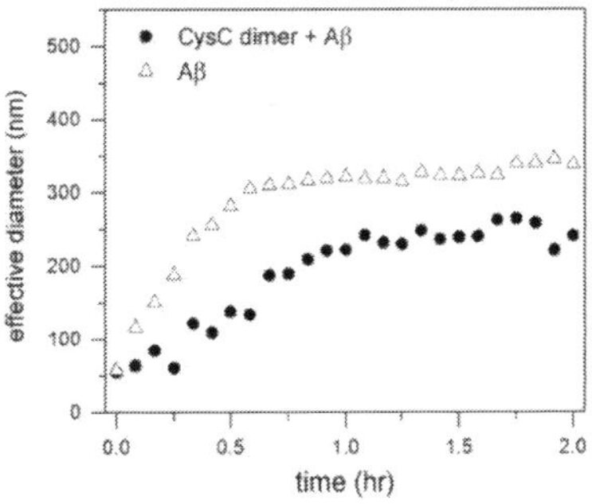


ThT time course study

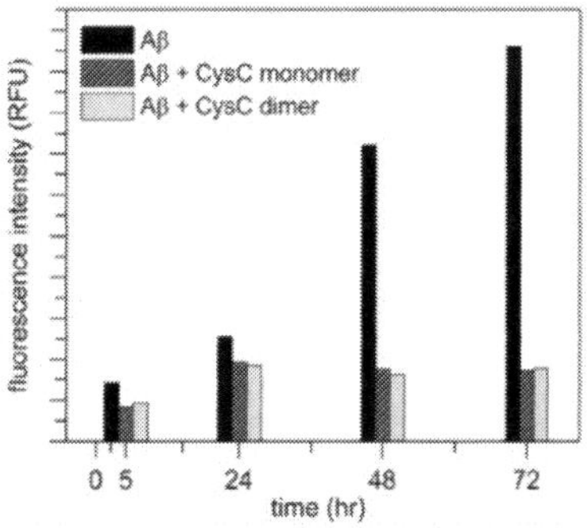


...and can reduce mean aggregate size

# Oligomers completely inhibit Aβ aggregation!

Dynamic Light Scattering (DLS)

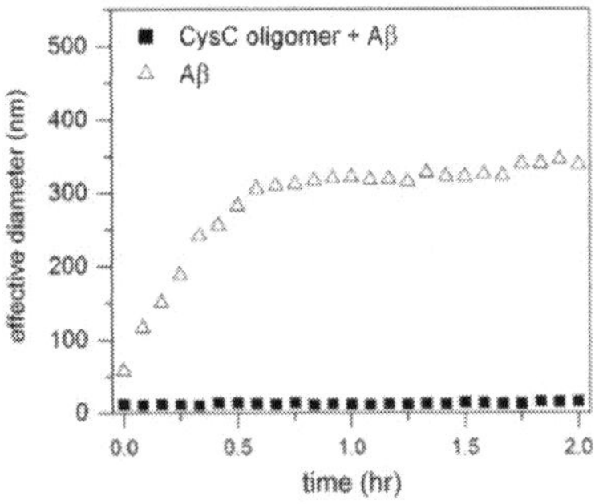


ThT time course study

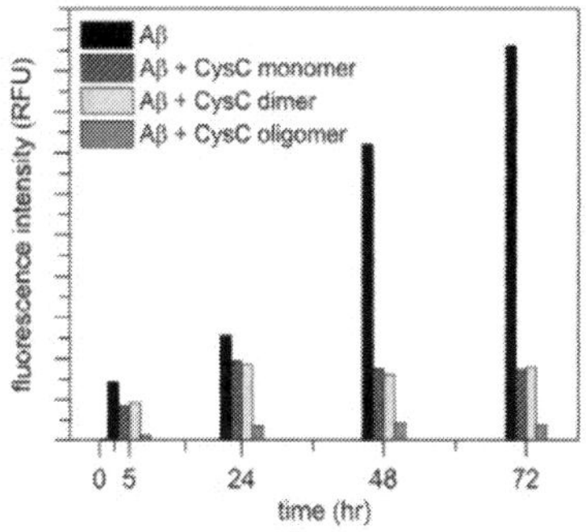

## Summary of CysC by multimer state

| | Monomer | Dimer | Oligomer |
|---|---|---|---|
| Structure | Native fold | Non-native (extended β-sheet interface) | Native secondary structure |
| ANS binding | - | - | Very strong |
| ThT binding | - | - | - |
| h-FTAA binding | - | Small | Medium |
| Inhibit Cathepsin B? | Yes | No | Yes |
| Aβ inhibition | Inhibits fibril formation | Inhibits fibrils & some amorphous aggregates | Total inhibition of aggregation |

## Concept diagram of CysC – Aβ interaction

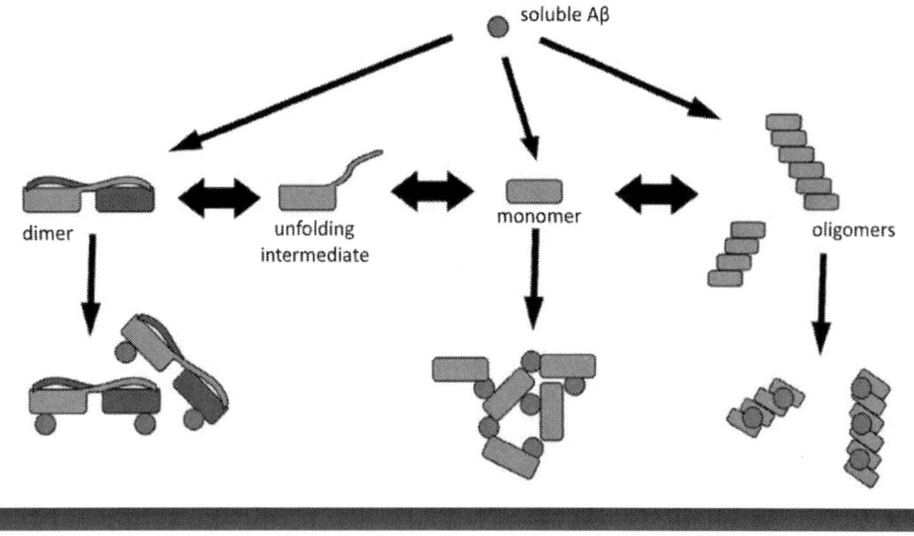

## Are CysC multimers relevant to neurological disease?

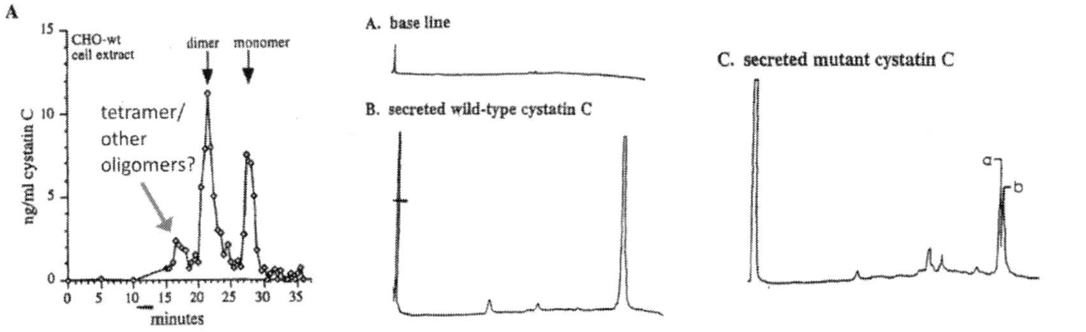


Normal cellular trafficking: extensive dimerization occurs

Secretion of wild-type CysC: possible oligomer population(s)?

Altered secretion profile of L68Q CysC: dimers & some oligomers likely present

For more information, see:
Benedikz & Rushbrook et al, 1999, Amyloid, 6:3, 172-182

## Future studies: regulatory network

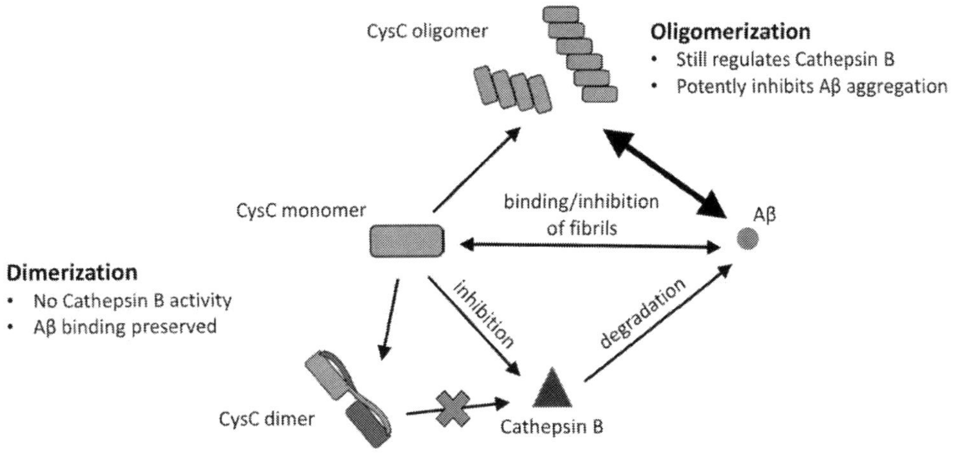

# Project support and thanks!

WISCONSIN
UNIVERSITY OF WISCONSIN–MADISON

Funding provided by:
- NSF CBET 1262729
- NIH Molecular Biophysics Training Grant

h-FTAA dye kindly gifted by K. Peter R. Nilsson, Linköping University, Sweden

Technical support from University of Wisconsin:
- Biophysics Instrumentation Facility, Dr. Darrell McCaslin
- Electron Microscopy, Randall Massey

# 426081 A New Class of Compounds Inhibiting the Amyloidosis of TTR(105-115): Insights from Experiments and Molecular Dynamics Simulations

**Abstract:**

Human transthyretin (TTR) is among the key proteins involved in serious amyloid diseases. It is a cerebrospinal fluid and serum protein, produced by the liver, which, in its regular function, transports thyroxin and retinol-binding protein. The misfolded mutants of this protein are involved in familial amyloid polyneuropathy (FAP), senile cardiac amyloidosis (SCA), and senile systemic amyloidosis (SSA), all very serious diseases [1,2,3,4]. A major class of inhibitors of TTR amyloidosis bind to the thyroxine (T4) binding pocket and interact specifically with a section of the TTR sequence, corresponding to residues 105−115, that is implicated in amyloidogenic propensity. The findings of a recent experimental study suggest that "stabilizing" inhibitors of the TTR 105-115 amyloidosis, may also directly disrupt amyloidogenic aggregation of TTR monomers through specific interactions with the exposed TTR (105−115) sequence [4]. Thus the discovery of novel inhibitors of amyloidosis by TTR 105-115 peptide fragments appears a promising direction for developing future therapeutics for diseases associated with the entire TTR protein.

In this work, we discovered a new class of molecules inhibiting the aggregation of TTR(105−115) peptides. A series of Resonance Raman and fluorescence studies were performed to investigate the inhibition effect through specific interactions of these molecules with the peptide. Replica Exchange Molecular Dynamics (REMD) simulations were introduced to provide further insights into the inhibition mechanism of the novel compounds. Our results reveal that the presence of inhibiting compounds in solution with TTR(105−115) peptides disfavors the formation of β-sheet structures of high complexity that can serve as structural units promoting the amyloid formation. We delineate the functional role of each group of the inhibitor with regard to the formation of interactions with TTR(105−115) peptides, and provide a comparison with existing molecules targeting the TTR(105−115) fragment and inhibiting the amyloid formation of the entire TTR protein. Our study offers significant insights for the design of novel molecules which can specifically target TTR(105−115) peptides and inhibit their amyloidosis, and can also potentially inhibit the amyloidosis of the entire TTR protein.

**References**

1. Lee KW, Lee DH, Son H, Kim YS, Park JY, Roh GS, Kim, HJ, Kang SS, Cho GJ, Choi WS. (2009) Clusterin regulates transthyretin amyloidosis. Biochem Biophys Res Commun 388:256−260.

2. Ando Y, Nakamura M, Araki S. (2005) Transthyretinrelated familial amyloidotic polyneuropathy. Arch Neurol 62: 1057−1062.

3. Pieridou G, Avgousti-Menelaou C, Tamamis P, Archontis G, Hayes SC. (2011) UV resonance Raman study of TTR(105-115) structural evolution as a function of temperature. J Phys Chem B. 115: 4088-4098.

4. Liang Y, Ore MO, Morin S, Wilson DJ. (2012) Specific disruption of transthyretin(105-115) fibrilization using "stabilizing" inhibitors of transthyretin amyloidogenesis. Biochemistry 51: 3523-3530.

# 413986 Effect of Oscillating Electric Field on Preformed Human Serum Albumin Fibrils

**Abstract:**

The stability and biological activity of proteins are highly correlated to their three dimensional structures with multi-level conformations. In the native folded state, proteins are soluble in aqueous solution. When the structure deviates from the native state, proteins have a propensity to aggregate. Such deviation can occur when proteins are subjected to certain specific biophysical conditions, where they tend to undergo self-aggregation to form polymeric species with different morphologies leading to the formation of insoluble amyloid fibrils. The extensive depositions of fibrils, associated with high cross β-sheet content in extracellular spaces of various tissues causing cellular damages, are linked to a number of neurological disorders such as Alzheimer's, Parkinson's, Creutzfeldt-Jakob disease, type II diabetes etc.[1] The possible remedy is removal of these toxic fibrillar deposits from affected tissues by developing inhibitors or disintegrating agents for the amyloid fibrils. Various external factors like temperature, pH, pressure, ionic strength, the presence of metal ions and chemical molecules have been found to play important roles in the formation and disintegration of amyloid fibrils. Many antibodies, aromatic drugs, and polyphenolic compounds have been found to destroy preformed fibrils or inhibit formation of amyloid fibrils. However, the uses of these compounds are limited due to their cytotoxicity.[2] Thus, it is critically important to develop an efficient analytical tool that enables high-throughput screening for these diseases by decreasing the beta sheet content. Proteins, having large electric dipoles, respond to external electromagnetic radiation. It is anticipated that the application of electric field may have a role in reducing the content of β-sheets in the protein and is the major focus of this study. Few simulation studies, regarding the effect of electric field on protein conformation are found in the literature but those are performed at very high electric field strength which is not practically feasible.[3] A recent experimental study reveals that static DC electric field of small strength ($8\times10^{6}$ V m$^{-1}$) can effectively disintegrate human serum albumin (HSA) fibrils.[4] Fibrillation of human serum albumin, an amyloidogenic model protein due to its tendency to aggregate in vitro, has been studied extensively under different experimental conditions. However, HSA being a natively α-helical protein does not form fibrils under normal circumstances. The HSA fibrils are prepared incubating the protein in presence of 50% (v/v) ethanol in 20 mM Tris-HCl buffer of pH 7.0 at 37°C for 24 h. Herein, we have studied the effect of the application of an AC electric field on preformed fibrils of HSA wherein the effect of oscillation frequencies are probed at constant voltage. For the purpose of application of electric field, an arrangement similar to electro wetting on dielectric (EWOD) setup is used. Oscillating electric fields of various frequencies (0-100 Hz) at a fixed voltage of $8\times10^{6}$ V m$^{-1}$ are applied on HSA fibril droplets for a period of 5 minutes. The disintegration process has been monitored by Thioflavin T fluorescence and circular dichroism spectroscopy. From the experimental observations it has been found that at a frequency 20 Hz, maximum reduction in β-sheet content is observed. This suggests that at a frequency of 20 Hz, the fibrillar structure is significantly disrupted due to the perturbation of various non-covalent (hydrophobic and electrostatic) interactions, responsible for the stability of fibrils. The application of AC voltage induced dynamic morphology change results in the generation of induced stress inside the HSA fibril droplets, and may be responsible for the reduction of the β-sheet content. The study may prove to be a precursor to a new methodology in the field of therapeutic treatment of neurological disorders and providing better understandings of the underlying mechanism of protein misfolding.

**References**

[1] Rauk, A. (2008). *Dalton Trans.*, 10, 1273-1282.

[2] Wersinger, C., Sidhu, A. (2003). *Neurosci. Lett.*, 342, 124–128.

[3] Lugli, F., Toschi, F., Biscarini, F., Zerbetto F. (2010). *J. Chem. Theory Comput.*, 6, 3516–3526.

[4] Pandey, N. K., Mitra, S., Chakraborty, M., Ghosh, S., Sen, S., Dasgupta, S., DasGupta, S. (2014). *J. Phy. D: Appl. Phys*, 47, 305401.

# 413683 The Fragmentation of Amyloid Fibrils Occurs in Preferential Positions Depending on the Environmental Conditions

**Abstract:**

Several different proteins and peptides readily self-assemble into highly organized fibrillar clusters that are characterized by supramolecular β-sheet structures. Such protein aggregates, also known as amyloid-like fibrils, have received a significant attention over the past decades due to both their connection with neurodegenerative disorders (e.g. Alzheimer's and Parkinson's) and their potential of being employed as bionanomaterials, thanks to their remarkable strength and rigidity.[1] The ambitions of combatting protein aggregation related diseases and of creating sophisticated materials, necessitate to thoroughly understand the process of amyloid fibril formation. It is commonly accepted that protein fibrillation undergoes a nucleated polymerization mechanism, consisting of a nucleation step followed by monomer addition events. Moreover, there is also an increasing evidence that fibril breakage plays a key role in reshaping the fibril length distribution, which may have crucial implications on biological functions and on biomaterial properties. In this frame, there is an increasing interest in unravelling the mechanisms of fibril fragmentation, which are still largely unknown.

The aim of the present work is to deepen the mechanism of amyloid fibril breakage as a function of temperature and agitation rate. We aim to clarify the dependence of the fibril breakage rate constant, both on the fibril length as well as on the fragmentation position along the fibril longitudinal axis. More specifically, we are interested in establishing a link between the environmental conditions and the three mainly accepted breakage mechanisms: central mechanism (i.e. fibrils preferentially break in their center), erosion mechanism (i.e. fibrils preferentially break at their ends), and random mechanism (i.e. fibrils break with the same likelihood at any position). To do so, we compare the predictions of a deterministic kinetic model with the time evolution of fibril distributions measured by AFM for three proteins (β-lactoglobulin, insulin, and β2-microglobulin) under various conditions of temperature and mechanical agitation,[2-4] employing own and literature data.

[1] Knowles, T.P.J.; Buehler, M. J., *Nat. Nano.* **2011**, 6, 469-479.

[2] Arosio, P.; Beeg, M.; Nicoud, L.; Morbidelli, M., *Chem. Eng. Sci.* **2012**, 78, 21-32.

[3] Xue, W. F.; Radford, S. E., *Biophys. J.* **2013**, 105, 2811-2819.

[4] Nicoud L.; Lazzari S.; Barragan D.B.; M. Morbidelli, *J. Phys. Chem. B* **2015**, in press

# 438312 Bioprocess Platform Design for Large Scale Production of Selenium Free and Selenium Rich Probiotic/Biotherapeutic Yeast Saccharomyces Boulardii

**Abstract:**

Probiotic organisms or biotherapeutic agent can be defined as live microorganisms which feed on or use in adequate amounts will beneficially affects the host by improving its intestinal microbial balance. The list of potential widely used probiotics includes many strains mainly belongs to *Lactobacillus, Enterobacter, Bifidobacterium, Bacillus,* and *Saccharomyces*. Nowadays *Saccharomyces boulardii* has been used in many countries either as preventive or therapeutic agent for diarrhea and also other GI disorders which cause by administration of antimicrobial agents. In addition, *S. boulardii* act as a shuttle liberating effective enzymes, proteins and tropic factors during intestinal transit that help to improve host immune defenses, digestion and absorption of nutrients. In this study, a special type of yeast adapted to dryness and high temperature stress were applied to increase the shelf-life of the product and to decrease the cost of downstream process by replacing freeze drying by spray drying. This work was focused on optimization of different cultivation parameters to increase cell mass production with minimal alcohol production. A maximal biomass of about 85 g/L was achieved by using exponential feeding strategy in stirred tank bioreactor. The scalability of this process was proven by cultivating the cells up to 70-L stirred tank bioreactor. Furthermore, studies were also carried out to produce selenium enriched *S. boulardii* in high cell density culture to increase the health benefit of this probiotic yeast and became as a potential source of organic selenium for food supplement industries.

# 437474 Flocculant-Aided Solid-Liquid Separation of Biomass Slurries

**Abstract:**

As biochemical conversion strategies transition away from biomass-to-ethanol pathways and into bio-based hydrocarbon fuels, the removal of residual biomass solids after saccharification (pretreatment and enzymatic hydrolysis) has become imperative. The residual solids are known to inhibit the growth and utility of fermentation media, limit oxygen transfer during aerobic fermentation, poison catalysts, and disrupt flow pathways within catalyst beds. In addition, continuous microorganism-driven sugar-to-lipid conversion often necessitates a solids-free, highly concentrated syrup-like sugar stream (for example, with *Lipomyces starkeyi*). As a result, removal of insoluble solids from hydrolyzate slurries is required prior to downstream purification, concentration, and upgrading. However, these solids are commonly deformable, porous, and asymmetric, have a density similar to water, and have a wide particle size distribution, making them extremely difficult to remove from solution.

Here we utilize polyelectrolyte-flocculating agents to enhance the solid-liquid separation of enzymatically-hydrolyzed (EH) corn stover slurries. The EH corn stover slurries are a two-phase system with 5–10 wt% insoluble solids (typically rich in lignin) suspended in an aqueous phase at pH 4–5. The undigested corn stover particles (23 μm mean diameter) displayed a negative zeta potential over a wide pH range (1–13), and were easily flocculated with low doses of polymeric cationic flocculants. Initial studies evaluated flocculant dosing and separation effectiveness using centrifugation. However, economics for a centrifugation process are not expected to be favorable. Therefore, the focus of this work was vacuum-driven bulk filtration aided by flocculation. Without flocculation, the solids form a compressible cake with low permeability. However, adding low doses of a cationic polyacrylamide flocculant generated large flocs (1 mm target size), which formed highly permeable cakes under low applied differential pressures. The use of flocculation increased the overall capacity of vacuum filtration (for the slurries tested) by ~50-fold. Unfortunately, this increase in filtration capacity came at the expense of increased wash water, where washing is used to remove the residual sugar retained in the solids cake bed.

# 437478 Bioreactor and Media Development for Amniotic Fluid Stem Cells

**Abstract:**

Human amniotic fluid stem cells (hAFSCs) are emerging as an important cell source for tissue engineering and regenerative medicine due to their easy accessibility and broad multi-potentiality. In clinical applications, a large number of hAFSCs and chemically defined media are required. To address these issues, the expansion of hAFSCs in 3-dimensional (3-D) polyethylene terephthalate (PET) scaffolds in a stirred bioreactor, and the effect of currculigoside and naringin as potential components of chemical defined media on osteogenic differentiation of AFSCs were evaluated. The results showed that 3-D PET scaffold with in vivo-like environment and a large specific surface area for cell adhesion promoted cell expansion (66-fold vs. 38-fold) compared to 2-D culture. A dynamic fibrous bed bioreactor (FBB) was used to expand hAFSCs to reach a high cell density of $3.2 \times 10^6$ cells/mL. The bioreactor-expanded cells maintained clonogenic ability, high levels of expression of characteristic stem cell surface makers, and very good multi-linage differentiation capabilities. Curculigoside and naringin stimulated alkaline phosphatase activity, calcium deposition, and expression of the osteogenic genes of hAFSCs during osteogenic differentiation in a dose-dependent manner (1–100 mg/mL). Concurrently, osteoclastogenesis of hAFSCs were inhibited by curculigoside. Moreover, the roles of Wnt/β -catenin and BMP signaling pathways were revealed during curculigoside and naringin treatments, respectively. In summary, this study demonstrated the feasibility of using the FBB to mass-produce hAFSCs, and the potential use of curculigoside and naringin for chemically defined media to induce hAFSC osteogenic differentiation for future applications in tissue engineering and regenerative medicine.

# 437476 An in-House Medium Development Platform: Development of a Chemically Defined Feed Medium for Monoclonal Antibody Production

**Abstract:**

An in-house medium development platform was developed and tested across different culture scales, ranging from shake flasks, automated microscale bioreactors (AMBR15$^{TM}$) and 2-L bioreactors integrated with high-throughput bioprocess analyzers. The platform was used to develop a prototype version of a chemically defined (CD) feed medium for fed-batch production of a recombinant monoclonal antibody in Chinese hamster ovary (CHO) cells. The medium design was conducted in two stages. First, studies were performed using a Plackett-Burman design to evaluate the statistical impact of grouped nutrient components on cell growth, productivity and viability. After two rounds of screening, elevated cell growth and viability were achieved with the prototype feed medium relative to a commercial feed medium control. Next, various titer enhancers were screened to improve culture productivity and process robustness. Feed formulation and feeding schedules were also optimized through a combination of multivariate DOE strategies in both shake flask and high throughput AMBR15$^{TM}$. After three rounds of optimization, a ~2.5 fold increase in titer was achieved with the feed medium, which slightly exceeded performance of the commercial feed medium. The results were confirmed in 2-L fed-batch bioreactors, showing comparable cell growth, viability and titer with the commercially available medium. The platform strategy will be leveraged for long-term media component screening and optimization.

# 437475 Scale-Down Modeling of Continuous Centrifugation for Mammalian Cell Culture Harvest

**Abstract:**

Disc Stack centrifuge has been widely used in the mammalian cell culture for continuous harvest process. It is often used as the primary clarification step to remove cell debris, cells and other impurities. Pilot scale disc stack centrifuge has been used to model the manufacturing scale; however its requirement for utilizing a significant amount of resources and materials makes the bench-top scale more appealing. Whisperfuge, a bench-top centrifuge, as well as a combination of shear device and batch centrifuge were studied to model the production and the pilot scale continuous centrifuges. Responses of centrate turbidity, Lactate dehydrogenase (LDH) and particle size indicated good reproduction of process scalability. The capability of generating a great amount of experimental data at a small scale efficiently renders greater process characterization and optimization for mammalian cell culture harvest operations. Furthermore, a small scale bench-top model could deliver material for filter sizing studies and could address potential scale up issues earlier during the process development.

# 437593 Integration of Raman Spectroscopy into Upstream Processes As a PAT (Process Analytical Technology) Tool for Real-Time Monitoring

**Abstract:**

Raman spectroscopy is used in chemistry to provide a fingerprint by which molecules can be identified and quantified. In the present case, the same principle for the determination of temporal profiles of key biochemicals was integrated into a *Pichia* fermentation process development. There are three stages involved in this study: 1) Evaluation of the feasibility in fermentation process development by building calibration models, and testing their performance. 2) Prediction of new batches under similar condition using a different Raman unit 3) Utilization of the model for different conditions within a fermentation process platform. Calibration models were developed using partial least-squares (PLS) regression employing concentration values obtained offline and spectra collected inline during the upstream processes.

Four compounds were investigated: glycerol, methanol and sorbitol, as the main biochemicals to sustain growth and productivity; and ammonium as the sole nitrogen source. Under routine development conditions, 14 different fermentation conditions, including changes of the strain, media components and process parameter set-points were used for both modeling and test. All models showed high $R^2$ and low error from cross validation. Additionally 4 independent batches using new Raman probes and analyzer were used for prediction with a remaining low error. Finally, models were tested on fermentation batches performed under platform process conditions but expressing a different product. Prediction showed low error proving the robustness of the models based on Raman Technology.

# 437479 Ultra High Density Perfusion for Enabling Automated Cell Banking

**Abstract:**

The typical process for manufacturing biologics begins with the thaw of a cell bank and the expansion of inoculum culture. Current technologies for the generation of these cell banks can be labor intensive, and include several steps that introduce a significant risk of contamination or operator error. We describe here the development of an ultra-high-density, controlled cell culture process that mitigates these risks, and can be used for the production of both standard- and high-density cell banks. Alternating tangential flow perfusion technology was implemented and challenges of oxygen delivery and carbon dioxide accumulation were addressed. The final process supported high-viability cultures with ultra-high cell densities. These cultures were subjected to a revised banking protocol, which involved in situ addition of DMSO and direct vialing, eliminating the need for pooling cultures, centrifugation, or re-suspension, and increasing process consistency. Cell banks generated in this manner showed comparable recovery from thaw as traditional cell banks, and were considered suitable replacements for the standard-density banks. High-density, large volume cell banks generated using this process can be thawed directly into a seed train bioreactor, thus avoiding shake flask culture, reducing the time required for inoculum expansion by up to two weeks, and lowering the cost of production.

# 413342 Exploring Alternative Materials to Fabricate Microfluidic Gradient Generators to Study Algal Growth and Migration

**Abstract:**

Algae play crucial roles in providing nutrients and food resources for the higher trophic levels in addition to serving as a promising biofuel source and an environmental hazard (in the form of harmful algal blooms, HABs). Current methods employed to evaluate algal growth, viability, and migration rely on batch culturing in shaker flasks that require large volumes of reagents, extended incubation times to achieve necessary biophase concentrations, and external counting and assessment. These flasks are also limited to uniform doses and concentrations of nutrients that do not accurately reconstitute the heterogeneous, dynamic environment in which the algae reside. Advantages of using a microfluidic approach over current methods include (i) the direct, real-time visualization of single algal species, (ii) a significant reduction in experimental time and reagent volume, and (iii) the ability to create heterogeneous or homogeneous cellular microenvironments.

This presentation highlights the development of a microfluidic gradient generator, fabricated from different materials, capable of both short- and long-term experimentation on algal growth and migration. One challenge in the design of a microfluidic device compatible with algae is the elimination of direct flow while allowing for the formation of a nutrient gradient in the culturing channels as algae are non-adherent cells. This was achieved in a device comprised of three separate fluidic channels fabricated into PDMS layered on top of an agarose slab. The two outer channels provided continuous flow of nutrients and, if necessary, a chemical gradient, while the center, 'flow-free', channel served as the experimental culture channel. To achieve a fluid tight seal in the channel, the PDMS/agarose device was enclosed in a custom built Plexiglas chamber. To overcome the need for this external chamber, we explored an alternative polymeric material using thiol-ene chemistry to generate the top and bottom components of the device. By tuning the polymerization chemistry of the thiol-acrylate, we were able to fabricate a rigid, PDMS-like top layer that could be chemically bound to a porous, hydrogel-like bottom layer to allow for 'flow-free' diffusion into the center channel. Nutrient diffusion and the stability of the chemical gradients of both materials were evaluated using both fluorescent microscopy and by COMSOL simulation. Finally, both devices fabricated from both materials were used to assess both the long-term growth dynamics and chemotactic response of the model algal species *Chlamydomonas reinhardtii* under varying nitrogen and phosphorus concentration. Ultimately, this technology allows for greater control and reduced experimental time and cost compared to current methods to assess algal growth and viability.

# 414068 Modular Cell Culture Platform with Passive Fluid Controls for GI Tract - Liver Tissue Co-Culture

**Abstract:**

**Modular Cell Culture Platform with Passive Fluid Controls for GI tract - Liver Tissue Co-Culture**

Mandy B. Esch,[1] Michael L. Shuler[2]

(1) Dept. of Biomedical and Chemical Engineering, Syracuse University, Syracuse, NY; (2) Dept. of Biomedical Engineering, Cornell University, Ithaca, NY

We have developed a microfluidic cell culture platform and used it to culture GI tract epithelial cells and primary liver cells together for 14 days. In the body, these tissues are responsible for the first pass meatbolism of nutrients and drugs. Both, GI tract epithelium and liver were scaled down from *in vivo* values by a factor of 110,000. Both tissues were perfused at physiologic fluid flow rates that were controlled via passive device elements. Representing the GI tract epithelium, Caco-2 cells maintained a transepithelial resistance (TEER) of 200 to 380 ½cm$^2$. Representing human liver tissue, a mixture of human primary nonparenchymal cells (fibrolbasts, stellate and Kupffer cells) and parenchymal cells (hepatocytes), maintained urea and albumin synthesis for the entire culture period of 14 days. The modular design enabled us to culture both tissues separate from each other in order to reach maturity before combining them. The device presents a low cost approach to culturing multiple tissue with ratios of tissue volumes and fluid flow rates that are of physiologic relevance.

**Introduction:** Patients who suffer from diabetes regularly inject insulin subcutaneously. If they took insulin orally, their stomach and intestine would break it down to such a high degree that even though any remaining insulin would be delivered directly to the liver, its concentration would be too low to be effective. In the case of insulin, the direct delivery to the liver would be an advantage, because that's where it's needed. Drugs that are needed at other organs, such as painkillers, are often additionally broken down in the liver. Considerable efforts go into developing drug formulations that can overcome these obstacles.[1] Animal models do not predict the first pass effects well,[2] and *in vitro* models of the human GI tract – liver unit would be of significant use when assessing new formulations of drug that seek to increase bioavailability.

**Materials and Methods:** We designed a modular device using Solidworks and printed it with a 3D Object printer from Stratasys (Israel). We calculated channel resistances so that medium flow rates reach physiologic values when tilted the device at an angle of 18¼. On separate tissue chips we cultured Caco-2 cells (for 16 days), and 150,000 primary human nonparenchymal cells (a mixture of fibrolbasts, stellate and Kupffer cells) and 250,000 hepatocytes (for 7 days). The two tissue chips contained chambers that were scaled down from human tissue volumes 110,000 fold. After maturation, we inserted the GI tract and liver chips into the multi-organ platform and operated the device for 14 days. We replaced 50% of medium with fresh medium dayly. We measured TEER, activity of CYP enzymes (CYP 3A4 and 1A1), and urea and albumin synthesis for 14 days.

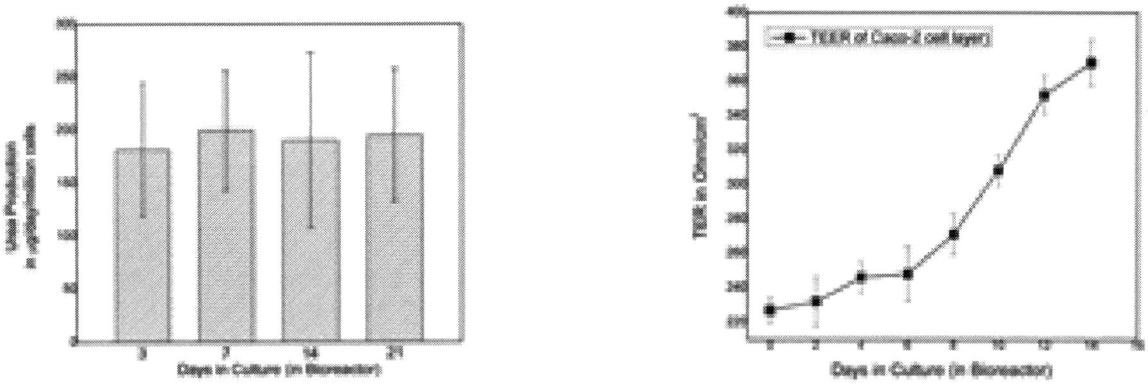


**Results and Discussion:** The devices operated for 14 days without loss of function of the tissues. The GI tract epithelium maintained TEER values between 200 and 380 ½cm$^2$ for the entire culture period (Fig.1 A). The liver tissue produced urea (Fig. 1B) and albumin (not shown) at rates we and others have previously observed under fluidic flow.[3] The cells also responded with CYP enzyme activity when stimulated with rifampicin or 3-Methylmethaxocine (results not shown).

**Figure 1.** A) TEER of Caco-2 cells for 14 days of co-culture. B) Urea synthesis of the liver tissue consisting of primary human liver cells [non-parenchymal cells (fibroblasts, stellate and Kupffer cells) and parenchymal cells (hepatocytes)].

**Conclusions:** We have developed a modular microfluidic device that operated without loss of tissue function for 21 days. We believe this device design could be useful when utilizing primary cells or stem cells to generate tissues for the purpose of combining them in order to simulate the first pass metabolism *in vitro*. The device design is expandable, meaning additional tissue modules could be designed and integrated in the future. **Ref.:** [1] S. Kalra, Diabetol Metab Syndr., 2010, 2, 66. [2] X.Gao, Pharmaceutical Research, 2006, 23(8), 1675. [3] Ebrahimkhani, ADDR, 2014, 69/70, 132.

# 418204 A Microtissue Array Device to Evaluate Effects of Multiwall Carbon Nanotubes on Lung Tissue Mechanics and Pulmonary Toxicity

**Abstract:**

Multi-wall carbon nanotubes (MWCNT) are widely used in many biomedical applications due to their excellent physical and chemical characteristics. However, the applications of MWCNT also raise health concerns. The small size and light weight of MWCNT allow them to enter the working environment as particulate matter of respirable sizes. Animal studies have shown that MWCNT cause inflammation, thickening and stiffening of the lung tissue, which leads to pulmonary diseases that are fatal and incurable. Therefore, understanding the effects of MWCNT on lung mechanics and pulmonary toxicity is important to the health safety in the applications involving MWCNT. However, due to the high cost and low throughput of the animal models, the progress in this field is very slow.

The bio-microelectromechanical systems (bio-MEMS) technology allows realistic modeling of 3D human tissue and high throughput testing of many samples in a single device, thus provides a potentially viable solution to screen the pathogenic factors. Here we report the development of a 3D human lung microtissue array device to evaluate the impact of two different types of MWCNT on lung mechanics and pulmonary toxicity.

Our device consists of arrays of PDMS microwells with each microwell containing a self-assembled microtissue hanging between a pair of micropillars. Using this device, we measured the changes in the mechanical properties and biomarkers related to the fibrogenesis in BEAS-2B human bronchial epithelial cells-populated microtissues. MWCNT (10-50um in length) and short-MWCNT (0.2-2um in length) with carboxyl modification (short-MWCNT-COOH) were selected. The cytotoxic effects of these 2 MWCNT were first evaluated by cell growth measurement and the reactive oxygen species (ROS) assay. BEAS-2B cells were treated with MWCNT at different concentrations (0, 50ng/mL, and 5ug/mL). At 48h post treatment, all treated groups showed the inhibition of cellular growth and generated significant ROS stress.

Next, the self-assembled BEAS-2B microtissues were used to study the biomechanical impacts of MWCNT. Almost all of the microtissues formed nice dog-bone shape between the micropillars and produced substantial contraction force at day 1 after initial seeding, except for the 5ug/ml short-MWCNT-COOH group, which was likely due to the cytotoxicity caused by overdose treatment of the short-MWCNT-COOH. The contraction forces continued to increase for all groups over a 3-day culture period and the 50ng/mL short-MWCNT-COOH treatment showed the largest contraction force and tissue width. However, the contraction force and tissue width of the 50ng/mL MWCNT treatment were closed to the untreated control group.

MicroRNA-21, a small non-coding RNA, has been found to mediate fibrogenic activation of lung fibrosis through the TGF-b signaling pathway, and thus represents a good biomarker for fibrogenic potential evaluation. Results from qRT-PCR showed significant up-regulation of microRNA-21 in 50ng/mL short-MWCNT-COOH treated BEAS-2B microtissues, which agreed with the high contraction force observed previously. Similarly, significant down-regulation of microRNA-21 was observed in 5ug/ml short-MWCNT-COOH treated microtissues, which was consistent with our contraction force results and probably due to the extensive cytotoxicity caused by such high concentration.

The 3D microtissue array device mimicked *in vivo* environment and allowed in-situ collection of mechanical properties of the model tissue under different treatment conditions. We demonstrated its application in investigating the effects of MWCNT on lung tissue mechanics and pulmonary toxicity. The applications of microtissue array may be extended to high throughput screening of potential pathogenic factors induced by pathogens and toxic materials.

# 425408 High-Density 3D CELL Culture and Analysis in an Integrated Microfluidic Platform

**Abstract:**

The development of robust high-throughput screening devises combining cell culture and detection systems is required for drug screening or for the applications of stem cells in therapy. In these fields, droplet-based microfluidics has recently shown avenues for controlled biological compartmentalization and high throughput cellular analysis, together with a significant reduction of the volume of reagents.

3D cultures have demonstrated improved biological functions of various mammalian cell types. In particular, the spheroids, which are tissue-like cellular aggregates, were reported to represent a physiologically relevant environment to mimic *in vivo* behavior. Various methods enable the formation of spheroids including hanging drops, low attachment wells etc. However, the application of these tools for high content screening is hindered by the limits on the number or the reproducibility of the spheroids. Moreover, these tools show limited compatibility with detection systems for in situ spheroid monitoring and analysis.

Here we demonstrate a microfluidic platform that integrates high-density spheroid formation, long-term culture and analysis. This platform is based on droplet techniques, with modifications that allow long term 3D culture at time scales relevant with drug or cellular microenvironment screening.

Nanoliter-sized aqueous droplets containing liquid agarose and hepactocytes were generated in fluorinated oil by flow focusing, in a microfabricated PDMS chip. The confined droplets were trapped in an array of 500 capillary anchors, through the reduction of their surface energy. After spheroid formation and hydrogel gelation, the oil phase was replaced with culture medium. The on-chip immobilized spheroids were cultivated and analyzed *in situ* during a 7 days culture period.

The individual spheroid formation kinetics was obtained by live imaging. Representative size and shape evolutions were determined and deviations from the mean behavior were identified. In situ live/dead, immuno-cytochemistry and BrdU staining were performed on-chip at various time points of the culture period. The analysis of over 10.000 spheroids indicated sustained viability and proliferation as well as improved expression of functional markers (i.e. intracellular albumin) compared to conventional 2D cultures. Moreover, the inter-spheroid analysis shed light on a significant correlation between albumin expression and morphometric parameters (i.e. spheroid index and size). The results were further validated by RT-PCR analysis on extracted spheroids from the chip. Furthermore, on chip intra-spheroids analysis was performed on over 100.000 cells. It was found a highly significant correlation between cellular organization (distance between cells) and the cell function.

Altogether, this study contributes to understanding the unique properties of 3D cultures by correlating micro-structural cues with the cell function. More importantly, this study demonstrates for the first time, the application of the droplet-based microfluidics into a robust and integrated long-term 3D culture and analysis platform, with potential applications in drug screening, stem cell and cancer research.

# 425762 A Microfluidic Single Particle Tracking Platform for Screening Ebolavirus Antivirals

**Abstract:**

Antiviral candidates are assessed based on their ability to prevent infection in animal models and cell lines, but these infectivity screens typically do not yield information on the mechanism of action of the potential drug. For example, the severity of the recent Ebola outbreak spurred the development of several antiviral strategies including antibodies and channel blockers, but it is not known what part of the viral life cycle these compounds are affecting. We present a platform that probes the ability of antivirals to block the entry of enveloped viruses. This platform provides us with information on how antivirals impact viral entry kinetics without the need for live Ebolavirus and microfluidics enable us to screen multiple compounds simultaneously.

Our platform has two configurations, both of which use supported lipid bilayers (SLB). In the first configuration, SLB are embedded with the receptors used for viral entry and mimic the endosomal membrane. Virus-like particles (VLP) containing the Ebolavirus fusion protein are fluorescently labeled then passed over the bilayers. Binding and fusion of individual VLPs is tracked using total internal fluorescence microscopy. In the second configuration, the SLB is embedded with viral fusion proteins to mimic the viral membrane and fluorescently labeled vesicles containing the receptor are passed over the SLB. In either configuration, the data from particle tracking is used to study the fusion between the virus-like and host cell-like membranes to determine the fusion kinetics. Measuring kinetic parameters in the presence and absence of various antivirals can provide us with insight into their mechanism of action and will allow us to determine whether or not potential Ebolavirus therapeutics impact the binding or fusion of the virus.

# 432123 Culturing Spatially Separated Anaerobes in Porous Microfluidic Culture Chambers

**Abstract:**

**Title: Culturing Spatially Separated Anaerobes in Porous Microfluidic Culture Chambers**

**Authors: Peter G. Shankles, M. J. Doktycz, J. Morrell-Falvey, M. Podar, S. T. Retterer**

Microbial communities are found throughout our environment and adapt unique architectures that enhance their survival and function. Their impact on our ecosystems and our individual health is tremendous. Current studies often rely on the use of non-culture based methods to study natural community composition. Such metagenomic and single-cell genomic techniques give rise to a more complete understanding of the individuals that comprise these communities and their metabolic potential. However, these techniques fail to capture the spatial dependencies and heterogeneities that shape community structure and function. Moreover, such techniques do not yield organisms for physiological studies, screening of specific microbe-microbe interactions, or the construction and testing of synthetic communities that can be used to recapitulate the function of natural communities.

The oral microbiome consists of over 1000 distinct species of microbes, and has a much higher culture rate than other natural environments. Nonetheless, only about half of the microbes catalogued using omics analysis have been successfully cultured. Even fewer of those associated with periodontal disease have been successfully isolated. Coevolution has caused the microbes associated with this disease to develop a complex system of relationships that are crucial to the growth of the microbes. As a result, the microbes do not grow when isolated individually. In order to culture uncultured bacteria from the oral microbiome, a co-culture approach, where bacteria are cultured together based on predicted dependencies, can be used.

In this study, we look at cultured strains of bacteria associated with periodontal disease in a microfluidic culture chamber to better understand the effects that co-culture has on growth of these bacteria. The goal of this study is to develop a platform through which sustainable culture of microbes associated with periodontitis can be achieved. To this end, we have developed a PDMS-based microfluidic device that incorporates culture chambers, media channels, and gas diffusion to allow for the culture of anaerobic bacteria. Multiple chambers are linked through micro- and nanofabricated pores that constrict the movement of microbes between chambers, but allow for cell-to-cell communication through molecular diffusion. The device is kept in an anaerobic state with a gas diffusion layer in the PDMS. $CO_2$ and $N_2$ are mixed at a ratio of 20:80 and flowed through the system. A thin, 100 mm separation, allows for gas diffusion that maintains an anaerobic environment. In earlier work, operation of the microfluidic culture chamber array, and transport across the porous membranes was characterized using fluorescent small molecules, nanoparticles and sender and receiver strains of *E. coli*. In the latter case, the production of AHL by the sender strains and transport of the small molecule signal across the connecting membranes was characterized by observing the fluorescence of the receiver strain in response to the diffusing AHL signal. Ongoing tests are intended to quantify the effects of co-culture on oral microbiome species such as *Synergistetes* and *Desulfobulbus*. Effects of co-culture on community health will be determined using these cultured microbes. In future studies, the demonstrated device can be applied to uncultured oral microbes as well as microbes from other environments.

# 432449 Microfluidic Cell Squeezing for Testing the Deformability of Tumor Cells

**Abstract:**

Metastasizing tumor cells must be able to squeeze through several narrow barriers, such as extracellular matrices and microvessels, on their way to secondary colony targets. During this process, the deformability of tumor cells is expected to play an important role. Microfluidics provides a unique opportunity to mimic in vivo cell squeezing and phenotype tumor cells based on their deformability. In this study, we developed a microfluidic device with arrays of microconstrictions. As cells flow through this multisqueezer device, we quantify cell sizes and phenotype them based on two deformability parameters – entry and transit time. We also determine the influence of channel wall friction on these two deformability parameters. We test the capabilities of the device with highly and weakly invasive prostate and breast cancer cells. We also assess the deformability of breast cancer cells that metastasize to the brain and those that do not.

We find that in general entry and transit times increase with cell size. Highly invasive prostate cancer cells CL1 have higher entry/transit times compared to lowly invasive LNCaP cells. A similar result is obtained for highly invasive breast cancer cells MB231 when tested against weakly invasive MCF7 cells. The multisqueezer device is also able to distinguish the breast cancer cells from pericardial effusion (HTB131) and brain metastasis (MB231BR). However, the device is unable to differentiate MB231BR (a subtype of MB231 cells that metastasize to brain only) and MB231 cells. These result suggest that the deformability metrics – entry and transit time – can distinguish at least some cancer cell lines. We hypothesize that differences in the frictional properties of MB231 and MB231BR might be more significant than differences in deformability allowing the MB231BR cells to cross the blood-brain barrier. We will test this hypothesis by coating the channel walls with polymers and results will be presented on the role of friction in tumor cell transport.

# 429090 Microtubules Can Bend By Dynein-Mediated Transport Towards Transiently Pinned Locations

**Abstract:**

In this talk I will outline progress we have made in combining optical microscopy with computer simulations to elucidate the mechanisms by which microtubules are bent in living cells. Microtubules have a persistence length of the order of millimeters *in vitro*, but inside cells they bend over length scales of microns. How these bends are formed is unclear, especially when the microtubule tips are far from the cell membrane. Experiments with LLC-PK1 epithelial cells, using microtubules marked by photo-bleaching show that loops develop primarily by plus-end directed transport of microtubule segments towards stationary pinning points along the length of the microtubule. The pinning points are transient in nature, and their eventual release allows the bends to relax. Experiments with cells where dynein and myosin were separately inhibited suggests that dynein is the primary force generator, able to bend microtubules against stationary pinning points. Simulations of microtubule dynamics accounting for filament mechanics and dynein forces predict the development of loops of similar size and shape to those observed in cells. Furthermore, simulations show that dynein-generated loops near a microtubule tip cause persistent rotation of the tip, leading to a change the direction of microtubule growth. Collectively, these results suggest a simple physical mechanism for the bending of growing microtubules by dynein forces accumulating at pinning points.

# 427219 Patterning of Wound-Induced Intercellular Ca2+ Flashes in a Developing Epithelium

**Abstract:**

The growth and patterning of developing epithelia result in tissues with spatially heterogeneous mechanical properties. Differential mechanical force distributions are increasingly recognized to provide important feedback into the control of an organ's final size and shape. However, the ability to quantitatively probe the mechanical properties of tissues is still in its infancy. As a second messenger that integrates and relays mechanical information to the cell, calcium ions ($Ca^{2+}$) are a prime candidate for providing important information on both the overall mechanical state of the tissue and resulting behavior at the individual-cell level during development. Still, how the spatiotemporal properties of $Ca^{2+}$ transients reflect the underlying mechanical characteristics of tissues is still poorly understood. Here we use an established model system for epithelial tissue, the *Drosophila* wing imaginal disc, to investigate how tissue properties impact the propagation of $Ca^{2+}$ transients induced by single cell laser ablations. Because experimentally manipulating mechanical properties is difficult to do without also impacting cellular differentiation, a computational model of calcium transients is employed to investigate how cell size and cell elongation (anisotropy) affect the spatiotemporal patterning of intercellular calcium flashes. We find that single cell ablations lead to intercellular $Ca^{2+}$ transients that show spatially non-uniform characteristics across the proximal-distal (PD) axis of the larval wing imaginal disc and that relative $Ca^{2+}$ flash anisotropy is principally explained by the local cell shape anisotropy. Further, $Ca^{2+}$ velocities are relatively uniform throughout the wing disc, irrespective of cell size or anisotropy. This can be explained by the opposing effects of cell diameter and cell elongation on intercellular $Ca^{2+}$ propagation. Thus, intercellular $Ca^{2+}$ transients follow lines of mechanical tension at velocities that are largely independent of tissue heterogeneity and reflect the mechanical state of the underlying tissue.

# 434514 An Inverse Problem Approach to the Analysis of Pluripotent Stem Cell Aggregation Dynamics in Stirred-Suspension Cultures

**Abstract:**

Stirred-suspension bioreactors are an attractive modality for the large-scale manufacturing of stem cell products. Key to the stirred-suspension cultivation of cells is their agglomeration which influences viability, proliferation and the stem cell differentiation trajectory [1]. A mathematical framework based on population balance equations (PBE) was applied to describe the temporal evolution of the size distribution of embryonic stem cell (ESC) aggregates cultured in a stirred-suspension bioreactor. Both cell proliferation and collision-induced aggregation were taken into account. Experiments were run at different agitation rates (60-100 rpm) and data on transient size distributions were collected. Coupling this information with the solution of the inverse problem [2] allowed for the extraction of aggregation kernels of ESCs cultivated in stirred suspension at different stirring speeds. The results obtained by forward simulation were in agreement with the experimental data supporting the validity of the computed kernels. The rate of change of the average aggregate size was greater at the intermediate stirring speed tested suggesting a trade-off between increased collisions and agitation-induced shear. More importantly, our study revealed that conditions at approximately the first 12 hours rather than the later stage (days 1-4) of cultivation determine the outcome of ESC culture. We conclude that the inverse problem framework can be applied to further our understanding of the dynamics of agglomeration processes transpiring in stem cell populations. Such frameworks are essential for the design and control of bioprocesses for the production of stem cell therapeutics.

References:

[1] J. Wu, M.R. Rostami, D.P. O. Cadavid, E.S. Tzanakakis. Oxygen transport and stem cell aggregation in stirred-suspension bioreactor cultures. *PLoS One* 9(7):e102486, 2014.

[2] H. Wright et al. Solutions of inverse problems in population balances. – I. Aggregation kinetics. *Comput. Chem. Eng.* 16(12):1019-1038.

# 426802 Amyloid Aggregation: Kinetics, Mechanisms and Models


**Abstract:**

Amyloidogenic proteins associate with each other through cross beta-sheet interactions to form fibrous ordered aggregates involved in a number of neurodegenerative diseases. The kinetics of aggregation depends upon two main factors: the primary sequence and environmental conditions. The aggregation kinetics also critically influences amyloid conformation and strain properties which govern disease patterns in mammals or phenotypic differences in yeast.

The process of amyloid formation follows a two-step pattern of initial nucleation corresponding to a lag phase followed by a fiber elongation phase to finally form a mature fiber. This behavior has been modeled empirically using mathematical functions such as the logistic function, which captures the near sigmoidal shape of the aggregation curve, and other kinetic models, for example the Finke-Watzky (F-W) 2-step mechanism which deconvolutes aggregation into slow continuous nucleation, with rate constant $k_1$, and typically fast autocatalytic growth, with rate constant $k_2$. Various other mechanisms have also been proposed to explain the amyloid aggregation process. However, many of them are not accompanied by proper kinetic analyses. This presentation will focus on the existing models for amyloid aggregation and compare mathematical models which have been used to fit amyloid aggregation data. Non-seeded and seeded aggregation of Sup35NM protein from *Saccharomyces cerevisiae* and Amyloid beta$_{1-42}$ in solutions of different ionic compositions and at different temperatures will be used to gain insights into the mechanisms of aggregation. Seeded aggregation is a complex process and depends on the seed characteristics, the monomer and recipient combination and the environmental conditions during aggregation. Further, the applicability of these models to the more complex cases of seeded aggregation will be discussed.



[1] Jonathan Rubin, Hasan Khosravi, Kathryn L. Bruce, Megan E. Lydon, Sven H. Behrens, Yury O. Chernoff, Andreas S. Bommarius.(2013), Ion-Specific Effects on Prion Nucleation and Strain Formation. *J. Biol. Chem.* 18; 288(42):30300-8 doi: 10.1074/jbc.M113.467829

# 422353 Isolation, Characterization and Analysis of the Osmotic Behaviour of hMSCs from UCB for Optimal Cryopreservation

**Abstract:**

Demographic studies have shown that population is ageing. As a consequence, degenerative events are increasing and the regenerative medicine market is growing rapidly. Within this context, stem cells possess an enormous therapeutic potential for regeneration and replacement of degenerated tissues. In particular, the ability to readily expand in culture, while maintaining a self-renewing phenotype, has made human Mesenchymal Stem Cells (hMSCs) a candidate for many cell-based therapies (Pittenger et al., 1999; Parekkadan et al., 2007). Unlike induced pluripotent stem cells and embryonic stem cells, adult hMSCs do not raise ethical and legislative issues, so that their use takes advantage of an increased likelihood for authority approval and public acceptance.

Even if bone marrow has been established as the primary source of adult hMSCs, due to the invasive nature of bone marrow aspiration, the identification of other abundant and reliable sources has nowadays become a priority. Regarding this, the successful isolation based on adherence capability to tissue culture plastic of hMSCs from peripheral sources, such as Umbilical Cord Blood (UCB), has been reported, even if, according to other contradictory studies, hMSCs in these biological samples were not found. In this framework, the ability to preserve these rare cells with high efficiency represents an even more crucial step in the regenerative medicine supply chain, since preservation now represents a core technology to bring cell-based products to market, on demand (Karlsson and Toner, 2000).

The principal preservation method consists of freezing the bio-specimens to cryogenic temperature in order to take advantage of the preservative power of the cold. If compared to the other preservation methods like maintaining the bio samples in continuous culture, cryopreservation has the benefits of affording long shelf lives, genetic stability, reduced microbial contamination risks, and cost effectiveness. The other side of the coin is that cryopreserved biological material can be damaged by the cryopreservation process itself. This damage ultimately translates into a reduced number of viable or functional cells, a loss that can be as high as 50% (Wang et al., 2011). This can be tolerated for some cell lineages, but it's unacceptable for others, as the hMSCs from UCB, whose collection and isolation is known to be difficult (Bieback et al., 2004). Even if, in principle, cell expansion/proliferation may solve the problem, an increased number of passages will inexorably lead these cells to lose their peculiar characteristics, and should be avoided (Lee et al., 2004).

Cryopreservation consists of cooling to sub-zero temperatures with or without a Cryo-Protectant Agent (CPA), storage, thawing and return to physiological environment for specific usages. Moreover, the steps of addition and removal of a CPA as DiMethyl SulfOxide (DMSO) which permeates through cells membrane into cytoplasm, needs to be taken into account when looking for optimal cryopreservation. A part from storage, all these different stages are potentially able to damage the cells due to the physical and/or chemical phenomena involved such as intra-cellular ice formation, excessive solutes concentrations and cell shrinkage or swelling. In general, due to the high number of trials actually required for experimental optimization of the cryopreservation process, mathematical modelling is considered a practical solution (Karlsson and Toner, 2000). To this aim, the osmotic behaviour needs to be first investigated in order to be able to predict the volume of residual intra-cellular water left by osmosis to form lethal ice or glass during the cryopreservation process, as well as to limit excessive cell volume excursions and solutes concentrations that might lead to the so-called *solution injury*(Fadda et al., 2010; 2011).

In this work, the hMSCs from UCB of three different donors were isolated by a density gradient centrifugation method, followed by plastic adherence of mononuclear cells. The isolation (20% success rate) was verified through phenotypic cytofluorimetric analysis, and adipogenesis/osteogenesis capability differentiations. The osmotic properties, namely inactive cell volume, water and CPA (DMSO) permeabilities, were determined by means of experimental runs carried out under hypertonic conditions (obtained with the addition of sucrose or DMSO to PBS, isotonic solution), at three different temperatures. To the best of authors' knowledge, these osmotic transport parameters have never been studied before for the hMSCs from UCB. Cell size was determined using an impedance measurement device (Coulter Counter), under equilibrium and dynamic conditions. Since the impedance measuring device does not discriminate single cells by debris or cells agglomerates, the measured cell volume distributions were filtered out from the data originally provided by the Coulter Counter through a novel data treatment, proposed in this work. Linear and non-linear regression analyses were carried out to determine the adjustable parameters by means of the salt-water sack model in the 2-parameters bi-compartimental version by Kleinahns (1998), as applied to a single-sized cell population (i.e. identical cells, with size equal to the average), as classically proposed for numerous cell lineages in the technical literature addressing cryopreservation. Basically, this model addresses a suspension of cells in a liquid, ideal solution, characterized by a given osmolality; the cells are supposed to act as a perfect osmometer in response of the ruling driving force, i.e. the difference between intra- and extra-cellular solute osmolalities, where only water and DMSO are assumed to permeate through cell membrane, while neglecting ionic transfer.

According to this model, in this work a rational parameter estimation was attempted by carrying out an ideal fitting procedure. It was found that the adopted model is not capable to simulate entirely the osmotic response for the cells under investigation. Specifically, only during swelling an apparent dependence of the so-called inactive cell volume from temperature and CPA concentration needs to be considered for hMSCs from UCB. Thus, these cells do not behave as a perfect osmometer, and show a peculiar osmotic response.

It may be concluded that, a regulatory volume system is activated for these cells, albeit only during swelling. This control system may be related to the action of ion pumps and transport channels, which are well-known in the literature for conditioning cells to return to their isotonic size, thus contrasting the lethal effect produced by osmosis (Hoffmann et al., 2009). In such a case, a more complex mathematical model than the standard salt-water sack model needs to be taken into account for capturing system behaviour. Specifically, resorting to the complex Goldman-Hodgkin-Katz model of permeant ions where quantities as membrane potential and ion permeabilities are introduced (Fernandez et al., 2013) may be considered. But, this demands a complex validation through direct comparison to data measured in well-designed experiments, which may be difficult to achieve. As an alternative, or in conjunction with this ionic transport mechanism for controlling cell swelling, the extrusion of permeant osmolites/solutes (produced inside the cells to the detriment of the inactive cell volume to osmosis) may be hypothesised. This would result in a smaller deviation from the standard salt-water sack model, simpler than taking into account the electro-diffusion of ions through cell membrane.

## References


Bieback, K., Kern, S., Kluter, H., Eichler, H., 2004. Critical Parameters for the Isolation of Mesenchymal Stem Cells from Umbilical Cord Blood. Stem Cells 22, 625–634.

Fadda, S., Cincotti, A., Cao, G., 2010. The Effect of Cell Size Distribution During the Cooling Stage of Cryopreservation without CPA. AIChE Journal 56 (8), 2173-2185.

Fadda, S., Cincotti, A., Cao, G., 2011. Rationalizing the Equilibration and Cooling Stages of Cryopreservation: The Effect, of Cell Size Distribution. AIChE Journal 57 (4), 1075-1095.

Fernandez, J.M., Di Giusto, G., Kalstein, M., Melamud, L., Rivarola, V., Ford, P., Capurro, C., 2013. Cell Volume Regulation in Cultured Human Retinal Muller Cells Is associated with Changes in trasmembrane potential. Plos One 8 (2), e57268.

Hoffmann, E.K., Lambert, I.H., Pedersen, S.F., 2009. Physiology of Cell Volume Regulation in Vertebrates. Physiological Reviews 89, 193–277.

Karlsson, J.O.M., Toner, M., 2000. Cryopreservation, in: Lanza, R.P., Langer, R., Vacanti, J., (Eds.), Principles of Tissue Engineering, 2nd Ed. Academic press, San Diego, pp. 293-307.

Kleinhans, F.W., 1998. Membrane Permeability Modeling: Kedem-Katchalsky vs a Two Parameter Formalism. Cryobiology 37, 271-289.

Lee, M.W., Choi, J., Yang, M.S., Moon, Y.J., Park, J.S., Kim, H.C., Kim, Y.J., 2004. Mesenchymal stem cells from cryopreserved human umbilical cord blood. Biochemical and Biophysical Research Communications 320, 273–278.

Parekkadan, B., Sethu, P. Van Poll, D., Yarmush, M.L., Toner, M., 2007. Osmotic selection of human mesenchymal stem/progenitor cells from umbilical cord blood. Tissue Engineering 13, 2465-2473.

Pittenger, M.F., Mackay, A.M., Beck, S.C., Jaiswal, R.K., Douglas, R., Mosca, J.D., Moorman, M.A., Simonetti, D.W., Craig, S., Marshak, D.R., 1999. Multilineage potential of adult human mesenchymal stem cells. Science 284(5411), 143– 147.

Wang, H.Y., Lun, Z.R., Lu, S.S., 2011. Cryopreservation of Umbilical Cord Blood-Derived Mesenchymal Stem Cells Without Dimethyl Sulfoxide. CryoLetters 32 (1), 81-88.

# 431170 Optimizing Surveillance Intervals and Surgery for Patients Diagnosed with Abdominal Aortic Aneurysm (AAA)

**Abstract:**

Abdominal aortic aneurysm (AAA) is when the diameter of the abdominal aorta is larger than 30 mm. The primary risk associated with AAA is aortic rupture which is fatal in 68-90% of cases. [1] Once a patient is diagnosed with AAA, the AAA is monitored via abdominal ultrasound. The rationale for the regular surveillance is that the risk of rupture is low for AAA less than 55 mm in size but increases dramatically in diameter larger than 55 mm.[2] Early surgery on patients with smaller AAA diameters (lower risk of rupture) has a higher mortality rate than taking no action. Despite numerous researches done about prediction of AAA size, there is a lack of a design which predicts the risk of surgery and rates of rupture and mortality at surveillances. This research tried to address the necessity of surveillance or surgery, rupture rate, and mortality rate in different time periods.

We applied Monte-Carlo simulation technique to a growth model based on Bayesian Analysis to simulate 10,000 hypothetical patients. [3] Using Cholesky decomposition matrix on the patient cohort data [1] kept the generated data correlated to the original data. Probability of each possible growth trajectory and cumulative risk of rupture is computed by Bayesian Analysis for each patient. Mortality and rupture rates are calculated individually applying Monte-Carlo simulation on meta-analysis papers. The risk of rupture increases in the patients with increase in the size and the mortality rate increases with the time. Generating all these predictions for numerous numbers of hypothetical patients and comparing the risk of mortality due to rupture versus the risk of mortality from surgery gives us a unique opportunity to analyze the effect of the surveillance and surgery decisions on patients' mortality. In this presentation we will utilize this mathematical technic to identify surgery and surveillance protocols to maximize the patients' survival.

**References:**

[1] Sherer EA, Bies RR, Clancy P, Norman PE and Golledge J. Growth of Screen Detected Abdominal Aortic Aneurysms in Men: A Bayesian Analysis. *Journal of Pharmacokinetics and Pharmacodynamics*, **1:**e12, 2012.

[2] Bown MJ, Sweeting MJ, Brown LC, Powell JT and Thompson SG. Surveillance Intervals for Small Abdominal Aortic Aneurysms; A Meta-analysis. *The Journal of American Medical Association*, **309:** 806-813, 2013.

[3] Shah BH, Borwanker JD and Ramkrishna D. Monte Carlo Simulation of Microbial Population Growth. *The Journal of Mathematical Biosciences*, **31:**1-23, 1976.

# 433480 Theory & Experiment for Tumor Growth, Regression & Metastasis

**Abstract:**

This paper posits a new mathematical model and complementary experiments for tumor evolution. Rather than focusing only on the how the mean tumor size evolves or on how the tumor size distribution changes due only to logistic growth and metastasis, as have been done in the literature, we model the effects of tumor growth, response to immunity and/or therapy and mitosis on the entire tumor distribution. These are chemical engineering methods that come from polymer modeling. The interaction of growth and immunity/ therapy leads to a novel diffusive term in tumor-size space that exhibits potentially patient-relevant new behavior. We will describe a fluorescent-melanoma-tumors-in-transparent-zebrafish model and planned experiments that should allow us to measure (possibly gender-dependent) model parameters for the fish system and to test model predictions to further our understanding of the fundamental science of tumor population progression. Insights from this integrated mathematical/ experimental program may help explain confounding human patient cohorts, suggest sources of gender disparities in melanoma and potentially lead to new treatment strategies.

# 425747 Particle Engineering of Cocrystals Using a Solvent-Free Approach By Spray Congealing

Wednesday, November 11, 2015: 12:30 PM
Ballroom D (Salt Palace Convention Center)

**Íris Duarte**[1,2], João F. Pinto[1] and **Márcio Temtem**[2], (1)iMed.ULisboa, Faculty of Pharmacy, University of Lisbon, Lisbon, Portugal, (2)R&D Drug Product Development, Hovione FarmaCiência SA, Loures, Portugal

**Particle engineering of cocrystals using a solvent-free approach by spray congealing**

Iris Duarte[1,2], Joao F. Pinto[2], Marcio Temtem[1*]

[1] Hovione Farmaciencia SA, Sete Casas, 2674-506 Loures, Portugal;
[*]mtemtem@hovione.com or +351 219 847 569

[2] iMed.ULisboa, Faculdade de Farmacia da Universidade de Lisboa, Av. Prof. Gama Pinto, 1649-003, Lisboa, Portugal

Around 90% of the new chemical entities emerging from the drug discovery phase present, at least, solubility constraints (BCS Class II+IV). Among the different solubilization strategies that have been applied to address this issue, the use of pharmaceutical cocrystals emerged as an alternative crystal-engineering platform to enhance the poor physicochemical properties of these challenging molecules [1].

Pharmaceutical cocrystals are stable multicomponent crystals, typically comprising two molecules, *e.g.* an API, or the salt of an API, and a coformer. The coformer can be an active or non-active ingredient (*e.g.* pharmaceutical excipients, vitamins, minerals, amino acids, other APIs) that has a favorable intermolecular interaction with the API, promoting hydrogen bonding, van der Wall forces or pi-stacking [2].

In general, and despite the great potential of cocrystallization, producing cocrystals with high efficiency at industrial scales is challenging due to the tight thermodynamic and kinetic crystallization windows necessary to achieve the desired crystalline form. Specifically referring to the pharmaceutical field, the wide application of cocrystals is still limited due, in part, to the scarcity of suitable large-scale production methods and lack of robust and cost-effective processes.

In order to address some of these challenges, spray congealing has recently emerged as new process for the preparation of cocrystals [3]. Spray congealing, which is a hybrid technology governed by similar principles of spray drying and hot melt extrusion, consists in feeding a molten mixture to an atomizer, which then breaks the liquid feed into small droplets, and those droplets are cooled thanks to a co-current stream of cooling gas (Figure 1).

The advantages of this method are the fact it can be conducted in the same apparatus of spray drying, with minor modifications, thus making the commercial scale of the process readily available; for being a solvent free technique, cocrystallization via spray congealing will also comply with green chemistry requirements and, most importantly, it can be used to produce "engineered" cocrystals (*i.e.* with the desired properties) in a particulate form, in a single stage operation without the need of any downstream processing.

The feasibility of using spray congealing to produce pharmaceutical cocrystals was successfully demonstrated using three model API:coformer systems (Table 1).

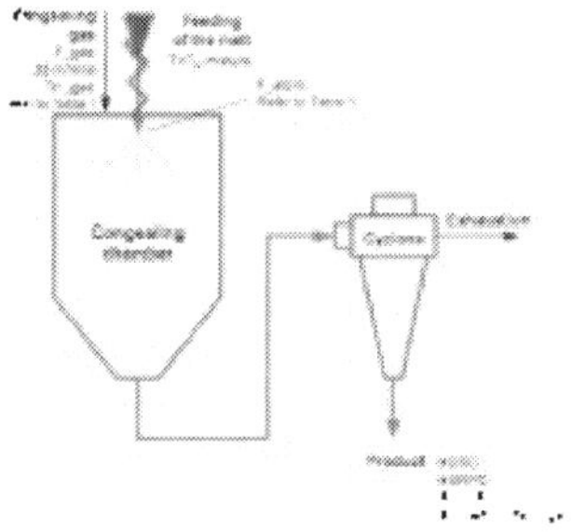


**Figure 1**. Schematic representation of the spray congealing set up used.

**Table 1.** Case-studies and respective process variables defined for the tests. The flow rate of the congealing (F_gas) was set at $0.35 m^3/min$.

| System | Molar Proportion | Process Variables | | Ref |
|---|---|---|---|---|
| | | ΔT [°C]* | F_atom [L/min]** | |
| Caffeine:Glutaric acid (CAF:GLU) | 1:1 | 0-50 | 11-20 | 4,5 |
| Caffeine:Salicylic acid (CAF:SAL) | 1:1 | 50 | 9 | 5 |
| Carbamazepine:Nicotinamide (CBZ:NIC) | 1:1 | 125 | 12 | 5-7 |

* ΔT: difference between the mixture melting temperature observed inside the beaker ($T_{M,mix}$) and the inlet temperature of the congealing gas ($T_{in,\ gas}$);

**F_atom: flow of the atomization gas.

All the final products obtained with spray congealing presented DSC thermal profiles and XRPD diffractograms characteristic of the respective cocrystals and different from the pure components or physical mixtures. As an example, Figure 2 shows the results obtained for the 1:1 Caffeine: Salicylic Acid system.

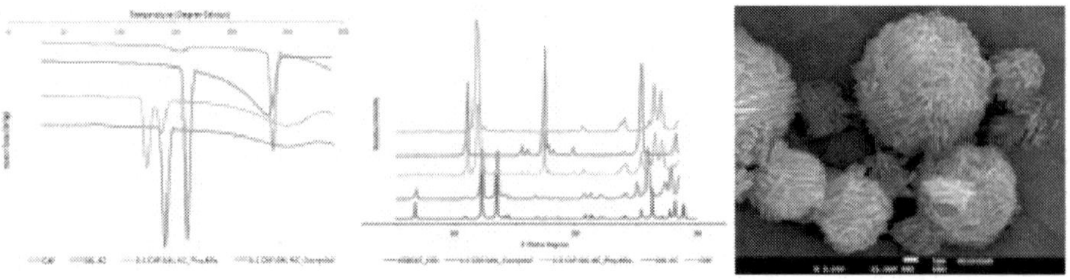

**Figure 2.** Results obtained for the 1:1 CAF:SAL system. From left to right: DSC, XRPD and SEM.

Moreover, and for one specific case-study, *i.e.* 1:1 Caffeine:Glutaric acid (CAF:GLU), a DoE with 2 parameters at 2 levels + 1 central point was conducted, to assess the effect of process variables on the final cocrystal particle and bulk powder properties. The parameters or process variables (*i.e.* ΔT and F_atom) were varied according to the ranges shown in Table 1.

According to Figure 3 it can be observed that particle properties (size, morphology, shape) can be adjusted, *in situ*, as part of the spray congealing process, without compromising cocrystal formation (DSC and XRPD data not shown).

Moreover, particle properties are known to affect the bulk behaviour of powders, which can impact subsequent downstream processing (*i.e.* blending, capsule filling, tableting, etc.) [8]. In order to evaluate this, bulk (*i.e*, compressibility and permeability) and shear properties of the powders produced were measured using a FT4 powder rheometer. Figure 4 shows the results from the permeability and compressibility tests at 15 kPa, and the shear cell test on samples that have been pre-consolidated at 9kPa normal stress.

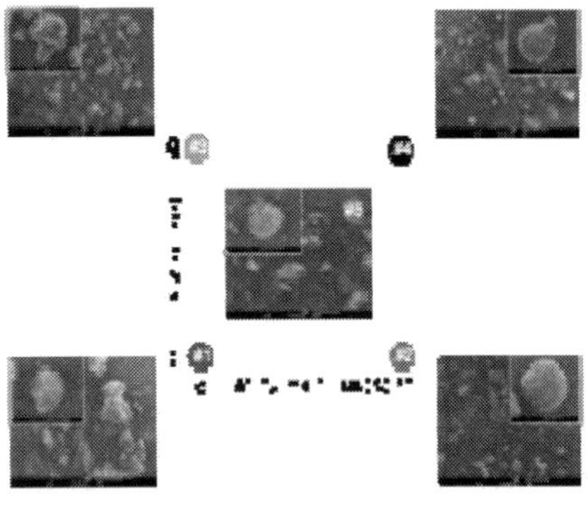

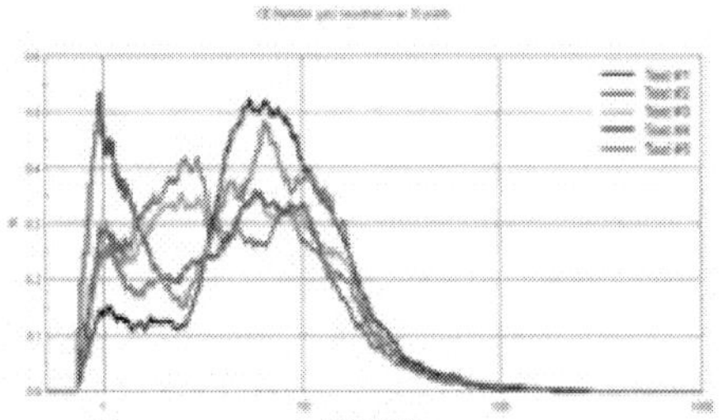

**Figure 3.** SEM results (right) and number-based circular equivalente diameter (CED) distributions (above) for the 1:1 CAF:GLU systems produced, when varying ΔT and F_atom.

Shear testing classified the powders as identical, whilst the permeability and compressibility tests show they are significantly different. The shear data suggest that the powders may behave similarly in a moderate to high stress environment, such as in a hopper, but it does not necessarily mean that they will behave similarly when aerated or flowing under low stress. The different tendencies of the permeability and compressibility results are clearly a consequence of the influence of multiple variables or particle properties (*e.g.* particle size, shape, surface properties, amount of fines, etc.) on the bulk powder rheology.

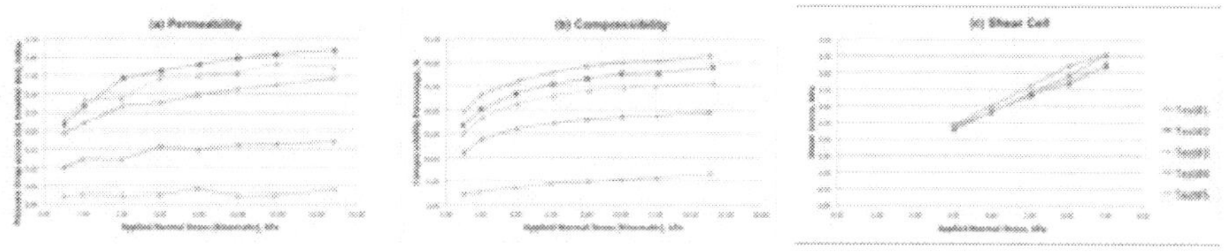

**Figure 4.** From left to right: permeability as a function of normal stress at constant air velocity of 2 mm/s, compressibility (*i.e.* % of bulk density change) as a function of normal stress and shear stress as a function of applied normal stress.

**References:**


[1] Blagden N. *et al.*, Pharmaceutical co-crystals - are we there yet?. *CrystEngComm*, 2014, 16, pp. 5753-5761;

[2] Qiao, N. *et al.*, Pharmaceutical cocrystals: An overview. *International Journal of Pharmaceutics*, 2011, 419, pp. 1-11;

[3] Duarte, I., *et al.* Synthesis and particle engineering of cocrystals. WO2015/036799A1 (2015);

[4] Trask *et al.*, Solvent-drop grinding: green polymorph control of cocrystallization. *Chem. Commun.*, 2004, 7, 890-891;

[5] Lu *et al.*, A rapid thermal method for cocrystal screening. *CrystEngComm*, 2008, 10, 665-668;

[6] Chieng *et al.*, Formation Kinetics and Stability of Carbamazepine - Nicotinamide Cocrystals Prepared by Mechanical Activation. *Cryst. Growth Des.,* 2009, 9(5), 2377-2386;

[7] Liu *et al.*, Improving the Chemical Stability of Amorphous Solid Dispersion with Cocrystal Technique by Hot Melt Extrusion. *Pharm. Res.*, 2012, 29, 806-817;

[8] Freeman R., Measuring the flow properties of consolidated, conditioned and aerated powders – a comparative study using a powder rheometer and rotational shear cell. *Powder Technology*, 2007, 174, pp. 25-33.

# 433246 A Case Study in Tuning API Powder Properties Via Crystallization to Meet Formulation Needs

**Abstract:**

Crystal form and powder properties are universal quality attributes which characterize every API. This presentation is a case study of one API that has the potential to form numerous polymorphs and a variety of crystal morphologies. Process development efforts were focused on obtaining the thermodynamically stable crystal form while targeting a narrow range of powder properties suitable for bioavailability and downstream drug product formulation. The work includes screening studies to select the appropriate solvent system and crystallization parameter optimization to control particle size and minimize agglomeration. Multiple technologies were employed to support this work, including FTIR, FBRM, wet milling, and dry milling. Understanding the impacts of crystallization parameters allowed for the generation of several lots of material with a variety of powder properties, which were evaluated for formulation behavior and bioavailability. The feedback and collaboration between drug substance and drug product process development teams enabled the selection of the desired target material properties (eg. particle size distribution, D90/D50 ratio) and formulation process (roller compaction or wet granulation) that would robustly lead to high quality drug to patients.

# 432520 Improving Dissolution Kinetics of Pharmaceuticals By Fluidized Bed Impregnation

**Abstract:**

According to the current Biopharmaceutics Classification System (BCS) adopted by the U.S. Federal Drug Administration (FDA), drug substances are categorized into four main classes, based on their solubility and permeability. Two of those classes (BCS classes II and IV) are characterized with low solubility. The trend in the past decade has been toward steady increase in the number of active pharmaceutical ingredients (APIs) belonging to BCS classes II and IV. It has been estimated that up to 60% of new chemical entities fall in the low solubility class, compared to 39% for marked drugs. This trend towards decrease in API solubility is undoubtedly one of the main hurdles for formulators today, as they continuously strive to develop novel formulations capable of effectively delivering the drug into the blood stream.

API solubility has a direct effect on the drug's bioavailability and therefore its efficacy. The absorption of BCS Class II and IV compounds and the resulting blood concentration-time profile depends on the rate-limiting step, which can be one of the following: permeability limited, dissolution limited and solubility limited. For poorly-soluble compounds, improving the rate (for dissolution limited) and the extent of dissolution (for solubility limited) will directly increase their bioavailability. There are two main techniques that are widely used today for improving dissolution rate and increasing solubility. The first technique involves particle size reduction as suggested by the Nernst-Brunner/Noyes-Whitney equation, where the dissolution rate for any API is proportional to the surface area available for dissolution. Solubility of any substance is a thermodynamic property, which among other environmental factors (solvent nature, pH and temperature) depends strongly on the chemical or physical structure of the substance itself (e.g. salts, crystalline or amorphous solids, solvates, co-crystals). Many variations of these two main techniques for improving dissolution rate and increasing solubility are employed into today's development or marketed drug formulations. The most common include various bottom up or top down approaches for size reduction and numerous means for producing amorphous materials (e.g. spray drying, freeze drying, hot-melt extrusion, polymer additives). Often these methods have major challenges and limitations associated with: difficulty in control or scale up, API stability, applicability to wide range of APIs, high cost, etc.

In this work we present a drug formulation method that is capable of increasing significantly the dissolution kinetics of a poorly soluble API (Fenofibrate), which shows a high degree of simplicity compared to many of the traditional methods. The method involves impregnation of the API into highly porous excipient in a fluidized bed (FB). Loadings of up to 40% were successfully formulated, showing significant improvement in the dissolution profile. Additional improvement in dissolution was achieved by mild milling of the impregnated drug product. Moreover, co-impregnation of the API with an FDA approved additive showed significant further improvement in dissolution profile, making our formulation comparable to one of Fenofibrate's patented market formulations (TriCor, 45mg). Other benefits of our impregnation method include: high blend uniformity independent on API nature or loading, excellent flow properties of the impregnated powder (no need for more additives), elimination of several unit operations from drug substance and drug product development.

# 423877 A Drop-on-Demand Manufacturing System for the Production of Amorphous Drug Products

**Abstract:**

Innovation is the key to success within the pharmaceutical industry. However, it has been largely sought through new drug discovery and development, while the development of more efficient manufacturing methods has received inadequate attention. Recently, as a result of encouragement from the regulatory authorities, the advent of globalization and the increasing awareness of environmental impact, the pharmaceutical industry has been reconsidering the way drug products are developed and manufactured. The US Food and Drug Administration (FDA) has promoted the Quality by Design (QbD) approach to increase process understanding and improve quality and efficiency while minimizing risk [1]. The Process Analytical Technology (PAT) guidance introduced by the FDA encouraged the monitoring of critical quality and performance attributes during processing with the goal of ensuring final product quality [2,3].

As part of this renewed emphasis on improvement of manufacturing and under the US National Science Foundation supported Engineering Research Center for Structured Organic Particulate Systems; we have developed a dropwise additive manufacturing process for solid oral dosage production. The process utilizes the drop-on-demand (DoD) inkjet printing technology for predictable and highly controllable deposition of active pharmaceutical ingredients (API) onto an edible substrate, such as a polymeric film or placebo tablet [4, 5]. This process uses fluid operations suitable for low volume production of personalized dosage forms. The advantages of liquid processing and reproducible production of small droplets create the special opportunity for the production of high potency, low dose drugs, which are difficult to produce with consistent quality via conventional powder processing methods.

Using the developed drop on demand manufacturing process for pharmaceuticals, different drug formulations including solvent based systems [4], i.e. solvent-polymer-API solutions, as well as melt based systems [5], i.e. polymer-API melts, can be printed. Melt-based printing applications eliminate the solvent evaporation step after drop deposition and thus allow on demand production of individual dosage forms with good control of drug solid state and morphology.

In this paper, we present the controlled production of amorphous dosages by the deposition of melt-based formulation using the DoD manufacturing system. Producing dosage forms in which the active ingredient is in amorphous forms can enhance the dissolution of low-solubility drugs. Amorphous drug forms are produced with alternative choice of polymers, which inhibit crystallization, along with suitable choice of operating conditions. Polymers are added to the printing material to help control drug morphology, material properties and the formulation composition [6]. Addition of crystallization inhibiting polymers, such as PVP, to the formulation, can result in the formation of dosage forms containing amorphous or semi-crystalline API. Operating conditions, i.e. crystallization temperature, are controlled using substrate temperature controller, which is used to achieve the desired product quality by influencing the dissolution behavior and therefore the bioavailability of the drug [7]. Since stability of amorphous drug forms can be challenging, design of crystallization temperature profile and precise control of substrate temperature profile are even more important than in the case of crystalline API formulations. The amorphous dosage forms are achieved by controlling the cooling profile of the deposits to solidify the drug in the amorphous form before crystallization occurs.

1. Yu LX, Amidon G, Khan MA, Hoag SW, Polli J, Raju GK, et al. Understanding pharmaceutical quality by design. AAPS J. 2014;16(4):771–83.
2. Food and Drug Administration CDER. Guidance for Industry PAT - A Framework for Innovative Pharmaceutical. 2004.
3. Simon LL, Pataki H, Marosi G, Meemken F, Hungerbu K, Baiker A, et al. Assessment of Recent Process Analytical Technology (PAT) Trends: A Multiauthor Review. Org Process Res Dev. 2015;19(1):3–62.
4. Hirshfield L, Giridhar A, Taylor LS, Harris MT, Reklaitis G V. Dropwise additive manufacturing of pharmaceutical products for solvent-based dosage forms. J Pharm Sci. 2014;103(2):496–506.
5. Içten E, Giridhar A, Taylor LS, Nagy ZK, Reklaitis G V. Dropwise Additive Manufacturing of Pharmaceutical Products for Melt-Based Dosage Forms. J Pharm Sci. 2015;104(5):1641-49.
6. Trasi NS and Taylor LS. Effect of Additives on Crystal Growth and Nucleation of Amorphous Flutamide. Crystal Growth & Design 2015; 12(6): 3221–3230.
7. Içten E, Nagy ZK, Reklaitis GV. Supervisory Control of a Drop on Demand Mini-manufacturing System for Pharmaceuticals. Comp. Aided Process Eng., 2014; 33: 535-540.

# 427189 Separation of Fine Oil Droplets from Oil-in-Water Mixtures By Dissolved Air Flotation

**Abstract:**

Dissolved air flotation (DAF) is a separation technique, often used after a primary gravity separator or filters, to enhance the quality of the wastewater before it can be released to streams, rivers and the sea. The main aim of the DAF investigations reported here was to measure the removal efficiency of oil droplets in the ranges of 15-80 μm from the oil-in-water mixtures. Several operating parameters were varied and two kinds of oil were used. It was decided that vegetable oil should be tested as it is relevant to the food industries. To compare the vegetable oil droplet removal efficiencies, investigations were also carried out using a mineral oil. Saline water in this experiment was made by adding 3.5% by concentration of NaCl, to mimic the average salinity of water produced from the oil wells. An oil-in-water measuring technique (FastHEX) and a droplet counting method (Coulter Counter) were used to measure the concentration and numbers of oil droplets in the water, respectively. The best oil droplet removal efficiency obtained from these experiments was found to be 96 % and it was for the removal of vegetable oil droplets from a saline oil-in-water mixture. The vegetable oil has a large and positive spreading coefficient ($14.6 \times 10^{-3}$ N/m), which facilitates the attachment of bubble and oil droplet and therefore produced higher removal efficiency than lamp oil that has a negative spreading coefficient ($-7.11 \times 10^{-3}$ N/m). The experimental results were subjected to dimensional analysis and the removal efficiency was found to be a function of several dimension and dimensionless groups. The experimental data tabulated in dimensionless form has been subjected to multivariable linear regression. The resulting correlation was found to have a root mean square error of 11.1 % but predicted oil droplet removal efficiencies greater than one and less than zero. An alternative mathematical formulation was devised that cannot predict efficiencies outside the range. Regression of this formulation, which had the same number of adjustable parameters as the linear regression, was successful with a root mean square error of 10.9 %.

# 419313 Modeling of Complex Bio Systems Using Batchreactor

**Abstract:**

Process simulation is widely used by process engineers in the design of new units as well as in operations of existing plants for process optimization, units troubleshooting or debottlenecking, plants revamping, performing front-end engineering analysis… It is used in nearly all process industries: chemicals, pharmaceuticals, petrochemicals, oil and gas, refining, specialty chemicals… However, bio-industries still do not use intensively process simulation. In these industries, microorganisms (fermentation), enzymatic reactions or reactive transformation of the plant material are involved. They are characterized by a great complexity of the raw material and the phenomena which are taking place in reactive unit operations. The number of components (often non-exhaustive for a plant material) and the number of reactions (several thousand for a given microorganism) limit the number of robust reactional models available in literature. As they most often operate in batch mode, bio-reactions are rarely presents and used in process simulation software.

BatchReactor® is a software dedicated to the simulation of batch chemical reactors [1], which provides the user with a detailed modeling of the reactor (heating/cooling system, condenser, mixing device…) and a reliable description of the production recipe. BatchReactor® allows modeling of complex systems by taking advantage of the power of Simulis® Thermodynamics (thermodynamic properties server) which offers an extensive set of thermodynamic models with a pure components database of more than 2,000 pure components (AIChE's DIPPR® database [2]). Additional new components can be easily added in database as for example microorganisms represented with a CHONPS formula. Moreover, thanks to the user "interpreted" kinetic rate model available through Simulis® Reaction (chemical reaction server), any kind of specific kinetic models can be used in BatchReactor®. This is essential to represent kinetic bio-reactions with often non-classical laws [3]. So, in BatchReactor®, stoechio-kinetic model can be coupled with thermodynamic and heat transfer models already available in the software to simulate the bio-reactor.

Several applications from various industries have been successfully simulated: food industry (tomato juice oxidation, brewing of beer, alcohol fermentation), pharmaceutical (production of thiolutine) and white biotechnology (production of polyhydroxybutyrate). A stoechio-kinetic model was necessary for all these applications but, when it was known, the strengths of BatchReactor appeared: accurate thermodynamic and transfer phenomena representation, recipe description, user-friendly interface.

References

[1] www.prosim.net

[2] R.L. Rowley, W.V. Wilding, J.L. Oscarson, N.F. Giles, DIPPR® Data Compilation of Pure Chemical Properties, Design Institute for Physical Properties, AIChE, New York, NY (2011)

[3] Ramon Portugal F., Fillon M., Meyer X.M., Pingaud H., Strehiano P., I.S.B.N. 2-87805-022-3, 159-170 (1997)

# 424888 Oxygen Transfer Rate Effect on Pigment Production By Penicillium Purpurogenum GH2 in Shake Flask Cultures


**Abstract:**

Fermentation processes have become a popular alternative to obtain natural food ingredients such as natural colorants. Filamentous fungi seem to be an attractive source of bio-pigments due to the wide range of colors that are able to produce at high yields. Production of pigments by filamentous fungi in submerged fermentation is affected by a large number of operating variables like pH, temperature, agitation and aeration rate that may interact strongly between themselves and with the specific characteristics and properties of growth medium and microorganisms, creating a unique environment resulting in a particular final yield of the process. Industrial application of fungal pigments will depend on the successful scaling up of the process. In aerobic fermentation, oxygen transfer rate (OTR) is a key parameter utilized on the design and scale up of microbial processes aiming for the highest productivity.

The presented study aimed to characterize the oxygen transfer conditions in shake flask culture and to analyze its effect on growth and pigment production by *Penicillium purpurogenum* GH2.

Fermentation experiments were carried out in Erlenmeyer flasks (125 mL) for 8 days at 30 °C. Oxygen transfer was controlled by varying agitation speed (200 to 250 rpm) and the medium volume in flask (25, 50 and 75 mL). Oxygen transfer coefficient ($k_La$) was determined without biological consumption of oxygen using the dynamic method. OTR at the evaluated agitation speeds and liquid volumes was estimated by OTR = $k_La$ ($C^*-C_L$) using the obtained $k_La$ values, where ($C^*-C_L$) varied approximately from 7.0-7.5 mg/L at the range of conditions studied.

Pigment production and growth were affected by the conditions tested. Pigment production increased as liquid volume decreased from 75 to 25 mL at both agitation speeds (200-250). The highest pigment production obtained at 250 rpm with 25 mL of liquid volume was 14.42 $OD_{500nm}$ which was 23% higher than the obtained with 25 mL at 200 rpm. Meanwhile the maximum biomass growth was obtained at 50 mL (3.36 g/L) at 200 and at 75 mL (3.37 g/L) at 250 rpm. The oxygen transfer coefficient $k_La$ decreased with the liquid volume and increased with the agitation speed (11.02-17.60 $h^{-1}$). Pigment production showed a linear correlation with $k_La$ (Pigment = 2.17$k_La$-23.73, $R^2$=0.98) and with OTR (Pigment = 0.32OTR-26.67, $R^2$=0.90). Biomass growth increased by decreasing $k_La$ from 17.60-11.59 $h^{-1}$ reached a maximum biomass 3.37 $gL^{-1}$ and decreased at a $k_La$ value of 11.02 $h^{-1}$.

The above results indicated that oxygen transfer rate in shake flasks depends on agitation speed and working volume used in submerged fermentation. The production of pigments by *Penicillium purpurogenum* GH2 is strongly affected by the oxygen supply in the medium. A direct correlation was found between pigment production and $k_La$ and OTR which represent a major finding as reference for future scale up of the process from shake flask to bioreactor.

# 435117 Experimental Orders on an Entrepreneurial Pace: Dairy Wastewater Characterization Control Using Sequenced Blending

**Abstract:**

A creative mindset must be adopted in the effort to create sustainable processes. Whether it be a process improvement to an existing technology or a novel application of sustainable processing, screening the effects through experimental orders often must be done in a fast pace environment with limited information. The case study presented is dairy wastewater control applying an entrepreneurial mindset to practical experimental orders.

Discharge specifications for effluents to the POTW have pH as a significant regulatory compliance component. Many small to medium cheese plant operations discharge directly to the POTW. Usually, cheese and milk products are produced in the same plant. There is both acidic and basic cleaning solution discharge streams. By using a selective blending control system the effluent from the plant to the POTW can be modified in process and in some cases water can be recovered for other operations. Cycling of the cleaning operations is not practical with the pressures upon production. The installation and use of pH monitoring and flow control systems can temper the blended discharge. pH curves can be determined to assess the capacities of both the basic and acidic fluids to accept volumes of the other for in process neutralization.

Cheese whey is acidic. Samples are in the pH 4 – 5 range. The high pH cleaning residue can accept cheese whey as an effluent constituent. Whey residue contains calcium which will precipitate as the pH is raised. The baseline pH variability of the current effluent treatment system can be established and translated to a bench lab titration. The cheese whey was characterized and added in a mass ratio to the available effluents then titrated similarly to the current system. The pH neutralization curves were established for each of the three loading levels. The family of curves were used to establish the mass balance expected at full scale Sample analysis was conducted at the nominal operating temperatures.

Typically high pH characteristic flow can accept up a mass of the cheese whey without dropping the pH below 10. This condition could occur frequently. A sequenced blending control scheme would control of the cheese whey addition should be programmed to occur only when the waste tank material has the capacity to accept the cheese whey. A second waste tank characteristic with a lower pH in the range of 7 – 9 was also observed. It did not have capacity to neutralize the added cheese whey. The sequenced blending control scheme will be needed to avoid this possible out of specification condition.

The addition of cheese whey may require additional caustic neutralization. This may add additional precipitates to the existing waste stream. The pH neutralization kinetics could be effected by the formation of additional precipitates. The mixing parameters can be validated by lab work in sample jars and beakers. These indicative results can then be run in a scalable mix system. Effects of mixing on the reaction kinetics can be difficult to observe in small lab volume samples.

# 428251 Chemical Mapping of Sweeteners and Additives in Commercial Chewing Gums Using Imaging FTIR Spectroscopy

**Abstract:**

**Chemical Mapping of Sweeteners and Additives in Commercial Chewing Gums using imaging FTIR Spectroscopy**

Ramazan KIZIL,

Istanbul Technical University, College of Chemical and Metallurgical Engineering, Chemical Eng. Dept. Maslak, 34469 Istanbul-Turkey. Contact: kizilr@itu.edu.tr

Determination of the structural organizations of chemical constituents, such as lipids, proteins and carbohydrates, within a food at the microscopic level is expected to significantly contribute to the modern food technology whose prime interest lies in development of controlled processes and operations to preserve, transform and create or destroy food structure (Aguilera, 2001). Food microstructure analysis has an ultimate goal of understanding structural hierarchies, forming naturally or manufactured, from images that depict spatial arrangements and interactions of chemical compounds that form flavor, color, texture and nutritional value. Developing new tool to correlation microstructure with food texture perception is expected to contribute to food industry in production of textually attractive food products (Wilkinson et al. 2000).

Vibrational spectroscopic tool such as Raman or infrared spectroscopy coupled with an optical microscope can be used to visualize structures in a form of image produced from chemical composition contrast. The technique is known to be chemical imaging and it is becoming a promising modality for diagnosis of cancer as a complimentary molecular histopathology tool. Another interesting application field of chemical imaging is pharmaceutical industry for which visualization of the distribution of active ingredients and excipients is prime interest.

Analyzing molecular vibrational transitions of atoms of a compound as a result of absorbance of IR radiation provides a strong basis for chemical structure determination of organic or inorganic substances. This analysis can be best performed by implementing a multiplexing detection of all wavelengths at the same time using a focal plane or linear array detectors. A classical FTIR system coupled to an optical microscope with a motion controlled stage and a multiplexing detector collects spectrum at a resolution set by the diffraction limit, which is around 5.5 micron for most commercial IR imaging systems.

Chemical imaging is a powerful tool for visualization of structure of a specimen from its functional groups in a form of chemical contrast map, representing the spatial distribution of chemical entities in the sample. FTIR provides the contrast necessary for enabling chemical structural imaging by means of the selectivity of vibrational absorption modes detected only at specific frequencies/energies. The spectrum at each pixel is referred to as fingerprint of chemical composition.

This talk covers experimental studies of determining the spatial distribution of sweeteners and additives in chewing gums to elucidate the chemical nature of developed microstructure in different brands. The samples were investigated in attenuated total internal reflectance (ATR) mode in order to obtain images at pixel resolution of 1.65 micron by oversampling approach and solid immersion lens (SIL) technology (Kazarian and Chan, 2010). Similar to oil immersion optics, ATR uses hemispherical Germanium element which has refractive index of 4.1. This allow to obtain images with 4 times enhanced resolution comparing to that of transmission or transreflectance modes of IR imaging. Three different brand commercial chewing gums were used to study microstructure imaging. ATR imaging necessitates a firm contact of the Ge element with the sample, so a little distortion of images due to contact pressure related stretch. Since all samples will be subjected to a similar stretch condition, the images offer a comparing realistic visualization of the microstructure. Chemical mapping strategy was followed to show the spatial distribution of sweetener and additives in the texture from their own molecular fingerprint. The main matrix was well represented in the images and inhomogeneous distributions of other components were mapped at a pixel resolution of 1.65 micron. The findings of the study and sensory evaluations suggests that controlling the microstructure can help improve the sensory quality and functional characteristics of chewing gums.

References:

Aguilera, J.M. 2005. Why food microstructure. Journal of Food Engineering. 67:3-11.

Kazarian, S.G., Chan, A. 2010. Micro-and macro ATR infrared spectroscopic imaging. Applied Spectroscopy. 64:135A-152A.

Wilkinson, C., Dijksterhuis, B.G., Minekus, M. 2000. From food structure to texture. Trends in Food Science and Technology. 11:442-450.

# 415742 Chemical-Genetic Inference of Antibiotic Interactions for Combination Therapies

**Abstract:**

Combination antibiotic therapies are being increasingly used in response to a global rise in drug resistance. Mapping synergistic and antagonistic interactions between antibiotics is essential for designing effective combinations that enhance potency and counter resistance. However, the large search-space of candidate drugs and dosage regimes makes such approaches highly challenging. Here, we present a computational approach that integrates publicly available chemogenomic data to predict dose-specific antibiotic interactions. We experimentally tested interactions for 171 antibiotic pairs in nine dose combinations in *Escherichia coli* to validate our approach. Our analysis revealed a core set of cellular pathways (e.g., central metabolism) and biophysical factors (e.g., drug lipophilicity) that are predictive of synergy and antagonism. Using an evolutionary approach, we modified our chemical-genetic model of *E. coli*, a Gram-negative bacterium, to accurately predict drug-drug interactions in the Gram-positive pathogen *Staphylococcus aureus*. This study provides a framework for chemogenomics-driven discovery and development of effective combination therapies.

# 431903 Dissecting the Effects of (5Z)-4-Bromo-5-(bromomethylene)-3-Butyl-2-(5H)-Furanone on Bacillus Anthracis Metabolism

**Abstract:**

Quorum sensing inhibitors or quorum quenchers hold great potential as therapeutics for treating infectious diseases. One such molecule, *(5Z)*-4-bromo-5-(bromomethylene)-3-butyl-2-*(5H)*-furanone (furanone) has been found to inhibit the quorum sensing circuit of *Bacillus anthracis*, the etiological agent of anthrax. However, the impact of furanone on the cellular system as a whole remains unclear. In this work, the affect of furanone on cellular metabolism was evaluated through an integrated analysis involving genome-scale metabolic modeling and experimental efforts. Specifically, a metabolic model of *B. anthracis* consisting of 1,003 metabolites and 1,049 reactions was used to analyze experimental results collected for the Sterne strain of *B. anthracis* when grown in the presence of furanone. It was found that although growth rate and glucose uptake rate both decreased in the presence of furanone, energy demand increased. It was postulated that the higher energy requirement was necessary to accommodate critical energy-dependent translocation processes of key nutrients. Additionally, the introduction of furanone appeared to alter the redox potential generation pathway, indicating that regeneration of NAD+ occurred by electron transfer to glycerol-3-phosphate rather than fermentation. The overall impact appeared to be indicative of an oxidative stress response.

# 422573 Discovery and Dissection of Metabolic Oscillations in the Nitric Oxide Response of *Escherichia coli* Under Microaerobiosis

**Abstract:**

Nitric oxide (NO•) is an important antimicrobial produced by the immune system to impair pathogens. Many pathogens, such as *Mycobacterium tuberculosis*, *Salmonella enterica*, and pathogenic *Escherichia coli*, harbor NO• detoxification and repair systems which have been linked to their virulence, and thus present an attractive target toward which novel anti-infective therapies could be designed. Unfortunately, the few compounds that have been found to inhibit major NO• detoxification systems in bacteria suffer from poor intracellular transport or severe toxicity toward humans, warranting a search for additional targets in bacterial NO• defense systems. The broad reactivity of NO• and its reaction products give rise to an expansive, complex reaction network, necessitating the use of a computational approach to effectively study the system in its entirety. To facilitate quantitative exploration of the *E. coli* NO• stress response, we previously constructed a kinetic model of the intracellular NO• biochemical network, including relevant processes such as enzymatic NO• detoxification, iron-sulfur cluster nitrosylation and repair, DNA deamination and repair, transcriptional regulation, and NO• autoxidation. Model simulations were validated with experimental measurements of [NO•], [$NO_2^-$], [$NO_3^-$], and [$O_2$] in *E. coli* cultures, and facilitated the discovery of a strong decrease in utility of the main aerobic NO• detoxification enzyme, NO• dioxygenase (Hmp), with increasing NO• delivery rate. Furthermore, Hmp was predicted to dominate cellular NO• consumption even into the microaerobic regime (35 μM dissolved $O_2$), which was confirmed experimentally.

Given that $O_2$ levels at *E. coli* infection sites *in vivo* range from anaerobic to microaerobic, we used the NO• model to investigate the dynamics of the *E. coli* NO• response network within this medically-important regime. Although the two major NO• detoxification systems in *E. coli*, Hmp and NorV, are known to dominate NO• consumption under aerobic and anaerobic conditions, respectively, little is known about their participation and dynamics within the microaerobic regime that lies between these two extremes. The model predicted a deficiency in the overall rate of NO• detoxification for $O_2$ environments between anaerobic (0 μM $O_2$) and the upper-end of microaerobic (50 μM $O_2$), which was attributed to an insufficient overlap of Hmp and NorV activity. Experimental measurements of NO• detoxification at these intermediate $O_2$ concentrations (5, 10, and 20 μM $O_2$) confirmed the predicted trend; however, the extent to which NO• detoxification was impaired was underestimated, and oscillations appeared in the [NO•] curves measured at 5 and 10 μM $O_2$. Efforts to reconcile the predictions with experimental measurements yielded a model that could quantitatively capture NO• dynamics throughout the microaerobic regime. An investigation of model parameters contributing to the oscillations predicted a mechanism involving competition between Hmp and respiratory cytochrome oxidases for $O_2$, which was confirmed experimentally through model-guided biochemical and genetic analyses. This demonstrates the utility of a modeling approach to quantitatively study the bacterial NO• defense network even under previously uncharacterized environmental conditions, and how it can facilitate the dissection of complex phenomena to yield novel insight about the quantitative interplay between network components.

# 432307 Identification of Gut Microbiota-Derived Metabolites As Liver Inflammation Modulators in Non-Alcoholic Fatty Liver Disease


**Abstract:**

Identification of Gut Microbiota-derived Metabolites as Liver Inflammation Modulators in Non-alcoholic Fatty Liver Disease

The human gastrointestinal (GI) tract harbors an estimated 10 to 100 trillion microorganisms, encompassing several hundred different species. Collectively referred to as the gut microbiota, the microbes perform a number of essential physiological functions, including the metabolism of complex carbohydrates [1], development of the immune system [2], and defense against pathogens [3]. Beyond the GI tract, gut-brain, gut-lung, and gut-liver associations have also been identified, highlighting the importance of the microbiota in host physiology. Consistent with the view that the microbiota is critical for whole body homeostasis, alterations in the intestinal microbiota composition (i.e. dysbiosis) have been correlated to various diseases, including obesity [4, 5], diabetes [6], and cancer [7].

Non-alcoholic fatty liver disease (NAFLD) has emerged as the most prevalent chronic liver disease in Western countries, paralleling the rise in obesity and the metabolic syndrome. It is characterized by steatosis, or fat deposition in the liver, which can remain a benign condition. However, in 10-20% of patients, NAFLD progresses from steatosis to non-alcoholic steatohepatitis (NASH), which can lead to cirrhosis and liver cancer. Seminal work by the Gordon laboratory [8, 9] and other groups demonstrated that dysbiosis could be a contributing mechanism for obesity and related diseases, including NAFLD, whose estimated prevalence in obese individuals exceeds 75% [10]. Direct, potentially causal involvement of the microbiota in NASH was highlighted by Le Roy et al. [11], who reported that high fat diet (HFD)-fed germ-free (GF) mice inoculated with bacterial isolates from feces of HFD-fed hyperglycemic mice developed steatohepatitis, whereas HFD-fed GF mice inoculated with bacteria from HFD-fed, but normoglycemic mice only showed mild steatosis. This suggests that the composition of the intestinal microbiota, and hence their metabolic products, could determine whether NAFLD progresses to NASH. While these and other recent studies have begun to build a strong case for potential mechanistic links between dysbiosis and NASH, the identities of microbiota-derived molecular mediators and their target pathways remain to be fully elucidated. The study described in this abstract utilizes both untargeted and targeted metabolomics to identify and characterize microbiota-dependent immunomodulatory metabolites that act on the liver to modulate liver inflammation in steatosis.

In previous work [13], we identified a panel of aromatic amino acid (AAA)-derived metabolites whose levels in the mouse intestine depended on the presence of the microbiota. Furthermore, several of these metabolites were found to activate the aryl hydrocarbon receptor (AhR), a transcriptional regulator of host-microbiota interactions in the intestine. To investigate a potential role for these metabolites in NAFLD, we compared their profiles in serum and liver samples from 14-week old male C57BL/6J mice raised for 8 weeks either on HFD or low-fat diet (LFD). The diets were sucrose calorie-matched and derived 60% and 10% of calories from fat, respectively. Metabolites were extracted from cold PBS-rinsed liver tissue and serum samples using a solvent-based method. Using an untargeted method, we identified 73 AAA-derived metabolites, including host- and microbiota-dependent metabolites as well as their methyl and hydroxyl derivatives. Interestingly, a large fraction of metabolites that were significantly elevated or depleted in liver and serum from HFD-fed mice (compared to LFD-fed mice) were the hydroxyl/methyl derivatives. Several of the metabolites that were consistently detected in all four sets of samples (liver and serum, HFD and LFD) were previously found to depend on the microbiota and activate the AHR [13]. We thus selected a pair of representative metabolites from this set, tryptamine (TA) and indole-3-acetate (I3A), for further analysis.

Absolute quantification of TA and I3A using targeted analysis confirmed that their levels were significantly elevated (ca. 2- and 3-fold, respectively) in the serum of LFD mice compared to HFD mice. We then investigated the ability of TA and I3A to modulate inflammation in hepatocytes using both human (HepG2) and murine (AML12) cell lines. Cultured HepG2 cells were treated with 1, 10, or 100 mM TA or I3A for 24 h, and then exposed to the pro-inflammatory cytokine TNFα for another 24 h in the presence of the metabolites. The expression of two anti-inflammatory genes (IL-4 and IL-10) and a pro-inflammatory chemokine (MCP1) were determined using qPCR. For all three markers, pretreatment with I3A or TA at the 10 mM dose significantly attenuated the pro-inflammatory responses elicited by TNFα treatment. These trends were in good agreement with the results obtained using the AML12 murine cell line.

Using HepG2 cells, we next investigated whether the AAA-derived, microbiota-dependent metabolites could modulate the impact of TNFα treatment on cholesterol and bile acid metabolism. Compared to vehicle control, TNFα treatment significantly increased both intra- and extracellular levels of several bile acids, including cholic acid (CA), glycocholic acid (GCA), and taurocholic acid (TCA). This profile was consistent with our findings that both CA and TCA were significantly elevated in livers of HFD-fed mice compared to LFD-fed mice. Pretreatment with either 1 mM TA or I3A significantly attenuated the TNFα–stimulated increases in bile acid levels, similar to the metabolites' effects on cytokine and chemokine gene expression.

To confirm activation of the AHR pathway, we utilized a hepatocyte reporter cell line, where binding of an AHR-ligand complex to the dioxin response element (DRE) leads to transcription of the reporter gene (GFP), resulting in fluorescence. Consistent with our previous findings [13], 24 h exposure to TA and I3A dose-dependently increased GFP expression in the reporter cells, indicating ligand activation of the AHR by these metabolites.

In summary, our results demonstrate that several AAA-derived microbiota metabolites are significantly depleted in a mouse model of diet-induced liver steatosis, and that these metabolites can act directly on hepatocytes to modulate inflammatory pathways. Our results also show that these microbiota metabolites are ligands for the AHR, providing a mechanistic link for the observed anti-inflammatory effects. Taken together, our findings support the hypothesis that dysbiosis of the gut microbiota could predispose the liver to inflammation in diet-induced steatosis through an altered microbiota metabolite profile. Prospectively, insights into the mechanisms underlying the link between microbiota dysbiosis and NAFLD could provide novel strategies to treat or prevent the progression of fatty liver diseases through the use of probiotics or postbiotics.

[1] T. Arora and R. Sharma, "Fermentation potential of the gut microbiome: implications for energy homeostasis and weight management," *Nutrition Reviews,* vol. 69, pp. 99-106, 2011.

[2] L. V. Hooper, D. R. Littman, and A. J. Macpherson, "Interactions Between the Microbiota and the Immune System," *Science,* vol. 336, pp. 1268-1273, June 8, 2012 2012.

[3] N. Kamada, G. Y. Chen, N. Inohara, and G. Nunez, "Control of pathogens and pathobionts by the gut microbiota," *Nat Immunol,* vol. 14, pp. 685-690, 2013.

[4] J. Shen, M. S. Obin, and L. Zhao, "The gut microbiota, obesity and insulin resistance," *Molecular Aspects of Medicine,* vol. 34, pp. 39-58, 2013.

[5] S. Greenblum, P. J. Turnbaugh, and E. Borenstein, "Metagenomic systems biology of the human gut microbiome reveals topological shifts associated with obesity and inflammatory bowel disease," *Proceedings of the National Academy of Sciences,* vol. 109, pp. 594-599, January 10, 2012 2012.

[6] J. K. Nicholson, E. Holmes, J. Kinross, R. Burcelin, G. Gibson, W. Jia, *et al.*, "Host-Gut Microbiota Metabolic Interactions," *Science,* vol. 336, pp. 1262-1267, June 8, 2012 2012.

[7] J. C. Arthur, E. Perez-Chanona, M. Mühlbauer, S. Tomkovich, J. M. Uronis, T.-J. Fan, *et al.*, "Intestinal Inflammation Targets Cancer-Inducing Activity of the Microbiota," *Science,* vol. 338, pp. 120-123, October 5, 2012 2012.

[8] R. E. Ley, F. Backhed, P. Turnbaugh, C. A. Lozupone, R. D. Knight, and J. I. Gordon, "Obesity alters gut microbial ecology," *Proc Natl Acad Sci U S A,* vol. 102, pp. 11070-5, Aug 2 2005.

[9] P. J. Turnbaugh, R. E. Ley, M. A. Mahowald, V. Magrini, E. R. Mardis, and J. I. Gordon, "An obesity-associated gut microbiome with increased capacity for energy harvest," *Nature,* vol. 444, pp. 1027-31, Dec 21 2006.

[10] S. G. Sheth, F. D. Gordon, and S. Chopra, "Nonalcoholic steatohepatitis," *Ann Intern Med,* vol. 126, pp. 137-45, Jan 15 1997.

[11] T. Le Roy, M. Llopis, P. Lepage, A. Bruneau, S. Rabot, C. Bevilacqua, *et al.*, "Intestinal microbiota determines development of non-alcoholic fatty liver disease in mice," *Gut,* vol. 62, pp. 1787-94, Dec 2013.

[12] W. R. Wikoff, A. T. Anfora, J. Liu, P. G. Schultz, S. A. Lesley, E. C. Peters, *et al.*, "Metabolomics analysis reveals large effects of gut microflora on mammalian blood metabolites," *Proceedings of the National Academy of Sciences of the United States of America,* vol. 106, pp. 3698-3703, Mar 2009.

[13] G. V. Sridharan, K. Choi, C. Klemashevich, C. Wu, D. Prabakaran, L. B. Pan, *et al.*, "Prediction and quantification of bioactive microbiota metabolites in the mouse gut," *Nat Commun,* vol. 5, p. 5492, 2014.

# 429442 Identifying Motifs Associated with Autoimmune Diseases Using Next Generation Sequencing Experiments

**Abstract:**

Autoimmune diseases are typically chronic, have major quality-of-life effects and afflict a large minority of the population. A hallmark of autoimmune diseases is the presence of antibodies that bind to "self" targets. Unfortunately, it has proven challenging to identify these autoantibodies and develop accurate diagnostic tests for their presence in many diseases.

We developed a method to Identify Motifs Using Next generation sequencing Experiments (the IMUNE method) capable of discovering disease-associated motifs. First, a large, bacterial-display peptide library is screened using magnetic and/or fluorescence-activated cell sorting to recover peptides bound by an individual's antibodies. Next generation sequencing technologies (e.g. Illumina NextSeq) are utilized to identify these millions of peptides. Because of the large diversities in the peptide library and in patients' antibodies, it is unlikely that this experimental approach alone will identify peptides bound by many disease patients but not control patients. Therefore, computational analysis is employed to identify patterns of amino acids that are statistically enriched in a cohort of patients compared to control patients, where a pattern is a series of specific amino acids possibly interspersed with undefined residues. Finally, these enriched patterns are clustered together to create the disease-associated motifs. We will present an overview of the experimental and computational methods, and show examples from Celiac disease and multiple sclerosis to highlight the method's ability to discover known and novel disease-associated motifs.

# 431408 Novel Approach to Target Endogenous KRAS Mutations By the Crispr/Cas9 Technology

**Abstract:**

**CRISPR/Cas9 is an enabling RNA-guided technology for genome targeting and engineering. Here we demonstrate that the lentiviral-mediated CRISPR/Cas9 technology can be used to specifically target single-nucleotide heterozygous mutations while exerting no aberrant effects on the wild-type alleles. Specifically, we used a novel gRNA-constraint-mediated method to target the heterozygous G13A activating mutation of KRAS in colorectal cancer cells and we show reversal of drug resistance to a MEK small-molecule inhibitor. Our study introduces a new paradigm in genome editing and therapeutic targeting via the use of gRNA to guide Cas9.**

# 433019 Rewiring Cellular Input-Output to Engineer Cell-Based Therapies That Interface Robustly with Human Physiology

**Abstract:**

Cell-based therapies have proven to be useful for treating a wide variety of diseases including autoimmune disease, infectious diseases, and cancer. A particularly promising frontier is the use of engineered cell therapies, wherein cells are programmed to carry out custom functions such as harnessing the immune system to find and destroy cancerous cells. Such therapies have now achieved robust clinical successes for some applications. However, controlling or modulating the activity of these therapies post-implantation remains both attractive, to enhance both safety and therapeutic efficacy, and challenging, since the tools necessary to engineer custom functional programs are limited. In particular, engineering cells to sense and recognize specific combinations of environmental cues would be desirable, but no such technology has yet been reported. Toward this goal, our lab has developed a platform for engineering novel protein biosensors, termed modular extracellular sensor architecture (MESA), for detecting exclusively extracellular cues. Upon binding extracellular ligand, MESA receptors release sequestered transcription factors from the cytoplasmic face of the plasma membrane, freeing the factor to regulate expression of an "output" gene or genes within the cell. Here we present three areas in which we have recently expanded upon the capabilities of the core MESA technology: (1) achieving novel ligand recognition by integrating modular ligand-binding domains, (2) integrating MESA with intracellular gene circuits that enable the cell to "process" sensory information by logical evaluation, and (3) rewiring cellular input-output by coupling extracellular sensing to intracellular modulation of endogenous gene expression.

First, as modular ligand binding domains, we used camelid single domain antibody fragments termed "nanobodies". Each nanobody's small size enables multiple nanobodies to recognize distinct, non-overlapping epitopes on a ligand of interest. Thus nanobody MESA are able to recognize monomeric or asymmetric ligand inputs. After demonstrating the feasibility of nanobody MESA, we integrated multiple such receptors into a logical processing circuit that enables the multiplexed logical evaluation of distinct protein ligand "inputs". Next, we demonstrated that human antibody-derived scFv could also be used as ligand-binding domains in MESA, and VEGF-MESA were thus designed to sense the tumor-associated cytokine, VEGF. In order to regulate endogenous genes, the MESA transcription factor domain was replaced with the dCas9 protein, which is able to act as a 'pseudo-transcription factor' by tethering transactivation domains to the dCas9 protein and providing small guide RNAs in trans to direct the Cas9 protein to a specific genomic locus. As proof of principle, we developed VEGF-MESA that induce expression of IL-2, a cytokine important for T-cell growth and proliferation. In principle, the dCas9 protein can be guided to any genomic locus of interest by simply delivering a different small guide RNA. Cells expressing VEGF-MESA exhibited minimal background signaling in the absence of ligand, and such cells showed significant increases in IL-2 gene expression upon exposure to ligand. Thus, this initial study demonstrates that MESA may be harnessed to effectively rewire cellular input-output behavior. This work lays the foundation for leveraging MESA receptors to boost the safety and efficacy of cancer immunotherapies as well as cell-based therapies for a range of clinical applications.

# 436637 Understanding Tumor Microenvironment's Metabolic Regulation of Cancer Cells

**Abstract:**

Tumor cells need appropriate nutrient for growth, invasion and energy metabolism. A growing number of studies indicate that metabolic dependencies of cancer cells are regulated by **tumor microenvironment (TME)** which rewires the cellular metabolism to promote tumorigenicity. Glutamine can play a critical role in cellular growth in multiple cancers. Recently, we uncovered a missing link between cancer invasiveness and glutamine dependence. Using isotope tracer and bioenergetic analysis, we showed that low invasive ovarian cancer (OVCA) cells are glutamine independent, whereas high invasive OVCA cells are markedly glutamine dependent. Here, we present a multidimensional approach, where we synergistically integrate *in vitro* stable isotope metabolomics, cell functional assays (invasion, migration, proliferation, 3D cocultures), mitochondrial bioenergetics, clinicopathological features (patient derived CAFs and stromal fibroblasts, and patient derived cancer cells), targeted therapies using in culture and *in vivo* models, to understand TME regulation of OVCA growth and metastasis. We will demonstrate a novel mechanism by which stromal cancer associated fibroblasts cells maintain tumor growth in OVCA cells.

# 431097 Refolding of Heterologous Proteins in the Extracellular Space after Secretion

**Abstract:**

Recombinant DNA technology enables the production of pure, folded heterologous proteins, where a gene coding for a protein of interest (POI) is expressed in a host cell. However, this process requires significant optimization for each new POI; heterologous protein production in bacteria is often a batch process, in which the cells are lysed and the POI is purified from the cellular milieu, sometimes requiring refolding. One strategy to simplify protein purification and enable continuous processing is to secrete the POI directly into the extracellular space. Toward this goal, we use synthetic biology and protein engineering techniques to optimize a bacterial strain for production and secretion of heterologous proteins. We engineer the type III secretion system (T3SS) of *Salmonella enterica*, a heterogeneous multi-protein structure that secretes proteins from the cytoplasm to the extracellular space in one step. In this work, we hypothesized that the secreted protein product would refold in the extracellular space. Recent structural data shows that the protein is unfolded and linearized as it is secreted through the "needle", which is 2 nm in diameter and 50 nm long. We use enzymatic activity as a proxy for protein folding, and demonstrate enzyme activity in the extracellular space, indicating that secreted proteins indeed refold after secretion. Genetic and chemical methods are used to probe the folded state of the model enzymes beta-lactamase and alkaline phosphatase and a single-chain variable fragment (scFv) from an antibody. Standard functional assays indicate that the enzymes and scFvs refold into functional forms after secretion and spontaneous form the requisite post-translational modifications (eg. disulfide bond formation, metal ion cofactor binding, multimerization). Recovery of folded heterologous protein in the extracellular space exhibits a strength of the use of a protein secretion chassis.

# 425236 Directed Co-Evolution of the Twin-Arginine Translocation Machinery for Enhanced Export of Recombinant Proteins

**Abstract:**

**Directed co-evolution of the twin-arginine translocation machinery for enhanced export of recombinant proteins**

May N. Taw[1,2], Mark A. Rocco[2], and Matthew P. DeLisa[2]

1. Microbiology Graduate Program, Cornell University, Ithaca, NY

2. Department of Chemical and Biomolecular Engineering, Cornell University, Ithaca, NY

*Escherichia coli* remains one of the preferred hosts for biotechnological protein production due to its robust growth in culture and ease of genetic manipulation. Often it is desired to secrete recombinant proteins into the periplasmic space as a means to simplify downstream purification and increase protein stability by avoiding proteolytic degradation in the cytoplasm. The predominant protein translocation mechanism in *E. coli* is the general secretory (Sec) pathway; however, a wealth of information has emerged over the past decade on an alternative, Sec-independent protein export mechanism known as the twin-arginine translocation (Tat) pathway. The hallmark of the Tat system is its unique ability to export already folded proteins to the periplasmic space. Our group and others have shown that the Tat machinery has an in-built quality-control mechanism that discriminates the folding state of substrate proteins, allowing only folded proteins to transit while rejecting misfolded substrates from the translocation cycle. However, despite its potential for localizing correctly folded proteins in the periplasm, translocation by the Tat system is approximately an order of magnitude slower compared to the Sec pathway. To address this issue, we engineered the Tat translocase, comprised of the TatABC integral membrane proteins, to hyper-secrete industrially relevant proteins into the periplasm using a directed co-evolution approach. This involved generation of a combinatorial *tatABC* library, followed by genetic selection of translocase variants capable of increased export of a single-chain Fv (scFv) antibody fragment specific for β-galactosidase called scFv13-R4. Following just a single round of mutagenesis and selection, we isolated five unique TatABC hyper-secretors that exhibited export efficiency that was greatly enhanced compared to the parental Tat machinery and also compared to Sec pathway. The selected mutants carried substitutions in TatB and TatC, but not TatA, implicating these proteins as key regulators of protein flux to the periplasm. We conclude that substrate "proofreading" is a rate-limiting step in the translocation cycle given that the quality-control mechanism of all hyper-secretors was found to be relaxed. Taken together, our results provide new insights on a poorly understood aspect of the Tat transport mechanism and at the same time furnish a set of optimized Tat translocases capable of high levels of recombinant protein export.

# 411674 The Potential of Small Molecules to Modulate Glycosylation By Media Design

**Abstract:**

**The Potential of Small Molecules to Modulate Glycosylation by Media Design**

David Brühlmann, Anaïs Muhr, Jürgen Hemberger, Markus Sauer, Henri Kornmann, Martin Jordan and Hervé Broly

Background and novelty
A large number of recent publications demonstrate the effect of cell culture media on post-translational modifications of recombinant proteins[1]. This study aims to extend the toolbox of media design beyond the commonly known media components. We identified and tested a large variety of cell culture compatible chemical components such as pigments and sugar derivatives in industrial relevant Chinese hamster ovary cell lines (CHO) expressing recombinant antibodies.

Experimental approach
The cells were cultivated in fed-batch mode cultures using shaking 96-deepwell plates, and process performance such as viable cell density, viability and product titer were monitored. Supernatants of each culture were purified and N-glycan analysis was performed by CGE-LIF. The findings were confirmed in 30 mL fed-batch shake tube cultures and 15 mL micro-scale bioreactors (ambr15$^{TM}$ system) operated at controlled pH and $pCO_2$.

Results and discussion
The addition of the components at the beginning of the culture exhibited significant changes of the glycosylation profile of the expressed protein. Furthermore, the adjustment of the levels and the supplement addition in the feed instead of the media allowed to fine-tune the effect of the components on glycosylation profile. Finally, the use of some of the tested supplements increased peak cell density to levels above 20 mio viable cells/mL and product titer up to 1.5-fold, while maintaining high viability throughout the culture. These results show that media design alone is sufficient to specifically modulate some of the essential protein quality attributes and to increase productivity, which circumvents the need of modifying the gene expression of the cell line.

[1] D. Brühlmann et al., Tailoring Recombinant Protein Quality by Rational Media Design, Biotechnology Progress [published online ahead of print April 10, 2015], doi: 10.1002/btpr.2089

## (570d) Engineering the N-Glycosylation Pathway in Pichia Pastoris for the Expression of Glycoprotein Hormones

Manoharan, S., Indian Institute of Science, Bangalore,India
Modak, J., Indian Institute of Science
Dighe, R.,

Originally presented on: 11/11/2015 16:09:00 - 16:27:00

Yeast cells serve as model systems for the study and understanding of eukaryotic cell functioning. They are also the ideal choice for the expression of proteins of therapeutic importance. Around 70% of the therapeutic proteins under preclinical and clinical development are glycosylated and hence an increasing demand for expression system with the ability to perform post translational modification. N-glycosylation is a post translational modification that occurs in the endoplasmic reticulum (ER) and the Golgi complex. The end product of the N-glycosyaltion is observed to be different among species. The N-glycan structure is important for proper folding and activity of the protein, its serum half-life and also defines its immunogenicity. Hence having an expression system with human like N-glycosylation is imperative.

The yeast *Pichia pastoris* is gaining increasing attention as a protein production host due to the ease of transformation, ability to perform post translational modification, growth on defined media to high cell densities and scalable fermentations, low concentration of host proteins secreted into the media enabling easy downstream processing and high expression rate. Many of the therapeutic proteins expressed in Pichia are currently in clinical trials. Yeast cells are known to have the high mannose type oligosaccharide chain attached to the proteins ($Man_{9-20}GlcNAc_2$). In contrast, the human N-glycosylation is of the mixed or hybrid type ($NANA_2Gal_2GlcNAc_2Man_3GlcNAc_2$).

Engineerring the N-glycosylation pathway in the yeast would involve introduction of the repository of the enzymes from the higher eukaryotic pathway and eliminating the non-human glycosylation reactions. The first step towards modification of the pathway involves the disruption of the endogeneous glycosyltransferase gene (OCH1) that initalizes the addition of excess mannose to oligosaccharide chain. Furthur heterologous glycosidases and glycosyltransferases have to be introduced. Given that these ezymatic reactions occur in a seqential manner with the product of one enzyme acting as the substrate of the next, proper targeting of these enzymes to the cisternae of Golgi complex is important. A host of enzymes from higher eukaryotic cells and a proper targeting sequence fused to the protein were used to create a host cell expressing proteins with human like N-glycans.

The host thus generated was used to express the glycoprotien hormones, Follicle Stimulating Hormone (FSH) and human Chorionic Gonadotropin (hCG). FSH and hCG controls the reproductive functions, gametogenesis and are important for the maintenance of pregnancy in humans. Both the hormones are heterodimeric molecules comprising of an α-subunit and a β-subunit in non-covalent association with two N-glycosylation sites on each subunit. Purified hormones are important for the study of stucture - function relationships, role of N-glycoslation in protein folding, treatment of in-fertile patients suffering from gonadotropin deficiency and in-vitro fertilization (IVF). In the present study we report the expression and purification of FSH and hCG under the costitutively expressing GAP promoter in the glycoengineered *Pichia pastoris*. The cells were grown to high densities of 540g/L in fermentor culture and the protein expression was found to be 51mg/L (FSH) and 37mg/L (hCG). The expressed proteins were purified by a two step chromatography and characterized by various biochemical assays. The purified hormones were found to have the modified glycan structure and showed activity in Radioreceptor assay with the cognate receptor. The recombinant hormones were administered to animals for testing their in vivo bioactivity and found to be biologically active. FSH activity was determined by the increase in ovarian and uterine weight in female rats and hCG activity was determined by the testosterone response in male rats and bonnet monkeys.

Abbreviations: NANA - Sialic Acid, Gal - Galactose, GlcNAc - N-acetylglucoseamine, Man - Mannose, GAP - glyceraldehyde-3-phosphate dehydrogenase

# 433566 Engineered $Ca^{2+}$-Binding Protein for Chemoenzymatic Tagging and Bioorthogonal Surface Conjugation

**Abstract:**

Calcium ($Ca^{2+}$)-binding proteins are involved in many biological processes, including vesicle release, cell proliferation, and apoptosis. One of the most widely studied $Ca^{2+}$-binding proteins, calmodulin (CaM), is a ubiquitous $Ca^{2+}$ sensor found in all eukaryotes that binds and activates more than 100 other proteins. In addition to playing a central role in intracellular signal transduction, CaM is used in affinity chromatography to purify both natural binding partners and proteins that have been engineered to carry a CaM-binding peptide. Previous studies of CaM biochemical function, which have focused on the binding of $Ca^{2+}$ to CaM and subsequent binding of CaM to downstream proteins, have generally employed established protein labeling strategies, often leading to a decrease in CaM activity. Here we report the engineering of CaM to display a recognition sequence for N-myristoyl transferase (NMT) that enables the covalent attachment of an azide fatty acid to the engineered CaM protein N-terminus. This chemoenzymatic tagging approach proceeds co-translationally in *E. coli* with greater than 95% of the expressed CaM labeled with the tag. We show that neither the NMT recognition sequence nor the presence of the azide tag decreases CaM activity. The presence of the azide group allows CaM to be conjugated to surfaces directly from complex cell lysate mixtures with the bioorthogonal azide-alkyne cycloaddition reaction (also known as click chemistry). The conjugation of azide-tagged CaM to cyclooctyne-functionalized resins provides a streamlined process for making CaM-affinity chromatography resins for rapid protein purification. Additionally, we show that azide-tagged CaM maintains activity when conjugated to the resin, and that azide-CaM resin outperforms traditional CaM resins for purification of a CaM-binding protein in a one-step procedure.

We expect our results to be translatable to the conjugation of other azide-tagged proteins to other platforms, such as protein microarrays and microtiter plates. Future work also includes optimization of the amount of azide-tagged CaM conjugated to the resin and purification of other CaM-binding proteins from multiple sources. Together, our methods and results have the potential to dramatically streamline the affinity resin manufacturing process.

# 433198 Post-Translational Azide Functionalization for Optimizing Immobilized Protein Properties

**Abstract:**

Site-specific azide functionalization of proteins allows for diverse chemical modification via azide-alkyne click chemistry and thus holds high potential for pharmaceutical and industrial protein applications. The introduction of azide groups benefits from post-translational techniques, which allow for precise temporal control of azide functionalization, preventing undesirable azide reduction reactions and, ultimately, improving the homogeneity of modified protein populations. One promising approach to post-translational azide functionalization entails the use of the enzyme, lipoic acid ligase, which covalently and site-specifically attaches an azide group to a short peptide recognition sequence known as the LAP sequence. We have found that this sequence, though traditionally used as an N- or C-terminal protein tag, may be inserted at numerous internal sites within the model protein, GFP, as well as within the industrially relevant enzyme, lipase A, without negatively affecting structure or function or significantly reducing expression levels. Azide attachment to these internal recognition sequences proceeds efficiently, with a high degree of temporal control and population homogeneity. Given the flexibility and robust nature of this approach, we have utilized internal LAP insertions within lipase for orientation-controlled immobilization to alkyne-functionalized silica nanoparticles. Specifically, we are observing the structural impact of different immobilized protein orientations on these nanoparticles through circular dichroism and coupling these assays with activity and stability measurements. Combined, these experiments allow us to identify azide attachment sites that yield optimal protein properties upon immobilization. Ultimately, exploring this approach to optimize protein function at material interfaces may yield a general strategy for designing enzyme-functionalized materials.

# 433864 Rationally Engineering Post-Translational Protein Pegylation for Improved Stability and Function

**Abstract:**

Years of work conjugating the biocompatible polymer polyethylene glycol (PEG) to proteins have already resulted in the development of FDA-approved PEGylated protein therapeutics that are more stable and have longer functional lifetimes than their un-PEGylated counterparts. The post-translational covalent attachment of PEG groups, or PEGylation, increases a drug's molecular weight, thereby reducing clearance from the body. PEGylation also shields the protein from proteases and can thermodynamically stabilize a protein's conformation. Polymer-protein conjugation is a challenging task, however; polymer additions may interfere with protein folding and disrupt the protein or enzyme function. Thus, the attachment site of PEG on a protein must be optimized to achieve maximum stability and activity. Using cell-free protein synthesis and the incorporation of non-canonical amino acids, we examine the optimal conditions for conjugating PEG to proteins and compare the effect of PEGylation at different sites. This work will provide a basis for the development of structure-based guidelines to aid in the rational selection of optimal PEGylation sites.

# 422817 Characterization and Modeling of Feeders: A Critical Component in Continuous Pharmaceutical Manufacturing

**Abstract:**

In recent years, continuous manufacturing technologies have been widely accepted in the pharmaceutical environment given the many advantages they offer to product quality. Some of the benefits continuous processes offer over batch counterparts include: enhanced process control, reduced solids handling, small equipment footprint, reduced scale up requirements, and more uniform processing [1]. These advantages make continuous manufacturing a desirable technology for many pharmaceutical manufacturers. Nonetheless, there are many challenges to this technology: higher initial cost, extended initial development time, lack of PAT tools available for monitoring continuous processes, and difficulty of implementation for low volume and dosage products [1]. This last challenge is primarily based on the difficulty to accurately feed small amounts of materials to the process and blend them efficiently. Research has focused on understanding the behavior of both feeders and blenders with the aim of improving upon these challenges [2, 3]. During the early phases of continuous manufacturing blending was studied more extensively given the legacy importance in batch processes (i.e., in batch it is one of most important unit operations). However, feeding has the most significant effect in continuous processing as its sets the mass balance for the entire process. Lack of feeder control and understanding, leads to lack of control of any downstream unit operation [4].

Feeding accuracy is of great importance in continuous processes given the stringent requirements for product composition. To this end, Loss-in-Weight (LIW) feeders under gravimetric control are commonly used. LIW feeders deliver powders by calculating the amount of material lost (i.e., fed out of the unit) and controlling the rate of material delivery accordingly [5]. This strategy provides the feeding accuracy required in pharmaceutical processes. LIW feeders generally consist of a hopper, a flow aid and a powder conveying system, all over a load cell that measures its weight. The hopper contains the material in the feeders and is the receptacle where refills are performed. The flow aid is located at the bottom of the hopper with the purpose of breaking any powder bridges and disrupt the statics to facilitate transport of material into the conveying system. The conveying system, often a screw due to its accuracy over other systems, is used to discharge the material out of the feeder. Previous research focused on studying LIW feeders to understand how these components affect the unit's performance (i.e., the output flow rate). The conveying system was found to be the most significant component on feeder performance, followed by the refill hopper refill strategy [2]. Material properties were also found to have a significant impact on feeding performance [6].

Feeders are often not considered in the absence of other processing equipment, therefore development of models for these units is not as extensive compared to other unit operations. Multiple articles focus on the use of Discrete Element Model (DEM) simulations to characterize the effect screw and hopper configurations have on the unit's performance [7, 8]. The results from these studies are often semi-qualitative in nature and do not allow for development of models that can be used for process modeling (e.g., flowsheet models and control strategy design). In this work we developed a semi-empirical model of feeder behavior using material properties as input parameters. This work is an extension of previously developed empirical models [5], with the goal of incorporating the behavior and effects of both the flow aid and conveying screws in the model.

The semi-empirical model was developed to simulate feeder behavior, primarily focusing on modeling screw speed and flow rate as a function of time. Frequency variations (i.e., noise) of these variables were characterized using Fourier series, whose coefficients were regressed from experimental data for individual components. Screw speed and flow rate were directly correlated in the model through the use of a feed factor, whose units are mass per screw revolution. Feed factor, defined as the maximum mass fitting in a screw flight, was modeled to account for changes in hopper material properties occurring due to material changes, flow aid systems, feeder emptying and refills. Material properties were characterized before and after the feeder in order to evaluate and model their relationship to screw speed and flow rate. Lowest feeding capacity (LFC) of the feeder was evaluated based on these findings and the developed model. The effect of flow aid systems on the mixing of powders inside the hopper was experimentally characterized using residence time distribution (RTD) experiments. Models to predict mixing behavior in the unit were developed using fundamental chemical engineering models (e.g., continuous stirred tank reactors (CSTR), plug flow reactors or any combination). The study was performed for multiple feeder types with varying the flow aid systems, hopper configurations, and conveying system locations. Operating conditions (e.g., flow rate, refill level) and powder properties were also varied to investigate the effects these parameters had on the mixing. An optimization framework for refill strategy with minimal trial-and-error was developed. The ultimate goal of this work was to develop a model that can be used to predict the behavior of LIW feeders for any material *a priori*, using the results from powder characterization test. Such model can lead to a significant reduction in development time and minimize product waste.

**References**


1. Rogers, A., A. Hashemi, and M. Ierapetritou, *Modeling of Particulate Processes for the Continuous Manufacture of Solid-Based Pharmaceutical Dosage Forms.* Processes, 2013. **1**(2): p. 67-127.

2. Engisch, W.E. and F.J. Muzzio, *Method for characterization of loss-in-weight feeder equipment.* Powder Technology, 2012. **228**(0): p. 395-403.

3. Vanarase, A.U. and F.J. Muzzio, *Effect of operating conditions and design parameters in a continuous powder mixer.* Powder Technology, 2011. **208**(1): p. 26-36.

4. Dai, J. and J.R. Grace, *Biomass granular screw feeding: An experimental investigation.* Biomass and Bioenergy, 2011. **35**(2): p. 942-955.

5. Boukouvala, F., et al., *An integrated approach for dynamic flowsheet modeling and sensitivity analysis of a continuous tablet manufacturing process.* Computers & Chemical Engineering, 2012. **42**(0): p. 30-47.

6. Falk, J., et al., *Mass flow and variability in screw feeding of biomass powders — Relations to particle and bulk properties.* Powder Technology, 2015. **276**(0): p. 80-88.

7. Owen, P.J. and P.W. Cleary, *Screw conveyor performance: comparison of discrete element modelling with laboratory experiments.* Progress in Computational Fluid Dynamics, an International Journal, 2010. **10**(5): p. 327-333.

8. Fernandez, J.W., P.W. Cleary, and W. McBride, *Effect of screw design on hopper drawdown of spherical particles in a horizontal screw feeder.* Chemical Engineering Science, 2011. **66**(22): p. 5585-5601.

# 425126 An Investigation of Mixing Performance of a Continuous Powder Mixing Process

**Abstract:**

Continuous manufacture of drug products is receiving increased attention both from academics and industry in recent years. Pfizer's approach to continuous manufacture, utilizing a Portable, Continuous, Miniature and Modular (PCM&M) technology is the focus of this presentation. In this work, the performance of a novel, vertical in-line powder mixing technology will be described. This mixer, which offers the independent selection of mixer shear rate, powder hold up mass and system mass throughput, is studied both experimentally and computationally to first characterize the mixing performance and secondly better understand how process parameters and equipment design can be manipulated to achieve the desired performance. The links between process conditions, powder flow regime, and mixing performance are illustrated with experimental data. Discrete element method (DEM) simulations are used to further characterize the mixer operation with respect to mixing performance and the extent of powder lubrication under a variety of conditions. Additionally, the models are used to characterize the coupled impact of gravimetric feeder variability and mixer performance (residence time distribution) on the variability in the outlet stream concentration. Together, these simulations also provide grounds to assess the influence of various design considerations on performance of the PCM&M equipment train.

# 422062 Comparison of Continuous High Shear Granulation Processes with Twin-Screw Wet Granulation Process


**Abstract:**

Wet granulation (WG) is an extremely prevalent unit operation in the manufacturing of drug product due to its numerous advantages. As particle size is enlarged in the process, WG can improve the handling and flow properties of powders, content uniformity and compression characteristics; and it also can minimize dust and ingredient segregation. These advantages could realize better drug release rate and tablets physical appearance. Currently, substantial attention has been attracted by continuous manufacturing (CM) because it can provide significant technical and business advantages relative to the conventional batch processes. CM methods have been demonstrated to be more robust and controllable, which can improve product quality and process efficiency. Furthermore, scale-up challenges can be overcome in CM by running the operation for a longer time, indicating the reduction of technique transfer risk and capital-intensive equipment requirement. In recent years, several continuous WG techniques have been developed. Most commercial continuous granulators are twin-screw extruders followed by fluid bed driers. In comparison, there have been very few studies so far on continuous high shear granulation, which resembles the batch high shear granulation in terms of dry powder mixing, binder addition and wet massing. In general, the granules produced by continuous fluid bed granulators or twin-screw extruders are different than those from high shear granulation due to the different shear and compressive forces that the material experiences in these systems.

The focus of the study is to quantify performance metrics (particle size distribution (PSD), bulk density, porosity, active concentration in granule, particle shape) of granules as a function of critical process parameters (throughput, impeller rotation speed and liquid to solid (L/S) ratio) for each formulation and granulator. Residence time distribution (RTD) studies are performed to understand the mixing dynamics which can aid in further understanding the relationship between input parameters, mixing and shear and the output granule properties.

Two continuous high shear granulators, namely Lödige CoriMix CM5 and Glatt GCG70, and Thermo twin-screw granulator/extruder (EuroLab 16-mm, 25:1 L/D) were examined. Two dosages formulations were used for experimentation, i.e., a high-dose drug formulation (15% -lactose monohydrate, 15% Microcrystalline cellulose and 70% semi-fine Acetaminophen) and a low-dose drug formulation (46% -lactose monohydrate, 46% Microcrystalline cellulose and 8% semi-fine Acetaminophen). Purified water served as the binder. A Design of Experiments (DoE) had been performed on each granulator with three levels of L/S ratio, rotation speed and throughput. Eyecon (InnoPharma) offline particle size analysis were used to quantify the median particle size and span of the output granules. Bulk density was measured from calculations involving the mass of the sample and the volume of particles (solid + liquid + pores). Scanning Electron Microscope images of the samples were taken to quantify porosity, active concentration and the shape of the granule. RTD studies, conducted by injecting colored tracer active particles and implementing a Near-infrared spectrometer, were used to evaluate the concentration of dye that was further correlated to the active concentration in the sample collected. The results allow for a quantitative understanding of the design space regarding the process parameters mentioned above, which can be used to identify an optimal operation condition to a specific formulation.

# 427602 Experimental Investigation Linking Granulation Performance with Residence Time and Granulation Liquid Distributions in Twin-Screw Granulation

**Abstract:**

Twin-screw granulation is a promising technique for the continuous production of pharmaceutical solid dosage forms based on wet granulation. In a short period, solid-liquid mixing must be achieved by arrangement of transport and kneading elements to produce granules with a particle size distribution appropriate for tableting. The residence time distribution and the solid-liquid mixing governed by the field conditions (such as location and length of mixing zones) in the twin-screw granulator thus contain interesting information about mixing and different granulation rate processes such as aggregation and breakage. In this study, the impact of process (feed rate, liquid-to-solid ratio and screw speed) and equipment parameters (number of kneading discs and stagger-angle) on the residence time, the granulation liquid-powder mixing and the resulting granule size distributions during twin-screw granulation were investigated. Residence time and axial mixing information was extracted from tracer maps and the solid-liquid mixing was quantified from moisture maps, both obtained by monitoring the granules at the granulator outlet using near infra-red chemical imaging (NIR-CI). The granule size distribution was measured by sieving. The screw speed most dominantly influenced the residence time distribution based responses, i.e. mean residence time and axial mixing. Increasing the screw speed caused reduction in mean residence time and an increase in axial mixing. Also, an interaction between screw speed and material throughput dictating the fill ratio caused a reduction in residence time and an increase in axial mixing when the throughput-force increased by the high fill ratio. The frequency and amplitude of perturbation in the mean moisture profiles (derived from the moisture maps) indicated that unlike residence time, the solid-liquid mixing is dominantly driven by the material throughput and the number of kneading discs. Increasing the number of kneading discs reduced the frequency of variation indicating reduction in the heterogeneity of liquid mixing, which is desired for a good granulation yield. Correlating these results with the particle size fractions, the low residence time and high axial mixing obtained at a high screw speed resulted in a reduced oversized (>1400 µm) fraction and an increase in yield fraction (>150 µm and < 1400 µm). However, increasing the material throughput, which had a similar effect as screw speed on the residence time and the axial mixing, resulted in a reduction of the yield fraction. This was due to the inferior solid-liquid mixing at high material throughput which ultimately negatively affected the granulation performance. This improper solid-liquid mixing at high throughput resulted in more oversized particles. Thus, a balance between material throughput and screw speed should be found to achieve the required granulation time and solid-liquid mixing for high granulation yield. The results from this experimental study improved the understanding regarding the interplay between granulation time and the axial and solid-liquid mixing responsible for the granulation yield after twin-screw granulation.

# 421449 NANEX: A Continuous Hot-Melt Extrusion (HME) Process for Nano-Based Formulations

**Abstract:**

The NANEX process is a continuous one-step HME process that converts stabilized aqueous nanosuspensions into a solid oral formulation with an increased solubility of poorly soluble drugs. A nanosuspension is fed to the extruder and the solvent is removed via devolatilization. Nanoparticles are embedded in the polymer matrix and solid nano-based pellets can be produced in a single step [1,2]. However, the amount of the active pharmaceutical ingredient (API) in a nanosuspension formulation is restricted to 20-30 % (w/w) making it hard to vary the API loading in the final extrudates.

Therefore, in the current study we focus on evaluating process parameters which impact the maximum amount of water that can be added during hot melt extrusion. Moreover, additional matrix materials that are suitable for incorporation of aqueous suspensions were investigated. Kollidon® VA64, Eudragit® E PO, AQOAT® (hypromellose acetate succinate, HPMCAS; in combination with plasticizer) and poly(ethylene glycol) 20000 (PEG 20000) were used. To assess whether the polymer could be mixed with water during the NANEX process and the added water could be removed completely via devolatilization (vacuum of 200 mbar), liquid (i.e., water) was fed to the hot melt extruder (MICRO 27 GL co-rotating twin-screw extruder, Leistritz Extrusionstechnik GmbH, Nürnberg, Germany) via a micro-angular gear pump in barrel 4, while adding the matrix material in barrel 1 and degassing of the added water in barrel 8. By adding water to the extruder at fixed throughputs and screw speeds, their influence on the maximum amount of water was examined. Clogging of the degassing unit and/or inclusions of moisture within the strands indicated the end-point. Additionally, the mean residence time at fixed process parameters was evaluated and residual moisture contents of the extrudates were measured via Karl Fischer titration.

The results demonstrated that the amount of water fed to the polymer melt is strongly dependent on the filling degree of the screw, which is a function of the throughput and the screw speed. The lower the filling degree was the higher was the maximum amount of added water. Similarly, the longer the residence time, the more water could be added. However, the quantity of added and removed water was not only related to the process parameters but also to the nature of the matrix material. It was found that polymers, which are miscible with water, could be processed via NANEX. For HPMCAS/plasticizer, 110%, and for PEG 20000 and Kollidon® V64 a maximum of 88% water could be added to the molten materials. Results of the Karl Fischer titrations clearly demonstrated that the excess water was removed entirely in the case of Kollidon® VA64 and HPMCAS/plasticizer. However, PEG 20000 extrudates exhibited increased mean residual moisture contents which can be attributed to process temperatures not exceeding 70 °C hampering the devolatilization of water at a vacuum of 200 mbar. To provide a remedy, a further drying step could be performed.

In summary, it was found that via the evaluation of the most important process parameters the API loading of nano-extrudates can be controlled easily and continuously without changing the complex and sensitive formulation of the nano-suspension.

1. Baumgartner, R.; Eitzlmayr, A.; Matsko, N.; Tetyczka, C., Khinast, J. and Roblegg, E. Nano-extrusion: A promising tool for continuous manufacturing of solid nano-formulations. Int J Pharm, 477, 1-11 (2014)

2. Khinast, J.; Baumgartner, R. and Roblegg, E. Nano-extrusion: a one-step process for manufacturing of solid nanoparticle formulations directly from the liquid phase. AAPS PharmSciTech, 14, 601-604 (2013)

# 427040 Continuous Manufacturing of Immediate-Release Tablets from Hot-Melt Extruded Pellets

**Abstract:**

Tablets are the most common solid dosage forms. In order to decrease variability in tablet quality many formulations are granulated prior to tableting. Compared to common continuous granulation methods, hot-melt extrusion (HME) offers the possibility to enhance or adapt certain physicochemical characteristics of the active pharmaceutical ingredient (API) [1]. This is achieved by the formation of amorphous dispersions or solid solution of the API in a polymeric carrier. Moreover, HME has the advantage of continuous processes, such as good process stability, permanent process control, small footprint and decreased manufacturing time.

Our aim is to develop a continuous process line for the production of immediate release tablets (85% API released within 30 min) for a poorly soluble API (BCS II). The line consists of (1) a hot-melt extruder, (2) a die-face pelletizer, (3) a continuous blender and (4) a tableting machine making tablets in a continuous way. The API solubility is increased by formation of an amorphous system via HME. A die-face pelletizer is used to directly produce almost spherical pellets for subsequent tableting.

Formulation screening was performed to select the most promising carrier substances to match the desired release profile and stability. Based on literature research and experience five promising carriers were selected for the screening. Melt rheology measurements were performed to estimate processability and thermal sensitivity. Additionally, table top extrusions (D=9 mm, L/D=20) were carried out to study the solid state after exposure to shear forces during extrusion. Based on this screening the best formulation was selected for HME runs (D=18 mm L/D=40) on technical lab scale. The extrusion setup was designed based on material information gathered during formulation screening. The temperature profile was selected based on the DSC and rheology data. Finally, the formulation was characterized regarding the mechanical behavior and surface properties.

Pure pellets and mixtures of tablets with compaction aids were subsequently tableted using a compaction simulator. Then the tablets were analyzed and compared to the target properties.

References:

[1] J. Breitenbach, "Melt extrusion : from process to drug delivery technology," *Eur. J. Pharm. Biopharm.*, vol. 54, pp. 107–117, 2002.

[2] C. R. Young, C. Dietzsch, and J. W. McGinity, "Compression of Controlled Release Pellets Produced by a Hot Melt Extrusion and Spheronization Process," *Pharm. Dev. Technol.*, vol. 1, pp. 133–139, 2005.

[3] M. R. Abbaspour, F. Sadeghi, and H. Afrasiabi Garekani, "Design and study of ibuprofen disintegrating sustained-release tablets comprising coated pellets.," *Eur. J. Pharm. Biopharm.*, vol. 68, no. 3, pp. 747–59, Mar. 2008.

# 426723 Process Intensification through Continuous Spherical Crystallization Using a Two-Stage Mixed Suspension Mixed Product Removal System

**Abstract:**

**Process intensification through continuous spherical crystallization using a two-stage Mixed Suspension Mixed Product Removal system**

Ramon Peña and Zoltan K. Nagy School of Chemical Engineering, Purdue University, West Lafayette, IN 47907, US

Of utmost importance in the crystallization of active pharmaceutical ingredients (APIs) in the pharmaceutical industry is to produce crystals of good physical, processing, and biopharmaceutical properties. The definition of good physical properties depends on what the end goal and the drug formulation that the crystals will be a part of, but often processing and biopharmaceutical properties are competing interests. In most cases, the crystallization process is tailored to improve downstream process efficiency rather than improve drug molecule efficacy in the human body. Herein a novel concept and method to help satisfy both processing and biopharmaceutical interests is proposed, which is based on performing crystallization and spherical agglomeration in a two-stage continuous mixed suspension mixed product removal (MSMPR) system. With the suitable operating mode, the system enables to decouple the nucleation and growth from the agglomeration mechanisms, while performing efficient continuous manufacturing of particles with desired properties. Decoupling will offer more degrees of freedom for the control of each mechanisms. This in turn provides the means by which properties in the nucleation/growth stage can be tailored to those of most biopharmaceutical benefit and efficacy (e.g. bioavailability, dissolution, morphology) while allowing the agglomeration stage to be tailored to produce spherical agglomerates of the most processing efficiency (e.g. filtering, drying, friability).

Continuous spherical crystallization (CSC) of a model drug compound was first carried out by Kawashima et al.[1] after Kawashima and Capes[3] realized that resulting products from spherical agglomeration had very good flow properties. In this work, a model continuous mixed suspension, mixed product removal (MSMPR) crystallizer consisting of one stage was fed an aqueous suspension. The critical parameters were identified as agitation rate, suspension feed rate, and bridging liquid feed rate[2]. Since Kawashima et al.[1] very little work has been done on CSC and few references make use of process analytical technology (PAT) tools. An abundance of work exists on spherical crystallization in batch operation. The first batch spherical crystallization (BSC) was carried out by Kawashima et al.[2] who crystallized salicylic acid in ethanol by pouring the solution into a water-chloroform mixture and found the critical parameters to be similar to that in a continuous system: agitation rate, temperature, bridging liquid content, and residence time. In this study a ternary solvent system was used. This consisted of good solvent to dissolve the drug, poor solvent to precipitate it, and bridging liquid to promote agglomeration of the precipitated crystals.

In this work, a novel two-stage continuous spherical crystallization system in which the nucleation and growth processes are separated (in the first stage) from the agglomeration mechanism (in second stage), enabling precise and independent control of the internal crystal size distribution (CSD) and the agglomerate size distribution (ASD). This can enable the simultaneous control of product properties (e.g. dissolution profile) and processing requirements (flowability, compressibility, etc.)

1. Kawashima, Y.; Kurachi, Y.; Takenaka, H. Preparation of spherical wax matrices of sulfamethoxazole by wet spherical agglomeration technique using a CMSMPR agglomerator. *Powder Technol.* **1982**, 32, 155–161.

2. Kawashima, Y.; Okumura, M.; Takenaka, H. Spherical crystallization: direct spherical agglomeration of salicylic acid crystals during crystallization. *Science.* **1982**, 20–21

3. Kawashima, Y.; Capes, C. An experimental study of the kinetics of spherical agglomeration in a stirred vessel. *Powder Technol.* **1974**, 10, 85–92.

# 437394 Keynote: CHO Cell Based Manufacturing - CHO Origin, Diversity, Genetics, Cloning, Metabolism - Past and Future

**Abstract:**

CHO cells are the most popular and most successful host system for the production of high value proteins for therapeutic use since 30 years. However, the genomic and phenotypic diversity of CHO cells is poorly understood and vastly underestimated. Knowledge about many aspects of this diversity has been known for decades, but is buried now in part in pre-internet publications rarely read or not accessible. Recent information, both from genomic sequence analysis and from phenotypic observations of hundreds of clonally derived cell lines enforce the notion that CHO cells are in fact highly dynamic in terms of their genetics, and their phenotypes are rapidly modified in culture. Most interestingly, any CHO population will respond efficiently towards "survival" while further enhancing their genetic and phenotypic diversity, each clone with a different type of response. The resulting genetic modifications, occurring in millions of cells of clonally or non-clonally derived cell lines, can and will happen within the timeframes of the 100 or so population doublings that are typically necessary, starting from a Master Cell Bank, when used for the manufacture of a given protein product under cGMP.

The talk will try to summarize the unique and remarkable history of Chinese hamster ovary cells, in culture now for more than 50 years. While doing so, both the phenotypic and genotypic features of the many academically well-studied cell lines as well as their genomic structures as they were analysed in the 60s-80s will be presented. Very recent genomic DNA sequence analysis of a few "known" host cell lines will be added to this discussion. Based on this, a case will be made to apply to CHO cells (and any other immortalized cell line) the notion of belonging to a class of replication competent organisms, termed "quasi-species". Population genetics theory and praxis predicts that such systems increase their genetic diversity when forced through narrow population bottlenecks – the cloning of cells being one of the most stringent of such bottlenecks. In conclusion, the talk will try to provide a better understanding of what CHO cells really are, how they evolve in the setting of product manufacture, and what one can do to enhance the robustness and quality of products derived from these remarkable cells.

# 437396 Green Biotechnology Research in Europe: From Environmental Protection to a Circular Bioeconomy

**Abstract:**

Thanks to a long tradition in the Life Sciences and Chemistry and also a strong and sustained manufacturing basis in the chemical and pharmaceutical industries, Europe continues to be at the forefront of Biotechnology R&D both in academia and in industry. It is currently estimated that the potential contribution of Industrial Biotechnology (IB) to gross added value worldwide hovers around 60 billion €. In this landscape Europe is heading the implementation of IB for fine chemicals and it is the world leading producer of enzymes (75%), while at the present time the share of European firms in R&D expenditure by leading companies worldwide represents nearly 70%. This long and consistent European dominance in Biotechnology has been evolving towards greener processes that meet the challenges of improved overall process sustainability with a strong focus on environmental protection and a trend towards insertion of more bio-based steps. This presentation will describe the recent efforts in European R&D actions, primarily exemplified by the multi-year research programs supported by the European Commission (7$^{th}$ Framework Program and Horizon 2020) to integrate individual strengths in academia and industry and provide a major boost on the road from fundamental discoveries to innovation. Several successful recently completed and on-going multi-national research projects will highlight the level of ambition and the challenges to be overcome, with a special focus on the transition from environmental biotechnology to resource management in order to make the circular BioEconomy a reality.

# 437393 Microdroplet Chemostats

**Abstract:**

We describe an automated microfluidic technology for manipulating and monitoring continuous growth of populations of bacteria in microdroplet chemostats. Classic chemostats facilitate the continuous culture of bacteria, yeast, and algae by replenishing liquids to maintain specific concentrations of cells and growth factors. Chemostats have facilitated a wide-range of studies, including microbial ecology, predator/prey dynamics, and the evolution of drug resistance. However, the consumption of large quantities of reagents and the significant operational challenges of traditional chemostats limit their use. Droplet microfluidics offer a unique solution to creating many parallel chemostats, each of minute volume. Compartmentalizing cells and nutrients in microdroplets of liquid can reduce the complexity and cost of operating many parallel chemostats. In addition, we characterize the dynamics of bacterial populations in microdroplet chemostats and cellular responses to a range of stable or changing antibiotic concentrations. The method could allow for parallel, long-term studies of microbial ecology, physiology, evolution, and adaptation to chemical environments.

# 437395 Scaling up and Down Cylindrical Orbshake Bioreactors: Engineering and Use of Osrs for High Density CHO Cell Cultures - 5 Ml to 2500 L

**Abstract:**

Impeller-free mixing of a liquid is possible: "Shaken, not stirred!" reflects this. However, technical parameters of shaking processes, such as vessel geometry, direction and frequency of vessel displacement, gas transfer, power input, shear stress impact, etc. have only been explored by a few groups recently. Mixing by stirring, in contrast, has been studied extensively over more than 5 decades and impeller driven technology remains industry gold standard, particularly when animal cells, such as Chinese Hamster Ovary (CHO) cells are cultivated in large-scale bioreactors. 5-10 ml suspension cultures of CHO were shaken "orbitally" for the first time by MdJ in the late 1990s in slim, cylindrical vessels (OrbShake Tubes). Today hundreds of thousands of these 50 ml plastic tubes, ventilated passively by diffusion through a sterility membrane and endowed with a conical bottom for easy harvest of cells after centrifugation, are used widely. Over more than 10 years now we established key engineering data and studied the culture performance of orbitally shaken cylinders from 300 ml to 50 L, to 250 L and eventually to 2500 L. We found superior cell growth and productivity in these reactors, solidified by very short mixing times, high gas transfer rates, and low shear stress. The talk will cover the essentials of our research over the last 15 years and will present surprising data of so far unpublished case studies with CHO cells and their performance.

# 437392 Towards Personalised Healthcare Engineering: A New Paradigm in Blood Disorder Treatment

**Abstract:**

Personalized medicine is a medical model that proposes the customization of healthcare, with decisions and practices being tailored to the individual patient by use of patient-specific information (initially genetics information) and/or the application of patient-specific cell-based therapies. The BioBlood project aims to deliver personalised healthcare through a "step change" in the clinical field of haemato-oncology. BioBlood represents an engineered bio-inspired integrated experimental/modelling platform for normal and abnormal haematopoiesis that receives disease & patient input (patient primary cells & patient/disease-specific data) and will produce cellular (red blood cell product) and drug (optimal drug treatment) therapies as its output. Herein, we will present the experimental platform, which is a novel three-domensional hollow fibre bioreactor capable of culturing normal and abnormal haematopoietic cells in the absence of exogenous growth factors by mimicking the structure and function of the bone marrow, alongside a population balance model (PBM) that is able to capture cellular heterogenity and in particular leukaemia heterogenity. The PBM, which is able to extract patient- and disease-specific information is linked to a pharmacokinetic/pharmacodynamic (PD/PD) model, which is the used to optimise chemotherapy treatment in a personalised manner.

# 430395 A Systems Biology Definition of the Core Proteome of Metabolism and Expression Is Consistent with High-Throughput Data

**Abstract:**

Finding the minimal set of gene functions needed to sustain life is of both fundamental and practical importance. Minimal gene lists have been proposed using comparative genomics-based core proteome definitions. A definition of a core proteome that is supported by empirical data, is understood at the systems level, and provides a basis for computing essential cell functions is lacking. Here, we use a systems biology-based genome-scale model of metabolism and expression to define a functional core proteome consisting of 356 gene products, accounting for 44% of the *Escherichia coli* proteome by mass based on proteomics data. This systems biology core proteome includes nearly 200 genes not found in previous comparative genomics-based core proteome definitions, accounts for 65% of known essential genes in *E. coli*, and has functional overlap with minimal genomes (*Buchnera aphidicola* and *Mycoplasma genitalium*) (AUC = 0.78). Based on transcriptomics data across growth conditions and genetic backgrounds, the systems biology core proteome is significantly enriched in non-differentially expressed genes, and depleted in differentially expressed genes. Compared to the non-core, core gene expression levels are also similar across genetic backgrounds. Furthermore, core genes exhibit significantly more complex transcriptional and post-transcriptional regulatory features (40% more transcription start sites per gene, >20% longer 5'UTR). Thus, genome-scale systems biology approaches rigorously identify a functional core proteome needed to support growth. This framework, discerned and validated using high-throughput datasets, facilitates a mechanistic understanding of systems-level core proteome function through in silico models; it de facto defines a paleome.

# 430216 Understanding Accessibility for Targeting Nucleic Acids in Living Cells Via a Thermo-Kinetic Model of in Vivo Oligonucleotide Hybridization

**Abstract:**

RNAs continue to be pursued as major regulatory targets to control cellular regulation for therapeutic and biotechnology applications. To this end, tuning of protein expression at the mRNA level by synthetic complementary anti-sense RNAs (asRNAs) has gained significant traction. A major requirement for this approach is the design of asRNAs that can effectively bind target mRNAs inside the dynamic environment of living systems. In designing these asRNAs, a major factor that is often excluded is the structure of the target RNA. To address this limitation, we used an in vivo experimental system (in vivo RNA Structural Sensing System-iRS$^3$) to study RNA-RNA interactions using a fluorescence reporter as a measure of RNA binding. Then, we derived a biophysical model that describes RNA binding and incorporates the structural accessibility of the target region. We used the model to investigate the importance of structural accessibility in the context of other kinetic and thermodynamic factors. This study demonstrates that thermodynamics of the interaction alone do not explain patterns of asRNA binding in living cells. Instead, our model shows that nucleotide accessibility within the binding region is a stronger indicator of the effectiveness of asRNA complementation. Collectively, our results suggest that binding of an asRNA to a highly structured target region seems to be driven predominantly by kinetic factors rather than thermodynamics. Therefore we conclude that accounting for nucleotide accessibility is key for designing asRNAs that can effectively target an RNA of interest.

# 434802 Interpreting Heterogeneity in the Response of Cells Expressing a Fluorescent Biosensor to Stimulation with Hydrogen Peroxide

**Abstract:**

Fluorescent, genetically encoded sensors of hydrogen peroxide have enabled visualization of perturbations to the intracellular level of this signaling molecule with sub-cellular and temporal resolution. Ratiometric sensors hold the additional promise of meaningful quantification of intracellular hydrogen peroxide levels as a function of time, a longstanding goal in the field of redox signaling. To date, studies that have connected the magnitudes of observed ratios with peroxide concentrations have either examined suspensions of cells or small numbers of adherent cells (~10). We examined the response of all cells in several microscopic fields of view to an identical perturbation and observed a striking degree of heterogeneity of fluorescence ratios from individual cells. The expression level of the probe and phase within the cell cycle were each investigated as potential contributors to the observed heterogeneity and found to be important. To aid in the interpretation of experimental observations, we incorporated the reaction of the reduced probe with peroxide and the reactions of the oxidized probe with glutathione and glutaredoxin into a larger kinetic model of peroxide metabolism. The predictions of the kinetic model suggest possible explanations for the experimental observations. This work highlights the importance of a systems-level approach to understanding the output of genetically encoded sensors that function via reversible redox reactions involving thiol and disulfide groups.

# 422511 A Kinetic Platform to Determine the Fate of Hydrogen Peroxide in Escherichia coli


**Abstract:**

Hydrogen peroxide ($H_2O_2$) is a toxic molecule utilized by the immune system in the first line of defense against infections. After being engulfed by phagocytic cells of the innate immune response, pathogens are exposed to reactive oxygen species (ROS) including superoxide ($O_2^{-}$•) and $H_2O_2$. $O_2^{-}$• cannot cross cell membranes, but $H_2O_2$ freely diffuses into the bacterium, where it can damage sensitive biomolecules, react with ferrous iron to form highly reactive hydroxyl radicals (•OH), or be detoxified by specialized enzymes. Accordingly, $H_2O_2$ detoxification is an important aspect of bacterial virulence, and deletion of one or more detoxification systems affects the ability of a variety of pathogens to establish or sustain an infection. While the importance of these $H_2O_2$ detoxification systems is known, a quantitative understanding of how they interact under different levels of oxidative stress is lacking. For instance, *Escherichia coli* utilizes an alkyl hydroperoxidase (AHP) and two separate catalases (HPI and HPII) to clear $H_2O_2$. AHP requires one NADH molecule per reaction cycle, affecting the maximum rate of clearance that can be achieved, whereas $H_2O_2$ is the only substrate required by HPI and HPII. The simultaneous activity of multiple detoxification systems and broad reactivity of $H_2O_2$ necessitates the use of quantitative kinetic modeling to understand how $H_2O_2$ distributes throughout its reaction network.

Thus, we have constructed an $H_2O_2$ detoxification network for *E. coli* consisting of 60 metabolites and enzymes, as well as 75 rate equations describing spontaneous and enzymatic reactions, transcriptional regulation, and degradation of key enzymes. Structural and parametric uncertainty with regard to enzyme degradation and the existence of an $H_2O_2$ gradient across the cell membrane led us to consider ten distinct structures while building the model. Using a strategic method of iteratively gathering data, optimizing uncertain parameters, and assessing comparative model performance, we found that only one model structure was able to consistently predict how $H_2O_2$ distributes amongst the different pathways. This observation suggested that bimolecular degradation of detoxifying enzymes is an important aspect of the $H_2O_2$ clearance network, and that under our conditions diffusion of $H_2O_2$ across the membrane was not limiting. Using a Monte Carlo method to explore the viable parameter space, we identified an ensemble of parameter sets that could all fit the data, and used this model ensemble to gain a better understanding of the importance of protein synthesis and reducing equivalents under oxidative stress. We have developed a platform to quantitatively explore $H_2O_2$ stress in *E. coli* which enhances understanding of the impact of mutations and environmental alterations on detoxification, and our method provides a template for translation to other organisms.

*Computational Modeling of Aspergillus Metabolism for Metabolic Engineering Purposes*

Yi Li, Lei Fang, Blaine Pfeifer
Department of Chemical and Biological Engineering
SUNY, University of Buffalo

University at Buffalo *The State University of New York*

# *Objective*

- Fuel production precursor
- *Aspergillus niger*-Industrial species
- Lignocellulosic feedstocks
- Sugar conversion
- Biomass utilization
- Building Block: Octatrienoic acid

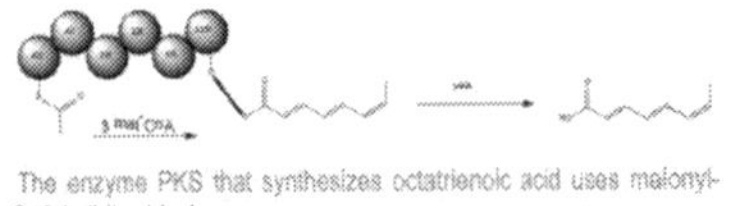

The enzyme PKS that synthesizes octatrienoic acid uses malonyl-CoA building blocks.

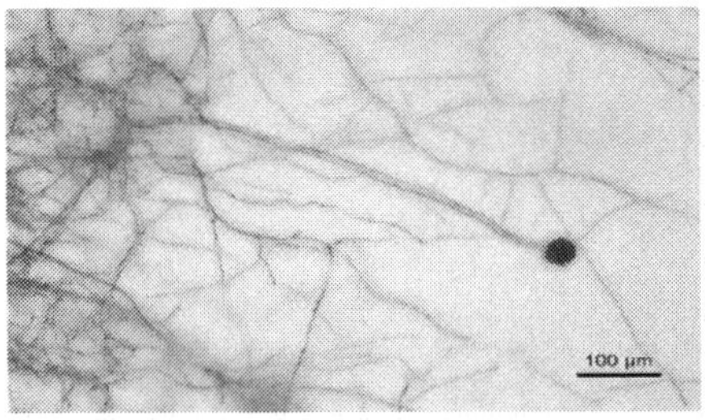


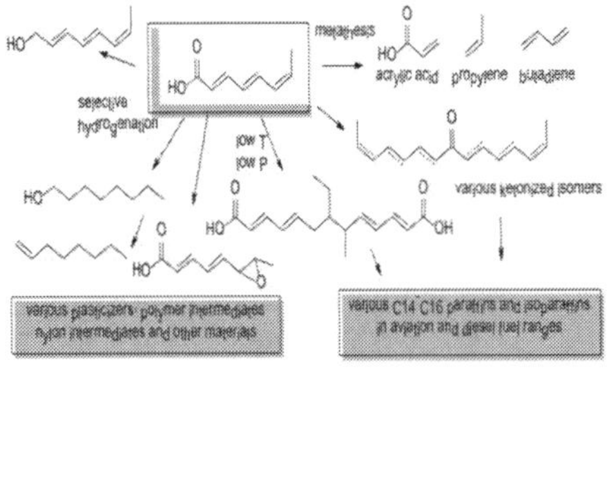


University at Buffalo *The State University of New York*

# Heterologous biosynthesis advantage

- Fungal metabolic pathway
- DOE's Joint Genome Institute (JGI)
- Polyketide synthesis
- Hydrocarbon backbone
- Robust genetic tool-reverse marker
- Highly tolerant to low pH
- Monomeric and polymerized sugars
- Relevant feedstock cornstover (Novozymes)

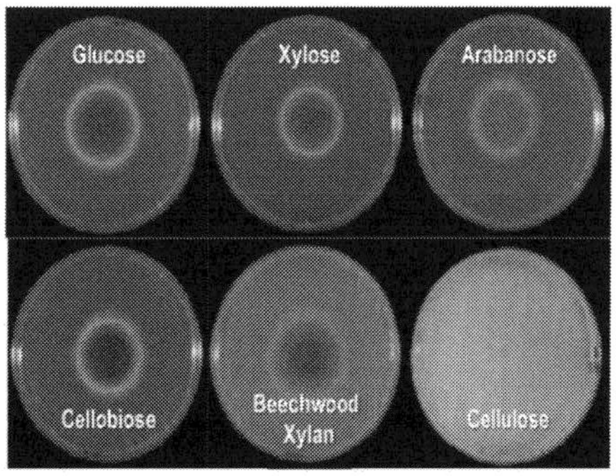


Growth of *A. niger* 1015 on various sugars similar to what is present in lignocellulosic hydrolysate (www.fung-growth.org).

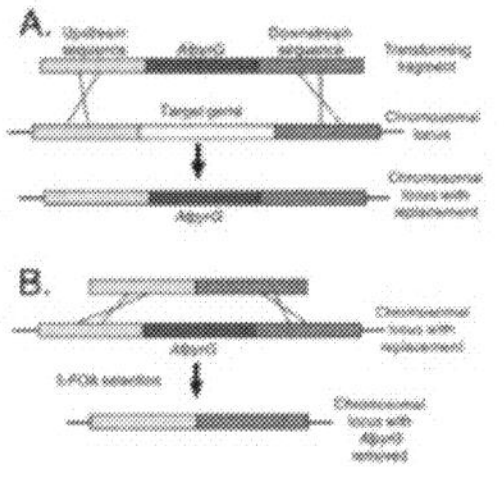


Gene deletion using a reversible marker.

University at Buffalo *The State University of New York*

# Introduction: Metabolic Modeling of A. niger

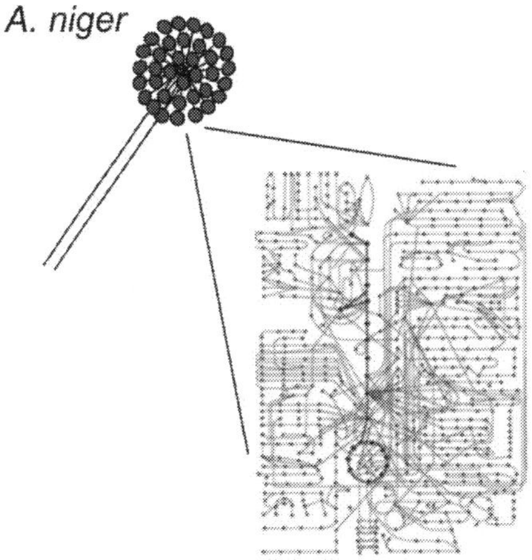


Project Goals: Identify experimental targets to improve the production of a desired metabolite using computational metabolic modeling.

Collaborators: Ken Brunno (PNNL); Berl Oakley (University of Kansas); Clay Wang (University of Southern California)

Computational Approaches: Flux Balance Analysis (FBA) and Elementary Mode Analysis (EMA)

Heterologous Reaction:

ACCOA[c] + 3 MALCOA[c] + 3 NADPH[c] <=> Oct[c] + 3 $CO_2$[c] + 4 COA[c] + $H_2O$[c] + 3 NADP[c]

University at Buffalo *The State University of New York*

# Flux Balance Analysis: Overview

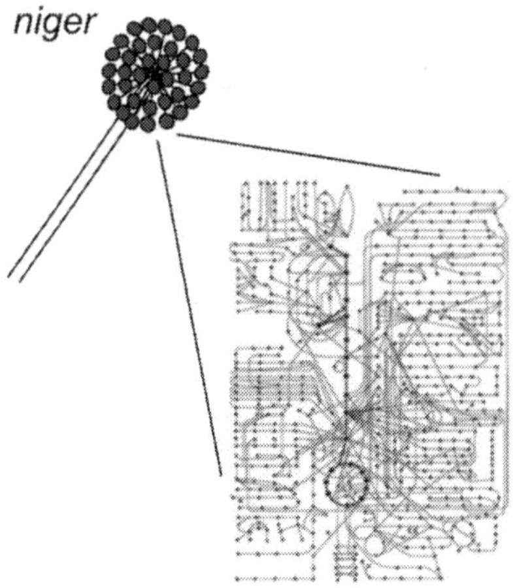


Aspects: Comprehensive; underspecified; more likely to generate non-obvious targets

Approach:
1) Select objective function;
2) Apply a second algorithm called Minimization of Metabolic Adjustment (MoMA) to identify gene deletion or over-expression targets.

Objective: Maximize target metabolite flux without sacrificing biomass formation.

Andersen M.R., Nielsen M.L., Nielsen *J. Mol Syst Biol.* 2008;4:178.

University at Buffalo *The State University of New York*

# Flux Balance Analysis: Description

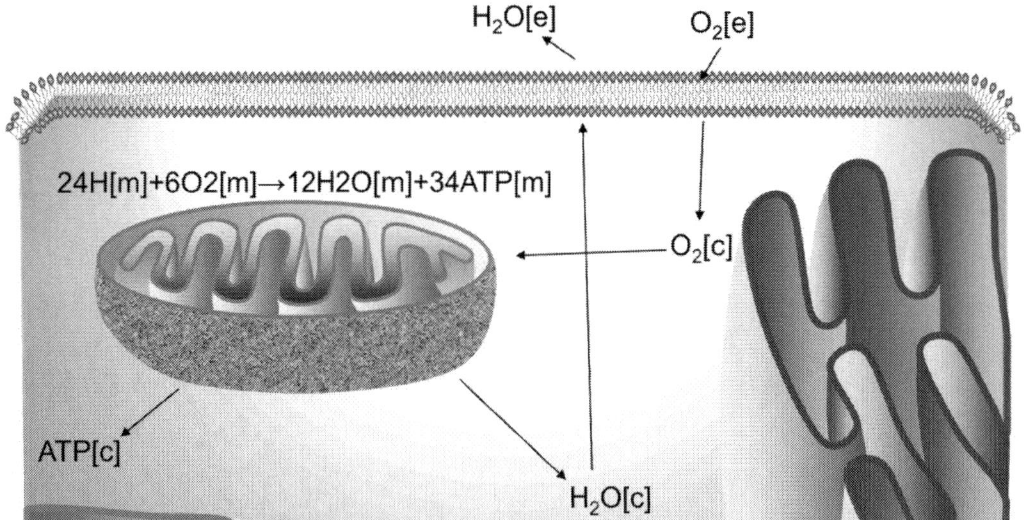


University at Buffalo *The State University of New York*

# *Flux Balance Analysis: Description*

**S**(stoichiometric matrix) of size **m**(metabolites)×**n**(reactions)

| S | RXN1 | RXN2 | RXN3 | RXN4 | RXN5 | RXN6 | ....... | v (rxn. flux) |
|---|---|---|---|---|---|---|---|---|
| $CO_2$[e] | 1 | 0 | 0 | 0 | 0 | 0 | ....... | $v_1$ |
| $H_2O$[e] | 0 | 0 | 0 | 2 | 0 | 0 | ....... | $v_2$ |
| $O_2$[e] | 0 | 1 | -1 | 0 | 0 | 0 | ....... | $v_3$ |
| ATP[m] | 0 | 0 | 0 | 1 | 0 | 0 | ....... | $v_4$ |
| ⋮ | ⋮ | ⋮ | ⋮ | ⋮ | ⋮ | ⋮ | ⋮ | ⋮ |

Positive coefficient for consumed metabolites
Negative coefficient for produced metabolites

University at Buffalo *The State University of New York*

# *Flux Balance Analysis: Description*

1. Mass balances on steady states : Sv=0

$$\left(\begin{array}{llllr} v_1+ & & & \ldots=0 & CO_2[e] \\ v_1+ & & +2v_4 & \ldots=0 & H_2O[e] \\ & v_2-v_3 & & \ldots=0 & O_2[e] \\ & & v_4 & \ldots=0 & ATP \\ & & & & \\ & & & \ldots n & \end{array}\right)$$

2. Flux constraints

Change flux upper/ lower boundaries

Examples: *E. coli* $CO_2$ transportation rate

$CO_2$[e] ⟷ $CO_2$[c]  $-2.1 \text{mmol gDCW}^{-1} h^{-1} < v_n < 0$

3. Objective function

Biomass=1·12 PROTEIN+0·56 RNA+LIPID+LPS+GLYC+PG+DNA+C1+PA+ATP[1]

University at Buffalo *The State University of New York*

1. Anke Kayser, etc. Metabolic flux analysis of Escherichia coli in glucose-limited continuous culture. I. Growth-rate-dependent metabolic efficiency at steady state. Microbiology, 2004

# *Flux Balance Analysis: Description*

Maximize the flux of objective function

Objective function: $z=a^*v_1+b^*v_2+c^*v_3+d^*v_4$ ....

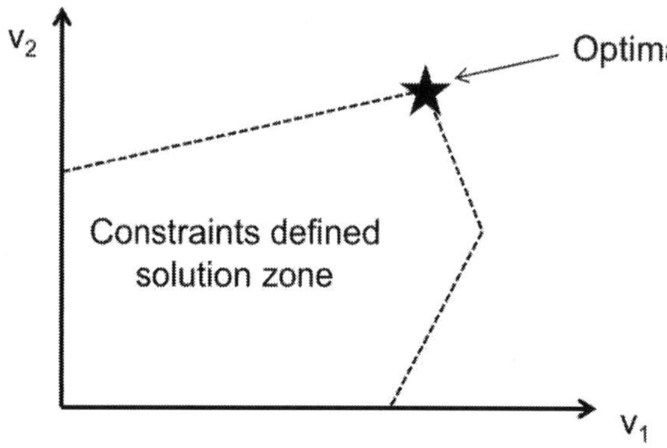




# *FBA: Minimization of Metabolic Adjustment*

Metabolic network perturbations:

Alter flux boundaries:

- Gene deletions: $v_i=0$
- Gene overexpression: $v_i=n^*v_i$

Minimize the Euclidean distance from a wild-type flux distribution

$*\min(v-w)^T(v-w)$; v altered network flux vector
w wild type flux vector

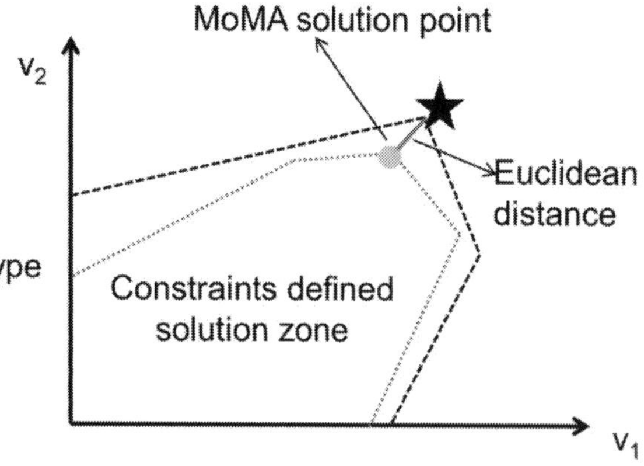


*Same constrains as FBA

# FBA: Phenotype Fraction

**The phenotype fraction** equally weights both the biomass flux and the flux of interest

$$f_{ph} = fbiomassfcom_{pound\ of\ interest} = (\frac{v_{biomass\ KO}}{v_{biomass\ wild}})(\frac{v_{compound\ of\ interest\ KO}}{v_{compound\ of\ interest\ wild}})$$

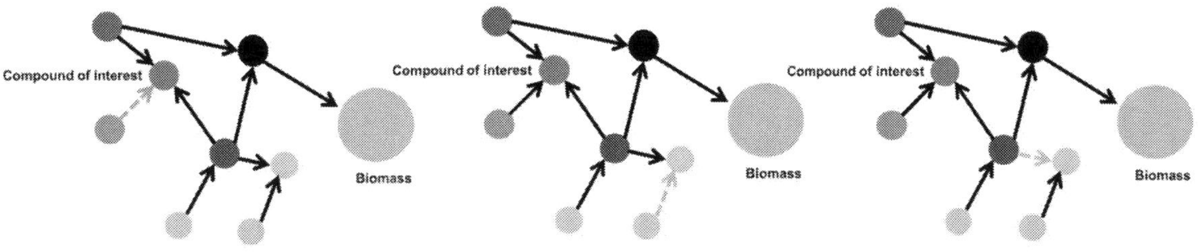


Scenario 1: fph<1 Scenario 2: fph=1 Scenario 3: fph>1



# Flux Balance Analysis: Results

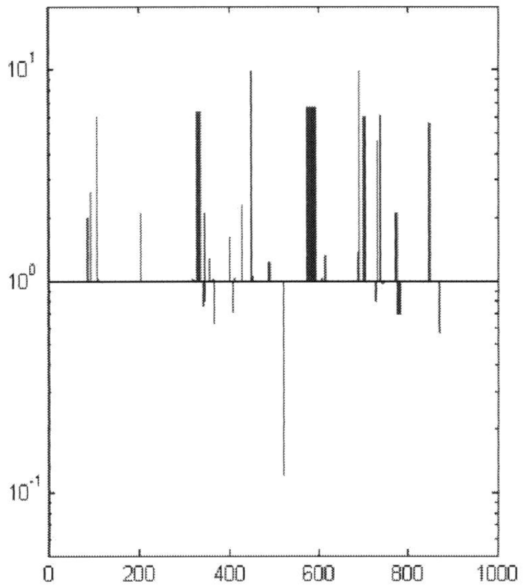


University at Buffalo The State University of New York

# *Flux Balance Analysis: Results*

**Table 1 Knockout targets for improving octatrienoic acid production**

| Enzyme | Reactions | fph | |
|---|---|---|---|
| Cytochrome c oxidase subunit I cox5 | 0.5 O2m[m] + 2 FEROm[m] + 4 HX_POm[m] -> H2Om[m] + 2 FERIm[m] + 4 HX_PO[c] | 6.2554 | Distribute ATP for biosynthesis of polyketide products |
| Ribose-5-phosphate isomerase | R5P[c] <=> RL5P[c] | 6.0951 | Disruption of the pentose affecting NADPH pools |
| Acetyl-CoA C-acetyltransferase, acetoacetyl-CoA thiolase | 2 ACCOA[c] <=> COA[c] + AACCOA[c] | 5.9645 | Possibly preventing the futile cycling of acetoacetyl-coA |
| Ubiquinol-cytochrome c reductase | 4 HX_POm[m] + 2 FERIm[m] + QH2m[m] -> 2 FEROm[m] + 4 HX_PO[c] + Qm[m] | 5.5773 | Distribute ATP for biosynthesis of polyketide products |
| Formaldehyde dehydrogenase | H2O[c] + NAD[c] + FALD[c] <=> FOR[c] + NADH[c] | 1.6118 | Increase NADPH for polyketide products |
| Malate synthase | H2Om[m] + GLXm[m] + ACCOAm[m] -> MALm[m] + COAm[m] | 1.2995 | Disrupting the glyoxylate cycle might decrease lipids/FAs |
| Glycerol 3-phosphate dehydrogenase M NAD c+ dependent | NADH[c] + T3P2[c] -> NAD[c] + GL3P[c] | 1.0449 | Reduce the amount of lipid synthesis |

University at Buffalo *The State University of New York*

# *Elementary Mode Analysis: Overview*

*A. niger*

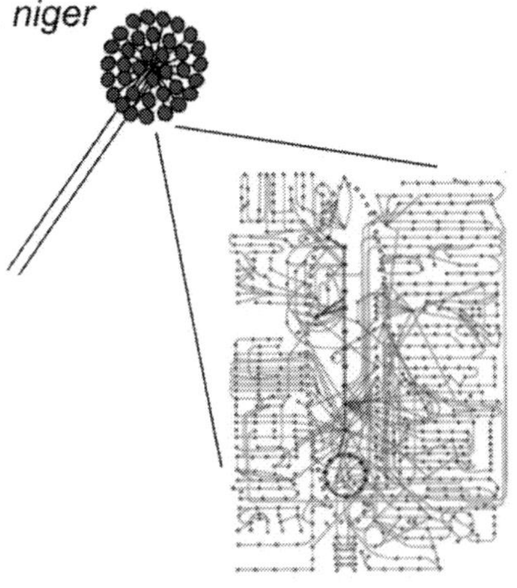

Aspects: Narrowed; does not require objective function (all plausible pathways ("modes") are evaluated)

Approach:

1) Pick limited number of conserved metabolic pathways;
2) Assess how pathway removal affects target metabolite levels.

Objective: Maximize target metabolite flux without sacrificing biomass formation.

University at Buffalo *The State University of New York*

# Elementary Mode Analysis: Description

Minimal set of enzymes at steady state
Optimization: Maximizing molar yields

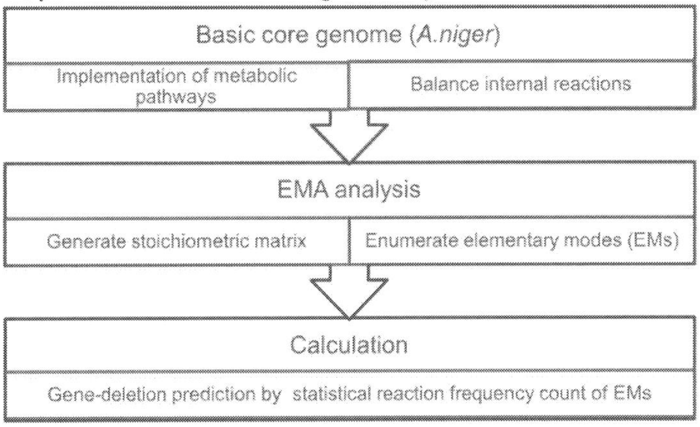


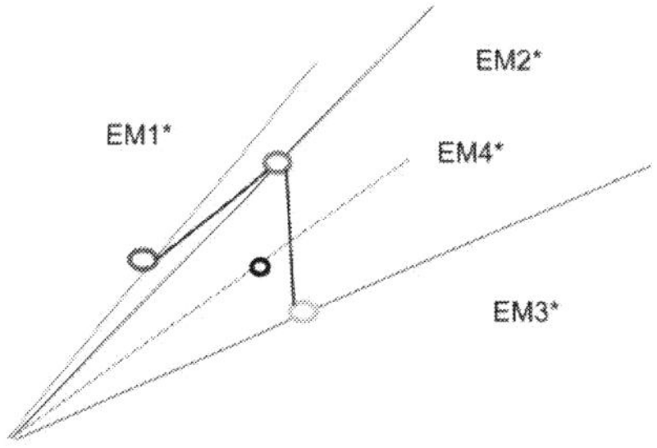


*A*: matrix with each row containing stoichiometric coefficients
*X*: Elementary flux
*b* : accumulation rates of extracellular metabolites

A*X=b

University at Buffalo *The State University of New York*

# Elementary Mode Analysis: Approach

| | Oct Model |
|---|---|
| **Number of reactions** | 68 |
| **Number of metabolites** | 53 |
| **Mean EMA Computation Times(s)** | 232.13 |
| **Number of EMs** | 174,555 |
| **Number of product –producing Ems(%)** | 13,908 (7.97%) |
| **Number of biomass-producing Ems(%)** | 7,302 (4.18%) |
| **Number of product- and biomass- producing Ems(%)** | 7,203 (4.13%) |

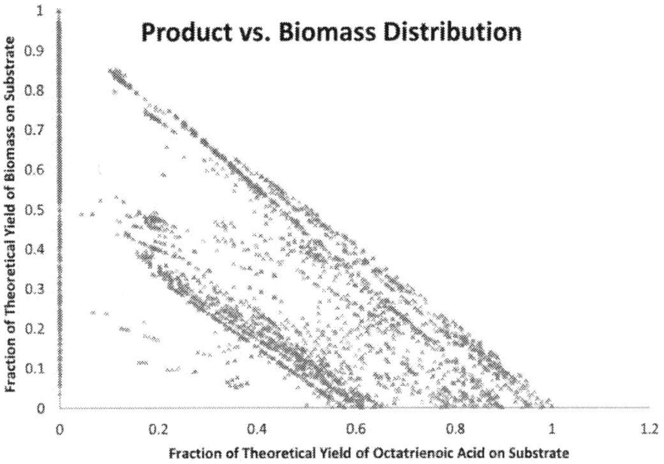


University at Buffalo *The State University of New York*

# Elementary Mode Analysis: Results

| Enzyme | Reaction | Frequency | Explanation |
|---|---|---|---|
| fumarate reductase | FUM + QH2 = SUCC + Q | 0.99 | TCA cycle<br>Growth-retarding reaction on glucose and glycerol* See paper |
| lactate dehydrogenase | PYR + NADH = LAC + NAD | 0.84 | Feasible in if A.niger has a functioning LDH |
| formate dehydrogenase complex | FOR = CO2 + H2_ext | 0.81 | Low risk and Feasible |
| acetaldehyde dehydrogenase | ACoA + NADH = ACA + NAD + CoASH | 0.80 | Higher risk but reduce ACoA conversion to ethanol |
| alcohol dehydrogenase ; ethanol dehydrogenase | ACA + NADH = ETOH + NAD | 0.80 | Increase certain acid production |
| acetate kinase | ACP + ADP = ACE + ATP | 0.71 | Down-regulation indicated oxalate increase |



# Acknowledgements


**Funding Sources**:
DoE

**Collaborators**:
Ken Bruno (PNNL)
Berl Oakley (KU)
Clay Wang (USC)
Mikael Andersen (DTU)

# 420198 Prediction of Peptide Solubility: An Experimental and Computational Framework

**Abstract:**

The design of peptides for therapeutic use has become an increasingly important and desirable approach for biomedical research. However, design of protein/peptide therapeutics is complicated by competing factors affecting the possible use of validated peptide agonists and antagonists, such as solubility, cell permeability, bioavailability, stability, and adverse side effects. As stark evidence for this fact, over 40% of drug candidates fail further pursuit because of adverse pharmacokinetics or bioavailability (1). For this reason, accurate methods for the prediction of peptide properties (e.g., solubility) based on amino acid composition alone, are highly desirable. In particular, amyloidogenic sequences are of significant concern for peptide solubility due to their role in the formation of fibrous aggregates.

In this study, we have utilized a combination of experimental and computational tools to investigate the factors leading to poor peptide solubility, with particular interest in developing predictive models of peptide solubility. We performed solid-phase peptide synthesis, using a Nautilus 2400 parallel synthesizer, to produce over 40 suspected amyloid peptides containing between 4 and 11 amino acids. The solubility of each peptide was measured using a modified version of the LYophilisation Solubility Assay (LYSA), as described in (2). SVM regression was performed based on predicted structure based features, including solvation free energy and solvent accessible surface area. Additionally, a novel docking algorithm, based on grey-box global optimization (3), was used to predict the structure of the optimal peptide aggregate. Finally, molecular dynamics simulations, as described previously (4), were also performed to evaluate the association affinity of each peptide.

1. J. Wang, T. Hou, Recent Advances on Aqueous Solubility Prediction. Combinatorial Chemistry & High Throughput Screening 14, 328–338 (2011).
2. D. Roy, F. Ducher, A. Laumain, J. Y. Legendre, Determination of the aqueous solubility of drugs using a convenient 96-well plate-based assay. Drug Development and Industrial Pharmacy 27, 107–109 (2001).
3. F. Boukouvala, M.M.F. Hasan, C.A. Floudas, Global Optimization of General Constrained Grey-Box Models: New Method and its Application to Constrained PDEs. Journal of Global Optimization, Submitted (2015).
4. J. Smadbeck et al., De Novo Design and Experimental Characterization of Ultrashort Self-Associating Peptides. Plos Computational Biology 10 (2014).

# 431652 An Information Theoretic Approach to the Model Selection Problem in Systems Biology

**Abstract:**

Data from biological systems can be modeled with a variety of differing mathematical descriptions. For instance, to model metabolic states one can use kinetic, constraint-based or cybernetic formulations to explain a given set of metabolic flux data. While arguments made from an intuitive standpoint may justify the selection of one model over another, the field of systems biology has yet to widely implement a universal scale to weigh the utility of models against one another. To address the problem of model selection, it is prudent to borrow arguments formulated from information theory. In this light, it becomes possible to distinguish models from one another by their ability to describe the regularity of a given system's output data. Capturing this regularity is in itself an act of compressing the data and therefore information theoretic methods are applicable. In this vein, one may apply the Minimum Description Length (MDL) principle to candidate biological models in order to compare how well these models describe regularity in data. Metrics derived upon MDL principles seek to identify a single model from a set of candidate models that best compresses data. This is accomplished through the consideration of both the model's likelihood of fit and its inherent complexity. The model that best compresses data from a process relative to other models can be said to be the most efficient and useful description of the system.

In this talk, MDL is employed to compare metabolic models on the basis of how well they compress metabolic flux data. Both static and dynamic data are considered for *S. oneidensis* and *E. coli* growing on lactate and glucose respectively. Metrics developed upon MDL principles including Akaike Information Criteria (AIC) and Bayesian Information Criterion (BIC) are applied in the analysis of a group of dynamic metabolic models to determine which model best compresses metabolic data through maximizing model fit and minimizing model complexity. This analysis identifies a point of diminishing returns in which additional model complexity provides little gain in describing data accurately for both nested and non-nested metabolic models. Starting with a comparison of flux predictions made by constraint based, kinetic and cybernetic metabolic models, this work develops a framework that intends to be extended to other model comparison applications in systems biology.

# 431635 Development of a Novel PAT for in-Line Measurement of Powder Dynamic Density inside Feed Frame of Rotary Tablet Presses

**Abstract:**

Consistent filling of the tablet press die with a uniform weight of powder has always been an empirical approach with little understanding of the powder flow pattern within the feed frame, especially at the last moment before powder is fed into the die and tablet potency is set. Compression is a complex unit operation comprising different regimes of powder flow (depending on the shear rate and packing fraction) which need to be fundamentally understood in order for this to be a predictable process.

In this work we studied powder flow (with application to die filling) in the feed frame of various rotary tablet presses by developing a novel in-line process analytical technology (PAT), the dielectric probe. This tool enabled us to capture powder dynamic density regime maps in different sections of a feed frame. The basic measurement principle of the probe is to estimate the physical properties of bulk solids (e.g. solid fraction or density) by standard correlation with the measured electrical properties (e.g. dielectric permittivity) through proper calibration for each formulation.

We investigated effects of press speed, feed frame speed, paddle design, and formulation properties on the die filling process by looking at dynamic density of powder and its fluctuations inside the feed frame. We discuss how these measurements can be used for faster and more effective set-up of the tablet press and to find and maintain the optimum operating condition throughout the entire compression. Furthermore, we show how these data are helpful in scale-up of the tablet press, especially for formulations that are sensitive to over-lubrication where it is important to maintain similar level of shear energy among different scales.

# 420074 Development of a Control Strategy for Continuous Manufacturing of Solid Dosage Forms

**Abstract:**

During the last years the advantages of continuous manufacturing have been increasingly demonstrated (and publicized) by academia, industry and regulators. Slowly, this is leading to a paradigm shift in pharmaceutical manufacturing, away from batch towards advanced continuous processing based on real-time quality assurance. As a result, different equipment suppliers are providing turn-key solutions for continuous manufacturing systems, ranging from powder feeding, to blending, granulation, tableting and/or coating. While several systems have been reported and studied in the literature, open questions still remain, especially with respect to system robustness, online measurement and process control.

In this work a control strategy for a novel industrial-scale production line for oral-dosage forms was developed. The objective of the control strategy is to consider all possible variability in raw material properties, process-settings, equipment design and environmental conditions. Constant product quality is maintained by proper equipment design and integration of unit operations, utilization of process understanding, a risk-based PAT strategy and appropriate control actions.

The presentation will highlight a successful implementation of an in-line monitoring system for critical quality attributes. In order to enable Real Time Release, PAT tools were selected with appropriate data acquisition systems to support control actions. This includes chemometric models for start-up and steady state. Out-of-spec materials are detected and eliminated at different stages of the process. All sensors are embedded in a PAT data management system to treat and register process data. Additionally, a concept for 100 % quality control of final tablets will be presented.

# 420103 Investigation of Segregation Effects in a Direct Compaction Process Via Multipoint NIR Monitoring and NIR Chemical Imaging

**Abstract:**

During the last years the application of near-infrared spectroscopy (NIRS) has become an established technology for in-line process monitoring of the API-content, as it is a fast and non-destructive method. Common methodologies have been further developed to multipoint applications allowing the timely aligned assessment of real time process data from different positions or even processes. Additionally, hyperspectral imaging (HSI) even allows measuring not only parts of powder streams or single tablets, but a large number of tablets. Even mapping of each single tablet becomes feasible and enables detection of API agglomerations.

In this study the evolution of API concentration within a direct compaction process is monitored in-line by means of a multipoint spectrometer using separate fibres at different locations within the press. In addition, a HSI unit assesses the quality of the final tablets. This approach enables the determination of content uniformity along the powder path and therefore, investigation of segregation effects along the processing route. Goal of the work is the evaluation, if those effects influence the final product quality and if different operating conditions affect the extent of segregation.

# 421775 Characterization of Tablet Core and Coating Roughness By Means of Optical Coherence Tomography


**Abstract:**

Although tablet coating is a well-established unit operation in pharmaceutical industry, the achievable quality of coating is still limited by the fact that it is a highly complex process depending on many parameters. Specifically for functional coatings slight changes of the coating equipment and process parameters may impact the physicochemical properties of the film and may consequently affect the coating quality. There are significant gaps remaining in the full scientific understanding of the process, which makes it a difficult task to produce tablets with an exact thickness and density and with only small tablet-to-tablet variations between batches or even within one single batch. The use of advanced quality control tools facilitates the understanding of physico-chemical changes of the film.

Such a quality control tool is optical coherence tomography (OCT) showing not only high potential for off-line quality control of film-coatings but also for in-line monitoring of coating processes. OCT is a high resolution imaging methodology to produce cross-sectional and three-dimensional (3D) images of film coatings in a non-destructive and contactless manner. This modality allows the direct measurement of the coating thickness under consideration of the refractive index of the coating. Beside the analysis of the coating thickness, it additionally facilitates the characterization of the tablet core and coating roughness.

An automated data evaluation algorithm extracts information about coating thickness as well as tablet core and coating roughness. This analysis can be carried out either on the basis of cross-sectional images or by using 3D data of samples. The focus in this study is on the analysis of 3D images of samples removed periodically from a pan coating process. A thickness map and profile maps of the tablet core and coating is computed from about 480,000 depth measurements per sample. This data enables among others the calculation of the arithmetical mean deviation, the root mean square deviation, the skewness and the kurtosis of the assessed profiles. Moreover, the impact of the tablet core roughness on the coating thickness is analyzed by correlating the tablet core profile and the coating thickness map. This analysis is further transferred to in-line OCT measurements allowing the investigation of core and coating roughness during the production. Such detailed information specifically emphasizes the high capability of the OCT technology to improve process understanding.

# 421912 In-Line Monitoring of a Pharmaceutical Fluid-Bed and Pan Coating Process By Optical Coherence Tomography

**Abstract:**

This study demonstrates a new in-line measurement technique for monitoring coating processes. Process monitoring is performed by an optical coherence tomography (OCT) sensor providing cross-sectional images of film-coatings in a non-destructive and contactless manner. Typically tablets or pellets are coated. Tablets are mostly coated in (perforated) pans or drums, whereas pellet coating is typically carried out in a fluid-bed. However, the monitoring of the coating growth as a function of process time is a key issue for both processes. This study shows results of coating thickness measurements from in-line OCT data for pellets and tablets coated in a fluid-bed and pan coater, respectively.

The coating thickness can be determined directly and automatically from OCT images and without the need of chemometric calibration models. Results of the in-line OCT measurements were validated by analyzing samples that were periodically removed from the process during the production. These samples were characterized off-line by OCT and dynamic image analysis in case of pellets or dimension measurements in case of tablets. Moreover, this methodology enables a multiple direct measurement of the coating thickness on individual pellets/tablets rather than providing the temporal and spatial average of a large number of pellets/tablets. This gives substantially more information about the coating quality, i.e., intra- and inter-particle coating variability, than standard quality control methods. The information about those parameters specifically emphasizes the high capability of the OCT technology to improve process understanding and to assure a high product quality. Specifically, the additional information as intra- and inter-particle coating variability as captured with the OCT sensor could be a valuable critical quality attribute (CQA). An in-line OCT system could act as a key tool for the development and control of coating processes within the process analytical technology (PAT) and quality by design (QbD) framework.

# 428833 Development and Demonstration of a Suite of Tools Aimed at Process Development Intensification and Efficient Experimentation Practices

**Abstract:**

Since its inception in early 2014, the Process Development Intensification Laboratory (iLab) within Chemical Process Development and Commercialization at Merck has been focused on the development of tools and techniques to enable acquisition of deeper, more fundamental process understanding in a resource sparing manner. This is predominantly centered around the concept of data intensification, specifically, around enabling more efficient data capture, data reduction and data analysis.

In this submission, we highlight the development and demonstration of novel, enabling technologies such as process modeling, bench-top automation, parallel processing, process analytical technologies and statistically designed experimentation. Furthermore, this submission uses a saponification deprotection reaction as a case study, highlighting the synergistic effects which can be realized through the application of many of the developed tools.

Specifically, we highlight the use of many of the tools the iLab has been employing, beginning with traditional DOE for selection of preliminary experimental conditions. We go on to discuss some of its limitations when the number of factors for investigation is high, prompting application of kinetic modeling for acquisition of deeper, more mechanistic process understanding. From there, it becomes apparent that alternative experimental techniques, specifically use of the Amigochem (a bench top automation tool combining a liquid handling robot for LC sampling and dilution with a parallel reaction station) offers an opportunity for building information-rich data sets rapidly. This enables kinetic profiling of the aforementioned DOE with minimal user intervention. We also demonstrated the use of probe-based sampling and dilution for unattended sampling of automated lab reactors. When coupled with online FTIR spectroscopy and chemometric techniques such as partial least squares regression, it becomes possible to gather compositional information and very detailed kinetic profiles rapidly in a wide variety of equipment types.

Simply collecting significant quantities of data is not always fruitful, however, unless that data can be transformed in some way to intimate process knowledge. We have also been developing means by which collected data can be used to build process models, and more importantly, how those process models can be used to drive future experimental design decisions. We have demonstrated two separate mathematical algorithms, one geared towards selection of optimal experiments to aid in model discrimination, and the other geared at selection of optimal experiments for kinetic parameter refinement.

Separately, many of the technologies developed by the iLab represent significant achievements in data collection, analysis and application in the development space. When combined as a holistic experimental design paradigm, there is significant synergy realized, and the complete toolkit stands to be truly transformative in the way in which process development is conducted in our area moving forward.

# 417087 Spectroscopy Assisted Process Design for Complex Multi-Phase Organic Reactions

**Abstract:**

Chemical reaction impurity control is an essential component of quality risk management in pharmaceutical drug substance manufacturing. Transformations involving reactive/unstable intermediates or multi-phase reaction media often present a challenge for conventional analytical techniques (e.g. HPLC, GC, NMR). During process development and plant execution of such processes, utilization of in-situ spectroscopic techniques offers a significant advantage from both sensitivity and time-of-analysis perspectives. In this work, two industrial case studies implementing spectroscopy will be discussed in the context of small molecule pharmaceutical intermediate laboratory process design and pilot plant execution.

In the first case study, process development for preparation of a triazole-based intermediate was performed utilizing Raman and Near IR spectroscopy. Such in-situ measurements were necessary as the reagents were both non-chromophoric and unstable using conventional off-line analytical methods. The data allowed for rapid analysis of kinetic information as well as mechanism elucidation. The approach was ultimately leveraged to define robust processing parameters (e.g. temperature, concentration, order of addition) for optimal impurity control. In the second case study, Near IR spectroscopy was utilized to monitor ammonia concentration in a telescope reductive amination process. Control of residual ammonia proved critical to limit formation of a des-halogenation impurity during the hydrogenation step. Application of Near IR to trigger additional vessel purges was successfully demonstrated in both the laboratory and pilot plant scale.

# 412815 Construction of Robust, Multi-Input Genetic Circuits Linked with Feedback Loops

**Abstract:**

**Construction of robust, multi-input genetic circuits linked with feedback loops**

Tatenda Shopera, William R. Henson, Young Je Lee, Kenneth Ng, Andrew Ng, and Tae Seok Moon

Washington University in Saint Louis

Saint Louis, MO

The ability to program living cells with novel functions has great potential to unlock important innovative solutions to many industrial, medical, and energy challenges. For example, artificial gene networks have been engineered in organisms to produce products ranging from pharmaceuticals to biofuels. However, many applications are yet to be fully exploited due to poor understanding of how robust, reliable, and predictable complex genetic networks are constructed. Thus, understanding the underlying design strategies implemented by natural biological systems to execute reliable behavior is essential for programming complex cellular functions. Our overarching goal is to determine the governing design principles of biological robustness by building simple genetic circuits from the bottom-up. Among these circuits, we have successfully constructed noise-tolerant bistable switches by combining multiple-protein interactions (regulatory cascades) and feedback systems using biological parts obtained from nature. Specifically, in this presentation, we will highlight how microorganisms were programmed to retain strong memory over a wide range of input signals in a predictable manner through engineering of the gene regulatory architecture.

CrossMark

# Discriminating direct and indirect connectivities in biological networks

Taek Kang[a,b], Richard Moore[a,b], Yi Li[a,b], Eduardo Sontag[c,1], and Leonidas Bleris[a,b,d,1]

[a]Bioengineering Department, University of Texas at Dallas, Richardson, TX 75080; [b]Center for Systems Biology, University of Texas at Dallas, Richardson, TX 75080; [c]Department of Mathematics and Center for Quantitative Biology, Rutgers-The State University of New Jersey, Piscataway, NJ 08854; and [d]Electrical Engineering Department, University of Texas at Dallas, Richardson, TX 75080

Edited by Wing Hung Wong, Stanford University, Stanford, CA, and approved September 10, 2015 (received for review April 12, 2015)

**Reverse engineering of biological pathways involves an iterative process between experiments, data processing, and theoretical analysis. Despite concurrent advances in quality and quantity of data as well as computing resources and algorithms, difficulties in deciphering direct and indirect network connections are prevalent. Here, we adopt the notions of abstraction, emulation, benchmarking, and validation in the context of discovering features specific to this family of connectivities. After subjecting benchmark synthetic circuits to perturbations, we inferred the network connections using a combination of nonparametric single-cell data resampling and modular response analysis. Intriguingly, we discovered that recovered weights of specific network edges undergo divergent shifts under differential perturbations, and that the particular behavior is markedly different between topologies. Our results point to a conceptual advance for reverse engineering beyond weight inference. Investigating topological changes under differential perturbations may address the longstanding problem of discriminating direct and indirect connectivities in biological networks.**

reverse engineering | synthetic biology | direct and indirect connectivities | human cells | nonparametric resampling

A focal point of systems biology is the reverse engineering of gene regulatory networks (1–5). The methods have shifted from intuitive inference of local connectivities to comprehensive analysis of large networks, involving heterogeneous data sets from high-throughput experiments and complex theoretical tools (6–10). Despite significant advances, a fundamental reverse engineering bottleneck is the ability to discriminate between direct and indirect connections. In a simple case, assuming three nodes in a cascade formulation, where an input node is activating an intermediary node which in turn is activating an output node, a reverse engineering algorithm may infer an activating edge from the input node to the output, even though there is no direct biological interaction.

Unfortunately, the limitation in correctly distinguishing the effects stemming from indirect connectivities is pervasive (11–13) and justifies the urgent need for new and reliable methods to eliminate spurious edges. Importantly, remedies to address this problem should not further muddle the interpretation by removing true network edges (14). A number of theoretical approaches have been proposed to overcome this hurdle (4, 15–18), but the ability to experimentally verify the conclusions drawn by reverse engineering tools remains paramount.

The majority of efforts to address the verification issue adopt in silico benchmark suites that are based on biological pathway approximations (19). Although these models do include a number of commonly observed topologies and have provided significant insights, they do not fully capture the complexity of the biological realm and the associated heterogeneity and intrinsic variability. On the other hand, engineered synthetic gene circuits are orthogonal to the endogenous pathways yet operate within the natural cellular context using the available resources. Thus, synthetic networks are a versatile platform for investigating specific connectivities and topological properties and can ultimately guide us to deriving fundamental insights about biological systems and pathways (20–23).

We previously proposed a strategy based upon using a synthetic gene network in human cells as a benchmark for reverse engineering validation and refinement (24). Here, we built three-node synthetic gene regulatory networks that incorporate direct and indirect connectivities and used them as benchmarks in human kidney cells. The first network is the type I coherent feed-forward loop (25, 26), where the origin node (X) activates the target node (Z) directly but also through an intermediate node (Y), with OR logic at the output (Fig. 1*A*). The second network is a cascade motif, where the origin node (X) regulates the target node (Z) indirectly via an intermediary node (Y) (Fig. 1*B*). More specifically, the node-to-node interactions are achieved through inducible transcriptional regulation. The origin (X) and intermediary nodes (Y) contain bi-directional promoter elements that drive the production of a fluorescent reporter protein and a transactivator unit, and the target node (Z) contains a unidirectional promoter for the production of a fluorescent reporter protein only. Each node produces a fluorescent reporter, which allows monitoring its state.

We commenced the experiments confirming the baseline behavior of the synthetic networks under boundary and control conditions. Subsequently, we systematically perturbed each network node using short interfering RNAs (siRNAs) (27); then, we collected and processed the flow cytometry measurements. Using these data, we performed network reconstruction via nonparametric single-cell data resampling followed by modular response analysis (4, 28). The reconstruction results reproduced the benchmark network topologies. Importantly, we identified divergent shifts in predicted interaction strengths under differential perturbations, a feature that can be critical toward discriminating between direct and indirect connectivities.

## Results

**Design and Assembly of the Benchmark Synthetic Regulatory Networks.** The first of the two networks is the type I coherent feed-forward loop (Fig. 1*C*). The plasmid for node X consists of a constitutively

### Significance

We used a combination of computational and theoretical approaches coupled to synthetic biology experimentation in mammalian cells to study direct and indirect connectivities in biological networks. After subjecting benchmark circuits to a range of perturbations, we recovered the edge weights using nonparametric single-cell data resampling coupled with modular response analysis. We discovered that inferred weights of specific network edges undergo divergent shifts under differential perturbations, and that the particular behavior is topology dependent. Incorporating this insight in the analysis of high-throughput experiments may provide a sought-after solution to a longstanding reverse engineering problem.

Author contributions: L.B. designed research; T.K., R.M., Y.L., E.S., and L.B. performed research; T.K., Y.L., and L.B. analyzed data; and T.K., R.M., Y.L., E.S., and L.B. wrote the paper.

The authors declare no conflict of interest.

This article is a PNAS Direct Submission.

[1]To whom correspondence may be addressed. Email: bleris@utdallas.edu or eduardo.sontag@rutgers.edu.

This article contains supporting information online at www.pnas.org/lookup/suppl/doi:10.1073/pnas.1507168112/-/DCSupplemental.

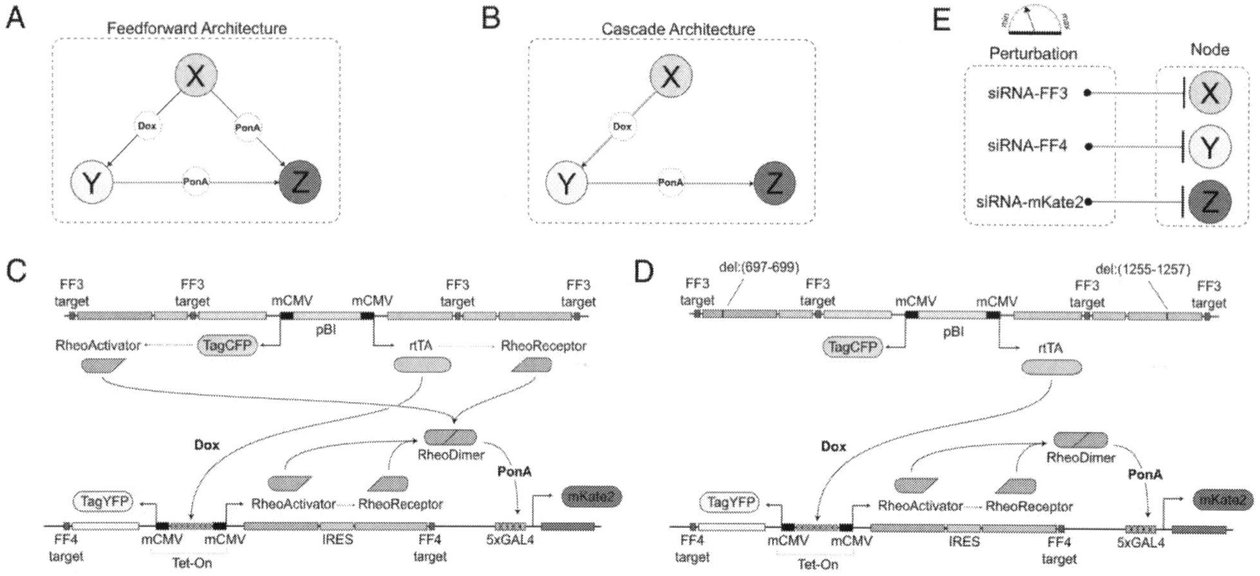


**Fig. 1.** The benchmark synthetic regulatory networks. (*A*) The first motif is a coherent feed-forward loop where node X regulates node Z in both a direct and an indirect manner. (*B*) The second motif is a cascade, where node X activates node Z only by activating node Y. (*C* and *D*) Detailed information about the synthetic gene networks. The activity of the three nodes X, Y, and Z can be quantified by the output fluorescent proteins TagCFP, TagYFP and mKate2, respectively. The constitutive bidirectional promoter of node X also transcribes rtTA for node Y induction and the RheoSwitch dimers for node Z induction. For the cascade motif, the translation of RheoSwitch dimer protein is prevented by nonsense mutation. In the presence of doxycycline, the constitutively transcribed rtTA induces transcription of RheoSwitch in node Y. When ponasterone A binds to the RheoSwitch dimer, the entire complex serves as a transactivator for the yeast Gal4 domain. Transcription at Gal4 domain results in production of mKate2 to indicate node Z activity. (*E*) Perturbation of each node in the system is performed siRNA. Nodes X and Y are perturbed by synthetic siRNAs (FF3 and FF4, respectively) with the targets located in the 3′ UTR of their corresponding targets. Node Z is perturbed by a custom siRNA that directly targets mKate2. IRES, internal ribosome entry site.

active bidirectional promoter flanked by the reporter fluorescent protein TagCFP and reverse tetracycline-controlled transactivator (rtTA) on either side. The regulatory unit rtTA serves as the transactivator of the tetracycline-inducible expression system (Tet-On) upon forming a homodimer and binding with the ligand doxycycline (Dox). Activation of the downstream target node Y, which consists of tetracycline response element (TRE) enhancer flanked by cytomegalovirus (CMV) promoters on either side, requires binding of active rtTA–Dox complex to the TRE enhancer. Thus, the activation of node Y by node X depends on doxycycline. The activation of node Y results in production of its fluorescent reporter TagYFP and a heterodimeric transactivator composed of the RheoActivator and RheoReceptor domains. The RheoActivator domain consists of a ligand binding domain fused with the viral transactivator VP16, whereas the RheoReceptor domain is a hybrid of insect hormone ecdysone receptor (EcR) fused to yeast GAL4 DNA binding domain for target binding specificity to GAL4 response element. After dimerization of RheoActivator and RheoReceptor, an EcR agonist such as ponasterone A induces conformational change to the RheoReceptor such that the heterodimer bound to GAL4 response element initiates transcription. In our synthetic network, RheoSwitch dimer activates node Z by initiating transcription of its reporter fluorescent protein mKate2. To achieve direct activation of node Z by node X, the node X produces the RheoSwitch proteins in addition to TagCFP and rtTA.

The second of the two networks is a cascade motif (Fig. 1*D*), where node X controls node Z exclusively through the activation of intermediate node Y. To implement this architecture, we modified the coherent feed-forward architecture by inserting a single base pair in both of the node X RheoSwitch genes to induce a nonsense frame-shift mutation. As the RheoSwitch heterodimer genes together constitute ~30% of total plasmid size we selected introducing

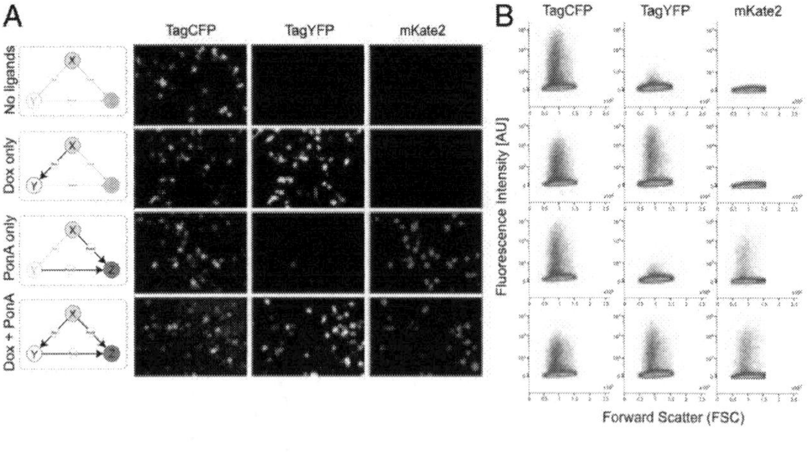


**Fig. 2.** Validation of the coherent feed-forward architecture. To validate the circuit behavior we tested all combinations of the two small molecules. The result analyzed by fluorescence microscopy (*A*) and flow cytometry (*B*).

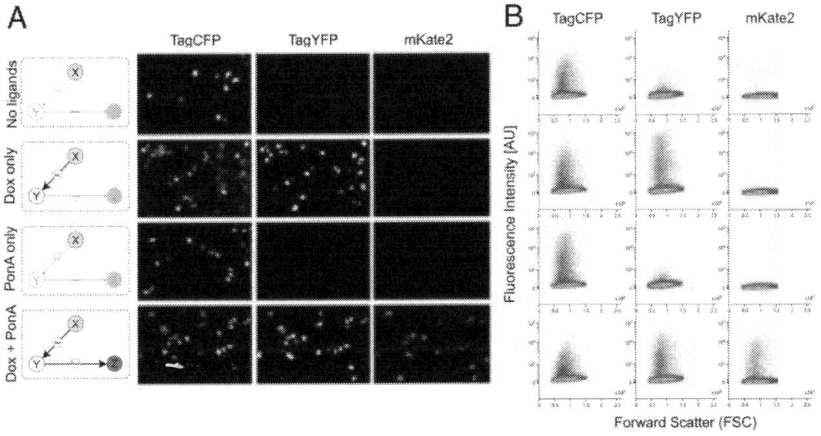


**Fig. 3.** Validation of the cascade architecture. To validate the circuit behavior we tested all combinations of the two small molecules. The result analyzed by fluorescence microscopy (*A*) and flow cytometry (*B*).

a mutation instead of complete excision to avoid possible discrepancies in transfection and transcription efficiencies due to differences in plasmid and cassette size.

During the design stage we opted for a simple yet effective means of perturbing the individual nodes via RNA interference (Fig. 1*E*). We use a set of siRNA with previously confirmed function (29) for X and Y and a custom siRNA for Z. More specifically, the siRNA-based suppression of node X is achieved through addition of an FF3 target into 3′ untranslated region (UTR) of each transcript produced by the constitutive bidirectional promoter. Similar to X, ubiquitous siRNA-based suppression of Y is made possible by inserting an FF4 target into the 3′ UTR of each transcript produced by the bidirectional TRE enhancer/CMV promoter. Node Z contains a single transcript (reporter protein mKate2), and its activity is modulated by a custom siRNA that directly targets the mKate2 transcript.

**Validation of the Synthetic Gene Network Behavior.** With the exception of node X which relies on a constitutive promoter, the activity of the synthetic networks depends on the presence of the appropriate ligand. In the cascade motif, Y requires doxycycline (Dox), and Z requires an EcR agonist such as Genostat or ponasterone A (PonA). In the type I coherent feed-forward loop, the requirement for activation of Y remains the same, whereas Z can be activated by the combination of Dox and PonA or PonA alone. To confirm these baseline conditions, the circuits were transfected in human embryonic kidney cell line (HEK293), the ligands were introduced at saturating concentrations, and measurements were performed using microscopy and flow cytometry ~48 h after transfection.

The microscopy measurements of the fluorescent outputs of both benchmark networks show that the inducible transactivators for both architectures function as desired with minimal leakage, thus confirming the designed circuit topologies. In the feed-forward loop,

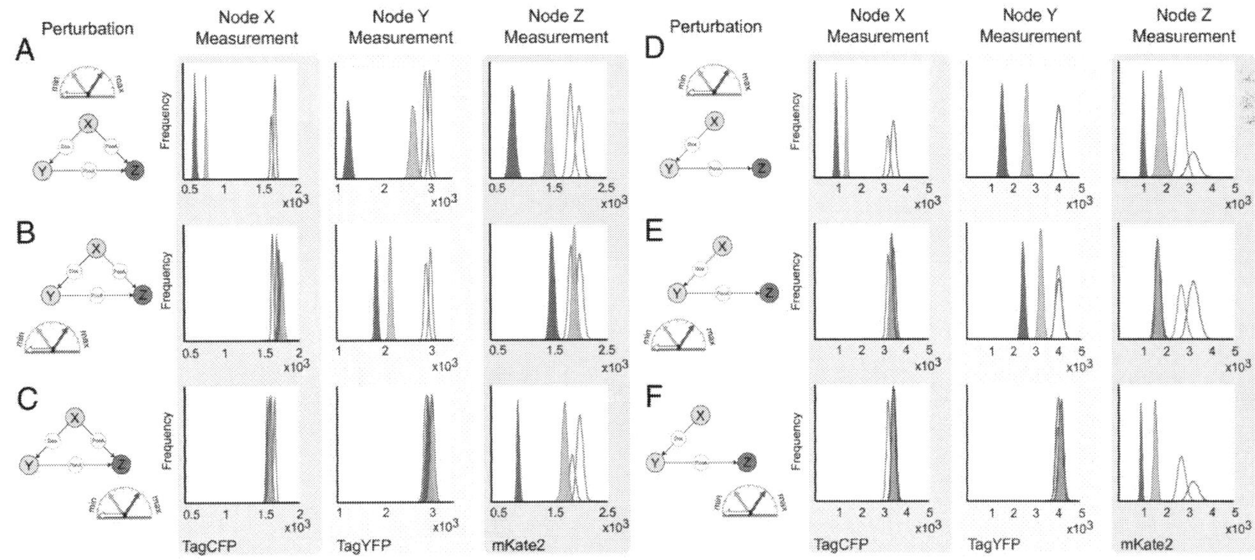


**Fig. 4.** Resampling the single-cell flow cytometry data after node-wise perturbations. Forty-eight h after siRNA perturbation, the expression level of the fluorescence reporters that represent nodes X, Y, and Z (TagCFP, TagYFP and mKate2, respectively) are measured using flow cytometry. To calculate the mean fluorescence of each population and the associated uncertainty, bootstrap resampling was performed. The resulting probability distributions of the resampled mean before perturbation (empty) and after perturbation (color-filled) are shown for the feed-forward loop (*A*–*C*), and the cascade (*D*–*F*). The colors of the peaks indicate the relative strength of the suppression applied (gray is used to indicate the low and purple the high perturbation). (*A*) The graphical representation of the X-node perturbation and the corresponding nodal responses using the feed-forward architecture. Probability distributions are composed of bootstrapped mean of the fluorescence reporters TagCFP, TagYFP and mKate2 (left to right) following perturbations to node X at two different siRNA concentrations. Color of the peak indicates the relative degree of suppression. (*B* and *C*) The graphical representation of the Y- and Z-node perturbations and the corresponding nodal responses using the feed-forward architecture. (*D*–*F*) Results from the same process using the cascade architecture.

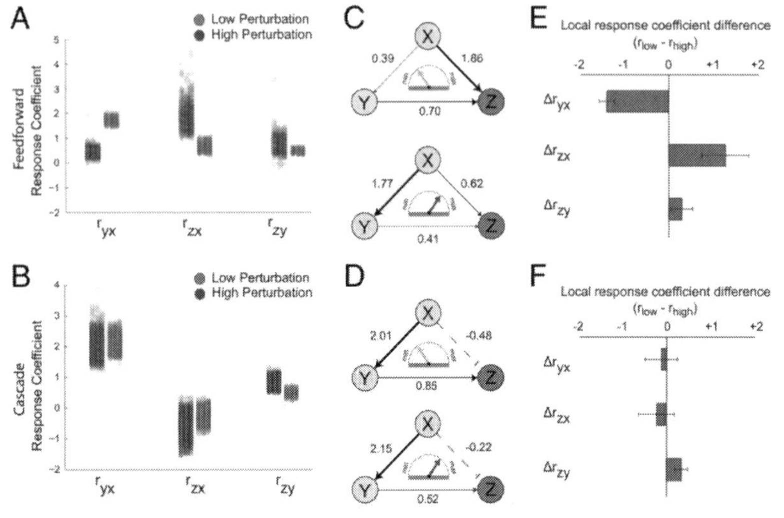


**Fig. 5.** Reverse engineering of the benchmark topologies using resampled single-cell data. (*A* and *B*) The complete reconstruction of the network with modular response analysis performed after two perturbations. For every set of subsampled means that make up the probability distributions, the MRA results along with the 95% confidence interval of the distribution are plotted as a 1D scatter plot. (*C* and *D*) The graphical representation of the reconstructed synthetic networks. (*E* and *F*) To probe the effect of response coefficient change due to perturbation magnitude shift we calculated the difference between coefficients of equivalent edges (*C* and *D*). The error bars were obtained using a propagation of error among the pair of local response coefficient distributions used to calculate this difference.

node X activity is represented by the constitutively produced fluorescent protein TagCFP and is observed regardless of the ligand conditions (Fig. 2*A*). The addition of doxycycline, which enables X-to-Y interaction by activating the synthetic transactivator rtTA, results in production of the TagYFP fluorescent protein (Fig. 2*A*). The activation of node Z is mediated by the active form of RheoSwitch dimer, which is produced by both nodes X and Y. Due to the constitutive activity of node X, PonA is sufficient to activate node Z in the feed-forward loop (Fig. 2*A*). These observations are confirmed by flow cytometry-based population measurements, which show TagCFP population in all scenarios and a distinct TagYFP population when Dox is present, and mKate2 when PonA is present (Fig. 2*B*). The same control experiments using the cascade network plasmid show an identical response to the ligand combinations except for the node Z activity (Fig. 3*A*). In the cascade motif, sequential activation of node X and node Y are necessary for node Z activation. Thus, mKate2 is only observed when Dox and PonA are present (Fig. 3*A*). Again, we confirm these observations with flow cytometry (Fig. 3*B*).

Subsequently, to probe the parameter space and general behavior of the circuits under perturbations we created mathematical models of our benchmark circuits (*SI Appendix*, *SimBiology Model*, Figs. S1 and S2). The kinetic parameters were selected from literature (21, 30). We performed sensitivity analysis of the output node Z protein concentration against the mRNA species of nodes X and Y (thereby emulating RNAi perturbation). We observe that, in the feed-forward loop, where node X activates node Z in a direct manner as well as an indirect manner, the cumulative sensitivity of the Z node protein to mRNA species of node X was always higher than that of node Y (*SI Appendix*, Fig. S3). Conversely, in the cascade, where node X only activates node Z indirectly through node Y, the production of node Z protein was more sensitive to the node Y mRNA. Based on the simulation results, we hypothesized that the topological differences of the examined architectures may yield divergent responses to different degrees of perturbation. This hypothesis points to an intriguing scenario where the properties and outcome of signal propagation after custom perturbation experiments can be exploited toward distinguishing direct from indirect connectivities.

**Modular Response Analysis.** An intrinsic difficulty in capturing direct interactions in a biological network is that any perturbation to a particular node may rapidly propagate throughout the network, thus causing global changes which cannot be easily distinguished from direct effects. Rooted in metabolic control analysis, Modular Response Analysis (MRA) uses steady state data obtained from node-wise perturbation to express the network in terms of pair-wise interaction sensitivities. To perform MRA (*SI Appendix*, *Modular Response Analysis Method*), we first calculate the global response coefficients (GRC) from experimentally measured responses to perturbations using $\Delta ln(x_i)$, where $x_i$ represents in our case the quasi steady-state measurement of fluorescent reporter obtained via flow cytometry. Once the functional modules (i.e., perturbation targets) of the target network have been selected, the experimental procedure consists of the following steps: (*i*) measure the steady-state $x_i$ corresponding to the unperturbed set of inputs $p_i$, (*ii*) perform a perturbation to each $p_i$ individually and measure the new steady-state $x_i'$, (*iii*) calculate the global response coefficients using the steady-state data, and (*iv*) convert global response coefficients to local response coefficients by inversion of the global response matrix.

In higher eukaryotes, perturbation can be achieved through the down-regulation of mRNA, and hence protein levels, using RNA interference (RNAi). This approach has shown to be successful in mapping the positive and negative feedback effects in the Raf/Mek/Erk MAPK network of rat adrenal pheochromocytoma (PC-12) cells (31). Using a variant of the MRA algorithm (32), the authors uncovered connectivity differences depending on whether the cells were stimulated by epidermal growth factor (EGF) or, alternatively, by neuronal growth factor (NGF).

We commenced the experimental reverse engineering process by performing a systematic perturbation of each benchmark architecture node. We first tested the efficacy of siRNA and calibrated the perturbation dosage against the feed-forward loop architecture plasmid (*SI Appendix*, Figs. S4–S6; quantitative RT-PCR results in *SI Appendix*, Fig. S7). As our goal was to find a range of siRNA concentrations that result in moderate yet distinct levels of suppression, we set our maximum siRNA concentration at the manufacturer-recommended dose of 5 pmol and tested five additional concentrations in decreasing magnitudes. To ensure consistency, we cotransfected each siRNA with the network plasmid and measured the circuit activity after 48 h via flow cytometry. Across all three nodes, each of the siRNA was most effective at suppressing the node for which it was designed to disrupt; at least 60% suppression was achieved with the highest siRNA concentration. We then selected the pair of siRNA concentrations that yield the largest difference in the activity of their respective target node, as measured by mean fluorescence level. Specifically, as illustrated in *SI Appendix*, Fig. S8*A*, we selected 1 pmol as "high" perturbation and 0.1 pmol as "low."

After selecting the perturbation magnitude, we performed a node-wise perturbation of the feed-forward circuit using the siRNAs that target each node supplemented with scrambled siRNA to control for the total mass. As before, we used saturating concentrations of the small molecule inducers and applied the predefined set of perturbations based on our calibration results. The three fluorescent reporter protein profiles indicate a response consistent with the benchmark network topologies, confirming the siRNA operation (*SI Appendix*, Fig. S8*B*). Down-regulation in the fluorescence reporter expression is observed in two cases. Directly by perturbing the actual node or indirectly by perturbing the upstream node responsible for its activation. For node X, a decrease in TagCFP is observed only after direct perturbation; for node Y, a decrease in TagYFP is observed after perturbation of nodes X and Y; and for node Z, a decrease in mKate2 is observed after perturbation of nodes X, Y, and Z (*SI Appendix*, Fig. S8*B*).

We then proceeded with the recovery of the network topology using population data (*SI Appendix*, *Reverse Engineering of the Benchmark Topologies Using Population Data*). For each set of perturbation responses, the global response coefficients were calculated based on the weighted mean fluorescence of gated populations. The pairwise sensitivity coefficients were then obtained via calculating the local response coefficient (LRC) (*SI Appendix*, Figs. S8 *C* and *F* and S9). To determine the significance of the recovered LRC, we performed error propagation using Monte Carlo simulations (31) rendering most of the predicted regulatory connections insignificant (*SI Appendix*, *Monte Carlo Simulation* and Fig. S10). Notably, the reverse engineering recovered a direct inhibitory connection between nodes X and Z for both perturbation magnitudes (*SI Appendix*, Fig. S8*F*), which may be attributed to mild retroactivity effects (33).

**Reverse Engineering of the Benchmark Topologies Using Resampled Single-Cell Data.** To increase our confidence in predictions we developed a technique based upon bootstrapping, an alternative to the sample statistics obtained from an aggregate population (i.e., mean and SD). Bootstrapping is a nonparametric resampling method (34) designed to estimate the confidence interval of a given statistic, and is particularly useful when the observed population distribution cannot be characterized by typical distributional assumptions such as normality (e.g., typical flow cytometry data). To obtain a bootstrapped mean we: (*i*) resample with replacement the dataset to the same number of times as the original population, (*ii*) calculate the desired statistic from each sample, (*iii*) repeat the process several times to form the probability distribution of the subsampled mean (*SI Appendix*, Fig. S11). For our analysis, we repeated the entire process 2,000 times to form the representative probability distribution of a fluorescent reporter expression.

Subsequently, we produced a unique panel for each of the three fluorescent reporters after perturbations to three different nodes, for a total of nine distinct panels for the feed-forward circuit (Fig. 4 *A*–*C*) and nine for the cascade (Fig. 4 *D*–*F*). Each frequency plot consists of four different probability distributions, from resampled means of fluorescent reporter before (empty) and after the two perturbations (filled with gray for low and purple for high). For every set of subsampled means that make up the probability distributions, we treated them as a unique instance of the perturbation response and fed these values to MRA to calculate the local response coefficients. The results, along with the 95% confidence interval of the distribution, are plotted as a 1D scatter plot and shown in Fig. 5 *A* and *B*. In this case, we were able to successfully recover all relevant regulatory connections of feed-forward loop (Fig. 5*C*) and cascade (Fig. 5*D*) networks with increased confidence. All of the reconstructed edges are included in *SI Appendix*, Fig. S12.

Using our prior knowledge of our network, we confirmed that the inferred connectivities are consistent with the network topology. Moreover, the inferred interaction strengths and distributions produced by the reverse engineering algorithm reveal features of the network that are not readily apparent and are in agreement with our in silico sensitivity analysis. Specifically, to probe the effect of response coefficient change due to perturbation magnitude shift we calculated the difference between coefficients of equivalent edges (Fig. 5 *E* and *F*).

For the feed-forward architecture we identified "perturbation-sensitive edges." In other words, we discovered edges that undergo distinctively large shift in interaction strengths under differential perturbations. In particular, we observe that for the feed-forward architecture, increasing the perturbation magnitude dramatically alters the inferred interactions arising from node X. Importantly, there is a noticeable reversal in the strengths of activation between node X to nodes Y and Z, whereas the interaction from node Y to node Z remain largely unchanged (Fig. 5*C*). Specifically, after a low perturbation, the recovered topology shows a prominent direct activation of node Z by node X, whereas the topology recovered after high perturbation shows a prominent activation of node Y by node X. We postulate that the low perturbation is buffered as it propagates through the intermediate node Y, therefore the direct connection between X and Z appears to be more important. Aligned with this observation, node Z is less sensitive to disruption of node Y in the context of the feed-forward loop (Fig. 5*B*).

Compared with the feed-forward circuit, we found that the reverse engineering of the cascade is robust to perturbation strengths. In fact, the recovered topology from two perturbation magnitudes are almost indistinguishable except for small decrease in the Y-to-Z interaction strength (Fig. 5*F*), despite the fact that the fluorescence reporter profiles clearly reflect the differences in perturbation magnitude (Fig. 4 *D*–*F*). In contrast with the feed-forward circuit, there is only one possible path of activation in the cascade motif thereby the presumed buffering effect is not critical.

To further explore the effect of differential perturbation on the reverse engineering results, we performed an additional experiment using three perturbation magnitudes (*SI Appendix*, Fig. S13). In this instance, we refer to the perturbation magnitudes as "low," "medium," and "high." We again observe the diverging trend of response coefficient values between the two architectures. In the feed-forward loop, each step-wise increase in perturbation magnitude affects the two edges originating from node X in contrasting manner, highlighting the activation from node X to node Y while reducing the weight of activation from node X to node Z (*SI Appendix*, Fig. S13 *A* and *C*). The recovered topologies of the cascade motif undergo little to no change over same perturbation magnitude intervals (*SI Appendix*, Fig. S13 *B* and *D*). Finally, to qualitatively probe our observations we developed a phenomenological model of the architectures (*SI Appendix*, *Phenomenological Model*). Using this model we analytically calculated the local response coefficients under low and high perturbations and we indeed confirmed the divergent shifts in interaction strengths.

## Discussion

Direct and indirect interactions are pervasive in all networks. The inability to disentangle these interactions hampers reverse engineering progress. Recent advancements in high-throughput approaches, combined with algorithm and methodological advances through a host of community-wide efforts (12, 14, 19, 35) have examined these aspects. In fact, attempts to fundamentally address the issue by recognizing and filtering out the effects of indirect interactions at a global scale have begun to surface (11). Meanwhile, parallel developments in synthetic biology (23) have endowed researchers with new tools that allow precise emulation of naturally occurring topologies (21, 22). Networks orthogonal to the cellular milieu can serve as a biomolecular topological "ground truth" (20, 24). Data gathered from benchmark synthetic circuits can complement and inform algorithms, and offer a unique opportunity to correlate topological properties to system identification.

The number of possible networks for a given set of nodes is large and it grows exponentially with the number of nodes, making impractical their exhaustive construction. Fortunately, recent research has uncovered that certain topologies appear more frequently than others. Those topologies were dubbed "network motifs (25, 36). The network topology does not specify the nature of the nodes, and indeed the expectation is that the network behavior will be invariant to the changes in the molecular nature of the nodes and the exact mechanism of the interactions between the nodes.

Here we constructed two synthetic networks that incorporate direct and indirect connectivities. We successfully engineered the benchmark architectures to be inducible with negligible leakage and amenable to simple perturbations to facilitate the reverse engineering analysis. After applying systematic perturbations and a combination of nonparametric single-cell data resampling and modular response analysis, we discovered response patterns that are markedly different between the two topologies.

Using the proposed methodology, individual nodes of a network can be perturbed from their steady-state using transcriptional or posttranscriptional inhibitors [e.g., TALEs/CRISPR (37, 38) or siRNAs]. The pre- and postperturbation steady states can be measured at the mRNA or protein levels, and fed into MRA to predict divergent LRC and accordingly the network structure. Beyond small-scale networks, although motifs are composed of relatively few elements, they are often embedded as "modules" (39–41) in large networks that exhibit complex behavior. The term "modular" in MRA indicates that the same theoretical tools, in principle, scale up to cover large networks that are connected through a small number of "communicating intermediaries" (4, 28).

To conclude, unraveling the complexity of biological networks is central to understanding biology. Our results point to a transformative opportunity in reverse engineering of biological networks. Taking into account inferred topological changes under differential perturbations may provide a solution to the longstanding problem of discriminating between direct and indirect connections.

## Methods

**Mammalian Cell Culture and Transfections.** HEK293 cell line was maintained at 37°C, 100% humidity and 5% (vol/vol) $CO_2$. Circuit plasmid transfection was performed with jetPRIME (Polyplus) in 12-well plates at a plating density of 200,000 cells. Transfection was performed 24 h after seeding, and each well received 10 ng of plasmid containing node X and 25 ng of plasmid containing nodes Y and Z, with 500 ng of cotransfection junk DNA and varying amounts of siRNA. Detailed information is provided in *SI Appendix, SI Methods*.

**Fluorescence Microscopy.** Approximately 48 h after transfection of network plasmid, fluorescence images of live cells were captured using an Olympus IX81 microscope. For ambient temperature control, the entire apparatus was housed in a Precision Control environmental chamber. The images were captured using a Hamamatsu ORCA 03 digital camera. Detailed information is provided in *SI Appendix, SI Methods*.

**Flow Cytometry.** All FACS experiments were performed 48 h after transfection with BD LSRFortessa. Data acquisition was performed using FACS Diva software and subsequent analysis with FlowJo (Treestar). The threshold fluorescence unit for selecting fluorescence-positive population was determined based on untransfected HEK293 cells (*SI Appendix*, Fig. S14). There was no compensation performed (*SI Appendix*, Fig. S15). Detailed information is provided in *SI Appendix, SI Methods*.

**Modular Response Analysis.** To obtain the pair-wise sensitivities between each node, we performed modular response analysis. In MRA, the local intermodular interactions, described by the local response matrix $r_{ij}$, are calculated from the global response matrix $R_{ij}$, which contains the observed change in the steady-state measurement of each node ($x_i$) due to the experimental perturbation ($p_j$). Because precise measurement of parameter perturbation size ($p_j$) is not possible in an experimental setting, the global response matrix is approximated by the fractional change of the steady states $R_{ij} \sim \Delta ln(x_i)$. After obtaining $R_{ij}$ from experimental data, we calculate the local response matrix $r_{ij}$ by solving $r_{ij} = -[dg(R_{ipj}^{-1})]^{-1} \cdot R_{ipj}^{-1}$. Detailed information is provided in *SI Appendix, Modular Response Analysis Method*.

**Resampling.** To estimate the 95% confidence interval of the obtained local response matrix, bootstrap resampling of the original flow cytometry population is performed. The steps for bootstrap resampling are as follows: From the original flow cytometry population, resample with replacement the same number of cells as the original population. Using the resampled population, compute the desired population statistic (mean), and then calculate the local response matrix using MRA. The bootstrapping and MRA process is repeated 2,000 times to create a distribution of local responses. The 95% confidence interval, which corresponds to values from 2.5th to 97.5th percentile of the calculated values, is used to estimate the error. The process is shown in *SI Appendix*, Fig. S11.

**ACKNOWLEDGMENTS.** This work was funded by the US National Institutes of Health Grants GM098984, GM096271, CA17001801, National Science Foundation Grant CBNET-1105524, and the University of Texas at Dallas. E.S. partially supported by Air Force Office of Scientific Research Grant FA9550-14-1-0060.

1. Csete ME, Doyle JC (2002) Reverse engineering of biological complexity. *Science* 295(5560):1664–1669.
2. Gardner TS, di Bernardo D, Lorenz D, Collins JJ (2003) Inferring genetic networks and identifying compound mode of action via expression profiling. *Science* 301(5629):102–105.
3. Khammash M (2008) Reverse engineering: The architecture of biological networks. *Biotechniques* 44(3):323–329.
4. Kholodenko BN, et al. (2002) Untangling the wires: A strategy to trace functional interactions in signaling and gene networks. *Proc Natl Acad Sci USA* 99(20):12841–12846.
5. Sachs K, Perez O, Pe'er D, Lauffenburger DA, Nolan GP (2005) Causal protein-signaling networks derived from multiparameter single-cell data. *Science* 308(5721):523–529.
6. Kholodenko B, Yaffe MB, Kolch W (2012) Computational approaches for analyzing information flow in biological networks. *Sci Signal* 5(220):re1.
7. Basso K, et al. (2005) Reverse engineering of regulatory networks in human B cells. *Nat Genet* 37(4):382–390.
8. Chen JC, et al. (2014) Identification of causal genetic drivers of human disease through systems-level analysis of regulatory networks. *Cell* 159(2):402–414.
9. Yeung MK, Tegnér J, Collins JJ (2002) Reverse engineering gene networks using singular value decomposition and robust regression. *Proc Natl Acad Sci USA* 99(9):6163–6168.
10. Tegner J, Yeung MKS, Hasty J, Collins JJ (2003) Reverse engineering gene networks: Integrating genetic perturbations with dynamical modeling. *Proc Natl Acad Sci USA* 100(10):5944–5949.
11. Feizi S, Marbach D, Médard M, Kellis M (2013) Network deconvolution as a general method to distinguish direct dependencies in networks. *Nat Biotechnol* 31(8):726–733.
12. Marbach D, et al. (2010) Revealing strengths and weaknesses of methods for gene network inference. *Proc Natl Acad Sci USA* 107(14):6286–6291.
13. Margolin AA, et al. (2006) ARACNE: An algorithm for the reconstruction of gene regulatory networks in a mammalian cellular context. *BMC Bioinformatics* 7(Suppl 1):S7.
14. Marbach D, et al.; DREAM5 Consortium (2012) Wisdom of crowds for robust gene network inference. *Nat Methods* 9(8):796–804.
15. de la Fuente A, Brazhnik P, Mendes P (2002) Linking the genes: Inferring quantitative gene networks from microarray data. *Trends Genet* 18(8):395–398.
16. Friedman N (2004) Inferring cellular networks using probabilistic graphical models. *Science* 303(5659):799–805.
17. Friedman N, Linial M, Nachman I, Pe'er D (2000) Using Bayesian networks to analyze expression data. *J Comput Biol* 7(3-4):601–620.
18. Pe'er D (2005) Bayesian network analysis of signaling networks: A primer. *Sci Signaling* 2005(281):pl4.
19. Prill RJ, et al. (2010) Towards a rigorous assessment of systems biology models: The DREAM3 challenges. *PLoS One* 5(2):e9202.
20. Cantone I, et al. (2009) A yeast synthetic network for in vivo assessment of reverse-engineering and modeling approaches. *Cell* 137(1):172–181.
21. Bleris L, et al. (2011) Synthetic incoherent feedforward circuits show adaptation to the amount of their genetic template. *Mol Syst Biol* 7:519.
22. Shimoga V, White JT, Li Y, Sontag E, Bleris L (2013) Synthetic mammalian transgene negative autoregulation. *Mol Syst Biol* 9:670.
23. Lienert F, Lohmueller JJ, Garg A, Silver PA (2014) Synthetic biology in mammalian cells: Next generation research tools and therapeutics. *Nat Rev Mol Cell Biol* 15(2):95–107.
24. Kang T, et al. (2013) Reverse engineering validation using a benchmark synthetic gene circuit in human cells. *ACS Synth Biol* 2(5):255–262.
25. Milo R, et al. (2002) Network motifs: Simple building blocks of complex networks. *Science* 298(5594):824–827.
26. Ma'ayan A, et al. (2005) Formation of regulatory patterns during signal propagation in a Mammalian cellular network. *Science* 309(5737):1078–1083.
27. Fire A, et al. (1998) Potent and specific genetic interference by double-stranded RNA in Caenorhabditis elegans. *Nature* 391(6669):806–811.
28. Sontag E, Kiyatkin A, Kholodenko BN (2004) Inferring dynamic architecture of cellular networks using time series of gene expression, protein and metabolite data. *Bioinformatics* 20(12):1877–1886.
29. Rinaudo K, et al. (2007) A universal RNAi-based logic evaluator that operates in mammalian cells. *Nat Biotechnol* 25(7):795–801.
30. Tigges M, Marquez-Lago TT, Stelling J, Fussenegger M (2009) A tunable synthetic mammalian oscillator. *Nature* 457(7227):309–312.
31. Santos SDM, Verveer PJ, Bastiaens PIH (2007) Growth factor-induced MAPK network topology shapes Erk response determining PC-12 cell fate. *Nat Cell Biol* 9(3):324–330.
32. Andrec M, Kholodenko BN, Levy RM, Sontag E (2005) Inference of signaling and gene regulatory networks by steady-state perturbation experiments: Structure and accuracy. *J Theor Biol* 232(3):427–441.
33. Del Vecchio D, Ninfa AJ, Sontag ED (2008) Modular cell biology: Retroactivity and insulation. *Mol Syst Biol* 4:161.
34. Efron B (1979) Bootstrap methods: Another look at the jackknife. *Ann Stat* 7(1):1–26.
35. Prill RJ, Saez-Rodriguez J, Alexopoulos LG, Sorger PK, Stolovitzky G (2011) Crowdsourcing network inference: The DREAM predictive signaling network challenge. *Sci Signal* 4(189):mr7.
36. Alon U (2007) *An Introduction to Systems Biology: Design Principles of Biological Circuits* (Chapman & Hall/CRC, Boca Raton, FL), p 301.
37. Moore R, et al. (2015) CRISPR-based self-cleaving mechanism for controllable gene delivery in human cells. *Nucleic Acids Res* 43(2):1297–1303.
38. Li Y, Moore R, Guinn M, Bleris L (2012) Transcription activator-like effector hybrids for conditional control and rewiring of chromosomal transgene expression. *Sci Rep* 2:897.
39. Kreimer A, Borenstein E, Gophna U, Ruppin E (2008) The evolution of modularity in bacterial metabolic networks. *Proc Natl Acad Sci USA* 105(19):6976–6981.
40. Bassett DS, et al. (2011) Dynamic reconfiguration of human brain networks during learning. *Proc Natl Acad Sci USA* 108(18):7641–7646.
41. Bullmore E, Sporns O (2009) Complex brain networks: Graph theoretical analysis of structural and functional systems. *Nat Rev Neurosci* 10(3):186–198.

# 425548 A Crispr-Cas9 Enabled Efficient and Scalable Method for High Throughput Chip-Seq Analysis

**Abstract:**

### A CRISPR-Cas9 Enabled Efficient and Scalable Method for High Throughput Chip-Seq Analysis

Genome-wide profiling for protein-DNA interactions is crucial to understand the transcriptional regulation networks. DNA-binding proteins play essential roles in many cellular processes to provide growing insights about cell developmental stages and disease states. Eukaryotic nuclear DNA interacts with multiple proteins, forming the complexes chromatins. Chromatins exist as intermediates between completely compacted chromosomes and nucleosomes. Chromatin immunoprecipitation (ChIP) enriches DNA fragments to bind a specific protein or a class of nucleosomes. Chromatin immunoprecipitation followed by deep sequencing (ChIP-seq) is a high throughput sequencing technique to detect the protein-DNA binding events and chemical modifications of histone proteins. Followed by recognition and direct sequencing, more comprehensive sequence information can be obtained in a high throughput fashion.

All organisms enable specific transcription regulation on transcriptional factors (TFs) to interact with regulatory regions. Eukaryotic TFs typically recognize short genomic sequence and the clustering motifs can be found multiple times in large genomes. ChIP-seq study on TFs will open a new avenue to discover new motifs as well as unravel genomic targets for potential therapeutics. The quality of any ChIP data is highly dependent on antibody. Specific antibodies are demanded for enrichment and the increase of signal contrast over background. However, commercial antibodies for TFs in human genome are very limited. To overcome this major limitation, we developed a new method for ChIP-Seq analysis by fusing the TFs with commonly recognized polypeptide protein tag, 3×FLAG-tag, to allow the track of TFs with antibody against FLAG sequence. Specifically, we combine the clustered regularly interspaced short palindromic repeats (CRISPRs) technology with microhomology mediated end joining (MMEJ) to insert an 3×FLAG-tag with screening markers to the ends of TFs. Current strategies for chromosomal integration typically rely on CRISPR mediated integration vis homologous recombination (HR) or non-homologous end joining (NHEJ). Scaling HR mediated integration on human genome is less feasible; and the reported NHEJ integration efficiency is low. MMEJ provides novel cellular repair mechanism with more precise integration over NHEJ. Using the CRISPR-MMEJ mediated tagging approach, we envision its power to improve and broaden ChIP-seq technology on mapping the human genome. This new ChIP-seq strategy can be applied to any TF targets in all organisms.

# 425814 Automated Metabolic Engineering of Yeast Cells Using a Biological Foundry

**Abstract:**

Given the complexity of biological systems, cellular reprogramming requires both identification of genetic targets on a genome scale and optimization of their expression in a concerted manner. Here we report the use of automated genome engineering to improve glycerol utilization and acetic acid tolerance phenotypes in *Saccharomyces cerevisiae*. Strains harboring multiplex genetic mutations, including both overexpression and knockdown modifications, were constructed and screened using the iBioFab (Illinois Biological Foundry for Advanced Biomanufacturing) platform. The evolutionary trajectories of improved mutants were also analyzed using next-generation sequencing. Our technology represents a prototype of industrialized biomanufacturing, where the design-build-test cycle is rapidly executed by a fully integrated robotic system.

# 426710 A Kinetic Model for Illustration of the Effect of Glutamine on the Apoptotic and Synthesis Pathways of Hybridoma Cells

**Abstract:**

**A Kinetic Model for Illustration of the Effect of Glutamine on the Apoptotic and Synthesis Pathways of Hybridoma Cells**

Denizhan Yilmaz, Satish J. Parulekar, and Ali Cinar

Department of Chemical and Biological Engineering

Illinois Institute of Technology, Chicago, IL 60616

(312) 567-3044 (voice), (312) 567-8874 (fax), parulekar@iit.edu

**Abstract**

Monoclonal antibodies (MABs) are important reagents used in biomedical research, in diagnosis and treatment of diseases. Monoclonal antibodies (MABs) derived from hybridoma cell cultures are used extensively in diagnostic assays and have found increasing use in therapeutic applications, affinity production systems, and in vivo imaging. The cell lines, hybridoma, are produced by fusing B cells (lymphocytes) from immunized animals with myeloma cells. The simplest approach for producing a MAB in vitro is to grow the hybridoma cells in batch and fed-batch cultures and recover and purify the MAB from the culture medium. The large scale production of MABs occurs via in vitro cultivation of hybridoma cell lines in bioreactors using techniques similar to those used for microbial cultivation. A cost-effective production of MABs requires an understanding of the effects of bioreactor process variables on the physiology of hybridoma cells.

The demand for MAbs has increased substantially over the last two decades due to their therapeutic and diagnostic applications. The development of methods for optimization of cell growth and cell productivity has become a crucial issue to enhance MAb yield in vitro production. The efficiency and performance of cell cultures depend on nutrient medium optimization and process monitoring and control strategies. Creating these strategies requires understanding of process dynamics affected by cell mechanism in culture environments. Mathematical models have been significant tools for combining cell physiology and engineering to predict the behavior of cell metabolism and optimize culture conditions by identifying process parameters that significantly impact cell growth and target metabolite productivity. In this work, we develop a hybrid model for antibody synthesis in hybridoma cell cultures. The kinetic model is comprised of an unstructured kinetic representation cell growth and death, uptake of key nutrients, glucose and glutamine, and accumulation of end-products of cell metabolism, lactate and ammonia and a structured kinetic representation of MAb synthesis. The structured portion of the hybrid model describes secretion pathways throughout cell growth and product synthesis considering the effect of glutamine depletion and ammonia accumulation on the antibody synthesis rate. Glutamine provides remarkable metabolic energy for cell growth and protein synthesis and is an important precursor of proteins and peptides, as well as amino sugars, purines and pyrimidines. It has been observed that glutamine depletion in the culture greatly influences the overall antibody synthesis rate and increases the apoptotic death rates of mammalian cells. When extracellular glutamine concentration is higher than a minimum required level, the translation rates and stability of mRNAs are high, leading to maximization of MAb synthesis rate per cell. The lack of glutamine in the culture medium activates a self-digestion system called autophagy for cellular energy. After autophagy period, cells enter a highly controlled and genetically defined terminal path known as apoptosis. Glutamine exhaustion increases cell death rate as per the programmed cell death pathway. Moreover, although glutamine is an apoptosis suppressor, feeding glutamine to the culture after cells entered the apoptotic pathway does not reduce cell death rate. Performance of batch, perturbed batch (intermittent addition of glucose and/or glutamine), and fed-batch cultures is simulated. The model predictions are in good agreement with experimental data reported in the literature for Immunoglobulin G (IgG) antibodies.

# 427425 Prediction of Double Substrate Microbial Growth Kinetics through Transcriptional Regulation: An Integrated Experimental/Modelling Approach on *Pseudomonas Putida Mt-2*

Wednesday, November 11, 2015: 5:25 PM
151D/E (Salt Palace Convention Center)

**Argyro Tsipa**, Department of Chemical Engineering, Biological Systems Engineering laboratory, Centre for Process systems Engineering, Imperial College London, LONDON, United Kingdom, Michalis Koutinas, Department of Environmental Science and Technology, Cyprus University of technology, Limassol, Cyprus, Efstratios N. Pistikopoulos, Artie McFerrin Department of Chemical Engineering, Texas A&M University, College Station, TX and Athanasios Mantalaris, Dept. of Chemical Engineering, Centre for Process Systems Engineering, Imperial College London, London, United Kingdom

**Prediction of double substrate microbial growth kinetics through transcriptional regulation: an integrated experimental/modelling approach on *Pseudomonas putida* mt-2**

Argyro Tsipa[a], Michalis Koutinas[b], Efstratios N. Pistikopoulos[c] and Athanasios Mantalaris[a*]

[a] *Centre for Process Systems Engineering, Department of Chemical Engineering, South Kensington Campus, Imperial College London, SW7 2AZ, London, UK*

[b] *Department of Environmental Science and Technology, Cyprus University of Technology, 30 Archbishop Kyprianou Str., 3036, Limassol, Cyprus*

[c] *Artie McFerrin Department of Chemical Engineering, University of Texas, Austin, TX 78712, United States*

*Pseudomonas putida* is a metabolically versatile soil bacterium and an industrially significant strain producing a series of fine and bulk chemicals, which has resulted in growing interest to understand specific metabolic pathways (Ballerstedt *et al.*, 2007). Among the several *P. putida* strains, mt-2 contains the TOL plasmid (pWW0), which specifies metabolic pathways for toluene and *m*-xylene degradation into Krebs cycle intermediates (Timmis, 2002). Successful activation of TOL relies mainly on the catabolic enzymes catalysing the degradation of the compound, and the promoters of genes subject to regulation by specific transcription factors (de Lorenzo and Perez-Martin, 1996). Therefore, **transcriptional regulation is a key step in the biodegradation process** acting as a controller regulating the appropriate metabolic cascades in response to the availability of specific substrate(s) (Diaz and Pieto, 2000).

The **TOL plasmid** is considered as a **paradigm of global and specific gene regulation** (Aranda-Olmedo et al., 2006). The transcriptional regulatory network of TOL constitutes four transcriptional units (*xylR*, *xylS*, *upper* and *meta*-operon) controlled by four promoters (*Pr*, *Ps*, *Pu* and *Pm*), respectively (Ramos et al., 1997). Entry of an effector in TOL provokes a cascade of regulatory events (Figure 1) initiated by *Pr* expression, which is followed by an increase in *Pu* expression. The latter controls the *upper*-operon that encodes for the enzymes transforming the inducer into an intermediate acting as environmental stimulus for XylS overexpression. XylS protein is produced constitutively by *Ps* and serves as the transcription factor for *Pm* activation. *Pm* controls transcription from the *meta*-operon, which encodes for the enzymes that catalyse the inducer further into Krebs cycle metabolites required for biomass growth.

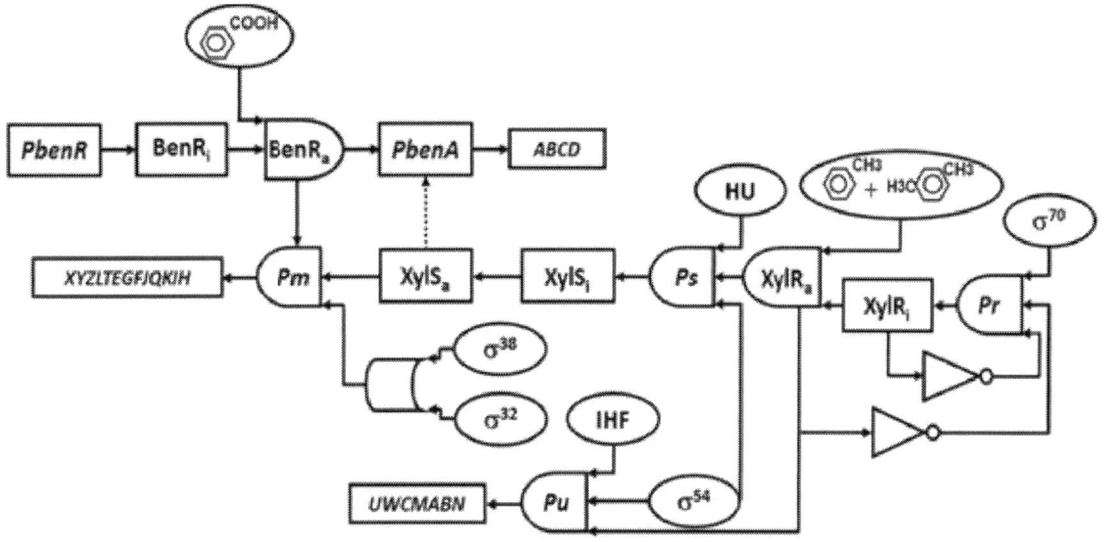


**Figure 1**: The regulatory network of TOL and *ortho*-cleavage pathways represented as logic gates of a genetic circuit (Weiss et al., 2003) SHAPE * MERGEFORMAT : input; SHAPE * MERGEFORMAT : output; SHAPE * MERGEFORMAT : AND; SHAPE * MERGEFORMAT : OR; SHAPE * MERGEFORMAT : NOT, i:inactive form of protein, a: active form of protein

Activation of ***Pu*** leads to the production of *upper*-operon enzymes catalysing the **transformation of *m*-xylene and toluene into methyl benzoate and benzoate**, respectively. The latter serves as **environmental stimulus of the *ortho*-cleavage pathway of the chromosome** activating *PbenR* and thus driving expression of the *benR* gene, which encodes for the **BenR protein** (Figure 1). This protein acts as **transcription factor for *meta*-operon activation in TOL, demonstrating a cross-talk between the two pathways** (Cowles *et al.*, 2000). However, induction with *m*-xylene leads into a metabolic conflict (Perez-Pantoja *et al.*, 2014) resulting in the formation of a dead-end product (Schmidt et al., 1985). Therefore, the *benR* gene of the chromosome is considered as an essential element of the catabolic process of toluene (Cuskey and Sprenkle, 1988). The presence of benzoate leads to a transcriptionally competent form of BenR regulating expression from *PbenA* and thus, transcription from the *ben*-operon of the *ortho*-cleavage pathway (Figure 1). Therefore, *ben*- and *meta*-operons are simultaneously activated by benzoate (Perez-Pantoja *et al.*, 2014) triggering the transformation of benzoate into catechol which is further catabolised into Krebs cycle intermediates (Burlage *et al.*, 1989).

The transcriptional kinetics of the TOL plasmid has been previously evaluated using *m*-xylene as the inducer of the pathway (Koutinas et al., (2010, 2011)). Furthermore, the specific transcriptional kinetics has been recently monitored upon toluene introduction in order to demonstrate the interplay between the TOL and the chromosomal *ortho*-cleavage pathways (Tsipa et al., 2015), suggesting interdependence of the two pathways through potential up-regulation of *PbenA* by the XylS protein. Nevertheless, the present study is the first to our knowledge to **evaluate the transcriptional kinetics of *P. putida* upon simultaneous induction with *m*-xylene and toluene**, although the specific pollutant mixture is commonly met in petrochemical waste streams. Both substrates are considered as predominant effectors of the TOL plasmid where *m*-xylene is strongly preferred by cells as compared to toluene (Duetz et al., 1998). At the genetic level the four promoters of TOL were stimulated first by *m*-xylene, and when the concentration was reduced below 0.5 mM the activity of the promoters was boosted due to toluene induction (Figure 2A, B, C, D). The expression of *PbenR* was maintained at a basal level and slightly decreased towards the complete depletion of both substrates (Figure 2E), while expression of *PbenA* was remarkably boosted only when toluene existed as the sole effector remaining in the culture (Figure 2F).

Taking into account the crucial role of promoters as controllers of biodegradation, a **mechanistic approach to describe *m*-xylene and toluene utilization as well as biomass formation in conjunction to the response of the TOL and chromosomal regulatory networks** is becoming a necessity. The performance of biodegradation in similar systems is typically monitored through Monod-type models. However, these traditional and widely used models are based on bulk measurements, ignoring the molecular interactions and gene regulation of the strain employed, resulting in limited predictive capability. Furthermore, prediction of double substrate microbial growth kinetics still remains a topic which has not been successfully realised in a wide range of conditions, constituting a major challenge in biochemical engineering. Herein, we present an **experimentally validated mathematical model** (using qPCR measurements), which is capable of capturing the **dynamic behaviour of the key promoters in the TOL and *ortho*-cleavage pathways**. The model's prediction is used to describe the activity of the enzymes responsible for substrates catabolism and biomass growth, followed by **prediction of specific growth, biomass formation and substrates utilization rates**. The application of the model in double substrate cultures exemplifies the importance of using transcriptional regulatory information for bioprocess prediction linking directly genetic and molecular mechanisms to macroscopic events. The developed model constitutes a novel mechanistic approach establishing an integrated framework for the prediction of microbial growth kinetics in environmental bioprocesses.

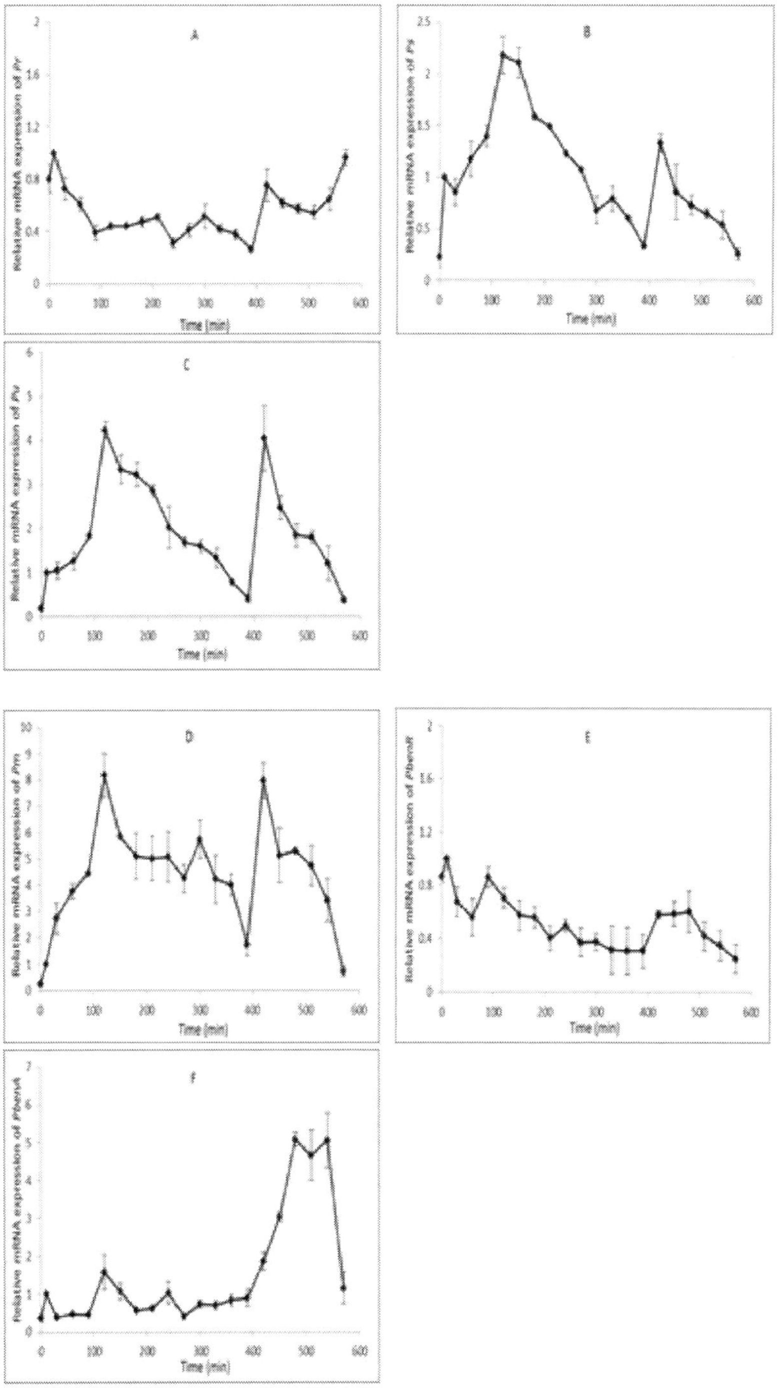

Figure 2: Relative mRNA expression of (A) *Pr*, (B) *Ps*, (C) *Pu*, (D) *Pm*, (E) *PbenR*, (F) *PbenA* promoters upon induction with *m*-xylene and toluene.

## References

- ADDIN EN.REFLIST Aranda-Olmedo, I., Marin, P., Ramos, J.L. and Marques, S., 2006. Role of the ptsN gene product in catabolite repression of the Pseudomonas putida

TOL toluene degradation pathway in chemostat cultures. Applied Environmental Microbiology, 72(11): 7418-7421.

-Ballerstedt, H. et al., 2007. Genomotyping of *Pseudomonas putida* strains using *P. putida* KT2440-based high-density DNA microarrays: implications for transcriptomics studies. Applied microbiology and biotechnology, 75(5): 1133-1142.

-Burlage, R.S., Hooper, S.W. and Sayler, G.S., 1989. The TOL (pWWO) Catabolic Plasmid. Applied and environmental microbiology, 55(6): 1323-1328.

-Cowles, C.E., Nichols, N.N. and Harwood, C.S., 2000. BenR, a XylS Homologue, Regulates Three Different Pathways of Aromatic Acid Degradation in Pseudomonas putida. Journal of Bacteriology, 182(22): 6339-6346.

-Cuskey, S.M. and Sprenkle, A.B., 1988. Benzoate-dependent induction from the OP2 operator-promoter region of the TOL plasmid pWWO in the absence of known plasmid regulatory genes. Journal of Bacteriology, 170(8): 3742-3746.

-de Lorenzo, V. and Perez-Martin, J., 1996. Regulatory noise in prokaryotic promoters: how bacteria learn to respond to novel environmental signals. Molecular microbiology, 19(6): 1177-1184.

-Díaz, E. and Prieto, M.A., 2000. Bacterial promoters triggering biodegradation of aromatic pollutants. Current Opinion in Biotechnology 11: 467–475

-Duetz, W.A. et al., 1998. Biodegradation kinetics of toluene, m-xylene, p-xylene and their intermediates through the upper TOL pathway in *Pseudomonas putida* (pWW0). Microbiology, 144: 1669-1 675.

-Koutinas, M. et al., 2011. Linking genes to microbial growth kinetics: an integrated biochemical systems engineering approach. Metabolic engineering, 13(4): 401-413.

-Koutinas, M. et al., 2010. The regulatory logic of m-xylene biodegradation by *Pseudomonas putida* mt-2 exposed by dynamic modelling of the principal node Ps/Pr of the TOL plasmid. Environmental microbiology, 12(6): 1705-1718.

-Perez-Pantoja, D., Kim, J., Silva-Rocha, R. and de Lorenzo, V., 2014. The differential response of the *Pben* promoter of *Pseudomonas putida* mt-2 to BenR and XylS prevents metabolic conflicts in m-xylene biodegradation. Environmental microbiology: 1-12.

-Ramos, J.L., Marques, S. and Timmis, K.N., 1997. Transcriptional control of the Pseudomonas TOL plasmid catabolic operons is achieved through an interplay of host factors and plasmid-encoded regulators. Annual Review of Microbiology, 51: 341-372.

-Schmidt, E., Bartels, I. and Knackmuss, H., 1985. Degradation of 3-chlorobenzoate by benzoate or 3-methylbenzoate-utilising cultures: 3-chlorobenzoate; mixed cultures; total degradation; incompatible substrates. FEMS Microbiology Letters, 31(6): 381-389.

-Timmis, K.N., 2002. Pseudomonas putida: a cosmopolitan opportunist par excellence. Environmental microbiology, 4: 779-781.

-Tsipa, A., Koutinas, M., Pistikopoulos, E.N. and Mantalaris, A., 2015. Transcriptional kinetics of the cross-talk between the ortho-cleavage and TOL pathways of toluene biodegradation in Pseudomonas putida mt-2

-Weiss, R. et al., 2003. Genetic circuit building blocks for cellular computation, communications, and signal processing. Natural Computing, 2: 47–84.

# 405396 Synthesis of Drug Intermediates By Monoprotection of Diols in C-Glycosides

**Abstract:**

Synthesis of Drug Intermediates by Monoprotection of Diols in C-GlycosidesADDIN IUCrAuthors

Tao Zhang and Zhijie Fang*

School of Chemical Engineering, Nanjing University of Science & Technology

Correspondence email: zjfang@njust.edu.cn

**Abstract:** Selective and per-O-esterification of carbohydrate derivatives using acyl chloride or anhydride and DMAP (4-dimethylaminopyridine), were first systematically reported. This was a common protocol that only one monoprotected product could be obtained by the screening of esterification reagents. For result confirmation, nuclear magnetic resonance study and theoretical calculation approach were set out to analyze and support the results i.e. 3-Ts (compound 6j) was the only per-O-substituted product when TsCl (4-toluene sulfonyl chloride) was used as esterification reagent. The research will be used as a quick and easy method to the application of synthesis of carbohydrate drugs.

# 407913 Degradation of a Fungicide Using a Microbiome : Importance in Organic Agriculture

**Abstract:**

In this paper the potential of a microbiome for degradation of prochloraz is studied. Concentrations of 100, 50, 10 and 5 ppm were considered, evaluating the rate of decrease of the concentration by HPLC-PDA chromatography. A decrease of 50% was observed, verifying that during the degradation process different populations were expressed microbiome. The effect of the initial concentration of the fungicide was analyzed, determining whether the degradative capacity is affected by this parameter. The operation of this microbioma allow its application in the agricultural fungicide contaminated areas in Mexico producing territories.

# 409344 Use of Lectin-Conjugated Magnetite Nanoparticles and Magnetic Force to Screen Mutant Cyanobacteria in a Single Cell Culture System

**Abstract:**

We developed a single cell culture system to screen mutant cyanobacteria by using magnetic nanoparticles (MNPs) and magnetic force. This system could perform a single cell isolation culture with high viability compared to conventional methods.

Cyanobacteria such as *Synechococcus elongatus* PCC7942 have the potential to produce biofuels directly from $CO_2$ [1]. To raise the productivity of biofuel, it is necessary to select the useful mutants such as strains which have high growth rate and are resistant to specific metabolic products. The general approach for screening of mutants involves single colony isolation in solid culture. However, it is inefficient for cyanobacteria with long doubling time and low colony-forming ratio due to the cell death by dehydration. Although we have previously developed the droplet culture system for single cell culture of cyanobacteria to solve the dehydration problem [2], the system was not suitable for a long term culture and a screening throughput of the system was low. Thus, more efficient single cell culture system for screening mutant cyanobacteria is still required.

In the present study, we applied the following cell patterning method established originally by us to a single cell culture of cyanobacteria in liquid. We have developed a magnetic force-based cell patterning method for mammalian cells and succeeded in analyzing cells at a single-cell level [3-5]. In this method, we fabricated magnetite cationic liposome (MCL) in which MNPs were coated by cationic liposome and MCLs were nonspecifically incorporated by mammalian cells. Then, the magnetically labeled cells can be arrayed by using a magnetic pin-holder device which has 6400 iron pillars (each 100 μm in width and 100 μm in length) with 150 μm intervals. In order to develop MCLs to label specifically PCC7942 cells, we focused on the lectin which can recognize specific sugar chain in the outer membrane of cells because the lipopolysaccharide exists on the outer membrane of cyanobacteria. We found that LCA (*Lens culinaris* Agglutinin) which recognize mannose as a sugar could bind to the surface of PCC7942 cells, and then developed the LCA-conjugated MCLs. The MCL labeled PCC7942 cells were magnetically patterned at a single cell level by iron pillars of the pin-holder device, and 1600 of single cell are enable to be arrayed in a square centimeter. As a result of culturing the patterned cells for 7 days, we observed colony formation from single cell in liquid medium and not observed any harmful effect of the LCA-conjugated MCLs to the cell growth. We achieved higher colony-forming ratio by the array culture method (78.4%) than that by the conventional solid culture method (4.8%). Furthermore, strains with different properties could be distinguished in the single culture system depending on their growth in the cell array. Finally, we could selectively pick up the target cells by a glass capillary operated by a micromanipulator and subsequently perform efficient isolation culture.

Thus, the developed system could overcome the problems of the conventional methods, such as dehydration of the cells, limitation of initial cell density, difficulty of long-term culture maintenance, and low throughput performance. It would be useful for screening mutant cyanobacteria.

## References

[1] T. Kusakabe, T. Tatsuke, K. Tsuruno, Y. Hirokawa, S. Atsumi, J. C. Liao and T. Hanai, *Metab. Eng.*, 20 (2013), 101-108.

[2] S. Arai, M. Okochi, T. Hanai and H. Honda, *Biotech. Rep.*, 4 (2014), 151-155.

[3] K. Ino, M. Okochi, N. Konishi, M. Nakatochi, R. Imai, M. Shikida, A. Ito and H. Honda, *Lab Chip*, 8 (2008), 134-142

[4] M. Okochi, S. Takano, Y. Isaji, T. Senga, M. Hamaguchi, H. Honda, Lab Chip, 9 (2009), 3378-3384

[5] S. Yamamoto, M.M. Hotta, M. Okochi, H. Honda, PLoS ONE, 9 (2014), e103502

# 410935 Development of an Automated in-Line Fraction Collection Device for Single-Sample, Multi-Peak 2D-HPLC Protein Characterization

**Abstract:**

The formation of aggregates during the manufacturing process of monoclonal antibodies is an important issue that can lead to problems downstream affecting the safety and efficacy of the drug product. Thus, analytical tools for monitoring protein aggregation during manufacturing would be a highly desirable component for implementing process analytical technology (PAT) and quality by design (QbD) in the biopharmaceutical industry. We have established an automated, off-line method for the two-dimensional (2D) chromatographic characterization of the multiple protein peaks that can result due to aggregation. This method utilizes a unique fraction collection device, herein termed Medusa, which was developed to enable the collection of up to twelve fractions between column one and two. Medusa consists of two electronically-controlled 12-position/13-port valve systems connected with 100 μl loops, and was placed in line with an Agilent 1200 series HPLC system configured for 2D chromatography. In our system, column one is Protein A and separation by size exclusion is column two. This method was challenged using an aggregated model IgG1 protein injected onto the first dimension column. Samples from multiple protein peaks of interest following purification from column one are fractionated by automated switching of the HPLC valve module to direct flow to the Medusa, and coordinated opening and closing of the Medusa's valves allowing for the isolation of these time-based fractions in the connected loops. Each sample can then be sequentially loaded onto the second chromatographic column by sequentially opening the Medusa valves. By placing the Medusa in-line with the 2D-HPLC in this way, we facilitate the timely analysis of multiple protein peaks from column one in a rapid, high-throughput manner.

# 417775 Effect of Transducer Position on Performance of an Enzyme Catalyzed Ultrasonic Reactor

**Abstract:**

Enzymes are biological catalysts in the form of protein which catalyze chemical reactions. They are highly active and specific and are capable of functioning at milder operating conditions. For industrial scale applications enzymes are immobilized on a carrier material (immobilized enzymes) in order to facilitate their separation from products and subsequent reuse. However, immobilization of causes a lowering of their activity owing to mass transfer resistance. Application of ultrasound (US) is found to be effective to overcome this limitation [1, 2]. In order to efficiently utilize ultrasound for industrial processes it is necessary to study the effect of different influencing parameters on performance of ultrasonic reactor. One of the key influencing parameters is the position of ultrasonic source. In context of the present work effect of ultrasonic source position on performance of ultrasonic reactor is investigated in a systematic manner by selecting an example reaction system. The results show that for a given energy input performance of reactor varies by varying the position of sound source. This observation is evident that a wrong positioning of the source in reactor will result in wastage of energy and thereby affecting the economy of the process.

[1] P.Schmidt; Ultrasonics, Vol. 25, PP 295-299; 1987
[2] Sang Hyun Lee; Process Biochemistry, Vol. 43, PP 1009–1012; 2008

# 412824 Engineering Escherichia coli for Production of 3-Dehydroshikimate and Its Derivative Aromatic Compounds

**Abstract:**

3-dehydroshikimate (DHS) is a key metabolic intermediate for aromatic compound production. 3-dehydroshikimate (DHS) and its related aromatic compounds could be important feedstocks for producing a variety of important industrial chemicals, fibers and plastics with great commercial value. In the previous study, *Escherichia coli* was always engineered to produce DHS by deleting *aroE* and overexpressing several key genes via plasmid under the control of inducible promoters. This would result in the genetically instability of the engineered *E coli* and the additional supplement of minor shikimate or aromatic amino acids in the fermentation medium for cell growth, thus affect the future industrial production of DHS. Hence, it is greatly necessary to construct the genetically stable strain as well as adaptable to simple synthetic media with any supplementation for higher and efficient production of DHS. Here, in this study, the synthetic regulatory parts with different strength were used for combinatorially regulating the different modules of DHS synthetic pathway for facilitating glucose transport, enhancing and balancing the supply of precursors, and blocking the degradation of DHS. These modifications further increased the production of DHS in the simple mineral salt media without supplementing shikimate or aromatic acids by about 40 fold. The metabolically engineered strain produced about 50g/L DHS in 7-L fed-batch fermentation with more than 30% yield (mol/mol) from D-glucose. With this chassis cell for DHS production, many other aromatic compounds, such shikimate, galliate quinate, protocatechuate, catechol, vanillate and vanillin were efficiently produced by redirect the flux based introducing heterogeneous enzymes. Our study laid important foundation for biomanufacturing of many useful aromatic compounds.

# 418831 A Diagnostics System for Monitoring Lupus Nephritis Based on Electrochemical Impedance Spectroscopy (Rapid Fire)

**Abstract:**

Millions of people worldwide suffer from the debilitating disease known as systemic lupus erythromatosus or SLE. Roughly half of these individuals will develop the potentially fatal kidney disorder lupus nephritis. In order to avoid this progression, maintaining SLE patient supervision through simple diagnostic testing is of the utmost importance. The following report details a novel biosensor platform for quick, non-invasive testing to assist researchers in identifying specific biomarkers of lupus nephritis for eventual patient monitoring.

The principal objective of our project is the development of a robust, low-cost, electrical biosensor capable of ultra-sensitive and highly selective protein biomolecule detection for monitoring lupus nephritis oxide platforms. We have designed a portable microelectrode sensor array platform for ultra-sensitive and highly selective detection of proteins. We have used a materials-based approach to overcome problems often associated with biosensors such as a lack of sensitivity and/or selectivity, dominance of signal noise, and adaptability of detection to a wide range of biomolecules. Point-of-care, rapid quantification of these biomarker molecules can help in various facets of disease detection, monitoring, and analysis. The key issues in protein detection have been achieving; (a) ultra-sensitivity in detection at concentrations (typically in lower pg/ml), and (b) selectivity in protein detection from complex solutions such as body fluids. We have designed a sensor for monitoring and tracking levels of vascular cell adhesion molecule-1 (VCAM-1) protein marker from patient urine samples in the range of 1fg/ml to 1μg/ml. We have demonstrated a great than 95% confidence in correlating the biosensor quantification of VCAM -1 with ELISA.

# 419250 Effect of Geometry of Confined Media on Folding and Stability of Proteins (Rapid Fire)

**Abstract:**

From catalyzing reactions to cell regulation, proteins represent one of the most important classes of complex molecules within living cells. Their folding and unfolding patterns from their primary to quaternary structures play a major role in determining their behavior and activity. The conditions that cause the misfolding of proteins also induce altered behavior in such molecules. One case in particular is the accumulation of toxic protein aggregates believed to affect the onset of neurodegenerative diseases such as Parkinson's and Huntington's. It has been shown that the three-dimensional structure of proteins is mainly dictated by their specific amino acid sequence. The transport properties and stability of proteins are, however, dependent on the local environment. Furthermore, while the behavior of such molecules is well understood under the bulk conditions, our understanding of their behavior under "crowded" cellular conditions is limited. The purpose of this study, therefore, is to determine the effects that high confinement representing a crowded medium, as may be seen in living cells, has on the structure, energy, and folding patterns of α-de novo-designed proteins for both attractive and repulsive nanopore walls.

Through the use of a united-atom force field the proteins are constructed by the protein intermediate resolution model (PRIME) via discontinuous molecular dynamics (DMD). DMD simulations have been carried out on the de novo-designed α-family of proteins with amino acid lengths of 9, 16, 23, and 30, with each amino acid chain consisting of 4 united atom beads. Three types of confined media have been considered: the space between a pair of flat walls, a cylindrical nanopore, and a spherical nanopore.

The DMD simulations for each amino acid length have been conducted using a range of nanopore sizes and temperatures for durations between 100 and 300 nanoseconds. The minimum nanopore radii wherein a fully folded stable protein can form in a slit and cylindrical nanopore are 1.28 and 1.31 nanometers, respectively, for all amino acid lengths. The respective values for a spherical nanopore are 1.91, 2.85, 3.87, and 4.88 nanometers for amino acid chain lengths of 9, 16, 23, and 30. The study conducted with a slit nanopore indicates that proteins with an α-helix native state are destabilized in the smallest nanopore size that does not affect the protein folding state. By slightly increasing the pore size, the optimal folding temperature is reached. Thus far, our studies indicate that increased confinement leads to enhanced instability with respect to both the folded and unfolded protein structures. We also show that for the α-helical proteins of various sizes in the cylindrical and spherical nanopores, the folding temperatures decrease with increased confinement for nanopore radii slightly larger than the minimum size. Furthermore, our present results reveal that in highly confined media the free energy surface becomes rough, and a new barrier for protein folding may appear close to the unfolded state.

# 419378 Development of PMMA Plate Immobilized with Fab Antibodies Possessing High Antigen-Binding Activity

**Abstract:**

Fab antibody is a recombinant protein which consists of Fd region of heavy chain (Fab H) and full-length of light chain (Fab L). Fab antibody can be produced at lower cost compared with conventional monoclonal antibody. Since Fab antibody is useful for economical diagnosis. In general, Fab antibody is prepared from IgG type antibody (Whole Ab) by enzymatic digestion, while it can also be produced as a soluble protein in recombinant *E. coli* cells. However, their productivity levels are generally low.

The aim of this study is development of production process for Fab antibody with high yield and high antigen binding activity by refolding. We prepared Fab antibody by mixing Fab H and Fab L fragments, both of which was separately produced in insoluble fraction of *E. coli* cells and refolded. In addition, we successfully prepared Fab antibody-immobilized poly (methyl methacrylate) (PMMA) plate by immobilization of mixture of PMMA-tag-fused Fab H (Fab H-PM) and Fab L (Fab L-PM). According to the results of sandwich ELISA, we successfully demonstrated preparation of PMMA plate immobilized with highly-activated Fab antibody. We further investigated preparation of Fab antibody-PMMA plate with a wide variety of antigen-binding specificities. By use of 5 different Fab H-PM and Fab L-PM as models, we tried to prepare total 25 types of Fab antibodies on the surface of PMMA plate. The results indicated that Fab antibodies on the PMMA plate show different binding affinity against the antigen, suggesting that they originally possess different antigen specificities. Thus, the developed method in this study will be useful for efficient preparation of antibody library on a PMMA plate with a wide variety of antigen-binding specificity.

# 420587 Metabolic Engineering of Saccharomyces Cerevisiae to Produce Long Chain and Medium Chain Fatty Alcohols from Xylose (Rapid Fire)

**Abstract:**

Metabolic Engineering of *Saccharomyces Cerevisiae* to Produce Long Chain and Medium Chain Fatty Alcohols from Xylose

Jiayuan Sheng, Weihua Guo, Xueyang Feng*

Department of Biological Systems Engineering, Virginia Polytechnic Institute and State University, Blacksburg, VA 24060

Fatty alcohols are important components of a variety of industries which have a 3 billion-dollar market globally[1]. Long chain fatty alcohols are mainly used in surfactants, lubricants, detergents, pharmaceuticals and cosmetics while medium chain fatty alcohols could be used as diesel-like biofuels[1,2]. We have previously engineered a *Saccharomyces cerevisiae* strain to produce over 1.1 g/L 1-hexadecanol from glucose by heterogeneously expressing a bird fatty acyl-CoA reductase (FAR) and optimizing the host metabolism[3]. In this study we introduced a fungal xylose utilization pathway, which is composed of xylose reductase (XR), xylitol dehydrogenase (XDH), and xylulose kinase (XKS), into our engineered *S. cerevisiae* to produce 1-hexadecanol from xylose, a non-food feedstock. Over 0.4 g/L 1-hexadecanol can be produced by using the native xylose utilization pathway. We further optimized the xylose utilization by combinational promoter engineering and evolutionary engineering, which improved 1-hexadecanol production titer by 200% (from 0.4 g/L to 1.2 g/L). We also successfully engineered *S. cerevisiae* to produce 1-decanol and 1-dodecanol by introducing the FAR enzyme into the peroxisome to hijack the medium chain fatty acyl-CoA generated from the beta-oxidation pathway. To our best knowledge, this is the first time that long chain and medium chain fatty alcohols were produced in yeast from xylose.

1 Steen, E. J. *et al.* Microbial production of fatty-acid-derived fuels and chemicals from plant biomass. *Nature* **463**, 559-562, doi:10.1038/nature08721 (2010).

2 Youngquist, J. T. *et al.* Production of medium chain length fatty alcohols from glucose in Escherichia coli. *Metabolic engineering* **20**, 177-186, doi:10.1016/j.ymben.2013.10.006 (2013).

3 Feng, X., Lian, J. & Zhao, H. Metabolic engineering of Saccharomyces cerevisiae to improve 1-hexadecanol production. *Metabolic engineering* **27C**, 10-19, doi:10.1016/j.ymben.2014.10.001 (2014).

# 419782 Development of ScFv-Immobilized Latex for Detection of CRP By Latex-Enhanced Turbidimetric Immunoassay

**Abstract:**

There are many immunoassays utilizing antigen-antibody reaction. Among them, latex-enhanced turbidimetric immunoassay for detection of C-reactive protein (CRP) has been widely used because rapid and quantitative detection is possible by just mixing human serum with latex beads immobilized with anti-CRP antibody. The pentameric antigen, CRP can be interacted with antibody molecules that were immobilized on the surface of latex, and then the latices are aggregated via CRP antigen. Consequently, turbidity of solution increased with CRP concentration. Therefore, the CRP concentration in serum could be quantitatively measured from the turbidity change.

However, there are some problems in the conventional antibody-immobilized latex for turbidimetric immunoassay. Production cost of whole antibody which has been used in this immunoassay is considerably high because it is produced by hybridoma cell or animals. Furthermore, remaining activity of antibodies in adsorption state might be significantly low due to suffering from conformational change after adsorption.

We investigated preparation of latex immobilized with single-chain Fv (scFv) antibody. ScFv is the fusion protein which consists of VH and VL domains via the flexible linker $(G_4S)_3$ and can be produced by recombinant *E. coli*. The material-binding peptide, PMMA-tag (DVEGIGDVDLVNYFEVGATYTFNK) which was originally-developed by this research group was also introduced to the C-terminus of scFv (scFv-PM). PMMA-tag has strong affinity toward poly(methyl methacrylate) (PMMA) and the other plastics.

ScFv-PM could be immobilized on the surface of latex and consequently, remaining activity of scFv-PM on the latex surface became higher than that of scFv without PMMA-tag immobilized, indicating that PMMA-tag prevented conformational change in scFv molecules after immobilization. Approximately 50 ng/ml of CRP was detected by use of the scFv-immobilized latices prepared in this study. Thus, scFv-immobilized latices is useful for economical and sensitive detection for CRP by latex-enhanced turbidimetric immunoassay

# 422094 Crispr-Based Self-Cleaving Mechanism for Controllable Gene Delivery in Human Cells

**Abstract:**

Controllable gene delivery via vector-based systems remains a formidable challenge in mammalian synthetic biology and a desirable asset in gene therapy applications. Here, we introduce a methodology to control the copies and residence time of a gene product delivered in host human cells but also selectively disrupt fragments of the delivery vehicle. A crucial element of the proposed system is the CRISPR protein Cas9. Upon delivery, Cas9 guided by a custom RNA sequence cleaves the delivery vector at strategically placed targets thereby inactivating a co-expressed gene of interest. Importantly, using experiments in human embryonic kidney cells, we show that specific parameters of the system can be adjusted to fine-tune the delivery properties. We envision future applications in complex synthetic biology architectures, gene therapy and trace-free delivery.

# 420590 Simulation and Comparison Between the Real and Supplied Oxygen Demand of a Fermentation Process

**Abstract:**

**Simulation and Comparison between the real and supplied oxygen demand of a fermentation process**

Serna Sebastián[1], Sánchez Julio C.[1], García Carlos A.[1], Cardona Carlos A.[A1]

[1] Instituto de Biotecnología y Agroindustria, Departamento de Ingeniería Química. Universidad Nacional de Colombia Sede Manizales. Cra. 27 No. 64-60, Manizales, Colombia

[A1] Tel.: +57 6 8879400 Ext 55354. Corresponding author E-mail: ccardonaal@unal.edu.co

Acetic acid bacteria (AAB) derive their energy from the oxidation of ethanol to acetic acid during fermentation. Acetic acid bacteria are airborne and are ubiquitous in nature. They are actively present in environments where ethanol is being formed as a result of fermentation of sugars. They can be isolated from the nectar of flowers and from damaged fruit. Other good sources are fresh apple, cider and unpasteurized beer that have not been filter sterilized. In these liquids, they grow as a surface film due to their aerobic nature and active motility [1].

This work analyzes by simulation the physic-chemical conditions of fermentation that affect both viscosity and oxygen solubility of acetic acid fermentation using *Acetobacter pasteurianus*. Concentration profiles of biomass, substrate (ethanol) and product (acetic acid) through the fermentation were tested [2]. *A. pasteurianus* was chosen due to available data (kinetic parameters and microorganism characteristics).

The aim of this analysis is to demonstrate the relation between concentrations of biomass, substrate and product with the viscosity of the fermentation media, in order to show the dependence of those with the oxygen supply. The real proficiency of fermentation was analyzed in terms of oxygen supply respect to oxygen demand of the microorganisms, and required agitation for full oxygen supply.

The viscosity of the fermentation is function of substrate, product and biomass concentrations, previously calculated by separate. Biomass viscosity was calculated using Vand's equation [3]. For the calculation of the viscosity of the ethanol and the acetic acid on an aqueous solution it was used the method of Teja and Rice [4], [5]. Mass transfer coefficient was calculated with the equation proposed for stirred fermenters containing non-coalescing non-viscous media [6]. The power dissipated and the amount of oxygen in the liquid was calculated with the model proposed by [7], and dissolved oxygen supply was calculated from [8].

It was concluded that viscosity of the fermentation solution and oxygen solubility are important variables to analyze. Both variables were directly related with the conditions that allow the proper functionality of the microorganism and therefore they are directly related with performance of the fermentation for acetic acid production. This analysis can be extended to any fermentation process. Results also reveal that agitation is an important variable on fermentation and affect directly the amount of available oxygen in fermentation media and therefore satisfy the microorganism oxygen demand.

Two cases were found, one in which the agitation and the oxygen supply exceeds the real requirements and another one doing the exact opposite, supplying less than the real requirements. In the first case we talk about over-costs and in the second one we talk about lower productivity. Therefore, the real equilibrium point between costs and productivity must be found within these two cases, with the aim of proposing an optimized performance of the fermentor.

Finally, this approach to the fluid dynamics of fermentation and its relation with the energetic demands opens a new perspective in which it is necessary to deepen the functioning of the associated processes of fermentation.

**BIBLIOGRAPHY**

[ SEQ [ * ARABIC 1] Guillamón, J & Mas, A. (2011) Chapter 9 - Acetic Acid Bacteria, In Molecular Wine Microbiology (pp. 227-255). Valencia, Spain: Elsevier Inc.

[2] Trcek, J., Toyama, H., Czuba, J., Misiewicz, A., & Matsushita, K. (2006). Correlation between acetic acid resistance and characteristics of PQQ-dependent ADH in acetic acid bacteria. *Applied microbiology and biotechnology*, *70*(3), 366-373.

[3] Vand, V. (1948). Viscosity of solutions and suspensions. I. Theory. *The Journal of Physical Chemistry*, *52*(2), 277-299.

[4] Teja, A. S., & Rice, P. (1981). A multifluid corresponding states principle for the thermodynamic properties of fluid mixtures. *Chemical Engineering Science*,*36*(1), 1-6.

[5] Teja, A. S., & Rice, P. (1981). The measurement and prediction of the viscosities of some binary liquid mixtures containing n-hexane. *Chemical engineering science*, *36*(1), 7-10.

[6] Van't Riet, K. (1979). Review of measuring methods and results in nonviscous gas-liquid mass transfer in stirred vessels. *Industrial & Engineering Chemistry Process Design and Development*, *18*(3), 357-364.

[7] Rushton, J. H., Costich, E. W., & Everett, H. J. (1950). POWER CHARACTERISTICS OF MIXING IMPELLERS. 1. *Chemical Engineering Progress*, *46*(8), 395-404.

[8] ENERGY, S., FIELDS, M. T. I. R., & AGEING, P. (1982). TRANSACTIONS REVIEW PAPER GAS-LIQUID MASS TRANSFER IN MICROBIOLOGICAL REACTORS. *Chemical Engineering Research and Design*, *60*.

# 422141 Reconfigurable Hybrid Interface for Molecular Marker Diagnostics and in-Situ Reporting (Rapid Fire)

**Abstract:**

Combinations of molecular signals such as transcription factors and microRNAs in cells are a reliable indicator of multi-gene disorders. A system capable of detecting these conditions in-situ may be used as a tool for diagnosis and monitoring of disease. Here, we engineer genetic circuits that sense endogenous levels of the androgen receptor (AR), the glucocorticoid receptor (GR), and the microRNA hsa-miR-21 (miR-21) in cervical cancer cells (HeLa). Furthermore, using the mediator molecule human chorionic gonadotropin (hCG), we interface the intracellular information to enzyme-linked immunosorbent assay (ELISA) test strips. We demonstrate that this hybrid genetic circuit and test-strip interface can accommodate combinatorial, low-cost, and in-situ reporting, a versatile profiling tool.

# 422181 Microneedle Mediated Vaccine Delivery As a Platform to Modulate in Vivo Release Kinetics

**Abstract:**

Administering vaccines with hypodermic needles results in patient discomfort, hazardous waste, and the storage of these vaccine solutions in refrigeration systems while en route to their destination can be prohibitively expensive. Microneedles, which have been shown to deliver subunit vaccines to the immune cells in the dermis, are currently being studied as a means to circumvent the aforementioned issues. An additional advantage of using microneedles is that they can be designed to achieve sustained release kinetics. These designs include silk, polysaccharide, polymer blend, and particle-based microneedles, and their release profiles follow Fick's law where the release rate is dependent on the amount of vaccine remaining within the implant. In contrast to these systems, we aim to design microneedles that exhibit exponential release kinetics since this would mimic how pathogen replication produces increasingly large quantities of antigen and Toll-like receptor ligands. We hypothesize that these microneedles that employ a degradable membrane fabricated with layer-by-layer techniques will elicit higher antibody titers than other microneedle designs and intradermal injections of the same antigen and adjuvant. These results could also provide a way to eventually generate even more complex release profiles, which could perhaps serve as both the prime and boost of a vaccine after administration with a single microneedle array.

# 423412 Site-Specific Immobilization of Nanobodies By Introducing Material-Binding Peptides

**Abstract:**

Antibodies having a strong affinity and specificity against a certain antigen, have been widely used in immunoassay and immuno-affinity chromatography. Since the conventional monoclonal antibodies (whole Ab) are often produced by mammalian cells or animal, their production cost is considerably high. Therefore, these operation costs using the whole antibody become high, while there is still space to reduce the cost. Recombinant antibody fragments such as single-chain Fvs and nanobodies (VHH) can be successfully produced by microorganisms such as *E. coli* cells. Furthermore, site-specific immobilization of these fragments can be attained by introducing our original material-binding peptides such as PS-tag, PMMA-tag and so on by this research group. Therefore, economical and high-performance antibody-immobilized supports may be prepared by combination these technologies. Here, we investigated production and immobilization of tagged recombinant antibody fragments for detection of target antigens with high sensitivity. In this study, anti-C-reactive protein (CRP) scFv from mouse and anti-green fluorescent protein (GFP) VHH derived from camel are used as models. We investigated availability of both PS-tag and PMMA-tag for site-specific immobilization of recombinant antibody fragments. When both of them with or without peptide-tags were overexpressed by *E. coli* cells, they were produced in insoluble fractions. After purification of antibody fragments followed by reconstitution of disulfide bonds, they were refolded by dialysis under the optimum conditions. Antigen-binding activities of antibody fragments in adsorption state were accessed by sandwich ELISA and SPR sensor equipped with polymer-coated Au sensor chip. Consequently, these peptide tags increased antigen-binding capacity of the antibody-immobilized solid supports and lower concentration of antigen was detectable, compared with antibody fragments without peptide-tags. Thus, recombinant antibody fragments fused with material-binding peptides will be useful for detection and/or separation of target substance with high efficiency.

# 424487 Enhanced Isoprenoid Production in *Saccharomyces Cerevisiae* through Xylose Utilization (Rapid Fire)

**Abstract:**

*Saccharomyces cerevisiae*, a workhorse yeast strain of the biotechnology and food industries, has been engineered for implementing economic processes for producing biofuels and chemicals. However, this yeast has limited capabilities for producing biomolecules which are derived from acetyl-CoA because of a rigid flux partition toward ethanol production during glucose metabolism. Even after introducing heterologous pathways for the biosynthesis of target molecules via acetyl-CoA, ethanol still remains as a major product. In this study, we aimed to investigate the effect of using glucose or xylose as a carbon source on the production of isoprenoids in engineered *S. cerevisiae*. To this end, we constructed efficient xylose-fermenting *S. cerevisiae* strains with the enhanced mevalonate pathway and compared the production of squalene from glucose and xylose. In order to construct an efficient xylose-fermenting *S. cerevisiae*, a xylose fermentation pathway from *Scheffersomyces stipitis* was introduced into *S. cerevisiae* and the promoter of transaldolase (*TAL1*) was substituted with a strong constitutive promoter in order to improve the xylose assimilating capability. The resulting strain was able to ferment xylose efficiently and rapidly. In order to increase metabolic fluxes toward the production of squalene, overexpression cassettes of a truncated HMG-CoA reductase 1 (*tHMG1*) and acetyl-CoA C-acetyltransferase (*ERG10*) were introduced into the xylose-fermenting *S. cerevisiae*. After cultivation on glucose and xylose, amounts of accumulated squalene in the engineered yeast were measured as an indicator of metabolic fluxes in the mevalonate pathway. Overexpression of *tHMG1* resulted in drastic increase in specific squalene content of the engineered cells regardless of a carbon source, but the positive effect was more outstanding under the xylose conditions as compared to glucose conditions. While overexpression of *ERG10* did not lead to any increase in squalene content regardless of carbon source, there was a synergistic effect on squalene production when both *tHMG1* and *ERG10* were overexpressed simultaneously. Specific squalene content in the engineered yeast was five-fold higher in the cells grown on xylose as compared to the cells grown on glucose. As a result, the *tHMG1* and *ERG10* co-overexpressing engineered yeast produced 472 mg/L of squalene with a productivity of 6.30 mg/L·h. These results suggest that the problem of the rigid flux partition toward ethanol production during the production of acetyl-CoA derived biofuels and chemicals can be bypassed through using xylose instead of glucose as a carbon source.

# 424834 Expression and Characterization of a Novel *Acinetobacter Baumannii*-Originated Alkaline Xanthine Dehydrogenase in *Escherichia coli*

**Abstract:**

Xanthine dehydrogenase (EC 1.17.1.4, XDH) catalyzes the oxidative metabolism of purines and aldehydes on CH or CH2 groups with $NAD^+$ or $NADP^+$ as the electron acceptor. XDHs have wide applications, such as in clinical diagnosis for enzymatic determination of inorganic phosphorus, 5`-nucleotidase and adenosine deaminase by coupling with purine-nucleoside phosphorylase and uricase. Here, we described cloning, expression and characterization of a novel alkaline XDH from *Acinetobacter baumannii* in *Escherichia coli*. A 4867-bp gene cluster consisting of 3 genes (*xdhA*, *xdhB* and *xdhC*) was cloned using PCR from the chromosomal DNA of *Acinetobacter baumannii* ATCC 19606. The 3 genes showed the highest homology of 91.4%~95.5% to the putative genomic genes of *Acinetobacter baumannii* retrieved in the GenBank, but only less than 52% homology to the experimentally confirmed XDHs genes of other species. The *xdhABC* gene cluster was functionally expressed in *E. coli* under the control of the *trc* promoter in pTrc99A plasmid, and the recombinant enzymes were purified to homogeneity by Ni-NTA affinity chromatography, anion exchange chromatography on Q-sepharose and hydrophobic interaction chromatography on phenyl-sepharose. The purified recombinant XDH was composed of two subunits with the respective molecular weights of 87 kDa(XDHB) and 56 kDa (XDHA), while the chaperone XDHC with a deduced molecular weight of 37 kDa was not confirmed by SDS-PAGE. The purified enzyme showed the maximum activity at pH 8.5 and high enzyme stability after incubation under the alkaline conditions (from pH 7.0 to pH 11.0) for 24 hours at 25. The optimum temperature was 40to 45, and the half-inactivation temperature () was 59. The enzyme exhibited a turnover number ($k_{cat}$) of $69s^{-1}$ and apparent Michaelis constant ($K_m$) of $25.3 \mu mol.l^{-1}$ for xanthine, resulting a catalytic efficiency ($k_{cat}/K_m$) of 2.74 $l.\mu mol^{-1}.s^{-1}$. These enzymatic properties were the best among the reported results. The high catalytic activity, thermal stability and alkaline stability of the recombinant XDH enable its application to be feasible.

# 425835 Engineering Transcriptional Regulation and Signal Peptides to Control the Catalytic Activity of Bacterial Microcompartments (Rapid Fire)

**Abstract:**

Bacterial microcompartments are subcellular bacterial organelles bound by a porous protein membrane. Enteric and other bacteria use these organelles to sequester certain metabolic pathways to enhance growth and pathogenicity. We aim to develop a tunable nanobioreactor system based on these protein structures to spatially organize heterologous biosynthetic pathways in bacterial hosts. To this end, we develop tools to control both the formation of the microcompartments and their enzymatic content.

By engineering the transcriptional regulation of the microcompartments, we can orchestrate their formation and exercise control over the loading of heterologous proteins to the nanobioreactor. We also engineer the N-terminal signal sequences responsible for enzyme targeting to the compartments, allowing fine control over the loading phenotype of the reactors. Our results have implications both for the controlled encapsulation of heterologous synthetic pathways and for *Salmonella* and *E. coli* pathogenesis *in vivo*, to which the bacterial microcompartments natively contribute.

廣西師範學院
GUANGXI TEACHERS EDUCATION UNIVERSITY
德才并育 知行合一

# Ying-zi He

College of chemistry and material science, Guangxi Teachers Education University, Nanning, China
email: hyz6929@163.com

廣西師範學院
GUANGXI TEACHERS EDUCATION UNIVERSITY
德才并育 知行合一

## ⑩Extraction Technology of Theabrownine from Liubao Tea and Its Liver Protection Research

**LiuBao Tea (Guangxi)**

**Puer Tea (Yunnan)**

广西师范学院
GUANGXI TEACHERS EDUCATION UNIVERSITY
德才并育 知行合一

# Abstract

- ［Objective］To study the extraction technology of theabrownine(TB) from Liubao tea，and its liver protection on acute hepatic injury induced by carbon tetrachloride ( $CCl_4$ ) in mice.
- ［Method］Orthogonal test was adopted to investigate effects of solid-liquid ratio，extracting temperature，time，times on yield of Liubao tea theabrownine，the technique was optimized，and the liver protection of theabrownine was analyzed.

广西师范学院
GUANGXI TEACHERS EDUCATION UNIVERSITY
德才并育 知行合一

- ［Result］The results showed that effects of extraction times and temperature on theabrownine yield reached significant level, while other two factors were not significant. The optimum extraction conditions were determined as follows: solid-liquid ratio 1∶30，extraction temperature 100 ℃，extraction time 80 min，extraction 4 times.

☐ [Result] Under this condition, the yield could be up to 22. 92%. Results of liver protection research showed that, following the treatment with TB, alanine aminotransferase( ALT) and aspirate aminotransferase (AST) activities and malondialdehyde (MDA) content were significantly reduced and superoxide dismutase (SOD) and glutathione peroxidase (GSH-Px) activities and the glutathione (GSH) content were obviously increased in the mice of treatment group.

廣西師範學院
GUANGXI TEACHERS EDUCATION UNIVERSITY
德才并育 知行合一

☐ [Conclusion] Liubao tea has abundant theabrownine, and TB have protective effects on acute hepatic injury induced by $CCl_4$ in mice, which will lay a solid foundation for deeply development of Liubao tea.

# 426406 Engineering Yeast Saccharomyces Cerevisiae for Assembly of Enveloped Influenza Virus-like Particles

**Abstract:**

Virus-like particles (VLPs) are nanoscale biological structures consisting of viral proteins assembled in a morphology that mimics the native virion but do not contain the viral genetic material. The possibility to manipulate the proteins contained within VLPs makes them an attractive tunable system for synthesizing predictable nanoscale structures that have numerous applications such as bioactive materials, biocatalysts, and vaccines. VLPs are especially amenable to vaccine development because they present viral proteins to the immune system in their near-native conformation without risk of infection. The FDA has approved VLP-based vaccines for human papilloma, hepatitis B, and hepatitis E viruses, and clinical trials for several other vaccines containing VLPs are currently underway.

The choice of the host expression system for VLP assembly requires careful consideration. For industrial and research applications, it is often advantageous to express protein in the most primitive host system possible to efficiently produce high yields of protein while minimizing the cost. However, primitive host systems such as *E. coli* may not be able to correctly fold more complex viral proteins. For this reason, VLPs containing glycoproteins from more complex enveloped viruses, such as influenza and Ebola, have been almost exclusively produced in insect, mammalian, and plant cells. To our knowledge, no attempt has been made to produce enveloped VLPs displaying authentic viral glycoproteins in a yeast-based expression system despite yeasts' ability to produce correctly folded glycoproteins of numerous enveloped viruses as well as their low cost and robustness in large-scale therapeutics production.

For the first time, we report that the yeast *Saccharomyces cerevisiae* is capable of assembling and producing enveloped influenza VLPs upon expression of hemagglutinin with or without the neuraminidase, M1, and M2 viral proteins from influenza subtypes H1N1 or H3N2. Budding of influenza VLPs off the yeast membrane was observed by examination of ultrathin sections of yeast cells using transmission electron microscopy (TEM). The influenza VLPs were shown to have key morphological and functional characteristics of influenza virus, including the presence of glycoprotein spikes on the VLP surface and the ability to agglutinate red blood cells. We are currently testing the effectiveness of vaccination with influenza VLPs in mouse models. These findings provide a proof-of-concept of using *S. cerevesiae* to produce complex enveloped VLPs for vaccine development. This system provides significant advantages over other higher eukaryotic systems because yeast cells can produce higher protein yields, are faster growing, are less expensive to maintain, and do not require the use of contaminating viral vectors.

# 426407 Development of a Lactococcal Vector for Antimicrobial Peptide Delivery in the Gut

**Abstract:**

The rise in antibiotic resistant pathogens has sparked an urgent need for new antimicrobial treatment options. Antimicrobial peptides (AMPs) are small proteins that exhibit micromolar to nanomolar activity against many pathogens of interest; however, their utility has been limited by their degradation in the gastrointestinal (GI) tract. We propose to overcome this delivery challenge by engineering probiotic bacteria that can survive passage through the upper GI tract and produce the desired AMPs at the site of infection. We are currently developing systems to target Vancomycin-resistant *Enterococcus faecium* and Vancomycin-resistant *Enterococcus faecalis* as these pathogens pose serious threats in health care. In our previous work, we engineered a powerful AMP expression system for *Lactococccus lactis* that was able to reduce *in vitro* counts of Vancomycin-resistant *Enterococcus faecium* (VRE) by approximately four orders of magnitude.[1] The work described here aims to build on this system by improving on three fronts; survivability of the probiotic in the gut, potency of the peptides, and decreased occurrence of resistance.

It is well-known that bile tolerance is a primary factor impacting bacterial survival in the gut. Consequently, we have now included a heterologous bile-resistance component in our AMP expression vector. We are currently testing these genes under different promoters to identify the most effective configuration to confer bile resistance both *in vitro* and *in vivo*. To improve the potency of our AMP expression system, we are exploring new AMPs exhibiting powerful activity both alone and in combination with our current peptides. Previously, we produced Enterocin A, Enterocin P, and Hiracin JM79.[1] In the work presented here, we demonstrate the successful production and secretion of Enterocin B from *L. lactis* using a fusion of Enterocin B with the lactococcal signal peptide Usp45. We are now testing the efficacy of these peptides individually and in combination to identify synergistic activity. To hinder resistance to our peptides, we are pursuing combinations of AMPs that rely on orthogonal targets; a method commonly used with traditional antibiotics to slow the development of resistance. We are thus working to identify peptide combinations that are not only effective at low concentrations, but also combinations that exhibit the rarest occurrence of resistance.

1. Geldart, K., Borrero, J., & Kaznessis, Y. N. (2015). A Chloride-Inducible Expression Vector for Delivery of Antimicrobial Peptides Against Antibiotic-Resistant Enterococcus faecium. *Applied and Environmental Microbiology*, *81*(11), 3889–3897. doi:10.1128/AEM.00227-15

# 425905 Metabolic Engineering of C. Tyrobutyricum for High n-Butanol Production By Rebalancing Carbon and Redox (Rapid Fire)

**Abstract:**

Biobutanol is a sustainable and environmentally friendly fuel that could be used as the substitute of gasoline. The acidogenic *Clostridium tyrobutyricum* had been engineered, ACKKO-*adhE2*, to produce butanol. However, our previous Omics study showed that the redox was not balanced and some enzymes involved in carbon metabolism had low expression in ACKKO-*adhE2*. In this study, both carbon engineering and redox engineering were performed to achieve high butanol production. The heterologous $NAD^+$-dependent formate dehydrogenase (*fdh*) was expressed in ACKKO-*adhE2* to synthesize a NADH producing pathway. The thiolase (*thl*) that directs metabolic flux from acetyl-CoA to acetocetyl-CoA was also overexpressed. Two mutant strains, ACKKO-*adhE2-fdh* and ACKKO-*thl-adhE2-fdh*, were constructed. The free-cell fermentation showed that butanol titer was increased to 19.03 g/L by ACKKO-*adhE2-fdh*. The ACKKO-*thl-adhE2-fdh* further improved butanol production, selectivity and yield. To obtain a stable strain, an advanced single plasmid was constructed by optimizing the promoter and terminator. This study demonstrated that the carbon and redox rebalance in *C. tyrobutyricum* enabled a high production of butanol.

# 426517 Impact of Nitrogen Removal Pretreatments on Algal Biomethane Production

**Abstract:**

Microalgal anaerobic digestion (AD) has been receiving increasing attention in the recent past owing to the versatility of algal biomass and the benefits of biomethane fuel. Algal methane production is known to be species dependent. The current work investigated the biochemical methane potential of 3 algal species namely *Chlorella pyrenoidosa* (fresh water green alga), *Tetraselmis chuii* (marine green alga), and *Phaeodactylum tricornutum* (marine diatom), which differ significantly in the cell wall structure and biochemical composition. The results indicated that biomethane yield was the highest for *P. tricornutum* (259.6 mL $gVS^{-1}$), followed by *C. pyrenoidosa* (234 mL $gVS^{-1}$) and the least for *T. chuii* (160.4 mL $gVS^{-1}$). Infact, the experimentally obtained biomethane yield was only about 33 % of the stoichiometric (theoretical maximum) yield for *T. chuii* and around 50-60 % of the respective stoichiometric yields for the other two species. These results clearly indicate a limitation of the AD process. Factors such as cell wall recalcitrance and the presence of inhibitory compounds such as ammonia significantly contribute to the limitation. Since algae have high nitrogen (and protein) content, they are capable of producing ammonia at inhibitory concentrations during AD process. Thus, pretreatments aimed at achieving nitrogen removal from the algal biomass may help in enhancing biomethane production from the residual biomass. Investigations are underway to validate this hypothesis for the 3 species listed above using chemical and hydrothermal protein/nitrogen removal pretreatments. Chemical protein/nitrogen removal was carried out by the alkali-acid method which involved solubilisation of proteins at alkaline pH followed by precipitation at acidic pH. Hydrothermal nitrogen removal was carried out by subcritical water extraction of algal proteins and amino acids at two different temperatures (125 °C and 160 °C) and 300 psi in a high pressure reactor.

# 427038 Screening and Characterization of Rabbit ScFv Antibodies Recognizing Cell Surface Antigen from ScFv-Displayed Phage Library

**Abstract:**

In cancer therapeutics, diagnostics in early stage of sickness such as canser is important and demand for methods to detect the certain antigen is considerably increased. In the present circumstances, cancer screening used is the focus of the organization of cancer was collected. These diagnostic methods are painful, costly, and risky to perform the analysis and diagnosis for patients, and need to conversion to technique of simple cancer screening such as blood test and urine analysis.

Cell surface proteins such as CD antigens and/or receptors sometimes are specifically expressed depending on cell species and morphology. Therefore, antibodies targeting these proteins result in valuable diagnostic agents and/or therapeutic drugs.

Here, we investigated screening and characterization of single-chain Fv antibodies specific to a certain membrane protein. The 2 kinds of oligo-peptides both of which was corresponding to a part of the target antigen, were designed and chemically-synthesized. These peptides were conjugated with carrier protein KLH, and were injected to rabbit. After extraction of total RNA from spleen, scFv-displayed phage library was constructed. After 4 rounds of biopanning procedure, antigen-binding activities of scFvs isolated were evaluated by enzyme-linked immunosorbent assay (ELISA). Furthermore, dissociation rate constant levels were also evaluated by SPR sensor equipped with peptide-immobilized sensor chip.

All of scFvs possessing the antigen-binding activities were ranked and top 5 clones were produced as soluble form by recombinant *E. coli* cells.

All the isolated scFvs reacted with the cells displaying target protein. These results suggested that scFv library prepared from rabbit has huge diversity and suitable for development of affinity molecules against target membrane proteins. The characteristics of scFv antibodies were also analyzed.

## (620bl) Conversion of Glucose to (2R, 3R)-Tartaric Acid By Sequential Whole-Cell Catalyzed Oxidation and Chemical Catalysis

Yuan, J., Zhejiang University
Wu, M., Zhejiang University
Yang, L., Zhejiang University
Lin, J., Zhejiang University

**Abstract** Development of bio-refineries has recently attracted increasing attention as a means to provide sustainable alternative solutions to reducing the dependence on petroleum resources and preventing environmental deterioration. (2R, 3R)-tartaric acid, an important natural existing hydroxyl carboxylic acid, is mainly used in pharmacy, food and dye industry. Currently, (2R, 3R)-tartaric acid on the market is produced by traditional technology from the cream of tartar or stereospecific hydrolysis of the *cis*-epoxysuccinic acid, derived from maleic anhydride, by *cis*-epoxysuccinate hydrolase. However, the former method is limited by the availibity of raw material and the latter is dependent on a relatively expensive petrochemical material. Therefore, new processes for the economical preparation of (2R, 3R)-tartaric acid from carbohydrate or renewable resource would be much more attractive. It is reported that glucose can be partially oxidized to 5-keto-gluconic acid (5-KGA) by several microbial cells, and then 5-KGA can be further transferred to (2R, 3R)-tartaric acid by chemical catalysis. But the efficiency of this process is still unsatisfactory. To obtain a robust whole-cell biocatalyst, *Gluconobacter oxydans* DSM2343 was metabolically engineered to improve the production of 5-KGA. First, the GOX1231 and GOX1081 genes, encoding gluconate-2 dehydrogenase and pyruvate decarboxylase, respectively, were eliminated via a clean deletion system to obtain the recombiant strain *G. oxydans* ZJU2. Then, the sorbitol dehydrogenase gene (GOX0855-0856, *sldAB*) was overexpressed in *G. oxydans* ZJU2 under a carefully selected promoter. The resulting strain, *G. oxydans* ZJU2/pB-$P_{0169}$-*sldAB*, produced 673 mM (about 131 g/L) of 5-KGA with a productivity of 2.08 g/(L·h) in 15 L fermentor. Further investigation to the metabolic pathway reveals that the electron transferring chain and its related coenzyme could be the crucial factors that hindered the conversion of gluconic acid to 5-KGA in this oxidation fermentation process. Hence, the gene cluster for the biosynthesis of coenzyme PQQ, *pqqABCDE* (GOX0983-0987), as well as *tldD* (GOX1104) gene, related to PQQ synthesis, and *cyoBACD* (GOX1911-1914) gene for terminal ubiquinol cytochrome $bo_3$ oxidase were fused into pUCpr plasmid and transferred into *G. oxydans* ZJU2/pB-$P_{0169}$-*sldAB*. The resultant recombinant strain *G. oxydans* ZJU2/pB-$P_{0169}$-*sldAB*/pUCpr-$P_{0169}$-*pqqABCDE*-*tldD*-$P_{0169}$-*cyoBACD* could produce 162 g/L 5-KGA within 64 h by using fed-batch fermentation and the productivity was 2.53 g/(L·h). At last, the transformation of 5-KGA to (2R, 3R)-tartaric acid was investigated with various transition metal catalysts in alkaline solution buffer in shaking flasks. After 72 h reaction, the molar conversion ratio and the selectivity with $Cu^{2+}$ as catalyst were 74.58% and 76.18%, respectively. The reaction mechanism of 5-KGA to (2R, 3R)-tartaric acid by $Cu^{2+}$ was also discussed.

# 427472 Sorting and Analysis of Yeast, Based on Replicative Age, Using High Reynolds Inertial Focusing (Rapid Fire)

**Abstract:**

In recent years, yeast has emerged as a valuable model in the study of aging. One of the main mechanisms of aging regulation in budding yeast is replicative aging, in which the mother cell ages with every budding event, by asymmetrically sequestering damaged proteins, thereby "resetting" the aging clock of the newly-emerged daughter cell. Although a number of tools already exist to both track individual yeast cells over time and to isolate successive generations of yeast, these tools are costly, time-consuming, and not easily scalable. Sorting based on fluidic forces confers the advantages of system simplicity, high-throughput work and low cost. Replicative age dependent separation of yeast requires characterization of the physical and biological properties of the cells, not previously done. An additional challenge is the size variance in each age group, followed by size overlap of different age groups. To withstand the challenge of replicative age dependent sorting of yeast, we coupled the unique phenomena of cross-section preferential concentration by Dean vortices with opposing shear induced lift forces, both occur in extremely high Reynolds numbers. We focused different size yeast to distant locations, to achieve biologically significant age dependent separation of the yeast despite the considerable age variance within each size binned group. Separation efficiency was monitored by yeast image processing, and replicative age dependent protein degradation was measured and analyzed.

# 427687 The Structure and Function of Photosynthetic Protein Complexes: An Experimental and Computational Study

**Abstract:**

Proteins mediate many biological functions and can possess a diverse set of features, all of which affect their interactions with other proteins, inorganic molecules, and surfaces. Gaining insight into proteins and their interactions is crucial to our understanding of the biological world, and could provide technological advancement in alternative energy, drug discovery, and a wealth of other areas. Membrane proteins are of particular interest, as they are responsible for mediating cell-environment interactions such as cellular recognition, signal transduction, and the transportation of ions and molecules across the membrane. Here, we are interested in studying photosystem I (PSI), a membrane integral protein involved in the photosynthetic cycle of plants and microorganisms that has been the subject of research for biorenewable energy conversion applications. We will present our work involving the use of a combination of small-angle neutron scattering and molecular dynamics simulation to study the structure and function of a PSI-detergent complex resulting from solubilization of this protein from its native membrane. We will also discuss current progress in studying the coupling of PSI with hydrogenase enzyme, which results in a unique complex capable of light-driven hydrogen production.

# 427866 Improving Thermotolerance of Industrial Saccharomyces Cerevisiae TSH3 through an Optimized Strategy for Genome Shuffling and Evolutionary Engineering

**Abstract:**

As a sugarcane-like non-food energy crop, sweet sorghum can accumulate high level of fermentable sugars within the stalks up to 18 wt%, and has superior agronomic advantages over sugarcane. Unfortunately, the potential of sweet sorghum is not fully harnessed as the feedstock for fuel ethanol production so far, due to quite low sugar recovery from juice pressing process that makes conventional liquid-state fermentation (LSF) not suitable for sweet sorghum ethanol production. Recently, advanced solid-state fermentation (SSF) developed by Tsinghua University was applied to fuel ethanol production using sweet sorghum, which exhibited a higher cost-competitive and lower environmental footprint over LSF, such as simpler process configuration and fewer sewage disposal. However, relatively low heat transfer efficiency of SSF at large scale always traps reaction heat within solid substrates during the process of fermentation, causing a high-temperature SSF, sometimes up to 40 . As a result, higher temperature of SSF inhibits cell metabolism, consequently reducing ethanol productivity. Meanwhile, among industrial ethanol producers, model strain *Saccharomyces cerevisiae* exhibits optimal growth between 28 and 30 , and there are also very few engineered *S. cerevisiae* that can grow with relatively high ethanol productivity at temperature higher than 41 . Therefore, all these factors mentioned above stymie the broad deployment of SSF at large scale, particularly in subtropical and tropical zones. In this study, a optimized genome shuffling and evolutionary engineering methods were adopted to improve thermotolerance of *S. cerevisiae* strain, TSH3, which was isolated from the sweet sorghum stalks and exhibits excellent ethanol fermentative capacity of SSF. Recombinant yeast strains were acquired by recursive DNA shuffling between the entire genome of *Trophcalis Candida* (CICC 32626) and *S. cerevisiae*. After two rounds of genome shuffling and screening, one potential recombinant yeast strain S2-1 was obtained. It was able to grow at 45 , and produce ethanol at 43 under ~100 g/l of glucose with 60.2% of theoretical ethanol yield during 48h. Using a long-term adaptation strategy, the recombinant yeast S2-1 produced ethanol more rapidly than the parental strain, and showed an improved ethanol yield (70%) at high temperature. In addition, the recombinant yeast S2-1 also obtained the capability of xylose utilization. Adaptation evolution work is still carried out and further improved strains might be obtained. In order to find out the key regulatory genes and networks of thermo-tolerance of *S2-1* strain, genome and transcriptome sequencing were carried out. Some interesting variation of genome profiles were found. Our study gives insight into the mechanism of thermotolerant yeast at genetic level, and might provide a novel strategy of rational genetic modification for obtaining engineering yeast strains that can make bioethanol production by SSF from sweet sorghum much more cost-competitive.

**Key words: genome shuffling, thermotolerance, *Saccharomyces cerevisiae*, ethanol, solid-state fermentation**

# 428548 Production of 3-Methylxanthine By Metabolically Engineered E. coli


**Abstract:**

**Production of 3-methylxanthines by Metabolically Engineered *E. coli***

Methylxanthines are naturally occurring purine alkaloids that are used in pharmaceutical preparations as diuretics, for asthma treatment, and as cardiac, pulmonary, and neurological stimulants. By using *E. coli*, metabolically engineered with N-demethylase genes as a biocatalyst, we were able to produce 100 mg of 3-methylxanthine by N-demethylation of theophylline at 30 °C and ambient atmospheric pressure. The conversion of theophylline is was 100%, 80% of which was 3-methylxanthine and 20% 1-methylxanthine.

It was previously reported that *Pseudomonas putida* CBB5 degrades caffeine via sequential N-demethylation to theobromine or paraxanthine, then to 7-methylxanthine, and further to xanthine. CBB5 contains five novel N-demethylase genes, *ndm*A, *ndm*B, *ndm*C, *ndm*D, and *ndm*E, which are responsible for caffeine degradation. Three genes, *ndm*A, B and D have been cloned, and expressed in *E. coli*. N-demethylase A (NdmA) and N-demethylase B (NdmB) are Rieske monooxygenases that catalyze position-specific $N_1$- and $N_3$- demethylations, respectively in the presence of NdmD. Both enzymes receive reducing equivalents from NADH via a redox-center-dense Rieske reductase, NdmD

Three *E. coli* strains was metabolically engineered using different combinations of *ndmA, B and D genes*. Strain pDdA which has one copy of each of NdmA and NdmD genes was grown overnight in 4 L super broth to produce 20 g wet cell paste. That cell paste was washed with potassium phosphate buffer pH 7.0 and used as a biocatalyst in 1.3 L reaction volume that contained 1 mM theophylline. The temperature and shaker speed used were 30 °C and 250 rpm respectively. Theophylline completely disappeared after two hours with 80% conversion to 3-methylxanthine as the major product. 3-methylxanthine was separated from post-reaction mixture by preparative chromatography by using BDS Hypersil C18 column (250 x 21.2 mm) as the stationary phase and 5% methanol as the mobile phase. The optimized solvent flow rate and the amount of post reaction mixture injected were found to be 2.5 mL/min and 25 mL respectively. 3-methylxanthine solution was concentrated by evaporation under vacuum at 70 °C and the resulting concentrated solution was freeze-dried. The purified product, 100 mg 3-methylxanthine was confirmed by LC-MS and NMR with respect to authenticity and purity.

# 427722 Soy isoflavones target amyloid-β oligomers associated with Alzheimer's disease (Rapid Fire)


**Abstract:**

Alzheimer's Disease (AD) is the most common neurodegenerative disease and the $6^{th}$leading cause of death in the US. One neurological marker of AD is the deposition of extracellular plaques composed of aggregates of the amyloid-β (Aβ) protein. As described by the amyloid cascade hypothesis, Aβ aggregation follows a nucleation-dependent pathway, beginning with protein monomer forming nuclei that grow into soluble aggregates and proceed to form the insoluble fibrils deposited in AD brain. In turn, these aggregate species are associated with increased intracellular reactive oxygen species (ROS) as well as cellular apoptosis. While such understanding of AD pathogenesis has increased significantly, AD remains the one disease in the top 10 that has no therapy, prevention, or cure. Among the most effective pharmaceuticals, 15% have been developed based on natural compounds or their derivatives. Thus, novel AD therapies can be developed and improved if natural compounds can be identified that effectively inhibit Aβ aggregation. Soy isoflavones (SIF), a subclass of naturally occurring polyphenols, are known for their beneficial effects in a myriad of diseases and cancers. While other subclasses of polyphenols have been studied for their ability to modulate Aβ growth and toxicity, SIFs remain unexplored. Here, four SIFS are studied for their ability to alter key steps in the Aβ aggregation pathway: genistein (GEN), genistin (O-GEN), daidzein (DEN), and daidzin (O-DEN). These compounds vary in hydroxylation as well as glycosylation on the fused center ring structure.

To examine the earliest stages of aggregation, Aβ oligomerization was induced by combining DMSO-solubilized $A\beta_{1-42}$ with a 10-fold excess of inhibitor followed by dilution into PBS. Oligomer formation was monitored via SDS-PAGE and Western blotting to quantify oligomer size. Additionally, oligomer samples were loaded onto APTES treated mica for AFM imaging. To study the overall process of aggregation, SEC-purified $A\beta_{1-40}$ was agitated with physiological salt in the presence of a 10-fold excess of compound and monitored using thioflavin T to detect aggregate β-sheet structure. Results were quantified as fold extension in lag time and reduction in equilibrium plateau. At terminal time points, samples were loaded onto copper grids for TEM imaging of fibril morphology. The antioxidant capacity of SIFs was evaluated using the OxiSelect™ Oxygen Radical Antioxidant Capacity (ORAC) assay. To confirm the antioxidant action of SIFs in cell culture, SH-SY5Y human neuroblastoma cells were treated simultaneously with SIFs and oligomers formed in the absence of SIFs. Cells were incubated with DCFH-DA, a probe that decomposes into fluorescent DCF in the presence of intracellular ROS, allowing for quantification of reduction in intracellular ROS.

When present during oligomerization, all SIFs reduced the amount of 25-100 kDa oligomers in size, while only GEN and O-DEN significantly reduced the amount larger 100-250 kDa oligomers. Additionally, GEN, O-GEN, DEN, and O-DEN increased the amount of monomer, trimer, and tetramer species. Together, these results suggest that SIFs keep small Aβ aggregates from progressing to larges sizes. In monomer aggregation, a corresponding effect was observed in lag extension. Through AFM and TEM, visual morphology changes were also apparent in oligomer and burblier aggregates, respectively. SIFs were also good antioxidants that were able to modulate Aβ-induced ROS.

In conclusion, all SIFs tested exhibit an effect on Aβ oligomerization and aggregation, as well as intracellular ROS, suggesting their potential as dual action therapies. Future work will determine the effect of SIFs on Aβ-induced cellular apoptosis to confirm that these drug-like compound properties have therapeutic potential.

# 427746 Hydrogel Ulva-Chitosan, Alternative Use of Algae As Bio-Injectable Material

**Abstract:**

**Hydrogel Ulva-Chitosan, Alternative Use of Algae Ulva as Bio-Injectable Material**

**Tania Kristanty [1] Farida Arisa [2]**

1. Student, Chemical Engineering Department, Universitas Gadjah Mada, 55281
2. Student, Chemical Engineering Department, Universitas Gadjah Mada, 55281

**Introduction.** Indonesia is a country that has large biodiversity but unfortunately the utilizations of that biodiversity are still sporadic. Wise use of biodiversity is important for sustainable development. The largest biodiversity is in the sea. For example, marine alga is the most abundant source of organic carbon in the biosphere, because its cell wall mainly consists of interconnecting polysaccharides. Ulva contains the functional acidic polysaccharide that has various functions. Beside that there is Chitin, a polysaccharide found in crab shells and shellfish which is being a marine waste. Chitin and chitosan(obtained by alkaline deacetylation of chitin) revealed that they possessed biological activities such as anti-bacterial effects. This research adds value to crab shell waste. One resource that is considered to have potential as a hydrogel material is Ulva lactuca.

Meanwhile, many researchers have reported various biological activities for ulvan that including anti-coagulant (Mao et al., 2009), anti-inflammatory (Leiro et al., 2010), anti-oxidant (Qi et al., 2006),immunostimulatory (Leiro et al., 2007), and anti-hyperlipidemic (Pengzhan et al., 2003) properties. However, despite the abundance and availability of green seaweeds or ulva and chitosan, relatively few studies have been performed on the molecular structures and the bioactivities of their sulfated polysaccharides. So, there is limited information available about that. Combination of both yields a hydrogel that can be use as a bioinjectable material because Chitosan has amine functional group (-NH2) which is highly reactive and positively charged, so it is able to bind to negative charge of Ulva lactuca.

In the medical world, surgery becomes an option in the medical world as a solution of the wound healing. However, the healing process will take a long time and not all operations are successful as expected and may pose a risk. For that, we need a new method in wound healing in order to minimize operating with the injection technique by the use of hydrogel as bio injectable material because it is capable of providing nutrients and place for the cell to live at the same place so that healing can take place. As injectable hydrogel material becomes very promising in the future as it has biocompatibility and easily applied without having to go through a complicated surgical procedure. (Bae, 2013).

Some previous researches have proof that ulva forms a hydrogel by combining it with another substance like Boron. However, the boron complex forms only at pH values over 9 (Lahaye et al., 1998). The ulvan-chitosan polyion complex-type gel was stable over a wide range of pH values. The reason for this may be that the polyion complex has many cross-linking groups (e.g., carboxyl groups and amino groups) in a single molecule. So, combining ulva and chitosan as a biohydrogel has side advantage.

**Experimental.** *Ulvalactuca*will be extracted by the combination of both alkali and hot water extraction. The aqueous solution of ulva extract will be mixed with acidic solution of chitosan to form hydrogel.Amino group on chitosan will react with polysaccharides from Ulva lactuca to form a hydrogel with the help of cross-linker (gluteraldehyde) which is intended to strengthen the bond of hydrogel.

Some tests should be done to know the characteristic and availability of this hydrogel as bioinjectable material. The tests consist of toxicity test, both in vitro and preliminary in vivo, which are held in the optimum condition of hydrogel, these test are on the process. The gel characteristic is obtained from the gelation ratio data or swelling ratio, and the gel morphology is obtained from SEM. To know the characteristic of ulva, there is moisture and ash content test.

**Results and Discussion.** Moisture content of Ulva: 19.33%; Ash Content of Ulva: 67.33%;

Table 1. Swelling Ratio Data

| Swelling Ratio (ulva:chitosan) | | 1:1 | 1:2 | 1:3 |
|---|---|---|---|---|
| Glutaraldehyde | Without | 32.58% | 34.48% | 52.63% |
| | With | 11.84% | 23.02% | 27.93% |

Thus, with the addition of chitosan, the swelling ratio is higher and the structure is better. With the addition of glutaraldehyde, the swelling ratio is lower but the structure is more rigid. Without the addition of glutaraldehyde, the swelling ratio is high, which means it can absorb a lot of nutrition, but its structure is weak. The specific function of ulva is coating the chitosan.

**Conclusion.** The idea to do research that utilizes the wealth of ulva and chitosan become hydrogel as bio-injectable material that has benefit to human.

**References.**

**Acknowledgement.** The authors thank to Yuni Kusumastuti, S.T.,M.Eng.,D.Eng. as supervisor of our research.

# 428383 Characterization and Reduction of Host Cell Protein Contaminants Via Mass Spec, Purification Optimization and Cellular Engineering

**Abstract:**

Host cell proteins (HCP) generated during cell culture processes to produce monoclonal antibody (mAb) therapeutics continue to be a contaminant in purified drug substance. Interactions between the HCPs and mAb products or purification matrices have made it difficult to remove these contaminants and result in significant downstream processing and ultimately increased production costs. Because we apply a platform cell culture process to our antibody programs, this provides a unique opportunity for characterization of the specific HCPs generated by our hosts and potential for engineering of the host cell line to reduce or eliminate expression of these proteins. To this end, we propose using mass spectroscopy to characterize the HCPs produced by our Chinese hamster ovary (CHO) host line expressing different platform antibody products. By comparing these products to material made using a null CHO host we will gain fundamental understanding of how different antibody expression affects the HCP profile. In addition, we will identify common HCPs which are non-essential to cellular function as candidates for cellular engineering, and perform proof of concept experiments to demonstrate reduction or removal of these HCPs in the culture supernatant. In parallel, studies optimizing purification conditions will be conducted to determine whether the process can be improved to reduce the HCPs without modification to the host cell line. We hope that this collaborative approach will not only strengthen our understanding of the cell's expression of HCP contaminants but also provide novel methods to remove some of the HCPs during downstream purification of monoclonal antibody products.

# 429182 Microcalorimetry Applied to the Study of Protein Stability (Rapid Fire)

**Abstract:**

Many different techniques are used currently to obtain a perfectly characterized, controlled and stabilized protein based drug. Amongst them calorimetry provides experimental data with a critical importance for the effect of the temperature on the unfolding and aggregation of proteins.

The microcalorimetry technique was used to study the unfolding and aggregation effects occurring in aqueous solutions of proteins. Simple data observation and comparison allows determining the most stabilizing formulation conditions, even in cases involving complex blends.

The thermally induced structural modifications of a protein do not only depend of its origin and extraction, purification conditions, but also of its environmental conditions like pH, salt content, or solvent type. The impact of formulation on the unfolding of RNase A was tested under different conditions. Tm temperature shifts to lower values in a presence of urea.

The β-lactoglobulin stability and aggregation was studied at different concentration, acidity and amount of NaCl. At pH 4 the protein is stabilized by adding sodium chloride. At neutral pH the mechanism is more complex and endothermic thermal effects are detected at high temperatures. They correspond to the depolymerization of the unfolded protein resulting in a formation of smaller peptides. The exothermic effects were attributed to Maillard reaction between the peptides and polysaccharide impurities contained in the material. In the presence of NaCl, an effect related to the protein aggregation is observed.

Additionally, the results from a thermal stability measurements of different gamma globulin proteins will be presented.

# 428900 A Hydrogel Platform to Understand Breast Cancer Metastasis to Bone Marrow

**Abstract:**


**A Hydrogel Platform to Understand Breast Cancer Metastasis to Bone Marrow**



Lauren E. Jansen and Shelly R. Peyton

Department of Chemical Engineering, University of Massachusetts Amherst



Metastasis accounts for 90% of cancer-related deaths. Breast cancer most frequently spreads to bone marrow tissue, resulting in painful osteolytic tumors that disrupt normal bone remodeling and production of lymphatic cells. While research efforts have provided genetic factors likely involved in breast cancer spread to bone, these findings are exclusive to the cancer cell and neglect contributions from the bone marrow tissue. We hypothesize that metastasis depends on both cancer cells and local cell and extracellular matrix (ECM) properties of the secondary tissue site [1-3]. I have designed a polymer hydrogel that mimics some aspects of human bone marrow tissue, which provides orthogonal control of scaffold stiffness, ligand presentation, and degradation. We have used this system to aid in isolating breast cancer cell phenotypes driven by local cells or matrix properties that may facilitate breast-to-bone metastasis.

Our system consists of a 4-arm star poly(ethylene glycol)-maleimide that encapsulates cells in a 3D microenvironment. Biochemically our system can be modified by crosslinking our scaffold with peptide sequences that degrade in the presence of cell-secreted enzymes, and by chemically incorporating peptides representing integrin-binding sites. Here, we have filtered for 12 peptide sequences that represent bone marrow tissue by capturing most of the unique integrin binding sites known to be present in marrow tissue[4]. We characterized the mechanics of porcine marrow using three different, but complementary techniques: rheology, indentation, and cavitation rheology[5]. Across techniques, the average Young's modulus of marrow is 4.4±1.0 kPa at physiological temperature. This stiffness and elasticity were recreated *in vitro* using our PEG hydrogel (20wt%, 4-arm 20 kDa PEG cross-linked with 1.5 kDa linear PEG). Both bone marrow and the PEG hydrogel followed a Hertzian model upon compression, validating the use of a PEG-based hydrogel for bone marrow tissue mechanics

Using a competitive binding assay, we have validated that hMSCs adhere to a select combination of these bone marrow peptides (peptides collagen I (GFOGER), fibronectin (PHSRN-RGD), osteopontin (SVVYGLR), and tenascin C (AEIDGIEL)) better than RGD alone. Additionally, an RNA profile (collagen type I, osteopontin, and osteocalcin) of hMSCs encapsulated in our hydrogel, with these same peptides and a tuned stiffness, showed that these cells are more comparable to hMSCs in native bone marrow tissue than hMSCs on tissue culture plastic. Using these hydrogels, we have screened for phenotypes across 4 patient hMSCs and 5 breast cancer cell lines. We found patient-specific phenotypes, like hMSC proliferation in response to some breast cancer soluble factors, and consistent phenotypes, such as increased breast cancer displacement in the presence of stromal cells. A high throughput bead based ELISA (Magpix) screened for 15 proteins, across 5 breast cancer cell lines and 4 human marrow-derived stem cells (hMSC), known to regulate breast cancer metastasis to the bone marrow. The highly expressed cytokines and growth factors have been isolated in both cell types and, to this end, we are working to identify the soluble factors and/or the tissue features driving these responses.

We have fundamentally quantified the mechanical and proteomic properties of bone marrow tissue, and used this knowledge to design an *in vitro* bone marrow model of the extracellular matrix. We have used this model to isolate phenotypes between cancer cells, local hMSCs, and bone marrow ECM that could be implicated in successful breast cancer cell adhesion and proliferation in secondary tissues. In sum, we demonstrate the benefit of using tissue-specific hydrogels to improve our understanding of breast cancer metastasis.



**References:** [1] Qian and Pollard, Cell. 2010. [2] Calvo, et al., Nature Cell Biology. 2013. [3] Coussens, et al. *Science.* 2002. [4] Uhlén et. al., *Science*. 2015. [5] Jansen et. al., *Submitted*. 2015. [6] Zhang et. al., *Cancer Cell*. 2009.

# 429194 *In Vivo* Continuous Evolution of Genes and Pathways in Yeast

**Abstract:**

Engineered microbial systems have the potential to solve pressing issues facing society, including the production of chemicals from renewable sources, degradation of environmental toxins, and the treatment of disease. However, the engineering efforts leading to these outcomes almost always require alterations to native host metabolism, often by modifying both enzyme activities and expression levels on a pathway scale. In the absence of detailed information about a biological system, directed evolution remains the most general method to generate improvements to performance, but is limited by low-throughput workflows and small library sizes, especially in workhorse eukaryotic systems such as yeast. Here, we establish an *in vivo* continuous evolution system in yeast by synthetically activating the endogenous retrotransposon Ty1 and optimizing the system to enable libraries of $10^8$ per liter per round of evolution by simply growing cells. This approach enables a rapid and generic increase in directed evolution throughput for synthetic parts, regulatory factors, single enzymes, and complete metabolic pathways. We demonstrate the utility of this approach by applying it to the engineering of the global transcriptional regulator Spt15p, the catalytic enzyme Ura3p, and a multi-gene xylose catabolic pathway, obtaining rapid and significant improvements to performance in each case. These improvements take place over days instead of weeks. Taken together, this work establishes a platform for rapid directed evolution in eukaryotes that is capable of identifying novel beneficial mutations to coding regions, regulatory elements, and entire pathways.

# 429691 Engineering Designer Glycopeptides As a Molecular Carrier for Cell Wall-Modifying Enzymes Expressed in Planta

**Abstract:**

Efficient conversion of lignocellulosic biomass to biofuels and biobased products has been hampered by recalcitrance of the plant cell wall to biochemical deconstruction. This project aims to leverage hydroxyproline (Hyp)-*O*-glycosylation – a process unique to plant cell wall glycoproteins – as an innovative technology for *de novo* design and engineering of novel "designer glycopeptide" (DGPs) that can facilitate plant cell wall reconstruction. This strategically designs Pro/Hyp-rich repetitive peptides, such as tandem repeats of "Ser-Pro" and "Ser-Pro-Pro-Pro-Pro" motifs as the DGP peptide backbone, which direct extensive Hyp-*O*-glycosylation in plants with arabinogalactan polysaccharides and oligoarabinosides, respectively. Such Hyp-*O*-glycosylated DGPs are proposed to be able to function as a *molecular carrier* for *in planta*-expressed cell wall-modifying (CWM) enzymes, targeting the enzymes to the cell wall matrix and stabilize the enzymes. Two types of DGP design, $(SP)_{32}$ comprised of 32 tandem repeats of "Ser-Pro" motif and $(SP4)_{18}$ comprised of 18 tandem repeats of "Ser-Pro-Pro-Pro-Pro" motif, was each engineered into tobacco plants as fusion with a reporter protein, enhanced green fluorescence protein (EGFP). The $(SP4)_{18}$ tag showed a better effectiveness on improving the accumulation of the tagged EGFP in cell wall matrix than the $(SP)_{32}$ tag. A thermostable CWM enzyme, E1 endoglucanase from *Acidothermus cellulolyticus*, was then engineered into tobacco as a fusion with the $(SP4)_{18}$ tag. The enzyme activity and subcellular localization of the enzyme, as well as phenotype and enzymatic digestibility of the plants were investigated. These DGP tags are also being engineered into the bioenergy crop−switchgrass to study their function as a molecular carrier in a monocot plant.

# 429730 Driving Potentials for Mass Transfer, Reaction Kinetics and Thermodynamics Define Syngas Fermentation to Produce Ethanol


**Abstract:**

Several acetogenic bacteria can produce ethanol from CO, $CO_2$ and $H_2$ supplied from synthesis gas. The production reactions are defined by the Wood-Ljungdahl pathway, and cell mass, ethanol, acetic acid and other chemicals are built from these inorganic molecules. Fermentation of the sparingly soluble gas substrates CO and $H_2$ requires robust gas to liquid mass transfer to sustain production through the pathway reactions. Further, the dissolved concentrations of CO, $CO_2$ and $H_2$ determine the thermodynamic state of the syngas fermentation and the kinetics of production.

CO inhibits hydrogenase activity in bacteria, so that uptake of $H_2$ concurrent with CO consumption in active fermentation indicates very low dissolved CO concentration ($c_{CO}$) and CO is mass transfer limited. Under mass transfer limitation ($c_{CO} \approx 0$) in a stirred tank fermenter (CSTR), the liquid film volumetric mass transfer coefficient for CO ($k_{L,CO}a$) is calculated from the observed inlet and effluent partial pressures of CO. Then $k_{L,CO2}a$ and $k_{L,H2}a$ derived from $k_{L,CO}a$ are used to calculate the dissolved concentrations $c_{CO2}$ and $c_{H2}$.

Observed conversion of CO and $H_2$ to acetic acid and ethanol confirms that each reaction in the biochemical pathway is thermodynamically favored and $DG_r < 0$. Concurrent uptake of CO and $H_2$ implies that the water gas shift reaction contained in the production pathway is in thermodynamic equilibrium, and the dissolved concentration $c_{CO}$ calculated using $c_{CO2}$ and $c_{H2}$ confirms the assumed mass transfer limitation. The dissolved gas concentrations ($c_{CO}$, $c_{CO2}$ and $c_{H2}$) in the water gas shift reaction represent the electrochemical half-cells for CO and $H_2$ oxidation, and define the electrochemical potential inside the cells at intracellular pH. All electrochemical reactions within the active cells can be characterized at this same intracellular pH and potential by the boundary condition $DG_r = 0$ assuming equilibrium of the reaction.

Reduction of free acetic acid through acetaldehyde to ethanol establishes the intracellular ratio of ethanol to acetic acid $c_{EtOH}/c_{HAc}$ as a thermodynamic parameter related to $c_{H2}$ and $c_{CO}/c_{CO2}$. The intracellular concentrations of ethanol and acetic acid cannot be measured directly. However, assuming that the acetate ion concentration is equal inside and outside the cell, as for assisted diffusion at equilibrium, intracellular $c_{EtOH}/c_{HAc}$ can be estimated from $c_{EtOH}$ and $c_{HAc}$ measured in the fermentation broth. Intracellular pH and potential can then be determined from measurements of the macroscopic fermentation.

Fermentation in a CSTR with continuous gas and liquid flow established a near steady state condition, while analysis using the derived mathematical model established a dominant thermodynamic determinant of ethanol versus acetic acid production with culture growth supported by either product. The CSTR was operated using automated process control for over 4000 hours, achieving an ethanol concentration of about 25 g/L. The ratio $c_{EtOH}/c_{HAc+Ac}$ was over 30, measured in the effluent fermentation broth representing moles of ethanol per mole of total acetic acid produced (free acetic acid plus acetate). Therefore, the ratio of $c_{EtOH}/c_{HAc}$ for free acetic acid in the extracellular broth was over 60 mol/mol at pH of 4.76. However, the model analysis using these conditions estimated ethanol to free acetic acid, $c_{EtOH}/c_{HAc}$, over 300 mol/mol inside the cells. The developed model is a tool for controlling agitation speed and gas feed rate to achieve efficient operation of syngas fermentation reactors with sustained conversion of CO and $H_2$ to the desired product.

# 429921 Optimizing Triacylglycerol Production Using the Crabtree Positive Fission Yeast, Schizosaccharomyces Pombe

**Abstract:**

The feasibility of biofuels as an efficient source of energy relies heavily on the efficiency of the organism to produce correct metabolites needed for biofuel production. Crabtree positive yeasts, which include *Saccharomyces cerevisiae* and *Schizosaccharomyces pombe*, direct a majority of available carbon towards ethanol, whereas traditionally considered oleaginous species such *Y. lipolytica* have minimal alcohol production. To date, the utilization of fission yeast as a biodiesel producing platform has yet to be fully investigated, so here we genetically engineered *S. pombe* to maximize triacylglycerol (TAG) synthesis. By overexpressing genes previously determined to enhance TAG yields as well as redirecting precursors of ethanol, we were able to achieve significantly higher triacylglycerol titers.

# 430592 Engineered Riboswitch: An Effective Way to Control Gene Expression in Cyanobacteria


**Abstract:**

Cyanobacteria are photosynthetic microbes that can derive energy from sunlight and carbon from atmospheric carbon dioxide, as well as other sources. In the past two decades, there were numerous efforts geared towards developing cyanobacteria as bioproduction platforms. The majority of these organisms are amenable to genetic manipulation, thus facilitating synthetic biology/metabolic engineering applications. The unique demands of living on light as a primary nutrient make the use of many established gene expression control systems difficult or impossible to apply. Among other approaches, researchers have used endogenous promoters such as *PpsbA2* and *PcpcB*, and synthetic riboswitches with limited/mixed success. Hence, developing an effective genetic toolkit to regulate gene expressions in cyanobacteria is of utmost importance. In this work, we combined a previously developed theophylline-responsive riboswitch with the T7 control system to drive expression of a yellow fluorescence protein (YFP) reporter in *Synechocystis* sp. PCC 6803. We then demonstrated superiority of this control system over available synthetic riboswitches and isopropyl-β-D-thiogalactopyranoside (IPTG) induced $lacI^{q}$-$P_{trc}$ promoter system. Thus, our work shows the utility of an engineered riboswitch system that can exploit the differences in light/dark life cycle and expedite biotechnological applications in cyanobacteria.

# 430048 Oligomer Formation of Toxic and Functional Amyloid Peptides Studied with Atomistic Simulations (Rapid Fire)

**Abstract:**

While amyloids have traditionally been associated with diseases such as Alzheimer's, it has recently been discovered that they play functional roles as well, including storage of peptide hormones. The differences between aberrant and functional amyloids remain unclear, but soluble oligomers formed during amyloid aggregation are known to be more toxic than the final fibrils. We have performed molecular dynamics simulations to study the aggregation of three different peptides: the amyloid-β peptide $A\beta_{25-35}$, which is associated with Alzheimer's disease, and two functional amyloid-forming tachykinin peptides, kassinin and neuromedin K. Although the three peptides have very similar primary sequences, the tachykinin peptides, unlike $A\beta_{25-35}$, form non-toxic amyloids. Our simulations suggest that the C-terminus charge is essential in controlling the aggregation process. In particular, if the kassinin C-terminus is not amidated, the kinetics slow down considerably. In addition, we observe that, in extended conformations, the monomeric peptides aggregate faster than the corresponding peptides in collapsed hairpin-like conformations.

# 431141 A Novel Soft Sensor Approach for Estimating Individual Biomass in Mixed Cultures (Rapid Fire)

Wednesday, November 11, 2015
Exhibit Hall 1 (Salt Palace Convention Center)
Kyle Stone, Chemical Engineering, Auburn University, Auburn, AL, Devarshi Shah, Chemical Engineering, Auburn Unviversity, Auburn, AL, Q. Peter He, Chemical Engineering, Tuskegee University, Tuskegee, AL, Jin Wang, Auburn University, Auburn, AL and Alejandro Zamora, University of Arizona, Tuscon, AZ

**A novel soft sensor approach for estimating individual biomass in mixed cultures**

Kyle Stone[1], Devarshi Shah[1], Q. Peter He[2], and Jin Wang[1]

[1]Department of Chemical Engineering, Auburn University, Auburn, AL, 36849, USA

[2]Department of Chemical Engineering, Tuskegee University, Tuskegee, AL, 36088, USA

<>Abstract:

Mixed cultures are ubiquitous in nature but their applications as biocatalysts in biochemical processes have been very limited. Until recently the application of mixed culture has been limited to wastewater treatment, but more diversified applications have started to emerge. For instance, Norferm Danmark A/S utilized a mixed culture of *Methylococcus* capsulatus (Bath) with heterotrophic bacteria to sustain growth on methane in a process to convert biomass into single cell protein for animal feed[1], and co-culture of xylose-fermenting strains have been paired with *Saccharomyces cerevisiae* to ferment lignocellulosic hydrolysate[2], [3]. Compared to single culture, there are several benefits of mixed culture. For example, a mixed culture can provide more versatile metabolic machinery due to a larger pool of collective genes, and tend to be more robust to different disturbances such as variations in feedstock composition and operation conditions. The main reason for the limited application of mixed culture approach is the difficulty associated with modeling, understanding and controlling of mixed culture.

To understand the metabolomic interactions of a mixed culture, quantitative analyses that can differentiate the dynamic properties of the mixed culture from those of monoculture are essential. One of the challenges in studying a mixed culture is to obtain the biomass concentration of each organism in the mixed culture. Cell count and fluorescent labelling have been used for microbial enumeration in mixed culture. But the former is time consuming and may not be accurate; the latter is also time consuming and may lack genetic tools and experimental protocols for some microbes. To address this challenge, we propose an absorbance spectra-based soft sensor to estimate individual biomass concentration in a mixed culture.

Soft sensor is commonly used in analytical chemistry and process industry to estimate variables (such as concentration) that are difficult to measure by using other easily obtainable measurements (such as absorbance spectra) [4], [5]. In this work, we propose to use absorbance spectra of the mixed culture broth, instead of the absorbance at a single wavelength, to estimate the biomass concentration of each individual strain. The proposed procedure is shown below:

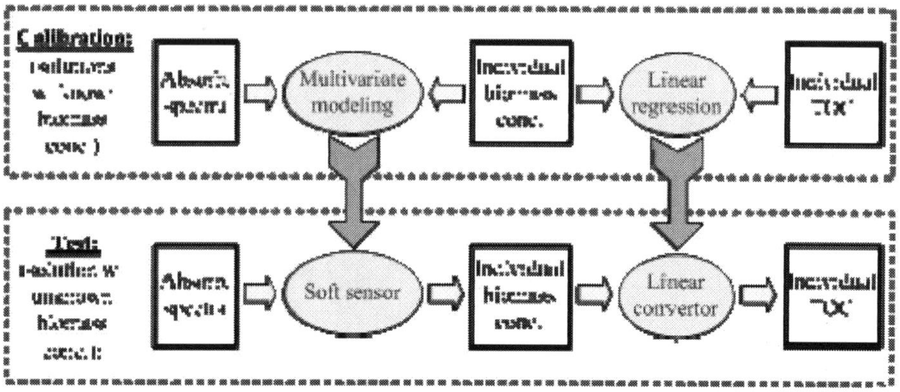


Case studies are carried out to show that the proposed soft sensor is highly accurate in estimating the individual biomass concentration from absorbance spectra of the mixed culture.

References:

[1] H. Bothe, K. M. Jensen, A. Mergel, J. Larsen, C. Jørgensen, H. Bothe, and L. Jørgensen, "Heterotrophic bacteria growing in association with Methylococcus capsulatus (Bath) in a single cell protein production process," Appl. Microbiol. Biotechnol., vol. 59, pp. 33–39, 2002.

[2] M. H. Kim, M. Liang, Q. P. He, and J. Wang, "A novel reactor for systematic investiagtion of co-culture systems," Bioresour. Technol., (submitted) 2015.

[3] M. H. Kim, M. Liang, Q. P. He, and J. Wang, "Simultaneous fermentation of glucose and xylose by co-culture in a novel bioreactor," in Proceedings of Sun Grant Initiative 2012 National Conference, 2012, pp. 97–101.

[4] H. J. Galicia, Q. P. He, and J. Wang, "Comparison of the performance of a reduced-order dynamic PLS soft sensor with different updating schemes for digester control," Control. Eng. Pract., vol. 20, pp. 747–760, 2012.

[5] H. J. Galicia, Q. P. He, and J. Wang, "A reduced order soft sensor approach and its application to a continuous digester," J. Process. Control., vol. 21, pp. 489–500, 2011.

# 431954 Dynamics of Aggregation of Proteins (Rapid Fire)

**Abstract:**

### Dynamics of Aggregation of Proteins

Size Zheng, Katherine S. Shing and Muhammad Sahimi

Misfolded proteins aggregate together and form large clusters. A normal protein typically has hydrophobic interior and shields itself with a hydrophilic exterior. When a protein misfolds, it exposes its inner hydrophobic portions to the outside, which may result in an inter-molecular interaction with other misfolded proteins, leading to aggregation. Protein aggregation is associated with a variety of human neurodegenerative diseases, including Alzheimer's, Parkinson's and prion disease. The aggregates' structural characteristics and molecular-level details are, however, hard to be studied by experimental methods and, thus, the mechanism of aggregations still remains poorly understood. But, advances in computational power and simulations, particularly, molecular dynamics (MD) simulations, provide the opportunity to simulate the aggregation process, hence helping us to unveil the secrets behind the phenomena and find the way to cure the diseases.

Previous works indicated that the proteins linked to neurodegenerative diseases, such as amyloid beta-protein to Alzheimer's disease and PrP prion protein to prion disease, only require a short portions of their entire chains, typically 4-10 residues in length, to begin aggregating. Such short portions contain high hydrophobic residues that are believed to drive the aggregation into fibrils. We have used discontinuous MD (DMD) technique to simulate the aggregation process of multiple 10-residue peptides into ordered fibrils, as the first step toward simulating the same in a crowded cellular environment. The DMD simulation uses coarse-grained peptides models and discontinuous stepwise potentials to provide dramatically faster simulation speed for big systems comparing with the classical continuous all-atom MD simulations, without sacrificing much of the necessary details.

We simulated multiple systems that have 8, 16 and 32 peptides. In each system we have simulated several cases at various temperatures, and investigated the characteristics of the aggregated protein clusters, including the number of hydrogen bonds and of the beta-sheets, the mean-square displacement and diffusion coefficient, as well as the radius of gyration, and the distribution of the cluster sizes. Interesting and physically relevant phenomena are observed, which will be reported.

# 431271 Engineer a Biosynthetic Platform to Branched Six- to Eight-Membered Lactones

**Abstract:**

Over the past several years, biological systems have been widely used for the sustainable production of several commodity chemicals, biofuels and more recently, for the production of bio-based polymers such as polylactic acid (PLA) and polyhydroxybutyrate (PHB). Although these bio-based polyesters show great potential, their stiff nature greatly limits their applications. As an alternative, substituted lactones have drawn great interest in the recent past as they generate a rubbery soft polymer block upon ring opening polymerization and further co-polymerization with hard PLA end blocks results in a triblock copolymer with mechanically tunable properties. Despite its numerous advantages, the ceiling temperature of six-membered lactone polymerization is around room temperature, making it unsuitable for hot melt polymerization. The goal of this project is to develop a biosynthetic platform in *Escherichia coli* for production of C6-C8 branched lactones. The advantage of higher lactones is that their higher ceiling temperature can allow high temperature polymerization reactions to be carried out directly in molds, thus avoiding the use of excess solvents.

# 432119 Engineering Antisense Transcription to Build Robust and Tunable Synthetic Genetic Switches

**Abstract:**

Antisense transcription has been extensively recognized as a regulatory mechanism for gene expression across all kingdoms of life. Growing evidence suggests the presence of non-coding *cis* antisense RNAs (AR) that regulate gene expression. Recent studies also indicate the role of transcriptional interference (TI) in regulating expression of neighboring genes due to traffic of RNA polymerases from adjacent promoter regions. While these mechanisms have been described to regulate gene expression in naturally occurring systems, the design principles for antisense transcription have not been explored yet. Here, we combine mathematical modeling and experimental approaches to engineer antisense RNA interaction, RNA polymerase traffic, relative promoter strength and promoter-to-promoter distance in a set of convergent promoters to develop a novel synthetic biology tool. We show that combination of TI and AR adds multiple-levels of regulation which allows such a system to have a highly tunable output in response to stimuli, ranging from a simple-first order response to biologically complex higher-order response such as bistable switch. Using a library of experimental constructs with engineered TI and AR by using face-to-face inducible promoters separated by carefully tailored overlapping DNA sequences we demonstrate tunable control of expression of a set of fluorescent reporter proteins. We show that even subtle changes in the physical arrangement of the convergent promoters can give rise to diverse gene expression patterns, which allows for design of highly-tunable synthetic genetic switches using minimum number of parts.

# 432168 Predictors of siRNA Functional Asymmetry (Rapid Fire)

**Abstract:**

RNA interference is a regulatory pathway that uses a small RNA trigger to direct gene knockdown through the destruction of mRNA containing complementary nucleotide sequences. The pathway can be initiated by exogenous short interfering RNAs (siRNAs), double-stranded RNAs with specific structures. For siRNAs to function properly they must be capable of base pairing with their target, meaning that one of the strands must be removed before forming the active complex. Proper strand selection is functionally critical to ensure targeting of only the intended mRNA target. The process of strand selection occurs through the orientation of the siRNA duplex within a multiprotein loading complex, followed by selection of one strand and destruction of the other. Once activated, this RNA-protein complex is known as the RNA induced silencing complex (RISC). However, not all siRNAs form highly active RISCs.

Focusing on known functional interactions of the proteins of the RNA interference pathway, we have characterized siRNA strand activity based on two criteria, the difference in the relative terminal hybridization stability ($\Delta\Delta G$) and the terminal nucleotide sequence classification (TN). Using a cell culture model, we investigated differences in relative siRNA strand activity, referred to as functional asymmetry, and differences in RISC loading. We found that the TN is a stronger predictor of siRNA functional asymmetry than $\Delta\Delta G$, but both characteristics do provide complementary information when predicting siRNA functional asymmetry. Specifically, $\Delta\Delta G$ has the strongest impact on functional asymmetry when siRNA strands have identical 5' terminal nucleotides. Our results also indicate that TN effects siRNA activity at the level of RISC formation and strand selection, similar to the literature, while, $\Delta\Delta G$ is less predictive of siRNA strand selection and therefore may have a larger influence on RISC activation and turnover. We will present our findings on siRNA functional asymmetry and the correlation of siRNA activity with siRNA strand selection.

# 432698 Mathematical Model of Nanoparticle Transport in Tumor Vessels Influenced By Radiofrequency Hyperthermia

**Abstract:**

The treatment for pancreatic and liver cancers has remained relatively unchanged for the last 40 years, with very little improvement in patient outcomes. Radiofrequency hyperthermia (RFH), or the use of electromagnetic fields to generate heat, combined with nanoparticle (NP) drug delivery has shown promise individually in the successful treatment of these two types of cancers. The question proposed by this work is how does RFH affect the transport and margination of nanoparticles? This question will be answered by focusing on mathematically modeling a rigid spherical nanoparticle moving through a blood vessel and the affect RFH has on it. The hypothesis is that RFH enhances NP margination in tumor blood vessels. Here I will present a literature review and preliminary model describing transport of a single, spherical nanoparticle in an electric field. The ultimate goal of my research is a model that will quantify the effect of RFH on NP margination and transport in a tumor blood vessel. Ultimately the results of this research can potentially enable the optimization and tailoring of treatment protocols to individual patients.

# 432306 Microbial Production of Chondrotin Sulfate

**Abstract:**

Chondroitin sulfate (CS) is an important glycosaminoglycan that predominantly presents in cartilage and surfaces of many cells and extracellular matrices. It has been widely used in treating osteoarthrithis due to its anti-inflammatory and chondro-protective properties. Given the importance of CS in medicine and as a dietary supplement, a traditional animal extraction method that is under potential risk of virus and/or prion contamination should be replaced by standardized bioengineering process. The capsular polysaccharide of pathogenic *Escherichia coli* K4 strain shares a similar structure to the CS precursor, chondroitin. Most of our effort has been focused on producing chondroitin or K4 polysaccharide. Overexpression of the biosynthesis pathway of K4 polysaccharide in a non-pathogenic production strain *E. coli* BL21*(DE3) results in ~2.4 g/L chondroitin in DO-STAT fed batch fermenter. Human Golgi chondroitin 4-sulfotransferase (C4ST1) catalyzes the transfer of a sulfo group to the 4-position of chondroitin. Functional expression of a soluble C4ST1 in *E. coli* BL21*(DE3) remains a challenge due to its sensitivity to the modification of N-glycosylation site. Recently, C4ST1 has been successfully expressed and purified resulting in a high level of conversion of chondroitin to CS type-A *in vitro*. The metabolic engineering of chondroitin biosynthesis and C4ST1 production into the same organism is underway to produce a bioengineered microbial CS as a promising alternative animal sourced CS.

# 432766 Microdroplet-Enabled Co-Cultivation and Characterization of Natural Microbial Communities

**Abstract:**

**Microdroplet-enabled co-cultivation and characterization of natural microbial communities**

Sida "Steven" Wang[1], Mark A. Burns[1,2], Xiaoxia "Nina" Lin[1,2]

[1]Department of Chemical Engineering, 3074 H.H. Dow, 2300 Hayward Street, Ann Arbor, MI 48109-2136

[2]Department of Biomedical Engineering, 1107 Carl A. Gerstacker Building, 2200 Bonisteel Blvd., Ann Arbor, MI 48109-2099

The majority of existing microbial species, in particular bacteria living in synergistic communities, have not been cultured and characterized in the laboratory. One important reason behind this "unculturability" is that conventional laboratory cultivation is aimed at pure cultures of individual species. This eliminates any potential interactions between different microbial species within a community necessary for growth of certain species. Another challenge comes from the sheer size and complexity of many microbial communities that require high-throughput cultivation, which conventional laboratory techniques cannot accomplish. We have developed a microfluidic platform utilizing aqueous-in-oil microdroplets for high throughput parallel co-cultivation of symbiotic microbial communities. We previously demonstrated the effectiveness of this cultivation platform using genetically engineered auxotrophic *E. coli* strains as our model system. In this work, we report new progresses in employing this conceptual and technological framework to isolate previously uncultured bacteria from natural microbial communities such as host-associated microbiomes.

Keywords: Droplet Microfluidics, Microbial Communities, Co-Cultivation, Unculturable Bacteria

# 432708 Dead-Box Helicase and Idp (intrinsically disordered domain) - Molecular Mechanisms By Single Molecule FRET (Rapid Fire)

**Abstract:**

Intrinsically disordered protein (IDP) motifs have a strong bias in their amino acid sequences, and also they are called as low complexity sequences (LCS). Interestingly, in DDX3 and DDX6 DEAD-box helicase family, ATP driven motor protein, has IDP, and the relationship between IDP and enzymatic function was not understood very well in the past. Here we show that three DEAD-box helicase, DDX3X, LAF1 and GLH-1 from human and C.Elegans has an interesting protein/RNA interaction as RNA enzymes. Those DEAD box helicases share few similarities - possess N-terminal IDP and a helicase core. Our results reveal that 1) they specifically bind single strand (ss) RNA and 2) dynamics of the RNP complex at certain condition. LAF-1 displays no unwinding of double stranded RNA, yet it promotes annealing of complementary ssRNA. ATP enzymatic reaction slows down the dynamics of RNP complex in LAF1. Series of truncation mutants in LAF1 reveal that the N-terminal IDP is responsible for providing the dynamic interaction in that. Together, we unravel the molecular mechanism how IDP motif can be regulated by DEAD-box helicase domain.

# 432828 Novel Tracking Peptide for Bio-Separation Processes

**Abstract:**

Development of a novel alternative visible spectrum tracking tool for bioseparation processes forms the motivation of this work. Current tracking tools such as spectroscopy (absorbance at 210 or 280 nm) or fluorescence are either non-specific or requires UV spectrum light. A novel peptide when fused to a recombinant product of interest is hypothesized to be useful for product tracking because of its small size and more importantly, visible spectrum signature. The poster focuses on the tracking features and discusses the methodology of its development. Chromatography is used as a representative unit operation of bioseparation. Breakthrough curves, often used for studying and optimizing column performance, were collected for model products that were extended by both the novel peptide and a fluorescent tag (control).

# 432845 Modeling Population Fluctuations in Nutrient Limited Environments (Rapid Fire)

**Abstract:**

*Trichodesmium erythraeum*, a colony forming filamentous cyanobacteria, is an integral part of the global nitrogen cycle; it is estimated that *T. erythraeum* fixes 42% of the biologically reduced nitrogen globally. Unfortunately, its metabolism and interactions with the environment and other marine organisms is poorly understood despite its ubiquitous presence in the ocean (*T. erythraeum* blooms can be so large that they can be seen from space). Detailed knowledge of this organism can provide insight into how nitrogen is made available in the ocean and how and why large blooms occur. Therefore, we have constructed a genome scale metabolic model of *T. erythraeum* which has been used to predict metabolic fluxes in a single cell. We will also present methods which allow this model to be extended to multicellular trichomes and colonies.

# 432904 Regulation of Carbon Metabolism and Lipid Accumulation in Yarrowia Lipolytica (Rapid Fire)

**Abstract:**

Regulation of Carbon Metabolism and Lipid Accumulation in *Yarrowia lipolytica*

Oleaginous yeasts are species that naturally accumulate high titers of lipid molecules[1, 2]. These are of particular interest because lipids and lipid derived molecules are attractive alternatives to petroleum based chemicals with applications ranging from fuels to food additives[3, 4]. The natural lipid accumulation, however, is known to occur when the growth media contains a high ratio of carbon to nitrogen. Furthermore, lipid accumulation in *Y. lipolytica* occurs only after a lag phase in which sugar is consumed but lipid is not produced. It has been shown that the deletion of the global AMPK type regulator Snf1 results in the abolition of the lipid accumulation lag phase as well as an increase in the amount of lipid produced[5]. We conducted studies on regulation of substrate transporters as well as enzymes in the lipid metabolism pathway to elucidate the mechanism by which lipid accumulation is controlled in *Y. lipolytica*.

1. Tai, M. and G. Stephanopoulos, *Engineering the push and pull of lipid biosynthesis in oleaginous yeast Yarrowia lipolytica for biofuel production.* Metabolic engineering, 2013. **15**: p. 1-9.

2. Ageitos, J.M., J.A. Vallejo, P. Veiga-Crespo, and T.G. Villa, *Oily yeasts as oleaginous cell factories.* Applied Microbiology and Biotechnology, 2011. **90**(4): p. 1219-1227.

3. Santos, T.M., A. Pinto, A.V. Oliveira, H.L. Ribeiro, C.A. Caceres, E.N. Ito, and H. Azeredo, *Physical properties of cassava starch–carnauba wax emulsion films as affected by component proportions.* International Journal of Food Science & Technology, 2014.

4. Moser, B.R., *Biodiesel production, properties, and feedstocks*, in *Biofuels*. 2011, Springer. p. 285-347.

5. Seip, J., R. Jackson, H. He, Q. Zhu, and S.-P. Hong, *Snf1 Is a Regulator of Lipid Accumulation in Yarrowia lipolytica.* Applied and environmental microbiology, 2013. **79**(23): p. 7360-7370.

# 433161 Construction of a Platform Saccharomyces Cerevisiae for the Production of Muconic Acid (Rapid Fire)

**Abstract:**

**Construction of a platform *Saccharomyces cerevisiae* for the production of muconic acid**

***Miguel Suástegui***[1], *John Matthiesen*[1], *Jean-Philippe Tessonnier*[1] & *Zengyi Shao*[1*]

[1] Department of chemical and Biological Engineering, Iowa State University

[2] Corresponding author (zyshao@iastate.edu)

Molecules deriving from the aromatic amino acid biosynthetic pathway have a great market opportunity. The chemical versatility of aromatic compounds can be exploited in various industries for examples for the production of polymers, pharmaceuticals, cosmetics, and nutraceuticals. In this work we focused on accumulation of 3-dehydroshikimate (DHS), a key aromatic compound used as precursor for the heterologous production of muconic acid. Flux balance analysis allowed constructing a platform *Saccharomyces cerevisiae* strain capable of accumulating the main precursors erythrose-4-phosphae (E4P) and phosphoenolpyruvate (PEP). Aided by 13-C metabolic flux analysis, the main rationale consisted on removing competing pathways, removing feedback inhibition, and overexpression of genes in the lower glycolytic node as well as in the non-oxidative part of the pentose phosphate pathway. The combination of these strategies and fermentation optimization allowed increasing the production of muconic acid around 5-fold of a previously reported *S. cerevisiae* strain. The biologically derived muconic acid was then electrochemically hydrogenated to 3-hexenedioic acid (3-HDA) at high conversion rates without the need for separation or cell clarification; potentially this molecule can be polymerized to obtain nylon-like polymers.

# 433251 Fabricating Surfaces for High-Throughput Sample Preparation and to Quantify Spatial Resolution in MALDI Imaging Mass Spectrometry (Rapid Fire)

**Abstract:**

Matrix assisted laser desorption/ ionization imaging mass spectrometry (MALDI – IMS) is powerful technology in medical, biological and material sciences. In a typical experiment, the tissue surface is coated with matrix, a small organic acid. Consequently, a laser pulse is directed on the tissue surface ablating and ionizing biomolecules. These molecules are separated based on their mass-to-charge ratio. The laser is rastered across the tissue surface to create molecular heat maps. Current techniques for studying molecular imaging such as immunohistochemistry and histology rely on specific targets requiring apriori knowledge of the biological specimen and monitor one molecular species at a time. In contrast, MALDI can detect hundreds of molecular species in a single experiment requiring no apriori knowledge of the tissue making it particularly useful for discovery.

Current instruments are limited to mass range of up to 20,000 m/z under typical operating conditions. Additionally, a MALDI spectrum consists of intact masses that tell us little about the identity of proteins. To address these challenges, the analytes within the tissue can be enzymatically digested and the resulting peptides can be identified using tandem mass spectrometry. Current methodology employs robotic spotters to deposit trypsin in spatially discrete regions followed by matrix deposition. This process requires many hours for a single tissue specimen and the spatial resolution is limited to 200 μm. We have developed MALDI targets pre-coated with trypsin and matrix, where tissue is mounted onto these coated surfaces.

The pre-coated targets are developed for both fresh-frozen tissues and formalin-fixed paraffin-embedded (FFPE) tissues. Gold slides are uniformly coated with alpha-cyano-4-hydroxycinnamic acid by solvent casting and subsequently spray coated with trypsin. Three micrometer tissue sections are sliced and thaw mounted on top of the pre-coated targets. The slide is then placed in a chamber saturated with N,N-diisopropylethylamine (DIEA) isopropylethylamine and water that converts the matrix into an ionic form with neutral pH of 7-8. After digestion for 2 hours at 50 °C, the sample is washed with 10% tri-fluoroacetic acid and analyzed with MALDI IMS. For FFPE tissue sections, the wax is first removed using a xylene wash and the rest of the process is similar.

We have successfully used pre-coated methods for protease digestion of fresh frozen tissues, reducing sample preparation time and improving spatial resolution 2-fold down to 100 μm. Thin tissue sections from rat brain were analyzed. Several proteins were identified including Myelin Basic Protein, PEP-19, Neurogranin and others. For each parent protein, several peptides were observed in the MALDI FT-ICR spectrum and matched to their theoretical mass to less than 1 ppm. Furthermore, each of the matched peptides for a specific protein was co-localized enabling high confidence in the identifications.

In addition to pre-coated surfaces, we are also fabricating synthetic patterns to evaluate instrument performance particularly spatial resolution. The standard reticle slides with a synthetic pattern of a laser desorbable compound can accurately measure laser spot sizes at multiple laser fluences, assess the positional accuracy of ablation spots on a target, and measure the reproducibility and accuracy of scans across the targets. This work also includes production of standard protocols and user friendly software tools to accomplish these tasks in the biological and clinical laboratory. Lines ranging from 10 – 100 μm are fabricated. These patterns are designed in computer-aided software and printed as masks that are used to fabricate photoresist masters using photolithographic techniques. PDMS molds are produced using soft-lithography.

The PDMS stamp is inked with a 2 mM solution of hexadecanethiol, and placed on a gold slide to selectively functionalize the gold surface yielding hydrophilic and hydrophobic regions. The substrate is immersed in a solution of crystal violet causing crystal deposition in pre-defined hydrophilic areas. The patterned surfaces are imaged using an Ultraflextreme MALDI TOF/TOF and FT-ICR instruments. The data are analyzed using MATLAB tools to quantify the spatial resolution as measured from the isolated spots on the reticle. Overall, the objective is to develop of a standard reticle to quantify MALDI imaging spatial resolution in the range of 10-100 um ± 10%.

We have also tested the patterned organic material to evaluate delocalization by different matrix application methods, comparing an excessively wet spray with a dry application process (sublimation). One slide was sprayed with sinapinic acid in 50% acetonitrile using a solvent sprayer. A second slide was coated with the same matrix by sublimation. The two slides were imaged using a Bruker Ultraflextreme MALDI TOF instrument with a raster step size of 50 μm. The ion images show that the solvent based spray application caused extensive delocalization while the sublimated sample preserved the pattern. This shows the importance of carefully controlling the surface wetness with solvent spray devices and the need for patterned slide to evaluate the quality of the matrix coating.

In summary, the pre-coated slides standardize the sample preparation for tissue specimen removing the burden of sample preparation from the end user. The reticle slides can evaluate the instrumentation performance particularly its spatial resolution. Combined these new methods can ensure standardized and reproducible performance so that MALDI IMS can be readily adapted in the clinical environment.

# 433623 Screening Signal Peptides for Heterologous Protein Secretion in *Geobacillus Thermoglucosidans*

**Abstract:**

*Bacillus* is widely used in industry for enzymes production because they are capable of secreting proteins at high concentration. We identified a related thermophile *Geobacillus thermoglucosidans* capable of secreting proteins at high titers and then engineered it to secrete heterologous protein. We examined the native capacity of three strains (*G. thermoglucosidasius* BGSC 95A1, BGSC 95A2, and C56-YS93) to secrete protein on different carbohydrate substrates. They were able to secrete grams per liter of protein on C5 sugars, which was much higher than on glucose and cellobiose. The highest yields are observed during growth on xylan with strain C56-YS93. We have also performed shotgun proteomics to determine and compare which proteins are secreted by strain C56-YS93 during growth on different substrates (xylan, xylose and cellobiose). Based on this analysis, the signal peptides (SPs) for highly secreted proteins were identified. Additionally, SPs were analyzed by prediction algorithms (SignalP 4.1) for strong signals and cleavage sequences. By using a thermostable amylase from *G. stearothermophilis* as a reporter, the top 25 SPs from prediction result were characterized for their potential for heterologous protein secretion.

# 433796 Enzyme Encapsulation Using Cell Free Protein Synthesis

**Abstract:**

### Enzyme Encapsulation using Cell Free Protein Synthesis

Seung-Ook Yang, Bradley C. Bundy

### Abstract

Nature has evolved to bear a micro-compartmentalization strategy for a number of various functions, such as protection and delivery of cargo molecules, regulation of chemical reaction, assisting proper folding of proteins, etc. These characteristics interest many biotechnologists to utilize viral and non-viral capsids to mimic nature for many applications. To date, a small number of viral capsids have been used to encompass enzymes to build nanoreactors. For example, Bacteriophage Q-beta's coat protein self-assembles into a 180-membered virus capsid. Recent studies show this capsid has an ability to encapsulate enzymes when co-produced. Substrates and product diffuse through pores in the capsid, while the encapsulated enzyme is protected and stabilized from biologically harsh conditions. Here we demonstrate the utility of a cell free protein synthesis (CFPS) environment for improved control over the encapsulation of enzymes. CFPS is *in vitro* protein synthesis where reagents for protein syntheses can be directly controlled, cytotoxicity is no longer an issue, and expression/folding rates can be tuned and engineered. Here we report the encapsulation of Lipase B from *Candida antarctica* in the Q-beta virus capsid using CFPS.

# 433849 Using Sacrificial DNA to Improve LET-Based Cfps Protein Yields

**Abstract:**

Using Sacrificial DNA to Improve LET-Based CFPS Protein Yields

Andrew Broadbent, Bradley C. Bundy

Brigham Young University, Provo, Utah, USA

AIChE 2015 Conference Abstract

Proteins—polymers of amino acids—are a major class of biomolecules whose myriad functions facilitate many crucial biological processes. Accordingly, human control over these biological processes depends upon the ability to study, produce, and modify proteins. One innovative tool for accomplishing these aims is cell-free protein synthesis (CFPS). This method, rather than using living cells to make protein, simply extracts the cells' natural protein-making machinery and then uses it to produce protein *in vitro*. Because living cells are no longer involved, scientists can freely adapt the protein production environment in ways not otherwise possible. However, improved versatility and yield of CFPS protein production is still the subject of considerable research.

This work seeks to improve CFPS protein yield by preventing the degradation of template DNA in CFPS. Among the advantages of CFPS is the option of using linear expression templates (LETs) in place of plasmids as the DNA template for protein production. Because LETs can be produced more quickly than plasmids can, using LETs greatly reduces the time required to obtain a DNA template for protein production. This renders CFPS a better candidate for high-throughput testing of proteins. However, LETs are more susceptible to enzyme degradation than plasmids are, which means that LET-based CFPS protein yields are lower than plasmid-based CFPS yields. This work explores the possibility of increasing the protein yield of LET-based CFPS by addition of sacrificial DNA, DNA which is not used as a protein-making template but which is degraded by the enzymes in place of the LETs.

# 432795 Ex Vivo Attenuation of Human Mesenchymal Stem Cell Senescence Via Metabolic Regulation


**Abstract:**

### Introduction

Cellular senescence is a form of permanent growth arrest due to consistent stress stimuli. While the decline of regenerative potential is a hallmark of aging, stem cell decline due to cellular senescence can disrupt tissue homeostasis and repair. As a primary cell source in cell therapy, human mesenchymal stem cells (hMSCs) isolated from elderly donor have reduced regenerative capacity associated with increased levels of senescence and senescence-associated changes. Moreover, this age related changes of hMSCs appear to be similar or even more pronounced during in vitro expansion. The lost of stem cell properties due to senescence/aging may reduce the therapeutic outcome, but more importantly may have deleterious effect on recipients after transplantation. Recent studies has suggested that in vitro senescence of stem cell is resulted from hyper-stimulation of nutrient sensing pathway from growth factors and nutrient that force cells into premature, and loss of quiescence. Understanding the metabolic regulation governing in vitro proliferation and quiescence has important implication in optimizing hMSCs properties while preserving their multipotent phenotype. Our prior studies have shown that glycolytic energy metabolism significantly regulates clonogenic MSCs proliferation and maintenance of multi-potency. The objective of this study is to investigate the metabolic regulation between hMSCs proliferation and quiescence during in vitro expansion, with the aim to develop a metabolic approach to preserve the multipotent state of hMSCs and consequently maintain their regenerative capacity.

### Materials and Methods

Standard hMSCs culture protocol was used to maintain hMSCs in continuous culture from passage 5 to passage 9. Metabolic modulators such as dehydroepiandrosterone (DHEA), 2-deoxyglucose (2-DG), and antimycin A (AmA), were used to selectively modulate the specific metabolic pathways and evaluate their regulatory roles.

### Results

The results revealed that cellular senescence of high passage hMSCs is significantly attenuated by inhibiting the rate-limiting enzyme of pentose phosphate pathway (PPP), glucose-6-phosphate dehydrogenase (G6PDH) via treatment of inhibitor DHEA at a concentration of 200uM for a period of 4 days. hMSCs treated with DHEA at passage 9 have significantly lowered staining and enzyme activity of senescent marker β-gal compared to hMSCs at the same passage without DHEA treatment. DHEA treated hMSCs retain regenerative capacity after continuous passaging, indicated by higher proliferation rate than cells without treatment, after being re-expanded into DHEA free culture medium. The attenuation effect of DHEA on hMSCs senescence is associated with cellular quiescence that DHEA treatment significantly decreased cell proliferation and growth, indicated by lower cell number fold increase and smaller cell size. The senescent attenuation effect by inhibiting PPP enzyme G6PDH is further associated with metabolic reconfigure towards a more oxidative status. DHEA decreased the mRNA level of glycolytic enzymes, including LDHA, and PKM2, while increased oxidative phosphorylation indicated by lower mRNA level of PDK1, higher cellular ATP content, and ROS level. Together, the results revealed the mechanistic connection between metabolic regulation and hMSCs in vitro quiescent state, and demonstrate that regulation of hMSCs metabolism as a strategy in preserving stem cell properties during expansion.

### Conclusions

The accumulative changes of hMSCs metabolic phenotype during expansion significantly influence hMSCs phenotypic properties, such as proliferation, quiescence, and multi-lineage differentiation. Understanding the metabolic mechanisms governing hMSCs fate decision can contribute to a better strategy to maintain stem cell properties in hMSCs expansion and application.


### References


Grayson, W.L., Zhao, F., Bunnell, B., Ma, T., Hypoxia enhances proliferation and tissue formation of human mesenchymal stem cells. Biochemical and Biophysical Research Communications, 2007. 358(3): p. 948-953.3.

Munoz, N., Kim, J., Liu, Y., Logan, T.M., Ma, T., Gas chromatography-mass spectrometry analysis of human mesenchymal stem cell metabolism during proliferation and osteogenic differentiation under different oxygen tensions. Journal of Biotechnology, 2014. 169: p. 95-102.

# 433187 Ectopic Expression of NANOG Restores the Actin Filamentous Organization and Contractile Capacity of Senescent Cells

**Abstract:**

Actin filamentous organization is critical for mediating cellular contraction. In the present work we hypothesized that cellular senescence decreases actin polymerization whereas ectopic expression of Nanog completely restores actin reorganization. To address this hypothesis, human hair follicle Mesenchymal Stem Cells and human dermal fibroblasts were isolated and transduced with a tetracycline regulatable vector that carries the Nanog gene. In this system the transgene is expressed only in the presence of the tetracycline analogue, Doxycycline. To induce cellular senescence, Early Passage (EP) carrying the Nanog gene were serially passaged in the absence of Doxycycline until they reached Late Passage (LP) or senescence. Subsequently Doxycycline was added to the culture medium (LP NANOG). Genome wide transcription analysis in EP, LP and LP NANOG revealed that upon senescence several components of the actin cytoskeleton pathway were affected. In agreement with this, we report that senescent cells failed to polymerize actin in respond to myogenic stimuli as evidenced by the complete absence of ACTA2 fibers as well as the weak F-Actin staining. The reduced filamentous actin was also associated with reduced contractile potential. Interestingly, Nanog expression in senescent cells restored the actin polymerization and the force generation capacity. To further elucidate the mechanism through which NANOG elicits its function, we studied the ROCK/ MRTF-A and the TGF-beta pathways, which are known to regulate the contractile phenotype. Interestingly, NANOG increased key components of the ROCK pathway such as ROCK1 and LIMK2, and activated the TGF-beta pathway by upregulating SMAD2. We also show that upon ROCK activation MRTF-A translocated into the nucleus and activated SRF-dependent myogenic gene transcription. Last but not least, inhibition of the ROCK and/or the TGF-beta pathway with either chemical inhibitors or shRNA abolished actin polymerization and myogenic differentiation. Taken together, our data provide novel insight into the mechanism of cellular senescence and propose a novel strategy to regain the lost properties, thereby increasing the potential of adult cells for tissue regeneration.

# 434205 Design, Selection and Optimization of Affinity Peptides for Applications in Bioprocessing (Rapid Fire)

**Abstract:**

Affinity ligands are employed in a range of applications as drugs, drug carriers, purification and imaging tags, biosensing agents, inhibitors of biomolecular interactions and modulators of crystal growth. The specific design and engineering of an affinity ligand depends on the desired application. Typically, the design of affinity ligands for different applications is guided by one or more end-points of affinity, selectivity and biological activity. While high-affinity and selectivity are often the governing design criteria, it is not always the case since certain challenges require additional criteria to be met. In such cases, affinity ligands often need to be tailored to simultaneously possess multiple functionalities and biochemical properties. Peptides are an attractive class of affinity ligands well-suited for such applications due to their amenability to rational design and modifications, ease of manufacture, low-cost, high stability and prolonged shelf-life. In this work, we will demonstrate the rational design and engineering of affinity peptides to meet one such challenge encountered in the bioprocessing industry.

Current bioprocessing strategies for the production of non-antibody based biologics are costly and require lengthy purification trains. While Protein A chromatography has emerged as the "gold-standard" step offering robust, high yield and high selectivity, affinity-based purification of monoclonal antibodies (mAbs), no affinity ligands currently exist for non-mAb based biologics. Further, non-affinity methods typically cannot match the process yields, effectiveness, reduced time and costs offered by affinity chromatography. By employing (and developing) a repertoire of design and screening strategies available in a biomolecular engineer's toolbox, our goal is to address this shortcoming and create, against any given biologic, robust, highly selective, low-cost, affinity peptides that possess all the desired properties essential for protein purification. These properties include not just high affinity for their target, but also high selectivity (in a milieu of host-cell proteins and impurities encountered with changing product titers and variable feed streams), sufficient dynamic binding capacity at the column scale (when immobilized on a support to create an affinity adsorbent), high elutability (to enable product dissociation and recovery), high stability (i.e. resistance to base and proteolytic cleavage) and good regenerability (for use over multiple purification cycles).

In this talk, we will focus on the design and optimization of affinity peptides for the purification of two key biologics – human growth hormone (hGH) and interferon α2b (IFNα2b). Using a diverse array of peptide design and screening strategies such as epitope mapping of natural binding partners, *de novo* design and *in silico* screening, mutational analysis, and cyclisation, we have designed libraries of peptides targeting both hGH and IFNα2b. We have tested these peptide libraries under different conditions and on various platforms (microarray screening and in-solution binding assays using fluorescence polarization, followed by batch-scale, resin screening experiments) to identify lead candidates that were carried forward for optimization at the column-scale. Some of the key results of this work are as follows:

- The designed peptide libraries exhibited a range of affinities, selectivities and elution characteristics for the target biologics, thus enabling the selection of the most optimal peptide ligands for the purification of recombinant hGH and IFNα2b.
- We found that short stretches of naturally-occurring amino acid sequences can be effectively re-designed to incorporate and display functionalities not intended for them by nature. For example, a peptide epitope involved in protein-protein complex formation in nature can be engineered by point-mutagenesis to display enhanced selectivity in hGH purification.
- Further, the rational insertion of specific amino acids at the front or tail ends of the same epitope yielded candidates with enhanced capacities and very different elution profiles for hGH. Interestingly, we also find that some peptides can be tailored to impart specificity in purification for a given biologic or a set of biologics.
- Finally, our results with a top-performing, hGH-binding peptide obtained after two rounds of design and optimization, demonstrate that affinity peptides can offer purification performances (good product recovery, effective DNA and host-cell protein clearance) comparable to those obtained in Protein A chromatography.

Our preliminary studies with the design and selection of IFNα2b-binding peptides have yielded several promising candidates that are now being optimized to provide enhanced purification performance. Based on the lessons learnt from these studies, our ongoing and future efforts will focus on engineering elutability in the ligands *a priori* and incorporation of non-natural amino acids and cyclisation to further optimize the specific molecular recognition and biochemical properties of these affinity peptides. Overall, this work demonstrates the potential impact and power of rational design of affinity peptides on downstream bioprocessing of therapeutic proteins.

# 434275 Converting a Thermostable Enzyme into a Binder of Explosive Molecules (Rapid Fire)

**Abstract:**

**Converting a Thermostable Enzyme into a Binder of Explosive Molecules**

Jennifer Haghpanah[1] and Scott Banta[1]

[1]Department of Chemical Engineering, Columbia University

The search for proteins that bind to small molecules is driven by the potential to develop sensors applications such environmental monitoring. Antibodies are generally utilized for molecular recognition; however, they can be difficult to generate and are generally unstable. By evolving a thermostable enzyme, (alcohol dehydrogenase (AdhD) from *Pyrococcus furiosus*) into a selective binding protein, we aim to overcome the problems associated with yield, purity, stability, and cost of antibodies. Cyclotrimethylenetrinitramine (Royal Demolition Explosive, RDX) is a harmful chemical and a common military-grade explosive that contaminates many environments around the world. The current detection methods for this explosive are expensive and time consuming. We are evolving the AdhD enzyme into a binder with affinity for RDX. Multiple AdhD libraries were generated by randomizing sites within and outside the cofactor binding pocket. Ribosome display techniques were employed to identify library candidates demonstrating an affinity towards RDX. To further investigate where the binding was taking place, we truncated the enzyme and confirmed the importance of the cofactor pocket for the binding event. After multiple rounds of selection, clones have been identified from several libraries and we are currently introducing diversity to our converged sequences to obtain a high-affinity binder. Promising candidates are being screened via enzyme-linked immunosorbent assay (ELISA) and isothermal calorimetry (ITC). The assay method ensures the mutants retain their thermal stability which will make them superior for the development of new biosensors for RDX detection.

# 434388 Engineering *Pseudomonas Fluorescens* for Efficient Bioseparation

**Abstract:**

*P. fluorescens* is slowly gaining popularity as a protein expression platform principally due to its low endotoxin potential and relative ease of cultivation. However, little is known about the proteome of this organism as it relates to downstream processing, as well as how targeted alterations can be used to mitigate the expression of nuisance proteins. To this end, this poster describes our effort to enhance the bioprocessing potential of *P. fluorescens* through rational host cell engineering. Host cell proteins were identified within the proteome that are retained strongly by ion exchange chromatography media, the data on which was used to guide recombineering efforts. Lack of an identified CRISPR system prompted our group to import elements of this mechanism for gene targeting in pseudomonas, used to delete or tune repression of proteins which were demonstrated to reduce column capacity and complicate gradient elution.

# 434398 A User Friendly Simulator for Antibody Producing Hybridoma Cell Cultures

**Abstract:**

**A User Friendly Simulator for Antibody Producing Hybridoma Cell Cultures**

Denizhan Yilmaz, Satish J. Parulekar, and Ali Cinar

Department of Chemical and Biological Engineering

Illinois Institute of Technology, Chicago, IL 60616

(312) 567-3044 (voice), (312) 567-8874 (fax), parulekar@iit.edu

**Abstract**

The demand for monoclonal antibodies (MAb), produced mostly from hybridoma cells, has proliferated vastly over the last two decades due to their therapeutic and diagnostic applications. The cultivation and development of methods for optimization of cell growth and cell productivity have become a crucial issue to enhance product yield in vitro production. The efficiency and performance of the cell culture systems depend on media optimization, process monitoring and control strategies. Creating these strategies requires understanding of process dynamics affected by cell mechanism in culture environments. Although the experiments provide data to discover underlying biological mechanisms and validate relationships between important cellular variables, acquiring necessary data on the dynamics of cells growth necessitates a broader amount of work. Moreover, the cost of these experiments accumulates with each trial, making these discoveries even more financially intensive. Mathematical models predicting cell growth behavior are becoming more necessary to reduce expensive trial and error experiments. Kinetic models are crucial for mammalian cell cultures as these not only reduce cost but also provide better insight into the process. Kinetic models provide better understanding of cell physiology for optimization and control of animal cell cultures. Demonstrating the most convenient experiments and pointing to strategies to improve process performance, simulation studies enable researchers to eliminate expensive and unnecessary time-consuming experiments. For this reason, we have developed a simulator for investigators without specific expertise in modeling to predict cell growth, cell death, target metabolite productivity, nutrition depletion, and waste product accumulation for different initial conditions, feeding strategies and bioreactor operation modes, such as batch, fed-batch, and perfusion (CSTR). The Matlab-based simulator is designed as a multifunctional tool incorporating five different kinetic models available in the literature for monoclonal antibody (target metabolite) production by mammalian cells. The user friendly graphical interface allows the user to specify and alter culture conditions - viable and non-viable cell densities, nutrient (glucose and glutamine) levels, and levels of inhibitory waste products (ammonia and lactate) of cell metabolism. For fed-batch and continuous cultures, the user can specify feeding strategies. The user can experiment by using different kinetic models, adjusting medium composition, and changing feeding strategies to observe their effects on cell growth and cell productivity.

# 434342 Synthesis of Well-Defined Protein-Polymer Conjugates Using Ligase-Mediated Site-Specific Modification and RAFT Polymerization

**Abstract:**

Advances in materials synthesis using proteins have allowed for the preparation of a variety of protein-polymer materials that find numerous applications in medicine, synthesis, and environmental monitoring. Protein-polymer conjugates, or bioconjugates, may be designed with linear and branched hydrophilic polymer such as PEG or may be combined with hydrophobic or crosslinked polymers resulting in insoluble supports in the form of gels, foams, fibers, wipes, and coatings. Synthesis of protein-polymer materials involves attachment of polymer chains or initiators for polymer synthesis to the reactive functionalities on the enzyme. Typically, protein modification approaches provide little control over the number and location of modification sites on the protein surface. Approaches that allow site-specific and bioorthogonal chemical modification of proteins are needed for the design of well-defined protein-polymer conjugates. We have used a novel approach to incorporate bioorthogonal reactive handles for site-specific protein modification that uses ligase-mediated post-translational modifications. This approach uses the enzyme lipoic acid ligase (LplA) to modify a protein, which contains a unique peptide tag (LAP), with clickable azide groups that may then serve as sites for polymer attachment. In this work we investigate the potential utility of designing uniform protein-polymer conjugates using the LAP/LplA system and RAFT polymerization. We have synthesized an alkyne terminated chain transfer agent; however, to improve the solubility of this RAFT agent a short macro-CTA of acrylamide was grown in solvent first. The macro-CTA was subsequently conjugated to the ligated GFP in a buffer solution using the copper catalyzed click reaction. The macro-CTA GFP conjugates were chain extended with functional monomers to create a library of bioconjugates with different insertion sites, chain lengths and chemical functionalities. These bioconjugates were characterized to determine the number CTA of attachments, polymer length, and stability.

# 434693 Investigating Neural Control of Epithelial Barrier Function of the Small Intestine with an in Vitro Engineered Model System

Wednesday, November 11, 2015
Exhibit Hall 1 (Salt Palace Convention Center)

**Marissa Puzan**[1], Daniel Ventre[2] and Abigail Koppes[1], (1)Chemical Engineering, Northeastern University, Boston, MA, (2)Therapeutic Innovations, Boston, MA

**Introduction:**

The epithelial cell barrier of the intestine controls the absorption of metabolites from the intestinal lumen. Improper function of this barrier can lead to inflammation and a 'leaky gut', where tight junctions do not form between enterocytes and toxins or bacteria may pass between these cells and into the body. This dysregulation plays a critical role in intestinal diseases such as irritable bowel disease (IBD), which affects up to 15% of the population of the United States (1).

Current *in vitro* models of the epithelial cell barrier typically consist of polarized epithelial colorectal adenocarcinoma cells (Caco-2) grown on a polymer transwell insert (2). Barrier function may be studied in various conditions based on the tight-junction resistance and flux through the transwell membrane. These models fail to consider the extensive neural network that lines the gastrointestinal tract (GI), the enteric nervous system (ENS), which modulates intestinal secretion and motility (3). Enteric glia, the support cells of the enteric nervous system, have also been reported to impact epithelial survival and maturation (4). Therefore, it is the hypothesis of this work that a co-cultured model consisting of both Caco-2 and enteric neurons and/or glia will provide a more realistic *in vitro* model of the epithelial cell barrier and will help to elicit the role of the enteric nervous system in barrier function regulation of healthy tissue and disease states.

**Materials and Methods:**

A mixed culture of enteric neurons and glia was isolated from 10 week old, adult black 6 mouse small intestine, (5) and seeded onto laminin coated glass cover slips in 24 well plates. Neurons were grown under standard incubated culture conditions in Neurobasal media supplemented with NGF and GDNF to promote neural survival in vitro.

At day 8 of neuron culture, Caco-2 were seeded onto 0.4 um PET transwells at approximately 60,000 cells/cm$^2$. The neurons remained on coverslips beneath the transwell. Control cultures were Caco-2 with no enteric cells, and enteric cells with no Caco-2. NGF/GDNF Supplemented Neurobasal media was used for both cell lines. After 5 days of coculture, TEER (transepithelial electrical resistance) measurements were taken with an EVOM from World Precision Instruments (in PBS, 37°C), and then both neurons and Caco-2 were fixed and stained. Enteric neurons were stained for Beta-III-tubulin to label neurites and nuclei (DAPI); Caco-2 were stained for ZO-1, a tight junction associated protein, and nuclei (DAPI). Cells were mounted on glass slides and imaged using an inverted Olympus fluorescent microscope.

**Results and Discussion:**

Enteric neurons and Caco-2 were cocultured with the Caco-2 in transwells and neurons on cover slips in the bottoms of the wells. The cells lines were kept spatially separate, but shared media and could thus interact through excreted metabolites and soluble proteins. Preliminary results show that the addition of enteric neurons to Caco-2 increases the transepithelial resistance and thus tight junction formation. After 5 days of culture, TEER measurements for the cocultured Caco-2 (Fig 1) were 270.8 +/- 3.9 ohms, while those of the Caco-2 alone were 238.6 +/- 10.0 ohms, a ~14% increase in a short duration culture. This indicates that soluble factors released from the enteric neurons impact the initial formation of tight junctions with this cell line.

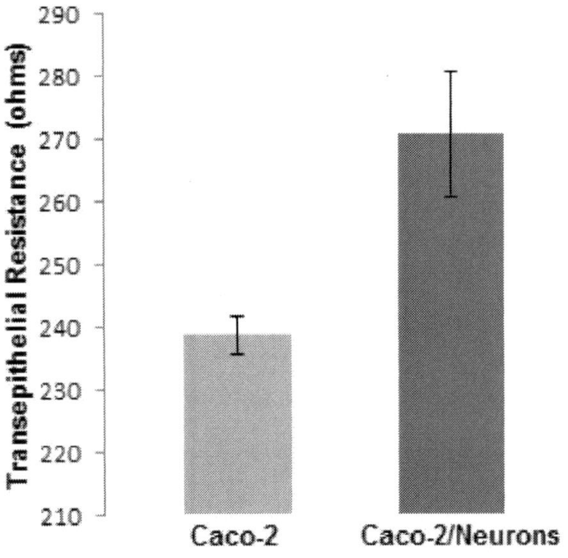


Figure 1: Transepithelial Resistance (TEER) for co-cultured Caco-2 and enteric neurons is 14% higher than the control after 5 days of culture

**Conclusions:**

These preliminary results imply a role of the enteric nervous system in the regulation of epithelial barrier function and flux across the epithelium. Further work will investigate the formation of tight junctions with coculture for an extended length of time (10-28 days) and determine if the addition of neurons and enteric glia to an *in vitro* model impacts transepithelial flux. Additionally, the epithelial cell line used in the coculture will be changed to include primary small intestinal stem cells (6) and/or mucous producing cells such as HT-29 (7) to determine if neural inclusion impacts barrier function. Neurite extension and functionality will be examined via neural tracing software and electrophysiology. This platform may provide mechanistic insight into the cross-talk between the epithelium and the enteric nervous system, which may lead to a more realistic pharmacological test bed that recapitulates the complex biological environment of the small intestinal niche.

**Acknowledgements:** The authors thank members of the ABNEL group and Northeastern University for their support. Authors also thank Dr. Rebecca Carrier for her insight into epithelial model systems.

**References:**

1. Quigley, E., Fried, M., Gwee, K. A., Olano, C., Guarner, F., & Khalif, I. (2009). Irritable bowel syndrome: a global perspective. WGO Practice Guideline.

2. Hidalgo, I. J., Raub, T. J., & Borchardt, R. T. (1989). Characterization of the human colon carcinoma cell line (Caco-2) as a model system for intestinal epithelial permeability. *Gastroenterology*, (96), 736-49.

3. Grundy, D., Schemann, M., (2006) Enteric Nervous System. *Curr. Opin. Gastroenterol.* 22. 102-110.

4. Gulbransen B., Sharkey, K. (2012) Novel functional roles for enteric glia in the gastrointestinal tract. *Nat. Rev. Gastroenterol. Hepatol.*

5. Smith, TH., Ngwainmbi, J., Grider, JR., Dewey, WL., Akbarali, HI. (2013). An in-vitro preparation of isolated enteric neurons and glia from the myenteric plexus. *J Vis Exp*. (78).

6. Sato, T., Vries, R. G., Snippert, H. J., van de Wetering, M., Barker, N., Stange, D. E., ... & Clevers, H. (2009). Single Lgr5 stem cells build crypt villus structures in vitro without a mesenchymal niche. *Nature,* 459(7244), 262-265.

7. Rousset, M. (1986). The human colon carcinoma cell lines HT-29 and Caco-2: two in vitro models for the study of intestinal differentiation. *Biochimie,* 68(9), 1035-1040.

# 434701 Evaluating Developmental Toxicity of Novel Molecules: A New Model System

**Abstract:**

It has been reported that bisphenol A (BPA) exposure is detrimental to human development, yet the toxic effects are not fully understood. BPA is widely used as an effective plasticizer and it provides desirable properties to materials in a variety of applications from water bottles to thermal resistant coatings on military vehicles. We recently developed an alternative monomer, bisguaiacol F (BGF), that maintains structural similarities to BPA and is developed from renewable resources. In this work, we evaluate the relative developmental toxicity of BGF and BPA using a new model system, the *Schmidtea mediterranea* (Smed) planarian. Smed are small, freshwater flatworms that possess a centralized nervous system, a sequenced genome, and extensive regenerative capabilities that enable the regeneration of a portion of, or all of, the central nervous system. Using an assessment of Smed cognitive ability throughout head regeneration, we found that BPA exposure induces a delay in the reacquisition of cognitive function. BGF exposure also induces a delay, but of considerably shorter duration. Furthermore, lethality testing showed that Smed are more resistant to BGF exposure than BPA exposure. These results suggest that BGF may reduce the developmental toxicity of BPA and that Smed planaria could provide a new model system for neurotoxicology and teratology investigations.

# 434796 A Method to Quantify Drug Uptake and Transport in 3D Multicellular Spheroids

**Abstract:**

P-glycoprotein (Pgp) is a regulator of the absorption, distribution, and excretion of clinically important drugs. In diseased tissues, such as the cells of a solid tumor, Pgp is often upregulated and increases the resistance to anti-cancer chemotherapeutics. Numerous inhibitors of Pgp have been identified using 2D monolayers of cells, but there has been poor clinical translation due to a lack of efficacy or unwanted side effects in vivo. In this study, we used three dimensional (3D) multi-layer tumor spheroids which mimic physiological barriers to drug transport in conjunction with a new algorithm to quantitatively study uptake and inward diffusion of fluorescent calcein. This analysis was applied to study the effect of Pgp efflux inhibitors (verapamil, loperamide, and cyclosporin A) in the 3D environment. While overall uptake of calcein was increased, a positional analysis revealed that samples treated with verapamil and loperamide only increased calcein concentration in the outermost layers of cells, and only samples treated with cylcosporin A had an increased calcein concentration at the spheroid core. This study shows that in addition to blocking drug efflux, Pgp inhibitors influence the transport of drug between cells and to the core of the spheroids. This method is amenable to many different cell types and may serve as a quantitative 3D model that more accurately replicates *in vivo* barriers to drug uptake and diffusion.

# 435192 Elucidate Robust Redox Metabolism of Clostridium Thermocellum (Rapid Fire)

**Abstract:**

Consolidated bioprocessing (CBP) is a potentially feasible route for sustainable production of bio-based fuels that condenses multiple steps of biomass degradation and sugar fermentation into a single step. *Clostridium thermocellum* is an anaerobic, gram positive, thermophilic bacterium that is capable of degrading cellulosic biomass directly into ethanol. Despite its growing popularity, the complete understanding of *C. thermocellum* central metabolism is still lacking with atypical glycolysis, incomplete pentose phosphate pathway and Krebs cycle, and complex redox pathways. In this study, a predictive metabolic model of *C. thermocellum* has been developed for metabolic flux quantification and rational strain design. We experimentally validated the model and investigated the range of phenotypes of *C. thermocellum* in response to significant perturbation of energy and redox pathways. The result revealed a complex, robust redox metabolism of *C. thermocellum*. By incorporating experimental data into the model, we identified redox bottlenecks hindering high-yield ethanol production in *C. thermocellum*. Furthermore, we show the model's rationale for why previous metabolic engineering strategies had low target ethanol yields, and provide different metabolic engineering strategies for reaching the target yield.

# 435199 Biotechnology of Phototrophs - from Metabolic Understanding to Scalable and Energy-Efficient Photobioreactors

**Abstract:**

Thanks to their sustainable capacity for solar energy conversion, phototrophic organisms like (micro/macro)algae and cyanobacteria represent almost one third of the total primary biomass that is being generated on the scale of our planet. At the same time they are by far the most important carbon dioxide consumers in the biosphere. Hence they are actively considered as potential bioeconomy actors for biofuels and high-value ingredients. The biotechnology of phototrophs is an emerging area with high potential to offer a paradigm shift in bioprocessing. This presentation summarizes our recent work ranging from metabolic understanding and exploitation of novel extremophilic microalgal strains to the rational design of scalable and energy-efficient photobioreactor systems. Photoheterotrophy conditions with glycerol addition to the new microalgal strain *Dactylococcus dissociatus* MT1 can simultaneously increase biomass production and reduce or eliminate the need for gas sparging. This behaviour is modelled mechanistically by linking the oxygen exchange between the cells and the culture medium as a function of glycerol concentration, cellular chlorophyll content, and photosynthetic efficiency to biomass productivity. The latter is greatest when the photosynthetic and respiratory pathways are nearly balanced. This implies that an intracellular recycling of $O_2$ and $CO_2$ increases biomass production efficiency and that a process set-point can be reached which eliminates the need to provide supplementary $O_2$ or $CO_2$. In the same strain, environmental stressors can induce the production of the valuable carotenoids canthaxanthin, adonixanthin and astaxanthin. Light intensity has a positive influence on the accumulation of the major carotenoid, canthaxanthin, similarly to salinity stress, while nitrate deprivation has more of an effect on lipid production. Nitrate depletion and salinity stress in combination increase both lipid and carotenoid accumulation. The growth and carotenogenesis phases can be initiated, reversed and repeated at regular intervals, thus opening the way to a scalable and sustainable bioprocess. The future establishment of a sustainable microalgal industry necessitates not only efficient and robust phototroph strains but large-scale, inexpensive and dependable photobioreactor systems. In addition to gaining further basic understanding on these organisms' metabolism, more technoeconomic interventions such as cutting down the construction and operational costs might emerge as the primary principle that should guide the development of future industrial photobioreactor systems. In this vein, we have designed a production platform which can generate algal biomass in an energy efficient and cost-effective manner while at the same time controlling the photonic input to the cell culture and maintaining aseptic conditions. This is a novel and controllable, yet simple, annular-plate airlift photobioreactor configuration with internal illumination and a high conversion efficiency of energy input to biomass.

# 435551 Improved Forms of Omega-PCR for DNA Plasmid Construction and Manipulation

**Abstract:**

The construction of DNA plasmids for metabolic engineering applications has long been a bottleneck in the discipline. It is either time-consuming and/or unsuccessful through traditional approaches in molecular biology or it can be cost prohibitive to use commercial cloning kits or custom DNA synthesis. This is particularly true as one tries to incorporate several genes, regulatory parts, and/or recombination sequences into a single DNA plasmid. In response to this molecular cloning bottleneck, useful solutions have often resulted in commercial ventures. Here, we describe the original development of omega-PCR (Ω-PCR) and our contributions of single-stranded and double-stranded versions of this technique (ds-Ω-PCR and ss-Ω-PCR). We also show how these techniques can be used for low cost, single tube, restriction enzyme-free cloning. The method is highly reliable and can be completed in less than 48 hours, even for the construction of DNA plasmids consisting of multiple parts. Here, we demonstrate the technique by inserting, removing, and editing multiple fluorescent reporter genes in a common plasmid backbone.

# 435588 Isolation of Caffeine-Degrading Bacteria from Soil

**Abstract:**

Caffeine (1,3,7-trimethylxanthine) is a natural compound frequently found in many common food and pharmaceutical products. Consumption of caffeine in low to moderate levels is not toxic to humans, but is lethal for many bacteria and insects. Caffeine infiltrates the environment through many methods, including human waste and plant decomposition. As a result, some bacteria have evolved to use caffeine as a nutrient source. Some of the metabolic intermediates formed by the bacterial degradation of caffeine are high-value chemicals that can be used in the food, cosmetics, and pharmaceutical industries.

Two pathways exist for the degradation of caffeine: *N*-demethylation and C-8 oxidation. During C-8 oxidation, caffeine is oxidized to 1,3,7-trimethyluric acid, which is further degraded to $CO_2$ and $NH_3$. In the *N*-demethylation pathway, caffeine is degraded through a series of *N*-demethylation reactions to form theobromine (3,7-dimethylxanthine), 7-methylxanthine, and xanthine. Xanthine then enters the normal purine catabolic pathway. In some bacteria, paraxanthine (1,7-dimethylxanthine) is also transiently formed during the first metabolic step by $N_3$-demethylation of caffeine, with subsequent degradation through 7-methylxanthine and xanthine.

Paraxanthine is a high-value compound that provides the stimulatory effects of caffeine with fewer side effects. In order to isolate a paraxanthine-producing strain of bacteria, we isolated several bacterial colonies from soil samples using an enrichment procedure. Soil samples were incubated in a modified liquid M9 media with 2.5% caffeine as the sole source of carbon and nitrogen. After three subcultures, bacteria were isolated on solid M9 + caffeine media. The isolates were characterized phylogenetically by 16S ribosomal RNA sequencing. Metabolism of each strain was determined by analyzing metabolites in growth media and resting cell assays by HPLC. Paraxanthine-producing strains were identified and selected for use in future studies.

# 435654 Constructing a Biosynthetic Pathway to 3-Hydroxy-2-Pyrrolidinone

**Abstract:**

The specialty chemical 3-hydroxy-2-pyrrolidinone is the subject of biosynthetic pathway construction and metabolic engineering due to its favorable selling price ($2,600/kg) and because there are no known routes for biological production. In order to produce 3-hydroxy-2-pyrrolidinone biologically, a new biosynthetic pathway was proposed, and enzyme promiscuity is now being engineered so that enzymes of the pathway can accept and convert non-native substrates. This is a significant challenge in biosynthetic pathway synthesis that will ultimately allow the production of new chemicals and provide more efficient routes to existing chemical targets. We are approaching the problem of biosynthetic pathway design using simultaneous computational and experimental approaches. In particular, molecular docking and dynamics simulations offer understanding of how native and non-native substrates interact with active site residues of an enzyme. This approach is used to suggest enzyme engineering strategies that may be required for an enzyme to accept and convert a non-native substrate. In addition, a new enzyme activity screening technique has been developed using Raman spectroscopy that can probe the activity of engineered enzymes without enzyme purification or the development of specific colorimetric assays. Here, we also describe how these methods can be implemented in a semi-high-throughput assay to screen enzyme mutants for biosynthetic pathway construction.

# 436052 Co-Immobilization of Cellulase Complex on Magnetic Nanoparticles Aiming to Sugarcane Bagasse Hydrolysis

**Abstract:**

Second-generation bioethanol produced from lignocellulosic biomass are a potential alternative to fossil fuels due to its renewability, low environmental impact, and limited employment of food crops. Compared to traditional chemical processes, the enzymatic hydrolysis of lignocellulosic biomass usually involves lower energy consumption, milder reaction conditions, substrate specificity and also achieving higher yields. High yields in the conversion of cellulose into glucose, used as substrate to bioethanol production, may be achieved by using cellulases, the enzymes able to hydrolyze β-1,4 linkages in cellulose chains, by the synergetic action of three main types: endoglucanases, exoglucanases and β–glucosidase. Although the advantages of second-generation bioethanol, its final commercialization must overcome primary costs, for instance, due to enzyme. A reduction in the cost of enzymatic hydrolysis of lignocellulosic biomass might be achieved by the development of a process that recover/reuse the enzyme. For this enzyme recycling process, some strategies have been proposed: enzyme ultrafiltration, enzyme re-adsorption, recycle for solids and liquids after hydrolysis, surfactant (or other enzyme coagulant agent) addition and enzyme immobilization. The use of magnetic materials as enzyme immobilization carries may be an interesting alternative since it is easy to separate, recover and reuse. The immobilized cellulases might be magnetically separated, recovered and reused in a new cycle of enzymatic hydrolysis. Therefore, the aim of this work was to evaluate the possibility of immobilizing cellulases on magnetic particles in order to enable enzyme re-use on the hydrolysis of sugarcane bagasse, a very abundant biomass in Brazil. The magnetic particles (iron oxide) were active by Glutaraldehyde at room temperature and pH 4.8. The enzyme (celullase from *Trichoderma reesei – Sigma Aldrich*) was maintained of citrate buffer (50mM; pH 4.8). The activity of cellulose was measured individually for each enzyme in the complex, as well the total celullase activity. Preliminary results showed the stability of the support prepared for the immobilization procedure, and the possibility of its magnetic recovery after the hydrolysis of sugarcane bagasse.

# 436506 A Novel Low-Volume Two-Chamber Microfabricated Platform for Evaluating Drug Metabolism and Toxicity (Rapid Fire)

**Abstract:**

To evaluate drug and metabolite efficacy on a target organ, it is essential to include metabolic function of hepatocytes, and to evaluate metabolite influence on both hepatocytes and the target of interest. Herein, we have developed a two-chamber microfabricated device separated by a membrane enabling communication between hepatocytes and cancer cells. The microscale environment created enables cell co-culture in a low media-to-cell ratio leading to higher metabolite formation and rapid accumulation, which is lost in traditional plate cultures or other interconnected models due to higher culture volumes. We demonstrate the efficacy of this system by metabolism of tegafur by hepatocytes resulting in cancer cell toxicity.

# 436588 Engineering Neurospora Crassa for Cellobionate Production from Cellulosic Biomass (Rapid Fire)

**Abstract:**

In recent years, carboxylic acids, such as lactobionic acid (LBA) have emerged as specialty acids due to their unique physiochemical properties. They are high value-added organic acids, with numerous applications that span the pharmaceutical, food, and cosmetics industries. Its sister chemical cellobionic acid is expected to have similar applications. In order to compete with petroleum-based processes for the production of carboxylic acids, the development of microbial processes utilizing low-cost substrates is essential. LBA is currently produced through chemical synthesis in an energy-intensive process requiring costly metal catalysts. Alternatively, LBA or CBA can be produced biologically by various bacterial and fungal strains using refined sugars as the substrate. Cellulosic biomass, which is available at low cost and in widespread abundance, is a potential alternative substrate for the bio-production of carboxylic acids. In this study, we report engineering *Neurospora crassa* strain for the production of cellobionic acid (CBA), directly from cellulose with exogenous laccase and redox mediator addition.

In this study, we report the production of cellobionic acid (cellobionate) directly from cellulosic biomass by a genetically engineered fungus strain Neurospora crassa. After deleting nine genes essential for cellobiose and cellobionate utilization and over-expressing laccase gene from *Botrytis aclada*, the engineered strain is able to convert pretreated poplar to cellobionate at high yield without addition of any exogenous enzymes.

# 440105 Bcr-Abl1 V304D Mutation in CML: Friend or Foe?

**Abstract:**

Our recent work revealed that some chronic myeloid leukemia (CML) patients failing imatinib therapy carry the BCR-ABL1 V304D mutation, and is associated with poor clinical outcomes. BCR-ABL1 V304D variant has rarely been described in clinical specimens, and its location within the BCR-ABL1 Kinase Domain (KD) appears to be in an area of the protein not involved in the direct binding with tyrosine kinase inhibitors (TKIs) such as Imatinib and Ponatinib. The present study integrates computational, structural, and molecular biology techniques to investigate the role of the BCR-ABL1 V304D mutation in kinase activity and TKI resistance.

Accordingly, in this work we used in silico experiments to probe whether, compared to the wild type (wt), the V304D mutation i) directly affects the kinase domain fold and/or thermodynamic stability and ii) if/how these effects might eventually interfere with BCR-ABL1 binding to the TKIs. Small angle X-ray scattering (SAXS), fluorescence, and circular dichroism spectroscopy are employed for structural characterization of the wt/mutated kinase domain and their interaction with TKIs. Cell proliferation, protein phosphorylation, and kinase activity assays are performed (with/without) the TKIs to assess the effect of this mutation on BCR-ABL1 KD, its enzymatic and transforming activities.

Our preliminary results show that BCR-ABL1V304D does not transform Ba/F3 cells, retains minimal kinase activity, and fails to activate downstream pathways. At the same time, the V304D BCR-ABL1 isoform partially protects cells from cytokine withdrawal and exhibits imatinib resistance. The SAXS-based structural model of the kinase domain mutant in complex with Ponatinib does not reveal major structural perturbations compared to the wild type protein. Molecular simulations on the SH2-linker-SH3-KD construct suggest a possible rationale for the TKI resistance seen in the clinic.

In aggregate, these results indicate that BCR-ABL1V304D results in kinase inactivation, pan-TKI resistance mediated by lower control of protein autoinhibition via perturbation of the SH3 binding domain and very poor prognosis.

# 440107 On the Mitochondrion's Secret Service: Genetic Knockdown and Ligand Binding Define the Role of the Chaperone Sigma-1 Receptor in Blocking Metabolic Stress Response in Cancer Cells

**Abstract:**

There is increasing interest in the biological pathways that determine the ability of cancer cells to adapt and survive to metabolic and micro-environmental stress and sustain tumor-initiating and metastatic properties. A better understanding of these pathways may lead to the identification of key nodes and targetable elements for development novel therapeutic strategies for cancer.

The sigma-1 receptor (Sig1R) is a ubiquitously expressed membrane-bound protein that acts as ligand-activated molecular chaperone. Sig1R is localized preferentially at the endoplasmic reticulum (ER) and has an important role in calcium signaling and cell survival in response to stress signals. In cancer cell lines and human tumors Sig1R is often up-regulated suggesting that it might play a relevant role in tumorigenesis, although detailed functional studies are missing.

In this study we examined whether Sig1R sustained proliferation, survival and tumorigenic properties of human cancer cells. Knockdown of Sig1R using small interfering RNA (siRNA) in human prostate and lung cancer cells had profound effect on proliferation, clonogenic capability and tumor sphere formation, indicating reversal of tumorigenicity and stem-like phenotype in absence of Sig1R.

Next, we tested a series of structurally diverse Sig1R ligands selected for high affinity and selectivity for the receptor in the attempt to discover pharmacological agents that could phenocopy the effects of the genetic knockdown on cancer cells. We identified various Sig1R ligands that behaved as antagonists of the receptor functions in cancer cells inhibiting clonogenicity and tumor sphere formation. Effective concentrations of the ligands were in low micromolar range (1-10 mM). At these doses, Sig1R antagonistic ligands almost completely suppressed clonogenic and tumor sphere forming capability of cancer cells. Furthermore, whereas cell proliferation and viability under standard culture conditions were minimally affected by Sig1R ligands, their effects were more pronounced under glucose starvation, a condition that causes metabolic stress in cancer cells. Both Sig1R knockdown and pharmacological antagonists led to impaired mitochondrial function, which was more evident under glucose starvation. Thus, the absence of functional Sig1R reduced mitochondrial activity and adaptability of cancer cells to metabolic stress.

These results demonstrate that the presence of functional Sig1R ensures survival and sustains tumorigenicity and stem-like properties of cancer cells by enhancing their metabolic flexibility. Highly selective Sig1R antagonists are promising leads for discovery of innovative therapeutic strategies and might represent effective candidates for pharmacological targeting of stem-like and tumor-initiating cells in human cancers.

# 437148 Engineering a Xylose Regulon in Saccharomyces Cerevisiae for Efficient Lignocellulosic Biomass Utilization (Rapid Fire)

**Abstract:**

*Saccharomyces cerevisiae* is one of the most widely used microbes for generating biofuels using lignocellulosic feedstock. Since this yeast cannot utilize xylose, the second most abundant sugar in lignocellulose, several metabolic engineering approaches have been attempted in the recent past for efficient xylose consumption. However, growth on xylose is still poor and economically unfeasible for biofuel production. We hypothesize that further enhancement in xylose utilization and biofuel production can be made by addressing the issue from a regulatory perspective rather than a metabolic perspective. Therefore, our work aims at developing a regulon (genome-wide regulatory infrastructure) to enhance the growth and biocatalytic fitness of this yeast for enhanced respiratory growth and advanced biofuel (non-ethanolic) production using xylose. To achieve this, we developed a protein sensor-actuator system that regulates the expression of aerobic xylose metabolism based on its concentration. The well-characterized galactose sensor protein, Gal3p, was engineered for increased affinity towards xylose through semi-rational protein engineering. Targets for mutagenesis were identified using *in silico* docking studies and high-throughput screens were developed based on G418 resistance and GFP fluorescence. In these screens, expression of *kanMX* and *gfp*, placed under *GAL* promoters, were triggered by a regulatory cascade initiated by the binding of Gal3p to xylose. Initial selection was performed in G418 selection plates and the strains that showed resistance were further quantitatively screened for increased fluorescence. The results reveal further scope for increasing xylose sensitivity in Gal3p mutant variants. Future work involves up- and down-regulating genes and pathways required for enhanced aerobic growth on xylose under the control of this synthetic regulatory system. Such a strain will act as a platform to engineer advanced biofuels using lignocellulosic biomass.

# 406810 Using Solvent NMR Relaxation and Diffusion Measurements As Aids in Fingerprinting Microemulsions

**Abstract:**

In part owing to their ease of formation and stability, the effective use of microemulsions has increased dramatically as major industrial applications have expanded in a variety of fields. One important example is as drug carrier systems for oral, topical, and parenteral administration; microemulsions offer the advantage of ease of manufacturing and scale-up, and improved drug solubilization and bioavailability.

Thorough characterization of any colloidal system is essential but presents distinctive problems in the case of microemulsions because they are clear and, further, since microemulsion domains exist at fairly specific concentrations of the three basic components, dilution is almost never an option. Hence, investigation by conventional light scattering is difficult, if not impossible. X-ray and neutron scattering and dielectric measurements have been used to provide insight to the structure and dynamics of such systems but such measurements are not routine and, certainly, are not suited to normal laboratory analysis.

Solvent NMR relaxation and diffusion measurements are non-invasive and non-destructive. Not only (micro)emulsions but also swollen micelles can be measured without dilution, stored under thermally controlled conditions (e.g. ICH guidelines) and re-analyzed to understand thermodynamic stability. Emulsion and micellar microstructure can be investigated with exquisite clarity via NMR.

With the advent of small bench-top low field NMR devices, the technique has become very useful as a routine aid in formulation of suspensions and emulsions. Using a 13MHz device, we will present data obtained on a model microemulsion system to illustrate both the utility and simplicity of the technique in fingerprinting not only the composition of both the preparatory oil and water phases but also the microemulsion system so-formed from them.

# 412759 The Challenge to Determine Particle Size and Specific Surface Area of an Agglomerated Drug Substance in the Late-Stage Development

**Abstract:**

Solid-state powder properties of a drug substance, alternatively called active pharmaceutical ingredient (API), such as particle size and specific surface area, play a critical role in late-stage pharmaceutical development. Specifically, surface area and particle size of APIs can impact bioavailability and dissolution of the solid oral dosage form. Further, these parameters are used to establish process robustness. Naturally, there is a need to determine particle size and specific surface area during API manufacturing. This presentation focuses on the particle size and surface area method development for an API that is agglomerated to enhance the flow of the material. Furthermore, this material is a non-stoichiometric hydrate that is particularly challenging in terms of method development. In particular, surface area measurements require an initial degassing of the material, which could alter the hydration state of this API. The difficulties and challenges to measure the primary particle size, agglomerate size, and specific surface area of this agglomerated drug substance will be presented. The effect of the agglomerated API and its hydration state on the particle size and surface area as well as their implications for dissolution and bioavailability will be discussed.

## (622q) Controlled Release of CGRP Antagonist Peptide By Temperature-Dependent Hydrogel Prolongs Analgesia in the Spared Nerve Injury-Induced Neuropathic Pain in Rats

Xie, X., Massachusetts Institute of Technology
Zhang, L. M., Sun Yat-Sen (Zhongshan) University
Xie, X., AfaSci Research Laboratories
Pascual, C., AfaSci Research Laboratories

Abstracts:

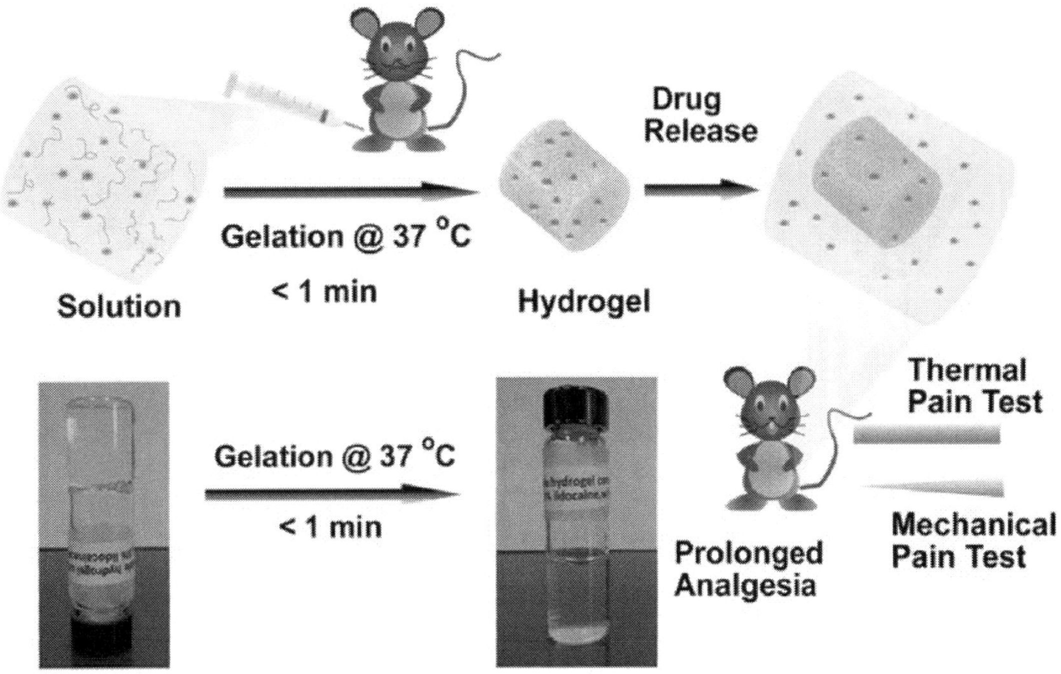


- - Abstract Text

# 414442 Biogas Production from Three Lignocellulosic Residues: Coffe Cut-Stems (CCS), Sugarcane Bagasse and Pinus Patula

**Abstract:**

**Biogas Production from three Lignocellulosic Residues: Coffe Cut-Stems (CCS), Sugarcane Bagasse and *Pinus patula***

Sanchez Julio C, Garcia Carlos, Cardona Carlos A

[1]Universidad Nacional de Colombia sede Manizales, Instituto de Biotenologia y Agroindustria, (+57) (6) 879400 ext 55354.

Biogas is a mixture of methane ($CH_4$), carbon dioxide ($CO_2$) and hydrogen ($H_2$) among others. It can be considered a form of green energy is gaining popularity worldwide, due to politic, economic and environmental reasons. Some countries in Europe have already applied biogas for its energy supply, as for example Germany that at the end of 2008 had approximately 4.000 agricultural biogas production units on its farms.

Biogas is produced in anaerobic digestion process which is carried out in four steps: hydrolysis, acidogenesis, acetogenesis/dehydrogenation and methanation. Each one of these steps is performed by different classes of microorganisms: acetogenic bacteria (hydrolyses substrates and synthetizes volatile fatty acids (VFA) mostly acetic acid), $H_2$-producing bacteria and methanogenic bacteria (break acetic acid into $CH_4$ and $CO_2$ or take hydrogen and carbon to form methane) [1].

Multiple sources of organic material can be useful for biogas production as they are residues of others processes such as municipal solid waste (MSW), crop residues, animal manure, forest residues or dedicated energy crops, which are mostly composed by lignocellulosic material (cellulose, hemicellulose, lignin and other substances) [2]. In Colombia case some available sources of lignocellulose considered as agriculture residues are coffee cut-stems (CCS) [3], *Pinus patula* residues and sugarcane bagasse [4].

The aim of this work is to present an approach for the sustainability assessment of biogas production from common agroindustrial residues for bioenergy production, specifically the biogas production from coffee cut-stems, *Pinus patula* residues and sugarcane bagasse. Residues were experimentally characterized by measuring klason lignin content, acid-soluble lignin content, holocellulose content, cellulose content and ash content. Pretreatment of lignocellulosic residues was done with particle size reduction and basic hydrolysis, fermentation was carried out in strict anaerobic conditions and constant temperature, biogas production rate was measure by volume displacement. Energy, economic and environmental simulations with experimental results were evaluated with Aspen Plus V8.4®, Aspen Economic Evaluator® and WAR® respectively [5],[6].

The obtained results lead to conclude that the studied lignocellulosic resources are a profitable option for bioenergy production, but these results dramatically depend of the scale of production.

**BIBLIOGRAPHY**

[1] Weiland, P. (2010). Biogas production: current state and perspectives. *Applied microbiology and biotechnology*, *85*(4), 849-860.

[2] Taherzadeh, M. J., & Karimi, K. (2008). Pretreatment of lignocellulosic wastes to improve ethanol and biogas production: a review. *International journal of molecular sciences*, *9*(9), 1621-1651.

[3] Triana, C. F., Quintero, J. A., Agudelo, R. A., Cardona, C. A., & Higuita, J. C. (2011). Analysis of coffee cut-stems (CCS) as raw material for fuel ethanol production. *Energy*, *36*(7), 4182-4190.

[4] Quintero Suárez, Julián Andrés (2011) *Diseño y evaluación de la producción de alcohol carburante a partir de materias primas lignocelulósicas*. Tesis Doctorado, Universidad Nacional de Colombia - Sede Manizales.

[5] Quintero, J. A., Moncada, J., & Cardona, C. A. (2013). Techno-economic analysis of bioethanol production from lignocellulosic residues in Colombia: a process simulation approach. *Bioresource technology*, *139*, 300-307.

[6] Sánchez, ". J., & Cardona, C. A. (2012). Conceptual design of cost-effective and environmentally-friendly configurations for fuel ethanol production from sugarcane by knowledge-based process synthesis. *Bioresource technology*,*104*, 305-314.

# 415649 Microfluidic Assembly of Multifunctional Polyplexes

**Abstract:**

The critical bottleneck to translation of human gene therapy remains the lack of safe and efficient methods for delivery of genetic material. Polymeric delivery agents have shown much promise. However, conventional polymeric vectors, which most often comprise plasmid DNA and a single polycation, cannot provide the multiple functionalities (e.g., binding of cellular receptors, intracellular trafficking steps, nuclear translocation, etc.) needed for efficient transfection. To achieve clinically relevant gene delivery activity with synthetic vectors will require more complex assemblies of DNA with multiple materials contributing the needed functions. Furthermore, construction of such multifunctional vectors will require new methods that provide controlled mixing and spatially defined arrangement of the DNA and polymers. We are developing microfluidic systems for construction of multilayered, multifunctional non-viral vectors. The devices bring solutions of plasmid DNA and polymer(s) into contact under laminar flow providing enhanced control of polymer/DNA stoichiometry and interaction times and, most importantly, allowing introduction of multiple materials in a sequential fashion. In this presentation, we will present design of a microfluidic device for layer-by-layer assembly of ternary polyanion/polycation/DNA polyplexes. We will compare physicochemical properties of the resulting polyplexes, including polyplex size, size uniformity and architecture, in comparison to polyplexes assembled via conventional bulk mixing. This novel polyplex assembly method is expected to provide the control and reproducibility necessary to enhance transfection efficiencies, facilitate investigation of gene transfer mechanisms, and ultimately enable clinical translation.

# 413498 Generation of Cardiac Cell Lineages from Human Pluripotent Stem Cells By Modulation of WNT Signaling Via Small Molecules

**Abstract:**

Human pluripotent stem cell (hPSC)-derived cardiomyocytes, endothelial cells and their progenitors are important tools for cardiovascular research and have significant therapeutic potentials. During the past 10 years, rapid methodological developments for creating de novo cardiac lineages have been made. However, most of these approaches require animal cells, fetal bovine serum, or various cytokines, limiting their application for large-scale cardiac lineage production for therapeutic applications. Here, we show that temporal Wnt signaling modulation is both necessary and sufficient for cardiomyocyte and endothelial cell specification from hPSCs under chemically-defined, albumin-free conditions. Firstly, we systematically optimized our previous GiWi protocol for cardiac differentiation, and found that RPMI basal medium alone is sufficient to support hPSC differentiation to cardiomyocytes with a purity of more than 90%. In addition, we developed a robust endothelial progenitor differentiation platform that uses a minimalistic medium, consisting of only DMEM and ascorbic acid, via modulation of canonical Wnt signaling pathway. These two fully defined platforms should facilitate large-scale production of xeno-free cardiomyocytes and endothelial cells respectively for both research and clinical applications.

# 419738 Interplay of Specific Trans- and Juxtamembrane Interfaces in the Regulation of Plexin-Semaphorin Signaling

**Abstract:**

Plexins are type I integral membrane proteins that serve as guidance receptors during angiogenesis, lymphangiogenesis, neuronal development, and zebrafish fin regeneration with a putative role in cancer metastasis. Receptor dimerization or clustering, induced through extracellular ligand binding but modulated in part by the plexin transmembrane (TM) and juxtamembrane (JM) domains, is thought to drive plexin activity. Previous studies indicate that isolated plexin TM domains interact through a conserved, small-x3-small packing motif, and the cytosolic JM region interacts through a hydrophobic heptad repeat, but the roles and interplay of these regions in plexin signal transduction remains unclear. In this study, we find disruption of the small-x3-small motifs in the *Danio rerio*plexin A3 TM domain enhances dimerization of the TM-JM domain in the AraTM assay by enhancing JM-mediated dimerization, whereas mutations to the cytosolic JM heptad repeat that disrupt dimerization do so even in the presence of TM domain mutations. However, mutations to the small-x3-small TM interfaces also disrupt plexin A3 signaling in a zebrafish axonal guidance assay, indicating the importance of this TM interface in signal transduction. Collectively, our results demonstrate that multiple TM and JM interfaces exist in the plexin A3 homodimer, and these interfaces independently regulate dimerization important in plexin A3 signal transduction. Furthermore, this work highlights the importance of JM interactions as well as how the interplay of TM and JM interactions can independently regulate receptor signal transduction.

This work was supported by the National Science Foundation (CBET-1227924) and the National Institutes of Health (P01 HL040387-26A1).

# 417074 β1-Integrin Surface Expression on Adipose Stem Cells Cultured in a Centrifugal Bioreactor with TGF-β3 and Cyclic Hydrostatic Pressure

**Abstract:**

**Introduction:** Articular cartilage (AC) is a smooth white tissue covering the articulating surfaces of bones to act as a lubricant and load distributer. Chondrocytes, the primary cell type in cartilage, remodel the tissue's extracellular matrix in response to mechanochemical cues from their environment. When injured or damaged, cartilage tissues have limited regenerative capabilities because of lack of vascularization. Unfortunately, damage to AC is common with millions of Americans suffering from the joint disease, osteoarthritis, every year. Current treatments for osteoarthritis are inadequate in restoring the structure or function of native cartilage, making AC a good candidate for tissue engineering. This project aims to culture human adipose-derived stem cells (hASCs) in our novel centrifugal bioreactor (CBR) with oscillating hydrostatic pressure and transforming growth factor-β3 (TGF-β3). These cues lead to differentiation of the hASCs into chondrocytes to form a cartilage tissue with desirable properties. In our work, we use atomic force microscopy (AFM) to measure the mechanical properties of the cell and cartilage tissue as well as to map the location and density of β1-integrins on the cell surface as a measure of mechanotransduction. We hypothesize that biomechanical and chemical stimulation will be sensed by the cell *via* mechanotransduction through β1-integrins and this will prompt cells to secrete extracellular matrix. We further hypothesize that this stimulation will improve the Young's modulus of cartilage tissues by increasing it to near native cartilage properties.

**Materials and Methods:** hASCs were cultured for 21 days in micromass, pellet and in the CBR with and without TGF-β3. Cells in the CBR were subjected to oscillating hydrostatic pressure. To determine tissue mechanics, a 10 $\mu m^2$ scan area was indented *via* AFM using gold-coated colloidal cantilevers in phosphate buffered saline. At least 256 points on each of 3 scan areas were indented per treatment group. Force-indentation profiles were fit to a Hertz model of contact mechanics to extract the elastic modulus of the engineered tissue. These values were then statistically analyzed to represent the mean and median Young's moduli for each treatment group. To determine the cell mechanical properties as well as the surface expression of β1-integrins on the cell, the tissue from each treatment group was digested with type II collagenase to isolate the cells. These cells were seeded onto poly-L-lysine coated glass slides for further AFM studies. A gold-coated sharp cantilever modified with a monoclonal antibody against β1-integrins was used to map the β1-integrin localization on the cell surface. As the cantilever probes the cell surface, specific adhesion interactions can be identified as single-molecule signatures that represent antibody-antigen binding. The location and sum of these binding events was recorded and compared across treatment groups.

**Results and Discussion**: CBR tissues grown with only oscillating hydrostatic pressure had Young's moduli 44-fold higher than the micromass or pellet static controls. With the addition of TGF-β3 to the CBR tissues, a 1.8-fold increase was observed, bringing the Young's modulus nearer to that of native cartilage. It was found that the average adhesion force for a single antibody-β1-integrin interaction was 80 pN, independent of treatment group. This fits the typical protein-protein interaction magnitudes found in the literature which are typically under 100 pN. Higher force magnitudes followed patterns indicating that multiple antibody-antigen interactions were present. The micromass and pellet cultures had two to five-fold larger β1-integrin counts compared to the cultures engineered in the CBR. We expect that this is a result of the cell responding to a lack of extracellular matrix (ECM). Because β1-integrins facilitate cell-ECM attachment, a deficient ECM may require more β1-integrins to attach to the cell. This would explain why static cultures with lower Young's moduli have more β1-integrins.

**Conclusions:** Cartilage tissues grown from hASCs had higher Young's moduli, nearing that of native cartilage, when cultured with oscillating hydrostatic pressure. The improvement in the Young's moduli was furthered by the supplementation with TGF-β3. The surface expression of β1-integrin is dependent on the tissue culture method and is lower in the CBR samples with oscillating pressure compared to static cultures. These data support the hypothesis that a larger number of β1-integrins are necessary when sub-optimal ECM is present.

**Acknowledgements:** This work was supported by an NSF EAGER grant 1212573, the NIH Protein Biotechnology Training Program 24280305, a NASA Space Grant, a WSU DRADS fellowship, and a Harold P. Curtis Scholarship.

# 418219 Ultrasound Assisted Pretreatment of Rice Husk and Plantain

**Abstract:**

** Corresponding author: ccardonaal@unal.edu.co

Lignocellulosic biomass such as rice husk and plantain pseudostem represents a potential source of sugars that can be used as raw material in biorefineries to produce chemical, biofuels, bioenergy, biomolecules and other kind of compounds [1], [2]. This type of biomass is arranged in a matrix composed by cellulose, hemicellulose and lignin. Reduction of polymerization degree and crystallinity are the principal objectives for lignocellulosic material pretreatment. The enhancement of the cellulose availability, surface area and lignin matrix destruction are allowed through hydrothermal, acid, alkaline and organosolv pretreatments. These conventional operations release inhibitory compounds affecting the fermentation yields and the microorganism performance. Some non-conventional operations such as ultrasound assisted pretreatment have been studied to enhance the pretreatment stage of the lignocellulosic materials [3], [4]

Ultrasonic technology is based on acoustic radiation phenomena in which mechanic energy form is transformed to chemical energy generating chemicals, physicals and biochemical energies [5]. Ultrasonic waves are the expression of this kind of energy they create pressure drops at frequencies beyond audible range between 20-1000 kHz. This phenomenon could generate bubbles that produce shock waves which create a flow patterns generating a hydrodynamic shear stress capable of inducing changes in the lignocellulosic material structure [6][7].

In this work different sonication times (10 to 60 minutes) and temperatures (25°C to 80°C) were evaluated for rice husk and plantain pseudostem pretreatment in 2% sulfuric acid medium. The ultrasonic irradiation at a frequency of 30 kHz was transferred through a cylindrical horn from a UP100H device. Rice husk and Plantain pseudostem were experimentally characterized by measuring moisture content (AOAC 928.09 method), Klason lignin content (TAPPI 222 om-83 method), acid-soluble lignin content (TAPPI 250UM-85 method) holocellulose content (ASTM Standard D1104 method), cellulose content (TAPPI 203 os-74 method), and ash content (TAPPI Standard T211 om-93 method). Then, the results were compared with the base case without ultrasound assistance. Later on, the pretreated samples were submitted to enzymatic hydrolysis to produce glucose by cellulases enzymes. The sugars and inhibitory compounds yields were determined through spectrophotometric analysis. All the experimental procedures were performed in the Biotechnology and Agribusiness Institute at Universidad Nacional de Colombia at Manizales.

With the purpose of understanding the effect of different composition profiles (after pretreatment) a biorefinery [8] [9] scheme to produce ethanol, citric acid, xylitol and energy considering the non-conventional and conventional pretreatment (Ultrasound assisted and diuted acid) were evaluated using the commercial software Aspen Plus V8.2 (ASPEN TECHNOLOGY INC). The economic evaluation was performed using the software commercial Aspen Process Economic Analyzer V8.2 (ASPEN TECHNOLOGY INC) taking into account the Colombian context. Finally, was evidenced the ecomonical improvement provided by the use of ultrasound assisted pretreatments in the biorefinery performance.

**References**

[1] J. Moncada, J. a. Tamayo, and C. a. Cardona, "Integrating first, second, and third generation biorefineries: Incorporating microalgae into the sugarcane biorefinery," *Chem. Eng. Sci.*, vol. 118, pp. 126–140, 2014.

[2] L. E. Rincon, J. Moncada Botero, and C. A. Cardona Alzate, *Catalytic Systems for Integral Transformations of Oil Plants through Biorefinery concept.*, First Edit. Manizales, Colombia: Universidad Nacional de Colombia sede Manizales, 2013.

[3] G. Ramadoss and K. Muthukumar, "Ultrasound assisted ammonia pretreatment of sugarcane bagasse for fermentable sugar production," *Biochem. Eng. J.*, vol. 83, pp. 33–41, Feb. 2014.

[4] E. V Rokhina, P. Lens, and J. Virkutyte, "Low-frequency ultrasound in biotechnology: state of the art.," *Trends Biotechnol.*, vol. 27, no. 5, pp. 298–306, May 2009.

[5] L. E. Robles Ozuna and L. A. Ochoa-Martínez, "Ultrasonido y sus Aplicaciones en el procesamiento de Alimentos.," *Rev. Iberoam. Tecnol. Postcosecha*, vol. 13, pp. 109–122, 2012.

[6] G. Muratov and C. Kim, "Enzymatic hydrolysis of cotton fibers in supercritical CO2," *Biotechnol. Bioprocess Eng.*, vol. 7, no. 2, pp. 85–88, Apr. 2002.

[7] C. Y. Park, Y. W. Ryu, C. Kim, A. Univezzity, and S. C. Dioxide, "Kinetics and Rate of Ezymatic Hydrolysis of Cellulose in Supercritical Carbon Dioxide," *Korean J. Chem. Eng.*, vol. 18, no. 4, pp. 475–478, 1994.

[8] Duarte, L.C.; Esteves, M.P.; Carvalheiro, F.; Gírio, F.M. 2007. Biotechnological Valorization Potential Indicator for lignocellulosic materials. Biotechnology Journal. 2:1556-1563.

[9] Carvalheiro F, Duarte LC, Gírio FM (2008) Hemicellulose biorefineries: a review on biomass pretreatments. Journal of Scientific & Industrial Research 67:849-864.

# 418350 Ethanol Production from Cocoyam (Xanthosoma sagittifolium). Design and Analysis

**Abstract:**

*Corresponding author:* ccardonaal@unal.edu.co

Xanthosoma sagittifolium (L.) Schott commonly known as cocoyam (Elephant Ear, Taro, and Malanga) belongs to the family of Araceas, originated in Central America and extensively grow in tropical regions [1]. The main producing regions in the world are West Africa, Asia and Oceania. Nigeria, China and Ghana are the leading countries in production with more than 1'300.000, 1'182.000, and 900.000 tons per year respectively [2, 3]. Currently in Colombia, this tropical plant is not extensively commercialized and is mostly grown around the land as a weed being the current uses mainly directed for feeding farm animals with an approximate production of 200 tons per year for farm animals feeding and only 20 tons per year for human food, sales and agroindustry [4]. However the use is limited to leaf and roots (tubers) are discarded. This discarded part contains about 15 and 39 percent of carbohydrates, 2-3 percent of protein and 70-77 percent of water. The potential of these roots can be used to obtain starches as value added product [5]. The average starch content of cocoyam is about 25% w/w [1]. That compared to others tubers such as potato (15%) [6] and yucca (18%) [7] can be considered a high value.

To understand the possibilities of this raw material, experimental work was done. The initial step of the process consisted in the pretreatment (including washing, peeling, grinding and gelatinization) of the plant in order to obtain the starch. Then, an enzymatic hydrolysis using α-amylase and glucoamylase enzymes was performed in order to convert the starch into sugars. The resulting stream was used for an alcoholic fermentation with*Saccharomyces cerevisiae* in a 1.5 L fermentor (Biotron LiFlus Autoclavable Fermentor of the brand Hanil Science Industrial) [8].

The experimental data obtained such as process conditions, conversion of starch and fermentation yield was used to assess the production process by means of the commercial software Aspen Plus V8.2 (ASPEN TECHNOLOGY USA). The economic evaluation was carried out using the commercial software Aspen Process Economic Analyzer V8.2 (ASPEN TECHNOLOGY USA). The environmental evaluation of the process was carried out using the software *WAR GUI (Chemical Process Simulation for Waste Reduction)* [9]. This environmental analysis is based in four main aspects which are: Human Toxicity Potential by Ingestion (HTPI), Terrestrial Toxicity Potential (TTP), Aquatic Toxicity Potential (ATP) and Acidification Potential (AP), correlating them into a total index, the Potential Environmental Impact (PEI) of process [10].

The obtained yields as well as the techno-economic and environmental analysis demonstrated that cocoyam can be an effective raw material in the production of ethanol. Additionally, compared to other starchy materials, cocoyam is not commonly used as a product for human food resulting in a process that does not affect food security.

**REFERENCES**

[1] Giacometti, D., & León, J. (s.f.). La agricultura amazónica caribeña. Consulted on March 28th of 2015. Available in: http://goo.gl/ayJzqu.

[2] Food and Agriculture Organization of the United Nations. (1990). Roots, tubers, plantains and bananas in human nutrition. Rome: FAO Italy.

[3] Onwueme, I. C., & Charles, W. B. (1994). Tropical Root and Tuber Crops: Production, Perspectives and Future Prospects. Rome: Food and Agriculture Organization of the United Nations.

[4] Gómez, M., & Acero Duarte, L. E. (2002). Guía para el cultivo y aprovechamiento del bore Alocasia macrorrhiza (Linneo) Schott. Convenio Andrés Bello: Serie Ciencia y Tecnología, 101: 43-76.

[5] United States Department of Agriculture. (s.f.). Germplasm Resources Information Network (GRIN). Consulted on March 28th of 2015. Available in: http://goo.gl/cfxIF6.

[6] Potatoes Goodness. (2014). Potatoes Goodness. Consulted on April 18th of 2015. Available in: http://goo.gl/M2e019.

[7] Rivera, M. A. (2012). Universidad Tecnológica de Pereira. Consulted on April 18th of 2015. Available in: http://goo.gl/5n6cI5.

[8] Cardona, C. A., Sanchez, O. J., & Gutierrez, L. F. (2009). Process synthesis for fuel ethanol production. CRC Press.

[9] United States Environmental Protection Agency. (2014). WAR GUI V1.0.17. Chemical Process Simulation for Waste Reduction: WAR.

[10] Quintero, J. A., Moncada, J., & Cardona, C. A. (2013). Techno-economic analysis of bioethanol production from lignocellulosic residues in Colombia: a process simulation approach. Bioresource technology, 139: 300-307.

# 419525 5 Fu (Fluorouracil) Loaded Sulfatedchitin Nanoparticles for Drug Delivery: Preparation, Characterization & Evaluation

**Abstract:**

5 Fu (Fluorouracil) loaded sulfatedchitin nanoparticles for drug delivery: Preparation, characterization & evaluation

TANVI JAIN[1,2], SUSHIL KUMAR[1] and P.K.DUTTA[2]

[1]Department of Chemical Engineering

[2] Department of Chemistry

Motilal Nehru National Institute of Technology, Allahabad 211004, India

***ABSTRACT***

**Background**: Now a days, natural polymers are much more attractive than man made polymers due to their excellent characteristics, such as biodegradability, biocompatibility, renewability, solubility and permeability. Chitin and its derivatives is being broadly utilized for the synthesis of nanocarriers for drug delivery applications as it has excellent properties of biocompatibility, nontoxic in nature. The main limitation of chitin is insoluble in water due to its intermolecular hydrogen bonds therefore there have been many reports on its biomedical applications. Sulfated chitin, a water soluble derivative showed various biomedical features like antibacterial properties, anticoagulant activity and also immunological activity etc.

**Experimental work**: In the present study, sulfatedchitin is prepared from chitin (Chitin derivative). After completion of reaction, solid state sulfatedchitin was obtained after dialysis against water for 2 days using freeze drier method. Then sulfatedchitin (SC) Preparation of sulfatedchitin nanoparticles is done by using cross linkers calcium chloride and ferric chloride using sonication method.

**Characterization & evaluation:** The prepared nanoparticles and drug loaded nanoparticles was characterized using FTIR analysis. FTIR spectra of desired samples have been carried out using Perkin Elmer Spectrum RXI-FTIR spectrophotometer with diffuse reflectance accessory in the wave-number range 400-4000 $cm^{-1}$.The particle size was analyzed using microtrac particle size analyzer. Drug release profile was also studied using conventional method by using model drug 5-flourouracil. Drug loaded nanoparticles of known weight were kept in 10ml buffer and put on a shaker at 37°C. 3 ml of the buffer was removed from the solution at regular time intervals and its absorbance recorded on Shimadzu UV spectrometer using 1.0-cm quartz cell at ambient temperature after 1, 2, 3, 4, 5, 6, 7 and 24 h at wavelength 262 nm for 5-Fu. The same amount of buffer was replaced in the solution at regular intervals and thus drug release profile (%) was calculated. The encapsulation efficiency was also calculated.

**Results and discussion**: The ITIR spectra of sulfatedchitin are presented, here new absorptions appeared at around 2564$cm^{-1}$ and 1738 $cm^{-1}$. Adsorption appeared at around 2564 $cm^{-1}$ that corresponded to the ----SH group which represents the formation of sulfated chitin. The size of sulfatedchitin (SC) nanoparticle using particle size analyzer is 81.8 nm which helps in skin permeation. The swelling behavior shows that the highest swelling ratio in acidic medium than neutral and basic medium. Polyelectrolytes like chitin/sulfatedchitin swells at pH below pKa. Drug release profile was also studied with the model drug 5-fluorouracil. Drug release profiles indicate that 5-fluorouracil loaded sulfatedchitin nanoparticles give a sustained release in 24 h. The entrapment efficiency was found to be 65.73%.

**Conclusion**: Sulfatedchitin has been successfully prepared from chitin and its nanoparticles has been synthesised and evaluated by and physiochemical evaluation and characterization like FTIR, Swelling behavior. Drug release profiles indicate that 5-Fu loaded sulfatedchitin nanoparticles sustain release in 24 hours.

***Key Words:*** *Nanoparticles, Drug Delivery, characterization, evaluation, 5 fluorouracil, sulfatedchitin*

**Acknowledgement:** TEQIP II grant, MNNIT Allahabad for providing financial support for the research project entitled" *Biochemical and bioengineering aspects of biopolymeric nanoparticles for targeted drug delivery*".

**References**

Zou,Y., Khor, E., (2009) Preparation of sulfated-chitins under homogeneous conditions. Carbohydr. Polym. 77, 516–525.

Dutta, P,K. , Dutta, J., Tripathi, V, S., (2004) Chitin and chitosan: Chemistry, properties and applications. *J .Sci. Ind. Res.* 63:20-31.

# 420263 Old Dog New Tricks; Electronic Platform to Quantify the Anti-Cancer Potential of a Cardiac Glycoside and Its Synthetic Analogue in Real-Time

**Abstract:**

Recent studies shows that digitoxin, a cardiac glycoside, exhibits strong anti-cancer effects with a high selectivity towards neoplastic cells when compared to normal cells. Previous research has also showed that the anticancer potency of digitoxin lays in its trisaccharide moiety with further manipulation of its structure leading to more potent analogues with different cellular mechanisms and possible clinical applications extended towards targeting lung, breast and leukemic cancers. In this research, we employed a new approach to characterize and quantify the cellular behavior of immortalized and tumorigenic human lung cells (BEAS-2B and H460 respectively) upon exposure to different concentrations of digitoxin and synthetic monosaccharide analogue (D6-MA) in real time. The approach relies on an electrical cell impedance sensing system (ECIS) used to monitor the morphological changes and cell-cell interactions in response to non-invasive electrical signals. Such measurements were further correlated with biophysical, microscopy and flow cytometry analyses to evaluate the cellular dynamics and the molecular pathways associated with exposure to the drug or analogue. Our results showed that both drugs target cancer cells in a time and dose-dependent manner by activating pro-apoptotic and anti-proliferative signaling cascades that result in strengthening cellular adhesion and sequestration of key regulatory proliferation proteins. Our study provides novel means to investigate the underlying anticancer mechanisms associated with exposure to natural or synthetic compounds and promises to help expedite the potential implementation of Digitoxin and its analogs as chemotherapeutic agents for cancer treatment.

# 420448 Vascular Differentiation from Embryonic Stem Cells in 3-D Auxetic Scaffolds

**Abstract:**

Auxetic scaffolds, i.e. scaffolds that can display negative Poisson's ratio, have unique physical properties and can expand transversally when axially strained. Auxetic materials have been used for bioprostheses and artery stents due to the enhanced compressive strength and shear stiffness. In tissue engineering, auxetic scaffolds could match both the elastic stiffness and the Poisson's ratio of the target tissue to better resemble native tissues and promote tissue generation. However, the influence of auxetic materials on the cellular fate decision in the local environment is unclear. In this study, auxetic polyurethane foams were fabricated and used to support cardiovascular differentiation from embryonic stem cells (ESCs). The scaffolds before and after auxetic conversion were compared. Mesoderm differentiation was induced with bone morphogenetic protein (BMP) 4 on the seeded ESCs. The expression of alkaline phosphatase (ALP), Oct-4 and Nanog were lower after four days of differentiation for the cells grown in auxetic scaffolds. Higher expression of vascular markers CD31 and VE-cadherin was observed for the cells from auxetic scaffolds compared to those from the scaffolds before auxetic conversion. Little influence on the expression of cardiac marker α-actinin was observed. The vascular cells secreted extracellular matrix (ECM) proteins vitronectin and laminin along the differentiation and expressed membrane-bound matrix metallopeptidase 9 (MMP-9), the enzyme that was involved in ECM remodeling. The examination of Yes-associated protein (YAP) expression indicated more cytoplasmic retention in the cells from auxetic scaffolds compared to those from regular scaffolds, suggesting that the auxetic scaffolds may affect the cellular organization and contraction. This study demonstrates a novel 3-D system that cellularizes auxetic scaffolds for vascular differentiation and provides a platform to study the influence of biophysical microenvironments on cellular differentiation from pluripotent stem cells. The outcome of this study has implications in regenerative medicine and drug discovery.

**(622n) Hepatic Differentiation of Controlled-Size Embryonic Bodies Derived By Inverted Colloidal Crystal Scaffolds**

Wang, Y., University of Michigan
Bahng, J. H., University of Michigan
Kotov, N., University of Michigan

Embryonic stem (ES) cells hold promising potential as a renewable cell source for tissue regeneration and cell-based therapy. ES cell-derived specific cell lineages were strongly influenced by the control of the microenvironment, such as the size of embryoid bodies (EBs). Several microscale engineering approaches have been developed to control the size and shape of EBs and to direct embryonic stem cell fate for cardiogenesis and neurogenesis. However, desirable microenvironment for directed differentiation of ES cells to hepatocyte-like cells are still unknown. Here, we used Inverted Colloidal Crystal with highly uniform porous structure to regulate EBs and differentiate them into hepatic fate. Murine ES (mES) cells were aggregated within the scaffolds, and their aggregate sizes were controlled by varying diameters of pores in scaffold (90 μm, 170 μm, 200 μm, 270 μm and 400 μm). Differentiation of murine ES cells into early hepatocytes was assessed by analyzing gene expression. We found that hepatic differentiation of ES cells were strongly regulated by the EB size, showing that EB equals to and larger than 179.6±26.9 μm maintained the functionality to differentiated to specific lineage and with the diameter of 201.7± 29.6 μm induced highest yield of hepatic differentiation where the optimal pore size of scaffolds is 270 μm.

# 424192 Migration of Schwann Cells in Micropatterned Channels

**Abstract:**

Schwann cell migration is important for both spinal cord and peripheral nerve injuries, as Schwann cells are responsible for remyelinating the regenerated axons as well as providing guidance for regrowing axons. However, although numerous studies have focused on designing transplantable scaffolds for bridging the damaged nerve, few have taken into consideration the migration of Schwann cells required for remyelination. Evidence suggests that filaments and channels hold promise in enhancing the directionality and efficiency of Schwann cell migration [1]. To date, studies have monitored single cell migration on grooved surfaces, however the migration speed of a single cell differs from a collection of cells, and given that cells in vivo migrate not as discrete cells but rather as cell populations [2], the latter would better inform on the scaffold design for tissue repair. Therefore to better optimize the design of transplantable scaffolds, we studied the influence of channel diameter on the migration behavior of large populations of Schwann cells.

Micropatterned polydimethylsiloxane (PDMS) channel surfaces with different channel sizes (50μm, 100μm, and 150μm) were prepared through photolithography, which served to mimic the varying channel sizes of transplantable scaffolds in vitro. Schwann cells were seeded onto PDMS substrates and monitored daily for cell migration. The average cell migration speeds were quantified and normalized. We hypothesize that the smaller channel size induces higher degree of contact guidance causing the cells to align in the channel direction and migrate faster. The experimental results support higher Schwann cell migration speeds on PDMS channels with decreasing channel sizes, which was most prominent in the 50μm channels with migration speeds of 2.3 times greater than on the 150μm channels. In conclusion, the data suggest that Schwann cell migration into the scaffold, required for nerve repair, could be manipulated by controlling the scaffold channel size.

**[1] E.C. Spivey, Z.Z. Khaing, J.B. Shear, C.E. Schmidt, The fundamental role of subcellular topography in peripheral nerve repair therapies, Biomaterials 33 (2012) 4264-4276.**

**[2] E.D. Miller, K. Li, T. Kanade, L.E. Weiss, L.M. Walker, P.G. Campbell, Spatially directed guidance of stem cell population migration by immobilized patterns of growth factors, Biomaterials 32 (2011) 2775-2785.**

# 427125 Quantitative Single Particle Tracking for Characterizing Virus Entry of H3 Influenza Strains

**Abstract:**

Class I membrane-enveloped viruses, such as influenza, infect host cells through the endocytotic pathway. For this class of viruses, one essential viral coat protein, hemagglutinin (HA), is responsible for both the attachment of the virus to the host cell (receptor binding) and the fusion of the viral membrane with the host endosomal membrane. Membrane fusion is necessary for genome release into the cytosol, with the timing of release being triggered by the increasingly acidic conditions during endosome maturation. Using single particle tracking (SPT) techniques, individual membrane fusion events can be observed under specific conditions, which gives detailed information regarding HA pH sensitivity, acid stability, and the rate and extent of membrane fusion. This approach provides a comparative way to characterize and distinguish fusion properties among virus strains. We used SPT to quantify the fusion properties of three H3 influenza strains: A/Aichi/68/H3N2 (X:31), A/Udorn/72/H3N2 (Udorn) and A/Brisbane/07/H3N2 (Brisbane). Among these strains, all utilize an H3-type HA protein for entry, but the X:31 and Udorn strains have been adapted to grow well in chicken eggs, while the Brisbane strain is a clinical isolate. Chicken egg culture is the standard method for the production of large quantities of virus for research purposes and vaccine production. However, typical vaccine efficacy is only about 70%. Our aim was to determine if any changes in the fusion process exist between egg-adapted and clinical strains that may provide some rationale for the reduced efficacy. We found that the rate of fusion for the most clinically relevant strain, Brisbane, is generally insensitive to decreasing pH, while fusion of the egg-adapted strains, Udorn and X:31, are strongly dependent on pH (and faster) as pH decreases. All strains exhibit similar acid stability (the length of time they remain fusogenic in an acidic environment) at higher pH's, but the egg-adapted strains become less acid stable at lower pH's. Thus, it appears that the egg-adapted H3 strains tested may have evolved to compensate for the faster HA deactivation at low pH with a commensurate increase in the rate of fusion and number of proteins facilitating fusion, relative to the Brisbane strain. The implications of these changes are in the timing of the release of the genome, which impacts its trafficking to the nucleus for reproduction. Thus, there may be some connection with fusion rates and properties with overall infectivity. With this single particle fusion assay, we now have the tools available to better quantify the membrane fusion process for better understanding of the fundamental steps in viral infection, which will enable development of antiviral drugs that specifically target the fusion process as well as improve vaccine strategies.

# 427655 Transport of Amyloid-β Across the Blood Brain Barrier By P-Glycoprotein (Rapid Fire)

**Abstract:**

Alzheimer's disease (AD) is the most common form of neurodegenerative disorder, affects over 5 million people, and is the 6th leading cause of death in the United States. AD is characterized by an accumulation of amyloid-β protein (Aβ) in the brain. Currently under study as a therapeutic approach is the transport of excess Aβ out of the brain through the single endothelial cell thickness of the blood-brain barrier (BBB). P-glycoprotein (P-gp), an ATP binding cassette transporter located on the apical side of the BBB, has been shown to transport monomeric and oligomeric forms of Aβ. While monomeric Aβ is inert, oligomeric Aβ exhibits neurotoxicity and leads to the formation of Aβ fibrils that accumulate as plaques in the brain. However, formation of oligomeric Aβ may be important to its clearance from the brain. To explore this hypothesis, this study sought to determine the Aβ assembly state (monomer, oligomer, fibril) that most effectively interact with P-gp.

An ATPase activity assay was used to quantify Aβ binding by P-gp. In this assay, when ligand binds to P-gp, ATP is hydrolyzed and inorganic phosphate ($P_i$) is released. Inverted vesicles containing P-gp (BD Biosciences) are incubated alone (negative control), in the presence of verapamil (positive control), or in the presence of Aβ prepared with different assembly states. Addition of MgATP initiates binding, which is halted for analysis by the addition of SDS. The concentration of $P_i$ is then measured via absorbance to quantify binding relative to the negative control.

To study monomeric Aβ, $A\beta_{1\text{-}40}$ (Anaspec) is purified using SEC. Aβ fibrils are formed by incubating the purified Aβ monomer under continuous agitation and in the presence of a high salt concentration. Sonicated Aβ fibrils are prepared by sonicating the prepared Aβ fibrils for 5 min. Aβ oligomers are prepared from $A\beta_{1\text{-}42}$(Anaspec) solubilized in DMSO and diluted into aqueous buffer. Oligomer sizes are then determined using SDS-PAGE and Western blot.

10 μM Aβ monomer and Aβ fibril exhibited negligible binding of P-gp, while 1 mM Aβ oligomer and 10 μM sonicated Aβ fibril effectively bound P-gp, as observed by $P_i$ concentrations elevated over the negative control. Moreover, Aβ oligomer exhibited the most pronounced binding, even at a 10-fold lower concentration. These results demonstrate that the size of the Aβ aggregate species plays a crucial role in the binding of Aβ to P-gp for transport. These experiments will be continued in MDR1 MDCK cells to ensure that this correlation is upheld in a cell culture model.

## (622e) Modified Cellulose Acetate to Sequester Acrolein for Neuropathic Pain Reduction

Muratori, B.,
Acosta, G., Purdue University
Shi, R.,

**Modified Cellulose Acetate to Sequester Acrolein for Neuropathic Pain Reduction**

### Introduction

The burden of spinal cord injuries (SCIs) is not only impacted by the loss of motor function but also by the persistence of neuropathic pain (NP) for decades after injury. Despite much recent research, effective analgesic treatments are not currently available for NP after SCI[1]. Evidence suggests that NP can be affected by exogenous factors such as cigarette smoke. Several case studies have reported that cigarette smoke exacerbates NP in people who have suffered an SCI[2,3]. However, lifestyle changes involving cessation of cigarette smoking may be difficult for a paraplegic due to the confinement, and depression in many cases, they experience from the paralysis. Further, this phenomenon of increased pain due to cigarette smoke may lead to the discovery or confirmation of theories regarding how NP develops and how it is intensified after spinal cord injury.

Cigarette smoke contains high concentrations of acrolein, a neurotoxin that has been implicated in playing a major role in the pathogenesis of SCI[4,5]. In particular, endogenous concentrations of acrolein are significantly elevated after spinal cord injury and acrolein inhalation in rodents, as well as after cigarette smoking in humans[6,7]. Acrolein is a direct agonist of the transient receptor protein ankyrin 1 (TRPA1) channel[8], which is a cation channel on sensory fibers that is crucial to the perception of thermal, chemical, and mechanical pain that has been shown to have increased mRNA expression after SCI[9]. Therefore, it's very likely that cigarette-based acrolein ingestion is an exogenous algesic factor contributing to NP. As such, reducing acrolein uptake is likely an effective way to curtail post-SCI pain.

The development of a filter to remove acrolein from cigarette smoke may be beneficial to smokers who may be unable to quit and also may provide useful information about the specific involvement of acrolein in the exacerbation of NP as opposed to the collective involvement of the many components of cigarette smoke. Current cigarette filters are composed of cellulose acetate and are mainly intended to capture particulate matter including tar. Previous studies have demonstrated the ability to perform surface modification cellulose acetate with 3-aminopropyltriethoxysilane (APTES)[10]. This study proposed that the stable surface attachment of APTES resulted a multilayer formation, with exposed amine functional groups. Amines have also been shown to actively scavenge acrolein in many *in vitro* investigations[11]. Therefore, it is reasonable to consider that chemically modifying a cellulose acetate filter would result in a reduction of acrolein in cigarette smoke.

The aim of this study is quantify the exacerbation of NP due to acrolein inhalation after injury and to chemically modify cellulose acetate in order to capture acrolein before entering the body to offer protection to those who may be unable to quit smoking. Spinal cord injury leaves the body susceptible to acrolein-mediated damage, and increased exposure to acrolein through cigarette smoke will result in sensory hypersensitivity. Our hypothesis is that this could be mitigated by eradicating acrolein through the use of an acrolein filter.

## Materials and Methods

***Cellulose Acetate Modification*** Prior to polymer modification, mass spectrometry (MS) analysis was used to confirm possible APTES-Acrolein interaction in a dilute IPA solution. Cellulose acetate films were modified using a procedure similar to one which was previously reported [10]. Briefly, 10% cellulose acetate films were suspended in anhydrous toluene to which APTES (1% vol, final) was added under flowing nitrogen. The reaction was allowed to proceed for 48 hours. X-ray photoelectron spectroscopy elemental analysis was used to confirm APTES deposition onto the cellulose acetate films. To deposit APTES on actual cigarette filters, the filters from 3R4F reference cigarettes were removed and opened to expose the fibrous cellulose acetate. The fibers were immersed in anhydrous toluene and the reaction was allowed to be carried out as previously described. Gas chromatography/mass spectrometry and changes in the reduction of pain thresholds during animal studies were used to quantify acrolein sequestering with the modified cellulose acetate filter during inhalation.

***Animal Studies*** In order to test the efficacy of the filters, animal studies were performed. Briefly, moderate contusion injuries were performed with an NYU-style impactor at the T-10 level on Sprague Dawley male rats in accordance with Purdue Animal Care and Use Committee protocols. Mechanical pain testing occurred every other day beginning 14 days post injury using von Frey filaments (0.01 to 15g) with the up-down method[12]. Acrolein inhalation sessions were performed using aerosolized acrolein (300 ppm) in combination with compressed air fed at flow rates that resulted in a final acrolein concentration of 1.5 ppm, which is in the range of acrolein concentrations in actual cigarettes. Inhalation of acrolein began 22 days after injury and proceeded until 36 days post-injury, with sessions occurring twice daily for 30 minutes. Urine collection occurred once weekly beginning two weeks after injury. Liquid chromatography with tandem mass spectroscopy (LC/MS/MS) was used to quantify the amount of 3-HPMA in urine[7]. Immediately after the 14 day inhalation period or one week after inhalation, animals were sacrificed and spinal dorsal horn, dorsal root ganglia, and paw skin tissues were harvested for RT-PCR and immunoblotting analysis for TRPA1 mRNA expression and acrolein-protein adduct quantification, respectively[9].

## Preliminary Results

In our preliminary studies, we have shown that acrolein inhalation does indeed decrease the paw withdrawal threshold of animals after SCI, indicating an increase in neuropathic pain. Specifically, Figure 1 shows that the paw withdrawal thresholds decreased by up to 80% for days 3-15 after the start inhalation period ($p<0.05$) and were not significantly different from the baseline results for days 17-21 ($p>0.05$). Further, in these animals, TRPA1 mRNA expression was increased during the inhalation period. This data supports the hypothesis that acrolein alone can exacerbate neuropathic pain after SCI. Acrolein's ability to both up-regulate and activate the TRPA1 channel make it a dual threat in the potentiation of neuropathic pain after SCI. To this end, removing acrolein, either endogenously through acrolein scavengers or exogenously through acrolein filters, could alleviate neuropathic pain for victims of SCI.

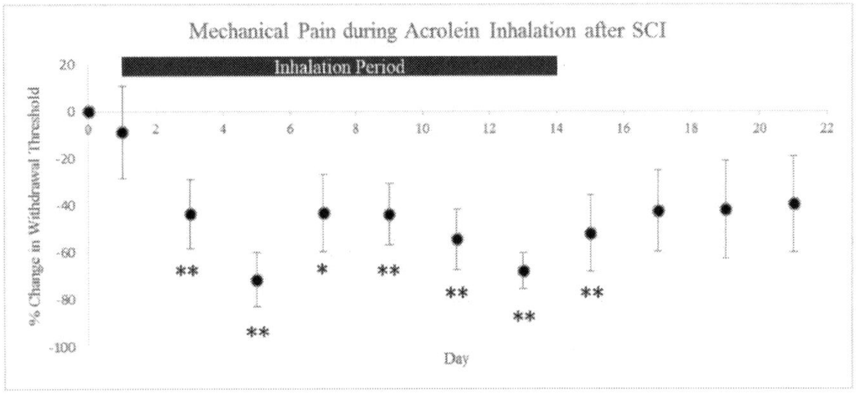


Figure 1: Change in paw withdrawal thresholds to mechanical stimuli from baseline over the course of 2 weeks of acrolein inhalation.

Preliminary mass spectrometry results indicate that there is covalent bonding between acrolein and APTES in a dilute IPA solution (.025 vol% APTES, 0.005 vol% acrolein). The spectrum for APTES (MW=222) incubated with acrolein (MW= 56) exhibited additional peaks at 278, 310, 366, and 398 m/z. A slight ion peak at 260 also existed in the spectrum. The expected structure of acrolein binding with APTES would have a molecular weight of 260. The additional peaks are a result of APTES-acrolein interactions in greater than a 1:1 stoichiometric fashion. The ability of acrolein to covalently bind to APTES indicates that the removal of acrolein with an APTES-modified cigarette filter is feasible. The sequestering of acrolein in this method could prevent the ingestion of acrolein and offer protection from increased neuropathic pain for those who have suffered spinal cord injuries.

## References


1. Hurlbert, R. J. Methylprednisolone for acute spinal cord injury: an inappropriate standard of care. *J. Neurosurg.* **93,** 1?7 (2000).

2. Richards, J. S., Jr, S. C. K., Ness, T. J. & Oleson, C. V. Effects of Smoking on Neuropathic Pain in Two People With Spinal Cord Injury. **28,** 330?332 (2005).

3. Botterell, E. H., Callaghan, J. C. & Jousse, A. T. Pain in Paraplegia. *Proceeding*

*R. Soc. Med.* **47,** 17?24 (1953).

4. Dong, J.-Z. & Moldoveanu, S. C. Gas chromatography?mass spectrometry of carbonyl compounds in cigarette mainstream smoke after derivatization with 2,4-dinitrophenylhydrazine. *J. Chromatogr. A* **1027,** 25?35 (2004).

5. Chen, P. X. & Moldoveanu, S. C. Mainstream Smoke Chemical Analyses for 2R4F Kentucky Reference Cigarette. *Contrib. to Tob. Res.* **20,** (2003).

6. Tully, M., Zheng, L., Acosta, G., Tian, R. & Shi, R. Acute systemic accumulation of acrolein through inhalation at a concentration relevant to cigarette smoke in mice. *Neurosci. Bull.* **In Press,** (2014).

7. Zheng, L. *et al.* Determination of urine 3-HPMA, a stable acrolein metabolite in a rat model of spinal cord injury. *J. Neurotrauma* **30,** 1334?41 (2013).

8. Bautista, D. M. *et al.* TRPA1 mediates the inflammatory actions of environmental irritants and proalgesic agents. *Cell* **124,** 1269?82 (2006).

9. Due, M. R. *et al.* Acrolein involvement in sensory and behavioral hypersensitivity following spinal cord injury in the rat. *J. Neurochem.* (2013). doi:10.1111/jnc.12500

10. Howarter, J. A. *et al.* Surface Modification of Polymers with 3-Aminopropyltriethoxysilane as a General Pretreatment for Controlled Wettability. 1128?1132 (2007).

11. Burcham, P. C., Kerr, P. G. & Fontaine, F. The Antihypertensive Hydralazine is an Effective Scavenger of Acrolein. *Redox Rep.* **5,** 47?49 (2000).

12. Chaplan, S. R., Bach, F. W., Pogrel, J. W., Chung, J. M. & Yaksh, T. L. Quantitative assessment of tactile allodynia in the rat paw. *J. Neurosci. Methods* **53,** 55?63 (1994).

# 428823 Integrin Alpha 6 and Calpain 2 Are Mechanosensitive Proteins in Breast Cancer

**Abstract:**

Cancer metastasis occurs when the primary tumor releases cells that travel through the vasculature and colonize distant tissues. Breast cancer spreads principally to the bone, brain and lung, however, it is not understood why there exists a clear preference to spread to these specific sites. Each secondary tissue has a distinct stiffness range and these differences alter gene expression, protein levels and activation, and cell phenotype, all of which impact the ability of cells to migrate, survive, and proliferate. We hypothesize that in breast cancer, mechanical forces could inform the tissue specificity of survival and growth at a secondary tissue site. To verify the role of tissue stiffness in metastatic cell fate, we used tunable biomaterials of varying stiffnesses and quantified genotypic and phenotypic characteristics of cells in each environment.

RNAseq implicated integrin α6 and calpain 2 as possible novel mechanosensitive genes and we examined their effects on cell adhesion and motility on synthetic PEG-PC hydrogel surfaces of varying stiffness. Integrin α6 is a surface receptor that binds specifically to laminin, which increases both the cell spreading rate and maximum area, as compared to collagen 1 alone, when on soft hydrogels. Gene expression levels show much higher expression of the B isoform at low stiffnesses. This variant of integrin alpha 6 has a splicing modification in the cytoplasmic domain, which is unique to a mesenchymal-like population and is a marker of cancer stem cells. Blocking integrin α6 function accentuates the stiffness dependence of cell spreading rate and area during adhesion on PEG-PC hydrogels. While integrin alpha 6 influences cell adhesion, calpain 2 is a protease that affects migration by controlling the disassembly rate of focal adhesions by cleaving structural proteins, such as talin. Calpain 2 inhibition decreases cell motility because cells are unable to release focal adhesions at the rear of the cell during migration. We are currently investigating the role of stiffness in mediating calpain 2 and we hypothesize that growth factor receptor activation (EGFR) accentuates the stiffness response due to receptor localization increasing the rate of focal adhesion turnover via cleavage of calpain 2. We have coupled genetic, protein and functional information to understand the mechanism by which stiffness cues initiate downstream signaling via integrin alpha 6B and calpain 2 to affect cell motility. We are working to connect the signaling roles of stiffness, EGF, and integrin binding to understand mechanosensitive cell adhesion and migration.

# 429657 Identifying Primo Vascular System inside the Lymphatic Vessel Using Hollow Gold Nanoparticle

**Abstract:**

Primo Vascular System (PVS) is newly recognized vascular system in the mammalian body. It was first discovered by Bonghan Kim in early 1960s in North Korea. In 2010, the system is officially named as the PVS by the group led by Kwang-Sup Soh, in South Korea. The PVS consists of primo vessels (PVs), primo nodes (PNs), and fluid flowing inside. The major composition of the fluid is various biochemicals and cells related to immune functions and special type of stem cells. PVS is present in the entire mammalian body, including inside the blood and lymphatic systems, on the surface of internal organs, on the brain tissue and continued through the spinal canal, and under the skin. Recent reports confirmed Kim's findings in 1960s on the role of the unique type of stem cells stored in the system performing the non-marrow based hematopoiesis and repairing damaged cells. These findings imply that the system is fundamental to maintaining mammalian lives. During cancer proliferation, the system also appears to have a role in the metastasis, and potentially storing cancer stem cells. The most important scientific claim on this system may be that it is to be the acupuncture meridian, although more studies are needed to done to confirm.

The PVS is very small (PV diameter, 20-50 mm; PN, 500-1500 mm) and translucent, highly challenging to identify it on/in the tissues with abundance of hemoglobin, which is a strong chromophore. Since biomarkers specific to the PVS is still not known, targeted contrast on its biomarker is still not possible. Harvesting a sufficient amount of PVS for further characterization has been, therefore, difficult. The Internal PVS subtype, one of the six PVS subtypes that Kim categorized, runs inside the blood and lymphatic vessels. Because these vessels are long, if it is successfully identified, the PVS harvested from them can potentially provide sufficient amount of samples. Out of two vessel types, the lymphatic vessel (LV) is a better candidate because its vessel wall is transparent and, with a help of a proper contrast agent, the PVS can be seen through its wall.

Over the years, several PVS scientists have attempted to develop techniques to identify PVS inside the LVs (intra lymphatic PVS; IL-PVS). The method currently used the most is using Alcian blue as the optical contrast agent, but this method is rather tedious in preparation and the repeatability is rather low even for the PVS experts. Our approach for developing a method for identifying the IL-PVS was utilizing the particular anatomical nature of the LV and IL-PVS within. The outer layer of a PV has pores in the size range of 2-5 mm, unlike blood or lymphatic vessels. Inside a PV, there are several sub-PVs and the interstitial space between the sub-PVs is filled with a polymer-like matrix. If nano-sized particles smaller than the PV pores but large enough to be trapped in the matrix are applied to the LV containing PVS, then the particles would be retained in the PVS and then generate the contrast by their colors.

Hollow gold nanospheres (HGNs) were found to be excellent for our purpose because of their variability in color and size, and good dispersity in bio-fluids. When a major LV is filled with the appropriately sized HGNs the spheres can get into the PVS inside the LV via the pores on its vessel wall. After a while the HNGs in the LVs are cleared by the natural flow of the lymphatic fluid and HGNs retaining PVS can be identified through the LV wall. The HGNs at a size range 50-125 nm, with a green to blue color range, allowed us to identify the PVS, within 15 minutes at a 95% success rate.

Currently, we are harvesting the PVS from the LVs for proteomics of PVS, as an effort to identify PVS specific protein biomarkers, which will be helpful for mapping the system in the body.

# 429641 Xeno-Free Differentiation of Human Pluripotent Stem Cells Toward Insulin-Expressing Pancreatic Progeny

**Abstract:**

Diabetes is a chronic disease caused by the loss of insulin-producing beta-cells due to autoimmune destruction (Type 1) or by insulin resistance and associated functional defects in beta-cells (Type 2). Human pluripotent stem cells (hPSCs) including embryonic stem cells (hESCs) and induced pluripotent stem cells (hiPSCs), can serve as a renewable sources for pancreatic beta-cells. Yet, stem cell-based diabetes therapies will rely on the efficient differentiation under xeno-free conditions of hPSCs to genetically normal and functional progeny resembling pancreatic endocrine cells. In this work, we employed both hESC and hiPSC lines aiming to develop a protocol for guiding stem cells along pancreatic cell lineages under defined conditions free of components of animal origin.

A stage-wise differentiation strategy was established to sequentially differentiate H9 hESCs and IMR90 hiPSCs into definitive endoderm, primitive gut tube, posterior foregut, and insulin-expressing pancreatic cells. Differentiation was induced in xeno-free media with growth factors involved in embryonic pancreas development. Different basal media and combinations of factors were screened and selected by standard design-of-experiments methodology based on the expression of stage-specific makers. The differentiating hPSC populations were analyzed by quantitative PCR, immunostaining and flow cytometry.

Human PSC monolayers were coaxed to definitive endoderm-like cells. Based on the choice of basal medium, a higher percentage of viable cells was obtained after 4-day differentiation with more than 60% cells co-expressing FOXA2 and SOX17. Subsequent induction led to the emergence of HNF1B$^{+}$/HNF4A$^{+}$ (foregut) cells after 4 days while over 30% of PDX1$^{+}$cells appeared on day 11 of the differentiation. These cells were further induced into cells displaying insulin and c-peptide when examined on day 19.

A protocol was established in this work for the efficient generation of insulin-expressing pancreatic progeny from hPSCs under complete xeno-free conditions and without the formation of embryoid bodies. These findings are an important step toward the establishment of strategies for the generation of clinically relevant pancreatic beta-cells for diabetes therapies.

# 430893 Modelling and Optimal Control of Non-Viral siRNA Delivery

**Abstract:**

Gene therapy gives great new hope to the patients with life-threatening diseases such as cancer and HIV (Al-Dosari and Gao, 2009). This approach can be defined as the insertion of genetic materials including RNA and DNA into selected cells in the body for obtaining a therapeutic effect by inhibiting gene expression associated with the pathogenesis, or correcting a genetic defect so as to obtain normal gene expression (Parra-Guillen et al., 2010). Vehicles used for gene delivery to transfer the genetic information into a cell are called vectors, which could be divided into two main groups: viral and non-viral (Liu and Huang, 2002). A successful approach depends on the effectiveness and safety of the delivery of therapeutic genetic materials (Wang et al., 2013). Thus, knowledge of the effects (pharmacodynamics) and kinetics (pharmacokinetics) of therapeutic agents in the body is critical (Urso et al., 2002).

In the current work, mathematical and computational modeling is used for analysis, optimisation and control of gene delivery. A pharmacokinetic model was developed using the experimental data reported in the literature to describe the transport of genetic material between the various compartments. The model development involves estimating the rate constants of siRNA translocation over the intracellular barriers. Parameter estimation is one of the important steps for developing high-fidelity mathematical models. This is based on minimizing an objective function (the error function) given by the summed square of the difference between the experimental data and the model predictions. Techniques that are used for solving the parameter estimation problems are chosen based on the model types of the system. The focus of this study is on systems involving ordinary differential equations (ODEs). The parameter estimation methodology is based on a simultaneous algorithm in which the solution of the model equations is the part of the overall optimisation problem. The dynamic model equations are converted into a system of algebraic equations using an Artificial Neural Network (ANN) transformation (Dua and Dua, 2012).

Then, the estimated rate constants of siRNA translocation are used to simulate the intracellular pharmacokinetic profile of siRNA delivery by non-viral nano-carrier (NC) to represent the intracellular endosome and cytoplasm exposure of siRNA. Following that, the developed models are used to obtain optimal gene infusion rates. The current work includes developing a pharmacodynamic model and providing a balance between two main and conflicting objectives: high efficacy and low toxicity. So, a mathematical formulation for optimal control of siRNA delivery was developed. Various practical constraints are also incorporated for the gene delivery problems. One of the important and practical constraints is on the time to take into account the cell multiplication so that the effect is manifested before the cell division takes place. Thus, a mathematical modeling and control framework to incorporate the time constraints for cell division has been developed.

**References**

AL-DOSARI, M. S. & GAO, X. 2009. Nonviral Gene Delivery: Principle, Limitations, and Recent Progress. AAPS Journal, 11, 671-681.

DUA, V. & DUA, P. 2012. A Simultaneous Approach for Parameter Estimation of a System of Ordinary Differential Equations, Using Artificial Neural Network Approximation. Industrial & Engineering Chemistry Research, 51, 1809-1814.

LIU, F. & HUANG, L. 2002. Development of non-viral vectors for systemic gene delivery. Journal of Controlled Release, 78, 259-266.

PARRA-GUILLEN, Z. P., GONZALEZ-ASEGUINOLAZA, G., BERRAONDO, P. & TROCONIZ, I. F. 2010. Gene Therapy: A Pharmacokinetic/Pharmacodynamic Modelling Overview. Pharmaceutical Research, 27, 1487-1497.

URSO, R., BLARDI, P. & GIORGI, G. 2002. A short introduction to pharmacokinetics. Eur Rev Med Pharmacol Sci, 6, 33-44.

WANG, W. W., LI, W. Z., MA, N. & STEINHOFF, G. 2013. Non-Viral Gene Delivery Methods. Current Pharmaceutical Biotechnology, 14, 46-60.

# 431558 Using the Fluorescence Lifetime As a Cytometric Parameter to Sort Unlabeled, Heterogeneous Populations of Cells

**Abstract:**

Time-resolved flow cytometers enable the measurement of the fluorescence lifetime as a singular parameter to count and sort cells. The fluorescence lifetime is the average time a fluorescent molecule spends in its excited state prior to relaxation to the ground state; it is a unique photophysical trait that, with time-resolved detection systems, can provide quantitative information for cellular assays. In the case of biomedical assays using proteins, polymers, fluorophores, chromophores, quantum dots, nanoparticles, and other targets for diagnostics, instruments that provide sensitive relaxation kinetic measurements may help bridge qualitative with quantitative information. Yet it is well known that to observe an excited state is difficult; radiative decay happens over nearly infinitesimal (e.g., nanoseconds to near-instantaneous) timescales. Moreover, it is non-trivial to capture heterogeneous time-resolved information from several different excitable molecules when excitation is observed from single cells and particles in fluidic states. Work by our laboratory has focused on making high throughput multi-exponential fluorescence decay detection possible in flow cytometry. In this contribution we demonstrate the utility of fluorescence lifetime measurements for cell sorting in a label-free fashion. That is, cells are detected based on their autofluorescence expression. The fluorescence lifetime of the autofluorescence specie, NADH, is measured and used as a sorting parameter. A range of mammalian cells were taken for this demonstration, including normal breast cells, breast cancer cells, and lymphocytes. We mix various populations of these cells to prove discrimination based on the NADH lifetime, which shifts distinctly when in a bound state during metabolic alteration (highly proliferative cell states vs. quiescent). The NADH lifetime shifts measured range from 1.1 to 2.4 nanoseconds, and the sorting purity ranged depending on the relative mixtures of the tumorigenic and non-tumorigenic cells. We aim to use label-free, lifetime sorting as a way to screen for circulating tumor cells, rapidly and without the need to pre-treat extracted blood cells from the body during this diagnostic procedure.

# 431672 Expansion of Human Mesenchymal Stem Cells in Fibrous Bed Bioreactor

**Abstract:**

### Introduction

Human mesenchymal stem cells (hMSCs) isolated from various adult tissues emerge as primary candidate in cell therapy for a wide range of diseases. The translation of hMSC into clinical application depends on the delivery of therapeutically competent hMSCs at the sufficient quantity. While most cell therapy products to date are based on autologous cells for immunologic compatibility, hMSCs have been used in allogenic transplantation as a potential "off-the-shelf" product. In the production of cells for allogenic transplantation, a scalable production process is crucial for the commercial scale cell production with required quality attributes. The conventional stirred-tank bioreactor provides the scalable possibility, but however has non-homogeneous hydrodynamic environment due to mechanical agitation. As hMSCs are highly sensitive and responsive to the mechanical forces, the potential impact of the biomechanical stimuli on hMSC phenotype and therapeutic potency is yet to be determined. Therefore, the present study investigates the feasibility of hMSC expansion in a 2.5 L CelliGen® 310 Bioreactor packed with Fibra-Cel® disks, providing the scalable potential with low and homogeneous shear stress microenvironment.

### Materials and Methods

A 2.5 L CelliGen® 310 bioreactor (Eppendorf) is used in the current study. The 2.5 L glass vessel is jacketed by a water bath layer and capped with a stainless steel head plate. Ports on the head plate are designed for pH and dissolved oxygen probes, seeding and sampling tube, and gas transfer. The vessel was autoclaved with 25.0 grams of Fibra-Cel® disks (Eppendorf) fixed within the packed bed. After sterilization, growth media was delivered into the vessel with pump first, and then hMSCs obtained from plate culture were seeded into the vessel. In the seeding process, 100 rpm agitation speed was used for 1 hour for even-distributed cell attachment. Temperature, pH, and mixture gas (5% CO2 and 95% air) were automatically controlled and monitored by the microprocessor control system. The agitation speed was adjusted to 50 rpm for cell expansion after seeding. 10% of working volume was replaced daily with fresh media and samples were collected before and after media change. The bioreactor operation was repeated twice using identical operation procedure.

### Result and conclusion

In our result, hMSCs on the surface of fibers were estimated to be exposed to a much lower shear stress of 0.54 dyne/cm$^2$ in average compared to the conventional stirred-tank bioreactors with suspended microcarriers, based on the dimension and packing density of the fibrous bed and the agitation induced pressure drop through the fibrous bed. After 9 days of expansion, a 9.2-fold increase in cell number with the population doubling (PD) time of 2.8 days (67.2 hours) was achieved. Specific glucose consumption and lactate production are 12.48 pmol/cell/day and 20.95 pmol/cell/day, respectively. Scanning electron microscope (SEM) and fluorescence DAPI staining images of harvested Fibra-Cel® disks show the uniform distribution of hMSCs on the fibers and the potential space for continuous cell growth. hMSCs harvested from the bioreactor maintained their properties based on the analysis of hMSCs surface markers, colony forming unit-fibroblasts (CFU-F) capacity, and multilineage differentiation ability. The results demonstrate the feasibility and the potential of the fibrous bed bioreactor for large scale hMSC expansion.

### References


1. T.W. Chiou, S. Murakami, D.I. Wang, W.T. Wu, A fiber bed bioreactor for anchorage dependent animal cell cultures: Part I. Bioreactor design and operations, Biotechnology and bioengineering, 37 (1991) 755-761.
2. F. Zhao, T. Ma, Perfusion bioreactor system for human mesenchymal stem cell tissue engineering: dynamic cell seeding and construct development, Biotechnology and bioengineering, 91 (2005) 482-493.
3. J. Kim, T. Ma, Perfusion regulation of hMSC microenvironment and osteogenic differentiation in 3D scaffold, Biotechnology and bioengineering, 109 (2012) 252-261.

# 432000 Delivery of Chemotherapeutic Drugs to Sensitize TRAIL Activity Against Gioblastoma Multiforme (GBM) Cells By Targeted MMP-Sensitive Peg-Hydrogel Microparticles

**Abstract:**

**Delivery of chemotherapeutic drugs to sensitize TRAIL activity against Gioblastoma Multiforme (GBM) cells by targeted MMP-sensitive PEG-Hydrogel Microparticles**

Targeting cancer cell surfaces and microenvironment via micro- and nano-range delivery vehicles is a promising approach for the treatment of cancer. These delivery vehicles improve the efficacy of chemotherapeutic drugs through functionalization and controlled release properties of delivery systems. TNFα-related apoptosis-inducing ligand (TRAIL) has been recently investigated as a therapeutic protein that induces cell death in human, particularly in cancer cells. Despite its success, some malignant cancer cell types can develop resistance to TRAIL-induced apoptosis. As a result, drug repositioning becomes more important to sensitize cancer cells to TRAIL- induced apoptosis.

In this study, we used integrin-targeting ligand functionalized and matrix-metalloproteinase (MMP)-sensitive poly (ethylene glycol) (PEG) hydrogels to synthesize microgels via water-in-water emulsion technique. Emulsions were prepared in dextran, where visible-light-induced photolymerization was used for microgel synthesis. The physicochemical properties of microgels were characterized via Dynamic Light Scattering (DLS) and Scanning Electron Microscopy (SEM).

The widely used FDA-approved drug, 03TR4, was used in combination with TRAIL. 03TR4 was encapsulated with PEG microgels and evaluated for its ability to sensitize GBM cells to TRAIL-induced apoptosis. Our results showed that cytotoxic potency of TRAIL could be improved with 03TR4, which suggests a cooperation between the drug and TRAIL. This study is the first report about the delivery of 03TR4 drug from a biocompatible system for effective treatment of cancer cells.

# 433481 The Relationship Between Oxidative Stress and Breast Cancer Cell Adhesion Using an Ischemia/Reperfusion Injury Model

**Abstract:**

Circulating tumor cells that metastasize from a primary tumor are the leading cause of cancer death in the United States. These circulating tumor cells adhere to the endothelial lining of the vascular system and extravasate through the lining to form deadly secondary tumors. Many pathways of oxidative stress can be found in pathological diseases, including one specific injury that leads to superoxide anion production. Ischemia, or the blockage of blood flow, can occur during extensive surgery in the removal of primary tumors. During an ischemic cascade, cells are introduced to a hypoxic environment and depleted of oxygen, which promotes purine nucleotide breakdown and produces xanthine and hypoxanthine. In the presence of xanthine oxidase, superoxide anions can be generated as a side reaction during the formation of uric acid. Superoxide anions can promote cellular dysfunction and even cell death. If there is a restoration of flow to the area subject to ischemia, reperfusion can allow for inflammatory responses to persist, permitting an even greater imbalance in the consumption of reactive oxygen species (ROS) and natural occurring antioxidants. If circulating tumor cells are within the system at the time of ROS generation, cancer cells may adhere to the endothelium more readily. The overall objective of this study is to investigate ROS generation under hypoxic conditions and to determine the role of this ROS generation in cancer cell adhesion to endothelial cells. Static studies were performed where a Hypoxanthine/Xanthine Oxidase (HXXO) injury model was used to mimic a hypoxic environment in the circulatory system and to determine its effect on the adhesion of MDA-MB-231 invasive breast cancer cells to human umbilical vein endothelial cells (HUVECs). ROS generation was measured using CellROX Green Reagent oxidative stress indicator to indicate superoxide anions and other general forms of oxidative stress present in the HUVECs. Endothelial cell monolayers that were exposed the HXXO injury model showed a 60% increase in ROS generation compared to monolayers not exposed to HXXO, providing confirmation that superoxide formation in endothelial cells can be induced in the presence of HXXO. In addition, MDA-MB-231 invasive breast cancer cells showed a significant increase in adherence to injured monolayers of endothelial cells correlating cellular dysfunction and increase in response to oxidative stress with the promotion of cancer cell binding. This positive correlation of ROS generation and cancer cell binding supports the use of antioxidant therapeutics to mitigate injuries that provide oxidative stress pathways and ultimately limit the occurrence of secondary tumor formation. Currently, a flow system using Ibidi $VI^{0.4}$ μ-slides is being used to induce flow over endothelial monolayers. Flow-regulated studies can be used to mimic the circulatory system physiologically and allow for endothelial cells to produce reactive oxygen species naturally by the physical stopping and starting of flow over endothelial monolayers. Results of ROS generation and cancer cell binding in this flow system will also be presented.

# 433443 Engineering Bioresponsive Nanoparticles for Drug Delivery

**Abstract:**

Nanoparticle penetration inside tumors is a critical step in the treatment of cancer disease. Although significant progress has been made in developing nanoparticle based drug delivery systems, the effects of the physical and chemical properties of nanoparticles on their transport behavior into tumor mass are largely unknown. In our lab, we work on characterizing the effects of nanoparticle geometry and its bioresponsive behavior in penetrating a solid tumor mass. We synthesize rod-shaped nanoparticles using bioresponsive polymers and anti-cancer drugs. Our previous results showed that rod-shaped nanoparticles enhance tumor targeting and uptake 2.5 times more than the conventional spherical nanoparticles. Bioresponsive drug nanoparticles release drugs 1.5 times more in the dense tumor mass than a non-responsive environment.Our results suggest the preparation of an efficient drug carrier system for NP design, drug release within the tumor.

# 433940 Role of Tryptophan in the Activity of IRE1Î± Protein Activation during ER Stress and the Induction of Epithelial-Mesenchymal Transition (EMT)

**Abstract:**

The endoplasmic reticulum (ER) is an organelle involved in protein synthesis, folding and post-translational modifications. Inositol requiring enzyme -1α (IRE1α), a type I transmembrane (TM) protein possessing both kinase and endonuclease functions, is the most conserved and well characterized ER stress sensor protein. IRE1α plays an essential role in maintaining ER homeostasis and its failure to reestablish homeostasis is associated with diverse diseases. ER stress is known to activate IRE1α by leading to dissociation of the BiP protein from the luminal domain of IRE1α.

Long chain free fatty acids (FFAs), especially palmitate (PA), are known to induce ER stress in many types of cells and contribute to diseases, including cancers. Cancer cells frequently exhibit alterations in fatty acid metabolism to sustain growth and proliferation, fulfill energy requirements and provide metabolites for anabolic processes. It has been observed that FFA levels are significantly elevated in the plasma of hepatocellular carcinoma (HCC) patients.

Recent studies implicate an involvement of PA in ER stress and diseases such as cancer, however the mechanisms that regulate ER stress and the unfolded protein response (UPR) (e.g. the IRE1α signaling) and, in particular, the role PA plays in this process is unclear. Recently we found that PA induces epithelial to mesenchymal transition (EMT), i.e. cells start to display more mesenchymal markers and phenotypes. In this study, we investigated wild-type (WT) IRE1α and IRE1α with a W457A mutation with respect to its splicing activity and migration ability to determine the functional importance of the tryptophan residue in activation of the ER stress pathway and its impact on EMT. The cells expressing WT IRE1α are more invasive and aggressive with an elevated resistance to apoptosis as compared with cells expressing the mutant IRE1α. Wound healing studies in mouse embryonic cells (MEF) IRE1-/- knockout cells transfected with IRE1α showed increased migration ability on treatment with PA as compared to IRE1α with a W457A mutation.

Studying the effect of fatty acids on the invasiveness and metastatic ability of cells through ER stress sensors like IRE1α provide insights into the biochemical changes induced by fatty acids that contribute to cancer progression.

# 434381 Impacts of Varying Hydrodynamics Conditions and Mass Transfer on Staphylococcus Aureus Biofilms and Their Resistance to Antimicrobial Agents

**Abstract:**

**Impacts of Varying Hydrodynamics Conditions and Mass Transfer on *Staphylococcus aureus* Biofilms and Their Resistance to Antimicrobial Agents**


Abiye Mekonnen and Patrick Ymele-Leki

Chemical Engineering Department, Howard University


Abstract

Biofilms occur when microorganisms attach to a surface and form a biolayer, which acts as a single organism. *Staphylococcus aureus* is one of the most common causes of bacterial biofilm infections. Growth of these biofilms and their ability to resist antimicrobial agents are the main cause of numerous tenacious and chronic bacterial infections. Bacteria within biofilms tend to be up to 1,000 times more resistant to antibiotics than free-floating bacteria. Biofilm development can be affected by several factors such as media concentrations, hydrodynamics conditions (fluid shear stress and shear rate), types of bacteria, and cell-cell communication (quorum sensing). This project aims to determine the isolated and combined effects of media concentration and shear rate on the application of antibacterial drugs. We hypothesized that the combined effect of media concentration and shear rate applied to the bacterial biofilm has statistically significant influence on the ability of antimicrobial drugs to degrade bacterial biofilms. To test this hypothesis, biofilms were grown under different media concentrations (2X, 1X, 0.5X, 0.1X) of TSB and fluid shear rates (ranging from 10 $s^{-1}$ to and 300 $s^{-1}$) conditions in the presence or absence of known antibacterial agents. The data found during this study suggest that: 1) Media concentration has significant impact on the growth of bacteria in a static system. Increasing nutrient strength, from 0.5X to 2X, corroborated with a significant increase in bacteria growth. 2) The growth of a biofilm at a constant shear rate was not dependent on media concentration under the conditions tested in a flow system. 3) At lower shear rates, biofilm growth curves overlap in the early stages of growth, regardless of media concentration. 4) The data also reveal that biofilms exposed to low media concentration continue to grow for much longer periods of time (> 2 hours) than biofilm grown with higher media concentration. 5) The previously mentioned differences in biofilm characteristics had significant impact on the action of a given antimicrobial agent and their dosage. This research can have different implications on combating the infections/contaminations caused by biofilms in both biological and industrial settings.

# 434359 Integrin-Growth Factor Receptor Co-Signaling during Breast Cancer Metastasis

**Abstract:**

Dissemination of cancer breast cancer cells from the primary tumor is an essential step in disease mortality. Dissemination requires invasion into the surrounding stroma, intravasation into the local blood or lymph vasculature, convective transport via circulation to distant tissues, extravasation out of the vasculature at the secondary tissue site, and, finally, growth in said secondary tissue. Many have noted that only a small fraction of cells (less than 1%) that are capable of invading into the surrounding stroma (the first step of metastasis) actually contribute to secondary tumors. Therefore, colonization and growth at a distant tissue site, such as the brain, bone, lung, and liver, is a rate-limiting step in the overall metastatic cascade.

We hypothesize that the ability of breast cancer cells to attach to their surroundings (via integrins) and respond to growth factors (via growth factor receptors) at these distant tissues dictates the ability of these cells to survive and grow. In this *in vitro* study, we created biomaterials that covalently and permanently present different extracellular matrix (ECM) proteins, such as collagen, fibronectin, and laminin, to cells, so that we may control integrin binding. We then studied the response of different breast cancer cell lines that span the disease spectrum to growth factors (epidermal growth factor (EGF), fibroblast growth factor 1 (FGF-1) and the insulin-like growth factor 1 (IGF-1)) on these ECM-controlled biomaterials. We quantified membrane extension (protrusion), morphological polarization, and spreading rate of SkBr3s (representing the HER2-overexpessing subtype of breast cancer), MCF-7s (representing luminal A), and MDA-MB-231s (triple negative). We found that SkBr3s were extremely sensitive to EGF stimulation on all three integrin-binding conditions, while FGF-1 stimulated membrane protrusion only on the fibronectin-treated surface, and SkBr3s were insensitive to IGF-1. MCF-7 cells did not show increases in membrane protrusion or spreading rate under any growth factor or ECM conditions. The MDA-MB-231s showed increases in membrane protrusion with EGF, FGF-1 and IGF-1 on collagen, and only with EGF, and FGF-1 on laminin. Thus far, our findings show that breast cancer subtype has some relationship to growth factor sensitivity and integrin-binding, and we are currently investigating whether this applies to other cell lines within these same disease subtype categories. We are now quantifying if these phenomenological observations correlate with growth factor receptor phosphorylation on the different integrin-binding regimes using a high-throughput bead-based ELISA platform. Coupled with our ongoing studies in tumor dormancy, we suspect that the ability of cells to respond to growth factors in different integrin-binding conditions is related to their ability to grow aggressively at certain tissues and not others. We propose this type of *in vitro* screening approach as a way to predict for tumor dormancy or relapse, and in the long-term, as a theranostic platform for patient prognosis.

# 435752 Analysis of Kinetics of a Serpin-Serine Protease Complex

**Abstract:**

Alpha-1-Antitrypsin (A1AT) Deficiency (A1AD) is a potentially debilitating inherited condition. It is associated with reduced levels of active A1AT being secreted from the liver thus resulting in an imbalance between A1AT and Human Neutrophil Elastase (HNE) in the lungs and predisposing those affected to early-onset emphysema. Further, certain genetic mutations lead to an accumulation of aberrant A1AT in the liver potentially leading to cirrhosis. No cure exists, and current approaches for diagnosis are insufficient as illustrated by the fact that A1AD is underdiagnosed and diagnosis is often delayed. In this work, a synthetic substrate (N-Succinyl-Ala-Ala-Ala-p-Nitroanilide), an enzyme-inhibitor system consisting of Porcine Pancreatic Elastase (PPE, an often-used surrogate for HNE) and a "normal" form of A1AT, and a mathematical model based on irreversible inhibition are used to understand rates of product formation and modes of inhibition. *Tian et al.* [Biochemistry, 21,1982] used basic conservation equations previously developed, with a modification for irreversibility, to describe enzyme inhibition for complexes traveling down an irreversible pathway but in which a reversible complex (*e.g.*, Michaelis-type) is formed along the way. That method goes on to describe approaches for exploring whether such inhibitors behave in competitive, non-competitive, or an uncompetitive manner, and the approaches reported there are used herein to characterize the A1AT-PPE system. The physiologic level of A1AT in the body varies between and within individuals and is associated with, but not totally diagnostic for, the development of lung and/or liver complications. It has been shown in this work that such modeling could be a useful tool to describe the interaction between A1AT and PPE, and as such, be important for the development of new diagnostic approaches for A1AD.

# 435250 Evaluation of Nano-Formulated Neuroprotective Against Traumatic Brain Injury

**Abstract:**

Repetitive traumatic brain injuries (TBI) can lead to progressive degeneration of brain tissue, resulting in chronic traumatic encephalopathy. Generation of oxidative stress (reactive oxygen species (ROS)), neuro-inflammation and mitochondrial dysfunction have all been implicated in the process of neurodegeneration associated with TBI. The loss of structure and function of neurons results in symptoms similar to neurodegenerative diseases such as Parkinson's, Huntington's and Alzheimer's. In this work we investigate the ability of nanoparticles to enhance the delivery of antioxidants to neuronal cells and protect against oxidative stress and impaired behavior from TBI.

Polyanhydride nanoparticles can be designed to elicit unique cellular responses by stimulating particle internalization and directing intracellular trafficking. By encapsulating the payload within the polyanhydride particles sustained release of these compounds over time can be achieved. Through modifications in the polymer chemistry, surface functionalization and particle size, the release kinetics of payloads and the fate of the particle within target cells can be tailored. In this work, the neuroprotective payload of interest is an anti-oxidant, mito-apocynin, which has been tailored to target mitochondria. The biodistribution of the mito-apocynin is evaluated through quantification of the compound in both target (i.e., brain) and off target tissues. Efficacy of the nano-formulated treatment is evaluated through a combination of cognitive function (i.e., locomotor activity) and neuronal loss.

# 419483 Cleaning in Place to Remove Allergenic Materials

**Abstract:**

Allergens are usually harmless substances. However, in susceptible individuals, they can trigger an adverse immune response, which can result in death. Consequently, individuals with food allergies need to be extremely careful of what they consume. Food labelling helps these individuals avoid the potentially serious consequences from consumption of foods they are allergic to.

Cleaning regimes have been developed to monitor soil and allergen removal, and understand the impact of varying cleaning regimes on their removal. A test section in a pilot scale cleaning rig is fouled with allergenic material, and cleaned at varying flow rates, temperatures, detergent concentrations, and cleaning times. Pressure, conductivity and electrical resistance tomography are used to monitor the dynamic process, and the cleaning fluid assessed at regular intervals to quantify the allergenic material present. By monitoring allergen levels, food manufacturers can increase the safety of products to allergenic individuals.

Bovine milk is used as the allergenic material, as it is easily available and cheap to obtain. Granulated milk is mixed with water to form a paste which is rolled to a controlled volume onto the test section, and inserted into the cleaning rig. The test section has a viewing section, enabling the volume of the fouling material to be observed during cleaning. This enables determination of when the test section is visually clean and consequently, measurements can be taken to determine if it is then also allergenically clean. An alternative test section also exists. This electrical resistance tomography test system provides a non-intrusive and non-invasive way to image the conductivity distribution within a vessel or pipeline through a number of measurements made from surrounding sensing elements.

Different sized areas of the test section were fouled and cleaning was conducted at different temperatures for the pre-rinse. During the pre-rinse, water is flushed through the test section until it is visibly clean. When the test section is visibly clean, a swab test is conducted on the test section to determine if more than 20 μg of milk protein is present in the swabbed area. An increase in temperature was found to reduce the cleaning time. However, higher temperatures increased the quantity of milk that would not remove from the test section during the pre-rinse phase. The volume of milk in the test section does not appear to affect the cleaning time. Cleaning time is determined by the area of milk covering the test section, as removal of small areas of milk are responsible for a significant quantity of the cleaning time. The results suggest that cleaning time is also determined by the adhesiveness of the milk paste to the test section. It was found that the conductivity of the pre-rinse water increased during cleaning. Conductivity measurements have also been used to determine the levels of cleanliness through calculations of the natural log of the root mean squared of conductivity.

In this study, an alkaline clean is conducted using a caustic based solution as the detergent. In this detergent phase, diluted detergent is circulated around the cleaning rig at a range of temperatures and flow rates. Samples of the cleaning fluid are removed from an outlet of the test section at regular intervals to quantify the allergenic material present using enzyme linked immunosorbent assay and bicinchoninic acid protein assay. The measured allergenic material in the fluid increases with cleaning time. The contaminated water is then drained from the rig and the rig filled with clean water again, for the final rinse.

Frequently, cleaning in place systems over clean food processing equipment and food manufacturers add precautionary labelling to their products. This study has deepened the understanding of the impact of various cleaning regimes on soil and allergen removal, preventing the need to over clean, therefore saving energy, time and money, and increasing process efficiency and ultimately eliminating the need for precautionary allergenic labelling.

# 422329 Techno-Economic and Environmental Assessment of Citric Acid Production By Aspergillus Niger (Rapid Fire)

**Abstract:**

Techno-economic and environmental assessment of Citric Acid Production by Aspergillus niger
Chacón Yessica1, Daza Laura1, Cardona Carlos A1**.
1 Universidad Nacional de Colombia sede Manizales, Instituto de Biotecnología y Agroindustria, (+57) (6) 8879400 ext 55354.
** Corresponding author: ccardonaal@unal.edu.co
Citric acid is an organic acid with high application in food industry as acidulant in soft drinks and confectionery [1]. In 2009, it was reported a world production of approximately 1,6 million ton [2]. Different fermentation techniques in which predominates submerged fermentation in different feedstocks of accessible cost (sucrose, glucose and cane molasses, among others) give high yields [3] [4]. In Colombia sucrose from sugar cane is used as substrate for citric acid production using the strain Aspergillus niger, based on the advantage of this microorganism to have an extra cellular mycelial bound invertase that hydrolyzes sucrose under acid conditions in the fermentation [1].

However submerged fermentation, different factors influence the production of other compounds reducing yields and obtaining oxalic or gluconic acids instead of citric acid. In this way the carbon source and other macronutrients as nitrogen, phosphorus and trace elements represent determinant factors in conidia growth and acid accumulation [5]. Carbon source in the medium should be adequate, since a low concentrations byproducts as the oxalic acid are generated and at high concentrations polyol can be obtained, leading to low yields of the interest product [4]. Nitrogen source leads to decrease pH, being ammonium sulfate of the most use because it increases the production of citric acid without oxalic acid formation. However, fermentation media with low levels of ammonium nitrate to prevent the formation of oxalic acid and without affecting the C/N ratio have been reported [2] [4]. High phosphorus contents easily lead to the formation of oxalic acid, hence low levels of phosphate are necessary. The most appropriate phosphate source use is potassium dihydrogen phosphate [2] [4]. Finally, different divalent metals (such as Zn, Mn, Fe, Cu and Mg) as trace elements inhibit the accumulation citric acid, while others (Mg, Cu) balance this effects and are necessary for the fungus growth and acid production [4] [6]. Fermentation process of A. niger for citric acid production is carried out at temperatures between 28°C-30 °C during 5-14 days with continuous agitation and tolerable aeration rate. High pH leads to oxalic and gluconic acids production due to the glucose oxidase enzyme activation [4] [5]. Low pH is advisable to performe this fermentation.

The aim of this work was to evaluate the citric acid production from glucose as carbon source using A. niger as microorganism. The study was carried out in the fermenter Liflus GX (BIOTRON, INC-HANIL). Biomass concentration, citric acid and total reducing sugars concentrations were followed using the techniques of dry weight, HPLC and the Saffran-Densted Method to 418 nm, DNS (3,5-dinitrosalicylic acid) Method to 540 nm, respectively. The techno-economic analysis was developed using the commercial software Aspen Plus V8.2 (ASPEN TECHNOLOGY USA) to simulate the process for obtaining the material and energy balances. These balances were use in the commercial software Aspen Process Economic Analyzer V8.2 (ASPEN TECHNOLOGY USA) to calculate the production cost per kg of citric acid considering the reported values in Colombia for tax rate (25%)m interest rate (17%), electricity and water costs [7] [8]. The study was performed for a period of ten years using the straight-line method for depreciation of capital. The environmental assessment was performed using the Waste Algorithm Reduction (WAR), developed by the Environmental Protection Agency (EPA) [9].

The results obtained allowed concluding the determinant factors in the citric acid production to reach an economic and environmental competitiveness of the most common production process in the field of biotechnology.

REFERENCES

[1] K. Kavanagh, "Antibiotics, Enzymes and Chemical Commodities from Fungi," in Fungi: Biology and Applications, L. Joh Wiley & Sons, Ed. 2005, pp. 129–134.
[2] K. Kirimura, Y. Honda, and T. Hattori, "Citric Acid," in Comprehensive Biotechnology, Second Edi., Elsevier B.V., 2011, pp. 135–142.
[3] J. A. Velásquez, D. Beltrán, L. Padilla, and G. Giraldo, "Obtención de ácido cítrico por fermentación con aspergillus niger utilizando sustrato de plátano dominico hartón ( musa aab simmonds ) maduro," RevistaTumbaga, vol. 5, pp. 135–147, 2010.
[4] H. S. Grewal and K. L. Kalra, "Fungal Production of Citric Acid," Biotechnol. Adv., vol. 13, no. 2, pp. 209–234, 1995.
[5] J. W. Deacon, "Introduction: the fungi and fungal activities," in Fungal Biology, Fourth Edi., Blackwell Publishing, Ed. 2006, pp. 1–15.
[6] M. M. Madigan, J. M. Martinko, and J. Parker, "Microbiología Industrial y Biocatálisis," in Brock. Biología de los microorganismos, 10th ed., PEARSON, Ed. 2004, pp. 957–985.
[7] J. A. Dávila, V. Hernández, E. Castro, and C. A. Cardona, "Economic and environmental assessment of syrup production. Colombian case," Bioresour. Technol., vol. 161, pp. 84–90, 2014.
[8] J. Moncada, M. M. El-Halwagi, and C. A. Cardona, "Techno-economic analysis for a sugarcane biorefinery: Colombian case," Bioresour. Technol., vol. 135, pp. 533–543, 2013.
[9] M. I. Montoya R., J. A. Quintero S., Ó. J. Sánchez T, and C. A. Cardona A., "Evaluación del impacto ambiental del proceso de obtención de alcohol carburante utilizando el algoritmo de reducción de residuos," Rev. Fac. Ing., vol. 36, pp. 85–95, 2006.

# 425979 Heterologous Biosynthesis and Diverse Applications for the Nonribosomal Peptide-Polyketide Siderophore Yersiniabactin (Rapid Fire)

**Abstract:**

Yersiniabactin (Ybt) is a mixed nonribosomal peptide-polyketide natural product natively produced by the pathogen *Yersinia pestis*. The compound enables iron scavenging capabilities upon host infection and is biosynthesized by a nonribosomal peptide synthetase featuring a polyketide synthase module. This pathway has been engineered for expression and biosynthesis using *Escherichia coli* as a heterologous host. In the current work, the biosynthetic process for Ybt formation has been improved through the incorporation of a dedicated step to eliminate the need for exogenous salicylate provision. Upon doing so, the compound was tested in parallel applications that highlight the metal chelating nature of the compound. In the first application, Ybt was assessed as a rust remover, demonstrating a capacity of ~40% compared to a commercial removal agent and ~20% relative to total removal capacity. The second application tested Ybt in removing copper from a variety of non-biological and biological solution mixtures. Success across a variety of media indicates potential utility in broad settings that include environmental and biomedical contexts.

# 426951 Construction of a Xylose-Induced System in Lactobacillus Brevis for Efficient Glutathione Production

**Abstract:**

**Abstruct**: Prioritization of sugar consumption is a common theme in bacterial growth. However, growth studies show that *Lactobacillus brevis* simultaneously consumes numerous carbon sources and appears to lack normal hierarchical control of carbohydrate utilization. Through this phenomenon, a new controlled expression system to target heterologous proteins to cytoplasm is described for *Lactobacillus brevis* CGMCC1306. It is based on the use of a xylose-inducible *lactobacillus* promoter, $P_{xylA}$. The capacities of this system to produce cytoplasmic was tested using a bifunctional glutathione synthetase gene (*gshF*) from *Streptococcus thermophilus*. Fermentation results showed that recombinant *L. brevis* CGMCC1306 (pNZ-$P_{xylA}$-gshF) harboring this system was able to accumulate GSH up to 36.9 mg/g cell dry weight in MRS medium with adding 10 g/L xylose and 0.6 g/L additional cysteine. Growth experiment with GFP as a reporter indicated that the induction of this system happened immediately in the present of xylose. These results confirm the potential of this expression system as an alternative and useful tool for the production of proteins of interest in *L. brevis*.

# Induction of Microbial Defense Mechanisms

**Elizabeth Frayne, PhD**

Frayne Consultants and University of Phoenix

## Introduction

Previous work has shown that thio-phosphate is universally taken up by cells and incorporated into nucleic acids leading to their nuclease resistance. The subsequent accumulation of mRNA in the cell results in competition for a limited number of ribosomes. Generally proteins derived from mRNAs that are less stable than average, preferentially accumulate at the expense of proteins derived from mRNAs that are more stable. This leads to a global shift in distribution of proteins in the cell. *Normally the level of those proteins increased is on the order of 5-20 fold* while for those proteins that are decreased, they are reduced by roughly the same amount. In some instances a protein can be elevated *greater than 40 fold, suggesting the activation of a regulatory switch* as a secondary response to the global stabilization of mRNA in the cell with the analogue. The present work is a description of some of those proteins induced to such a high degree in E. coli and Bacillus subtilis. The results suggest that several defensive proteins were activated and thus the analogue may be useful a useful tool in identifying defensive proteins in microbes.

## Objectives

1) *To compare the global shift in distribution of proteins in the secretome of B. subtilis with that of the E. coli proteome*
2) *To identify proteins increased to levels substantially above the average or expected amount*
3) *To characterize those proteins induced to exceptionally high levels suggestive of a regulatory response being activated with the use of the phosphate analogue*

## Materials & Methods

Cells were grown in minimal media or a semi-synthetic media depleted of bactopetone, using thio-phosphate as the sole source of added phosphate. Thio-phosphate was purified to biological grade. Secreted proteins from B. subtilis were TCA ppt. prior to analysis on 2D polyacrylamide gels. Total cellular protein was isolated from E. coli and used in 2D gel analysis. Protein spots were measured as a percent of the total stained mass using computer analysis of gel images undertaken with laser scanning and SameSpots software (Non-Linear Dynamics). The MW and pI of selected protein spots was used to identify proteins in TagIdent (SIB, ExPasy resource portal). The compute pI/MW tool was also used to generate pI and MW values for known secreted proteins with or without a signal sequence by entering the appropriate name for the gene from UniProt (2002-2014). The analysis was limited to those proteins with a pI between 4 and 7.

## Results with B subtilis

Secreted proteins from *B. subtilis* were analyzed during late log cultures (6-8 hrs) from cells grown with or without the phosphate analogue. The ppt. proteins were separated via 2D polyacrylamide gel electrophoresis (pH 4-8 IEF gradient) and the spots quantitated via scanning densitometry. 626 of the protein spots in the control gel and 785 in the treated gel had values of at least .01% of the total protein mass detected. 18% of the detected spots had at least a 5 fold difference and 35% at least a 3 fold difference in expression levels. The average fold increase was 8.1 but ranged from 3 to 77 fold. The average fold decrease was 6.6 but ranged from 3 to 29 fold. A total of 229 proteins were induced at least 3 fold and 75 inhibited at least three fold. In general, those secreted proteins with an initial mass of 0.05% or less were enhanced while those proteins with a mass greater than 0.25% were reduced. Shown below is the difference gel generated by overlaying a 2D gel image of proteins from cells grown with thio-phosphate over a control 2D gel image.

Fig. 1. Difference gel of B. subtilis secreted proteins. The green represents proteins that were enhanced in the thiophosphate gel and the magenta indicates those reduced in comparison with a control gel. White appears where spots had similar intensity.

16 secreted proteins were potentially identified and most of these were previously known to be secreted such as as Ybdn, Ggt, YusC, XynD. Several putatively secreted proteins were identified with signal sequences YdjN, Yrak, YkuE, and YkyA (all decreased) as well as two without signal sequences Ysdc and YopN (a prophage protein with presumably an atypical transport mechanism). Two proteins were elevated more than 40 fold namely, Ybdn enhanced 77 fold and YopN enhanced 48 fold. Ybdn has recently been shown to be a class III bacteriocin.

Vpr, Ybdn, MsmX, YusC, and Ysdc all of have reported mRNA half-lives less than the cellular average of 7.5 min. (Note, longer mRNA half-lives are detected when micro-array studies are done). These findings are consistent with the ability of thio-phosphate to lead to the preferential accumulation and translation of mRNAs with half-lives less than the average of total mRNA.

## Results with E. coli

Total cellular protein was isolated from E. coli grown with or without the phosphate analogue and separated in 2D polyacrylamide gels. Computer scanning of 2D gels for E. coli detected 815 spots and of these spots, 28% showed significant changes. 125 spots were increased an average of 5.5 fold while a total of 104 spots were decreased an average of 4.9 fold. In general those spots with an initial mass of .038% or less were enhanced while those with an initial mass of .176% were reduced. The results were similar to what was found for secreted proteins in B. *subtilis*.

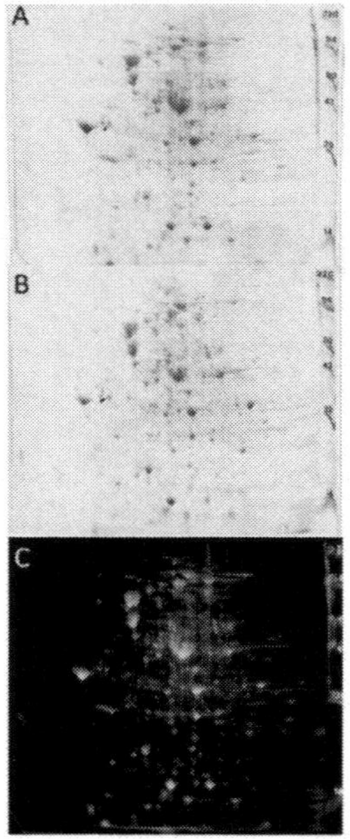


Fig. 2. 2D gel analysis of E. coli proteins, using 200 ug of total cellular protein from either cells grown in normal media in (A) or cells grown in thio-phosphate containing media (B). Each were separated by a pH 4-8 IEF gradient and then run on a 10% polyacrylamide gel and stained with Coomassie blue. Arrow indicates tropomysoin standard at MW 33,000 and pI of 5.2. (C) Difference gel of gels in A and B. Green represents proteins which are enhanced in the thiophosphate gel and the red indicates those reduced. White appears where spots are of similar intensity.

Two proteins were identified that were increased more than 40 fold, namely theT1R1 type one restriction enzyme EcoR124II R protein, increased 150 fold and a putative diguanylate cyclase increased 41 fold.

## Conclusions

Exposure of two diverse bacterial species to a phosphate analogue triggered the upregulation of several proteins involved in protective responses. In Bacillus a protein with a MW of 28,624 and a pI of 6.6 was found to be upregulated 77 fold as judged by quantitative analysis of 2-D protein gels. This protein corresponds to the recently identified bacteriocin, YbdN, which has a MW of 28,699 and pI of 6.79. While the mechanism by which YbdN kills bacteria is not known, other class III bacteriocins are known to attack the cell wall or disrupt the cell membrane. Another, possibly protective protein induced 48 fold in Bacillus, is the YopN prophage derived protein. This protein may prevent superinfection with another bacteriophage. In E. coli a protein of ~120,000 MW and a pI of 6.08 was upregulated 150 fold. This protein corresponds to the T1R1 restriction enzyme otherwise known as EcoR124II R protein. Restriction enzymes are well known for their role in defending bacteria and archaea from viruses. Overall, the accumulation of these proteins suggests the activation of a regulatory switch with the analogue via mRNA stabilization and/or a response to physical changes of the modified nucleic acids themselves. The approach taken may be useful in identifying other bacteriocins, microcins, colicins, and defensive proteins in a variety of microbes. The use of the analogue also greatly facilitates the detection of minor proteins on 2D gels.

## References

1. E. Frayne. Microbial production of nuclease resistant DNA, RNA, and Oligo Mixtures. *US Patent 7,125,982, related patents, and patents pending.* Washington DC: Patent and Trademark Office (2006).
2. Frayne, E. (2013). Global Alterations in the Secretomes of *Bacillus subtilis* and S. *cerevisiae*. *Proceedings of the Annual Meeting of the American Institute of Chemical Engineers*, San Francisco, CA.
3. SIB, ExPASy portal (2015). TagIdent. Retrieved from http://web.expasy.org/tagident
4. Burgess-Cassler, A., Johansen, J., Santek, D., Ide J., and Kendrick N., (1989). Optimization of 2D Gel Transblotting for Host Cell Protein Analysis. *Clin. Chem. 35*: 2297.
5. Jamal, M. T. and Mudarris, M. S. A. (2010). Separation of YbdN bioactive protein from Bacillus subtilis isolated from the Red Sea algae Sargassum sp. with bioactivity against antibiotic resistant bacterial pathogens. *JKAU: Mar. Sci. 21*: 53-64.
6. Tjalsma H, et al. (2004). Proteomics of protein secretion by Bacillus subtilis: separating the "secrets" of the secretome. *Microbiol. Mol. Biol. Rev. 68:207–233.*
7. Hambraeus, G. et al. (2003). Genome-wide survey of mRNA half-lives in *Bacillus subtilis* identifies extremely stable mRNAs. *Mol Gen Genomics 269*: 706-714.
8. Han, M.J. and Lee, S.Y. (2006). The Escherichia coli proteome: Past, present and future prospects. *Microbiol. Mol. Biol. Rev.* 70:362-439.

## Contact

efrayne@fraynconsultants.com, efrayne@email.phoenix.edu

# 423406 Techno-Economic and Environmental Assessment of Supercritical Pumpkin (Curcubita maxima) Extraction

Wednesday, November 11, 2015
Exhibit Hall 1 (Salt Palace Convention Center)

**Ashley Caballero**, Instituto de Biotecnologia y Agroindustria, Universidad Nacional de Colombia, Manizales, Colombia, Angela Idárraga, Instituto de Biotecnología y Agroindustria, Universidad Nacional de Colombia, Sede Manizales, Manizales, Colombia and Carlos A. Cardona, Instituto de Biotecnología y Agroindustria, Departamento de Ingeniería Química, Universidad Nacional de Colombia, Manizales, Colombia

**Techno-economic and Environmental Assessment of Supercritical Pumpkin (*Curcubita maxima*) extraction**

Caballero Ashley S[1], Idárraga Ángela M[1], Cardona Carlos A[1**].

[1] Universidad Nacional de Colombia sede Manizales, Instituto de Biotecnología y Agroindustria, (+57) (6) 879400 ext 55354.

[**] Corresponding author: ccardonaal@unal.edu.co

## ABSTRACT

The pumpkin (*Curcubita maxima*) is a vegetable native of Central and South America widely spread in the world, which is used due to its nutritional and medicinal properties. The pumpkin is mainly composed by aminoacids, vitamins, fat, fiber, minerals and sugars[1].

The Supercritical Fluid Extraction (SFE) is a mass transfer operation which occurs above the solvent supercritical point having advantages of high selectivity, and high purity extracts. The SFE is applied for chemical process, food, pharmaceutical, oil industry and waste removal [2]. The aim of this work was to obtain antioxidants and phenolic compounds from pumpkin seeds, pulp, and peels. The antioxidants are very important in pharmaceutical and food industry to reduce oxidative stress in cells and treat human diseases[3]. This process was performed through supercritical extraction technology as an alternative to conventional soxhlet extraction, and microwave-assisted extraction process.

The raw materials were dried and ground at 45°C to avoid the degradation of thermolabile compounds. The extractions were performed using supercritical CO2 which increases yields and avoids degradation of components[4] and ethanol as co-solvent[5][6]. The three different pressures (200, 250 and 300 bar) were evaluated at 50°C, which was ideal to avoid degrade phenolic compounds[7]. The total antioxidant activity and total phenolic content of the samples were determined by Folin-Ciocalteu test and DPPH method[7][8][9].

Techno-economic assessment at the propused processes was carried out employing Aspen Plus V8.2 for generating the mass and energy balances, while the economic assessment was performed using the commercial software Aspen Process Economic Analyzer V8.2. Environmental assessment to determine the potential environmental impact was made using Waste Reduction Algorithm (WAR) per kilogram of product.

As a result of this prefactibility study, the potential of pumpkin to obtain antioxidant compounds through supercritical extraction technology was shown.

## References

[1] T. Kikuchi, M. Takebayashi, M. Shinto, T. Yamada, and R. Tanaka, "Three new multiflorane-type triterpenes from pumpkin (Cucurbita maxima) seeds," *Molecules*, vol. 18, pp. 5568–5579, 2013.

[2] A. M. Velasquez, "La tecnología de fluidos supercríticos , un proceso limpio para el sector industrial *," *Prod. + Limpia*, vol. 3, pp. 98 – 104, 2008.

[3] I. Cerón, J. Higuita, and C. Cardona, "Capacidad antioxidante y contenido fenólico total de tres frutas cultivadas en la región andina," *Vector 5*, vol. 5, no. 2010, pp. 17–26, 2011.

[4] Ž. Knez, E. Markočič, M. Leitgeb, M. Primožič, M. Knez Hrnčič, and M. Škerget, "Industrial applications of supercritical fluids: A review," *Energy*, vol. 77, pp. 235–243, 2013.

[5] J. P. Maran and B. Priya, "Supercritical fluid extraction of oil from muskmelon (Cucumis melo) seeds," *J. Taiwan Inst. Chem. Eng.*, vol. 47, pp. 71–78, 2015.

[6] L. Miscibles, "C o m e n t a r i o s s o b r e s o l v e n t e s y s o l u b i l i d a d e s d e s u s t a n c i a s o r g á n i c a s," 2008.

[7] M. R. Rover and R. C. Brown, "Quantification of total phenols in bio-oil using the Folin-Ciocalteu method," *J. Anal. Appl. Pyrolysis*, vol. 104, pp. 366–371, 2013.

[8] W. Brand-Williams, M. E. Cuvelier, and C. Berset, "Use of a free radical method to evaluate antioxidant activity," *LWT - Food Sci. Technol.*, vol. 28, pp. 25–30, 1995.

[9] R. M. F. Vargas, M. S. T. Barroso, R. G. Neto, R. Scopel, M. A. Falcão, C. F. Da Silva, and E. Cassel, "Natural products obtained by subcritical and supercritical fluid extraction from Achyrocline satureioides (Lam) D.C. using CO2," *Ind. Crops Prod.*, vol. 50, pp. 430–435, 2013.

Tecnológico de Monterrey

15AIChE Annual Meeting, Salt Lake City, UT

CIDPRO Centro de Investigación y Desarrollo de Proteínas

# ALKALINE PEANUT PROTEIN EXTRACTION ASSISTED WITH MICROWAVES (MAE)

Ariana Ochoa-Rivas, Jessica González-Ruiz, Cristina Chuck-Hernández

Centro de Investigación y Desarrollo de Proteínas (CIDPRO), Escuela de Biotecnología y Ciencias. Instituto Tecnológico y de Estudios Superiores de Monterrey, Edificio CEDES PB, Av. Eugenio Garza Sada 2501, Monterrey, Nuevo León, 64849. A00793563@itesm.mx

## INTRODUCTION

Peanut is a leguminous with high protein content (~28%) and with an excellent amino acid profile, used mainly for oil extraction, yielding a protein-rich (~55%) by-product (DPF) mainly used for animal feed, providing an excellent opportunity to produce high-protein food ingredients (protein isolate) with a better economic and social benefit [1]. The traditional method used for protein extraction from vegetable sources is alkaline solubilization followed by acid precipitation, but when this method is used for protein extraction in DPF, the final protein yield is low because of the changes in protein structure during oil extraction, leading to unavailability of molecules [2]. Therefore, the pursuit of an alternative extraction process for DPF is desirable. In the last years the use of novel methods in food industry has arisen, standing out microwave because of their convenience from economic point of view and their efficient energy use, but the information available about the use of this technology in plant protein extraction is scarce [2]. Thus the aim of this research was the evaluation of microwave-assisted protein extraction (MAE) of peanut and the purity assessment of the final high-protein product (PPI).

## METHODOLOGY

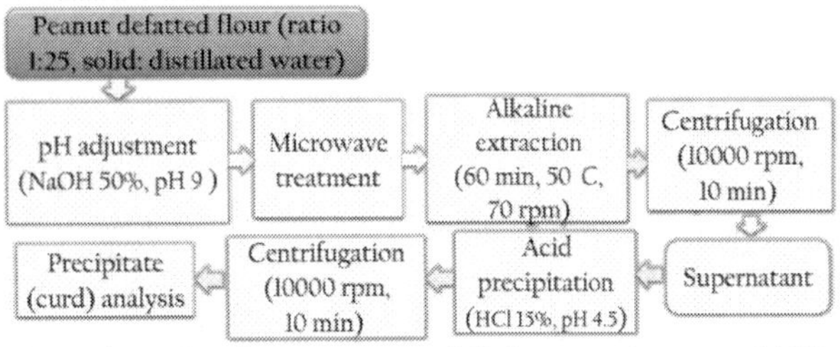


Fig. 1. Description of microwave assisted alkaline extraction process (MAE).

The microwave assisted alkaline extractions were made using DPF as raw material (54% of protein) and a commercial microwave (LG, MS1142GW, Mexico) varying: A) power (145, 290, 435 and 580 watts) and B) processing time (2, 4, 6 and 8 min). After the process, the curd was stored under refrigeration for later weight, protein (Kjeldahl) and moisture (oven method, 105 °C) determination. Then extraction yield (mass of protein in PPI/mass of protein in DPF *100) and purity (mass of protein in PPI/mass of PPI *100) were calculated.

## RESULTS AND DISCUSSION

The main effect of microwaves was detected in extraction yield, because the higher was the microwave power and treatment time, the higher was the extraction yield (see Fig. 2). The final curds had a high purity (>95%), with no difference among treatments.

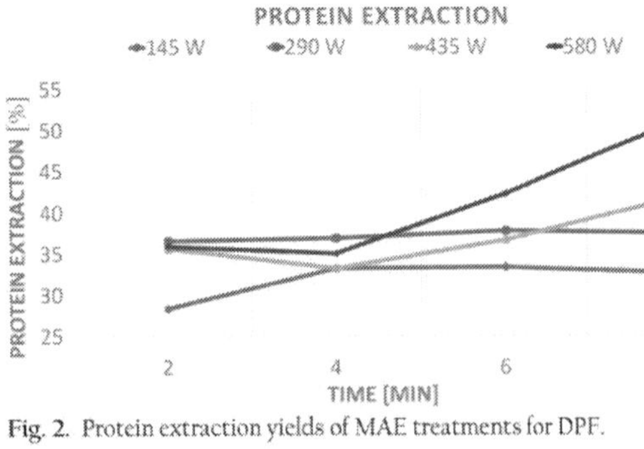


Fig. 2. Protein extraction yields of MAE treatments for DPF.

A higher time (10 min at 580 watts, with an augment of 63.63%) and power (725 watts for 8 and 10 min, with an augment of 59.52 and 76.93% respectively) were also evaluated. Despite their values were higher, the effect of the increment in time and power was minor (an augment with a minor slope) and the temperatures reached values near to the boiling point, therefore not higher conditions were tested.

In order to evaluate the effect of these treatments in other vegetable sources, soy and chickpea flours (obtained without aggressive heat treatments) were also tested (see Fig. 3).

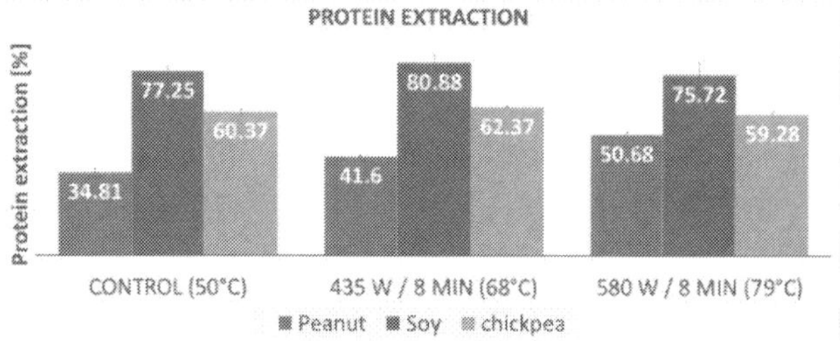


Fig. 3. Graphic comparison of the effect of MAE in the protein extraction from different sources (peanuts, soybean and chickpea).

Results showed that there is no effect of MAE in soy and chickpea extraction yield. As soy and chickpea storage proteins constitute 65–80% and 62% respectively of the total seed protein [3][4], control treatments extracted the total storage proteins contained in the flours without the need of the microwave treatment.

Bath and direct heating treatments (to 90 °C) were also tested, to probe if the effect of microwave treatment is related only with the increase in the temperature of the DPF dispersion.

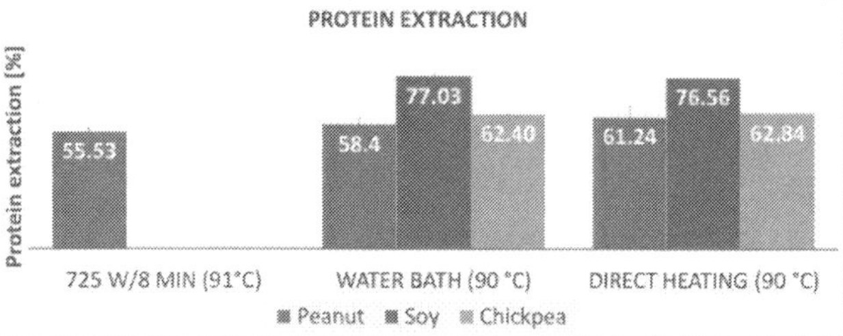


Fig. 4. Water bath and direct heating treatment (90 °C)

The effect of MAE on protein extraction is due to the rise in the temperature and when temperature is increased, the protein extraction increased too (see Fig. 4).

## CONCLUSION

- Microwave is a an efficient technology that can be used to improve peanut protein extraction from the by-product of oilseed industry that are subject to aggressive heat treatments during oil extraction.
- The effect of MAE on protein extraction is due to the increase in the temperature.
- Microwave is recommended to increase temperature because of the high heating velocity.

## REFERENCES

[1] Zhao, X., Chen, J., & Du, F. (2012). Potential use of peanut by-products in food processing: a review. *Journal Of Food Science And Technology-Mysore*, *49*(5), 521-529.

[2] Gayol, M., Pramparo, M., Nepote, V., Fernandez, H., & Grosso, N. (2013). Optimization of the protein concentration process from residual peanut oil-cake. *Grasas Y Aceites*, *64*(5), 489-496.

[3] Matina, M. (2009). Compositional, functional and sensory properties of protein ingredients. Graduate Theses and Dissertations. Paper 10482.

[4] Singh, D K and Rao, A S and Singh, R & Jambunathan, R. (1988). Amino acid composition of storage proteins of a promising chickpea (*Cicer arietinum L*) cultivar. *Journal of the Science of Food and Agriculture*, 43 (4), 373-379.

# 427104 The Effects of Different Bioprocessing Parametrs on Pectinase Production By Aspergillus Niger from Orange Peel Waste (Rapid Fire)

**Abstract:**

Pectinase is a heterogeneous group of enzymes that have a polysaccharides substrate and breaks down the pectin component which found in the plant cell wall into simple sugar and galacturonic acid. The pectinases or usually known as pectinolytic enzymes were extensively used in food industries that involve degradation of plant materials to fasten the fruit juice extraction process. Among different biofactories of pectinases, filamentous fungi such as *Aspergillus niger* were best known for the production and secretion of pectinase. Therefore, the objective of this research was to develop industrial culture medium and a cultivation strategy for the production and secretion of pectinases in a semi-industrial scale 16-L bioreactor by *A. niger*. In this study, the effect of medium composition on the production and secretion of pectinase was studied through the one-factor-at-a-time (OFAT) method and statistical approach. Therefore, the pectinase production from the new optimized production medium when pectin from orange peel as the carbon source which produced more than 64.83 % of pectinase when compared to unoptimized medium. Followed by batch cultivation in the semi-industrial scale 16-L bioreactor under uncontrolled pH yielded 51.35 % more pectinase compared with the cultivation under controlled pH of 5.5. Finally, the fed-batch cultivations employing a constant feeding mode under uncontrolled pH with the initial pH cultivation condition of 5.5 under two different feeding solutions which were full medium compositions of the production medium and concentrated sucrose. The higher pectinase yield was produced when feeding with concentrated sucrose about 429.95 U $mL^{-1}$ of pectinase activity compared to the full medium feeding only 138.27 U $mL^{-1}$. These all together will lead to the development of industrial process for pectinase production in semi-industrial scale bioreactor. Therefore, *A. niger* is best cultivated in pH-uncontrolled condition with DO consistent at 30 % level using the constant feeding strategy of concentrated sucrose.

# 427172 Optimization of Growth Medium and Bioprocessing Condition in Pilot Scale Stirred Tank Bioreactors for Production of Levan By Bacillus Subtilis (Rapid Fire)

**Abstract:**

Levan is extracellular polysaccharides of fructose and it can be found in plants and microbial products. Recent studies of levan were motivated for its role as antitumor, antioxidants, antiviral and anticancer agents with fibrinolytic action and were produced from various types of microorganisms. This study focused on cultivation of *Bacillus subtilis* to optimize production medium to produce levan in the shake flask levels including effect of bioprocessing parameters when cultivation in the semi-industrial scale 16L- bioreactor. Recent studies reported that the highest amounts of levans are produced from the production medium with commercial sucrose as the carbon source with minor quantity of alcohol precipitate. The un-optimized medium yielded the cell mass of 1.945 $gL^{-1}$ and 13.5 $gL^{-1}$ of levan from the shake flasks studies. and after medium optimization using statistical approach the cell mass increased to 2.26 $gL^{-1}$ and the production of levan also increased to 22.1 $gL^{-1}$. This showed increment of levan production up to 38.07% from statistical optimized medium compared to un-optimized medium. Batch cultivations in 16-L stirred tank bioreactors were carried out under uncontrolled pH and controlled pH condition by using the new optimized production medium generated from statistical tool. The new formulations were composed of (g $L^{-1}$): Sucrose, 130.0, Yeast extract, 6.3, $K_2HPO_4$, 5.0, and $MgSO_4.7H_2O$, 0.2. The result showed that bioreactor under uncontrolled pH was favorable compared to controlled pH condition in obtaining the highest amount of cell mass and levan production which gave about 5.06 $gL^{-1}$ and 26.65 $gL^{-1}$, respectively. This was almost 49.34% higher than those obtained in shake flask culture. However the yield of levan per cell mass was 5.64g $L^{-1}$ compared to 11.54 g $L^{-1}$from batch cultivation in the bioreactor due to the low cell mass production of *B. subtilis*. In conclusion, this study proposed a reliable industrial production medium and aneffective cultivation strategy for the cell mass and levan production in semi- industrial scale bioreactor.

**(623m) Improving Thermostability of Cis-Epoxysuccinate Hydrolase from Rhodococcus Opacus By Semi-Rational Designs and Directed Evolution**

Qiao, P., Zhejiang Uinversity
Wu, M., Zhejiang University
Zhu, L., Zhejiang Uinversity
Zhang, Y., Zhejiang Uinversity
Yang, L., Zhejiang University
Lin, J., Zhejiang University

*Cis*-epoxysuccinate hydrolases (CESH), belonging to soluble epoxide hydrolases (sEHs, EC 3.3.2.10), catalyze the reaction of hydrolyzing the oxirane ring of *cis*-epoxysuccinate (CES), adding one water molecule onto it and synthesizing L-(+)-tartaric acid. The first report about biologically preparing L-(+)-tartaric acid was carried out by Miura and his colleagues in 1977, and the similar enzymes with the same function were found in various kinds of bacterial afterwards. The gene sequence (GenBank DQ471957) of *cis*-epoxysuccinate hydrolases from *Rhodococcus opacus* was deposited in 2007 and the catalytic properties of it were also studied thoroughly. Though has been industrially utilized, the low thermal resistance of CESH has caused extra production costs in continuously maintaining low temperatures and repeatedly preparing new biocatalysts. Low temperatures lead to low reaction rate, as well as low solubility of substrate and product, thus reduce the overall efficiency and increase the consumption of water and electric power. Consequently, improving the thermal resistance of CESH will have a positive influence to the biologically industrial production of L-(+)-tartaric acid. Because of the fact that there is still no exact information about the 3D structure and catalytic mechanism of CESH, we employed vitro protein engineering strategies including semi-rational designs and directed evolution to stabilize the CESH from *Rhodococcus opacus*. After screening 10,000 colonies from directed evolution, we obtained mutant Q122R whose half-life time at 50°C ($t_{1/2,50°C}$) increased by 1.75-fold and the $T_{50}^{15}$ value (temperature under which the original activity of enzyme would decrease 50% in 15min) of it increased by 5.5°C. As was manifested in our work, the directed evolution was effective, but the screening of the mutant library was random. Thus, we turned to sequence-based semi-rational redesign and computational enzyme redesign. We compared the sequence of CESH with four thermostable enzymes from haloacid dehalogenase-like hydrolases (HAD) superfamily (GenBank cl21460), and predicted seven mutants. After analyzing the thermal resistance of those proposed mutants, we obtained F26V and I83R whose $t_{1/2, 50°C}$ increased by 2.47- and 0.21-fold and the $T_{50}^{15}$ value of them increased by 4.5 and 0.5°C individually. After constructing the homology model of CESH and carrying out simulated mutagenesis, we proposed ten mutants within which the mutants D8K and S90R turned out to be positive substitutions. The $t_{1/2, 50°C}$ of mutants D8K and S90R ascended by 1.98- and 0.35-fold and the $T_{50}^{15}$ value of them increased by 5.1 and 1.0°C respectively. To maximize the effects of the positive substitutions, we introduced site-saturated mutagenesis into amino acid residues Asp8, Phe26, Ile83, Ser90, and Gln122. After screening over 2,000 colonies from the site-saturated mutant library, we isolated mutant F26W whose $t_{1/2, 50°C}$ increased by 3.38-fold and the $T_{50}^{15}$ value of it increased by 7.2°C. At last, an integrated mutant whose substitutions included D8K, F26W, I83R, S90R, and Q122R was obtained, the optimum reaction temperature of it extended from 50°C to 60°C, the $t_{1/2, 50°C}$ increased by 33.49-fold and the $T_{50}^{15}$ value increased by 20.0°C. As far as we know, the final mutation obtained in this study is the most thermostable CESH.

# 433584 Solubility of Vanillin and Vanillic Acid in Supercritical Carbon Dioxide

**Abstract:**

Vanilla (*Vanilla planifolia* Andrews) has been highly prized for its aroma, and it is now considered the most important flavoring in food and perfume industries. The development of vanilla aroma takes place during processing of green vanilla beans to produce cured vanilla beans, during which the green and odorless fruits become black and highly aromatic pods. Vanillin and vanillic acid are components of vanilla aroma which is obtained from a natural fermentation process.

Since most of the components that provide the flavor are volatile and thermo-labile, extraction and purification methods of vanilla extracts determine the overall quality of the final product. Actually, the main extraction method for vanilla oleoresin is multistage lixiviation, or leaching at 80°C employing ethyl alcohol, but if the process does not have a good control of temperature, degradation of vanilla may occur and a loss of quality extract exists.

Supercritical extraction with $CO_2$ seems to be a good alternative. Due to carbon dioxide's relatively low critical temperature ($T_c$= 304.4 K), it is especially suitable for processing thermo-labile compounds. Information about solubility of solids in supercritical fluids is of great importance because supercritical fluids have several advantages over organic solvents about physical properties. Accurate experimental solubility data are fundamental to test theoretical, empirical or semi-empirical models. Semi-empirical models based on density are common approaches used to correlate experimental solid solubilities in supercritical fluids.

The aim of this work is to present new experimental solid solubilities of vanillin and vanillic acid in carbon dioxide, at 313, 323 and 333 K an pressures up to 30 MPa, based on the static-analytical method. The solubility data were correlated with the Mendez-Santiago and Teja model.

* Corresponding author: e-mail: **lgalicial@ipn.mx** (Luis A. Galicia-Luna)

# 435062 A Comparative Study of the Effects of Specific Growth Factors and Extracellular Matrix Proteins on Neurite Outgrowth (Rapid Fire)

**Abstract:**

Nerve damage due to injury and disease is a significant problem affecting millions of people. Current treatment options are limited and pose many problems, including low effectiveness and the possibility of infection. New technologies are being developed which seek to build upon and enhance the body's natural mechanisms of regeneration and repair. Tissue engineering is a relatively new field involving a wide range of applications, from engineering more robust tissues to stimulating healing of damaged tissues. The focus of this work is on stimulating nerve growth in dorsal root ganglia neurons. The effects of a combination of growth factors and extracellular matrix components on cellular development are analyzed in 2D and 3D culture conditions. Specifically, the effects of nerve growth factor, fibronectin, osteopontin, tenascin R, and tenascin C, alone and in combination, are discussed in relation to neurite outgrowth in dissociated dorsal root ganglia neurons. The goal is to determine conditions conducive to optimal nerve growth as measured by neurite length and density of proliferation as visualized by bright-field and fluorescence microscopy.

# 436709 Operational Mode Selection for Succinic Acid Production (Rapid Fire)

**Abstract:**

The optimal process mode to improve the performance of succinic acid production from biodiesel glycerol is assessed through the use of a predictive macroscopic model which is used for experimental design. Typically, biochemical process design is performed in laboratory scale, where different parameters are examined and optimised in batch mode before scaling up. Experimental cost can be significantly decreased by performing a limited number of key experiments, dictated through optimisation of a detailed model, that can predict the system's behavior in different conditions (e.g. different nutrients concentration, volume or process mode).

Succinic acid is one of the top-value added chemicals [3] and its production from glycerol, which is the major by-product of the bio-diesel production, is the bacterial fermentation under study. The system's behaviour has been examined based on a double substrate limitation model [1, 2] in batch mode. The developed model was proved to be useful for designing purposes as it takes into consideration process parameters that influence the extra-cellular availability of dissolved $CO_2$, and therefore the production of succinic acid [3, 4, 5], as well as parameters accounting for the effect of scaling up the system.

Having computed the effective ranges, in terms of yield and productivity, of medium composition and process parameters for each scale, process optimisation is performed at the bench scale for 4 different process modes (batch, fed- batch, continuous, continuous with cell recycle). In all cases, the maximization of yield and productivity is sought and the initial conditions along with the feeding strategy (flow rate, feed concentration, time intervals) and the recycling strategy (filtrate & retentate rate) where applicable, comprise the degrees of freedom. The model is validated against experimental results and the optimal process mode is assessed. The possibility that different process modes might be more suitable at different scales is also examined by applying the suggested methodology in pilot scale.

References

[1] A. Rigaki, C. Webb, C. Theodoropoulos, Chem. Eng. Transactions. 2013; 35: 1033 - 1038.

[2] A. Vlysidis, M. Binns, C. Webb, C. Theodoropoulos, Biochemical Engineering Journal 58 -59.2011;1-11

[3] T. Werpy, G. Peterson, US Department of Energy.2004; DOE/GO-102004-1992.

[4] M.J. Van der Wef, M.V. Guettler, M. K. Jain, J.G. Zeikus, Arch Microbiol. 1997; 167: 332-342

[5] Y. Xi, K. Chen, J. Li, X. Fang, X. Zheng, S. Sui, P. Wei, J Ind Microbiol Biotechnol. 2011; 38:1605-1612

# 438533 High Strength Acetic Acid Production in Acetobacter pasteurianus by Transcriptional Analysis and Enhancement of Alcohol Respiratory Chain Run

**Abstract:**

**Background:** Nowadays, achieving high strength vinegar fermentation is still important to vinegar producers. And the molecular mechanism of acetic acid production and resistance is hotspot in this area. We disclose and improved by transcriptional analysis and enhancing alcohol respiratory chain (ARC) run of *Acetobacter pasteurianus*.

**Method:** A better strain *Acetobacter pasteurianus* FS2-4 which is improved from HS-1 was got through a combination of modified UV irradiation and nitrous acid mutation after acetic acid shock. Under a batch culture in a 7L fermentor, the bacteria are harvested when $OD_{600}$ is 1.0. The transcript profile about acetic acid production and resistance of *Acetobacter pasterurinaus* is examined by qRT-PCR. Twenty genes related to tricarboxylic acid (TCA) cycle, acetic acid production and resistance system are chosen. Based on above analysis, the ARC was chosen to be enhanced to improve the production.

**Results:** The maximum acetic acid yield of *Acetobacter pasteurianus* FS2-4 increased 26.6% and the initial acidity tolerance increased from 20g/l to 50 g/l. The transcriptional result shows that acetic acid produciton and resistance system is up regulagted while TCA cycle is down regulated. Among acid resistance systems, it shows kdpA, clpB, ompA are enhanced and fapA,B, frbC, purE are down regulated while aatA, Gro EL, Gro ES aren't changed. Except suc B, all other gene in TCA cycle are down regulated. And we also found acetification rate showed strong positive correlation with ARC activity. So, activity of ARC enzymes, ubiquinone biosynthesis and oxygen availability were taken into account to improve ARC run. Results disclosed that acetification could not be fully inspired until the above three aspects being balanced properly. Therefore, to achieve high strength fermentation, some precursors of alcohol respiration associated factors were added along with an appropriate aeration rate for repeated batches. Obviously, average acetification rate was enhanced to 2.21±0.02 g/L/h with the optimized method, while the original rate was only 1.78±0.02 g/L/h.

**Acknowledgments**

This work was supported by a grants from National Natural Science Foundation of China (No. 31301540), the National High Technology Research and Development Program of China (863 Program) (No. 2012AA021201).

# 411010 Redox Properties of Food-Derived Polyphenols

**Abstract:**

Polyflavanols (procyanidins), anthocyanins, and phenolic acids are known naturally-occurring antioxidants. Polyphenolic structures of these antioxidants enable one-electron transfer reactions to quench radicals; the reaction is reversible if the antioxidant forms stable semiquinone radicals. Because of its aromatic, polyphenolic structures, procyanidin extracted from sorghum reportedly have several-fold greater peroxyl radical-quenching ability than phenols and Trolox. Limited information is available on the chemical constituents responsible for the antioxidant activity of sugarcane extracts. For example, kokuto, "black sugar" in Japanese, is produced by evaporating water from sugarcane juice at elevated temperature. Like sugarcane molasses, kokuto contains phenolic acids, phenyl glycosides, phenyl propanoid glycosides, and phenyl propanoids. The objective of this study was to understand the relationships between the structure (by solid-state NMR and fluorescence/UV-visible spectrophotometry) and redox property of naturally occurring antioxidants: grape seed, evaporated sugarcane juice, and sorghum grain. Fully oxidized and reduced forms of sugarcane, sorghum, and grape seed extracts were prepared using Fremy's salt. Solid and liquid-state NMR techniques were then utilized to investigate the structural components responsible for the redox reactivity. Iron (III) reduction kinetics was examined, and compared with DPPH, acid-butanol, and vanillin assays.

# 424355 In silico Analysis of Anaerobic Methane Oxidation By Methanosarcina Acetivorans

**Abstract:**

There is a renewed industrial interest in capturing methane, major constituent of greenhouse and shale gas. Anaerobic oxidation of methane (AOM) has been evaluated to be more efficient in terms of carbon conversion of methane to valuable products. However, there has been limited success in isolation and characterization of anaerobic methanotrophs due to the fact that they are often found in consortia of organisms and, thus, are not easily cultured in lab. Interestingly, trace methane oxidation has been demonstrated in the methanogenic archaeon *Methanosarcina acetivorans*. While the pathways describing the metabolism of native substrates of this organism have been well established, relevant pathways for methane oxidation remain unexplained. This talk is aiming at addressing thermodynamically feasible routes for anaerobic methane oxidation by *M. acetivorans* leading to proposal of optimized pathways for methane (co)utilization and candidate biofuel production. To this end, an updated genome-scale metabolic model of *M. acetivorans* is introduced (iMAC868 containing 868 genes, 840 reactions, and 712 metabolites) by integrating information from two previously reconstructed metabolic models (i.e., iVS941 and iMB745), modifying 17 reactions, adding 19 new reactions, and revising 64 gene-protein-reaction (GPR) associations based on newly available sources. The new model establishes improved predictions of growth yields on native substrates and is capable of correctly predicting the dispensability for 27 out of 28 gene deletion mutants. The iMAC868 model enables prediction of thermodynamically feasible pathway for co-utilization of methane and bicarbonate in the presence of ferric, nitrate, sulfate, and manganese dioxide as external electron acceptors. We find that the reversal of the acetate to methane pathway is the prominent route for AOM by *M. acetivorans*, which produces acetyl-CoA, the precursor for the production of ethanol, 1-butanol, and iso-butanol. We also find that the methylotrophic pathway must carry a minimum mandatory flux to ensure thermodynamic feasibility. The implications of acetate and $CO_2$ production, and the interplay between externally supplied electron acceptors and various by-products are discussed. This effort paves the way in informing the search for thermodynamically feasible ways of (co)utilizing novel carbon substrates in archaea, in general, and in *M. acetivorans*, specifically, since the archaeon possess a diverse carbon utilization capability and its genetic tools are well studied among archaea.

# 424627 Genome Scale Loss-of-Function Screening in *Saccharomyces Cerevisiae* Using Crispr/Cas9 System

**Abstract:**

The baker's yeast, *Saccharomyces cerevisiae*, represents a useful host for heterologous production of therapeutic proteins, biofuels and value-added chemicals. Resistance to harsh industrial fermentation conditions, such as low pH, high temperature or inhibitory products, is thus a desired trait. Despite the great effort in engineering high-performing yeast strains, only minimal success has been achieved, largely due to the low efficiency and throughput of yeast genome engineering. Previously, we developed a homology-integrated CRISPR/Cas (HI-CRISPR) system capable of knocking out single genes in *Saccharomyces cerevisiae* with near 100% efficiency. Here we sought to exploit this system for rapid, genome wide loss-of-function screening to improve complex yeast phenotypes. A total of 24765 homology integrated crRNAs were designed to target 6458 open reading frames (ORFs) including essential genes, with an average of 4 crRNAs targeting each ORF. The crRNA library was synthesized on chip, pooled and cloned into an ultrahigh copy number plasmid already expressing Cas9 and tracrRNA. A yeast mutant library with single gene disruption can be rapidly generated in approximately 6 days. Coupled with next generation sequencing, our method is suitable for both positive and negative screening. Furthermore, iterative rounds of genome wide mutagenesis can be conducted using the same plasmid library to enrich multiple mutations into a single cell, enabling the previously unfeasible screening of yeast mutant libraries with multiple knock-outs. We envision our method to be a useful tool for yeast metabolic engineering as well as understanding yeast genetics.

# 430632 New Yeast Strains Optimized for Metabolic Engineering for Biobased Products with Reduced Ethanol Production

**Abstract:**

The yeast *Saccharomyces cerevisiae* is a favored industrial organism due to its robustness, low nutritional requirements, high tolerance to low pH, suitability for large-scale fermentation, and ease of genetic manipulation. However, this yeast preferentially converts fermentable sugars into ethanol over any other products of interest. Therefore, yeast strains engineered to produce other products are suboptimal due to their production of ethanol as a major byproduct that significantly diminishes the yield of the desired product. Strains have been developed to overcome this central challenge in yeast metabolic engineering by deleting either pyruvate decarboxylases, or alcohol dehydrogenases involved in ethanol formation, which dramatically reduces ethanol formation. However, if these enzymatic activities are completely eliminated, the resulting strains are unable to grow on glucose, a preferred industrial feedstock. Restoration of growth of these strains on glucose can be achieved through directed evolution, but the resulting strains have impaired fitness and are difficult to work with. We have developed a new yeast strain that produces almost no ethanol, has improved fitness when growing on glucose, and overcomes many of the limitations that exist in currently available strains. This optimized strain can be engineered to produce different fuels or chemicals of interest from glucose, with significantly enhanced yields.

# 431916 Instating the Wood-Ljungdahl Pathway for Carbon Dioxide Fixation into the Solventogenic *Clostridium Acetobutylicum*


**Abstract:**

Atmospheric $CO_2$ has increased largely due to the burning of our depleting fossil fuel resources. Decreasing $CO_2$ emissions has become an increasingly appealing topic, and much research has been dedicated to understanding the biological $CO_2$ fixation route naturally found in acetogens. Acetogenic bacteria capture and utilize $CO_2$ along with electron sources such as $H_2$ or CO using the Wood-Ljungdahl pathway. This pathway has been shown to be the most energy efficient pathway for the production of both biomass and metabolic products [1]. Of particular interest is the production of butanol, a great alternative to current transportation fuels. Most acetogens primarily produce acetate as a primary fermentation product, and butanol titers, if at all exist, are much lower than those reported in *Clostridium acetobutylicum* [2, 3]. Therefore, we have investigated $CO_2$ sequestering enzymes found in the Wood-Ljungdahl Pathway by heterologous expression of these complex proteins in *C. acetobutylicum*. This presentation will focus on the expression of the two core carbon sequestering enzymes that form the CODH/ACS complex (carbon monoxide dehydrogenase and acetyl-CoA synthase) through in vivo and in vitro studies and assays. This will allow us to directly study activity and determine the feasibility of combining carbon sequestration with solvent production. The combination of the native solvent producing and the heterologous carbon sequestration abilities aims to create a strain capable of producing chemicals from cheap and abundant, renewable carbon resources.



[1] Fast, A. G. & Papoutsakis, E. T. Stoichiometric and energetic analyses of non-photosynthetic $CO_2$-fixation pathways to support synthetic biology strategies for production of fuels and chemicals. *Curr. Opin. Chem. Eng.*1, 1–16 (2012).

[2] Papoutsakis, E. T. Engineering solventogenic clostridia. Curr. Opin. Biotechnol. 19, 420–9 (2008).

[3] Bruant, G., Lévesque, M.-J., Peter, C., Guiot, S. R. & Masson, L. Genomic analysis of carbon monoxide utilization and butanol production by Clostridium carboxidivorans strain P7. *PLoS One* 5, e13033 (2010).

# 433174 Deregulating Escherichia coli Non-Growth Metabolism for High Productivity Chemical Synthesis without Carbon Loss to Growth

**Abstract:**

Substantial feedstock carbon (e.g. glucose) is converted to biomass during cell growth and is unavailable for product formation. Product synthesis is commonly growth-coupled, so a dilemma arises: continual growth production strategies are necessary for productivity, but these methods limit product yields and introduce processing/separation challenges. Ideally, carbon fed to non-growing cells would be solely and rapidly converted to product. However, *E. coli* cells entering stationary phase naturally (e.g. nutrient limitation) have drastically reduced overall metabolism. Therefore, deregulating non-growth metabolism is essential to achieving efficient stationary phase chemical production.

We have evaluated the effect of metabolic gene deletions coupled with different nutrient limitations on central carbon metabolism. We have identified promising non-growth conditions that generate *E. coli* with rapid glucose uptake rates and organic acid secretion profiles that would be ideal catalytically active, non-growing cells. We will use metabolomics to identify regulatory bottlenecks in central carbon metabolism for future strategies to increase non-growth metabolism and divert carbon to valuable products. This work should be applicable to improving the yields and productivities for the biosynthesis of a wide range of different chemicals and fuels.

# 432293 Improving the Production of Mvl in E. coli By Directed Evolution

**Abstract:**

**Improving production of bMdVL in *E. coli* by directed evolution**

**Yi-Shu Tai, Jilong Wang, Kechun Zhang**

b-Methyl-d-valerolactone (bMdVL) is an attractive monomer which can be polymerized to amorphous polyester with a low $T_g$ at -58 C. Previously, we developed a total biosynthetic pathway of bMdVL from glucose in engineered *E. coli*. However, the low activities of downstream enoate reductase, OYE2 or YqjM, in the pathway have limited the production titer. To address this challenge, we have developed directed evolution platforms to screen enzymes with improved activities. The first platform exploits a *Ralstonia* strain which we isolated from soil samples. This strain can use bMdVL as the carbon source to grow, but cannot use the precursor, anhydro-bMdVL (AMVL), for growth. Therefore, when anhydro-bMdVL was fed to the *Ralstonia* strain containing a library of enzyme mutants, the improved variants can be selected. In the second platform, we use toxicity of AMVL for screening. We found that 15 g/L of AMVL would significantly inhibit cell growth. Cells harboring more active enoate reductase can consume excess AMVL and restore cell growth. Overall, these selection platforms enable us to improve the activity of the pathway enzymes for bMdVL. This can further reduce the production cost of this novel biobased monomer used for soft polymer.

# 436420 Engineering Redox Homeostasis and Aldehyde Detoxification to Improve Lipid Production in Y. Lipolytica

**Abstract:**

*Yarrowia lipolytica*, an oleaginous yeast, can naturally accumulate large quantity of neutral lipids using a variety of carbon sources. Previous metabolic engineering efforts working on the acyl-CoA related pathways (Tai M *et al*, Metabolic engineering, 2013) have resulted in efficient triacylglyceride producers by increasing the carbon flux towards malonyl-CoA and sequestrating fatty acyl-CoAs in neutral lipids. One major obstacle for efficient production of lipids in *Y. lipolytica* is pertinent to the unique nitrogen starvation conditions which predispose the cell in an unfavorable physiological states and lead to relatively low productivity and yield. We have recently identified an oxidative stress defense mechanism to regulate lipid biosynthesis in Y. lipolytica. By scavenging reactive oxygen and aldehyde species, we were able to optimize the cell physiology and morphology and achieved very high level of oil content and lipid productivity in our previously engineered *Y. lipolytica* lipid overproducers. The strategies reported in this study represent a promising solution to develop a yeast biorefinery platform that potentially upgrades low value carbons to high value commodity chemicals.

# 433065 Single-Cell Transcriptional Analysis Reveals Principles of Neuronal Phenotype Organization and Plasticity

**Abstract:**

Single-cell analysis has shown consistently that a significant amount of transcriptional heterogeneity exists within and across differentiated cell-types. This transcriptional heterogeneity is pervasive across post-mitotic neurons in the brain and conflicts with the traditional expectation that a neuronal phenotype consists of functionally identical neurons that respond uniformly to synaptic and neuromodulatory inputs. High-throughput "-omic" level analysis suggests that transcriptionally heterogeneous neurons, within a neuroanatomical phenotype, function within a complex molecular organization in the brain. We aim to determine the functional relevance of this transcriptional heterogeneity (if it exists) in order to gain insight into how variability contributes to neurogenic control of physiological systems in health and disease.

To this end, we analyzed the transcriptional responses of individual neurons in distinct brain nuclei to specific physiological perturbations. We generated a high-dimensional gene expression dataset measured from hundreds of single neurons collected from the nucleus tractus solitarius (NTS), a brainstem nucleus involved in regulating cardiovascular homeostasis. NTS neurons were collected from rats undergoing an acute hypertensive challenge in order to measure their transcriptional responses to this blood pressure perturbation. We used multivariate analytical techniques such as principal component analysis and multidimensional scaling to analyze the high-dimensional gene expression dataset representing the transcriptional states of neurons, and identified a gradient of neuronal subtypes that aligned with synaptic input-types corresponding to blood pressure changes and to cardiovascular demands involved in homeostatic regulation. Also we used our recently developed fuzzy logic-based gene network modeling methodology to identify distinct gene networks that correspond to these neuronal subtypes and which drive heterogeneous gene expression across NTS neurons.

Additionally, single-cell analysis of the suprachiasmatic nucleus (SCN), a hypothalamic brain nucleus that regulates circadian rhythm, provided an alternative perspective of the organization of cell-interaction networks. Our analysis of hundreds of SCN neurons responding to a perturbed circadian cycle revealed a previously unidentified neuronal population, characterized by mRNA expression of pituitary adenylate cyclase-activating peptide (PACAP), a peptide typically thought to be an input to the SCN, rather than something produced endogenously. In addition to the traditionally defined SCN neuron-types that express arginine vasopressin (AVP) or vasoactive intestinal peptide (VIP) in a region-specific manner, the presence of the PACAP-expressing neuronal phenotype indicates that the SCN cellular interaction network is more complex than previously understood. Moreover, functional gene expression behavior across single SCN neurons correlated more strongly with receptor expression for AVP, VIP, and PACAP rather than with the mRNA expression of the neuropeptides themselves. This receptor-based correlation suggests that gene regulation driving the transcriptional state of SCN neurons is governed by inputs and coincides with the input-driven behavior observed in the NTS.

In conclusion, analysis of transcriptomic data of single neurons from distinct brain nuclei has revealed unexpected molecular organizational structures within neuroanatomical phenotypes. These organizational structures suggest that transcriptional heterogeneity reflects an adaptive response to inputs and provide a more complex perspective of how adaptive gene regulation drives neuronal state within and across phenotypic populations.

1. Park J, Brureau A, Kernan K, Starks A, Gulati S, Ogunnaike B, Schwaber J, Vadigepalli R. 2014. Inputs drive cell phenotype variability. Genome Res.
2. Park, J., Ogunnaike, B., Schwaber, J., & Vadigepalli, R. (2014). Identifying Functional Gene Regulatory Network Phenotypes Underlying Single Cell Transcriptional Variability. *Progress in Biophysics and Molecular Biology*, *117*(1), 87–98.

Research Support: NIH (F31 AA023143-01 NIGMS R01GM083108, NHLBI R01HL111621, NIAAA T32 AA007463)

# GLOBAL SHIFT IN THE CYTOSOLIC PROTEINS OF HUMAN HEK293 CELLS AND E. COLI INDUCED WITH A PHOSPHATE ANALOGUE CORRELATES WITH MRNA STABILITY AND RELATIVE PROTEIN ABUNDANCE

Elizabeth Frayne, PhD
Faculty UOP

## INCORPORATION OF THIO-PHOSPHATE INTO CELLS

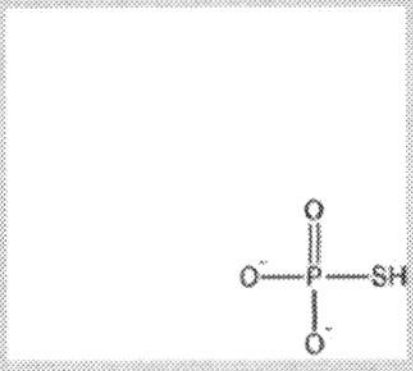

Thiophosphate can potentially enter cells via **GLYCOLYSIS** or mitochondrial oxidation

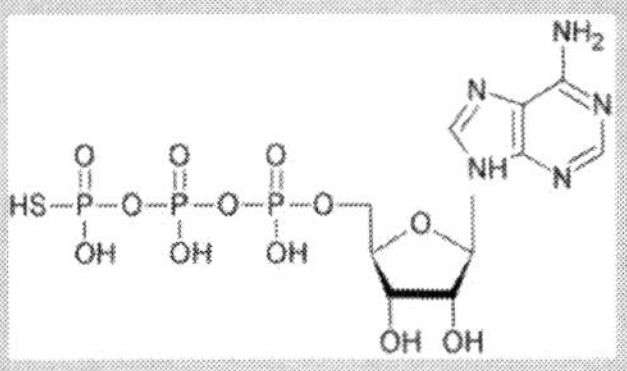

To be incorporated into RNA and DNA it must enter nucleotide pools such as ATP shown for use in nucleotide synthesis

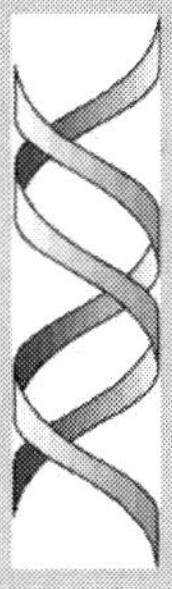

Modified nucleotides are then used to synthesize RNA and DNA polymers with phosphorothioate linkages

# MODEL FOR PHOSPHATE ANALOG

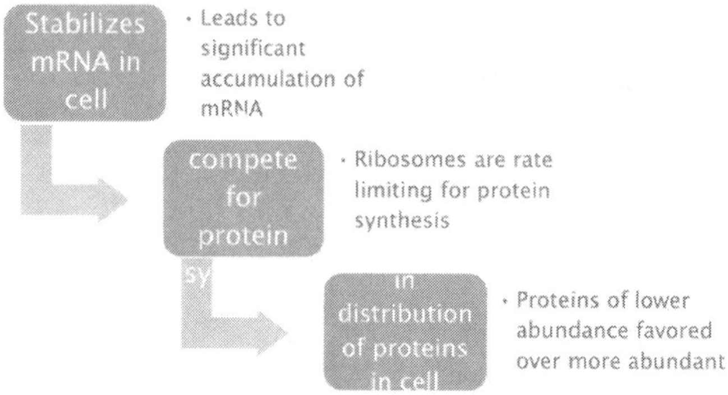


# PROTEIN SECRETION IN SACCHAROMYCES CEREVISIAE

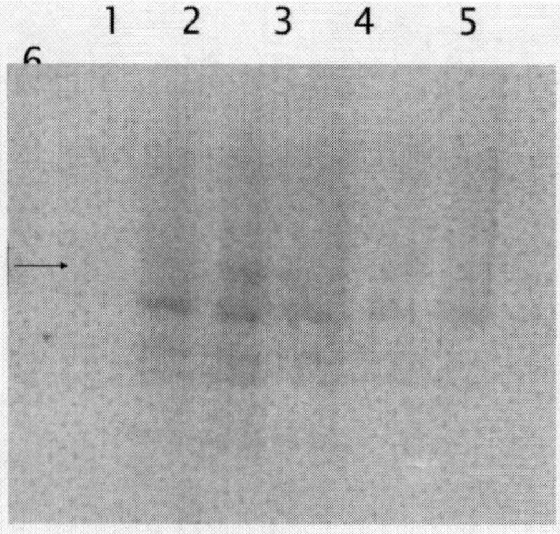

## ECOLI 2D POLYACRYLAMIDE GELS

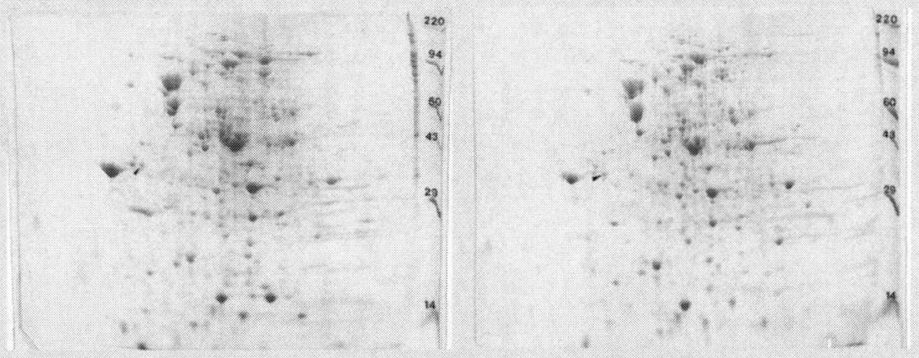


## ECOLI 2D DIFFERENCE GEL

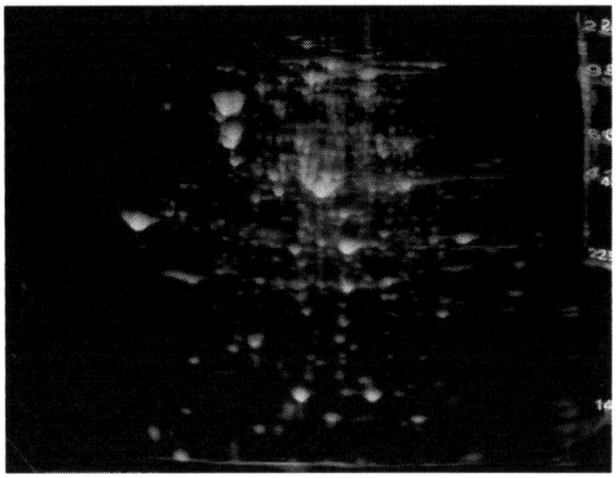

# HEK 293 CELLS

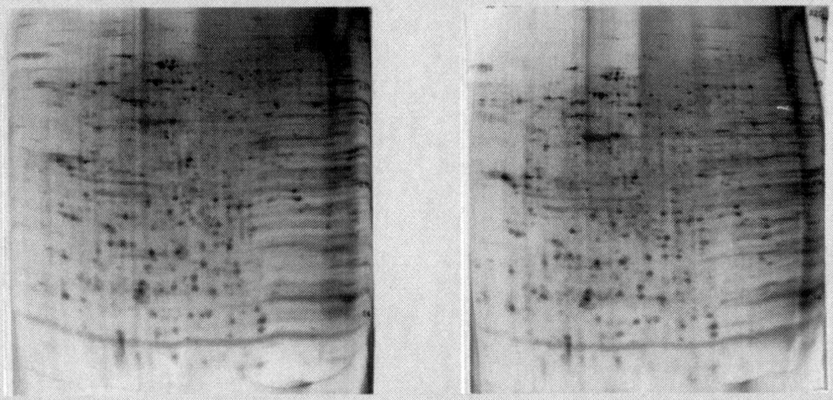

# HEK 293 CELLS 2D DIFFERENCE GEL

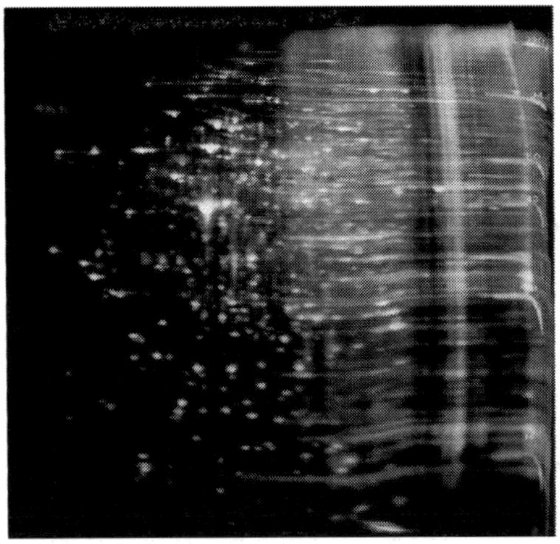

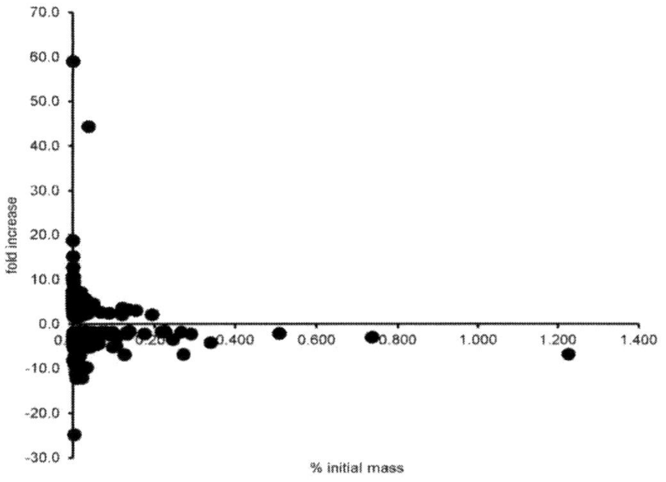


| ORGANISM | % of total protein spots changed; % increase; %decreased | Average % initial mass of proteins increased | Average % initial mass of proteins reduced | Average reported half-life of mRNA for organism | Average fold increase; range | Average fold decrease; range |
|---|---|---|---|---|---|---|
| Bacillus subtilis secretome | 35; 26; 9 | .04 | .169 | ~3.2 min. | 8.1; 3-77 | 6.6; 3-29 |
| Escherichia coli total soluble protein | 28; 15; 13 | .038 | .176 | ~6.8 min. | 5.5; 1.7-150 | 4.9; 1.7-32 |
| Saccharomyces cerevisiae secretome | 74; 21; 52 | ND | ND | ~22 min. | ND | ND |
| HEK 293 human cell line total soluble protein | 18; 7.3; 10.8 | .014 | .049 | ~10 hrs | 4.9; 1.7-59 | 3.2; 1.7-25 |

# APPROACH TO IDENTIFY PROTEIN SPOTS

**In silico gel**

- Theoretical translation of entire genome for a given organism projected on a 2D gel

**TagIdent SIB Ex Pasy**

- MW & pI observed compared to database of predicted proteins for a given organism

**Computation of candidate proteins**

- MW & pI predicted
- MW & pI observed

| Ecoli Gel MW; PI | Putative Ecoli Gene | Putative Gene MW; PI | Reported half-life | % initial mass | Relative Activity |
|---|---|---|---|---|---|
| 120,533; 6.1 | T1R1 type 1 restriction enzyme | 119,657; 6.08 | unknown | .000 | > 150 |
| 122,299; 6.1 | YegE probable diguanylate cyclase | 123,888; 5.8 | unknown | .000 | > 41 |
| 88,355; 5.9 | ThrA | 89,120; 5.47 | 2.1 min | .035 | > 2.3 |
| 48,352; 5.3 | ThrC | 48,579; 5.3 | 2.1 min | .009 | > 7.0 |
| 16,659; 5.4 | OmpX | 16,382; 5.30 | 2.6 min | .021 | > 3.2 |
| 39,693; 5.0 | HisC | 39,360; 5.01 | 3.6 min | .003 | > 19.1 |
| 27,364; 6.3 | OmpR | 27,353; 6.04 | 6.1 min | .007 | > 3.5 |
| 25,628; 5.0 | PhoP | 25,535; 5.1 | 5.7 min | .076 | > 1.8 |
| 34,538; 5.6 | Fmt | 34,037; 5.56 | 5.9 min | .012 | > 4.7 |
| 38,047; 5.9 | GalE | 37,265; 5.89 | 4.3 min | .120 | > 1.9 |
| 39,176; 4.9 | GlaM | 38,190; 4.84 | 6.2 min | .004 | > 3.1 |
| 35,025; 5.9 | AccA | 35,110; 5.76 | 12.6 min | .026 | < 3.2 |
| 16,347; 5.8 | Eco | 16,099; 5.94 | 11.2 min | .039 | < 5.5 |

| HEK 293 Gel MW; PI | Putative Human Gene | Putative Gene MW; PI | Tissue Specificity | % initial mass | Relative Activity |
|---|---|---|---|---|---|
| 178,271; 4.3 | Nestin | 177,439; 4.34 | neuronal | .001 | > 4.3 |
| 140,638; 5.5 | Neurabin | 140,303; 5.39 | neuronal | .000 | > 10.1 |
| 55,473; 5.5 | Alpha-internexin | 55,390; 5.34 | neuronal | .000 | > 7.4 |
| 91,583; 6.2 | NPAS2 HLH transcription factor | 91,7790; 6.35 | neuronal | .001 | > 5.4 |
| 49,017; 5.8 | Inwardly rectifying potassium channel | 49,500; 5.82 | neuronal | .001 | > 3.5 |
| 123,730; 5.6 | Alpha 2 delta calcium channel subunit isoform 1 | 129,735; 5.54 | neuronal | .001 | > 4.0 |
| 22,092; 5.4 | HIP2 ubiquitin conjugating enzyme | 22,406; 5.53 | neuronal | .024 | > 4.9 |
| 47,248; 5.7 | E3 ubiquitin ligase | 46,987; 5.75 | ubiquitous | .025 | > 2.2 |
| 54,766; 7.6 | E3 ubiquitin ligase isoform 2 | 54,339; 7.39 | ubiquitous? | .001 | > 10.6 |
| 55,473; 5.5 | Hsp60 | 55,390; 5.4 | ubiquitous | .000 | > 7.4 |
| 46,642; 6.5 | GSA alpha subunit trimeric G | 47,154; 6.48 | ubiquitous | .031 | < 2.4 |

# EXPERIMENTAL APPLICATIONS

1) Allows for a way to predictably manipulate the stability of mRNA in cell as well as to assess the stability of a given mRNA indirectly by fold change in protein

2) Allows for a way to predictably shift the distribution of proteins in the cell and study related cell functions

3) Enables a way to perturb a system to identify coordinated responses and/or to measure system responses

4) Enables a way to identify regulatory responses by identifying departures from global trends

5) Enables a way to assess function of whole cell systems as well potential cellular interactions

# WHAT HAS BEEN LEARNED?

## Bacteria & yeast

- Enhances protein secretion in yeast but not the Bacillus bacterium
- Global shift similar for secreted proteins in Bacillus and cytosolic proteins in E. coli
- Possible defensive response in bacteria with activation of DNA restriction enzyme system and a protein antibiotic (bacteriocin type III)
- At least for E. coli competition for ribosomes may not be only factor

## Human cells

- Enhances tissue differentiation of HEK 293 cells promoting a neuronal identify
- Evidence of conserved mechanism of action from bacteria to humans

# 430945 Proteomics and Factorial Experimental Design Enabled Novel Insights into Sulfur Metabolism and Microbe-Mineral Interactions in a Model Phototrophic Bacterium

**Abstract:**

Sulfur is a major byproduct of crude oil and natural gas refining, where it is estimated that over 1.5 million tons of waste sulfur were produced in 2014 alone. In developing strategies for managing this waste sulfur, an understanding of the environmental microbes that interact with and transform insoluble forms of sulfur is essential. Currently, mechanisms of microbial attachment to and transformation of insoluble sulfur are poorly understood. 'omic' analyses have the potential to provide insight into these mechanisms, but challenges with culturing these and other environmental microbes can obstruct these high-throughput approaches.

The model organism of the phototrophic green sulfur bacteria, *Chlorobaculum tepidum*, both produces and degrades (oxidizes) extracellular, insoluble sulfur globules ($S^0$) as an obligate intermediate of its energy metabolism. Thus, *Cba. tepidum* provides a platform to study microbial production and degradation of insoluble sulfur, as well as microbe-sulfur interactions, within a single system. Through the application of factorial experimental design and proteomic analyses, this work seeks to obtain novel insights into microbial interactions with insoluble sulfur.

To address known challenges with growth variability, a systematic approach to reproducibly culture and characterize *Cba. tepidum* under a range of conditions was first established. These studies, which spanned a factorial space of $S^0$-synthesizing and $S^0$-degrading states and a range of light fluxes (the energetic landscape), provided unexpected insights into *Cba. tepidum* sulfur and energy metabolism. For example, at low light flux, kinetic limitations on electron donor oxidation were observed which are not obvious consequences of current electron transport models for *Cba. tepidum*.

Proteomic analyses identified over 40 proteins that associate with purified biogenic $S^0$, including uncharacterized proteins and proteins involved in sulfur oxidation pathways. Beyond an improved mechanistic understanding of microbe-sulfur interactions, these protein identifications inform efforts to functionalize nanoparticles and other surfaces for biocompatibility.

# 425158 Omics Guided Rationally Metabolic Engineering of Clostridium Tyrobutyricum

**Abstract:**

The objective of this study was to achieve high butanol production by rational metabolic engineering of *Clostridium tyrobutyricum*. Our proteomics analysis showed that the expression of thiolase (*thl*) that converts acetyl-CoA to acetoacetyl-CoA and the butanol dehydrogenase (*bdh*) that catalyzes the formation of butanol from butyryldehyde had strong correlation with butanol production. These enzymes regulated carbon metabolism or/and redox balance in the metabolic pathway of butanol production. Then the homogenous *thl* gene and heterologous *bdh* gene from *C. acetobutylicum* were overexpressed in butanol producing mutant of *C. tyrobutyricum*. Both single- and double-vector gene expression systems with optimized promoter were developed to rationally engineer *C. tyrobutyricum*. The butanol fermentation showed that the redirection of carbon in the central metabolic pathway greatly improved the butanol titer to > 20 g/L and selectivity > 0.90 g/g. This study revealed that the Omics guided metabolic engineering is an efficient cell engineering strategy to achieve high biobutanol production.

# 417508 Understanding the Mechanism of Multi-Drug and Extensively-Drug Resistant Mycobacteria: Insights into Transport Processes


**Abstract:**

Emergence of antibiotic resistance is a major problem especially in the control of tuberculosis (TB) globally. *Mycobacterium tuberculosis*, the causative agent of TB develops resistance to multiple drugs simultaneously. Strains resistant to first-line anti-TB drugs, isoniazid and rifampicin, are referred to as multidrug-resistant TB (MDR-TB) and those resistant to fluoroquinolones and any one of the second-line injectable drugs in addition to isoniazid and rifampicin resistance are referred to as extensively drug-resistant TB (XDR-TB). In order to develop strategies to counter multiple drug resistance it is important to have in-depth knowledge of various mechanisms that makes the organism resistant to multiple drugs. In the current study, we use a combination of gene sequencing, expression analysis, biochemical studies and microscopy to understand the varied survival mechanisms that makes the mycobacteria tolerant to various anti-TB drugs.

For this purpose, we have created three *in vitro* random mutants in *Mycobacterium smegmatis*, a model mycobacteria, which can sustain a very high drug concentration in comparison to the wild type (WT) strain. The mutants were selected by exposing the wild type strain to any one of the three drugs, rifampicin, isoniazid and norfloxacin as selection pressure. Interesting, two of the finally selected strains were MDR strains with resistance to both rifampicin and isoniazid while the third mutant exhibited characteristics of XDR strain and was resistant to rifampicin, isoniazid, amikacin and many fluoroquinolones.

To understand the mechanism of drug resistance in these mutants, we sequenced the drug targets to check for mutations that may lead to drug insensitivity. Further, measurement of intracellular drug levels along with extensive transport studies using fluorescent tracer confirmed that the uptake or efflux of drugs in the mutants was effected either constitutively or induced in the presence of drug. Moreover, these results were confirmed at molecular level by performing the gene sequencing of major efflux pumps and porins, along with real-time PCR studies. The morphology of mutants was also altered, from rods to coccid or elongated rods, in response to high drug concentrations, indicating that alteration of cell morphology could be one of the mechanisms involved in multi-drug or extensively-drug resistant mycobacteria.

Thus, this study demonstrates the involvement of multiple mechanisms in development of multi-drug and extensively-drug resistant tuberculosis.

# 434760 Modulation of Transcriptomic Response of Streptomyces Coelicolor Towards Ciprofloxacin with Altered Nutrient Composition

**Abstract:**

**Modulation of transcriptomic response of *Streptomyces coelicolor* towards ciprofloxacin with altered nutrient composition**

Minal Patkari, Sarika Mehra*

*Department of Chemical Engineering, Indian Institute of Technology Bombay, Mumbai-400076*

Antibiotic resistance is a major concern in the field of medicine as bacteria in the environment can come in close contact with one another and spread the resistance to many of the clinically important antibiotics. Some non-pathogenic bacteria harbour genes for intrinsic resistance, which are likely to transfer to pathogens. Study of intrinsic resistance mechanisms in non-pathogenic organism is, therefore, as important as a similar study in pathogens. Understanding the multiple ways in which bacteria develop resistance is important to devise effective solutions and to develop novel antimicrobials. Study of intrinsic resistance mechanisms in non-pathogenic organism is, therefore, as important as a similar study in pathogens. An extensive study on transcriptomics of *S. coelicolor* showed altered gene expression in presence of ciprofloxacin in Supplemented Minimal Medium (SMM) (Patkari & Mehra, 2013). The susceptibility for ciprofloxacin is decreased by four fold in a nutrient rich medium referred to as R5. This indicates that the growth conditions may also determine the response of bacteria towards particular antibiotic. Therefore, susceptibility of a bacterium to an antibiotic is dependent on the expression of various genes responsible for protection against stress conditions. This study aims to analyse the transcriptomic response of *Streptomyces coelicolor* towards ciprofloxacin treatment in rich medium as compared to minimal medium. A detailed transcriptomic study is performed at sub-inhibitory and inhibitory concentrations of ciprofloxacin to understand the mechanisms underlying the elevated resistance. The intracellular drug levels in the two media are also compared.

The response at sub-inhibitory concentration is mainly a stress response showing activation of most of anti-oxidant genes and few genes from nucleotide biosynthesis pathway. However, when the cells were exposed to a higher concentration of drug (MIC), upregulation of additional genes involved in protection against the stress and DNA damage were observed. The up-regulated genes belong to a variety of functional classes. Many genes from the DNA repair pathways and nucleic acid and protein metabolism are differentially regulated in both rich R5 medium and in supplemented minimal medium. Thus, the concentration of ciprofloxacin to induce SOS response is almost 8-folds higher in cells grown in richer R5 medium. We compared the basal expression of drug transporters which was found to be higher in rich medium cells as compared to cells grown in minimal medium. This data justifies the lower intracellular drug levels in rich medium cells which was almost 12 fold less in cells grown in minimal medium at ciprofloxacin 80 μg/ml. When exposed to inhibitory concentration of drug, cell attempts to survive by activating various mechanisms that assist in damage control. Thus, in this work we have compared the global response of *S. coelicolor* towards ciprofloxacin to establish the role of gene expression modulation with the nutrient supplements in resistance.

**Keywords:** *S. coelicolor*, ciprofloxacin, gene expression, antibiotic resistance

# 434252 Characterization of Protein-Metabolite Regulatory Interactions Via Complementary Parallel Approaches

**Abstract:**

Computational metabolic modeling plays an increasing role in both metabolic engineering and identification of therapeutic targets to combat disease. The construction of such models relies primarily on understanding of the network architecture conferred by large-scale characterization at the genomic, transcriptomic, proteomic, and increasingly, metabolomic levels. Regulatory interactions between proteins and metabolites, however, are only sparsely characterized by comparison, despite being a common phenomenon with great functional importance in cellular metabolism.

Available evidence suggests that current knowledge of these interactions encompasses only a fraction of their actual occurrence; therefore, a broad understanding of protein-metabolite regulatory interactions should confer substantial benefit in the development of predictive metabolic models. Efforts at characterization of this level of regulation face the challenge of capturing interactions that are often as weak as micromolar- to millimolar-level $K_D$s. A systematic methodology for discovery and characterization of these protein-metabolite regulatory interactions is necessary for broad-scale understanding.

The model organism *Saccharomyces cerevisiae* provides an excellent subject of study for the development of this methodology, as it is an industrially important organism and has been the subject of many previous mathematical modeling efforts. Novel regulatory data for its metabolism will therefore be immediately useful for the improvement of such models. Here, we present our work in developing a multi-pronged strategy for the characterization of protein-metabolite regulatory interactions, which can be generalized to study this level of regulation in other organisms of interest. We have developed an *in vitro* binding assay as a medium-throughput screen for protein-metabolite binding interactions; cocktails of metabolites are incubated with known proteins, washed, and eluted prior to GC-MS analysis for enriched metabolites to identify putative regulatory pairs. These interactions may then be validated and further characterized using an *in vitro* reaction assay, also with GC-MS based analysis. We additionally present the *in vitro* reaction assay itself as a means of investigating putative regulatory interactions that are predicted by any other method. In parallel, small-molecule microarrays provide a high-throughput platform for broader-range protein-metabolite interaction discovery at the cost of sensitivity. Small-molecule microarrays have previously been applied to the discovery of strong-affinity binders for putative drug targets; we have worked toward adapting this strategy to qualitatively probe for weaker, endogenous interactions. Arrays of covalently-attached yeast metabolites are synthesized and probed with tagged, known proteins to identify protein-metabolite binding pairs, which may then be characterized via *in vitro* reaction assay.

The diversity of binding affinities across the scope of protein-metabolite regulatory pairs provides both challenge and opportunity; employing complementary discovery methods in parallel will allow throughput to be balanced with sensitivity. Data gathered via these complementary approaches will aid in the eventual construction of a systems-level model of *S. cerevisiae* metabolism.

# 430148 Development of Synthetic-Hybrid Systems for Enhanced Gene Expression Control

**Abstract:**

**Development of synthetic-hybrid systems for enhanced gene expression control**

Brittany Forkus and Yiannis Kaznessis

Controllable gene regulatory systems that exhibit new biological phenotypes can be developed by arranging naturally occurring regulatory genes in synthetic DNA sequences. In our work, we have built several synthetic DNA promoter systems by stitching together various components of the common and well-studied tetracycline, luminescence, and lactose operons.

Previously we designed simple logical AND gates by shuffling *tetO* and *lacO* operator sites in a single transcriptional unit. We then built ProteOn, the first inducible activator transcription factor, by fusing the tetracycline repressor with the activating domain of the luxR protein.

In this presentation we will discuss the most recent system we built, which implements inducible activation of protein expression with positive-feedback. We call this new system ProteOn+. We characterized this system extensively by expressing green fluorescent protein (GFP), and using flow cytometry techniques. We compared ProteOn+ to commercially available promoter systems and found that ProteOn+ elicits many-fold increases in gene expression than commonly used DNA promoters. We believe that ProteOn+ adds significant value to the current synthetic biological-toolbox for bacterial gene expression.

In this presentation we will also discuss applying ProteOn+ in the development of antibiotic technologies. We have engineered probiotic *E. coli* strains to express antimicrobial peptides from ProteOn+. We assess the strength of this synthetic DNA promoter and its utility in the delivery of antimicrobials to treat gastrointestinal infections.

# 413127 De Novo Design of Heat-Repressible RNA Thermosensors in E. coli

**Abstract:**

Inducible gene expression systems are powerful tools for optimizing metabolic pathways or generating logical behavior in an organism. However, in most systems, chemical inducers must be added exogenously, leading to extra costs and potential barriers to scale-up. Expression systems that are induced by environmental changes avoid the need for costly chemical additions and allow bacteria to respond directly to growth conditions. The overall goal of this research is to develop small, synthetic, heat-repressible RNA thermosensors. These thermosensors are located in the 5' UTR upstream of the Shine-Dalgarno site, where they contain a cleavage site for ribonuclease E, an enzyme native to *Escherichia coli* that binds at the cleavage site and degrades the mRNA. At low temperatures, the recognition site is sequestered in a stem-loop structure, turning on gene expression. At high temperatures, the stem-loop unfolds and exposes the recognition site, and thus the mRNA is degraded, turning off gene expression. Synthetic heat-repressible RNA thermosensors were designed and tested *in vivo* by varying length, number of recognition sites, size of the loop, and predicted melting temperature of the stem. The RNA thermosensors worked as expected, with high expression levels at low temperatures. Several experiments were performed to support the hypothesized mechanism, and to demonstrate that these thermosensors respond exclusively to temperature. Furthermore, these thermosensors were shown to function in a multi-input genetic circuit. Finally, design guidelines are presented, based on the behavior of the 24 designed thermosensors. These heat-repressible thermosensors are small, do not require expression of protein regulators, and can function independently of expensive chemical inducers. In addition, they are relatively simple to design and can be implemented in the optimization of a range of metabolic processes.

# 434835 Retargeted Bacterial Regulatory RNAs for Metabolic Engineering

**Abstract:**

In metabolic engineering we genetically alter cells to enzymatically produce specialty chemicals such as drugs and fuels. RNA is among the most flexible and potent molecular-genetic regulatory molecules that can be used to engineer these cells, as gene sequences can be edited to create tailored RNA circuit elements with high specificity and limited off-target effects. To optimize gene expression for specific processes we have created a synthetic genetic system for engineering protein translation control by RNA. The metabolic engineering application of this work is to improve *n*-butanol yield and selectivity in biofuels fermentation. In bacteria, small regulatory RNAs (sRNAs, ~50- 250 nucleotides) natively govern gene expression by binding protein-coding mRNAs and enhancing either their expression (at translation) or degradation/turnover. The sRNAs are ubiquitous in bacteria and act by specific sRNA-mRNA base-pairing interactions with single or multiple mRNA targets. Altering sRNA sequences permits targeting of desired mRNA transcripts and thus tailored protein (enzyme) expression. To systematize retargeting of our sRNA, we have created a quantitative gene control assay comprising one sRNA (*E. coli* DsrA) that acts on two reporter gene fusions of the native mRNA targets of DsrA (*rpoS*-GFPuv and *hns*-mCherry). Each mRNA and the sRNA are under separate transcription-repression control activated by small inducer molecules, and expression of both mRNAs can be read out in separate fluorescent channels during growth *in vivo* (96-well plate format). We utilized this inducible, scalable expression testbed system for three main experiments. First, we re-targeted DsrA by altering one of its two pairing sequences to bind with and regulate translation of mRNAs of interest in the *n*-butanol synthesis pathway, specifically acetate or butyrate kinases (to alter pathway flux and improve fermentation selectivity) or a hydrogenase (to improve n-butanol yield by NAD/NADH redox balance). Newly targeted mRNA sequences were used to make a new reporter gene fusion with one of the two fluorescent proteins (*e.g.* mCherry). Second, in order to determine the extent and basis of partitioning of the sRNA between mRNA targets and thus predict likely sRNA dosage and response, we de-targeted the DsrA by diminishing its interaction ($K_d$) with one mRNA target (*hns*) while retaining DsrA native pairing with the other target (*rpoS*). Finally, to more clearly delineate those sRNA structural features relevant to its mRNA strand targeting, we used this system to assess mutated but functional structures of DsrA stem-loop 2. These experiments fine-tune target mRNA translation to optimize metabolic pathway flux, inform the design of sRNAs to target specific mRNA transcripts, and create a generalizable toolset for characterizing and retargeting sRNAs in bacteria.

# 415736 Discovery and Design of Novel Regulatory Noncoding RNA in Bacteria

**Abstract:**

The work here describes the identification of new regulatory noncoding RNAs in bacteria and their potential use in synthetic biology applications. Produced during futile attempts by the RNA polymerases to clear the promoter region and transition to transcriptional elongation phase, the synthesis of such "abortive" RNAs have been known for decades in all kingdoms of life. However, until the recent identification of abortive RNA in the bacteriophage T7 with transcriptional antitermination function, they had been considered only as "junk" RNAs. Functional abortive RNAs have not, however, been identified to exist endogenously in more complex biological systems such as bacteria. Using an *in silico* predictive approach, we evaluated whether these abortive RNAs might be able to serve biological functions in *E. coli*. Abortive RNAs generated from 3,780 transcriptional units (obtained from RegulonDB) were used as query sequences within their respective transcription units to search for possible binding sites. Sites that fell within known regulatory features, based upon data obtained from RegulonDB and WebGeSTerDB, were then ranked based upon the standard free energy of annealing (reported by UNAfold) of the transcript to the binding site. Matches with a sufficiently high standard free energy of annealing were then explored further as likely sites for abortive RNA-mediated regulation. We will discuss one such result – the uncovering of a potentially novel abortive RNA-mediated regulation of an alcohol dehydrogenase in *E. coli*. Future experimental validation of this and other loci may reveal the pervasive physiological role of this new class of noncoding RNAs in *E. coli*. We also discuss the potential application of using such functional abortive RNAs in a design-based approach to regulating gene expression in synthetic systems.

# 428854 Integrating Multiple Omics Studies to Uncover Underlying Post-Transcriptional Regulatory Networks

**Abstract:**

An important challenge in gene regulation studies is to identify the targets and effects of global regulators. A single global regulator can often control hundreds of targets and impact multiple levels of cellular process (transcription, translation, metabolite profile, etc). In our studies, we explore these key issues using the *Escherichia coli* carbon storage regulator (Csr) system as a case study. The Csr system is a well-studied posttranscriptional regulatory system that responds to various extracellular stresses. The system has four major regulatory elements, two proteins, CsrA and CsrD, and two small RNAs, csrB and csrC. CsrA, the main effector of the system, can affect a wide variety of cellular mRNAs and differentially regulate their translation via multiple mechanisms. Many CsrA targets have been suggested and studied in the literature, but further characterization for the majority of these targets is lacking and novel targets likely remain unknown; this largely limits our ability to exploit this powerful system for any engineering application. Furthermore, the experiments performed to identify CsrA targets have been performed in a highly traditional manner, at a single time point while ignoring potential dynamic network reconfigurations. To address these knowledge gaps, we performed a time-resolved of transcriptomic, proteomic, and pull-down study of the Csr response in *E. coli* to environmental stress in various genetic backgrounds. Using this new approach, we discovered that signature patterns on the proteomics, transcriptomics, and pull-down assays can be used to confidently *predict* a more complete set of real Csr targets. Additionally, we have experimentally validated and characterized these predicted omics targets using a combination of fluorescence and traditional biochemical approaches. As a result of these studies, we identified a number of novel Csr targets for which we have predicted and tested potential mechanisms of control. This large scale studies provides new insights on how to use global omics studies to identify new regulatory targets and suggest possible mechanisms of action for these targets.

# 434002 Reconstructing Anaerobic Microbiomes from the 'Bottom-up': New Techniques to Decipher Interwoven Metabolism

**Abstract:**

Anaerobic microbiomes are among the most diverse microbial communities on earth, yet the mechanisms that control community structure and stability are poorly understood. For example, the rumen microbiome within large herbivores consists of four interdependent microbial populations (bacteria, protozoans, fungi, and methanogens) that work together to drive crude biomass into sugars and fermentation waste products. However, the interdependency of these microbes and their impact on microbial metabolism has been difficult to characterize since the vast majority of these microbes cannot be cultured. To characterize interwoven metabolism of anaerobes, we have focused on an interdependent sub-population within the rumen microbiome – anaerobic fungi and their associated hydrogenotrophic archaea (methanogens). In this community, gut fungi hydrolyze cellulose-rich biomass in the herbivore rumen, and a syntrophic partnership with methanogens allows fungi to thrive by siphoning hydrogen to methane. Our goal is to understand this natural syntrophy, and mimic it to engineer synthetic, stable anaerobic consortia that funnel crude biomass to sustainable chemicals.

To construct anaerobic consortia from the "bottom-up", we have overcome culture challenges by isolating anaerobic fungi with dependent methanogens from herbivore fecal materials. These environmental co-cultures establish a simplified system to model the cooperative action of the anaerobes, and our prokaryotic/eukaryotic systems are stable in culture > 60 weeks. Using RNA-Seq, we have modeled the regulatory patterns for fungal genes from the *Piromyces, Neocallimastix, and Anaeromyces* genera during hydrolysis of reed canary grass and discovered several conserved regulons of novel genes that govern function, which are catabolite repressed. Addition of methanogens (e.g. *Methanocorpusculum*) during biomass breakdown shows that fungal-methanogen syntrophy drastically accelerates cellulose and lignocellulose breakdown by the fungi. Furthermore, the excess sugar hydrolysates and metabolites from an *Anaeromyces/Methanocorpusculum* co-culture enable metabolic linkage of non-native facultative anaerobes to the consortium. Strains of *S. cerevisiae* and *E. coli* were engineered to produce Flavin based fluorescent protein (FbFPs) as a reporter of co-culture growth on excess sugars (5-8 g/L) left over from fungal cellulose breakdown. We will discuss stability of these synthetic co-culture platforms, and strategies we have used to optimize production of n-butanol by an engineered *E. coli* strain in co-culture with anaerobic fungi. As anaerobic fungi are not yet genetically tractable, our strategy offers a path forward to make value-added products directly from crude lignocellulose by compartmentalizing lignocellulose breakdown and production in co-cultured microbes.

# 424701 The Role of Modeling, Simulation and Optimization Approaches in the Design of Pharmaceutical Production Processes

**Abstract:**

Achieving cost-effectiveness while maintaining good manufacturing practice (GMP) is an emerging topic in pharmaceutical manufacturing. In this presentation, the role of modeling, simulation and optimization (SMO) approaches is explored by showcasing two recent research examples on active pharmaceutical ingredients (APIs) and excipients, respectively.

The first example is on optimizing the purification process of thermally unstable APIs. The target here was an industrial process for sterilizing API as the raw material of a Parenterals (or injectable) product. The process consists of dissolution of crude API with water by heat exchanger, sterile filtration, crystallization and drying, and as a product sterile API crystal is obtained. In the heat exchanger, the crude API needs to be dissolved completely while degradation reaction of the API occurs in the liquid phase. This dissolution process of thermally unstable API was modeled with the overall yield as the objective function, and the flow rate and the temperature in the heat exchanger as design variables. The process model also considered a couple of characteristics such as particle size distribution, non-linear temperature profile in the heat exchanger, and sterility requirements by GMP. The simulation result indicated an optimal set of the design variables as well as certain ranges where the yield decreases dramatically due to the degradation or non-dissolution.

The second example is on the multiobjective optimization of the use process of pharmaceutical excipients, i.e., reception, storage, testing, and application of the material. As an example, benzyl alcohol, an oxidation-sensitive material, was investigated, and a model was created, which defines the amount of oxidized impurities, and the costs associated with testing as the quality and economic objectives, respectively. The process was modelled and simulated, considering container volume and purchase frequency as the design variables, and also the use of Raman spectroscopy as an alternative to conventional manual testing. The multiobjective evaluation indicated options on the Pareto frontier of the design variables as well as the effectiveness of the use of Raman spectroscopy.

In these two examples, the SMO approaches showed its supportive role for shifting the design concept of pharmaceutical production processes from "by-testing" to "by-design". The goal of the design can cover not only a single aspect but also to multiple ones, such as quality, economy, supply stability or risks regarding environment, health and safety.

# 431772 Modeling of Drug Substance Manufacturing in a Multi-Purpose Plant

**Abstract:**

This work presents different modeling strategies to address the manufacturing of drug substances in a multi-purpose plant. In multi-purpose plants the process equipment needs to have flexibility to tackle a wide range of chemical and physical operations, often with multi-phasic systems, and a broad range of solvents, temperature and pH. It is not uncommon that such process equipment in multi-purpose plants are not fully adequate to specific highly demanding chemical processes (eg a given impeller may be adequate for intense local mixing, but not fit for axial mixing of solids). Hence, it is crucial to use adequate modeling tools and methodologies to optimize the process performance as much as possible considering the equipment constraints; such strategies are aligned with the Quality by Design initiave being promoted by the regulators.

The case-study to be shown comprises the scale-up of the manufacturing of two distinct drug substances from the laboratory to the same full scale reactor. The kinetics of both processes is assessed in detail in the laboratory scale through the fitting of the transient differential equations to the analytical data (using Dynochem); the intrinsic kinetics of both processes is determined through decoupling of the fluid dynamics portion of the mass transfer resistance in the kinetic constants using both Dynochem and CFD (Computational Fluid Dynamics) - this is made possible since the laboratory reactor geometry, impeller and baffle system are relatively simple. In constrast, the geometry, impeller and baffle system of the full scale reactor are not simple, so this is addressed differently. The fluid dynamics of both processes in the full scale reactor is assessed solely with CFD, which enables the determination of the local mass transfer coefficient distribution. It is then shown how the use of the local mass transfer distribution along with the intrinsic kinetics of each process enables an improved understanding of the scale-up to the full scale.

# 426939 Systematic Retrofitting Methodology for Pharmaceutical Drug Purification Processes


**Abstract:**

Pharmaceutical production processes are gaining research attentions in the field of CAPE/PSE. In this work, we present a systematic method for retrofitting pharmaceutical manufacturing plants, together with an industrial case study.

The method consists of five tasks: (I) collection of process understanding/data, (II) creation of process model, (III) adaptation of the model for optimization, (IV) optimization and (V) task outcome interpretation. In Task I, an overview of the process is obtained by performing Path Flow Decomposition (PFD), which determines the sources and sinks of the all substance in a batch, and thus facilitates the creation of a model. In Task II descriptive mathematical models are created for every unit, which will then be evaluated for feasibility in Task V. The aim of Task III is to transform the descriptive model into an analytical model, which relates the optimization variables to the objective function. In parallel to identifying variables to be optimized, analysis is performed on constraints, especially on the pharma-specific, GMP and product quality aspects. All these information are considered in Task IV, i.e., optimization, where MINLP plays a central role. Task V runs simultaneously to all the other tasks, evaluating the feasibility of the outcomes and the necessity to iterate corresponding Task(s), which resembles iterative nature of the design.

The methodology was applied to a manufacturing plant of a solid drug, where crude crystal is purified through the process of Dissolution-Filtration-Crystallization (DFC). In Task I, a trade-off was identified: the higher the process temperature, $T_H$, the more degradation of the crude crystal; the lower the $T_H$, the more amount of undissolved crystal. The process flow rate, $F$, affects also the overall yield. In Task II, process model was created with considering measured data, e.g. particle size distribution (PSD) during the dissolution for the heat exchanger. The descriptive model was adapted for optimization by incorporating accurate information, e.g., liquid-wall-liquid heat transfer. We were able to model reactive-dissolution of crystals with PSD inside the heat exchanger, which is the secondary novelty of the work. As Task IV, three alternative process layouts were considered. The insertion of coolers before and after the filtration unit enabled a substantial improvement, lowering the degradation reaction and thus increasing the yield and the net present value. The feasibility of the model was supported by the result of the performance of lab scale experiments on a plant with the same new layout.

The methodology is a practical decision-supporting tool for the retrofitting of pharmaceutical purification processes, and its application is not only limited to the DFC processes. The novelty lays in its integrative nature, indeed the combination of different technics, e.g., information gathering in GMP environment, mathematical modelling and multiobjective optimization, and decision making, in the field of pharmaceutics. The methodology can show its weakness in case the data measurement is difficult, however, elucidation of trends and behaviours inside the plant can still serve as a basis for process retrofitting.

# Model based approach to develop a ring closure reaction with high product selectivity

Marimuthu Andiappan, Adam McFarland, Jared Fennell

Small Molecule Design and Development (SMDD)

Eli Lilly and Company, Indianapolis, USA

AICHE Annual Meeting 2015

1

## Outline

- Goal – Ring closure reaction with high product selectivity
- Characterization of reaction: Flow NMR, Hypothesized mechanism
- High Selectivity (high yield, low impurity) criteria
- Kinetic Model prediction
- Proof of concept experiment

2

# Ring closure reaction - Example

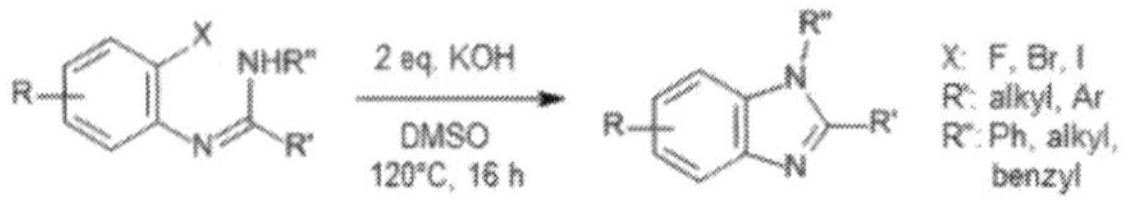


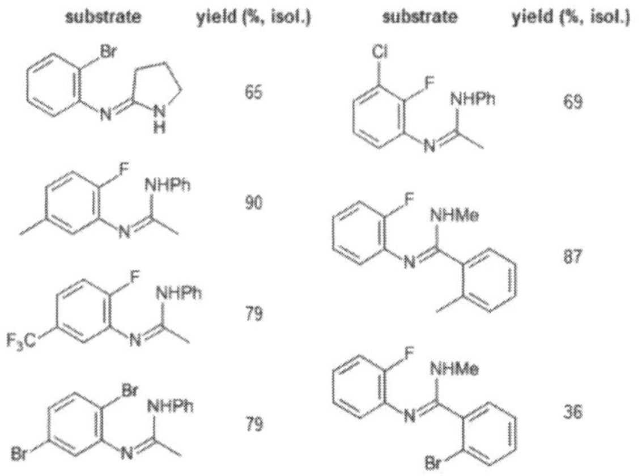


- In-situ product selectivity, impurity levels and isolated product yield depends on the nature of substrate
- How to design reaction with high product selectivity?

H. Baars, A. Beyer, S. V. Kohlhepp, C. Bolm, *Org. Lett.*, **2014**, *16*, 536-539

3

# Our study: Ring closure reaction in toluene

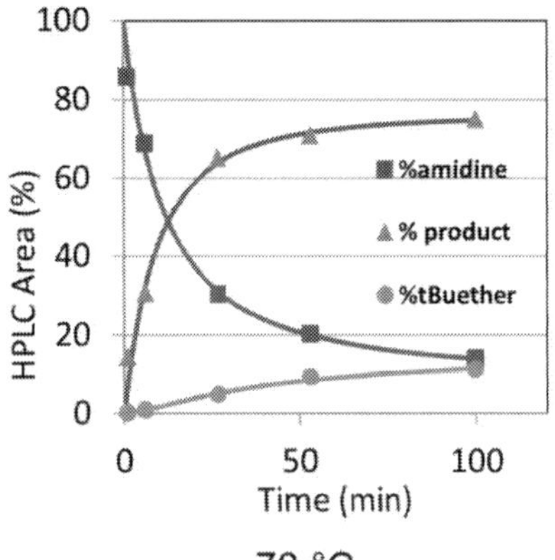


70 °C

- Desired reaction

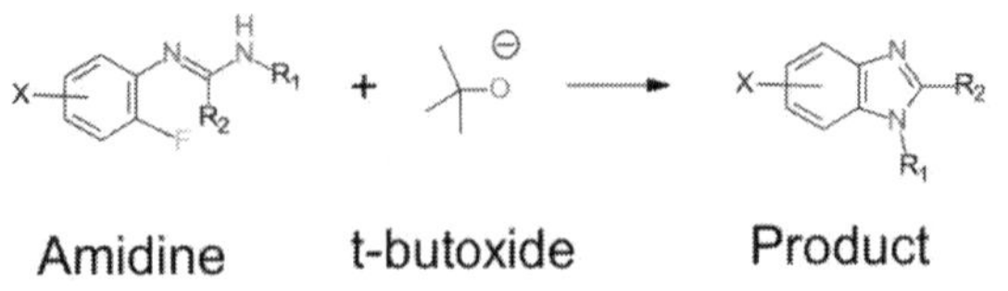


4

# Our study: Ring closure reaction in toluene

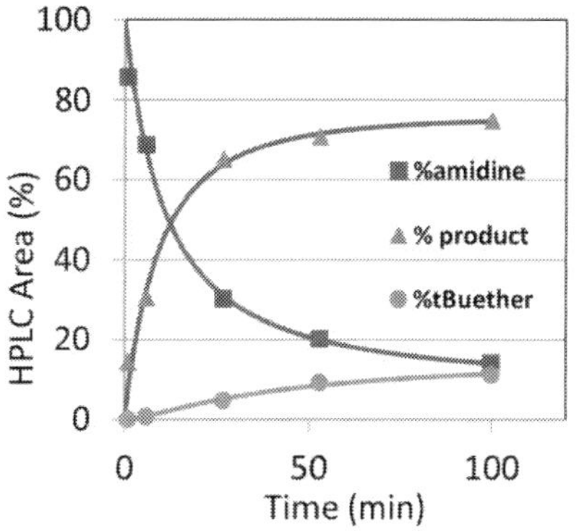


70 °C

- Problem: t-butyl ether impurity formation due to competing reaction

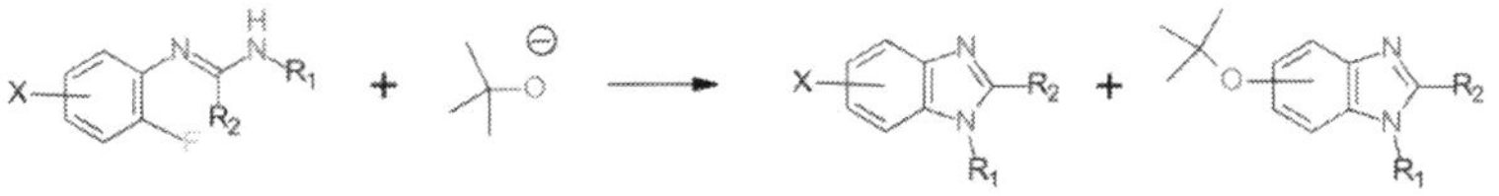


Amidine t-butoxide Product t-butyl ether impurity

5

# Our study: Ring closure reaction in toluene

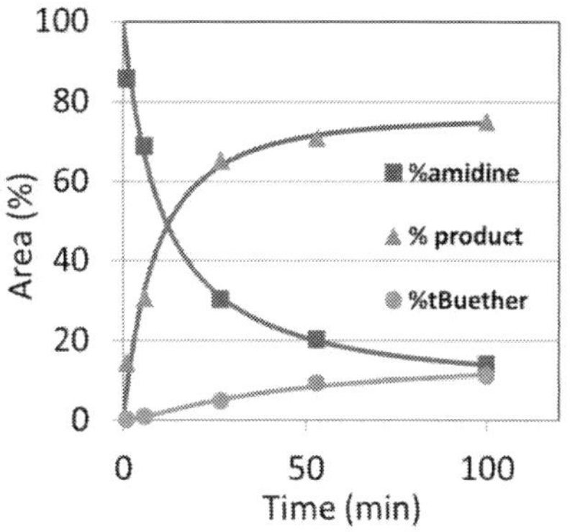


70 °C

- Acceptable level of impurity in isolated product: Not more than 0.5%

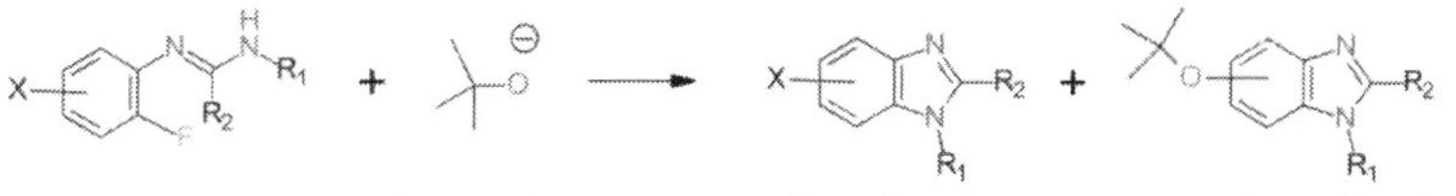


Amidine t-butoxide Product t-butyl ether impurity

6

# Our study: Ring closure reaction in toluene

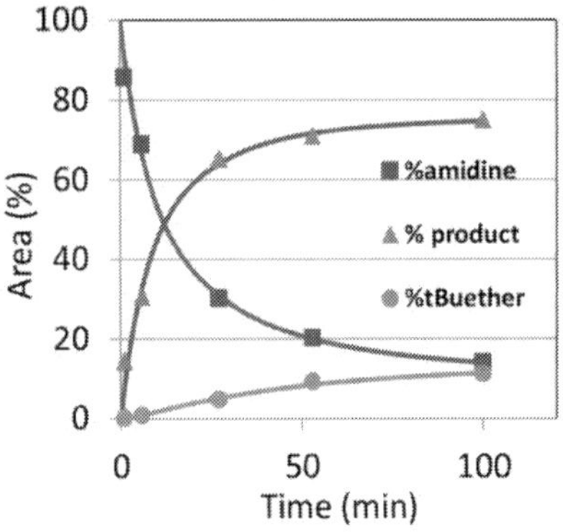


70 °C

- Can we design reaction with impurity level less than 0.5% at EOR using Mechanistic approach?

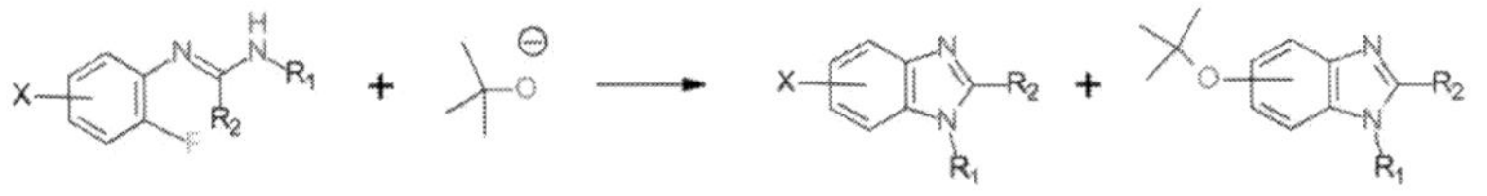


7

# Flow NMR - Ring closure reaction at 25 °C

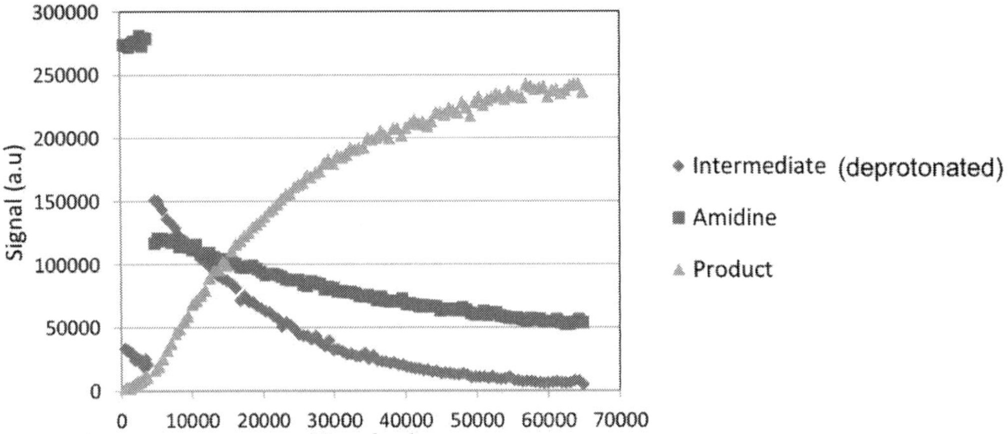


Signal from Proton NMR

- Equilibrium between amidine and deprotonated intermediate (equilibrium depends on t-butoxide concentration)
- Product forms from deprotonated intermediate through irreversible reaction

8

# Mechanism – Production formation

**Overall Reaction**

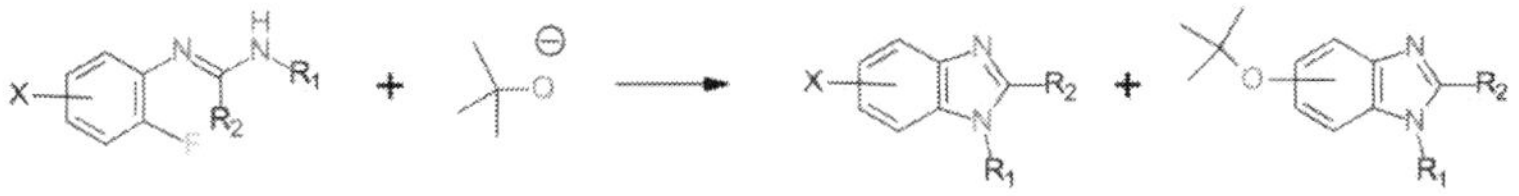

Amidine t-butoxide Product t-butyl ether impurity

**Step 1**

Amidine t-butoxide intermediate t-butanol

**Step 2**

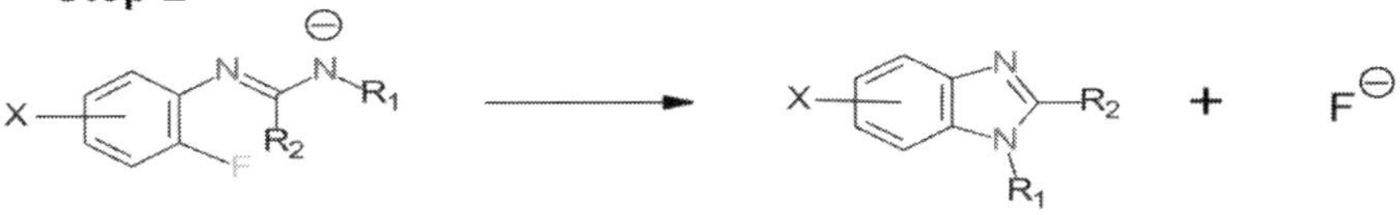

Intermediate Product

9

# Possible mechanisms for t-butyl impurity formation

**I. t-butyl impurity formation from product**

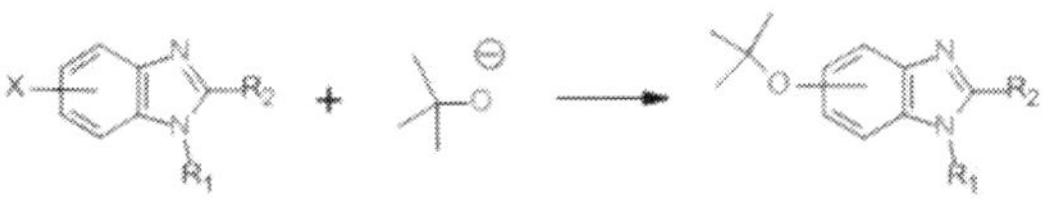

Product t-butoxide t-butyl ether impurity

**II. t-butyl impurity formation from amidine substrate**

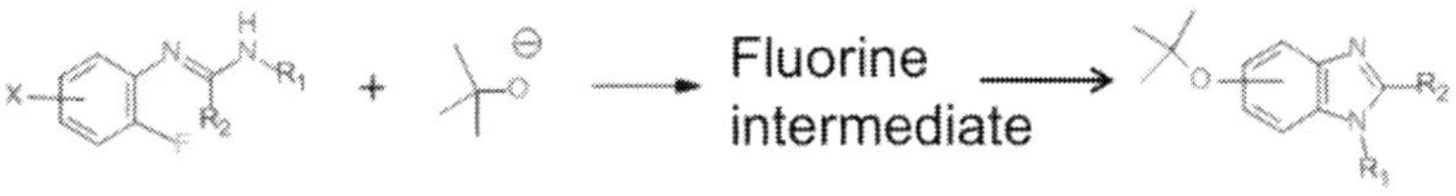


Amidine t-butoxide t-butyl ether impurity

**III. t-butyl impurity formation from deprotonated amidine intermediate**

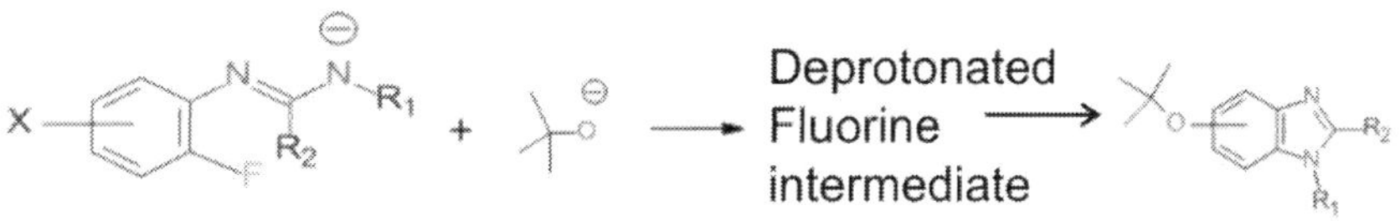


Deprotonated intermediate t-butoxide t-butyl ether impurity

10

# Flow NMR: Reaction at 50 °C, Impurity formation from product or amidine?

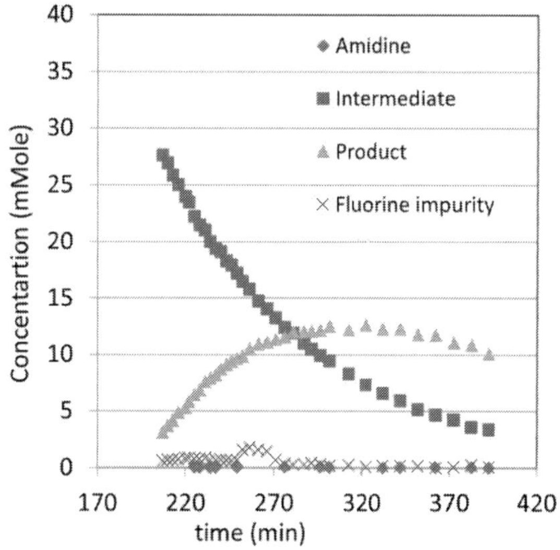


Signal from Fluorine NMR

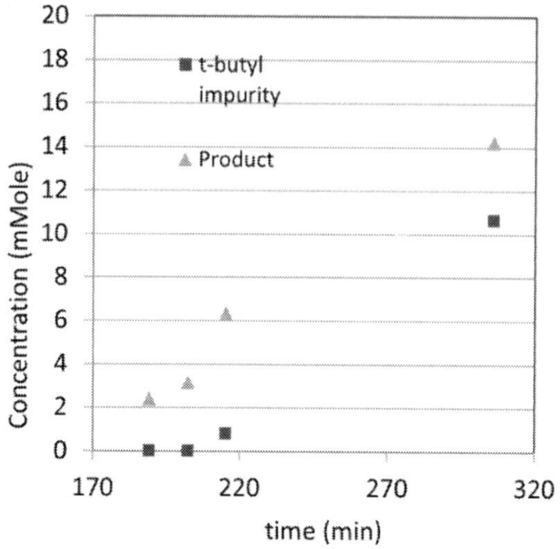


Signal from Proton NMR

- ~10 equiv. of t-butoxide: Transient Fluorine impurity (t-butyl ether impurity could form from amidine) – Disappears to t-butyl impurity

11

# Flow NMR: Reaction at 50 °C, Impurity formation from product or amidine?

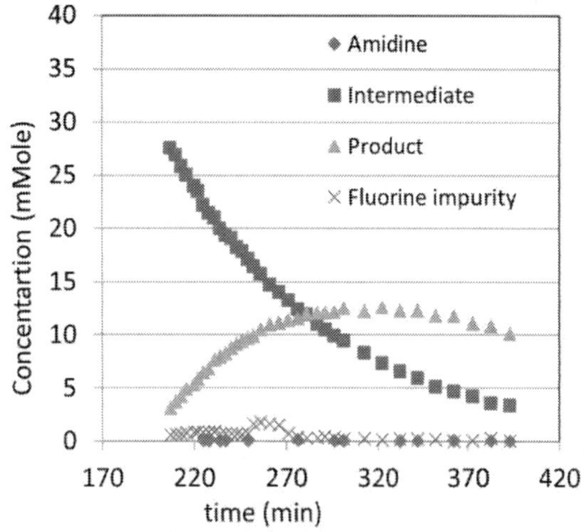


~10 equiv. t-butoxide

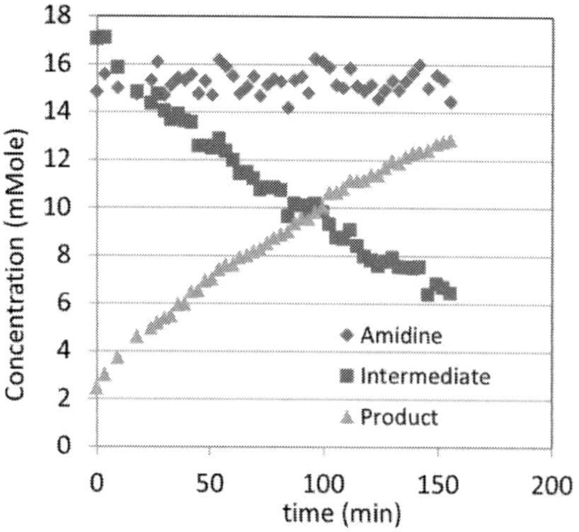


~2 equiv. t-butoxide

- ~10 equiv. of t-butoxide: Transient Fluorine impurity (t-butyl ether impurity could form from amidine) – Disappears to t-butyl impurity
- ~2 equiv. of t-butoxide: No significant transient Fluorine impurity (t-butyl ether impurity formation from amidine may not be significant)

12

# Mechanism (Hypothesized)

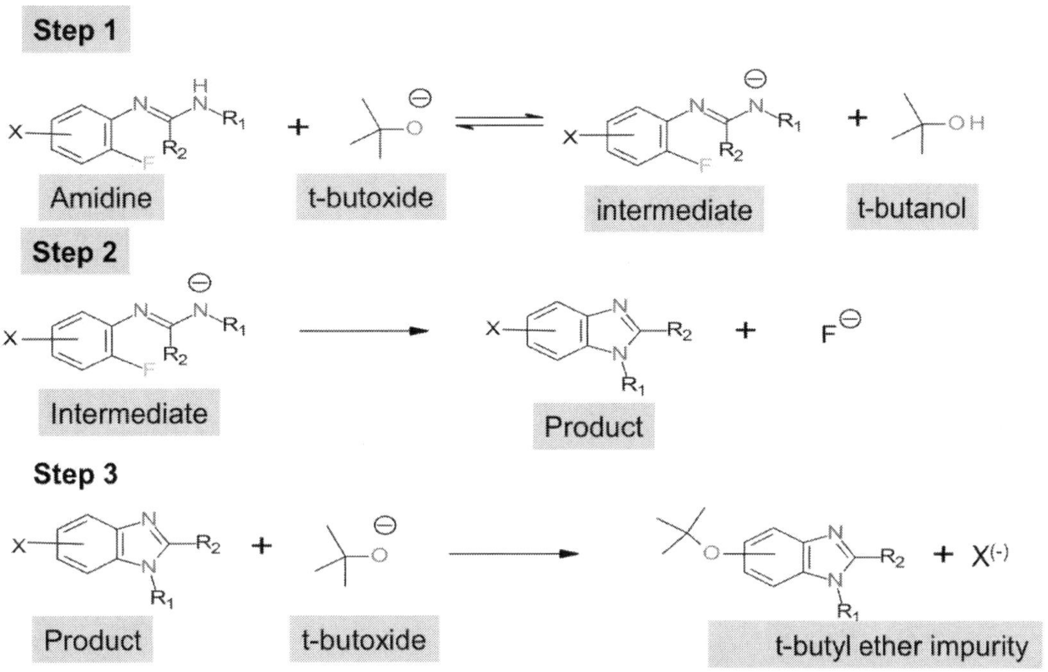


13

# High Product Selectivity Criteria

- Reaction 1
  - Amidine + t-butoxide ⇌ Intermediate + t-BuOH **(Keq)**
- Reaction 2
  - Intermediate ⟶ product + byproduct2 **($k_2$)**
- Reaction 3
  - product + t-butoxide ⟶ impurity + byproduct2 **($k_3$)**

- High product selectivity criteria, Selectivity=$r_p/(r_p+r_{imp})$

  **$k_2 K_{eq}$ [Amidine] [t-butoxide]/[t-butyl alcohol] >> $k_3$ [Product] [t-butoxide]**

- Controlling selectivity
  - I. Metering addition of t-butoxide may not be help
  - II. Dilution may help
  - III. Removing t-BuOH from solution may help

14

# Model build in DynoChem

- $K_{eq}$ is calculated using Marvin software

- Reaction 1
  - Amidine + t-butoxide ⇌ Intermediate + t-BuOH (**$K_{eq}$ = 0.001)**
- Reaction 2
  - Intermediate ⟶ product + byproduct2 (**$k_2$**)
- Reaction 3
  - product + t-butoxide ⟶ impurity + byproduct2 (**$k_3$**)

15

# Exp. vs. Model (fitted in DynoChem)

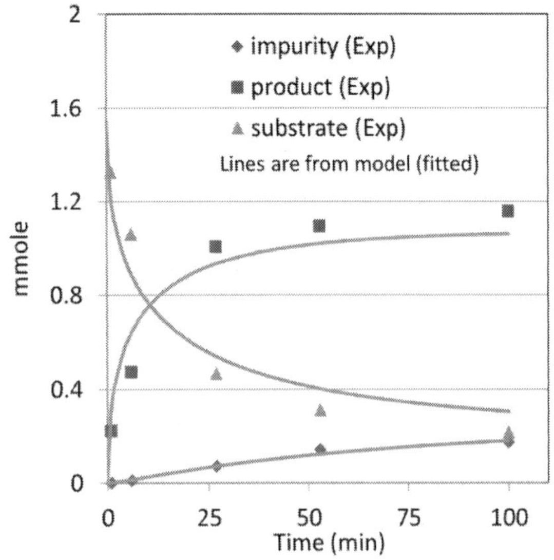


70 °C

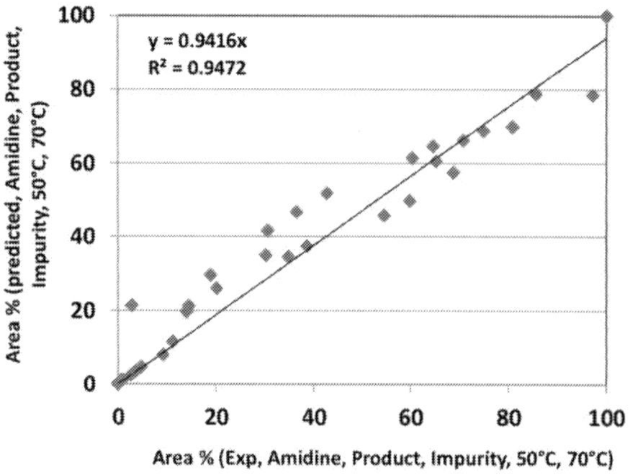


Overall fit

- Mechanistic kinetic model is reasonably well fitted to experimental data

16

# Exp. vs. Model (fitted in DynoChem)

| | Final Value | Units | Confident Interval (%) |
|---|---|---|---|
| Rxn2.Ea> | 129.682 | kJ/mol | 17.724 |
| Rxn3.Ea> | 106.257 | kJ/mol | 11.785 |
| Rxn3.k> | 7.14E-4 | L/mol.s | 18.023 |
| Rxn2.k> | 5.001 | 1/s | 33.343 |

- Reaction 1
  - Amidine + t-butoxide ⇌ intermediate + t-BuOH (**Keq=0.001**)
- Reaction 2
  - Intermediate ⟶ product + byproduct2 (**$k_2$**)
- Reaction 3
  - product + t-butoxide ⟶ impurity + byproduct2 (**$k_3$**)

17

# Prediction from Model : I. Effect of metering addition of t-butoxide

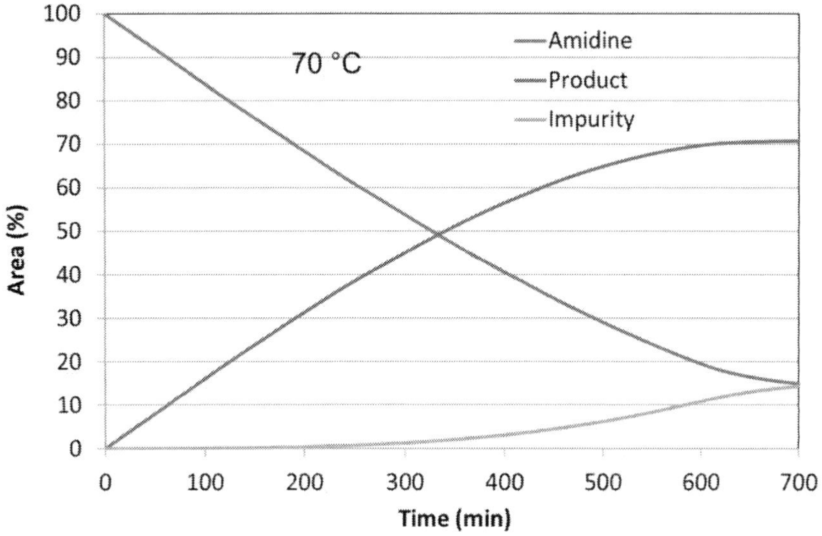


- Metering addition of t-butoxide through 600 mins – Impurity level is similar to batch reactor

18

## Prediction from Model : II. Dilution effect

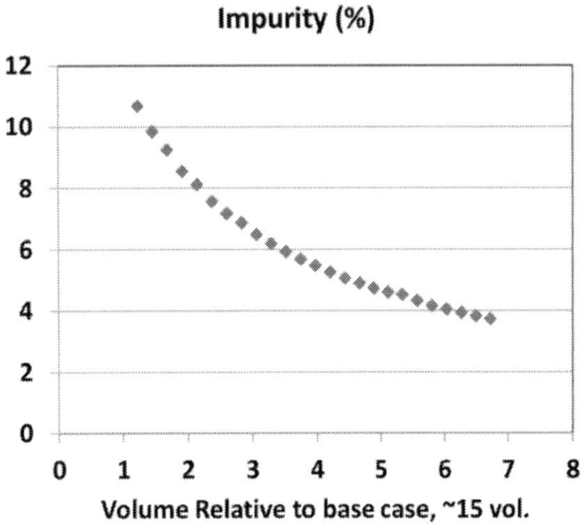


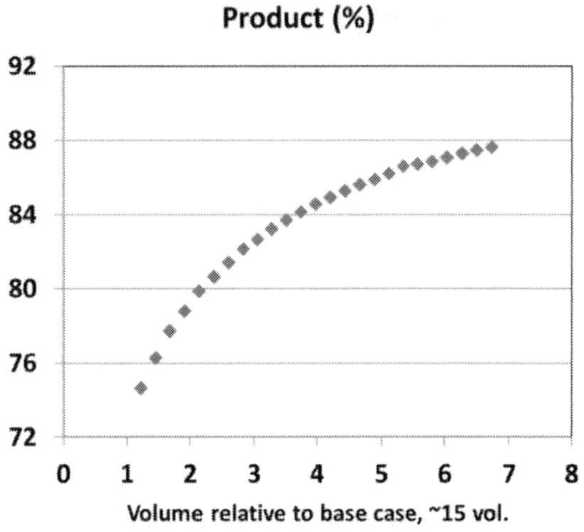


$k_2 K_{eq}$ [Amidine] [t-butoxide]/[t-butyl alcohol] >> $k_3$ [Product] x [t-butoxide]

- Super-dilution may be required – is not practically possible in the plant
  - ~100 vols. to minimize impurity less than 4%
  - ~ 300 vols. to minimize impurity about ~ 1%

19

## Prediction from Model: III. Effect of t-BuOH removal (removal rate= 0.05 mole$L^{-1}Sec^{-1}$)

$k_2 K_{eq}$ [Amidine] [t-butoxide]/[t-butyl alcohol] >> $k_3$ [Product] [t-butoxide]

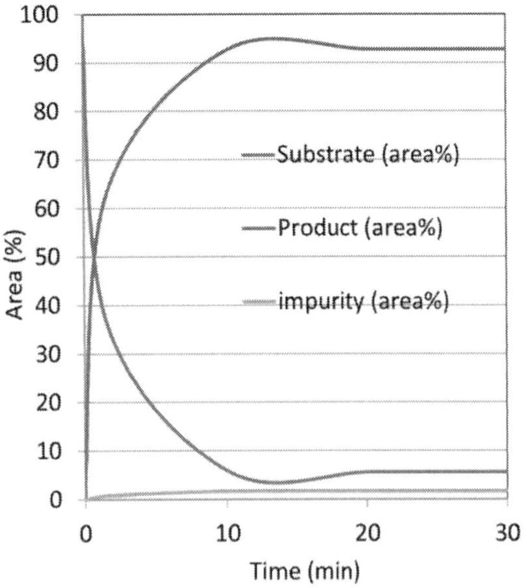


With removal of t-butyl alcohol

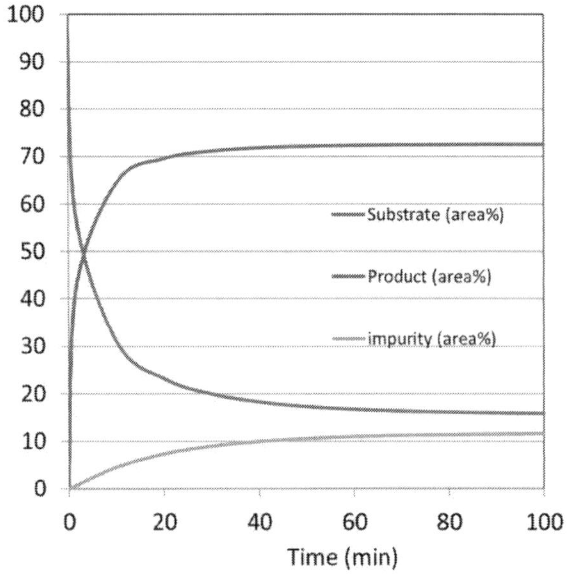


Without removal

20

# Binary VLE of t-BuOH and toluene at 1 atm (DynoChem NRTL)

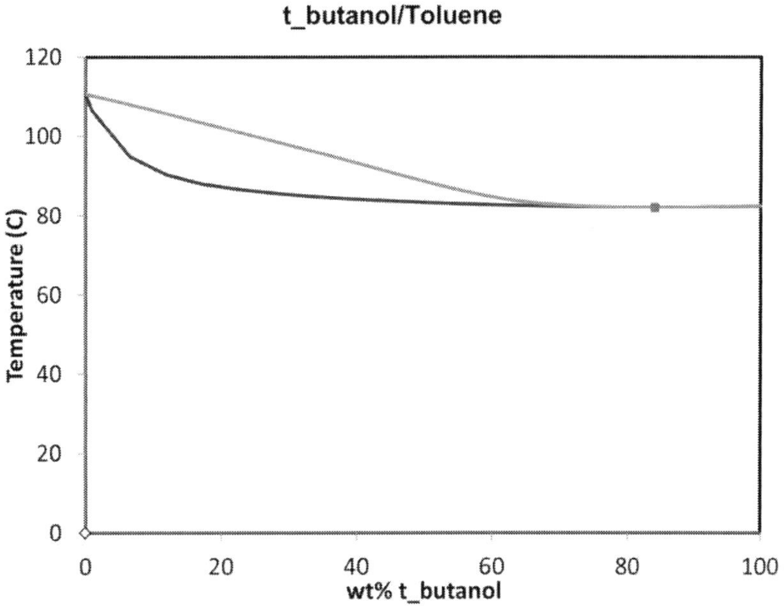


- Distillation can be an option to remove t-BuOH

21

# Reactive distillation

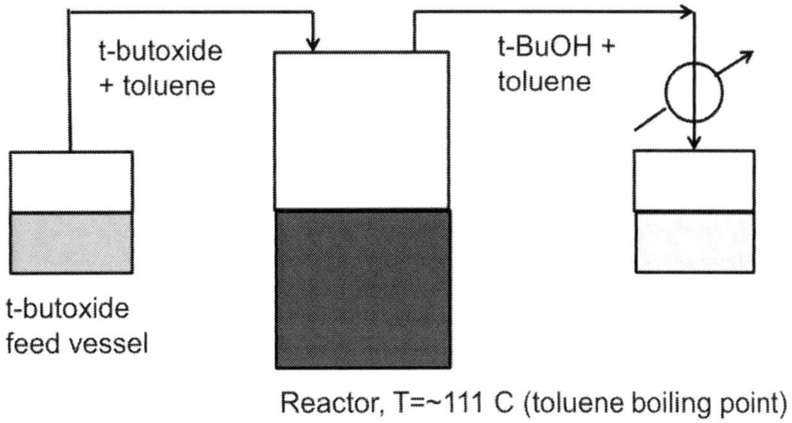


- Fed-batch reactive distillation
  - t-BuOH removal rate ≈ t-BuOH formation rate
  - Control rate of reaction: t-butoxide (in toluene) addition rate
  - Constant volume reactive distillation, distillation rate = t-butoxide (in toluene) addition rate

22

# Proof of concept experiment - Reactive distillation

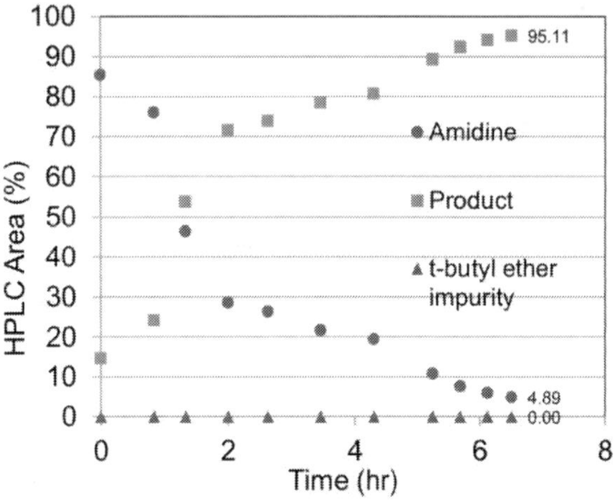


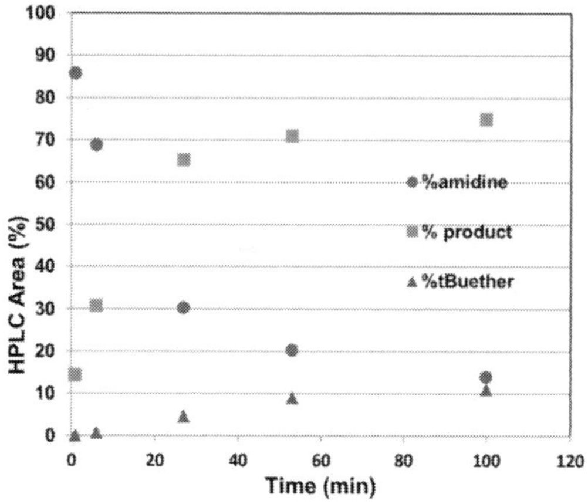


Reactive distillation at ~111 °C

Reaction at 70 °C (No distillation)

- Reaction rate - controlled by metering addition of t-butoxide in toluene (+ little bit THF)
- No significant t-BuOH in the reaction samples
- No significant t-butyl ether impurity : t-BuOH removal effect

23

# Summary

- Mechanistic model based approach led to
  - Ring closure reaction with extremely high selectivity (in-situ, EOR) for the possible use in the future campaign
- The approach is applicable for other reactions to minimize impurity formation from similar competing reaction pathways.

24

# Acknowledgement

- Matt Embry
- Stan Kolis
- Doug Kjell
- Jeremy Merritt
- Shankar Vaidyaraman
- Shekhar Viswanath

25

**Thanks**

**Questions ?**

26

# Proof of concept experiment - Reactive distillation (continuation)

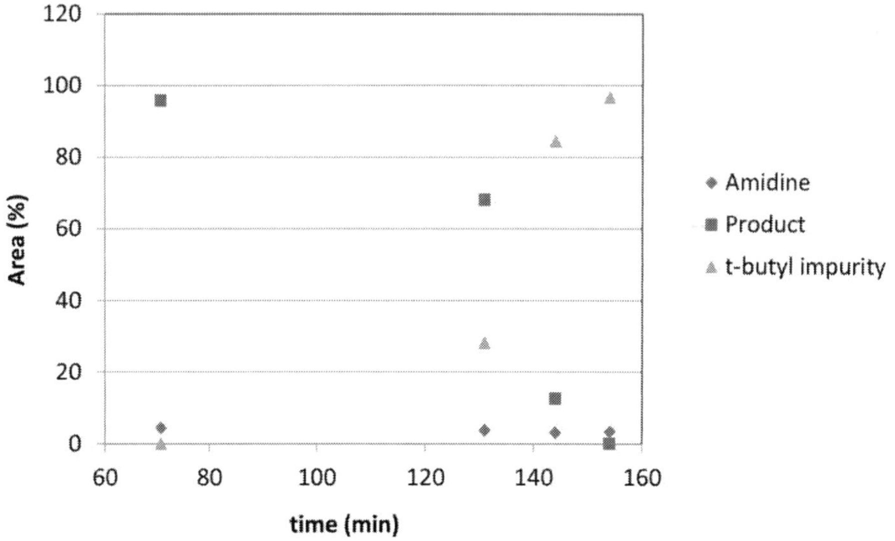


Reactive distillation at ~111 °C (continuation)

- Confirms t-butyl ether impurity formation from product and t-butoxide

27

# Proof of concept experiment - Reactive distillation, reason for the trend

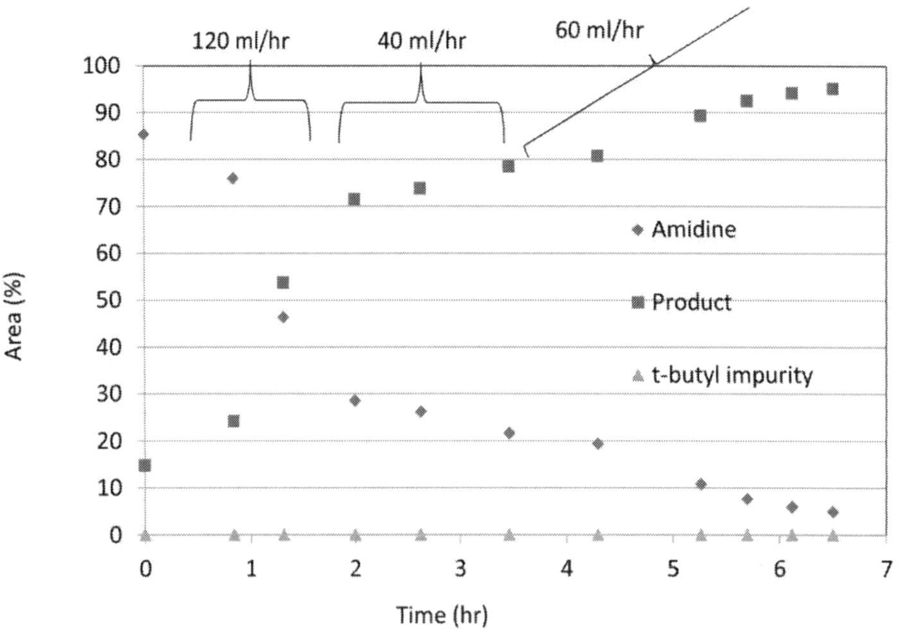


28

# Proof of concept experiment 2

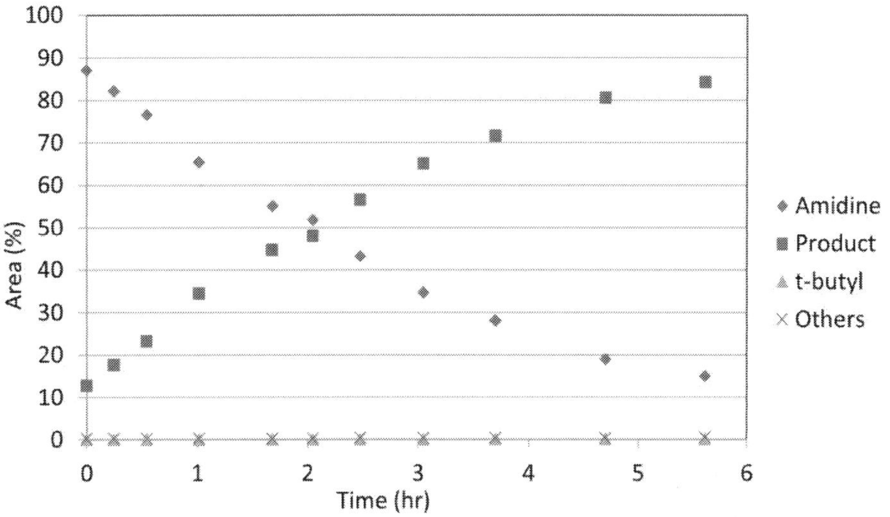


Reactive distillation at ~111 °C

- ~1 equiv. of t-butoxide (in toluene+THF): ~constant addition rate

29

# Proof of concept experiment 3

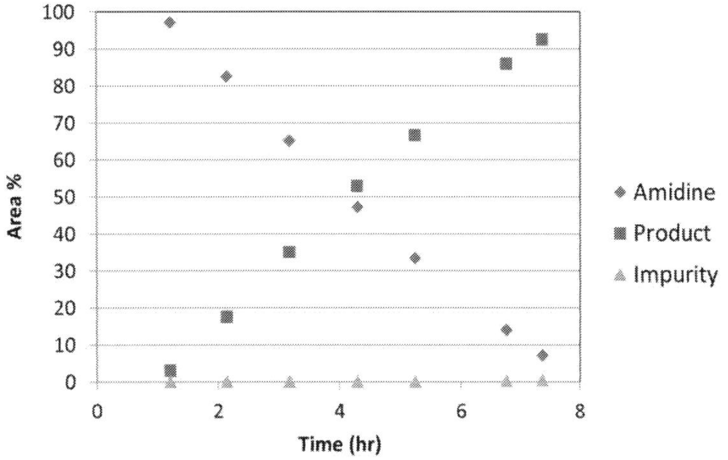


Reactive distillation at ~111 °C, Tj=115 C

- Initial Amidine: 1 g, Distillation working volume: ~20 ml, Tj=115 C
- K t-butoxide (in toluene+THF): 1 ml/hr addition rate (total: 6-7 ml)

30

# 430528 Simulation of Biopharmaceutical Bioreactor Using an Integrated CFD-Population Balance Modeling Approach

**Abstract:**

Since mid-1980's the specific productivity and density of cells in animal cell cultures have been increased several folds. These have been achieved via improving vectors, host cell engineering, selection, gene amplification and cell line screening [1] and optimum design of medium and feeding strategy and process engineering [2]. The necessity of integration of these goals has been established and partially addressed but a comprehensive bioreactor model in which extracellular and intracellular phenomena are coupled is still missing.

To represent the effect of agitation, aeration and feeding strategy on productivity, growth and viability of cells, the model of a large-scale bioreactor has multiple compartments which capture fluid dynamics, cell source variability, interphase mass transfer, gas-liquid interaction, and cell metabolism. The turbulent multiphase flow inside the bioreactor has been successfully modeled using *k-e* models of eddy viscosity provided in commercial CFD software packages [3]. The output of the CFD model is local velocity fields, shear and energy dissipation rates. The most important effect of shear on cells is lysis. To take into account the effect of this phenomenon on the viable cell density a population balance model can be implemented and integrated in which the death kernel includes lysis in addition to other causes of death. The merit of this type of modeling is in the prediction of the average behavior of a group of individuals showing randomness. Therefore it has been a trusted approach to model population of cells and to compute the distribution of cells over cellular states [4]. Due to low solubility of oxygen in water, high densities of cells quickly consume all of it in a saturated culture and produce enough carbon dioxide that it can have inhibitory effects. As a result addition and removal of gases are inevitable parts of operation of a large-scale fermenter. Dissolved oxygen concentration is calculated through estimation of gas-liquid mass transfer coefficients under the operating condition. The gas phase volume fraction can be calculated using gas hold-up correlations available in the literature or from the CFD with an additional Eulerian or discrete phase. The energy dissipated due to bubble rupture is usually higher than that in the bulk of the tank. So gas-liquid interaction directly affects the number of viable cells. Population balance modeling has the capability to capture the stochasticity involved in breakage and coalescence of bubbles and therefore is a good candidate for modeling the gas phase. Another advantage of this approach to capture the effects of bubbles is the more precise calculation of the gas-liquid interphase area; hence a better estimation of cell entrapment rate in the surface of bubbles. Besides lysis, bioreactor hydrodynamics also affects distribution and number of surface receptors, production of specific proteins, certain metabolic processes and can induce apoptosis [5]. Furthermore, nutrients and oxygen accessibilities to cells as well as the toxic effects of the material discharged from cells should also be taken into account. So a connection to cell metabolism seems necessary which can be built via a single cell model. Animal cells contain a large number of molecules so the main decision is the level of the complexity of the single cell model. Metabolic engineering and flux analysis can provide a framework to measure the importance of certain biochemical reactions and their effects on the systemic behavior of cells.

In summary, the model is composed of three main parts which represent bioreactor hydrodynamics, cells population and cell metabolism. The first component consists of a CFD coupled with a population balance model of bubbles. The result of solving this sub-model is nutrients and dissolved oxygen distributions, velocity fields and shear. The cells population model needs this information for calling the single cell model and calculation of its kernels; birth, death, growth and transition. The three components together connect protein production and viable cell density to operational parameters such as aeration and agitation rates and feeding schedule.

References

1. Wurm, F.M., *Production of recombinant protein therapeutics in cultivated mammalian cells.* Nature Biotechnology, 2004. **22**(11): p. 1393-1398.

2. Xie, L.Z. and D.I.C. Wang, *High cell density and high monoclonal antibody production through medium design and rational control in a bioreactor.* Biotechnology and Bioengineering, 1996. **51**(6): p. 725-729.

3. Schmalzriedt, S., et al., *Integration of physiology and fluid dynamics*, in *Process Integration in Biochemical Engineering*. 2003. p. 19-68.

4. Sidoli, F.R., S.P. Asprey, and A. Mantalaris, *A Coupled Single Cell-Population-Balance Model for Mammalian Cell Cultures.* Industrial and Engineering Chemistry Research, 2006. **45**(16): p. 5801 - 5811.

5. Shuler, M.L. and F. Kargi, *Bioprocess Engineering*. 2 ed. 2002: Prentice Hall.

# 436072 Systematic Design and Optimization of Fines Removal in Mixed-Product Mixed-Suspension Crystallizer

**Abstract:**

Fines removal in mixed-product mixed suspension crystallizer (MSMPR) via controlled growth and dissolution is systematically studied in this work. The problem is formulated as a multi-objective optimization, with the fitness functions defined as (1) the ratio of fines particle to the total number of particles (or the "fine index"), (2) the particle mean size, and (3) yield. The decision variables are the temperature set-points of each MSMPR stage, in which the optimization is carried out under different operating constraints, including seeded vs. unseeded operations, size-dependent vs. size-independent crystallization systems, different number of stages and different residence times. The results demonstrate that there exist clear trade-offs between the different objectives and that the various operating constraints significantly dictate if the obtained Pareto solutions are suitable for experimental implementations.

# 432317 Characterization of Lab and Plant Reactors Using Fast Competitive Reactions and a Predictive Model

**Abstract:**

Scale-up of reaction and crystallization conditions often involves assumptions about equivalence of mixing between lab and plant. Scale-down of process for lab studies requires similar assumptions. A question that often arises is how small can the lab equipment be, so that the results translate to scale. With modern lab synthesis workstations providing good control of addition rates and temperature in reactors that resemble their larger scale equivalents, scale-up and scale-down work sometimes takes place in lab systems with volumes as low as 100 mL. It is possible using mixing-sensitive test reactions to quantify the mixing environment in vessels at any scale and to thereby identify lab and plant conditions that are more quantitatively equivalent.

The fast competitive reaction system involving neutralization and ester hydrolysis (a so-called 'Bourne Reaction', ref 1) has been used to study mixing characteristics in several common lab reactors used for routine development of chemical processes. The final product distribution after running the reaction in fed batch mode reflects the history of relative mixing and reaction rates in the plume or reaction zone that forms near the feed location.

Most previous applications of this technique have been in larger vessels where a critical feed time can be identified and longer feed times represent micromixing-controlled conditions. It is possible using a predictive model containing both kinetics and mixing parameters to estimate the product distribution as a function of feed time and mixing energy dissipation rate. This latter parameter, often referred to as epsilon, with units of W/kg, is of interest for a wide range of mixing phenomena. Using the predictive model together with measured product distribution, it is possible to estimate the corresponding epsilon. This represents a useful level of vessel characterization, in addition to the more widely known heat and mass transfer characteristics (UA and kLa).

We will present and discuss results for different reactors and operating conditions and also how best to use predictive models to design and analyse the experiments, which require a good dynamic range of selectivity to be informative. We will compare mixing performance at operating conditions that are commonly used in the lab and plant. Finally, we will deduce some guidelines towards obtaining equivalent mixing between lab and plant reactors.

1. Investigation of micromixing in stirred tank reactors using parallel reactions, John R. Bourne , Shengyao Yu, Ind. Eng. Chem. Res., 1994, 33 (1), pp 41–55

# 420196 Case-Studies of Enhancing Manufacturing Flexibility for a Commercial Biologic

**Abstract:**

During the manufacturing process of a commercial recombinant molecule, CHO cells are cultured through the inoculum expansion in shake-flasks and seed bioreactors followed by inoculation of the production bioreactor. During primary recovery, cell culture is centrifuged and the centrate is passed through a series of depth filters and sterile filters prior to transfer to downstream operations. Three case-studies will be presented to illustrate typical study workflows carried out to support the manufacturing of a commercial biologic moiety. The first case-study entails the validation of three optional, additional passages during inoculum expansion. Evaluation of process parameters against acceptance criteria derived from the in-process control strategy for inoculum expansion will be described. The second case-study details how a selective pressure agent used in inoculum expansion, and originating from an alternate vendor, was evaluated by comparison to the same compound originating from the primary approved vendor. The third case-study discusses a process qualification study for changing out a depth filter and a membrane filter during primary recovery. These case-studies successfully improved manufacturing flexibility allowing full utilization of capacity while handling several multi-product campaigns.

# 432326 The Confined Impeller Stirred Tank (CIST): A New Mixing Device Designed to Properly Scale-Down Local Mixing Effects in Industrial Applications

**Abstract:**

**The confined impeller stirred tank (CIST): a new mixing device designed to properly scale-down local mixing effects in industrial applications**

Marcio B. Machado, Suzanne M. Kresta

University of Alberta, Edmonton AB T6G 2V4

Scaling-down and -up an industrial application remains challenging. The effectiveness of these industrial processes is improved if efficient dispersion and dissolution of the additives is achieved. It is known that current bench scale tests do not always represent the industrial scale behavior. One of the most critical steps for a process scale-down or scale-up is to have a fully turbulent flow regime in the bench scale mixing vessel, because if the flow falls into transitional flow, the results are not reproducible. This work presents the confined impeller stirred tank, as shown in Figure 1, which is a mixing device designed to provide uniform intensity of mixing over the entire volume of the tank. The CIST is a 1 L tank filled with five impellers. Such configuration confines the flow. The tank has a diameter of T = 7.6 cm and a height H = 3T. Mean velocity profiles were measured at different axial positions of the tank using a Laser Doppler Velocimeter. Several impeller geometries, impeller diameters and fluids were used in order to check their effect in the flow. This tank is able to sustain fully turbulent flow throught the entire vessel at the Reynolds numbers less than 3000 while conventional stirred tanks require Reynolds numbers up to 300 000 to achieve fully turbulent flow at the surface. Having fully-developed turbulent flow at the surface, where the chemicals are usually added, is ideal for decreasing the dissolution time, avoiding mesomixing effects and providing reproducible bench scale experimental data which can be used to specify local mixing conditions in the industrial plant. These results have implications both for scale-up and for many industrial applications with surface feed or with dip pipes in the top third of the tank. A methodology to scale-down or –up an industrial process based on the total mixing energy provided to the system is also presented.

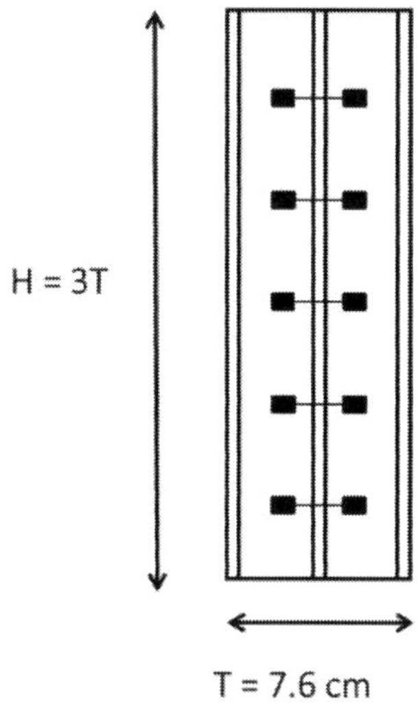


Figure 1: The confined impeller stirred tank (CIST) operating with a set of five Rushton turbines

# 429039 A Hybrid Experimental-Computational Method to Scale-up Powder Dissolution in Viscous Solvent: An Industrial Approach

**Abstract:**

Dissolution of solid particulates in a liquid solvent is an important process in several pharmaceutical operations. Our application relates to topical ointment manufacture where a particulate ingredient must be dissolved in a viscous liquid. From a lab-scale process, we seek to scale-up this operation to pilot scale, and finally to commercial scale.

In this work, we developed a hybrid experimental-computational approach to simulate and scale-up the dissolution process. A population balance model (PBM), implemented in-house, served as the framework to track the size and distribution of particles as they evolve with time. Experimental dissolution kinetics data generated from lab-scale reactor vessels (Mettler-Toledo OptiMax-500) was used to calibrate the mass-transfer rate law. Our PBM framework also incorporates the effect of fluid-particle flow on the dissolution rate—the flow behavior for varying agitation conditions at different scales was characterized by performing a series of computational fluid dynamics flow simulations. The calibrated PBM framework was then used to execute "virtual experiments" in lieu of expensive, actual experiments at the larger scales.

The initial particle size distribution is an inherent source of uncertainty in our process since size distributions of the solute may vary lot-to-lot. The solute powder also is prone to *in-situ* agglomeration when added to the viscous solvent —the extent of agglomeration is random and unpredictable. Therefore, we quantitatively captured this size-distribution uncertainty in our modeling framework and developed confidence intervals, instead of point estimates, for the scaled-up dissolution time predictions.

Our work showcases a successful example where confidence-based model predictions have guided process development and reliably addressed scale-up of an industrial pharmaceutical operation.

# 419741 Scale-up of Agitated Drying: Using Agitator Motor Power Draw to Control Changes in Powder Properties

**Abstract:**

Scale-up of agitated drying operations to produce Active Pharmaceutical Ingredients (APIs) with the desired powder properties can be challenging. Traditional laboratory drying equipment has not successfully reproduced the degree of agglomeration or attrition observed at scale. The development of new lab-scale tools that can reproduce the magnitude of forces observed during agitation of bulk powders has provided insight into the factors that influence changes in powder properties during API manufacturing. Several recent reports have shown that agitator torque and the total work per unit mass imparted on the drying cake through agitation can be correlated to changes in API particle size. Factors such as API characteristics, batch size, dryer geometry, residual solvent content, and the defined agitation protocol impact the agitator torque, work input, and extent of changes in powder properties during drying. This work builds off these previous reports by using agitator motor power draw, which can be easily measured at both laboratory and pilot plant scale, to compare agitated drying studies conducted at different scales. The application of power measurement to define and implement control strategies for agitated drying of two APIs will be described. In the first case, the power measurement was used to identify an upper limit on agitation and avoid breaking up agglomerates during the drying process. For the second API, the agitator power draw was used to define a lower limit for agitation time to ensure sufficient breakage of needles occurred to provide material with the target bulk density.

# 424868 Rational Scale-up of Hot Melt Extruders: 1D Predictive Modeling Based on 3D Smoothed Particle Hydrodynamics

**Abstract:**

**Rational Scale-Up of Hot Melt Extruders: 1D Predictive Modeling based on 3D Smoothed Particle Hydrodynamics**


Josip Matić[*,**], Andreas Eitzlmayr[*], Johannes Khinast[*,**]

*Institute for Process and Particle Engineering, Graz University of Technology, Inffeldgasse 13/III, 8010 Graz, Austria

**Research Center Pharmaceutical Engineering, Inffeldgasse 13/III, 8010 Graz, Austria

Email for correspondence: khinast@tugraz.at


**Introduction**

Hot melt extrusion (HME) is a continuous manufacturing process primarily using co-rotating twin-screw extruders (TSE). The potential to increase the bioavailability of poorly soluble drugs, the steady product quality of a continuous process, the increased efficiency as well as a reduction of operational costs led to the specific interest of pharmaceutical industries in HME. In order to effectively predict the impact of screw configuration and process parameters on the HME process, an improved understanding of flow and mixing in individual screw elements as well as in the entire extrusion process is of high interest.

In our previous work[1], we showed that the smoothed particle hydrodynamics [2] (SPH) simulation method provides significant benefits for extruder screw simulations. In contrast to computational fluid dynamics (CFD), SPH inherently accounts for convective mixing and free-surface flows, allowing a detailed investigation of flow and mixing in fully filled and partially filled extruder screw elements. In order to provide correct boundary conditions for the complex screw geometries, a novel wall interaction method was developed [3].

**Methodology**

Due to the high computational expense of spatially resolved flow modeling, simplified modeling approaches are still required for the modeling of entire extrusion processes. One-dimensional (1D) models consider the spatial dependencies only along the screw axis and thus, provide the benefit of comparably low computational costs, suitable for the industrial use in design, optimization and scale-up of extruders. Following the idea of a continuous stirred tank reactor cascade with back-mixing, as shown by Choulak et al. [4], a 1D extruder model was developed [5], [6] in order to predict profiles of filling ratio, pressure, specific energy input, material temperature and mixedness along the screw axis. Due to its strong simplifications, the 1D approach requires pre-computed data about flow and mixing in individual screw elements, which were obtained from the spatially resolved SPH model. This combination of the three-dimensional (3D) SPH approach for the detailed investigation of individual extruder screw elements and the 1D approach for the description of process variables along entire screw configurations results in a comprehensive tool, which increases the process understanding, supports design, optimization and scale-up, and reduces the need for experiments.

**Results**

An extruder scale-up scenario was analyzed based on proposed scaling laws for process parameters as throughput and screw speed which result in similar conditions at different length scales. Our results show the impact of the extruder dimensions (scale) on filling ratio, pressure, specific energy input, temperature and mixedness.

**References**

[1] A. Eitzlmayr, G. Koscher, and J. Khinast, "Co-Rotating Twin-Screw Extruders: Detailed Analysis of Conveying Elements Based on Smoothed Particle Hydrodynamics. Part 1: Hydrodynamics." Chemical engineering science, Oxford, p. In Press, 2015.

[2] J. J. Monaghan, "Smoothed particle hydrodynamics," *Reports Prog. Phys.*, vol. 68, no. 8, pp. 1703–1759, Aug. 2005.

[3] A. Eitzlmayr, G. Koscher, and J. Khinast, "A novel method for modeling of complex wall geometries in smoothed particle hydrodynamics," *Comput. Phys. Commun.*, vol. 185, no. 10, pp. 2436–2448, May 2014.

[4] S. Choulak, F. Couenne, Y. Le Gorrec, C. Jallut, P. Cassagnau, and a. Michel, "Generic Dynamic Model for Simulation and Control of Reactive Extrusion," *Ind. Eng. Chem. Res.*, vol. 43, no. 23, pp. 7373–7382, Nov. 2004.

[5] A. Eitzlmayr, J. Khinast, G. Hörl, G. Koscher, G. Reynolds, Z. Huang, J. Booth, and P. Shering, "Experimental Characterization and Modeling of Twin-Screw Extruder Elements for Pharmaceutical Hot Melt Extrusion," *AIChE J.*, vol. 59, no. 11, pp. 4405–4450, 2013.

[6] A. Eitzlmayr, G. Koscher, G. Reynolds, Z. Huang, J. Booth, P. Shering, and J. Khinast, "Mechanistic Modeling of Modular Co-Rotating Twin-Screw Extruders," *Int. J. Pharm.*, vol. 474, no. 1–2, pp. 157–176, 2014.

## (661g) Designing for Compaction in Tablets with Spray Dried Intermediates

Ploeger, K., Merck & Co, Inc.
Temtem, M., Hovione FarmaCiência SA
Cordeiro, P., Hovione
Aghajanian, A.,

Originally presented on: 11/12/2015 10:30:00 - 10:50:00

**Title:** Designing for Compaction in Tablets with Spray Dried Intermediates

**Authors:** Kristin Ploeger (kristin.ploeger@merck.com), Marcio Temtem (mtemtem@hovione.com), Paula Cordeiro (pcordeiro@hovione.com), Angela Aghajanian (angela.deese@merck.com)

**Abstract**:

In an environment with increasing numbers of poorly-soluble drug substances, spray drying is used to enhance the solubilization of some of these molecules. Inclusion of this unit operation in the manufacture of oral solid dosage forms results in a new challenge, where the compactability for the formulation is frequently impacted in three separate unit operations (spray drying, roller compaction, and tabletting). The interconnection between the unit operations requires a Quality by Design approach to development, where impact of process parameters across multiple unit operations is considered since the design space in one unit operation has the potential to restrict the design space in the others.

Approaches for designing the spray dried particle attributes will be discussed. The combination of modeling and simulation tools with process understanding and prior knowledge were crucial to deliver a commercially viable process. Moreover, the description of the process and its design space through meaningful and scale-independent parameters (some of them, such as relative saturation of the drying gas and mean droplet size, accurately estimated by modeling tools) were used in the benefit of the scale-up (Figure 1). Corrections between process parameters and material attributes will be

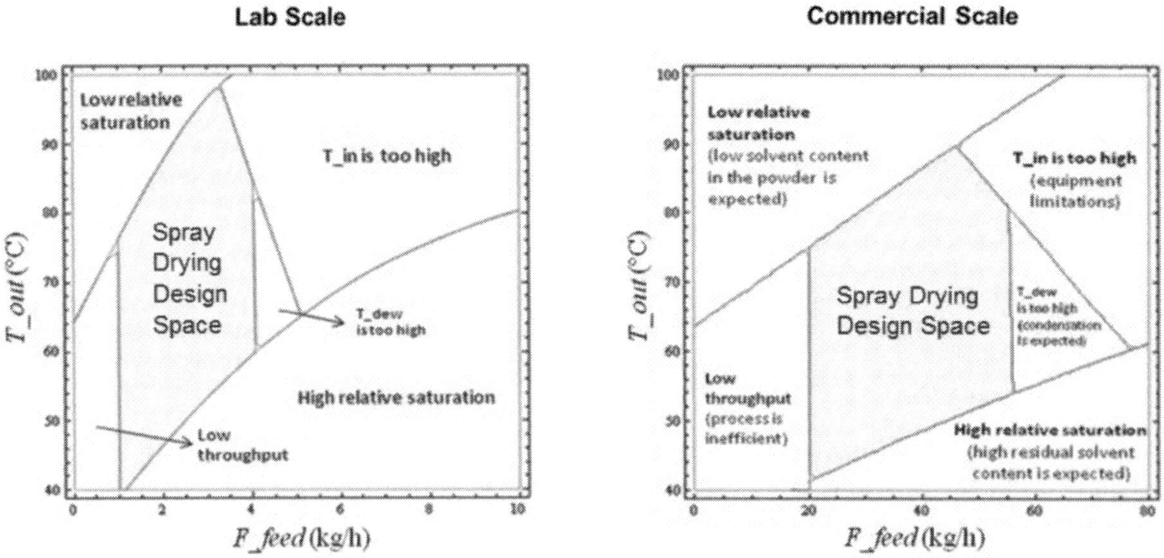


presented.

Figure 1 ? Schematic representation of the lab and commercial thermodynamic design space

Spray dried particle attributes were evaluated to determine which, if any, could be a surrogate measure of particle compactability. A multivariate study of compactability across the three unit operations was performed (Figure 2), to evaluate if enhancement of compactability in the spray dried particles could permit more aggressive roller compaction conditions, without impact to the tabletting process or final attributes (such as disintegration, dissolution, and friability). While traditional studies of roller compaction typically include independent variables like roll pressure, this work will discuss the potential benefits (and challenges) of designing the DoE with a parameter like tensile strength (analogous to the practice of designing tabletting DoEs with tablet hardness as a parameter).

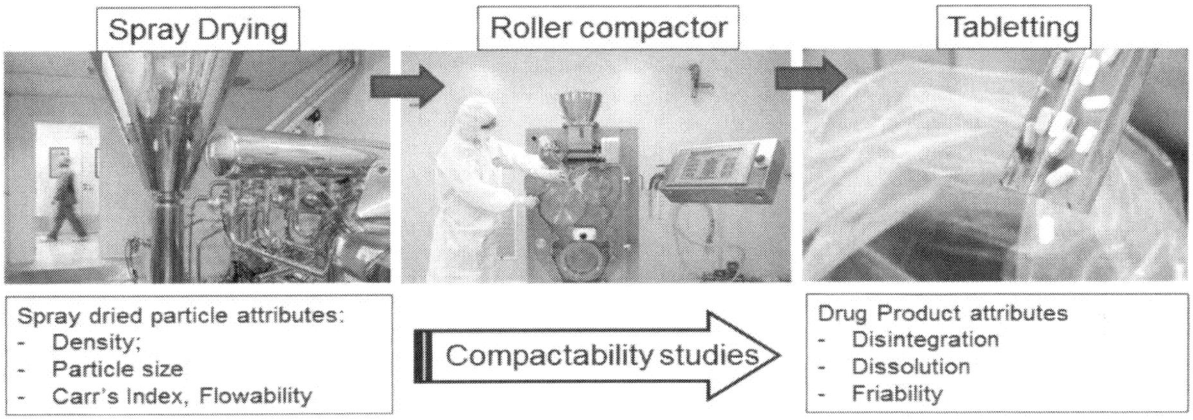


Figure 2 ? Schematic of the proposed experimental protocol (Images ? Hovione copyright)

# 424217 Tuning Phage Display and Yeast Surface Display for Panning for Novel Ligands on Tumor Cells

**Abstract:**

**Tuning Phage Display and Yeast Surface Display for Panning for Novel Ligands on Tumor Cells**

Lawrence A. Stern, Ian A. Schrack, Aakash Deshpande, Sadie M. Johnson, and Benjamin J. Hackel

Yeast surface display and phage display have proven to be effective tools in the evolution and screening of ligands with novel binding activity. Generally, selections for binding activity are carried out using immobilized recombinant domains of target molecules. While this method often provides ligands with apparent high affinity and specificity for the target of interest, translatability to true cellular target *in vitro* is a common concern. Direct selections on mammalian cell surfaces can be carried out either exclusively or in combination with soluble target-based selections to further direct toward evolution of ligands with affinity for true cellular target. Further, this method can be applied to simultaneous discovery of molecular targets and new binding ligands using poorly characterized cell types. This study aims to optimize direct cellular selections using multiple formats of phage display and yeast surface display by using an epidermal growth factor receptor (EGFR) model system that encompasses multiple tiers of affinity and cellular target expression, and to directly compare these methods for applicability to binder selection from mock libraries. Cell-based selections with standard yeast surface display can be successful with up to mid-nanomolar affinity, yielding 15-30% of inputted binders with enrichment ratios reaching $96 \pm 84$. However, this level of enrichment requires at least one million targets per cell, with yields and enrichment ratios decreasing dramatically for a 10-fold decrease in targets per cell. One hypothesis for this decrease in enrichment is a lack of accessibility of relevant epitopes due to steric hindrance by glycosylation and other membrane-bound proteins on cell surfaces. The impacts of displayed protein linker length, receptor epitope, and receptor identity will be discussed. Mathematical modeling of phage display selections highlights the importance of (a) binding strength, which is being evaluated with various intrinsic affinities and four modes of valency – monovalent pIII, bivalent pIII using a previously validated antibody hinge - leucine zipper fusion approach (Lee, *et al.*, *J. Immun. Meth.* 2004), avid pIII from phage vector, and multivalent pVIII via pVIII engineering – and (b) fraction of phage displaying ligand, which is being evaluated using the aforementioned display approaches. Western blotting techniques are used to quantify displaying fraction. Mock library selections are used to quantify enrichment. Optimal conditions, and their consistency with mathematical modeling, will be discussed as well as ongoing efforts in target-ligand co-discovery.

# 431998 Stabilization of Xylanase By Its Random Insertional Fusion to a Thermophilic Host Protein Domain


**Abstract:**

Protein stabilization has been of continuous interest to researchers and is an intense field of study with various applications in pharmaceutical sectors and energy industries. In particular, improved thermal stability of an enzyme enables its application at high temperature, offering many advantages such as low microbial growth, high substrate solubility and an enhanced reaction rate. Our previous study has demonstrated that a target enzyme domain (guest) may be stabilized by inserting into a stable scaffold (host) protein domain. It was also found that stability conferred was a direct influence of the host protein as well as the point of insertion. In the present study, to systematically explore all potential insertion sites of a host protein for guest enzyme stabilization, we facilitated a combinatorial approach. To this end, we engineered a MuST transposon in order to construct insertional fusion libraries, where xylanase from *Bacillus circulans* (BCX) as a guest domain protein was randomly inserted into maltose binding protein from a hyperthermophilic archaeon *Pyrococcus furiosus* (PfMBP) as a host domain protein. The random insertion libraries were screened based on xylanase activity and stability to identify enzymatically active and thermally stable PfMBP-BCX insertional fusion complexes. Our results indicate that thermal unfolding of the BCX domain inserted into PfMBP was reversible in contrast to irreversible thermal unfolding of the wild-type BCX. We also found that while thermodynamic stability of the BCX domain within the selected PfMBP-BCX fusion proteins remained similar, its kinetic stability during prolonged thermal denaturation was substantially higher when compared to the wild-type BCX and another control, an end-end fusion complex of PfMBP-BCX. In summary, our results demonstrate that combinatorial insertional fusion can serve as an important molecular platform for protein stabilization.

# 428048 Co-Evolution of Affinity and Stability of Alzheimer's Domain Antibodies Using Yeast Surface Display

**Abstract:**

An attractive method for designing lead antibodies is to mimic natural protein-protein interactions by grafting key recognition peptides into the antibody binding loops. We are using this approach to design single-domain antibodies that are specific for amyloid-forming polypeptides such as the Alzheimer's Aβ peptide. To further improve binding properties of these lead antibody domains, we have used random mutagenesis and yeast surface display to select for mutations that enhance the binding affinity of antibodies with grafted amyloidogenic (hydrophobic) peptide segments in their binding loops. Interestingly, traditional selection methods yield domain antibodies with improved affinity and reduced stability. The highest affinity antibody domains obtained using such methods were strongly destabilized both on the surface of yeast and as autonomous domains. However, we find that stable antibody domains with improved affinity are reliably identified by replacing the display antibody that recognizes an epitope tag at the terminus of both folded and unfolded antibody domains with a conformational ligand (Protein A) that recognizes a discontinuous epitope on the framework of folded antibody domains. Importantly, selecting only for improvements in one property (e.g., stability) often resulted in reductions in the other property (e.g., affinity). Our findings highlight the importance of simultaneously optimizing both affinity and stability during the selection process to improve the rapid isolation of well-folded antibody fragments with specific binding activity.

# 436951 The Distribution of Fitness Effects of Indels in TEM-1 Beta-Lactamase

**Abstract:**

Insertions and deletions (INDELs) are an important source of genetic diversity in molecular evolution, both in nature and in the laboratory, yet a comprehensive picture of the fitness effects of INDELs is lacking. Taking advantage of innovations in mutagenesis techniques, selection schemes, and deep-sequencing technology, we created an extensive library of single codon INDELs in the *TEM-1* beta-lactamase gene, which confers resistance to penicillin antibiotics, including ampicillin (Amp). The ability of *E. coli* cells expressing the TEM-1 protein to grow in the presence of Amp is a direct reflection of the protein's cellular properties (e.g., $k_{cat}$, $K_m$, protein abundance), making TEM-1 beta-lactamase a convenient and well-studied model protein for sequence/structure/function studies and molecular evolution experiments. Much work has been done to characterize TEM-1, including a comprehensive map of the distribution of fitness effects of missense mutations. Our study helps to further characterize the fitness landscape of TEM-1, and adds to molecular evolution studies in general by offering an extensive picture of the distribution of fitness effects of single codon INDELs within a gene. We used PFunkel mutagenesis and inverse PCR to construct two libraries: one designed to contain all possible single codon insertions, and one with every possible single codon deletion. Using a band-pass selection scheme, each library was partitioned into sublibraries based on relative fitness, as determined by Amp resistance. Barcoded sublibraries were deep-sequenced to determine the fitness effects of individual INDELs. The distribution of fitness effects of these INDELs in TEM-1 and the relationship of fitness effects to structural features will be presented and discussed.

# 433405 Protein Engineering As a Tool to Improve the Surfactant Activity of OmpA Protein from Escherichia coli

**Abstract:**

The study and investigation of new products to replace those that are harmful to the environment has become very important these days in many fields. Surfactants are constituents mostly derived from petroleum, capable of reducing the surface tension on interfaces. Their slow degradation and high contamination in nature make the applications of these substances a huge concern. This is the main reason why bio-surfactants are really attractive now days. These molecules are obtained from different microorganisms during the occurrence of natural processes. Research has been growing in this matter, aiming to satisfy the necessities of effective surfactants that are not aggressive to the environment. We demonstrated in previous studies the surface tension activity of the transmembrane protein OmpA from *Escherichia coli [1]*. Here, we carried out DNA shuffling for OmpA in order to improve its interface activity. To select the best candidates we evaluated surface tension changes when exposing protein mutants to water-dodecane emulsions and the time elapsed until destabilization. Moreover, additional underpinnings behind the potential mechanisms for destabilization of the mutants were analyzed by obtaining the drop size distribution profile in time.
References

[1] S. M. A. Seguro, A. P. Macías, D. C. Pinto, W. L. Vargas, M. J. Vives-Florez, H. E. C. Barrera, O. A. Álvarez y A. F. G. Barrios, «Escherichia coli's OmpA as Biosurfactant for Cosmetic Industry: Stability Analysis and Experimental Validation,» *Advances in Computational Biology*, vol. Volume 232, pp. pp 265-271, 2014.

# 434992 Directed Evolution of Self-Cleaving Intein Purification Tags with Improved Cleavage Kinetics

**Abstract:**

The lack of a purification platform technology for non-mAb (non-monoclonal antibody) proteins represents a primary impediment to product and process development in the biopharmaceutical industry. Advances in mAb-based technologies have been greatly bolstered by the development of protein A affinity chromatography into a generalizable platform for the purification of IgGs. Nevertheless, to date, no such platform exists for non-mAb products. Purification of non-mAb proteins has historically been problematic and costly, with each product purified using a uniquely optimized combination of chromatographic steps. The use of affinity purification tags represents a potential purification platform for non-mAb proteins, but the high costs associated with the tag removal step have greatly diminished the industrial relevance of such technologies. Self-cleaving intein purification tags circumvent the complications accompanying tag removal; however their use is currently limited to bacterial expression due to premature cleavage (leading to product loss) of the intein tag at mammalian cell culture conditions. This work aims to develop a CHO-suitable self-cleaving intein.

We have developed a yeast surface display-based platform for the directed evolution of inteins that have slower, more desirable, cleavage kinetics at the temperature and pH of CHO cell culture. Further screening experiments with the selected mutants have indicated that some of these mutant inteins also appear to retain full cleavage activity under purification conditions. This makes them ideal candidates for use in mammalian cell expression due to the enhanced pH-and-temperature sensitivity of the cleavage reaction. These mutants could also serve as the template/basis for future rounds of mutagenesis that may lead to the development of inteins possessing cleavage kinetics that are even more drastically pH-and-temperature dependent.

# 424768 Induced Pathway Optimization By Antibiotic Doses in Saccharomyces Cerevisiae

**Abstract:**

*Saccharomyces cerevisiae* has increasingly been engineered as a cell factory for efficient and economic production of fuels and chemicals from renewable resources. Here we report a new and simplest synthetic biology approach, named induced pathway optimization by antibiotic doses (iPAD), to improve the performance of multi-gene biosynthetic pathways in *Saccharomyces cerevisiae*. First, plasmids with step-wise increased copy numbers were constructed by engineering the expression level of selection marker proteins, including both auxotrophic and dominant markers. More importantly, the copy number of the plasmids with engineered dominant markers (5-100 copies per cell) showed a positive correlation with the concentration of antibiotics supplemented to the growth media. Based on this finding, iPAD was developed for modular optimization of multi-gene pathways by different combinations of antibiotic concentrations. To demonstrate this approach, iPAD was applied to optimize the lycopene and *n*-butanol biosynthetic pathways, with the production of lycopene and *n*-butanol increased by 10- and 100-fold, respectively. Finally, the pathways optimized by iPAD were integrated to the chromosome to increase the strain stability and eliminate the requirement of antibiotic supplementation, by taking advantage of the iPAD and CRISPR-Cas9 technologies for multiplex pathway integration.

# 429390 Novel Gut Fungal Sugar Transporters for Improved Bioprocess Efficiency

**Abstract:**

Lignocellulosic materials are one of the most abundant feedstocks for the economical production of biofuels and other value-added chemicals. Many sugars derived from such sources, however, remain unused due to poor substrate uptake. For instance, *Saccharomyces* lacks xylose-specific transporters, severely limiting its ability to effectively use these pentose sugars in engineered production pathways. In contrast, anaerobic fungi from the guts of large herbivores are able to digest a wide array of lignocellulosic feedstocks and thrive on the resulting sugars. Here, we pursued a systems-level strategy integrating transcriptomics and phenotype to mine this diverse phylum for novel sugar transporters for improved bioprocessing. A panel of gut fungi isolated from a horse, sheep, and goat, representing three distinct genera, were examined with RNAseq to generate *de novo* transcriptomes containing hundreds of putative sugar transporters. These transporters were classified on the basis of homology into families and a putative substrate was assigned. These predictions were subsequently validated through phenotypic growth studies and sugar utilization assays to reveal dozens of novel sugar transporter candidates for biotechnology. In this talk, I expand on this workflow for general transcriptome mining and discuss some of the more promising candidates.

# 429640 Constructing a General Bioinformatics Pipeline to Identify Regulatory Mechanisms That Improve Ethanol Tolerance in Zymomonas Mobilis

**Abstract:**

*Zymomonas mobilis* has been identified as a promising cellular factory for biofuels due to its efficient, natural production of and tolerance to ethanol. Recent discovery of ethanol-responsive small regulatory RNAs (sRNAs) in *Z. mobilis* suggested the potential of exploiting these elements for strain engineering. As global controllers of gene expression, sRNAs represent powerful tools for engineering complex phenotypes. However, mechanistic analysis of these regulators in bacteria lags far behind their high-throughput search and discovery; this makes it difficult to understand how to efficiently identify sRNAs that could be used to engineer a phenotype of interest. In this study, we use a forward systems approach to first predict sRNAs that impact ethanol tolerance in *Z. mobilis* using large-scale transcriptomics and proteomics profiles of *Z. mobilis* under ethanol stress. The effect of the bioinformatically predicted candidates were experimentally characterized by building overexpression strains and this led to the successful uncovering of several sRNAs that could be manipulated to enhance ethanol tolerance. Using these ethanol-related sRNAs, we then performed traditional genetic and biochemical approaches to identify a variety of mRNA targets and pathways that were being regulated under conditions of enhanced tolerance. In this way, we have demonstrated application of a novel bioinformatics pipeline to accelerate the discovery of specific pathways and extract insightful regulatory mechanisms that could be further optimized to enhance a given complex phenotype. This work represents the first application of a de novo sRNA engineering strategy in non-model *Z. mobilis* that is of relevance to biofuel technologies.

# 430342 Facilitating Unconventional Yeast Engineering for Biorenewables Production

**Abstract:**

Many of the 1800 known yeast species have highly unusual metabolic, biosynthetic, physiological and fermentative capacities that are not possessed by model yeasts such as *Saccharomyces cerevisiae*. Here we report our recent success in establishing the first stable expression platform for engineering *Scheffersomyces stipitis*, one of the species with the highest native capability for xylose fermentation. The current expression plasmid is extremely unstable, being lost within 2 days of cultivation. With the perfect integration of *in silico* prediction and library-based screening, we were able to isolate all eight centromere sequences from *S. stipitis* genome in a monthly manner. The identified centromeres significantly enhanced the stability of the episomal vector, resulting in homogenous expression of target proteins and significant increase in production of valuable biochemicals. Isolating a functional centromere in an efficient manner by far is still a missing piece from literature. Immunoprecipitation with specific kinetochore protein antibodies is the predominant method, but suffers from the extensive experiment procedures and the prerequisite of antibody that needs to be developed for each species. We are currently investigating the generalizability of our hybrid approach in rapid centromere identification in a broad spectrum of unconventional yeasts. In contrast to the previous developments in this field that were fairly scattered and short of systematicness, we are ambitious to elucidate design principles and establish platform technologies to enable rapid functional modifications of a series of high-potential yeast species.

# 433236 Optimization of the Escherichia coli Methyl Erythritol Phosphate Pathway

**Abstract:**

Isoprenoid pathways present attractive solutions for the production of biofuels, biomaterials, and biopharmaceuticals. In the current study, we report the production of isopentenol in *Escherichia coli* via the non-mevalonate pathway by the stable expression of the pyrophosphohydrolase enzyme NudB fused with N-terminal Flag tag. The expression of deoxy-xylulose-phosphate (DXP) synthase gene *dxs*that condenses pyruvate and glyceraldehyde 3-phosphate into the MEP pathway was limited by the formation of insoluble inclusion bodies and cell lysis. Upon optimization of DXS expression and solubility, DXP concentrations and culture physiology were significantly enhanced thereby improving the isopentenol titers in the culture.

The carbon-efficient non-mevalonate pathway is limited by the flux imbalance of its intermediate metabolites due to the technical issues related to the post-translational modification of its enzymes with iron-sulfur clusters in *E. coli*. Due to this, the intermediate metabolite methyl erythritol cyclopyrophosphate (MECpp) of the MEP pathway which is utilised by the enzyme IspG, an Fe-S protein; was found to be effluxed into the culture supernatant, thereby resulting in significant carbon loss. Since anaerobic conditions have been observed to be better redox mediators for higher IspG activity, when isopentenol-producing *E. coli* strains were cultivated under anaerobic conditions, the specific productivity of isopentenol improved to 6 mg/L/ODunitcells as compared to 4 mg/L/ODunitcells under aerobic conditions. Supplementation of Fe-S cluster assembly proteins by the expression of the ISC operon further improved the isopentenol production levels. Through these multi-step optimizations, 64 mg/L isopentenol was obtained from 20 g/L glucose in *E. coli*cultures.

The balance of the MEP-intermediate fluxes via the active expression of Fe-S cluster proteins IspG and IspH would be critical for further optimization of isopentenol production. To achieve this, we have further explored the expression of apoproteins IspG and IspH in *Escherichia coli* and their maturation into their holo-forms. The Fe-S cluster assembly proteins were supplemented by the co-expression of the ISC (Iron Sulfur Cluster) operon with the target apoproteins. We further investigated the expression of four proteins CyaY, ErpA, RIC and NfuA for their possible role as iron-donors or Fe-S cluster assembly/repair. It was interesting to observe that ErpA expression enhanced the labile iron and labile sulfide content in purified IspG as well as IspH. When the effect of these helper proteins were further studied under anaerobic culture conditions, the RIC protein which has been reported to be involved in repair of iron centres significantly improved the labile iron content in IspG and IspH. The increased iron and sulfide content in the purified target proteins was also accompanied by higher intensity bands on Western blots from NATIVE protein gels of the crude cell extract, clearly indicating higher level of expression of soluble IspG and IspH.

These studies also serve as model proteins to explore biogenesis of Fe-S cluster proteins that are involved in a variety of biological roles including electron transfer, catalysis, sensor and regulatory mechanisms, protein stabilizations etc. Further studies are being conducted to quantify the active Fe-S cluster in the target holo-proteins and investigate their effect in the in-vivo cultures and balance the fluxes of the MEP pathway.

# 433625 Kinetic Modeling of Lignin Biosynthesis in Arabidopsis thaliana for Improved Biofuel Production


**Abstract:**

**Kinetic Modeling of Lignin Biosynthesis in *A. thaliana* for Improved Biofuel Production**



Rohit Jaini[1]*, Longyun Guo[2], Peng Wang[2], Rachel McCoy[2], Clint Chapple[2], Natalia Dudareva[2], John A. Morgan[1,2]

[1]*School of Chemical Engineering, Purdue University, West Lafayette, Indiana 47907,* [2]*Department of Biochemistry, Purdue University, West Lafayette, Indiana 47907*



Lignin is a hetero-phenolic polymer formed as a result of radical polymerization of monolignols namely coniferyl, syringyl and *p*-coumaryl alcohols – products of the phenylpropanoid pathway. It imparts structural strength, vascular integrity and pathogen resistance to plants. Although lignin is essential for plant sustenance, it renders valuable lignocellulosic biofuel feedstock recalcitrant to chemical, mechanical and biological treatment. Several attempts to manipulate lignin monomer content and composition, to improve forage digestibility and saccharification efficiency, have revealed a lack of understanding of the dynamics and regulatory properties of monolignol biosynthesis. We are developing a kinetic model to elucidate the underlying mechanisms regulating carbon flux through the phenylpropanoid network. Kinetic parameters for the model will be obtained by non-linear least squares optimization of metabolomics data from *Arabidopsis thaliana* stem tissue fed with [ring-$^{13}C_6$]-phenylalanine, the pathway precursor.

As a first step, we developed a novel and comprehensive analytical method based on liquid chromatography coupled to tandem mass spectrometry (LC-MS/MS) for quantifying intermediates of the phenylpropanoid pathway. The experimental procedure entailed extraction of metabolites from *A. thaliana* stem tissue using 75% (v/v) methanol in water as a solvent. The extract was then analyzed using reverse phase LC-MS/MS by electrospray ionization in the negative ion mode with multiple reaction monitoring (MRM) of two ions (Q1–parent/Q3–daughter) per compound. The chromatography method was improved by tuning for chromatographic parameters such as buffer concentration, flow rate, pH of mobile phase, and column temperature. Transgenic Arabidopsis plants with knockdown or overexpression of the ferulate 5-hydroxylase gene were utilized to validate the analytical method. Next, isotopic feeding experiments with three different concentrations (100, 300 μM & 1 mM) of [ring-$^{13}C_6$]-phenylalanine were conducted to obtain kinetic parameters for the entry reactions to the phenylpropanoid network, starting at phenylalanine to *p*-coumaroyl CoA.

The analytical method allowed detection and quantification of 12 of the 17 compounds of the pathway for which standards were available in wild type tissue. Signal suppression due to matrix effects and analyte losses during sample preparation and extraction were accounted for by standard spiking studies to accurately quantify the pathway metabolites. A set of kinetic parameters were identified that simultaneously fit fractional label and pool sizes of the endogenous intermediates obtained using the analytical method. The model showed good agreement with experimental data allowing reliable parameter estimation. Model validation was conducted using an independent data set obtained on a *pal1/pal2* knockout line. Our study has laid foundation to extend the kinetic model to the entire metabolic network and shows promise as a predictive tool to facilitate rational metabolic engineering of plants for improved biofuel production.

# 434103 Transforming Our Understanding of Algal and Plant Lipid Metabolism with a High-Throughput Whole Genome Functional Genomics Strategy

**Abstract:**

Microalgae have been of recent interest for use as a biofuel feedstock because they can produce large quantities of lipids, contain little recalcitrant biomass, and do not impact the food supply. Lipid metabolism is central to algal/plant biology, human nutrition, and biofuel production; however, details on how it is regulated, what genes are involved but currently unknown, and how these metabolic pathways can be optimized for bioenergy production remain unknown.

To address these questions we are using a newly developed functional genomics strategy to identify all single gene mutations that alter cellular lipid metabolism in *Chlamydomonas* at a genome scale, a first for a unicellular eukaryotic phototroph. A genome-wide, barcoded insertional mutant library has been created and genomic insertion sites mapped. A Fluorescence-Activated Cell Sorting (FACS) pipeline has been used to isolate mutants with increased and decreased triacylglyceride (TAG) content. Mutants identified in the screen have been characterized for total TAG content, fatty acid distribution, and lipid class abundance. These mutants are revealing insights into the functional relationship between specific genes and lipid metabolism in algae. We are pursuing genes involved in the regulation of lipid metabolism, in particular, genes involved in the switch from basal TAG production to hyper-accumulation during nitrogen deprivation, or so called 'lipid triggers'. This whole genome functional genomics approach to studying algal lipid metabolism is yielding insights that can inform metabolic engineering strategies to maximize biofuel production in these organisms.

# Relationships between $CO_2$ Consumption and Co-Product Formation during Bubble Column Photobioreactor Cultivation of the Diatom *Cyclotella* Under Light-Limited and Light-Saturated Growth Conditions

Altan Ozkan and Gregory L. Rorrer

Oregon State University
School of Chemical, Biological, and Environmental Engineering

Paper 424694, Session: Food, Pharmaceutical & Bioengineering
AIChe Annual Meeting, November 12, 2015

Emerging Frontiers in Research and Innovation (EFRI) program, Award Number 1240488

## Diatoms

- Most species rich group of algae
- Eukaryotic, single cellular, photosynthetic microorganisms
- Sizes ranging from 1 to 500 µm
- Cell walls (frustules) composed of amorphous silica
- Many high rate biomass and lipid producing strains

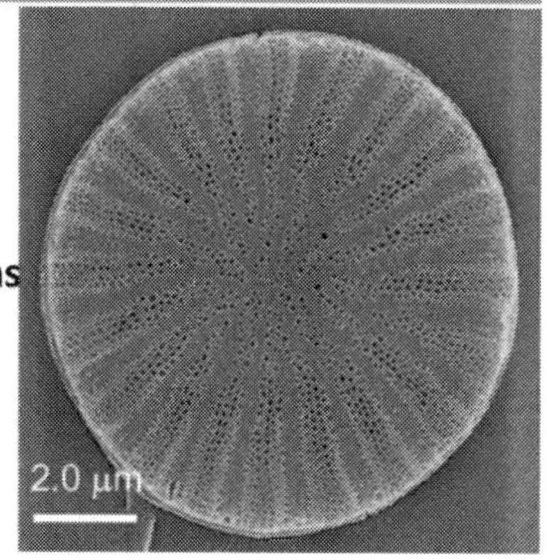


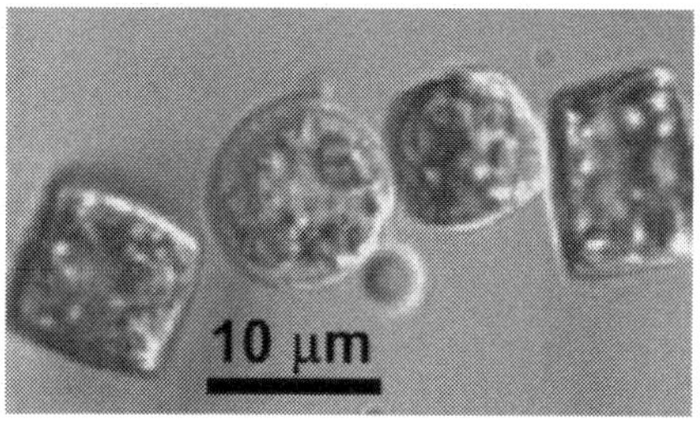


1

# Experimental Set-up and Process Parameters

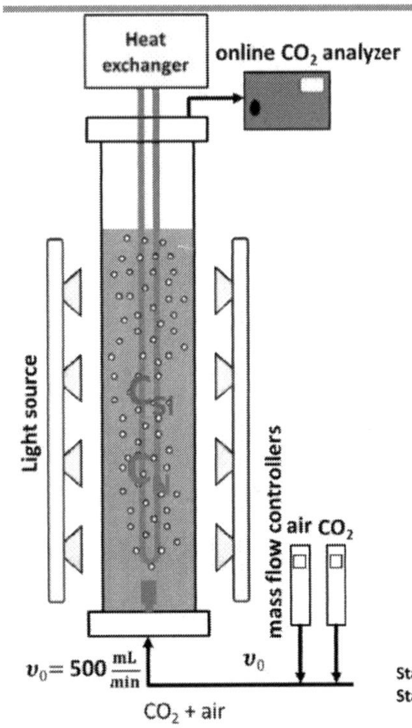


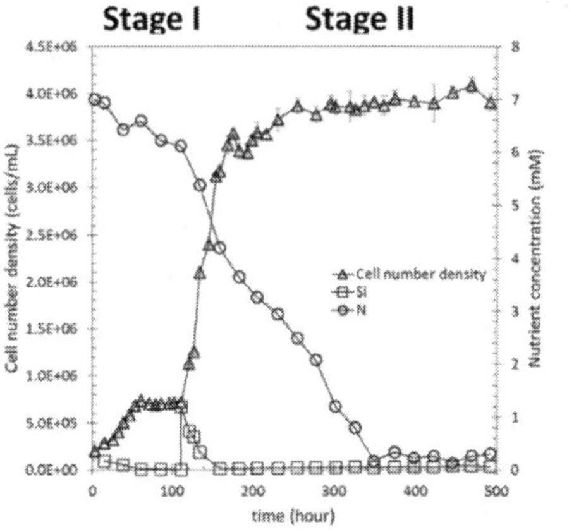


| Parameter | Stage I | Stage II |
|---|---|---|
| $C_{Si}$ (mmol/L) | 0.3 | 1.2 |
| $C_N$ (mmol/L) | 7.0 | 6.2 |
| $I$ (μmol/m$^2$-sec) | 36 | 70/220/371/525 |
| $v_0$ (mL/min) | 600 | 600 |

**Stage I:** Cell growth to silicon (Si) starvation at 7x10$^5$ cells/mL

**Stage II:** Continued cell 4x cell growth to a fixed cell number density of ~4x10$^6$ cells/mL



# $CO_2$ Consumption and Fate

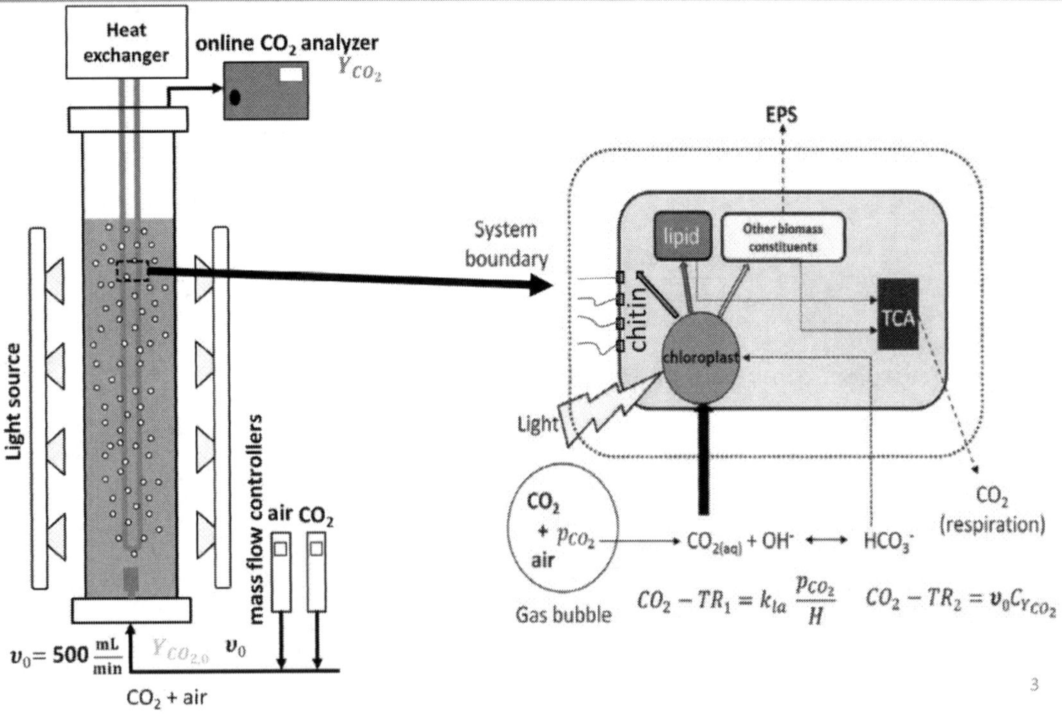

# $CO_2$ at Reactor Inlets and Outlets

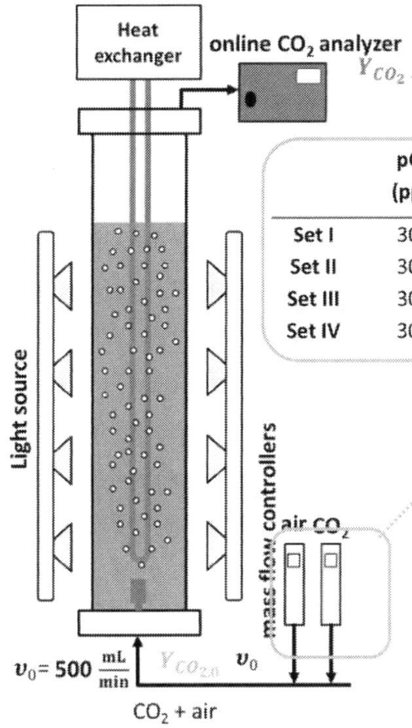


| | $pCO_2$ (ppm) |
|---|---|
| Set I | 3000 |
| Set II | 3000 |
| Set III | 3000 |
| Set IV | 3000 |

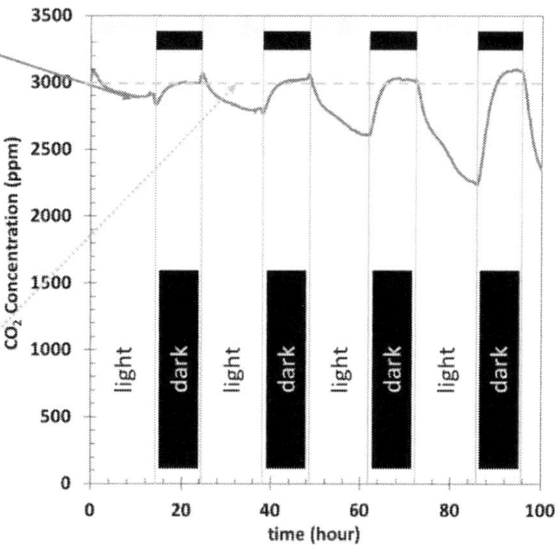


4

# Estimation of $CO_2$ Consumption

**Inlet & Outlet $CO_2$**

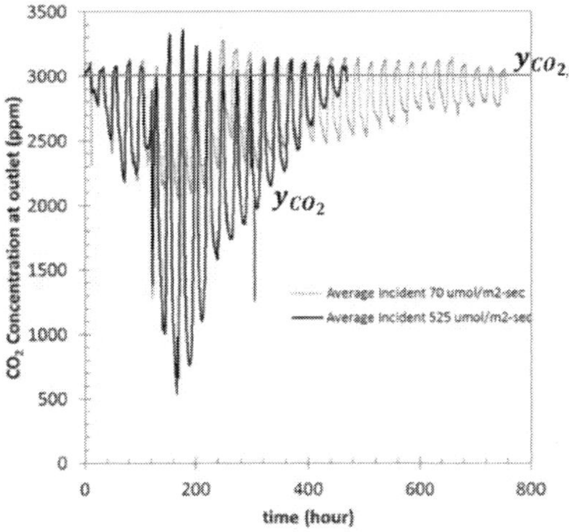


**Cumulative $CO_2$ Consumed**

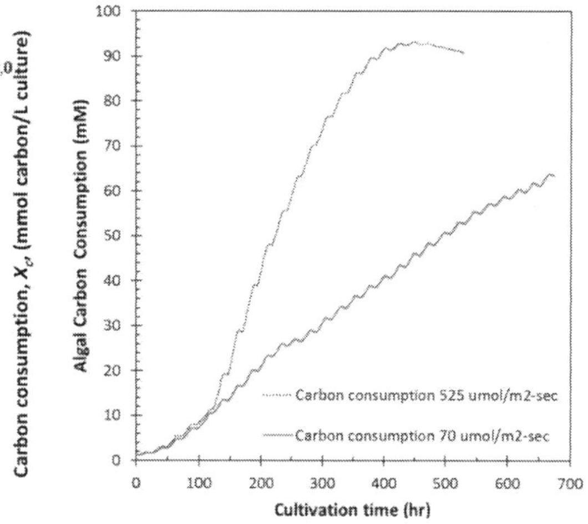


**$CO_2$ consumption from outlet gas**

$$X_C(t) = X_{C,i} + \frac{v_0}{V}\frac{P}{RT}\int_{t_i}^{t_{i+1}} \left(y_{CO_{2,0}} - y_{CO_2}\right) dt$$

5

# Estimation of $CO_2$ Consumption

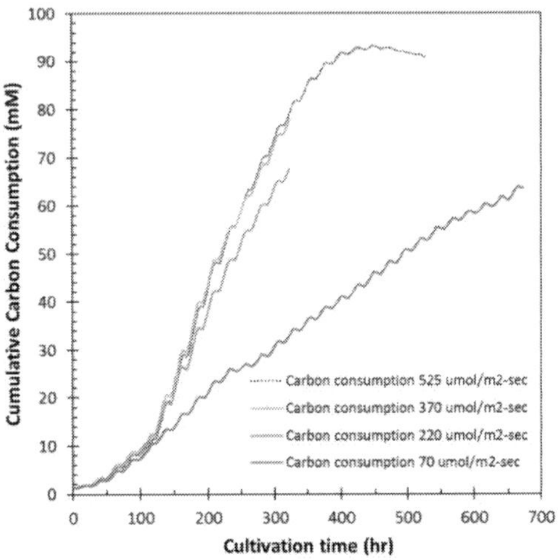


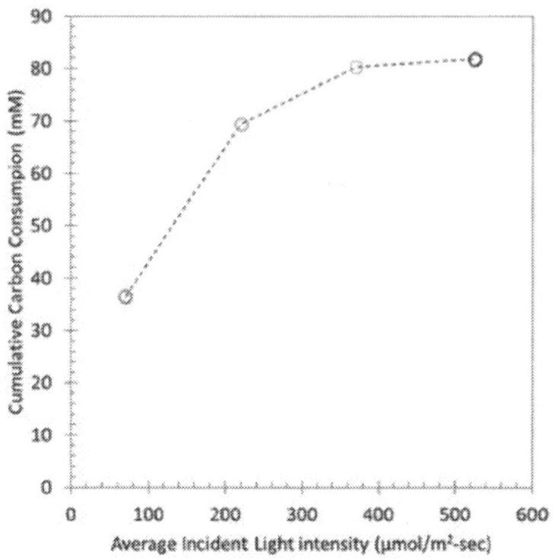


**$CO_2$ consumption from outlet gas**

$$X_C(t) = X_{C,i} + \frac{v_0}{V}\frac{P}{RT}\int_{t_i}^{t_{i+1}}\left(y_{CO_{2,0}} - y_{CO_2}\right)dt$$

6

# Comparison of Cell Division and Nutrient Consumption at Low and High Light Intensity

**Incident 70 µmol/m²-sec**

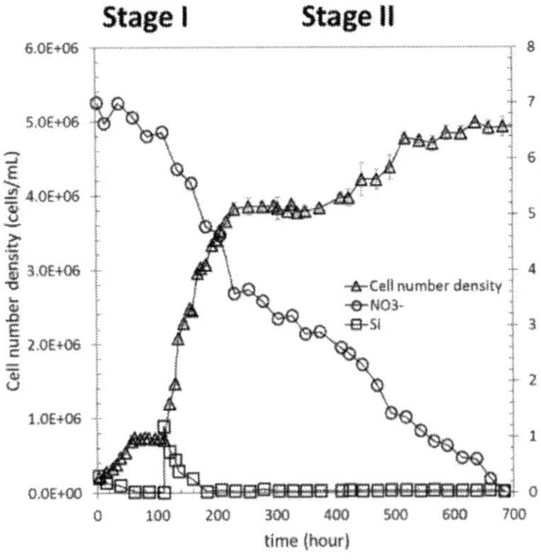


**Initiation of Stage II**

Silicon addition = 1.2 $\frac{mmol}{L}$

**Incident 525 µmol/m²-sec**

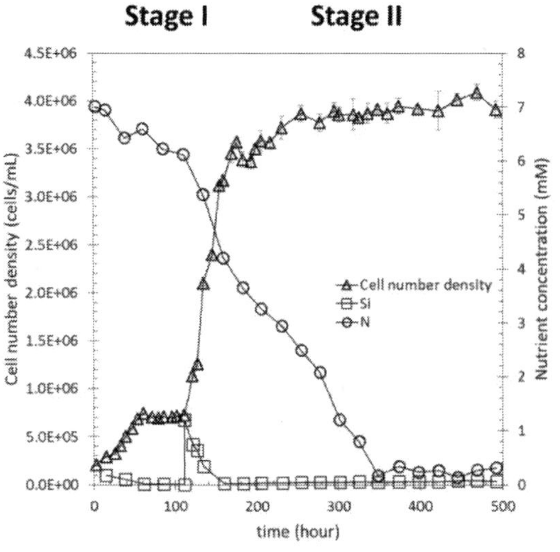


**Initiation of Stage II**

Silicon addition = 1.2 $\frac{mmol}{L}$

7

# Stage II-Biomass and Lipid Production

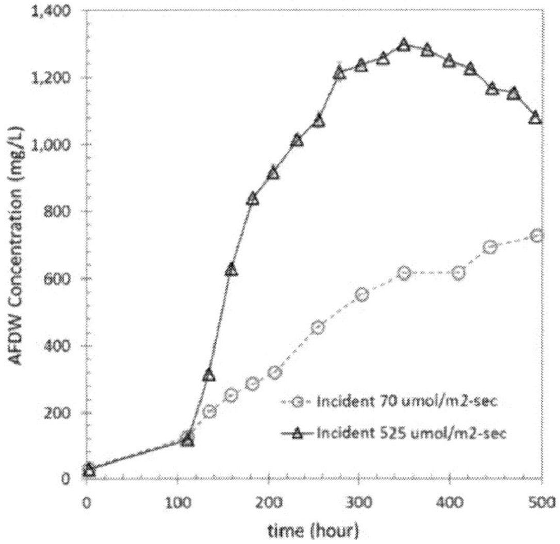


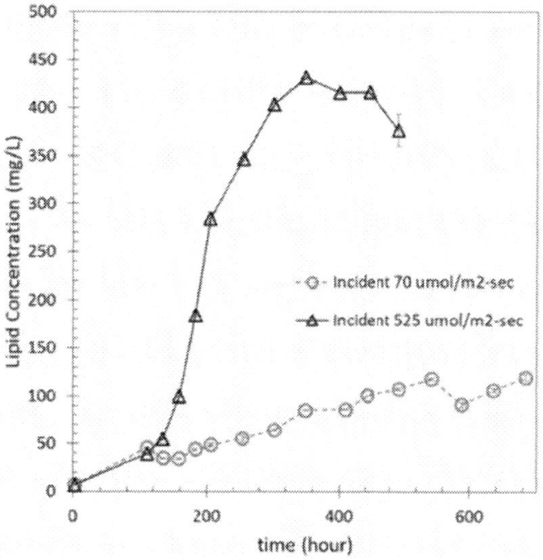


8

# Stage II-Chitin Production

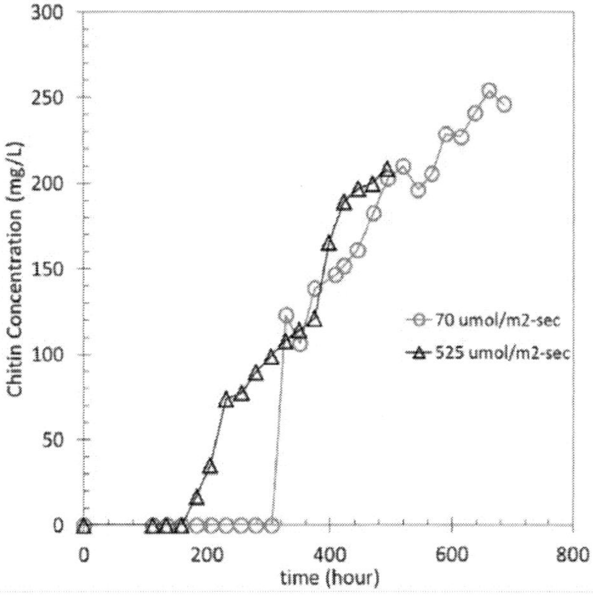


9

# Conclusions

**Summary:**

- The photosynthetic diatom *Cyclotella* was cultivated to silicon starvation at average incident radiation ranging from 70 to 525 µmol/m$^2$-sec.

**Major Outcomes:**

- $CO_2$ consumption, biomass production, and lipid production continued after silicon depletion and final cell number density achieved.
- Carbon consumption rate plateaued after an average light intensity of 371 µmol/m$^2$-sec.
- Lipid productivity increased with increasing light intensity.
- Chitin productivity did not follow the increase in biomass production rate, i.e. increased average incident light intensity, and decreased with increased production rate of lipids.

10

# 429019 On-Line Monitoring of Protein Product Formation in Mammalian Cell Culture System


**Abstract:**

Recombinant protein therapeutics are the most important new therapeutic entities developed in the past twenty years. In the processes of cell line development for producing these proteins, on-line and *in situ* monitoring the product titer is crucial. In this work, we report a novel on-line monitoring method to predict the monoclonal antibody (mAb) titer in a Chinese hamster ovary (CHO) cell bioreactor based on mass balance of amino acids. By measuring a few parameters beforehand and conducting simple optical readings *in situ*, we were able to promptly predict the mAb titer in bioreactors at any specific fermentation time with high accuracy. This rapid and low-cost method could be extremely useful for clone selection and process development in the bio-pharmaceutical industry.

# 434841 Application of Scale-Down Modeling for Improving Robustness of a Manufacturing Process By Supplementation of Limiting Nutrients

**Abstract:**

Representative scale-down models of manufacturing processes allow for troubleshooting and improvement of the cell culture process at scale. In this work, we applied scaling parameters such as volumetric flow rate for oxygen transfer and $CO_2$ stripping, mass transfer coefficient ($k_La$) for oxygen delivery, and feed strategy alignment between scales to create a representative 5L scale-down model of a manufacturing process at the 15,000L scale. Multivariate analysis (MVA) was implemented to compare the process performance between bench scale and manufacturing scale, and univariate analysis was used for product quality comparison. The confirmed scale-down model was statistically similar to two Biogen manufacturing facilities in process performance and product quality. Further, the scale-down model was applied to understand two growth phenotypes observed in manufacturing. Two critical nutrients were identified to be depleting in the higher cell density growth phenotype, leading to a faster drop in cell viability. A simple feed supplementation did not improve process performance. Subsequently, methods of delivering higher levels of the limiting nutrients in the form of a stock solution addition during the process or upfront supplementation in the feed formulation proved to increase the process performance. The current work shows the successful application of a scale-down model developed systematically, to improve the robustness of an existing process in manufacturing.

**(682e) Genomics Based Methodology of Cell-Culture Media Formulation for Improved Bio-Therapeutic Productivity and Quality Consistency**

Hemlata, H.,
Yoon, S., University of Massachusetts Lowell

Originally presented on: 11/12/2015 13:58:00 - 14:20:00

Biosimilar drugs are emerging very fast as the patents of innovator's drugs are expiring. Media formulation development has to be carried out for each biosimilar, which takes up a significant amount of time and generally, a random approach is taken to meet comparability requirement and improve the product titer. A targeted approach with genome information of host cell as a function of media composition can provide useful information to explain the product titer variability and comparability. A 2-step correlation model explaining relationship between cell-culture productivity and quality, and media compositions taking gene expression as intermediate attributes can provide a precise and robust platform. Experimental design was conducted for different media compositions having different productivity enhancer components to get a wide range of product titer. Gene expression data obtained from the above cell culture samples shows significant differences among these samples. This is an on-going work. Expected outcome of this work is the development of a novel mechanistic model based on gene expression data, which will potentially decrease the time for media formulation and is expected to be applicable to different types of cell clones.

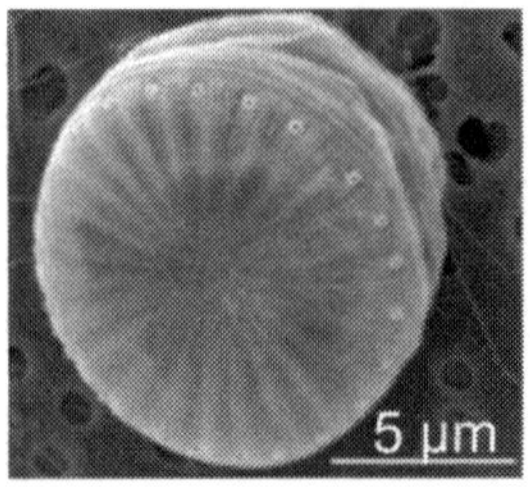


# Control of chitin and lipid production in the photosynthetic diatom *Cyclotella sp.* by co-limitation of silicon and nitrate in batch and fed-batch cultivation

Omar Chiriboga (speaker) and Gregory Rorrer.
School of Chemical, Biological and Environmental Engineering
Oregon State University, Corvallis – OR

Nattaporn Chotyakul, Xulei Wu, Bob Durst and J. Antonio Torres.
Department of Food Science and Technology
Oregon State University, Corvallis - OR

Paper 430145, Session: Food, Pharmaceutical & Bioengieenering Division
2015 AIChE Annual Meeting, November 12, 2015

EFRI 1240488

# Diatoms – Biomineralizing Photosynthetic Algae with a Biosilica Shell

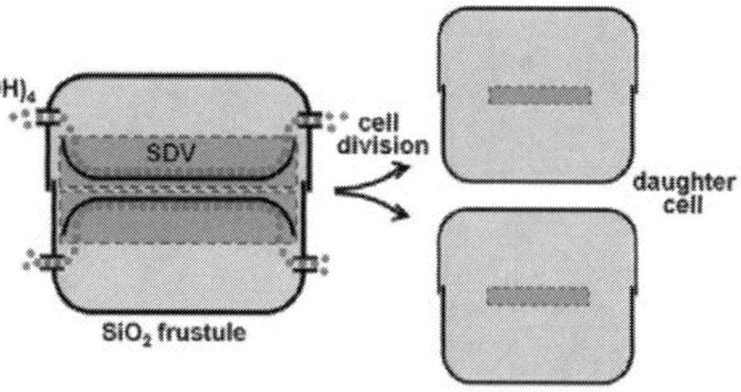


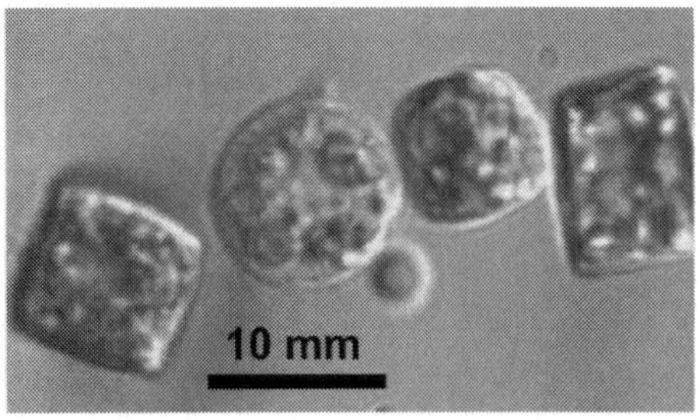


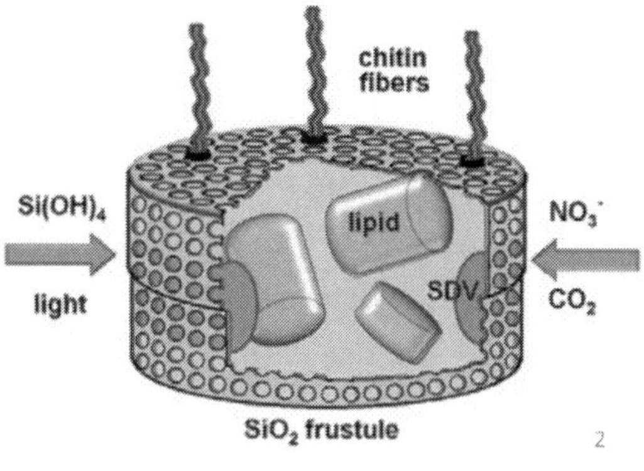


2

# Chitin and Lipid Production in Diatoms

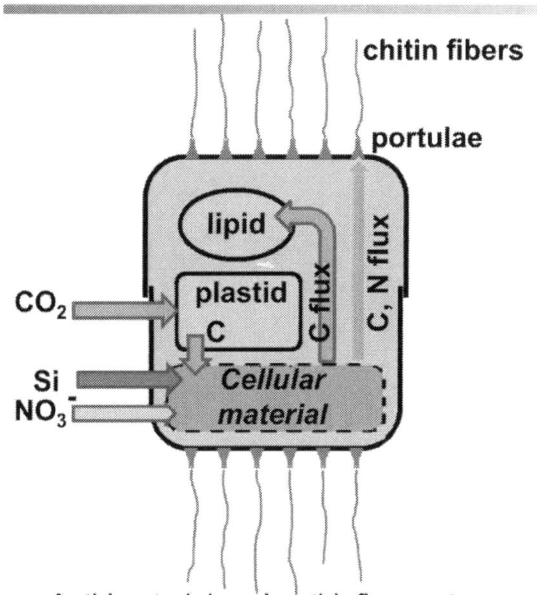


chitin

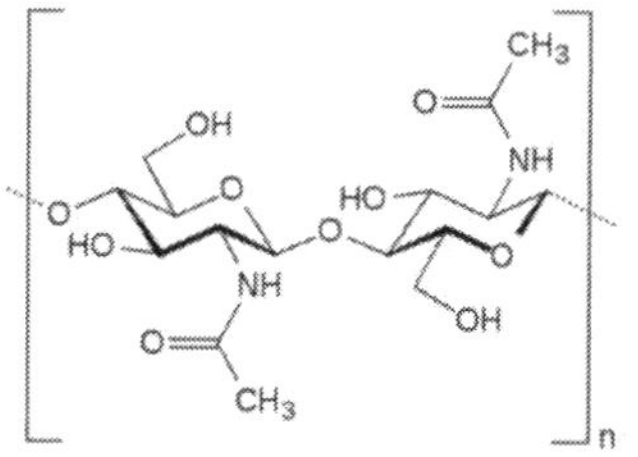


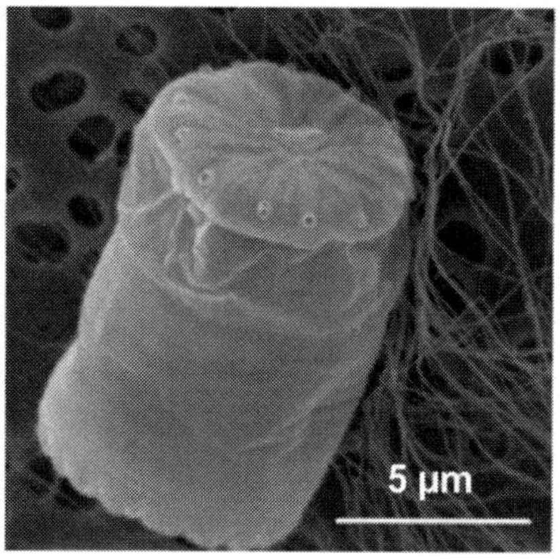


- Anti-bacterial and anti-inflammatory
- Used in: cosmetics, food preservation, paper making, wastewater treatment
- Glucosamine: deacetylated monomer of chitin

3

# 5 Liter Bubble Column Photobioreactor

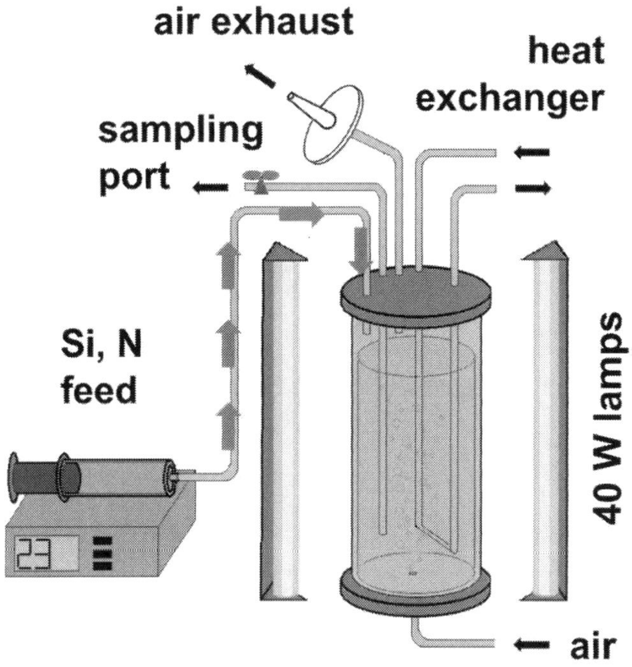


**Medium Delivery Conditions**

| Nutrient | Stage I | Stage II |
|---|---|---|
| Silicon | 0.8 mM | 0.8 mM |
| Nitrogen | 1.0 mM | 1.0 mM |
| | **Batch** | **Fed-Batch** |

Medium Addition Rate: silicon and nitrate delivery rate, to a fixed amount of Si + N delivered on Stage II

4

# 5 Liter Bubble Column Photobioreactor

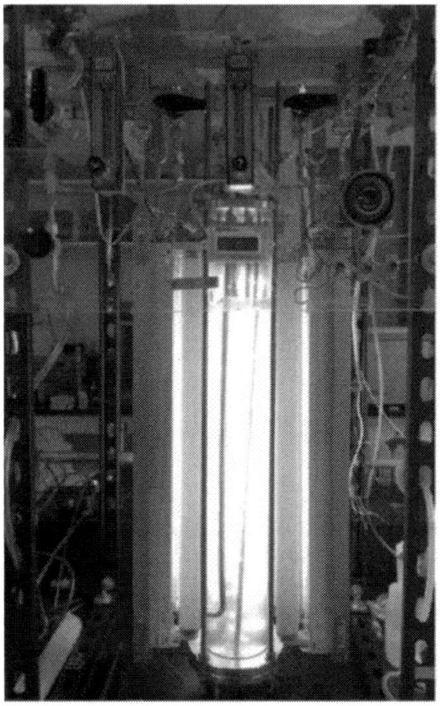

| Process parameters | Value |
|---|---|
| Initial cell density | 1e5 (cells/mL) |
| Initial pH | 8.1 |
| Air flow | 1.4 (L/min) |
| Temperature | 20 (°C) |
| Light Intensity | 150 ($mE/m^2s$) |
| Photoperiod (L/D) | 14/10 (hr) |

5

# Lipid and Chitin Production in the Photosynthetic Diatom *Cyclotella sp.*

Objective: Identify a cultivation strategy that selectively directs N to chitin, and lipid production.

Presentation outline

- ❑ Cultivation strategy of *Cyclotella sp.* in a 5 L bubble column PBR
- ❑ Comparison of nutrient consumption and chitin production
- ❑ Lipid production at different nutrient delivery rates
- ❑ Allocation of lipid vs. chitin production at low and high medium delivery rates

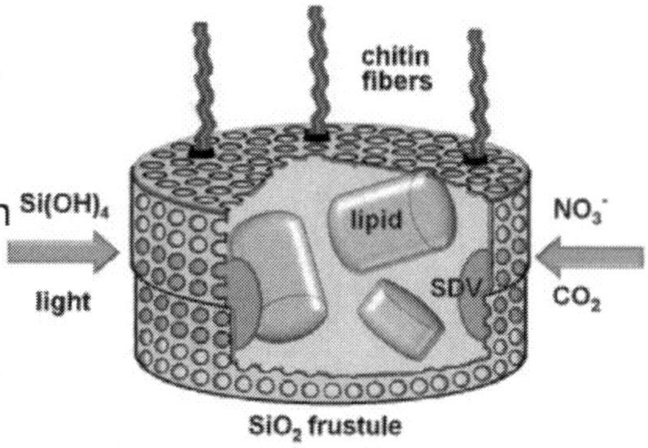


6

# Two Stage Photobioreactor Cultivation Process

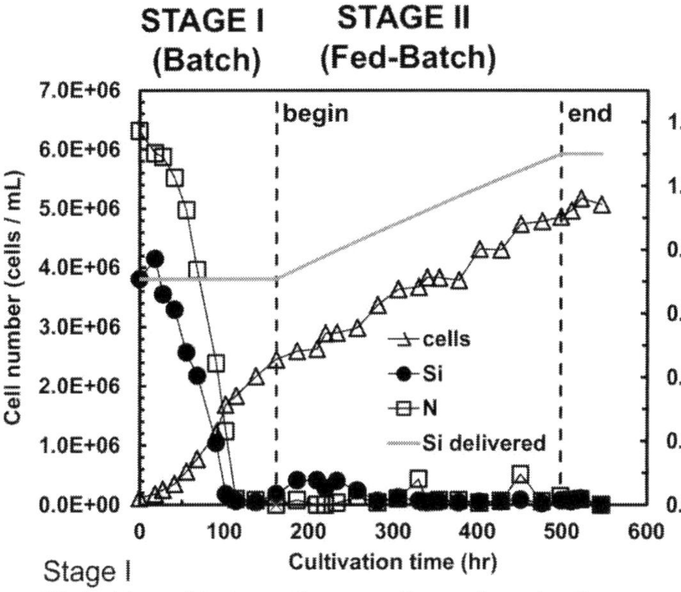


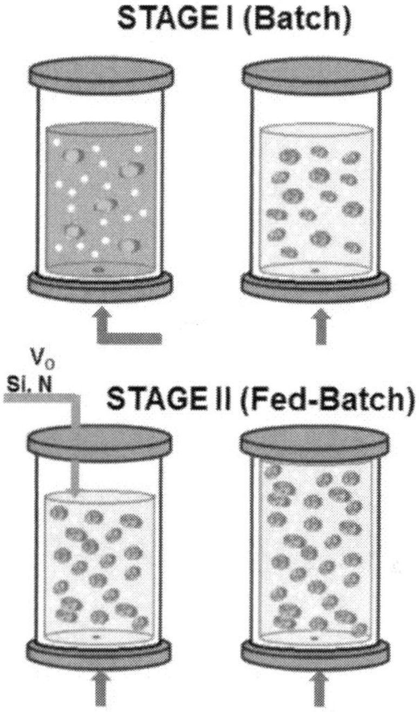


Stage I

- ❑ Achieve N starvation – cell synchronization

Stage II

- ❑ All nutrients delivered are consumed
- ❑ Cell production limited by N, Si co-addition

7

# Two Stage Photobioreactor Cultivation Process

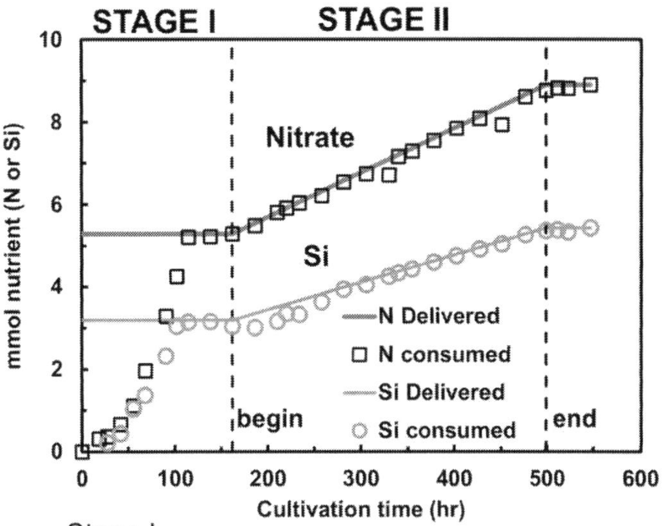


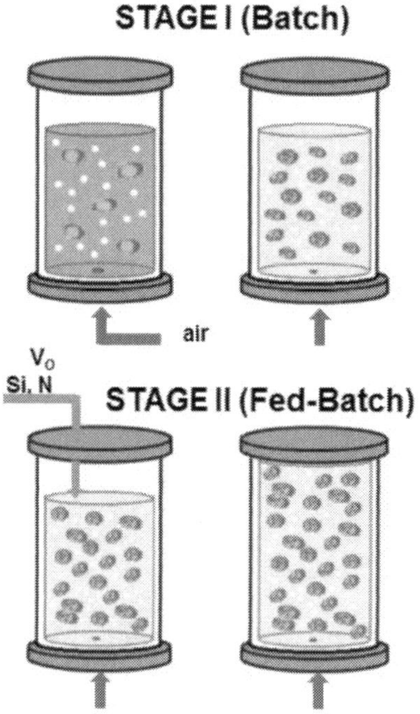


Stage I

- ❑ Achieve N starvation – cell synchronization

Stage II

- ❑ All nutrients delivered are consumed
- ❑ Cell production limited by N, Si co-addition

8

# Two Stage Photobioreactor Cultivation Process

STAGE I STAGE II

N delivered stage II = 3.7 mmol

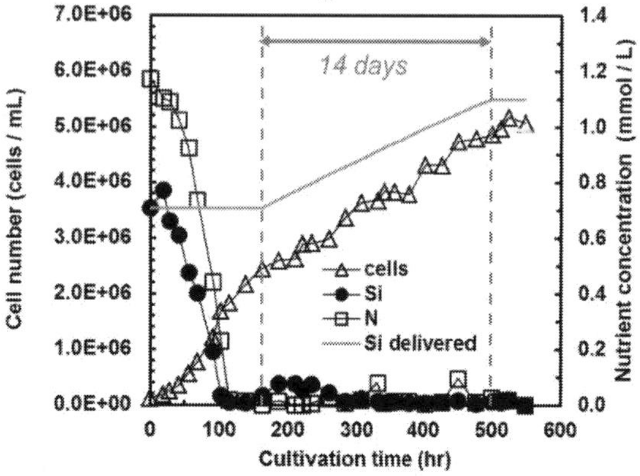


Stage II
**Low** Medium Addition Rate
($v_o$ = 1.26 mL/hr over 14 days)

STAGE I STAGE II

N delivered stage II = 3.7 mmol

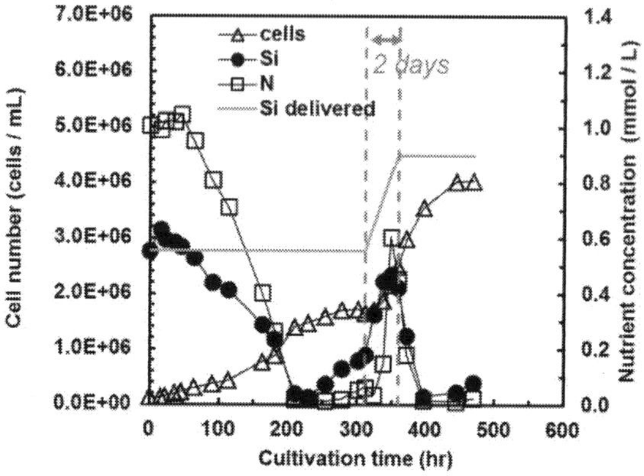


Stage II
*High* Medium Addition Rate
($v_o$ = 1.26 mL/hr over 2 days)

9

# Two-Stage Cultivation: Nutrient Consumption & Chitin Production

mg chitin / L culture

STAGE I

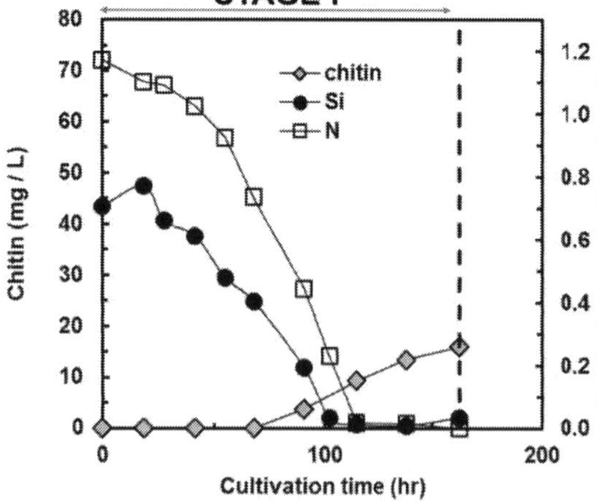


mg chitin / L culture

STAGE I STAGE II

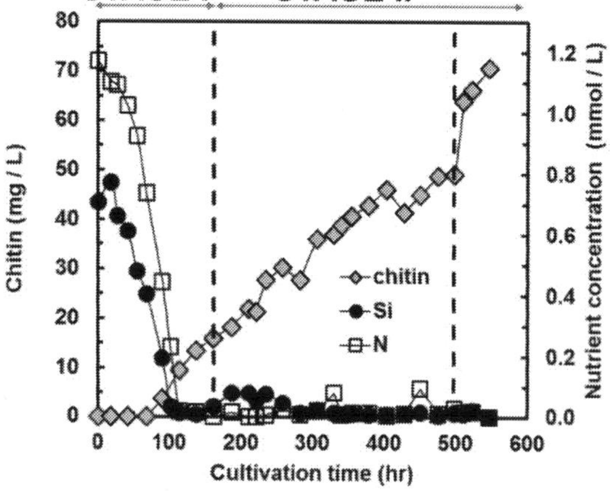


**Stage I:** chitin production induced at silicon starvation
**Stage II:** chitin production maintained while all Si+N delivered is consumed

10

## Nitrate Delivery and Incorporation into Chitin

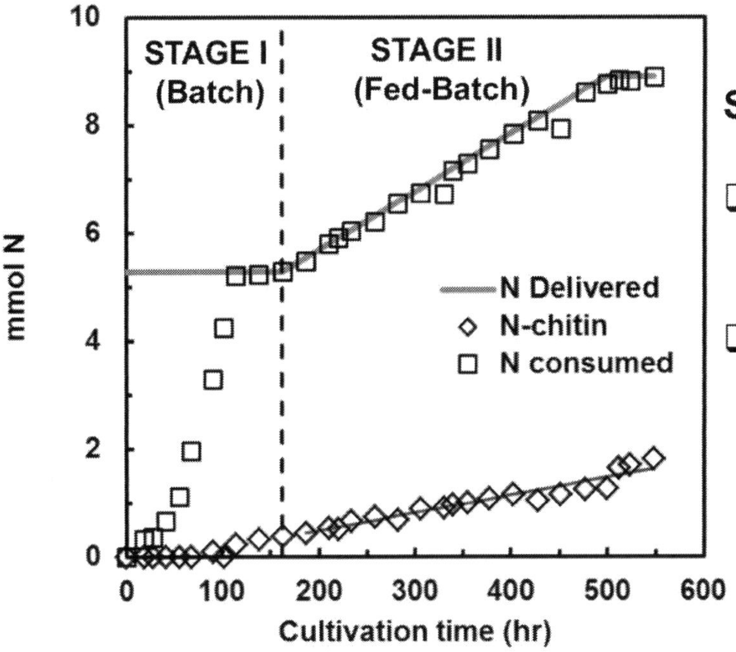


**Stage II:**

- All nitrate delivered is consumed
- Chitin production is approximately linear in Stage II

11

## Stage II: Specific Growth Rate & Nitrogen and Silicon Delivery Rates

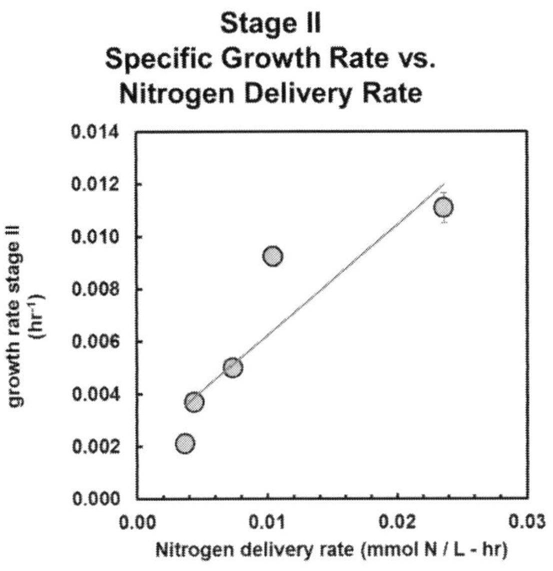


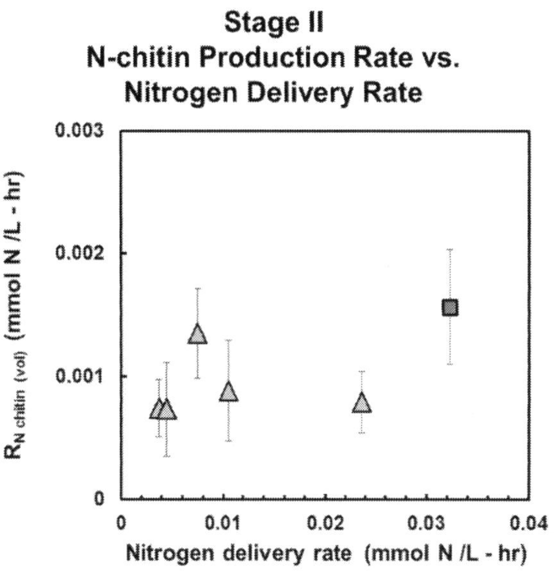


12

# Stage II: Nitrogen Allocation into Chitin at Different N Delivery Rates and Growth Rates

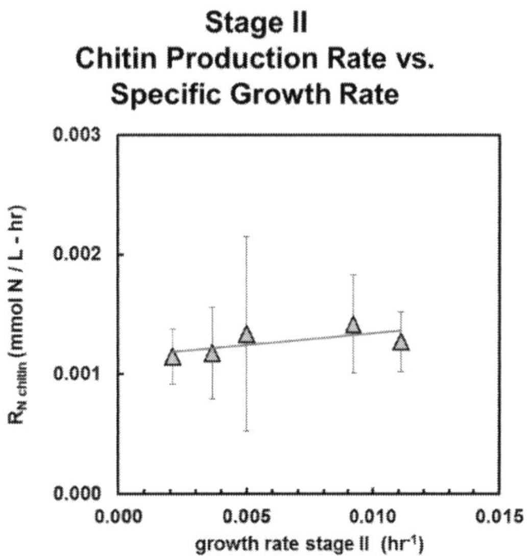


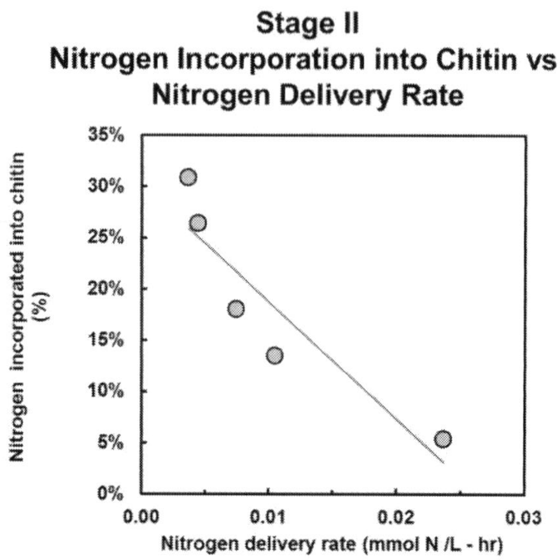


$$\% \text{ NI} = \frac{\text{Rate Chitin-N produced}}{\text{Rate Nitrate-N delivered}}$$

13

# Chitin is Produced According to Nitrogen Delivery Rate

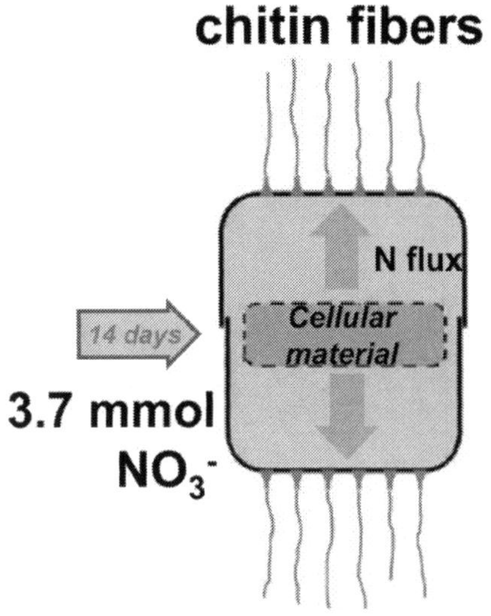


**Low** Medium Addition Rate

- ❑ Selective for chitin production

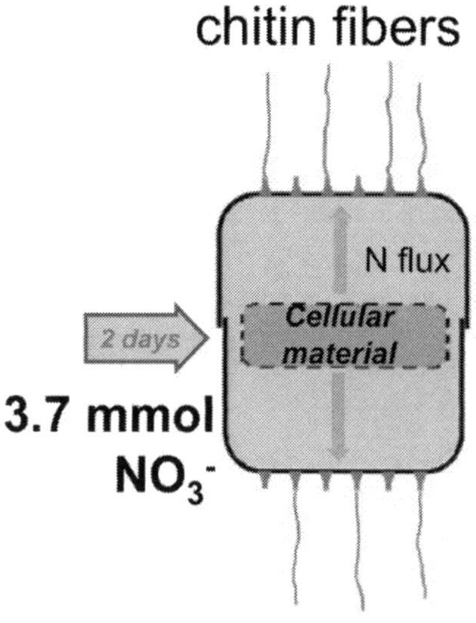


*High* Medium Addition Rate

- ❑ Chitin production rate maintained, but ratio of N diverted was reduced

14

# Lipid Production vs. Stage II Medium Delivery Rate

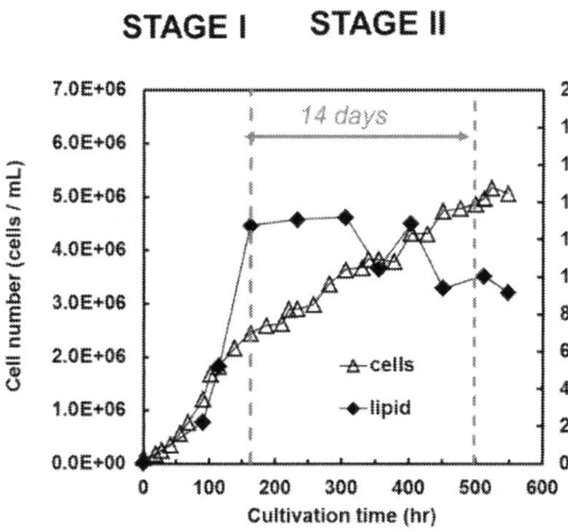


Stage II
**Low** Medium Addition Rate

- ❑ Lipid production shutdown

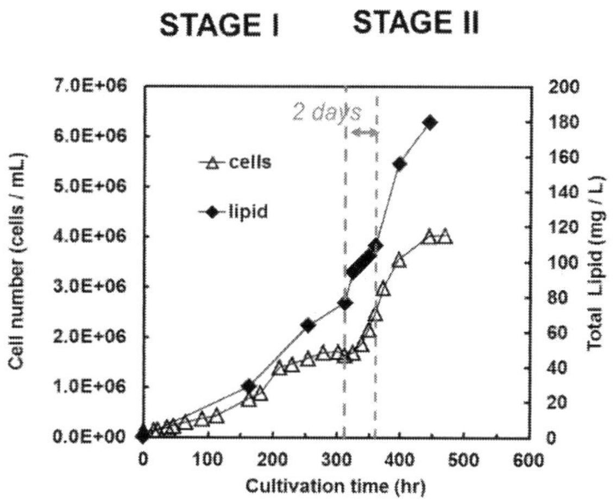


Stage II
*High* Medium Addition Rate

- ❑ Lipid production maintained during cell production

15

# Proposed Control of Lipid vs. Chitin production in Stage II

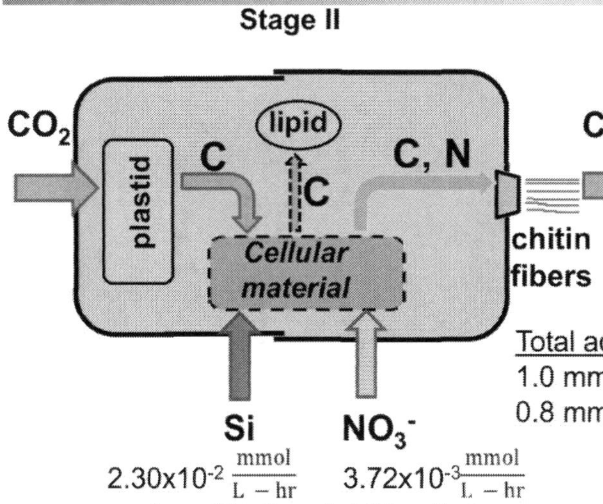


Total addition
1.0 mmol/L N
0.8 mmol/L Si

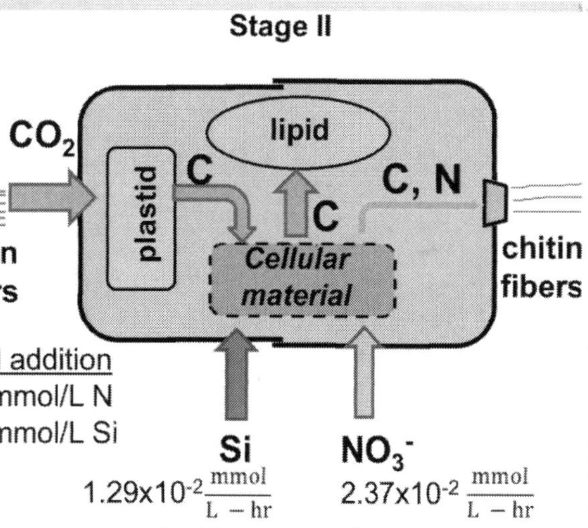


**Low** Medium Addition Rate

- ❑ Lipid production shutdown
- ❑ Selective for chitin production
- ❑ Cell production rate limited by Si + N addition rate

*High* Medium Addition Rate

- ❑ Lipid production maintained during cell production
- ❑ Chitin production rate maintained, but reduced
- ❑ Cell production rate not limited by Si + N addition rate

16

# Rate of Addition of Nutrients (Si+N) Divert the Production of Chitin and Lipids in *Cyclotella sp.*

Concluding Remarks:

Two-stage culture system studied:

- Stage I – batch culture to silicon (Si) and nitrate (N) starvation.
- Stage II – perfusion of nutrient medium under Si+N limited growth condition

- Chitin production was linear with time during Stage II

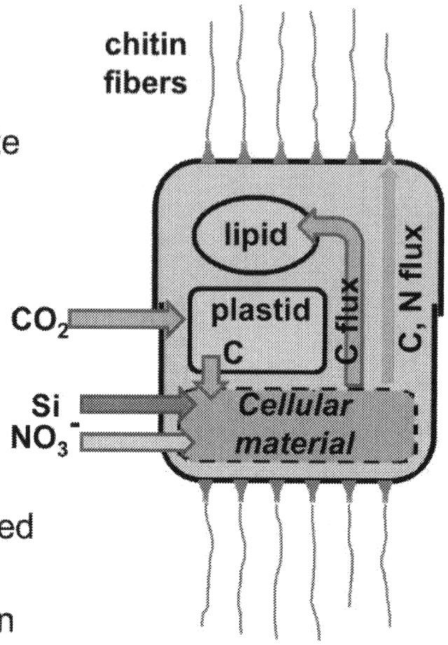


Increasing Stage II Si+N nutrient flow rate changed patterns of lipid and chitin product formation

- Low rate for 14 days shut down lipid production but maintained chitin production
- High rate for 2 days promoted both lipid and chitin production

17

# 435172 Fed-Batch Production of Novel Antifungal Peptide

**Abstract:**

Development of recombinant platform for the production of bioactive peptides is particularly important since current solid phase synthesis is difficult and sometimes uneconomical for products greater than 40 amino acids in length. These limitations impair the depolyment of longer peptides for therapeutic applications and provide an important challenge to tackle, namely efficient production. Our group has been investigating the use of recombinant technology and fed-batch fermentation to produce a novel antifungal peptide that shows broad spectrum activity. This presentation will summarize our efforts, and include a description of the strategies used to construct an effective recombinant strain, an examination of the conditions necessary for efficient fermentation and bioseparation, and an assessment of efficacy when compared to its solid phase homolog.

# 408475 Molecular Mobility As a Tool for Understanding the Impact of Polyvinylpyrrolidone (polymer) and Tpgs (surfactant) in Crystallization Kinetics of Amorphous Celecoxib

**Abstract:**

Formulating drugs (active pharmaceutical ingredients) as amorphous solid dispersions is an alternative for improving the solubility of poorly water soluble drugs. An amorphous solid dispersion ideally consists of the amorphous drug homogeneously distributed and thermodynamically stable within a polymer matrix. In addition to the polymer, the formulation may contain other materials such as surfactants to improve processability, dissolution and flow behavior, among others. Understanding the crystallization tendency of such compounds is crucial when assessing it's feasibility for a solid dispersion formulation, as re-crystallization of the drug is an undesired effect and may lead to potential market withdrawal of a commercial drug product. Celecoxib is a poorly water soluble (BCS class II) non-steroidal anti-inflammatory drug. When compared to other BCS class II compounds, it has relatively low molecular mobility (high viscosity) and crystal growth rate upon cooling from the undercooled melt state [1,2]. The presence of surfactants or polymers has demonstrated to significantly increase or inhibit the radial crystal growth rate of Celecoxib, respectively. Interestingly but not necessarily advantageous, the combination of both, results in a complex behavior where the polymer is able to reduce the growth rate acceleration caused by the surfactant resulting in an increase or decrease of the overall crystal growth rate depending on the miscibility of Celecoxib and the surfactant [3]. Even though, there is a large number of surfactants and polymers available for amorphous solid dispersions, few studies elucidate the impact of all of these in the crystallization kinetics of drugs. While the impact of surfactants and polymers in the crystallization growth rates of pure amorphous Celecoxib has been studied and reported, this impact analyzed in terms of the molecular diffusivity is a knowledge still not available. Crystallization of a drug is a diffusive process where the nucleation rate is dictated by the fluidity of the liquid (viscosity). Fundamental knowledge may be obtained when the crystallization tendency (or lack thereof) is analyzed in terms of the "glass forming ability" of the liquid, *i.e.*kinetic fragility. Here, the liquid's viscosity with temperature may follow Arrhenius or Vogel-Tamman-Fulcher (VTF) dependence, depending on the relaxation of the liquid molecules. It is commonly known that glassy liquids at high temperature follow Arrhenius dependence (linear), however upon lowering the temperature deviates from Arrhenius and follows VTF law (non-linear). Interestingly, some speculate that the VTF behavior transitions back to an Arrhenius dependence at low temperatures [4]. The viscosity is basically the resistance of a fluid to flow under an external applied stress, it is then related to the molecular relaxation and hence mobility of a material. We have characterized the rheological properties of Celecoxib with small amounts of PVP (polymer) and/or TPGS (surfactant) with the purpose of understanding the impact on the molecular diffusivity of the Celecoxib during crystallization from the undercooled melt state. We expect these results to complement the knowledge available of molecular mobility of Celecoxib with regards to thermodynamic quantities for solubility prediction [5]. Additionally, in general for when assessing a drug as a candidate for a solid dispersion formulation during developmental stages of a drug product. Amorphous samples where prepared using a cryogenic mill. Binary and ternary mixtures consisted of 10 wt.% TPGS or 10 wt.% PVP dispersed in a 90wt.% Celecoxib, and ternary mixtures of 10wt.% PVP, 10 wt.% TPGS and 80 wt.% Celecoxib. Samples were characterized in an Anton Paar MCR 301 series rheometer, using 20mm diameter parallel Peltier Plates. Temperature dependent Dynamic Viscosity was measured while cooling the sample from melting point at a cooling rate of 10°C/min. We measured the dynamic rheological properties of the Celecoxib samples. The dynamic viscosity data was analyzed in terms of molecular mobility, Activation Energy ($E_a$), and fragility (Strength Parameter) (D) of the drug near the glass transition ($T_g$) and melt temperature ($T_m$), this was possible by analyzing the viscosity data with Arrhenius and Vogel-Tamman-Fulcher (VTF) equations. The temperature dependent dynamic viscosity curves resulted in overall viscosity increase upon cooling from the melting temperature, which was expected due to the nature of the measurement. The rate at which the viscosity increased with temperature varied with sample. Generally speaking, pure Celecoxib and Celecoxib/TPGS resulted in a small degree of undercooling compared to Celecoxib/PVP and Celecoxib/PVP/TPGS, this due to sample solidification during measurement. Pure Celecoxib resulted in an increase in viscosity of approximately one decade at roughly 10°C degrees of undercooling; at this point the viscosity suddenly increased dramatically almost three orders in magnitude. This sample could not be measured more than 30°C degrees of undercooling. Celecoxib/TPGS resulted in a similar behavior as pure Celecoxib, with a reduced degree of undercooling, as the sample showed a marked increase in viscosity near the melt temperature. Celecoxib/PVP resulted in a significant increase in the degree of undercooling (~95°C) as the dynamic viscosity could be measured up to the glass transition temperature of the sample with an increase of roughly one decade of viscosity every 10°C in temperature. Interestingly, almost 10°C before the glass transition, the viscosity increase rate with temperature decreased significantly. Near the glass transition temperature, this sample showed roughly a viscosity increment of seven orders of magnitude. Similar behavior was observed for Celecoxib/PVP/TPGS, having similar viscosity values as the Celecoxib/PVP sample near melt temperature, as the glass transition approached during cooling, the Celecoxib/PVP/TPGS sample showed a decrease of one decade in viscosity compared to Celecoxib/PVP. Celecoxib/PVP/TPGS viscosity increase rate also was reduced near the glass transition. Analyzing the viscosity data in terms of diffusivity indicates that the molecular mobility of the Celecoxib is impeded by the presence of the PVP, this occurs by the hindering of local arrangements of the Celecoxib molecules within a crystal lattice (crystallization inhibition). This was evidenced by the significant increase in viscosity and degree of undercooling compared to pure Celecoxib. This induced hindering of Celecoxib molecules allowed for viscosity measurements near the glass transition temperature. On the other hand, the TPGS acted as a plasticizer where miscibility between TPGS and Celecoxib aided in the molecular mobility of Celecoxib molecules assisting therefore the crystallization near the melting temperature. The combination of Celecoxib with PVP and TPGS resulted in interesting results, generally a decrease of molecular mobility similar to the effect of PVP alone, however the plasticizing effect of TPGS was able to increase the mobility, hence lower viscosity values near the glass transition compared to the effect of PVP. The viscosity curves showed mostly linear dependences with regard to temperature, however some deviation from linearity were observed most significantly for the samples containing PVP and PVP/TPGS. Fitting the viscosity curves with the Arrhenius and VTF equations evidenced these dependence. The hindering of local arrangements of the Celecoxib molecules (crystallization inhibition) by PVP was evidenced by the increase in activation energy of Celecoxib/PVP (115.96kJ/mol) compared to pure Celecoxib (92.28 kJ/mol) near the melt temperatures, whereas TPGS decreased the activation energy necessary for crystallization to occur (60.52kJ/mol), evidencing thus the plasticizing effect of TPGS. Interestingly, the combination of PVP and TPGS was not capable of reducing the activation energy (98.89kJ/mol), as significantly as PVP alone complementing the similar viscosity values near the melt temperature. Deviations from the Arrhenius equation were observed as the degree of undercooling progressed (10°C-80°C), evidenced by a decrease in the $R^2$ value of up to 12% for the Celecoxib/PVP/TPGS sample, compared with $R^2$ = 10% of Celecoxib/PVP sample. These deviations could be successfully corrected with the VTF model, increasing the $R^2$ value to 99.9% for both samples within the same temperature range. A marked VTF-to-Arrhenius transition near the glass temperature was not observed for these two samples within the temperature range studied, however Arrhenius equation resulted in higher $R^2$values compared to the VTF model (10% difference between both). VTF fitting demonstrated that all samples had a calculated strength parameter (D) of less than 10, confirming the kinetic fragility (crystallinity) of the liquids. However, no correlation could be observed between the impact in strength parameter, and the crystallization tendency and molecular mobility. The impact on the crystallization of Celecoxib by PVP (polymer) and TPGS (surfactant) was studied in terms of molecular mobility. Over all it was demonstrated that the polymer decreases the mobility of the drug, whereas the surfactant acts as a plasticizer. PVP hinders the mobility of Celecoxib molecules inhibiting to some degree the crystallization of the drug, evidenced by the significant increase in measured degree of undercooling, dynamic viscosity and calculated activation energy. The surfactant acted as a plasticizer aiding the mobility of the drug allowing therefore the crystallization of the drug to occur at earlier, and lowers the activation energy necessary for crystallization to occur. The dynamic viscosity of the Celecoxib with polymer and/or surfactant follows Arrhenius temperature dependence near the melting point (10°C of undercooling), however the VTF model results in higher agreement as the undercooling degree increases.

# 413132 Process Development for a Co-Crystal of Telaprevir (VX-950) and 4-Hydroxybenzoic Acid

**Abstract:**

When development work was initiated, the existing process for generation of a co-crystal of telaprevir (VX-950) and 4-hydroxybenzoic acid (4-HBA) was uncontrolled precipitation from a solvent mixture. Anecdotally, there was the existence of a "magic" co-crystal batch that had superior dissolution performance and had been generated from the uncontrolled process. For a time, technical work was driven by this knowledge and there was an ill-fated attempt to reproduce this batch. However, when it was determined that the precipitation led to co-crystal with variable properties and performance, focus shifted to obtaining a more fundamental understanding of the crystallization. The objective for developing the process was then to design a controlled crystallization that would lead to co-crystal material with reproducible physical properties and dissolution behavior. There is a significant difference in solubility of the API and the coformer which led to difficulty in choosing an appropriate solvent system. This presentation will review the candidate solvent systems and modes of crystallization that were attempted. After choosing dichloromethane and methyl *tert*-butyl ether as the solvent system, an equilibrium phase diagram was generated for the solvent composition at the seed point. This phase diagram indicated that the process was at risk to produce co-crystal with excess VX-950. By slightly changing the parameters in the crystallization, it was found that only co-crystal would be formed. Further work involved optimization of the solvent composition at the end-point to improve yield and minimize agglomeration. Development and understanding of the phase diagram was essential to reliably producing VX-950/4-HBA co-crystal.

# 416357 Primary Nucleation Under High Shear Conditions

**Abstract:**

The influence of external energy on the kinetics of primary nucleation has been investigated under high shear conditions. It is shown in literature, that shear-induced molecular alignment and in particular agitation-enhanced cluster aggregation are mechanisms that appear to deserve further attention [1]. To study this phenomenon further, the induction time has been measured at different supersaturation, and levels of energy input, at various scales using magnetic stir bar and rotor-stator based IKA wet mills. Since there is a distribution of induction time possible, a statistical view is proposed for the value of induction time at a given condition [2]. There is an overall tendency in the experiments that the primary nucleation is promoted by increased input of shear imparted due to mixing. In small vials agitated by magnetic stir bars and in a larger scale reactor with high shear wet mill, the induction time was found to decrease with increasing wet mill tip speed to the power 0.66, the induction time is inversely related to the shear rate. By fitting the parameters of the classical nucleation theory to experimental data, it is shown that the results can be explained as an influence on the pre-exponential factor. Interfacial tension is estimated between API and solvent. An empirical equation is proposed to quantify the energy input from wet milling, and to predict the induction time.

References:

1. Liu, J.; Rasmuson, A.C. Cryst. Growth Des., 2013, 13 (10), pp 4385–4394
2. Kulkarni, K.A.; Kadam, S.S.; Meekes, H.; Stankiewicz, A.I.; ter Horst, J.H. Cryst. Growth Des., 2013, 13 (6), pp 2435–2440

*AbbVie contributed to the design, research, and interpretation of data, writing, reviewing, and approving the publication. Huayu Li, Moiz Diwan and Fredrik Nordstrom are AbbVie employees at the time that they worked on this project.*

# 417728 Microfluidic Spherical Crystallization and Co-Formulation of Drugs and Excipients – Controlling Morphology and Polymorphism

**Abstract:**

**Microfluidic Spherical Crystallization and Co-formulation of drugs and excipients – controlling Morphology and Polymorphism**

Eunice W. Q. Yeap, Ambika Somasundar, Denise Z. L. Ng, Yiji Wang, and Saif A. Khan*

Department of Chemical and Biomolecular Engineering,

National University of Singapore, SINGAPORE

There is considerable interest in the creation of "designer" pharmaceutical microparticles with controlled morphology and polymorphism, as drug product intermediates for facile formulations. Emulsion-based crystallization is an attractive route for controlling crystal size in downstream processing of active pharmaceutical intermediates (API), yet crystals formed in batch processes show limited control over key attributes like size and shape.[1] We circumvent these limitations, by coupling capillary microfluidics based emulsion generation, with evaporative crystallization for the production of monodispersed spherical API agglomerates. This platform has allowed us to produce exquisitely designed spherical agglomerates (of drug or drug mixtures), formed in uniform size, composition, morphology and polymorphism.[2-3] We study how precise control over key attributes of drug or drug-excipient microparticles can be achieved by a combination of process and formulation parameters.

In this work, we demonstrate the co-formulation of model hydrophobic API drugs – ROY and carbamazepine – with excipient, ethyl cellulose (EC). We generate monodisperse emulsion drops containing the API-excipient mixture, and subject these droplets to evaporative crystallization in thin films of continuous phase. We find that the co-formulation of API with EC not only shortens the API crystallization process, but also allows for the direct creation of highly monodisperse spherical microparticles containing API crystals and macromolecular excipient, with nearly complete and tunable control over both morphology and polymorphic outcome. Specifically, we control polymorphic selection of both ROY and CBZ in the API-excipient microparticles, by implementing different film thicknesses for the evaporative crystallization; the film thickness mediates the rate of solvent evaporation from the droplets, controlling the kinetics of nucleation and crystallization. We obtain two completely specific and different polymorphic outcomes for each drug when crystallized in the presence of EC under different film thickness. In summary, our methods pave the way for the synthesis and formulation of novel drug product intermediates that bridge the considerable gap between primary and secondary pharmaceutical manufacturing.

**Word Count:** 308

**CONTACT**

* S. A. Khan, Tel: +65-6516-5133; saifkhan@nus.edu.sg

[1] Chadwick, K.; Davey, R. J.; Mughal, R.; Marziano, I., Org. Process Res. Dev. 2009, 13, 1284 – 1290.

[2] Toldy, A. I.; Badruddoza, A. Z. M.; Zheng, L.; Hatton, T. A.; Gunawan, R.; Rajagopalan, R.; Khan, S. A., Cryst. Growth Des. 2012, 12, 3977 – 3982.

[3] Leon, R. A., Badruddoza, A. Z. M., Zheng, L., Yeap, E. W., Toldy, A. I., Wong, K. Y., Hatton, T. A. Khan, S. A., Cryst. Growth Des., 2015, 15 (1), pp 212–218

# 423568 Morphology Evolution of Quadrangular Dendrites of Erythritol

**Abstract:**

The change of the morphology of erythritol was investigated in this paper. Offline tools such as scanning electron microscope (SEM), XRD, DSC and optical microscope and online tools such as PVM and FBRM were used to test the effects of the factors including temperature, additives, solvents, cooling rate and so on. It was found that when the erythritol solution is added the same anti-solvent, its habit changes from dendritic structure to octahedron with the rising of the initial temperature, and the transition temperature changes with the kind of the anti-solvent. A possible explanation of this phenomenon has been proposed on the basis of experiments. The dendritic erythritol is the early stage forms of the octahedron. The morphology of erythritol experiences a transformation process of dendritic—snow-like with four corners—octahedron. The influences of other factors on this phenomenon have also been discussed in detail.

# 414966 Implementation of a Reduced System Model to a Continuous Direct Compression Manufacturing Process

**Abstract:**

The continuous direct compression manufacturing process consists of several loss-in-weight (LIW) feeders, a horizontal mixer, a surge hopper, and tablet press equipped with a near infrared (NIR) probe at the die-filling station of the tablet feed frame. A system-wide deterministic dynamic model was used to fit dispersion coefficients within the various elements for a given formulation/throughput using the feed frame NIR to measure the system response. This partial differential non-linear algebraic system model consisting of ~8,000 equations is sufficiently complex that it cannot be solved real-time with the current technology. This necessitates a second, reduced system model based upon the original deterministic model and uses thousands of digital experiments to create a set of response surfaces. The result is a data-driven version of the model that can be executed real-time using a PAT controller (e.g. SynTQ) to assess the impact of feeder variability on the final product within the distributed control system (e.g. DeltaV). The real-time reduced system model provides prediction of the feed frame concentration, the time when a disturbance will reach the die-filling station, and the time when the disturbance will clear the system.

# 427926 Flowsheet Modeling for Oral Solid Drug Product Manufacturing


**Abstract:**

QbD is a systematic approach to pharmaceutical development based on scientific knowledge and risk management [1]. Continuous manufacturing is a promising alternative and important for the implementation of QbD because it reduces the need of scale-up studies [2]; it requires smaller equipment size and thus minimizes the plant footprint [3]; and it decreases human factor using automated operation and thus decreases labor cost [4]. To facilitate the development of continuous pharmaceutical manufacturing, each unit operation must be well understood in terms of the effect of different material properties and operational conditions on the final product quality attributes. This goal can be facilitated by the utilization of computer-aided process systems engineering (PSE) approaches. Flowsheet modeling is an efficient way to understand manufacturing process dynamics. An integrated flowsheet model can be used in the design, analysis, and optimization of a chemical process [3]. Simulation results can help identify the possible process integration requirements, study multiple designs and test control strategies. In pharmaceutical process modeling, a computationally efficient equation-oriented process simulator (gPROMS) has been used to develop models for unit operations and build flowsheet models [3] [4].

Among different modeling methods, residence time distribution (RTD) theory has been commonly used in developing pharmaceutical unit operation models. RTD is defined as the probability distribution of the time that the particles stay inside one or more unit operations in a continuous flow system [5]. With RTD, process changes (e.g., material changes, process disturbances) can be traced throughout the system. This helps determine disturbance dissipation along the process and thus help decide when to collect final products with desired quality. Consequently mathematical models based on RTD theory are very useful for modeling the continuous pharmaceutical manufacturing process.

We focus on building validated integrated flowsheet model with RTD-based unit operation models for direct compression and granulation lines. Major unit operations such as feeders, comil, blenders, transfer pipes, feed frames, tablet presses, and granulation equipment were modeled and properly integrated. These flowsheet models are used to simulate and predict the dynamics of entire continuous direct compression processes given input such as material properties and operational conditions. Moreover, the behavior of different processes in response to disturbances can also be simulated. Flowsheet models can also be easily customized. Users can drag and drop necessary unit models and connect them based on their process design or to test different designs and different processing strategies. Graphic user interfaces (GUIs) have been implemented in our flowsheet simulations to provide users with an icon-based interface (i.e., no exposure to model code) and set parameters for each unit. Multiple process configurations can be easily set with GUIs, which makes it convenient to run different simulations.

The objective of this presentation is to highlight the use of flowsheet modeling tools in the pharmaceutical tablet manufacturing process and to demonstrate their applications for process design. Because the process models can have thousands of differential and algebraic equations (DAEs) and integral expressions, they can be difficult to solve with deterministic solvers. In these cases, surrogate modeling methodologies are adopted to address the challenges. However with the high-dimension problems, the application of surrogate modeling is limited. For example, Kriging method is computationally expensive for high dimensional problems because it needs to numerically solve a maximum likelihood estimation problem to fit the model [7]. In contrast, Radial Basis Functions (RBF) is a promising alternative because it can provide simple and efficient model to approximate the original model [8]. We focus on solving high-dimensional design problems by using RBF-based surrogate modeling strategies. We introduce flexibility analysis for the design problem to quantify the ability of a process to maintain feasible operation in the presence of inherent variability or external disturbances [6]. The final goal is to achieve a more robust design for the continuous pharmaceutical manufacturing process.

**References**

[1] Lionberger, Robert A., Sau Lawrence Lee, LaiMing Lee, Andre Raw, and X. Yu Lawrence. "Quality by design: concepts for ANDAs." The AAPS journal 10, no. 2 (2008): 268-276.

[2] Leuenberger, Hans. "New trends in the production of pharmaceutical granules: batch versus continuous processing." European journal of pharmaceutics and biopharmaceutics 52, no. 3 (2001): 289-296.

[3] Boukouvala F, Niotis V, Ramachandran R, et al. An integrated approach for dynamic flowsheet modeling and sensitivity analysis of a continuous tablet manufacturing process[J]. Computers & Chemical Engineering, 2012, 42: 30-47.

[4] Rogers A J, Inamdar C, Ierapetritou M G. An integrated approach to simulation of pharmaceutical processes for solid drug manufacture[J]. Industrial & Engineering Chemistry Research, 2013, 53(13): 5128-5147.

[5] Gao Y, Muzzio F J, Ierapetritou M G. A review of the residence time distribution (RTD) applications in solid unit operations[J]. Powder Technology, 2012, 228: 416-423.

[6] Biegler L T, Grossmann I E, Westerberg A W. Systematic methods for chemical process design[J]. 1997.

[7] Regis R G. An initialization strategy for high-dimensional surrogate-based expensive black-box optimization[M]//Modeling and Optimization: Theory and Applications. Springer New York, 2013: 51-85.

[8] Shan S, Wang G G. Metamodeling for high dimensional simulation-based design problems[J]. Journal of Mechanical Design, 2010, 132(5): 051009.

# 427688 A Multi-Fidelity Mechanistic Modeling Framework for Wet Granulation Processes: Characterization and Technology Transfer

**Abstract:**

Although wet granulation processes are widely used in a range of industries, they are typically designed empirically and operated inefficiently. In order to improve these processes, predictive modeling tools are desired for process design, validation, and optimization. Recent advances have expanded the mechanistic understanding of these processes, enabling quantitative descriptions to be formed using population balance modeling techniques.

Although they share a common name, wet granulation processes in practice are operated using a range of configurations, resulting in different process outcomes. Further, various mechanisms drive wet granulation processes, including wetting, drying, particle flux, agglomeration, drop nucleation, breakage, consolidation, and layering. The dominant phenomena are often governed by equipment configuration and process conditions, and different types of granulators exhibit different behavior.

Consequently, there is no single population balance model for wet granulation processes. Instead, custom models must be configured to meet the needs of the application. Some considerations in model development include population balance dimensionality and grid resolution, representation of particle composition, use of compartmental models.

Oversimplified models are unable to capture realistic behavior and provide little additional insight into process outcomes, only representing idealized cases. In contrast, overly complex models are inefficient, and numerical challenges can mask the true results. These high-fidelity models can often be simplified by identifying and eliminating irrelevant details. Despite the broad range of wet granulation models developed in academia and industry, a unifying framework is needed to ensure consistency, establish best practices and facilitate technology and knowledge transfer.

In this study, various alternatives for modeling wet granulation are discussed. These configurations are unified in a flexible modeling framework that encompasses varying levels of fidelity. Through this framework, population balance dimensionality, compartmental configuration, and governing rate mechanisms can be customized to simulate a variety of wet granulation processes. Case studies of twin screw and fluid bed granulation processes are presented to demonstrate the implications of model alternatives and establish best practices for model-based design and operation of efficient and robust wet granulation processes.

# 425191 Prediction of Lubrication-Based Tensile Strength Reduction Using a Compartment Modeling Approach

**Abstract:**

In the production of pharmaceutical tablets and capsules, a lubricant such as magnesium stearate is frequently added to the powder blend or granulation to reduce friction between the powder and the tablet press or encapsulator components. While the presence of the lubricant is generally necessary to improve manufacturability of the dosage form, if the lubricated blend is exposed to excessive shear strain during processing and handling prior to tableting or encapsulation, adverse effects on quality attributes of the final dosage form may be observed. These effects can include an increase in wetting contact angle, a slowdown in disintegration and/or dissolution, and a reduction in tensile strength.

In this work, the extent of lubrication in a powder feeding process is studied experimentally and with a compartment model approach. Since a placebo blend is used to study this phenomenon, the tensile strength is the quality attribute used to quantify the extent of lubrication. Here, a compartment modeling approach is proposed to model the powder flow pattern and predict the lubrication-based tensile strength reduction that may occur during the production of a large batch. Parameter estimation is conducted through the use of a separate experiment utilizing an input step change in powder feed from undyed to dyed material. Model predictions are compared with experimental results as well as predictions from a separate discrete element method (DEM) modeling approach. Finally, additional predictions from the compartment model are made to assess the impact of various processing modifications.

15AIChE
Annual Meeting, Salt Lake City, UT

# Finite Element Simulations on Strength and Impact of Pharmaceutical Tablets

Jie Ren, Jerry Klinzing, Jennifer Cislo, Adam Procopio

Merck & Co.

P Public

MERCK
Be well

## Tablet Chipping

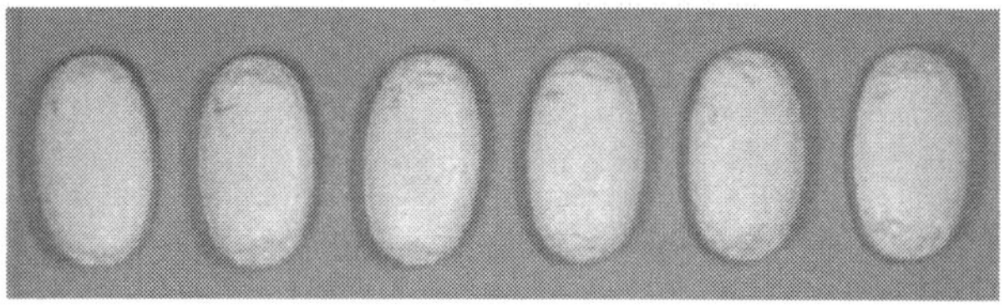

- Tablet chipping: among the most common modes of failure during solid oral drug development.
- Usually occurs during the post-processing steps of tablet manufacture – film-coating, packaging, transportation, patient handling
- In order to prevent/mitigate chipping, compressed tablets need to meet hardness and friability criteria.
- Hardness testers and friability testers

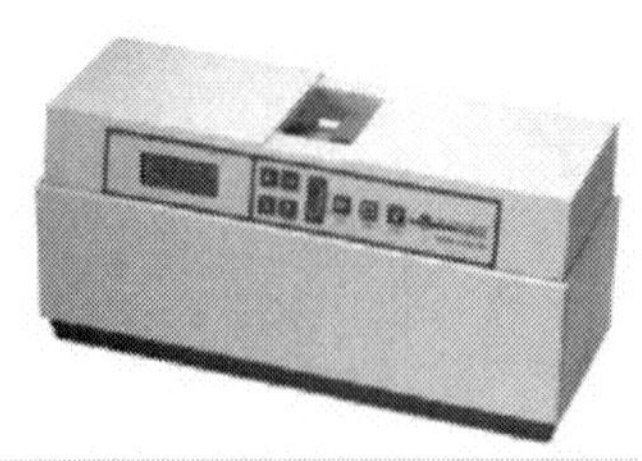

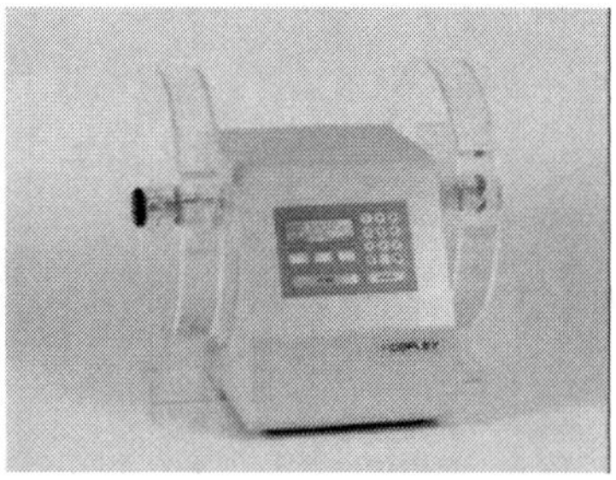

P Public

MERCK
Be well

# Limitation of hardness & friability testing

- Hardness and friability values can be hard to interpret
  - Hardness can be different for different shapes, even with the same material property.
  - Friability reveals weight loss after repetitive drops; not a direct reflection of elegance.
  - Neither hardness nor friability testing represent the real failure mode in a film-coating drum.

**Hardness test**

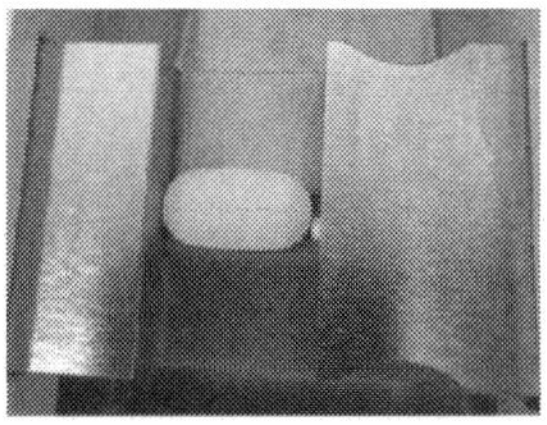

- Tablet geometry
- Break force (hardness)
- Tablet material property

→ ? → Will the tablet break in friability test, and then, will it break during film coating?

**Friability test**

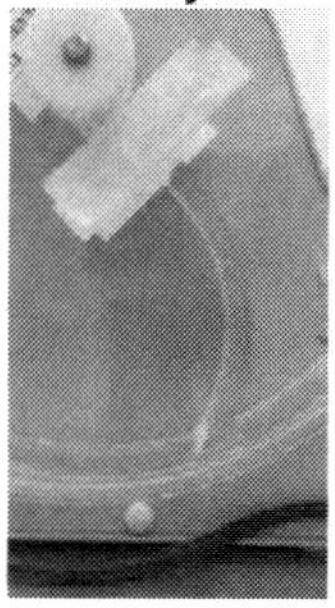

Ground-up understanding needed on the strength and impact response of tablets!

MERCK *Be well*

Public

# Tablet Hardness vs. Tensile Strength?

Which shape is ideal?

| Shape | Kp Value |
|---|---|
| | 67 |
| | 44.8 |
| | 47.5 |
| | 41 |
| | 41 |

Compute Tensile Strength!

Pitt's equation:

Round Flat Tablets

$$T.S. = \frac{2P}{\pi D t}$$

Round Concave Tablets

$$T.S. = \frac{10P}{\pi D^2}\left(2.84\frac{t}{D} - .126\frac{t}{W} + 3.15\frac{W}{D} + .01\right)^{-1}$$

Other shapes?!

No analytical solution!

FEM Simulation

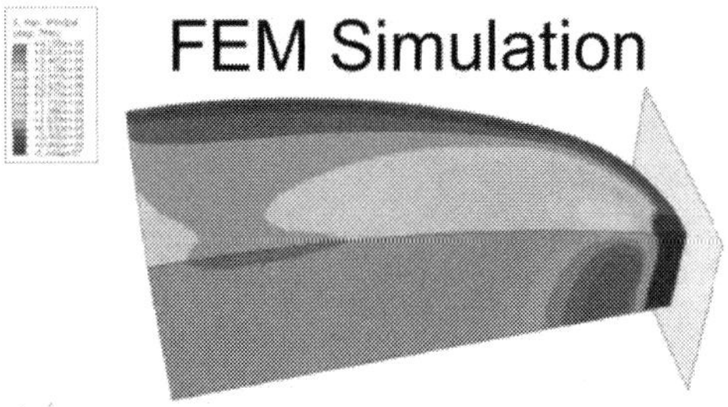

MERCK *Be well*

Public

# Calculate tensile strength: diametrical test

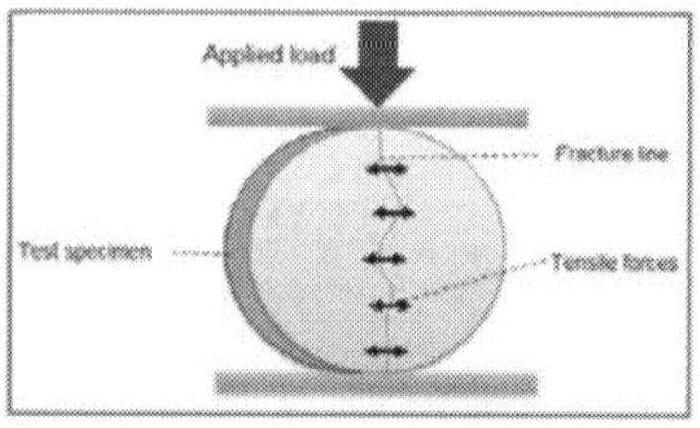


- Simulate 1/4 of a tablet based on symmetry
- Rigid plate moves towards the tablet at constant speed

Stress distribution along Z direction (transverse tc compression direction)
This is the tensile stress that pulls the tablet apart

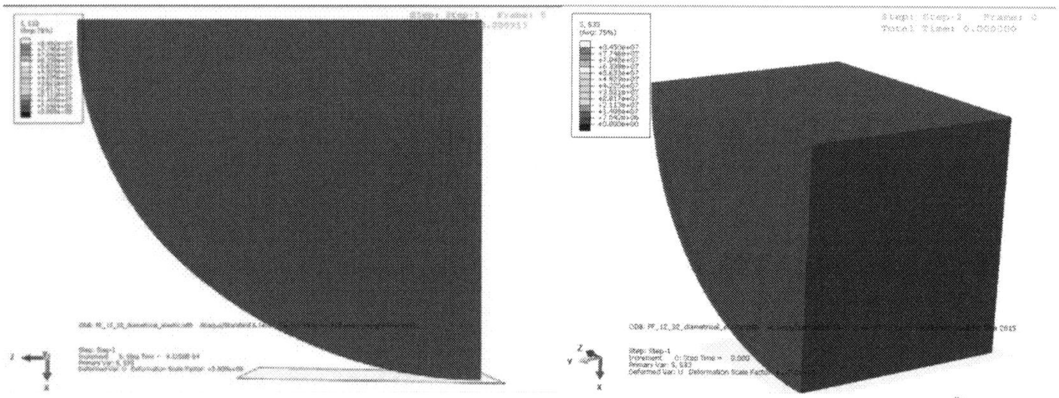

Public

MERCK *Be well*

# Calculate tensile strength: diametrical test

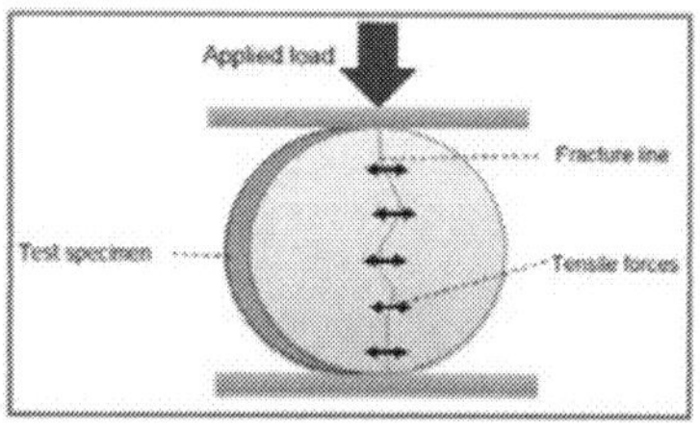


- Simulate 1/4 of a tablet based on symmetry
- Rigid plate moves towards the tablet at constant speed

Stress distribution along Z direction (transverse to compression direction)
This is the tensile stress that pulls the tablet apart

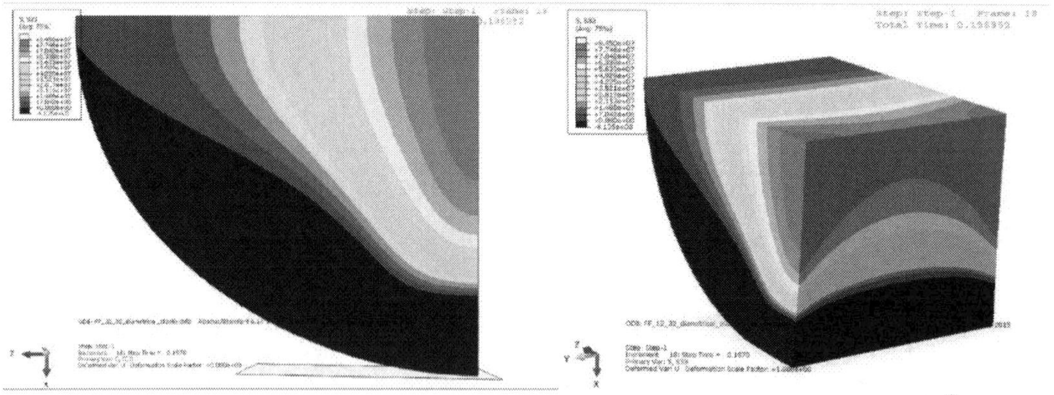

Public

MERCK *Be well*

## Diametrical test: Material Property Matters

- Compare Linear Elastic and Elastoplastic materials
  - Linear Elastic (typical for brittle materials: Lactose, Dicalcium Phosphate)
    - $\rho$ = 1.2g/cc, E = 3.65GPa, $\nu$ = 0.25; frictionless contact with the plate
  - Elastoplastic (typical for plastic materials: Microcrystalline Cellulose)
    - All the above properties, plus yield strength $\sigma_y$ = 36.5MPa.

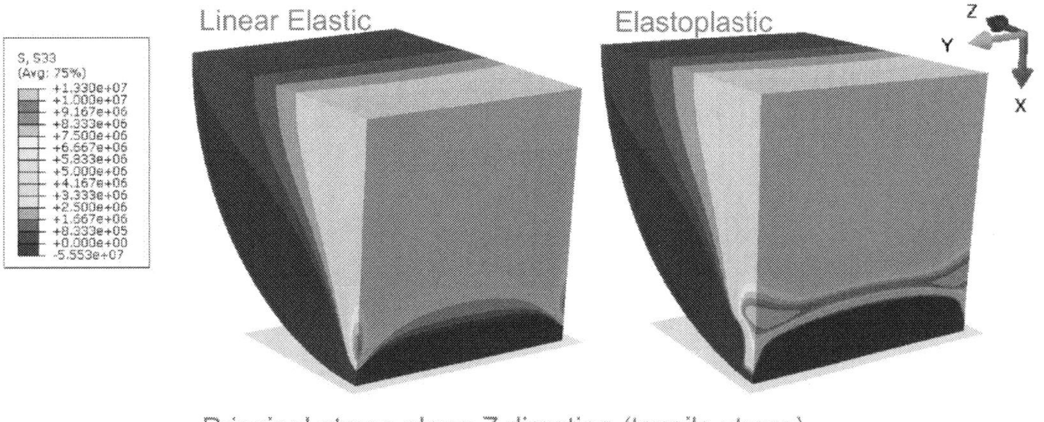


Principal stress along Z direction (tensile stress)

MERCK *Be well*

Public

## Diametrical test: Material Property Matters

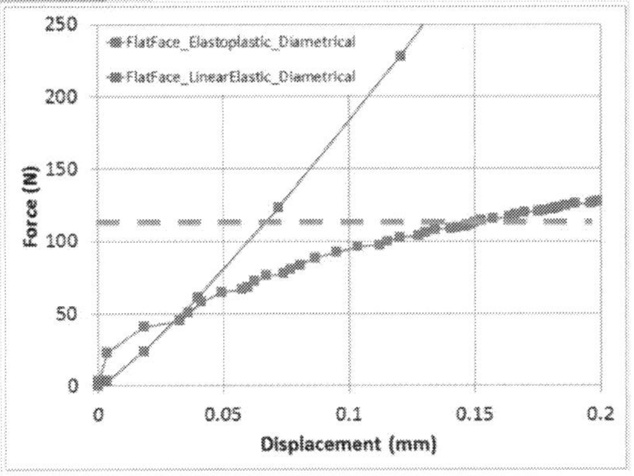


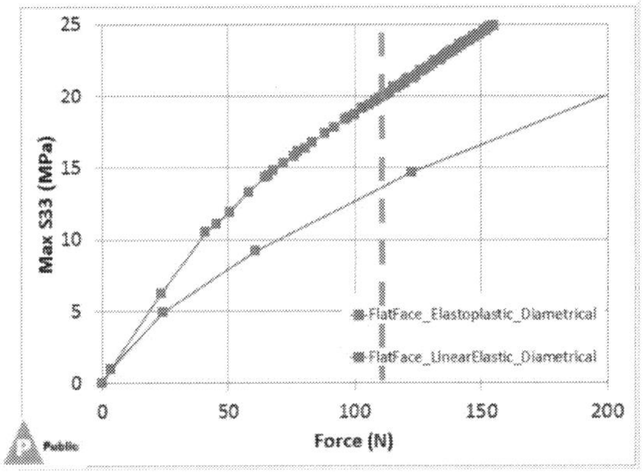


Big difference between Linear Elastic and Elastoplastic materials

- Linear Elastic – typical for brittle material: Lactose, Dicalcium Phosphate
- Elastoplastic – typical for ductile material: Micro-crystalline Cellulose

Force vs. Displacement:

It takes an elastoplastic tablet much larger deformation to reach the same diametrical force

Max Tensile Stress vs. Force:

Same diametrical force generates higher maximum tensile stress inside an elastoplastic tablet

Recall Pitt's equation: Round Flat Tablets

$$\text{T.S.} = \frac{2P}{\pi D t}$$

Green dashed line: experimental break force of a typical 330mg tablet

MERCK *Be well*

Public

# Diametrical test: Shape Matters

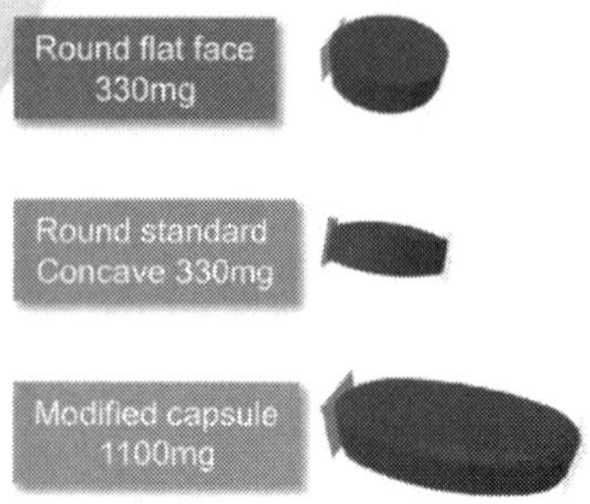


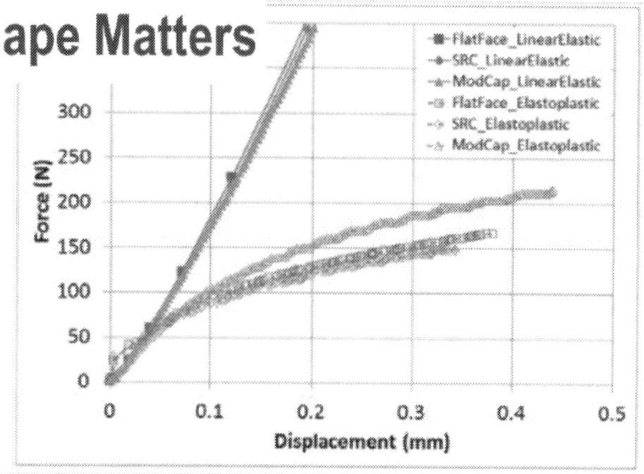


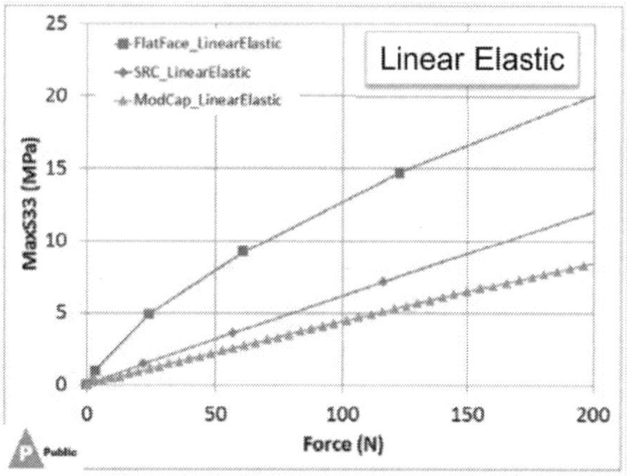


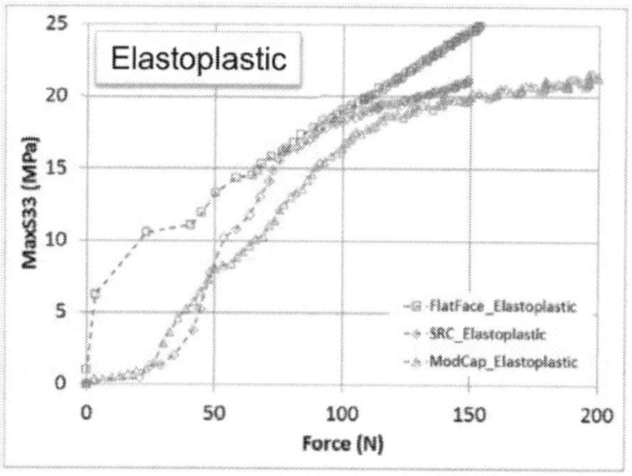


Public

# Diametrical test: Stress Distribution

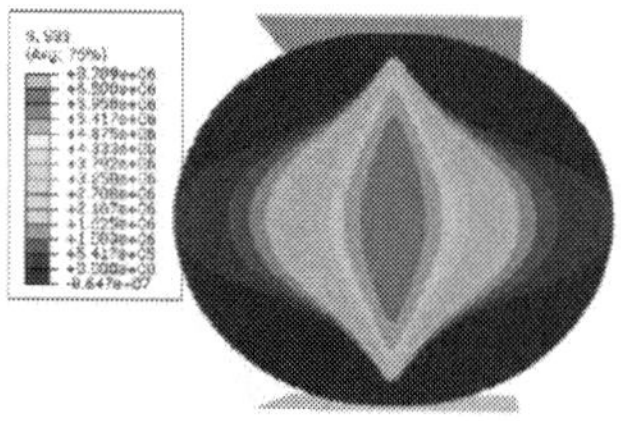

*All shapes are simulated with Linear Elastic material

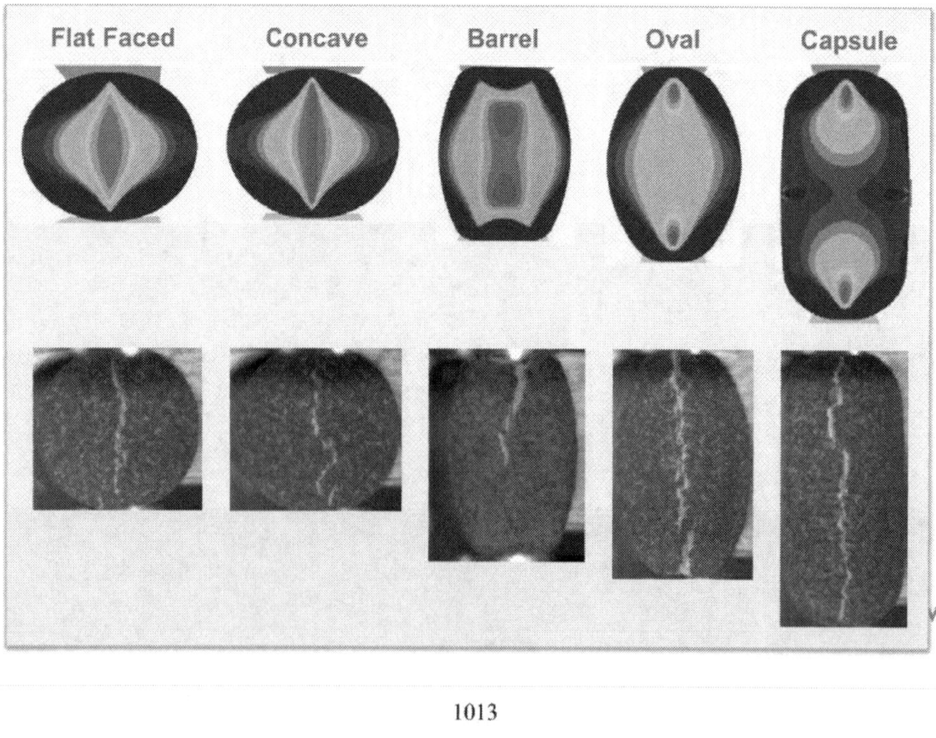


Public

MERCK
Be well

## Diametrical test: from Hardness to Tablet Strength

Tablet geometry + Material property (E, v, $\sigma_y$) → FEM model → (Break force (hardness) from experiment) → Threshold tensile stress

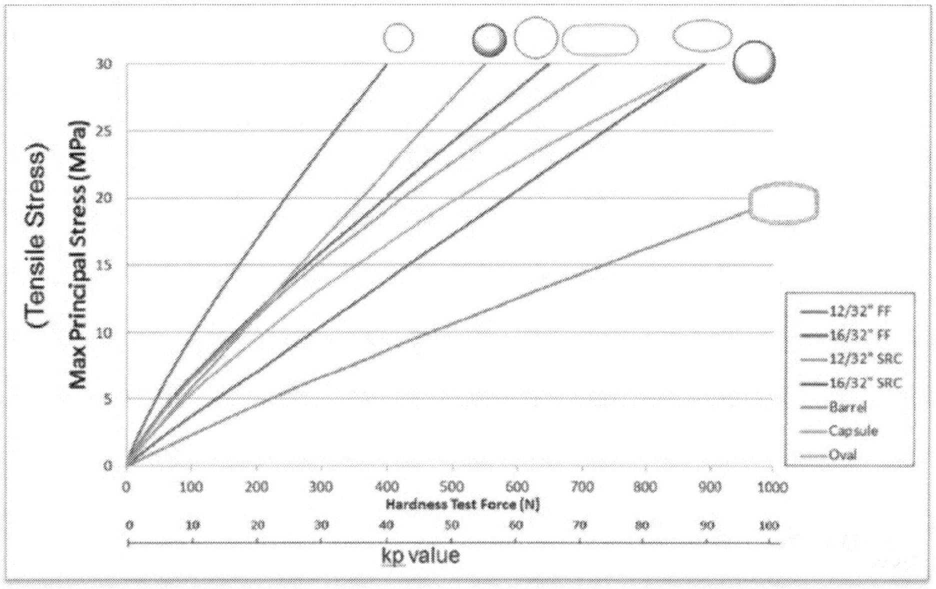


Public

## Dynamic impact response?

Hardness / Tensile Strength → ? → Dynamic impact response

Hardness Testers or Instron machines: ~0.4mm/s ($\dot{\varepsilon} \sim 10^{-2}$ $s^{-1}$) test speed

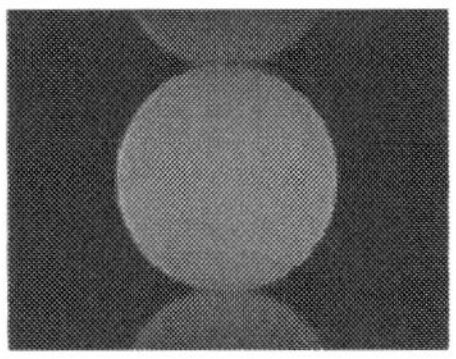

A tablet dropped in friability drum (156mm drop height): 1.75m/s ($\dot{\varepsilon} \sim 10^{2}$ $s^{-1}$)

A tablet dropped from 1m high: 4.4m/s

### Drop test in FEM

We model a tablet impacting a hard surface

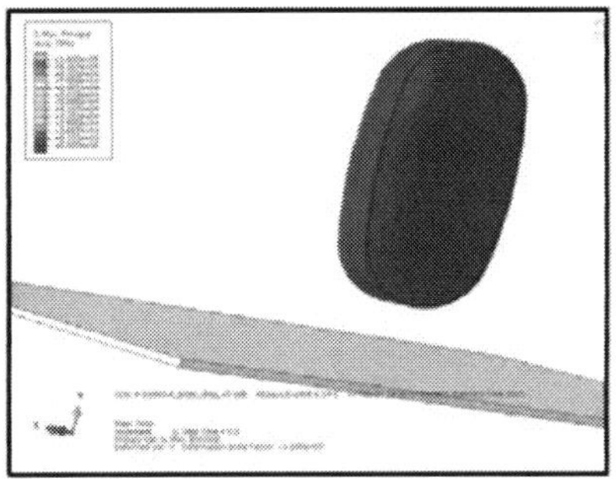

...with very fine spatial/temporal resolution:

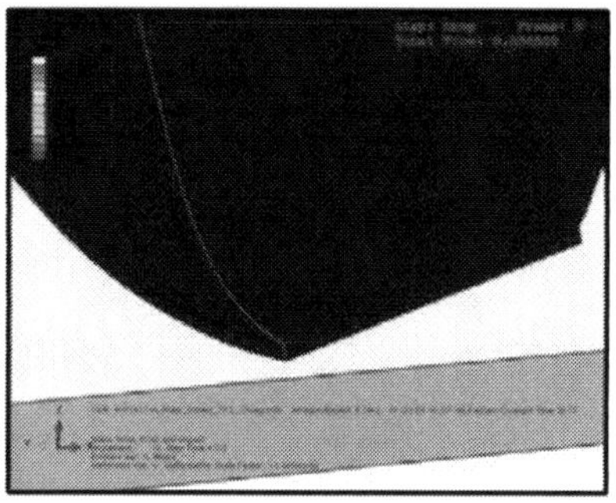

Public

# Drop test: Critical Factors

**Impact Angle**

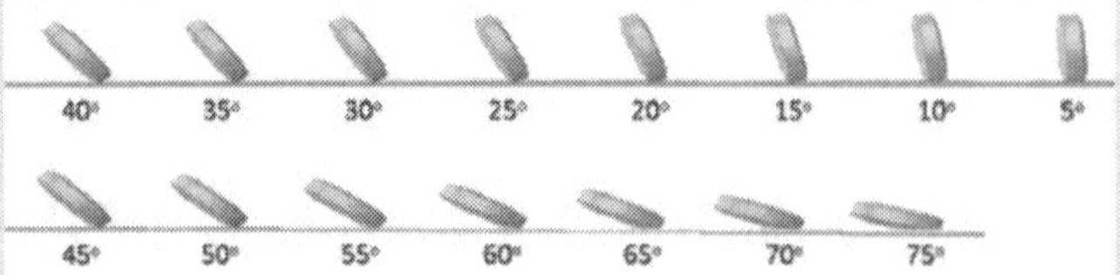


**Material Property**

- Linear Elastic
- Elastoplastic

**Impact Speed**

0.5m/s (0.013m drop height – friability drum) ⟶ 4.5m/s (4.5m drop height – film coating / patient handling)

**Tablet Shape**

Flat Faced (330mg) SRC (330mg) Mod. Capsule (1100mg)

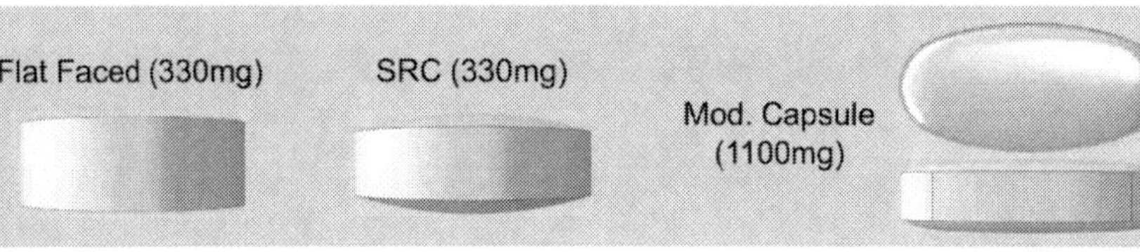

Public

MERCK Be well

# Drop test: Impact Angle

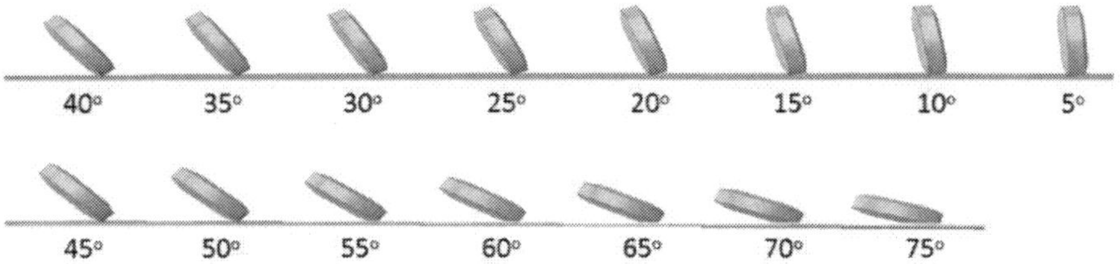


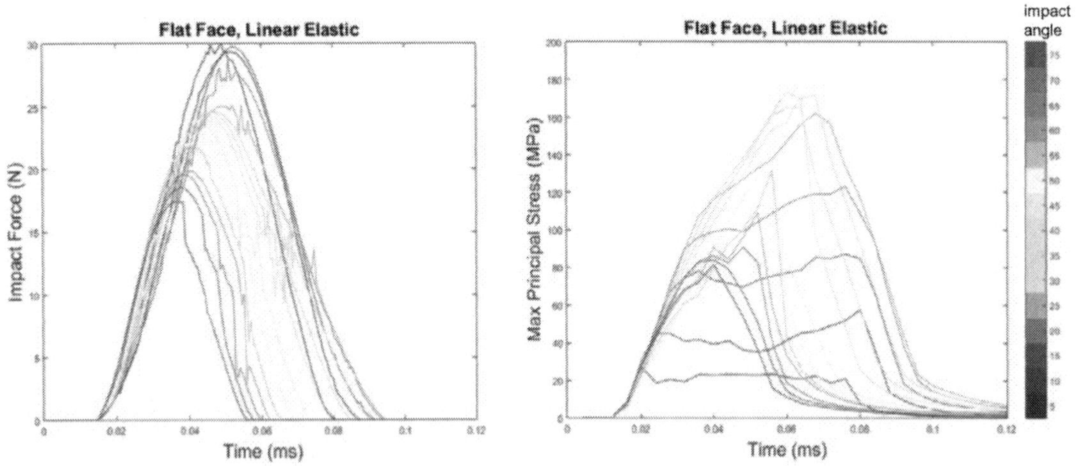


1.75m/s impact speed – typical for friabilily tester

Public

MERCK Be well

# Drop test: Impact Angle

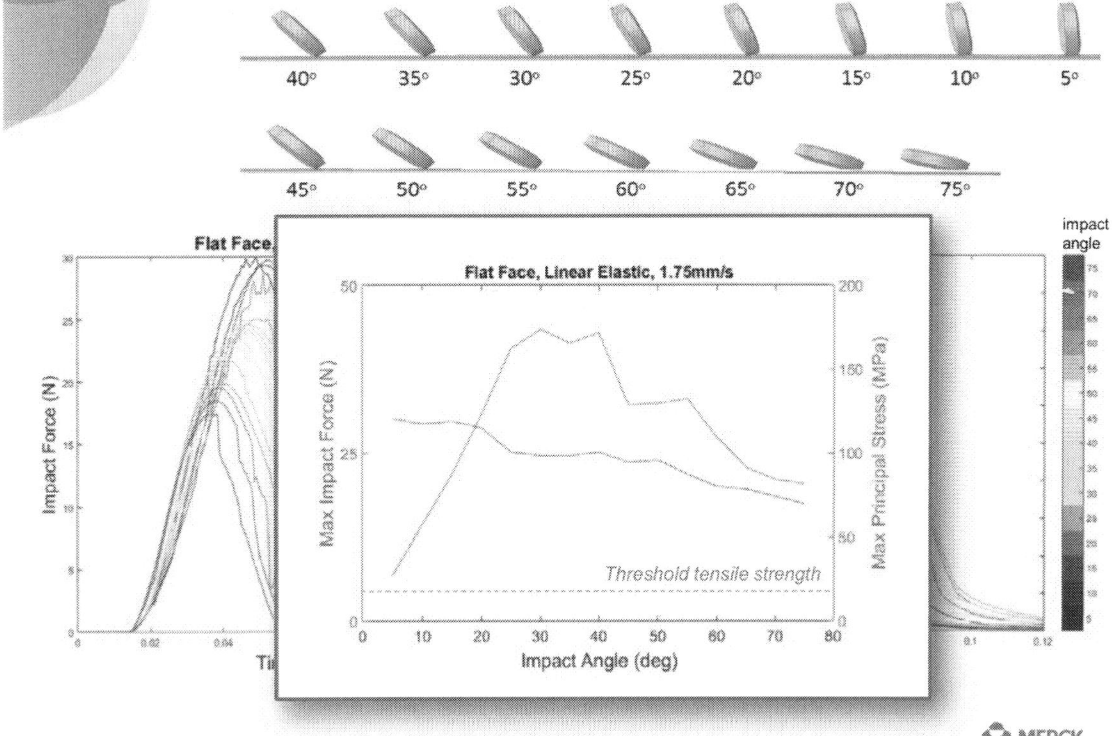


Public

MERCK
Be well

# Drop test: Material Property

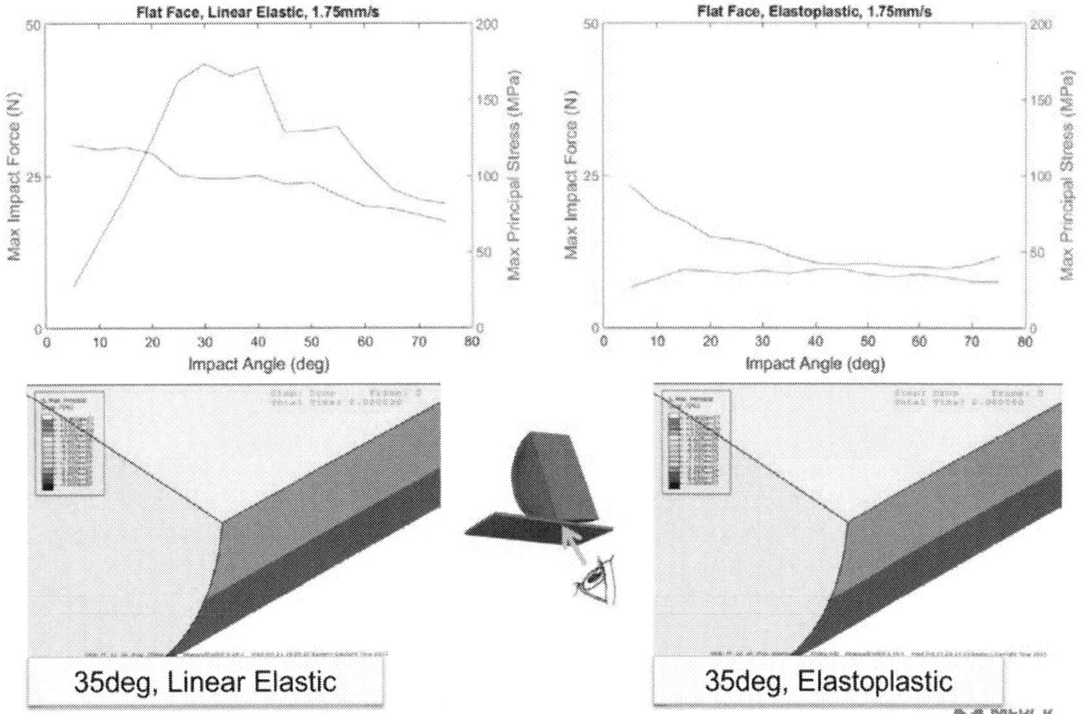


MERCK
Be well

Public

# Drop test: Stress Distribution near Contact

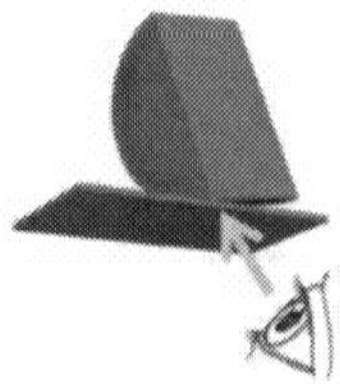

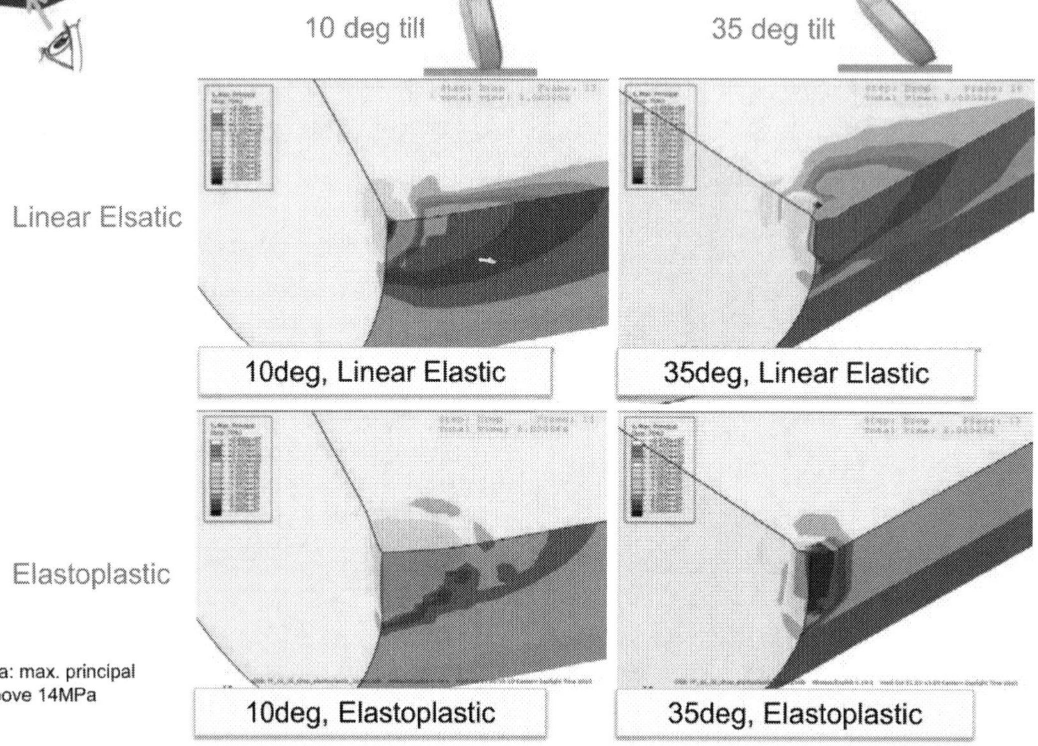


Gray area: max. principal stress above 14MPa

Public

# Drop test: Impact Speed

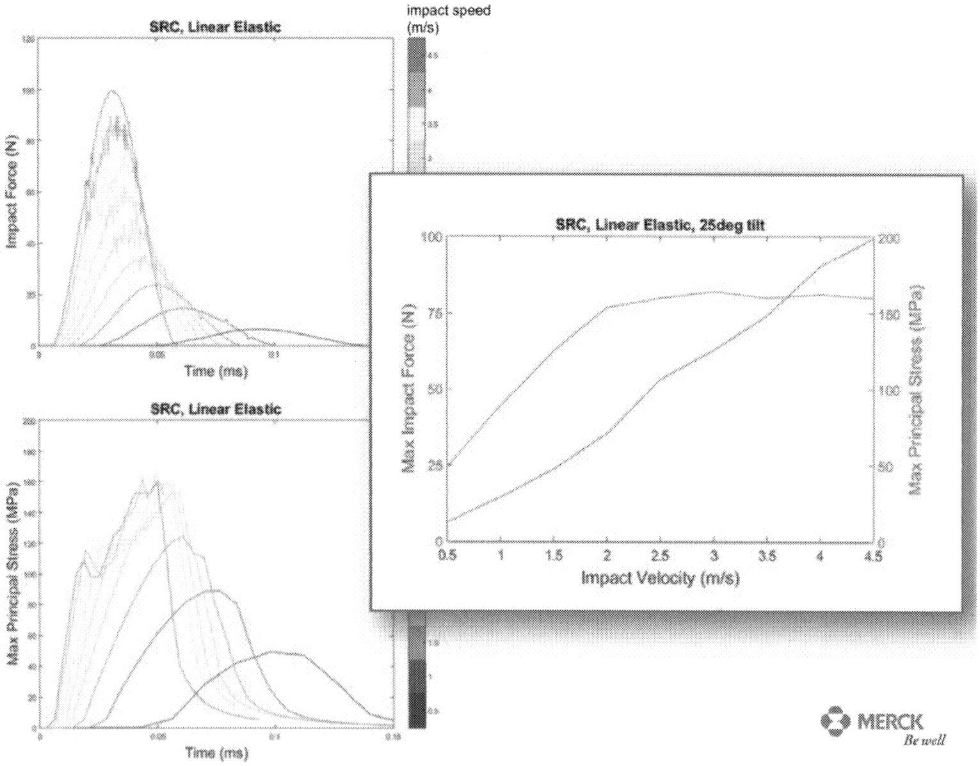


Public

MERCK
*Be well*

# Drop test: Tablet Shape

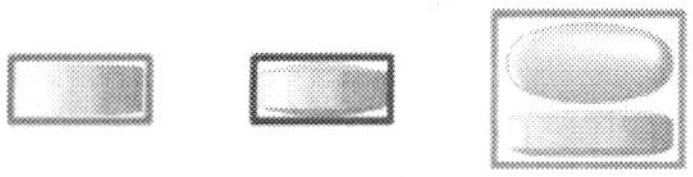

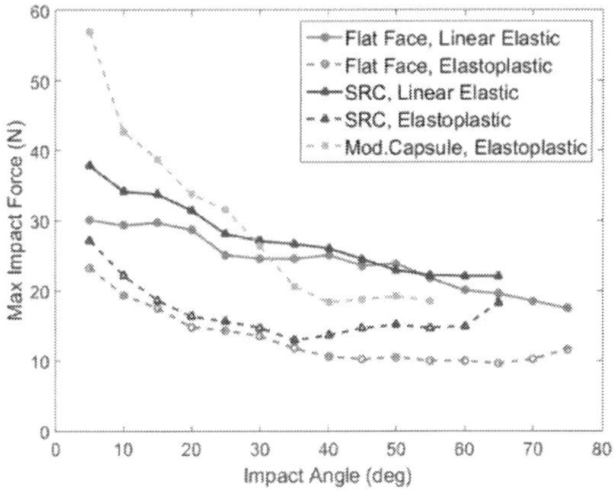


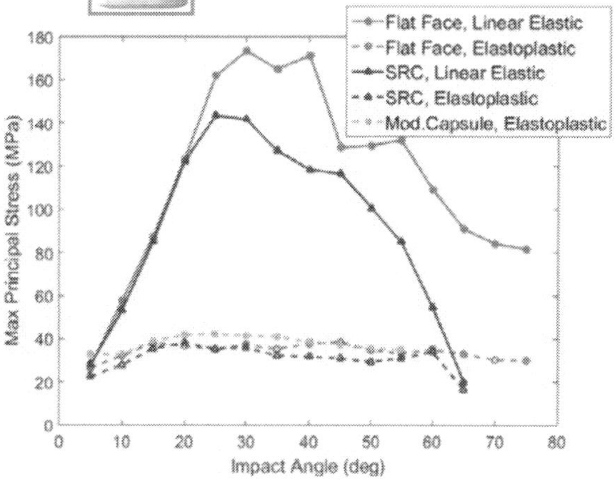


- Tablets with different shape => different angle dependencies of max impact force and max principal stress

Also:

- Linear Elastic material: higher max impact force; max Principal Stress peaks at 20-40deg
- Elastoplastic material: lower max impact force; max Principal Stress does not change with impact angle

Public

MERCK Be well

# Conclusion

- We are now a step forward in predicting impact damage of pharmaceutical tablets
  - There are still liabilities: simplified material model, no fracture dynamics
- Next step: experimental confirmation of the impact modeling

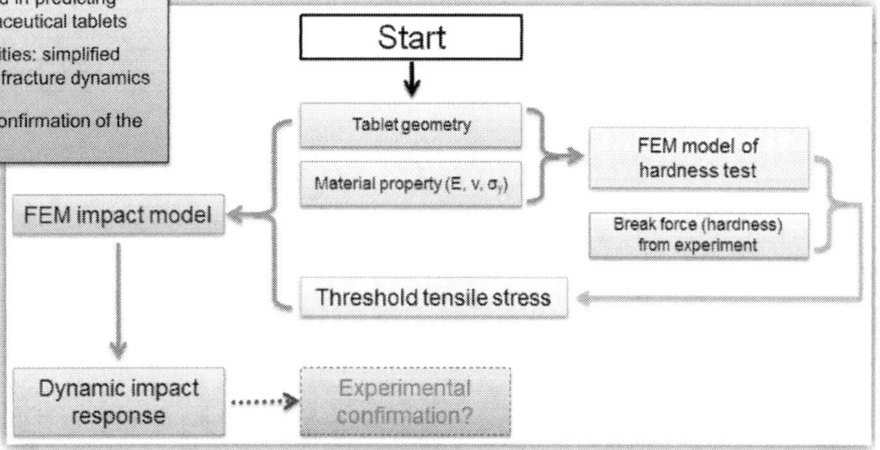


Hardness test

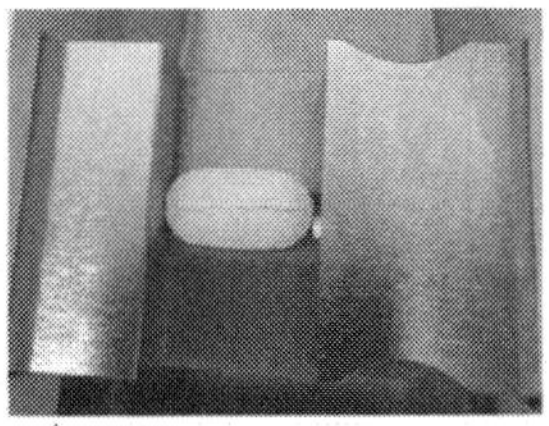

Impact test

Friability test

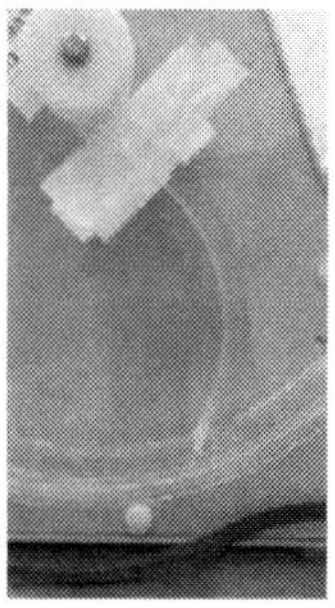

Chipped tablets

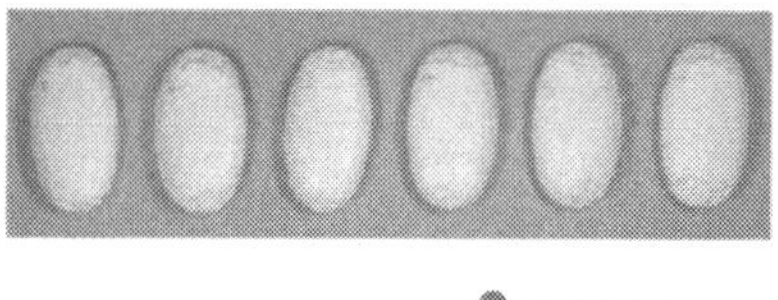

Public

MERCK Be well

## Questions?

Public

MERCK
Be well

## Diametrical test: Shape Matters

Round flat face
330mg

Round standard
Concave 330mg

Modified capsule
1100mg

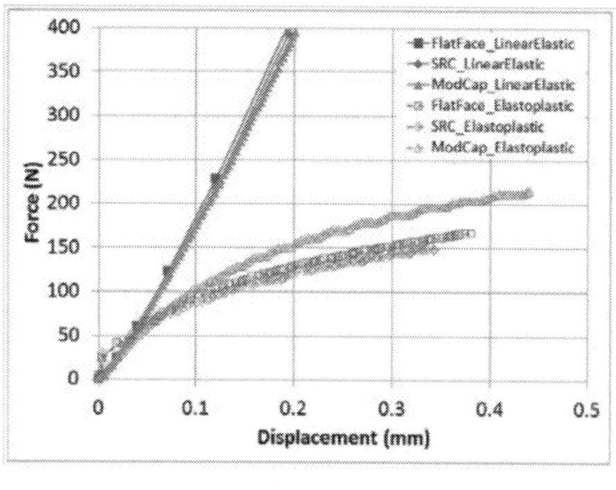


Public

MERCK
Be well

# Drop test: Impact Speed

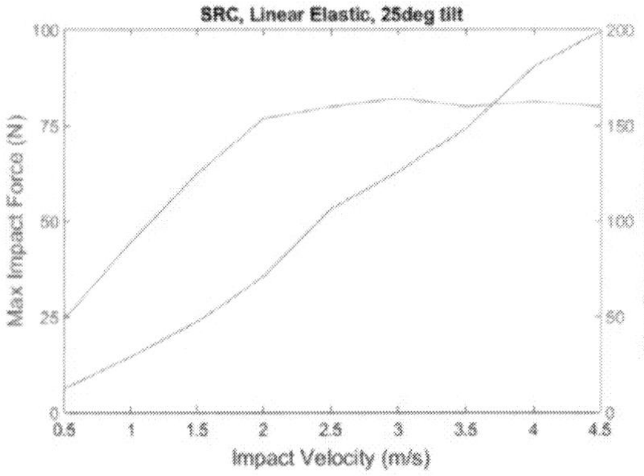


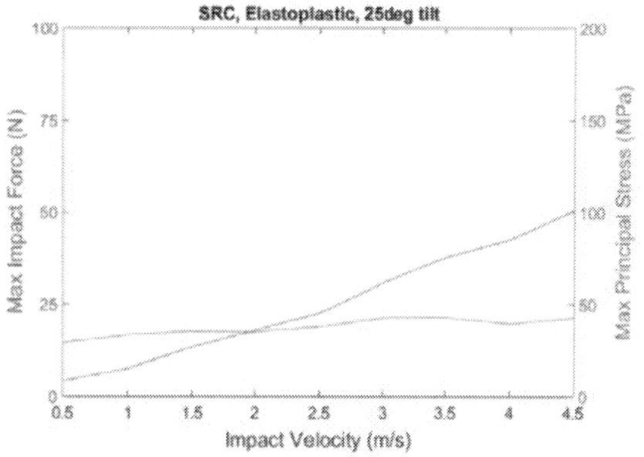


Public

MERCK Be well

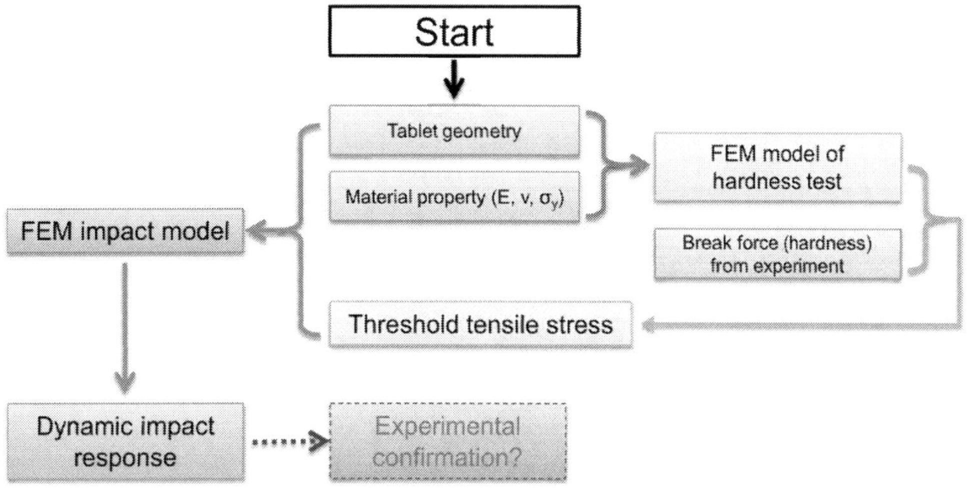


Public

MERCK Be well

# Impact Breakage

Tablet tensile strength or Hardness are currently measured by Hardness Testers or Instron machines
...with speed ~0.4mm/s ($\dot{\varepsilon}\sim10^{-2}$ s$^{-1}$)

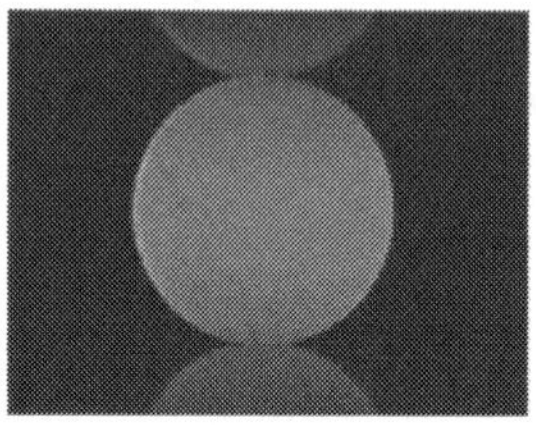

A tablet dropped in friability drum (150mm drop height): 1.7m/s ($\dot{\varepsilon}\sim10^{2}$ s$^{-1}$)

A tablet dropped from 1m high: 4.4m/s

Tablet tensile strength or Hardness are currently measured by Hardness Testers or Instron machines
...with speed ~0.4mm/s ($\dot{\varepsilon}\sim10^{-2}$ s$^{-1}$)

Drop test in FEM

We model a tablet impacting a hard surface

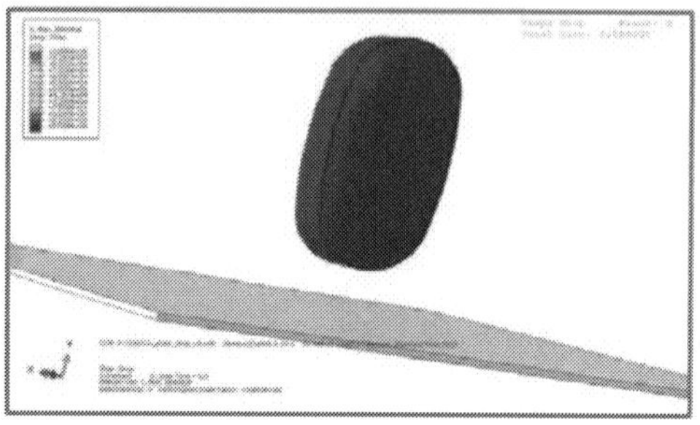

...with very fine spatial/temporal resolution:

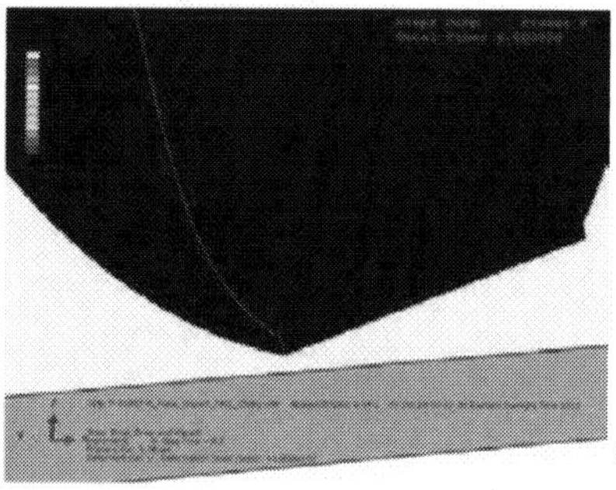

MERCK
*Be well*

Public

# Drop test: Material Property

Linear Elastic

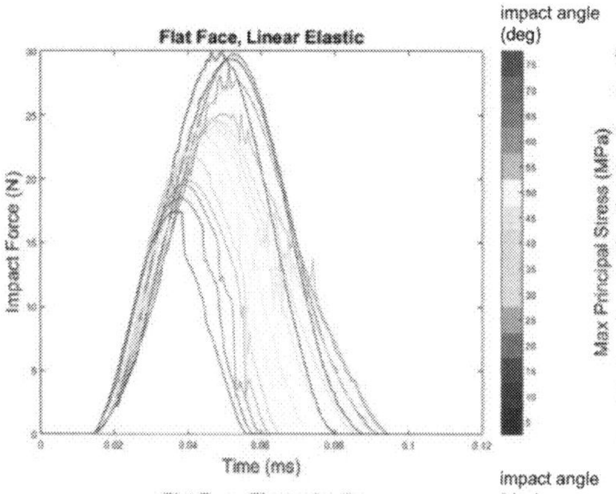


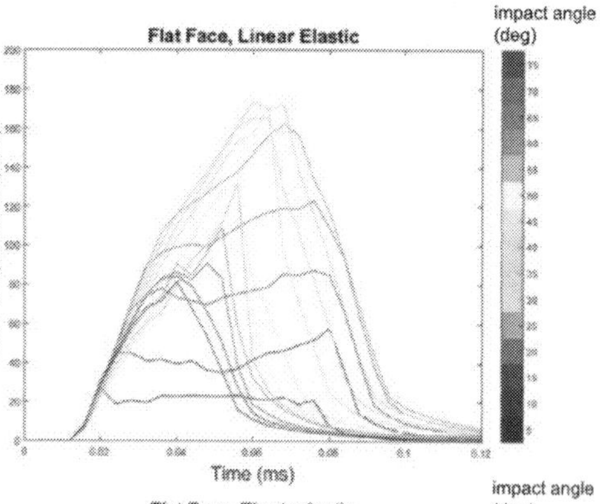


Elasto-plastic

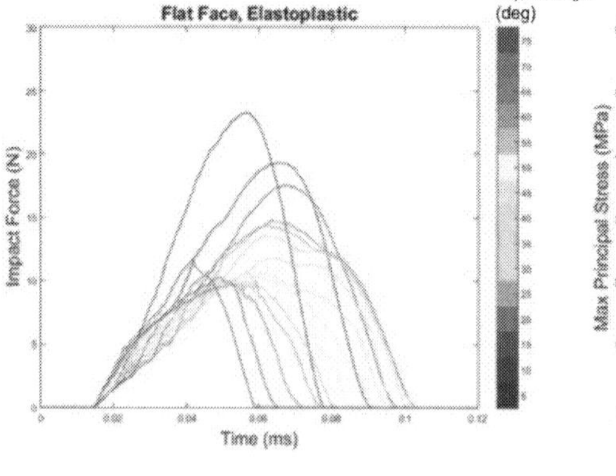


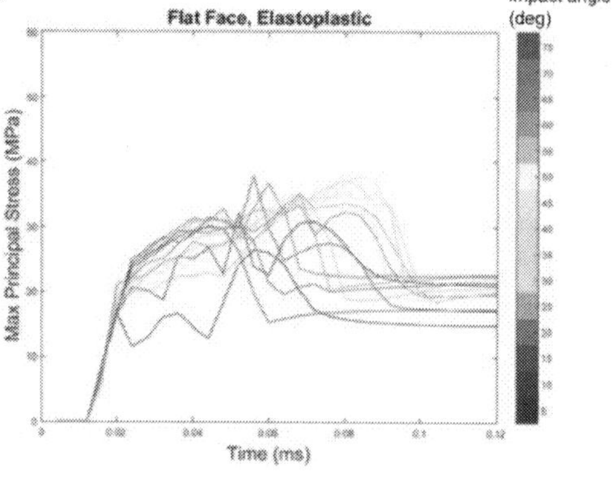


Public

# Drop test: Impact Speed

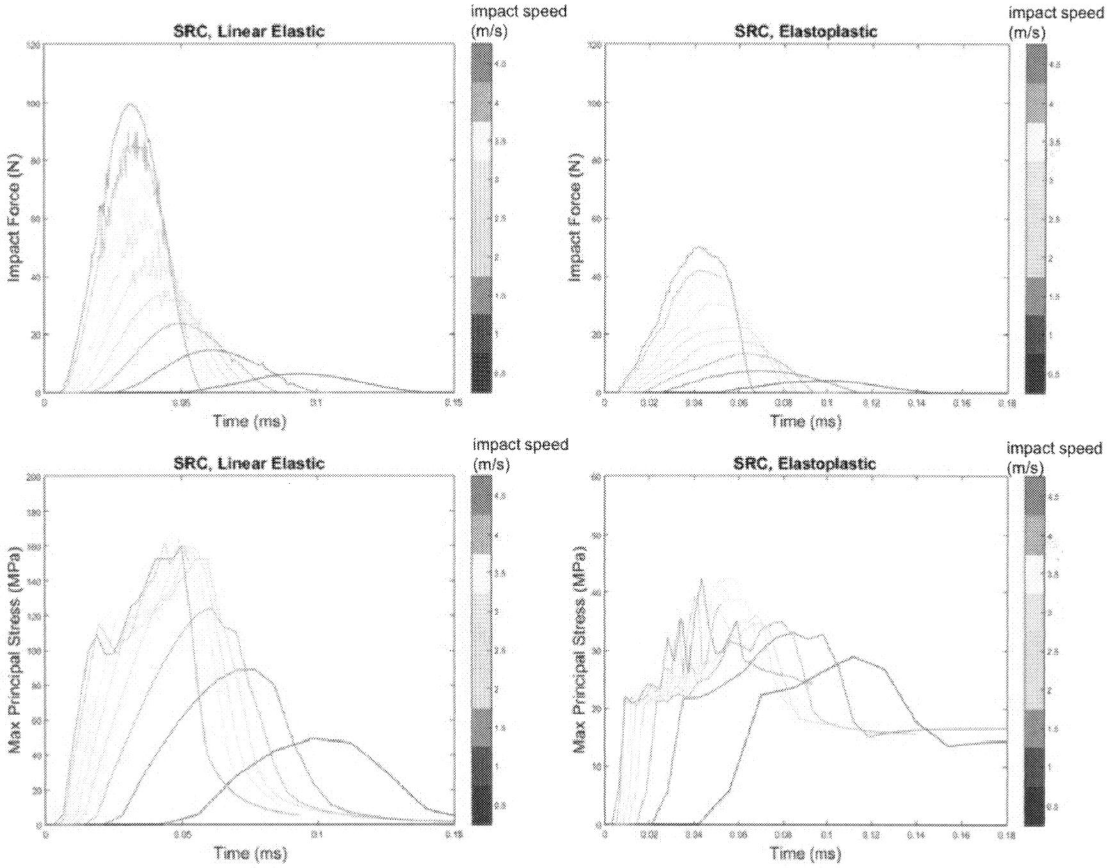


Public

# 425643 Dynamic Modeling and Advanced Control of Tablet Press


**Abstract:**

The tablet press is one of the most critical but least understood unit operation involved in both batch and continuous tablet manufacturing process. Various operations that take place simultaneously in a tablet press, including flow in a chute or a hopper, feeding of powder blend/granules into dies via feed frame, and tablet compression, are characterized by complex phenomena that are difficult to model. Moreover, the control system of existing tablet presses has not changed much in several decades and needs to be improved significantly. For example, with current state of the art techniques and tools, the tablet weight and the tablet hardness can not be controlled simultaneously in real time and raw materials and process variability can not be mitigated proactively. Powder raw material characteristics, feeder level, shear history, upstream milling and blending operation, and powder transport can all have significant impact on powder bulk density, flowability, and lubricity and thereby on critical tablet quality attributes. Therefore, dynamic modeling to understand tablet press performance, and an advanced control system is highly desirable. The Rutgers continuous pharmaceutical manufacturing platform provides an opportunity to improve the tablet press control system by allowing integration of meaningful real time measurements with feed-forward and feedback control systems.

In this work, dynamic models of tablet press, including a detailed feed frame model as well as a compression model, are developed and applied for the design of an advanced control system. The feed frame operation can have significant impact on powder mixing and die filling and thereby on critical quality attributes such as tablet weight, hardness, dissolution and API potency. One of the objectives of feed frame modeling was therefore to provide a predictive tool to understand the complex mixing and segregation phenomena occurring at this stage and to understand tablet die feeding behavior. The integrated tablet compression model will then allow studying the effects of feed frame speed and raw material properties on critical quality attributes. The tablet press model is integrated with the direct compaction line to study the effects of feeding, milling and blending effects on critical quality attributes. The developed mechanistic dynamic model has been employed to design an advanced tablet press control system and to evaluate its performance prior to implementation to the pilot-plant situated at Rutgers University. First, the feedback tablet weight and hardness control loops are decoupled so that these variables can be controlled independently and simultaneously. Then a feed-forward control loop is integrated with the feedback control loops to take the proactive mitigation actions based on powder bulk density variability.

The objective of this presentation is to demonstrate the dynamic model of tablet press, and to illustrate its applicability for design and evaluation of advanced tablet press control system.

# 419773 Data Driven Predictive Modeling of Formulation and Process Conditions of Pharmaceutical Spray Dried Intermediates

**Abstract:**

**Data driven predictive modeling of formulation and process conditions of pharmaceutical spray dried intermediates**

Elise Miller*, Gregory M. Troup, Justin Moser

Merck & Co., Inc., Formulation Sciences, West Point, PA

* E-mail: elise.miller@merck.com, gregory.troup@merck.com, justin.moser@merck.com

The pharmaceutical industry has experienced a shift in the physicochemical properties of pre-clinical drug candidates rendering them less soluble and bioavailable. Spray drying is used to enhance solubility and oral bioavailability by stabilizing the amorphous phase of a drug in a polymer matrix. However, the interaction of formulation and process variables is extremely complex. To fully simulate the process from atomization to final particle attributes based on first principles would require multiple computationally heavy models that still do not adequately capture formulation aspects.

In this work, a series of data driven models is developed and operationalized to relate formulation and process inputs to final spray dried particle properties. A user-friendly simulation tool will be showcased that enables *a priori* prediction of spray dried intermediate (SDI) particle properties as a function of formulation and process parameters. It combines a set of common formulation and process inputs, thermodynamic calculations of outlet process conditions, and SDI particle attributes in partial least squares projection to latent structure (PLS) models. The tool enables simultaneous evaluation of multiple hypothetical formulations and process conditions on the predicted physical attributes of particle size, bulk density, and tapped density. The multi-Y PLS models utilize data from 54 historical spray drying batches. The variables include SDI composition, solution composition, process inputs, process dependent variables, and SDI particle attributes. This exploratory analysis shows that the complexities of the spray drying process are readily captured and that the multivariate model results are consistent with process fundamentals. Then, separate optimized PLS models are developed for the key quality attributes: particle size, bulk density, and tapped density. The resulting PLS models have R2 values greater than 0.8, and Q2 values greater than 0.6, with root mean square estimation errors that are between 8% and 12% of the range. The optimized PLS models allow for prediction of these particle attributes from a limited set of formulation and process inputs, allowing proposed Design of Experiments to be simulated to ensure that a meaningful range of particle attributes can be achieved. Further, the user interface incorporates multivariate diagnostics to access the similarity between the proposed hypothetical batches and the model calibration data. The diagnostics also highlight input parameters that are outside of historical operating ranges. The tool has added utility as a training platform for scientists new to the technology and requires no modeling expertise.

# 425262 Improvement of KM of Ammonia and Mechanistic Insight of Amine Dehydrogenases through Rational and Data-Driven Protein Engineering

**Abstract:**

Recently, amine dehydrogenase (AmDH) has been developed to catalyze prochiral ketones to chiral primary amines, with the addition of ammonia and the cofactor NADH. Three AmDHs have been developed so far: leucine amine dehydrogenase (L-AmDH) [1], phenylalanine amine dehydrogenase (F-AmDH) [2] and a chimeric amine dehydrogenase (cFL1-AmDH) [3]. One of the limitations of the AmDH family is the high $K_M$ for its ammonium substrate ($K_M$: 650 mM). This work will investigate the improvements of the AmDH binding to ammonia through 1) protein engineering of the F-AmDH and 2) adapting reaction conditions.

The protein engineering work will mostly investigate mutations within the active site of the F-AmDH, with a few variants into the secondary shell. The recently solved crystal structure of the F-AmDH helped to guide the rational protein engineering of the AmDHs. Over 80 active variants were expressed and characterized and the $K_M$ of ammonia was improved to two-fold to 350 mM. In addition, specific activity of the F-AmDH was improved 2.5-fold by opening up space within the binding pocket. Various reaction medium conditions were investigated such as pH and ammonium buffers (acetate, bicarbonate, citrate, chloride, formate, phosphate). Changes to $K_M$ and specific activity of the F-AmDH were observed in various buffers. Lastly, substituents on the model substrate of the F-AmDH, para-fluoro-phenyl acetone, were investigated and will be presented with this work.

[1] Abrahamson, M. J.; Vazquez-Figueroa, E.; Woodall, N. B.; Moore, J. C.; Bommarius, A. S., Development of an Amine Dehydrogenase for Synthesis of Chiral Amines. Angew Chem Int Edit 2012, 51 (16), 3969-3972.

[2] Abrahamson, M. J.; Wong, J. W.; Bommarius, A. S., The Evolution of an Amine Dehydrogenase Biocatalyst for the Asymmetric Production of Chiral Amines. Adv Synth Catal 2013, 355 (9), 1780-1786.

[3] Bommarius, B. R.; Schurmann, M.; Bommarius, A. S., A novel chimeric amine dehydrogenase shows altered substrate specificity compared to its parent enzymes. Chem Commun 2014, 50 (95), 14953-14955.

# 415877 Barrel-Stave Pore Formation for Four De Novo Designed Antimicrobial Peptides

**Abstract:**

The interactions of four *de novo* designed and four natural antimicrobial peptides (AMPs) with supported lipid bilayers (SLBs) were measured and used to gain insight into the mechanism of the four *de novo* designed AMPs. These interactions with a Gram-positive bacterial-membrane mimetic SLBs were monitored using a quartz crystal microbalance with dissipation (QCM-D). Protegrin-1 and indolicidin operated through the same toroidal pore formation mechanism with a SLB comprised of 3:1 mole ratio 1-palmitoyl-2-oleoyl-*sn*-glycero-3-phosphocholine:1-palmitoyl-2-oleoyl-*sn*-glycero-3-phospho-(1'-*rac*-glycerol) (POPC:POPG) lipids. Magainin-2 inserted into the same SLB with minimal mass addition and caused little SLB disruption, while α-defensin-1 did not measurably interact. Using these results as references and known mechanisms from the literature for the natural AMPs, the four *de novo* designed AMPs were found to disrupt the SLB via a 4-stage barrel-stave pore formation mechanism. The more potent AMPs were able to access further stages of the mechanism at lower AMP doses. The discovery that the *in vitro* disruption of bacterial-membrane mimetic SLBs correlated with AMP potency against *Mycobacterium tuberculosis* cells suggests that this QCM-D method could be used as a relatively fast *in vitro* screen for AMP efficacy.

# 426293 Toward Design of Antibody Binding Loops That Mediate Conformation- and Sequence-Specific Recognition of Amyloid-Forming Polypeptides

**Abstract:**

Antibodies with conformational and sequence specificity for protein aggregates linked to disorders such as Alzheimer's and Parkinson's disease are valuable for a wide range of applications. We are investigating to what extent such antibodies can be designed by mimicking the natural process of protein aggregation. Our approach is to graft peptide segments that mediate amyloid formation into the complementarity-determining regions (CDRs) of single- and multidomain antibodies. Here we have investigated the impact of inserting polar mutations at the edges of a large hydrophobic Aβ peptide segment in CDR3 on the solubility and binding of the corresponding domain antibodies. We find that antibody expression and solubility are strongly enhanced by introducing multiple negatively charged or asparagine residues in CDR3, while other polar mutations are less effective. Moreover, Aβ domain antibodies with negatively charged CDR mutations show significant preference for recognizing Aβ fibrils relative to Aβ monomers, while the same domain antibodies with other polar CDR mutations recognize both Aβ conformers. We observe similar behavior for a domain antibody grafted with a large hydrophobic peptide segment from islet amyloid polypeptide (IAPP) that contains negatively charged mutations in CDR3. These findings highlight the sensitivity of antibody binding and solubility to residues at the edges of CDRs, and provide guidelines for designing other grafted antibody fragments with hydrophobic binding loops.

# 431203 Engineering the Beta Roll Peptide for Biotechnology Applications

**Abstract:**

**Engineering the Beta Roll Peptide for Biotechnology Applications**


Beyza Bulutoglu[1], Kevin Dooley[2] and Scott Banta[1]

[1] Chemical Engineering Department, Columbia University, New York, NY

[2] Center for Engineering in Medicine, Harvard Medical School, Boston, MA



In nature, there are many proteins and peptides that are natively or intrinsically disordered. Often these peptides gain structure upon interaction with other proteins or molecules. Using protein-engineering tools, we have explored one of such peptide: the β-roll. The β-roll is a unique intrinsically disordered, allosterically–regulated peptide motif. This peptide is isolated from the repeats-in-toxin (RTX) domain and is intrinsically disordered in the absence of calcium. In calcium rich environments, the peptide binds $Ca^{++}$ ions and folds into a β-roll secondary structure composed of two parallel β-sheet faces. We have extensively characterized this calcium responsive RTX domain and evaluated its potential as a cross-linker for hydrogel formation and as an alternative scaffold for biomolecular recognition.

By rationally designing the two faces of the folded beta helix to contain leucine residues, we have created an environmentally–responsive cross-linking domain, genetically fused to an α-helical leucine zipper, capable of self-assembly only in the presence of calcium. We characterized this "double-face" leucine-rich RTX domain and evaluated its mechanical and biophysical properties using various techniques such as circular dichroism and michrorheology. To further investigate its cross-linking capability, we constructed concatemers of this β-roll with maltose binding protein (MBP) and demonstrated that the engineered β-roll peptide can mediate calcium-dependent proteinaceous hydrogel formation without the need for other cross-linking moieties.

In addition, we proposed that the β-roll peptide would be a suitable binding scaffold combining molecular recognition with a simple elution mechanism. This will be of great interest in the bio-separation area as well as other areas of biotechnology, where binding events can be controlled by benign environmental perturbations. Engineering the β-roll results in an allosterically–regulated scaffold, where the folding function can be decoupled from the binding function. Thus, this peptide can operate as a protein switch for biomolecular recognition, which can be mediated by simply changing the calcium concentration, allowing control over the binding behavior of the molecules.

For these purposes, we constructed a library containing mutated residues in eight positions, which form the beta sheet face upon folding. This way, one face of the folded β-roll is fully randomized. Using the ribosome display technique, we selected new β-roll peptides showing affinities for various targets including lysozyme and the Fc region of an IgG antibody. Isothermal titration calorimetry (ITC) was utilized to quantitatively determine the thermodynamic parameters of interactions in solution between the mutants and their targets. The latest results on the molecular recognition capabilities of the candidates obtained via this method will be presented.

# 427170 Genetically Encoded ATP Sensor Based on Magnetic Resonance Imaging

**Abstract:**

In this work, we report the development of the first genetically encoded sensor for noninvasive detection of adenosine triphosphate (ATP) using magnetic resonance imaging. Despite the critical relevance of ATP in cellular metabolism, neurotransmission, and as a clinical biomarker for neurodegenerative diseases and cerebrovascular injury, there is currently no method for noninvasive imaging of ATP *in vivo*. In particular, existing sensors for ATP imaging are based on fluorescence [1], which renders them unsuitable for imaging in preclinical animal models owing to the limited tissue penetration that is achievable using visible light. With a view towards enabling truly noninvasive molecular imaging of ATP, we developed an ATP sensor that is based on a bacterial glutamine synthetase. Glutamine synthetase produces potent MRI contrast owing to the presence of high spin manganese ions in the enzyme [2]. Here, we demonstrate that the contrast can be modulated by the binding of ATP to the enzyme, which serves as a molecular framework to develop an MRI-based contrast agent for ATP. We engineered glutamine synthetase to abrogate enzymatic activity while retaining ATP-dependent MRI contrast, thereby developing an enzymatically silent variant (known as $GS_{ATP}$) as a sensor for quantitative detection of ATP. We characterized relaxivities of $GS_{ATP}$ *in vitro* at magnetic field strengths of 1 T and 7 T and demonstrate an approximately 2-fold difference in relaxivities between the ATP-bound and ATP-free states of the sensor that results in appreciable ATP-dependent contrast on T1 and T2-weighted MR images. Finally, we demonstrated dynamic imaging of ATP using $GS_{ATP}$ to monitor enzymatic degradation of ATP by apyrase. Overall, $GS_{ATP}$ produces significantly stronger ligand-dependent contrast compared to existing genetically encoded MRI biosensors [3] and provides a platform for the development of other noninvasive MR sensors based on high spin manganese binding metalloproteins.

References:

1. Imamura, H. **et al**. *PNAS*, 106:15651-15656 (2009)
2. Villafranca, J.J. **et al**. *Biochemistry*, 15:536-543 (1976)
3. Shapiro, M.G. **et al**. *Nature Biotechnology*, 28: 264-270 (2010)

# 434630 Development of Biosensors to Identify New Chemicals Against Dengue Fever Vectors

**Abstract:**

The development of a bacterial biosensor containing a nuclear hormone receptor (NHR) has proven to be an effective alternative to current compound screening strategies that rely on hormone-mimicking properties. NHRs comprise one of the largest classes of protein targets because they can bind to a wide variety of endocrine disruptors in addition to their native hormones. Our allosteric biosensors can detect compounds that target NHRs and distinguish between their positive or negative effects on the receptor's activity. When this design is extended from human to insect NHRs, an opportunity arises to screen compounds for their potential role as insecticides against infectious disease vectors such as *Aedes aegypti*, the dengue fever mosquito. This project seeks to construct a nuclear hormone binding sensor in *E. coli* to favor the identification of those compounds that disrupt the endocrine system of this target insect. By incorporating the ligand-binding domain of *Aedes aegypti's* NHR 96 into our *E. coli* biosensing plasmid, we are able to detect its response to CITCO, a chemical compound known to be the NHR96 native ligand. After a thorough optimization of parameters such as incubation time, temperature, and dilution in a full factorial experimental setup, we seek to attain a reproducible growth phenotype assay for this novel biosensing system. The use of domain deletion, linker engineering, and point mutations can help improve the biosensor's ligand-response ability. Furthermore, the translation of the biosensing protein in an *E. coli* cell-free system will allow the test compounds to come directly in contact with the biosensing protein as it is made, further increasing the sensitivity and efficiency of the assay.

# 434417 Discovery and Engineering of Cytochromes P450 from Plant Secondary Metabolism

**Abstract:**

Plant natural products are an excellent source of anti-cancer drugs and other therapeutics. For example, the defense compound indole-3-carbinol from broccoli (*Brassica oleracea*) arrests reproductive cancer cell lines[1], while brassinin from Chinese cabbage (*Brassica rapa*) increases immune response to cancerous cells[2]. To produce these complex, small molecules on an industrial scale, we must first identify and engineer the proteins which synthesize them. Crucifers like *B. rapa* and *B. oleracea*use specific oxidizing and tailoring enzymes to create these tryptophan-derived therapeutic natural products: cytrochromes P450 (P450s). P450s possess a unique capacity for multi-step catalysis, performing not just site specific oxidation, but further tailoring of oxidized intermediates. This "catalyst within a catalyst" characteristic make P450s powerful frameworks for engineering. If structural elements causing tailoring activity are known, they may be altered, or swapped into another P450, towards the creation of new molecules. Although work has been done to understand P450 oxidation, little is known about how P450s, especially those from plants, do this specific tailoring. Because plant P450s are difficult to isolate in large numbers, the best approach for analyzing and engineering these enzymes is rational design of a small mutant library based on structural models and simulations.

Such models exist for the P450s CYP71A12 (A12) and CYP71A13 (A13) from *Arabidopsis thaliana.* These enzymes share 89% sequence identity but tailor molecules differently, making them an ideal case study. Both enzymes dehydrate and hydroxylate the same tryptophan-derived substrate, while only A13 promotes a unique sulfur addition[3]. In this study, we analyzed protein model simulations to identify structural elements correlated with substrate processing of A13. These regions were swapped from A13 into A12 in chimeric proteins which were heterologously expressed and assayed *in vitro*. We found distinct regions and residues responsible for specificity towards both products and the overall stability of the protein. In this work we also describe efforts to diversify the library of tryptophan-derived therapeutics by the discovery of P450s from non-model plants including *B. rapa*and other crucifers. By analyzing the structure function relationships of these enzymes and by identifying new classes of P450s, we are expanding the library of novel therapeutics derived from plant natural products.

1. Hsu, J.C., *et al. Biochemical pharmacology* **72**, 1714-1723 (2006).

2. Banerjee, T. *et al. Oncogene* **27**,2851–2857 (2008).

3. Klein, A.P., *et al. Angewandte Chemie International Edition* **52,** 13625-13628 (2013).

# 414209 Living Reverse Osmosis Membranes: A Model for the Beneficial Use of Engineered Biofilms

**Abstract:**

Reverse osmosis membrane systems, increasingly used for developing new water sources and recycling water, are a key technology to ensure water sustainability. However, as with all membrane water treatment processes, biofouling, the build-up of microbes and their polymeric matrix, clogs these systems and reduces their efficiency. Realizing that a microbial film is inevitable, we engineered a beneficial biofilm that prevents membrane biofouling by limiting its own thickness by sensing the number of its cells that are present via a quorum-sensing circuit; the beneficial biofilm also prevents biofilm formation by deleterious bacteria by secreting nitric oxide, a general biofilm dispersal agent. In addition, the beneficial biofilm was engineered to produce an epoxide hydrolase so that it efficiently removes the environmental pollutant epichlorohydrin. Thus, we have created a living reverse osmosis system that simultaneously reduces biofouling while providing a platform for biodegradation of persistent organic pollutants.

# 426431 Development of Synthetic-Hybrid Systems for Enhanced Gene-Expression Control

**Abstract:**

**Development of Gene-Regulatory Control Tools for *Salmonella* Reduction in Poultry**

Brittany Forkus and Yiannis Kaznessis

The development of antibiotic resistance in foodborne pathogens poses a major challenge to current healthcare efforts as it limits our potential treatment options. The rise and spread of these resistant strains is believed to be largely attributed to the overuse of antibiotics, particularly in livestock. Gastrointestinal illnesses caused by these pathogens affect millions and resistant infections yield higher rates of mortality, morbidity, and increased hospital costs when compared to infections caused by antibiotic-sensitive strains. *Salmonella*infections, in particular, are primarily derived from poultry sources and yield the highest number of foodborne infections, hospitalizations, and deaths in the U.S, making it a rising concern to limit the number of resistant strains that make their way into food products.

We propose the development of an antibiotic alternative for livestock purposes to reduce *Salmonella* carriage in poultry. We have re-engineered the probiotic *E.coli* strain, Nissle 1917, to produce and secrete the antimicrobial peptide, Microcin J25 (Mccj25). Antimicrobial peptides are essentially the antibiotics of the bacterial kingdom and are natively produced by bacteria to ward off competitors and to modulate bacterial populations, but from an engineering standpoint, they serve as an untapped reservoir of antibiotic activity. The engineered Nissle we have developed constitutively produces high titers of Mccj25. *In vitro* results of this engineered strain demonstrate remarkable antagonistic activity against the leading foodborne strain, *Salmonella* enteritidis (SE). In our work, we have run pilot studies to observe the effect of our engineered probiotic on young, male turkey poults challenged with SE. We hypothesize that our engineered probiotic could aid in the reduction of SE in turkey poults by an antibiotic-free mechanism.

# 430615 Towards a Genetically-Engineered Bacterium for Gastrointestinal Wound Healing

Thursday, November 12, 2015: 1:10 PM
150G (Salt Palace Convention Center)
Chelsea Virgile[1], Eunkyoung Kim[1], Gregory F. Payne[1,2] and William E. Bentley[1,2], (1)Fischell Department of Bioengineering, University of Maryland, College Park, College Park, MD, (2)Institute for Bioscience and Biotechnology Research, University of Maryland, College Park, MD

Bacteria are often thought of as "good" bacteria (e.g., commensals, probiotics) or "bad" bacteria (e.g., pathogens). Synthetic biology enables the augmentation of biosynthetic capabilities and retooling of regulatory structures in the creation of cells with unprecedented ability to make products. One can also, however, think of the cell as the product – a cell that operates in a noisy environment to execute non-native tasks. There have been several recent reports of the rewiring of bacterial cells to function as conveyors of therapeutics. The engineering and rewiring of the bacteria such as *E. coli* into 'smart' bacteria potentially allows for a broad range of applications, from the treatment of wounds, the elimination of pathogenic strains, to the delivery of vaccines, particularly in the GI tract.

'Smart' bacteria must survey their environs, respond to the appropriate cues, while at the same time not respond to inappropriate cues. They must stay on task. Our approach comprises electrochemistry and synthetic biology for the creation of a biological 'test track' for ensuring the appropriate design and testing of engineered bacteria. We are engineering bacterial motility for response to wound-generating signals such as hydrogen peroxide and reactive oxygen species. Specifically, we have placed motility enzyme CheZ under the control of the hydrogen-peroxide-responsive *oxyR* promoter so that the "run" in the tumble and run scheme of bacterial movement is externally regulated. Engineered cells exploit pseudotaxis for directional swimming.

In addition, we engineer the surface of these bacteria for the display of therapeutic agents including human proteins and enzymes. Specifically, we have expressed human tissue transglutaminase (htTG) on the outer surface of *E. coli* for the crosslinking of proteins containing exposed lysine and glutamine residues. We exploit the AIDA (AIDA-1) system for presentation of the htTG on the outer surface using the motives shown below. For therapeutic application, tissue transglutaminase would then potentially aid in wound repair. Our presentation will highlight both successes in engineering "smart" bacteria using this methodology as well as limitations based on system design and the parameter space for use.

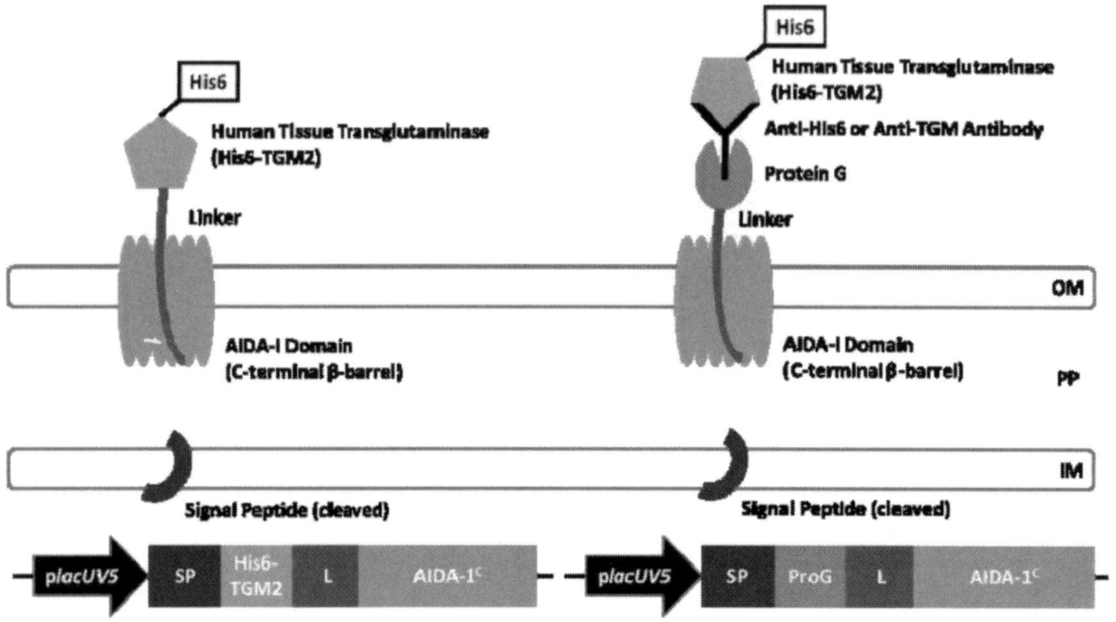
His6
Human Tissue Transglutaminase
(His6-TGM2)
Linker
OM
AIDA-I Domain
(C-terminal β-barrel)
IM
Signal Peptide (cleaved)
placUV5
SP
His6-TGM2
L
AIDA-1C
His6
Human Tissue Transglutaminase
(His6-TGM2)
Anti-His6 or Anti-TGM Antibody
Protein G
Linker
AIDA-I Domain
(C-terminal β-barrel)
PP
Signal Peptide (cleaved)
placUV5
SP
ProG
L
AIDA-1C

# 432244 Perturbing Bacterial Adaptive Resistance Using Crispr-Cas9 Mediated Gene Targeting: Emerging Applications in Synthetic Biology

**Abstract:**

The rate at which new antimicrobial resistances are emerging at the clinical level has surpassed our ability to develop novel therapeutics to combat them. Recent studies have suggested that, when exposed to antimicrobials, bacteria enter an "adaptive resistance" state by exploring multiple pathways sampling a dynamic gene regulatory space. We investigated these adaptive pathways by controllably up-regulating and down-regulating the expression of genes known to be involved in bacterial tolerance. We employed emerging synthetic biology techniques to investigate gene regulatory networks involved in controlling adaptive resistance. Using deactivated CRISPR-Associated Protein 9 (dCas9), we designed a library of synthetic genetic devices to activate and inhibit native gene expression of stress-response networks. Here we show that in the presence of these synthetic constructs, significant control over bacterial fitness can be achieved during adaptation to sub-minimal inhibitory concentrations of a range of toxins, including disinfectants and antibiotics. The range of adaptive phenotypes observed allowed us to identify individual genes important in the evolution of bacterial tolerance towards specific stress conditions. Moreover, we employed this approach to design constructs targeting multiple genes simultaneously, and observed predominant negative epistasis with severe loss in fitness. We ultimately discern genes responsible for the development of bacterial tolerance, providing direction for novel therapeutics which can impede intrinsic adaptive pathways leading to antimicrobial resistance. The techniques presented in this work offer a novel approach for both impeding the intrinsic adaptive pathways leading to antibiotic resistance and re-sensitizing antibiotic resistant pathogens to traditional therapies employed at the clinical setting.

# 429577 Harnessing Synthetic Biology to Uncover & Engineer Plant Natural Product Biosynthesis in Yeast

**Abstract:**

Plant natural products are valuable compounds with diverse therapeutic activities whose biosynthesis is poorly understood due to the challenges in functionally identifying and characterizing pathway enzymes. While the recent discovery of plant gene clusters offers the potential to accelerate pathway discovery in plants, current enzyme characterization strategies limit the efficiency and accuracy of subsequent pathway elucidation. We describe an alternative approach, fueled by recent methodological advances in synthetic biology, for the rapid elucidation of complex plant biosynthetic pathways through the step-wise reconstruction of plant gene clusters in *Saccharomyces cerevisiae*. We demonstrate this approach for the full elucidation of one of the most complex gene clusters identified to date. Our work not only offers novel insight into several of the enzymatic steps in the pathway, but also provides yeast strains engineered to produce previously inaccessible pathway intermediates and novel derivatives, thereby providing platforms for advancing drug discovery. The synthetic biology-based approach offers several advantages over traditional elucidation strategies based on gene silencing in the plant host and *in vitro* enzyme characterization, including characterization of pathway metabolites in a clean host background, direct supply of pathway intermediates for structural validation, and simultaneous development of scalable synthesis platforms for intermediates and novel derivatives.

# 433414 Composability and Design of Parts for Application to Large-Scale Pathway Engineering in Yeast

**Abstract:**

Yeasts and fungi can be powerful platforms for production of antibiotics and other small molecules. Yet, rewiring yeasts *a priori* for optimal production is not possible, necessitating an iterative design-build-test-learn cycle. Within this framework, large-scale combinatorial libraries can be used to rapidly learn design rules to guide subsequent iterations. To enable this process, standards and methods for achieving rapid assembly of varied levels of expression, while preserving genetic stability, must be developed.

To this end, the first part of the current study advances *S. cerevisiae* parts library design, characterization, and analysis. Employing high-throughput techniques and automated liquid handling, the expression of over a thousand natural and synthetic yeast transcription units under different growth conditions can be analyzed via flow cytometry. With this approach, the combined effects of promoters and terminators on gene expression can be modeled. Our initial analyses point to a well-defined expression space that can be modeled with a simple second-order expression.

The second part of the current study demonstrates how parts characterization on this scale enables construction of combinatorial libraries of multi-gene pathways without repeating parts within a design. Coupled with standardized assembly techniques, a library of nearly two hundred variants of a six-gene pathway were constructed in a matter of weeks. The pathways were subsequently assayed for performance and, using statistical analysis, bottlenecks and optimal relative expression levels for the pathway were determined.

With these advances, we demonstrate a large-scale design-build-test-learn cycle that can be generally applied to any multi-gene system in yeast.

# 436429 Rapid Evolution of Regulatory Elements for Multiplexed Combinatorial Metabolic Engineering

**Abstract:**

Optimizing microbial cell factory to produce targeted chemicals often requires the expression of multiple enzymatic steps. It is desirable to coordinate the expression of these genes to prevent the accumulations of toxic intermediates/byproducts and maximize the carbon flux towards the target pathway. To meet this challenge, recent advances in metabolic engineering have spurred great interests to explore precise and combinatorial approaches to control and optimize biosynthetic pathways for production of fuel and pharmaceutical molecules. Resembling of the engineering principle, most of the current combinatorial pathway optimization follows a rather intuitive framework – an iterative cycle of design, characterization, construction and testing. Despite great success to improve the production of an array of metabolites, the current approach suffers from tedious trial-and-error work that obviously restricts our ability to realize the full potential of a specific biosynthetic pathway. To bridge this gap, we have invented an evolutionary approach that efficiently incorporates single strand regulatory DNAs (ssDNAs) and rapidly tuned the expression of multiple genes at the same time. As a case study, we have simultaneously evolved the promoter, operator and terminator regions of a five gene pathway and resulted in significant titer improvement. Combining with sophisticated DOE (design of experiments) principles, this approach opens up new avenues for quantitative multiplexed metabolic engineering.

# 421921 Improved Functional Screening of Metagenomic Libraries Via Heterologous Sigma Factor Expression

**Abstract:**

The microbial diversity present in nature is an enormous stock of genetic material that remains largely unexplored. With fewer than 1% of organisms successfully cultured in the laboratory, this enormous genetic diversity can be harvested best in the form of metagenomic libraries. A major limitation in functional screening of heterologous genomic or metagenomic libraries is the low expression of heterologous genes in screening hosts, such as *Escherichia coli*. To overcome this limitation, here we engineer *E. coli* strains that are capable of recognizing heterologous promoters by expressing heterologous sigma factors. Among several sigma factors tested, the primary sigma factor RpoD from *Lactobacillus plantarum* (Lpl) is able to initiate transcription from many diverse sources of DNA. Using a promoter GFP-trap concept, we show greatly improved transcription initiation from DNA libraries with heterologous and metagenomic DNA (sourced from *Lpl*, *Bacillus subtilis*, *Deinococcus radiodurans*, *Clostridium pasteurianum*, *C. acetobutylicum*, and soil samples), thus enlarging the genomic space that can be functionally sampled in *E. coli*. We show that screening fosmid-based heterologous (Lpl) genomic library in an *E. coli* strain with a chromosomally integrated Lpl *rpoD* enables the identification of Lpl genetic determinants imparting strong ethanol tolerance in *E. coli*. Strand-specific RNA-seq analysis confirms increased expression of heterologous genes in the engineered strain. Metagenomic and heterologous libraries were screened for functional traits in the wild type and engineered strains.

# 434504 Rapid Engineering of Bacteria with Multiplexed Editing, Testing and Tracking Approaches

**Abstract:**

Multiplexed genome design and editing strategies can produce targeted libraries containing billions of combinatorial mutations on laboratory timescales. When combined with high-throughput screens or selections, such strategies enable metabolic engineering at scales orders of magnitude beyond prior approaches. A major challenge has been the inability to efficiently and dynamically map such combinatorial library diversity; thus preventing the adoption of more sophisticated engineering strategies that learn from the recursive mapping of a large number of designs onto targeted performance metrics. In this talk, I will discuss approaches to engineering and tracking combinatorial designed genetic libraries. Specifically, I will cover the development and application of a high-throughput genotyping approach where distal genomic sites targeted by genome-editing methods are copied and assembled into individual DNA constructs that are compatible with next-generation sequencing strategies, thus enabling a ~$10^4$ fold increase in combinatorial library tracking capabilities. We subsequently use this multiplexed editing and testing approach to engineer organisms towards alcohol and biofuel production. Additional data will be presented tracking population diversity and mapping growth selection dynamics towards rapidly engineering alcohol tolerant organisms.

# 428093 Controlling Central Carbon Metabolism for Improved Pathway Yields

**Abstract:**

**15C13 Advances in Metabolic Engineering**

**Controlling Central Carbon Metabolism for Improved Pathway Yields**
Sue Zanne Tan & Kristala L. J. Prather

Microbial fermentation has the potential to produce valuable compounds, such as pharmaceuticals, fragrances, and fuel-like compounds, from simple and inexpensive starting materials. One of the challenges in manufacturing molecules with microbes is achieving sufficient titers and yields, such that the overall process can be economically feasible. Some common methods to improve pathway yields include overexpression of pathway enzymes, deleting competing pathways, and engineering redox and energy balances. However, these methods are not always successful due to the interconnectivity of metabolic pathways and the finite nature of cellular resources. Alternative to these traditional approaches is to control and regulate metabolic pathways in a dynamic manner. This method enables flux balance between growth and production, such that production is maximized while still maintaining high cell viability. As a proof-of-concept, we have successfully constructed a metabolite valve in *Saccharomyces cerevisiae* that provides the control of central carbon metabolism and redirection of carbon flux into a heterologous pathway. Using the tetracycline transactivator protein (tTA) system, we saw a 50-fold increase in yields of a model compound, gluconate. By varying different induction times and inducer concentrations, we obtained strains with different growth rates and gluconate titers, demonstrating the control of glucose flux either to growth or production. In addition, the successful redirection of carbon flux away from glycolysis significantly reduced the accumulation of ethanol, even in the presence of high glucose concentrations.

# 426014 Development and Application of Genetic Tools for an Industrial Strain of Solventogenic Clostridium

**Abstract:**

The production of ABE (acetone, butanol, ethanol) by the anaerobic bacteria *Clostridium* is a bioprocess capable of converting various renewable substrates into high value solvents and fuels. While ABE fermentation by *Clostridium* has been performed on an industrial scale since the early 1900s, several significant technical challenges have prevented this process from being economically competitive with the conventional production of solvents and fuels from petroleum. Among these challenges is the lack of genetic tools available for manipulating the powerful metabolism of many of these organisms. While progress has been made in developing genetics for a handful of strains over the last 25 years, a vast majority of known solventogenic *Clostridium* strains remain genetically intractable. Here, we discuss our development of a genetic transformation method for the industrial butanol hyperproducer *Clostridium saccharoperbutylacetonicum* strain N1-4. While the wild type strain has been used on the industrial scale in Japan since the 1940s, we have demonstrated, for the first time, the use of rational metabolic engineering to expand the range of high value solvents produced in this strain. While the final applications of these genetic tools are important, we believe that discussing the successes and failures encountered while developing this genetic transformation method will help others in approaching genetics in other intractable species of *Clostridium*.

# 419759 Reversing Methanogenesis to Capture Methane for Liquid Biofuels

**Abstract:**

**The world is rapidly basing much of its economy on methane deposits such as those of the Marcellus Shale in the United States. Although the promise of energy independence is enticing, many of these sites are remote and significant amounts of the methane are released into the atmosphere. This unintended release is detrimental since methane is a potent greenhouse gas, so it is imperative to convert these energy resources efficiently into liquid fuels that may be more readily transported. Since aerobic oxidation of methane is less efficient, we focused on anaerobic processes to capture methane, which are accomplished by anaerobic methanotrophic archaea (ANME) in consortia. However, no pure culture capable of oxidizing and growing on methane anaerobically has been isolated. Here we show methane can be consumed during anaerobic growth, and further converted to liquid fuels, by combining archaeal pathways. Specifically, *Methanosarcina acetivorans,* an archaeal methanogen, was metabolically engineered to take up methane, rather than to generate it. To capture methane, we cloned the DNA coding for the enzyme methyl-coenzyme M reductase (Mcr) from an unculturable archaeal organism from a Black Sea mat into *M. acetivorans* to effectively run methanogenesis in reverse. The engineered strain produces primarily acetate, and our results demonstrate that pure cultures can grow anaerobically on methane. Also, we anticipate that our metabolically-engineered strain will provide insights into how methane is cycled in the environment by Archaea as well as will possibly be utilized to convert remote sources of methane into more easily transported biofuels via acetate.**

# 430594 Development of an E. coli Co-Culture for the Improved Conversion of Phenylpropanoic Acids to Catechins in Vivo

**Abstract:**

The field of metabolic engineering has equipped scientists with the ability to harness the power of biology to produce a wide variety of difficult to synthesize plant natural products at high titers, yields, and productivities. The high-titer microbial synthesis of these compounds is possible through the careful optimization of both internal genetic elements and external fermentation conditions. Although, recent advances have shown highly efficient production to be possible, many systems still result in non-industrially feasible titers. This less than desirable production is, in part, due to the high metabolic burden of expressing extensive extrinsic pathways in a single microbial strain. This work attempts to reduce the metabolic stresses associated with expressing the 6-gene catechins pathway by separating the burden of pathway over-expression across two *E. coli* strains in co-culture. Using this method, we demonstrate up to a 400-fold improvement in the titer of green tea catechins over previous literature reports. These results highlight the potential for the use of co-culture methods to further improve microbial production of plant natural products in *E. coli*.

清华大学化学工程系邢新会教授实验室

# Engineering of Mevalonate-Synthesis Platform from Methanol in *Methylobacterium extorquens* AM1

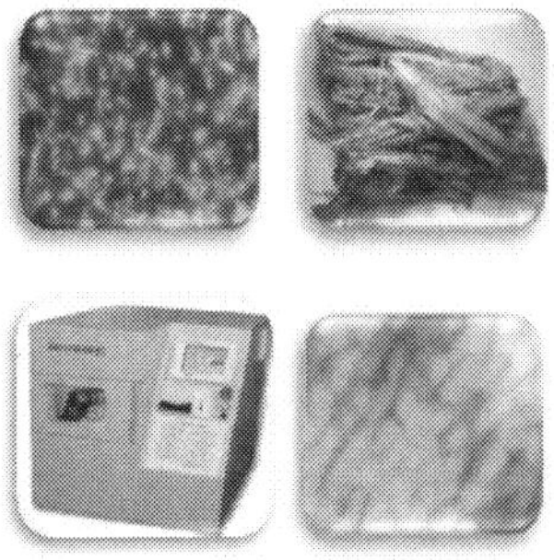

**Speaker:** Wei-Fan Liang

**Supervisor:** Prof. Xin-Hui Xing

## Microbial Cell Factory

Chemical synthesis

Microbial biosynthesis

**Advantage:**

- Pollution avoidance ;
- High substrate specificity, reduce the formation of byproducts;
- Some compounds with complex structures are difficult to synthesize by chemical synthetic routes

**Microbial Cell Factory Products**

- Energy: Ethanol、Butanol、Fatty acid esters、Alcohols、Alkanes
- Chemicals: Organic acids、Rare sugars and sugar alcohols、PHA、1,4- Butanediol, Vitamins and amino acids、Taxol

**Value-added natural Products**

- Alkaloids
- terpenoids
- Flavonoids
- polyketide
- nonribosomal peptides

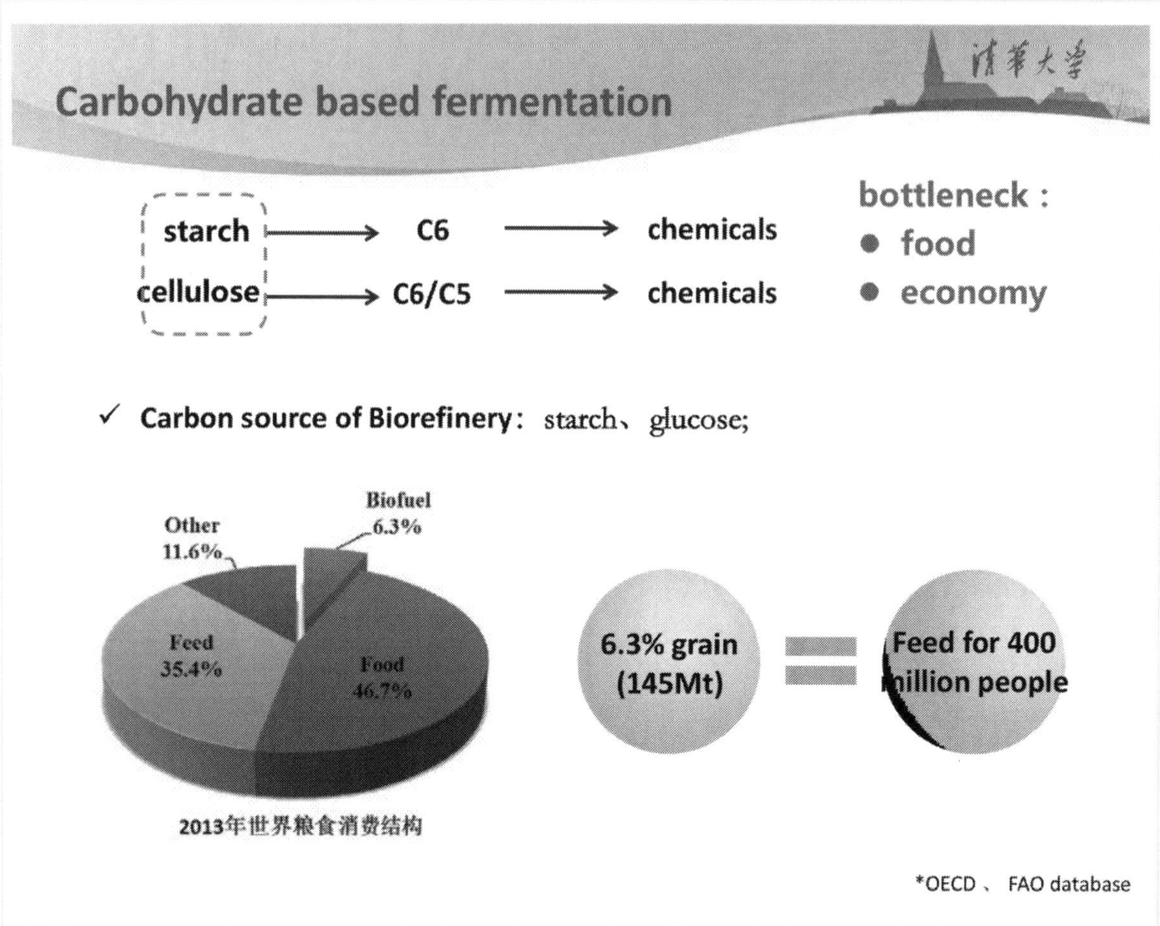
Carbohydrate based fermentation
starch → C6 → chemicals
cellulose → C6/C5 → chemicals
bottleneck :
food
economy
Carbon source of Biorefinery：starch、glucose;
Biofuel 6.3%
Other 11.6%
Feed 35.4%
Food 46.7%
2013年世界粮食消费结构
6.3% grain (145Mt)
Feed for 400 million people
*OECD 、 FAO database

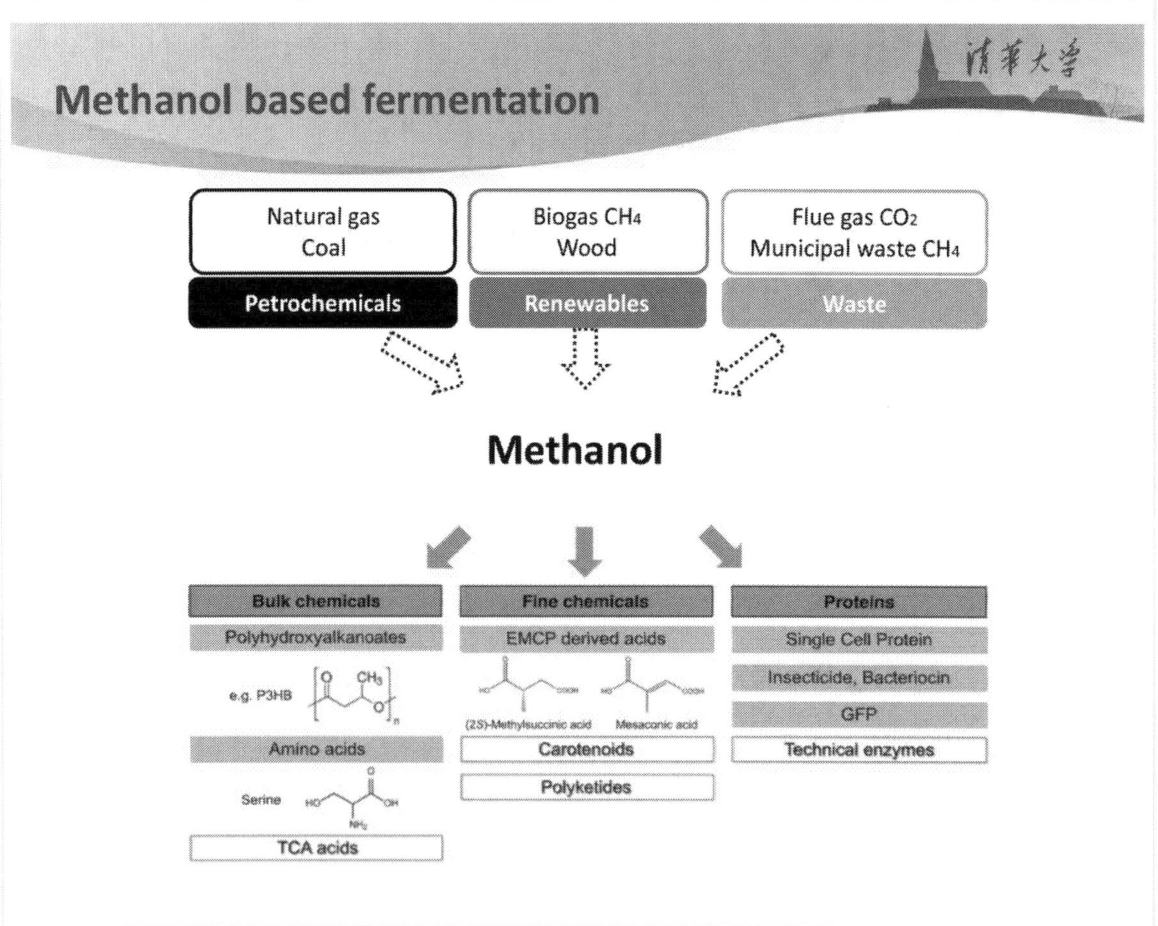
Methanol based fermentation
Natural gas
Coal
Petrochemicals
Biogas CH4
Wood
Renewables
Flue gas CO2
Municipal waste CH4
Waste
Methanol
Bulk chemicals
Polyhydroxyalkanoates
e.g. P3HB
Amino acids
Serine
TCA acids
Fine chemicals
EMCP derived acids
(2S)-Methylsuccinic acid
Mesaconic acid
Carotenoids
Polyketides
Proteins
Single Cell Protein
Insecticide, Bacteriocin
GFP
Technical enzymes

# Methylotrophy

**In nature, methylotrophic microorganisms can utilize methanol as their sole source of carbon and energy.**

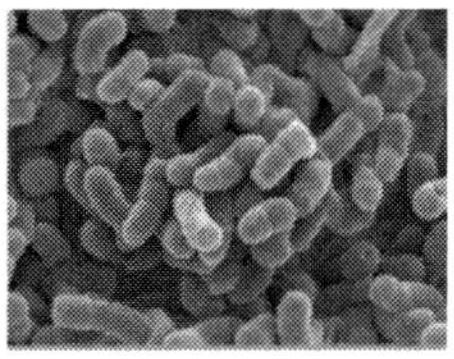

*Methylobacterium.extorquens* AM1

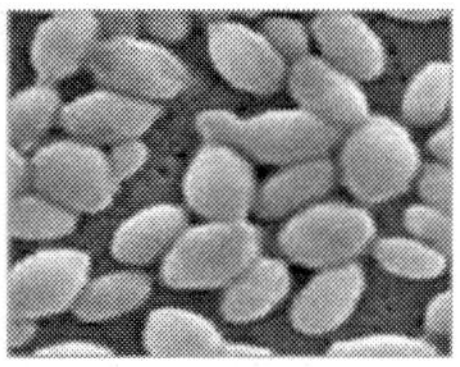

*Pichia pastoris*

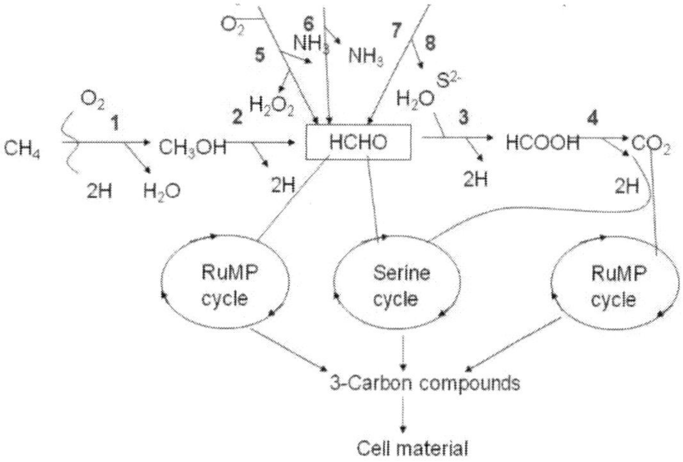


# Methanol based Microbial Cell Factory

**coal/natural gas**
**Bio-methane**
**$CO_2$**

**Methanol**

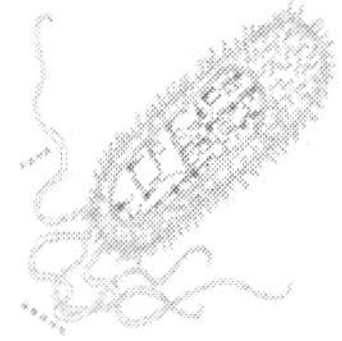

**Value added compounds**

**Methanol**

- ✓ **Non-food**
- ✓ **Sustainable supply**
  capacity 129Mt
  production 66Mt
- ✓ **Economy**
  methanol: 1900-2400yuan/ton
  starch: 3000-4000yuan/ton

**methylotrophy**

- ✓ **Model strain**
- ✓ **Methanol substrate**
- ✓ **High carbon conversion rates**
- ✓ **High cell density cultivation**

## *Methylobacterium extorquens* AM1

- **Gram-negative bacteria， α-proteobacteria ;**
- **Use methanol as sole carbon and energy source ;**
- **Best studied methylotroph with high carbon conversion rate;**

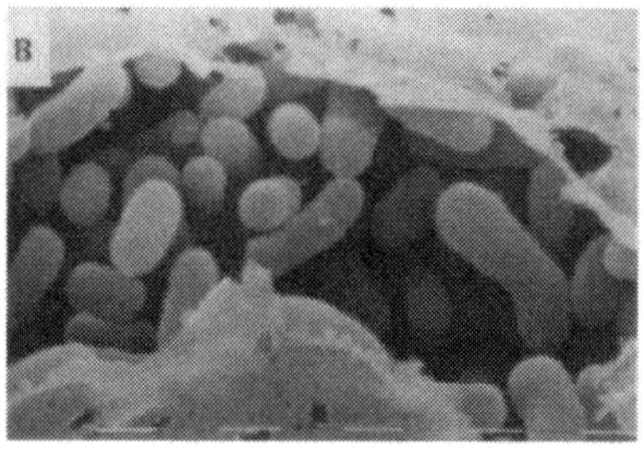

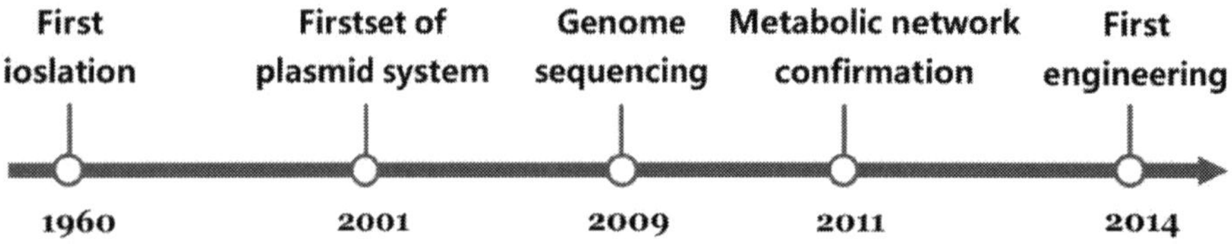


**PHA copolymers (2014.01)**
**dicarboxylic acids (2014.01）**
**Butanol (2014.11)**
**Humulene (2015.09)**

## *M. extorquens* AM1
## Metabolic network

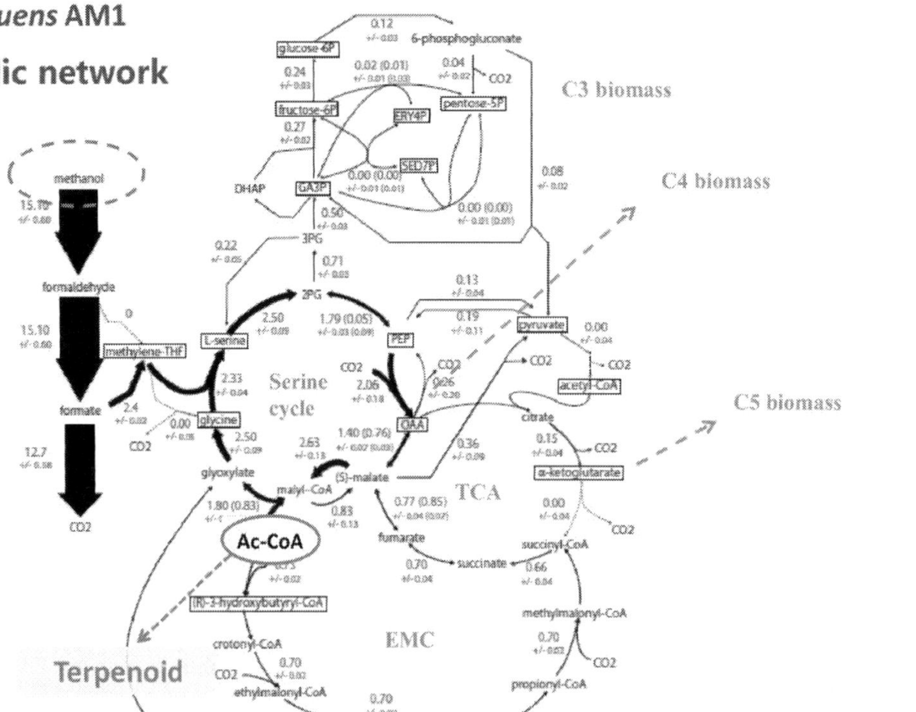


Mary et al，2011

*M. extorquens* AM1

## Metabolic network

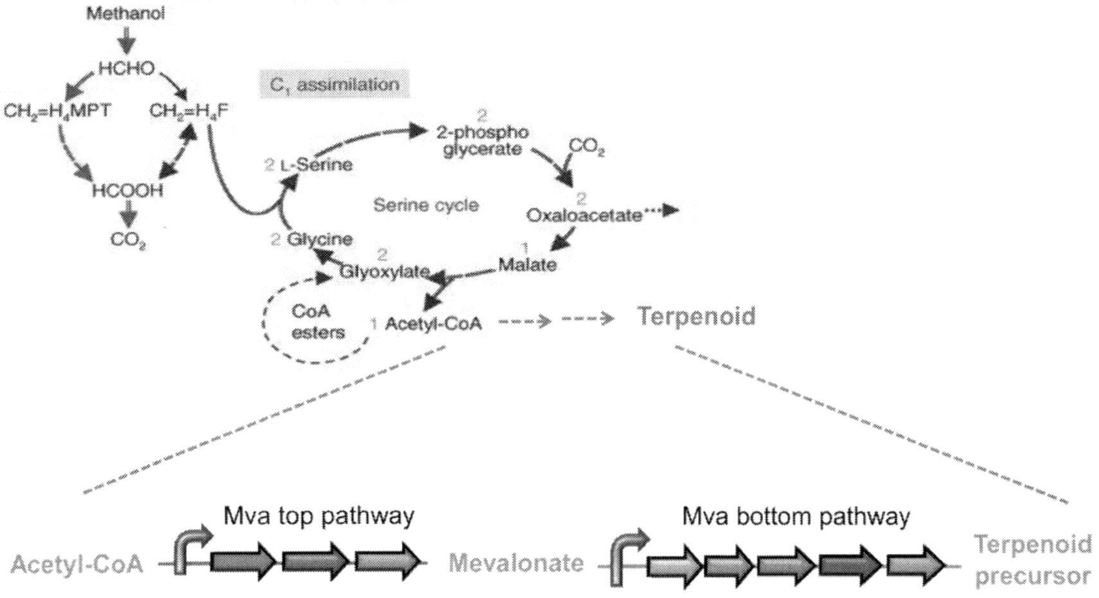


*M. extorquens* AM1

## Metabolic network

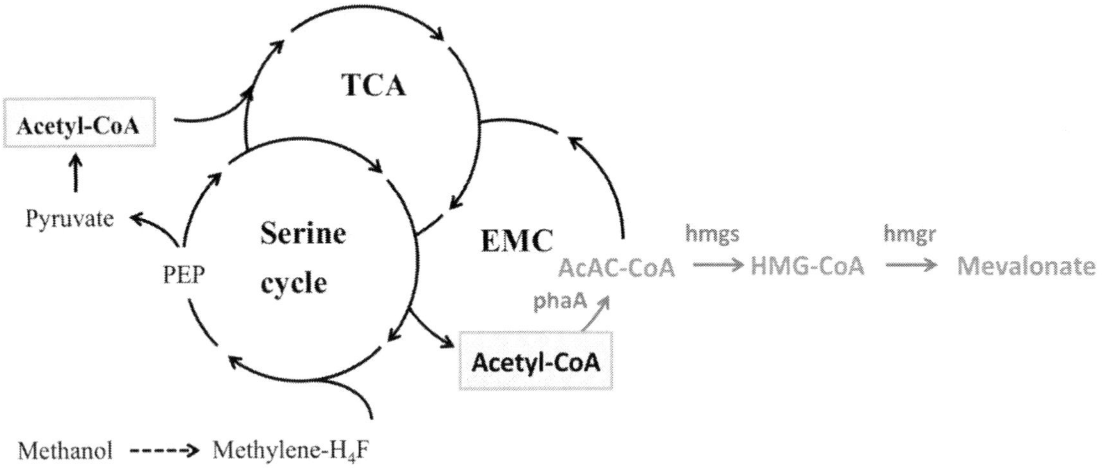

清華大學

# Part Ⅰ: Pathway engineering of MVA Top pathway

清華大學

## Gene selection

Kinetic constants of HMGSs and HMGRs used in this study

| Source and reference | HMGS | | | HMGR | | |
|---|---|---|---|---|---|---|
| | $K_m^{Ac\text{-}CoA}$ (μM) | $K_m^{app\ AcAc\text{-}CoA}$ (μM) | $V_{max}$ (eu/mg)[a] | $K_m^{HMG\text{-}CoA}$ (μM) | $K_m^{NADPH}$ (μM) | $V_{max}$ (eu/mg)[b] |
| *E. faecalis* | 350 | 10 | 10 | 20 | 30 | 2.0 |
| *B. germanica* | 15.3 | <0.5 | 40 | — | — | — |
| *T. cruzi* | — | — | — | 13±2 | 43±3 | 28±3 |

• 1eu means quantity of enzyme HMGS needed for conversion of 1 μmol AcAc-CoA per minute
• 1eu means quantity of enzyme HMGR needed for conversion of 1 μmol NADPH per minute

Operons in our work

| Operon | ***hmgs*** | ***hmgr*** | **Strain** |
|---|---|---|---|
| MVE | *E. Faecalis* | *E. Faecalis* | **AM1-mve** |
| MVH | *B. Germanica* | *T. cruzi* | **AM1-mvh** |

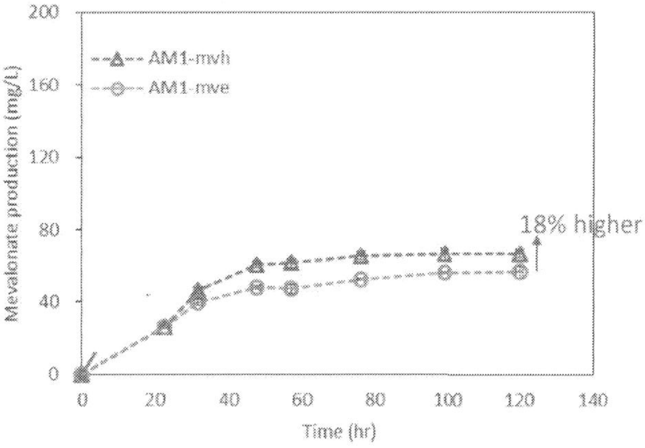


Mevalonate production of engineered strains with different operons

| Operon | ***hmgs*** | ***hmgr*** | **Strain** | Titer |
|---|---|---|---|---|
| MVE | *E. Faecalis* | *E. Faecalis* | **AM1-mve** | 56 mg/L |
| MVH | *B. Germanica* | *T. cruzi* | **AM1-mvh** | 66 mg/L |

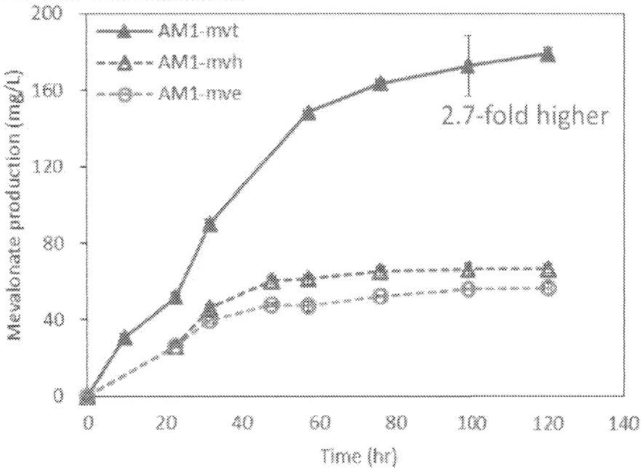


Mevalonate production of engineered strains with different operons

| Operon | ***phaA*** | ***hmgs*** | ***hmgr*** | **Strain** | Titer |
|---|---|---|---|---|---|
| MVE | —— | *E. Faecalis* | *E. Faecalis* | **AM1-mve** | **56 mg/L** |
| MVH | —— | *B. Germanica* | *T. cruzi* | **AM1-mvh** | **66 mg/L** |
| MVT | *Ralstonia eutropha* | *B. Germanica* | *T. cruzi* | **AM1-mvt** | 180 mg/L |

## phaA : RBS engineering

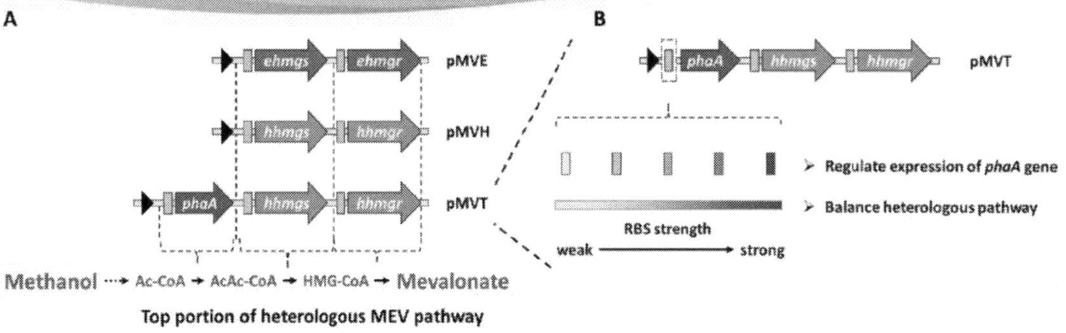


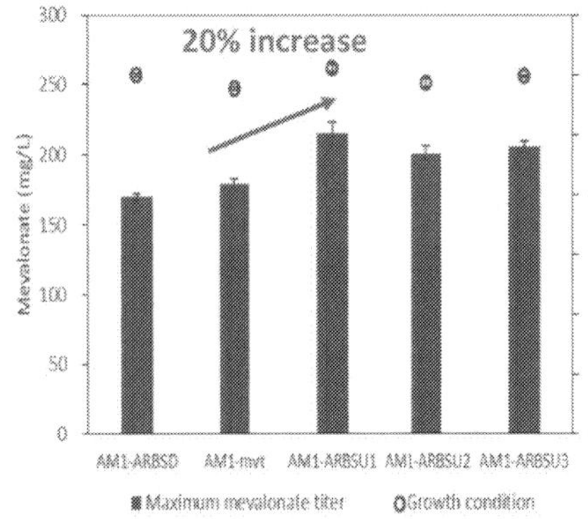


| Engineered Strain | RBS | TIR |
|---|---|---|
| AM1-ARBSD | pCM110 | 22.87 |
| AM1-mvt | $P_{alkB}$ | 93.02 |
| AM1-ARBSU1 | segB | 259.87 |
| AM1-ARBSU2 | gp46 | 768.76 |
| AM1-ARBSU3 | gp4 | 10410.21 |

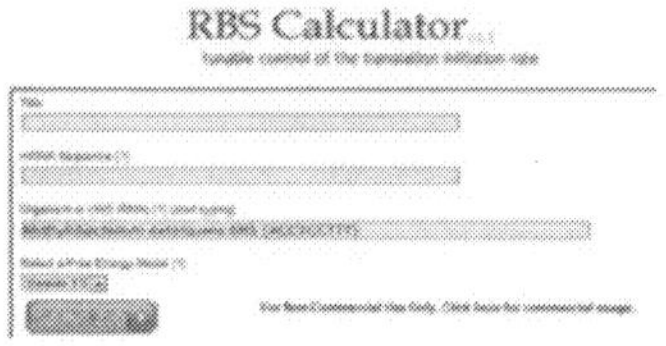


Salis,2015

## HMGR : RBS engineering

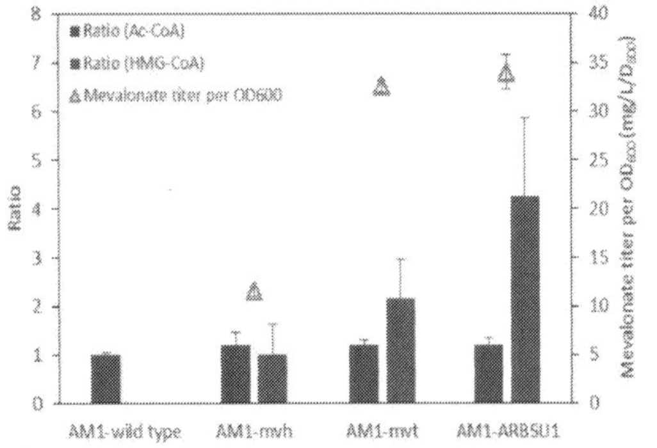


*In vivo* metabolites in AM1-ARBSU1

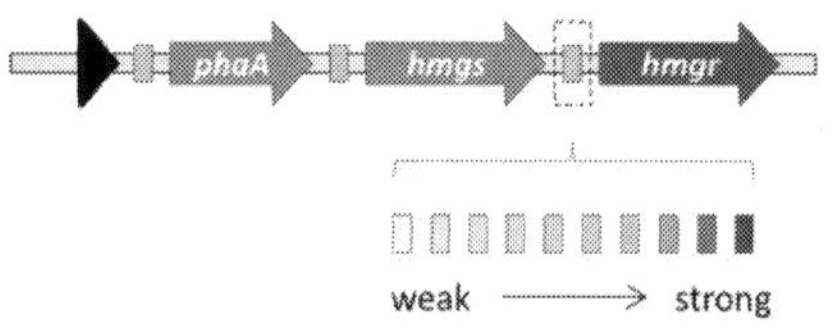


| Engineered Strain | TIR | Source |
|---|---|---|
| HMGR-RBS-1 | 4 | RBS calculator |
| HMGR-RBS-2 | 84 | |
| HMGR-RBS-3 | 3698 | |
| HMGR-RBS-4 | 7728 | |
| HMGR-RBS-5 | 48199 | |

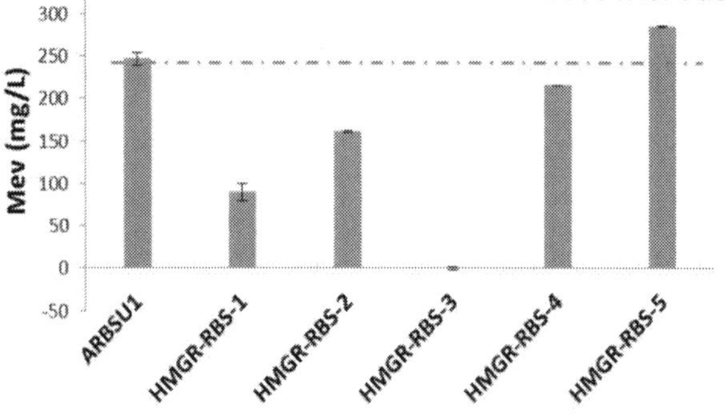

## Genome integration

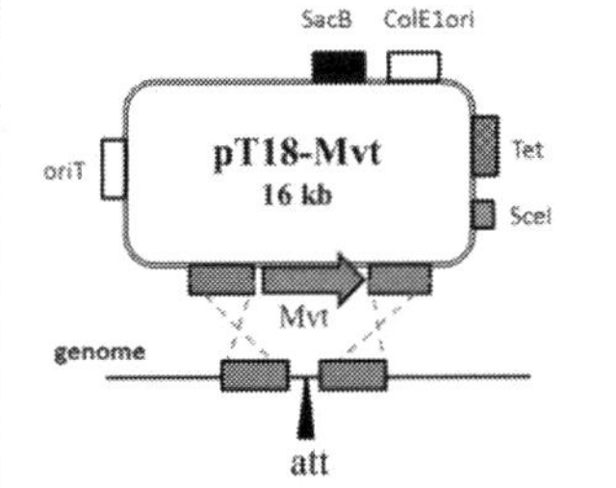


| Verification | AM1 Wild type | AM1 1st | AM1 2nd |
|---|---|---|---|
| SacB | -- | 1.4kb | -- |
| GFP | -- | 2.6kb | 2.6kb |
| ::GFPK | 3.1kb | 13kb(--) | 5.5kb |

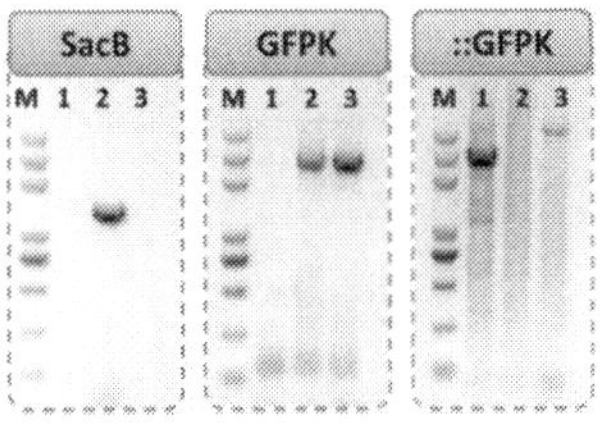


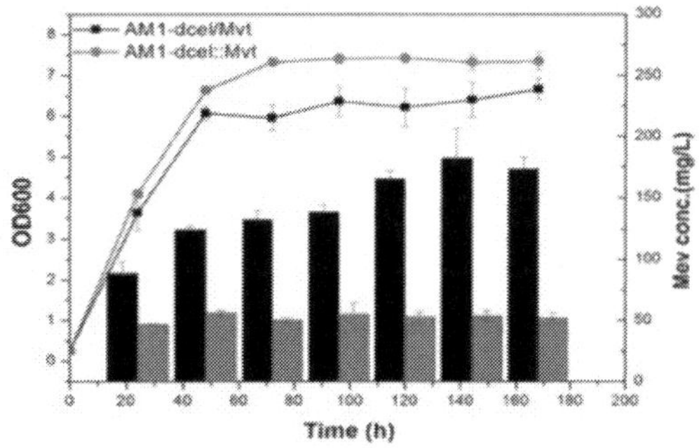


- Mvt pathway was integrated into genome (insertional expression);
- $OD_{600}$ was higher than origin, but mevalonate production was only 1/5 (50mg/L);
- Genome is one copy and plasmid is five copy number;

## Next work : Multicopy integration into genome

**Cre-loxP system from P1 phage**

1. Insert loxP sequence into genome;
2. Co –electroporation of Mva pathway with loxP and cre plasmid;
3. Screening with high concentration of antibiotic;

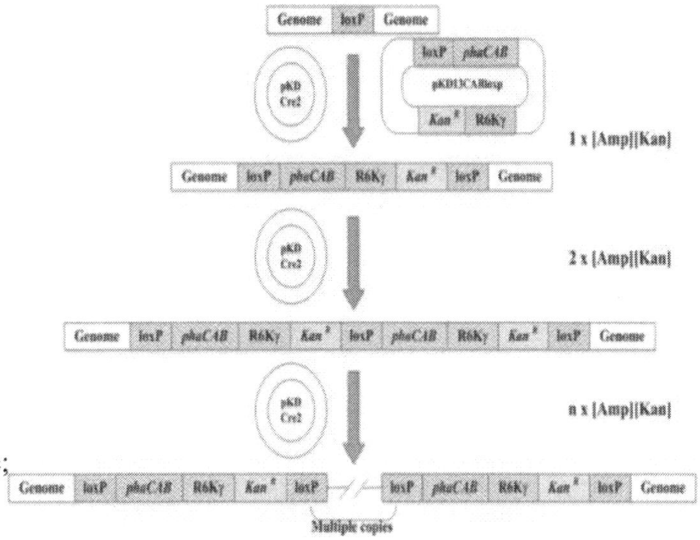


| Kanamycin (50ug/mL) | 1X | 1.5X | 2X | 3X | 4X | 5X | 6X | 7X | 8X |
|---|---|---|---|---|---|---|---|---|---|

**AM1::Kana**

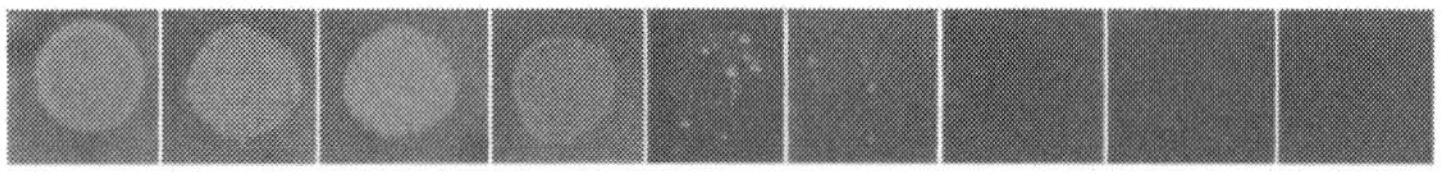

## Part II: Regulation of Acetyl-CoA accumulation

## *qscR* regulation

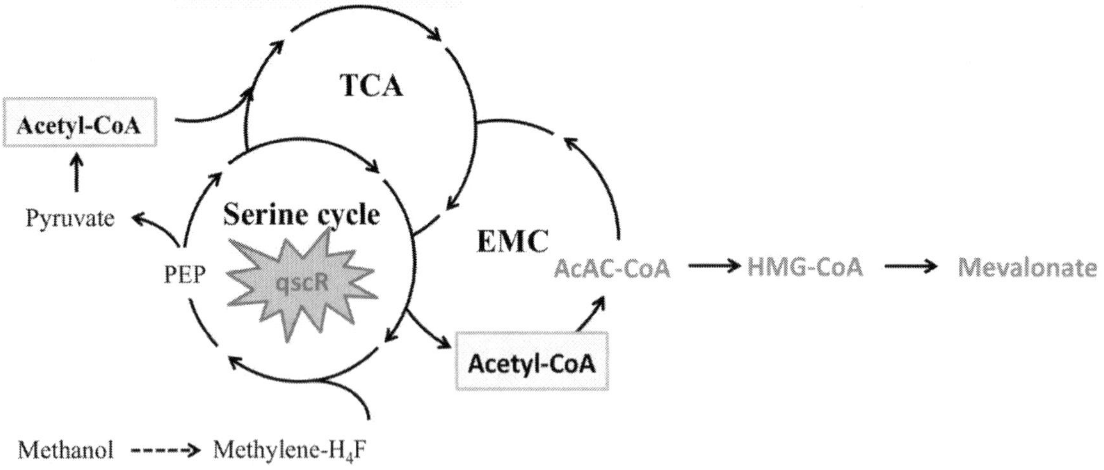


**qscR library (error prone PCR)** → Mevalonate biosensor → **High throughput screening**

## Biosensor

**Mev-sensor is based on arabinose operon**

**AraC-Mev（T24L,H80L,Y82L,H93R）**

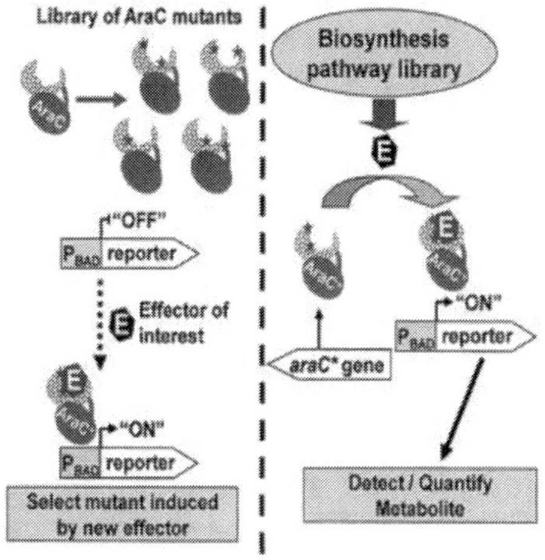


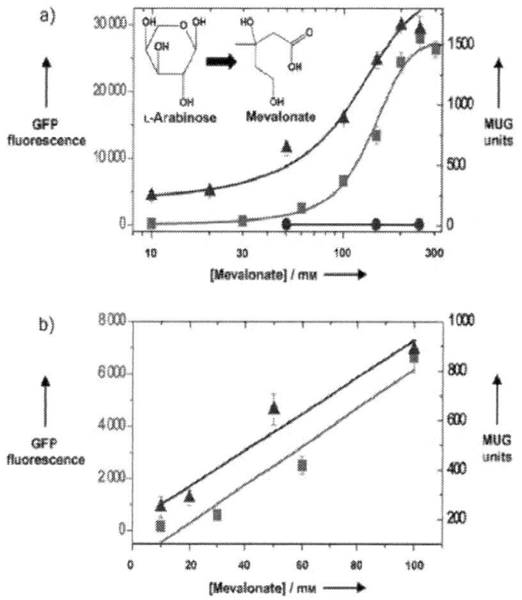


Patrick, 2010

## Mev-sensor mechanism ( Bidirectional Regulation )

**No mevalonate:**

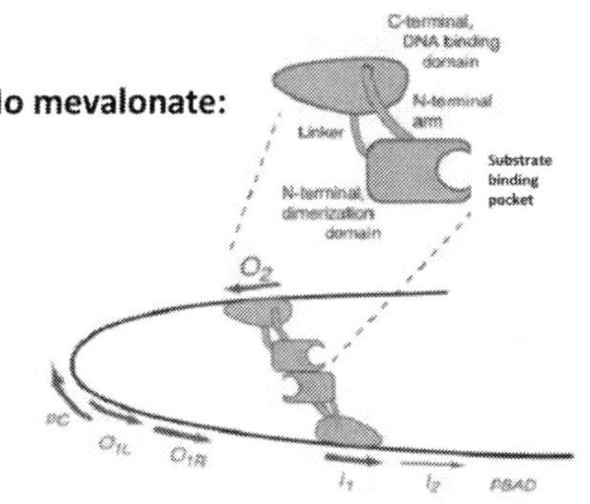


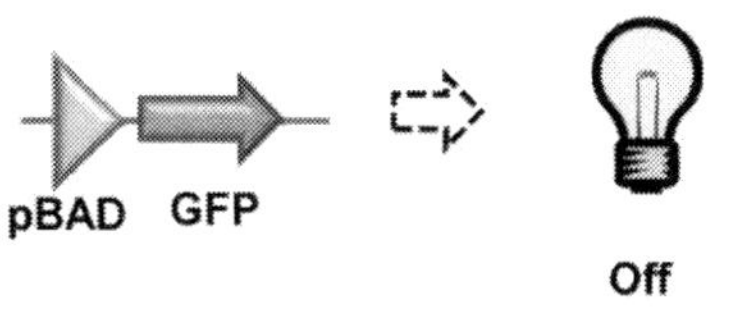


**Add mevalonate:**

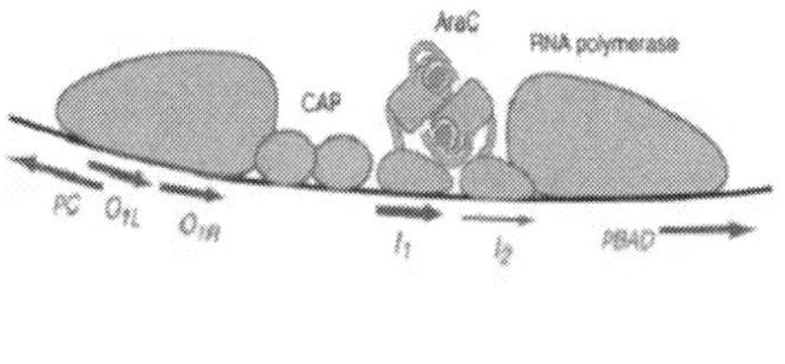


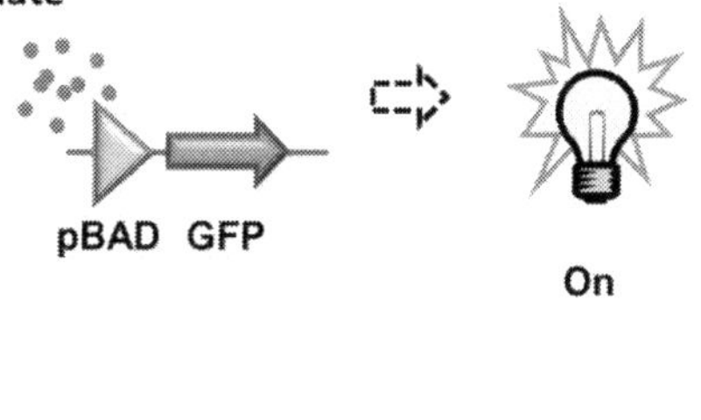

## Mev-sensor reconstruction

### Difficulty：cross-species construction of functional element

***M.extorquens*** **AM1：α-proteobacteria;**

***E.Coli*** **：γ-proteobacteria;**

- **Most promoters of E.coli can not work in *M.extorquens* AM1;**
- **Common replicons of E.coli can not work in *M.extorquens* AM1;**

Original sensor
Bidirectional regulation

Reconstructed sensor
Unidirectional regulation

## ① modeling

**Relationship between araC concentration and fluorescence:**

$$\begin{cases} [Mev\cdot araC]=K_1[Mev][araC]; \\ [DNA\cdot araC]=K_2[DNA][araC]; \\ \\ [DNA]_{Total}=[DNA]+[DNA\cdot araC]; \\ [Mev]_{Total}=[Mev]+[Mev\cdot araC]; \\ [araC]_{Total}=[araC]+[Mev\cdot araC]+[DNA\cdot araC] \\ \\ [Fluorescence]=[GFP]=K_3[Mev\cdot araC]\ [DNA] \end{cases}$$

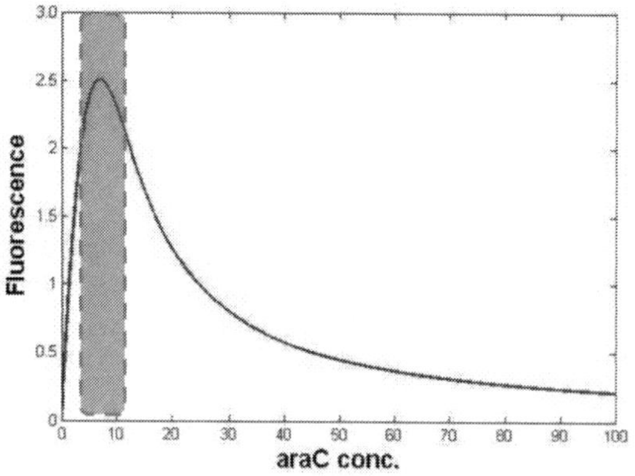


Set $K_1=K_2=K_3=1$, $[DNA]_{Total}=1$, $[Mev]_{Total}=10$;

- Reconstructed Mev-sensor was araC expression dependent model;
- araC concentration had great influence on fluorescence output.

## ② design

Principle： low araC expression

1. Express araC by replacing Pc promoter with a functional promoter；
2. Weaken araC expression by RBS;
3. Reduce leakage；
4. Other optimization;

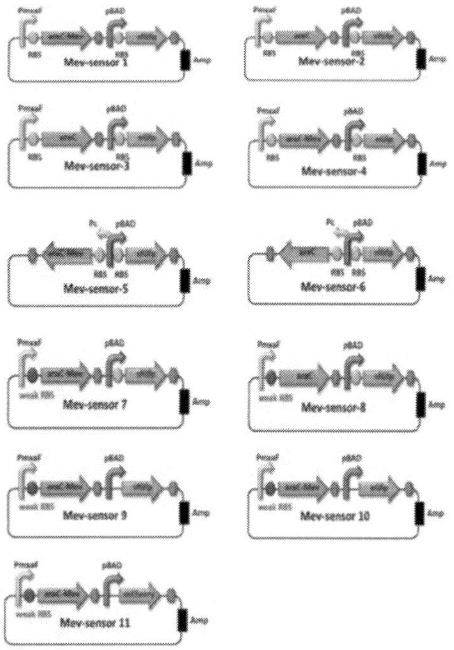

**RBS calculator**

| RBS | Translation Initiation Rate (au) |
|---|---|
| RBS-1 | 358.88 |
| RBS-2 | 126.75 |
| RBS-3 | 25.98 |
| RBS-4 | 6.74 |

## ③ result（plasmid）

**Intracellular response evaluation ：mevalonate was insertional expressed, and detected RFU/OD600.**

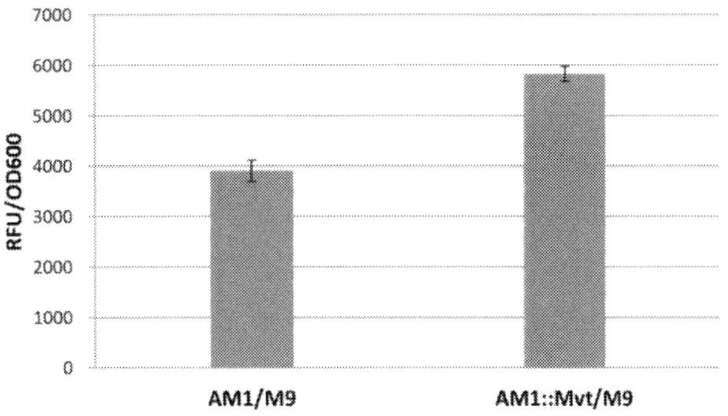


- Mev-sensor-9 can response intracellular mevalonate and induction ratio was 1.60;

## ④ result ( genome )

**Insertional response**

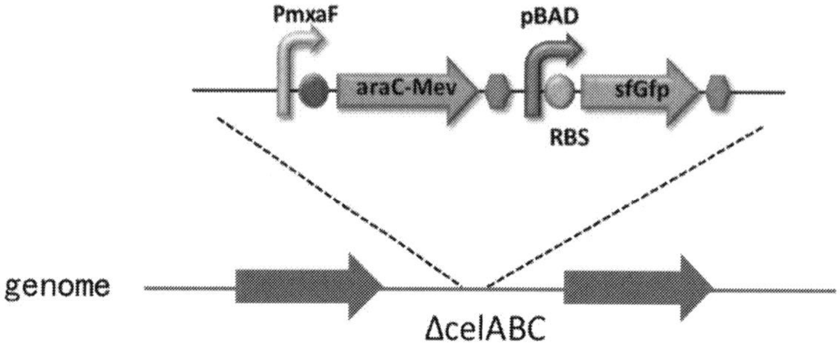


**araC** ↓ **Fluorescence** ↑

**DNA** ↓ **Fluorescence** ↓

## ④ result ( genome )

**Insertional response**

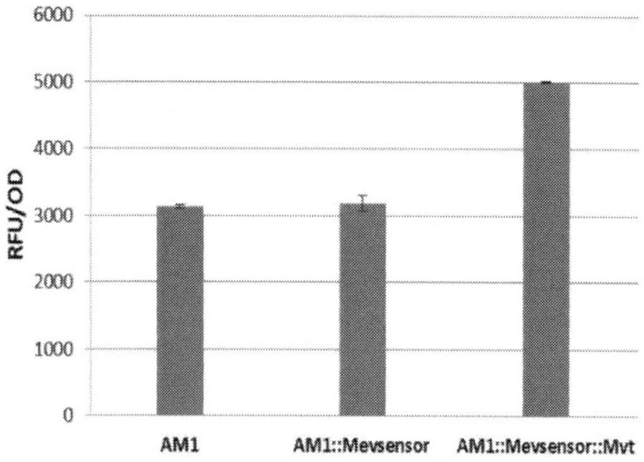


- AM1Δatt::MvtΔcekABC::Mevsensor (S1) can response intracellular mevalonate and induction ratio was also 1.60;

## Next work : High throughput screening

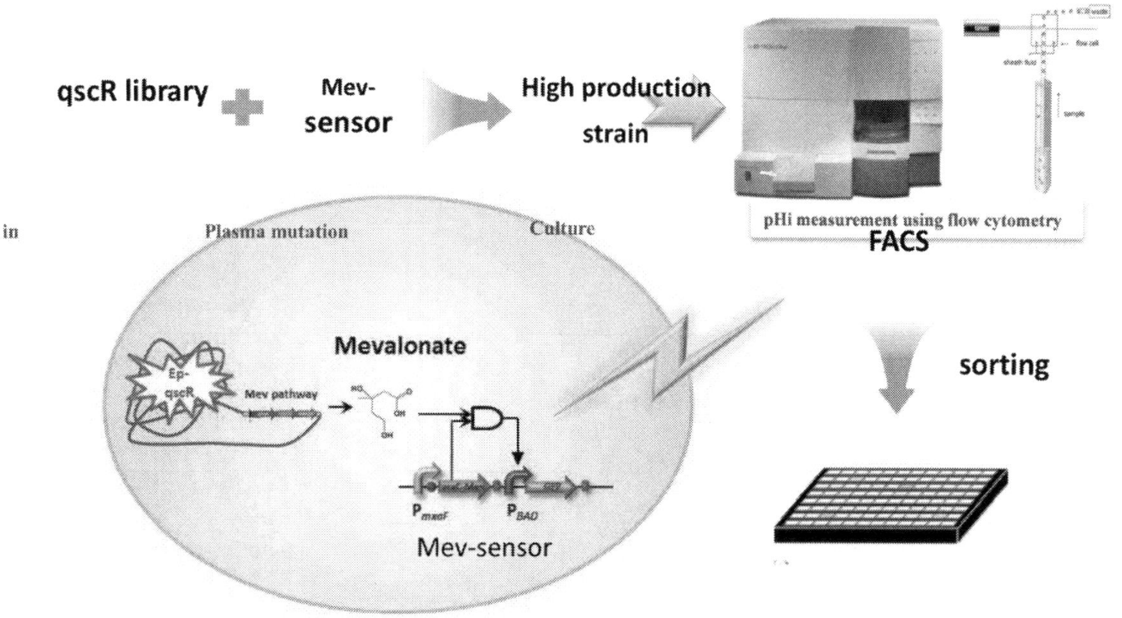


# Thank you ！

# 435301 Designing De Novo Overall Stoichiometric Conversions and Intermediate Metabolic Reactions

**Abstract:**

Existing computational tools for the *de novo* pathway assembly, either based on mixed integer linear programming techniques or graph-search applications, generally only find linear pathways connecting the source to the target metabolite. The overall stoichiometry of conversion along with alternate co-reactant (or co-product) combinations is not part of the pathway design. Therefore, global carbon and energy efficiency is in essence fixed with no opportunities to identify more efficient routes for recycling carbon flux or redox equivalents closer to the thermodynamic limit. In this work, we introduce a stoichiometry and pathway design tool that first identifies an optimum overall stoichiometry by exploring exhaustively co-reactant/co-product combinations while meeting thermodynamic feasibility. Following the selection of the overall reaction, intervening reactions can be identified that link the chosen reactants and products in the desired stoichiometric ratios. Requirements on the minimality of size of the network and negativity of the free energy of change of each individual reaction step can also be imposed to help rank-order all identified designs. Implementation for recent pathway design studies identified non-intuitive designs with improved efficiencies. Specifically, multiple alternatives for non-oxidative glycolysis are generated and non-intuitive ways of co-utilizing carbon dioxide with methanol are revealed for the production of $C_{2+}$ metabolites with higher carbon efficiency.

# 431422 Integrative Analysis of Gene Co-Expression and Metabolic Networks in Microbial Communities: Organizational Properties of a Unicyanobacterial Consortium

**Abstract:**

Biological networks often show remarkable robustness by preserving their functionalities against perturbations. Robust networks are known to possess specific topological characteristics such as being scale-free, meaning that they can tolerate random disruption of genes (nodes), but are vulnerable to targeted removal of highly connected genes (hubs). Case studies in the literature have reported such topological signatures in gene regulatory, protein-protein interaction, and metabolic networks of single organisms, but, as yet, little is known regarding general organizational principles of microbial communities. The main reason for this is the lack of reliable network models for complex communities. Here, we set out to identify the structural properties of a unicyanobacterial consortium by viewing it as a network of genes. For this purpose, we collect a compendium of gene expression data of this consortium across multiple time points and conditions and reconstruct a compartmentalized gene co-expression network. Using a platform previously developed by our group, we then combine the reconstructed gene co-expression network with a community-scale metabolic network. This integrative analysis offers an opportunity to reveal the interrelationship between topological and functional roles of genes across multiple interacting organisms.

# 417510 Spatiotemporal Metabolic Modeling of Clostridium Ljungdahlii for Renewable Alcohol Production from Synthesis Gas

**Abstract:**

One of the most promising routes to renewable liquid fuels and chemicals is the fermentation of synthesis gas (syngas; mainly comprised of $H_2$/CO/$CO_2$) streams by specialized bacteria to synthesize desired products such as ethanol and 2,3-butanediol. The most commonly studied syngas fermenting bacterium is *Clostridium ljungdahlii*, and a genome-scale reconstruction of *C. ljungdahlii* metabolism has recently been published. While commercial development of syngas fermentation technology is underway, many research problems must be addressed to further advance the technology towards economic competitiveness. Commercial development efforts are currently focused on bubble column reactors due to their superior gas-liquid mass transfer characteristics and enhanced operational flexibility.

In this presentation, we describe a spatiotemporal metabolic model of syngas bubble column reactors that combines the *C. ljungdahlii* genome-scale reconstruction with multiphase transport equations that govern convective and dispersive processes within the spatially varying column. Our solution procedure involves spatial discretization of the partial differential equation model followed by numerical integration of the resulting system of ordinary differential equations with embedded linear programs using DFBAlab, a MATLAB code that performs reliable and efficient dynamic FBA simulations. We show that the model is able to predict biomass, dissolved gas, and secreted ethanol and acetate concentrations in a spatially and time resolved manner. To expand the range of metabolic byproducts that can be synthesized, we utilize the metabolic reconstruction and the OptKnock framework to identify gene knockouts that lead to the production of 2,3-butanediol and butanol as well as the overproduction of ethanol. These knockout strategies are further screened using our syngas bubble column reactor model to access their relative performance under realistic fermentation conditions. The simulation results provide new metabolic engineering targets for *C. ljungdahlii* that can be experimentally tested with the continued development of genetic engineering tools.

In addition to providing new insights into bottlenecks to biochemical production in syngas bubble column reactors, the study established a new paradigm for formulating and solving genome-scale metabolic models with both time and spatial variations.

# 421173 Hybrid Agent-Based Framework Towards Virtual Cell Culture Bioreactors

**Abstract:**

Manufacturing complex therapeutic proteins requires unique operational conditions and highly specialized process knowledge to obtain consistent product quality and productivity. Optimization and control of these processes are challenging due to complexity involved in the interactions between mammalian cell metabolism and the bioreactor environment. These interactions become even more critical for large-scale bioreactors where cellular and fluid phases are heterogeneous.

Understanding hydrodynamic influence on culture viability and productivity is crucial. Mammalian cells are highly sensitive to shear stress and its damaging effect. In order to reduce the shear exposure of cell, mammalian bioreactors often operate with low agitation speeds that may cause further increase in the presence of spatial gradients and mass transfer limitations. This in term may further impact cell cycle of the individual cells, leading to sub-optimal growth.

Due to the complexity and heterogeneity involved in the culture environment, conventional mechanistic modeling efforts are often incomplete when used to describe the interactions of cell physiology and environmental conditions. Hybrid computational models provide a better approach for studying mammalian cell culture bioreactor processes where environmental factors and cellular interactions may be coupled. Agent-based modeling is a natural choice to account for the heterogeneity in the cell population where agents (cells) take action based on changing dynamics of their immediate vicinity. These models provide more realistic description of process when coupled with computational fluid dynamics (CFD) models. An ABM was previously developed to simulate individual mammalian cell behavior and its cell cycle regulation in response to dynamic bioreactor conditions (Bayrak et al., 2015). In the current study, this model has been further developed and combined with a CFD model to account for the mass transport of nutrients and oxygen and shear stress on individual cells.

Reference

Bayrak, E. S., Wang, T., Cinar, A. & Undey, C. Computational Modeling Of Fed-Batch Cell Culture Bioreactor: Hybrid Agent-Based Approach. International Symposium On Advance Control Of Chemical Processes, June 7-10 2015 Canada.

# 422442 A Model-Based Strategy for Understanding and Improving Recombinant Protein Glycosylation in CHO Cells

**Abstract:**

Glycoprotein therapeutics represent an increasing sector of the pharmaceutical industry. Glycoproteins are large complex molecules, which are heterogeneous in terms of their glycan moieties. This heterogeneity arises from a sequence of complex post-translational enzymatic modifications that take place in the ER and Golgi apparatus and play an important role in drug safety, efficacy and half-life. It has been shown that the process of protein glycosylation can be partially driven by the intracellular availability of nucleotide sugars (NSs), which are the co-substrates to the glycosylation reactions. It is also known that the availability of the NSs is affected by process conditions including cell culture mode, temperature, dissolved oxygen tension, nutrient availability and addition of precursor molecules [1,2].

In this work, we describe a model-based platform for understanding the culture factors that affect recombinant protein glycosylation in Chinese hamster ovary (CHO) cells, the workhorse of the biopharmaceutical industry. Our platform constitutes a multi-scale model describing the process at three different scales and through different approaches. Specifically, the higher scale model is a cell culture dynamics (CCDyn) model, which describes cell growth and the variation of the nutrient and product concentrations over time in the extracellular medium. The specific production and consumption rates ($q_i$) in the extracellular medium are used as inputs for an intracellular compartment (cytosol) model.

The cytoplasmic compartment has been described following different approaches. The first approach, termed the Jedrzejewski model [3], consists of a semi-structured purine and pyrimidine synthesis network, which describes the synthesis of nucleotides based on the concentrations of extracellular metabolites, and a structured and mechanistic representation of the NSD synthesis pathway, based on the murine metabolic network as depicted in the Kyoto Encyclopaedia of Genes and Genomes and depending on the concentrations of sugars in the extracellular medium. The latter model follows the bottom-up approach because the kinetic pathway has been described in its

completeness according to the individual enzyme data as found in BRENDA.

On the contrary, the second approach known as the del Val model follows a top-down modelling method because the kinetic pathway has been reduced through the adoption of a macroscopic reaction scheme (macroreactions). Both models describe the temporal profiles of nucleotides as well as NSDs concentrations in the cytoplasm in response to extracellular feeding. They both are theory-based models that are based on biophysical equations that require many, sometimes difficult-to-measure kinetic parameters.

For this reason, the intracellular compartment has been described also through a fully determined dynamic metabolic flux model (dMFA) which describes, in a parameter-free manner, the most significant metabolite fluxes in the cytoplasm comprising a more reduced version of a genome-scale model. The metabolic flux network consists of 55 material balances and 55 metabolite fluxes. Eight NSs and outlets for host cell protein N-linked and O-linked glycans as well as mAb glycosylation have been considered. The outputs are the intracellular metabolite fluxes varying with respect to time as a function of changes in the extracellular environment.

The lowest scale model is the del Val glycosylation model [4] describing the N-linked glycosylation process of the antibody heavy chain through the Golgi apparatus viewed as a plug flow reactor. Given the concentrations or fluxes of the NSDs from the higher scale model in the cytosol, this model is capable of reproducing glycosylation profiles of commercial mAbs.

Our platform has been validated against experimental data from two IgG-producing CHO cell lines representing all aforementioned scales. The platform allows one to capture the effect of different amino acid feeding strategies [5], as well as that of hexose and nucleotide precursor additions to culture media and their impact on NSD concentrations in the first place, and consequently on the product glycoform. Finally, based on this platform, we present the computational design of novel strategies for improving the glycan profile of antibody products and their experimental implementation.

**References**

1. del Val, I. J., Kontoravdi, C. and Nagy, J. M. (2010), Towards the implementation of quality by design to the production of therapeutic monoclonal antibodies with desired glycosylation patterns. *Biotechnol. Prog.*, 26: 1505–1527.
2. Sou, S. N., Sellick, C., Lee, K., Mason, A., Kyriakopoulos, S., Polizzi, K. M. and Kontoravdi, C. (2015), How does mild hypothermia affect monoclonal antibody glycosylation? *Biotechnol. Bioeng.*, 112: 1165–1176.
3. Jedrzejewski, P. M., del Val, I. J., Constantinou, A., Dell, A., Haslam, S. M., Polizzi, K. M., and Kontoravdi, C. (2014), Towards Controlling the Glycoform: A Model Framework Linking Extracellular Metabolites to Antibody Glycosylation. *Int. J. Mol. Sci.*, *15*: 4492-4522.
4. del Val, I. J., Nagy, J. M. and Kontoravdi, C. (2011), A dynamic mathematical model for monoclonal antibody N-linked glycosylation and nucleotide sugar donor transport within a maturing Golgi apparatus. *Biotechnol. Prog.*, 27: 1730–1743.
5. Kyriakopoulos, S. and Kontoravdi, C. (2014), A framework for the systematic design of fed-batch strategies in mammalian cell culture. *Biotechnol. Bioeng.*, 111: 2466–2476.

# 431013 Metabolic Flux Analysis Using Isotope Labeled Metabolites to Understand High Lactate Phenotype in a Fed-Batch Cell Culture Process

**Abstract:**

Empirical approaches have been widely used in cell culture development to identify key variables that improve cell culture process productivity. These approaches, although useful and in some cases successful, do not lead to an understanding of the underlying mechanism and can result in unintended consequences. Here we report a chemically defined fed-batch antibody production process using CHO cells, which is reproducible with respect to growth and titer but variable in its metabolic phenotype, particularly with respect to glucose consumption and lactate production rates. In this work we investigate the pathways underlying the variability in the metabolic phenotype by profiling the levels of major extracellular and intracellular metabolites and their mass isotopomers at various phases of the cell culture process.

The insights obtained from the isotopic labeling experiments were used to update and refine a previously developed kinetic model, which was then fitted to the metabolite time course data of the culture. Our results to date show good agreement between the simulated and experimentally measured metabolite profiles as well as overall culture viability and product titer.

# 427968 Application of 13C Flux Analysis to Identify High-Productivity CHO Metabolic Phenotypes

**Abstract:**

Industrial bioprocesses place high demands on the intermediary metabolism of host cells to meet the biosynthetic requirements for maximal growth and protein expression. Identifying host cell metabolic phenotypes that promote high recombinant protein titer is a major goal of the biotech industry. $^{13}C$ metabolic flux analysis (MFA) provides a rigorous approach to quantify these metabolic phenotypes by applying isotope tracers to map the flow of carbon through intracellular metabolic pathways. We have previously performed $^{13}C$ labeling experiments and MFA to characterize the metabolism of a IgG-expressing DHFR host (Amgen) during four separate phases of a fed-batch culture designed to closely represent industrial process conditions. We found that peak specific growth rate during early exponential phase was associated with high lactate production and minimal citric acid cycle (CAC) flux. Conversely, we found that lactate metabolism switched from net production to net consumption as the culture transitioned from peak growth to peak IgG production. During stationary phase when IgG production peaked, energy was primarily generated through CAC and oxidative phosphorylation, which was associated with elevated oxidative pentose phosphate (OPPP) flux. Overall, we found that an oxidative metabolic phenotype corresponded with peak IgG production, whereas peak cell growth was characterized by a glycolytic metabolic state.

We have recently examined ten CHOK1SV (Lonza) clones cultured in 3-liter fed-batch bioreactors, to assess their metabolism during stationary phase. Six of the clones used the GS system to express one of three different IgGs. Four of the clones were genetically manipulated to be apoptosis-resistant by expressing Bcl-2Δ. Hierarchical clustering was performed to assess correlations amongst the measured fluxes of all nine clones. The three non-producing control lines CHOK1SV, Bcl-2Δ, and CHOK1SV-GS clustered together. Likewise, the six antibody producing clones clustered together and were separated by host background (Bcl-2Δ or CHOK1SV). The lactate dehydrogenase (LDH) flux was most closely associated with specific IgG productivity: as IgG productivity increased, lactate production decreased. Additionally, elevated CAC fluxes corresponded strongly with increased specific productivity. Taken together, these studies indicate that oxidative metabolism is enhanced in high-producing CHO cell lines. This presentation will discuss the central metabolic trends observed among both GS and DHFR expression systems as a means to provide potential metabolic engineering strategies to further enhance IgG productivity and titer of industrial CHO hosts.

# 427565 Development of a Reactive Slurry-Slurry Approach to Improve Crystallization Process Performance and Scalability

**Abstract:**

The crystallization of a potent compound, hereafter refered to as compound A for the sake of confidentiality and currently in development at GlaxoSmithKline proved to be challenging in terms of obtaining acceptable yield, consistent and desirable API powder properties, and a good process robustness and scalability. In order to address these issues, the salt formation crystallization was redesigned and a relatively novel approach was developed that allowed significant improvement in yield (from ≈70% to ≈95%), consistent API powder attributes within the desired range and an improved process manufacturability. This presentation describes the approaches and the process design of the crystallization recipe which utilizes a seed bed of the base as mitigating factor to prevent spikes in supersaturation with respect to the salt. The new process design resulted in a relatively novel crystallization path, which is a direct result of the solubility behaviors of both the base and the salt. The process was scaled-up successfully in the pilot plant to manufacture material for clinical supply.

# 427844 Effect of Seed Type on the Resistance of API Agglomerates to Agitated Drying


**Abstract:**

Control of the Active Pharmaceutical Ingredient's (API) powder properties is often necessary to ensure successful drug product development and manufacture. This is, in part, accomplished by designing a crystallization process that can deliver consistent batch-to-batch powder properties. When an API forms agglomerates during the crystallization, the resulting powder often exhibits enhanced flow properties and bulk density, which are of benefit to the formulation. However, development of a crystallization to consistently produce agglomerates that can withstand downstream processing, particularly agitated drying, represents a significant challenge. Weak agglomerates that are prone to break during agitated drying increase the risk of batch-to-batch powder property variability and scale dependence.

This presentation describes the development of a crystallization process that consistently yields API agglomerates which are robust to agitated drying. During development, the impact of several factors on agglomerate formation and strength was evaluated, and a mechanism of agglomeration was proposed. These studies showed that seed type has a strong impact on agglomerate properties and, subsequently, on the extent of agglomerate breakage during drying. Seeds with varying properties were generated via wet milling, jet milling, and hammer milling, and evaluated in the crystallization on laboratory and pilot scale. Filter, conical, and helical dryers were used on pilot and manufacturing scale to investigate the propensity of agglomerates made from the different seed types to withstand agitation. Results from the studies enabled definition of a robust manufacturing process and powder property control.

# 429782 Continuous Nanoparticle Production Using Focused Acoustics


**Abstract:**

### Continuous nanoparticle production using focused acoustics

Many drugs currently in development are either water insoluble or lipophilic, thus creating formulation challenges to administer these drugs at a high dose. In some situations, nanoparticles are an appropriate formulation strategy as a result of their intrinsic properties such as small shape and size (e.g., radius) and large surface area to volume ratios. Nanoparticles can be generated by grounds up crystallization, which is typically processed in batch mode. One basic approach of crystallization is to vary the solubility of the solute of interest with different methods, such as vary the ratio to non-solvent vehicle (anti-solvent).

Low frequency, unfocused ultrasound has previously been demonstrated for improving crystallization kinetics. However, unfocused ultrasound has efficiency limitations due the inability to focus the energy, creating issues in thermal management such as non-uniform mixing, and uncontrolled thermal fields. Additionally, these unfocused solutions are not able to provide sufficiently tight, uniform nucleation control at the chemical solvent boundary interface that is required for nanoparticle production. The Covaris Adaptive Focused Acoustics™ (AFA) technology is a highly controlled mechanical based process that provides a practical tool to precisely control both the nucleation formation and the rate of crystal growth. The high energy efficiency of AFA avoids excessive heat generation, reducing potential thermal damages to the sample and allowing greater control over thermally dependent chemical reactions. Moreover, the precisely controlled energy delivery improves the reproducibility of the sample processing for higher quality control.

As a model of nanoparticle formation, a Solvent:Anti-solvent crystallization method with an active AFA energy field was tested in different batch volumes as well as in a continuous flow format. Cinnarizine was used as a model

th

compound. Nanoparticles were measured using a Malvern Mastersizer 3000, and particle sizes were reported for 10 percentile, 50$^{th}$ percentile, and 90$^{th}$ percentile populations (d10, d50, and d90 populations respectively).

Initial studies were performed in triplicate on batch scale processing of 2 mL samples. Parameters evaluated include:

| | |
|---|---|
| Solvent | Dimethylacetamide (DMA) |
| Drug solvent concentrations | 50 mM, 125 mM, 250 mM |
| Stabilizer | 0.2% PVP30 and 0.25 mM SLS |
| Stabilizer percentage | 0.2% 1.5%, 3.0% |
| Solvent:Anti-solvent ratio | 1:29, 1:14, 1:6.6 |

For comparison, a control was run at each setting. The control was mixing 2 mL samples via a magnetic stir bar in a 20 mL scintillation vial for the same time duration of 5 minutes.

In all cases, the particle sizes of the AFA-processed samples were lower and more reproducible than the control. A typical comparison is shown in the following tables:

| | |
|---|---|
| S220x | Covaris Focused-ultrasonicator |
| 300 W | Peak Incident Power (PIP) |
| 50% | Duty Factor (DF) |
| 1,000 | Cycles Per Burst (CPB) |
| 12x24 | 2 mL process vessel |
| 300 | processing time (seconds) |
| 6°C | water bath temperature |

Table 1. Results from the stir bar method (Control).

| **125mM-Control** | | | | |
|---|---|---|---|---|
| **Ratio** | **d10-Control** | **d50-Control** | **d90-Control** | **Span** |
| 1:6.6 | 9.0 | 29.9 | 1730.0 | 57.5 |
| 1:6.6 | 8.0 | 26.8 | 66.5 | 2.1 |

| | | | | |
|---|---|---|---|---|
| 1:6.6 | 8.0 | 26.5 | 64.2 | 2.1 |
| Average | 8.3 | 27.7 | 620.2 | 20.6 |
| SD | 0.45 | 1.54 | 784.70 | 26.1 |

All values are in microns

Table 2. Results from the AFA method.

| **125mM-AFA** | | | | |
|---|---|---|---|---|
| **Ratio** | **d10-AFA** | **d50-AFA** | **d90-AFA** | **Span** |
| 1:6.6 | 0.0406 | 0.122 | 0.325 | 2.335 |
| 1:6.6 | 0.0406 | 0.122 | 0.333 | 2.387 |
| 1:6.6 | 0.0407 | 0.123 | 0.334 | 2.395 |
| Average | 0.041 | 0.122 | 0.331 | 2.372 |
| SD | 0.000 | 0.000 | 0.004 | 0.027 |

All values are in microns

d50 particle size populations for the control sample averaged 27.7 microns while the same method processed with AFA yielded average d50 particle size populations of 122 nanometers.

A continuous manufacturing process was evaluated with the following settings:

Drug solvent concentration 125 mM (46 mg/mL)

Solvent flow rate 800 μL/min

Anti-solvent flow rate 10 mL/min

Stabilizer 0.2% PVP30 and 0.25 mM SLS

Stabilizer percentage 0.2%

Solvent:Antisolvent ratios 1:12.5 (10 mL/min anti-solvent, 800 μL/min solvent)

**With 125 mM at a 800 μL/min solvent flow rate, the production rate was 36.5 mg/min.**

Higher total acoustic power input (Peak Incident Power) yielded both smaller particle sizes and repeatable results. At PIP values of 500 Watt, particle size distributions are listed in the following table.

Table 3. Particle size distributions at PIP = 500 W.

| | Sample 9 | Sample 10 | Sample 11 | Sample 12 |
|---|---|---|---|---|
| d10 | 0.085 | 0.091 | 0.098 | 0.090 |
| d50 | 0.180 | 0.191 | 0.202 | 0.187 |
| d90 | 0.415 | 0.409 | 0.423 | 0.399 |
| Span | 1.834 | 1.663 | 1.611 | 1.654 |

All values are in microns

Control of particle size was demonstrated by varying PIP. Particles sizes at 300 PIP are listed in the following table.

Table 4. Particle size distributions at PIP = 300 W.

| | Sample 5 | Sample 6 | Sample 7 | Sample 8 |
|---|---|---|---|---|
| d10 | 2.15 | 1.79 | 1.89 | 1.78 |
| d50 | 3.48 | 3.23 | 3.03 | 2.86 |
| d90 | 5.55 | 5.01 | 4.95 | 4.73 |
| Span | 0.98 | 1.32 | 1.01 | 1.03 |

All values are in microns

The current study demonstrates the technical feasibility of using Adaptive Focused Acoustics (AFA) to control nucleation and crystal growth for continuous flow Solvent:Anti-solvent crystallization. Nanoparticles were successfully and reliably produced in a continuous flow process with d50 particle size populations below 200 nm. By using the AFA system to control the crystallization process, a choice of nano- or micron-sized crystals can be easily generated with a simple change of PIP process parameter. In addition, the isothermal, reproducible AFA process enables control of fluid shear effects on secondary nucleation to minimize crystal size, accelerate the overall rate of crystal formation, and improve yield. The effect of AFA on the morphology of the crystal is currently under investigation.

# 430222 Crystal Growth Inhibition of a Poorly Water-Soluble Drug Using Polymeric Additives

**Abstract:**

A large percentage of active pharmaceutical ingredients (APIs) in the drug development pipeline demonstrate poor aqueous solubility as crystalline solid forms. This poses a problem for oral therapeutics due to decreased bioavailability. Utilizing the amorphous form of the drug rather than the crystalline form is an effective method to alleviate solubility issues, due to its greater overall free energy. Much emphasis has been placed on methods to stabilize the inherently unstable amorphous form and prevent crystallization. One such method utilizes polymers to act as a mechanical barrier, occupying growth sites. The effectiveness of this method, however, varies for each polymer-drug combination.

In this work, we study inhibition of crystal growth of the drug felodipine in solution using a variety of polymers. Using atomic force microscopy (AFM), we can determine the fractional surface coverage as well as the conformation of adsorbed polymers on crystalline felodipine. These results reveal insights into the mechanisms by which polymers can effectively inhibit crystal growth, as well as elucidate important polymer properties that impact their effectiveness

# 430469 Dynamic Modeling and Control of Multi-Stage Slug-Flow Crystallization

Thursday, November 12, 2015: 4:40 PM
Ballroom D (Salt Palace Convention Center)
Michael L. Rasche[1], Mo Jiang[2] and Richard D. Braatz[2], (1)Department of Chemical and Biomolecular Engineering, University of Illinois at Urbana-Champaign, Urbana, IL, (2)Department of Chemical Engineering, Massachusetts Institute of Technology, Cambridge, MA

The design of robust pharmaceutical crystallization processes has been of high interest in academia and industry for the last decade (Nagy and Braatz, 2012). Continuous-flow tubular crystallizers have shown potential for high reproducibility and process efficiency at low capital and production cost (e.g., Alvarez & Myerson, 2010; Lawton et al., 2009; Vacassy et al., 2000). A recent advance that combines the advantages of continuous and batch crystallizers is the air/liquid slug-flow crystallizer (Eder et al., 2010; 2011; 2012; Jiang et al., 2014; 2015) which has unique advantages including narrow residence time distribution, no stirrer generating secondary nucleation, and easy post-crystallization separation. The potential application of slug-flow crystallization in the final stage of pharmaceutical manufacturing is the focus of this design study.

Similar to batch crystallization, the design of a robust slug-flow crystallization process comes from a good understanding of how the supersaturation profile is affected by the design variables, such as the method and rate of supersaturation generation, and the corresponding implementation. For example, supersaturation should be minimized so as to avoid impurity incorporation and secondary nucleation (e.g., Jiang et al., 2012; 2014), which is important for slug-flow crystallization with its possibly short residence time (on the order of minutes) (Eder et al., 2012; Jiang et al., 2014) and with fast heat transfer associated with the large surface area-to-volume of tubular crystallizers.

Unlike past studies that involve mathematical modeling or analysis of similar continuous crystallizers (e.g., Eder et al., 2010; Kubo et al., 1998), this study only analyzes an experimental demonstration (Jiang et al., 2014) but also investigates the effect of and optimizes numerous design variables (e.g., number of heat exchangers, length of tubing in each heat bath/exchanger, mass flow rate through the heat exchangers) while minimizing the total equipment/material use in the design. This presentation describes the first dynamic simulation of slug-flow crystallization for a series of counterflow, concentric shell-and-tube heat exchangers.

The method of moments is combined with the numerical method of lines with a finite difference approximation to reduce the model described by four partial differential equations (balances on population, solute, and energy in both the shell and tube) to a system of ordinary differential equations. The system is to be designed so that, under normal operating conditions, the temperature drop in each heat exchanger produces an environment to minimize the maximum supersaturation, to promote molecular purity, while applying a constraint that ensures a nearly maximum yield.

Operating so that the exit concentration is near the minimum concentration provides the desired yield. The optimal crystal size is specified by the upstream nucleation rate. Excessive length of the tubes should be avoided, as that would introduce a time delay in the process dynamics that is undesirable for closed-loop control. The control objective is to minimize the maximum supersaturation in the system by manipulating the flowrate and inlet temperature of cooling water in each heat exchanger to suppress the effects of process disturbances.

Optimal steady-state spatial profiles for the four-stage slug-flow crystallizer are shown in Figure 1. The supersaturation reaches the minimized maximum value in less than 5 meters, and returns to that maximum once in each heat exchanger. The drop in supersaturation near the outlet also occurs in less than 5 meters.

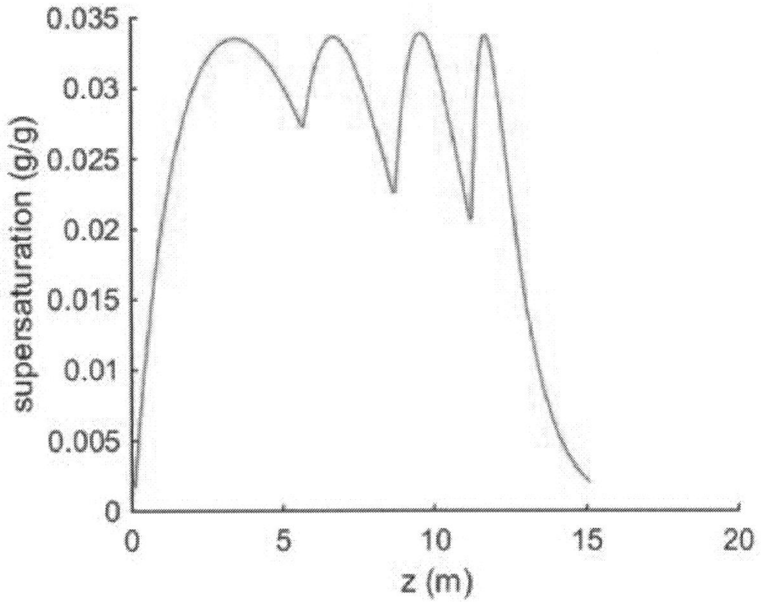


Figure 1: Supersaturation profile under steady, optimal conditions in the slug-flow crystallizer.

Investigating the dynamic response of the measured and product quality variables to disturbances, manipulations, and uncertainties provides a great deal of information and insight not available from steady-state models. The solute concentrations in the slug as it leaves each heat exchanger ("exit concentrations") are observed to be highly sensitive to cooling water temperatures and variation in the growth kinetics. The cooling water flowrates have relatively low effectiveness as manipulated variables for rejection of some disturbances. Increasing the cooling water flowrate or decreasing the cooling water temperature decreases the exit concentrations. The model shows that the effect of most input disturbances on concentration, yield, and average particle size are dampened by the process design automatically.

The response of maximum and average supersaturation is of vital importance as the driving force for growth as well as nucleation. While this system is designed to minimize secondary nucleation, supersaturation levels must be kept under control to avoid the incorporation of impurities into the crystalline matrix. Cooling water temperature and flowrate are seen to have a large effect locally. Inlet conditions for the slug stream have a decreasing effect with each subsequent heat exchanger. Persistent sources of relatively large deviation are again variation in the growth kinetics.

Outlet temperatures are also compared and evaluated in terms of value as measurements for feedback control, due to the relative ease and accuracy at which temperature can be measured. Not surprisingly, the greatest effect by far on the outlet slug stream temperature is the temperature of the cooling water.

Manipulating only the cooling water flowrates is shown to be insufficient to reject process disturbances that result from the deviation of growth kinetic parameters from nominal values. Additionally, without a tight control of the inlet temperature of cooling water, manipulation of the cooling water flowrates will be insufficient to suppress that resulting deviation as well. Changes in the inlet temperature and concentration for the slug stream are less problematic; the system design is inherently robust to disturbances which occur far upstream. The robustness allows the on-line measurement of particle size, a relatively expensive and inaccurate endeavor, to be eliminated from the design of a control algorithm; particle size is not observed to be affected greatly by upstream disturbances.

In an effort to increase the output response to changes in manipulated variables, the cooling water inlet temperature should be a manipulated variable rather than allowing it to be a disturbance. That is, a lower level regulatory control system should be implemented so that the cooling water inlet temperature is controlled, with the setpoint to the cooling water inlet temperature being a manipulated variable of the higher level control system. The extra degrees of freedom from controlling not only the cooling water flowrates, but also the cooling water temperatures, greatly increase the controllability of the slug-flow crystallization process. From a design standpoint, sensitivity analysis indicates the tubing length should be increased beyond nominal design conditions to serve as a buffer to disturbances.

Several control strategies motivated by the dynamic analysis are investigated for the multivariable control of the slug-flow crystallizer (e.g., Figure 2). Neglecting particle size measurements for the reasons mentioned above, there are two measurements of interest that can be taken at the outlet of each heat exchanger: temperature and solute concentration. There are also two manipulated variables for each heat exchanger: cooling water flowrate and cooling water temperature. A variety of factors are considered in selecting the best control strategy in terms of the number of measurements, number of controllers, and which controllers receive which measurements (and control which manipulated variables). Of the eight possible measurements, the four temperature measurements are inexpensive and accurate while four measurements of solute concentration would be relatively expensive and difficult. The design of this process does not provide evidence that intermediate solute concentration measurements would be useful process control elements, and it is unlikely that the final implemented control systems design would contain those measurements. With respect to the eight manipulated variables, the inlet temperature of cooling water could be altered by applying energy to a thermal bath or, perhaps more likely, combining multiple feeds from multiple baths at Y-mixers. The cooling water temperature displays a greater effect on process dynamics than the cooling water flowrate, but incompressible fluid flowrates have the advantage of being easily and quickly controlled. After comparing the pros and cons of different control strategies, an optimal control system design will be described.

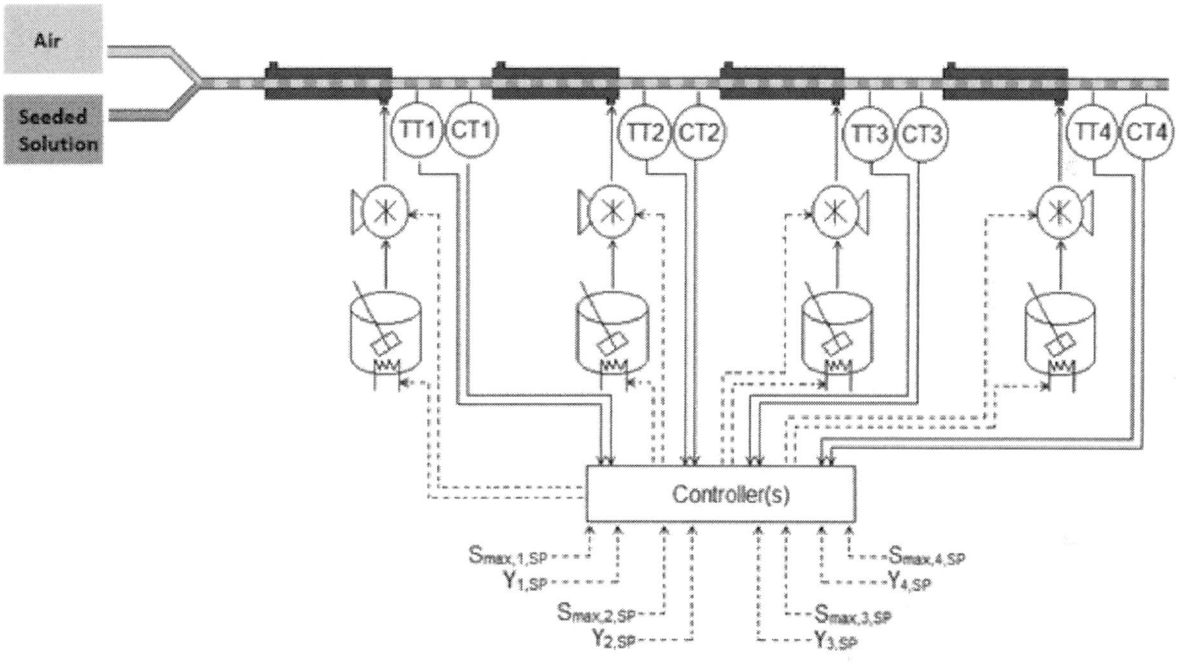


Figure 2: Example control strategy for the slug-flow crystallizer, which uses the complete set of eight measured and eight manipulated variables.

**References:**


Alvarez, A.J., Myerson, A.S., 2010. Crystal Growth & Design, 56, 349-369.

Eder, R.J.P., Radl, S., Schmitt, E., Innerhofer, S., Maier, M., Gruber-Woelfler, H., Khinast, J.G., 2010. Crystal Growth & Design, 10(5), 2247-2257.

Eder, R.J.P., Schmitt, E.K., Grill, J., Radl, S., Gruber-Woelfler, H., Khinast, J.G., 2011. Crystal Research and Technology, 46(3), 227-237.

Eder, R.J.P., Schrank, S., Besenhard, O., Roblegg, E., Gruber-Woelfler, H., Khinast, J.G., 2012. Crystal Growth & Design, 12(10), 4733-4738.

Jiang, M., Zhu, Z., Jimenez, E., Papageorgiou, C.D., Waetzig, J., Hardy, A., Langston, M., Braatz, R.D., 2014. Crystal Growth & Design, 14(2), 851-860.

Jiang, M., Papageorgiou, C.D., Waetzig, J., Hardy, A., Langston, M., and R. D. Braatz, R.D., 2015. Crystal Growth & Design, 15(5), 2486-2492.

Kubo, M., Kawakatsu, T., Yonemoto, T., 1998. Chemical Engineering Research & Design, 76(6), 669-676.

Lawton, S., Steele, G., Shering, P., Zhao, L., Laird, I., Ni, X.-Y., 2009. Organic Process Research & Development, 13(6), 1357-1363.

Nagy, Z.K., Braatz, R.D., 2012. Advances and new directions in crystallization control, Annual Review of Chemical and Biomolecular Engineering, 3, 55-75.

Vacassy, R., Lemaitre, J., Hofmann, H., Gerlings, J.H., 2000. AIChE Journal, 46(6), 1241-1252.

## (730f) Polymorphism in the Development and Scale-up of Reactive-Crystallization of an Active Pharmaceutical Ingredient

Zhang, H., Sunovion Pharmaceuticals Inc.
Saranteas, K., Sunovion Pharmaceuticals Inc.
Originally presented on: 11/12/2015 17:00:00 - 17:20:00

**Polymorphism in the Development and Scale-up of Reactive-crystallization of an Active Pharmaceutical Ingredient**

Haitao Zhang, Kostas Saranteas

Chemical Process Research & Development, Sunovion Pharmaceuticals Inc., 84 Waterford Drive, Marlborough, Massachusetts 01752, U.S.A.

### Abstract

Polymorphism is widely observed in pharmaceutical compounds and continues to be an important issue in drug development because of its impact on the physicochemical properties of drugs. It is well recognized that polymorphism and solvate formation affect the various pharmaceutically important physicochemical properties, such as stability, solubility, dissolution rate, crystal habit (shape), tableting behavior. Crystallization is a crucial step in the pharmaceutical industry in terms of the understanding and further control over the polymorphism during the manufacturing of active pharmaceutical ingredients (APIs).

Herein is described the observed polymorphic behavior of the API SEP-363856-01 in the reactive-crystallization process. This API is crystallized out from isopropanol (IPA) as the HCl salt through the addition of HCl/IPA to an IPA solution of the SEP-363856 freebase. As the API crystallizes it displays two distinct morphologies, the first a block like crystal (Form A) and the second a needle like crystal (Form B). Single crystal x-ray diffraction has shown Form A has a monoclinic crystal system while Form B has an orthorhombic crystal system. The experimental results indicate that the supersaturation control during the addition of HCl/IPA is critical in terms of the polymorphism control. Polymorphism transformation of kinetically preferred Form B to thermodynamically stable Form A is observed during a solvent-mediated phase transition in isopropanol. It has been shown that the transformation of Form B needles to Form A blocks is influenced by the quantity of Form B present and is accelerated with increasing temperature. A controlled HCl/IPA addition strategy has been proposed and successfully implemented in the lab development and scale-up manufacturing campaign. With this, the generation of unstable form B crystal can be avoided in the first place.

Solid phase polymorphism transformation is observed as well. The isolated samples of Form B needles held in the laboratory for a period of time yield a more rod like morphology, however XRPD analysis shows this morphology is consistent with the Form A blocks. Single crystal x-ray diffraction has shown Form A has a monoclinic crystal system while Form B has an orthorhombic crystal system.

# 433636 Structure-Function Design Principles at Nucleotide Resolution for Crispr Small Guide RNAs

**Abstract:**

The recent discovery of the Clustered Regularly Interspersed Short Palindromic Repeats (CRISPR) system in prokaryotes has triggered a rapid expansion in our ability to edit genomes and control gene expression. In one example, the engineered type II CRISPR system from *S. pyogenes*, Cas9 (CRISPR associated protein 9) binds to a synthetic small guide RNA (sgRNA) that targets Cas9 to a specific DNA sequence for cleavage [1]. This ability to target Cas9 by only changing part of the sgRNA has led to the widespread use of the Cas9-sgRNA system for genome editing [2]. Further engineering of Cas9 has also converted this system into an RNA-guided platform for sequence-specific control of gene expression [3], making the Cas9-sgRNA system one of the most versatile biotechnology platforms for cellular manipulation currently at our disposal. Thus, there is great interest in understanding the design principles behind the type II CRISPR system to facilitate its further engineering. In this work, we explore the structural requirements of the sgRNA for binding dCas9 (catalytically dead Cas9) and targeting DNA sequences for regulation in *E. coli*. We find that nucleotide flexibility is critical for function within specific regions of the sgRNA. This work is uncovering new design principles of sgRNAs that will allow further engineering of this system to make new, more sophisticated regulators from the type II CRISPR system.

1. Horvath, P., and Barrangou, R. (2010) CRISPR/Cas, the immune system of bacteria and archaea. *Science 327*, 167–170.
2. Sander, J. D., and Joung, J. K. (2014) CRISPR-Cas systems for editing, regulating and targeting genomes. *Nat. Biotechnol. 32*, 347–355.
3. Qi, L. S., et al. (2013) Repurposing CRISPR as an RNA-Guided Platform for Sequence-Specific Control of Gene Expression. *Cell 152*, 1173–1183.

# 421539 A Synthetic-Biology Approach to Rapidly Characterize Novel Crispr-Cas Systems

**Abstract:**

CRISPR-Cas systems have rapidly transitioned from curious defense systems to powerful and versatile genetic tools. A central driver has been the ease in designing synthetic CRISPR RNAs that guide the system's Cas proteins to bind and cleave target DNA. The only design rules are ensuring the spacer portion of the CRISPR RNA is complementary to a target sequence and selecting target sequences flanked by a protospacer-adjacent motif (PAM). Once directed to the target sequence, the system can be utilized for a range of applications that include genome editing, gene regulation, sequence-specific antimicrobials, and many others. To date, these applications have overwhelming relied on the CRISPR-Cas system from the bacterial pathogen *Streptococcus pyogenes*. However, thousands of other CRISPR-Cas systems have been found in nature and offer distinct attributes and capabilities. The challenge is rapidly characterizing these systems in order to generate the next generation of CRISPR technologies.

We developed a simple genetic circuit in *Escherichia coli* to rapidly identify the PAM, an essential design feature that varies widely between systems and has proven difficult to predict a priori. The designed circuit relies on CRISPR-Cas systems modified for gene repression. As part of the circuit, CRISPR-based repression is converted into the upregulated expression of a fluorescent reporter through an inverter gate. By introducing libraries of random sequences in the location of the PAM, functional PAM sequences can be rapidly identified using fluorescence-activated cell sorting followed by next-generation sequencing. We applied this screen to a collection of Type I systems and Type II systems—the two types that require PAMs and target DNA. The screen reproduced known PAMs for well-characterized systems and revealed PAMs for never-before characterized systems. The screen also revealed flexibility in the spacing between the target sequence and PAM, suggesting that CRISPR-Cas systems have a greater propensity for off-target effects than previously realized. Overall, our synthetic circuit offers a simple and rapid means of characterizing novel CRISPR-Cas systems, with a step toward bringing the remarkable diversity of these natural defense systems to the forefront of biotechnology.

# 433810 Design Principles of a Robust Genetic Switch Using Antisense Transcription

**Abstract:**

Antisense transcription has been extensively recognized as a regulatory mechanism for gene expression across all kingdoms of life. Growing evidence suggests the presence of non-coding *cis* antisense RNAs (AR) that regulate gene expression. Recent studies also indicate the role of transcriptional interference (TI) in regulating expression of neighboring genes due to traffic of RNA polymerases from adjacent promoter regions. While these mechanisms have been described to regulate gene expression in naturally occurring systems, the design principles for antisense transcription have not been explored yet. Here, we combine mathematical modeling and experimental approaches to engineer antisense RNA interaction, RNA polymerase traffic, relative promoter strength and promoter-to-promoter distance in a set of convergent promoters to develop a novel synthetic biology tool. We show that combination of TI and AR adds multiple-levels of regulation which allows such a system to have a highly tunable output in response to stimuli, ranging from a simple-first order response to biologically complex higher-order response such as bistable switch. Using a library of experimental constructs with engineered TI and AR by using face-to-face inducible promoters separated by carefully tailored overlapping DNA sequences we demonstrate tunable control of expression of a set of fluorescent reporter proteins. We show that even subtle changes in the physical arrangement of the convergent promoters can give rise to diverse gene expression patterns, which allows for design of highly-tunable synthetic genetic switches using minimum number of parts.

# 433631 Rewriting the Language of Biology: Unnatural Amino Acids Incorporation at Sense Codons Using Cell-Free System

**Abstract:**

Although 20 canonical amino acids are life's building blocks, today's applications (e.g. biocatalysts, biotheraputics, site–specific conjugation) often require unique chemical functionality beyond those amino acids. Therefore, site-specific incorporation of multiple unnatural amino acids (uNAA) has been a major aspiration of protein engineering and synthetic biology. While there are many promising efforts to incorporate uNAA, the amber stop codon is the predominant coding target, due to the fundamental constraints imposed by the genetic code. In theory, the great redundancy in genetic code has potential to be exploited for uNAA, but such extensive rewriting is extremely difficult in biological systems.

To address this challenge, we propose a cell-free synthetic biology approach to rewrite the genetic code and incorporate multiple uNAAs simultaneously. The open nature of the cell-free system enables us to modify the system and reengineer the genetic code to emancipate some codons for incorporating uNAAs. In fact, there is potential to emancipate up to 43 codons, since only 21 of 64 codons are required for translation of 20 standard amino acids and translation termination. To that end, we removed all *E. coli* endogenous tRNA and then repopulated the system with a designed minimal set of synthetic tRNA made *in vitro*. By importing orthogonal sets of tRNA/synthetase from different organisms, we were able to emancipate codons and demonstrate the incorporation of uNAA at sense codons.

# 434245 Development of a Modular, Small Molecule Responsive, Transcriptional Regulator Platform

**Abstract:**

Our ability to engineer microbial factories has accelerated as new technologies have enabled rapid mining of biological diversity and high throughput synthesis of novel ensembles of enzymatic pathways. To date, however, it remains challenging to monitor the presence of metabolites in individual cells, outside of the situations in which a natural metabolite-responsive transcriptional regulator has evolved and been identified. Such naturally occurring biosensors have proven useful in applications including both screening of high-dimensional libraries and engineering feedback control to optimize biosynthesis. To help address this need for novel biosensors, we have developed a modular and general approach for engineering novel transcriptional regulators that are actuated by the presence or absence of metabolites of interest – the Small molecule Responsive Transcriptional Regulator (SmaRTR) biosensor platform. The central goal of this approach is to develop a systematic strategy for converting proteins that bind small molecules into proteins that bind a target DNA sequence in a manner that depends on whether those small molecules are present. In our initial evaluation of this strategy, we utilized the BCR-ABL1 zinc finger protein as a model DNA-binding domain. We built and characterized a library of promoters including BCR-ABL1 binding sites at various positions to elucidate which promoter design features are most important for achieving effective zinc finger mediated-transcriptional repression. Then, we evaluated several strategies for converting the well-characterized maltose binding protein (MBP), a model of a ligand-binding protein, into a novel maltose-actuated transcription factor that regulates gene expression in a ligand-dependent manner. Our investigations using this model system resulted in several robust and functional novel biosensors, establishing the fundamental feasibility of the SmaRTR approach. We also elucidated protein and promoter engineering strategies that may ultimately be broadly applicable to developing a range of engineered biosensors. Ultimately, this technology may be harnessed to generate quantitative insights into metabolite dynamics in living cells, provide new tools for high throughput screening, and enable improved biomanufacturing through implementation of metabolic feedback control.

# 425166 Synthetic Extracellular Sensing Circuit By Intein-Mediated Reconstitution of Yeast Mating Factor

**Abstract:**

Increasing demands for alternative fuels such as bioethanol has driven the development of more efficient processes. In particular, the use of microbial consortia is a promising approach to dramatically reduce costs and to improve fuel production directly from renewable biomass. However, regulation and coordination of different populations within the consortia remain challenging since currently available intracellular sensing circuits are incapable of responding to the two key substrates, cellulose and hemicellulose, that are presented exclusively in the extracellular medium. To meet these challenges, we present a new synthetic extracellular sensing circuit to control yeast behaviors by reconstitution of extracellular protein sensors. These synthetic fusion proteins are designed to trigger split intein-mediated reconstitution of functional α-factor peptide analogues in response to nutrient conditions. The reconstituted α-factors bind onto Ste2p, a native yeast G protein-coupled receptor, and activate the yeast mating response and transcriptional induction. By integrating this native mating signaling pathway with heterologous gene expression cassettes, we can control timing and level of induction and direct phenotypic behaviors by applying our split intein fusion proteins in yeast cultures. This synthetic control strategy will enable regulation of yeast functions in a wide range of applications, including consolidated bioprocessing.

# 432043 New Tools for Controlling Gene Expression in Drosophila Melanogaster Embryos Using Hammerhead Ribozymes and Short Open Reading Frames

**Abstract:**

There are many situations where it is desirable to finely tune gene expression levels, including the study of both native and synthetic gene networks. Two methods that are currently being explored are hammerhead ribozymes and short upstream open reading frames. These two systems work through very different mechanisms. The hammerhead ribozyme leads to mRNA degradation through autocatalytic self-cleavage, resulting in changes in both mRNA and protein levels (Yen et al. 2004, Auslander, Ketzer & Hartig 2010). The sequences directly upstream of the ribozyme can be altered to adjust the ability of the ribozyme to be properly folded and thereby altering the amount of mRNA degradation (Yen et al. 2004, Auslander, Ketzer & Hartig 2010). In the short upstream open reading frame approach, three upstream open reading frames (uORFs) encoding two amino acids were placed upstream of the open reading frame encoding the protein of interest. Only a small number of ribosomes are able to "leak" past the uORFs to the ORF for the protein of interest (Ferreira, Overton & Wang 2013). A different number of uORF and altering the surrounding sequences of these uORF can be used to alter the amount of ribosomes that "leak" through and translate the downstream ORF, producing the protein of interest (Ferreira, Overton & Wang 2013). Both of these systems have been previously tested in single-cell systems to various extents (Auslander, Ketzer & Hartig 2010, Ferreira, Overton & Wang 2013). We further validated these systems and showed their ability to regulate gene expression in a multi-cellular system, namely the *Drosophila melanogaster* embryo. We were able to demonstrate the tuning of gene expression of EGFP within the embryo at both the protein and mRNA level for the hammerhead ribozyme and at the protein level for the uORFs. Our results can be used to tune protein or mRNA level of endogenous or exogenous genes in *Drosophila melanogaster* or other multi-cellular system to a known level based on the insertion of a ribozyme or uORF.

References:

Auslander, S., Ketzer, P. & Hartig, J.S. 2010, "A ligand-dependent hammerhead ribozyme switch for controlling mammalian gene expression.", *Mol Biosyst,* vol. 6, no. 5, pp. 807-814.

Ferreira, J.P., Overton, K.W. & Wang, C.L. 2013, "Tuning gene expression with synthetic upstream open reading frames", *Proceedings of the National Academy of Sciences of the United States of America,* vol. 110, no. 28, pp. 11284-11289.

Yen, L., Svendsen, J., Lee, J., Gray, J.T., Magnier, M., Baba, T., D'Amato, R.J. & Mulligan, R.C. 2004, "Exogenous control of mammalian gene expression through modulation of RNA self-cleavage.", *Nature,* vol. 431, no. 7007, pp. 471-476.

# 429113 Hybrid Cell Culture Models for Efficient Process Development

**Abstract:**

Cell culture processes are dynamic systems governed by a large number of influential and highly interacting process variables. Despite the stringent control of several process variables, considerable variation in the process outcome is often observed. Moreover, generalized deterministic models for dynamic process prediction based on the impact of the process variables on the product quantity and quality attributes do not exist. According to the Process Analytical Technology (PAT) initiative, multivariate data acquisition and analysis should be implemented to achieve advanced process knowledge and to gain effective control on risk in manufacturing. While several studies have shown that such multivariate statistical models provide promising possibilities for process monitoring and prediction based on the process history, they are very limited regarding their capability for the support of experimental design.

This work demonstrates how the flexibility of statistical models combined with the formalized bioengineering knowledge of simplified deterministic models can improve the efficiency in cell culture process development. Based on a rigorous case study, the results obtained with the hybrid approach will be compared to the ones from the purely statistical and deterministic models to visualize its promising capability of providing significant process information using the least number of experiments. Different kinds of hybrid cell culture models will be discussed and the possibility to incorporate such models into a sequential experimental design methodology to define the process design space will be presented.

# 420830 The Impact of Glycation on Protein Color

**Abstract:**

Health Authorities expect that color of liquid formulations of monoclonal antibodies be monitored and controlled. Subcutaneous drug formulations often require high drug substance concentrations which may result in more intense product color. A potential mechanism of coloration is through the Maillard reaction, where reducing sugars in the cell culture media react with the secreted protein product in a series of non-enzymatic reactions that result in formation of glycated protein and advanced glycation end products (AGEs). We varied the glucose feeding strategy and cell culture media across multiple cell lines to study and delineate the effects. Presence of AGEs was also measured using mass spectrometry. We observed good correlation between the glucose fed to the culture and the level of the glycated protein. However, the glucose feeding strategy tested did not significantly impact color, and we observed no strong correlation between glycated protein and color. Changes in the levels of the AGEs for different culture conditions will also be discussed.

# 433606 Increase Cell Culture Development Value Using the Modular Automated Sampling Technology (MASTTM) Platform in an Integrated Bioprocess Lab Environment

**Abstract:**

Biopharmaceutical companies are continually striving to fundamentally understand and optimize cellular performance within their bioreactors. While on-line process analytical technology (PAT) tools like dielectric and Raman spectroscopy are helping to provide insight into cell processes that impact titer and antibody quality, the integration of these data with off-line measurements such as cell density, viability, and titer remains an elusive goal.

Bend Research Inc., in collaboration with Pfizer Inc. and other major biopharmaceutical companies, is advancing the Modular Automated Sampling Technology (MAST$^{TM}$) platform and integrating this system with automated analytical and data retrieval systems. This approach allows application of novel experimental methods providing greater insight into the bioreactor environment. In this process, aseptically collected bioreactor samples are delivered to multiple analytical devices with resulting data being automatically retrieved and curated. This data is presented at a graphical user interface (GUI) for real-time data analysis and predictive model development. Early studies with the MAST sample delivery system coupled with our data-processing software increased system observability and real-time process understanding by the end user.

This presentation describes how Bend Research has used the MAST platform as a cornerstone of an integrated lab environment. We will describe how the MAST system works and the features that set it apart from other autosampling solutions. Applications will be presented where MAST facilitated the use novel development approaches yielding enhance process understanding

From this new level of insight into the bioreactor, scientists can obtain enhanced data-driven guidance for key activities like optimizing process operation. With this new technology, the bioprocess industry can make major advances toward advanced real-time testing, predictive control, and overall enhanced bioprocess design and operation.

# 428887 Application of Raman Spectroscopy in Cell Culture Process Development

**Abstract:**

In recent years, the European Medicines Agency (EMA) and the United States Food and Drug Administration (FDA) have stressed the need of greater process understanding and control in the area of pharmaceutical development, manufacturing and quality assurance, all which is summarized in the Process Analytical Technology (PAT) initiative. Therefore, in bioprocesses, and particularly in cell cultures, multivariate data acquisition and analysis are likely to provide versatile and significant real-time information. Lately, Raman spectroscopy has been characterized as a promising tool to detect multiple process variables simultaneously.

This work presents several cases studies, where Raman spectroscopy was successfully applied for the quantification of significant variables in monoclonal antibody (mAb) cell culture processes. These studies include online monitoring of fed-batch cultures, offline analysis of (partially spiked) cell culture samples as well as the utilization in perfusion cell culture processes. Besides the evaluation of those applications of Raman spectroscopy, the central role of multivariate spectra processing techniques including pretreatment, spectral range selection and data reduction will be discussed with regards to robust calibration and prediction of the components of interest. Finally, the role of those spectral data with regards to overall process prediction, monitoring and control will be highlighted within the scope of all acquired online and offline data in bioprocesses.

# 411442 Density Surrogate for Fine Tuning Dissolution of an Extended Release Tablet to Increase Manufacturing Process Robustness

**Abstract:**

Tablet dissolution can be affected by several factors including raw materials, formulation and processing conditions. For a tablet with a fixed formulation, such as that of a commercial drug product, there remain few levers by which to improve tablet critical quality attributes (CQA). Tablet dissolution is one such known critical quality attribute to be affected by tablet hardness, which can be controlled by varying the compression force, typically in a directly proportional relationship. Beyond a certain threshold of compression force value, however, hardness reaches a maximum value and plateaus. Continuing to increase compression force after this threshold value could lead to loss in tablet cohesion due to high internal stresses or eventually tooling and equipment damage. Near the plateau threshold, there is a space over which tablet hardness may not adequately capture the tablet properties controlling its dissolution. This work explores this space for a polymer matrix extended release tablet where improved dissolution release robustness is desired. A parameter of weight/thickness ratio is suggested as a density surrogate to fine tune dissolution performance near and within the hardness plateau region.

# Engineering Proteins Targeting Tumor Biomarker Mesothelin with Applications in Cancer Diagnostics and Therapeutics

**Allison Sirois**
Katia George
Dr. Sarah Moore
Department of Biological Sciences
Department of Biochemistry
Picker Engineering Program
Smith College

## Cancer is a global health concern

**Estimated New Cases**

| Males | | | Females | | |
|---|---|---|---|---|---|
| Prostate | 220,800 | 26% | Breast | 231,840 | 29% |
| Lung & bronchus | 115,610 | 14% | Lung & bronchus | 105,590 | 13% |
| Colon & rectum | 69,090 | 8% | Colon & rectum | 63,610 | 8% |
| Urinary bladder | 56,320 | 7% | Uterine corpus | 54,870 | 7% |
| Melanoma of the skin | 42,670 | 5% | Thyroid | 47,230 | 6% |
| Non-Hodgkin lymphoma | 39,850 | 5% | Non-Hodgkin lymphoma | 32,000 | 4% |
| Kidney & renal pelvis | 38,270 | 5% | Melanoma of the skin | 31,200 | 4% |
| Oral cavity & pharynx | 32,670 | 4% | Pancreas | 24,120 | 3% |
| Leukemia | 30,900 | 4% | Leukemia | 23,370 | 3% |
| Liver & intrahepatic bile duct | 25,510 | 3% | Kidney & renal pelvis | 23,290 | 3% |
| **All Sites** | **848,200** | **100%** | **All Sites** | **810,170** | **100%** |

Siegel, R.L. *et al.* (2015) CA *Cancer J Clin. 65,* 5-29.

# Traditional therapy relies on cell cycle arrest, targeted therapies inhibit specific molecules

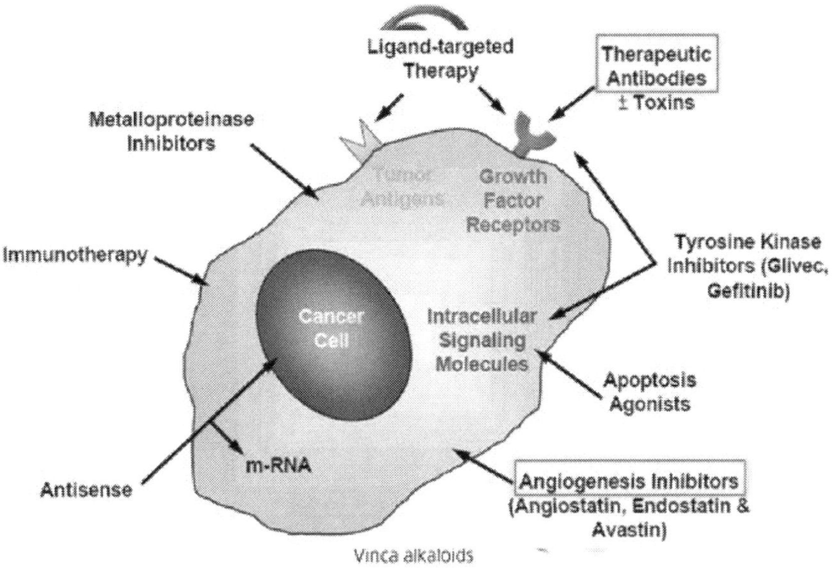


Gerber, D.E (2008). *Am Fam Physician*. 77, 311-9

# Molecular imaging obtains information in a non-invasive manner

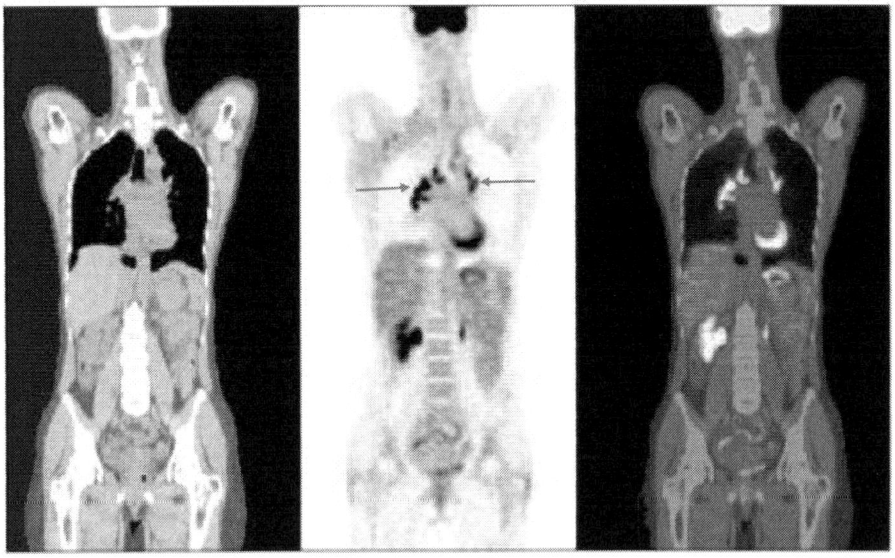

Computed Tomography Positron Emission Tomography PET/CT Fusion

UPMC (2015)

# Mesothelin is overexpressed in many cancers and interacts with tumor biomarker MUC16

MUC16 (CA125)

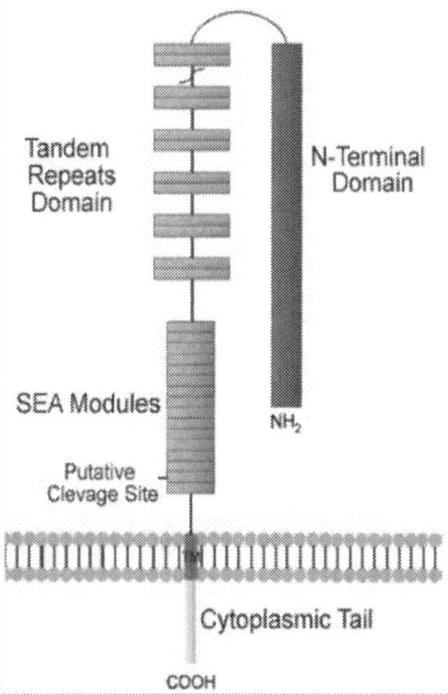


Mesothelin (MSLN)

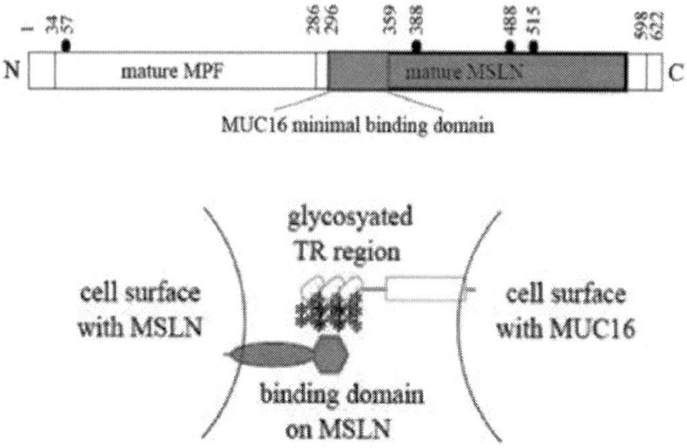


S.A. Farghaly. (2012) *InTech*

# Project Goals

- We propose to engineer proteins targeting tumor biomarker Mesothelin to interfere with the MSLN-MUC16 interface.
- These proteins will have applications as both a molecular diagnostic and targeted therapeutic

## Yeast surface display engineers proteins for enhanced binding affinity and stability

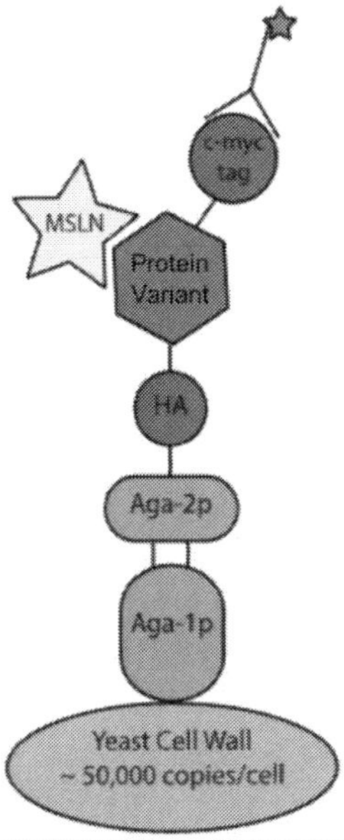


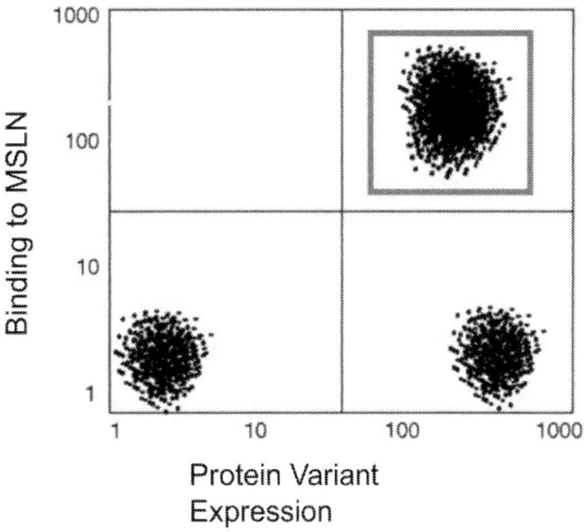


## Fibronectin is a widely used alternate scaffold for engineering novel proteins

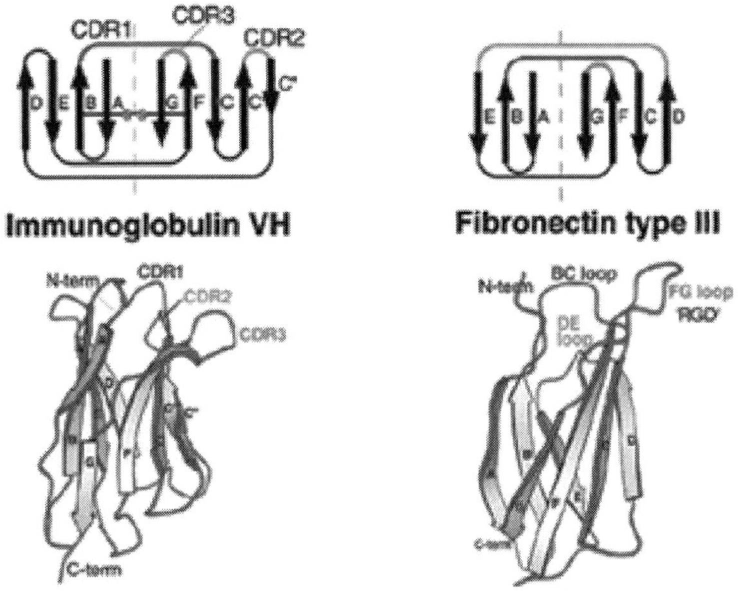


Koide, A. and Koide, S. (2007) *Methods Mol Bio.* 352, 95-109

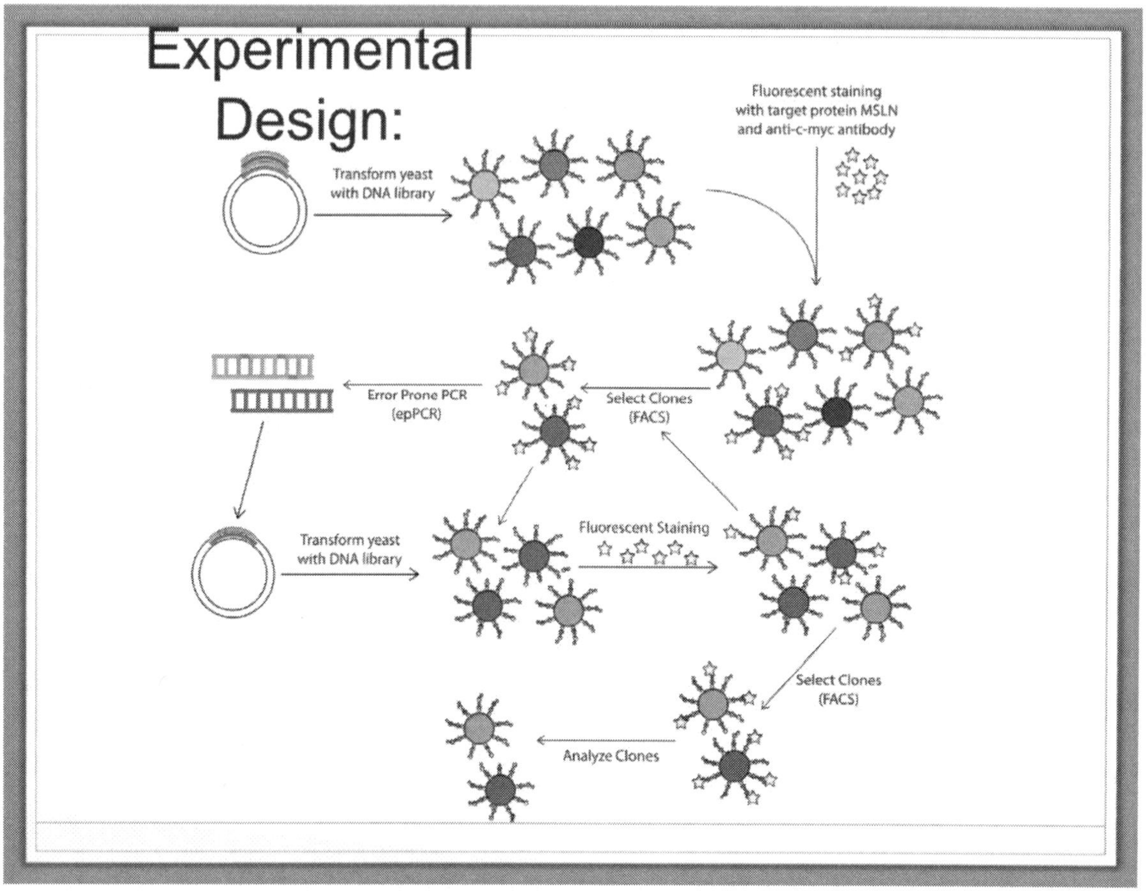
Experimental
Design:
Transform yeast with DNA library
Fluorescent staining with target protein MSLN and anti-c-myc antibody
Error Prone PCR (epPCR)
Select Clones (FACS)
Transform yeast with DNA library
Fluorescent Staining
Select Clones (FACS)
Analyze Clones

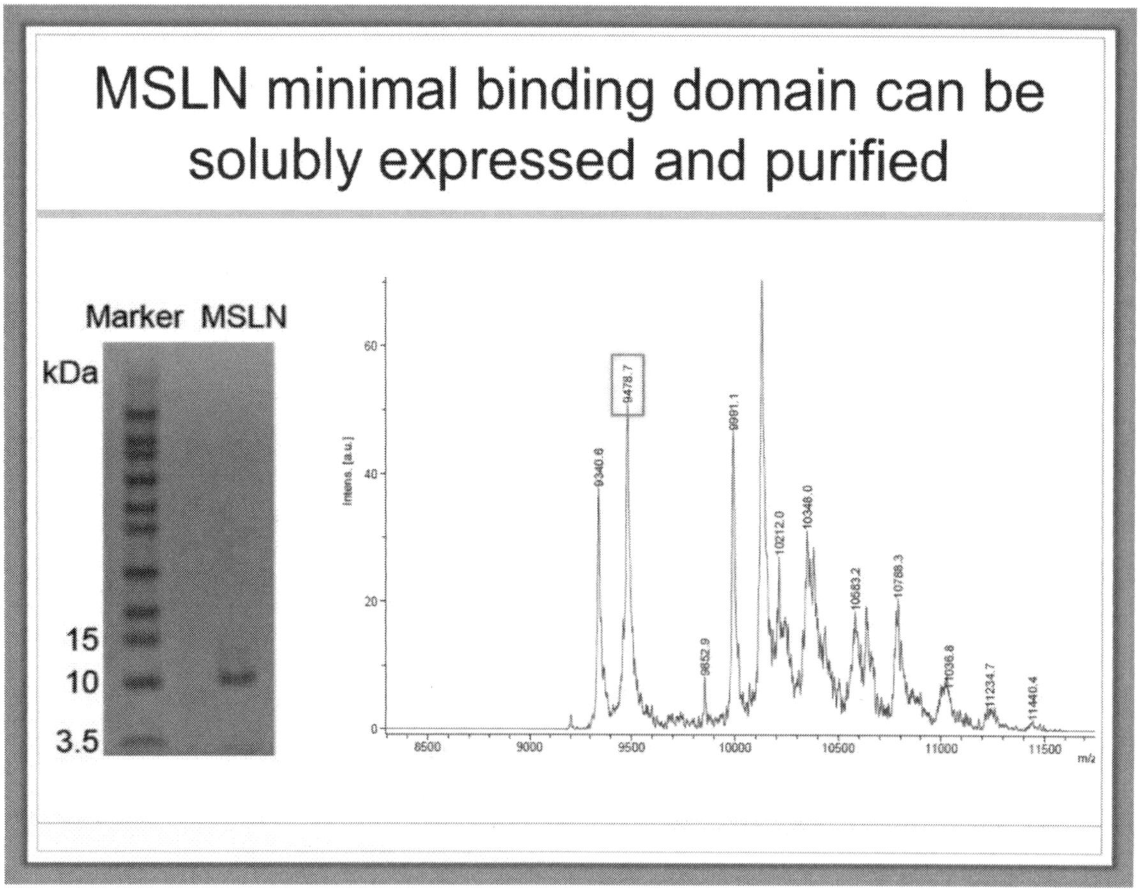
MSLN minimal binding domain can be solubly expressed and purified
Marker MSLN
kDa
15
10
3.5
Intens. [a.u.]
9340.6
9478.7
9852.9
9991.1
10212.0
10348.0
10583.2
10788.3
11036.8
11234.7
11440.4
8500
9000
9500
10000
10500
11000
11500
m/z

# MSLN minimal binding domain can be displayed on the surface of yeast

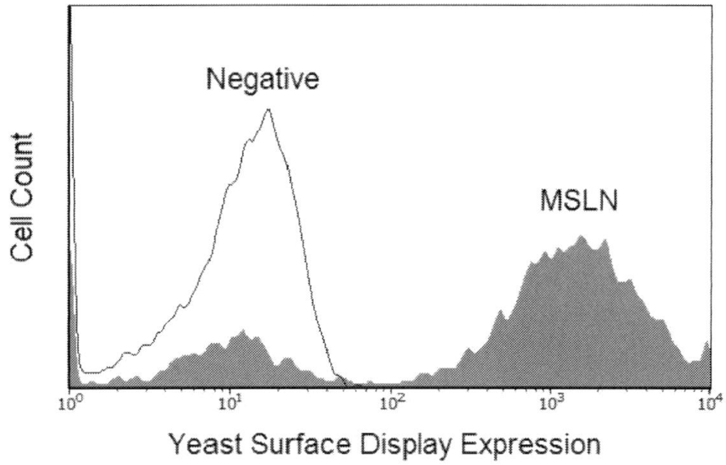


# OVCAR-3 cells express MSLN and MUC16 on their surface

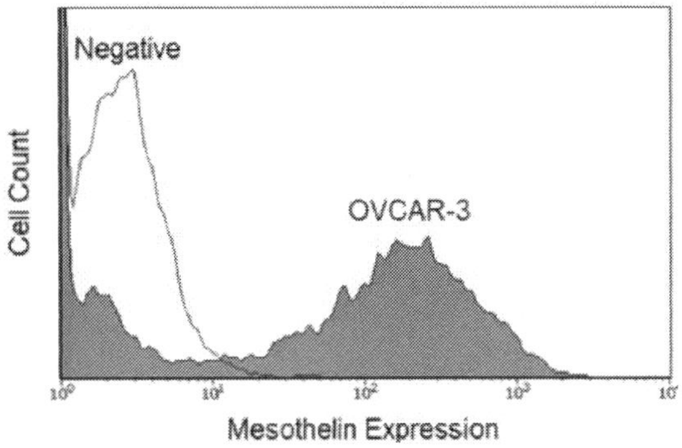


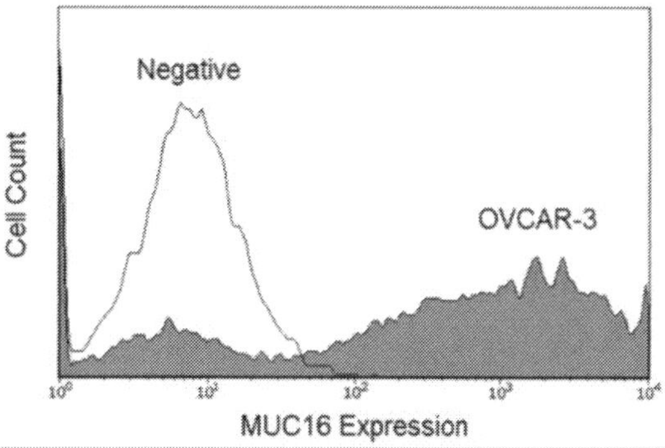

# Recombinant MSLN binds to OVCAR-3 cells expressing MUC16

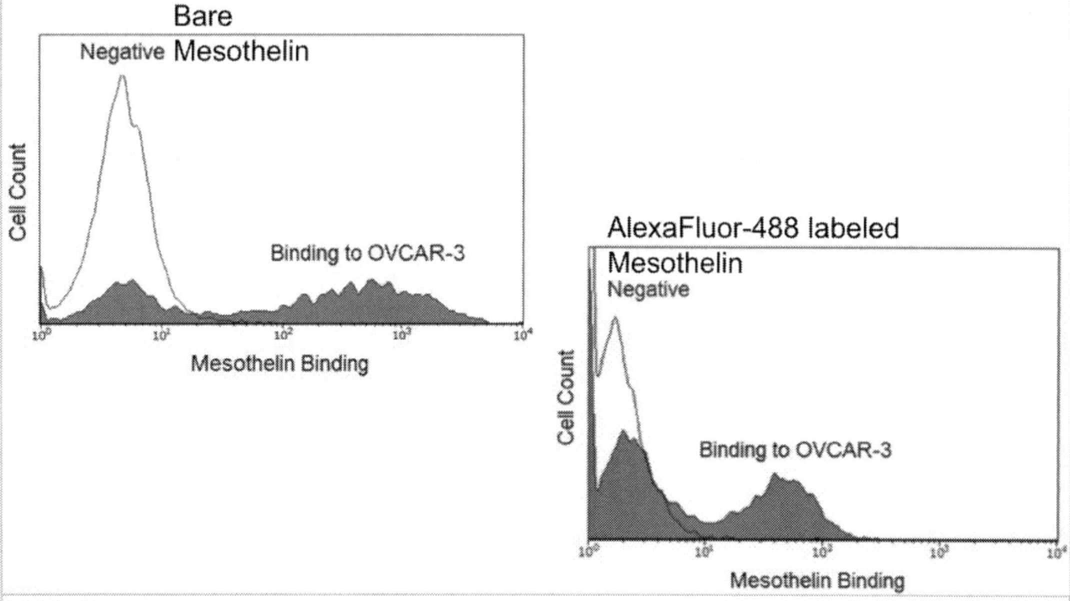


# Experimental Design:

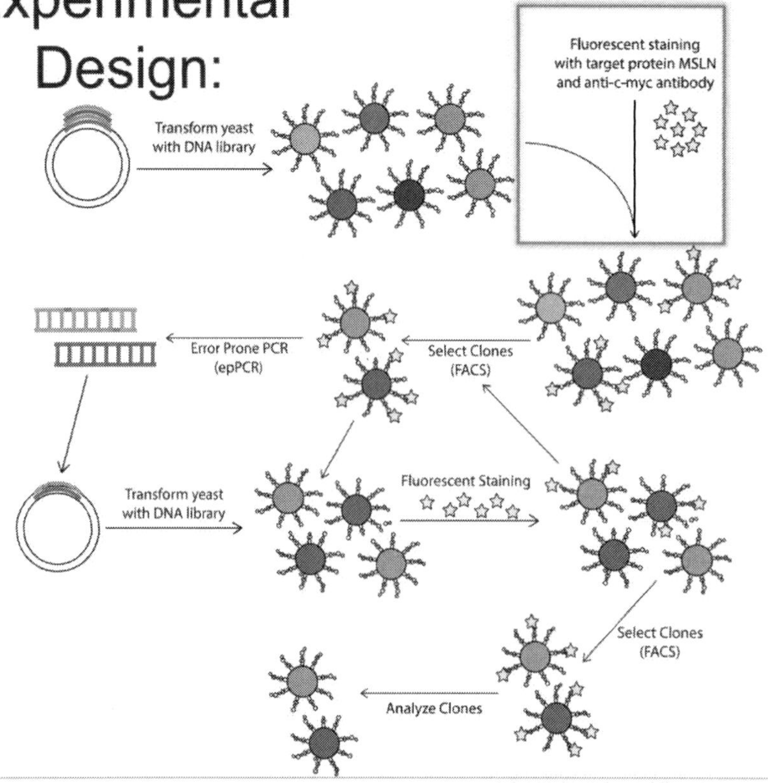

# Initial sorts show increasing MSLN binding

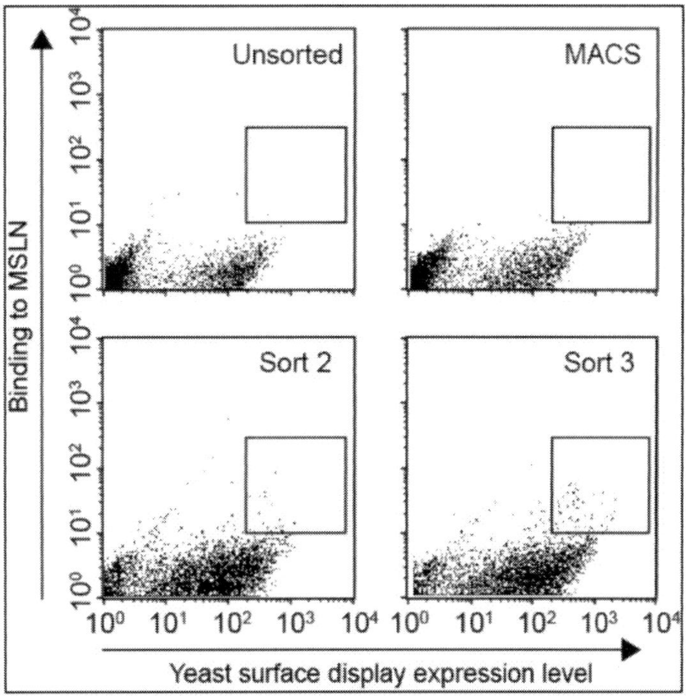


# Re-mutagenize to re-establish binding to MSLN

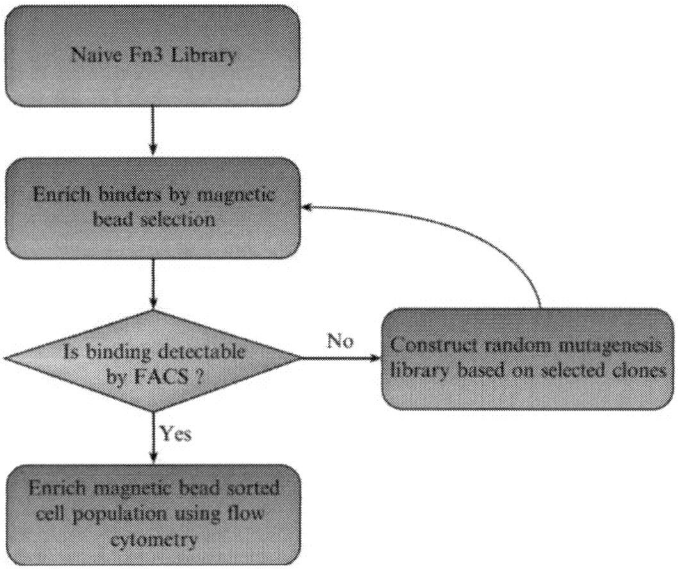


Chen, T.F., *et al.* (2013) *Methods Enzymol.* *523*, 303-26

# Mutations introduced at low levels to gene and at high levels to loops

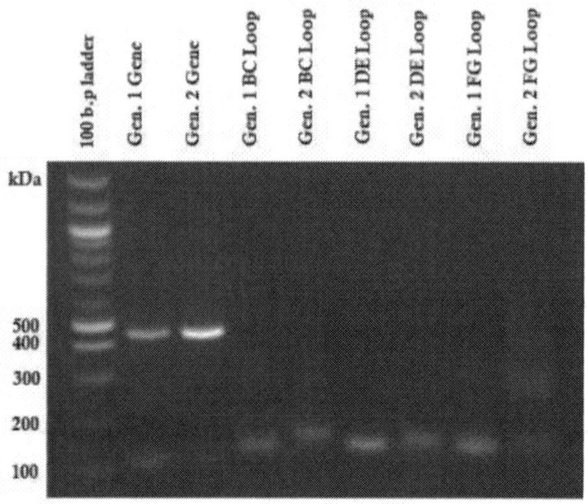


"Generation 3" FN3 α MSLN Library

# Experimental Design:

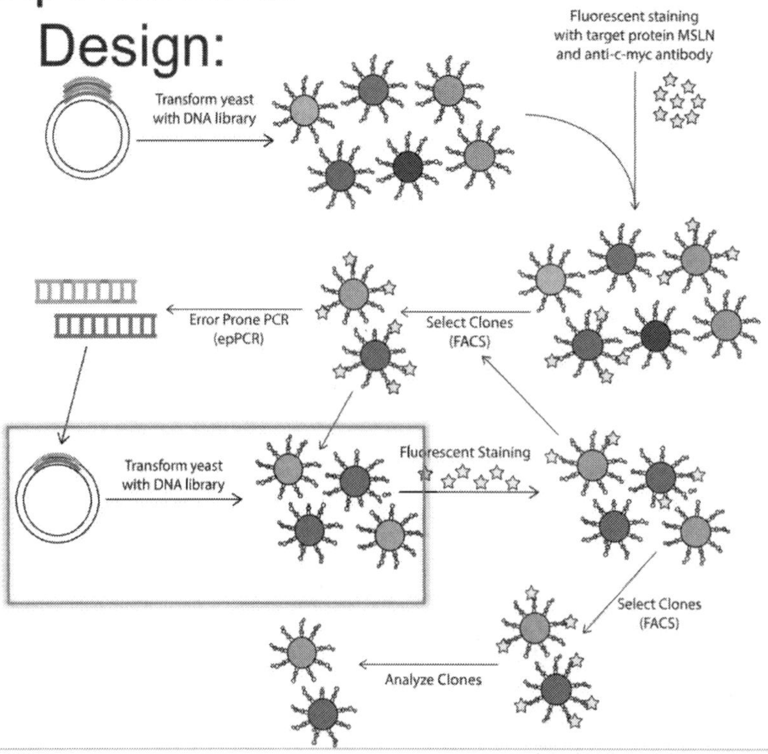

# 3° FN3 α MSLN library demonstrates full length clones and bind MSLN

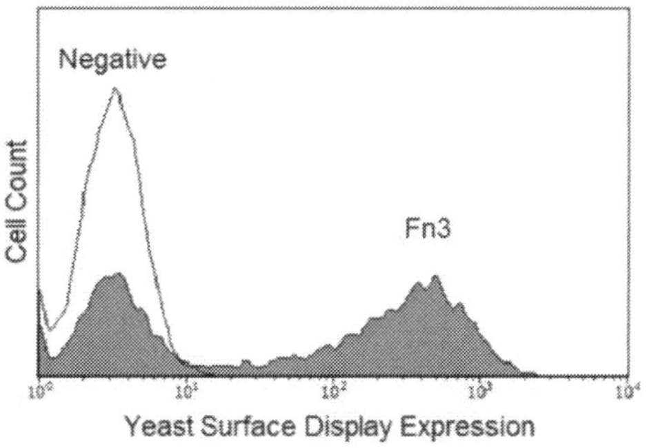


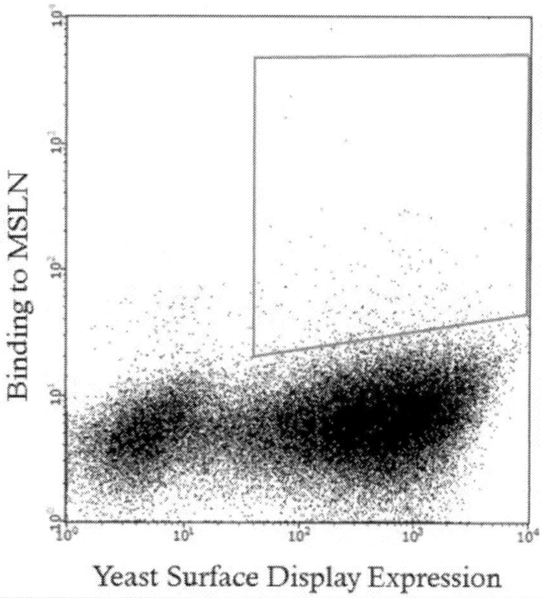


# Conclusions

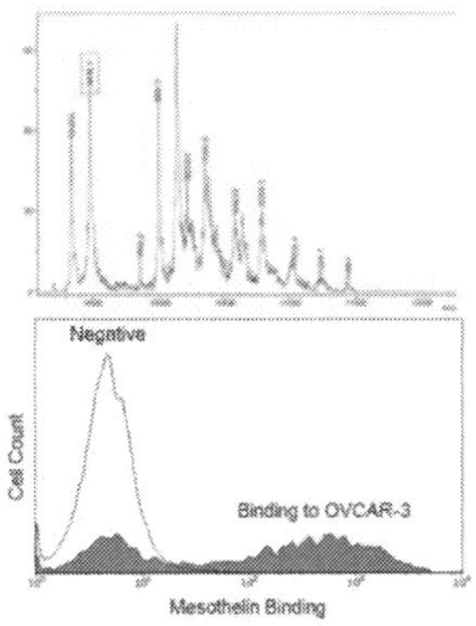


Produced MSLN minimal binding domain that is properly folded and binds MUC16

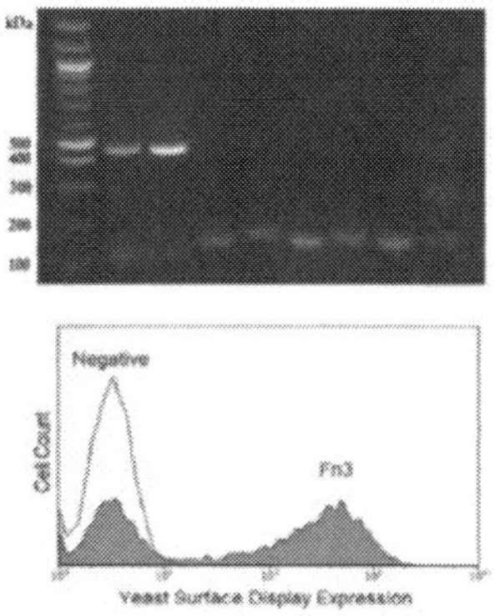


Introduced diversity by error prone PCR and transformed into yeast

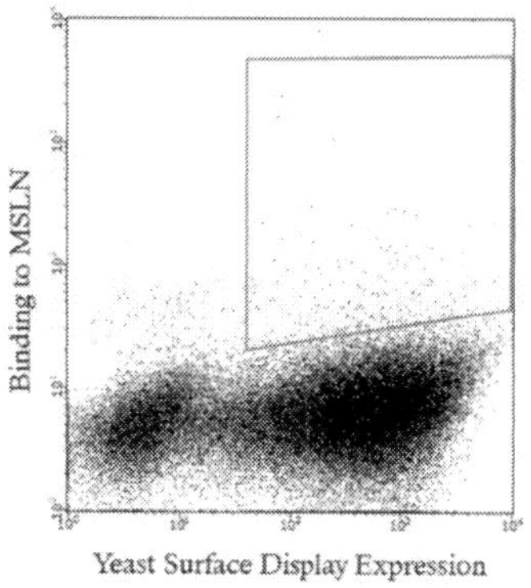


Initially selected for Fn3 variants demonstrating MSLN binding

# Future Work

- Continue sorting through 3° FN3 α MSLN library
- Solubly produce, purify and characterize engineered proteins
- Cell binding assays
- Tumor models in nude mice

# Acknowledgements and Funding

Sarah Moore
Katia George
The Moore Lab
Dr. Dane Wittrup, MIT
Dr. Benjamin Hackle, UMN
Dr. Stephen Eyles, UMass Amherst
Dr. Amy Burnside, UMass Amherst

Smith College
The Schultz Foundation
NIH/NCI R15 CA198927-01

Thank You!

Questions?

# MSLN promotes tumor cell survival and proliferation through multiple pathways

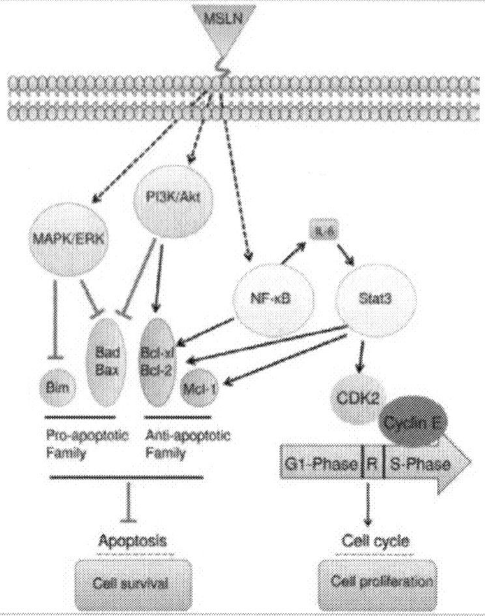


# MSLN promotes pancreatic tumor invasion and migration

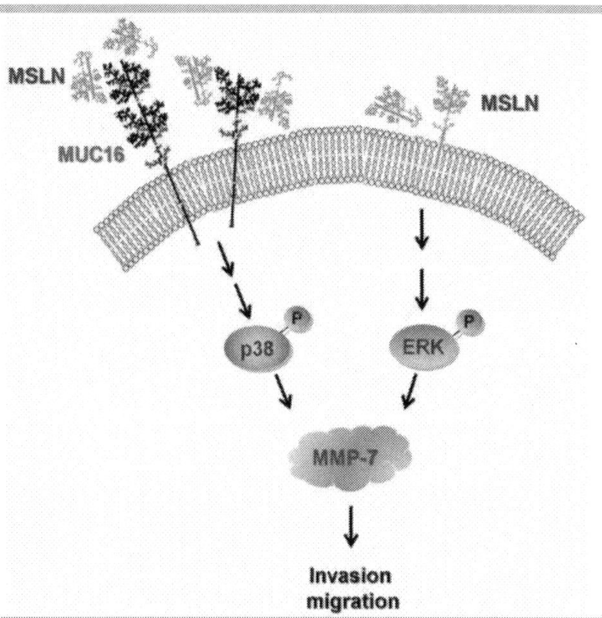

# Vector map of yeast surface expression pCT vector

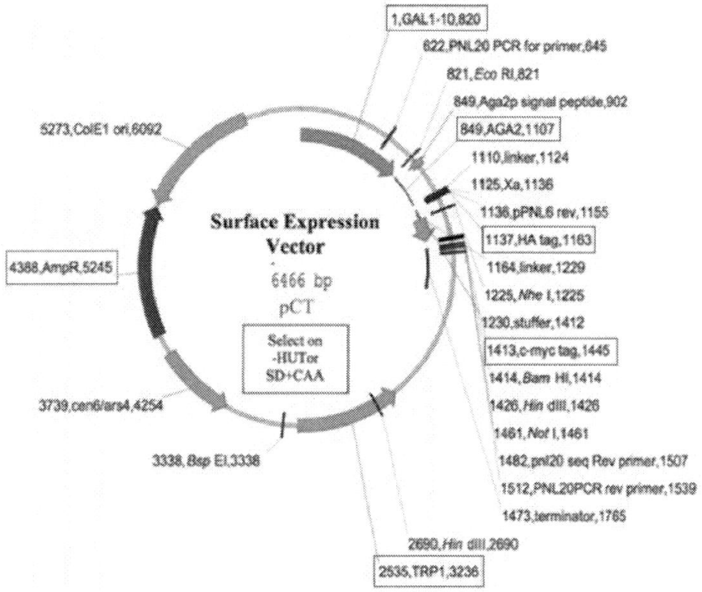


Chen, T.F., *et al.* (2013) *Methods Enzymol. 523,* 303-26

# FACS separates cells into sub-population based on fluorescent labeling

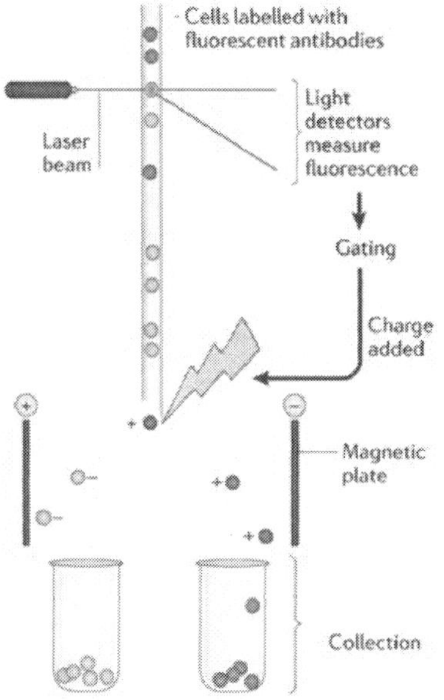

# Antibodies: Natural biomolecular scaffold for basic science and medicinal applications

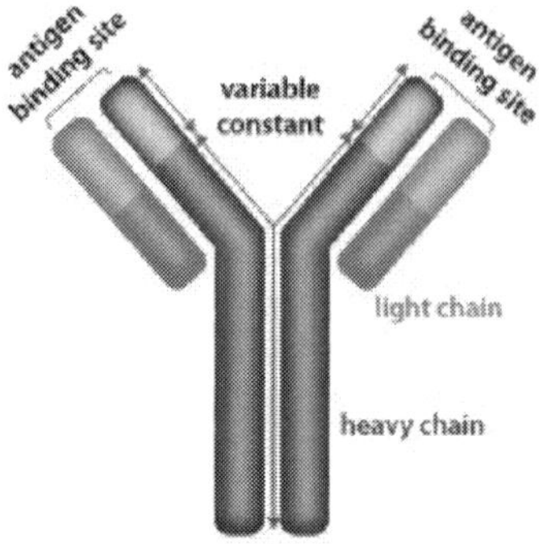


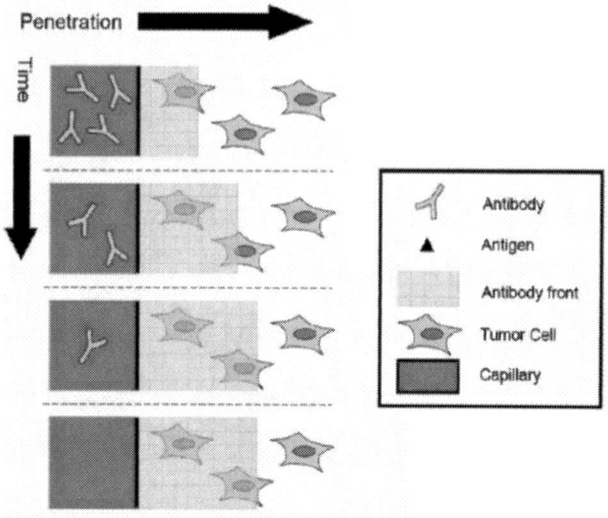


Thurber, G.M., *et al.* (2007) *J Nuc Med.* 48, 995-9

# Biotinylated MSLN no longer binds MUC16 on OVCAR-3 cells

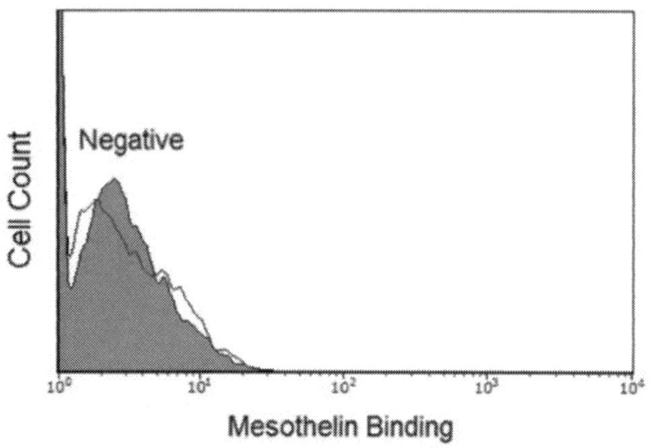

# Initially analyzed clones bind nonspecifically to Streptavidin

```
                BC Loop                        DE Loop                        FG Loop
WT Seq.   LISWDAPA-VTVRYYRITYGETGGNSPVQEFTVPGSKSTATISGLKPGVDYTITVYAVTGRGDSPASSKPIS
Seq. 1    LISWYYTN-FYVIYYRITYGETGGNSPVQEFTVPRPGSTATISGLKPGVDYTITVYAVTYRYAY---SHPIS
Seq. 2    LISWYYQHPHYARSYRITYGETGGNSPVQEFTVPDAHCNATISGLKPGVDYTITVYAVTSRMDGYYFSSPIS
```

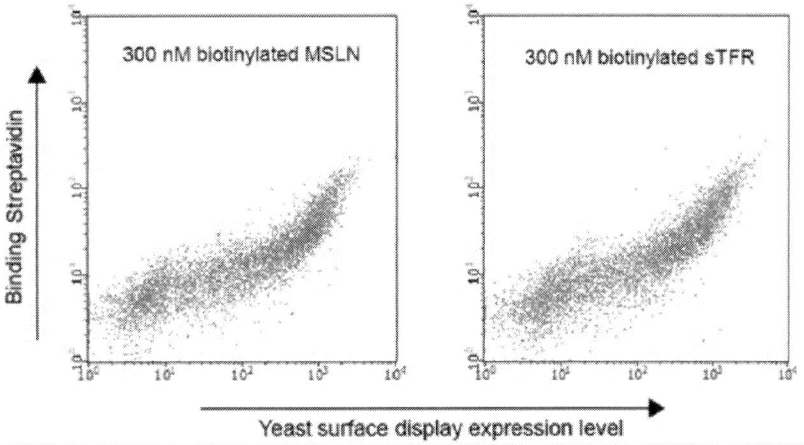


# Analysis of 3° Fn3 α MSLN demonstrates unique sequences

```
                BC Loop                        DE Loop                        FG Loop
Fn3 WT    LISWDAPA-VTVRYYRITYGETGGNSPVQEFTVPGSKSTATISGLKPGVDYTITVYAVTGRGDSPASSKPIS
20_Seq    LISWSNS---ALDYYRITYGETGGNSPVQEFTVPYT-TAATISGLKPGVDYTITVYAVTCTRVVS--SPPIS
18_Seq    LISWCSRH-SCAEYYRITYGETGGNSPVQEFTVPDASATATISGLKPGVDYTITVYAVTVRDLFF--SSPIS
19_Seq    LISWSTL--GYSCYYRITYGETGGNSPVQEFTVPEP-STATISGLKPGVDYTITVYAVTHRDT-D--SGPIS
11_Seq    LISWPYLSFPGSCYYRITYGETGGNSPVQEFTVPACKSTATISGLKPGVDYTITVYAVTLTGHC---SYPIS
13_Seq    LISWLYH--CDVRYYRITYGETGGNSPVQEFTVPGSSVAATISGLKPGVDYTITVYAVTLTGHC---SYPIS
16_Seq    LISWPSP--CYLYYYRITYGETGGNSPVQEFTVPMLYSPATISGLKPGVDYTITVYAVTDPY--W--SYPIS
15_Seq    LISWSDPYCNCAPYYRITYGETGGNSPVQEFTVPQYLSVATISGLKPGVDYTITVYAVTGRC--R--SLPIS
12_Seq    LISWNPSHDSYLVYYRITYGETGGNSPVQEFTVPGSFFIATISGLKPGVDYTITVYAVTVRR--I--SHPIS
14_Seq    LISWNLSYDLYSVYYRITYGETGGNSPVQEFTVPGSYHGATISGLKPGVDYTITVYAVTSKCVCY--SCPIS
```

# 424551 Covert Cancer Therapeutics: Engineering T Cells to Interrogate Intracellular Tumor Antigens

**Abstract:**

Adoptive T-cell therapy has shown tantalizing promise as a cancer treatment strategy, with several clinical trials reporting that T cells expressing chimeric antigen receptors (CARs) can eradicate tumors in patients with relapsed disease. However, CAR-T cells rely on receptor-mediated recognition of surface-bound antigens that are seldom tumor-exclusive, resulting in severe on-target, off-tumor toxicities that have led to patient deaths in clinical trials. There is a growing consensus that the lack of suitable antigens poses a major obstacle to the broad application of engineered tumor-targeting T cells. The ability to overcome T cells' reliance on surface antigen presentation and interrogate intracellular disease signatures would significantly expand the pool of detectable tumor markers and improve tumor-targeting specificity. Here, we present a novel strategy to reprogram T-cell–mediated cytotoxicity to interrogate intracellular disease signatures. We have engineered a switchable form of the cytotoxic protein Granzyme B (GrB) that is produced and delivered by T cells into target cells, but becomes active if and only if a tumor-associated protease is present inside the target cell. As a proof of concept, we have developed a GrB switch that responds to Sentrin-specific protease 1 (SENP1), an oncoprotein known to be overexpressed in prostate, pancreatic, and thyroid oncocytic tumor cells. We demonstrate that this GrB switch, termed cytoplasmic oncoprotein verification evaluator and response trigger (COVERT), is efficiently expressed and packaged by human T cells and properly trafficked to the immunological synapse between T cells and target cells. Furthermore, we show that COVERT is produced as an enzymatically inert protein that is activated by SENP1 in a dose-dependent manner. Finally, we describe designs to adapt COVERT into a modular platform technology that will expand the repertoire of candidate target antigens. We envision that COVERT can be utilized in combination with existing CAR technology to improve the tumor-targeting precision of cell-based immunotherapy.

# 425371 Protease-Activatable Viruses for Cardiovascular Disease Therapy

**Abstract:**

Cardiovascular disease (CVD) is the leading cause of death around the globe today. Gene therapy is an emerging technology that may be used to treat and cure a wide range of pathologies, including CVD. However, a major issue most researchers face is targeting specificity of gene vector delivery. One hallmark of cardiovascular disease (as well as most human cancers and neurological disorders) is increased protease activity at the affected site. In particular, matrix metalloproteinase-9 (MMP-9) plays a major role in the progression of CVD. Our lab has recently engineered a protease-activatable virus (PAV) based on adeno-associated virus serotype-9 (AAV9) that delivers genetic cargo to diseased cells only upon MMP-9 proteolysis. Briefly, a negatively charged tetra-aspartic acid ($D_4$) peptide "lock" was inserted throughout the virus protein capsid to prevent the virus' interactions with its primary cell receptor, galactose. This peptide lock is flanked by two protease-cleavable peptide sequences, which, upon MMP-9 proteolysis, release the $D_4$ lock and allow subsequent cell binding and internalization of the virus. A major advantage of our approach compared to other targeting approaches, such as targeting peptide insertion, is several diseased cells lack specific receptors that identify them as unique. Instead, PAVs recognize the diseased, MMP-rich tissue microenvironment and deliver genetic cargo only to areas that require therapy. We have already created and characterized over a dozen PAV variants that display varying specificity and kinetic activation for MMP-9. Our goal is to develop specialized PAVs that recognize the specific biomolecular protease signatures of CVD for targeted therapeutic gene delivery to the myocardium.

# 432039 Targeted Enzyme Prodrug Therapy for Prostate Cancer – a Comparative Study of L-Methioninase, Purine Nucleoside Phosphorylase, and Cytosine Deaminase

**Abstract:**

Prostate cancer is the most common non-skin malignancy and the second leading cause of cancer-related death in American men, yet remains essentially incurable. Since the introduction of prostate specific antigen (PSA) screening, the lethality of prostate cancer stems not from a lack of early detection but more commonly from the failure of loco-regional therapies creating a need for improved systemic therapies. Currently, most single-agent anticancer drugs face challenges due to increased multidrug resistance, pharmacokinetic limitations, and restricted clinical dosage or frequency of administration due to cytotoxicity in non-cancerous tissues.

Enzyme prodrug therapy shows promise for the treatment of solid tumors such as prostate cancer, but current approaches lack effective/safe delivery strategies. To address this, we have developed three enzyme-containing fusion proteins targeted via annexin V (AV) to phosphatidylserine exposed on the tumor vasculature and tumor cells, using the enzymes L-methioninase (MT), purine nucleoside phosphorylase (PNP), or cytosine deaminase (CD). In enzyme prodrug therapy, the fusion protein is allowed to bind to the tumor before a nontoxic drug precursor, a prodrug, is introduced. Upon interaction of the prodrug with the bound enzyme, an anticancer compound is formed, but only in the direct vicinity of the tumor, thereby mitigating the risk of side effects while creating high intratumoral drug concentrations. Prostate cancer is especially suited to this targeting strategy as prostate carcinomas have been shown to have approximately twice the vascular density of healthy prostate tissue. The applicability of these enzyme prodrug systems to treating prostate cancer has remained unexplored. Additionally, target availability may increase with the addition of low-dose docetaxel treatment to the enzyme prodrug treatment. Therefore, in this study we examined the binding strength and the cytotoxic efficacy of these enzyme prodrug systems on human prostate cancer cell line PC-3 and on human endothelial cells grown to mimic the tumor vasculature. We also examined the effect on PC-cells of low-dose docetaxel on the cytotoxic efficacy for the three enzyme prodrug systems.

Enzymatically active fusion proteins were recombinantly produced in *Escherichia coli* and purified. The binding to cells of all three fusion proteins was relatively strong, with dissociation constants of 1.5 nM and lower for both the endothelial and PC-3 cells. In the *in vitro* cytotoxicity studies, the system using the MT-AV fusion protein performed the best, giving greater than 95% killing of endothelial cells grown to mimic the tumor vasculature and near complete killing of the PC-3 cells. In the presence of only the selenomethionine prodrug, there was a slight or no effect on these cells. Low-dose docetaxel treatment was found to increase the cytotoxic effect of the annexin V-targeted therapeutics for the PNP-AV and CD-AV systems. In conclusion, the enzyme prodrug system using MT-AV appears to be the most promising as an effective treatment of prostate cancer.

# 432263 DNA Aptamer Switch for Enzymatic Prodrug Activation Against Cancer

**Abstract:**

Targeting-moiety conjugated prodrugs have more cancer-specific targeting and reduced background toxicity over traditional chemotherapy. One promising strategy is to conjugate the targeting moiety with a prodrug activating enzyme, termed suicide enzyme, instead of the prodrug itself. The suicide enzyme will catalytically turnover the prodrug into its toxic form, thus providing greater cancer killing than a direct prodrug conjugate. Here, we employ yeast cytosine deaminase (yCD), which converts the non-toxic prodrug, 5-fluorocytosine, into the toxic chemotherapeutic, 5-fluorouracil. To guarantee low nonspecific activity, we employ a split version of yCD which is inactive until the two yCD fragments are brought into close proximity.

We utilize HaloTag technology for efficient covalent linkage of DNA to split yCD proteins and use cancer-specific DNA aptamer switches to control the reconstitution of split yCD. The binding of the DNA aptamer to its cell surface protein target induces a structural shift. This shift exposes single-stranded sequences for hybridization and allows for the split yCD-DNA to be brought together and regain activity. DNA targeting moieties make this strategy modularly adaptable to various targets without disrupting the split yCD design. The switchable aptamer means the system is turned off unless it is bound to a cancer cell, upon which the suicide enzyme is turned on for cancer killing by amplified prodrug activation.

# 413347 Generation of Protease-Inhibiting Monoclonal Antibodies By Novel Paratope Design


**Abstract:**

As extremely important signaling molecules, proteases precisely control a wide variety of physiological processes, and thus serve as one of the largest families of pharmaceutical targets. Most therapeutic protease inhibitors currently in clinical use or under developments are peptides or their chemical compound mimics. Considering that 500-600 proteases have been identified in human genome, specificity is highly desired for any protease inhibition therapy. However, the reaction mechanism is highly conserved among the same class or family of proteases, thus achieving target specificity has been recognized as the main hurdle for protease inhibitor developments. The crucial importance of specificity has been highlighted by the extensive studies on inhibition of matrix metalloproteinases (MMPs) using zinc-chelating compounds as a strategy for treating cancer. Although pre-clinical results were promising, previous attempts focusing on development of chemical compound inhibitors, *e.g.,* hydroxamates, targeting broad-spectrum MMPs all failed in clinical trials due to severe side effects. With these aspects, antibody-based inhibitors are emerging as very attractive therapeutic agents due to their exclusive specificity given by a large antigen-antibody contact surface.

Inspired by the fact that a large proportion of antibodies isolated from camels and llamas can bind to active pockets and inhibit enzymatic reactions, we analyzed the sequences and known structures of inhibitory antibodies, and hypothesized that convex shaped paratopes encoding by ultra long CDRs are inhibition-prone. Large synthetic human antibody libraries ($>10^9$ variants) carrying extended CDR-H3 ranging from 23 to 27 aa were constructed, and subjected for isolation of antibodies targeting MMP-14. Notable, dozens of inhibitory antibodies have been isolated, which accounted for > 50% of identified binding clones. CDR-H3 sequences of these inhibitory clones showed dramatic patterns with central histidines flanked by arginine and lysine rich sequences. These residues presumably contribute to chelate the catalytic zinc and interact with negatively charged surface of active sites of MMP-14, respectively. Among isolated high affinity inhibitory clones, several showed exclusive selectivity toward MMP-14 over MMP-2/-9. Particularly, 3A2 Fab was a competitive inhibitor exhibiting a binding affinity of 4.85 nM by SPR analysis and an inhibition potency of 25 - 50 nM. Gelatin zymography and type-I collagen degradation using fibrosarcoma cell lines demonstrated its inhibitory function at physiological environments. We believe our novel inhibitory antibody discovery approach has great potentials to be applied for many biomedically important but challenging targets for conventional antibody techniques

# 433868 Cell-Free Protein Synthesis: A Dynamic Platform for Rapid Screening and Production of Cytotoxic Cancer Biotherapeutics

**Abstract:**

Cancer, the second leading cause of death in the United States, claims the lives of 7.6 million people worldwide every year. Protein-based anticancer therapeutics are promising alternatives to traditional chemotherapies. However, many such proteins pose challenges to recombinant *in vivo* expression, such as cytotoxicity and aggregation, which make them difficult, time-consuming and expensive to express and screen. In this work, we demonstrate how the open nature of cell-free protein synthesis (CFPS) can be harnessed to overcome these challenges. We established an *E. coli*-based cell-free protein synthesis method to produce and directly assay a cytotoxic anticancer therapeutic, onconase, as a model protein. This protein is difficult to express *in vivo* because of inclusion body formation and its ribonucleic activity which targets three main components of the translation process: mRNA, tRNA, and rRNA. We overcame these obstacles by directly adding tRNA at specific time intervals to the CFPS environment. As a result, production yield increased 56-fold to >1.8 mg per ml with more than 95% solubility. We also established an optimized method for tRNA purification, which reduced the cost of tRNA required by 13-fold. Finally, we demonstrate that our CFPS-produced onconase can be assayed for anticancer activity directly without any prior purification. The ability to directly assay for therapeutic function drastically reduces the time and cost required for protein engineering and selection.

# 431023 Can R&D be Standardized? Implementation of a Recipe Framework Supporting Lab to Patient Execution and Analytics

**Abstract:**

The benefits of a recipe framework have been recognized in manufacturing organizations for several years, but implementation in a development environment brings different challenges – and measurable benefits to drive appropriate scientific and engineering knowledge. To accomplish this task in a development environment a significant amount of structure needs to be implemented in the underlying laboratory systems, ensuring data can be captured with appropriate context. In this presentation, we will outline how we are implementing a flexible S88 recipe model across the organization to enable process definition, drive execution with context, and provide visualization of experiments/batches to compare this contextualized data across sites and scales. Detailed examples of the approach, resulting analysis tools, and challenges encountered, will be provided. This presentation will also outline how this framework can help a research and development organization easily analyze scale up and manufacturing processes, and enable transfer of process knowledge through development and commercialization.

# 414967 Enabling Rapid Elemental Analysis of Highly-Automated, Parallel Screening Studies for Pharmaceutical Process Development

**Abstract:**

In the pharmaceutical industry, utilization of lab automation for parallel, statistically designed experiments to optimize reagent and processing parameters allows for accelerated development of chemical processes. The resulting generation of comprehensive, high-fidelity data sets provides in-depth knowledge which ultimately informs a robust chemical manufacturing process. For a given chemical step that may require downstream removal of a metal catalyst, the use of X-ray fluorescence (XRF) enables rapid elemental analysis of samples for pharmaceutical process development. Together, automated parallel experimentation and XRF technology have proven to be a valuable partnership for the high-throughput analysis of large sample arrays for obtaining process knowledge and advancing pharmaceutical process development. Herein, we exhibit a case study where the combination of highly-automated design of experiments (DoE) studies and XRF allowed for rapid metals analysis and, subsequently, significant improvements in processing a late stage asset to remove trace metals.

# 423647 Pharmaceutical Process Development Using Computer Aided-Methods and Tools

**Abstract:**

A majority of pharmaceutical production systems involve batch processes, which have functioned well for pharmaceutical industries as well as the regulatory bodies. Batch vessels provide the flexibility of handling different unit operations in a multipurpose plant. However, batch processes are not very efficient for product quality assurance and have a number of drawbacks. Over the last decade the regulatory bodies required the pharmaceutical companies to demonstrate more process understanding. Moreover, financial drivers require profit expectations to be met due to increasing R&D cost and high competition [2]. Consequently, the pharmaceutical industry is expected to look for opportunities to rapidly evaluate the different alternatives for process improvement. To achieve this objective, Process Systems Engineering (PSE) methods and tools can play an important role, for example, (a) apply computer aided-methods and tools that are mature for other industries (such as chemical and petrochemical) also in solving problems in the pharmaceutical industries, (b) in the evaluation and implementation of alternative solutions and/or designs and (c) to evaluate opportunities for continuous manufacturing [1-2].

An integrated computer-aided framework for the development of pharmaceutical processes has been developed to tackle key research areas for pharmaceutical processes (such as continuous processing, separation and reaction processes, solvent selection/recycle/optimization and process intensification/integration [3]). The objective of the framework is to improve pharmaceutical process operation/design and to evaluate opportunities for continuous manufacturing through better process understanding of pharmaceutical processes by using systematic model-based methods. Computer aided-methods and tools are important for such a framework as they can assist in solving various problems such as property prediction, mixture separation (ternary and binary azeotropic separations, separation sequence and type), solvent selection and process modeling and in providing data through knowledge libraries. The developed framework consists of four main parts: reaction mechanism identification, reaction analysis, separation synthesis and design and finally process simulation, evaluation, optimization and operation.

To extend the applicability of the framework a solvent swap method has been developed and implemented in a computer-aided tool. The solvent swap method is based on the method proposed by Gani et al. 2006 [4] and is treated as a special solvent selection problem with additional criteria consideration. The objective of the method is the quick and reliable identification of swap solvents which are suitable for the following process step and it can increase the process efficiency (minimize solvent and product loss and reduce energy demands).

The application of the proposed integrated framework with the added features will be highlighted through the BHC (Boots and Hoechst-Celanese) patented synthesis pathway for ibuprofen and the production of region-someric products through Heck reactions [5]. The improved BHC synthesis of ibuprofen consists of three reaction steps [6] and it is used to verify the proposed integrated framework. Finally, the application of the solvent swap method using the computer-aided tool for both examples will be illustrated.

References:

[1] K. Plumb, 2005. Chem. Eng. Res. Des., 83, 730.

[2] K.V. Gernaey, A. E. Cervera- Pardell, J. M. Woodley, 2012. Comput. Chem. Eng., 42, 30

[3] C. Jiménez-González et al., 2011. Org. Process Res. Dev. 15(4), 900.

[4] R. Gani, M. Jones, I. Powell, J. H. Atherton, J. L. Cordiner, 2006. Chem. Eng. 30.

[5] R. L. Hartman, J. R.Naber, S. L. Buchwald, & K. F. Jensen, 2010. Angew. Chem. Int. Ed. Engl., 49(5), 899.

[6] Elango, V. Murphy, M., Smith, B.L., Davenport, K.G., Mott, G.N., Zey, E.G., Moss, G.L. (Hoechst Celanese Corp.), 1990. European Patent 4.981.995

**Accuracy of Micro Powder Dosing and Feeding via a Vibratory Sieve-Chute System**

**M.O. Besenhard[1,2], E. Faulhammer[1,2], S. Fathollahi[1,2], G. Reif[2], V. Calzolari[3], S. Biserni[3], A. Ferrari[3], S.M. Lawrence[4], M. Llusa[1], A. Paudel, J.G. Khinast[1,2*]**

1: Research Center Pharmaceutical Engineering (RCPE) GmbH, 8010 Graz, Austria
2: Graz University of Technology, Institute of Process and Particle Engineering, 8010 Graz, Austria
3: MG2, Via del Savena 18, I-40065, Pian di Macina di Pianoro, Bologna, Italy

4: GlaxoSmithKline, New Frontiers Science Park, Third Avenue, Harlow, Essex CM19 5AW, UK

* Corresponding author:

*khinast@tugraz.at*

**Abstract:**

This presentation describes our current studies on powder dosing [1] and feeding via a vibratory sieve-chute system, i.e., the MG2's Microdose®. Figure 1 shows the operating principle and parts of the set-up. The chute is tilted at a fixed angle of 5° and sieves with 5-10 holes of e.g. 0.7mm in diameter are fixed on its top. The powder was discharged from the sieve into the chute and the capsule body using gravity. Vertical vibration occurred with a broad range of frequencies and amplitudes. The fill weight ($fw$ $fw$) during dosing events was recorded via a capacitance sensor, which had two parallel electrode plates encompassing the capsule body. The electrical field and the capacitance varied depending on of the powder quantity in the capsule. Due to this scale, fill weights could be recorded dynamically facilitating the analysis of filling characteristics, i.e., the fill rates and their robustness.

First, the range of frequencies and amplitudes was screened for settings that facilitated reasonable (no blocking of spilling in the sieve) fill rates for three lactose powders, serving as model substances. The filling characteristics were studied in detail within this operating space.

The filling rate was determined by averaging the slopes ( $\Delta fw(t)\ \Delta t\ [mg\ s]$ $\Delta fw(t)/\Delta t\ [mg/s]$ ) of $30$ $30$ filling profiles. The higher the probabilities of an unsteady filling behavior (as shown in Figure 5 b) or fluctuations in the filling rate are, the higher is the standard deviation ([mg/s]) of the evaluated slopes. Therefore, the latter can be used to characterize the robustness of the fill rate, which is of prime importance for selecting the settings for continuous (micro) feeding using such a system.

The results reveal similar operating spaces for all powders and the fill rate robustness varied distinctly in the operating space. Figure 2 depicts the results for one of the studied powders. Utilizing the gained knowledge on filling characteristics micro dosing studies have been performed, dosing capsules with 2.5 mg. These studies compared accuracy of dosing via a constant frequency and amplitude or a constant frequency but the amplitude manipulated by a PID controller. The results reveal that dosing was the most accurate when using constant settings, since the usage of PID control caused inhomogeneous powder beds on the chute, as shown in figure 3.

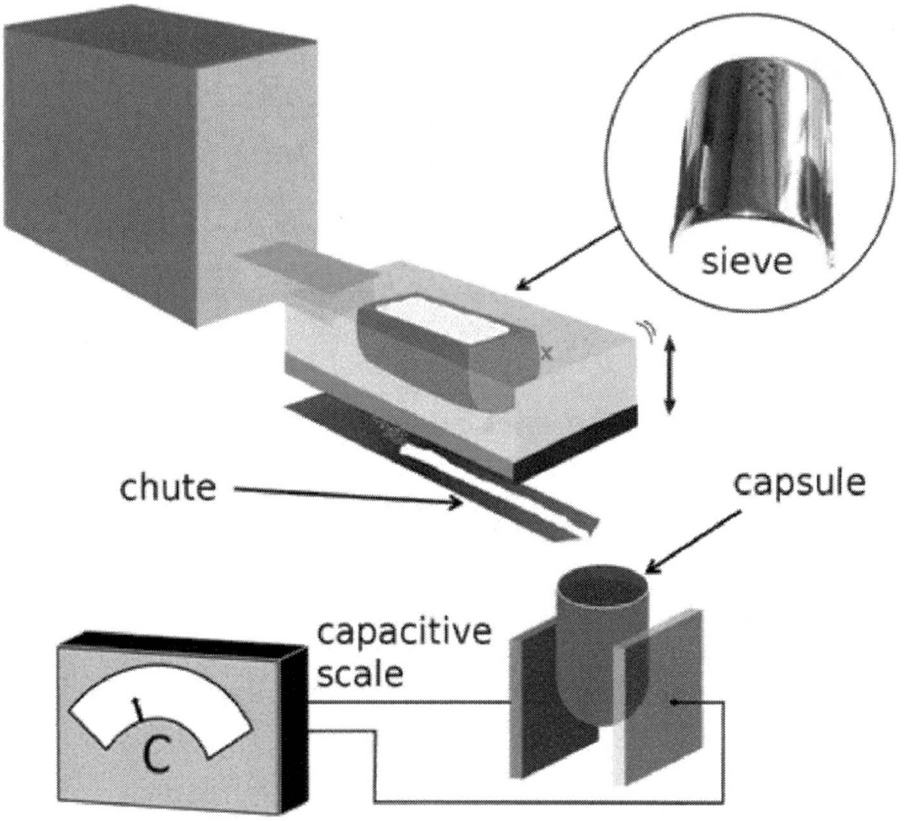


Figure SEQ figure * ARABIC 1: Principle of powder dosing via a vibratory sieve-chute system. The inlet shows a photo of the sieve used, and the red x depicts the position of an acceleration probe used to study the oscillation patterns.

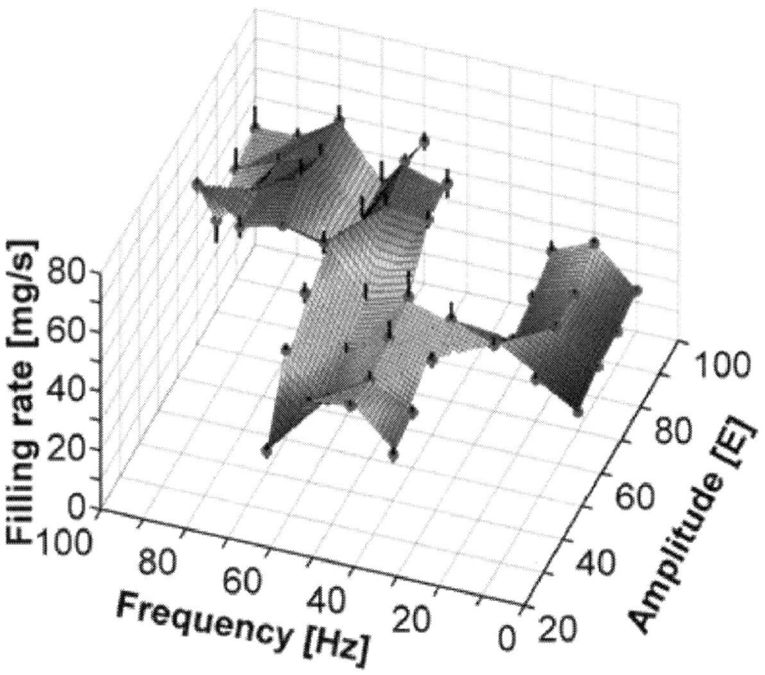


Figure 2: Filling rate and its standard deviations (vertical bars) for InhaLac 230 in the amplitude and frequency settings of the operating space.

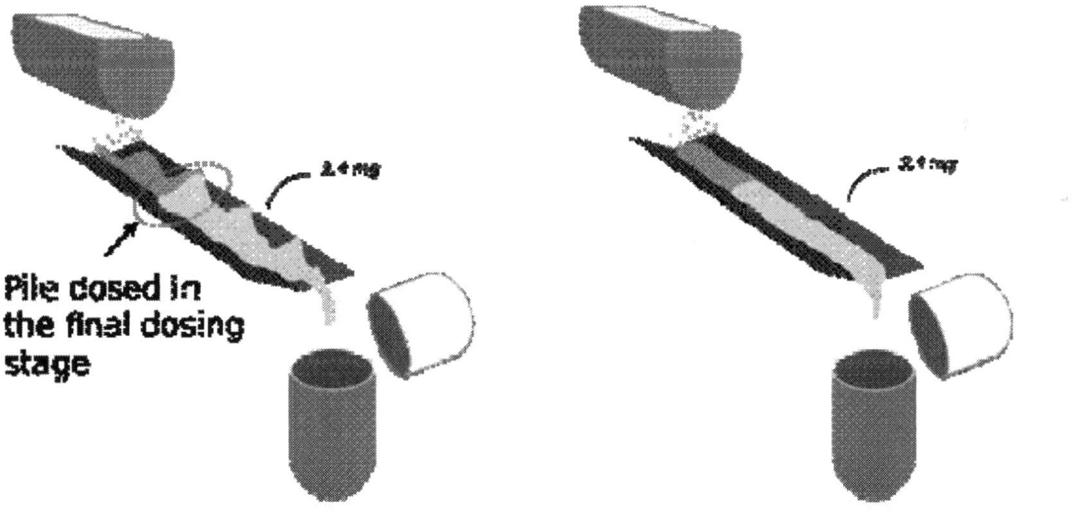


Figure 3: Powder bed in the chute during a PID controlled dosing (left) and dosing using a constant amplitude (right).

[1] Besenhard, M.O., Faulhammer, E., Fathollahi, S., Reif, G., Calzolari, V., Biserni, S., Ferrari, A., Lawrence, S.M., Llusa, M., Khinast, J.G. (2015). Accuracy of Micro Powder Dosing via a Vibratory Sieve-Chute System. *European Journal of Pharmaceutics and Biopharmaceutics,* accepted.

# 431162 Chromatographic Tools to Predict the Stability of MAbs for Faster Identification of Therapeutic Candidates

**Abstract:**

**Chromatographic Tools to Predict the Stability of mAbs for Faster Identification of Therapeutic Candidates**

S. Hedberg*, J.Y.Y Heng*, D.R. Williams* and J. Liddell**

*Surface and Particle Engineering Laboratory
Department of Chemical Engineering
Imperial College London, SW7 2BY,
United Kingdom

**Research and Development
Fujifilm Diosynth Biotechnologies
Billingham, TS23 1LH, United Kingdom

## ABSTRACT

Protein-protein molecular interactions are known to be involved in protein solution aggregation behaviour and are a common issue for the manufacturing of therapeutic proteins such as mAbs. Much effort has been employed to gain a better understanding of aggregation, however the mechanisms leading to protein aggregation are still not fully understood. The osmotic second virial coefficient ($B_{22}$) is a fundamental physiochemical property that describes protein-protein interactions solution, which can be a useful tool to predict aggregation propensity of proteins.

One way of predicting aggregation propensity is self-interaction chromatography (SIC), which recently have shown to be a promising tool for better understanding of phase behaviour of proteins. Another technique, cross-interaction chromatography (CIC), has shown to be an even more high-throughput technique than its predecessor with possibly the same capabilities. This work consists of two experimental studies with therapeutic mAbs to improve SIC and CIC as a tool to predict protein aggregation.

The first part includes a 10 times scale-down study of therapeutic mAbs from laboratory scale macro-columns to micro-scale columns, which will enable the determination of $B_{22}$ for the individual protein as well as the cross-virial coefficient, $B_{23}$, between two proteins. Micro SIC and CIC uses only a few milligrams of mAb in order to obtain a complete formulation study. The results from the first part of the study proved to give good comparable results between the micro and macro scales enabling the use of micro SIC for $B_{22}$ determinations.

The second part of this work presents an extensive formulation study of mAbs, varying pH and salt, as well as the presence of different stabilisers as well as different external factors known to induce aggregation. The $B_{22}$ and $B_{23}$ values determined from the formulation study are then correlated with aggregation data obtained from size-exclusion chromatography. It was shown that over all test conditions, good correlations could especially be found between $B_{22}$ and aggregation rate.

In summary this work hopes to lead to a better insight into the mechanisms of aggregation by looking at the extent of tailing of the peaks in the chromatograms and comparison with conformational studies.

## REFERENCES

1. Dumetz, A.C., Snellinger-O'Brien, A.M., Kaler, E.W., and Lenhoff, A.M. (2007) *Patterns of Protein–Protein Interactions in Salt Solutions and Implications for Protein Crystallization.* Protein science. **16**(9): p. 1867-1877.

2. Quigley, A. (2012) *Protein Aggregation Behaviour and the Second Virial Coefficient,* Ph.D. Thesis, Imperial College London.

3. Tessier, P.M., Sandler, S.I., and Lenhoff, A.M. (2004) *Direct Measurement of Protein Osmotic Second Virial Cross Coefficients by Cross-Interaction Chromatography.* Protein science. **13**(5): p. 1379-1390.

4. Jacobs, S.A., Wu, S.-J., Feng, Y., Bethea, D., and O'Neil, K.T. (2010) *Cross-Interaction Chromatography: A Rapid Method to Identify Highly Soluble Monoclonal Antibody Candidates.* Pharmaceutical Research. **27**(1): p. 65-71.

# Advancements of Accelerated Stability Studies and Modeling through Use of Nonlinear Kinetics and Bootstrapping

GlaxoSmithKline

**Don Clancy, Neil Hodnett, Rachel Orr, Martin Owen, and John Peterson**

**Primary corresponding author email: Donald.J.Clancy@gsk.com**

## What is Accelerated Stability? – 4 Steps

- Conduct a short term designed experiment at elevated temperature with varying humidity.

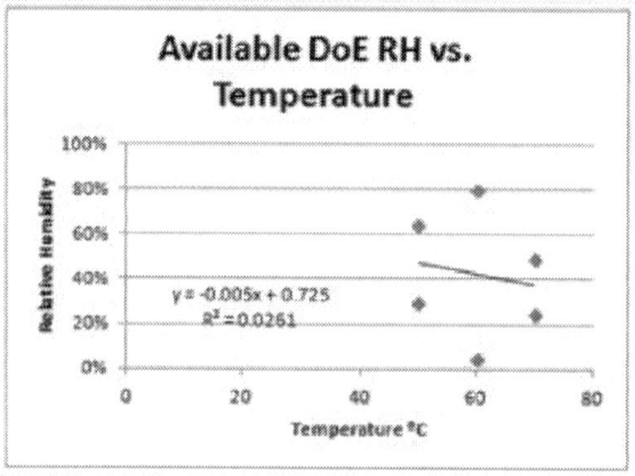


- Fit kinetic models for impurities and physical changes using short term data under accelerated conditions.
- Develop a long term model to predict container humidity and moisture content for a typical medicine bottle or blister pack.
- Extrapolate kinetic models from short term studies and integrate them with the long term humidity model to predict long term stability.

# Background and Current Available Approach

- Currently, the most widely published approach for modeling drug product degradation based on accelerated stability data is that described by Waterman (Waterman, 2012), (Adami & Waterman, 2004). This is based on the modified Arrhenius equation shown in Equation (1):

$$ln(k) = ln(A) - \frac{Ea}{R \cdot T} + B \cdot (RH) \quad (1)$$

Where "A", "Ea", and "B" are fitted terms, "R" is the gas constant, "T" is temperature in Kelvin, and "k" is defined as the change in impurity with respect to time.

Rearranging into a typical kinetic rate expression:

$$\frac{dy}{dt} = A \cdot exp\left(\frac{-Ea}{R \cdot T}\right) \cdot exp(B \cdot RH) \quad (2)$$

Makes the implicit assumption the rates of reaction is an exponential function of RH, which implies moisture affects molecular mobility similar to temperature.

# Background Continued - Isoconversion

- Equation 1 is further reduced to predict time to failure (t.alpha), or "isoconversion."

$$ln\left(t_{\alpha}\right) = \frac{Ea}{R \cdot T} - B \cdot RH - C \quad (3)$$

- This approach often requires significant extrapolation or interpolation of data to a failure limit to estimate the time till failure for each condition.

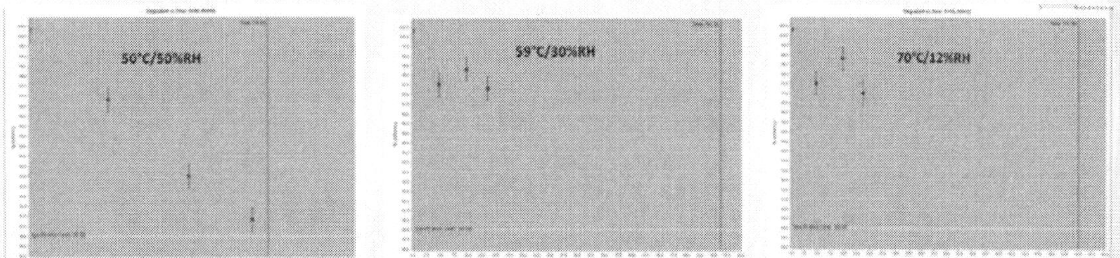


Drug Product "N" Isoconversion Time Estimations from ASAP Prime as reported by Waterman.

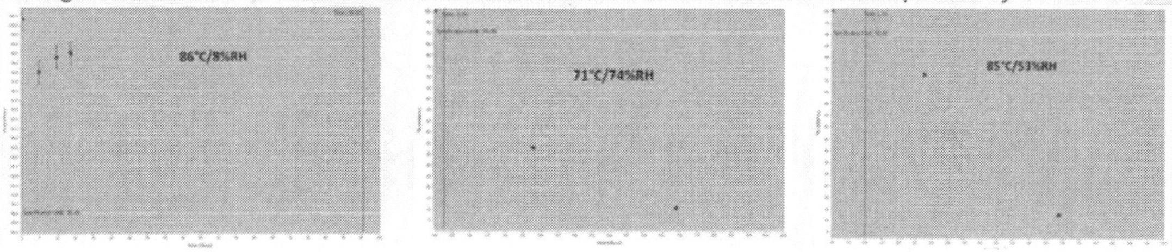

## Kinetic Model Development With a Different Starting Assumption

Change in reaction conversion (called alpha in this analysis) as a solid API reacts with water is typically related to the "activity" of the reactants, raised to a power, as:

$$\frac{d\alpha}{dt} = k \cdot \left[ H2O_{activity} \right]^{N1} \tag{4}$$

The "k" term in these equations is an Arrhenius factor relating an exponential relationship between temperature "T", the activation energy of the reaction "Ea" and a gas constant "R". Expanding on the Arrhenius term for equation (5) yields (7):

$$\frac{d\alpha}{dt} = K1 \cdot exp\left(\frac{-Ea}{R \cdot T}\right) \cdot RH^{N1} \tag{7}$$

Where, K1 is a pre-exponential term called the "frequency factor." For simplicity of nomenclature and units in the models described in this report, the activation energy divided by the gas constant (Ea/R) will be simplified to a fitted parameter "K2".

$$\frac{d\alpha}{dt} = K1 \cdot exp\left(\frac{-K2}{T}\right) \cdot RH^{N1} \tag{8}$$

## Kinetic Model Development - Continued

Fitting Equation 8 can be challenging as K1, and K2 are correlated in an exponential relationship. A way of reducing this exponential correlation between K1 and K2 is to scale the exponent to a value of 1.0 at the average temperature in the study (Tref) by modifying equation 8 to create equation 9.

$$\frac{d\alpha}{dt} = K1 \cdot exp\left(\frac{-K2}{T} + \frac{K2}{T_{ref}}\right) \cdot RH^{N1} \tag{9}$$

What is actually fitted to the experimental data is the integrated form of the kinetic models, as the final impurity value at time "t" is what is observed. Integrating (9) with respect to time yields (10):

$$\alpha = K1 \cdot exp\left(\frac{-K2}{T+273.15} + \frac{K2}{Tref}\right) \cdot (H)^{N1} \cdot \Delta t + C \tag{10}$$

Where "C" is the value of alpha at time zero, and "H" is the humidity expressed as either RH or Vapor pressure. The type of humidity dependence will need to be tested within the experimental analysis as two alternate models.

## Accounting for the Potential Curvature With Respect to Time

| Model Type | Description | Typical Application | Nonlinear Model (Integrated Form) Used for Fitting: | Limitations | Terms Needed to Fit |
|---|---|---|---|---|---|
| A | Reaction slows exponentially with time | Reaction rate proportional to a limiting reagent that is close to being consumed | $\alpha = C + K3 \cdot \left(1 - exp\left(-K1 \cdot exp\left(\frac{-K2}{T+273.15} + \frac{K2}{Tref}\right) \cdot H^{N1} \cdot \Delta t\right)\right)$ | | 4 |
| B | Reaction slows down with time | Diffusion | $\alpha = K1 \cdot exp\left(\frac{-K2}{T+273.15} + \frac{K2}{Tref}\right) \cdot (H)^{N1} \cdot \Delta t^{N2} + C$ | N2<1 | 4 |
| C | Reaction is linear with time | Zero order reactions | $\alpha = K1 \cdot exp\left(\frac{-K2}{T+273.15} + \frac{K2}{Tref}\right) \cdot (H)^{N1} \cdot \Delta t + C$ | | 3 |
| D | Reaction speeds up with time | Power Law Induction Reactions | $\alpha = K1 \cdot exp\left(\frac{-K2}{T+273.15} + \frac{K2}{Tref}\right) \cdot (H)^{N1} \cdot \Delta t^{N2} + C$ | N2>1 | 4 |
| E | Reaction speeds up exponentially with time | Nucleation, or reaction rate dependent on product concentration | $\alpha = exp(C) \cdot exp\left(K1 \cdot exp\left(\frac{-K2}{T+273.15} + \frac{K2}{Tref}\right) \cdot (H)^{N1} \cdot \Delta t\right)$ | | 3 |

**Table 1** Kinetic Equations Available in the Accelerated Stability Dashboard

- Each of these 5 possible curvature types with respect to time is tested with either relative humidity or vapor pressure in the "H" term.
- All models are fitted and then compared.
- Data with large extents of conversion (>15%) may need an Avrami transform if exponential.

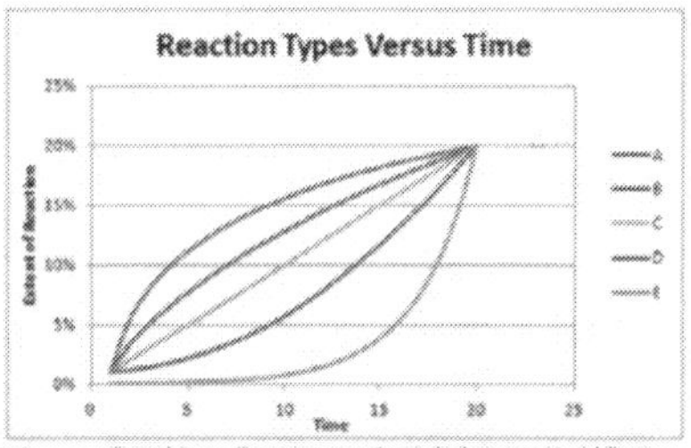


**Figure 1** Typical Curve Shape Associated with Each Kinetic Model Type

## Scientific Rational for Initial Model Selection

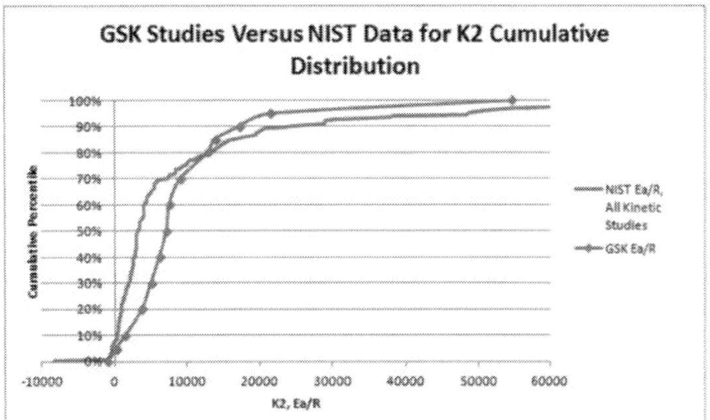


**Figure 2** Cumulative Distribution of Ea/R for GSK studies Compared to All Organic Reactions in NIST Data Set

- To help decide if a potential kinetic model might be real, the activation energy K2 value is compared to the known potential distribution of activation energies to assess how unusual it is.
- Models with extremely unusual activation energy values are eliminated from further consideration.

## Typical Power on Humidity Term (N1)

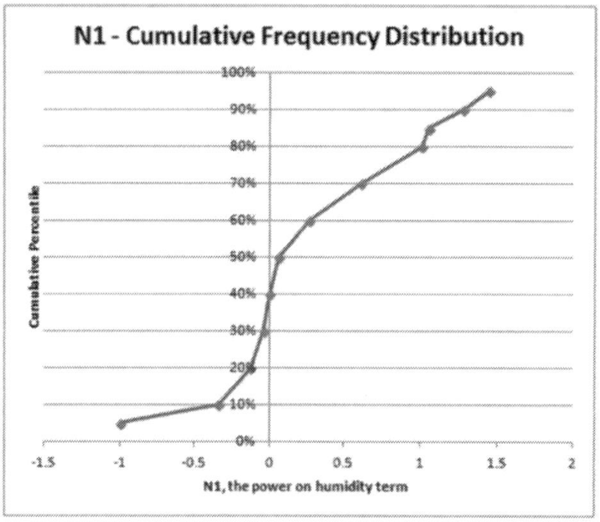


**Figure 5** Cumulative Percentile Plot of N1 for GSK Accelerated Stability Studies

- From 120+ impurity analysis, the cumulative frequency distribution for the humidity power (N1) is found to typically fall between -1 and 1, which is the opposite of an exponential curve with respect to RH.
- There is a long tail to the right with N1 values as high as 6 being noted for some unusual hydrolysis reactions in the top 5% of the distribution.

## Example Kinetic Model Selection in GSK Software within JMP

**Potential Kinetic Models and Parameters**

| Model Type | K1 | K2 | K3 | N1 | N2 | nParams | RMSE | Score | AICc | BIC |
|---|---|---|---|---|---|---|---|---|---|---|
| RH Linear Time Kinetic Model | 0.69 | 8374 | - | 3.34 | - | 4 | 0.1115 | - | -22.964 | -21.742 |
| AH Linear Time Kinetic Model | 118 | -8801 | - | 3.34 | - | 4 | 0.1115 | - | -22.966 | -21.744 |
| RH Accelerating Kinetic Model | 4.4 | -17419 | - | 11.19 | - | 4 | 0.3514 | - | 25.237 | 26.459 |
| AH Accelerating Kinetic Model | 1 | 1000 | - | 1 | - | 4 | 1.3584 | - | 82.024 | 83.247 |
| RH Decelerating Kinetic Model | 0.33 | 12601 | 3.7 | 3.35 | - | 5 | 0.0221 | - | -88.206 | -87.939 |
| AH Decelerating Kinetic Model | 57.33 | -4588 | 3.7 | 3.34 | - | 5 | 0.0221 | - | -88.3 | -88.032 |
| **RH Power Model** | 1.9 | 7500 | - | 3 | 0.5 | 5 | 0 | - | -848.163 | -847.896 |
| AH Power Model | 193.48 | -7937 | - | 3 | 0.5 | 5 | 0.0001 | - | -312.928 | -312.66 |

**Table 2** Fit of All Possible models to Data Set Created by an RH Model, with K2 Value of 7500, and N2 Value of 3.0, as applied to a typical Accelerated Stability study

- Potential kinetic models fit with highly unusual K2 and N1 parameters are first eliminated, as shown in red.
- Of the remaining models, the model with the best (lowest) Bayesian Information Criterion (BIC) is then chosen.

## What Can Go Wrong in Accelerated Stability? Nitrazepam Stability (1977)

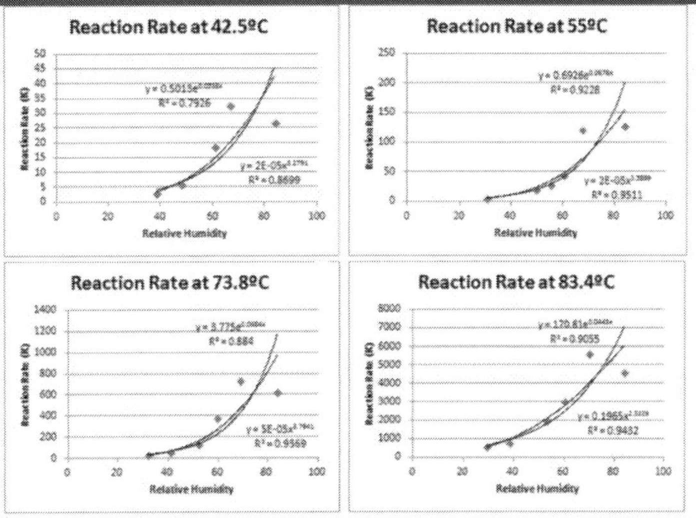


**Figure 6** Data Visualization for Reaction Rate for Disappearance of nitrazepam at varying temperatures

The analysis of nitrazepam stability data from Kesselring's and Genton's 1977 paper (Kesselring & Genton, 1977) is of particular interest as it's an unusual case due to high level of humidity sensitivity and is a good example of what can go wrong in an Accelerated Stability study.

This study also has historical significance as it is referenced in Waterman's papers to support the hypothesis that the kinetics of degradation for drug products can best be modeled as an exponential function of relative humidity.

## Analysis of Nitrazepam Stability

The data in the paper is expressed as reaction rate "K" which is the change in nitrazepam purity with respect to time. Our standard model given as equation 11 is simplified to equation (11) to predict K.

$$\frac{dy}{dt} = K = K1 \cdot exp\left(\frac{-K2}{T+273.15} + \frac{K2}{335.83}\right) \cdot (RH)^{N1} \tag{11}$$

Waterman's equation from equation 2, the "RH Exponential Model" is expressed as equation (12) using the standard nomenclature developed in this paper.

$$\frac{dy}{dt} = K = K1 \cdot exp\left(\frac{-K2}{T+273.15} + \frac{K2}{335.83}\right) \cdot exp(B \cdot RH) \tag{12}$$

- Since the Nitrazepam data set was the primary reference in prior work to the hypothesis that reaction rates are proportional to the exponential of RH, both equation forms will be tested.

## Comparison of Model Fits

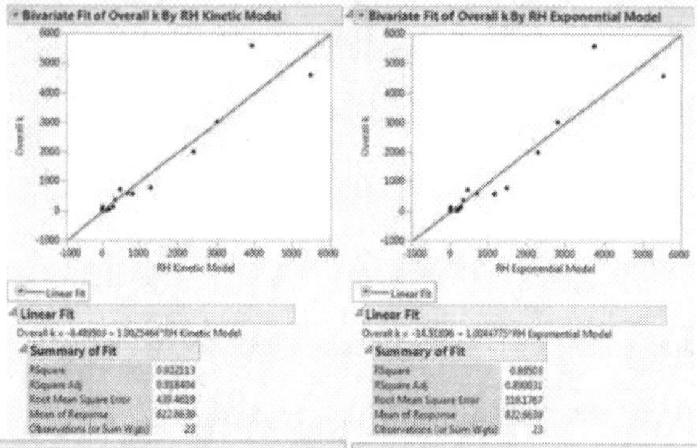


Solution (Left)

| SSE | DFE | MSE | RMSE |
|---|---|---|---|
| 4056908.012 | 20 | 202845.4 | 450.38361 |

| Parameter | Estimate | ApproxStdErr |
|---|---|---|
| K1 | 0.0167416035 | 0.02574611 |
| K2 | 26757.744805 | 5939.75581 |
| N1 | 1.8220013208 | 0.27078848 |
| C | 0 | 0 |

Solution (Right)

| SSE | DFE | MSE | RMSE |
|---|---|---|---|
| 5469488.3438 | 20 | 273474.42 | 522.94781 |

| Parameter | Estimate | ApproxStdErr |
|---|---|---|
| K1 | 4.8687759394 | 6.08294001 |
| K2 | 26807.829412 | 6997.3303 |
| C | 1e-100 | 0 |
| B | 0.0285954924 | 0.00467036 |

**Figure 7** Fits for Linear Kinetic Model as Derived versus Waterman's Exponential RH Model (Right)

- This data does not support the hypothesis that this reaction has an exponential rate with respect to humidity. A typical kinetic equation fits the data slightly better.

## Dropping Highest Temperature Data Due to Change in Activation Energy

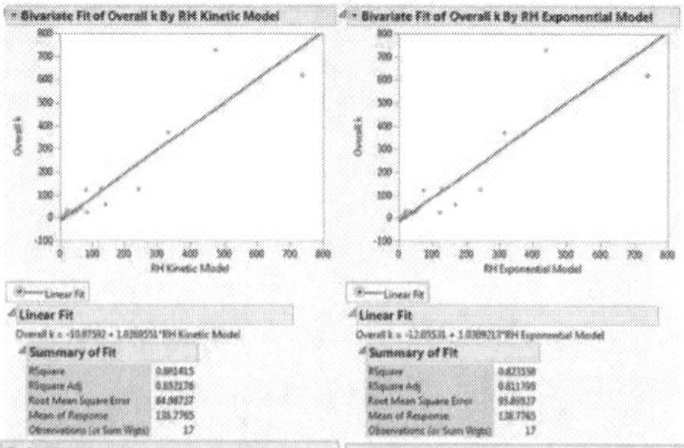


Solution (Left)

| SSE | DFE | MSE | RMSE |
|---|---|---|---|
| 109627.14446 | 14 | 7830.5103 | 88.490171 |

| Parameter | Estimate | ApproxStdErr |
|---|---|---|
| K1 | 0.0106708776 | 0.02166786 |
| K2 | 10458.983857 | 2823.98606 |
| N1 | 2.2911430108 | 0.4702406 |
| C | 0 | 0 |

Solution (Right)

| SSE | DFE | MSE | RMSE |
|---|---|---|---|
| 139482.01185 | 14 | 9963.0008 | 99.814833 |

| Parameter | Estimate | ApproxStdErr |
|---|---|---|
| K1 | 14.846054948 | 9.50787154 |
| K2 | 10364.598576 | 3210.1629 |
| C | 1e-100 | 0 |
| B | 0.0348019883 | 0.00768587 |

Figure 8 Fits for Normal Kinetic Model by Equation 11 (Left) versus Waterman's Exponential Model (Right) for Overall Disappearance of nitrazapem, excluding data at highest temperature in study

- If the activation energy markedly changes as the highest temperature data is dropped, then likely there was a phase change, form change, or some other physical change impacting the results. Here the estimate changed from K2 value of 26,800 to 10,500.

- **RECOMMENDATION:** Get DSC and/or TGA data up to 90 degrees prior to the study.

## Final Model Fit – Dropping Highest Humidity Data

**Solution** (Left)

| SSE | DFE | MSE | RMSE |
|---|---|---|---|
| 3620.9308018 | 11 | 329.17553 | 18.143195 |

| Parameter | Estimate | ApproxStdErr |
|---|---|---|
| K1 | 7.6732252e-8 | 8.45784e-8 |
| K2 | 10836.303303 | 770.225387 |
| N1 | 5.1846287097 | 0.26325205 |
| C | 0 | 0 |

**Solution** (Right)

| SSE | DFE | MSE | RMSE |
|---|---|---|---|
| 4957.4557942 | 11 | 450.6778 | 21.229173 |

| Parameter | Estimate | ApproxStdErr |
|---|---|---|
| K1 | 0.6966261868 | 0.22797375 |
| K2 | 10723.079571 | 899.058984 |
| C | 1e-100 | 0 |
| B | 0.086179662 | 0.00482877 |

Figure 9 Parameter Fits for Normal Kinetic Model by Equation 11 (Left) versus Waterman's Exponential Model (Right) for Overall Disappearance of Nitrazapem, excluding data at highest temperature and highest humidity in study

- Dropping highest humidity in this study also results in significant changes in the estimate of the humidity sensitivity. Likely some change in physical behavior above 75% RH.
- Leads to conclusion that experimental designs need to have the flexibility to test for changes in humidity sensitivity at extremes in humidity.
- Finally, while the impact of RH may have looked exponential in this case, it fits standard kinetics consistently better. High values of N1 as high as 6 have been observed for some unusual hydrolysis reactions.

## Initial Experimental Designs - Hexagon

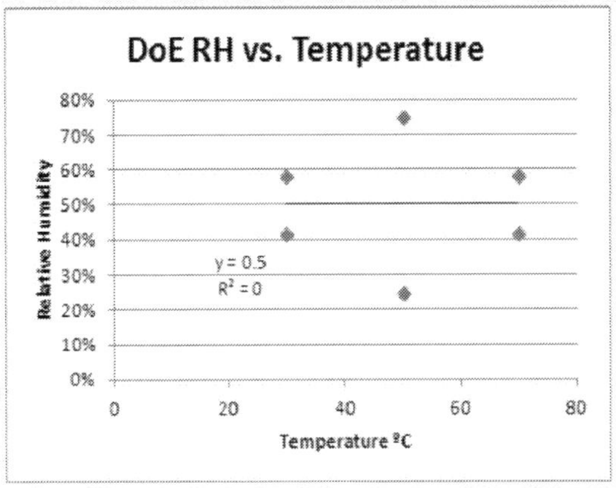


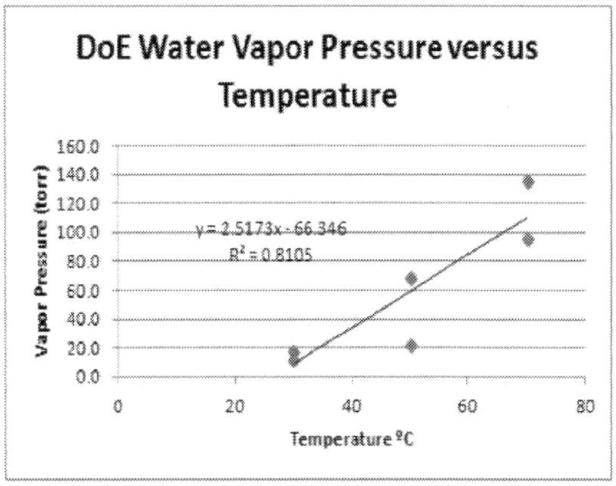


- A mathematically "optimized design" for fitting kinetic models looks like a square, and only tests a low and high humidity value in the design. This is a risky design due to the observed potential changes in kinetics at low or high humidity.
- A Hexagon is an attractive design, as it allows for 3 levels of temperature, and 3 levels of humidity within the design.
- Problem with this design is the potential correlation between temperature and vapor pressure makes fitting vapor pressure based models prone to error in parameter estimation.

## Typical Initial Experimental Design - Offset Hexagon

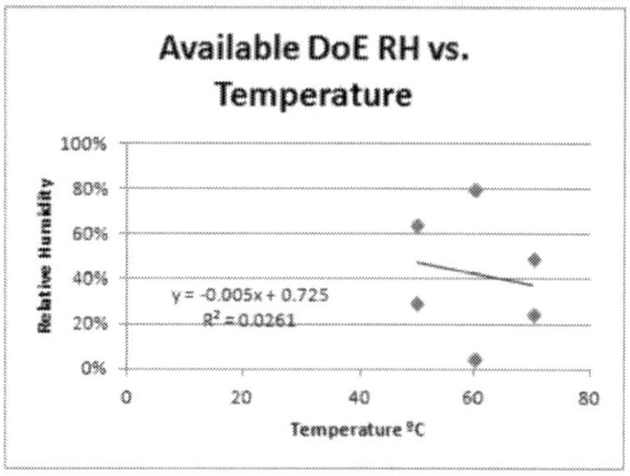


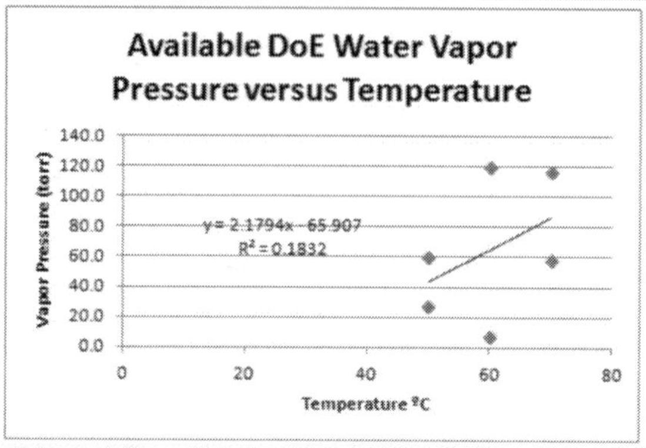


- Skewing the run points slightly reduces the resulting correlation between vapor pressure and temperature in the study without compromising the RH based design.
- Allows either RH or vapor pressure to be fit as the humidity term within the kinetic equations.

## Estimation of Relative Humidity in Long Term Storage

- Moistures permeate into plastic storage bottles, following a standard permeation model, same process as polymer membranes:

$$Flux = P \cdot \left(\frac{SA}{thickness}\right) \cdot \left(VP_{exterior} - VP_{interior}\right) \cdot \Delta t \quad (3)$$

Where

"Flux" is the water vapor transmission rate,
"P" is the permeability coefficient,
"SA" is the surface area,
"thickness" is the wall thickness,
"$VP_{exterior}$" is the water vapor pressure on the exterior of the bottle or capsule film,
"$VP_{interior}$" is the water vapor pressure in the interior of the bottle or capsule film and,
"$\Delta t$" is the change in time.

Permeability coefficients follow an Arrhenius relationship with temperature, and can be calculated at different temperatures using equation 4 (Abdel-Bary, 2003).

$$P = P0 \cdot exp\left(\left(\frac{Ea}{R}\right) \cdot \left(\frac{1}{273.15 + T_{ref}} - \frac{1}{273.15 + T}\right)\right) \quad (4)$$

# Solve for Equilibrium RH on Each Day

- Need to experimentally measure and define models for:
  - Moisture permeation of bottles at varying temperatures.
  - Moisture versus RH relationship of desiccant
  - Moisture versus RH relationship of tablet
- With known initial water content, and known permeation of water for a given day, four equations and four unknowns can be solved to find the equilibrium RH on each day.
  - Water Fraction in air
  - Water fraction in tablet
  - Water fraction in desiccant
  - Overall mass balance

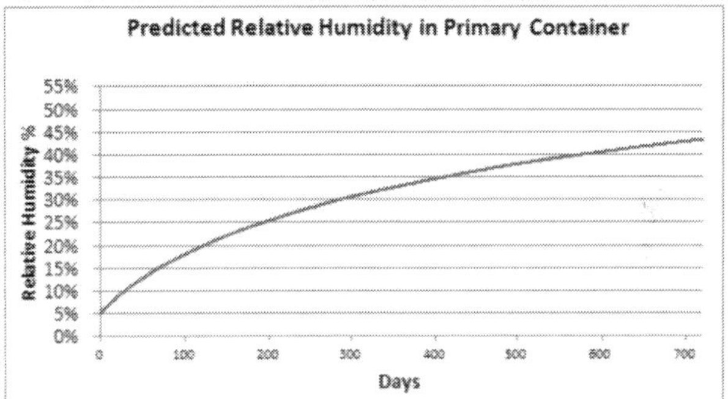


# Estimation of Uncertainty in Prediction – Bootstrapping Approach

- The kinetic models can then be integrated over time with the estimated humidity curve to create a predicted line.
- To estimate the error of the prediction, the error in the kinetic model fit is assumed equivalent to the error in the experimental data.
- Each raw data point is perturbed with random noise equivalent to the error of the model. A new model is generated and plotted for the long term prediction.
- Process is repeated several thousand times to generate a distribution in the prediction.

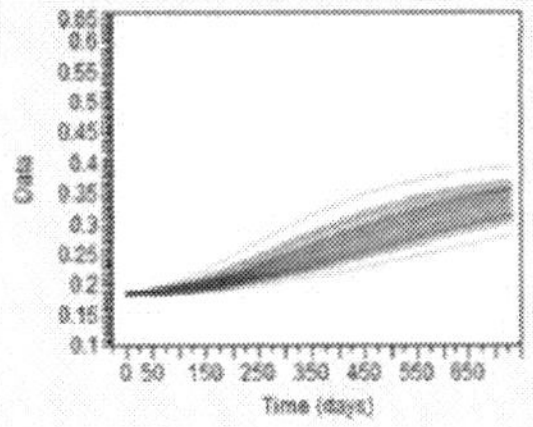

50 Bootstraps

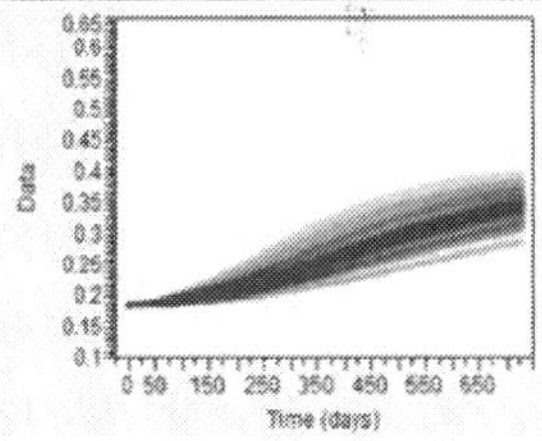

100 Bootstraps

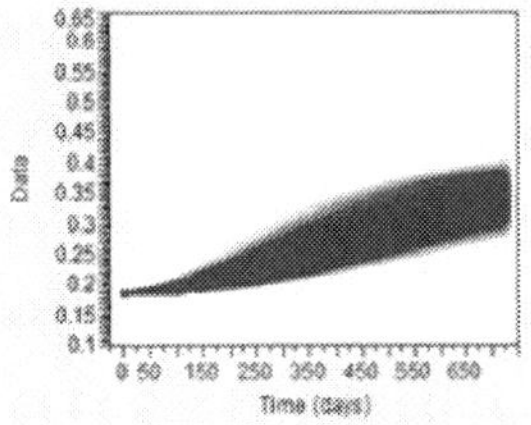

500 Bootstraps

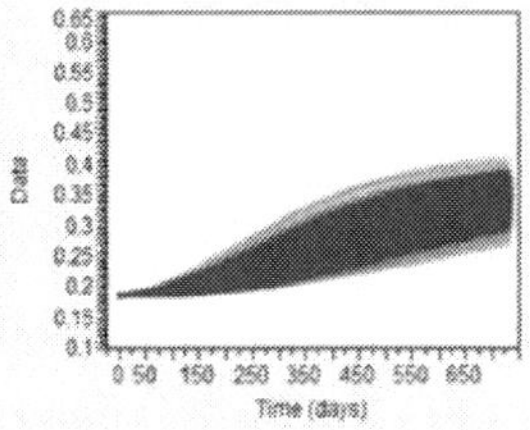

1000 Bootstraps

# Example 1 from Drug Product 81:

| | Specification Limit | | 35 | 35 | 35 | 34 | 60 | 80 |
|---|---|---|---|---|---|---|---|---|
| | Assay/Imp | | Impurity | Impurity | Impurity | Impurity | Impurity | Impurity |
| Temperature | Relative Humidity | Time | Imp101 | Imp966 | Imp545 | Imp799 | Imp283 | Imp102 |
| °C | % | days | parts per 10000 | parts per 10000 | parts per 10000 | parts per 10000 | parts per 10000 | parts per 10000 |
| 71.3 | 77.6 | 4 | 6.7 | 14.8 | 9.3 | 7.7 | 38.2 | 13.2 |
| 79.3 | 27.7 | 4 | 8.3 | 13.5 | 9.5 | 6.9 | 36.8 | 14.0 |
| 61.8 | 39.9 | 7 | 7.4 | 6.0 | 9.4 | 6.1 | 34.1 | 13.1 |
| 71.5 | 51.3 | 7 | 7.5 | 12.9 | 9.7 | 7.2 | 35.6 | 13.7 |
| 71.3 | 77.6 | 7 | 7.2 | 13.2 | 8.8 | 8.8 | 37.8 | 14.1 |
| 79.3 | 27.7 | 7 | 9.8 | 16.3 | 9.9 | 7.5 | 37.7 | 14.0 |
| 61.9 | 6.6 | 10 | 9.2 | 5.7 | 9.4 | 6.0 | 34.9 | 13.8 |
| 61.8 | 39.9 | 10 | 8.1 | 6.9 | 9.4 | 6.4 | 34.6 | 14.1 |
| 71.5 | 51.3 | 10 | 8.2 | 14.6 | 9.7 | 7.8 | 36.1 | 13.4 |
| 50.4 | 74.8 | 14 | 6.8 | 6.0 | 9.5 | 6.6 | 34.4 | 13.7 |
| 61.9 | 6.6 | 14 | 10.1 | 5.5 | 9.6 | 6.5 | 34.6 | 14.5 |
| 50.4 | 74.8 | 10 | 6.9 | 5.4 | 9.4 | 6.6 | 34.3 | 14.3 |
| 61.8 | 39.9 | 0 | 5.2 | 2.0 | 9.5 | 5.8 | 29.8 | 13.1 |

**Figure 13** Drug Product 81 Accelerated Stability Experimental Data

- Just eyeballing this accelerated data, surely Impurity 966 is the worst one?
- Eyeballing again, that impurity 102, can't possibly be a risk, it's nowhere near the specification limit.
- Lesson #1: Don't believe that eyeballing data is a good approach. Don't stop doing an analysis due to little response in the data.

# Example #1: Comparisons of Model Predictions to Long Term Data for Drug Product 81

- Impurity 102 had a low activation energy, did not accelerate substantially at higher temperatures, but ended up the most critical impurity at lower temperatures.
- Impurity 966 was low risk with high activation energy, and showed little to no increase at low temperatures.
- Points out a problem with the isoconversion approach – low activation energy reactions.

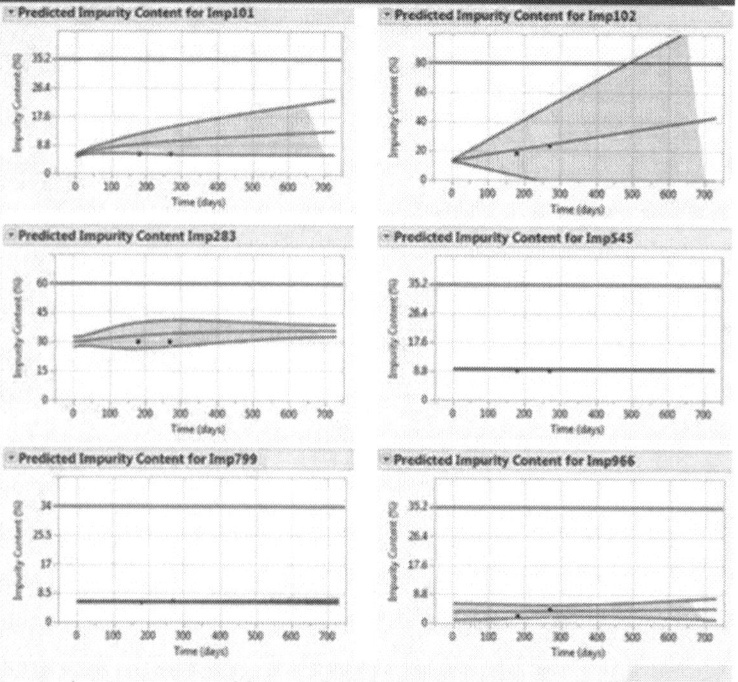


Drug Product 81 impurity models overlaid with long term data at 30 deg C, 65% RH in HDPE Bottle with Desiccant

# Example #2 – Drug Product "P"

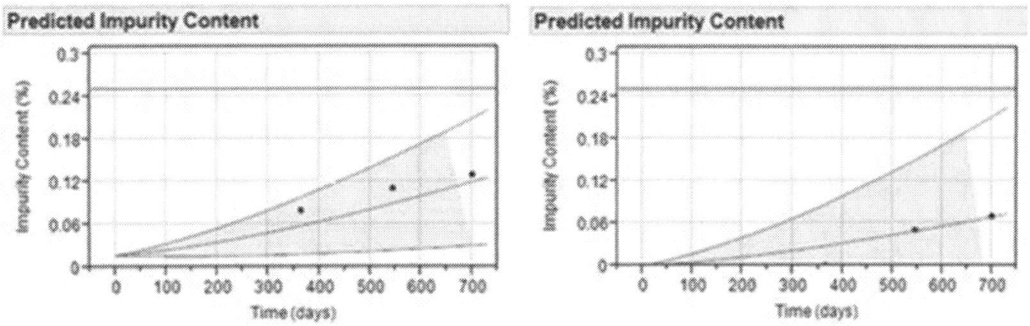


**Figure 15** Drug Product P Impurities 0.94 (left) and 0.11 (right) models overlaid with long term data

•Predicted passing for both impurities of concern, and this proved true.

# Example 3: Drug Product S

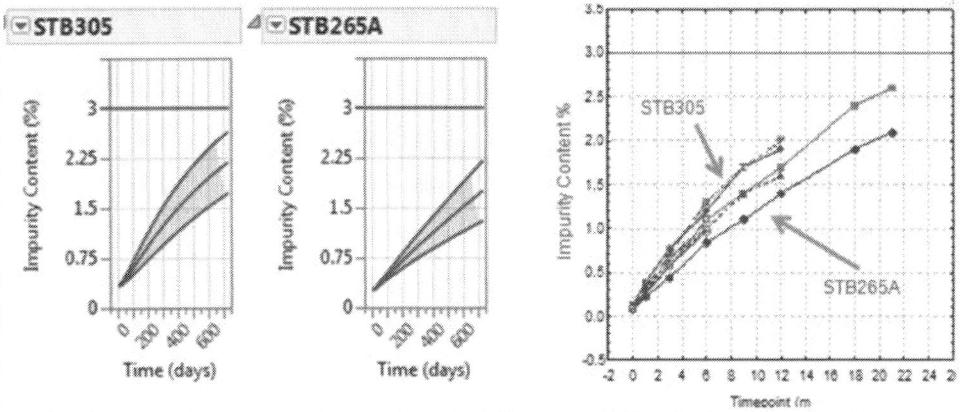


**Figure 16** Drug Product S Batches 305 and 265A in Accelerated Stability Prediction (left and center) Compared to Long Term Observation (Right)

- Multiple formulations were compared in accelerated stability prediction (left), then run in long term stability (right).
- Accelerated stability predicted which formulation was going to have the most stable results.

# Example 4: Desolvation of DMSO From an API

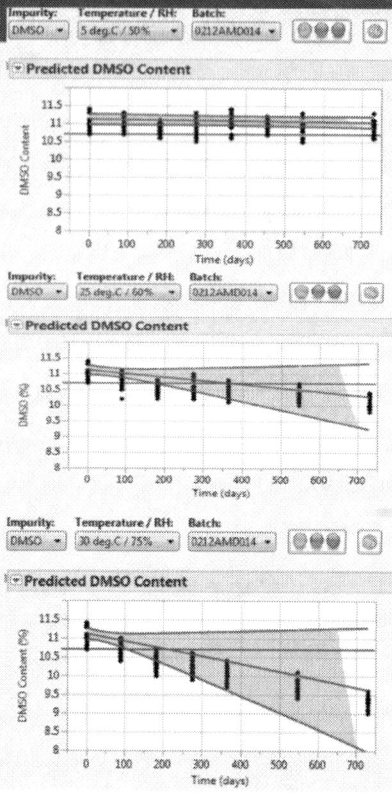


- Accelerated stability modeling can be used to predict some physical phenomenon, such as desolvation.
- Cannot predict other physical mechanisms, such as loss of free DMSO that became unbound from the API prior to the experiment start.
- The accelerated experiment must be run with similar DMSO vapor pressure as in final package in long term stability.

# Example 5: Salt Disproportionation to Free Acid

- Can also predict salt disproportionation, and the subsequent nucleation and growth of free acid crystals.
- Note that this phenomenon was exponential in time, and followed an Avarami curve. What's predicted here is the Avarami transform of the data: $Y^*=\ln(Y/(100-Y))$

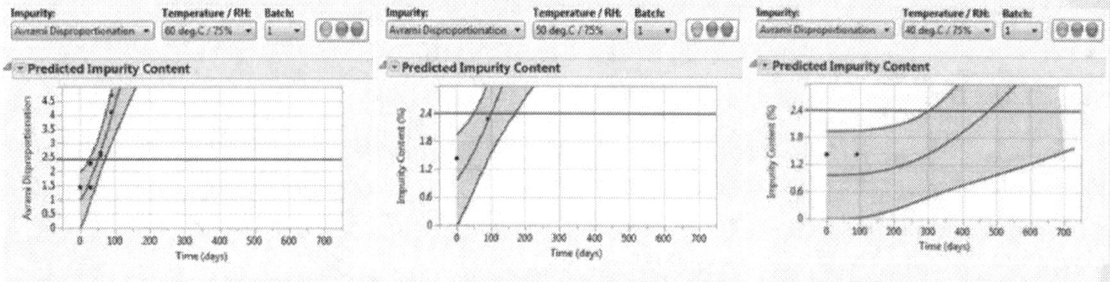


- The important prediction that the 30 degree 65% RH condition with desiccant will remain flat and undetectable, has proved true for the first year so far.

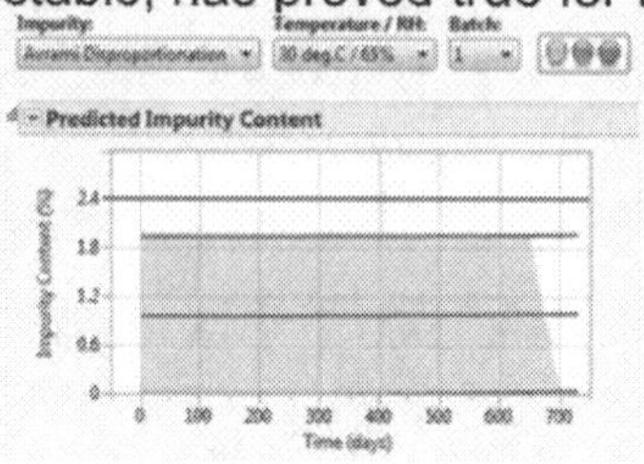

# Example 6: Biological Product Main Isotherm Impurity

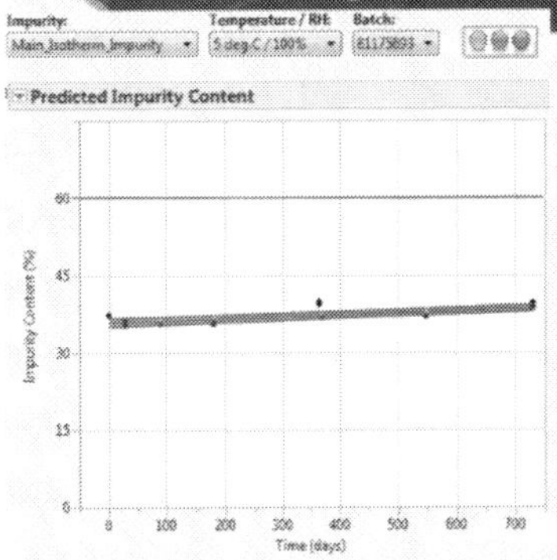


5 degrees

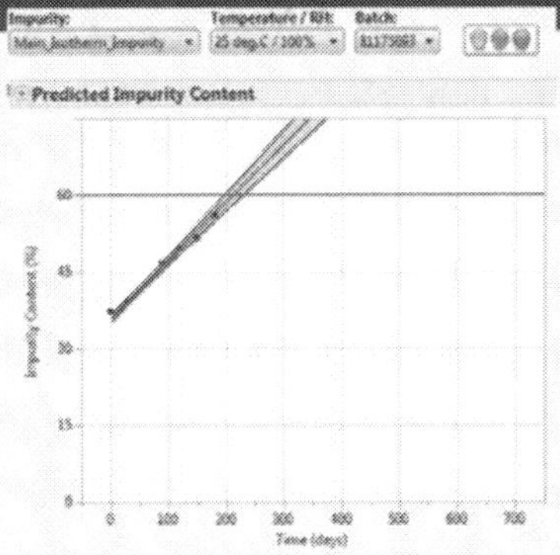


25 degrees

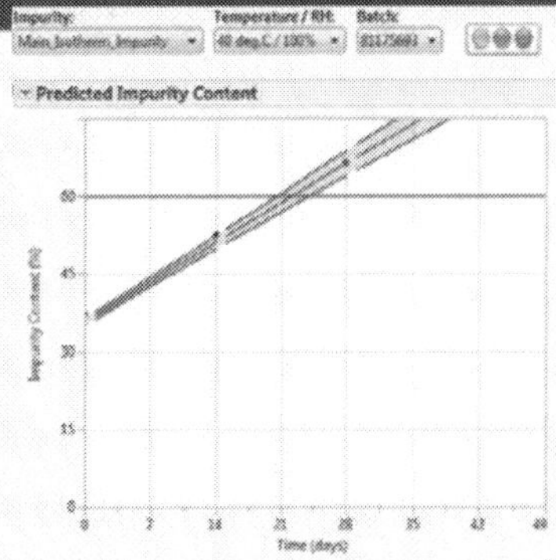


40 degrees

- Main isotherm impurities, and caliper reduced impurities are often well modeled as an Arrhenius model, linear with time, with no humidity impact since it's in liquid.
- Need lower temperatures in the study. Multiple years of shelf life at 5 degrees, reduces to 21 days at 40 degrees.
- Cannot predict below the freezing points generally around -15 degrees.

# Example 7: Drug Product "N" Done in Both ASAP Prime and GSK Methodology

| | | Specification Limit | 91.5 | 10 | 2.406846103 |
|---|---|---|---|---|---|
| | | RRT | impurity | Impurity | Impurity |
| Temperature | Relative Humidity | Time | Assay | Conversion | Avrami_Assay_Loss |
| °C | % | days | % a/a | % a/a | % a/a |
| 50 | 50 | 0 | 100.3 | 0.7 | -0.349720122 |
| 50 | 50 | 4.8 | 96.8 | 4.2 | 1.477558048 |
| 50 | 50 | 9.8 | 93 | 8 | 2.161962565 |
| 50 | 50 | 13.7 | 90.8 | 10.2 | 2.428848952 |
| 59 | 30 | 4.8 | 97.5 | 3.5 | 1.288031107 |
| 59 | 30 | 9.8 | 98.3 | 2.7 | 1.020348263 |
| 59 | 30 | 13.7 | 97.3 | 3.7 | 1.345654347 |
| 70 | 12 | 4.8 | 97.5 | 3.5 | 1.288031107 |
| 70 | 12 | 9.8 | 98.8 | 2.2 | 0.810480272 |
| 70 | 12 | 13.7 | 97 | 4 | 1.426703899 |
| 71 | 74 | 2.8 | 37.8 | 63.2 | 5.129115715 |
| 71 | 74 | 6.9 | 10.5 | 90.5 | 6.76909511 |
| 86 | 8 | 4.8 | 97.5 | 3.5 | 1.288031107 |
| 86 | 8 | 9.8 | 98.3 | 2.7 | 1.020348263 |
| 86 | 8 | 13.7 | 98.5 | 2.5 | 0.941354701 |
| 85 | 53 | 2.8 | 72.5 | 28.5 | 3.681438042 |
| 85 | 53 | 6.9 | 12.5 | 88.5 | 6.572394425 |

- Drug product "N" has an assay that declines with time.
- Since the extent of conversion is high (>15%) and exponential in time, an Avrami transform is used to linearize the data first.

Conversion = 101-Assay

Avrami Assay Loss = ln(100*(Conversion)/(101-Conversion))

# Example 7: ASAP Prime Analysis

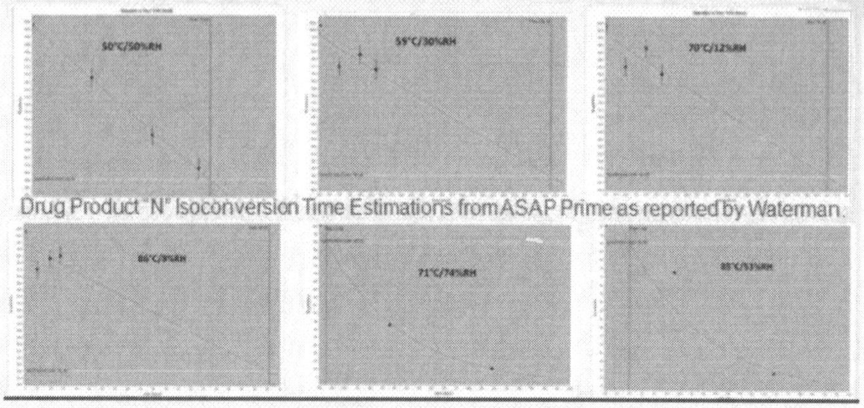

Drug Product "N" Isoconversion Time Estimations from ASAP Prime as reported by Waterman.

| Temperature (°C) | %RH | Isoconversion Time (Days) | SD (Days) |
|---|---|---|---|
| 50 | 50 | 14.6 | 1.2 |
| 59 | 30 | 56.4 | 18.4 |
| 70 | 12 | 55.0 | 17.5 |
| 71 | 74 | 0.3 | 0.1 |
| 85 | 53 | 1.0 | 0.0 |
| 86 | 8 | 95.6 | 44.8 |

Table 2 Calculated Isoconversion Time for Experimental Design Conditions

# Example 7: ASAP Prime Predictions

| | |
|---|---|
| ln A | 18.7±3.1 |
| $E_a$ | 15.6±2.0 kcal/mol (65.3±8.4 kJ/mol) |
| B | 0.103±0.008 |
| $R^2$ | 1.00 |

Table 3 Parameters for Moisture Modified Arrhenius Equation

Predictions were then made for shelf life in PVC blisters.

| Condition | Measured Shelf-life (mos.) | ASAP*prime*™ Shelf-life (mos.) with PVdC |
|---|---|---|
| **25°C/60%RH** | 17.9 (13.4-32.5) | 11.2 (9.2-13.6) |
| **30°C/65%RH** | 8.2 (5.7-11.3) | 6.2 (5.3-7.3) |
| **30°C/75%RH** | 4.8 (4.4-7.6) | 5.2 (4.5-6.1) |
| **40°C/75%RH** | 1.1 (0.8-1.3) | 2.0 (1.8-2.2) |

Table 4 Comparison between the measured long-term storage shelf-life and ASAP*prime*™ predicted shelf-life (to 91.5% potency)

- Due to the exponential assumption, an exponent cannot approach zero reaction rate at near zero humidity. The limit of an exponent of a small number is 1.
- As a result it tends to substantially overestimate reaction rates of humidity sensitive compounds extrapolated to low environmental humidity. These PVC blisters started with an RH of 6%.
- In this case the predicted failure time of the product in PVC blisters was 11 months, where the actual was approximately 18 months at 25 degrees C.

# Example 7: GSK Methodology Prediction of RH in Blisters

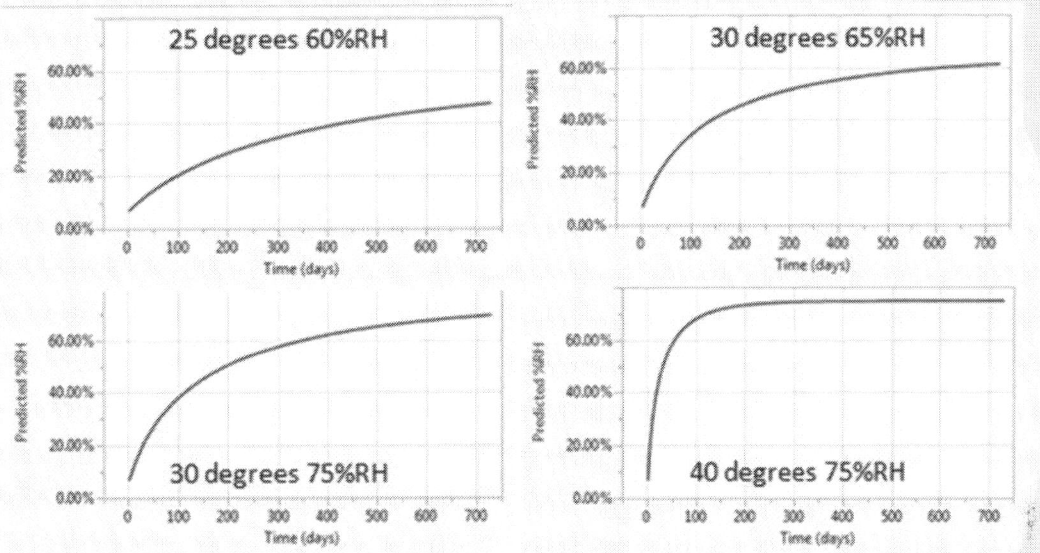


# Example 7: Model Selection and Long Term Predictions versus Actual Long Term Data

**Avrami_Assay_Loss Model Selection**

| Model Type | K1 | K2 | K3 | N1 | N2 | nParams | RMSE | Score | AICc | BIC |
|---|---|---|---|---|---|---|---|---|---|---|
| RH Linear Time Kinetic Model | 1.74 | 6562 | - | 2.61 | - | 4 | 0.6414 | - | 44.037 | 42.749 |
| AH Linear Time Kinetic Model | 45.74 | -6632 | - | 2.59 | - | 4 | 0.6424 | - | 44.089 | 42.801 |
| RH Accelerating Kinetic Model | -188386.9 | 15388 | - | -3 | - | 4 | 3.0955 | - | 97.556 | 96.268 |
| AH Accelerating Kinetic Model | -1 | -7500 | - | -3 | - | 4 | 3.197 | - | 98.654 | 97.365 |
| RH Decelerating Kinetic Model | 0.82 | 7344 | 6.82 | 3.67 | - | 5 | 0.4607 | - | 36.375 | 32.974 |
| AH Decelerating Kinetic Model | 80.08 | -11170 | 6.82 | 3.63 | - | 5 | 0.4608 | - | 36.378 | 32.977 |
| RH Power Model | 3.77 | 4843 | - | 2 | 0.51 | 5 | 0.4789 | - | 37.688 | 34.287 |
| AH Power Model | 45.39 | -5216 | - | 1.97 | 0.51 | 5 | 0.4786 | - | 37.665 | 34.264 |

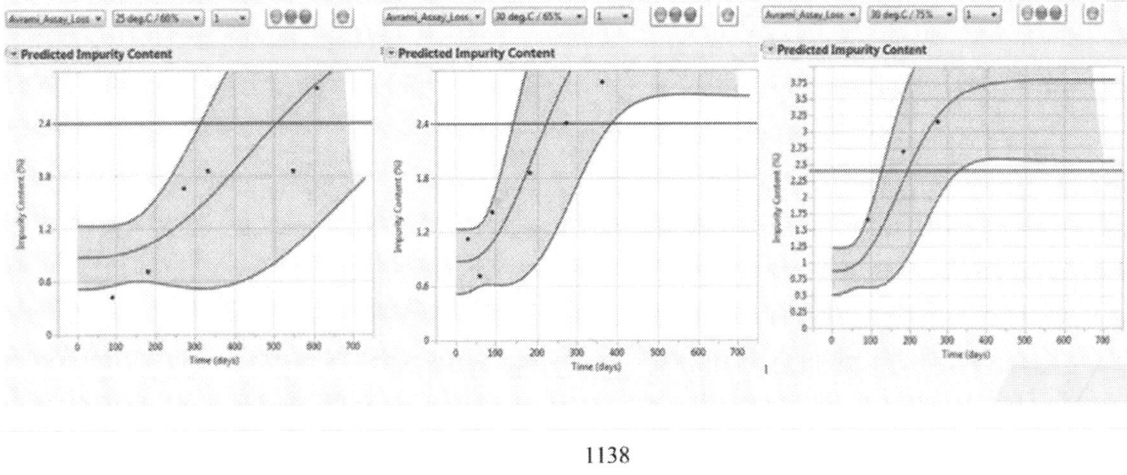

## Example 7: Summary of Comparison of Results Between ASAP Prime and GSK's Methodology for Drug Product "N"

| Storage Condition | Observed Time to Failure (months) | ASAP Prime Isoconversion method Prediction | GSK's Prediction |
|---|---|---|---|
| 25 degrees 60% RH | 17.9 | 11.2 | 16.8 |
| 30 degrees 65% RH | 8.9 | 6.2 | 7.4 |
| 30 degrees 75% RH | 5.3 | 5.2 | 6.3 |
| 40 degrees 75% RH | 1.3 | 2 | 1.5 |
| | | | |
| | | | |
| 25 degrees 60% RH | percent error | 37.4% | 6.3% |
| 30 degrees 65% RH | percent error | 30.2% | 16.7% |
| 30 degrees 75% RH | percent error | 1.2% | 18.8% |
| 40 degrees 75% RH | percent error | 52.0% | 12.5% |
| | | | |
| | Average percent error | 30.2% | 13.5% |
| | Slope actual vs. Predicted | 1.84 | 1.10 |

- ASAP Prime's average error in prediction was 30% for this example.
- Methods described in this presentation resulted in an average of 13% error compared to actual long term data for this example.

## Conclusions

- At GSK, we've observed that stability does not appear to be exponentially affected by increasing RH. Typically curvature is closer to cubed root or square root rather than exponential with power terms typically between -1 and 1, and the majority of data falling within -2 and 2 as would be typically expected for the order of reaction in reaction kinetics.

- We've chosen to pursue an approach using kinetics rather than focus on the isoconversion approach as it allows us to improve the logistics of doing a large number of accelerated studies.

- The aspiration of this model development would be to predict long term data ranging up to two years from a short 2 to 4 week accelerated stability study. The approach we've developed has thus far shown good fit with long term trends in available data. The number of studies is currently greater than 80, and impurities or physical phenomenon modeled greater than 300.

# 425091 Hydrodynamics and Mass Transfer Effects in Mini Vessel Dissolution Testing Systems

**Abstract:**

In the pharmaceutical industry, dissolution testing is routinely conducted to evaluate the *in vitro*performance of solid dosage forms in pharmaceutical development, determine the dissolution behavior of formulations and provide an optimization of drug release dosage forms, and as a quality control tool to insure that solid dosage forms have consistent dissolution properties. The USP Dissolution Testing Apparatus 2 is the device most commonly used for this purpose. In a typical test, a tablet is added to the vessel and samples are manually withdrawn from the vessels over time and analyzed for their drug content.

The USP 2 drug dissolution testing method is not always appropriate, especially for dissolution testing involving very small tablets or when not enough drug substance may be available for proper dissolution testing in the USP 2 Apparatus, especially during the initial stage of drug development. Small-volume dissolution testing apparatuses (mini vessel systems) can be advantageously used instead of the standard USP 2 system to overcome these limitations. The most common of these mini vessel systems is similar in shape and operation to the USP 2 apparatus, but it only requires a small volume vessel (200 mL) and a mini paddle impeller.

Before mini vessel systems can gain wide acceptance in industry, their hydrodynamics needs to be studied, and their operating conditions must be defined in order to compare the velocity distribution and mass transfer coefficient in a mini vessel system with those in the standard USP2 system.

Very few studies on mini vessel systems are available. Therefore, the objective of this work was to quantify the hydrodynamics and drug mass transfer rates in mini vessel systems, and determine the operating conditions under which mini vessel systems can be reliably used instead of USP 2 system for dissolution testing purposes.
The velocity distribution in a minivessel was obtained here using Computational Fluid Dynamics (CFD) at different agitation speeds. The predictions were compared to the velocity vector maps obtained using Particle Image Velocimetry (PIV). Four horizontal iso-surfaces were selected to be studied in detail. The CFC predicted and PIV measured flow patterns were found to be in good agreement. As expected, the 3D flow velocity magnitude increased as the agitation speed increased. However, the velocity in the inner core region below the impeller always remained very low.

The tablet-medium mass transfer process was additionally modeled, and the mass transfer coefficients at four different agitation speeds were obtained and compared to the standard system. The mini vessel system has a flow pattern similar to that of the standard USP 2 system. It was found that similar velocities at the tablet surface (on the iso-surface at y=7mm, i.e., at the vessel bottom) in the mini vessel system and in the USP 2 system were obtained when the mini vessel was operated a 100 mL, 125 rpm and the USP 2 system was operated at 900rpm and 100rpm, indicating that these two systems should result in similar dissolution profiles when operated under these conditions.

# 427772 Stability By Design (SbD): Utilizing Material Science Principles to Predict Physical and Chemical Stability of Pharmaceutical Solids

Friday, November 13, 2015: 8:30 AM
Ballroom D (Salt Palace Convention Center)

**Michael Brunsteiner**[1], Corinna Gressl[1], Johannes G. Khinast[2], Amrit Paudel[3], Adrian L. Davis[4], Margaret Landis[5], Klimentina Pencheva[4], Garry Scrivens[4] and Geoff Wood[6], (1)RCPE, Graz, Austria, (2)Research Center Pharmaceutical Engineering (RCPE), Graz, Austria, (3)Advanced Products and Delivery, RCPE, Graz, Austria, (4)Pfizer Worldwide R&D, Sandwich, Kent UK, Kent, England, (5)Pfizer Worldwide R&D, Groton, CT, Groton, CT, (6)Pfizer

**Stability by Design (SbD): Utilizing material science principles to predict physical and chemical stability of pharmaceutical solids**

*M. Brunsteiner [1], C. Gressl[1], J.G. Khinast[1], A. Paudel[1]*

*Adrian L. Davis [2], Meg Landis [3], Klimentina Pencheva [2], Garry Scrivens [2], Geoffrey P.F. Wood [3]*

*[1]Research Center Pharmaceutical Engineering, Graz, Austria*

*[2] Pfizer Worldwide R&D, Sandwich, Kent UK*

*[3]Pfizer Worldwide R&D, Groton, CT, USA*

Ensuring, both, physical and chemical long-term stability of the drug product is one of the key requirements in successful drug formulation development. While significant knowledge exists on solution-state stability of pharmaceuticals, the predictive knowledge of solid-state physical and chemical stability in solid oral dosage forms is limited. Here we present two examples of our research towards achieving stability by design (SbD) for solid oral dosage forms through the use of *in-silico* methods to accelerate and streamline formulation development.

As a model for chemical drug degradation we consider Carvedilol (CAR), a compound used in the treatment of congestive heart failure. The compounds secondary alcohol has been shown to undergo esterification when formulated with citric acid (CA)as an excipient.(1) The free base form of CAR can crystallize in at least two different polymorphic forms. Based on structural and energetic results from Molecular Dynamics (MD) simulations of the crystal surfaces of two different CAR polymorphs in contact with vacuum or citric acid, we calculated a range of descriptors, including surface energies and exposed reactive surface areas. A thorough analysis of these descriptors and their relative magnitudes for different crystal surfaces of the CAR polymorphs allows us to investigate the relative importance of degradation of molecules on crystal surfaces versus degradation of molecules that are highly mobile due to surface defects and partial amorphization. Accordingly, two different crystalline forms (included in MD simulation) and an amorphous form of CAR were generated of two different size ranges. Following the comprehensive solid-state and particulate characterization, CAR-CA blends prepared using different solid-states and sizes of CAR were subjected to accelerated condition of 50°C. The comparison of the experimental data on kinetics and extent of CAR-CA interaction products formation with the descriptors derived from MD simulations will provide the predictability of chemical stability by the computed descriptors.

As models for physical drug stability, we consider a series of API-polymer combinations that have been used to produce amorphous solid dispersions (ASD), (2) a commonly used strategy in the formulation of poorly soluble drugs. We performed extensive MD simulations of ASDs at various drug loads and calculated descriptors such as relative hydrogen boding propensities, free volumes, API mobilities, and API-polymer intermolecular interactions. A comparison of the obtained results with existing literature data for the relative stabilities of the different ASDs shows that both kinetic and energetic factors can play a role in determining the stability of ASDs. We suggest criteria that can be determined *in-silico* and used to estimate the relevance of these factors for a given API-polymer combination.

References

1. Larsen, J., Cornett, C., Jaroszewski, J. W., and Hansen, S. H. (2009) Reaction Between Drug Substances and Pharmaceutical Excipients: Formation of Citric Acid Esters and Amides of Carvedilol in the Solid State., Journal of Pharmaceutical and Biomedical Analysis 49, 11–17.

2. Eerdenbrugh, B. V., and Taylor, L. (2010) Small Scale Screening to Determine the Ability of Different Polymers to Inhibit Drug Crystallization upon Rapid Solvent Evaporation, Molecular Pharmaceutics 7, 1328–1337.

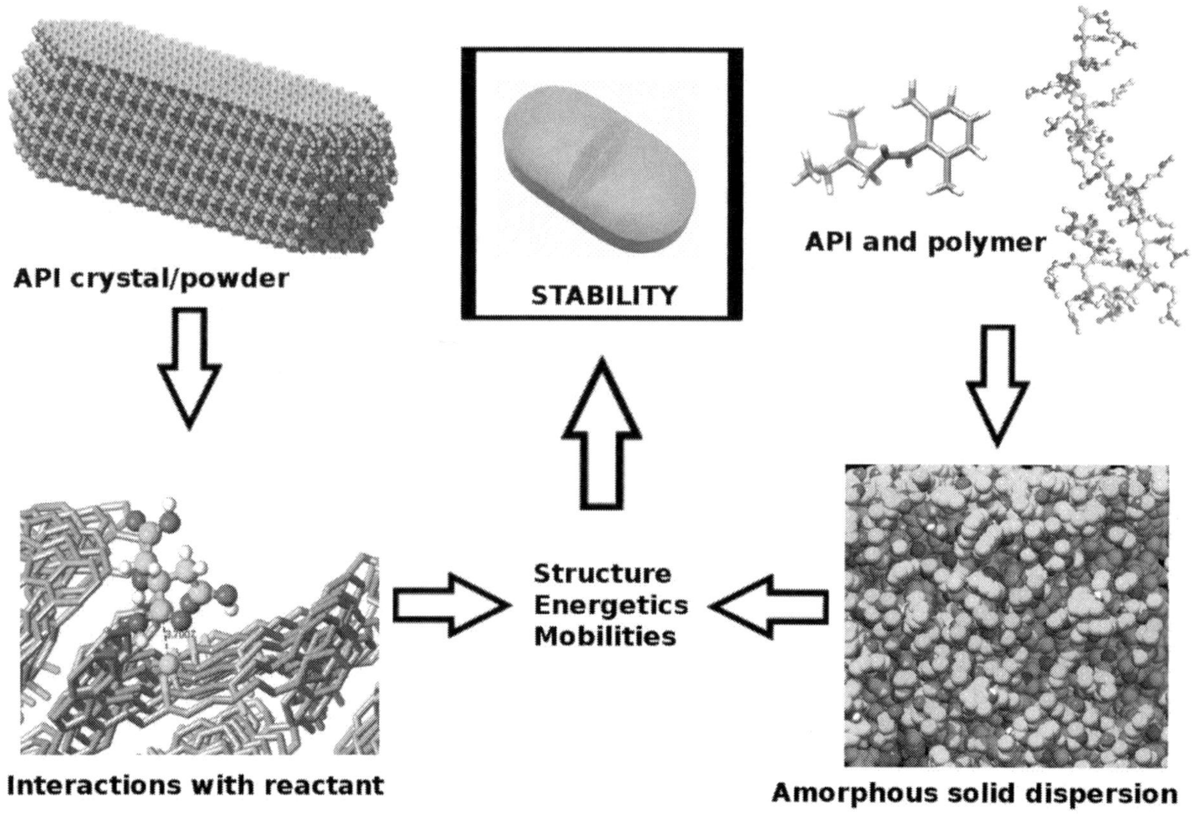
API crystal/powder
STABILITY
API and polymer
Interactions with reactant
Structure
Energetics
Mobilities
Amorphous solid dispersion

# 425319 Exploring Polymorph Free Energy Landscapes with Hamiltonian Reweighted Molecular Dynamics

**Abstract:**

The presence of multiple stable crystal structures in solid organic materials can limit their commercial viability when two or more observable polymorphs exhibit markedly different physical properties. Crystallography researchers estimate that 50% of all drug molecules exhibit polymorphism, but only 1% of the materials with a well-characterized crystal structure in national crystal structure databases contain a similar level of data for an alternative polymorph structure.

Current computational methods for predicting polymorphic behavior generate many possible crystal structures and rapidly evaluate the lattice energies of these candidate structures to determine the most likely to be observed experimentally. However, lattice energies alone do not fully capture the stability of systems that are at room temperature and sample a range of configurations due to thermal motion. In addition, *ab initio* quantum mechanical models or at minimum polarizable force fields are likely necessary to represent the physical system, but the extensive sampling required in molecular dynamics can currently only be performed with classical point-charge force fields.

In this work, we present an atomistic simulation approach to evaluate the free energy difference between crystal polymorphs in both computationally cheap and expensive Hamiltonians. Molecular dynamics simulations are conducted in a classical point-charge all-atom potential designed to overlap with the expensive Hamiltonian. Multistate Hamiltonian reweighting is then employed on these configurations post simulation to estimate the free energy difference in more expensive force fields without any additional sampling. We therefore access computationally demanding free energy differences by avoiding direct simulation within the polarizable potentials.

This simulation technique is applied to solid phase benzene, which is known to have multiple distinct polymorphs. We reweight to polarizable potentials and the simulation results indicate that the exact polymorph stability is strongly sensitive to the choice of Hamiltonian. Furthermore, the successful Hamiltonian reweighting technique demonstrates that the relevant phase space of polarizable Hamiltonians in this system can be approximated using only point-charge models. These preliminary results represent a significant step toward the ultimate goal of tractable thermodynamic calculations using more expensive many-body energies.

# 431785 Molecular Modeling to Design and Control (M2DC): A First Principles Approach to Polymorph Prediction and Crystallization Unit Design

**Abstract:**

Crystallization remains one of the most important industrial separation processes to date. Despite its wide spread use, a fully fundamental understanding of the nucleation process is lacking. The classical nucleation theory (CNT) rate expression, formulated by Gibbs, Weber, and Volmer, at the beginning of the $20^{th}$century, remains the main framework against which experimental evidence is compared, despite being overly simplistic, and failing to capture data well in a range of systems. Skepticism towards the degree to which thermodynamic bulk properties used in the CNT rate expression can be even applied to nano scale clusters is warranted. Rather, with the advent of modern supercomputers, the possibility to relax the suit of CNT assumptions in favor of fully atomistic calculations provided via molecular dynamics is becoming a more and more appealing alternative. From a solid-state physics perspective, polymorphism, the ability of a molecule to exist in multiple crystalline phases, remains a thorn in the side of both industrial crystallization experts, and molecular modelers alike. Landscape energy minimization techniques, although elegant in their own right, fail to include standard design variables such as solvent, temperature, or pressure. Molecular dynamics, which encapsulates the aforementioned design variables naturally, has suffered from time scale challenges associated with activated rate processes, such as nucleation. However, with the advent of modern supercomputers, and HPC hardware such as xeon phi coprocessors, these types of simulations are becoming more and more feasible. The crystallization community would more than like molecular simulations to become primetime, because in the absence of in silico polymorph specific nucleation rates, expensive solvent screening experiments must be performed. Furthermore, the question of "have we truly found all possible forms" is never truly answered by such experiments.

In this work, we investigate the potential of atomistic information provided by hybrid monte carlo molecular dynamics (MCMD) to be used to not only predict polymorphism as a function of operation conditions, but also the degree to which polymorph specific solubility, and metastable zone curves can be calculated. These calculations will serve as inputs into control models, allowing a fully atomistic design of a crystallization unit. As microfluidic devices provide a means to measure homogeneous nucleation rates using minimal reagents and in minimal time, they will serve as a benchmark for determining the degree of accuracy of our homogeneous nucleation rate calculations prior to further scale-up for industrial applications.

# 412913 PAT for Form Monitoring during Drying

**Abstract:**

**PAT for Form Monitoring During Drying**

M. Dimitri, S. Allwein, P. Mudipalli, and M. Christie

Teva Branded Pharmaceuticals, Inc.

Chemical Synthesis Center

383 Phoenixville Pike

Malvern, PA 19355

A.I.Ch.E. Annual Meeting 2015

Session: "PAT for Process Understanding and Elucidation of Fundamental Phenomena in Drug Product/Substance Development"

Abstract: Controlling and monitoring polymorphic form during drying of active pharmaceutical ingredients (APIs) is increasingly an area of interest particularly for specialized drying, such as humidified drying. The use of PAT instruments is often powerful in understanding the drying kinetics of an API, identification of the final polymorphic form in situ in the dryer, and determination of drying end point. This study examines the development and use of Raman Phat Probe technology to monitor a humidified drying process of an API and examines some of the key scale-up parameters involved. For this API, three different known polymorphs are possible during drying, with one polymorph being irreversible once formed. In addition, online analysis was determined to be critical as the API is extremely sensitive to relative humidity and form conversion occurs rapidly, making offline analysis challenging. This study demonstrates the use of raw spectra and chemometric analysis in monitoring and defining the drying kinetics of the two main forms of interest. The example of the use of this technology that is being presented shows its applicability both for drying process development and for confirmation of form at scale (ie, in situ in the dryer).

Keywords: Drying, Phat Raman, Form Monitoring, PAT

# 443949 Molecular Modeling of Additive Effects on Calcium Pyrophosphate Crystals

**Abstract:**

Calcium pyrophosphate (CaPP) deposition disease is a rheumatologic condition caused by the formation of calcium pyrophosphate crystals in the joints. CaPP crystals form in joint cartilage as the results of an unknown metabolic disorder. Eventually, CaPP crystals are ejected from the cartilage into the surrounding synovial fluid where they continue to grow. CaPP deposition disease becomes increasingly common with age, affecting approximately a quarter of people over 85 years old.[1] Despite its prevalence, there are no treatments to inhibit the growth or formation of CaPP crystals. Mechanistic understanding of how additives interact with crystal faces is important for the development of additives to inhibit CaPP crystal growth.

We use atomistic simulations to study the effects of additives on calcium pyrophosphate crystal faces. Accurate modeling of additive effects on CaPP crystal faces requires a molecular force field that can reproduce the experimental crystal structures of CaPP. We develop a new force field for pyrophosphate using the functional form of the CHARMM force field. Starting with parameters in the CHARMM General Force Field, we refit bond, angle, and dihedral parameters to match ab initio simulation data. Finally, the non-bonded parameters were optimized to minimize the root mean squared deviation between the experimental and force field derived crystal structure of the thermodynamically stable $Ca_2P_2O_7$Ÿ$2H_2O$. The resulting force field accurately reproduces the crystal structure of all solved hydrated CaPP forms.

Simple salts have been shown to affect polymorph selection and crystallization of CaPP. We study the effects of $Ca^{2+}$, $Mg^{2+}$, and $Na^+$ on CaPP, by calculating their binding free energies to CaPP surfaces using adaptive biasing force and free energy perturbation simulations. Ion binding energies are calculated for fast growing CaPP crystal faces for crystals grown homogeneously and heterogeneously on model biological surfaces. Positive ions that bind more strongly to fast growing crystal faces than $Ca^{2+}$ may inhibit crystal growth. We find that $Mg^{2+}$ binds more strongly than $Ca^{2+}$ to the fast growing faces of β-$Ca_2P_2O_7$Ÿ$4H_2O$, while $Ca^{2+}$ binds more strongly than $Mg^{2+}$ to the fast growing faces of the thermodynamically stable $Ca_2P_2O_7$Ÿ$2H_2O$.

[1] Guerne, P. A., and R. Terkeltaub, "Calcium pyrophosphate dihydrate crystal deposition: epidemiology, clinical features, diagnosis, and treatment," Chapter 21 in Terkeltaub, R., ed., "Gout and Other Crystal Deposition Arthropathies", Elsevier Heath Sciences, Philadelphia, PA (2012).

# AUTHOR INDEX

# AUTHOR INDEX

# AUTHOR INDEX

# AUTHOR INDEX

# AUTHOR INDEX

# AUTHOR INDEX

# AUTHOR INDEX

# AUTHOR INDEX

# AUTHOR INDEX

# AUTHOR INDEX

# AUTHOR INDEX

# AUTHOR INDEX

# AUTHOR INDEX

# AUTHOR INDEX

# AUTHOR INDEX

# AUTHOR INDEX

# AUTHOR INDEX

# AUTHOR INDEX

# AUTHOR INDEX